The CRC Press Advanced and Emerging Communications Technologies Series

Series Editor: S. Zamir

Data and Telecommunications Dictionary, Julie K. Petersen

Handbook of Sonet Technology and Applications, Steven S. Gorshe

The Telecommunications Network Management Handbook, Shervin Erfani

Handbook of Communications Technologies: The Next Decade, Rafael Osso

ADSL: Standards, Implementation, and Architecture, Charles K. Summers

JavaBeans Handbook , Ricardo Devis

Protocols for Secure Electronic Commerce, Ahmed Sehrouchni and Mostafa Hashem Sherif

After the Fireworks: Business and Technology Strategies for Surviving Y2K Hits, Bhuvan Unhelkar

Web-Based Systems and Network Management, Kornel Terplan

Intranet Management, Kornel Terplan

DATA & TELECOMMUNICATIONS DICTIONARY

Julie K. Petersen

STANDARD-SETTING 820-PAGE TECHNICAL REFERENCE WRITTEN IN PLAIN LANGUAGE. GENEROUSLY ILLUSTRATED WITH MORE THAN 300 CHARTS AND DIAGRAMS.

CRC Press
Boca Raton London New York Washington, D.C.

Library of Congress Cataloging-in-Publication Data

Petersen, Julie K.
Data & telecommunications dictionary / Julie K. Petersen.
p. cm. -- (Advanced and emerging communications technologies)
Includes bibliographical references (p.).
ISBN 0-8493-9591-7 (alk. paper)
1. Telecommunication--Dictionaries. 2. Data transmission systems--Dictionaries. 3. Computer networks--Dictionaries. I. Title.
II. Title: Data and telecommunications dictionary. III. Series.
TK5102.P48 1999
621.382'03—dc21 98-46077
CIP

Direct all inquiries to CRC Press LLC, 2000 Corporate Blvd., N.W., Boca Raton, Florida 33431.

Text design and layout by the author.
Illustrations and photos by the author unless otherwise credited.

International Standard Book Number 0-8493-9591-7
Library of Congress Card Number 98-46077
Printed in the United States of America 1 2 3 4 5 6 7 8 9 0
Printed on acid-free paper

Preface & Acknowledgments

Creating a technical reference for a field which changes by the minute is a particular challenge and this book would not have come about without the assistance and support of a great number of dedicated librarians, museum curators, and museum volunteers, who gave graciously and generously of their time and energy. They patiently answered my numerous and persistent questions and pointed the way to original and, in some cases, rare references. Special thanks go to the Bellingham Antique Radio Museum, which provided a provocative and educational look into the development and construction of historic and present-day communications technologies, and permitted me to photograph the fascinating collection. Several local independent book stores deserve thanks for providing leads to turn-of-the-century early editions on telegraph and telephone technology: Michael's Books, Eclipse Books, Henderson Books, and Village Books.

Old books, historical artifacts, diagrams, and schematics are more important to the study of contemporary electronics than many people realize. Since the development of semiconductors, technology has become 'invisible,' that is, you can no longer see what a component does or how it works by simply looking at it, or taking it apart and putting it back together. Yet, to work in design and engineering-related fields still requires an understanding of the underlying physical and relational concepts. For this reason, a number of basic historic devices and schematics are included here to give a better understanding of the background and inter-relationships of electronics and the roles they play.

Sources

In the course of developing this dictionary, I discovered that many references are derivatives of each other and do not go back to original sources, instead perpetuating common lore that lacks verification and accuracy. For the *Data & Telecommunications Dictionary*, I made special efforts to seek out the origins of various telecommunications inventions, discovering that Morse probably didn't develop Morse Code, that the Wheatstone bridge was described but not invented by Charles Wheatstone, and that printed circuit boards were invented at least 12 years earlier than previously acknowledged. These stories give life to what might otherwise be a dry reference, and a few more important antecedents and surprises are similarly documented in these pages.

Dictionaries tend to be ethnocentric. Language is partly responsible for this, as it is difficult for most authors to access or read foreign references. I was aware, while writing this book, that there are undoubtedly foreign contributions to technology, particularly from Asia, an ancient culture, that might be commonly overlooked and not acknowledged. The Web helped. It enabled me to locate information on early inventions and cultural contributions from Russia, the Orient, India, and the West Indies that were not documented in any of the written references I had so carefully and persistently ferreted out. We can look forward to more of these unsung heroes coming to light as the global populace comes online. Their stories are inspiring and intriguing and, even if their contributions didn't reach the mainstream of western culture in terms of human endeavor, they have made important contributions that are worth acknowledging.

Format

In reviewing other references, I noted that most dictionary authors avoid the time and effort required to find out which entries are spelled in upper or lower case, choosing to begin every entry with a capital letter. This may save work, or satisfy some aesthetic need, but it does not fulfill one of the functions of a reference, which is to provide you, the reader, with correct spellings, or at least to suggest the conventional spellings of the various terms. In the telecommunications industry, there is a lot of variation and, in some cases, no definitive answer, but there are discernible patterns, so I have endeavored to find reputable sources and to provide good guidelines. In general, protocol names and commercial products or commercially derived technologies begin with capital letters; software commands, generic technologies, and general concepts are spelled with lower case letters.

In the same vein, American authors tend to Americanize the spellings of European organizations (organisations), inventors, and commercial products. This reference seeks to respect the names assigned by foreign countries (e.g., Fibre Channel rather than Fiber Channel). There are a few exceptions. In some cases an Americanized version is used so often it becomes an ad hoc standard, and may be used here somewhat interchangeably, depending upon the context, e.g., ISO is expressed by many Americans as International Standards Organization when it is actually the International Organization for Standardization.

Recent Technologies

In surveying other computer references I discovered that some authors convey the mistaken impression that Microsoft invented everything. Without taking any credit away from the best-known commercial software vendor, in fact, most of the significant computer technologies were developed long before Microsoft was founded, by pioneers at Bell Laboratories, the IBM research labs, Xerox PARC, Dartmouth College, Carnegie-Mellon, and many small independent design and development firms. I have taken pains to credit Joseph Henry, the Lorimer brothers, Gary Kildall, Alan Kay, Jonathan Postel, and many lesser-known contributors, so that this reference reflects a more accurate slice of history. As such, it credits the Kendak-1 as the first microcomputer, three years prior to the introduction of the Altair. I felt readers would appreciate this attention to accuracy. I only wish there had been more time to delve into the 'real' history of computing and telecommunications.

Many references steer clear of documenting recent technologies, particularly satellite projects, new products of significance, important protocols, and organizations developing and supporting them. I now understand why. No sooner did I list a new development, when another stepped in to take its place. I further discovered that it takes as long to research a definition of a new technology (you essentially have to learn it before you can describe it) as it does to research an entire page in many types of prose books. To deal with obsolescence, I have tried to present information in its broader context, since the concepts remain, even if the mechanics change over time. It is hoped this approach will extend the useful life of this reference, and that over time it will retain its value because it documents not just current products, but the overall development and evolution of the industry.

Special Thanks

The author would like to thank Saba Zamir for suggesting the topic and Mimi Williams for her patience and assistance. Finally, special thanks go to Dan Suslo, Craig Butler, and Paul Carpentier, who went out of their way to offer support and assistance during the final and most difficult parts of assembling this extensive project. Your efforts are remembered and greatly appreciated.

JULIE PETERSEN
JULY 1998

About the Author

Julie Petersen is a technical consultant, author, software designer, and internationally recognized computer artist living in the Pacific Northwest. In her few spare moments she enjoys music, outdoor activities Aikido, and Go. She has been a professional in the microcomputing industry since 1980 when personal computers were a rare sight in retail outlets. She remembers how much microcomputing and telecommunications have changed.

> "The computing domain was male, born from the engineering and computer science fields. I remember searching in vain for a woman's washroom in an engineering building at the University of British Columbia where some of the terminals were installed. When I visited a Radio Shack store to replace a DC converter circuit on my modem card, I was invisible to the sales clerks who jumped in startled confusion every time I asked for assistance. Yet, despite a general lack of support, those were exciting times. In creating an outlet for their intellectual curiosity, creativity, and common interests, society's undercurrent of techies and geeks had unleashed technology with revolutionary repercussions for all humankind."
>
> "The early computer hobbyists were hungry for software, any software. In those days, 'commercial software' was a cassette tape packaged in a zip-top bag with a two-page, nine-pin dot matrix printout. An investment of $6,000 was enough to become a developer. One of my friends and colleagues programmed an early computer game that earned enough to finance him for the rest of his life. In stark contrast, most commercially successful games are now created by teams of 20 to 150 people with development and marketing costs of $2- to $40-million."
>
> "In the late 1970s, if you wanted to use a computer, you had to learn to fix it and to program it. My unreliable system taught me basic electronics. I learned to diagnose common system problems, to upgrade chips, and to distinguish between hardware and software malfunctions. Most computer users now have very little understanding of the design and functioning of their systems. In many ways this is good, as a computer should be a tool, not an albatross demanding constant care and attention. In other ways it results in buyers who are at risk of being cheated. When users buy defective computer tools, it's hard to determine if the problem is the product or their use of the product. This results in a Wild West mentality in terms of lack of accountability and product support on the part of some developers and vendors. Consumers can protect their investments by insisting that products be fixed *before* they are enhanced. The reality is that an upgrade often has as many bugs as the original. With greater awareness and advocacy on the part of consumers, I think this will change."
>
> "One of the greatest benefits of telecommunications technologies is that they can open a portal to the rest of the world. I was living in a beautiful but small and conservative town, and I was lonely for conversation about technology and philosophy. My modem made it possible to find people with common interests 24 hours a day and my sense of isolation vanished. Most computer bulletin boards were completely open in the early 1980s; passwords were rarely required. The BBSs freely offered email, chat, and online games. Eminent scientists and celebrities conversed with friends and fans. Vandalism was rare. By the mid-1980s, computers had become mainstream, and many of the unspoken ethical codes respected by early users were disregarded. Passwords were added, boards were shut down, VIPs hid behind aliases, chat lines were discontinued, newsgroups were moderated; the atmosphere of voluntary responsibility and open cooperation gradually gave way to regulation and commercialization. We can mourn those days or we can strive to celebrate and enhance the aspects of the technology that improve the quality of our lives. If we seek to promote universal access, telecommunications can be used to engender tolerance and understanding, to bring together people of all sizes, shapes, colors, and cultures. It can be a fantastic resource for our combined knowledge of medicine and cosmology, a boon to the bedridden, a shot in the arm for distance education, and an option for those who prefer telecommuting to 15-mph stop-start traffic on overcrowded highways. It is my hope that this reference will help readers understand the overwhelming profusion of technical terms and telecommunications inventions so we can focus our efforts on developing the life-enhancing aspects of the technologies, and evolve them in the most positive sense of the word."

How to Use
The Data & Telecommunications Dictionary

General Format. There are two sections to this reference: a main alphabetical body, with numeral entries at the end, after Z; and appendices with various charts, an extended section on ATM, and a timeline of telecommunications inventions and technologies.

Entries. Dictionary entries follow a common format, with the term or phrase in **boldface**, followed by its abbreviation or acronym, if there is one. Pronunciation is included in cases where it may not be obvious. Alternate names (e.g., William Thompsom, aka Lord Kelvin) are cross-referenced. The body is included next, with multiple definitions numbered if there are several meanings for a term. Finally, where appropriate, there are cross-references, RFC listings, and URLs included at the end, in that order.

Abbreviations. In many cases, the term and its abbreviation are described together so the reader doesn't have to look up abbreviated references to understand a particular entry; for example, cathode ray tube will often be followed by *(CRT)* so the words and their commonly used abbreviations become familiar to the reader.

Web addresses. Web addresses, based on Uniform Resource Locators (URLs) are listed for nonprofit, not-for-profit, charitable, and educational institutions and, in a few rare instances, for commercial enterprises with particular relevance for telecommunications or with substantial educational content on their Web sites. For the most part, commercial URLs are not listed. They can often be guessed (http://www.companyname.com/) or otherwise easily located through Web search engines listed in Appendix C.

RFCs. Request for Comments (RFC) documents are an integral part of the Internet, and extremely important in terms of documenting the format and evolution of Internet protocols and technologies. For this reason, RFC references are listed with many of the Internet-related references and can be found in numerous RFC repositories online. There is also a partial list of significant or interesting RFC documents included in Appendix E.

Diagrams and charts. Illustrations are included as close to the related definition as was possible in the space provided. Extensive listings of the various ITU-T Series Recommendations are included because they are the standards on which most Internet technologies, telecommunications standards, and commercial products are built.

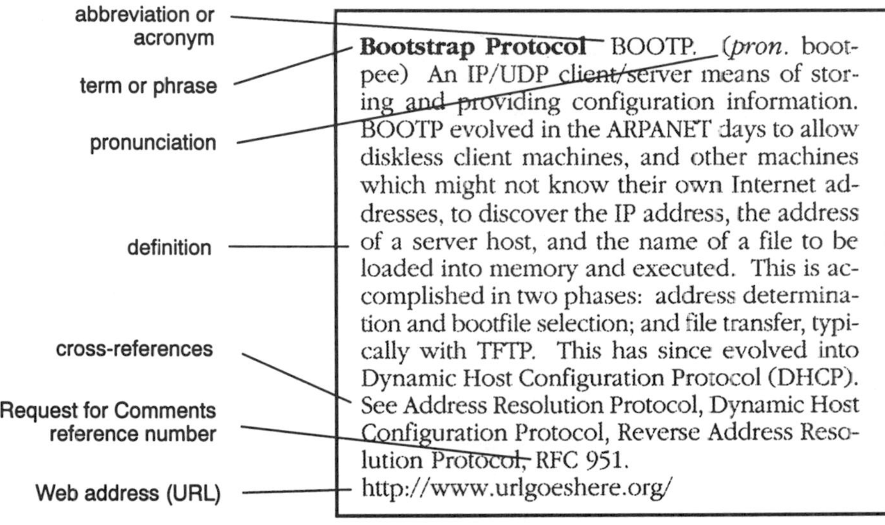

Contents

a 1. Symbol for acceleration 2. Symbol for anode. See anode. 3. Abbreviation for area. 4. Abbreviation for atto-. See atto-.

A 1. Symbol for acoustic velocity. See acoustic velocity. 2. Abbreviation for ampere. See ampere. 3. Symbol for gain. See gain.

A & B bit signaling A data signal bit used in data transmissions to convey a signal state. These bits may be conveyed as an in-band *robbed bit* from various frames to signal line status. For example, *A bits* are used in voice communication implemented over T1 to indicate outbound call signaling, with *B bits* as mirrors to the A bits. A & B signal bits are commonly carried in each sixth frame.

A battery 1. A low voltage battery originally used to provide current to filaments or cathode heaters in electron tubes, now commonly used for small electronic appliances such as cameras, calculators, pen lights, etc. See battery. 2. An *air cell* A battery is a nonrechargeable wet cell with carbon electrodes providing an average power of 2.0 volts. See talk battery.

A Block A Federal Communications Commission (FCC) designation for a Personal Communications Services (PCS) license granted to a telephone company serving a Major Trading Area (MTA). This grants permission to operate at certain FCC-specified frequencies. See band allocations. See frequency chart below.

FCC-designated Communications Frequency Blocks

Block	Carrier	Frequency	Paired Frequency
A Block	30 MHz	1850-1865 MHz	1930-1945 MHz
B Block	30 MHz	1870-1885 MHz	1950-1965 MHz
C Block	30 MHz	1895-1910 MHz	1975-1980 MHz
D Block	10 MHz	1865-1870 MHz	1945-1950 MHz
E Block	10 MHz	1885-1890 MHz	1965-1970 MHz
F Block	10 MHz	1890-1895 MHz	1970-1975 MHz

A carrier *a*lternate *carrier*. A designated nonwireline competitive telephone cellular service carrier which is not the established local wireline carrier (B carrier). See B carrier.

A channel In a stereo system, the designation for the left audio channel, usually connected to the left speaker.

A interface, Airlink interface A frequency-translating interface for over-the-air communications to enable Cellular Digital Packet Data (CDPD) to be deployed over AMPS. The A interface connects the Mobile End System (M-ES) to the Mobile Data Base System (MDBS). See B interface, C interface, Cellular Digital Packet Data, D interface, E interface, I interface.

A link See access link.

A minus, A- The negative polarity of a voltage source as, for example, the negative terminal of an A battery.

A News An early UUCP-based news-reading software program released in 1979. It was developed by Tom Truscott, Jim Ellis, and Steve Bellovin, with enhancements and modifications a year later by Steve Bellovin, Steve Daniel, and Tom Truscott. It inspired many subsequent programs for reading Internet discussions. See newsgroup, USENET.

A plus, A+ The positive polarity of a voltage source, as, for example, the positive terminal of an A battery.

A port In a Class A, dual-attachment (dual ring) Fiber Distributed Data Interface (FDDI) token-passing network, there are two physical ports, designated PHY A and PHY B. Each of these ports is connected to both the primary and the secondary ring, to act as a receiver for one and a transmitter for the other. Thus, the A port is a receiver for the primary ring and a transmitter for the secondary ring. The dual ring system is configured to provide fault tolerance for the network.

Port adaptors can be equipped with optical bypass switches to avoid segmentation, which might occur if there is a failure in the system and a station temporarily eliminated.

FDDI ports can be connected to either single mode or multimode fiber optic media, providing half duplex transmissions. LEDs are commonly used on port adaptors as status indicators. Optical bypass switches may in turn be attached to the port adaptors. See dual attachment station, Fiber Distributed Data Interface, optical bypass, port adaptor.

A-0 A programming language developed by Grace Hopper in the early 1950s.

A-1 time An atomic time scale established by the U.S. Naval Observatory. The origin is set at 1 January 1958 zero hours Universal Time with a second unit equal to 9,192,631,770 cycles of cesium at zero field. See atomic clock, Universal Time.

A-law A pulse code modulation (PCM) coding and companding standard used outside North America as the CEPT standard. A-law is commonly used for encoding speech by sampling the audio waveforms. See E carrier, Mu-law, pulse code modulation, quantization, U-law.

A-scope A specialized radar tracking screen which displays only the range of objects being detected, appearing as illuminated blips from left to right. See B-scope.

A/B switch 1. A dial or switch with two settings for controlling sources of input and output to a circuit. A/B/C and A/B/C/D switches are also common. See A/B switchbox, switcher. 2. A setting on various appliances allowing the user to select between two modes. 3. In cellular communications, an A or B setting on most new transceivers which designates whether the user wants to use a wireless or wired connection when roaming.

A/B switchbox A very common, usually passive, connection-routing device selected by a switch and providing receptacles or sockets for various connectors. In computing, A/B switchboxes are commonly used to switch a serial communications line between a printer and a modem or facsimile machine, or between two different printers, such as a laser printer and a pen plotter. These serial boxes commonly have 25-pin D connectors on the back, one for the input, which may be from the computer, and two for the output, which may be a printer and modem. A gender changer or converter (e.g., nine-pin) is sometimes needed to connect the selected cable. On the front, there is usually a blank space for labeling the connection.

A/B/C and A/B/C/D switchboxes are also common. A *crossover* switchbox is similar to a straight switchbox, but provides multiple input and output combinations, and usually has four or more ports on the back for attaching the input and output connectors.

Switchers are similar to switchboxes, and are frequently used in live broadcasts and video editing to select among various video sources (cameras, VCRs) and computer-generated signals. Video switchers typically use RCA and BNC connectors to accommodate standard video cables. See switcher.

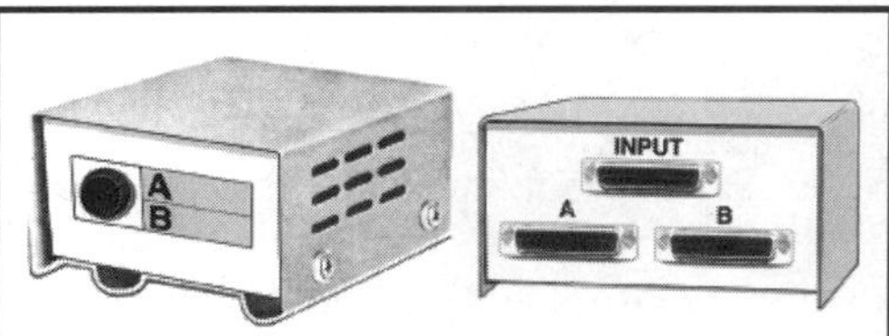

A/B switchbox front and back. Passive switchboxes are commonly used to interconnect computers with various peripherals. For example, a serial cable leading to a modem could be plugged into the input connection, and A and B could each be connected to a different computer to share a modem. Another configuration is to connect a computer to the input connection, and attach A to a modem and B to a printer, so that a computer with one serial port can alternately use two periperhals. Most common switchboxes have female 25-pin D connectors, as shown here.

A/D analog to digital.

A/D conversion Conversion of analog to digital signals, often for transmission over data networks or for sampling by computer applications such as speech or voice recognition software or music sequencing and editing software. The advantage of converting to digital format is that many types of processing can be applied to the information, including image or sound editing, sequencing, compression, encryption, error-correction, and more. Some simple examples include:

Microphones can be used as analog input devices to a computer to capture voice or music which can be digitally represented as sound 'samples.'

Analog videoconferencing cameras can capture still frame images of a caller and send them to the computer or video processing unit for transmission over phone lines or the Internet in digital form. They can then be displayed on a computer monitor at the receiving end. See audiographics.

Modems can receive analog phone signals and demodulate them to serial transmission signals to transfer data with file transfer protocols.

A/UX A 32-bit Unix operating system designed for the Apple Macintosh computers, derived from AT&T's UNIX (when capitalized, UNIX refers to the registered trademark of X/Open Company), BSD, with full POSIX compliance and System V Interface Definition (SVID) compliance. A/UX provides The X Windows System, sh, csh, and ksh.

A4 An international paper size standard, commonly used in Europe, corresponding to 210 mm *x* 297 mm (8.27" *x* 11.69"). Similar to this in North America is *letter sized* paper which is 8.5" *x* 11".

AA See Automated Attendant.

AAAC all aluminum alloy cable. See ACSR.

AAAI American Association for Artificial Intelligence. A nonprofit organization founded in 1979 to advance education in and scientific understanding of thought and intelligent behavior and their embodiment in machines. http://www.aaai.org/

AAAS American Association for the Advancement of Science. Descended from the Association of American Geologists and Naturalists, the AAAS was formed with a broader mission in 1848 to promote the development of science and engineering in the United States. http://www.aaas.org/

AABS See Automated Attendant Billing System.

AAC 1. abbreviated address calling. 2. Aeronautical Administrative Communications.

AACS See attitude and articulation control subsystem.

AAL ATM adaptation layer. See asynchronous transfer mode, and see appendix for details and diagrams.

AAMOF An abbreviation for "as a matter of fact" which is used in email and on online public forums. See AFAIK, IMHO.

AAP Applications Access Point.

AAPI Audio Applications Programming Interface.

AAPT American Association of Physics Teachers.

AARP See AppleTalk Address Resolution Protocol.

AAS authorized application specialist.

ab- A prefix commonly used with names of practical electrical units in the centimeter-gram-second (CGS) electromagnetic system, e.g., abampere.

AB roll, A/B roll In editing, a system configuration in which information coming from two separate components is alternately merged or selectively merged from various portions of the two inputs, sometimes with special effects, and recorded on another system, as in video or music editing. AB roll editing in video allows the assembly of cuts, fades, rolls, inserts, page turns, etc.

abacus An ancient computing device, derived from boards with sand that were used as accounting and counting aids. Native Americans used a type of abacus using grains as counters. One familiar form of Chinese abacus consists of beads resting in grooves or strung on wires within a frame. The beads are separated by a bar into upper and lower regions. In the version shown in this diagram, the values of the beads are assigned from botton to top, and right to left. In the first row each lower bead represents one, while in the upper row, each bead represents five. From there, each row of beads increases in value by a factor of ten compared to the previous row.

This common Chinese abacus is one of many arrangements of boards, beads, and frames. Variations are still used in commerce in many Asian and Eastern European regions.

ABAM A Western Electric (now Lucent Technologies) cable designation for 22 AWG, 110 ohm, individually shielded, twisted pair cable typically used in central office trunk line, circuit line, T1, and T1 to E1 channel service installations.

abampere, ab-ampere In the centimeter-gram-second (CGS) system, an absolute unit for current. Since the abampere is often too large for practical convenience, current is described instead in terms of amperes (one-tenth of an abampere). See ampere.

abandoned call See call abandons.

abandoned call cost An economic calculation intended to determine the amount of revenue lost. Abandoned call cost estimates are used by businesses whose customers order products or services through the telephone, or whose inquiries lead to sales later on. It's impossible to know how many of the calls would have generated revenue and how many would have been completed later, but business owners may benefit from rough estimates based on the number of abandoned calls times the percentage of anticipated sales resulting from those calls. See call abandons.

Abbe condenser A simple type of two-lens condenser invented by Ernst Abbe. It is used in photomicrography, where sufficient lighting is important. The condenser is located below the stage of a microscope so it can collect, direct, and spread light up onto the object being examined and recorded. It aids visibility in high magnification environments.

Abbe, Ernst (1840-1905) A German mathematician and physicist who worked at Zeiss Fabrications developing a number of optical theories, and inventing a variety of optical condensers and metering instruments.

abbreviated address calling In data network information routing, calling an address with fewer than the normal number of characters, usually from a table or file in which abbreviated address codes are stored. Similar to speed dialing or abbreviated dialing on phone networks.

abbreviated dialing AD. 1. A feature of a phone which allows a short dialing sequence to replace a longer one. The abbreviated sequence can be programmed in and associated with a longer number, then, when the shorter sequence is dialed, the system connects to the associated phone number. Also known as speed dialing. 2. A priority telephone service using special grade circuits, in which two or more subscribers can connect calls with fewer than usual dial tones.

ABC 1. arbitration bus controller. 2. See Atanasoff-Berry Computer. 3. automatic bass compensation. A circuit that increases the amplitude of bass notes to create more natural sound at low volumes. Used especially for playing back music recordings. 4. automatic bias control. See bias. 5. Automatic Bill Calling. A billing method for coin phone calls which is being superseded by calling card billing. 6. automatic brightness control. A circuit which senses ambient light levels and adjusts a display automatically in order to optimize the levels for the viewer.

ABCD bits In network systems, a method for signaling using robbed bits, which are then used to provide status information. The number of bits robbed depends upon the system. In Extended SuperFrame systems, four bits, designated ABCD, are utilized. See Extended SuperFrame, robbed bits.

abend *ab*normal *end*. Abnormal or premature termination of a task or process, one which cannot be handled by available error recovery mechanisms. In workstation level computers, abend problems with applications software are usually handled by the operating system so that the system itself does not crash, and there are usually mechanisms for killing individual processes that are locked or hung. System-level abend problems on well-tuned networks are actually relatively rare. Some, not all, of the microcomputer single-tasking systems, and less robust task-switching or multitasking systems, experience abend problems that may require a system reboot. (As a point of interest, this dictionary was created on a workstation that hasn't crashed in over three years of 24-hour a day operations.) See abort.

aberration 1. Deviation from expected shape, behavior, or path. 2. Failure of an image to coincide point-by-point with its original, as in a television image or facsimile.

ABIST autonomous built-in self test. The capability of a system to automatically run built-in diagnostic routines.

ablation 1. Removal of a part. 2. The process of removing parts, such as small holes, grooves, or pits in order to encode information on a medium. Many computer storage media are recorded by ablating thin layers of plastic or metal, e.g., optical media such as compact discs.

ABM See asynchronous balanced mode.

ABME asynchronous balanced mode extended.

abnormal propagation In broadcast transmissions, an undesirable effect of atmospheric or ionospheric changes that interferes with signal integrity.

abort 1. Stop prematurely, cut off. 2. To terminate the transmitting or receiving of a message in progress. 3. To stop a software program or process in progress, often abruptly. An abend may be one type of abort, but *abort* more often signifies a situation in which a process is cleanly or voluntarily terminated without compromising system operating functions. 4. To terminate user access through a network or during a login, usually due to detection of unauthorized access or tampering.

Above 890 decision A 1959 decision of the Federal Communications Commission (FCC) which allowed private construction and use of point-to-point microwave links. Thus, private companies, especially in remote locations, could utilize frequencies above 890 Mhertz for communications with oil rigs, power plants, gas pipelines, research stations, etc. The decision came about partly because of changes in technology, which made it less expensive and easier to use the higher frequency ranges for communications. This resulted in pressure to make these capabilities more widely available. Microwave Communications Inc. (MCI) was the first private commercial carrier service to take advantage of the Above 890 decision. See Telecommunications Act of 1996.

ABR 1. See available bit rate, cell rate. 2. autobaud rate. Early modems had to be individually matched to the same baud rate in order to communicate successfully with one another, but since the mid-1980s, when 1200 baud transmissions were common, most modems have incorporated autobaud capabilities in which the called modem and the calling modem negotiate a common speed and then commence with user communications. Autobaud capabilities have been a great boon to bulletin board systems (BBSs) which must accommodate users calling in on a variety of types of computers and modems.

abrasion resistance A measure of the ability of a material to resist surface wear and tear. This is expressed in various ways, depending upon the industry and the type of material.

abs *abbrev.* absolute value. See absolute value.

ABS See Alternate Billing Services.

abscissa Typically the horizontal axis or X-axis in a standardized coordinate system.

Absent Subscriber Service, Vacation Service A service offered by local telephone carriers that retains the absent subscriber's phone number at a reduced rate so the subscriber will get the number back later, and provides a standard recorded message to any people who call while the subscriber is away.

absolute address In computer programming, the actual address in which a unit of data is stored (in contrast to a pointer to its storage location). 2. The binary address which directly designates a storage location.

absolute altitude Altitude described relative to the surface of the Earth, as distinguished from altitude measured relative to sea level.

absolute coding Machine level coding, that which can be processed directly by a computer.

absolute delay The time interval between two synchronized transmission signals from the same or different sources.

absolute error 1. A means of expressing a deviation from a standard or expected value in terms of the same units as the units of the value. In statistical population distributions or other scatter distributions, this is a common way of indicating a deviation. 2. The absolute value, that is, the value without regard to sign, equal to the value of the error.

absolute gain In antennas, the gain in a given direction when compared against an isotropic reference antenna, typically expressed in decibels. If a direction for the antenna is not specified, then radiant energy in all directions is assumed and gain measured along a selected axis. See isotropic antenna.

absolute position Position on an agreed-upon coordinate system, e.g., a system with a point of origin defined as the center of the mass of the earth (geocentric).

absolute scale Kelvin scale.

absolute standard An assigned mass of one unit applied to a specified particle or object so that it can be used as a reference guideline.

absolute value A numerical notation and corresponding mathematical concept of the magnitude of a value without respect to its sign. Thus, the numeral -5 without respect to sign is written 5.

absolute zero The lowest point in an absolute temperature scale system, zero degrees Kelvin; the low point at which there is thought to be no molecular activity and thus no heat energy, which can also be expressed as -273.15 degrees centigrade or -459.67 degrees Fahrenheit. The Kelvin scale is named after William Thompson (Lord Kelvin).

absorption 1. The process by which particles penetrate and are subsumed by matter. 2. Penetration of a substance or wave into another substance. A sponge will absorb water and vegetation will absorb radio waves. 3. Dissipation, as of a wave, into another material as a result of its interaction with the other material. Sometimes this is desirable, as in sound-editing studios. See acoustics. 4. The process by which particles entering matter are reduced, or reduced in energy, as a result of interaction with that matter. 5. The reduction of energy that occurs as particles pass through or into another substance as a result of interaction with that substance. In radio wave frequencies, absorption tends to occur more readily at the highest frequencies, e.g., microwaves. Absorption can also be used as a means to add information to a signal. See absorption modulation.

absorption band The radiant energy of a range of electromagnetic waves or frequencies which are absorbed by a substance.

absorption current Current flowing into or out of a capacitor after its initial charge or discharge.

absorption fading Slow fading of transmission waves due to various absorption factors along the path. Complete fading or significant dissipation is known as absorption loss. Depending upon the transmission medium, degree of loss is sometimes expressed in terms of decibels over distance.

absorption loss The portion of a transmission that is lost due to interaction with another material through partial reflection or complete absorption into the material. This interaction may cause the conversion of energy into other forms, such as heat.

absorption modulation A means of modulating the amplitude of a wave, such as a radio carrier wave, using a variable-impedance device. See amplitude modulation.

absorption wavemeter An instrument for measuring frequency or wavelength and sometimes the amplitude of the harmonics of that frequency by absorbing energy from the circuit being tested. When absorption is at its maximum, the wavemeter is 'tuned' to the corresponding frequency of the circuit. This instrument is often used in conjunction with antenna systems.

Abstract Syntax Notation One ASN.1. A data definition language defined in 1988.

ABT Advanced Broadcast Television.

abuse numbers A database of phone numbers which are known to be inappropriate for outgoing calls (i.e., numbers not associated with typical business transactions). Some suppliers provide an option to track and highlight any calls to specified abuse numbers so they can be readily identified on billing statements.

ABX See Advanced Branch Exchange.

AC 1. See Authentication Center. 2. See alternating current.

AC bias In recording processes, a technique of adding a high frequency to aid in linearizing the record head.

AC power phone Most small residential phones operate off current from the phone line, but if the phone has extra features, such as electronic displays and speakerphones, or if it is a multiline business phone system, then dedicated alternating current (AC) from a wall socket is generally used. Battery systems also exist, typically for backup power in the event of failure of the AC source. Private branch phone systems can consume a significant amount of power if many calls are being processed and may require power from both the phone switching cabinet (through the line) and from an AC power source serving the phone console.

AC ripple Undesired modulation in an alternating current (AC) circuit. Filtering may be employed to reduce or eliminate ripple.

AC to DC converter A device for converting alternating current (AC) to direct current (DC). The current that comes from most wall sockets is AC current, but many devices including answering machines, feature phones, modems, etc. require DC current and will include a converter attached to the power cord or incorporated into the device.

It is unwise to swap around these power converters, as they have widely varying specifications. Most will list the voltage and amperage on the converter, and some will list the corresponding voltage and amperage on the device itself (usually on the underside). Installation of incorrect converter cords can damage sensitive electronic devices. If the device is NOT labeled, it is prudent to mark it as soon as you take it out of the box, with a felt pen or label, so that if the converter and the device get separated from one another, you can correctly match them up again.

AC/WPBX Advanced Cordless/Wireless Private Branch Exchange.

ACA 1. See American Communications Association. 2. Australian Communications Authority. http://www.austel.gov.au/ 3. Automatic Circuit Assurance.

ACADEMNET A Russian academic network.

ACAR aluminum conductor alloy-reinforced. See ACSR.

ACARD 1. Advisory Council for Applied Research and Development. U.K. advisory organization superseded in 1987 by ACOST. See ACOST. 2. Acquisition Card Program.

ACB 1. Annoyance Call Bureau. 2. Architecture Control Board. 3. ATM Cell Bus. 4. automatic callback.

Accademia del Cimento A group consisting of Florentine experimenters, founded in 1657 by the students of Galileo, dedicated to carrying on scientific study. Based partly on the writings of W. Gilbert, the participants carried out important early experiments in magnetism and contributed the use of the pendulum as a tool for studying the attractive properties of materials. See barometer.

Accelar routing switch Commercial switcher/router device from Bay Networks which makes switching decisions based on Internet Protocol (IP) addresses embedded in the local area network (LAN) switch hardware, without proprietary protocols or appended bits. See IP switching.

accelerated aging, accelerated life test A design and diagnostic technique that involves subjecting a process, material, or mechanism to short-term conditions that simulate long-term use and environmental influences. Accelerated conditions simulate factors such as weather, movement, mechanical stress, chemical exposure, use, etc.

Accelerated Graphics Port AGP. A dedicated graphics video slot incorporated into motherboards or PC interface cards which supports Intel AGP product specifications. AGP systems can access system memory in conjunction with a dedicated graphics frame buffer. AGP supports several modes at different clock speeds. This standard is superseding older video formats like PCI buses which may be incompatible with devices increasing in popularity, such as DVD.

accelerating electrode A device in an electron tube, such as a cathode ray tube, which increases the velocity of the electron beam.

acceleration The expression of a change in velocity over time. Acceleration is commonly expressed in meters per second per second. An international standard value for acceleration due to gravity on a free-falling object in a vacuum has been established as 9.807 meters per second per second.

acceleration voltage In a cathode ray tube, the accelerating potential which controls the average velocity of electrons that are directed toward the imaging surface from an electron gun. The voltages are tuned in conjunction with the magnetic coil through which the electrons pass to create the *sweep* and image *frames* that help build the picture on the tube.

accelerator A system, process, chemical, organic substance, or device which acts on something to speed it up. Accelerators are used in many areas including, but not limited to: studies of elementary particles, chemical reactions, transmission circuits, and computer systems.

accelerator board, accelerator card A peripheral card designed to fit into a computer slot that increases the speed of the system, usually by increasing the CPU speed, or by taking over some of the more demanding of the CPU's functions, such as graphics manipulations.

accentuation 1. Intensification, emphasis. 2. In transmissions, the emphasis of a particular channel or frequency, often to the exclusion of others. Accentuation is found in the high frequencies in FM transmitters.

Acceptable Use Policy AUP. A license or purchase agreement which sets out limitations, restrictions, and acceptable uses which are binding to the purchaser or receiver. For example, a number of freely distributed network software programs stipulate that they may not be used or sold for commercial purposes.

acceptance period A period, usually of a few weeks, during which a product or service is evaluated by the receiver as to its conformance to the agreed-upon specifications. This is more commonly a stipulation of custom installations, than of off-the-shelf products. Acceptance differs from a warranty in that it applies mainly to initial configuration during the ramp-up or installation period, whereas a warranty may cover other factors, and last several months or years after purchase and installation are complete.

acceptance test A test, which usually follows installation, that demonstrates that the product or services purchased conform to the agreed-upon specifications. An acceptance test may be contractually required by the purchaser before making final payments on the purchase.

access 1. *n.* The point through which a circuit or communications device is entered, or the point at which the communications process is entered and initiated. 2. *v.* To gain entry into a circuit or communications device. Phones are generally accessed by dialing a number, although an *access code* may be required on a secure system. Dialing '9' first to obtain an outside line is a common access procedure. Account codes are sometimes used to assign billing to specific departments or individuals. Access codes may be used by installation or maintenance technicians to initiate services or procedures not available to the subscriber. Secured computer systems are accessed by *logging in* with or without a password. See access code.

access arm The positioning mechanism which supports a read/write head for reading from or writing to magnetic or optical storage media. On a computer hard drive, the access arm moves across the disk and positions the head directly to within thousandths of an inch of the area of magnetic particles to be read (or written). See seek time.

access carrier An interconnect agreement through which a carrier can gain access to the services and network facilities of another carrier.

access charge 1. The charge made for access to a computer system or network. An access charge may be assessed on a periodic basis or per time or volume of use. Internet Service Providers typically charge flat monthly rates, although some will assess extra charges for storage, peak-time connects, or access to chat areas, or special online services. 2. The Modified Final Judgment (MFJ) which broke up the Bell system included the rationale and stipulation that users should be able to choose a long-distance carrier, thus changing the way in which long-distance access charges were structured. Compensatory restructuring resulted in two categories of access charges: Customer Access Line Charges (CALCs), and Carrier Access Charges (CACs). The first applies to local phone loops and varies according to the subscriber (residential or business), and the characteristics of the service. The latter applies to service providers connecting to the local exchange circuits and varies according to factors such as distance. Adjustments and modifications in order to implement the many changes have subsequently occurred. See Telecommunications Act of 1996.

access code One or more characters which must be entered in order to obtain use authorization to a system such as a phone or network. Access codes are generally used for security, monitoring, and billing purposes. They can also be used by technicians to set up a system for use with specified features and, more recently, to program a telephone system. Some typical telephone access code implementations include 1. dialing codes to access an outside line, or to dial a long-distance number (dialing '9' is common), 2. dialing an access code to bill the call to a particular line or department, 3. dialing a code to obtain authorized use on a privileged system.

access control 1. A physical or virtual control point, gateway, or other filter or security system which selectively allows data to pass through according to general or specific parameters which may include priority level, data characteristics, sender, receiver, etc. 2. The policies, procedures, and system configurations which control security or utilization of resources. Access control operates on many levels, including building access, system access, applications access, network access, device access, and computer operations access. See access code.

access control field Information in the header of a synchronous multimegabit data service (SMDS) cell which provides access to a shared bus, which in turn provides access to the SMDS network.

access control list ACL. A list, table, or database which provides a reference for various levels of security within a system. It can be as simple as a piece of paper with names in the hand of a doorway security guard, or as sophisticated as a tiered database of levels of security for different people and processes on a computer network. On bulletin board systems, many access control lists consist of a series of flags for each user which can be toggled individually by the sysop to control user access to services such as chat, email, doors, downloads, etc.

access control method A system for controlling access to systems, processes, or devices on a network. A variety of general guidelines, and specifically defined systems for particular types of networks and protocols have been developed. Access control can be set up "by user," "by workstation," "by application," "by file," "by network," or a combination of these, which may be hierarchical. Examples of particular types of access control on specific types of networks include *carrier sense multiple access* (CSMA) on Ethernet systems, and token passing schemes on IBM Token-Ring networks. See Media Access Control.

access coupler A connecting device used between physical network segments, such as fiber optic cable legs, to allow signals to be passed on to the next leg. Access couplers are sometimes used in conjunction with relays and amplifiers, depending upon the type of signal and the distance.

access delay In a packet-switched network, this provides a performance measure for polling systems, measuring from the time of arrival of a data packet to the time it is retransmitted.

access group A group of accounts or individuals who have specific defined levels and types of privileges within a system which may be different from individual privileges and from other groups. For example, on a private branch phone system, a group of managers may be designated as having access to long-distance lines, or outside lines, whereas a new employee may be assigned to a group with limited privileges until an evaluation period has passed. On a computer network, an access group may have certain read and write file privileges which differ from individual privileges and the privileges of other groups. Thus, they may be permitted to run only certain applications, look at certain directories, etc. according to settings established by the system administrator.

access line 1. The physical link between the subscriber box and the local telephone switching center. From the subscriber box to the telephones is considered inside wiring and may be installed by the subscriber or, for a fee, by the phone company. See local loop. 2. In BBSs, the line through which the caller accesses the BBS modem. There may be multiple lines, sometimes with different baud rate capabilities. Historically, BBSs have been accessed through phone lines, but more and more, BBSs are interfacing with the Internet to provide online access through telnet. 3. In frame relay systems, a communications circuit that connects a frame relay device to a frame relay switch.

access mechanism A device for moving and positioning an access arm, usually on random access read and/or write media.

access method Logical guidelines established by International Business Machines (IBM) in the 1960s for input and output access to computing resources, particularly those which are shared, as on local area networks (LANs). By consolidation of instructional sequences in common procedures, functions, and subroutines, the overall structure can be simplified.

access software provider This is defined in the Telecommunications Act of 1996, and published by the Federal Communications Commission (FCC), under SUBTITLE A—Telecommunications Services as:

"... a provider of software (including client or server software), or enabling tools that do any one or more of the following:

(A) filter, screen, allow, or disallow content;

(B) pick, choose, analyze, or digest content; or

(C) transmit, receive, display, forward, cache, search, subset, organize, reorganize, or translate content."

access tandem switches Specific types of switches which are used to connect End Offices to Interexchange Carrier (IXC) switches, or to interconnect central office (CO) switches.

access time 1. The interval between a signal or instruction to access information or a device, and the time it takes to successfully retrieve that information, or interact with the device. Depending upon the system, the access time may or may not include the time it takes to *display* the requested information to the user. For example, in a database query on a computer system, the access time may be two seconds to search and retrieve a long list of names and addresses, but it may take twenty additional seconds to fully display all the listings, and the *access time* may not include the display time, or may include the display time for the initial information, but not the time during which the software may be building additional viewable information below a scrolling window. Access time is described in terms of units appropriate to the average time and device involved. For example, access time for a process may be described in CPU *clock cycles*, for data access on storage media, by *seek time* in *milliseconds,* or *fetching* in *nanoseconds*. 2. In magnetic storage devices, the interval during which the access mechanism, once information is requested, moves across the medium to the desired location and successfully reads the data.

access unit AU. 1. In Token-Ring networks, a wiring concentrator that connects the end stations. The AU provides an interface between the Token-Ring router interface and the end stations. Also known as Media Access Unit (MAU). 2. In many X.400-based commercial software applications, the AU works in conjunction with mail servers to provide synchronization between post offices and other services, such as directories, address books, etc.

access.bus A serial communications bus topology protocol developed jointly in the mid-1980s by Digital Equipment Corporation and Philips Semiconductors, for connecting peripheral devices such as mice, keyboards, card readers, scanners, etc. to computers through a four-wire serial bus. See Universal Serial Bus.

accounting server A software application,

sometimes operating from a dedicated, secured computer, which monitors network usage, stores the information, and may assess charges for usage based on CPU time, real time, time of day, department, or some other measure appropriate to the type of use.

ACCS Automatic Calling Card Service.

accuracy 1. Degree of conformity to a stated or observed value considered to be optimal or correct. See calibration. 2. Precision. 3. Degree of freedom from error.

AC/DC, AC-DC An electrical appliance that can operate on alternating current (AC) or direct current (DC).

ACD See Automatic Call Distribution.

acetate A cellulose acetate chemical compound that was once commonly used as a coating for storage media such as audio recordings. This application has been superseded by various types of magnetic storage and compact discs.

ACF Advanced Communication Function.

achromatic 1. Uncolored; unmodulated; neutral; black and white; grayscale. 2. In the visual spectrum, lightwaves which are not dispersed or singled out according to a particular wavelength.

ACIA asynchronous communications interface adapter. A data formatting device that translates signals between the computer and a peripheral such as a modem.

ACK See acknowledge.

acknowledge, acknowledgment ACK. A message or signal from the receiver to the sender confirming receipt of data, or accurate receipt of data. In handshaking, ACK sometimes also signifies that the receiver is ready for further data. ACK and NACK are commonly used on bidirectional communications systems, which can only transmit in one direction at a time. See negative acknowledge.

ACL 1. See access control list. 2. Applications Connectivity Link. 3. Association for Computational Linguistics.

aclastic Having the property of not reflecting light.

ACM 1. See Association for Computing Machinery. 2. Automatic Call Manager. An administrative and operations system that handles inbound and outbound calls integrated with a database. Telemarketing, teleresearch, and collection agencies make use of these types of systems. 3. Address Complete Message. A call setup message in ATM networking which is returned to indicate that the address signals required for routing the call have been received by the called party. The ACM is sometimes sent in conjunction with other routing messages.

ACO 1. Additional Call Offering. 2. alarm cutoff. A switch which suppresses an audible alarm, while not affecting a corresponding visual alarm.

ACOnet Austrian Academic Computer Network. An ATM-based Austrian research network funded by the Austrian Ministry of Science, Transport, and Art. ACOnet interconnects about a dozen universities and provides international links to other countries through EBS-Vienna. http://www.aconet.at/

acorn tube A very small vacuum tube named for its squat, rounded shape, which has electrodes leading directly through the glass on several sides. It was developed for use at extremely high frequencies. While most vacuum tubes have been superseded by transistors and other modern electronics, there are still some high frequency applications where vacuum tubes are practical.

ACOST Advisory Council on Science and Technology. A U.K. organization which superseded ACARD in 1987, ACOST is a government advisory and coordinating body on policy and research.

acoustic Relating to sound phenomena, the science of sound, and biological structures and nonbiological apparatus for generating, conveying, controlling, or apprehending sound. In music, an acoustic instrument is one which does not require electronic enhancement or modification, but rather depends upon its physical structure and the surrounding medium to create and convey the desired sounds at the desired intensity.

acoustic coupler Any device that is designed to interface with an audio sending and/or receiving circuit to provide amplification or conversion between analog and digital audio signals. The coupler is usually designed to exclude extraneous noise that could interfere with a signal, and may be a self-contained unit or a peripheral interfaced with another system.

Acoustic couplers that resemble large suction cups (sometimes called suction cup modems) were incorporated into early modems to provide a way to interface telephone handsets with computers. The coupler was designed so that the outbound modem signal played into the mouthpiece microphone and the inbound signal played from the earpiece speaker into the modem. See acoustic modem.

acoustic feedback See feedback.

acoustic model In software applications, a means to apprehend and interpret sound input, such as speech, by breaking it down into smaller units and then using those together to build a 'picture' or interpretation of input combined from these units into larger words and speech patterns.

Early attempts at speech recognition were hit-and-miss, and very person-specific, but new programs can transcribe speech into text with a useful degree of accuracy up to about 70 words per minute. Computerized speech used to be characterized by very flat, mechanized sounds, but with faster processors and better methods and sound samples, natural sounding voices can be generated. Many automated phone voice applications now use speech generation for messages, queries, and instructions. See phonemes, speech recognition.

acoustic modem A modulating/demodulating computer peripheral which converts the digital signals created by a computer into audible tones that can be coupled with the transmitting end of a telephone handset or other audio transmissions device so they can be sent through an analog phone line. The device then converts the audible tones generated by the other end of the transmission back into digital signals for the computer to interpret. The modem is usually attached to the computer by means of an RS-232 (EIA-232) interface, although some acoustic modems designed for the early Apple computers were connected through the joystick port.

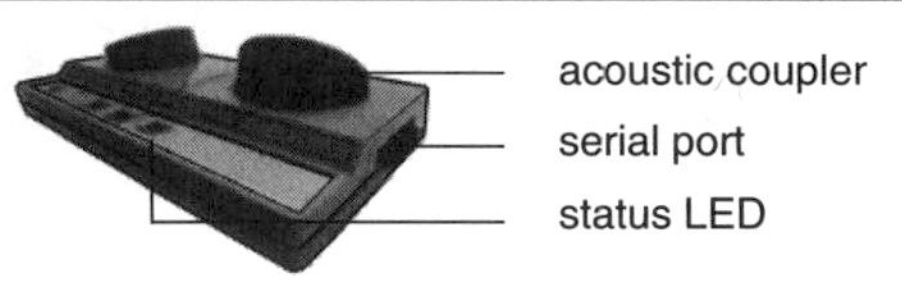

Acoustic couplers were designed so that the handset of a telephone could rest in the rubber sound shields. Since the sizes and shapes of handsets began to vary at about the same time that acoustic modems were distributed, the physical connection was often less than satisfactory and stray room noises could interfere with transmission. This type of coupler has been superseded by direct connect modems and is now useful only in specialized situations (as when using a payphone).

Acoustic 300 baud modems were prevalent on personal computer systems in the late 1970s. These were gradually superseded by direct connect modems in the 1980s. Acoustic modems have many limitations. They tend to be bulky, as they need sufficient shielding around the transmitting and receiving electronics to prevent the tones from crossing over and interfering with one another. They are subject to interference from external noises. They only work well with old-style phone handsets. The newer, flatter ones do not get sufficient shielding or contact with the couplers to transmit clean tones, and they do not generally employ any sophisticated data compression capabilities, resulting in slow transmission speeds. See acoustic coupler, direct connect modem.

acoustic telegraph Messages conveyed by sound, such as bells ringing specific tones or sequences (still used in many European churches), drumbeats, horns, or shouts that are passed from one person to the next (still used on large sailing vessels, railroad lines, or other areas where other means of distance communication are not available).

In the early days of electrical telegraphs, a number of inventors were seeking ways to use tones to convey more information over a single line than was possible with the simple on/off system that was gaining widespread use. Experiments in trying to send tones led to the invention of the telephone, and so the technology leap-frogged over the acoustic telegraph. Basic telegraphic systems were used for a long time concurrently with the evolution of telephones.

acoustic velocity (Symbol - a) The speed of sound (technically, *velocity* is the rate of motion in a direction).

acoustical Doppler effect A characteristic of a sound when there is motion that causes the object emitting the sound to be moving further away or closer in relation to the listener. The perceived pitch of the sound changes from the pitch that is being emitted. As an example, imagine standing by a railroad track when a train goes by, blowing its whistle. Although the effect is not dependent on the pitch of the whistle being constant, imagine that the train whistle is a constant note at one pitch. As the train moves toward the listener, the perceived pitch of the whistle rises, and as it passes and moves away from the listener, it falls. This effect occurs due to the characteristics of the sound waves traveling through air relative to the rate of change and the distance of the wave from the listener. It can readily be heard on musical instruments that short strings vibrate at a higher pitch than long strings of the same thickness and materials. Now imagine ripples in a pond from a stone thrown in the center. The ripples close to the source are higher, those farther from the stone are shallower, that is, the amplitude of the wave diminishes as it moves away from the source. Similarly, as the train moves through the air, and is oriented differently with respect to the listener, the characteristics of the sound waves change, resulting in different pitches reaching the ear.

Acoustical Society of America ASA. A scientific society founded in 1928 after an initial meeting at the Bell Telephone Laboratories in New York. It began publication of its professional journal in 1929. ASA merged with three other societies in 1931 to form the American Institute of Physics. ASA has been involved in research, development, promotion, and standardization efforts in the field of acoustics. http://asa.aip.org/

acoustics 1. The art and science of sound production, transmission, and reception. 2. The sound-carrying capacity, in terms of quality, fidelity, and loudness, of an environment such as a concert hall or recording studio. See anechoic.

acoustics, architectural The art and science of the propagation of sound in enclosed structures (concert halls, museums, classrooms, auditoriums, etc.), or circumscribed environments (amphitheaters, stadiums, landscaped areas such as parks, etc.). Reverberant sound fields are created, studied, and adjusted to suit the needs of the structure. Concert halls and theaters are designed to carry the fullness and dynamic ranges of music or voice to as many listeners as possible within the building, while still minimizing noise that might disturb activities outside the building.

acoustics, engineering The art and science of sound control in electronic structures. This includes amplification, propagation, dampening, and the harnessing of sound to carry information, as in data and broadcast transmissions.

acoustics, musical The art and science of sound control and propagation through acoustical and electronic instruments (many electronic sounds are created by digitizing the sounds of acoustical instruments). This involves the careful study of sound in relation to materials, shapes, resins (which help wood to retain its resilience and vibratory qualities), sound transfer through sound boards and bridges, and much more. Much of the history of musical acoustics is based on the subjective observations of individuals with good craftsmanship and good musical ears, but in recent years scientific instruments have given us additional tools with which to craft instruments and propagate musical sounds. See patch, sampling.

acousto-optic modulation A technique which can be used quite effectively for color control, dimming, and blanking in laser light beams. The beam is shone through an acousto-optic crystal. The modulation is applied with electrical impulses to the crystal to affect the intensity of the beam. Three beams can be used, red, green, and blue, as in a cathode ray tube, to provide color modulation. This is known as polychromatic acousto-optic modulation. In other applications of acousto-optics, they can be used to tune filters. See modulation.

ACP See activity concentration point.

acquisition 1. The gathering, receipt, and possession of data. 2. The process of orienting toward and acquiring data, that is, seeking a source; setting up the necessary protocols; aiming an aerial, scanning network inputs, or broadcast frequencies; and receiving the transmission.

acquisition and tracking A data detection or receiving system such as radar, which seeks out a signal, locks in on it, and orients toward the source of the signal while receiving.

acquisition time The time required to seek out and lock on to the source of the desired signal. Commonly used in microwave transmissions such as radar and satellite communications.

ACR 1. allowed cell rate. In ATM, an available bit rate (ABR) service parameter which describes the current allowable sending rate in cells per second. See cell rate. 2. attenuation to crosstalk ratio.

Acrobat An Adobe Systems commercial page layout software application used to create documents containing text and graphics. This format is popularly used to distribute documents on the Web. The software is available as a reader/writer for viewing, creating, and editing Acrobat format files, and as a reader (viewer), downloadable from many FTP sites. A number of word processing and desktop publishing programs will export Acrobat format files and the resulting files may be several times smaller than the original. Acrobat files are characterized by a *.pdf* file name extension.

acronym A word that is formed by taking the first letter or letters from each of a number of successive words in a phrase or compound term. Examples include scuba (self-contained underwater breathing apparatus), radar (radio detecting and ranging), and BASIC (Beginner's All Purpose Instruction Code).

ACS 1. See Advanced Communication System. 2. automatic call sequencer. A simple form of automated phone call handler which hands off calls to available agents.

ACSE See Association Control Service Element.

ACSL Advanced Continuous Simulation Language.

ACSR aluminum conductor steel-reinforced. Although aluminum is light and a good conductor, aluminum cables with steel cores tend to be bulkier and heavier than copper.

ACT 1. See Applied Computer Telephony. 2. See Authorization Code Table.

ACTA See America's Carriers Telecommunications Association.

ACTAS See Alliance of Computer-Based Telephony Application Suppliers.

actinism A property of radiant energy in the X-ray, ultraviolet, and visible parts of the spectrum to promote chemical changes.

ACTIUS See Association of Computer Telephone Integration Users and Suppliers.

activation fee, setup fee In many communications services, there is an activation or 'setup' fee associated with starting a new account. This fee covers the service provider's administrative costs of installing the account and providing the new user with operating instructions, passwords, etc. Sometimes providers will waive activation fees in order to attract new subscribers.

active communications satellite A communications satellite which employs transponders (a type of repeater) or other means of amplifying and forwarding (relaying) a signal, usually with the frequencies shifted so the uplink and downlink transmissions do not interfere with one another. Unlike the larger passive satellites launched in the 1960s, newer active satellites can amplify a signal without the extra bulk needed in earlier systems. Virtually all current satellites are active.

active jamming The deliberate interposition of signals intended to disrupt communications such as radio or radar transmissions.

active line A communications channel that is currently being used. While no human communication may be taking place, if it is a data line, there may nevertheless be meaningful activity on the line, such as computer processes interacting with one another.

active lines In a television image, those lines which are visible to the viewer at any one time. Since a frame consists of many sweeps of the beam, only some of the possible lines may be seen by the viewer at any one time, but because the transition is so fast, the image is perceived as continuous. Those lines which are not active are *blanked*. See blanking, frame, scan line.

active matrix display Usually a liquid crystal display (LCD), active matrix is a means of brightening an electronic display by adding transistors to individual elements to maintain the image between successful scans or refreshes of the screen. Thus, the screen appears to refresh more quickly and gives a crisper, more contrasting appearance that aids in legibility. Color laptops frequently incorporate this technology and active matrix screens are gradually replacing passive matrix screens.

ActiveX Descended from Microsoft's Object Linking and Embedding (OLE), but intended to run over the Internet and to compete with Sun's Java, ActiveX provides a means to utilize animation, sound, and interactive elements in Web documents. ActiveX components are somewhat similar to browser plugins, or Java applets. See ActiveX Controls.

ActiveX Controls Microsoft ActiveX controls are interactive objects created individually by developers which can be embedded in various Web-related applications. ActiveX controls can be programmed in a variety of languages, including Visual BASIC, Java, or C++. They can then be readily shared with other programmers. A number of commercial vendors of authoring and page layout display systems have incorporated ActiveX Controls into their software.

activity concentration point ACP. A place in which there is a high traffic load, that is a focal point for higher activity than is ordinarily found in other locations in the system.

activity reports Automated usage logs that are generated by computing devices and usually accessible through a file or printout. Activity reports can provide information about times and types of use, errors, and sometimes transmitter/recipient information. Activity reports are commonly available on facsimile machines, high-end printers, and some electronic photocopiers. There are disadvantages to the glut of information and statistics that can quickly and easily be generated by electronics. There aren't sufficient hours in the day to evaluate all of it, nor storage space to keep hardcopy versions.

Some corporations now require their employees to wear sensors which monitor and record all their movements and activities within the premises. There are individuals understandably concerned about this excess detailed electronic monitoring of employee actions, and its implications for privacy and abuse. They are further concerned about what happens to these reports, and who can see them once the employee has left the company, and whether employees have opportunities to assess their accuracy and interpretation.

ACTRIS A Swiss joint telecommunications research project.

ACTS 1. Advanced Communications Technologies and Services. 2. Advanced Communications Technology Satellite. 3. See Association of Competitive Telecommunications Suppliers. 4. Automatic Coin Telephone Service. An automated system for handling payphone traffic, it directs the user on how much money to insert, handles calling card calls, provides diagnostic and tuning information to technicians, etc. In areas without ACTS service, calls are handled by TSPS operators.

actuator A mechanical or electromechanical positioning mechanism used to aim an antenna so it can remotely or automatically scan the arc of a satellite.

ACUTA The Association of College and University Telecommunications Administrators. http://www.acuta.org/homepage.html

AD See administrative domain.

ADA 1. A high level, structured, data-typed programming language, somewhat like an extended Pascal, developed by and mandated within the Department of Defense, but not popular outside of this circle. It has been criticized by some programmers as being cumbersome and difficult to use. The language is named after Ada Lovelace, the technically astute daughter of Lord Byron. There have since been variations on ADA, including ADA++, which in turn has been superseded by ADA 95. See Lovelace, Ada. 2. Average Delay to Abandon. The average length of call duration for a caller held in a queue who hangs up before being connected with the callee.

ADACC Automatic Directory Assistance Call Completion.

Adams, Scott Founder of Adventure International in 1978, Adams created a microcomputer games empire by developing and marketing a series of text adventure games that were wildly popular in the early 1980s. Over the last twenty years, the availability of entertainment products like the Scott Adams games has strongly influenced users to purchase computers and network cards for multiplayer games.

adapter A device to connect one type of component, system, or connector to other components, systems, or connectors to provide physical and electronic compatibility on each end of the connection. An adapter is used when the two connections do not naturally couple with one another. Related to adapters are connectors, which are most often small passive devices, simply passing information or current through, while adaptors tend to be combined with active, signal-processing or enhancing components, or with gender changers, extenders, or splitters.

adapter card See peripheral card.

adaptive antenna array A series of antennas grouped and arranged so the combination of antennas provides enhanced reception or transmission over individual antennas. An antenna array can be configured to monitor signals, or signal conditions, or to use input from other sources, such as computers, and to adapt to them as appropriate. For example, in a directional antenna array, if the signal shifts due to movement on the part of the sending antenna, the array may be able to move or swivel to optimize the signal level (as in elliptical satellite orbit communications).

adaptive communication A communications system which incorporates intelligence and feedback mechanisms to optimize signal or data transfer. In telephony, a cordless phone may automatically switch channels to find a better signal if the current one deteriorates. In the telephone switching system, a phone call may be routed through another trunk if congestion is detected. In computer network systems, the system may reroute packets if one of the hops in a journey changes or becomes unavailable.

adaptive differential pulse code modulation ADPCM. An ITU-T standard for voice digitization and compression in which sample rate speeds are related to the variation in the samples, thus using fewer bits than pulse code modulation (PCM), which is commonly used in digital voice coding, if the sample speeds are slow. An analog voice can be carried on an up-to-32 Kbps channel. This can be used over digital networks such as frame relay systems.

adaptive routing A system of routing in networks that utilizes intelligence in addition to information in routing tables, to establish best routes, fastest routes, or alternate routes in the case of obstructions in the usual paths. See hop-by-hop routing.

ADAS See Automated Directory Assistance Service.

ADB Apple Desktop bus.

ADC, A/DC 1. analog-to-digital converter. 2. automated data collection, automatic data collection.

ADCA Automatic Data Collection Association.

ADCCP Advanced Data Communication Control Procedures. A bit-oriented, ANSI-standard communications protocol related to High Level Data Link Control (HDLC).

ADCIS Association for the Development of Computer-based Instruction.

Adcock antenna A transmitting/receiving antenna with two or more vertical conductors arranged so that the pickup is minimized in the horizontal wires. Adcock antennas can be arranged in arrays to provide directional transmitting/receiving; one such array system resembles the configuration of the number five on a throwing die.

ADCU Association of Data Communications Users.

add-on 1. More commonly known as three-way calling, or add-on conference, a telephone subscriber feature which permits the connection of a third phone into an ongoing conversation. This is usually accomplished by putting the conversation on hold, calling the third party, and returning to the initial call with the third party linked into the call. 2. See applications processors, peripheral device.

address A locator, usually in the form of a number, of a position in memory or other storage medium, such as a hard drive or floppy diskette. A telephone number is a unique address on a phone system, used in establishing a connection. An email address is a unique identifier used in the transmission, receipt, and storage of electronic messages over a network. There are directories on the Web which store the email addresses of specific individuals or companies on the Internet, or which can retrieve a name and address, given a specific email address. The individuals whose addresses are listed are not necessarily aware of the fact. See address, MAC; ego surfing; electronic mail.

address, Internet An Internet address, or Internet Protocol (IP) number, is a unique host name identifier on the Internet. IP addresses can be expressed as numbers, *255.0.0.0*, or as a full DNS name, *www.crcpress.com*. A registration process is required to obtain a unique address on the Internet. See Domain Name Service, InterNIC.

address, MAC A Media Access Control (MAC) address is a device address on a network. See MAC address, Media Access Control.

address filtering Decision-making on a network as to which data packets will be permitted to continue. For example, a filter evaluates the source and destination media access control (MAC) address and compares this against any specific restrictions or instructions that have been set up for the system. On a general level, address filtering can be used to keep out messages from unwanted sources, such as bulk commercial mail senders, and to reject messages to local destinations which may no longer exist, or which may be restricted. See firewall.

address resolution AR. On the Internet and local area networks (LANs) using ATM, the conversion of an Internet Protocol (IP) address or local address into its corresponding geographical/physical address. This may be done in stages, through a discovery process, with the layer address being sought first, and other parts of the address, such as a media access control (MAC) address, being resolved at a more local level. This hierarchical approach can streamline the amount of information that needs to be processed and carried initially, and provides the flexibility to reorganize machines, switches, and routers at the local network level.

Address resolution is done by broadcasting from the sender to a number of nodes at the general destination and then responding to a specific destination, once information has been sent back from the appropriate end station to show where it is. See address, Address Resolution Protocol, MAC address, Media Access Control.

Address Resolution Protocol ARP. A protocol used to systematically, dynamically discover the low level physical network system which corresponds to an Internet Protocol (IP) address for a given host. ARP is used over physical networks that can handle broadcast packets (not all networks have a broadcast layer) to all the hosts, or the relevant hosts, on the system. By broadcasting to a general destination and then evaluating the responses by the local hosts, the specific address can be discovered and resolved without all the information about all possible destinations being stored at the originating system. See address, address resolution, MAC address, RFC 826.

address space 1. There are a number of definitions for address space, that is data storage locations, as it relates to computer memory. Computer operating systems handle addressing in various ways. For example, some OSs cannot put a large block of information across noncontiguous segments of available memory. In these systems, if the system has been running for some time, with many processes writing to memory, they may 'bog down' due to memory fragmentation. Others can map the memory and use the noncontiguous spaces provided they are not too small or too numerous. Systems also vary in their ability to 'clean up' memory once an application has been closed down, or a process completed, and may not release the memory to other programs or processes. On these systems, it may be necessary to reboot at intervals to clean up the system. Philosophies differ. Some feel the OS should clean up the system, others feel the programmer responsible for the application should clean up when the application is terminated. The second, cooperative system may not work as well in environments where software is unnaturally terminated (aborted rather than closed down normally). Address space is also related to hardware configuration. Some systems can only address memory in chunks of limited sizes, up to a fixed amount. Others can dynamically map the memory space up to a relatively large sizes (e.g., 256 Mbytes of RAM). Address space on specialized devices, such as digital cameras, may be limited to the size of a memory storage card or internal components. 2. The total amount of available memory that may be directly used, which may be broken up by other data, or across memory segments. 3. A single contiguous segment of memory which may be directly used.

addressable point On a computer display, any point that can be directly written or read by the system. Unknown to many consumers, many of the older microcomputer systems could not directly read or write every point which is displayed. Some could address only odd or even points in the horizontal or vertical directions, limiting the power and flexibility of graphics imaging and painting programs. For example, some systems could only draw lines (and thus the edges of windows and dialog boxes) on every other line, so programmers could not precisely position the boxes, and had to make the widths divisible by two (or one side would disappear). Systems with these limitations are no longer common.

addressee The intended recipient of a written message or data communication. See email.

addressing In computer programming and operations, a means of keeping track of stored information so it can be accessed in the future as needed.

ADF 1. See automatic direction finder. 2. automatic document feeder. A built-in or optional device on a printer, photocopy machine, facsimile machine, or scanner that holds a sheaf of paper, usually unattached single sheets, and feeds these pages individually through the machine. Some machines have a series of paper trays for different sizes or types of paper, and can cycle through the trays as needed, or automatically select the paper size.

ADIO Abbreviation for analog/digital input/output.

adjacent Near; next to; directly before or after; beside. Having a shared border, contiguous with. If something is adjacent, then no other device or process of the same kind is between it and that to which it is adjacent. For physical devices, the adjacent entities may or may not be physically touching or connected by cables or other means.

adjacent channel interference Due to demand, broadcast spectrums are subdivided into narrow bands in order to accommodate many channels. When broadcast channels are adjacent, the signal from one may interfere with the ones close by. Most people have experienced this type of interference in AM car radios; as they move farther from the signal of the current selected station, adjacent stations (or stronger stations) may be heard over the desired station. For this reason, some of the better radios are equipped with adjacent channel selectivity circuitry which rejects the transmissions of adjacent channels to provide cleaner reception.

adjunct 1. Something which is additional to, or joined to, something else, but which is not essentially part of it. 2. Assistant, aide, associate. 3. A peripheral device which enhances a system, without being essential to its basic operation, such as a computer microphone, joystick (gamers would argue that this is essential), modem, telephone headset, etc.

Adjunct System Application Interface ASAI. A set of AT&T technical specifications for the controlling of private branch exchange (PBX) systems by computers.

ADK application-definable keys.

ADM 1. adaptive-delta modulation. 2. add/drop multiplexer.

administrative domain AD. The group of network hosts, switches, and routers and their interconnections managed by a specified administrative authority, such as a system administrator on a small network, or a network control center for a larger network.

admittance (Symbol - Y or y) In an electrical circuit or material, a measure of the facility with which the current flows through the circuit or material. Admittance is rather whimsically expressed in *mho* units, which is *ohm* spelled backward, since ohms are used to express impedance, the reciprocal of admittance. Contrast with impedance.

ADN See Advanced Digital Network.

Adobe Systems Incorporated A California and Seattle-based company, Adobe is best known for PostScript, Acrobat, Pagemaker, and Illustrator, products which are aimed at the large number of home and professional publishers and graphics users. See Acrobat, PostScript.

ADONIS Article Delivery Over Network Information Systems.

ADP automated data processing.

ADPCM See adaptive differential pulse code modulation.

ADQ Average Delay in Queue. A measure of the average time a caller waits before a call is processed or handled by an agent. It is important to keep this time a short as possible, to discourage the caller from hanging up or negatively perceiving the service.

ADR 1. achievable data rate. 2. aggregate data rate. 3. analog to digital recording. 4. ASTRA Digital Radio. Radio based on the ASTRA European satellite system.

ADRMP autodialing recorded message player. An automatic dialer which plays a recording to the person who answers the phone to keep him or her on the line until an agent can take the call. These are used by telephone solicitors and collection agencies. ADRMP systems are disliked by many callees who consider it intrusive to pick up the phone and be connected to a recorded message and asked to wait.

ADRT approximate discrete Radon transform. A mathematical technique used in situations where substantial redundancy is expected or encountered. See discrete cosine transform, Fourier transform.

ADS 1. advanced digital system. 2. See AudioGram Delivery Services. 3. automated data systems.

ADSL See asymmetric digital subscriber line.

ADSL Forum An international association of ADSL professionals formed in 1994 to promote and disseminate information about asymmetric digital subscriber line (ADSL) services, fast communications over copper wires. The Forum provides technical and marketing information, including conferences and analysis of ADSL-related technology. http://www.adsl.com/

ADSP AppleTalk Data Stream Protocol.

ADSTAR Automated Document Storage And Retrieval.

ADSU ATM Data Service Unit.

ADT abstract data type.

ADTV Advanced Definition Television.

ADU asynchronous data unit.

advance ball A mechanism for supporting and steadying a cutting stylus (mounted just ahead of it) which is used for recording media such as phonograph platters. Most information storage is now done by rearranging magnetic data, or creating pits, rather than by cutting grooves into a physical medium. See acetate.

ADVANCE Project A project of the European Community Telework Forum (ECTF) to stimulate and coordinate leading global telework development throughout Europe, in conjunction with other organizations committed to this goal. The stimulation of new types of businesses, particularly small businesses, and the support of existing businesses are key goals of the project. See European Community Telework Forum, telework.

advance replacement warranty A type of warranty return/replacement service in which the replacement device or component is shipped prior to the returned item so the user can continue using it until the problem is corrected or the unit replaced. This is valuable if it is an essential component and partial use of it while awaiting a replacement is better than no use at all. It's important to check billing policies on ARWs because some companies will bill a credit card until the return unit is received, and then apply a credit, all of which may be prone to error and confusion if not monitored carefully.

Advanced Branch Exchange ABX. Not in common usage, but a phrase which has been used to distinguish traditional voice-only telephone exchange branches from those providing newer integrated voice/data capabilities.

advanced common-view ACV. A time-referencing technique used to transfer frequencies and times of various of the standards which contribute to Coordinated Universal Time.

Advanced Digital Network ADN. A commercial leased line 56 Kbps digital phone subscriber service.

Advanced Intelligent Network AIN. A telephone services architecture based around Signaling System 7 (SS7), and possible future versions of SS7, which is intended to integrate ISDN digital capabilities and cellular wireless services into a personal communications system (PCS). The AIN grew out of the Intelligent Network (IN) initiated by Bell Communications Research (Bellcore) in 1984. It can dynamically process calls by evaluating 'trigger points' through the call handling process.

Currently a newer technology to AIN is in development, called Information Network Architecture (INA), that may coexist with AIN or eventually supersede it. See Information Network Architecture, Intelligent Network, Personal Communications System.

Advanced Mobile Phone Service AMPS. An analog cellular system utilizing frequency modulation (FM) transmissions. AMPS was the first standardarized cellular phone service (1983) using the 800 MHz to 900 MHz frequency range, which is still the predominant type of cellular system in the world. NAMPS (Narrowband Analog Mobile Phone Service) is an interim enhancement to AMPS, which uses *frequency division* as a way of sectioning the bandwidth a tradeoff that increases calling capacity, but may also increase interference. See AMPS, cellular phone, mobile phone, cell, cluster, roaming.

Advanced Network and Services ANS. A nonprofit organization founded jointly by the National Science Foundation, Michigan Education and Research Infrastructure Triad (MERIT), IBM, and MCI in September 1990 to develop a gigabit network to benefit American education and research. Initially ANS planned two independent networks running over the same system of physical lines. Various issues emerged as controversial, such as corporate access, cost of operations, and use of the MCI backbone topology, which was criticized as being insufficiently robust and lacking in redundancy.

Advanced Peer-to-Peer Networking APPN. A distributed networking system which is now included in the Systems Network Architecture (SNA) developed by IBM. APPN workstations are dynamically defined to reduce the need for extensive changes when the network is reconfigured. APPN provides optimization of routing between devices, direct communication between users, direct remote station communication, and transparent sharing of applications over the network.

Advanced Peer-to-Peer Networking+ APPN+, APPN Plus. An enhanced IBM APPN which includes faster throughput, dynamic rerouting and congestion control, and other features to make it competitive with TCP/IP. See Advanced Peer-to-Peer Networking.

Advanced SCSI Programming Interface ASPI. A SCSI host adapter-independent programming interface released by Adaptec in the late 1980s. ASPI permits multiple device drivers to share a disk controller by providing a consistent device driver interface. Typically developers have had the burden of supporting many different host adapters, writing several, sometimes dozens of individual device driver definitions and programs for their users. The user then has to either install and load them all, or search through them at installation time, trying to locate the right device driver for the hardware peripheral, often a time-consuming, hit-and-miss process.

With ASPI, vendors can make their products ASPI-compatible, so software can talk to the hardware without a lot of extra files or hit-and-miss installation effort on the part of users. While there are similar systems from other vendors, this is one of the more popular ones.

advanced telecommunications capability This is defined in the Telecommunications Act of 1996, and published by the Federal Communications Commission (FCC) as:

" ...without regard to any transmission media or technology, as high-speed, switched, broadband telecommunications capability that enables users to originate and receive high-quality voice, data, graphics, and video telecommunications using any technology."

advanced TV, advanced television ATV. A generic category for television broadcasts that supply better audio and/or video properties than are generally associated with the traditional NTSC system in North America. Various means of digital manipulation at the broadcast or receiving ends can result in better picture viewing or sound without changing the underlying broadcast format, while others require a completely different way of sending and encoding a signal. High Definition Television (HDTV) is a type of advanced TV.

ADVENT The first major geostationary satellite, launched by the U.S. Department of Defense. It included a directional antenna, and a three-axis stabilization system.

advocacy Providing support, promoting acceptance or use of. Advocacy has played a prominent role in the evolution of the Internet, and in computer development and marketing. Because the interests of promoters of universal access for technology are not always consistent with the goals of government or big business, many users' groups and advocacy forums have been established online. Almost every USENET newsgroup devoted to a specific computer platform has an advocacy section for the discussion of issues related to platform design, acceptance, and use.

AE 1. acoustic emission. 2. Application Entity.

.ae Internet domain name extension for the United Arab Emirates.

AEA 1. American Electronics Association. 2. American Engineering Association.

AEC acoustic echo canceller.

AECS Plan Aeronautical Emergency Communications System Plan. A voluntary system of communication established and organized for the provision of emergency communications to the U.S. President and federal government representatives.

AECT Association for Educational Communications and Technology. An organization committed to providing communication among professionals with a common interest in using technology for education.

AEEM Aerospace Engineering and Engineering Mechanics.

AEGIS Advanced Electronic Guidance and Instrumentation System.

AEP AppleTalk Echo Protocol.

aerial This term for conductive wires or structures is derived from the fact that most of them are suspended from poles, towers, or other aerial structures high enough to provide safety from interference and electrical hazards. Sometimes aerials are distinguished as signal receivers, and antennas as signal senders. And sometimes the opposite distinction is made, so there isn't a lot of consistency in usage. Since insect antennas can be considered as 'receiving' units, it might make sense to call the receiver the *antenna*. Because of the lack of standardization of the terms, and because many of the same concepts of design and construction apply to both sending and receiving structures, this dictionary groups most of the information on aerials and antennas under the heading of antenna. See antenna.

aerial cable Transmission receiving circuits which are strung through the air, typically supported by utility poles to keep them out of reach since many carry hazardous levels of current. Contrast with buried cable.

aerial distribution Aerial cabling configuration, with wires running through the air among buildings and poles. Various insulators and amplifiers or repeaters are used in many cable installations to protect signals from interference or to extend them over distance. See distribution frame.

aerogram 1. A European term for correspondence which is lightweight, and intended to be transported on planes. Charles Lindbergh, the pilot, was one of the early pioneers of air mail service. Many countries of the world have high postal rates for air mail, based on weight. Aerograms are often written on very light, fine paper, or even on paper which folds into its own envelope, rather than requiring a separate envelope. 2. A letter sent by airwaves, as through radiotelegraphy. Radiograph is an early name for a radio telegraph.

aeronautical broadcasting Various government and commercial services providing information to the aeronautics industry, especially regarding meterological conditions.

aerospace Space consisting of: 1. the Earth's atmosphere above ground, the region in which much broadcasting and travel occurs, and 2. the space beyond, into which satellites, probes, and spacecraft are launched. Also, the industries and sciences concerned with space travel and communications.

AES 1. Application Environment Standard, Application Environment Service. 2. atomic emission spectroscopy. 3. Audio Engineering Society.

AES 90 The first large-scale commercial word processor, introduced by a Canadian company, Automatic Electronic Systems, in 1972. The system featured magnetic storage, with commands such as "Memorize" for saving a file, built around a custom microprocessor. Word processing on the AES system did not become widespread until the mid-1980s, as the business world was slow to make the transition from typewriters to word processors.

AESS Aerospace and Electronics Systems Society.

AEW 1. aircraft early warning. 2. airborne early warning. This includes not only aircraft, but other airborne objects such as missiles and probes.

.af Internet domain name extension for Afghanistan.

AF audio frequency. A spectrum of wavelengths which can be heard. For humans this is from about 30 hertz up to about 20 kilohertz, although the upper level declines to about 16 to 18 kilohertz by adulthood.

AFAIK An abbreviation for "as far as I know" used in email and on online public forums. See IMHO.

AFAST Advanced Flyaway Satellite Terminal. A family of commercial, modular, portable satellite terminals operating in the C-, Ku-, and X-band frequencies, from California Microwave, Inc. (CMI).

AFC 1. Advanced Fibre Communications. 2. See automatic frequency control.

AFCEA Armed Forces Communications and Electronics Association.

AFE 1. See analog front end. 2. antiferroelectric.

affiliate In the Telecommunications Act of 1996, published by the Federal Communications Commission (FCC), the term affiliate has a specific meaning as follows:

"... a person that (directly or indirectly) owns or controls, is owned or controlled by, or is under common ownership or control with another person. For purposes of this paragraph, the term 'own' means to own an equity interest (or the equivalent thereof) of more than 10 percent."

See Federal Communications Commission, Telecommunications Act of 1996.

affine redundancy A phrase attributed to Michael Barnsley, who used it to describe the characteristics of fractals in terms of their self-similarity, and their likelihood of looking more like parts of themselves, than parts of other things. See fractal.

affirmative In voice communications where signals are weak or noise is present, a synonym for 'yes' which is intended to be clear and unambiguous.

AFI Authority and Format Identifier. In ATM, part of the network level address header.

AFIPS American Federation of Information Processing Societies. A national organization of data processing societies which organizes the National Computer Conference (NCC).

AFK An abbreviation for "away from keyboard," used especially on public chat channels on the Internet, and on BBS chat rooms to indicate someone with his or her hands full who can't type (eating, reaching for a book), or who is away for a while to answer a door, a phone, to help a child, or visit the washroom. AFK messages should be taken with a grain of salt, as the person may have stepped out to go to the store, or attend a class, and become caught up in other activities that keep him or her away from the keyboard for extended periods. See also BRB.

AFM 1. Adobe Font Manager. 2. Adobe Font Metrics. 3. antiferromagnetism.

AFMR antiferromagnetic resonance.

AFNOR Association Français Normal. The national standards organization of France.

AFOSR Air Force Office of Scientific Research.

AFP AppleTalk Filing Protocol. A network file protocol which allows file sharing over an AppleTalk network.

AFS See Andrew File System.

AFT Automatic Fine Tuning. See automatic frequency control.

afterimage A visual image that may appear in pale outline or as a complementary color if an object is viewed for some time without moving, after the source of the image has changed or disappeared. See persistence of vision.

AFTRA American Federation of Television and Radio Artists. A trade organization representing performers which was founded in 1937. http://www.aftra.org/

AFV audio-follow-video. In many broadcast systems, audio and video are recorded and/or transmitted separately. In AFV, the audio signals are automatically routed with their associated video signals.

.ag Internet domain name extension for Antigua and Barbuda.

agate line In typography, a unit of measurement typically used by periodicals to communicate column sizes for articles and ads. 14 agate lines = 1". Agate line measurements may cross columns, since display ads and images for headline stories are often two or more columns wide.

AGC 1. AudioGraphic Conferencing. ITU-T terminology related to transmissions protocols for multimedia. See audiographics. 2. See automatic gain control.

AGCOMNET A U.S. Department of Agriculture voice and data communications network.

aged packet In packet-switched networks, a data packet which has exceeded a prespecified parameter such as node visit count or elapsed time. Aged packets may be handled in a number of ways, depending upon their nature and the configuration of the network. They may be discarded, assigned a different priority, or returned to the originator.

Agency of Industrial Science and Technology AIST. A Japanese organization which is part of the Ministry of International Trade and Industry (MITI) which superintends research laboratories which are known for technological innovation.

agent 1. Representative, broker, one who acts in place of, or on the authority of another. 2. One who handles customer inquiries and procures services or products for that customer, often through other firms. Real estate agents serve as liaisons between sellers and buyers, and travel agents are liaisons between airlines and travellers. Many long-distance providers are agents who procure services through other companies or through leased lines rather than by installing their own physical equipment. 3. On networks, software agents are frequently used in client/server transactions to gather, organize, or exchange information according to security and priority levels usually established by the server. 4. On computers, in a general applications sense, agents are products (such as utilities or plugins) that do long, tedious or complex tasks, in conjunction with, and generally on behalf of, server software or user applications.

aggregate bandwidth In a stream carrying more than one communication through some system of multiplexing, the aggregate bandwidth is the total combined bandwidth.

aggregation The bringing together or combining of physical, data, or radiant waves as in cables or transmissions. Aggregation typically refers to bringing together in terms of proximity, usually without a merging of information or electrical characteristics. However, some types of data are aggregated through an interleaving process, while still keeping individual portions true to their origins. Multiplexing is often used in conjunction with, or as a means of, aggregation. Agents sometimes aggregate, that is 'bundle,' services for consumers. Cable companies sometimes aggregate certain types of stations into 'package deals' for cable subscribers.

aggregate transmission The multiplexing of the transmissions of large numbers of users over a network backbone.

aggregator A service agent, broker, or liaison who coordinates negotiations on behalf of a block of subscribers, usually to get reduced rates. Billing is done by the service provider once the service has been established or facilitated by the aggregator.

aging 1. *v.t.* A process of storing materials until their properties become essentially stable, or reach a desired set of characteristics. 2. *v.i.* The characteristics of a material or process over time under a certain set of conditions. This may be an improvement, a deterioration, or simply a change.

agonic In magnetism, an imaginary line connecting all points on the earth where the magnetic declination is zero. See declination, isogonic, magnetic equator.

AGP See Accelerated Graphics Port.

AGT AudioGraphics Terminal.

AGU 1. address-generation unit. 2. Automatic Ground Unit.

Ah ampere-hour.

AHT Average Handle Time. A call management phrase that describes the amount of time it takes, on average, to take a call, talk to the caller, and handle the caller's needs at the end of the call. For example, on a typical sales call, it may take a minute to connect with the desired person, fifteen minutes for the call, and twenty minutes after the call to log the caller's feedback and arrange to have a sales brochure sent to the caller.

.ai Internet domain name extension for Anguilla.

AI 1. Airborne Interception radar. 2. See artificial intelligence.

AIA 1. Aerospace Industries Association. 2. Application Interface Adapter. A software utility which converts client function calls to standard SCSA messages.

AIEE American Institute of Electrical Engineers. This was consolidated with IRE to form the IEEE, an influential body of engineering professionals. See IEEE.

AIFF Audio Interchange File Format.

AIIM Association for Information and Image Management.

Aiken, Howard Hathaway (1900-1973) A Harvard student who proposed a calculating machine, a forerunner of computers. Aiken received financial support in the 1940s from the President of International Business Machines (IBM), Thomas J. Watson, to build the Automatic Sequence Controlled Calculator, more commonly known as the Mark I. This was later followed by the Mark II and Mark III computers. See Harvard Mark I.

AIM 1. amplitude intensity modulation. 2. See Ascend Inverse Multiplexing protocol. 3. See Association for Interactive Media. 4. ATM inverse multiplexer.

AIN See Advanced Intelligent Network.

AIOD Automatic Identified Outward Dialing. A multiline phone option which records the extension number of the originating phone in order to facilitate billing. AIOD is especially common for the identification of long-distance calls. AIOD leads are terminal leads used to transmit this information to the phone carrier.

AIP ATM Interface Processor. A Cisco Systems commercial router network interface (ATM layers AAL3/4 and AAL5) for reducing performance bottlenecks at the UNI.

AIR 1. additive increase rate. In ATM, a traffic flow control available bit rate (ABR) service parameter which controls cell transmission rate increases. See cell rate. 2. Airborne Imaging Radar. 3. All India Radio.

air blown fiber See blown fiber.

air bridge In electronics, a suspended interconnect, usually of metal.

air capacitor, air condenser A capacitor/condenser whose dielectric is air.

air cell A type of electrolytic wet cell once widely used in phone applications. It consists of separate cells connected in order to increase the voltage. Polarization is reduced because oxygen from the air combines with hydrogen from the carbon electrode to form water. These early cells had a useful life of a thousand or so hours, and required sufficient ventilation. See dry cell, wet cell.

air column A channel of air, usually with certain size specifications or sound characteristics, within a piece of equipment, instrument, or chamber. Air column cables sometimes employ air as a dielectric, thus enabling a lighter, more flexible cable than one with a solid dielectric. See air-spaced coaxial cable.

air conditioning Running air through a system to alter its characteristics to make it suitable for people, equipment, or both. An air conditioner can affect temperature, humidity, and ion balances. Air conditioners are often used to cool work rooms in hot climates, and to cool equipment that generates heat but may be damaged by the heat if the air temperature is not kept down. Many large supercomputing installations require cooling, and chip manufacturing plants condition the air to keep it free of dust, smoke, and other particles.

air core transformer A type of transformer designed to overcome some of the limitations of iron core transformers. At the higher frequencies used by broadcast communications, various problems such as the eddy effect and the skin effect will interfere with transmissions. Thus, air core coils and transformers, carefully tuned, can overcome some of these problems by eliminating the core.

air gap A region of air through which an electrical spark or magnetic current travels, as in spark gaps in gasoline engines.

air miles Accumulated miles traveled on a particular airline, or under a special program. Air miles programs are marketing tools used by airlines and their affiliates to calculate bonus trips, class upgrades, and other customer incentives. The term air miles has become very broad, since air miles can now be accumulated in a variety of ways, including through long-distance calls, hotel stays, rail trips, etc.

air time Time spent online, broadcasting, or engaged in two-way, or multiple connect wireless conversation. A measure of accumulated air time is used by service providers as an accounting tool for scheduling, billing, and time management on shared systems.

air-spaced coaxial cable A type of cable assembly design which incorporates air as a dielectric in order to minimize the loss of signal. Since there is no way to suspend the central core exactly in the middle of the column of air, air-spaced cables require 'spacers' which are usually of some type of plastic, inserted at intervals over the length of the cable, sufficiently far apart to let the air do its job (and to prevent moisture from entering), and sufficiently close together that a twist or bend in the cable doesn't allow the inner core to make contact with the next layer. See coaxial cable.

airbrush A painting tool that combines a fine spray nozzle with an air compressor to create fine gradations and details. In computer imaging, a functionally similar tool has been incorporated into paint programs to allow the creation of subtle blends and misty effects. The density of the individual dots, and the shape and extent of the spray area are usually configurable.

aircraft earth station A mobile satellite transceiving station which, instead of being stationed on the ground, is installed on board an aircraft.

AIRF Additive Increase Rate Factor.

airline miles See air miles.

airplane dial A type of rotary dial common on old radio systems which, when turned, moves a needle-like indicator back and forth in an arc, or straight line according to a marked gauge, similar to the gauges seen in airplane cockpits. Airplane dials are often used along with 'sliders' on analog systems, and pushbuttons on analog/digital systems.

airtime, air time 1. The time during which a specific broadcast is active (airs). 2. Time allocated to a specific broadcast, whether or not it is used. 3. The time spent on a radio phone call. This information is frequently used in billing calls, as in cellular phone systems. Unlike wired systems where toll-free numbers or busy numbers are not billed, many wireless services bill for the amount of time the call is online, regardless of whether it is connected to a toll-free or local callee.

AIS 1. See alarm indication signal. 2. Automatic Intercept System. 3. Association for Information Systems. 3. Automated Information System.

AIST See Agency of Industrial Science and Technology.

AISTEL Associazione Italiana Sviluppo delle Telcomunicazioni. http://www.venus.it/aistel/

AIT 1. assembly, integration and testing. 2. Atomic International Time (more correctly known by TIA). See International Atomic Time. 3. Automatic Identification Technology.

AITS Australian Information Technology Society.

AIX Advanced Interactive Executive. An International Business Machines (IBM) implementation of Unix.

AJ anti-jam. A communications signal which is structured so that it is resistant to jamming or interference.

AJP American Journal of Physics.

AKA also known as. 1. Alias, handle, nickname. 2. False or fraudulent name.

.al Internet domain name extension for Albania.

Al aluminum.

AL Adaptation Layer. See ATM in appendix.

ALA American Library Association. A governing body and support group for American Librarians. The ALA provides member services, workshops, conferences, and administrative support. The author received help from many librarians in the creation of this dictionary.

ALAP AppleTalk Link Access Protocol.

alarm Warning signal, a signal to indicate an error or hazardous situation. Alarm signals are generally designed with flashing lights, or raucous noises to attract immediate attention. In electronic equipment, alarms are signaled by various messages, flashing elements, or sounds and may or may not indicate the priority level, and possible location or cause of the problem.

alarm indication signal, alarm indicating signal AIS. 1. In ATM, a signal transmitted downstream, that signifies that an upstream failure has occurred. 2. Blue signal, blue alarm. A signal which overrides normal traffic during an alarm situation.

ALASCOM A commercial, regional communications service, consisting of satellite earth stations, fiber optic, and microwave links serving the state of Alaska.

albedo A ratio of the amount of electromagnetic radiation reflected by a body, to the amount incident upon it. This reflectance may be described in the context of a portion of the spectrum (as the visible spectrum) or of the whole spectrum. A concept used in telecommunications in relation to satellites and other celestial bodies. Complementary to absorptivity. Often expressed as a percentage.

ALBO automatic line buildout. In data transmissions, a means of automatic cable equalization.

ALC 1. automatic level control. 2. automatic light control.

ALDC adaptive lossless data compression.

ALE 1. Application Logic Element. 2. Atlanta Linux Enthusiasts. 3. Automatic Link Establishment.

alert signal, alerting signal A transmission signal designed to gain the attention of an administrator or user. In computer networks, alert signals signify many things, such as imminent shutdown of a system, talk requests, new user logins, newly arrived email, etc. On telephone networks, alert signals are often used to indicate an incoming call.

Alexanderson alternator A high-frequency generator designed by E.F.W. Alexanderson which was used in pioneer transatlantic communications. One of the historic uses of the Alexanderson alternator in broadcasting was at the Fessendon station which, in 1906, broadcast Christmas music to surprised and delighted listeners.

Alexanderson antenna A vertically polarized wired antenna used for low frequency (LF) and very low frequency (VLF) transmitting and receiving. Not commonly used above amplitude modulation (AM) frequencies.

Alexanderson, Ernst F. W. (1878-1975) A pioneer developer of radio alternators in the early 1900s. GE had been contracted by Fessendon to develop a high frequency alternator for his pioneer radio station in 1904. Ernst Alexanderson was assigned to the project and achieved this significant engineering feat. He was involved in some of the early television development that was occurring in the 1920s, and demonstrated a home television receiving unit. The Alexanderson alternator and Alexanderson antenna are named after him. See Alexanderson alternator.

A 200-kilowatt Alexanderson motor used for radio frequency alternation for the Radio Corporation of America (RCA) in New Jersey. Scientific American Monthly, October 1920.

Alford antenna A rectangular loop antenna, with each of the corners slightly infolded toward the center to lower impedance at the nodes.

ALGOL *Algo*rithmic *L*anguage, *Al*gebraic *O*riented *L*anguage. A computer programming language developed in the 1950s by P. Naur, and others, for manipulating mathematical algorithms. C is said to be evolutionarily descended from Algol (with an intervening language called B).

algorithm A procedure consisting of a finite series of steps, defined to solve a problem or execute a task. The solution to the problem does not necessarily have to be known to create an algorithm to seek out a solution, or a path toward a solution. Logical/mathematical algorithms are widely used in the computing industry. The algorithm itself may not have a fixed number of steps, since an algorithm can be designed to be self-modifying, but the initial tasks, as set out by a programmer, for example, are finite. See brute force, heuristics.

ALI 1. ATM Line Interface. 2. automatic location identification, automatic location information. A feature of enhanced 911 emergency systems which automatically provides information on the source of the call from a database.

alias *n.* 1. Pseudonym, assumed name, substitute or alternate name. 2. On operating system command lines, a short, easily remembered label for a longer, harder to remember label or command. Most systems will allow users to set up aliases at boot-up time, or in a file that can be reread while the system is running, to update the aliases. On Unix systems, a convenient alias is *ll* in place of *ls -la*. It's easier to type and displays more information in the subsequent directory listing, including permissions, file size, etc. 3. On Macintosh systems, there is a menu command to alias a filename. When selected, it causes an extra icon to appear, matching the original, under which the user can modify the name of the application, if desired, to better remember its function. This can be placed on the Desktop (or anywhere that's convenient), in place of the original icon which may be buried several folders deep, or have an obscure name. When double-clicked, the alias then finds the original and launches it on behalf of the user. 4. Online, many users will assume an alias identity, known as a handle, or nickname, in order to present a friendlier, more interesting, or more obscure face to others. 5. In computer imagery, a visual artifact consisting of rough, staircased edges. This may result from low sampling, or from low resolution in the output device. See aliasing.

aliasing 1. In imaging, a visual artifact which causes rasterized images to take on a staircased effect when displayed or translated into resolutions that are too coarse to clearly resolve the image. See antialiasing. 2. In audio, a frequency distortion that occurs in sampling, when the sampling rate and the frequency interact in undesirable ways. A filter can sometimes reduce distortion.

aligned bundle A bundle of fibers or wires, in which the relative positions of each of the ends at one end are retained at the other end. In fiber optic transmissions, the bundling alignment is quite important to the quality of the transmissions.

ALIT Automatic Line Insulation Testing.

All Call Paging A capability through which a spoken message can be broadcast through a phone system, to all speakers and phones on that system. See hoot'n'holler.

all dielectric cable A cable consisting of dielectric materials, that is, insulating materials, and which has no metal conductors as in most conventional cables.

all number calling Most people are now familiar with phone addresses consisting entirely of numbers, but in older phone systems in many regions of North America, a unique phone ID consisted of two letters, usually indicating the region or neighborhood, followed by five numbers. Thus, the number 525-1234 would have been called Larch 51234, Ladysmith 51234, LA 51234 or something to that effect. This was a somewhat more poetic and easy to remember system than the current all number system. All number calling was instituted to provide more numbers, as human populations and the demand for phone lines increased. In most areas, all number calling was in place by the 1960s. Since numbers are difficult for many people to remember, companies often request 'gold numbers,' that is numbers that correspond to letters which spell out the name of the company or some aspect of its service.

All Trunks Busy ATB. A tone indicator or recording to the caller that all trunks in a specific routing group are unavailable. The tone sequence sounds like a fast busy signal.

all-wave antenna A multipurpose antenna designed to broadcast and/or receive a wide range of frequencies. All-wave antennas may include a number of different types of receiving structures on one basic supporting structure, and may achieve even further enhanced reception by being adjusted (i.e., it may tilt or rotate manually, or electronically on servos).

ALLC Association for Literary and Linguistic Computing.

Allen, Paul (1953-) The renowned cofounder of Microsoft Incorporated, Paul Allen was Bill Gates' teenage Seattle high school friend and business partner, and coauthored a number of early programming projects with Gates. Together they founded Traf-O-Data around 1972 and worked on commercial programming contracts. Allen discussed a number of ideas for creating and selling microcomputers with Gates, but Gates didn't get as fired up about hardware ideas as he did about software, and these ventures were not aggressively pursued. After graduation, Gates went to Harvard and Allen took a job with Honeywell in Boston.

When Allen was in Harvard Square, he learned of the Altair computer from the January 1975 issue of Popular Electronics magazine. He and Gates discussed the article together and conceived the idea of developing a BASIC interpreter for the MITS Altair. Apparently others had called MITS with a similar idea. Allen and Gates had previously had access to computer code for a BASIC interpeter when they were working at Dartmouth, so Allen developed a simulation environment for the 8080 processor on a DEC PDP computer, and he and Gates developed a BASIC interpreter in a few weeks. Allen flew to the MITS offices in New Mexico to demonstrate the software and, remarkably, it worked the second time they fed the tape through the machine. Gates and Allen moved their business to Albuquerque to work in cooperation with MITS, and Allen became their VP of Software.

Allen always had an interest in hardware as well as software, and when the transition from 8-bit to 16-bit machines was taking place, Allen originated the idea of the Apple SoftCard emulator, to straddle the two markets, and to provide greater flexibility while Microsoft moved into 16-bit software development.

The single most important alliance in Microsoft history was the contract to develop an operating system for International Business Machines (IBM), under controversial and competitive circumstances with Gary Kildall, the developer of the CP/M operating system. The text-based QDOS system, based on a mid-1970s manual on Kildall's CP/M, was their flagship to success. They purchased QDOS, developed by Tim Paterson, and developed it into PC-DOS for IBM, and MS-DOS for Microsoft. Later Paul Allen left Microsoft to pursue other interests, including investments in a number of ventures, and in 1994 he founded the Paul Allen Group to monitor the performance of the various companies in which he has significant investments. See Altair; Gates, William H.; Microsoft BASIC, Microsoft Incorporated.

Alliance of Computer-Based Telephony Application Suppliers ACTAS. A trade organization established to promote the distribution and development of computer-based telephone applications and standards. ACTA is associated with the Multimedia Telecommunications Association (MMTA).

Alliance for Telecommunications Industry Solutions ATIS. An organization of industry professionals from North America and World Zone 1 Caribbean service providers. ATIS was initially the Exchange Carriers Standards Association (ECSA) in 1983, when it was created as part of the Bell System divestiture. It became ATIS in 1993. ATIS is concerned with a variety of issues ranging from telecommunications protocols and interconnection standards to general administrative operations of systems among competing carriers. ATIS has cooperated on many projects with the U.S. Federal Communications Commission (FCC). See Committee T1. http://www.atis.org/

alligator clip A long-nosed metal pressure clip with small teeth on the inner surface of the clip for holding fast small objects or wires. Often two clips are mounted on a firm base to make them free-standing. Commonly used in electronics to hold wires and various components, especially for soldering, or for establishing temporary connections.

allocate To apportion or earmark for a specific purpose. Resource allocation is an important aspect of computer and network operations. Memory, storage space, CPU time, and printers are queued and prioritized as part of the allocation process. Allocation is also essential to broadcasting and two-way radio communications, as there are only a limited number of frequencies available, and these must be carefully administrated to avoid interference, and to maximize the number of regions in which they can be reused.

Allouette-I Canada's first research satellite, launched in 1962, to study radio communications in the northern reaches and the ionosphere. Three years later, the Allouette-II was launched. See ANIK.

alloy A combination of a metal or metals with nonmetals, or of metal with metal, accomplished through the intimate fusing or amalgamation of the components. Alloys are often created to combine the better qualities of the components. For example, blending gold with a stronger metal provides the greater malleability and beauty of gold with the greater durability of the alloyed metal.

ALM 1. airline miles. See air miles. 2. AppWare loadable module. 3. automated loan machine. This is a type of commercial access point, similar to withdrawal ATMs, in which financial services in the form of quick loans can be negotiated through an automated teller machine.

almanac 1. Publication containing astronomical and meteorological data useful for navigation/positioning technology. 2. File detailing satellite orbits, and related atmosphere and time information.

alnico An iron alloy with *al*uminum, *ni*ckel, and *co*balt, sometimes with various combinations of cobalt, copper, and titanium added. It is commonly used to make permanent magnets, used in many electronics components including speakers, motors, meters, etc.

ALOHA A method of radio wave transmission in which transmission can occur at any time. This means many transmissions may happen simultaneously, and may cause interference, but sometimes it's a practical way to deal with unusual situations. The basic idea is to send out a signal, see if there's a response, and if there isn't, send again. Pure ALOHA and slotted ALOHA are variations. Pure ALOHA is very much a free-for-all, and has been used for packet radio communications since the early 1970s. It has a low capacity rate, usually only about 18%. In slotted ALOHA, the transmissions are slotted according to time access, which may provide about double capacity of pure ALOHA. The name is derived from a failing satellite whose use was donated to researchers in the South Pacific. Since capacity outstripped demand, the loose ALOHA method fitted the circumstances.

ALOHANET An experimental frequency modulation (FM) transmission in which data frames are broadcast to a specific destination, developed by the University of Hawaii. See Aloha, packet radio.

alpha channel A portion of a data path, usually the first 8 bits in a 32-bit path, which is used with 24- and 32-bit graphics adapters to control colors. Popular paint programs like Adobe Photoshop allow the contents of alpha channels to be individually manipulated to create special effects.

alpha testing In-house testing of software or hardware. In software alpha testing, employees attempt to find and eradicate all the bugs, flow control, and user interface issues that can be determined by internal staff. See beta test.

Alphanet Telecom A new Internet protocol-based long-distance company based in Toronto, Canada. Phone, fax, and data transmissions will be jointly available as IP-based calling services leased through private carriers.

alphanumeric A set of characters comprising the upper and lower case letters of the English alphabet from A to Z, and the numerals 0 to 9. On some devices, lower case letters may not be included.

alphanumeric display A very common, usually inexpensive type of display on consumer appliances and electronics in which basic letters and numbers, and sometimes a few symbols, can be seen well enough to be understood for simple tasks. Alphanumeric displays are commonly based on liquid crystal diode (LCD) or light emitting diode (LED) technology. Alphanumeric displays are used in digital clock radios, microwaves, calculators, music components, handheld computers, and many other items.

ALPS automatic loop protection switching.

Altair 1971 is the year that commercial distribution of personal computers really began, when the Kenback-1 Digital Computer was advertised in the September issue of Scientific American magazine. Thereafter, a steady trickle of assembled units and kits, including the Micral, Mark-8 and Scelbi-8H foreshadowed the growing interest in hobbyist and home computers. None of these predecessors to the Altair sold in significant numbers. Then, at the end of 1974, Edward Roberts' MITS company introduced the Altair 8800, which was featured on the January 1975 cover of Popular Electronics magazine. This exposure was a significant boost to its visibility. International Business Machines (IBM) released the 5100, the first significant corporate competitor, the same year, but it failed to catch on and was quickly forgotten.

The Altair was designed by Roberts, William Yates, and Jim Bybee. The introductory price for the first three months was $395 for the kit, and $650 for a fully assembled unit. Programming was accomplished by means of small dip switches on the front of the computer; if the power was interrupted, the programmer had to start all over again and the available 'memory' was infinitesimal by today's standards, only 256 bytes. It featured an 8-bit Intel 8008 central processing unit (CPU) and room for the addition of up to 15 peripheral cards. Later Altair-compatible buses incorporated Intel's upgrade to the 8008, the 8080, which was significantly faster.

Through marketing, a little luck, and the growing interest of electronics hobbyists, the Altair line was the first to successfully capture the hearts and imaginations of computer pioneers, and Micro Instrumentation and Telemetry Systems (MITS) sold more than 40,000 units by the time the company was sold in 1978.

The Altair was a history-making invention; in spite of its modest capabilities, it became wildly popular with insightful hobbyists who grasped its potential and significance. The Altair bus, more commonly remembered as the S-100 bus, was quickly copied and a number of clones, most notably the IMSAI 8080, began to appear. MITS set to work adding to their product line, creating a Motorola-based version, the Altair 680. The mass market computer had been born and the industry quickly shifted into high gear, with far-reaching changes to society.

Paul Allen and Bill Gates, friends not long out of high school at the time of the release of the Altair, provided MITS with a BASIC interpeter just in time for it to be included with fully assembled versions of the machine, thus launching Microsoft Incorporated, the world's best-known software company. Steve Wozniak, inspired by the little kit computer, designed his own computer circuit board and, with Steve Jobs, formed Apple Computer, Inc., another of the world's most successful computer hardware/software companies. See Alto, Intel MCS-4, Kenbak-1, IMSAI 8080, LINC, Mark-8, Micral, MITS, Scelbi-8H, Sol, SPHERE System, STPC 6800, TMS 1000.

The Altair 8800 was available assembled or as a kit from MITS, a New Mexico-based company. It was introduced late in 1974 and was prominently featured in the January 1975 issue of Popular Electronics.

The November 1975 issue of Popular Electronics featured an article on building the Motorola MC6800-based Altair 680 by Edward Roberts and Paul Van Baalen. This was likely a strong factor in introducing the Motorola MC6800 CPU to hobbyist hardware designers.

Altair 680 A Motorola and American Micro-Systems, Inc. 6800 CPU-based computer from MITS, the same company which released the Intel 8008-based Altair a little less than a year earlier. The Altair 680 was featured in an article in the November 1975 issue of Popular Electronics as having a built-in TTY interface, a capacity of 72 program instructions, and room for up to five interface cards. The 680 was intended to appeal to hobbyists who liked the architecture of the MC 6800, and who were looking for a smaller, less expensive kit to build. The 680 was less than a third of the size of the Altair 8800 and much less expensive to build. See Altair.

Altair bus The original data bus that was developed by MITS for the Altair computer line. Later vendors changed the name to S-100 bus, and it became common in many different computers in the late 1970s and early 1980s.

Altair Users Group, Virtual There is a Virtual Altair Users Group on the Internet, comprised of hobbyists who still build, repair and operate Altair computers. One of the participants, Tom Davidson, hosts an excellent Web site with schematics and circuit board images. http://hyperweb.com/altair

AltaVista One of the significant World Wide Web search engines on the Internet, AltaVista draws from one of the largest Web database catalogs online. http://www.altavista.com/

alternate access carriers A telephone service vendor other than the Local Exchange Carrier (LEC) can be authorized under competitive Federal Communications Commission (FCC) guidelines to provide alternate access.

Alternate Billing Services ABS. Telephony services which allow collect or bill-to-another-number services to callers, especially applicable to long-distance calls.

Alternate Mark Inversion AMI. Line transmission code used for T1 and E1 lines in which successive *marks* alternate in polarity (negative and positive).

alternate route AR. An alternate data or telephone communications route selected when the initial choice is unavailable due to load, or a break in the path. In telephony, sometimes called *second-choice route.*

alternate routing In both circuit switching and packet switching network systems, there are times when the initial attempt to trace and complete a transaction between a sender and the destination is unsuccessful. This can be due to high traffic, intermediary switching links being down, the destination not being available, etc.

In most circuit switching implementations, the transmission cannot go through until an end-to-end connection is set up, dedicating an established path to the call, so alternate routing to find another way to connect the requested call, must take place before any data (or voice, in the case of a phone call) can be sent. In telephony, alternate routing usually involves locating a less-busy trunk.

In contrast to circuit switching, packet switching does not require the establishment of an end-to-end connection before data can be sent; it can be sent regardless of whether it is known whether the destination is reachable or available. Packet switching is used in dynamic environments where it is not known, or cannot be known, which routing nodes may be available, which route is most efficient, and whether the destination is online at any particular time. The packets are sent by various means, usually through hop-by-hop systems, and the packets from an individual message may be broken up and sent through different routes if a bottleneck or break occurs in the original path. At the destination, separated packets are reassembled, and there are usually several attempts to 'deliver' the information before it is returned (in the case of most email) or abandoned (in the case of low-priority data). To facilitate alternate routing, packet system routers may have extensive routing tables listing a wide variety of connections within that 'region' of the network. See router.

alternate use AU. The capability of a communications system to switch from one mode of service to another, e.g., between data and voice. See alternate voice/data.

alternate voice/data AVD. A transmission system which can be used for voice and data over one line, by alternating the services as needed, usually switched manually, as between voice through a telephone, or data through a modem. Some modems are equipped with speaker-phone capability to allow switching between voice and data, and further to detect the mode of an incoming transmission in order to switch to the correct mode automatically. More flexible and sophisticated systems are always being developed, and some success with newer, faster modems has been achieved to allow simultaneous voice/data communications. See simultaneous voice/data.

alternating current AC, ac. A very commonly used form of electrical current with a periodically reversing charge-flow with an average value of zero. Unlike direct current (DC), alternating current (AC) varies continuously in its magnitude. For the supply of electricity to businesses and residences, it is set to reverse about 50 to 60 times per second, depending upon regional electrical codes.

Voltages in North America are supplied as 120 plus or minus about 10% for regular wall outlets, and to 220 for heavy duty outlets (for dryers, stoves, etc.). Voltages in Western Europe are set to 220.

Alternating current is typically used in commercial and residential power circuits leading to wall sockets, whereas direct current is typically used in battery-operated devices and sensitive electronic components. The large converters/transformers attached to the power cords of small components like modems convert the AC power from the wall circuit to DC power compatible with the component. Given the greater sensitivity of electronic components, plugs now commonly have one wide leg and one narrow leg, to correspond with wider and narrower holes in newer wall or extension cord sockets. The wider and narrower pins correspond to the different characteristics of the wires to which they are connected, with one being a hot 'live' wire, and the other being a neutral or 'grounded' wire.

Much of early communications technology was based on direct current (DC) as a power source. Telephones had 'talking batteries' and 'common batteries.' These batteries were large, leaky wet cells, which were inconvenient if moved or exposed to fluctuating temperatures. Surprisingly, Thomas Edison was opposed to alternatiing current for the power supply for communications circuits, and hotly contested the concept with Nikola Tessla. More than fifty years after the invention of the telegraph, AC power for telegraph systems was still considered a novel idea, but the shortage of batteries, and their high cost, provoked French and Swiss engineers to experiment with AC generators, as described in the *Annales des Posts, Télégraphes et Téléphones* in September 1919. Eventually the advantages of AC power were better understood, and its use became common.

See B battery; direct current; ground; impedance; surge suppressor; talking battery; Tessla, Nikola.

alternator An electronic or electromagnetic device for producing alternating current.

altimeter An instrument for measuring altitude. Altimeters are essential in aviation, and not uncommon in cars. There are also handheld and wrist-worn altimeters for researchers, forest service workers, and mountain climbers. Many altimeters are based on the same principles as barometers, using relative pressure (usually set initially at sea level) as a means of showing changes in pressure, and hence altitude. Since pressure changes, altimeters must be adjusted.

Global Positioning System (GPS) consoles can provide information on location and altitude which is derived from satellite data and is not dependent on air pressure.

Radio altimeters, developed in the late 1930s at Bell Laboratories, indicate distance above ground by reflecting frequency modulated (FM) radio signals off the surface of the Earth, and measuring the reflections. See barometer, Global Positioning System.

Alto A pioneering computing system developed at the Xerox PARC laboratories around 1973. The Alto was the inspiration for the graphical user interface incorporated into the Macintosh line of computers, and later into Microsoft windowing software. Some argue that the Alto was the first microcomputer, but that honor really belongs to the Kendak-1 (1971), and the commercially successful Altair (1974), since the Alto was never available to the general public in its original form, and its price tag was thousands of dollars. Nevertheless, many of the revolutionary graphical user interface ideas that filtered out to the commercial world were developed and implemented on the Alto. See Altair; Kay, Alan; Macintosh computer; Microsoft Windows; Xerox PARC.

ALU arithmetic and logic unit. An integral part of most computer processors' logic architecture, for performing operations.

aluminum A silvery, dull, malleable, light, inexpensive metallic element with good electrical conductivity and resistance to oxidation. Aluminum is somewhat brittle, but is still commonly used in cables, antennas, reflectors, and other communications-related structures.

AM 1. access module. 2. active messages. 3. active monitor. 4. See amplitude modulation.

AM broadcasting Transmission through amplitude modulation technologies on approved AM frequencies with the appropriate AM broadcasting license. In the United States, AM stations are spaced at 10 kHz intervals, ranging from 540 to 1700 kHz. See amplitude modulation, band allocations, broadcasting, FM broadcasting.

AM/VSB amplitude modulated vestigial sideband. See modulation, sideband.

amalgamation 1. *v.t.* The uniting, consolidation, or merging of. Amalgamation of metals may be done to reduce the effects of chemical deterioration. 2. *n.* Blend, composite, alloy, mixture.

AMANDA Automated Messaging and Directory Assistance.

amateur bands Frequency spectrums set aside by regulatory authorities for the use of amateur radio operators. These are geographically subdivided, with some ranges designated for international use. Not all countries permit broadcasts by amateurs, licensed or unlicensed.

amateur call sign A set of identification characters licensed to amateur broadcasters by a regulating agency such as the U.S. Federal Communications Commission (FCC). Call signs in the U.S. indicate the country and region of the licensee.

amateur radio operator, ham radio operator A radio broadcasting hobbyist permitted to transmit radio signals over specific frequencies. Most licenses that permit amateur transmissions require that the operator be licensed and fulfill certain requirements. In the U.S., also often called *ham radio* operator.

amateur satellite service A radio communication service using space stations on earth satellites for the purposes of the amateur radiocommunications service. See amateur service, AMSAT, OSCAR.

amateur service Radiocommunications services for the purpose of self-training, intercommunication, and technical investigations carried out by amateurs, that is, licensed, or otherwise authorized persons interested in radio technique solely with a personal, educational, nonpecuniary aims.

amateur television ATV. Black and white or color image broadcasts through amateur radio frequencies, with or without accompanying sound broadcast. Some amateur enthusiasts prefer to use ATV to mean fast scan TV over amateur bands, and SSTV for slow scan image transmission. See slow scan television.

amber A very light, transparent or semi-transparent, warm golden substance arising from the fossilization of tree resin from pine trees that have been extinct for millions of years. Amber floats in water, and occasionally washes up on the coasts of Europe after storms, intermingled with kelp and other natural debris. Sometimes insects can be found imbedded in the amber, preserved for centuries. Amber can be highly polished, and has been used for jewelry for thousands of years.

The chief importance of amber to telecommunications is its static electrical properties, which can be observed by rubbing amber with a cloth (or on your hair) and using it to attract small fragments of tissue paper. In fact, the Greeks observed this property, and Plato recorded "... the wonderful attracting power of amber ..." in his *Timaeus* dialog. The Greek word for amber is elektron.

ambient *n.* Environment, atmosphere, mood, surroundings.

ambient noise, room noise The general acoustic noise level of an environment, usually measured in decibels. The ambient noise in terminal rooms with printers or other equipment may be sufficient to cause hearing loss over time. Technicians who work long hours with high speed printers should wear ear protection.

AMDM ATM multiplexer/demultiplexer.

America Online AOL. A large, commercial Internet Services Provider (ISP) that provides access to the Internet, and AOL-specific forums, news, email, and other features.

America's Carriers Telecommunications Association ACTA. A U.S.-based trade organization, representing commercial long-distance vendors (nondominant inter–exchange vendors). Of significance is the fact that ACTA has lobbied the Federal Communications Commission (FCC) to bar long-distance digital telephony over the Internet. The focus of ACTA is to provide representation for its members to various legislative and regulatory bodies, and to further the business activities of its members.

American Bell Telephone Company In 1875, the Bell Patent Association was formed by Alexander Graham Bell with investors willing to finance his telegraphy research. Two years later, in 1877, The Bell Telephone Company was formed by Bell, who included his associate, Thomas Watson. The company was formally incorporated in Massachusetts in 1878. Theodore N. Vail was hired as the General Manager and had a long association with the company and its successors. In 1878, the Bell Telephone Company and the New England Telephone Company were consolidated into the National Bell Telephone Company. Then, in 1880, American Bell Telephone Company was incorporated. In 1881, American Bell purchased Western Electric Manufacturing Company, and was made into Western Electric Company, the equipment manufacturing arm of American Bell.

American Bell was the parent of the American Telephone and Telegraphy Company (AT&T). AT&T was established in New York as a subsidiary in 1885 for handling long-distance calls. These two were then merged into AT&T in 1899. See AT&T; Vail, Theodore N.; Western Electric Company.

American Communication Association ACA. http://www.americancomm.org/

American Institute of Electrical Engineers AIEE. Formed as a result of growing electrical development in the 1800s, and the International Electrical Exhibition in 1884, to represent the profession and develop standards for the industry. Norvin Green, president of Western Union Telegraph Company, was the first president. Alexander Graham Bell and Thomas A. Edison were among the first six vice-presidents. AIEE was presented The Clark Collection in 1901 by Schuyler Skaats Wheeler. The Clark Collection was one of the world's great libraries of electrical technology. Andrew Carnegie further donated $1.5 million for AIEE premises. AIEE was merged with the Institute of Radio Engineers (IRE) in 1963 to form the IEEE. See IEEE, Institute of Radio Engineers.

American Mobile Satellite Corporation AMSC. A commercial provider of seamless mobile communications services across North America under the SkyCell trademark. Hughes Communications is the largest shareholder, joined by AT&T Wireless, Singapore Telecom, and Mitel Corporation. A variety of services are marketed to government agencies, emergency organizations, and major corporations. AMSC is permitted to provide domestic mobile satellite services (MSS) in the upper L-band.

American Morse Code, Railroad Morse A system of dots and dashes used to represent characters for distance communications, probably developed by Alfred Vail, while working with Samuel Morse. Due to the fact that American Morse includes some characters with internal spaces, which can be confusing to some, it is not often used, International Morse code being preferred. See Morse code.

American National Standards Institute ANSI. A significant U.S. private sector, nonprofit, standards-promoting body based in New York. ANSI was founded in 1918 by a group of engineering societies and government agencies. The ANSI Federation contributes the enhancement of global competitiveness of U.S. businesses, by promoting the development and support of consensus standards and conformity assessment systems. http://web.ansi.org/

American Public Power Association APPA. A national American service organization representing local or publicly owned electric utility companies.

QST has been published for its members by the American Radio Relay League since 1914.

American Relay Radio League ARRL. Founded in 1914 by Hiram Percy Maxim, with assistance from Clarence D. Tuska and a number of their colleagues, the ARRL is now a worldwide organization with almost 200,000 members, headquartered in the United States. Tuska was a youthful tinkerer and radio hobbyist when he met Tuska. The ARRL name is derived from the way in which amateurs, constrained to certain power levels and frequencies, would cooperate by relaying messages from one person to another in order to send over greater distances or difficult terrain.

The ARRL cooperates with various radio groups and governing authorities like the International Telecommunications Union (ITU) and the Federal Communications Commission (FCC). Its members have contributed to many of the technological milestones in communications history, including the pioneering of frequencies which were originally thought to be 'useless' (and hence were assigned to amateurs). More recently, amateur radio enthusiasts have cooperated in satellite communications projects with AMSAT. The ARRL monthly publication QST has been available for more than 80 years. See International Amateur Radio Union. The ARRL's call letters are W1AW. http://www.arrl.org/

American Sign Language Ameslan. A system of gestures to indicate words, phrases and letters without speech over short distances, or if televised, over long-distances. Ameslan is a specific 'dialect' of sign language. When you see "Closed Captioned for the Hearing Impaired" on the corner of a television image, it indicates that sign language interpretation of the speech content of a program is available.

Used primarily for instruction and communication among the deaf, sign languages are also useful in environments with a lot of noise, where earphones are used, or where quiet is important (e.g., recording studios). Ameslan is a very expressive and often aesthetic mode of communication which employs a characteristic grammar and syntax different from English. Gestural sign languages are more universal than spoken or written languages because they incorporate many intuitive and cross-cultural modes of physical expression.

Speech through signing, in general, is more readily learned by many children than spoken language, with some children learning to communicate simple concepts through basic gestures at six to 12 months of age. Signing can also be used to communicate with learning disabled or autistic children, some of whom may have difficulty acquiring spoken language skills.

While signing is primarily visual, it is also useful for individuals who are both blind and deaf. Helen Keller, who was befriended by Alexander Graham Bell, is the most well-known example of a blind and deaf person who learned to communicate with her tactile senses, through signed letters spelled into her hand, or by her into the hand of another.

American Standard Code for Information Interchange ASCII. An important, alphanumeric 7-bit (128-character) communications standard which is widely used around the world for the transmission of textual messages. ASCII is a simple system, used on telegraph systems and computers. It doesn't support formatting attributes such as bold, italic or underline, and it is primarily useful for English and western European languages.

ASCII often functions as a lowest common denominator for textual communications since it is supported by most electronic mail, word processing, text editing, and desktop publishing programs, which may otherwise be incompatible. Differing formats are often resolved through ASCII translation and conversion. See ASCII for a chart showing the characters, control characters, and hex, decimal, and octal values for each. See ASCII (for chart), EBCDIC.

American Telephone and Telegraph Company See AT&T for an explanation of the company's origins, history, and technologies.

American Voice Input/Output Society AVOIS. A not-for-profit organized dedicated to the promotion of information regarding speech applications technologies. Speech applications include voice recognition, speech recognition, and speech generation, and all of these are now important input and output capabilities of computer systems. http://www.avios.com/

American Wire Gauge, Brown & Sharpe Wire Gauge AWG. A standardized wire diameter measuring system, exclusive of covering, for non-ferrous conductors. With a range from 1 to 40, lower numbers denote thicker wires, higher ones, thinner wires. Generally, for a specific material, the current-carrying capability increases as the diameter of the wire increases, and the AWG number decreases.

Since heavier wires are usually more expensive, consumers tend to purchase the thinnest wire that will accomplish the task at hand. It's important to get wire that not only is adequate to carry the current desired, but that is strong enough to bend and stretch, especially around connectors, panels, punch-down blocks, etc. If the wire breaks at the connection point, it's not very useful. See Birmington Wire Gauge.

Ames Research Center ARC. A research organization dedicated to creating new knowledge and technologies within NASA's areas of interest. ARC was formed in 1939 by the U.S. National Advisory Committee on Aeronautics (NACA), which became part of National Aeronautics and Space Administration (NASA) in 1958.

Ameslan See American Sign Language.

AMHS Automated Message-Handling System.

AMI 1. See alternate mark inversion. 2. analog microwave link

Amiga computer A remarkable system developed by Jay Miner (hardware), Carl Sassenrath, R.J. Mical, and others in the early 1980s. The Amiga was well-equipped for its time with full serial, parallel, and joystick ports, full-color graphics, the ability to run multiple screens simultaneously in different resolutions, NTSC video compatibility, built-in 4-channel (16-voice) stereo sound, fast graphics display with coprocessing chips, and a Motorola MC68000 CPU chip running at 7.15909 MHz with 32/16-bit internal/external addressing. When it was released in 1985, this broad range of built-in features was uncommon.

The Amiga had a fully pre-emptive multitasking operating system (working smoothly in only 256 kilobytes of memory) which came with both a graphical user interface (GUI) and a text command line interface, both available for use at the same time. It helps to remember that in 1985 most personal computers lacked peripheral ports and employed single-tasking, monochrome graphics, and command line interfaces for prices ranging from $3,000 to $6,000. The Amiga 1000 offered everything built in, including monitor and sound, for under $2,000. The only other computer at the time significantly competitive with the Amiga was the Atari ST. Other Amiga models, including the 2000, 3000 and 4000, and updates to the OS were released over the next several years, followed by a new type of product from Commodore, the CD32.

The original name of the Amiga was the Lorraine, developed by a small company which was purchased by Commodore Business Machines in the fall of 1984. Commodore shipped the Amiga 1000 a year later with limited memory and no expansion slots. These compromises were presumably to keep the price down, although Jay Miner, the hardware developer, and 'Father of the Amiga' envisioned the Amiga 1000 with a megabyte of memory and several expansion slots.

The first of the Amiga line of computers, the Amiga 1000 was released in August 1985. It featured pre-emptive multitasking, built-in serial and parallel ports, a Motorola MC68000 CPU with coprocessor chips, two mouse/joystick ports, composite or RGB color graphics up to 640 x 400 pixels (more in overscan mode), and two-channel (16-voice) stereo sound.

At first the Amiga was hotly competitive with the Atari 520ST, which also supported good sound and graphics. Eventually, approximately five million Amigas were sold worldwide, many configured by NewTek as Video Toaster systems, thus launching the desktop video industry. In the mid-1990s, the Amiga was acquired by Amiga Technologies in Germany, and later resold to an American firm.

The Amiga is historically significant not only for providing the first viable platform for desktop video, but for its many capabilities that have subsequently been incorporated into other systems, showing the prescience and desirability of its design and features. Even a decade after its release, most personal computers lacked many of the Amiga's early capabilities, in spite of faster CPUs and other advances in technology. Recently the Amiga rights have been acquired by Gateway. See Amiga CD32; Commodore Business Machines; Mindset computer; Miner, Jay.

Amiga CD32 This MPEG 68EC020-based computer/game console released in 1993 was the last product promoted by Commodore, Inc. before they were bought out by Amiga Technologies. The small, sub-$350 unit was capable of playing standard MPEG, CDTV, and CD-I video CDs off the shelf and could function as an Amiga computer with an optional infrared keyboard. The CD32 supported 24-bit color, at resolutions up to 1280 *x* 512 pixels, and 8-bit stereo sound, with 18-bit CD-DA. The CD32 came with a game controller and built-in CD player. It was one of the first commercial systems to provide MPEG capabilities at 352 *x* 288 at 25 fps in PAL mode, and a little less in NTSC mode. S-Video, composite video, and radio frequency (RF) output were provided. See Amiga computer.

AMIS Audio Messaging Interchange Specification.

AML 1. Actual Measured Loss. 2. ARC Macro Language.

AMLCD active matrix liquid crystal display. See active matrix display.

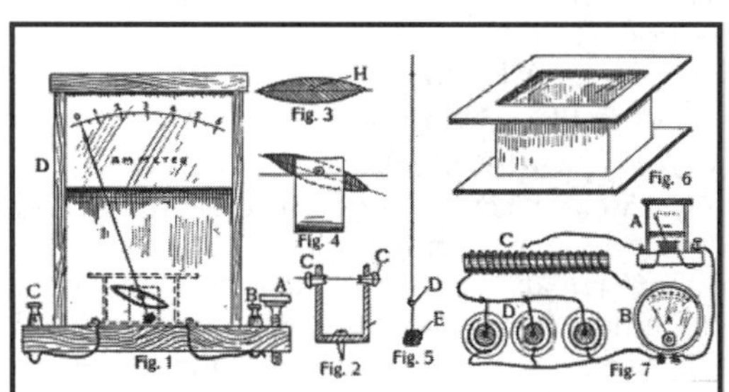

Historic drawings are often useful for understanding basic mechanics and electronics. The above diagram shows the basic structure and components of a historic ammeter. The ammeter is descended from the galvanometer. Popular Mechanics, May 1907.

ammeter, ampere meter An instrument for measuring the flow of electric current in alternating or direct current in ampere units. In communications circuits, where current may be very small (below one ampere), milliammeters and microammeters are used.

When used as a measuring and diagnostic instrument, an ammeter is connected in series with a circuit to measure the current as it passes through. If the total current is above the range of the ammeter, or is such that it might cause damage to the sensitive instrumentation, part of it may be predirected through a shunt connected in parallel. See ampere, shunt.

AMN Abstract Machine Notation.

ampacity The current-carrying capability, in amperes, of a circuit or cable. Typically ampacity is specified in product descriptions to indicate various types of cable assemblies, which may collectively consist of various combinations of wires and insulating materials.

AMPAS Academy of Motion Picture Arts and Sciences.

ampere, amp (Symbol - A) A unit of measurement of flow of electric current named after André-Marie Ampère. It is a practical meter-kilogram-second unit of electric current equivalent to a flow of one coulomb per second, or to the steady current produced by one volt when applied across a resistance of one ohm.

The international ampere was traditionally expressed as the steady current which will deposit silver at the rate of 0.001118 grams per second when flowing through a neutral silver nitrate solution.

The accepted scientific definition has since been replaced by a SI unit of electric current defined as a constant current which, in two straight parallel infinite conductors of negligible cross section placed one meter apart in a vacuum, would produce a force between conductors of $2 x 10^{-7}$N m-1. See volt, watt.

Ampère, André-Marie (1775-1836) A French physicist and mathematician who described and developed terminology for the nature of electricity. He also sought in 1820 to formulate a combined theory of magnetism and electricity following some of the investigations of H. C. Ørsted.

In 1826, Ampère published an important paper, the "Memoir on the Mathematical Theory of Electrodynamic Phenomena, Uniquely Deduced from Experience" in which he described electrodynamic forces in mathematical terms. Many later experimenters built on Ampère's ideas, and his discoveries led to the development of magnet-moving coil instruments. The ampere unit of measure of electric current is named after him. See ampere, galvanometer.

André-Marie Ampère was inspired by the discoveries of Ørsted, and worked with Arago to follow them up. Together they further investigated electrical and magnetic forces from which Ampère sought to formulate a unified theory to explain these phenomena.

Ampère's rule Based on his discoveries in electromagnetism, André-Marie Ampère described a method for determining the direction in which a magnetic needle orients itself when in the vicinity of a current of electricity. See left-hand rule, right-hand rule.

ampere-second A unit of electric charge flowing past a point in a current-carrying wire per second with a constant current of one ampere. Thus, amperes times seconds equals coulombs. See coulomb.

amplifier A device or system that increases the magnitude or intensity of a phenomenon such as sound. This is accomplished in electronics through an increase in power, voltage, or current. Amplifying a signal doesn't necessarily make it louder, bigger, brighter, etc. than the original. The effect of amplification at the receiving end, or at a transfer point, may increase the signal that is received above its characteristics at the point it is received, but not necessarily above the original. Some systems are intended to increase the signal above the level of the original, as in public address systems and blow horns. Amplification systems seek to minimize the possible amplification or introduction of noise in the signal, while increasing the meaningful parts of the signal. See regenerative relay.

amplitude 1. The measure of the magnitude or extent of some property, movement, or phenomenon. 2. The magnitude of variation in some changing quantity from an established value such as zero, or from its extents. See amplitude modulation. 3. In a diagrammatic representation of a wave, the measure of the magnitude from the highest point in the waveform, to the lowest.

amplitude distortion Assuming a fundamental wave in a steady-state system, an undesirable condition in which the outgoing waveform differs from the incoming waveform sufficiently to affect the perception or informational content of the signal.

amplitude equalizer Corrective electronics, usually passive, designed to compensate for less than desirable amplitudes over a range of frequencies. Equalizers are used in audio recording and playback, especially for music.

amplitude fading In an amplitude modulated carrier wave, fading is the attenuation of the amplitude across frequencies, more or less uniformly.

amplitude modulation AM. A very common means of adding information to a carrier wave. A basic wave carries no information. By varying or *modulating* the amplitude in a predetermined way, signals can be created which can be reconstructed as data, sound, or images at the receiving end of the transmission. This system was adopted in the early telegraph systems, and is familiar in the form of AM radio broadcasts. AM radio typically requires about 10 kilohertz of bandwidth, and is more subject to noise than frequency modulated (FM) radio. Designation of AM radio frequencies is under the jurisdiction of the Federal Communications Commission (FCC), and the they have changed from time to time. In 1993, the FCC increased the upper limit of the AM band from 1605 kHz to 1705 kHz. Once frequency modulation (FM) was developed by Armstrong, it was thought that its superiority would overshadow amplitude modulation, but AM radio stations are still common decades later.

One of the simplest ways to modulate is to create intervals of current which are either on or off, as in Morse code telegraph communications, and some types of binary computer signaling. Most computer modems use amplitude modulation and demodulation to convert from digital computer transmission signals to analog telephone transmission signals, and back again at the receiving modem.

Various types of amplitude modulation have been developed, and other nonamplitude modulation techniques exist, such as frequency modulation, in which the frequency of the signal, rather than its amplitude is varied. See absorption modulation, frequency modulation, modem, modulation, quadrature phase shift keying.

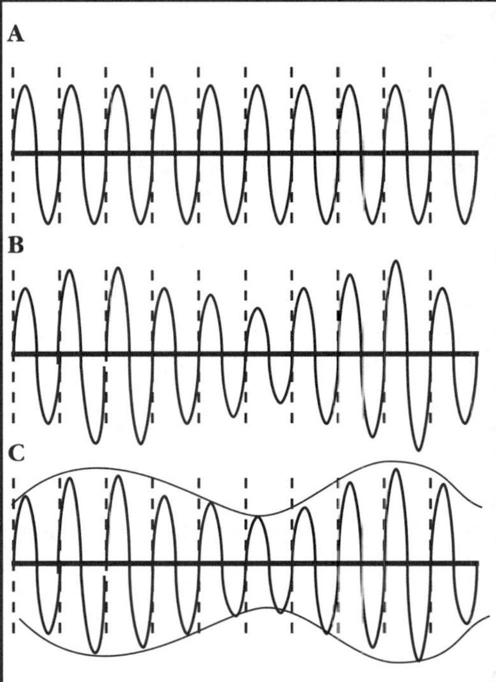

The top diagram (A) shows an unmodulated 'carrier' signal. The middle diagram (B) shows the signal modulated so that the amplitude varies through time. The bottom diagram (C) shows the 'modulation envelope' which conveys useful information, such as the magnitude of the modulation.

amplitude separation In television transmissions, the separation of the incoming signal into a video component and a synchronization signal component.

amplitude shift keying, intensity modulation, on/off keying ASK. A basic type of modulation that employs a constant-frequency signal, with two different signal levels used to represent binary values. In its simplest form, one state is represented by the lack of presence of the carrier, and the other by the presence of the carrier at a constant amplitude, hence on/off keying (OOK).

AMPS Analog/Advanced Mobile Phone Service. An analog mobile phone service first implemented in 1976 in the U.S. and Korea. It uses the same bandwidth as a landline voice channel, but is modulated onto a frequency modulated (FM) carrier using frequency division multiple access (FDMA). This system is slowly giving away to digital systems which offer greater features, and more call security. See cellular phone, DAMPS, NAMPS.

AMR anisotropic magneto-resistance.

AMS 1. Account Management System. 2. American Meteorological Society. 3. Attendant Management System.

AMSAT The Radio Amateur Satellite Corporation. A global organization of amateur radio operators who share an active interest in building, launching, and communicating amateur radio signals through non-commercial satellites. AMSAT was founded in 1969 as a result of the 1961 Project OSCAR satellite launchings. AMSAT was established as a not-for-profit educational organization to foster amateur participation in space research and communication.

The following chart shows some of the earlier satellite projects, following the initial OSCAR series.

AMSAT now consists of a number of loosely affiliated organizations around the world, some bearing the AMSAT name with extensions, working together through cooperative rather than formal arrangements.

Many early launchings have piggy-backed as secondary payloads on weather satellites. More recently, AMSAT satellites have shared launch vehicles with other commercial and scientific craft.

Many AMSAT enthusiasts are highly skilled technicians, and their knowledge and expertise have contributed to developing new technologies, in cooperation with a number of agencies, including the European Space Agency (ESA). See OSCAR.

AMSC See American Mobile Satellite Corporation.

AMSC-1 A commercial satellite, operating in the L-band frequencies, owned by American Mobile Satellite Corporation. AMSC-1 provides voice, data, facsimile, paging, and other mobile communications services, particularly commercial transport companies. Communication is through satellite phones, or cellular/satellite

Satellite	Launch Date	Tech. Details	Notes
Phase II Satellites - developmental, low-orbit, operational, longer lifespan.			See OSCAR.
Phase III Satellites - operational, high elliptical orbit, longer lifespan.			
AMSAT-PHASE 3A	12 Dec. 1961	10 lb., beacon, 22-day orbit Non-rechargeable batteries.	Initiated by a U.S. west coast group. U.S. Air Force launched.
AMSAT-OSCAR 10	2 Jun. 1962	Better coatings.	Similar to OSCAR I, with some improvements. Coatings provided better temperature control.
AMSAT-OSCAR 13			
AMSAT-PHASE 3D			
Phase IV Satellites - operational, high geostationary or drifting geostationary orbit, long lifespan.			
AMSAT-PHASE 4A	12 Dec. 1961	10 lb., beacon, 22-day orbit	Initiated by a U.S. west coast group.

hybrid phones.

AMSS Aeronautical Mobile Satellite Service.

AMTOR *am*ateur *t*eleprinting *o*ver *r*adio.

AMTS Automated Maritime Telecommunications System.

.an Internet domain name extension for Netherlands Antilles.

AN access network.

analog 1. Relating to, similar to, linear, continuous with. 2. Circuits or devices in which the output or transmission varies as a *continuous function* of the input. Here are two examples which are commonly given to explain the distinction between analog and digital display and selection systems:

1. Time Piece Displays. Most analog watches have continuously sweeping minute and hour hands that move through a 360 degree arc through the action of internal rotating gears. Contrast this to a digital watch which stays on a one-minute setting until the next has been reached, and then 'flips over' the display to the next minute in discrete units.

2. Dials and Buttons. In older AM radios, the turning of an analog radio dial will move the station pointer in a continuous path through the various frequencies, and the transitions can be heard as the signals from various stations get stronger and weaker. In newer car radios, a push-button digital system is often used (sometimes in conjunction with an analog dial) to store the locations of preferred radio stations. Pushing the buttons 'jumps' to the desired stations without passing through the intervening frequencies.

Most traditional phone conversations occur as analog transmissions over copper wires, but newer digital systems are being produced, especially for mobile communications, and phone over Internet. When computers were first remotely accessed over analog phone lines, it was necessary to convert the digital signals from the computer to analog signals through modulation. With the growing availability of ISDN, end-to-end digital transmissions are possible. See digital, ISDN, modem.

anamux analog multiplexer.

ANC All Number Calling.

anchor 1. Something which serves to steady or hold, such as a guy wire or stake. 2. In hypertext programming, an element which serves to link to related information. The anchor delimits the two ends of the hyperlink, designated with a tag as follows:

```
<A link tag="location">link text</A>
```

ancillary charges Charges for optional or value-added services.

AND Automatic Network Dialing.

Anderson bridge A device, usually employing a galvanometer, that measures reactance in order to determine capacitance or inductance by balancing against a frequency standard.

Andreessen, Mark Andreesssen developed the first version of Mosaic, the precursor to the Netscape Navigator browser, in early 1993 while at the National Center for Supercomputing Applications. He was working with the Software Development Group developing for Unix. In 1994, he joined forces with Mark Bina, some of his colleagues at the University of Illinois, and developers from Silicon Graphics to form Mosaic Communications. They essentially rewrote the code, as the new Mosaic company didn't have the rights to market the version developed at the University. The company also had to change its name, so as not to infringe on the University rights to "Mosaic" as a tradename. The new company was called Netscape Communications and is now well-known for creating the Web browser known as Netscape Navigator.

Andrew File System AFS. A distributed file system named for Andrew Carnegie and Andrew Mellon which grew out of a collaboration between Carnegie-Mellon University and International Business Machines (IBM).

anechoic An environment without noise, or without significant noise. Sound recording rooms are designed to echo as little as possible, with thick, porous materials resembling foam egg crates absorbing the sound so it is prevented from reflecting back to the recording equipment. Speakerphones work better in anechoic environments. See acoustics.

angle brackets < > Symbols very commonly used in programming code as delimiters or arithmetic operators. These are best known as *greater than* (>) and *less than* (<) symbols. In HTML, the angle brackets delimit markup tags, e.g., *<p>* signifies a paragraph marker.

angle of arrival The angle between the Earth's surface and the center of a radiant beam from the antenna to which it is radiating.

angle of beam The predominant range of direction of radiant energy from a directional transmitting antenna.

angle of deflection See angle of divergence.

angle of divergence In a cathode ray tube (CRT), a description of the spread or divergence of an electron beam from an imaginary center position for that beam as it travels from the cathode to the coating on the inside surface of the front of the tube. A well-focused beam should spread as little as possible. Higher amplitudes tend to result in higher divergence. A perfectly straight beam has an angle of divergence that equals zero.

angle of incidence The angle that a radiant beam (or line) contacting with a surface makes perpendicular (normal) to that surface at the point of incidence.

angle of radiation The angle between the Earth's surface and the center of a radiant beam from the antenna from which it is radiating.

ångstrøm, angstrom (Symbol - AAU, Å) A unit of measurement of length named after Anders J. Ångstrøm. Ångstrøm applied this unit to the measurement of wavelengths when mapping the sun's spectrum. It is now also used to express atomic and molecular dimensions. It can be expressed as one ten-billionth of a meter, or one-tenth of a nanometer, or 1×10^{-8} centimeters.

Ångstrøm, Anders J. (1814-1874) A Swedish scientist who researched the solar system and radiant waves. The angstom unit of measurement is named after him.

angular misalignment loss In systems which use optical beams, a misalignment of fibers, mirrors, or connecting pieces resulting in the loss of beams that deviate from the desired path.

ANI See Automatic Number Identification.

ANIK The first domestic communications satellite, launched in 1972 by Telesat Canada, ANIK was fully operational by 1973. Circuits on the satellite were leased to Radio Corporation of America (RCA) until RCA had its own satellite. ANIK is actually a series of satellites, ANIKs C, D, and E were built in Canada's David Florida Laboratory (DFL) facility. The Canadian Broadcasting Corporation was the first television broadcast station in the world to use satellite broadcasting of their shows, utilizing ANIK in 1972. See Canada Space Agency.

animate To bring to life, to give movement to, to move to action, to manipulate so as to simulate the effect of movement.

animation, cell To create the illusion of movement through the rapid sequential presentation of a series of cells. These are individual still frames that are similar to one another except in small details, drawn on cellophane, or another transparent material, so that background images and other frames can be subimposed or superimposed. Each cell is photographed once or twice, depending upon the speed of the movement, and the number of images needed. The human visual perception system functions in such a way that such a series of still frames presented at about 24 to 40 frames per second is perceived as movement. Humans are not able to individually resolve or distinguish each frame at those speeds. Film and computer animation models are based on this characteristic of perception. See frame, persistence of vision.

anisotropic Direction-dependent electrical or optical properties, e.g., polarized antennas.

anneal To use heating and subsequent cooling to alter the properties of a substance (such as glass or wire), to make it stronger, less apt to crack or tear, or to fuse it with associated substances. Wires can be annealed to make them more durable.

announcement 1. The message that plays on an answering machine when the machine accepts an incoming call. 2. A message sent from a system administrator on a network to users, usually to let them know that the system may be shutting down temporarily for backups or maintenance. 3. A general message or page over a public address (PA) system.

annular ring A ring inserted around a hole as a support structure to hold a connection or wire, or to serve as an indicator. Small annular rings are used in printed circuit boards. Slightly larger annular rings are sometimes used on cables to indicate connection points.

annunciator An intercept device that indicates (with light or tone) the state of a circuit for information or diagnostic reasons. Information revealed by the annunciator may be as simple as the fact that the phone is ringing, or more sophisticated, as in the state of a specified piece of equipment elsewhere on the line.

anode (Symbol - P) 1. The positive terminal of an electrolytic cell. 2. The negative terminal of current-providing cell or storage battery. 3. In a system of moving electrons, as in an electron tube, the direction to which the electrons flow, or are attracted, originating from an cathode, and sometimes passing through a controlling grid. The anode is sometimes in the form of a thin plate of metal. See cathode.

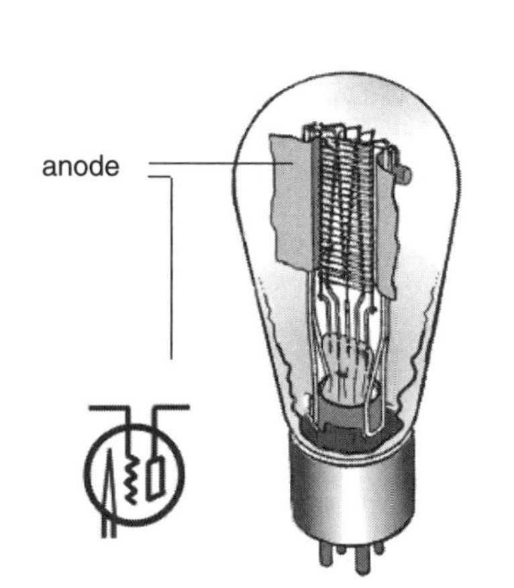

On the left is the symbol for a three-element electron tube. On the right is a tube drawn so the different elements can be seen behind the thin metal plate which is the anode, next to the grid (resembling a fine Venetian blind). The anode functions to attract the electrons emitted by the cathode (the filament, in this case).

Anonymous Call Rejection ACR. An optional telephone subscriber service in which a call made to the subscriber which is 'blocked,' that is, one that doesn't show up on a Caller ID system, is rejected. A message is then played, advising the caller to disable call blocking and dial again.

anonymous FTP A configuration of an FTP archive site that provides limited public access to users without the assignment of individual passwords. When you log into an FTP site, you will be prompted for a username. Type "anonymous" or "ftp" (in a text window, the command must be typed in lower case); you will then be prompted for a password, to which you respond with your full email address.

Assuming you have responded correctly to the prompts, and the system is set up for anonymous FTP, you will now have limited access to file directories, downloads, and perhaps uploads on the system. Many vendors are now using FTP sites to distribute demonstration versions of their software, and to dispense upgrades and technical support documents. A sample ftp login is illustrated under the entry for ftp. See Archie, FTP, File Transfer Protocol.

anonymous remailer An electronic mail transit point which deliberately obscures the identity of the poster in order to ensure his or her privacy. These remailers have been known to provide protection to email senders from war-torn countries who are providing information, or asking for assistance, and wish to protect their personal safety and anonymity. Anonymous remailers are occasionally used for illegal purposes, or to harrass people on the net, but generally, anonymous servers provide an important service. Refugees from political persecution have sometimes used them, and a number of celebrities on the Internet, who wish to maintain their privacy, use anonymous remailers to post to public newsgroups.

ANS 1. See Advanced Network and Services. 2. answer.

ANSI See American National Standards Institute.

ANSII IISP ANSI Information Infrastructure Standards Panel.

Answer Back A signal (light or tone) that indicates the called party is ready to accept a call or transmission, or which acknowledges receipt of a transmission. See ACK, Answer Supervision.

Answer Back Supervision See Answer Supervision.

Answer Supervision A verification system consisting of an electrical signal which provides information between the local phone company and a long-distance service as to the successful connect status of a call. The signal is transmitted through the long-distance connection to make sure the call has been answered by the callee, and billing timing is initiated. In the past, long-distance calls were billed on an averaged wait-time-to-connect billing system without actual verification of the connection, and in fact, some small long-distance services still do it that way, and initiate billing after a specified number of rings, before the called party answers.

answering jack In manual switchboard systems, the jack that is used by the operator to connect the circuit for the *incoming* call. This is then connected with the calling jack to put the call through to the subscriber.

ANT Access Network Termination.

antenna In its simplest form, a passive conductive device for transmitting and/or receiving signals, chiefly broadcast signals from radio, television, and radio phones. Most antennas for use with longer wavelengths are constructed from wires and metal cylinders or rods. Most antennas for use with very short wavelengths (microwaves) are designed as parabolic dishes.

A simple, vertical, one-quarter wavelength conductive wire can function as an antenna, if it is mounted where there are transmission waves, and is connected at one end to a receiving device, like a radio. Most mobile 'whip' antennas are of this kind, with maximum transceiving capabilities oriented along a horizontal plane, without much vertical capability. They are commonly seen on cars and trucks.

Antennas can be designed to transmit selectively, or in combination, various ground, directed, reflected, or ionospheric waves. Transmitting antennas tend to be placed high, to transmit unobstructed in all directions, while receiving antennas tend to be focused, to capture just the desired transmission, with as little interference as possible from others.

Antennas are mounted in many places: on components, on rooftops, on mountaintops, in orbit, and on moving vehicles. They vary widely in shape, from thin rods, to branched, tree-like structures, to monuments like the Eiffel Tower in Paris, France.

An antenna generates two types of fields, electrostatic (along its length), and magnetic (associated with the antenna's current). Antennas come in many sizes, from rabbit ears on old TVs, to high poles with guy wires in the yard of a CB radio enthusiast. Generally the higher and broader the antenna, the greater its range or scope, although there are exceptions to this general rule, based on the shape, and the frequencies involved. The Eiffel Tower was used by Lee de Forest as an antenna for sending a historic transatlantic radio broadcast. The orientation, length, and shape of an antenna will affect the type of frequency it can draw or transmit, and its signal strength. A radio antenna, for example, is commonly designed so that its length is some multiple (e.g., double), or division (one-half, or preferably, at least one-quarter) of the radio wave frequency.

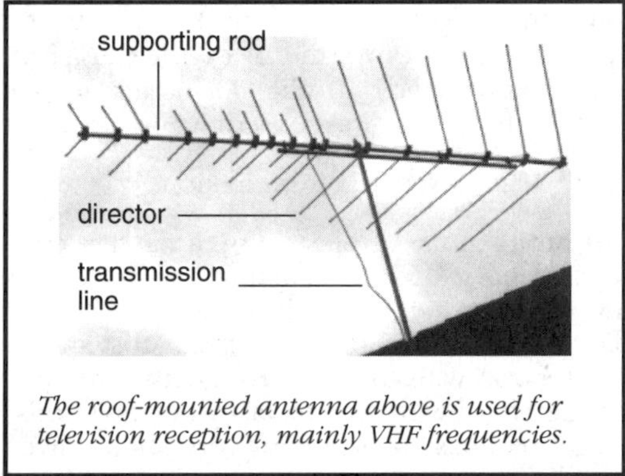

The roof-mounted antenna above is used for television reception, mainly VHF frequencies.

There is no one type of antenna which is best for all frequencies, since its shape must be optimized in relation to the length of the waves it is transmitting or receiving. Some types of transmission, such as broadcasts from satellite cable stations or pulses from distance stars, must be captured with devices, such as parabolic antennas, that focus the waves. Due to their importance to telecommunications, this dictionary includes many listings under individual types of antennas. See also ground wave, Hertz antenna, ionospheric wave, isotropic antenna, Maxwell's equations, Marconi antenna, polarization, radio wave, satellite antennas, waveguide.

antenna effect Improperly shielded loop antennas, or those in which the loop is incorrectly constructed, or too closely spaced, may lose the benefit of the structure and behave as though they were simple whip antennas instead.

antenna gain An expression of the effectiveness or power of a signal from an antenna, usually selected at the point of its maximum radiation, when compared to a standard such as an isotropic antenna. Gain is commonly expressed in decibels. Gain is the greater power of transmission of a beam in a particular direction, as compared to a reference standard. See isotropic antenna.

antenna impedance A ratio, at a specified point, of voltage to current such that impedance equals voltage divided by current. The impedance of any antenna will vary along its length according to a variety of factors. See resonant frequency.

antenna lobe, antenna pattern A 2D or 3D diagrammatic description of the direction angles and numbers of radiating patterns (or receiving patterns) of a specific type and configuration of antenna. The name is derived from the fact that waves tend to spread out in a more-or-less rounded or circular pattern, hence creating lobes in the diagram. Sometimes these are compared against a hypothetical isotropic antenna. The antenna pattern of a directional antenna and that of general-direction antennas can be quite different.

antenna noise bridge A diagnostic device for determining the complex impedance of an antenna system. It is placed in series between the antenna feed line and its receiver.

antenna polarization A number of polarization structures and schemes to maximize the effectiveness or versatility of an antenna for different uses. This polarization may be linear or rotating circular. Once polarized, an antenna transmits and receives with the same polarization (unless, of course the antenna is reoriented between transmitting and receiving modes). Polarization is employed in ground-wave antennas, but is less effective for ionospheric waves (sky waves). See polarization.

antenna stacking An arrangement of antennas in a vertical plane, one above the other, with a common transmission line, to improve gain and horizontal directivity.

antenna tuning The process of maximizing transmitting or receiving capabilities; if you're trying to do both with one antenna, sometimes the result is a compromise. This can be done through structure, by adjusting the sizes and positions of the various parts, and by orientation, by adjusting the angle and direction of the antenna. Even the degree of overlap of the tubing in dipole or Yagi-Uda antennas can be important. Since antenna structures are tied to the length of the wavelengths concerned, structure is quite important. In directional antennas, such as parabolic antennas, computerized servos are often used to make small adjustments, and can be programmed to track a satellite in its orbit. See waveguide.

anthropomorphize *v.t.* To confer or ascribe human form or characteristics to something which is not human. Cinema robots tend to be anthropomorphized, to give them appeal. People often anthropomorphize through projection by attributing human thoughts and motives to non-human creatures such as pets. See android, bot.

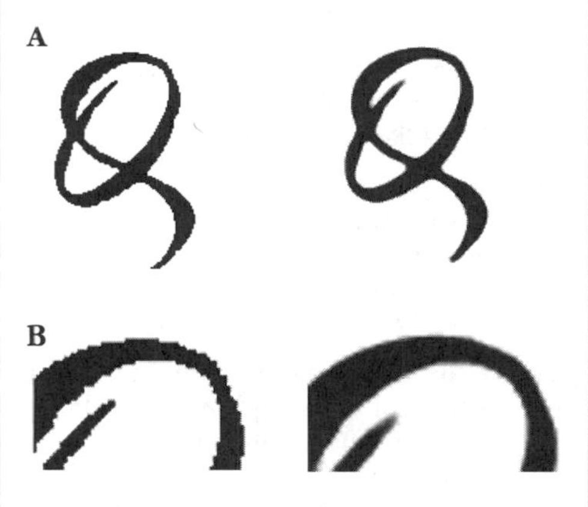

The raster image on the left is aliased, that is, the edges are rough, with a 'staircased' effect. The effect is even more noticeable when the image is scaled up (B). By using gray tones that are intermediate between the white background and the black letter, the image can be 'antialiased' so the contours appear smoother.

antialias *v.t.* In computer imaging, to add picture elements to create an illusion of gradual transitions between otherwise jagged or sharp transitions. Aliasing (also called staircasing or 'jaggies') may occur at sharp tonal changes in a grayscale image, or at line boundaries in a monochrome image.

In low resolution raster images in grayscale or color, it is possible to use intermediary tones or colors between dark and light areas to reduce the effect of aliasing, providing the illusion that the shape or object is smooth. In sound reproduction, the same principles can be applied to 'smooth out' a rough sound transition which may occur due to low quality components, or sound recording technologies, or digital sound sampled at low resolutions.

anticurl A property of some printing media which allows them to pass through a roller or other paper feed mechanism without curling up, or a feature of some printers and facsimile machines, which feeds the paper through a straight paper path, or one which has a double bend, so the paper doesn't curl after exiting.

antinode In a standing wave in an oscillating body, the point of maximum amplitude between the nodes on either side.

Antique Wireless Association, Inc. AWA. Founded as a not-for-profit in 1952, operating from New York State, the AWA is committed to research, preservation, and documentation of the history of wireless, and administrates the Antique Wireless Association Electronic Communications Museum.

antistatic A specialized tool or material that resists the buildup of static charges or which gradually dissipates a charge rather than sending out a quick spark. There are antistatic wrist bracelets and antistatic mats for people who work on electronics, and antistatic packaging for the storage and shipping of sensitive electronic components. See static.

antistuffing flap An antitheft, antivandalism mechanism in a coin operated vending machine like a payphone that prevents unauthorized users from blocking the coin return chute and then coming back later to extract any coins caught in the block from interim users.

antivirus program A software program intended to detect and disable computer 'viruses,' that is a software program designed to vandalize or penetrate a system without the consent or knowledge of the user. Some virus checkers run as background tasks, and monitor any new files which are copied to the system. If a known virus, or unusual program, is detected, the user is alerted, and the software attempts to disable the 'intruder.' It is almost always advisable to run good antivirus software, particularly if software is being downloaded from bulletin boards, the Internet, or other public file archives. It is also a good idea to do so on any networked computer that shares file access with other computers. See virus.

ANU-NEWS A complete news system designed for Digital Research VMS systems by Geoff Huston. It supports standard DEC screen-oriented reading, posting, replies, and support of NNTP in a way similar to other full news systems. The software is freely distributable.

anycast In IPv6, the proposed successor to IPv4, the primary protocol used for the Internet, anycast is related to communications between devices within a group, with the host device passing on some of the responsibility for routing updates to the closest member of a group.

anywhere fix The capability of a receiver to begin position calculations without an initial approximate location and approximate time, used in Global Positioning Systems (GPS).

.ao Internet domain name extension for Angola.

AO acousto-optic.

AOCS attitude and orbit control system. See telemetry.

AOL America OnLine. A large, commercial Internet services provider which requires the use of proprietary software to access the service. Basic monthly rates are available, although a number of the premium services are charged additionally, per access, or per minute.

AOM acousto-optic modulator. See acousto-optic modulation.

AOR Atlantic Ocean Region. A longitudinal regional designation for geostationary satellites.

AOS 1. Alternate Operator Services. See Operator Service Providers. 2. Area of Service.

AOSSVR Auxiliary Operator Services System Voice Response.

AOTF acousto-optic tunable filter.

AP 1. action potential. 2. aiming point. This can be used to direct an antenna beam. 3. application program. 4. Applications Processor. An AT&T telephone add-on which provides more options. 5. array processor. 6. Associated Press. A commercial association with a long history of using long-distance communications services to gather and disseminate news.

Apache A freely distributable HTTP server for Unix systems, developed in the mid-1990s. It is the most prevalent server on the Internet.

APAD asynchronous packet assembler/ disassembler.

APAN Asia-Pacific Advanced Network Consortium. This organization was established in 1997 to carry out research and development in advanced networking applications and services in the Asia-Pacific region.

APaRT See Automated Packet Recognition/ Translation.

APC 1. adaptive-predictive coding. 2. advanced process control. 3. Association for Progressive Communications.

APCC The American Public Communications Council, affiliated with the North American Telecommunications Association (NATA).

APD avalanche photodiode.

APDU Application Protocol Data Unit.

aperiodic antenna A circuit or antenna structure which does not have a tendency to vibrate within the range of frequencies to which it is tuned.

aperture 1. An opening or hole, usually for controlling the admission of waves or particles, as in cameras and telescopes. The size of the opening, and the speed with which it can be opened or closed, may be adjustable. 2. In a one-way antenna, the portion of the plane surface, perpendicular to the direction of maximum radiation, through which the major portion of the radiation passes. See aperture antenna.

aperture antenna An antenna characterized by a lens, horn or reflector used as an 'aperture' or directed region through which the majority of the radiant energy passes.

aperture distortion Aberrations in the focus, size, or shape of an image recorded through an aperture. Faults in an aperture, such as shape, orientation, perforations, jamming, speed of opening, etc. can cause undesirable effects.

aperture grill A focusing mechanism inside a cathode ray tube (CRT) which is analogous to a shadow mask that helps target a beam on the inside coating of the monitor. An aperture grill consists of fine, aligned wires, and is said to have advantages over conventional shadow masks. See shadow mask.

aperture mask A thin grill or perforated sheet control mechanism that is commonly mounted inside an electron tube such as a color cathode ray tube. The aperture mask is used to more precisely control and single out the electron beam, or portion of a beam, which is intended to be passed through the mask to the display. See shadow mask.

aperture ratio In optics, especially photography, the ratio of the useful diameter of a lens to its focal length, the reciprocal of the *f*-number. See aperture, *f*-stop.

API application programming interface.

APIC advanced programmable interrupt controller.

APL Abbreviation for A Programming Language which is a high level computer programming language.

apoapsis The point of greatest separation between two orbiting bodies. See apogee.

apogee The highest or most distant point, such as the apogee of Earth's orbit, that is, the point at which it is farthest from the sun. The apogee of an orbiting artificial satellite is the point at which it is most distant from the Earth (which can be described in more than one way, but is usually from the center of Earth's gravitational field, or the center of an elliptical orbit). See apoapsis, geostationary, orbit. Contrast with perigee.

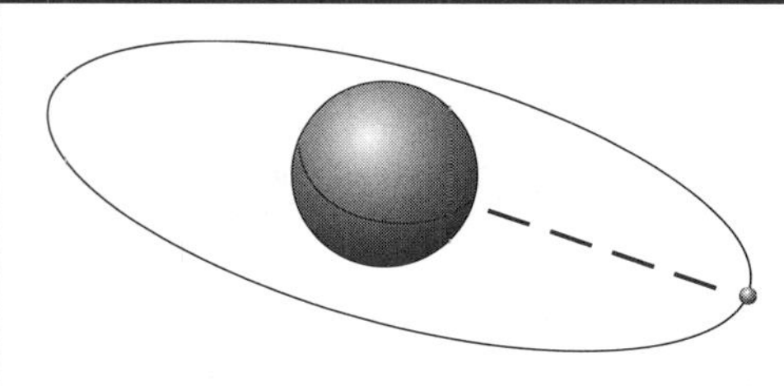

The two common types of orbits used in satellite communications are elliptical and geostationary. Elliptical orbits are sometimes planned so that the satellite spends a greater part of the orbit over specific land masses, rather than over the more plentiful oceans.

app See application.

APPA See American Public Power Association.

APPC advanced program-to-program communications.

APPC/PC An International Business Machines (IBM) application that implements advanced program-to-program communications (APPC) on a personal computer.

append Add, affix, subjoin. It is very common in software programming to add the contents of a list, table, or file to another. Append is used most commonly to indicate additions to the *end* of a file; if the additions are in the middle of a file, or spread through various parts of the file, the term *merge* is used. See adjunct.

Apple IIGS computer A 65C816-based 2.8-MHz 16-bit addressing computer in the Apple II line, released in fall 1986 by Apple Computer, Inc. Unfortunately, in its support of the earlier 6502-based Apple IIs and competition from the Atari 520ST and the Amiga 2000, the Apple IIGS, in spite of its attractive features, good graphics, and fantastic 32-oscillator Ensoniq sound chip, was dubbed 'too little too late' and didn't attract enough support to sustain the machine.

Apple Computer, Inc. A microcomputer hardware and software company located in Cupertino, California. Apple Computer was founded in 1976 by Steven P. Jobs and Stephen G. Wozniak, with Mike Markulla providing early business plan and financing support and Arthur Rock providing venture capital. Steve Jobs is known best for his marketing presence and administration tasks; Steve Wozniak is rememberd for hardware design and computer-related technical tasks. Their initial product, leading up to the formation of Apple Computer, was a *blue box* designed to gain unauthorized access to long-distance lines, after which Wozniak developed a microcomputer circuit board, much like the original Altair kit, and this became the original Apple I computer. The Apple I was little more than a circuit board with neither case nor keyboard, yet the entrepreneurs sold about four dozen to excited hobbyists. They soon followed up with the Apple II at the West Coast Computer Faire in 1978.

Both Wozniak and Jobs had a strong commitment to providing computing services to education. Evidently the alliance of the young entrepreneurs was successful because Apple grew from those small beginnings to be one of the most significant microcomputer developers and retailers of the 1980s and 1990s, particularly with its Macintosh line, introduced in 1984 (following the less successful introduction of the Lisa a year before). Paired with the Laserwriter printer, the Macintosh launched a desktop publishing revolution. The subsequent PowerMac and G3 lines provided fast processors at lower prices than previous systems.

When sales flattened out and doom-sayers predicted the demise of the company, Apple responded by launching the *iMac,* a powerful, portable, individualist computer with an upbeat design and appeal similar to that of the Volkswagen 'Bug' in the 1960s. The *iMac* evidently attracted more than loyal Macintosh users, with 16% purchased by new computer owners or those who had previously used other brands.

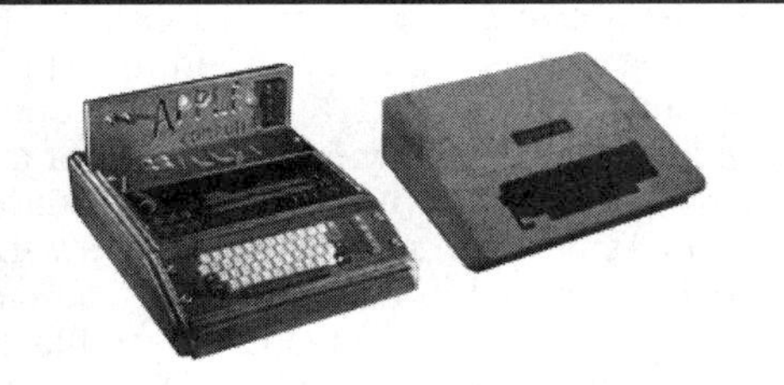

On the left is a composite image of the original Apple computer next to the popular commercial version of the Apple][on the right, which was especially popular in schools and homes. The top flap allowed access to the components.

Apple Computer went public in 1980 and forged new directions, pioneering the graphical user interface developed at Xerox PARC, and incorporating the point-and-click style of interaction into the Lisa computer in 1983. The Lisa was ahead of its time, and under-appreciated. It did not sell well, probably due to the steep price tag. However, most of the characteristics of the Lisa showed up over the years in the Macintosh line, introduced in 1984, which eventually began to sell very well, after a slow start with the cute, but limited *Little Mac* which had a small black and white screen and a single floppy drive.

Apple Computer continues to market computers and software, continually bringing out new desktop models and laptops, and now offering two operating systems, MacOS and Rhapsody. See Jobs, Steven P.; Macintosh; Wozniak, Stephen.

AppleTalk A proprietary computer network protocol developed by Apple Computer, Inc., which functions independently of the layer on which it runs. Implementations vary, and include 1. LocalTalk and similar (230 to 300 Kbps), commonly used among printers, Macintosh computers, and emulators; and 2. EtherTalk (10 Mbps), which provides broader multiplatform communications.

AppleTalk Address Resolution Protocol AARP. A protocol in the AppleTalk networking protocol stack which maps a data link address to correspond to a network address.

AppleTalk Control Protocol ATCP. A means for configuring, enabling, and disabling AppleTalk Protocol modules at both ends of a point-to-point (PPP) link. ATCP uses the same basic packet exchange mechanism as the Link Control Protocol (LCP). See RFC 1378.

AppleTalk Remote Access ARA. A capability which allows remote access, through two or more Macintosh computers, connected through an AppleTalk network, to share a serial device on the remote system, usually a modem. In other words, if there is only one phone line and one modem, and four computers attached to the network, ARA can be set up so any one of the people using the computers without a modem can access the modem through the other computer (one at a time) as though it were attached to the local machine.

application, applications program A catch-all designation for computer software programs, especially high-level ones intended for end-users, such as databases, spreadsheets, word processors, graphics programs, telecommunications programs, programming tools, etc.

application framework The basic logical structure in an object-oriented development environment. When software is being designed, there is often a pre-existing set of assumptions within which the user interacts with the computer. For example, when a user sees something on a screen that looks like a button, he or she will expect something to happen when it is clicked, or double-clicked, depending upon the system, and the experience of the user.

These basic assumptions are cultural and experiential, and are important in the design of software. If the software interface is obscure, or too radical to be understood, it may not be of practical use. A certain degree of consistency, immediacy, and familiarity are important factors.

By using an application framework, not only will the user be presented with a consistent set of stimuli and tools, but the programmer will have a context within which to create the software. The framework exists at several levels, at the user interface level, at the applications design level, and at the lower levels in which the parts, components, interactions, and processes are created.

In an object-oriented programming environment, it is easier to apply a framework, and to work within a framework, when shared objects, classes, and other programming primitives and structures are being used and reused. For this reason, most of the thinking about application frameworks has arisen in object-oriented programming environments, such as those utilizing Smalltalk, C++, and various graphical interface builders such as the NeXTStep Interface Builder, or Apple Computer's MacApp. See application generator.

application generator AG. A software program which greatly facilitates the development of software applications code by providing a set of tools to describe the program, leaving the details to the software. It's a way of automating programming, and taking out many of the drudge activities and details that are easy to mistype when coding in text with an editor.

This type of programming approach wasn't prevalent on desktop computers until Power Windows was released for the Amiga 1000 in 1986. It was one of the earlier microcomputer application generators, allowing the user to essentially 'draw' the application as though using a paint program, placing buttons and icons, windows, and other structures where they were needed. Colors and logical relationships could then be dynamically adjusted with the mouse, and then presto! select build and it would automatically generate C, BASIC, or Assembler code. The code could then be edited and changed as needed.

With this type of programming environment, the programmer doesn't have to worry about counting pixels, about guessing what the interface will look like, or about writing reams of C code before even the smallest activity can take place on the screen. This is a very good idea. NeXTStep incorporated a very nice interface builder about two years later, which took aesthetics, utility, and logical linking to object-oriented structures several steps beyond Power Windows, and facilitated graphical creation of windows, menus, tables, buttons, and much more, providing a fast and easy way to create an interface and connecting structures that were consistent with the NeXT *application framework*. An hour with the interface builder could easily equal two days of coding by hand with a text editor. In the 1990s, other desktop systems began to come out with interface builders, an idea that has great practical value, especially as object-oriented environments become more prevalent.

Some authoring systems also function as application generators, as do some programmable databases. If the software front-end that allows authoring and database configuration without programming, also provides an option to save out code that can be accessed and manipulated, usually with a text editor, and linking to operating system or program structures, then it is a form of application generator. See application framework.

application program A very broad, generic term for almost any computer program which interacts at the user level. That's not to say that system administrators don't use applications; they do. They just happen to be more technical applications aimed at a technical user level. Short, specific applications that do a single task or a small number of tasks are sometimes called utilities, such as a disk utility for formatting disks, or a conversion utility for changing a TIFF file to a BMP file, or a copy utility for duplicating disks.

Application Program Interface API. An interapplication, or intervendor interface that provides a somewhat standardized means of allowing programs to talk and work together.

Application Software Interface ASI. A means of working within a common application interface for provision of ISDN-related digital telephony services. See North American ISDN Users Forum.

application-specific integrated circuit ASIC. A computer chip or small, specialized circuit designed to enable or enhance a specific type of application. As examples, ASIC video cards have been designed to drive specialized monitors, ASIC modem cards provide functionality to specialized or enhanced modems, ASIC daughterboards sometimes provide hardware support to rendering and ray-tracing applications.

applications processor A computerized system that can be integrated with a phone system to add functionality. These may include voice mail, Automated Attendant, Call Detail, networking (packet switching), and others. See peripheral device.

Applications Technology Satellite program ATS. A series (ATS-1, ATS-2, etc.) of satellite launchings carried out by the U.S. National Aeronautics and Space Administration (NASA) to test payloads and study space. Five of these craft in three configurations were manufactured by Hughes between 1966 and 1969. See chart on following page.

Applied Computer Telephony ACT. A commercial product from Hewlett-Packard Company for integrating voice and data analysis technologies on HP systems. The system is used in conjunction with private branch phone exchanges to record, handle, and evaluate call-related transactions. See Hewlett-Packard.

APPN See Advanced Peer-to-Peer Networking.

APS 1. Advanced Photo System. A new format for color photography media and finishing being promoted by the larger film suppliers. 2. Automatic Protection Switching. A network switching technique which varies from system to system. In some, it is a device which automatically switches over from a primary to a secondary circuit if excessive error conditions are detected on the primary circuit. For Synchronous Optical Networks (SONETs), APS is defined in ANSI T1.105-1995. In SONET, APS carries the signaling bytes associated with establishing and releasing the protection of the optical facility.

APTS Association of Television Stations.

.aq Internet domain name extension for Antarctica.

AQL acceptable quality level. An industry-established confidence level.

aquadog A colloidal suspension of graphite in water used as a coating inside electron tubes, especially cathode ray tubes (CRT), to facilitate the conduction of electrons away from the screen.

.ar Internet domain name extension for Argentina.

Ar argon.

AR Automatic Recall.

ARA See AppleTalk Remote Access.

Aragon, Dominique In 1820, Aragon described how an artificial magnet could be created by winding a coil around a piece of iron or steel that was carrying an electrical current. Soon after, electromagnets were developed.

ARAM audio RAM. A low-cost, low-grade integrated memory chip which is good enough for digital answering machines, and other inexpensive consumer products.

ARB all-routes broadcast.

arc A very bright discharge across a gap in a circuit. See arc lamp; Aryton, Hertha Marks.

ARC See Ames Research Center.

arc converter A device which is used to convert direct current (DC) into undamped or continuous wave radio frequency (RF) signals. This technology was used in early radiotelegraphy. Many aspects of global radio communications in the early 1910s were based on this technology. See Poulsen arc.

arc lamp A type of electrical lamp which takes advantage of a characteristic of electrons to jump a gap in a circuit and produce an intense light under certain circumstances. It was invented by Sir Humphrey Davy in the early 1800s, and came into widespread commercial use in the late 1870s when it was incorporated into street lamps.

Archie Network archiving software developed by Peter Deutsch, Alan Emtage, and Bill Heelan. Named for the word *archive*, Archie is an Internet query tool which tracks the contents of anonymous ftp sites. It was introduced at McGill

Application Technology Satellite Program

Satellite	Launched	Notes
ATS-1	1966	Spin-stabilized synchronous altitude. Electronically despun antenna. Stationed over the Pacific Ocean. Successfully photographed Earth. Provided a Presidential communications link for recovery of Apollo 11.
ATS-2	1967	Gravity gradient stabilized. Insufficient thrust resulted in an elliptical orbit and it lost orbit after only 880 days.
ATS-3	1967	Synchronous orbit. Mechanical despun antenna, color camera which photographed tornados in 1968 and an eclipse of the Sun in 1970.
ATS-4	1968	Gravity gradient stabilization in synchronous altitude. Failed to reach intended orbit and lost orbit in 1968, two months later.
ATS-5	1969	Synchronous orbit. Gravity gradient booms for stabilization didn't deploy correctly, but some of the experiments were successful. It was retired in 1984.

University, Canada, in 1990. Archie allows users to retrieve a list of FTP locations by submitting file search criteria to an Archie server. See also Anarchie, Veronica.

Archimedes (~287-212 BC) A Greek mathematician and inventor born in Sicily who made contributions to our understanding of volume and displacement, and created the mathematical treatise "Measurement of the Circle" in which he described the calculation of the ratio of a circle's circumference to its radius. See Archimedes's principle.

Archimedes's principle A body immersed in fluid is buoyed up by a force equal to the weight of the fluid displaced. (This principle is humorously alluded to by actor Gary Oldman in a bathtub in the Cinecom Entertainment movie production of "Rosencrantz and Guildenstern are Dead.")

architect *n.* One who designs a layout or topology, such as a building layout, circuit board architecture, network routing system, etc. The architect frequently is also the one who drafts the technical drawings associated with the layout, and may or may not check electronics codes, building codes, or other regulations associated with the design.

architecture The design and layout of a process, system or facility. The architecture involves the overall plan and topology, in addition to the relationships and interconnections between the individual parts. It may also include the *direction* of information paths, or movement within the system. Good architectures usually try to incorporate, or at least balance, flexibility, robustness, efficiency, and scalability, whether it be the design of a building, or of a microprocessor chip.

archival A format, medium, or protective system designed to facilitate preservation. Archival papers and plastic sleeves are acid-free, or free of plastics that may change the information or degrade rapidly. Archival data storage formats are nonvolatile (magnetic media such as video tapes, audio tapes, floppy diskettes, etc. are not very stable over time and may be damaged by proximity to magnets) and resistant to damage and degradation. In the data industry, archival file formats are as important as the materials on which they are stored, as the information is useless if it can no longer be read or deciphered.

archive *n.* 1. A repository of records or files. A backup or duplicate of information made to preserve or prevent loss in compressed or uncompressed form. An archive generally contains information which needs to be kept over time, for one reason or another (legal, historical, etc.), but which is seldom or ever accessed. See anonymous FTP, Archie, archival, FTP.

archiver A term for software tools that are designed to store files in such a way that they take a minimum of space, and can be retrieved, reconstructed, and viewed at a later date. Software archivers often include compression algorithms and switches to allow an archived file to be scanned for header information without decompressing it. The degree of compression possible is very dependent on the interaction between the type of compression algorithm, and the type of data being compressed. Common software archivers include zip, lharc, Stuffit, and tar.

archiving The process of storing information, compressed or uncompressed, encoded or not encoded, such that it can be accessed and viewed at some future date, if needed. Archiving involves selecting a storage format, medium and location, and carrying out occasional or scheduled consolidation and organization of the objects or information. One of the big issues with archiving, besides space, is the development of efficient search and retrieval methods that make it possible to find a desired piece of information in a vast amount of data. See archival, archive, database, FTP.

ARCnet *A*ttached *R*esource *C*omputer *net*work. A popular pioneering local area network (LAN) developed by Datapoint Corporation in 1977 for use with thin coaxial cable. Incorporating a modified Token-Ring passing scheme, ARCnet provides high-speed baseband communications at 2.5 Mbps with either a bus or star topology. ARCnet became standardized as ANSI 878.1. Although not as widespread as it once was, ARCnet has been upgraded to include transmission over copper twisted pair wire and fiber optic cables.

Arcstar The brand name for the Nippon Telegraph and Telephone Corporation's (NTT) global services including NTT Worldwide Telecommunications Corporation, NTT Europe, ntta.com and Asian branches. These include managed bandwidth, frame relay and Internet Protocol (IP) virtual private networks (VPN). See Nippon Telegraph and Telephone Corporation.

ARD 1. advanced research and development. 2. See Automatic Ring Down.

Ardire-Stratigakis-Hayduk algorithm ASH. A lossless compression algorithm, named after its creators at Western DataCom, which was developed between 1990 and 1993. It was intended for use over synchronous data communications with varying media characteristics.

Unlike asynchronous transmissions protocols, framed data can contain a very large number of bits, and does not have to be timed with start and stop bits. ASH provides a means to provide good compression ratios on various types of traffic in a multiuser network.

ASH incorporates some interesting concepts from the world of artificial intelligence. By using pattern-matching and predictive algorithms, data not yet transmitted and non-identical strings can be processed and evaluated. As part of the compression methodology, ASH uses an Occurrence Optimized Codebook (OOC) for 'fast-cache' access to commonly occurring tokens and strings. ASH safeguards against data expansion and latency. A patent is being sought for the ASH technology. See Lempel-Ziv-Welch.

ARDIS A commercial packet-switched nationwide wireless data communications service developed in the mid-1980s by Motorola and International Business Machines (IBM) and now owned by Motorola. It was originally developed for field technicians, and is appropriate for short messages and quick database lookups for a variety of applications. ARDIS is somewhat similar to CDPD except that it is a data only service. It is used for wireless faxing, and realtime messaging with any Internet address worldwide. ARDIS can be accessed through laptops, and personal data assistants. See RAM Mobile Data.

ARE All Routes Explorer. In ATM, a means of sending a transmission through all possible routes.

area code A three-digit code in a phone number that designates the region. See North American Numbering Plan (NANP).

area code restriction A service which selectively denies telephone calls to specified area codes. This is not a blanket restriction as in some long-distance call blocking services.

Area of Service AOS. The geographical area supported by a vendor, carrier, or service provider.

Arena 1. The name of an HTML3 browser from the World Wide Web Consortium (W3C) designed as a proof-of-concept demonstration tool for HTML+ ideas preceding HTML3.

ARES Amateur Radio Emergency Service.

ARF Alternative Regulatory Framework.

argon laser A type of gas laser which primarily uses argon gas. This type of laser can be used to produce green and blue light. It is similar to a krypton laser, except that it produces a little more light. Argon is one of the most common types of laser, used in laser light shows, and is sometimes combined with krypton gas to produce an argon-krypton hybrid. Argon lasers are typically water-cooled.

ARI Automatic Room Identification. Used especially in the hotel/motel industry.

ARIB See Association of Radio Industries and Businesses.

Ariel 1 The first satellite launched by the United Kingdom to study the ionosphere in April 1962. This satellite's orbit decayed in 1976.

ARIES 1. Angle-Resolved Ion and Electron Spectroscopy. 2. The name of a commercial satellite service. See Constellation Communications, Inc.

ARINC Aeronautical Radio, Inc. An organization founded in the 1920s, which coordinates airline industry telecommunications activities, communications, and information processing systems.

ARISE Advanced Radio Interferometry between Space and Earth.

ARISTOTELES Applications and Research Involving Space Technologies/Techniques Observing The Earth's Fields from Low Earth Orbiting Satellites. A joint project of NASA and ESA carrying out scientific research data gathering on the Earth's gravity and magnetic fields. It is equipped with a Global Positioning Service (GPS) receiver, gradiometer, and scalar magnetometer.

ARL Association for Research Libraries.

armature A portion of a magnetic circuit which typically consists of a wound conducting material such as wire encircling a core, which is moved within a strong magnetic field to create current. If the armature is revolved, the wound material interacts with the magnetic lines of force, in a sense, cutting in and out, and the current generated by this interaction can be drawn out. The arrangement is used in generators and alternators, where the current may be drawn out by brushes. See coil.

armor 1. Defensive or protective covering. 2. A type of heavy-duty waterproofing or other shielding used especially in underwater or underground installations. 3. Heavy shielding to protect facilities, equipment, and personnel from radiation or chemical contamination. 4. In computer technology, heavy security measures taken to keep users off a system, which may range from inaccessible facilities to extra software measures taken to discourage unauthorized users and intruders.

armor-plated A physical, or administrative high security system which the administrators deem to be virtually impenetrable. Highly resistant to access or tampering. Bullet-proof.

armored cable Any cable which is wrapped in a strong, environment-resistant, or vandalism-resistant covering, usually of wound metal. Armored cables are sometimes used to chain costly equipment, like computer terminals, to walls or work desks, to prevent theft. Armored cables are sometimes used on telephone book holders and handsets in public areas and penal institutions (to prevent the handset or the cord itself from being cut away and stolen).

Armstrong, Edwin Howard (1890-1954) A gifted American researcher who invented the superheterodyne circuit. In the 1920s, he contested Lee de Forest in the invention of regeneration, but de Forest won the suit. After many years of painstaking research against conventional wisdom and predictions of mathematicians and engineers, Armstrong proposed a method of wave modulation which varied the frequency rather than the amplitude of a wave. Armstrong waged a long and tragic legal battle with RCA over his patents to the FM technology, which after his death, were upheld in his favor. During World War II, Armstrong furthered the art of radar transmission by suggesting the use of FM signals, rather than the short pulse radar bursts which were used at the time. His ideas are now commonly incorporated into frequency modulated (FM) radio, television, and radar transmissions. See frequency modulation, heterodyning.

AROS Amateur Radio Observation Service.

ARP See Address Resolution Protocol.

ARPA Advanced Research Projects Agency. See Defense Advanced Research Projects Agency.

ARPANET Advanced Research Projects Agency Network. The historical basis of the Internet, ARPANET was originally discussed by the ACM in 1967, presented to ARPA the next year, and put into operation in 1969. The first widespread demonstration of ARPANET occurred at a Computer Communications conference in 1972, and a year later ARPANET traffic had grown to millions of packets of data transfer per day. By 1975, the ARPANET had been transferred to the Defense Communications Agency (later the Defense Information Systems Agency).

In 1982, ARPA endorsed TCP/IP as its protocol suite. In 1983, ARPANET split into MILNET and ARPANET (mandated to use TCP/IP), which together formed the Internet. Each was given a network number, and gateways were installed to provide packet forwarding between them. ARPANET was officially discontinued in 1990, largely due to the evolution of the Internet. See ARPA, BITNET, IANA, NSFNET, NEARNet, SPAN.

ARQ See automatic retransmit request.

array A type of data organization structure commonly used in programming. An array consists of an ordered list or matrix of information, which can be visualized as 2D or 3D tables of information contained in 'cells' which often have common characteristics, such as the size of the data cell (although the data in the cells may vary in length). Arrays form the structural basis for many types of data bases, including tables, and lists. Many software programs have built-in array-handling functions to automate common ways in which arrays are manipulated.

array antenna 1. An antenna with a number of directing, reflecting, or other elements arranged in a more-or-less regularly spaced, often symmetrical pattern. See antenna, Yagi-Uda antenna. 2. One antenna in an array of antennas which are organized and connected in such a way as to significantly boost power, range, and performance. These powerful antenna systems are used for picking up weak signals as in astronomy and military applications.

ARRL See American Relay Radio League.

ARRN Amateur Radio Repeater Network.

article A distinct section of text, often numbered for reference and clarity, as in legal documents, policy statements, association bylaws, and the charters of online discussion groups.

articulation Clear utterance or playback of sounds. The measure of the degree to which the reproduced or transmitted sounds are understood by a listener. When people say "Testing, testing, ..." on a sound system, they are testing not only whether the circuit is operating, and its volume, but its articulation, to see if the recipients can make out the content of the message. Articulation does not have to be high fidelity to be understood. Articulation is related to the ability of the recipient to perceive and understand the message, rather than the degree to which it perfectly matches the original. This is an important aspect of data communications as well. When conversations are converted from analog to digital, through a process called sampling and quantization, it is important to determine how much of the information is needed in order for the communication to be understood by the recipient. This

ASCII (Character and Control Codes)

Oct	Dec	Hex	Name	
000	0	0x00	NUL	
001	1	0x01	SOH	
002	2	0x02	STX	
003	3	0x03	ETX	Control-C
004	4	0x04	EOT	
005	5	0x05	ENQ	
006	6	0x06	ACK	
007	7	0x07	BEL	
010	8	0x08	BS	backspace
011	9	0x09	HT	tab
012	10	0x0a	LF	line feed, newline
013	11	0x0b	VT	
014	12	0x0c	FF	form feed, NP
015	13	0x0d	CR	carriage return
016	14	0x0e	SO	
017	15	0x0f	SI	
020	16	0x10	DLE	
021	17	0x11	DC1	XON, Control-Q
022	18	0x12	DC2	
023	19	0x13	DC3	XOFF, Control-S
024	20	0x14	DC4	
025	21	0x15	NAK	
026	22	0x16	SYN	
027	23	0x17	ETB	
030	24	0x18	CAN	
031	25	0x19	EM	
032	26	0x1a	SUB	
033	27	0x1b	ESC	escape
034	28	0x1c	F	
035	29	0x1d	G	
036	30	0x1e	RS	
037	31	0x1f	US	
040	32	0x20	space	space
041	33	0x21	!	exclamation mark
042	34	0x22	“	double quote
043	35	0x23	#	number sign, hash
044	36	0x24	$	dollar
045	37	0x25	%	percent
046	38	0x26	&	ampersand
047	39	0x27	’	apostrophe/quote
050	40	0x28	(	open parenthesis
051	41	0x29	)	close parenthesis
052	42	0x2a	*	asterisk, star
053	43	0x2b	+	plus
054	44	0x2c	,	comma
055	45	0x2d	-	minus
056	46	0x2e	.	period, full stop
057	47	0x2f	/	oblique stroke
060	48	0x30	0	zero
061	49	0x31	1	one
062	50	0x32	2	two
063	51	0x33	3	three
064	52	0x34	4	four
065	53	0x35	5	five
066	54	0x36	6	six
067	55	0x37	7	seven
070	56	0x38	8	eight
071	57	0x39	9	nine
072	58	0x3a	:	colon
073	59	0x3b	;	semicolon
074	60	0x3c	<	less than
075	61	0x3d	=	equals
076	62	0x3e	>	greater than
077	63	0x3f	?	question mark
0100	64	0x40	@	commercial at sign
0101	65	0x41	A	
0102	66	0x42	B	
0103	67	0x43	C	
0104	68	0x44	D	
0105	69	0x45	E	
0106	70	0x46	F	
0107	71	0x47	G	
0110	72	0x48	H	
0111	73	0x49	I	
0112	74	0x4a	J	
0113	75	0x4b	K	
0114	76	0x4c	L	
0115	77	0x4d	M	
0116	78	0x4e	N	
0117	79	0x4f	O	
0120	80	0x50	P	
0121	81	0x51	Q	
0122	82	0x52	R	
0123	83	0x53	S	
0124	84	0x54	T	
0125	85	0x55	U	
0126	86	0x56	V	
0127	87	0x57	W	
0130	88	0x58	X	
0131	89	0x59	Y	
0132	90	0x5a	Z	
0133	91	0x5b	[	open square bracket
0134	92	0x5c	\	backslash
0135	93	0x5d	]	close square bracket
0136	94	0x5e	^	caret
0137	95	0x5f	_	underscore
0140	96	0x60	‘	back apostrophe
0141	97	0x61	a	
0142	98	0x62	b	
0143	99	0x63	c	
0144	100	0x64	d	
0145	101	0x65	e	
0146	102	0x66	f	
0147	103	0x67	g	
0150	104	0x68	h	
0151	105	0x69	i	
0152	106	0x6a	j	
0153	107	0x6b	k	
0154	108	0x6c	l	
0155	109	0x6d	m	
0156	110	0x6e	n	
0157	111	0x6f	o	
0160	112	0x70	p	
0161	113	0x71	q	
0162	114	0x72	r	
0163	115	0x73	s	
0164	116	0x74	t	
0165	117	0x75	u	
0166	118	0x76	v	
0167	119	0x77	w	
0170	120	0x78	x	
0171	121	0x79	y	
0172	122	0x7a	z	
0173	123	0x7b	{	open curly bracket
0174	124	0x7c	|	vertical bar, pipe
0175	125	0x7d	}	close curly bracket
0176	126	0x7e	~	tilde
0177	127	0x7f	delete	delete

information can be applied to compression and decompression systems for speeding up transmissions. See fidelity, intelligibility.

artificial intelligence AI. Intelligent insights or behaviors attributed to an entity, usually a machine, which is not traditionally perceived by humans as having the capability to think in ways that involve problem-solving, insight, and other 'uniquely' human characteristics. The field of artificial intelligence has spawned many useful approaches, languages, techniques, and programming algorithms. Expert systems, neural networks, robotics, vision systems, and natural language processing all have their origins in AI research. People interested in artificial intelligence come from a diverse range of backgrounds.

Many of the origins of artificial intelligence, as they apply to computers, occurred in the 1950s, particularly as the result of the work of researchers like A. Turing, J. McCarthy, and N. Wiener. In some ways, Ada Lovelace was far ahead of her time, in the early 1800s postulating thinking machines that could create music or art. See expert systems; Lovelace, Ada; robotics.

Artron *ar*tificial neu*ron*. The familiar name for an electronic 'neuron' used in a maze-running robotic mouse in the early 1960s. See Melpar model.

Aryton, Hertha Marks (1854-1923) A British physicist, inventor, and author who investigated electricity, particularly electric arcs. She was the author of "The Electric Arc," which became a standard textbook on the subject. Aryton also was awarded a patent for the invention of an instrument which was used for dividing a line into any number of equal parts.

.as Internet domain name extension for American Samoa.

AS See autonomous system.

as is A term applied to products that are bought and sold with no implied or stated warranties. Condition can be somewhat ascertained by inspecting and trying the equipment, but there is no way to know the completeness, remaining useful life, or technical functionality of the equipment. See fair, good.

AS&C Alarm Surveillance and Control.

ASA See Acoustical Society of America.

Ascend Inverse Multiplexing AIM. An in-band networking protocol from Ascend Communications, which manages interconnections between two remote inverse multiplexers.

ascender In typography, the portion of a character that extends above the main body of a typestyle. That is, the upper stem in the letters *b* or *t* are called ascenders. See descender.

ASAI See Adjunct Switch Application Interface.

ASAPI Advanced Speech API. An open, cross-platform speech applications programming interface developed by AT&T.

ASCA Advanced Satellite for Cosmology & Astrophysics.

ascending node Intersection of a satellite's orbital plane with the earth's equatorial plane.

ASCII (as-kee) American Standard Code for Information Interchange developed by the American National Standards Institute (ANSI). Also known as ASCII International Telegraph Alphabet 5. The most widely used computer character set encoding scheme currently employing seven bits, thus making a total of 128 possible characters/symbols. ASCII is suitable predominantly for English language communications. Since it is very limited in its letters and symbols, many extensions to ASCII have been incorporated into key mappings on various computers to include western European characters. Sometimes called extended ASCII (even though the extensions aren't standardized), these eight bit encodings provide 256 possible characters, but the higher 128 characters are not usually compatible across platforms. See EBCDIC, Unicode.

ASCII editor A text editing tool which handles basic, simple characters standardized as ASCII text, which are cross-compatible and transferable over almost all seven-bit systems. Since the ASCII standard does not support style attributes (bold, underline, oblique, etc.), it cannot be used for extensive formatting. Due to the limitations and simplicity of its character set, ASCII editors are very fast. They are also good for writing computer source code, which typically needs speed and compatibility more than style tags.

If you require style tags and indentation for text formatting, and want to transfer the documents across applications or platforms, the best supported format which includes them is Rich Text Format (RTF), also known as Interchange Format (developed by Microsoft and supported across its products). It's not 100% compatible across platforms, but it's pretty close and can be read and written by most word processors (with import and export menu options). Another good format for transferring more complex documents is generic Adobe PostScript, which includes not only text and style support, but image positioning, layout effects and more, or its younger cousin, Adobe Acrobat.

When designing Web pages which have downloadable files, there should be more than one format available. If all Web site managers were to include these three: an ASCII version, an RTF version (which can be read into virtually any popular word processor), and a PostScript version, then the needs of low-end and high-end users would be well met, and at least one of the files would be accessible to virtually everyone using the Web.

ASDSP application-specific digital signal processor. See digital signal processor.

ASE Application Service Element. An element of an application layer protocol in the Open Systems Interconnection (OSI) layered network model. It is combined with other elements to form the complete protocol. See Open Systems Interconnect.

ASH See Ardire-Stratigakis-Hayduk.

ASI 1. Adaptive Speed Leveling. A U.S. Robotics modem term for adjusting the speed of a transmission, depending upon line conditions, to optimize the transfer of data. 2. Advanced Study Institute. 3. artificial sensing instrument. 4. Astronomical Society of India.

ASIC See application-specific integrated circuit.

ASK See amplitude-shift keying, modulation.

ASN 1. Abstract Syntax Notation. See ASN.1 2. See Autonomous System Number.

ASN.1 Abstract Syntax Notation 1. An ISO/ITU-T standard machine- and implementation-independent language defined in 1988 for the description of data, to facilitate the exchange of structured data among applications programs on a network. ISO 8824, ITU TS X.208.

ASP 1. Abstract Service Primitive. In ATM, an implementation-independent description of user/provider interactions, as defined by the Open Systems Interconnection (OSI). 2. Adjunct Service Point. A network feature of peripherals which are designed to respond intelligently to processing requests. 3. administrative service provider. SCSA term. 4. analog signal processing. 5. AppleTalk Session Protocol. 6. See ATM switch processor. 7. Attached Support Processor. 8. Association of Shareware Professionals.

aspect ratio The relationship of the proportions of the width to the height, usually of a rectangular form. A two-to-one aspect ratio, for example, is commonly represented as 2:1. The aspect ratios of televisions and monitors are similar, but cinematic films which are shown in theaters with panoramic screens, have a much greater width to height ratio. This is why letterboxed films have a dark strip on the top and bottom to preserve the full width of the image. Unletterboxed films have been modified to remove part of the picture from the sides.

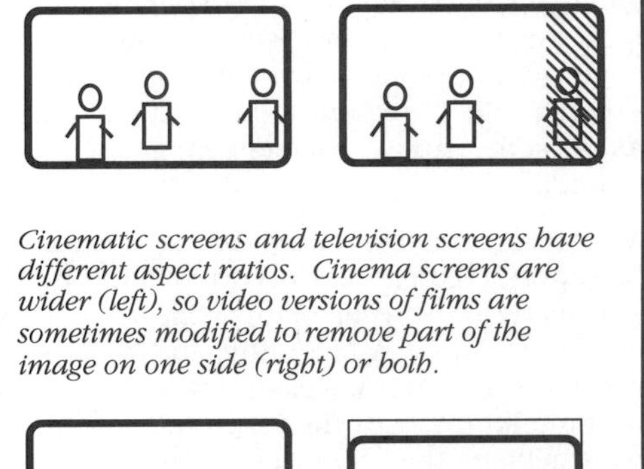

Cinematic screens and television screens have different aspect ratios. Cinema screens are wider (left), so video versions of films are sometimes modified to remove part of the image on one side (right) or both.

Letterboxing, on the other hand, reduces the size of the entire image, without removing information, preserving the panoramic aspect ratio and all the picture elements.

ASPI See Advanced SCSI Programming Interface.

ASQ Automated Status Query.

ASR 1. Access Service Request. A request sent to a Local Exchange Carrier for access to the local circuit. 2. Airport Surveillance Radar. 3. Automatic Send/Receive. A system that can send and receive messages unattended. 4. Automatic Speech Recognition.

assembler A program which converts symbolic assembly language program code into machine instructions that can be directly executed by the computer CPU. On early microcomputers, in the 1970s most serious programming was done with an assembler.

assembly language A low level symbolic computer language which structurally, and mnemonically, fits somewhere between machine code and higher level languages such as C, BASIC, Java, and Perl. Languages like BASIC and Perl are typically run in interpreted mode (although compilers exist for almost everything, if you really want one). When compiled and assembled, C and assembly language are converted into machine language, which typically consists of the binary digits one and zero, and is very difficult (for normal folks) to read and debug.

By coding in assembly language (which is also difficult for most folks) and then using an assembler, a software utility to convert to machine code, the program can often be optimized to run faster, and may be more difficult to reverse-engineer. See symbolic code.

assigned numbers A sequential numbering system administrated by IANA to organize and assist in the search and retrieval of Request for Comments (RFC) documents. See IANA, Request for Comments.

Association Control Service Element ACSE. An International Standards Organization (ISO) application layer service for establishing a connection, which is part of the Open Systems Interconnection (OSI) model.

Association of Competitive Telecommunications Suppliers ACTS. A Canadian-based association which represents telecommunications equipment manufacturers and suppliers in order to support and encourage market competition.

Association of Computer Telephone Integration Users and Suppliers ACTIUS. A trade organization in the United Kingdom which promotes awareness and acceptance of computer-telephone integration (CTI) technology through campaigns and educational programs.

Association for Computing Machinery ACM. A well-known association of more than 80,000 computing professionals in over 100 countries participating in the exchange of ideas, discoveries and information in many areas of academia, government, and industry. The ACM was founded in 1947. http://www.acm.org/

Association for Interactive Media AIM. A Washington, D. C.-based nonprofit trade association dedicated to promoting consumer confidence and government support of interactive media products and related technologies. http://www.interactivehq.org/

Association of Radio Industries and Businesses ARIB. A research and development organization, headquartered in Tokyo, Japan, which studies radio waves and developing radio systems and industries in telecommunications and broadcasting, in order to promote public welfare. A number of committees work under ARIB, including the Infrared Communications Systems Study Committee.

Association for Women in Computing AWC. A not-for-profit professional association founded in 1978 to promote the advancement of women in computing. http://www.awc-hq.org/

ASSP 1. acoustics speech and signal processing. 2. application-specific standard product. An integrated circuit designed for a specific application.

ASSTA Australian Speech Science and Technology Association.

Assured Link A telephone link meeting certain minimum transmission, loss (5.5 dB in the 300- to 3000-Hz bandwidth range), and service standards for a communications circuit for voice grade analog signals and sometimes one-way digital signals. See Basic Link.

astatic galvanometer A device developed by William Thompson (Lord Kelvin) in 1858 to overcome the limitations of earlier instruments that were subject to interference from the Earth's magnetic field. Unlike previous galvanometers employing one needle, the astatic galvanometer uses two needles, each with a separate coil. The needles are oriented so that the north and south poles effectively cancel out the Earth's magnetic interference. See galvanometer.

ASTC Australian Science and Technology Council.

ASTER Advanced Spaceborne Thermal Emission Reflectance Radiometer. A Jet Propulsion Laboratory (JPL) satellite imaging instrument project. Since 1998, ASTER has been obtaining moderate to coarse detail maps of Earth's temperature, emissivity, reflectance, and elevation characteristics. The satellite that carries ASTER, the EOS AM-1, is part of NASA's Earth Observing System (EOS). See Earth Observing System. See the NASA ASTER Web site for information on ASTER's progress.
http://asterweb.jpl.nasa.gov/

ASTRAL Alliance for Strategic Token-Ring Advancement and Leadership. A vendor-supported organization formed in 1997 to support migration to High Speed Token-Ring LAN technology. The group prepared a draft standard for 100 Mbps Token-Ring transmissions. See High Speed Token-Ring.

Astrolink A commercial global satellite communications service scheduled to come online in 2001. Astrolink International Limited is an independent Lockheed Martin venture. Lockheed Martin has been active in global communications frequency utilization conferences, and on various ATM- and ITU-T-related technical and standardization committees and working groups.

Astrolink is targeted at providing multimedia applications over virtual private networks (VPN) with a focus on secure transmissions and connectivity between private and public networks.

The Astrolink system consists of nine geostationary satellites, five to provide global coverage, four to come online later. They will be operating over Ka-band frequencies with approximately 6 Gbps capacity per satellite using continuous beam uplinks and multifrequency TDMA.

astronomical unit AU. A unit of length defined as the distance from the Earth to the Sun, a measure which is generally given as about 149,579,000 to 149,599,000 kilometers. The variation in the unit comes about due to the variation in the Earth's movement in relation to the Sun, and the measuring system and criteria used to establish the distance.

astronomy The science of the celestial bodies, including but not limited to, their magnitudes, relationships, evolution, motions, history, and constitutions.

ASU application-specific unit.

ASV Air-to-Surface-Vessel radar.

asymmetric 1. Not symmetric, lopsided, irregularly proportioned, unbalanced, one-directional.

asymmetric Digital Subscriber Line ADSL. Installation involves hooking up a splitter at the subscriber premises, running separate lines to the phone and computer, and installing a special modem and software on the computer. Copper wire lines also are not optimal for ADSL, as the common bridge taps and load coils can interfere with ADSL signals. As a result, various alternatives have been proposed. See Digital Subscriber Line for a fuller explanation and chart, G.lite, UAWG.

asymmetric transmission A transmission channel in which information flows more readily (faster) in one direction than the other, or moves primarily in one direction or the other at any one time, or in which a greater volume of information flows in one direction or the other. There are many instances in which information typically flows more in one direction than another, as in interactive TV, where most of the time the user is observing and not transmitting, but may just make an occasional request for a specific movie or file. The medium itself may not be inherently asymmetric. For example, a data upload over a modem is primarily one-way, but the line capacity is two-way, and the direction can be easily switched when uploading. The slower channel, or the one with a lower volume capacity may be called the *back channel*.

asymmetrical compression In data compression techniques, there are some types of files that can be compressed faster than they can be decompressed and some that work the other way around. In designing compression algorithms, sometimes optimization in one direction or the other is preferred. In creating animation sequences, it is usually very important that they decompress and play quickly, otherwise the illusion of motion is lost. However, it is usually not a problem if the compression takes longer than the decompression because the computer can handle that while the user is working on drawing the next frame.

asymmetrical modem A modem designed to favor the transmission of the bulk of the data in one direction over the other. This is appropriate in situations where most of the communication is one-way, as in managing an archive site, where downloads typically outnumber uploads thousands-to-one. This is a way of optimizing performance over limited data circuits. See asymmetric.

asynchronous Not synchronous. A concept which applies across many areas of telecommunications, in which the timing of the information being received and transmitted is not predefined, and may be unpredictable, as in many modem communications and interactive radio communications. This type of communication typically requires some means of indicating the starting and stopping points of the transmission. There are various schemes for handling this, from verbal cues ("Roger"), to start/stop bits, and various handshaking signals.

asynchronous balanced mode ABM. In an International Business Machines (IBM) Token-Ring network, a service in the logical link control (LLC) at the SNA data link control level that allows devices to send and respond to data link commands.

asynchronous transfer mode ATM. A high-speed, cell-based, connection-oriented, packet transmission protocol for handling data with varying burst and bit rates. ATM is a commercially significant protocol due to its flexibility, and widespread use for Internet connectivity. ATM evolved from standardization efforts by the CCITT (now ITU-T) for Broadband ISDN (B-ISDN) in the mid-1980s. It was originally related to Synchronous Digital Hierarchy (SDH) standards.

ATM allows integration of LAN and WAN environments under a single protocol, with reduced encapsulation. It does not require a specific physical transport, and thus can be integrated with current physical networks. It provides Virtual Connection (VC) switching and multiplexing for Broadband ISDN, to enable the uniform transmission of voice, data, video and other multimedia communications. See appendix for more details and diagrams on ATM and the ATM adaptation layers.

AT 1. Access Tandem. 2. AudioTex.

AT, PC/AT Advanced Technology. The common name for a series of 80286-based personal computers introduced by International Business Machines (IBM) in the mid-1980s. This model was released about a year later than the Apple Lisa, at about the same time as the Apple Macintosh, and about a year before the Amiga 1000, Apple IIGS, and Atari ST computers. This is historically significant in the development of user interfaces, as most of the competing computers were evolving graphical user interfaces (GUIs) and included built-in serial ports and sound cards, while most of the AT systems were text-oriented (primarily MS-DOS), with sound and various interface cards optional. The IBM AT and licensed 'clones' from other manufacturers were purchased primarily by business users, in part because the IBM name was well-known in the business industry, and also because IBM had a decades-old traditional of providing service and repair options to business owners. Some of the chief software products used on the AT were spreadsheets and word processors.

AT commands, Hayes Standard AT Commands A very simple control and reporting language built into Hayes Microcomputer Products, Inc. modems, and Hayes-command-compatible modems from other manufacturers. Originally modems were 'dumb' devices, that is, they had no significant memory or algorithms incorporated into the device to process commands or data from the computer. Hayes introduced 'smart' modems in the early 1980s which could process a limited command set and enhance the utility of modems. This instruction set has since been incorporated into almost every make and model of computer modem, usually with enhancements by individual manufacturers.

The AT command set allows computer control of a modem, and provides a way for the modem to report information back to the computer software. The AT stands for "attention" and is a way of alerting the modem that there is an instruction set following the "AT" which is to be acknowledged or executed. When you run a telecommunications program through your modem, the software is talking to the modem with AT Commands along the path provided by the serial cable that typically connects the modem to the computer. If your software can be set to interactive mode, you can type the AT commands directly to your modem and see what happens. The AT commands are usually listed at the back of the manual that comes with a modem.

Many modem manufacturers have included supersets of the basic Hayes command set to provide control of proprietary or enhanced features specific to their products, so AT commands usually include most or all of the Hayes commands, and additional ones as well.

AT commands fall into a number of categories. There are commands for querying the status of the phone line, for querying the status of the modem, and for carrying out operations such as dialing, setting the transmission speed, setting the number of redials, setting the length of wait periods, etc.

Here are a few common and handy AT commands. Assume that "AT" is included at the front of each series of commands and that some of the following require additional information after the command (such as a number to be dialed) or allow optional parameters to modify the settings (indicated by *n*).

Sampling of common modem commands.

Command	Description
AT	attention (followed by one or more of the following commands)
A	answer
DT*n*	go off-hook and dial tone
DP*n*	go off-hook and dial pulse
H*n*	set hangup process
L*n*	set speaker volume level
M*n*	set speaker options
W	wait for dial tone
X*n*	tone detection and reporting
+++	escape, go to command mode

Modems contain a number of 'registers' in which information is stored, often in the form of a toggle (true or false) or integer setting. Thus, setting the register to *zero* signifies one thing, and setting it to *one* or another integer, when appropriate, signifies another. Thus, AT S0=0 sets the 'S' register to zero. Since register S0 determines how many rings to AutoAnswer, setting it to zero effectively turns off AutoAnswer. AT S0=1 instructs the modem to AutoAnswer after it detects one ring. If you are running a computer bulletin board, or a friend is calling to send you a file over the phone line, AutoAnswer can be turned *on* (or you can type ATA "attention, answer" when you hear the phone ring). Remember to set AutoAnswer *off* when you are finished transmitting, or the next voice caller may get a nasty modem-blast to the ear. (Some modems have enhancements that allow them to autodetect whether the incoming call is voice or data, and react accordingly so this doesn't happen.)

AT commands can be combined. You don't have to type AT in front of each individual instruction. For example, you might wish to initialize your modem, and dial out as a single string of commands as follows:

```
AT S0=0 M1 DT 555-4321 W DT 123
```

attention; set autoanswer to zero rings; set speaker to be on (M1) during establishment of call (so you can hear dial tone and dialing) and off during connection (so you don't have to hear the modem sounds); dial tone mode 555-4321; wait for dial tone; dial tone mode 123 (to dial an inner extension).

AT&T American Telephone and Telegraph Company. A company established almost 150 years ago to create practical commercial applications from the early telegraph and telephone patents filed in the 1870s, primarily those of Alexander Graham Bell, and Elisha Gray. Some of the patents became the property of the Bell System, and some served simply as competitive motivation to implement the new ideas and technologies.

The American Telephone and Telegraphy Company (AT&T) began as a long-distance subsidiary of the American Bell Telephone Company in 1885. In 1899 the two companies were again merged into one under the AT&T name. In the 1900s, AT&T was reorganized, becoming a holding company, the parent of the Bell companies and Western Electric. In the ensuing years, several additional reorganizations occurred, some voluntary, some mandated by U.S. justice authorities.

In the early 1900s, there was a period of substantial change in the phone industry, since the original Bell patents, protected for a term of 17 years were expiring, and independent companies were entering the phone market in substantial numbers. This resulted in independents collectively holding almost half of the phones until, by 1913, AT&T was again the majority holder due to mergers and acquisitions, and was legally restrained from acquiring any more independents. AT&T was also mandated to permit independents to use the AT&T toll lines.

Further regulation was implemented in the Communications Act of 1934 in which the Federal Communications Commission (FCC) was established, and given jurisdiction over the telephone and broadcast industries. In 1956, the U.S. government and AT&T entered into an agreement that AT&T would offer only phone-related services, and not engage in common carrier communications such as computer network services. AT&T was further required to license Bell patents for royalties to interested applicants.

A number of antitrust suits ensued in the 1970s charging AT&T with monopolistic practices, and there were calls for divestiture resulting in divestiture proceedings in the 1980s. During this same period, AT&T was granted limited permission to engage in computer-related services.

While its political history was undergoing many ups and downs, the researchers in the Bell Laboratories provided an enormous amount of research and development in telephone technologies, beginning in the late 1800s and early 1900s. AT&T researchers developed the first two-wire telephone circuit, which is still in use today, the first practical transistor, and many other inventions that are in broad use.

See Bell, Alexander Graham; Bell Laboratories; Bell System; Carty, John J.; Kingsbury Commitment; Modified Final Judgment; Vail, Theodore.

AT&T TeleMedia Connection A Microsoft Windows-based videoconferencing product from AT&T Global Information Solutions, which provides video, audio, file transfer, and application-sharing utilities over ISDN. It uses various ITU-T H Series and Q Series Recommendations standards and encoding techniques.

Atanasoff, John Vincent (1903-1995) American physicist and inventor, in the 1930s, of a vacuum tube calculating device which foreshadowed some of the characteristics of the famous ENIAC computer which was operational in the post World War II years. In 1939 he pioneered a binary logic computer called the ABC, or

Atanasoff-Berry Computer. See Atanasoff-Berry Computer.

Atanasoff-Berry Computer ABC. A pioneering binary, direct logic computer with a regenerative memory, designed and built by J. V. Atanasoff with assistance from his graduate student, Clifford E. Berry. After two years on the drawing boards, it was prototyped in 1939. It is significant not only for its historic place in the early history of computers, but also because it was designed with a separation between memory and data processing functions. The electricity needed to keep the memory refreshed, so the information wasn't lost, was provided by rotating drum capacitors.

Atanasoff had been working on the ideas that led up to the ABC since 1935, and related that the idea for the ABC came to him in a roadhouse in 1937 after he and his graduate students had developed a calculator for complex mathematics manipulation. Punch cards, which had been developed to store information for electromechanical devices in the late 1800s, were used to enter data into the ABC. Much of the information about Atanasoff's invention did not come to light until a long court battle in the 1970s between Sperry Rand and Honeywell. See ENIAC.

Atari Corporation A significant games and computer company established in 1972 by Nolan Bushnell. Atari shipped the first computer game which gained wide commercial acceptance. "Pong" was a simple monochrome game with a ball and two paddles, a form of electronic table tennis that became wildly popular. Atari continued developing games but also subsequently introduced a number of microcomputers, including the Atari 800 and the Atari 520ST. The 520ST had a graphical interface, built-in MIDI and was competitive with the Amiga for the home market in the mid-1980s.

ATB all trunks busy.

ATCP See AppleTalk Control Protocol.

ATCRBS air traffic control radar beacon system.

ATD 1. asynchronous time division. 2. Attention Dial. A modem command from the Hayes set that instructs a modem to dial the number following the command. Often a *T* or *P* will precede the number to indicate whether to dial as a *tone* or *pulse* signal. For example, *Attention Dial Tone* would be ATDT 555-1234. 3. advanced technology demonstration.

ATDRSS Advanced Tracking and Data Relay Satellite System.

ATG address translation gateway. A Cisco Systems DECnet routing software function for routing multiple, independent DECnet networks.

Athena project, Project Athena A project of the Massachusetts Institute of Technology (MIT) Computer Science Lab begun in 1984. The goal was to take various incompatible computer systems, and develop a teaching network that could utilize the different resources of each in a consistent manner. The development of The X Windows System originated from efforts to provide a graphical user interface (GUI) for Athena. See X Wndows System.

ATIS See Alliance for Telecommunications Industry Solutions.

ATM 1. See asynchronous transfer mode. 2. Automated Teller Machine. Any automated walk-up or drive-up console system designed to carry out many of the transaction activities (usually financial) which have historically been handled by human bank tellers. Typical ATM functions include deposits, withdrawals, payments, transfers, and balance inquiries. ATMs are intended to provide services 24 hours a day, or an option to those who prefer automated services. Early critics of ATM systems proclaimed that they wouldn't last, that "People will never prefer conducting a transaction with a machine rather than with a human." The naivete of this statement is evidenced by the current proliferation of ATMs and some banks are promoting this trend by charging consumers for visits to tellers. ATMs are networked to a central system, if free-standing or off-site; or may be directly linked to the local network when attached to the building with which it is associated.

Walk-up Automated Teller Machine, sometimes colloquially called a 'cash machine.'

ATM-PON asynchronous transfer mode passive optical networks. A type of optical distribution network, being promoted as a means to implement large-scale, full-service subscriber telecommunications services. See fiber to the home.

ATM Adaptation Layer See asynchronous transfer mode in Appendix B for information and diagrams.

ATM cell The basic unit of information transmitted through an ATM network. An ATM cell has a fixed length of 53 bytes, consisting of a 44- or 48-byte payload (the information being transmitted), and a 5-byte header (addressing information) with optional 4-byte adaptation layer information. Interpretation of signals from different types of media into a fixed length unit of data makes it possible to accommodate different types of transmissions over one type of network. See asynchronous transfer mode, see Appendix B for details and diagrams.

ATM cell rate In ATM networks, a concept that expresses the flow of basic units of transport used to convey data, signals, and priorities. Common cell rate concepts include leaky bucket and cell rate margin. See ATM Cell Rate Concepts Table.

ATM endpoint In an ATM network, the point

ATM Cell Rate Concepts

Abbrev.	Name	Notes
ACR	allowed cell rate	A traffic management parameter dynamically managed by congestion control mechanisms. ACR varies between the minimum cell rate (MCR) and the peak cell rate (PCR).
CCR	current cell rate	A traffic flow control concept that aids in the calculation of ER, and which may not be changed by the network elements (NEs). CCR is set by the source to the available cell rate (ACR) when generating a forward RM-cell.
CDF	cutoff decrease factor	Controls the decrease in the allowed cell rate (ACR) associated with the cell rate margin (CRM).
CIV	cell interarrival variation	Changes in arrival times of cells nearing the receiver. If the cells are carrying information which must be synchronized, as in constant bit rate (CBR) traffic, then latency and other delays that cause interarrival variation can interfere with the output.
GCRA	generic cell rate algorithm	A conformance enforcing algorithm which evaluates arriving cells. See leaky bucket.
ICR	initial cell rate	A traffic flow available bit rate (ABR) service parameter. The ICR is the rate at which the source should be sending the data.
MCR	minimum cell rate	Available bit rate (ABR) service traffic descriptor. The MCR is the transmission rate in cells per second at which the source may always send.
PCR	peak cell rate	The PCR is the transmission rate in cells per second which may never be exceeded, which characterizes the constant bit rate (CBR).
RDF	rate decrease factor	An available bit rate (ABR) flow control service parameter which controls the decrease in the transmission rate of cells when it is needed. See cell rate.
SCR	sustainable cell rate	The upper measure of a computed average rate of cell transmission over time.

at which a connection is initiated or terminated. See asynchronous transfer mode.

ATM endpoint address A location identifier which is functionally similar to a hardware address in an ATMARP environment, although it need not be tied to hardware. See asynchronous transfer mode, ATM endpoint.

ATM Forum, The An international nonprofit organization founded in 1991 to further the evolution and implementation of asynchronous transfer mode (ATM) technology as a global standard. The Forum provides educational information on ATM, and specifications and recommendations to the ITU-T based on standards of interoperability between vendors, with consideration to the needs of the end-user community. The ATM Forum is a membership-by-fee group which includes a number of technical committees to discuss and report on specific issues such as signalling, traffic management, emulation, security, testing, and interfacing. See asynchronous transfer mode, UNI. http:/www.atmforum.com/

ATM hardware address The individual IP station address. See asynchronous transfer mode, ATM endpoint address, Internet Protocol.

ATM Link Enhancer ALE. An error-correcting mechanism for satellite communications developed by COMSAT. The Header Error Control (HEC) specified for asynchronous transfer mode (ATM) is suitable for transmissions carried through low error rate media such as fiber optic cables. It becomes inadequate, however, in bursty transmissions environments such as wireless networks, particularly those which are satellite-based. To compensate for this limitation, COMSAT developed an ALE module which is inserted in the data paths before and after the satellite modems to isolate ATM cells from burst errors. This module allows selective interleaving of ATM cells before they are transmitted through the satellite link, thus providing a lower bit error rate (BER) and an improved cell loss ratio (CLR). See asynchronous transfer mode, cell rate.

ATM Models There are a variety of types and implementations of ATM networks including Classical IP, LANE, IP Broadcast over ATM, and others. See asynchronous transfer mode, ATM Transition Model, Classical IP Model, Conventional Model, Integrated Model, Peer Model. See Appendix B for details and diagrams.

ATM slot A time indicator for the duration of one cell, usually described in microseconds. This will vary depending upon the cell-carrying medium. In ATM, one use of the term *slot* is to describe delay in switch performance. See asynchronous transfer mode, ATM cell.

ATM switch processor ASP. A modular component from Cisco Systems which provides cell relay, signaling, and management processing functions. It includes an imbedded 100-MHz MultiChannel Interface Processor (MIP) R4600 RISC processor, with ATM access to the switch fabric, to provide high call setup rates and low call setup latencies. It includes an Ethernet port and dual serial ports.

The ASP works in conjunction with a field-replaceable feature daughtercard which supports advanced ATM switch functions, including intelligent packet discard, dual leaky bucket traffic policing, and available bit rate (ABR) congestion control mechanisms. See asynchronous transfer mode.

ATM traffic descriptor A list of network traffic parameters, such as cell rates and burst sizes, and, optionally, a Best Efforts indicator, within an asynchronous transfer mode (ATM) virtual connection. This information is used to determine traffic characteristics, and to allocate resources. See asynchronous transfer mode, BEC, cell rate, PCR, SCR.

ATM Transition Model A model lying between the Classical IP Models and the Peer and Integrated Models. See ATM models.

ATMARP ATM Address Resolution Protocol. See asynchronous transfer mode.

atmosphere 1. Ambience, mood, feeling about a location or room. 2. A gaseous mass enveloping a celestial body.

atmosphere, Earth's The gaseous envelope surrounding the Earth which provides breathable air, moisture, weather variations, protection from the sun's radiation, especially the ultraviolet rays, and particles which deflect radiant energy that can be harnessed for telecommunications. Atmospheric pressure at sea level is approx. 14.7 pounds per square inch, with local weather variations, and decreases somewhat uniformly as altitude increases. Barometers are used to measure atmospheric weather, and barometric altimeters indicate altitude through changes in pressures in the atmosphere.

The atmosphere has been divided into three main regions. From the surface going away from the Earth, they are the troposphere, stratosphere, and ionosphere.

atomic clock An instrument devised in the 1940s for precise timing and synchronization, it is now particularly important in the U.S. Global Positioning System (GPS), and many scientific research applications.

An atomic clock uses the frequency associated with a quantum transition between two energy levels in an atom as its reference. It exploits the unique frequency characteristics of photons in a given transition. In the early 1990s, atomic clocks were further improved with the introduction of a Hewlett-Packard cesium-beam clock which was more rugged and more stable than previous models. Coordinated Universal Time reporting centers make use of atomic clocks for establishing an international time reference.

atomic number A number characteristic determined by experimentation, since the atom is too small to be seen by any natural or microscopic means. This number is used to represent an element in a periodic table and describes electrons in relation to the protons in a neutral atom.

ATS See Applications Technology Satellite program.

ATT See Automatic Toll Ticketing.

attachment In data communications, a note or file that is attached to the end of an existing file, or a communication. Commonly binary files are sent with email as attachments, because the message text part of many email systems cannot transcribe or transmit 8-bit binary code. An email binary attachment allows you to send a picture, sound file, Adobe PostScript document, or other non-text transmission.

attachment unit interface, autonomous unit interface AUI. Certain cables and connectors used to attach equipment to Ethernet transceivers.

attack time The time it takes for a signal or sound to go from its initiation to its full volume or power. On a violin, for example, it's the time interval from the moment the bow begins to move and a consistent note achieves its full volume and tone. The attack time on an electronic system is the time it takes from the initiation of a pulse or signal, or power on action, until the system reaches its intended activation threshold, output, or throughput level. See decay time.

attempt An effort to initiate or establish a communications connection. In some systems that are billed on a flat rate or per-call basis, attempts are not billed. In other systems, such as those that bill by air time, the attempts are charged by the minute or second whether or not the call is connected.

attenuation The decrease between the power of the initial transmission, and its power when received, usually expressed as a ratio in decibels. Loss in power can result from distance, transmission lines, configurations, faults, and weather. See gain.

attitude and articulation control subsystem AACS. A spacecraft guidance system employed on the Cassini spacecraft mission to permit dynamic control of rotation and translation maneuvers. The AACS uses star and sun sensors to establish reference points for determining its position. The main engine and smaller engines are used for propulsive maneuvers. Sensors estimate attitude and rate of both the base body and the articulated platforms. A series of vectors, kinematically propagated in time, aids the system in determining motion of various bodies in relation to the base frame. The AACS works in conjunction with the command and data subsystem (CDS), which is the main processor on the craft. The CDS receives RF signals from Earth, and sends information and control parameters to other systems, such as the AACS, accordingly. See Cassini.

atto- (Symbol - a) Used as a prefix to represent a very minute quantity, one quintillionth of, 10^{-18}. See femto-.

ATV 1. See advanced TV. 2. See amateur television.

audible ringing tone An audible signal transmitted to the calling party to let the caller know that the called number is ringing. See busy signal.

audible sound Sound waves which are recognizable to the ear of a particular species. Audible sound ranges vary from species to species, with humans hearing generally between the ranges of 20 to 20,000 hertz. The upper ranges tend to drop off during the teenage years, and decline gradually throughout a person's lifetime. Illnesses, very sudden loud noises, protracted loud noises, and sustained low level noises can have profound negative effects on a person's hearing. See audio, sound.

audio Pertaining to sounds, primarily those within range of human perception, from frequencies of about 20 to 20,000 hertz (the upper range especially tends to diminish as people get older). The comfortable hearing range varies in loudness from a few decibels to about 100 decibels. At volumes near and above 160 decibels, permanent hearing damage is almost certain. Sudden loud sounds, frequent exposure to loud sounds, or even long-term exposure to medium level sounds can damage the sensitive structures associated with hearing.

The types of sounds most commonly used for communication are speech and music. Most hearing is done with the ears, although some people augment their understanding of auditory information by reading lips or sensing physical vibrations through their fingers or bodies. Helen Keller was known for 'listening' to symphonies through a sensitive sound board placed in the symphony hall under her chair. Many deaf or hard-of-hearing people use their fingertips pressed against the larynx of a speaker to aid in sensing auditory vibrations.

While humans can hear a broad range of frequencies, not all these frequencies are used in human speech. We can detect pitches up to about 18,000 to 20,000 hertz, but don't utter sounds that high in conversation. Thus, telephone and other speech circuits typically are not designed to transmit the full hearing range of frequencies, and will be optimized for those frequencies which are associated with the information being transmitted.

audio tape A type of magnetic storage medium used for audio recordings. Most audio tapes are small, so they can be used in portable tape decks or car stereos, with playing times ranging from 10 minutes to 120 minutes. Common music tapes are 30, 45, or 60 minutes per side for a double-sided tape. Some audio tapes are designed as a continuous loop with the tape ends fused for continuous playing. Video tapes are sometimes used as high quality audio tapes. Computers in the late 1970s and early 1980s used 'computer' and audio tapes for recording data, and the same is still true for tape backup systems today.

With good recording equipment, the quality of video tape audio recording comes close to that of CD sound. Eight-track tape cartridges were introduced in the early 1960s and were popular for a few years. Cassette tapes were introduced soon after eight-tracks and eventually superseded them. As CD players become less expensive and more prevalent, audio tapes are less favored for portable applications.

audio-on-demand AoD. Audio services provided to a user on request. AoD is one of the earliest services-on-demand (SoD) systems implemented in the telecommunications industry. In the days of operator-managed telephone services, some imaginative service providers realized that they could place a phone at the switchboard center near a radio or gramophone player and play music for the subscriber on request. It was an unsophisticated system, but the concept was timely, and the same idea is now being implemented with digitally automated technologies in the form of video-on-demand, and other custom request services. See services-on-demand, video-on-demand.

AudioGram Delivery Services ADS. A Nortel subscriber telephone service option which allows callers who get a busy signal or no answer to their ring to leave a message that will be delivered to the callee at a later scheduled time. Essentially it's a phone line answering machine service.

audiographics A system suitable for distance learning, in which remote computer screens are shared as a conference and lecture interactive medium for dynamically sharing images, video, and text. Electronic Classroom, written by Robert Crago for the Macintosh, is an example of this type of application, designed to work over public switched telephone networks (PSTNs). Audiographics is sometimes called *telematics.* Some people like to make a distinction between audiographics, which is the transmission of still images and sound, and videoconferencing, the transmission of motion video and sound. With improvements in technology, the distinction is blurring. See whiteboarding, electronic; videoconferencing.

audiometer An instrument for measuring hearing acuity, invented in 1880 by Alexander Graham Bell.

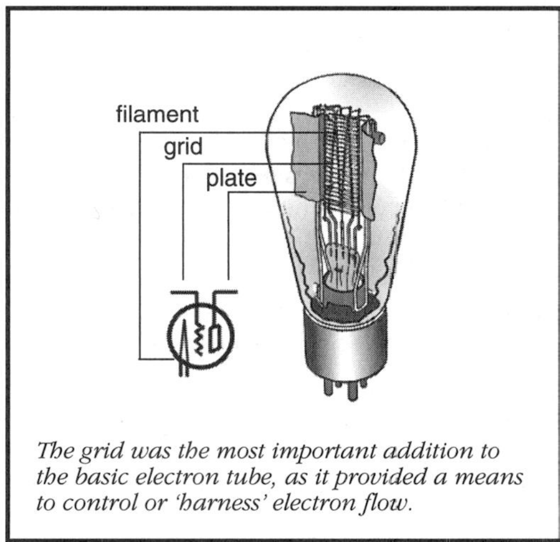

The grid was the most important addition to the basic electron tube, as it provided a means to control or 'harness' electron flow.

Audion An extremely significant early invention, evolutionarily descended from simple *flame detectors*, that led to the three-element vacuum electron tube developed and patented by American inventor L. de Forest. The Audion was a tantalum lamp with an evacuated glass globe sealed around a filament and plate. A simple wire bent in a zigzag pattern became a grid, providing control over the flow of the electrons from the filament to the plate in a way that had not been previously achieved.

This *triode* electron tube's control grid represented breakthrough technology which de Forest sold to AT&T at the bargain price of $50,000. It was used for decades throughout the electronics industry until it was superseded by the transistor. Repeater devices based on the Audion enabled long-distance telephony. Interestingly, like many inventions through history, the inventor himself didn't understand the detailed mathematics/physics behind *why* the Audion worked. This created problems in manufacturing. The only way to know if the tube was a good one was to test it, and the sensitivity varied from tube to tube. Edwin Armstrong was one of the few early people who grasped some of the physics associated with its functioning. The term Audion was originally trademarked. See de Forest, Lee; Edison effect; electron tube; flame detector.

auger A tool designed for boring, or a bit which fits into a drill designed to make large bore holes, that can be used for wiring installations.

Augustine, Saint A philosopher who authored *De civitate Dei* (The city of God) in 428 AD. This important record of western knowledge includes observations of magnetic phenomena.

AUI See Attachment Unit Interface.

AUP See Acceptable Use Policy.

aural Heard or perceived through the ear, auditory. See acoustic, sound.

aurora Solar flare, a nuclear effect from the sun which can sometimes be seen by its influence on the Earth's upper atmosphere. The ionization that results causes the undulating light shows we know as the aurora borealis and aurora australis.

Aurora 1 A regional communications satellite in geostationary orbit over Alaska.

authenticate To establish the identity and authorization status of a user, device, process, or data which is seeking entry to a system, or seeking to negotiate a transaction. Authentication is used at network access points, such as gateways and firewalls, in electronic transactions such as purchases or contracts, and at password prompts to systems, servers, and applications. Data is often authenticated to see whether it has been altered during transmission, and email messages may be authenticated before being sent or received. See certificate, Challenge-Handshake Authentication Protocol, Clipper Chip, encryption, Pretty Good Privacy.

authenticator In packet networking, the end of the link which requires authentication, and specifies the authentication protocol to be used in the link establishment phase. See Challenge-Handshake Authentication Protocol.

authoring The process of using authoring systems to create computer software or multimedia presentations without a large expenditure in time in learning to program. The rationale for authoring is to free the creator from technicalities to concentrate on content and flow. Authoring involves developing a scenario, providing content, and putting them together in an interactive environment according to how the author wishes the user to interact with the software or presentation. Some authoring languages are very similar to English or BASIC. Others use graphical interfaces to allow the user to create the scenarios, and inter-relate the program building blocks with a minimum of programming or text entry.

authoring system, authoring language A high-level computer programming language designed to be quickly and easily learned and used such that professionals (e.g., teachers) with expertise in their specific subject fields, but without programming experience, can develop computer software such as courseware and multimedia presentations.

authorization Permission to use a product or service, or to gain entry to an area or structure. More and more, computerized means are being used to assign, track, and administrate access to secure areas, or within corporate premises. Authorization can be monitored through video systems, magnetic cards, retina or fingerprint scanning, visual recognition of faces, passwords, and voice recognition.

authorization code A code which must be entered into a telecommunications system to gain access to the service, or specific features of the service, or to generate statistical records of use. It is used for security or efficiency monitoring, and frequently used on touchtone phone systems to permit long-distance calls, or to gather departmental data. When used for security, the code is often typed before the desired number, although in cases of departmental billing or monitoring, it may be required after the number has been dialed.

Authorization Code Table ACT. A lookup table for determining whether a phone call is authorized on the list and should be permitted to ring through. If a number is not found on the list, it is called an *unmatched call* and may be rejected or forwarded to someone in authority.

authorized agent A person authorized to resell, release, or represent the product or services of a company in a somewhat cooperative, independent manner with state restrictions. It is common for large telephone carriers, both landline and wireless, to permit authorized agents to resell or repackage phone services such as long-distance services. In some cases, the agent is a software developer or Internet Services Provider (ISP) who works in conjunction with the phone company to provide digital value-added services.

authorized user A person or entity authorized by the company or provider to use a service.

auto answer The capability of a telecommunications receiving device to automatically detect and respond to an incoming transmission. Facsimile machines, BBS modems, and answering machines are examples of devices with auto answer capabilities. Sophisticated systems can be configured to auto answer a voice call and, by using Caller ID, to display the person's file on a computer monitor. See Caller ID, Caller Name, bulletin board system, auto dial.

auto attendant, automated attendant An automated voice system, designed to provide a 24-hour a day substitute for an operator or receptionist, it answers incoming calls and plays a recorded message to the caller providing a number of touchtone options, or selections from a touchtone-activated menu. Different systems can transfer calls to humans or voice-mail systems, perform transactions, provide information, initiate a faxback transmission, or initiate a fax tone for those with manual fax machines. The better systems allow you to go to submenus without waiting for the current recording to end, and will give you an easy way to return to the main menu. Auto attendants are used by banks, mail order companies, information service companies, and others. See voice mail, Automatic Call Distribution.

auto baud See autobaud.

Auto Busy Redial A surcharge phone service, multiline subscriber feature, or consumer phone feature in which the last number called can be redialed continuously until a connection is made. The system recognizes a busy signal, hangs up and redials. There is a similar feature in most telecommunications software which is used to connect to BBS or Internet services which can cycle through a list of numbers, trying each one in turn, or which continuously attempts to connect with a specific number. The software can often further be configured to dial at specific intervals, or for a specific duration of time. The Auto Busy Redial service is useful when combined with auto dial for voice communications. See auto dial

auto dial 1. A phone or phone system feature in which a short code has been assigned to a longer number to allow the longer number to be dialed automatically with fewer keystrokes. Also known as speed dial. See abbreviated dialing.

auto dial capability A software/hardware applications feature for dialing a phone number through a modem and setting it up for voice, rather than data communication. It's very handy for dialing from a laptop, a cellular laptop link, or from a database on a desktop computer. Some phone solicitors use auto dial in conjunction with phone listings to maximize the number of call connects. Be aware that there are strict regulations governing the use of automated procedures for phone solicitations. See auto answer.

auto discovery, auto mount An automated process whereby a system or a network server is alerted to a new device on the system and can gain sufficient information about its operating characteristics to bring it online and make it available to users. Device tables and databases are sometimes used to make this possible, and manufacturers are creating more devices that signal their presence and include electronically accessible information about the brand, model, capacity, and attributes.

auto start 1. The capability of an emergency power system to detect when electricity falls below a certain crucial level and start up standby generators to provide continuous service. 2. The capability of a computer system to restart or *reboot* after a power outage, or power fluctuation sufficient to take down the system.

autoanswer See auto answer.

autobaud The capability of a modem to detect the incoming baud rate and adjust its transmission speed and handshaking to match the rate to establish a connection. Useful in 24-hour a day, unattended services like BBSs.

AUTODIN *Auto*matic *Di*gital *N*etwork. A global communications network of the U.S. Department of Defense.

Automated Attendant See Auto Attendant.

Automated Attendant Billing System AABS. In telephony, a system in which the caller dials collect and long-distance calls with the aid of an automated voice prompting system which seeks authorization from the called party, connects or rejects the call, and bills accordingly. Most telephone services in North America have become automated in this way with the use of speech recognition and synthesized operator-assist voices.

Automated Directory Assistance Service ADAS. A telephone service in which a speech recognition system is used to get information from a caller who has requested directory assistance. It requests the location and name, and then provides the phone number, or patches the call to an operator who provides the phone number.

Automated Interchange of Technical Information CALS. A data interchange standard.

Automated Packet Recognition/Translation APaRT. A Cisco Systems technology that allows automatic network configuration and translation, e.g., Ethernet clients and a CDDI or FDDI server, so that workstations or switches do not have to be individually configured. APaRT recognizes and, if necessary, translates specific data link layer encapsulation packet types.

Automatic Call Distribution ACD. A multiline phone capability or service which automatically manages and routes incoming calls to assigned lines. If there are no available lines at the moment the call is received, it is placed on hold, and may be configured to play a recording such as "Your call is important to us, please stay on the line and your call will be answered in the order received." ACD systems can put the party on hold and play a recording, or they can be quite sophisticated, and can perform significant traffic direction and business transactions.

Mail-order companies, airlines, and other high phone-traffic businesses utilize ACD systems, although smaller companies are starting to use them as they become less expensive. An ACD system detects and answers incoming calls, searches a database for instructions on how to handle the call, responds to the call (as with a recording), and reroutes it appropriately as human operators become available. The routing itself can be programmed to the subscriber's needs, with a number of options available: *Uniform* distributes calls evenly, *Top-down* distributes the calls according to a list in the same order each time, so that calls go to the top of the list first, and work their way down, and *Specialty* distributes calls according to the callee who most appropriately can handle the call. ACDs can also be used to gather statistical data on the number of calls received, and how they are handled in order to fine-tune the system, and to respond to the business needs of the subscriber to improve call handling or change it as the need arises. See Centrex, private branch exchange.

automatic callback A subscriber option usually found on private branches in which a caller can key in a code or press a button for automatic callback if she or he has encountered a busy signal on an extension line. When the line is freed up, the caller's phone and the callee's phone both ring so that the connection can be made.

automatic calling unit ACU. A device used to automatically dial numbers (a modem and the appropriate software can also do this) in order to save a human operator the time and inconvenience of dialing a lot of calls. This type of system is used by fundraisers, telemarketers, researchers, and others who make frequent calls to a predetermined list of numbers. On computer systems, ACU software is sometimes coupled with database directory programs or address books.

automatic circuit assurance ACA. A diagnostic feature of private branch exchanges which aids in locating possible trunk malfunctions. The system keeps a record of call performance as significant changes might signal a trunk problem or failure. When a potential problem is detected, an attendant is notified.

automatic dialer, autodialer A device designed to save time by allowing a user to program a short sequence to represent a longer number. When the short sequence is keyed in on a phone console, the phone checks memory and if the sequence is found, dials the corresponding longer number. See speed dialing.

automatic direction finder ADF. An antenna which usually rotates or arcs back and forward and continuously monitors signals until it finds a strong one or one with specified desired characteristics. It then locks on to that signal, or provides the direction or frequency information on some type of output device such as a monitor or dial.

Automatic Electric Company A company supplying automatic telephone switching systems based on Strowger technology, co-founded in 1901 by Almon B. Strowger, a mortician who reportedly wanted an automatic exchange because human operators were diverting business to his competition. This company was able to compete by installing working systems quickly, according to customer specifications, and was the largest company supported by telephone company independents. It was directed by Alexander E. Keith.

Originally most of the Automatic Electric systems were three-wire systems which used two wires plus the Earth as the 'third wire' return path conductor for the transmission. Later, they developed a two-wire system. In 1955, it was merged into General Telephone and Electronics (GTE). See Strowger switch.

automatic exchange A central telephone switching office in which calls from subscribers are automatically routed to the callee through mechanical, electromechanical, or electronic switching. There are still a few operator-assisted exchanges around, mostly in remote locations or third-world countries, but automatic exchanges are found in most developed nations. The history of automated exchanges is interesting. Besides the economic motivation of not having to pay wages to operators, one of the early switching systems was designed by a mortician because he was apparently concerned that operators were channeling calls to his competition.

automatic exclusion Once a call has been answered, subsequent stations, nodes, or consoles are excluded from having access to the line.

automatic frequency control AFC. 1. Periodic sampling of a frequency modulated (FM) signal to focus the receiver on the approximate center of the transmission band. This came into widespread use in the 1930s. 2. A device that can seek out a particular frequency or monitor the incoming frequency to keep the tuning accurate. AFC is common on FM receivers and other devices that must maintain operations within a very narrow range.

automatic gain control, automatic volume control AVC. A circuit designed to sense the level of incoming sounds and adjust their volume. This can serve two common purposes: to increase the dynamic range of the sound, by making quiet sounds quieter, and loud ones louder; or to condition the sound, by making the volume more consistent (e.g., by quieting down the loudest sounds, and strengthening the quietest sounds) when incoming signals are fluctuating more than is desired. Volume conditioning is widely incorporated into sound receivers with tuners, as the signal coming through an antenna can vary significantly due to varying broadcast characteristics and weather.

automatic hold A convenience in which the operator of a multiline telephone console or switchboard can switch between active call lines without having to push a hold button. This saves operator time and prevents caller frustration as the operator can't disconnect the caller by forgetting to press the hold button.

automatic level control See automatic gain control.

automatic light control A feature on many different types of cameras, in which the camera will adjust the settings to changes in lighting without manual metering or intervention by the user. This is particularly prevalent on small automatic cameras and many camcorders. Sometimes backlighting, high contrast, and other lighting situations can be preset with buttons, so that a general ambience is made known to the camera, but the final settings are still automated. Professionals prefer automatic systems only if they also have a manual over-ride for tricky lighting situations.

automatic line hold See automatic hold.

Automatic Number Identification ANI. 1. An identifier that provides the calling number to the callee, provided the callee has the service and equipment to display the information. It was historically distinguished from Caller ID by the number of rings within which the information was sent to the caller, but that distinction is disappearing. See Caller ID, Signaling System 7. 2. A multifrequency signaling parameter by which a long-distance carrier receives the caller's number from the local carrier for billing purposes.

automatic privacy A feature on some multiline systems that automatically locks out the ability of other people who pick up a phone to select the line already in use. These systems may also include a 'release' button that allows others to pick up the line and join the conversation.

automatic recovery If there is a power outage or other problem that interrupts a phone system, bulletin board system, network, etc., automatic recovery is the ability of that system to power up to operating status and to recover as many of the original operating parameters and files as possible, as well as to retain the information which may be retained in memory that is of importance to continued operations.

Automatic Redial A surcharge phone service which allows a caller to recall the most recent previously dialed number and dial it again by inputing a short code instead of rekeying in the whole number. This is handy if the line was busy the first time it was called. Many business and consumer phones now have a redial button, thus decreasing demand for this service.

automatic rerouting The capability of a system to route a transmission through another leg, hop, or path when the original, or expected path is not available. Dynamic routing in large systems often works this way. Large distributed systems where the physical and virtual pathways change constantly usually function with automatic rerouting. In some systems, such as Fiber Distributed Data Interface (FDDI) alternate routing is supplied in a dual ring system in which the port adaptor and various ports are quickly reconfigured to the secondary systems to prevent loss of data and connectivity.

automatic retransmit request, automatic repeat request ARQ. In its simplest form, as in CB radio, a verbal request for the sender to repeat a message that did not come through clearly or completely.

More standardized, automated ARQ systems exist, including one in which characters are sent in groups of a set length, and the sender waits for an *acknowledged* (ACK) or *not acknowledged* (NAK) signal (or no response) before retransmitting or continuing. In some systems, such as amateur radio communications, ARQ is called *mode A*.

In high-speed data transmission, error-detection fields are built into the data and used as check fields by the recipient. As in broadcast ARQ systems, an acknowledge (ACK) or not acknowledge (NAK) is transmitted to the sender and the sender responds accordingly.

Automatic Route Selection A phone service which automatically seeks and selects the most economical circuit (usually the least expensive carrier) for the path of an outgoing call. See Least Cost Routing.

automatic sounder A type of historic telegraph device which created audible clicks of the incoming transmission which could be heard and interpreted by the telegraph receiving operator. The term was also applied to a sounding device used for the teaching of telegraphic sending and receiving skills. When contact was made between the arm and the stop, the sounder's circuit was closed. When the contact was broken, it was open. See sounder.

automatic switching system Various types of mechanical and electrical telephone switching systems became prevalent after the invention of the Strowger switching system. These automatic systems allowed the control of telephone circuitry in such a way that a call was connected by dialing a code rather than by asking a human operator to manually patch through the call. A number of large and small telephone switching manufacturers, including AT&T/Bell and the Lorimer brothers, created automatic switching equipment in the early 1900s, a trend that continued until the 1970s. By the 1980s, very few manual systems were in use except in rural or specialized situations, as they had been superseded by electronic switching systems.

automatic volume control AVC. A circuit in a radio receiver which is designed to prevent loud blasts from strong transmitting stations when the user moves the dial through the various stations. It can be disconcerting to tune through several weak stations, and then hit a strong one that assaults your ear drums. AVC was designed in the 1930s to prevent these sudden gains and dramatic volume changes, and most systems now incorporate this feature.

automatic wakeup Any timing device set to create an alarm, or other alert to wake up a person. These can include a clock radio, alarm clock, bell, computer programmed sound file, rooster crow, or telephone signal.

autonomous switching A Cisco Systems router feature which allows the ciscoBus to independently switch packets, without interrupting the system processor, to provide faster packet processing.

Autonomous System Number ASN. A common administrative routing setup identifier, that is routing through a collective numbered common domain. The ASN designates a system under common operations control, using common routing protocols, with the various routing tables dynamically maintained.

AUU ATM User-to-User.

AUUG Australian Unix User Group.

AUXBC auxiliary broadcasting.

availability 1. The amount of time during which a telephone or network system is available for handling calls. This is expressed as the ratio of denied calls to attempted calls. See reliability. 2. In Global Positioning Service (GPS), the period of time during which a particular location, within *angle of elevation* parameters, has sufficient satellites to make a position fix.

avalanche noise In semiconductor junctions, a situation in which sufficient high-voltage energy is generated by some carriers such that others are physically impacted.

avatar 1. An embodiment in human form. 2. An electronic image or other embodiment of an individual which is computer-generated and holds some 'essence' or presence of individuality or actuality beyond that of a photographic image or scan. A concept in computer gaming, and especially in virtual reality simulations.

AVC See automatic volume control.

AVD See alternate voice data.

Average Busy Hour The hour in a day during which the most traffic is carried on the system. This information is important for configuring and tuning a system to handle traffic efficiently.

Average Speed of Answer ASA. The average time it takes for an operator or automated system to answer a call, usually measured in seconds. Used for system configuration, statistical, and staff training and management purposes.

AVHRR advanced very high resolution radiometer. A device for sensing passive radiation emitted from the earth and its atmosphere. Used on orbiting satellites.

aviation channels A set of broadcast frequencies set aside for aviation purposes.

AVIOS See American Voice Input/Output Society.

AVRS automated voice response system. A system designed to respond to voice commands without the intervention of a human operator. This type of system is often used by banks and mail order companies. See speech recognition, voice recognition.

AVSSCS Audio/Visual Service Specific Convergence Sublayer. A convergence protocol for transmitting video over AAL-5 using available bit rate (ABR) services. See asynchronous transfer mode.

AWA See Antique Wireless Association.

AWC See Association for Women in Computing.

AWG See American Wire Gauge.

awk An interpreted computer language common on Unix systems, developed by *A*ho, *W*einberger, and *K*ernighan (who is also an author of C) with C-like syntax. See Perl.

AX.25 A communications protocol designed for packet radio communications that operates at the link layer level. It is based on the ISO Open Systems Interconnection Reference Model (OSI-RM). Since its introduction, it has generally been superseded by NET/ROM, a more flexible means of transmission. See Open Systems Interconnection.

axial leads Leads on a component which are arranged to protrude along a linear axis in a common plane. In other words, they stick out the ends rather than out the sides.

axial ratio In elliptically polarized radiant energy, the ratio of the major axis to the minor axis.

axis 1. A reference, orientation, or vector in a coordinate system, typically depicted as a line when graphed. 2. A primary direction or line of motion. 3. An imaginary or implied line around which other elements appear to be oriented, as for example, the vertical axis of a tree trunk, or the horizontal axis of a sea/skyscape. 4. A drawn line, usually straight, used as a reference in a graph or chart. 5. The longitudinal center of a wire or cable.

axis of rotation A straight line around which a body or representation is symmetrically aligned, or around which it rotates.

axis ray, axial ray A beam that travels along the axis (the longitudinal center) of a fiber optic cable. See fiber optic cable, waveguide.

AZERTY A designation for a computer keyboard or typewriter keyboard used in some European countries such as France. The letters represent the first six top left letters directly below the number/symbol keys. See QWERTY.

azimuth 1. A geometric arc used in navigation and astronomy which is calculated, for example, between a fixed point on the horizon, and clockwise through to the center of a specified object. 2. A horizontal direction calculated from the angular distance between the direction of a fixed point, such as a navigational heading, and the direction of the object (boat, spacecraft, etc.). 3. A specific arc described in relation to a fixed point and a moving object or radiating transmission such as a rotating storage medium (drive, tape, etc.), or antenna. 4. The horizontal direction of a celestial point from a reference terrestrial point, expressed as an angular distance.

available bit rate ABR. Sometimes called Best Effort service. A class of service (Cos) defined by the ATM Forum which utilizes bandwidth on an availability basis for the transport of bursty data traffic. ABR is a traffic flow control designator designed to approximate the traffic characteristics of existing local area network (LAN) protocols. See cell rate, constant bit rate, variable bit rate.

azimuth-elevation mount A common type of antenna mount that permits two types of rotation for adjusting horizontal orientation (azimuth) and elevation. This type of mount is frequently used with parabolic antennas that work best when focused precisely toward highly directional beams. See parabolic antenna, microwave antenna, polar mount.

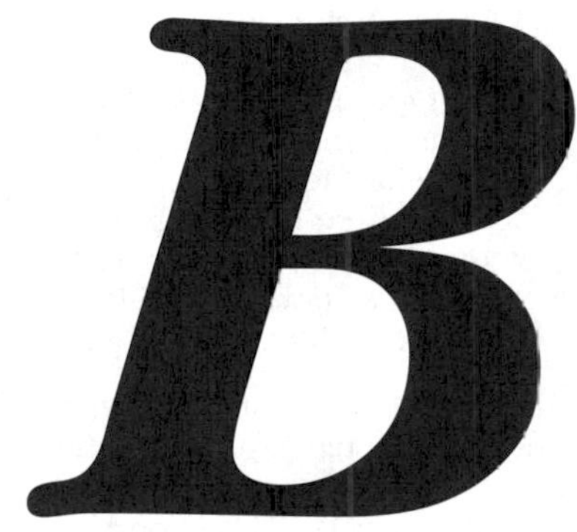

B 1. Symbol for magnetic flux. 2. Abbreviation for brightness, as on a computer monitor or TV picture tube setting. 3. Symbol for byte, a unit of data consisting of eight bits.

B battery A low voltage source of direct current (DC) power, historically used to provide power to the plate, or anode, in electron tubes, or to relays in a communications circuit. B batteries ranged from about 24 to 130 volts, with 48 volts common in communications circuits. The early B batteries were wet cells, often consisting of a matrix of 1.5 volt cells combined. Later dry cells, with two leads replaced the more cumbersome wet cells. See battery.

On the right is a somewhat cumbersone older wet B battery, with the individual cells connected in series. Later, dry B batteries, like the one on the right, superseded wet cells. These batteries are from the Bellingham Antique Radio Museum collection.

B Block A Federal Communications Commission (FCC) designation for a Personal Communications Services (PCS) license granted to a telephone company serving a Major Trading Area (MTA). This grants permission to operate at certain FCC-specified frequencies. See A Block for a chart of designated frequencies for Blocks A to F.

B Carrier A local wireline cellular telephone communications carrier. This is the designation for the local phone company. See A Carrier.

B channel 1. In a stereo system, the designation for the right audio channel, typically connected to the right speaker. 2. bearer channel. A channel in a circuit-switched ISDN connection with bidirectional data transmission capability. For a fuller description, see ISDN.

B interface An interface used in Cellular Digital Packet Data (CDPD) which is deployed over AMPS. The B interface connects the Mobile Data Intermediate System (MD-IS) to the Mobile Data Base System (MDBS). See A interface, C interface, Cellular Digital Packet Data, D interface, E interface, I interface.

B minus, B- A negative terminal on a B battery. A negative polarity in a vacuum tube anode.

B News A UUCP-based news reading program that evolved from A News in the early 1980s. Mark Horton and Matt Glickman rewrote the software to accommodate a greater volume of postings, and to add functionality. This came to be known as B news, which was first introduced in 1982. See A News, USENET.

B plus, B+ A positive terminal on a B battery. A positive polarity in a vacuum tube anode or voltage source in an electronic transistor. See B battery.

B port In a Class A, dual-attachment (dual ring) Fiber Distributed Data Interface (FDDI) token-passing network, there are two physical ports, designated PHY A and PHY B. Each of these ports is connected to both the primary and the secondary ring to act as a receiver for one, and a transmitter for the other. Thus, the B port is a transmitter for the primary ring and a receiver for the secondary ring. The dual ring system is configured to provide fault tolerance for the network.

FDDI ports can be connected to either single mode or multimode fiber optic media, providing half duplex transmissions. LEDs are commonly used on port adaptors as status indicators. Optical bypass switches may in turn be attached to the port adaptors. Optical bypasses are provided to avoid segmentation which might occur if there is a failure in the system, and a station is temporarily eliminated. See dual attachment station, Fiber Distributed Data Interface, optical bypass, port adaptor.

B-911 A telephone emergency response system which has a subset of the capabilities of a full 911 system. Most notably, it doesn't include Automatic Location Information (ALI).

B-CDMA Broadband Code Division Multiple Access. InterDigital Communication Corporation's commercial wireless local loop TrueLink product is designed to provide enhanced broadband phone features through CDMA technology. InterDigital is collaborating with Siemens AG and Samsung Electronics Company, Ltd. in developing the proprietary B-CDMA technology. See CDMA.

B-DCS See broadband digital cross-connect system.

B-frame bidirectionally predictive-coded frame. In MPEG animations, a picture which has been encoded into a video frame according to both past or later frames in the sequence using predicted motion compensation algorithms. See I-frame, P-frame.

B-ICI B-ISDN (Broadband-ISDN) InterCarrier Interface. 1. A specification defined by The ATM Forum for the connecting interface between public ATM networks, for the support of user services across multiple public carriers. 2. An ITU-T standard for protocols and procedures for broadband switched virtual connections (SVCs) between public networks.

B-ICI SAAL Broadband Inter-Carrier Interface Signaling ATM Adaptation Layer. A signaling layer which permits the transfer of connection control signaling, and ensures reliable delivery of the protocol message. See asynchronous transfer mode, SAAL, AAL5.

B-ISDN Broadband ISDN. See ISDN for an introduction to ISDN concepts. B-ISDN was designed to meet some of the demands for increased speed and enhanced services on primary ISDN lines. This was geared to the needs of commercial users. It has since evolved into a strategy for delivery for many new telecommunications services including teleconferencing, remote banking, videoconferencing, interactive TV, audio, and text transmissions. Broadband ISDN is intended for services that require channel rates greater than single primary rate channels (i.e., voice at 64 kbps). B-ISDN services can be broadly organized as is shown in the following chart:

Category	Example activities
messaging, data	paging, electronic mail, data files (images, sound, formatted documents).
conversation	telephone, conferencing, audiographics, videotelephone, videoconferencing
interactive	distance education, services-on-demand, Web browsing, retrieval services such as news, stocks, etc.

The essential characteristics of B-ISDN services were approved in the I-series Recommendations by the ITU-T in 1990. These developed into broader standards, and specific recommendations for implementation including network architecture, operations, and maintenance.

Recommendations for ATM to be the switching infrastructure for B-ISDN contributed to the formation of the international ATM Forum which promotes commercial implementation of ATM and related technologies.

Physical layer transmission for B-ISDN is accomplished through the Synchronous Optical Network (SONET) system. See I Series Recommendations.

B-LT broadband line termination.

B-MAC Broadcast Master Antenna Control.

B-NT broadband network termination.

B-picture bidirectionally predictive-coded picture. In MPEG animations, a picture which is to be encoded according to both past or later frames in the sequence using predicted motion compensation algorithms. Once it is encoded, it is considered to be a B-frame.

B-scope A radar screen which displays information on range and bearing in rectangular coordinates. See A-scope.

B-TE Broadband Terminal Equipment. An equipment category for broadband ISDN (B-ISDN) connecting devices which encompasses terminal adapters and terminals.

B8ZS binary/bipolar eight-zero substitution. A line-code substitution technique to guarantee density in network transmissions independent of the data stream, used on T1 and E1 lines. The zeros can be replaced at the receiving end to restore the original signal.

Babbage, Charles (1791 or 1792-1871) An English researcher who contributed a great deal to the theory and practice of computing and conceived his now-famous 'analytical engine' by 1834. While Babbage's ideas for computers could not be easily built with technology available in the 1800s, the basic ideas were sound and have stood the test of time. There is a crater on the moon named after Charles Babbage. Ada Lovelace collaborated with him in his work.

babble Crosstalk from other communications circuits, and the noise resulting from such crosstalk. The term generally implies a number of noise sources combined.

babble signal A deliberate transmission consisting of composite or otherwise confusing signals to obscure the intended transmission from unwanted listeners. A babble signal may be used as a type of jamming mechanism to deliberately interfere with other transmissions. See frequency hopping, jam signal.

BABT British Approvals Board for Telecommunications. A telecommunications regulatory organization.

baby monitor An audio or audiovisual monitoring device which is used to keep track of the well-being and activities of a baby or small child in another room or out in a yard. Low power frequency modulated (FM) signals are often used for these devices which are similar to small cordless phones or intercoms.

BAC binary asymmetric channel.

back bias 1. A technique for restoring the environment in a vacuum tube which may have been altered by external forces, by applying a voltage to the control grid. 2. A means of feeding a circuit back on itself before its point of origin or contact. One important application of this technique has been the creation of regenerative circuits in electron tubes, an important milestone in radio signal amplification. Regeneration was developed independently by E. Armstrong and L. de Forest and hotly contested in a patent suit. 3. When used in the context of semiconductors, back bias is sometimes more commonly called reverse bias. It refers to an external voltage which is used to reduce the flow of current across a *pn junction*, thus increasing the breadth of the depletion region.

back door A security hole that is accessible without going through the normal login/password procedure. A back door may be deliberately left by the developers or maintainers of a software application or operating system in order to gain entry later, sometimes much later. Back doors have legitimate uses for maintenance and configuration, but are sometimes abused by disgruntled ex-employees or employees engaged in embezzlement or other illegal or unauthorized activities. See back porch.

back electromotive force, back EMF An electromotive force opposing the main flow of force in a circuit.

back end 1. A program which sends output to a particular device or front end. See client/server. 2. The final step in a transparent (to the user) task or process. 3. In networking, the manner in which a lower layer provides a service to the one above it. 4. In electronics, the final production stages of assembly and testing.

back haul See backhaul.

back lobe In a directional antenna, there is a main lobe and there may be additional lobes, one of which extends backward from the direction of the channeled signal, called a back lobe.

back porch 1. On a computer system, a file access point to the system or an application with limited privileges which may not be publicly announced or which may have a group password. In other words, there may be some files available to certain employees that may not be generally accessible by all employees. It's like a meeting place on a friendly neighborhood porch where people are welcome to visit as long as they don't go inside the house and disturb the privacy of the home owners. This environment is somewhat like an anonymous FTP environment in that users of the 'porch' do not have full privileges or access to all parts of the system. A back porch differs from a back door in that it is a circumscribed, known area, with limited privileges. A back door, on the other hand may provide full privileges, and is often not known to anyone but the person who programmed the software. 2. In video broadcasts, the portion of a composite picture signal which is between the edge of the horizontal synchronization pulse and the edge of the associated blanking pulse.

back projection A means of presenting information on a visual display system by illuminating, or otherwise activating the display elements from behind. In its broadest sense, most TV and computer screens are back projection systems. However, a further distinction can be made that a projection system implies a larger display system, as would be used in a seminar, theater, or lecture hall, environments which are traditionally equipped with front projection systems (film projectors, slide projectors, etc.). In these environments, back projection screens are less common.

One of the main advantages of a back projection system is that the audience and various speakers, can stand or sit directly in front of the display without affecting the image with shadows. Back projection also tends to show up better in rooms where there is ambient light, sufficient for people to take notes. The main disadvantage of such a system is that it usually requires specialized equipment for both the projection and the display screen, whereas films and slides can be shown on many types of surfaces, including a light-colored wall.

back scatter See backscatter.

backboard A mounting board often used to provide a sturdy surface to support electrical panel boxes, punch-down blocks, or other threading or wiring equipment that needs a firm backing, and wouldn't be secure if mounted on plaster, wallboard, or some other brittle surface. Sometimes equipment is preinstalled and tested on a backboard, so it can be assembled lying down in a convenient position, sometimes off-premises, and then quickly mounted where desired.

backbone A primary ridge, connection link, or foundation, generally pictured as longitudinal with branches. See 6bone, Mbone.

backbone data circuit A main data communications circuit, usually of national distribution, from which there are secondary branches. The term was originally used to describe key USENET/email sites, but is now used more generally. A backbone is sometimes defined in terms of the speed of communications, and primary nature of the data, and is sometimes considered that part of the circuit which customarily carries the heaviest traffic. A backbone can connect a mainframe with local area networks (LANs) or individual terminals; or individual systems with peripherals such as modems, printers, video cameras, etc. Bridges, routers, and switches perform a variety of traffic control and direction functions within the system. More regional, medium-sized installations, as at universities and large corporations, are called campus backbones.

Backbones can generally be categorized into three types: distributed backbones, utilizing multiple routers; collapsed backbones, with a configuration switching hub generally contained within a single building complex; and hybrid backbones which include collapsed backbones in individual building complexes, interlinked with FDDI distributed backbones. See campus backbone.

backbone radio circuit In packet radio communications, a packet-radio bulletin board system (PBBS) that provides automatic routing services for a number of users.

background Behind the scenes, active or visible, but not commanding the attention of the user. See background process.

background communication Data communication that occurs while other user actions are taking place, that is, it carries on in the background without intruding on other activities. For example, a user may be using a word processor while a file is uploading or downloading in the background. Single-tasking systems don't do this. Background communications are characteristic of multitasking systems; and some task-switching systems, which will 'time-splice' the processor between the two activities.

background music A system used over telephone lines, especially on *music hold* systems, and public address (PA) systems, which broadcasts stock music, specialized radio stations, or regular radio stations through the telephone receiver, or PA speakers. Some phone systems are equipped with a jack receptacle for plugging in the music source. The biggest advantage of music hold systems is that the listener knows he or she is still on hold, and hasn't been cut off.

background noise Ambient noise, environmental noise, noise without significant meaning. If background noise levels are too high, they can interfere with communications. There are now digital systems, such as cellular phones, which can selectively screen out background noise and increase the clarity of a transmission from a noisy environment. This has both industrial and social communication advantages.

background process, background task A computer program which operates or waits in the background, that is, not in immediate sight or use of the user, at a lower priority level, becoming active quickly when needed or brought to the foreground, or when other processes are idle. On data and phone systems, system operations, archiving, cleanup of temporary files, print spooling, diagnostics, etc. are frequently run as background tasks, and may function primarily on off-peak hours or when more CPU time is available.

backhaul *v.* In telephone and computer network communications, to send a signal beyond a destination and then back. For example, a phone call from Seattle to north San Francisco may be routed through Palo Alto and back to San Francisco. Backhauling can occur for a number of reasons, including cost, availability, and traffic levels. Backhauling may also happen in companies with a number of branch offices. A call to one branch may, for various business reasons, be backhauled to another, in order to serve the caller's needs. In Internet usage, backhauling is quite common. For example, in some cases it may be cheaper to Telnet to an ISP in a distant city with better rates and services, and then access ftp sites, chat channels, or other services by backhauling, perhaps even to the originating city, than to call out from a more limited local service.

backhaul broadcasting In cable broadcasting, to bring back a signal (haul) from a remote site (such as a hot news flood zone, or big sports event) to the local TV station, or network head station for processing, before being distributed to viewers.

backoff *n.* A retransmission delay which may occur when a transmission cannot get through to its destination, due to an interruption, collision, a medium already in use, etc. If a transmission fails, rather than trying again immediately, the sending or interim system may wait momentarily before retransmitting. The retransmission interval may be random, or may be set within a certain range by backoff algorithms incorporated into the protocols being used. Backoff (one word) is the noun form, back off (two words) is the verb form.

backplane, backplane bus 1. In desktop computers, the physical connection between a data bus and power bus (both of which are usually on the motherboard), and an interface link or card (which are usually inserted into slots). See bus. 2. In phone exchanges, the high-speed line and power sources which connect individual components, often through circuit board slots. The speed and quantity of transmissions through the exchange are in large part determined by the capacity of the backplane. See bus.

backpressure, backpressure propagation In a network, the information that is being transferred is always accompanied by data about the information, and on large networks, about its progress from source to destination. In hop-by-hop routing, there is network communication about the location and subsequent routing as well. This overhead can sometimes add up, if there is congestion on the network, and may propagate upstream to form backpressure.

backscatter Backscatter is a phenomenon in which radiant energy is propagated in a reverse direction to the incident radiation, sometimes in a diffuse pattern. Backscatter usually happens when the radiant energy comes in contact with an object, particles, or various projections in an uneven terrain, or when it encounters outer boundaries or particles in a transmissions medium, as in fiber optic cables.

Sometimes back scattering is useful, and sometimes it is undesirable. In radar, the returning deflected signals from when a radar signal hits a target and is reflected back to the sensing device are used to track the location and movement of the target. In directional antenna assemblies, backscattering of signals to the rear of the antenna may cause interference. See zone of silence.

backscatter, ionospheric In the E and F regions of the ionosphere, where many radio waves are 'bounced' from the sender to the receiver, generally at the angle at which the wave hits the ionized particles, there is typically a certain amount of backscatter that happens in which some of the waves are propagated back in the direction from which they came. This may cause interference to the original signal, or may result in the transmission being heard by receivers near the transmitting station, although the signal is generally weak. See E layer, F layer, ionosphere.

backslash A character symbol also called the *virgule* which is commonly found on computer keyboards. The backslash is rarely used on most systems, with the exception of Microsoft's MS-DOS in which it is used for the commonly-needed directory listings to indicate a subdirectory.

backup An alternate resource in case of failure or malfunction of the primary resource. The alternate may be identical (or as close as possible) to the original, as in data archives; or may be a substitute which is just sufficient for short-term functioning, as in a backup light source or power supply.

backup file A copy of computer data in a separate file or partition from the original file, or, preferably, on a storage device separate from the original (tape, floppies, magneto-optical discs, etc.). Better yet is a copy at another location.

Software developers will sometimes keep copies of important releases in a safety deposit box off-site in case of theft, water damage, or fire to the main premises or equipment. Since backup archives do not necessarily reflect all the changes that have occurred, it is wise to create backups on-site every hour, or every day, and off-site backups once or twice a week, at least! Most companies avoid buying backup peripherals, citing cost as a deterrent, yet the cost of a backup device is small compared to the cost of recovering from hours or days of lost productivity and serious data loss. A redundant array of inexpensive disks (RAID) is a dynamic means of providing data striping and mirroring, that protects the information without using a separate backing up schedule. For people who hate separate backup procedures, RAID might be the answer. See archiving, RAID.

backup link A secondary link which may not typically carry traffic, or may only carry overload traffic, unless there is a failure in the primary link, in which case it becomes available for transmission until the fault is corrected. See alternate routing.

backup ring On Token-ring networks, a second ring is often set up to provide a backup in case of failure of the first ring. Depending upon the setup, the system may switch automatically or may need to be switched manually. See Fiber Distributed Data Interface.

backup server A server system expressly designated to automate the handling of data protection tasks. The server can be configured to back up certain machines, directories, or files at predetermined times, or when processing overhead from other tasks is low. A backup server is usually configured with drivers for a number of backup devices such as tape drives and magneto-optical disks and may be secured against fire or public access to protect backed up data.

backward channel A channel in which the transmissions are flowing in the direction opposite to the flow of the majority of the data, usually the informational data. Some interactive systems are designed so that control signals and queries flow through the back channel, while the majority of the data flows through the forward channels, as in video-on-demand. Thus, the system can be optimized to accommodate faster data flow rates in the forward direction. Some simplified Internet access systems are designed this way, with a modem or other connect line set for faster data rates for downloading, and slower data rates for querying as, for example, for Web browsing.

backward compatibility The ability of a system to run legacy (older model) programs, or to support older equipment. For example, 1.4 megabyte floppy drives are usually backward compatible with 770 kilobyte floppy diskettes, that is they can read, write, and format the older, lower capacity floppy diskettes. Similarly, a new version of a word processing program may be able to read and write data files created by an older version of the software.

Backward Explicit Congestion Notification BECN. In frame relay networking, a flow control technique that employs a bit set to notify an interface device that transmissions flowing in the other direction are congested, and congestion avoidance procedures should be initiated by the sending device for traffic moving in the direction opposite to that of the received frame.

backward learning An information routing system based on the assumption that network conditions in one direction will be symmetric with those in the opposite direction. Thus, a transmission moving efficiently through a path in one direction, would assume this to be an available, efficient route in the other direction as well.

backwave In radiotelegraphy, an undesirable interference heard between code signals.

Bacon, Roger (~1220-1292) An English philosopher, scientist, and a member of the Franciscan Order. In 1265 he completed an encyclopedic document of the knowledge of the time entitled "Opus majus."

BACP See Bandwidth Allocation Control Protocol.

bad block In magnetic storage that is segmented into blocks, a section which has experienced write or read failures. Some operating systems will map out bad block sectors on a diskette or hard drive so they will not be addressed or used, and will continue to format the remaining good parts of a disk. This is one of the reasons why the amount displayed for the usable portion of a disk differs from the total storage capacity of the disk.

Baekeland, Leo (1863 - 1944) A Belgium-born American inventor of modest birth whose mother encouraged him to get a good education to improve his opportunities in life. He had an agile mind, and emigrated to the United States to pursue his interests and professional connections. He is responsible for the invention of Bakelite, the first synthetic polymer, and Velox, a new type of photographic paper. See Bakelite.

baffle A device to direct sound, and to prevent sound waves from interfering with one another. A baffle consists of a series of carefully spaced corrugations that provide a longer path within a limited amount of space. It can be constructed of wood, metal, or synthetics, and works by lengthening the air path along the diaphragm through which the sound waves travel and by reducing interaction among them. This is commonly used in speaker systems to maximize the clarity of the sound.

BAFTA British Academy of Film and Television Arts.

bag phone *slang* See transportable phone.

Bain, Alexander (1811-1877 [dates approximate, as reports vary]) A Scottish chemist and clockmaker who developed an electrochemical paper tape recording system in the mid-1800s, suitable for telegraphic signals, at about the same time Samuel Morse was developing a somewhat similar system. The Bain system worked reasonably well except in situations with high noise on the line, which would create spurious marks on the tape.

Bain received a patent for his version in the 1840s which was contested by Morse, but which was sufficiently different to hold up in court. Morse subsequently bought out the Bain systems and converted them to his own.

Baird, John Logie (1888-1946) Although historical research makes it clear that a number of people independently developed different aspects of television reception and display, in the late 1800s, John Baird, a Scottish inventor, is attributed with being one of the earliest successful experimenters in this area. He was able to transmit a two-tone image of a face onto a small television screen in 1926, and finally by 1932, had developed a practical system for broadcasting images.

Baird used some of the principles of the Nipkow disc to develop his system. A light-sensitive camera was placed behind a perforated rotating disc, just as Nipkow had placed light-sensitive selenium behind a perforated rotating disc. The Baird system could only display a crude 30-line image at a frame rate a little less than half of that used now, but the 'proof of concept' technology launched an industry that is still going strong. See Nipkow, Paul.

Bakelite The development of Bakelite in 1907 revolutionized industrial production and heralded the 'age of plastic.' Inventor Leo Baekeland created this first synthetic polymer with a trademarked mixture of phenol, formaldehyde, and coloring agents. This new material was hard, and acid-, heat-, and water-resistant. It was quickly put to use in thousands of industrial products as a non-corrosive coating, and chemical binder for composite materials. Bakelite also provided new means to create colorful, water-resistant, moldable household products, dials, and even jewelry. Many early telephones and radios used Bakelite in their construction. See Baekeland, Leo.

Bakelyzer It looks like a B-movie adaptation of a Jules Vern diving bell on wheels, but it actually is a floor-standing iron pressure cooker devised by Leo Baekeland to mix simple organic chemicals into his versatile Bakelite synthetic resin. See Bakelite.

Bakken Library and Museum A museum located in Minneapolis, Minnesota which houses a collection of about 11,000 books, journals, and manuscripts focusing on the history of electricity and magnetism, and their applications in life sciences and medicine. The collection focuses on 18th through 20th century works, including those of Franklin, Galvani, Volta, and well-known contributors. In 1969, the collection of historical electrical machines was added to the activities of the museum, including several Oudin and D'Arsonval coils, and many electrostatic generators. http://bakkenmuseum.org/

balance *v.t.* To equalize, to counterbalance, to bring into harmony or equipoise, to offset in equal proportion, to arrange such that opposing elements cancel one another out or are of comparable weight, size, construction, value, strength, or importance. Balancing is commonly done in electrical circuits to equalize loads, or to diagnose the location of breaks or interruptions in a line. Stereo volume is usually balanced to equalize the volume or perceptual evenness of the left and right channels.

balanced bridge A bridge circuit in which the measured output voltage is equal to zero. Bridge circuits are sometimes used diagnostically to seek out and measure unbalanced circuits in order to detect a break or anomaly in the wiring. See Wheatstone bridge.

balanced circuit A circuit in which the electrical properties are symmetric and equal with respect to ground.

balanced configuration A point-to-point High Level Data Link Control (HDLC) network configuration with two combined stations.

balanced line An electrical circuit consisting of two conductors which have matched voltages at any corresponding point along the circuit, and which have opposite polarities with respect to ground. It is not uncommon to use more than one line to carry related transmissions or a 'split' transmission, especially in newer multimedia applications. By matching voltages and setting opposite polarities, it is possible to reduce the probability of crosstalk and interference, resulting in cleaner signals.

balanced modulation Modulation is a means of adding information to a carrier signal. In the early days of broadcasting radio waves, experimenters sought ways to manipulate or reduce the amount of bandwidth that was needed to carry the desired information. It has been found in amplitude modulation (AM) using electron tubes, that the control grids of two tubes can be connected in parallel, and the screen grids connected for push-pull operation such that the sidebands are singled out for transmission without the carrier. See amplitude modulation, modulation, single sideband.

balcony A small ledge or platform for aerial jobs used by film crews, antenna technicians, or utility wire workers.

bale, bonfire A signal fire, one of the oldest means of communicating over distance, and one which could be used at night. In the 1400s in Scotland, a simple signal code, using one to four bales, was established by an act of Parliament.

ballast Something which provides stability. In an electrical circuit, a device which stabilizes a current or provides sufficient voltage to start up a mechanism (such as a fluorescent bulb) or transmission.

balloon help A computer graphical user interface system which provides information to the user when requested in the form of text in a cartoon speech balloon shape hovering near the associated logical elements. On the Macintosh system, the balloon help can be turned off and on in the upper right corner. If on, balloon help is activated if the mouse pointer is held in one position for a length of time near an item for which help is available. This is helpful when it's needed and annoying when it isn't.

balun balanced/unbalanced. A small, passive transforming device used to match impedance on unbalanced lines that are connected together, such as twisted-pair cable and coaxial cable, so the signal can pass through the differing types of lines. As with many interface devices, there may be some signal loss through the balun. See bazooka.

BAN 1. base area network. 2. See Billing Account Number.

banana plug A simple, small, slightly curved, single-lead, spring-tip jack that slips easily in and out of its plug, yet exerts a low resistance against the plug in order to hold it in place.

band 1. The range of frequencies between two defined limits, usually expressed in hertz (Hz). See bandwidth. 2. A group of electronic tracks or channels. 3. A group of channels assigned to a particular broadcast spectrum, e.g., UHF (300 to 3,000 MHz). See chart of regulated band designations. 4. The range or scope of operations of an instrument. 5. An AT&T designated WATS Service Area.

band allocations Frequency ranges for radio wave communications have to be shared, and devices communicating on similar frequencies can have devastating effects on one another. For this reason, the frequency spectrum is allocated and regulated in order to maximize use of the available spectrum, and also to designate waves suitable for different types of activities. In the U.S., this information is contained in the Federal Communications Commission (FCC) Table of Frequency Allocations and the U.S. Government Table of Frequency Allocations, which together comprise the National Table of Frequency Allocations. Other organizations such as the ITU have tables as well. The values in the tables change, and what is represented in the Frequency Allocations and Common Uses chart is a very generalized overview to provide a basic understanding.

When new frequencies are available, they may be allocated to amateur or specialized uses, or auctioned for commercial use. When a user stops using an allocated frequency, it is reassigned. The available spectrum ranges have been established for various types of communications and some regions are unlicensed. Some of the more interesting unlicensed uses have been listed in the Frequency Allocations chart.

band center. The computed arithmetic mean between the upper and lower frequency limits of a band. This measure can be used to adjust modulation, to constrain it, or to provide the maximum possible amplitude range for an amplitude modulated (AM) signal.

band splitter A multiplexer which subdivides an available frequency into a number of smaller independent channels, using time division multiplexing (TDM) or frequency division multiplexing (FDM). See bandpass filter.

band, citizens See citizens band radio.

band-elimination filter BEF. A resonant circuit filter with a single, continuous attenuation band, in which the lower and higher cutoff frequencies are neither zero or infinite.

band-stop filter, band-rejection filter A resonant circuit filter for locking out a specified range, or ranges, of transmissions according to their frequency ranges.

banded cable Two or more cables physically held in proximity to one another (aggregated) with metal or plastic straps or bands.

bandpass The range of frequencies that will pass through a system without excessive weakening (attenuation), expressed in hertz. See bandpass filter.

B

Frequency Allocations and Common Uses

Name	Range	Notes
AM band	535 - 1605 kHz	Amplitude modulation, used commonly for radio broadcasts.
Videoconf.	around 24 MHz.	Certain local videoconferencing systems.
Mobile	various	Frequencies around 48 MHz are used for consumer outdoor mobile intercom units.
Radar	10 - 200 MHz	Imaging radar applications.
Amateur	50 - 54 MHz 144 - 146 MHz	Amateur radio use. Frequencies allocated for amateur use are frequently changed as the FCC often puts a higher priority on commercial users. This is in spite of the fact that amateurs have contributed a great deal to radio communications technology.
FM band	88 - 108 MHz	Frequency modulation, used commonly for radio broadcasts, and some low power FM transmitters (intercoms, bugs).
SAR	141 MHz	Synthetic Aperture Radar for environmental sensing and image processing.
Radar	300 MHz +	Approx. lower end of radar for remote sensing applications.
USDC	824 - 894 MHz	U.S. Digital Cellular FDMA and TDMA cellular phone services.
A-F block	1850 - 1910 MHz 1930 - 1975 MHz	Personal Communications Services (PCS) A to F block licenses granted to phone companies serving MTAs.
UPCS	1890 - 1930 MHz	Unlicensed Personal Communications Services (PCS).
S-band	2310 - 2360 MHz	Frequencies sensitive to terrain, making them unsuitable for some types of transmissions.
U-NII	5150 - 5350 MHz	Unlicensed National Information Infrastructure wireless communications, including PCS.
P-band	.22 - .39 GHz	Experimental radar. SAR.
C-band	4 - 8 GHz	Microwave frequencies, more specifically 3.40 to 6.425 GHz. Satellite - larger antennas. VSATS. Incumbent telephony operations (2. 0 GHz). Experimental radar. SAR.
L-band	1 - 2 GHz	Experimental radar. SAR.
X-band	8 - 12.5 GHz	Dedicated for use by the U.S. military for satellite communications. SLAR.
Ku-band	10.95 - 14.5 GHz	This has now been subdivided into fixed satellite service (FSS) at 11.7 to 12.2 GHz, and broadcasting satellite service (BSS) at 12.2 to 12.7 GHz. VSATs.
K-band	18.5 - 26.5 GHz	Satellite applications with smaller antennas, radar.
Ka-band	26 - 40 GHz	Satellite applications with smaller antennas, radar.
Q-band	36 - 46 GHz	Satellite, radar.
V-band	40 - 75 GHz	Radar band.
W-band	75 - 110 GHz	Radar band.

Frequency Range Designations

ITU	Designation	Abbrev.	Frequency	Wavelength	Typical or Example Uses
2	extremely low	ELF	30 - 300 Hz	10 Mm - 1 Mm	
3	ultra low	ULF	300 - 3000 Hz	1Mm - 30 km	
4	very low	VLF	10 - 30 kHz	30 km - 10 km	
5	low	LF	30 - 300 kHz	10 km - 1 km	Facom distance measurement and navigation
6	medium	MF	300 kHz - 3 MHz	1 km - 100 m	AM radio
7	high	HF	3 - 30 MHz	100 m - 10 m	CB radio
8	very high	VHF	30 - 300 MHz	10 m - 1 m	TV channels, FM radio, land mobile radio (cellular), ISM, LAWN, amateur radio
9	ultra high	UHF	300 MHz - 3 GHz	1 m - 100 mm	TV channels, CB radio, land mobile radio (cellular), PCS, radar
10	super high	SHF	3 - 30 GHz	100 mm - 10 mm	Satellite, amateur satellite, U-NII bands, radar
11	extremely high	EHF	30 - 300 GHz	10 mm - 1 mm	Satellite
12	tremendously high	THF	300 - 3000 GHz		

Note: In the above frequency ranges, the lower limit is exclusive, the upper limit inclusive.

bandpass filter A device with a resonant circuit, often used in conjunction with frequency division techniques, which recognizes and selectively allows control of frequencies, letting through those which are desired. A band-reject filter is complementary in the sense that it recognizes and selectively screens out a range of frequencies in order to form a 'blackout' area within the full spectrum of available frequencies. See band splitter.

bandspread tuning A means of spreading a band of frequencies over a wider area in order to adjust the tuning more precisely. This is most commonly found in shortwave radios and more than one set of dials may be used, with differently spaced tickmarks on the tuning gauges to aid in adjusting the settings.

bandwidth 1. The extent of a range of frequencies between the minimum and maximum endpoints, typically measured in hertz (cycles per second). Technically, the term bandwidth is associated with analog systems. In recent years, it has been more loosely applied to mean 'data rates' in digital systems and hence, is sometimes expressed in bits per second (bps). 2. The range of the frequency required for the successful transmission of a signal. This may range from a few kHz for a slow-scan or sideband signal to 100 kHz for a frequency modulated (FM) signal. That is not to say that the bandwidth of a signal necessarily takes up the entire range of the band that may be designated for its use. See band spectrum table. 3. In a cathode ray tube (CRT) device, the speed at which the electron gun can turn on and off. 4. The capacity to move information through a social, data, or physical system. 5. A numerical expression of the throughput of a system or network.

bandwidth allocation, bandwidth reservation In a network, the process of assessing and allocating resources according to flow, priorities, type, etc. This allows priority administration of the network traffic when congestion occurs.

Bandwidth Allocation Control Protocol BACP. In ISDN, a protocol which provides mechanisms for controlling the addition and removal of channels from a multichannel link.

bandwidth augmentation 1. Adding additional frequencies or channels to an existing bandwidth range. 2. Replacing existing physical transmissions media with broader bandwidth media in a system where the data transmission is capable of broader bandwidth, but the physical media caused a bottleneck due to its inherent limitations (e.g., replacing copper wiring with fiber optic).

bandwidth compression Techniques for increasing the amount of data that can be transmitted within a given frequency range. The increased demand for broadband applications such as video has motivated technologists to find more efficient ways to use existing transmissions media, resulting in better compression schemes and better management of the direction of transmission in bidirectional systems.

Compression can be very medium-specific. For example, in sending voice, blanks between words may be removed; in sending images, white pages may be compressed or eliminated; in sending complex multicolored images, lossy formats such as fractal compression or other lossy compressions such as JPEG may be used.

bang *colloq.* ! Exclamation point. 1. A common symbol used in many programming languages. For example, in C it represents a logical not. 2. Although its use is diminishing, it was at one time used in email addresses to designate a break between portions of an address, where a dot (period) is now typically used. Here is an example of a 'bang path':

```
{uunet,ucbvax}!galileo.berkeley.edu!username
```

bank A row or matrix, usually of similarly sized or configured components or data cells. Individual units in a bank are often interrelated, by shape, function, or electrical contact. In its simplest sense, a physical bank does not necessarily have connections between individual cells, but may just appear similar, and be mounted in rows and columns. Banks may also be electrically related, either by induction, physical connections between the cells themselves, or by temporary electrical connections that occur when a bar drops down over a bank of cells, or a brush passes over the bank.

Many large-scale telecommunications devices and junctions are set up in banks. Punchdown blocks at switching centers are set up in banks, often on racks or panels. Memory banks can be physical rows of memory chips in a circuit board. Large Internet Access Providers (IAPs) may have banks of hundreds or thousands of modems, and their associated phone wires. See bank switching.

bank switching A method of extending access to banks of components, such as memory chips, beyond the extents of any of the individual components, especially in situations where the operating system or microprocessor can only address a limited amount of memory at one time. By paging or swapping between banks, the virtual memory capacity is extended beyond the default physical memory or operating system (OS) or central processing unit (CPU) capacity. Bank switching may result in slower memory access.

banner A clearly visible, often graphic representation that heralds a new section in a printout or other text or image communication. Its purpose is to demarcate the end of one communication or section, and the beginning of the next one, especially in situations where many people are sending information through the same queue, such as a print queue; or to identify the type of communication, or the person to whom it belongs. The title page on a fax document is one type of banner.

bantam tube A squat electron tube with a normal-sized base which was once commonly used in small appliances or battery-operated mobile appliances like portable radios. Modern transistors have since made most types of vacuum tubes obsolete. See acorn tube.

bar code Identification, information, and management code designed for scanning by a laser reading device. Black and white visual bar codes are familiar identifiers on consumer products. They assist the checkers in entering prices and adjusting inventory databases. Bar codes are frequently inserted on letters that have been optically scanned. Less familiar is the fact that information is stored via bar codes onto video and audio discs.

bar generator A device used to generate horizontal or vertical bars on an output device to determine and adjust linearity.

Barclay insulator A type of early glass utility pole insulator invented by John C. Barclay. See insulator, utility pole.

bare metal, down to the bare metal The essentials of a machine or system. The low level systems functions. Programming 'down to the bare metal' usually means programming in assembly or machine language.

bare wire A wire without any kind of protective or insulating cover. The ends of insulated transmission wires are usually stripped of their covers to provide bare wire for a good electrical contact at a circuit junction.

Barge In A surcharge phone service, or feature of a multiline subscriber service, which allows someone (hopefully in authority) to barge into specified lines and interrupt a call in progress. It is a privilege that should not be used indiscriminately, but may be important in emergency situations. See buttinsky.

barge out To abruptly leave a call in progress.

barium ferrite Barium, a silver-white, malleable substance in combination with iron produces a substance that can be used in magnetic recording media. Methods of synthesizing barium ferrite nanoparticles through precipitation and spray pyrolysis are being studied.

Barkhausen-Kurtz oscillations In a vacuum tube, oscillation of electrons by means of electrodes and the grid. See klystron.

barometer An instrument designed to measure atmospheric pressure. It is one of the tools commonly used to evaluate and predict weather patterns. Barometers are incorporated into a number of other instruments as well, most notably traditional altimeters. Newer altimeters sometimes incorporate Global Positioning Service (GPS) capabilities.

Barometers were important instruments in early studies of magnetism, particularly in Italy where members of the Accademia del Cimento used barometers in the 1660s to provide an airless environment to study whether the attractive properties of various substances were dependent on air. Unfortunately, the difficulties of creating a vacuum, and manipulating the materials within the small area hindered them from making any definite conclusions. See Toricelli, Evangelista.

The two outer barometers are antique mantle and wall barometers. The center barometer is a portable barometer/altimeter combination for hiking and mountaineering.

barometric light A phenomenon observed at least as early as the 1600s, in which a glow or flash appears above the mercury in a barometric tube if it is moved quickly or shaken. The phenomenon is similar to that exhibited by neon light, although this was not known at the time. See barometer.

barrel distortion A type of visual aberration in which the outward corners of an image are contracted inward. This may happen on a convex or concave surface (depending upon whether it is backlit or frontlit) and is noticeable on older, more highly curved monitors and television screens. The opposite of barrel distortion is pincushion distortion. See keystoning.

barretter, barreter A device whose resistance changes in relation to temperature. The hot-wire barretter was devised in 1901 by R. Fessenden to improve the technology which was then being used to detect radio waves. The technology was incorporated into voltage regulating devices consisting of a wire filament connected to the circuit in series contained within a gaseous envelope. In conjunction with a waveguide, a barretter can be used to measure electromagnetic power.

BARRNet Bay Area Regional Research Network. An association of university campuses and government research centers in the San Francisco area. See BBN Planet.

Barton, Enos Melancthon (1840s-early 1900s) Barton cofounded Western Electric with telegraph/telephone pioneer Elisha Gray. The business partners first established Gray & Barton in 1869, which became Western Electric Company in 1872. This is the spiritual forerunner of today's Lucent Technologies. Graybar Electric Company, Inc. was spun off from Gray & Barton in 1925 in order to provide electrical distribution. See Gray & Barton; Graybar Electric Company, Inc.

base Bottom, lower support portion. In insulators used in utility poles, the base may be smooth or may have drip points, little knobs for the distribution of streaming moisture.

base film A substrate used to hold magnetic particles as in audio and video tapes. The materials used for base film vary but generally have the characteristics of flexibility, resistance to wear, and affinity for holding the magnetic coatings that are applied to their surfaces.

Base Information Transport System BITS. A U.S. Air Force military network system. Information about BITS is published on the Web as an aid to outside contractors planning and installing military information infrastructures.

base insulator A large support and insulating structure used on transmissions towers to insulate the tower from the ground.

base memory *jargon* The first block of memory, consisting of 640 kilobytes, in the older Intel-based desktop computers.

base rate 1. The basic rate without options or value-added services. 2. The basic charge per minute for measured service.

base station A main transmitting and/or receiving station or central switching station, often one which serves as a junction between wireless and wireline communications paths, or between broadcast signals and cable subscribers.

baseband 1. A simple type of transmission in which the signal is sent without altering it, as by modulation, and which does not require demodulation through modems to alter the signal at its destination. A transmission which is not segmented by frequency division (multiplexing). This basic signal is centered on or near the zero frequency. See sideband. 2. A one-channel or one carrier-frequency data network such as Ethernet, which is alternately shared by the various peripherals, such as computers and printers, or allocated as requested. See narrowband, Token-Ring. Contrast with broadband.

baseband modem This phrase is an oxymoron, since a baseband signal is one which has not been modulated, and thus doesn't require a modem to demodulate it, but baseband 'modems' do sometimes provide an interface device with some simple translation capabilities, and may resemble standard modems, hence the name. Sometimes better termed a short haul modem, as it is suitable for short distances.

baseband repeater A common main station repeating system used to retransmit a signal, and in some cases drop out selected channel groups, e.g., voice channels, before retransmission. Over long distances with several legs, heterodyne repeating, which is less subject to loss or distortion from modulation and demodulation, may be used in conjunction with baseband repeating to exploit the better properties of each method. See baseband, heterodyne repeater.

baseboard raceway A cable conduit or pathway along wall baseboards. The raceway may run along the baseboards, or be built into them, usually behind, so they won't be seen. Thus, 'exterior' wiring can be installed and hidden without making holes in the walls.

baseline 1. In coordinate systems, a scale, often the horizontal X axis, that establishes a reference structure on which related data can be depicted. 2. In radar, a line displayed to show the track of a scanning beam. 3. In typography, an imaginary line extending through a font (horizontal in Roman and Cyrillic fonts) used for alignment. Desktop publishing software is not entirely standardized, some programs treat the baseline as the bottom edge of nondescending letters, and others treat it as the first unit beneath the bottom edge of these letters. 4. In a Global Positioning System (GPS), a baseline is a pair of stations for which simultaneous data have been collected.

bash, bash shell Bourne-again shell A popular, powerful, practical sh-compatible Unix command interpreter shell released in the late 1980s by the Free Software Foundation. Bash is based on the Bourne shell, with some features from Korn (ksh) and C (csh) shells.

BASIC Beginner's All-purpose Symbolic Instruction Code. An English-like programming language designed at Dartmouth College in the early 1960s to satisfy a need for a programming language that was more comprehensible, and quicker to learn than FORTRAN, and lower level languages with rather obscure commands and symbols.

There are numerous implementations of BASIC, some of which can be compiled, but the majority are run as interpreted languages. The program instructions were originally entered as text on a command line, but more recent versions include graphical user interfaces which allow some parts of the code to be generated automatically by the software.

Historically, BASIC had an important role in bringing microcomputers out of the hobbyist domain into home and business markets. The Altair is the most significant early computer that could be purchased with a BASIC interpreter, and various BASIC compilers and interpreters have been available for virtually all microcomputer systems since that time. The first complete systems introduced in the 1970s with BASIC interpreters came, for the most part, without software, because hardly any software had been written at that time. The only way to get any use out of the computers was to learn to program them, and many people got their start in programming on these early machines. It is now not necessary to have programming skills to use the majority of software available on the market.

basic cable service The base service offered by a television cable company, consisting of a cable feed to the premises, and broadcasting of a specific package of programs. The cable company's programming provisions, signals and public, educational, and government access channels are government regulated under the Cable Act.

Basic Control System BCS. An interrupt-driven computer satellite control system.

Basic Encoding Rules BER. Standardized rules for data encoding that provide support for the abstract syntax description language described in Abstract Syntax Notation One (ASN.1).

Basic Exchange Radio Telecommunications Service BERTS. A system developed in the 1980s to provide wireless services through radio signals to standard local telephone loops, especially to rural areas or for emergency services.

basic information unit In packet networking, a unit of data and control information consisting of a request/response header (RH) and a following request/response unit (RU).

Basic Link A voice grade circuit providing certain specified standard levels for services, transmission, and loss (8 dB in the 300 to 3000 Hz bandwidth). See Assured Link.

Basic Rate Interface BRI. There are two basic types of ISDN service available: BRI, and PRI. BRI is an ISDN service consisting of two bidirectional 64 Kbps bearer channels (B channels) for voice and data, and one delta channel (D channel) for signaling or packet networking at 16 Kbps or 64 Kbps. It requires two conductors through a U Loop, from the carrier to a terminator (NT1) at the customer premises. Except in the rare cases of extremely long phone lines with load coils, most existing phone lines can be used for BRI without significant changes to the actual wire. BRI is aimed at residential and small business users. See ISDN.

base station 1. In mobile communications, a fixed station within the transceiver system. See cellular phone. 2. In Global Positioning Systems (GPS), a receiver established in a known location to provide reference data for differentially correcting rover files. Baseline data can be correlated with position data from unknown locations collected by roving receivers to improve accuracy.

Basic Trading Area BTA. An organizational designation for wireless telecommunications in which the United States is subdivided into almost 500 basic trading areas (BTAs) which are collectively grouped into Metropolitan Trading Areas (MTAs). The BTAs are used by the Federal Communications Commission (FCC) as a basis for assigning PCS wireless phone systems.

basket winding A technique of winding a wire coil, or other filamentous conducting material, such that the paths of the various turns of the winding do not touch except at junctions where they may cross. Basket winding is used in applications where a long length of wire, or a greater degree of surface area, is desired in a small amount of space. Basket wound antennas and other devices can be quite aesthetic, resembling arabesque. See basket winding tuners.

basket winding tuners Various types of basket winding with fine thread-like materials were commonly used in old radios to act as frequency tuners. The windings were of many shapes, cylindrical, circular, somewhat spherical, and varied in complexity from a dozen turns or so, to many hundreds of turns in an intricate pattern in many layers. By varying the shapes, sizes, and the thickness of the wires, different frequencies could be selected. A radio often came with a selection of basket wound tuners with electrical contacts on the base that could be 'plugged in' as needed.

bass In audio, a low pitch, deep tone.

bat switch A small, narrow switch, usually a toggle switch that is rounded on the end and tapered more finely at the point of attachment, roughly in the shape of a baseball bat. This type of switch is common in the aviation and audio industries where quick toggle adjustments are needed and many components are crowded for space in a small area.

batch An assortment of data or objects grouped to be processed during a single run of a program or process. See batch file, batch processing.

batch file A data file designed to save time by grouping, storing, and allowing the user to quickly execute computer commands that are frequently used. Batch files are a convenient way to store configuration parameters, frequently used groups of commands, a list of applications which are executed one after another, commands intended for deferred execution, scripts launched from Web pages, and startup commands for a computer system. ".BAT" is a familiar extension given to batch files on MS-DOS-compatible systems. Many systems provide job control languages (JCLs) or a variety of scripting languages for the quick creation of batch files. Perl is an excellent multiplatform programming tool for creating batch files for use on the Web and off. Batch command files have traditionally been created with text editors, but graphical tools can be used as well. See batch, batch processing, JCL, Perl, Java.

batch processing Deferred or off-line processing of an assortment of data, programs, or objects handled during a single program or process run. Unless there is a fault condition, batch processing usually assumes once the job is initiated, it will run undisturbed and unattended. Email is often handled as a batch process, e.g., your Internet Services Provider may wait a specified period of time before posting a group of messages to your account rather than posting each one as it is received. Payroll accounts are often run as batch processes, as are many data collection programs, such as weather testing, astronomical observations, etc.

It is not uncommon for batch processes to run as background tasks, executing while users continue to use the system for other applications. Deferred batch processes can be scheduled to run during times when network access is low, thus not sapping the resources of the system when many users are online. Batch processes can also be used to schedule transmissions, such as facsimiles, during hours when phone rates are low. See batch, batch file, real-time processing.

battery A group of two or more cells connected together in such a way that they produce a direct electric current (DC). While historians believe battery power may have been used for electroplating by the Parthians as early as the third century BC, the first significant records of modern battery experiments date from the work of C. A. Volta.

Battery-generated electricity was widely used in industrial applications and telegraph and telephone communications in the early 1900s. Edison was a strong proponent of DC current and received a lot of opposition from Tesla and Westinghouse, who were advocating alternating current.

Batteries are used widely in portable devices, and as emergency or backup power for systems whose main power source is alternating current. The chart on the following page describes a few common batteries, and interesting historical antecedents. See B Battery, cell, storage cell, talking battery.

battery, rechargeable A direct current (DC) power source. A rechargeable battery is designed to readily have its power restored, usually through consumer-priced battery chargers, or through an alternating current (AC) transformer attached between the wall socket and a battery-charging device. Rechargeable batteries are commonly used on palmtops, laptops, camcorders, etc. Most of them need to be fully discharged before being recharged, or a 'memory effect' results in only a partial charge. Some can be 'trickle-charged' when plugged into an outlet and in use. Larger rechargeable batteries for consumer electronics usually supply from two to five hours of charge.

battery, storage A type of battery which, once charged, will hold that charge for a practical amount of time without constant electrical refreshes from another source, such as alternating current (AC). Car batteries are a type of common storage battery that are recharged when the engine is running. They have a useful life of about three to five years if not completely discharged too often (by leaving lights on, for example). Storage batteries are often used as backup systems for alternating current (AC) systems. See battery, rechargeable.

battery backup A direct current (DC) backup system which kicks in if something happens to the primary power system. For example, many phones now have memory storage for names and numbers, or extra features like speakerphones or text display which require more power than is provided by the current coming from the phone line. These phones usually have a battery to help power the extra features which also functions as a backup battery to protect the contents of the electronic phonebook, if the phone line power is interrupted by a power failure, or if the phone is disconnected and moved from one location to another.

Most computer systems have small batteries on the motherboard to protect the contents of certain types of chips that hold information such as configuration parameters. Lithium batteries are commonly used and should be replaced every 5 to 7 years or so.

Many microwaves, clock radios, and VCRs have backup batteries so that the time is not lost during a power outage. If your appliance flashes 12:00 after a power failure, it probably doesn't have a backup battery.

baud, baud rate A unit signifying a rate of transmission of data indicating the modulation rate, named after French engineer J.M.E. Baudot. The term is commonly used in connection with modem data transfer rates (e.g., 9600 baud), although it originated from telegraph signaling speed in the 1920s. Note that the rate of transmission is not necessarily equal to the rate of acquisition of the data. Line interference, handshaking, error correction, and other factors can cause the actual rate of data receipt to be less than the raw transmission speed applied to the amount of data sent.

Baudot, J. M. Emile (1845-1903) A French engineer and inventor, Baudot made many contributions including a means, in the 1870s, to

Sampling of Battery Cell Types

Type	Developer	Notes
Daniell battery	J. F. Daniell	A chemical battery used in early telegraph systems, ~1 volt.
Edison cell		A variable storage nickel hydrate (positive) and iron oxide (negative) cell with an electromotive force lower than that of a lead cell. An older battery that was used in automobiles due to its ruggedness.
gravity/crowfoot cell		A voltaic wet cell for providing small currents at a constant emf.
Grove battery		Primary, 1.96 volts per cell. Zinc, platinum. Used by Morse.
A battery		Historically used as 'talk batteries' in telephone installations and as low voltage batteries for electron tube filaments. Now commonly used for cameras, calculators, and other small portable appliances.
B battery		Historical provider of low voltage power to the plates (anodes) in electron tubes and to communications relay circuits.
C battery		Introduced in the early 1920s, C batteries provided bias voltage to electron tubes for the control of the grid circuit and were often used in conjunction with B batteries to extend the life of the B battery. C batteries are now commonly used with small portable devices.
silicon solar	Bell Labs	Announced by Bell in 1954, there are now a number of variations on this technology from different developers. Kyocera introduced a multicrystal silicon solar battery in 1996 that has a conversion efficiency rate of 17.1%, very good in the solar industry.
lithium nouveau		A battery chemistry based on lithium polymer which may provide longer life for power-hungry mobile phones, laptops, etc.

insert synchronization signals between baseband signals so time division multiplexing (TDM) could be used to combine signals into a 'bundle.' He developed the Baudot code for telegraphic communications.

Baudot code, Murray code A data code used in asynchronous transmissions, named for its inventor J. M. Emile Baudot. It was widely incorporated into teletypewriter communications beginning around 1870. Baudot code was based on a marks and spaces character-representation scheme employing five equal-length bits to symbolize upper case letters. A simple method of reversing the polarity of the line was in use for about half a century before it was superseded by frequency shift keying (FSK modulation techniques). The character set was very limited and eventually standard codes such as EBCDIC and ASCII superseded Baudot code except for specialized communications, as for the hearing impaired. See ASCII; EBCDIC; Baudot, J.M. Emile.

Bauschinger effect Straining a solid body beyond its yield strength in one direction decreases its yield strength in other directions.

bay 1. Harbor, indentation, arced enclosure. 2. An opening in a rack or panel into which modular components can easily be inserted. A patch bay has a series of regular openings designed to securely hold modular components while still providing easy access, ease of configuration, and swapping in and out as needed. 3. Part of an antenna array.

bayonet base A type of jack or jack-like base, as on a bulb, which has a small projection on one side, and slips into a receptacle with a turn so the projection catches within a small trough and secures the inserted object.

bazooka A device for isolating an outer conductor from other surfaces and connecting an unbalanced line to a balanced line. Bazookas are commonly used on the ends of coaxial cables that are to be connected to two-wire lines (e.g., copper twisted pair). See balun.

.bb Internet domain name extension for Barbados.

BB 1. See baseband. 2. See broadband.

BBC 1. Broadband Bearer Capability. A bearer class field which is part of an initial address message. See ISDN. 2. See British Broadcasting Corporation.

BBG Basic Business Group.

BBN Bolt, Beranek, and Newman, Inc. A high-technology company in Cambridge, MA which developed, maintained and operated the historically significant ARPANET, and later, the Internet gateway, CSNET CIC, and NSFnet NNSC. See BBN Planet.

BBN Planet A subsidiary of Bolt, Beranek, and Newman, Inc., which operates a national Internet access network. See BBN.

BBS See bulletin board system.

BBT broadband technology.

Bc See Committed Burst Size.

BC 1. backward compatible 2. beam coupling. 3. binary code. 4. broadcast.

BCC 1. Bellcore Client Company. 2. See block check character.

BCD binary coded decimal. 1. A system wherein each decimal digit is coded into a four bit word. 2. A system wherein each octet within an ATM cell has each bit set to one of two allowable states, i.e., one or zero.

BCNU An abbreviation for "be seein' you" used in email, and postings on the Internet. See AFAIK, IMHO.

BCOB Broadband Connection Oriented Bearer. In ATM networks, information in the SETUP message that indicates the type of service requested by the calling user.

BCRS See Bell Canada Relay Service.

BCS 1. basic control system. 2. Batch Change Supplement. 3. Boston Computer Society. 4. British Computer Society.

.bd Internet domain name extension for Bangladesh.

BDCS Broadband Digital Cross-Connect System. A SONET DCS used for cross-connecting various signals.

BDF block data format.

BDT See Telecommunications Development Bureau.

.be Internet domain name extension for Belgium.

Be 1. See Excess Burst Size. 3. Be, Inc. A computer software company founded by Jean-Louis Gassée, head of R&D at Apple Computer during the Apple II years. Be developed and released the BeOS (Be Operating System) in 1997. It is a fast, integrated database, multiplatform OS which is aimed at the audio and graphics/video computer using markets.

BE 1. base embossed. A designation for glass and ceramic utility pole insulators which have embossings on the lower edge, usually of the size, company, and/or patent date. 2. Bose-Einstein.

beacon 1. A signal, locator, or guidance beam or tone. 2. A transmitter which aids in monitoring radiant energy propagation.

beacon alert An alert frame in a Token-Ring or Fiber Distributed Data Interface (FDDI) device signaling a serious problem, such as a physical interruption of the signal or Media Access Unit (MAU). The frame includes information on the location of the break, or the station that is down.

beam 1. *n.* A ray, shaft, or other directed energy or illumination, as an electron beam or light beam. 2. *v.* To direct or aim, as in a broadcast beam.

beam antenna An antenna which transmits and/or receives within a narrow, confined directional range.

beam divergence 1. As a beam travels through the air, various factors may cause it to spread out. This may result in attenuation or dispersion of the signal strength of a transmission over distance. 2. The path of a beam may progressively move away from the axis of the original trajectory, resulting in divergence.

beam power tube An electron tube with a beam that is directed and concentrated in certain specific directions by a special electrode. Used, for example, in radio frequency (RF) transmitters.

beam splitter A device which produces two or more separate beams from one incident beam. Mirrors and prisms are commonly used to direct or split light beams.

bearer channel See B channel.

beat Short percussive tone, one instance of a repetitive sequence, a reaction from the impact of one object or process on another, the interaction of two different frequencies when certain portions of their cycles interact. See beat frequency, beat reception, zero beat.

beat frequency The frequency resulting when two different frequencies beat together on a nonlinear circuit. The beat frequency is equal to the difference between the two separate frequencies, typically expressed in cycles per second. See heterodyne.

beat reception The combining of two different frequencies, usually the external, incoming frequency, and an internally generated frequency which are then easier to amplify or otherwise condition as a single frequency than the incoming frequency would be by itself. See heterodyne. For contrast, see zero beat reception.

beating A wave phenomenon that occurs when two or more periodic waves of different frequencies combine to form a periodic amplitude pulsation. In audio, this can quite often be heard and felt as an undulating pulse by those in listening range. See beat frequency, heterodyne.

Beaufort notation In meteorology, a code which is used for indicating the state of the weather.

Beaver Falls Glassworks A lesser-known utility pole glass insulator fabrication company founded by William Modes in Pennsylvania in 1869. See insulator, utility pole.

BEC 1. See Best Effort Capability. 2. Bose-Einstein condensation.

BECN See Backward Explicit Congestion Notification.

beehive insulator A type of early glass utility pole insulator, characterized by its beehive shape. See insulator, utility pole.

beeper Colloquial term for a portable device that alerts the user with an audible tone. Commonly applied to pagers, a beep signifies that there is a message awaiting the user, or some action to be taken as a result of the alert signal.

Beilby layer A microcrystalline or amorphous layer that is formed on the surface of metals by polishing.

Being There A consumer-priced, Macintosh-based videoconferencing product from Intelligence at Large, which provides video, audio, whiteboard, and file sharing utilities over

Bell Serial Communications Standards

Standard	Speed	Notes
Bell 103	300	Asynchronous full duplex communications standard for transmitting at speeds up to 300 bps over publicly switched telephone networks (PSTNs). This standard was commonly used with computer modems in the late 1970s, but was superseded by Bell 212 in the early 1980s.
Bell 212	1200	An AT&T asynchronous full duplex communications standard for transmitting at speeds up to 1200 bps over publicly switched telephone networks (PSTNs). This standard was commonly used with computer modems in the early 1980s, but was superseded by Bell 201 in the mid-1980s.
Bell 201	2400	Asynchronous full duplex communications standard for transmitting at speeds up to 2400 bps over publicly switched telephone networks (PSTNs). This standard was commonly used with computer modems in the mid-1980s. Many other vendors began entering the modem manufacturing/standards industry at this time.
Bell 208	4800	Asynchronous full duplex communications standard for transmitting at speeds up to 4800 bps over publicly switched telephone networks (PSTNs). This standard did not particularly catch on in consumer markets. Many users leapfrogged from 2400 bps to 9600 bps as vendor participation and competition for faster speeds increased in the mid-1980s.

AppleTalk and TC/IP local area and wide area networks.

bel (Symbol - B) A unit of relative power or strength of a signal, which is not commonly used because it is so large. It is used in conjunction with amplitude, usually by its tenth measure, the *decibel.* Named after Alexander Graham Bell. See decibel.

Belden A major commercial manufacturer of communications media which has been responsible for influencing cable standards for many telecommunications systems.

bell 1. Audio device, often of resonant hollow metal, designed to emit sound when struck, vibrated by a column of air, or vibrated through electrical stimulation. 2. A digitally reproduced simulation of a physical bell, created either by sampling a physical bell and playing back the sound, or by analyzing the type of sound wave patterns produced by a physical bell and simulating them mathematically. A computer requires built-in electronics or a peripheral sound card in order to send sounds to a speaker, especially if it's good quality 16-bit stereo sound. Digital music synthesizers often have a wide array of bell sound patches from which to choose. 3. A phone bell that is activated by line current from the switching office to indicate that there is an incoming call on the line.

Bell asynchronous standards A series of full duplex standards which has been developed by AT&T. These were widely supported by other manufacturers in the late 1970s and early-/mid-1980s. Other vendors and standards bodies began competing with the Bell standards, most notably Hayes, in the early 1980s. The V Series Recommendations by the ITU-T are now the dominant formats. The Bell standards shown in the following chart are of interest.

Bell Atlantic A holding company which was created as a result of the AT&T divestiture in the mid-1980s. See Bell Operating Company.

Bell Canada The Canadian arm of the Bell system, Bell Canada is now a member of the Stentor consortium, along with BC Tel Ltd. and others. It is a major telecommunications carrier and supplier of telecommunications equipment in Canada.

Bell Canada Relay Service BCRS. A 24-hour service which allows TTY users, who may be hearing impaired, to talk to one another, or to a hearing person with the help of specially trained operators. The TTY equipment can signal up to 60 words per minute.

As an example of the service, the subscriber calls the BCRS operator and provides his or her name and number, and the number of the person to be called. The operator requests billing information and then places the call. The operator then acts as a translator, conveying a text message by voice to the hearing callee, and a voice message by text to the hearing impaired caller.

The call is kept confidential by the operator, and no record of the conversation is retained. BCRS services are billed at the same rate as normal phone charges.

Bell Communications Research Bellcore. An organization established as a result of the AT&T divestiture to provide a variety of central administration, training, standards, documentation, and quality services to the regional Bell companies who fund Bellcore, and their subsidiaries. It is roughly equivalent to the Central Services portion of the pre-divestiture AT&T organization.

Bell Laboratories, Bell Telephone Laboratory, Bell Labs The research arm of the Bell system responsible for many important discoveries and the development of thousands of telecommunications technologies and devices over the decades. The labs were established as a combined effort of the Western Electric Company, and the AT&T engineering departments in 1907. This grew to be the largest industrial research organization in the U.S., and in 1925 the engineering department of Western Electric was incorporated as Bell Laboratories, with the head office in New York City.

In 1941, headquarters were moved to Murray Hill, New Jersey and larger plants were later established in Denver and Atlanta. Smaller field stations and satellite labs were regularly established over the years in many parts of the United States. In 1934, AT&T's research division was merged into Bell Laboratories.

Bell Labs Museum An online resource sponsored by Lucent Technologies. You can visit the images and historical references at the Website. http://www.lucent.com/museum/

Bell Operating Company BOC. This is defined in the Telecommunications Act of 1996, and published by the Federal Communications Commission (FCC), as

"... any of the following companies: Bell Telephone Company of Nevada, Illinois Bell Telephone Company, Indiana Bell Telephone Company, Incorporated, Michigan Bell Telephone Company, New England Telephone and Telegraph Company, New Jersey Bell Telephone Company, New York Telephone Company, U S West Communications Company, South Central Bell Telephone Company, Southern Bell Telephone and Telegraph Company, Southwestern Bell Telephone Company, The Bell Telephone Company of Pennsylvania, The Chesapeake and Potomac Telephone Company, The Chesapeake and Potomac Telephone Company of Maryland, The Chesapeake and Potomac Telephone Company of Virginia, The Chesapeake and Potomac Telephone Company of West Virginia, The Diamond State Telephone Company, The Ohio Bell Telephone Company, The Pacific Telephone and Telegraph Company, or Wisconsin Telephone Company; and

(B) includes any successor or assign of any such company that provides wireline telephone exchange service; but

(C) does not include an affiliate of any such company, other than an affiliate described in subparagraph (A) or (B)."

See Federal Communications Commission, Telecommunications Act of 1996.

Bell speak *colloq.* A phrase to describe the substantial body of telephone jargon which grew up over the decades within the Bell system, particularly among technicians and scientific researchers.

Bell System The original holders of the Bell telephone patents formed by Bell, Sanders, and Hubbard in 1877, and incorporated in 1878, less than 15 years after the invention of the telephone. The company thrived and grew under the management of Theodore N. Vail. Since the term of exclusivity granted by a patent lasts only 17 years, the expiry of the Bell patents resulted in the founding of thousands of new independent phone companies. These gradually were merged and consolidated into the Bell system. In a 1984 court decision, divestiture of the American Telephone and Telegraph company (AT&T) removed the distinction between the Bell company and independent phone companies.

Bell Telephone Company of Canada Inc. Established in 1880, it began by providing service to the larger centers in eastern Canada, most of which were interconnected within about 10 years. Bell Canada is under the jurisdiction of the Canadian Radio Television and Telecommunications Commission (CRTC).

Bell Telephonic Exchange The first telephone exchange in Ohio State.

Bell, Alexander Graham (1847-1922) A Scottish-born American inventor, who was one of the original founders of the National Geographic Society. He emigrated to Canada with his family and subsequently found employment in Boston. He studied aviation, electricity, fresh water distillation, etc., but is chiefly credited with the invention of the telephone. In fact, a number of independent inventors in Europe and America were working on ways to transmit tones, and in some cases, voice over telegraph lines. Three significant contemporaries of Bell who achieved success in some of these technologies were Philip Reis, Elisha Gray, and Thomas Edison.

Alexander Graham Bell continued to think of himself as a 'teacher for the deaf' long after he became famous and financially independent from his technological inventions. He and Helen Keller were friends. Many of the members of his family were known as excellent orators.

In March 1876, two years after working out the original concept, Bell and his assistant Watson reported having transmitted Bell's spoken voice over electrically charged wires, a story that caught the world's imagination and ushered in the telephone age. Bell's patent was filed just hours before a caveat to file was entered at the same office by Elisha Gray, for a similar harmonic telegraph. Bell's talking phone was not demonstrated publicly until some time later. One interesting historical note is that Bell's patent did not mention the use of a liquid medium, yet the rudimentary telephone that was later demonstrated did use a liquid medium, a detail which was mentioned in the Gray caveat.

In 1882, Bell was granted United States citizenship.

Bell achieved enormous financial success and could have ceased working at a young age, had he so chosen, but he continued to research aeronautic kites, hydrofoils, and various technologies to aid the deaf, most notably the audiometer. For much of his life, he listed his occupation as 'teacher of the deaf.'

While the invention of the telephone was not as revolutionary as the telegraph, it was a significant evolutionary step that personalized distance communications in a way not previously possible. See audiometer; Bell System; Edison, Thomas Alva; Gray, Elisha; telephone.

Bellcore See Bell Communications Research.

Bellingham Antique Radio Museum A diverse, well-selected collection of over a hundred years history of antique radio and electrical technologies, including a Tesla coil, Nipkow disc, Leyden jars, static generators, phonographs, and all sorts of makes and models of crystal detectors and historic radios. Located in Bellingham, WA. Curator: Jonathan Winter. http://www.antique-radio.org/radio.html

BellSouth Corporation A large regional holding company which was created as a result of the AT&T divestiture in the mid-1980s. It is comprised of Southern Bell Telephone and South Central Bell Telephone Company, and a number of other companies. BellSouth is cooperating with Nippon Telegraph and Telephone to provide large-scale integration of residential fiber multimedia telecommunications services by 1999. See Bell Operating Company, fiber to the home.

benchmark 1. A very specific expression of performance based on agreed-upon test criteria. 2. A criterion expression, often numeric, against which other systems or processes are compared. Benchmarks are so specific that it is hard to translate benchmark performance scores to real-life computing situations, and their validity is often hotly contested. See benchmark test.

benchmark test A criterion test for evaluating the performance of a system, often applied to the speed of processing. Although benchmark tests may be straightforward for simple electronic components, they are sometimes used to evaluate the system performance of complex systems, which are difficult to measure in objective units. For example, a computer with a 40-MHz CPU will perform more slowly on benchmark tests than a 200 MHz RISC chip CPU, yet a word processor running on one system may have the same apparent speed to the user as one running on a faster system due to many factors such as load on the system, user interaction, software optimization, address bus bottlenecks, amount and type of memory, etc. In a broad sense, benchmarks cannot be said to provide definitive performance measures, but they are nevertheless often established as a best-efforts way of comparing and contrasting systems with significantly different construction and characteristics. Yet, even these are often considered "better than nothing" performance indicators. See Dhrystone, Rhealstone, Whetstone.

bend loss In cabling, attenuation caused by bends and twists in the wires or fibers. At each bend there is a tendency, especially in optical fibers, for the signal to want to continue to radiate in the same direction, resulting in slight losses as the cable curves.

bend radius In cabling and cabling enclosures, a description of the bend tolerance of a certain material at a certain radius, often under a certain pulling force. This measure is important for manufacturing, for selecting types and sizes of parts, and for installing pulleys, cables, and wires. See bend loss.

Benedicks, Manson An American researcher who investigated the electromagnetic-influencing properties of germanium crystals in the early 1900s and found that they could be used to convert alternating current (AC) to direct current (DC).

bent pipe A description for a communications conduit, path, or transmissions medium which reflects an incoming signal back out at an angle, usually between 20 and 70 degrees, thus following a path that resembles a bent pipe. This is a very common configuration for Earth-satellite/satellite-Earth transmissions and for radio transmissions which are bounced off the ionosphere.

BeOS An object-oriented, multitasking, fast, nonlegacy, microcomputer operating system developed by Be, Inc. under the leadership of Jean-Louis Gassée. Gassée is well-known for his previous contributions to R&D at Apple Computer, when the Apple II line was being developed. BeOS is aimed at multimedia audio and visual applications.

BeOS was introduced to developers late in 1995 along with the Be computer. About a year later, Be, Inc. discontinued the hardware, to concentrate on software development, as their operating system software is able to run on several hardware platforms by various vendors.

BEP Back End Processor.

BER 1. See Basic Encoding Rules. 2. See bit error rate.

Berkeley Internet Name Domain BIND. A popular implementation of the Internet domain name service (DNS) originally developed and distributed by the University of California in Berkeley. There have since been numerous commercial implementations of BIND.

Berkeley Standard Distribution BSD. A family of Unix-style operating systems originally released in 1974, adapted to the Digital VAX and PDP-11, and now widely ported to many systems. BSD was further developed by Bill Joy and others at the University of California in Berkeley, who released it in 1978. Joy subsequently wrote the well-known *vi editor*, and cofounded Sun Microsystems.

BSD flourished with the development of the ARPANET, the forerunner to the Internet, and the Computer Systems Research Group (CSRG) enhanced BSD with 32-bit addressing, virtual memory, and a fast file system supporting long filenames. They further introduced BSD Lite which was BSD without the licensed AT&T code, which could be freely distributed. The CSRG disbanded in 1992, and the community at large adopted BSD and developed FreeBSD. See FreeBSD, Unix, UNIX.

Berners-Lee, Tim Tim Berners-Lee of CERN has gained a spot in the history books with his Web project proposal introduced in March 1989, and his demonstration of World Wide Web software in the winter of 1989. The rapid acceptance and growth of the Web is a tribute to the viability of this concept. Prior to that, Berners-Lee developed Enquire, in 1980, a hypertext system that no doubt formed the seed for his Web project. See World Wide Web.

One of the earlier removable computer storage media, called Bernoulli drives, were named after Daniel Bernoulli, a mathematician.

Bernoulli, Daniel (1700-1782) Bernoulli was a Netherlands-born Swiss mathematician, who pioneered the principles of fluid dynamics, now used broadly in aeronautics, electronics, and other fields. The Bernoulli principle is derived from his writings in "Hydrodynamica" which described basic properties of fluid pressure, density, and flow.

Bernoulli box An electronic, high-speed data mass storage and retrieval device based on technology pioneered by Daniel Bernoulli.

Bernoulli-Euler law In a homogenous bar, the curvature of its central fiber is proportional to the bending movement.

Bernoulli's Theorem in a Field of Flow At every point in a steadily flowing fluid, the sum of the pressure head, the velocity head, and the height is constant.

Berry, Clifford E. A collaborator with J. V. Atanasoff in the development of one of the pioneering digital computers in the 1930s. See Atanasoff-Berry computer.

BERT bit error rate test/tester; block error rate test/tester. A diagnostic device which is used to test data integrity by transmitting a known pattern of bits, and evaluating the subsequent bit error rate (BER), usually on a cable segment. See bit error rate.

BERTS See Basic Exchange Radio Telecommunications Service.

bespoken, bespoke Custom-made, made to order, made by engagement, requested item.

Best Effort See available bit rate.

Best Effort Capability A capability offered on some ATM networks which tries to provide transmission, but provides no guarantees of throughput. Might be used between two routers. See ATM traffic descriptor, RFC 1633.

beta 1. In electronics, the current gain of a bipolar transistor in a grounded-emitter amplifier. 2. (Symbol - B) quartz. 3. A version of software which is mostly complete, has been in-house tested, but requires wider input and trials from testers and users outside the company. See beta test.

BETA Business Equipment Trade Association.

beta site A location or group of people designated to test and use a piece of nearly completed, internally tested software in working conditions more nearly like those in which the software will eventually be used.

beta testing The in-house process of testing and using a nearly complete software or hardware product to try to determine if there are still bugs, or problems of usability, consistency, continuity, and ergonomics. Beta testing can take months or years, depending upon the state of readiness and the complexity of the software. Some developers use automated 'monkeys,' programs which systematically 'climb through' the software, to identify bugs or flow problems. This is a good process to use in conjunction with human testing.

In the author's experience, as much as 85% of software may be commercially introduced without sufficient beta testing. Because computing is confusing to the average user, users are hesitant to complain, thinking the fault is in their use of the product rather than the product itself, and sometimes this is true. But, upper managers often insist the product has to ship (whether it's ready or not) in order to generate revenues to stay in business even though, in most cases, it's bad economy. The cost of testing and correcting bugs before the product ships is almost always lower than the cost of repeated patches, upgrades, technical support, and loss of business due to consumer dissatisfaction when firefighting and corrections are done after the product ships.

If cars were sold with the same number of defects as many software products, consumer rights organizations would boycott the manufacturers. The author of this dictionary has contributed many hundreds of hours to beta testing and once found more than 300 bugs in two days of testing in a software product that the manufacturer insisted was 'complete, ready to ship, and absolutely bug-free.' There are responsible software houses which engage in extensive testing and quality control and their efforts should be recognized and rewarded. For those that don't, caveat emptor or get your money back. See alpha testing, upgrade.

BETRS See Basic Exchange Telecommunications Radio Service.

bezel The rim or edge of a tool or piece of equipment, often angled or sloped. On a computer monitor, the edge around the cathode ray tube.

Bézier curve A mathematically generated curve to delineate contours, especially in vector drawing programs. The Bézier curve is commonly used in illustrations and in the design of type styles due to its aesthetic qualities. Sometimes called B-spline.

.bf Internet domain name extension for Burkina Faso.

BFI Bad Frame Indicator. A packet networking error condition alert.

BFO beat frequency oscillator.

BFOC bayonet fiber optic connector.

BFT See binary file transfer.

.bg Internet domain name extension for Bulgaria.

BG, BGND background.

BGP See Border Gateway Protocol.

.bh Internet domain name extension for Bahrain.

BHLI Broadband High Layer Information. An ATM information element that uniquely identifies an application (or session layer protocol of an application). BHLI is implemented differently, depending upon whether the codepoint is user-specific, vendor-specific, or ISO.

.bi Internet domain name extension for Burundi.

bias 1. Expected or consistent deviation, inclination of outlook. 2. Deviation from expected value, systematic error. 3. In an electron tube, the fixed voltage which is applied between the control grid and the cathode.

bias distortion A device with inconsistencies or aberrations in the linearity of the signal. In finely tuned equipment, this is usually an undesirable property.

bias stabilization A means of controlling the bias in a circuit so that it does not fluctuate. Heat or signal variations can throw off bias resulting in damage to components. See bias, bias distortion.

biasing To apply a small amount of positive or negative stimulus to a circuit, as in an electron tube, to shift it in one direction or the other.

BIB Backward Indicator Bit, a status bit used in Signaling System 7 (SS7).

BIBO bounded input, bounded output.

BICEP See bit-interleaved parity.

biconical antenna A balanced broadband antenna which resembles a bowtie in the sense that it has two metal cones mounted in the same axis, that meet at the narrow ends where the feed line is attached. The orientation of the assembly affects its polarity. It is suitable for transmissions in the VHF range.

BIDDS base information digital distribution system.

bidirectional printing Any printing mechanism that prints in both directions, e.g., printing in horizontal sweeps while moving vertically down a sheet of paper from one line to the next. Unlike traditional typewriters which do not strike the paper during the carriage return line feed, a computer printer or memory typewriter will type as the carriage moves from left to right, and continue to print as the carriage moves back from right to left. This enables faster printing.

bifurcated routing A routing technique which splits data traffic so that it continues through multiple routes (technically it would be two routes, as *bi*furcated means split into two branches).

BIG See broadband integrated gateway.

big-endian Stored or transmitted data in which the most significant bit or byte precedes the least significant bit or byte. Many file incompatibilities between computer systems, in which the file formats are otherwise almost identical, are due to platform conventions which store the data in big-endian or little-endian form.

BIGA Bus Interface Gate Array.

BII base information infrastructure.

bilateral antenna An antenna whose maximum transmitting or receiving poles are diametrically opposite, that is 180 degrees apart in a plane.

Bildshirmtext [picture screen text] A German interactive videotext system from the German Bundespost. It is similar to the French Telecom Minitel service, except that the German Bundespost did not provide the terminal free. See Minitel.

Bill Of Materials BOM. A list or table of the specific items and quantities required to produce a given project, job, or output from a process. In engineering and CAD, the BOM is often printed on the plan or blueprint in the lower right corner.

billboard array antenna An antenna array which resembles a billboard in that it uses a large sheet metal reflector behind the stacked bipole arrays.

Billing Account Number BAN. An identifier that allows telephone carriers to bill individual customers, or each of multiple accounts belonging to the same customer.

billion In North America and France, one thousand million (10^9 - 1,000,000,000). In the U.K. and parts of Europe, one million million (10^{12} - 1,000,000,000,000).

BINAC Binary Automatic Computer. A joint project of J. Mauchly and J. P. Eckert who founded the Eckert-Mauchly Computer Corporation. The BINAC was unveiled in 1949, and was historically significant not only as a successor to ENIAC, but for its ability to store data on magnetic tapes rather than on tape or punch cards. See ENIAC.

binaries, binary files Files which have been compiled or assembled into machine-readable codes, usually 8-bit executable files, that are inscrutable to most human beings. Source code is higher level code (as in BASIC, C, or Perl) that can more easily be read and modified by a programmer. Binary files can be edited directly with a hexadecimal (base 16) editor. See attachment.

binary, base 2 A system of numeric concepts and numerals representing quantities in terms of ones and zeros with the smaller units on the right. Thus, two in the base 10 decimal system is written as "2." The same quantity expressed in binary is "10" or "00000010" with the one in the 'twos' position, second from the right. The columns from right to left are thought of as the 'ones column,' 'twos column,' 'fours column,' 'eights column,' etc. so that a digit in a specific column indicates the presence or absence of that amount. Thus, the following numeral in binary: 0011010 can be transposed to decimal by adding its values: 0+0+16+8+0+2+0 = 26.

In electronics, binary values can be variously represented by pulses of unequal length, by amplitudes of specified magnitudes, by power on or off conditions, or by different tones.

Because most computers are two-state systems, the binary number system is used for programming and storage of data. Thus, zero and one can represent states such as 'on' or 'off,' 'yes' or 'no,' etc.

binary file transfer BFT. Binary files are those which have been translated into a base 2 system to be more readily used by a computer. Binary files are usually 8-bit encoded so they cannot be readily transported, without conversion, over 7-bit systems that typically use ASCII (or EBCDIC on some older systems). Due to the encoding, they cannot be directly read by (most) humans, or directly edited by most text editors, although a hexadecimal editor is sometimes used to make limited changes in binary files.

Binary file transfers are usually accomplished by encoding the file into ASCII with a utility such as BinHex, transmitting the file, and re-encoding it at the destination. Most email clients now do this automatically as 'file attachments.' Multipurpose Internet Mail Extension (MIME) is one of the protocols used with mail clients to transparently handle the encoding and decoding of attachment files.

binary signaling A common form of digital modulation.

binaural Related to two sound sources or two sound receiving sources, as human ears. Since humans are accustomed to using two sound sources to distinguish the quality and directionality of sound, monaural music tends to sound somewhat flat. Thus, stereo (binaural) sound systems have evolved to provide a more natural representation of sound.

BIND See Berkeley Internet Name Domain.

bind triangle In an International Business Machines (IBM) SNA implementation, a session setup message sequence.

binding post In electrical installations, a screw terminal with a corresponding nut around which 'U'-shaped lugs or wrapped wires can be wound and secured with the nut. Sometimes there are two nuts, close together so the wire can be secured between the two screws. Binding posts tend to be used in temporary circuits, or in small installations. In medium- and large-scale telephone installations, mounting blocks and punchdown tools are much faster.

BinHex A very useful software archiving/translation tool which can be used to convert an 8-bit binary file into a 7-bit ASCII file through run-length encoding, so that it can be handled by 7-bit systems that may use different protocols, but which understand ASCII. Many email clients use BinHex internally to handle binary attachments to text messages. At the receiving end, the file must be converted back to its original form before it can be executed or otherwise used as originally intended. BinHex is a very widely used application on many platforms, but is especially prevalent in Unix and Macintosh environments.

bionics A term coined for a computing conference in 1960, sponsored by the U.S. Air Force to describe 'unit of life' neural network computing applications. The Air Force was particularly interested in intelligent guidance systems because aircraft and other technologies were becoming too fast and sophisticated to be handled by humans unaided. See intracellular electrode, neural network.

BIOS basic input/output system. A system in read only memory (ROM) on some Intel-based desktop computers, which supports the central processing unit (CPU) by supplying access to a variety of input/output devices, such as serial ports, joysticks, monitors, keyboards, etc. As these peripherals are basic to the functioning of the computer, they are frequently used and loaded from ROM into RAM for fast access as a system comes online during the powerup sequence.

BIP bit interleaved parity. In ATM, a method used at the physical layer to monitor the error performance of the link. A check bit or word is sent in the link overhead covering the previous block or frame. Bit errors in the payload will be detected and may be reported as maintenance information.

biphase coding A networking bipolar coding scheme in which clocking information is carried in the synchronous data stream without separate clocking leads.

biphase shift keying BPSK. A simple type of modulation scheme used in digital satellite transmissions. In BPSK, each phase of the carrier wave is shifted once with each complete cycle, with a shift indicating the change of the value (from one to zero or zero to one).

bipolar 1. Having two mutually opposing or repelling forces, characteristics, or viewpoints. 2. Having two poles. 3. A circuit with both positive or negative polarity, or alternating between positive and negative polarity. 4. In electronics, a type of construction prevalent in integrated circuits (ICs). 5. A device having both majority and minority carriers. 6. Having electromagnetic characteristics alternating between two poles. 7. A type of signaling in digital transmissions in which a binary value represents a signal amplitude of either polarity, and no value represents zero amplitude.

bipolar receiver A type of telephone receiver used extensively in the Bell System. It improved on earlier technology by using new magnetic alloys, and employing a different acoustical system for the diaphragm. See ring-armature receiver.

bipolar signal A signal with two nonzero polarities, that is it can represent two states or three states in a binary coding scheme. See bipolar.

bipolar transistor A semiconductor commonly used in oscillators, switches, and amplifiers.

bird 1. A feathered, warm-blooded vertebrate with specialized forelimbs of the class Aves. 2. *colloq.* Something which resembles a bird in shape or ability to stay aloft, as a blimp, plane or satellite.

birdie A lightweight cable or wire installation accessory device. Once the conduit has been installed for a wiring/cabling installation, a birdie, attached to the wire by a long lead, can be blown through the conduit with a compressed air tool, and the wire subsequently pulled through with the birdie as a lead to make it easier. See pulling eye, snake.

Birmington Wire Gauge A gauge standard for describing the diameter of iron wires (nonferrous wires are generally described with American Wire Gauge). The thinner the wire, the higher the number from 1 to 20, exclusive of the coating. See American Wire Gauge.

bis Second, update, revision, encore. In the V Series recommendations of the ITU-T related to telecommunications, 'bis' is used to indicate a second version or update to a previously numbered standard. This was probably substituted for a revision number to prevent confusion between the series number and revision level. Similarly, 'ter' is used to designate three, or third.

BIS border immediate system.

BISDN See B-ISDN.

BISSI Broadband Inter-Switching System Interface.

BIST built-in self-test.

Bisync (*pron.* bye-sink) Binary Synchronous Communication Protocol. A character-oriented serial network protocol that was developed in the 1960s, at a time when International Business Machines dominated the network market. This is now mainly supported as a legacy protocol.

bisynchronous transmission A transmission that can flow in two directions on the same line or channel, usually at the same time. Traditional wireline telephones are bisynchronous, whereas some types of radios or intercoms only transmit in one direction, or in one direction at a time.

bit binary digit. A basic unit of digital information with two (bi-) states. Many schemes for signaling binary states have been developed: on/off (early telegraphs), high/low, one or zero (mark or space, data bits), etc.

bit bucket *slang* A mythical container into which unwanted or unused code, email, or other computer files are discarded or lost. Deleting a file sends it into the bit bucket. Losing a file unaccountably is another bit bucket casualty.

bit error A fault situation in which the value of an individual bit is changed by transmission or data interpretation errors.

bit error rate BER. A measure of transmission quality, usually expressed as a ratio of error bits to total bits received. A high bit error rate does not necessarily result in a faulty transmission. Error-detecting and correcting algorithms are incorporated into most current transmissions protocols. However, a high BER may result in slower transmissions, smaller packets, a higher percentage of retries, and perhaps even the necessity to connect several times to complete a file transfer, for example.

Bit Oriented Protocol BOP. A network control protocol functioning at the data link layer. This is one of various bit-oriented protocols typically used for synchronous transmissions.

bit pipe A telephone circuit which is used to transmit digital data packets.

bit robbing A process of commandeering bits in a transmission for something other than their usual purpose. Extra bits may be robbed to convey signaling information, especially if the signals are only occasionally needed. See robbed-bit signaling.

bit stuffing See zero bit insertion.

bit-interleaved parity BICEP. In ATM networks, an error monitoring method implemented at the physical (PHY) layer. The link overhead contains a check bit or word for the previous frame to flag errors.

bit-oriented Data communications which can encode control information in small single-bit data units.

bitmap A pixelated image. An image (picture or font) which is represented by discrete dots or pixels on a monochrome display, which is typically white, green, or amber (technically a multicolor image is called a pixmap). The displayed image does not necessarily reflect the format of the image file. For example, a vector file may be represented as a bitmap or pixmap image on a printer or computer monitor, or a pixmap file may be converted to a continuous tone image when printed to a dye sublimation printer.

On the printed page, bitmap images of moderate complexity appear somewhat coarse and grainy to the human eye at resolutions of 300 dots per inch or less. Above 600 dpi most fonts and curved lines appear reasonably smooth, and at 1200 - 2400 dpi the resolution of a bitmap is high enough to satisfy most professional applications. Monitors display bitmaps and pixmaps at much lower resolutions than those of printouts, usually about 75 pixels per inch, with color, and human perception, assisting in making even this low resolution sufficient for most daily applications. See raster, pixels, vector.

BITNET Because It's Time Network. An international, cooperative, academic network established in 1981 by Ira H. Fuchs (City University of New York) and Greydon Freeman (Yale University). It began as a cooperative project at the City University of New York, with Yale as the first outside connection through a leased telephone line. It ran originally on IBM mainframes and Digital VAXes communicating through EBCDIC formats. From there, it spread across the U.S. and became international when it was joined by the European Academic and Research Network (EARN) in 1982.

BITNET promoted the noncommercial exchange of research and education information among its participants and was organized as a nonprofit corporation in 1987. By 1992, BITNET included almost 1,500 organizations in 49 countries and for a while was the world's largest academic network. By 1993, the number of participants was declining due to the rapid growth of the Internet. More recently it was merged with The Computer+Science Network (CSNET) to form the Corporation for Research and Educational Networking (CREN).

BITS See Base Information Transport System.

bits per second bps. A very common means of describing data transmission per second unit of time. A megabit per second, or Mbps, represents a million bits per second. Common consumer modems operate at data rates of about 9600 to 28,800 bits per second. T1, fiber lines, and other higher speed protocols and media function at much higher rates.

BIU See basic information unit.

bj Internet domain name extension for Benin.

BL bit line.

black and white monitor A specific type of monochrome monitor in which the displayed image is roughly white, depending upon the brightness adjustment, and all lit parts of the monitor are displayed at the same intensity, rather than each pixel being varied in intensity from the others. See monochrome monitor. Contrast with color monitor, grayscale monitor.

black body A theoretical body which absorbs all incident light with no reflection, and consequently appears 'black' (without light) at all wavelengths.

black box 1. A registered trademark of The Black Box Corporation of Pittsburgh, PA. 2. *colloq.* A device whose internal workings are obscure or obscured. That is, the outside may not indicate what is inside, or how it works. 3. A device that is used by a lay person without technical knowledge of its construction or functioning. 4. A type of clandestine phone interface device used in a central office to gain unauthorized access to phone services by obscuring the fact that a long-distance call had been answered. See blue box, red box.

black box design A design model for inputs and outputs which function independently of the various ways the internal components might be configured. For example, a converter or transformer for matching two types of signals might be specified, with leeway given to a manufacturer as to the best way to implement and build the hardware itself.

black hole 1. A theorized invisible (thus perceived as dark) region in space with a small diameter in relation to its intense gravitational field that is proposed to result from the collapse of a massive star in which the escape velocity equals the speed of light. 2. *colloq.* A fictional area into which things disappear when they can't be found, and those looking for them are sure they should be 'right there.' 3. In networks and computer systems in general, data that went in and apparently didn't come out, or which ended up in a metaphorical repository for lost data. It may also be described as having disappeared into the ether, into the bitstream, or into the bit bucket.

black jack When Samuel Morse won a contract from the U.S. Congress in the 1800s to build a telegraph line from Washington, DC to Baltimore, he initially tried to install the lines underground, alongside railroad tracks. There were problems with the line, however, and the wires were subsequently suspended from poles. To this day, millions of miles of communications cables are installed on utility poles. Since the poles were subject to weathering, they were coated with creosote, a preservative derived from coal tar, and came to be called black jacks.

black level A reference level on a display device corresponding to the lowest possible luminance setting, which typically appears as black, or nearly black, depending upon the characteristics of the display device.

black matrix tube A cathode ray tube in which black fills the spaces between color phosphors on the inside front coating of the tube. The greater contrast between the lit phosphors and the small surrounding area results in a picture that appears to have crisper, brighter colors.

black recording In recording systems employing amplitude modulation, black recording is the correlation between the maximum power of the transmission and the maximum density of the recording device. In recording systems employing frequency modulation, black recording is the correlation of the lowest frequency received and the maximum density of the recording medium. The phrase applies to various wired and wireless facsimile machines, printers, electronic photocopiers, etc. See black transmission.

black transmission, AM In an amplitude modulated (AM) transmission of an image, a black transmission means that the greatest divergence in amplitude in the signal represents the black tones, and the narrowest divergence represents the lightest tones (or no tone at all). In white transmission, the opposite is true.

black transmission, FM In a frequency modulated (FM) transmission, a black transmission means that the lowest frequency corresponds to black and the highest frequency corresponds to white, or no tone, and in a white transmission the opposite relationship is used. Black transmission concepts in general can be applied to image scanners, facsimile machines, photocopiers, etc. See black recording

Blacksburg Electronic Village BEV. An interesting cooperative effort to create an electronically-linked community, which includes citizens, university members, and the Bell Atlantic phone company. Blacksburg Village participants are interconnected to one another and to the Internet through modems installed in homes, businesses, and educational institutions. http://www.bev.net

Blake transmitter A very early telephone transmitter designed so the diaphragm varied the strength of an already established current from a battery, rather than generating energy by means of electromagnetic induction, as in earlier models, thus producing a stronger sound.

blank 1. A transmissions gap, one in which no signal or no data is coming through. 2. A space or nonprinted area on paper. 3. A spacer on a Web page to format a blank area, such as a blank image file, a <PRE> (preformatted) tag, or   (nonbreaking space) tag will create blanks in HTML documents that are used on the Web. 4. An advancing key on a teletype, typewriter, typesetter, label-maker, or other device such that an unprinted area is established. The word 'space' is often used interchangeably with blank.

blanking In a display device, such as a cathode ray tube (CRT), the interval during which part or all of the display is suppressed. Blanking can be used to suppress artifacts from the display sweep of an electron gun. The sweeping of a screen is often done in a repeated zigzag, but the part of the transmission intended to be seen is displayed on the 'straight' sweep (usually the horizontal sweep), and the beam drops to the next line as it moves back to the other side, ready to sweep again. It is somewhat like the line feed and carriage return on a typewriter; as the print head goes back and down to the beginning of the next line, it shouldn't make marks on the page. A blanking pulse is used for blanking. See blanking pulse.

blanking interval, blanking time The period during which a display is suppressed, usually to allow an electron gun to return to the next display position. See blanking, cathode ray tube, frame, sweep.

blanking pulse A mechanism for suppressing a display, usually on a cathode ray tube. This is sometimes accomplished by means of a positive or negative square wave. A series of pulses can be combined to create a blanking signal that is synchronized with the sweep.

bleed *n.* Extra image elements that extend beyond the significant portion, the portion which is viewed or printed. Bleed is often employed in the printing industry as a way to avoid fine white lines around the edge of an image when it is cropped. Suppose you have a business card with the picture of an airplane extending off the edge of the page. If you bleed the end of the plane 1/8" of the edge of the paper, then when the cards are printed and cut, there's no guesswork as to where the edge needed to be, and there is no chance of a telltale white line on the edge marring the image.

Printing jobs that require bleed are usually priced higher than regular printing jobs because of the extra paper size that may be needed to accommodate the larger image size, and the trimming which is done to adjust the size of the paper to the size of the image. When choosing a color printer for your computer, you may have to specify oversize paper capabilities if you need printouts with bleed. For example, if you need to print an image that is a full 8.5" *x* 11", you can't do it on most 8.5" *x* 11" printers, because of the nonprinting margin areas. Thus, you may need a printer that will hold letter extra size paper, or legal (8.5 *x* 14), or perhaps even tabloid (11 *x* 17) in order for the image to cover the entire area desired. See tiling.

blend In a paint program, to intermix colors or shapes so they merge gently into one another. This is used for shading and overlay effects. Alpha channels, masks, and other transparency effects are sometimes used to blend elements into one another.

blind transfer, cold transfer Transfer of a call without seeking the identity of, or receiving information about the caller.

blind zone A zone where there are no transmission signals. A skip zone is one type of blind zone. See zone of silence.

blinking A regular periodic or other intermittent visible increase and decrease in intensity, or on-off illumination display, bulb, or liquid-emitting diode (LED). Blinking is commonly used as an attention-getter or fault-condition alert.

Bloch's theorem of superconductivity The lowest state of a quantum mechanical system, in the absence of a magnetic field, can carry no current.

block check character BCC. An error-checking technique developed for early teletypewriters in which a control character is appended to blocks for longitudinal checking and CRC. In packet networking, as a packet is assembled, the data is processed to create a BCC, which is then incorporated into the packet, checked at the receiving end, and acknowledged (ACK) or not acknowledged (NACK) if it does not match. The data can then be resent until the BCC matches or until the process is stopped.

block diagram A type of visual communications aid which uses simple shapes to symbolize objects, functions, relationships, conditions, and processes. A flow diagram or flow chart is a type of block diagram in which specific shapes have been standardized to have certain meanings within the context of the diagram. Rectangles, diamonds and arrows are commonly used.

block transfer The process of moving data in a block, instead of in individual bits. Double buffering, in which a screen of information or block of data is built in the background, and then instantly presented or displayed by transferring it from one area of memory to another, is a type of block transfer commonly used to reduce delays. Various file transfer protocols make use of block transfer techniques, often reducing the size of the block if many errors are occurring.

blocking 1. Preventing entry/exit or transmission through. 2. Holding until time or space is available, as in a queue, or until the data, person, or object can be turned back. 3. A circumstance in which a call cannot be completed (the exchange may be overloaded, or the line busy). See grade of service, call abandons. 4. Deliberate exclusion of certain parties from certain numbers (such as prevention of 900 calls, long-distance calls, etc.) 5. In business, an illegal practice which prevents others from engaging in fair competition. 6. In vacuum tubes, creating very high negative grid bias to lower the plate current to zero.

blocking capacitor, blocking condenser A device in a circuit which blocks direct current (DC) while permitting alternating current (AC) to pass through.

blocking probability A performance measure describing the likelihood of data, or a user, being rejected.

bloom On a cathode ray tube (CRT) display device, the tendency of a phosphor excitation level to create a 'halo' effect of extra light that spreads out beyond the area being targeted. This tends to happen at higher intensity levels with lighter colors.

blooper 1. Goof, embarrassing error, bungle. 2. In transmissions through a regenerative relay, an unwanted signal created by the relay which is not part of the desired transmitted communication.

blower 1. *colloq.* A device to blow a current of air or gas, e.g., a fan used for ventilation or cooling. 2. A speaker or blowhorn.

Blowfish A 64-bit (8 bytes) encryption algorithm developed by Bruce Schneier that has become the basis for a number of encryption schemes, including Kent Briggs' Puffer, Harvey Parisien's VGP, and Philip Zimmermann's PGP. Blowfish uses a variable length key up to 448 bits in length. It may not be sold outside the U.S. due to federal export restrictions. See Pretty Good Privacy.

blown fiber A fiber optic installation system designed by British Telecommunications PLC that allows faster, more flexible installation and reconfiguration of fiber optic cable systems by literally blowing the fiber lines into a grouped tube cable 'hose'. This type of system is often combined with point-to-point modular connectors that eliminate splicing. Since splicing is an exacting job in fiber optic installations, this is a great convenience.

blown fuse A fuse which has broken its connection, due to some electrical abnormality on the circuit on which it is installed, which might have endangered other links in the system. Fuses typically cannot be reused and must be replaced with another with the appropriate voltage. A blown fuse will sometimes show a blackened area inside the glass. Circuit breakers have superseded fuses in many types of electrical wiring, but the phrase has remained, and is often used to indicate a tripped circuit breaker. See circuit breaker, fuse.

BLSR Bidirectional Line Switched Ring.

BLU basic link unit. A generic term used in a variety of types of networks.

Blue Book standard A document published by Digital Equipment Corporation (DEC), Xerox Corporation, and Intel Corporation in 1980 to provide information on the Ethernet protocol standard, version 1.

blue box *colloq.* A small hand-held device designed to emit tones in the same frequencies as touchtone telephones, often used in the 1980s for connecting long-distance calls illegally through direct tones rather than dialing. Typically a connection was established through normal means, usually through a toll free 800 number and then the blue box was used to disconnect the remote ringing, without actually disengaging from the long-distance connection. A new connection could then be established within about a 10 second window, by punching in appropriate operator tones from a keypad on the blue box. (Some individuals have even learned to reproduce some of these tones by whistling, without having to use a blue box.) Newer systems can move these tones out of band, and use more sophisticated monitoring and tracing of suspected connections to reduce the possibility of abuse.

Blue boxing probably originated in the very early 1960s, and the Bell System first apprehended a blue box user in 1961. The myth that blue boxing is done almost entirely by young college students is refuted by a report by AT&T that almost half of those caught stealing phone services with blue boxes are businessmen, many of them wealthy, along with a number of doctors and lawyers.

blue gun In a color cathode ray tube (CRT) using a red-green-blue (RGB) system, the electron gun which is specifically aimed to excite the blue phosphors on the inside coated surface of the front of the tube. Sometimes a shadow mask is used to increase the precision of this process, so the green and red phosphors are not affected, resulting in a crisper color image. See shadow mask.

blue pages A convention in telephone directories in which government listings are printed on pages with a blue background to distinguish them from residential and business listings. Online directories of government information and email addresses are now sometimes called blue page listings.

blue wire A color designation used by IBM to indicate patch wires used to correct design or fabrication errors in situations where it is not practical to recreate the board with the corrections. See purple wire, red wire, yellow wire.

BLV/BLI An operator call wherein the caller requests information about the busy status of a line, or requests an interruption of a call on an Exchange Service.

.bm Internet domain name for Bermuda.

BM burst modem.

BMEWS Ballistic Missile Early Warning System. A U.S. government long-range warning and tracking radar network designed to detect missile fire along the northern approaches.

BMP, .bmp A file extension standing for bitmap, often expressed as a three-character file name extension to maintain backward compatibility with operating systems which can't use a longer file extension. Bitmap files are raster graphics files. While .bmp is used by some as a generic file extension name for any type of raster graphics file, it also has a specific meaning for a standardized file format. See raster.

.bn Internet domain name extension for Brunei Darussalam.

BNC bayonet nut connection, bayonet navy connector. A common quick-connect bayonet-locking connector, especially for coaxial cables for network or video transmissions. It is intended to provide a constant impedance. The outer shell has a small bayonet the inserts in a helical channel in the receptacle to aid in holding and aligning the connector. See connector.

BNCC base network control center.

.bo Internet domain name extension for Bolivia.

BO branch office.

board See printed circuit board.

Boardwatch A good *prosumer*-level print and Web publication dealing with the telecommunications industry, particularly the Internet. http://www.boardwatch.com/

bobtail curtain antenna A phased array, bidirectional, vertically polarized wire antenna, intended for high-frequency transmitting and receiving.

BOC See Bell Operating Company.

body The main informational portion of a communication, sometimes sandwiched as a block between headers and trailers. Sometimes called the payload. In a picture file, the body is the portion that carries the information about the image. In a word processed document, the body is the portion that contains the informational text and accompanying illustrations. For contrast, see header.

body, type In typography, the main portion of the shapes that constitute a character set (typestyle). The portion from which ascenders and descenders originate. Sometimes called x-height.

BOF 1. Birds of a Feather. An informal, often ad hoc, focused discussion group among people with similar interests, which may develop into a special interest group (SIG) or larger working group or organization. 2. Business Operations Framework. A document which describes a specific area of operations within a firm. This may also constitute or include statements of policy regarding the described operations.

Boggs, David Along with Robert Metcalfe, the co-developer and co-builder of the first Ethernet systems in 1973. See Ethernet; Metcalfe, Robert.

Bohr's correspondence principle In an atomic system, the behavior of the electrons must increasingly approach that predicted by classical physics the higher the quantum number of the orbit.

bolometer An instrument for measuring the intensity of radiant energy through a thermal-sensitive resistor, a type of actinometer.

Boltzmann constant (Symbol - k) The ratio of the universal gas constant, R, to the Avogadro constant, Na. Named after Ludwig Boltzmann.

Boltzmann, Ludwig (1844-1906) An Austrian experimenter who built on the ideas of James Clerk-Maxwell, studying electromagnetism, thermodynamics, and statistical mechanics. Boltzmann demonstrated a number of Maxwell's predictions, confirming them, and published his results in 1875. The Ludwig Boltzmann Institute for Urban Ethology in Vienna is named after him.

BOM 1. Beginning of Message. In ATM, an indicator contained in the first cell of an ATM segmented packet. 2. bill of materials.

bond *n.* 1. To join, adhere, or unite into a combined unit or system. Bond usually implies a semi-permanent or permanent adherence, as opposed to wrapping a wire, which would not be considered a bond. A bond is often accomplished with a bonding agent such as glue, weld, or solder.

bond, electrical To form an electrical connection by joining two conductive surfaces, usually metal, to provide a low-resistance path for the circuit. In electronics, wires are often bonded to a small metallic pad on a circuit board. See bonding.

bonding 1. Joining two or more items with adhesive, weld or solder. In PC boards, there may be a bonding pad on the board, or on a chip for the express purpose of providing sufficient space and electrical contact for a potential bond (usually solder). 2. An inverse multiplexing specification described by BONDING. See BONDING.

BONDING Bandwidth On Demand Interoperability Group. A set of protocols, known as the BONDING specification, developed by a consortium of data communications consultants and suppliers. BONDING arose from efforts to create a standardized inverse multiplexing protocol in order to improve interoperability among multiplexers from various vendors.

The BONDING specification describes a number of modes of interoperability for switched networks, so a sideband signal can be subdivided into multiple 56 Kbps or 64 Kbps channels, and recombined at the receiving end.

bong A tone transmitted through a phone line to indicate to the listener that additional information is required. The information is usually entered through a touchtone key pad.

bookmark 1. A place holder. 2. A computer virtual address, often identified as a file name, or Web site, stored and displayed with an application, like a database, or browser, in order for the user to quickly select and jump to the desired location without having to look up or type in lengthy instructions. Bookmarks are often displayed in a menu list on a pull-down or pop-up menu for quick selection. See hot list. 3. A place-holder in a word processed, or desktop-published file which allows a user to mark and quickly relocate a desired position in the text.

Boole, George (1815-1864) An English-born mathematician, son of a maid and a shoemaker, Boole set up a school at the age of only 19. He taught himself mathematics and began publishing his ideas, introducing Invariant Theory. His 1854 publication "The Laws of Thought" introduced mathematical concepts applicable to computing operations, and earned him the sobriquet of "father of symbolic logic." Boolean logic is named after him.

Boolean expression A type of expression often used in programming to control binary relational operations that may be executed or may express true or false. Boolean algebra in a broader sense in set theory involves the intersection and union of sets and elements of sets. It also provides a practical means for implementing logic in digital computers. Boolean algebra works readily on binary computing systems.

boom *n.* 1. Vertical spar, beam, pole, or suspended piping. 2. In video, a vertical bar, rod, or other support for microphones, cameras, or other equipment that need to be suspended near a source with a minimum of visual obstruction. 3. The horizontal supporting rods for many common antennas from which there may be secondary protruberances to increase transmission or reception.

boom mike A microphone secured to the end of a pole to extend it toward the source of the sound while keeping it out of the view of the camera.

boom pole A long pole with a spike on the end used by crews of *boomers*, early telephone line installers, to guide a long telephone pole into a deep hole. A boom pole is sometimes called a pike pole.

boomer *colloq.* Telephone line installer. The name is derived from the boom poles which installers used to hand guide telephone poles into their holes before machinery for this job became prevalent.

boot *abbrev., v.* To start, to power up, to get a machine going. See bootstrap.

boot *v.* Derived from *bootstrap*, which is further derived from the phrase "pulling yourself up by your bootstraps." This term aptly describes how a computer has to launch its basic lower processes so it can recognize its own hardware and capabilities in order to further launch the higher level processes.

Bootstrapping is the process during which a computer brings the basic essential hardware and software online to the point where it can recognize devices, input from the user, network configurations, basic operating parameters, etc. When a machine is rebooted or *warm* booted from a power on condition, it rereads the boot ROM and resets the basic operating parameters without actually powering off the system. Of course, when you bring up a system from a power off state, it also has to boot, sometimes called a *cold boot*. Stable operating systems rarely crash or hang, but there are some microcomputer operating systems which do crash, and rebooting is often the only way to bring the system back into operating mode. See device drivers.

boot ROM A read-only computer memory chip usually located on the motherboard as an essential part of a basic system. This chip provides the minimum necessary information for bringing the computer hardware online, and may include diagnostic routines that test systems before bringing the whole system up.

In simple terminals, the boot ROM may include all basic operating software needed, or as is the case on most self-contained desktop systems, it may include only the essentials and will seek a floppy diskette, hard drive, or other boot information for further instructions and parameters for launching the operating system, device drivers, and sometimes user applications.

On many Intel-based desktop computers, the information for accessing devices may be transferred to the BIOS during system startup. See BIOS.

BOOTP See Bootstrap Protocol.

Bootstrap Protocol BOOTP. An IP/UDP client/server means of storing and providing configuration information. BOOTP evolved in the ARPANET days to allow diskless client machines and other machines which might not know their own Internet addresses, to discover the IP address, the address of a server host, and the name of a file to be loaded into memory and executed. This is accomplished in two phases: address determination and bootfile selection; and file transfer, typically with TFTP. This has since evolved into Dynamic Host Configuration Protocol (DHCP). See Address Resolution Protocol, Dynamic Host Configuration Protocol, Reverse Address Resolution Protocol, RFC 951.

BOP See Bit Oriented Protocol.

Border Gateway Protocol BGP. An interdomain gateway routing protocol which is superseding Exterior Gateway Protocol (EGP). BGP is used on the Internet. See Exterior Gateway Protocol, RFC 1163, RFC 1267, RFC 1268.

Border Gateway Protocol Version 4 BGP4. A version of BGP which uses route aggregation to reduce the size of routing tables, and which supports Classless Inter-Domain Routing (CIDR).

Border Node In ATM, a logical node in a specified peer group, which has at least one link that crosses the peer group boundary.

bot, 'bot A term frequently used on the Internet to describe software 'robots' which manage tasks on behalf of users and operators, especially in Internet Relay Chat (IRC) channels. Since IRC is an interactive social medium, these software programs have frequently been given personalities by their respective programmers, and thus take on anthropomorphic characteristics not usually attributed to applications programs, hence the term bot instead of application. See avatar, robot.

bottleneck A term for a point in a system, device, or transmission that slows the rate of communication below the capabilities of other links in the system. For example, a computer may have a CPU capable of 64-bit processing trying to send the information through a 32- or 16-bit data bus. As another example, you may have a fast serial card and an ISP with a T1 line, but your 9600 baud modem creates a bottleneck, limiting the upper speed of the transmission of data. Bottlenecks may be a constant limitation of a system, or may be a limitation that only occurs during times of peak traffic, such as rush hour automotive bottlenecks.

bottom line A phrase frequently used in finance to describe a variety of 'end price' conditions. Some examples include the ultimate price of a system once all the essential parts have been tallied up; the no-frills, basic price without options; what is left after all the expenses have been factored in, in other words, the net profit. Outside of finance, bottom line is metaphorically used to ask "What will it take to make this happen?"

bounce *v.* 1. To rebound, to come back, to deflect off of. 2. In electronic transmissions, if data doesn't reach its intended destination and is routed back to the sender, it is said to have 'bounced.' This may happen when email is sent to an address which no longer exists.

bounce, broadcast 1. In broadcast transmissions, if a signal hits a physical impediment, it may bounce, sometimes causing a zone in which there is interference in the transmission or no transmission at all. In other instances, the physical characteristics of the Earth and the ionosphere and the position of repeaters or satellites, etc., may be used to selectively to deliberately bounce a signal in order to direct it. See ionosphere, Moon bounce. 2. In visual media such as television broadcast displays, bounce is an undesirable and unexpected variation in the brightness of the image.

Bourseul, Charles A Belgian-born French researcher who described, but apparently never followed up, a means of transmitting speech electrically through wires.

Bower-Barff process A process in which metal (iron or steel) is heated to red heat and then treated with super-heated steam in order to reduce vulnerability to corrosion.

Boyle, Robert (1627-1691) A British physicist and chemist who developed pumps which could create near vacuums. Boyle subsequently observed that sound required a medium for its transmission. He also did numerous experiments on atmospheric pressure and discovered an important relationship between gas and pressure in 1662. In 1675, he published a treatise on electricity and observed that the attractive properties of amber did not require the presence of air. Boyle's law is named after him. See barometer; Boyle's law; Hauksbee, Francis.

Robert Boyle conducted numerous experiments on the properties of gases and the behavior of electricity in the presence or absence of gases.

Boyle's law, Marriotte's law At a constant temperature, the volume of a definite mass of gas is inversely proportional to the pressure such that the product of the volume (PV) is constant.

BP 1. bandpass. 2. base pointer. 3. bypass.

BPAD Bisynchronous Packet Assembler/ Disassembler.

BPDU Bridge Protocol Data Unit. In ATM, a message type used by bridges to exchange management and control information.

BPI bytes per inch. In recording, a measure of the amount of data which can be stored on a medium. This is usually used to describe data density in media that are long and narrow, like magnetic recording tape. See superparamagnetic.

BPON Broadband Passive Optical Network.

BPM Beam Position Monitor.

BPS See bits per second.

BPSK binary phase shift keying. See phase shift keying.

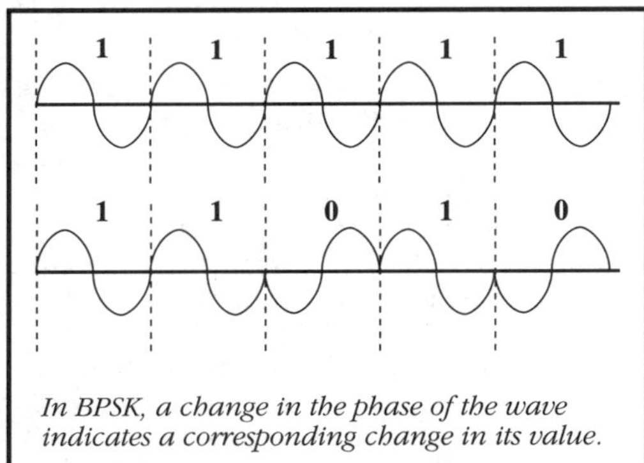

In BPSK, a change in the phase of the wave indicates a corresponding change in its value.

.br Internet domain name extension for Brazil.

BR 1. beacon receiver. 2. Bureau of Radiocommunications.

Bragg reflector A technology used in diode lasers which allows very fine control over the focus of the beam. A Bragg reflector is also called a 'grating' reflector, due to the ridges used to direct the beam that change along their lengths. Bragg reflectors are being researched as a means of increasing throughput of data transmissions in existing cable installations. By finer focusing of the beams and multiplexing, capacity may be improved on fiber channels. See quantum cascade laser.

braid A fibrous or filamentous, long, tubular intricately-woven structure usually of plastic or fine metal which forms a covering over a conductive or insulating core in a layered cable.

brain fart *jargon* Well, it sounds rude, but the phrase is used so frequently in programming circles, that it's appropriate to include it here. A brain fart is a temporary lapse in doing or describing something which is usually self-evident to the person suffering, and suddenly realizing, the lapse. On receiving an answer, it is often followed by a slap on the head and the expostulation, "Duh, I knew that!"

Brainerd, Paul Brainerd founded Aldus Corporation in 1984, the year after the introduction of the Apple Lisa computer and the year before the release of the Apple LaserWriter printer. Aldus specialized in graphics applications, particularly for vector drawing and desktop publishing. Macintosh computers and Aldus software quickly became favorites with print industry service bureaus. The Aldus Corporation was one of the few developers that created some really good, quick, intuitive user interfaces. Good interface design is a rare talent in the software development industry. Aldus PageMaker and Aldus Freehand, developed by the Aldus Corporation, were acquired by Adobe Systems and Macromedia.

branch 1. A junction point from which there is more than one path along which to continue. 2. An instruction in a computer program which, when evaluated, can lead to more than one possible destination for execution of the next step, depending upon the condition. 3. A substation, subsidiary office, or other facility which is a satellite of, or auxiliary to, the main operations.

branch circuit In a wiring installation, a separate circuit which, if damaged or tripped, doesn't affect the other branch circuits. This divides the power so the main circuit is not overloaded. On a circuit breaker panel, the branch circuit is a constellation of appliances and sockets wired to a particular breaker.

branch feeder In an electrical distribution system, a cable which connects between the main cable and the subscriber distribution system, as between a phone switching center's main cable and a business distribution closet.

branching 1. Dividing, splitting into two or more paths or sections. 2. A hierarchical structure often used for database creation, search, and retrieval. 3. Branching electrical distribution systems for electrical installations and data networks.

branching filter 1. A device for separating or combining separate frequencies when used in conjunction with a guiding structure for the wave. 2. In computer networking, a software utility for selectively routing data into several paths or files based on specified characteristics.

branding A marketing strategy that involves advertising, repetition, and other means to associate an action, product, or service with a particular brand name in the minds of consumers and potential consumers.

Branly detector A device created in 1890 by Édouard Branly, consisting of a small, glass, metal-filled tube with a short wire inserted to make contact with the metal filings. When connected between a power source and a meter, current didn't pass through the glass unless a spark was discharged. The spark caused the filings to cohere, and thus act as a conductor.

This on/off quality of the Branly detector was very useful to the development of radio.

Branly, Édouard A French inventor who devised the Branly detector in the late 1800s, a cohering device that contributed to the development of radio, and which was adapted by G. Marconi. Branly also investigated the transmission of nerve impulses. See Branly detector, coherer.

Braun, Karl Ferdinand (1850-1918) A German researcher who discovered in the 1870s that certain minerals had a property of one-way conductivity of radiant energy; they could function, in a sense, as a one-way gate. Braun was awarded a Nobel prize in Physics in 1909, along with G. Marconi, for his contributions to wireless telegraphy. See crystal detector.

BRB An abbreviation for "be right back" used especially on public chat channels on the Internet, and on BBS chat rooms to indicate someone who has to briefly answer a door, phone, help a child, or visit the washroom. See also AFK.

BRCS Business and Residence Customer Services.

breadboard A board with numerous attachment points, often in a grid, which permits the prototyping of circuits. Breadboards often resemble a nest of colored worms, as they are frequently hand-wired with a lot of crisscrossing conductors with temporary attachments. Breadboards are handy for concept design, testing, teaching, temporary circuits, and convincing the boss that you have a good idea that will work. See proof of concept.

break Willful or inadvertent interrupting or stopping of a process, transmission, or broadcast. On computer terminals, a break can be sent in many instances with Ctrl-C, or Esc, depending upon the software.

break in 1. *v.* Interrupt, or take control of, a circuit or process. This may be from human intervention, systems intervention, or through an automated system. See Barge In, buttinsky. 2. *v.* Gain illegal entry to a system. See back door, hacking, Trojan horse.

Break key A specialized key included on some computer keyboards that permits a one-keystroke interruption of the current task, assuming the software supports, and correctly interprets, the input from the keystroke. Break keys are included since some of the common ways to interrupt tasks involve combination keystrokes, such as Ctrl-C, and hitting one key is easier, especially for less experienced computer users. See break.

break out box See breakout box.

breakdown potential, breakdown strength Dielectric strength, the maximum voltage that can be tolerated without breakdown.

B

breakdown voltage 1. The voltage at which an insulator or dielectric breaks, or at which ionization and conduction occurs in a gaseous environment. 2. The voltage which needs to be applied in a device to jump a gap (in air).

breaker 1. In electrical installations, a point in a circuit, usually a junction installed in series between the main electrical source and a branch circuit, in which excessive voltage is detected and the circuit 'tripped' or broken in order to prevent overload, electrical fires or damage to appliances or the main panel. 2. In radio communications, anyone who drops in on a channel to communicate when others are already engaged in conversation. 3. In public online forums, a person who breaks into an ongoing conversation or thread with irrelevant or unkind and unwanted comments.

breakeven point The point at which the cost of a service, product, or operation equals the revenue generated by that activity or product. In telecommunications, breakeven points are often evaluated when choosing a service, such as cellular, or Internet Services Provider, wherein a low monthly cost plus surcharges is balanced against a higher monthly charge with fewer surcharges.

breaking strength 1. An industrial measure of the force needed to break a specific structure or material. Used in structural and safety design, and selection and installation of appropriate wires and cables.

breakout An exit point for electrical conductors (wires, cables, etc.) along the length of the circuit, between the endpoints of the circuit. Breakouts can be used for additional installations, or testing and diagnosis of the circuit. Breakouts are usually covered or capped in some way to prevent interference with the circuit, and shock or fire hazards.

breakout box A diagnostic instrument used to tap into an existing circuit to evaluate its functioning, to see whether individual lines within a group are correctly connected, and, in some cases, to detect which signals are being transmitted. Breakout boxes often have indicators such as tones or light-emitting diodes (LEDs) to help assess the line, and may include jumpers to temporarily switch connections. One useful application of a breakout box is to test a serial circuit, since computer equipment manufacturers don't consistently follow the RS-232 or RS-423 specifications. In the case of a temporary need for a null modem cable (e.g., for transferring information from one computer to another through telecommunications software), a breakout box can be used to cross the transmit and receive lines. See breakout.

Brewster's law The refractive index of a beam of light is equal to the tangent of the angle of polarization, with the result that the sum of the angle of incidence and angle of polarization is equal to a right angle. The angle of maximum polarization is known as the Brewsterian angle, and varies as the substance varies.

BRI See Basic Rate Interface.

brick slang A somewhat pejorative term for any handheld mobile equipment which is heavy or cumbersome, as a heavy laptop, mobile phone, or intercom radio.

bridge *n.* 1. A link that provides a connection across a physical or conceptual gap. This link may or may not be intended to affect the quality or format of the objects or information crossing the gap. 2. In networks, a device that handles communications between separate local area networks (LANs) which may or may not use the same protocols. In frame relays, a bridge encapsulates LAN frames and feeds them to a frame relay switch for subsequent transmission. It also receives frames from the network, strips the frame relay frame, and passes the LAN frame on to the end device. With increasing technological sophistication, the distinction between bridges and routers is lessening. See brouter, router.

bridge, acoustic In acoustic instruments, the bridge elevates and spaces the strings, and transfers vibrations to the body (soundbox) of the instrument. In electronic instruments, the bridge elevates and spaces the strings, and transfers vibrations to the body of the instrument where they are, in turn, converted into electrical signals, usually by a device called a 'pickup.'

bridger switching A technique for improving low return transmissions, as in cable networks, by sequentially turning on and off each leg of the distribution circuit. This is sometimes used in conjunction with high pass filters. While improvements in reliability can be attained in this way, it is at the cost of greater complexity, and hence, greater expense.

bridging clip A small piece of conducting apparatus used to connect nearby terminals, contacts, or other circuit elements that are close together, either for the purpose of changing a circuit (usually temporarily), or for testing it.

Bright, Charles Tilson (1832-1888) An English inventor and chief engineer for the Magnetic Telegraph Company. Bright was the first to undertake underground cable installation with gutta-percha as an insulating material. When Edison was installing the historic Washington-Baltimore line in the 1830s, problems with insulation and ground-breaking caused the line construction to be changed from underground to overhead, so Bright's success with an underground line was important. Later, Bright installed the first cables to be laid in deep water, first with a shallower line across the Channel in 1851, and two years later a deepwater line between England and Ireland. Further lines around the world followed.

brightness The level of luminosity, or amount of illumination of a surface or display medium. Luminance is used to describe the lightness or brightness component of a television broadcast signal. Brightness across the visible spectrum is not equally perceived for different colors. See contrast.

British Broadcasting Corporation BBC. A television broadcast provider since the late 1920s, when it began its first experimental television transmissions, the BBC began widespread public broadcasting from London in 1936.

brittle A quality of a physical substance lacking elasticity, one that is vulnerable to breakage. Substances may be brittle in one set of circumstances and not another, e.g., electrical components or connectors may be vulnerable below or above certain operating temperatures.

brittle software Software which may function correctly, but which may not be 'user-proof' in the sense that it may break or crash easily when the user makes unanticipated selections or inputs; or which may not be robust in the context of the operating system. A database that is easily corrupted, or which does not transparently save data is said to be brittle.

broadband A band of frequencies wide enough to be split into narrower bands, each capable of individual use for a variety of transmissions, or for transmissions by a variety of users. Broadband transmission requires suitable hardware and cabling, capable of quickly transmitting and receiving a large amount of information. Fiber optics are often used. The entire breadth of the band is not necessarily used for transmissions, depending upon supply and demand. Also, there may be gaps between bands to prevent interference. Cable

TV is a ubiquitous example of broadband transmissions where the band is split into all the different channels to which the recipients have subscribed. As in many broadcast media, broadcast technologies tend to be one-way, or mostly one-way, but with the increased demand for interactivity, more two-way communications over broadband are being developed. See baseband, telecomputer, wideband.

broadband digital cross-connect system B-DCS. A digital cross-connect system which accepts a variety of optical signals and is used to terminate SONET and DS-3 signals. B-DCS accesses STS-1 signals and switches at this level, and is appropriately used as a SONET hub for routing and other functions. See wideband digital cross-connect system.

broadband integrated gateway BIG. A component of HFC (Hybrid Fiber Coax) networks which converts an ATM transmission into a signal that can be transmitted over the HFC. Working in conjunction with a connection management controller (CMC), the BIG strips information from ATM cells, and orders and addresses them for further transmission. See connection management controller, HFC.

Broadband ISDN See B-ISDN.

Broadband Telecommunications Architecture BTA. An architecture introduced by General Instrument for multimedia networking.

broadband terminal adapter BTA. A data communications device that interfaces a broadband ISDN (B-ISDN) connection to other terminal equipment which is not directly compatible with B-ISDN.

broadcast *v.* To transmit sound, images, or data over distance, in the context of more-or-less simultaneous receipt by a larger audience. Transmission can occur through a variety of media, over airwaves, satellite links, wire or fiber, or a combination of these. Reception can often be enhanced with antenna, cable, or satellite hookups. Radio, television, and Internet chat channels are common broadcast channels.

Commercial and high power broadcasting is regulated. In the United States, the Federal Communications Commission (FCC) is the primary regulatory body and has jurisdiction over the allocation of broadcast frequencies. In Canada, the Canadian Radio Television and Telecommunications Commission (CRTC) handles many of the same functions.

In most of North America, very low power broadcasting is permitted without a license, otherwise it wouldn't be possible for people to use cordless phones, baby monitors, and wireless intercoms without being licensed. Generally these low power broadcasts are limited to a signal strength of 250 microvolts per meter, as measured three meters from the transmitter for FM transmissions, and 0.1 watts on a maximum three meter antenna for AM transmissions. This effectively limits the broadcast distance to 100 feet or so for FM and a couple of blocks for AM.

Traditional broadcasts are typically in the range of 535 to 1605 kHz.

Commercial entertainment broadcasts are often financed by revenues from sponsors which are aired in the form of commercials. Since this is a revenue model which has been successful for quite some time in the television and radio industries, it is not surprising that many broadcasters are turning to the same ideas in designing information to be viewed over the Web. In contrast to commercial stations, however, the Web is far less regulated, and has many more participants, and it will be interesting to see how it evolves over the next several years. See television, radio, multicast, narrowcast, unicast.

broadcast list 1. On computer networks, a list of users to whom broadcast messages are sent, usually by a system operator (sysop), or other privileged administrator. See broadcast message. 2. On fax machines, a list of recipients to whom the same fax will be sent.

broadcast medium In some networks, a physical layer capable of supporting broadcast messages.

broadcast message 1. A message sent out to a group of users (or all users) on a computer network. A common broadcast message informs users that the system is about to shut down in five or 10 minutes. This allows users to save work, close files, and finish up before being logged out. On networks, broadcast functions are usually only available to those with system privileges, as it is a capability that is easily abused. One apocryphal example is the Cookie Monster message which flashed onto users' screens and demanded a cookie. If there was no response, or the user continued working, the Cookie Monster would become more and more demanding until it was requesting a cookie so often that it was no longer possible to use the terminal. The solution was to type the word cookie, but most disgruntled users didn't know that. See broadcast list. 2. A message broadcast over a public broadcast medium, such as a news flash, or Emergency

Alert System (EAS) message. 3. A message broadcast over a paging or public address (PA) system.

broadcast over network In ATM, data transmissions to all addresses or functions.

Broadcast Pioneers Library An education and research resource in the Hornbake Library at the University of Maryland, College Park. The collection includes correspondence, books, film and video, periodicals, historic photographs, scripts, and transcripts. More information is available online through the Pioneers' Web site. http://www.lib.umd.edu/UMCP/LAB/

broadcast standards Established in the late 1930s, professional standards still exist as important guidelines for ethical business practices in the broadcast industry.

broadcast storm A broadcasting clamor which is excessive, and which overrides other communications due to the number of messages being simultaneously transmitted. In radio broadcasts, a broadcast storm may occur due to a sudden emergency, if a number of operators simultaneously try to call for help, or get messages through at the same time. In end-to-end systems, such as analog telephone systems, broadcast storms as such don't occur (except perhaps, in a different sense, on a party line), because excessive calling will result in a fast busy being sent to the caller, indicating that no trunks are available, rather than in many people talking at once. In data networks, however, a broadcast storm can occur as a fault condition in which some process goes wild and starts broadcasting to all workstations and disrupting user interactions and work. This may occasionally be deliberate vandalism, in the form of a virus, on unsecured networks.

broadcasting satellite service BSS. One of two divisions into which Ku-band satellite broadcast services have been split. BSS operates in the 12.2 to 12.7 GHz range. The other is fixed satellite service (FSS). See ANIK, Ku-band.

broadside array antenna A phased array (with harnesses) of antennas with the maximum radiation directed perpendicularly to the plane which holds the driven elements. This antenna arrangement can be configured as a billboard antenna by adding a reflecting sheet behind the array. See billboard antenna.

Broadway The internal development code name for The X Window System 11 Release 6.3 (X11R6.3) from The Open Group. See X Window System 11 Release 6.3.

broker An entity (person or business) which assists in negotiating contracts for services or equipment on the part of a user, or other equipment purchaser, for specialized or bulk services, with the intent of reselling these services to consumers, sometimes at discount rates. If equipment, the materials are often 'drop-shipped' to the customer directly from the supplier, without going through the hands of the broker. A service which is frequently brokered is a merchant card (a card which allows a merchant to accept credit card purchases). If you go to your local bank for a merchant card, that bank may not be the primary provider of the card and will sometimes go through a broker, who then shops around for a primary lender. Everyone along the line gets a small fee, or percentage, for services, depending upon the arrangements. Travel agents are also brokers, sometimes getting discount rates for their customers by purchasing a block of seats on a flight, and reselling those seats at competitive rates.

bronze An alloy which consists primarily of copper, with tin, and occasionally other elements, added.

Brooks' law Adding manpower to a late software project makes it later. From Frederick P. Brooks, author of "The Mythical Man-Month," a much-quoted provocative book about the engineering development culture.

brouter bridge router. Combination devices which function as links between different networks. The combination of a bridge and a router provides the physical and logical connections between networks which may or may not have different protocols, and routing tables to facilitate the efficient transmission of information to the desired destination. A brouter typically performs its functions based on information in the data link layer (bridging), and the network layer (routing). See bridge, router.

Brown & Sharpe Wire Gauge See American Wire Gauge.

Brownian movement, Brownian motion Botanist R. Brown observed in early 1827 that pollen grains suspended in water were in a continual state of agitated motion. This motion has been widely observed for small particles suspended in fluids. It is said that the molecules of the suspension medium continually buffet the particles, resulting in the characteristic movement. Einstein later provided a mathematical explanation of Brownian motion.

brownout 1. A situation in which power is partially, but not completely lost. Some companies use an industry-specific definition for a brownout, usually based on a relative or specific drop in voltage. Complete loss of power is called a blackout. 2. In cellular systems, a security precaution used by some companies to prevent fraudulent use. When brownout is in effect, there may be roaming areas in which a subscriber's system will not function.

browser 1. An object-oriented software development tool for inspecting a class hierarchy. 2. A software utility for displaying and traversing files and directories.

browser, Web A historically important software application designed to make it easy to access the resources of the Internet through a graphical browser. The software was originally developed for NeXTStep by Time Berners-Lee in 1989. It became commercially prominent in the early 1990s, and developed into an easy tool for accessing, traversing, and displaying files on the World Wide Web. The browser interprets the HTML tags that are used to describe the Web page, and displays the results on the user's system. This simple, accessible, means of finding information on the Internet has resulted in an explosion of interest and participation, increasing from a handful of users in 1989, to more than 30 million in 1998.

Browsers typically download HTML pages onto the local drive, so they can be more quickly redisplayed when the user moves back through previously viewed pages, and further permit the transfer of files through ftp. Some browsers incorporate email functionality as well, or will launch the email utility of choice when an email anchor is selected on a Web page. There are many browsers from which to choose; common ones include: Lynx (text browser), AWeb, OmniWeb, Microsoft Explorer, and Netscape Navigator. Navigator is open source software. See HTTP, HTML, SGML, FTP, World Wide Web, Java, applets, Internet.

browsing Searching or scanning through data for information, or to get a general feel for the format or contents of a body of information. The information may take a variety of forms: text, files, directories, images, sounds, etc. See browser.

brush A conductor that provides an electrical connection between a motor and its power source.

brush, graphics A computer graphics software tool that mimics a traditional paintbrush, or rather, an unlimited box of paintbrushes, since an electronic 'brush' can be configured in a variety of shapes and sizes, with a number of added special effects, depending upon the application. A brush tool can be used to draw a new image, or can be used to modify an existing image by overlaying colors, or textural effects.

Deluxe Paint, released by Dan Silva on the Amiga in 1985, has one of the best basic brush tools ever designed. Fractal Design Painter is a more recent program with excellent brush tools.

brute force 1. A problem-solving method that involves trying every possible combination and permutation. This method is only practical for small problems of limited scope, and is usually unwieldy for larger, or more complex problems. Sometimes it is used in conjunction with other problem-solving methods such as heuristics. 2. A programming approach that involves reliance on a system's basic capabilities and processing power, rather than on efficient algorithms and elegance of design and concept. A brute force application generally does not run quickly on legacy systems.

brute force attack An attack on a security system by using every possible combination, password, login name, or other entry data, usually generated automatically by the computer software. This type of attack is usually easily detected, and is often not very effective.

.bs Internet domain name extension for Bahamas.

BS 1. back scatter. 2. band signaling. 3. base station. 4. beam splitter.

BSAM Basic Sequential Access Method.

BSCC BellSouth Cellular Corporation. A corporation serving about 10% of the U.S. wireless market, formed in 1991.

BSD See Berkeley Software Distribution.

BSE 1. back-scattered electrons. 2. Basic Service Element. 3. Basic Switching Element.

BSF bit scan forward.

BSI British Standards Institution. A United Kingdom standards body which provides input to various international standards associations, including ISO and ITU-T.

BSL British Sign Language.

BSMS Broadcast Short Message Service.

BSMTP Batch Simple Message Transfer Protocol.

BSP 1. Bell System Practice. Bell internal policies and procedures for creating instructional manuals for the servicing, support, and operation of phone equipment. 2. Byte Stream Protocol.

BSR 1. bit scan rate. 2. bit scan reverse.

BSS 1. Base Station System. 2. See broadcasting satellite service. 2. Business Support System.

BSVC Broadcast Switched Virtual Connections.

.bt Internet domain name extension for Bhutan.

BT 1. British Telecom. 2. Burst Tolerance. In asynchronous transmissions method (ATM) connections supporting variable bit rate (VBR) services, BT is the limit parameter of the GCRA. See cell rate.

BT cut crystal A type of crystal with vibratory qualities that makes it suitable for crystal radios.

BTA 1. See Basic Trading Area. 2. See Broadband Telecommunications Architecture. 3. See broadband terminal adaptor.

BTag Beginning Tag. In ATM, a one-octet field of the CPCS_PDU used in conjunction with the Etag octet to form an association between the beginning of a message and end of a message.

BTBT band-to-band tunneling. Direct transfer of electrons from filled valence band (VB) states to empty states, or recombination of electrons with holes in the valence band.

BTE 1. Boltzmann Transport Equation. 2. broadband terminal equipment.

BTI British Telecom International.

BTL Bell Telephone Laboratories.

BTM Broadband Transport Manager. In telephony, a transport for long-distance portions of a connection.

BTN Billing Telephone Number, Billed Telephone Number. In some situations, the number billed may be one of several associated numbers, but for simplicity, all the calls are billed to one. This is sometimes done with extension numbers. In other situations, the main number used may be different from the number to which the calls are billed, again, usually to simplify accounting, or billing statements.

BTRL British Telecom Research Laboratories.

BTS 1. Base Transceiver Station. In mobile communications, an end transmission point. 2. bit test and set.

BTU 1. basic transmission unit. 2. British Thermal Unit.

BTW An abbreviation for "by the way" frequently used in email and on online public discussion groups. See AFAIK, AFK, IMHO.

bubble memory A type of nonvolatile memory, that is, it doesn't have to be constantly refreshed to keep the data. Bubble memory, as used in computers, consists of a thin layer of material that has magnetic properties. A magnetic field is used to manipulate a circular area such that the diameter becomes smaller, forming a bubble.

Buchmann-Meyer effect The sound track on the rotating disc of a record forms a pattern when light is reflected from it allowing the lateral velocity to be determined.

buffer 1. A circuit or device designed to separate electrical circuits one from another. 2. A physical or electronic storage device designed to compensate for a difference in the rate of use or flow of objects or information. Generally a buffer is intended to increase speed of access and efficiency. In a computer, a buffer is often used as a storage area for frequently accessed information, so the software doesn't have to constantly access slower storage devices such as a hard drive, if sufficient fast access chip memory (e.g., RAM) is available. Cut and paste functions make use of a buffer. Data in a buffer tends to be temporary and volatile. See cache, frame buffer, RAM disk.

buffer box See Logical Storage Unit.

buffer condenser A condenser installed in an electronic circuit to provide protection to other components by reducing excessive voltages, especially surges.

buffer memory, buffer storage Electronic memory, usually RAM, used for information storage, particularly for applications programs which make use of chunks of information which are frequently recalled. See buffer, cache.

bug *n.* A small, concealed listening device used in surveillance and espionage. Placing a covert bug in a room, or on a phone line, is almost always illegal. The term is also used in conjunction with small, hobbyist transceiver projects which may be used for electronics education, wireless intercoms, child monitors and other legitimate uses. See wire tapping.

bug rate In software testing, the rate at which bugs are located. As the rate decreases, the software may be moved through stages of the testing and revision cycle, as from alpha to beta versions. Software is usually distributed to beta testers outside the company when all the major bugs have been located and fixed. See bug.

bug, software/hardware A software or hardware error which adversely affects operations or user interaction. Grace Hopper is credited with relating the first story about a computer bug that was found by a technician, and for preserving the bug itself in a log book. This story has long been a part of hacker lore as the origin of the term 'bug' in computer technology. The bug in the story apparently was moved to the Smithsonian Institution in the early 1990s (after an earlier unsuccessful attempt to have it accepted), but was not immediately exhibited. However, there are earlier anecdotes about bugs in industrial settings that indicate the term may go back decades, if not hundreds of years.

Removing bugs from software (debugging) is an art form, and not all programmers who are good at writing code are good at finding and correcting bugs. Unfortunately for developers, removing one bug often introduces one (or more) elsewhere. Unfortunately for consumers, some commercial software vendors release products knowing they are full of bugs, and there are no specific regulations prohibiting it. Because computer technology is technical, the user may not know whether a problem is from bugs, or from incorrect use of the software.

Another unfortunate aspect to bugs is that companies often combine software enhancements with bug fixes, and sell the new product as an 'upgrade' with no guarantee that it is more robust than the previous version (sometimes it is less so). This is like buying a $15,000 car with a faulty engine, and having the manufacturer refuse to fix it, instead advising you to pay $5,000 to upgrade to next year's model, and when you do, you find that the engine's been fixed, but the axles are defective, and the car has racing stripes that you didn't want in the first place. This situation in the software industry won't change until consumers stop buying substandard software and 'enhanced' upgrades, and support instead the more responsible vendors who provide patches for bugs separately from releases of 'enhanced' versions.

bug, telegraph A telegraph lever which, depending upon its position, can be used to send dots or dashes to partly automate transmission.

build *n.* 1. An increase in diameter of a line or object attributable to insulating materials. 2. In software development, the process of combining, compiling, or linking code so as to 'build' an application.

bulb The sealed glass enclosure for an incandescent or fluorescent lamp. Bulbs provide protection for the gaseous environments, and the delicate filaments that they enclose. See Edison, Thomas Alva; lamp.

bulk encryption Simultaneous encryption of a group, or set of communications, such as multiple data messages, or multiple channels on a broadcast medium.

bulk eraser, bulk degausser A electromagnetic device designed to save time by clearing the data from a large number of floppy disks at one time. By rearranging the particles on the physical disk, the electronic information is destroyed. This is handy for recycling the diskettes, or for providing a measure of security with data that needs to be destroyed. It is wise to keep magnetic storage media away from computer monitors, which have magnets, or you may inadvertently erase or damage the data on them. A large-scale pirate software vendor that was apprehended in Vancouver, B.C. is rumored to have had a bulk eraser in a storage cabinet wired to a button under the service counter to destroy evidence in the case of a police raid. See diskette.

bulk storage Media on which large amounts of electronic data can be stored. The amount of storage that constitutes 'large' keeps increasing. In the mid-1970s, a tape holding 100 kilobytes was considered bulk storage! In the mid-1980s, a writable optical disk holding 600 MBytes was bulk storage. Now hard disks and tapes holding 4 GBytes or more are being bundled with consumer machines.

bulletin board system BBS. The precursor to the Internet, BBS systems are typically individual computers set up for public or private modem access, by a number of users, on which there are shared files, mail, and chat services. The administrator is usually called the SysOp (System Operator). In the late 1970s and early 1980s, it was extremely rare for a BBS to be password-protected; there was open access to all. Unfortunately, persistent abuse has made this type of BBS almost extinct. In the mid-80s, there were still many BBSs running on TRS-80s, Color Computers (CoCo), Commodore 64s, Apple IIe's, and Amigas, with only 5 or 10 MBytes of hard drive storage for the entire system. BBSs have since become more sophisticated, offering credit card payment options, and increasingly are being linked to the Internet through telnet. See FidoNet.

bump contacts Small conductive lumps on electronic circuits to enhance electrical contact, such as those which allow chips to contact with terminal pads.

bunch stranding A technique used to combine wires so they fit tightly together, with individual strands retaining the same directional relationship to one another to form a stranded wire. Stranded wire is useful in situations where flexibility is desired, or in which the electrical properties of the wires are influenced by proximity to others.

bunching An alternating convection-current effect in an electron stream caused by velocity modulation. Bunching is quantified as a parameter based on the relationship of the depth of velocity modulation to the absence of modulation. Used in electron tubes to generate ultra-high and microwave frequencies. See klystron.

Bundesamt für Zulassungen in der Telekommunikation BZT. A German telecommunications approval authority established in the early 1980s.

bundled 1. Combined products or services, sometimes from a variety of manufacturers, offered at a combined price. Phone and cable companies often have bundled or 'packaged' deals, such as regular telephone service and Caller ID-related services offered at a flat rate, or movie and educational channels combined. Software products are often bundled with computer systems. Operating systems are almost always bundled with computers, often along with various productivity applications, demo programs, and clipart libraries. 2. Individual wires or cables combined or interwoven to form a bundle for ease of handling and installation.

Bunsen cell A type of cell devised by R. W. von Bunsen which was an adaptation of an early wet cell in which the positive electrode was suspended in a bath of contained nitric acid (separated from the outer electrolyte solution) so that hydrogen would be oxidized and the cell depolarized. See dry cell; wet cell; von Bunsen, R.W.

buried cable An underground cable installation which cannot be altered without disturbing the soil, or accessing the cable under the soil through some entrance point. Buried cables are more aesthetic, as they do not clutter the landscape with utility poles and wires, but may be less easily accessed to make changes or repairs. See Call Before Digging.

burn-in A diagnostic preliminary operation, sometimes at high temperatures, to test devices and circuits in order to identify those likely to fail. In electronics, many problems will show up early, in the first few weeks of operation, or under stress from heat and humidity. Sometimes called early failure period.

burn-in, monitor An undesirable ghost image on the coating inside a cathode ray tube (CRT) monitor resulting from the persistent display of the same image or similar images. Monitor burn-in can be prevented by turning the monitor off when not in use, and using screen savers which darken the screen completely, or second-best, which move around small, low-contrast images. Many commercial products called screen savers do not save your screen. If a bright, still image covers most or all of the screen, it's not a screen saver, no matter what it is called, and persistent use will cause burn-in.

burnout A condition in which a person's physical and psychological resources are severely depleted and stressed. Symptoms vary, but may include: lethargy, anxiety, shivering, vertigo, headaches, stomach upsets, insomnia, and apathy. Factors contributing to burnout can include: overwork, lack of variety, low pay for the type of work done, lack of appreciation, poor time management, lack of rest and exercise, poor working conditions, long hours spent with a demanding public (such as sales or telesolicitation). See a health care professional if you think you might be suffering from burnout.

burn rate The rate at which a seed or startup company uses its initial cash resources in the process of making the company financially viable.

burst *n.* 1. Sudden increase in signal strength. See surge. 2. A color burst, or reference burst, is an oscillator phase reference in a color broadcast receiver. 3. In printing, the separation of continuous-feed or multipart pages into individual sheets, usually along a perforation. See burster. 4. In a frame relay network, a sporadic increase in a circuit where the total bandwidth is not continually in use. 5. In data communications, a sequence of more-or-less contiguous signals that are treated as a unit according to a predetermined set of criteria.

burst error, burst noise 1. A data burst sufficient to garble or interrupt data transmission. For example, scratches may result in a burst error on a laserdisc, audiodisc or magneto-optical storage disc. Some systems have software error-checking which will minimize the negative effects of a burst error, or which will make a 'best guess' as to what the data should have been. 2. In audio transmissions, noise and interference that substantially exceed the ambient noise, or the level of the desired transmission, e.g., sudden pops and clicks on an old vinyl recording.

burst mode A data transmissions mode in which the data is sent faster than usual, sometimes due to a device temporarily monopolizing the transmissions channel. See caching.

burst pressure The maximum pressure which a device or mechanism can tolerate before rupturing. This phrase is often used in reference to liquid or vapor conductors.

burst sequence In color video transmission, a mechanism for improving the stability of color synchronization by controlling the polarity of the color burst signals.

burst transmission A transmission in which the signal is intentionally sent in a group at significantly higher speeds than is usual, with as much informational content as possible. For example, in radio transmissions, the information may be sent in a burst at 50 or 100 times the normal speed, and then played to the listener at normal listening speeds. Burst transmission is a technique for getting more information transmitted in less time.

burst transmission, isochronous In a data network, where many different devices may be operating at different speeds, burst transmission may be used to resolve some of those speed differences in order to operate the network more efficiently.

burster In printing, a device that speeds and facilitates the separation of continuous feed or multipart pages into individual sheets, usually along a perforation. This type of equipment is sometimes incorporated into a multiple-capability machine (burster-trimmer-stacker) which also trims the pages to remove the tractor-feed strips, and stacks the paper.

bursty data Data that comes in spurts; it is often unpredictable. The nature of the data is important in the design of network traffic flow control procedures and protocols. See variable bit rate.

bursty information Information which alternates between intervals of low transmission and short bursts of high transmission incorporating a lot of data. Print jobs tend to be bursty, with long periods of idleness, and short periods where it seems as if everyone on the network submits a job to the print queue at the same time (e.g., just before lunch break).

bus *n.* 1. An uninsulated conductor such as a wire, bar, or printed metal patch on a circuit board, intended to provide an electrical contact point for adjoining conductors or devices. Commonly used in telephone and computer circuits. See backplane, edge connector, expansion slot. 2. A category of standards which facilitate the compatibility and interconnection of consumer electronics products. 3. One type of computer architecture in which a series of computer processing units (CPUs) are interconnected. 4. In its most basic sense, an uninsulated solid or hollow conducting bar or wire.

BUS Broadcast and Unknown Server. In ATM networks, this server handles data sent by an LE client to the broadcast Media Access Control (MAC) address, all multicast traffic, and initial unicast frames which are sent by a LAN Emulation (LANE) client.

bus, data The data pathways internal to a computer that permit the transfer of data within the system and to peripherals associated with the system. Common buses include address and data buses which may or may not match the capacity of the CPU. Bottlenecks are possible at the bus if the information reaching the bus is greater than its physical or logical carrying capacity (e.g., a 16-bit address bus on a 32-bit CPU).

Bus Interface Gate Array BIGA. A Cisco Systems technology for allowing a Catalyst 5000 to receive and transmit frames from the packet-switch memory to the Media Access Control (MAC) local buffer memory, independently of the host processor.

bus mastering A capability of a peripheral device to take over functions and transfer data through a computer's system bus. This is incorporated into PCI video cards, for example, so the card can directly access system memory, and to provide other performance improvements. Bus mastering is a means of direct memory access (DMA) processing.

bus topology A network topology in which individual nodes are connected to a single communications line which is terminated at either end. Like a ring topology, this arrangement is fragile in that if one node or system goes down, it affects the entire network. See star topology, topology.

Bush, Vannevar (1890-1974) An American engineer and writer who devised the product integraph in the mid-1920s. This device was a semiautomatic machine for solving problems. He later invented the differential analyzer, a mechanical apparatus for solving problems, evolved from his earlier experiments. He was president of the Carnegie Institution of Washington from 1939 to 1955.

bushing 1. A cylindrical lining in an opening that aids in sizing the hole, insulating it, or providing a path, as for wires. See ferrule. 2. A cylindrical utility pole insulator with external ribs at one end. This type of insulator was typically used on high voltage leads.

busy 1. A system which is in use. For example, a printer may be busy processing data or handling a current job. 2. A telephone which is in use or off-hook, or a telephone trunk which is at capacity and can't finish processing the call. See busy signal, fast busy.

busy back A signal that communicates the call status back to the caller. See busy signal.

busy hour That hour during a specified period (day, week, etc.) in which the greatest volume of traffic is carried, as on a phone or computer network. Busy hour traffic volume and characteristics provide important data about the capacity, equipment and switching needs of a network. The busy hour characteristics can be used to determine the amount of variation from low to peak usage times, the expected maximum requirements of the system, and other administrative and operational parameters. On phone networks, busy hour times have changed (become later in the day), probably as a result of the greater use of off-peak hours for facsimile transmissions and Internet usage.

busy lamp, busy light A lighted indicator on a device that shows it is currently active, or in use. The busy lamp is often paired with a second lamp that shows that the device is powered on. Thus, a printer may have two lights next to each other, one to show power is going to the printer, and the second, to show that data is currently being sent to the printer (this second lamp often flashes). Busy lamps are often included on modems to indicate whether data is being transmitted or received (usually shown as two different lamps). This is very useful, as it helps the user to know at a glance, if a transmission is coming through. On a phone console, the busy lamp is a useful indicator of which lines are currently in use, so you can pick up on a different line and not cut in on someone else's conversation.

busy season In any service or product distribution, there are usually low periods and peak periods. In the airline industry, the peak periods are Thanksgiving, Christmas, other holidays, and summer vacation time. During these peak periods, there are greater demands on communications systems, and the computer systems that support them. Thus, software and equipment have to be designed and configured to handle busy times without loss of business, or substantial slowdowns.

busy signal, busy tone A regularly recurring signal (beep) on a phone line, transmitted to a caller to indicate that the trunk is not available (fast beeps), or that the called party is on the line and cannot be connected with you unless they have a service such as Call Waiting. If so, and they wish to interrupt their call in progress to talk to the second caller, they can do so by typing in a code, and then return to the original call, if that person is still waiting. See Call Waiting, Audible Ringing Tone.

busy test In telephone networks, a diagnostic technique for determining the availability (or disability) of an expected service, such as a successful transmission through a newly installed line.

butcher *colloq.* An individual who takes a hack-and-slash approach to his or her vocation, without first verifying the necessity or wisdom of those actions. In electronics or networking, a butcher is someone who cuts or reconfigures circuits without checking at each step to see if that's the right course of action, or if the lines are reactivated as appropriate.

Butler antenna A type of array antenna in which the feed lines include combined junctions for more than one beam.

butt joint A wire connection or splice in which the ends of the conductors are butted up against one another and combined by soldering, welding, etc.

butt-in, butt-set See buttinsky.

butterfly capacitor A tuning device with good selectivity resembling a butterfly or bowtie, used for very high frequency (VHF) and ultra-high frequency (UHF) transmissions.

Butterstamp telephone An early innovation, a telephone which combined the receiver and transmitter into a single unit, as is still used, except that it was necessary to turn the instrument around to alternately listen or speak. An improved version was soon developed which could be used to listen or speak without turning the handpiece around.

buttinsky *slang* A type of telephone system which includes a transceiver worn on the hips that is used to 'butt in' on conversations, that it, is permits a technician to break in on or monitor a call. It is used in diagnostics and installation.

button caps Portable key sleeves designed to fit over individual buttons on programmable pushbutton phones. These may be plain (for hand labeling), transparent, or preprinted, to indicate the newly programmed function of the key.

buttoned up Sealed, completely closed.

buzz *v.* 1. *colloq.* To catch the attention of someone at a distance, or out of eyesight. To call for a quick chat on the phone. 2. To press a buzzer.

buzz test To perform diagnostic testing of the continuity of a circuit by placing a buzzer on one end and then sending a current from the other end to see if the buzzer rings. Often used when the far end is out of eyesight, but not out of hearing. See buttinsky.

buzzer A signaling device, usually electromechanical, which makes a raucous noise when the circuit is completed, at an attention-getting volume and frequency. Buzzing noises can also be generated on computer systems. Buzzers have many uses; they can catch a person's attention, indicate a fault condition, or provide a tone that can be used as a diagnostic tool when tracing a circuit. See tone generator, tone probe.

buzzer leads The connecting posts or wires attached to a device which are intended for the connection of a buzzer.

.bv Internet domain name extension for Bouvet Island.

BVR beyond visual range. Something which is outside human sight, or in some contexts, out of sight of human vision with binoculars. In a very general sense, it can mean something distant, or something obstructed. In telecommunications contexts, it is more often used to indicate objects, communications means, or antennas which require 'line of sight' distances or unimpeded pathways to be effective.

.bw Internet domain name extension for Botswana.

BW See bandwidth.

BX cable A type of cable used in electrical wiring consisting of insulated wires enclosed within a flexible metal tube.

B/Y signal In a color television signal, this is one of the three primary signals (of RGB), providing blue (B) when combined with a luminance signal (Y).

.by Internet domain name extension for Belarus.

bypass *v.* To shunt around the normal path, to provide an alternate path, often for temporary installation or diagnostic purposes.

Byron, Ada Augusta Lord Byron's daughter. See Lovelace, Ada Augusta.

byte A unit of data that is bigger than a bit (binary digit) and smaller than a word. The relationship of bits to bytes and bytes to words varies according to the system, but in most computing, a byte is said to consist of eight bits. File sizes and storage on smaller media are usually displayed by the system in terms of bytes or kilobytes. Many character sets encode each character within a byte of data. International character sets tend to use two-byte encoding schemes to accommodate the much larger number of letters and symbols. In Internet protocol documents, it is more common to see the term 'octet' (8 bits) in use, instead of the term byte, presumably because its meaning is more explicit. See kilobyte, megabyte.

Byte magazine One of the first popular small computer systems journals, founded in 1975, Byte magazine is still publishing in 1998, while many other computer magazines have come and gone. One of the most popular features of Byte magazine in the 1970s and 1980s was Steve Ciarcia's circuit cellar, an electronics column for hobbyists. At around the same time that Ciarcia left Byte for other activities, the tone of the magazine changed, and it became more IBM-system-oriented, and "Small Systems Journal" was removed from the masthead.

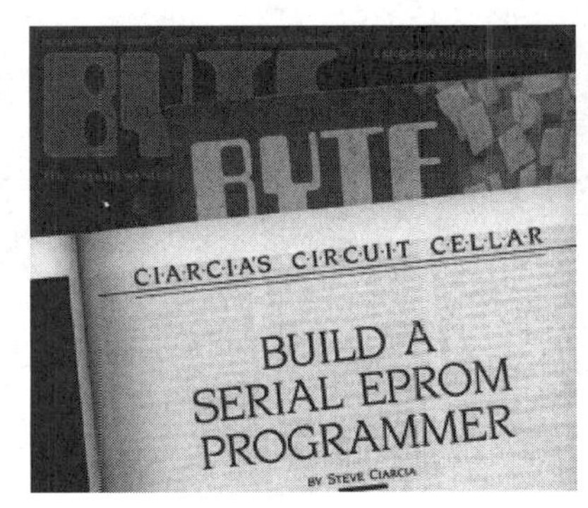

Byte magazine has been a popular source of industry news and articles for software and hardware engineers for two decades. Particularly popular in the 1980s was Steve Ciarcia's "Circuit Cellar" which explained how to construct hobbyist electronic gadgets.

bytes per second Bps. Many people abbreviate this bps, which causes confusion because bits per second is conventionally abbreviated bps. Bytes per second is a description of transfer or transmission rates in bytes over time. A byte is eight bits, or one octet. Data frame format sizes are sometimes described in terms of bytes or octets.

.bz Internet domain name extension for Belize.

BZT See Bundesamt für Zulassungen in der Telekommunikation.

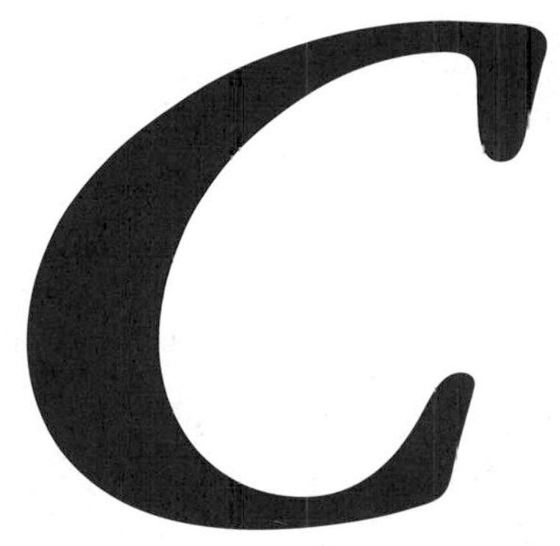

C 1. Symbol for capacitor or capacitance. 2. Symbol for Celsius or centigrade. 3. Symbol for the velocity of light.

C minus, C- Symbol for the negative terminal of a C battery. The connecting point at which the negative terminal of a grid bias voltage source is connected, as in a vacuum tube circuit.

C plus, C+ Symbol for the positive terminal of a C battery. The connecting point at which the positive terminal of a grid bias voltage source is connected, as in a vacuum tube circuit.

C++ A high-level programming language, a superset of C, developed primarily by Bjarne Stroustrup at Bell Laboratories. In endeavoring to be backwardly compatible with C, it is criticized by some as being unwieldy. Nevertheless, it is widely used in commercial software development and many programmers like it. See C language.

C battery A type of power cell first introduced in the early 1920s. Historically, C batteries supplied bias voltage to electron tubes which were used to control a grid circuit. C batteries are still commonly used in small flashlights, portable radios, and many small, portable, electronic appliances.

C Block A Federal Communications Commission (FCC) designation for a Personal Communications Services (PCS) license granted to a telephone company serving a Major Trading Area (MTA). This grants permission to operate at certain FCC-specified frequencies. See A Block for a chart of frequencies.

C interface An interface used in Cellular Digital Packet Data (CDPD) which is deployed over AMPS. The C interface connects the Mobile Data Intermediate System (MD-IS) to the Intermediate System (IS). See A interface, B interface, Cellular Digital Packet Data, D interface, E interface, I interface.

C jack The USOC Federal Communications Commission (FCC) code for a flush or surface mounted jack.

C language A sophisticated, fast, widely used, medium-high level programming language developed by Dennis Ritchie at Bell Laboratories in the early 1970s. The name was chosen because C was descended from B. Originally written 'C' the quotes are now usually omitted. C became widely distributed with Unix, and was used for programming the Amiga computer in the mid-1980s. By the late 1980s, universities were teaching both C and Pascal as basic skill sets and C has become widely used in the commercial software development industry. The chief advantages of C are power and flexibility. Its chief disadvantages are the logistics of keeping track of pointers and memory allocation and the many pages of code that are needed to accomplish basic tasks. The popular introductory book on C was written by Kernighan and Ritchie. See C++.

C lead In communications lines utilizing three wires, where one is positive and one is negative, the third wire or C lead may be used as a ground and can be manipulated to connect or release the circuit. In telephony, the control provided by the third wire is useful on trunk circuits.

C News A UUCP-based news-reading program developed from the earlier A News and B News programs. Along with InterNetNews (INN), C News superseded B News. The networks were changing, TCP/IP had been introduced, along with Network News Transfer Protocol (NNTP), and times they were a'changing. C News, written by Geoff Collyer and Henry Spencer of the University of Toronto, were released in fall

1987. C News was faster, more reliable, and supported a bigger article database than previous versions. A port of C News for AmigaOS was developed by Frank Edwards. See A News, B News, USENET.

C-band A portion of the electromagnetic spectrum which is used extensively for the transmission of communications signals, especially those to and from satellites. The C-band extends from 4 GHz to 8 GHz with satellite uplinks in the 5.925 GHz to 6.425 GHz range, and downlinks in the 3.7 to 4.2 GHz range. Downlink and uplink frequencies are different in order to reduce interference between received and sent signals. C-band transmissions require relatively large receiving antennas, making them less popular for consumer services than Ku-band. See band allocations for chart.

C-Scope In radar, a screen that displays bearing and elevation information relative to the center of the region being scanned.

C-stock Bell Telephone jargon for refurbished telephone equipment.

C/A Code, civilian code, S-code In the Global Positioning System (GPS), a Clear/Acquisition Code in which the carrier wave is modulated with a sequence of pseudo-random, binary, biphase signals to provide civilian locational information transmissions. This information is at a lower resolution level than the classified government GPS transmissions.

C64 See Commodore 64.

C7 The European analog to the North American Signaling System 7 (SS7), which is similar but not directly compatible. C7 is widely deployed for digital phone communications. See Signaling System 7.

.ca Internet domain name extension for Canada.

CA Call Appearance.

CAB See Canadian Association of Broadcasters.

Cabeo, Niccolo An Italian scholar and experimenter who recorded a number of electrical observations and wrote the first Italian treatise on magnetism in 1629.

cable Wire, fiber or other conductive material in single or multiple (bundled) strands used for the transmission of light, heat, electricity, or data. Although the terms *wire* and *cable* are often used interchangeably, some technicians make a distinction based on the bundling. If it is a single core, it's called a wire, if it is a combination of layers of two or more separately insulated wires, or if it is a fiber optic bundle, it's called a *cable*.

In telecommunications, the speed and quantity of data which can be conducted along a cable varies greatly with the materials which are used in its manufacture. Traditional phone lines are usually copper wires, while cable television broadcasting uses fiber optic cable with greater speed and bandwidth. The computer industry, with its demands for simultaneous transmission of data, sound, and video, has greatly increased interest in high-speed, high-bandwidth cable media. Wireless services such as cellular phone and individual satellite modem transceivers provide an option to physical cable links. See conductor, conduit, fiber optic.

cable access, PEG By regulation, cable television broadcast providers must set aside and reserve channels for use by the public, educational institutions, and government entities (PEG). The cable company is limited from exercising editorial or content control over these public and government programming channels.

Cable Act Of 1984 An Act of the U.S. Congress broadly deregulating the cable TV (CATV) industry. This significantly reduced the Federal Communications Commission's (FCC's) jurisdiction in this area. In 1992, the Act was partially repealed and further shaped by the Cable Reregulation Act of 1992.

cable assembly A pre-assembled cable, ready for installation (e.g., fitted with jacks, or other relevant attachments).

cable bay An installation setup designed to accommodate many sets of cables in rows or a matrix, to facilitate easy access, identification, and maintenance.

cable core The inner, conductive portion of a sheathed or insulated cable.

cable coupler A hardware connector used to complete the circuit between similar or dissimilar cables with the same electrical characteristics.

Cable Deregulation Act of 1992 In 1984, an Act of the U.S. Congress broadly deregulated the cable TV (CATV) industry. This significantly reduced the Federal Communications Commission's (FCC's) jurisdiction in this area. In 1992, the Act was partially repealed and further shaped by the Cable Reregulation Act of 1992; rates were mandated to be lowered by an FCC-prescribed percentage in 1993 and again in 1994.

cable diameter An important indicator of the size and other properties of a cable. The diameter of a wire or fiber optic cable can dramatically affect its transmission properties, weight, flexibility, cost, and ease with which it can be interconnected with other components. It may also influence the distance over which it can carry a signal. See American Wire Gauge, Birmington Wire Gauge, fiber optic, multimode optical fiber, single mode optical fiber.

cable drop The subscriber connection segment of a wired *cable access* installation. The cable drop is generally the section that originates at the *cable tap* on a utilities pole and ends at the subscriber's television, or at a connector fed through the subscriber's wall to which the subscriber can hook in the television.

Cable/Information Technology Convergence Forum CITCF. An organization representing the cable industry which was established to further communication between vendors and cable industry professionals.

cable loss An important property of a cable's transmission characteristics over distance. Loss of signal, or *attenuation*, is the gradual diminution of the signal to the point that it is no longer useful or can no longer be detected at the receiving end. Loss may be due to many factors that often occur together including interference, the construction and materials used in the cable, the number of connections, the proximity of other conducting surfaces, the thickness, distance, weather conditions, etc.

Some types of transmissions can only be carried over a short distance. For example, external SCSI devices usually can only be used with cables up to about six feet in length, and shorter cables are recommended. With Fibre Channel technology, the distance for SCSI devices can be extended enough to use a separate maintenance room for aggregating the devices. For many types of optical fiber installation, the transmission distance may only be a couple of kilometers. See attenuation, multimode optical fiber, single mode optical fiber.

cable map A diagrammatic record of the type and location of cables in a distribution system. For decades, hand-drawn cable diagrams were used. More recently, computer-generated diagrams and databases are sometimes used to keep track. To aid with local management, cables are frequently color-coded, marked, or bundled, and this information may or may not be redundantly recorded on the cable map.

Cable maps are particularly important in institutions and business complexes serving many rooms and buildings, and in submarines and ships which have substantial numbers of cables running through corridors and walls. In data networks, the physical cables between routers, switchers, and workstations are sometimes diagrammed in the routing software, displayed as a color-coded schematic on-screen. These applications may have two types of maps included in them, a physical cable map, and a virtual connection map, one of which is overlaid on the other and each of which may be managed somewhat differently.

cable modem A device which permits computer access to high-speed transmissions via a broadcast cable network and subscriber service. Although fiber cables can handle two-way communications, most implementations tend to be asymmetric, giving more time to downstream data (since most information is downloaded from source to user). Cable affords broadband, round-the-clock, fast access to data services like the World Wide Web, while keeping phone lines free. Cable modems allow users to download information about 20 times faster than ISDN modems, and about 80 times faster than 28.8 bps phone line modems. 360k cable modems began shipping in 1997. See ASDL, ISDN.

cable riser Cable that is installed and supported vertically, usually through walls and ceilings, in order to reach upper floors in multiple-story buildings.

cable run A conduit, or other piping or pathing system which provides a means to thread cables or which constitutes the path of the cables. Cable conduit runs are used for a number of reasons. In some cases, they make it possible to add cable later on, if the full cabling requirements are not known at the time the run is installed. They may provide extra insulation or fire protection and they may be more aesthetic, enclosing a bundle of cables that might otherwise be distracting or unsightly.

cable tap In *cable access* installations, the tap is the physical connection, usually on a utilities pole, to which the subscriber line is attached.

cable television, community antenna system CATV. A television broadcast system that transmits licensed television programs and local programs to subscribers over a wired network, usually over fiber optic cable. Cable

TV, that is, television broadcast delivered over wire, was established in Europe in the early 1930s, less than a decade after the viability of the television medium was first demonstrated. In North America, the distances between communities was much further, and cable TV was slower to develop. Satellite transmission options to cable television are becoming more widely available. See basic cable service.

Cable Television Relay Service Station CARS. A television relay station is a transceiving point between the original broadcaster and the subscriber. This may be a building facility, an unstaffed tower relay, a mobile relay, or other point at which the transmission is received, sometimes processed, and then retransmitted. For example, the broadcast may be sent over airwaves to the local relay station, which may be a local cable TV supplier. Once the various broadcasts are received, the local station subsequently sends the signals to the subscribers through physical cables. This way there is only one powerful antenna needed to serve the local area (otherwise each subscriber would need an antenna, the way it was before cable TV became available). This also gives the local station the capability of transmitting only those stations which the subscriber may desire, or which fit the payment package arranged with the local station.

cable vault An enclosed area, often in a basement with extra fire-proofing, which encloses a large number of cables leading into a building or distribution frame. Cable vaults are usually used in situations where the type or number of cables pose extra risks in terms of fire or electrocution, and where access by only trained personnel is desirable.

cablehead The point at which a land cable and marine cable are joined. It may be indicated with a sign (if you live near the ocean, you may have seen these signs).

cableway A hole, slot, or other opening in a component unit or work surface which allows cables to be fed in from behind or underneath. Most computer desks now include slots and holes for cables. Some even put slots into the drawers, so a printer can be installed in a drawer, and pulled out only when needed.

CAC 1. Customer Administration Center. A phone console used for maintenance and diagnosis of a multiline phone system. 2. See call admission control, connection admission control.

cache *v.* In the traditional sense, to cache something is to put it away, or hide it, in a secure place for later use. In computer terminology, the sense of putting it away for later retrieval is retained, but ease of access is also implied. Thus, to cache is to store information in a readily accessible, fast access location, as in RAM on a computer, so it can be retrieved quickly when needed. Many systems are specifically configured with cache memory, while others may use the hard drive as a cache location, which is not as fast, but still may be effective in certain circumstances. Information is cached by an applications program for items that are often consulted or executed. This speeds up operations for priority activities. See cache, cache memory, RAM disk.

cache, cache memory *n.* A high-speed electronic memory buffer used in computing to increase apparent processing speed by more effectively managing resources. The cache storage is usually within a designated amount of random access memory (RAM) and thus is volatile, (although in its most generic sense, a hard drive would be a suitable cache device for a slow sequential storage medium like a tape drive). A hard drive controller card may itself include a cache.

The effectiveness of a cache depends upon a variety of factors, including the size of the cache, the ability of the software to utilize it, the types and variability of operations being done, the design of the caching logic, and the speed of the microprocessor. Since RAM access is faster than hard drive access, efficiency can be increased by storing frequently accessed information in the cache memory, where it can be written and retrieved more quickly than from disk. Information which is not found in the cache, may then be added to the cache for future reference. In networking, a cache can be used to store frequently accessed information (often the locations or contents of data files or applications) in order to serve it more quickly to users, as it is requested. FATs and hash tables may be stored in the cache to increase file access speed. BIOS device-controlling functions may be loaded from read only memory (ROM) into cache memory during a startup sequence. See cache, cache hit, cache miss.

cache controller In some computer architectures, a circuit that is specifically included to administrate the storage, organization, and retrieval of cached information. This may be incorporated into a specialized chip.

cache hit A situation in which the data which was sought in a cache access was found, and it is not necessary to access the slower storage medium (usually a hard drive), resulting in faster retrieval of the desired information. See cache, cache miss.

cache miss A situation in which the data which was sought in a cache access was not found and is consequently sought on the slower storage medium (usually a hard drive). It may subsequently be stored in the cache for future reference. See cache, cache hit.

caching Putting information in a storage area where it can quickly be retrieved when needed. It is a means of speeding up effective and perceived performance of a system. Disk caching and memory caching are two ways to speed up access to frequently used commands, device drivers, or frequently accessed data.

caching, data entry A means of speeding up data entry by retaining previously inserted information so that it can be reused or overwritten for subsequent entries. In other words, sometimes it's more efficient to edit or retain the data in the field from the previous entry, than to type it in from scratch.

CACM Communications of the Association for Computing Machinery.

CAD 1. See computer-aided dispatch. 2. See computer-aided design/drafting.

caddy A protective case for storage media, especially magneto-optical discs. A caddy may be used for storage, but often it is a necessary adjunct to the disc in order to fit in the insertion mechanism of a disc player. The caddy may in some instances provide write protect/write enable mechanisms.

cadence A rhythmic measure or beat. In telecommunications, many signals are identifiable by a pattern of tones and silences. Cadence has implications for telegraph, radio, and telephone communications where many audio signals are coded to particular rhythms (Morse Code, distinctive ringing, international variations in rings, and busy signals). People can learn to distinguish different types of data communications by pitch and cadence, as fax tones differ from data tones, and data tones vary further according to baud rate.

cadmium A bluish-white, malleable, ductile, metallic, non-corrosive element which is commonly used in protective coatings and platings.

CADS Code Abuse Detection System.

CAE 1. See Common Applications Environment. 2. computer-aided engineering.

CAFA computer-aided financial analysis.

cage antenna A multiwire antenna (imagine a ring of horizontal parallel wires somewhat constricted in the middle, resembling a cage) similar to a dipole antenna, configured to improve capacity and reduce loss.

CAI 1. See computer assisted instruction. 2. common air interface. An international interface standard defined to provide interoperability between wireless handsets and compatible networks.

CAL computer-aided learning, computer-assisted learning. See computer-assisted instruction.

CALC See customer access line charge.

calculator A device for facilitating the speed and accuracy of mathematical computations. Early calculators could handle only simple arithmetic functions, while current ones include storage, automation, and programming capabilities for doing complex computations.

The calculator is the forerunner of the general purpose computer, in fact the early computers were very large, very powerful calculators, and their history runs hand-in-hand. Handheld calculators were devised in the late 1960s and became widespread in the early 1970s, costing about $200 for a very simple arithmetical calculator. As prices came down, calculators superseded slide rules and abacuses for quick computations.

The most celebrated early microcomputer was developed in 1974 by MITS, a company that was producing scientific calculators. With competition from bigger companies such as Texas Instruments, MITS needed a new source of revenue, and developed the Altair. Since that time, calculators have been incorporated into many devices, such as cash registers and wrist watches, and even some computer keyboards. Current calculators include graphical displays, square root computations, multiple memory registers, and programming languages like Forth. They are more powerful than computers from two decades ago. See abacus.

calendar routing An administrative method for directing inquiries according to the time of year, week, or day. Used especially in industries where inquiries are cyclic (travel industry), or where availability of personnel to assist callers is cyclic.

calibrate 1. To set, align, or mark a measuring or timing instrument according to an accepted standard. 2. To ascertain, record, or correct variations in a measuring or timing instrument with reference to another, or to an accepted standard.

California Education and Research Federation Network CERFnet. This research and education network was founded in the late 1980s by General Atomics, with aid from a National Science Foundation grant, and grew to be a national backbone by the early 1990s. CERFnet joined with other nets in 1991 to form the Commercial Internet Exchange (CIX). In 1996, the Teleport Communications Group Inc. (TCG), one of the largest competitive local telephone companies in the U.S., acquired CERFnet to provide Internet services to corporate and institutional clients.

CERnet is based on ATM and SONET architectures, with each Local Access and Transport Area (LATA) served by at least two backbone nodes.

call 1. *v.i.* To attempt to contact, or succeed in contacting another party or entity. A unit of virtual or human communication across some type of communications medium, or at some distance. 2. *n.* A unit comprising a successful communication through some type of communications medium, or at some distance, between two or more parties or entities who are more or less simultaneously in contact, frequently with a 'give-and-take' character to the contact. Human participants in a call are generally called *parties* to the call. 3. *n.* In networking, a communications association between a user and a network entity, or between two or more users across the network.

call abandons, abandoned calls Calls which are terminated by the originator before completion of the intended contact. For telephone calls, reasons for abandoning calls include fuzzy connections, wrong numbers, answering machines, being put on hold, obnoxious hold music (not all hold music is obnoxious, just some of it), even more obnoxious hold commercials, interruptions (children, doorbells), transfer to the incorrect person or department, etc.

Since any call connection in progress impacts on system capacity, abandoned calls have to be considered when structuring and managing a system. If a high proportion of abandoned calls occur after a human operator has made verbal contact with the caller, it's important to determine and evaluate the reasons and take corrective measures to increase call completion. See abandoned call cost.

call accounting system A system of recording the type and quantity of calls on a system. This was originally recorded manually, and operators of public phone systems had elaborate card systems on which to record calls, particularly long-distance calls. Now accounting has been computerized, and the system can constantly monitor call volume, number of connects, number of abandons, peak hours, trunk allocation, and other statistics related to economics in general and call billing in particular. When used in private branch systems, it can further be used to track agent activities, length of calls, departmental use, etc. and integrated with revenue and customer databases to give an overall picture of the role of the phone calls within the company's business. See call card.

call admission control, connection admission control CAC. The set of actions taken by a network during a call setup, or re-negotiation, to evaluate whether to accept or reject a connection or re-allocation request, based partly on the ability to supply Quality of Service (QoS). See crankback.

call announcement A feature in a telephone system in which an operator or other agent announces the call to the callee before connecting the call.

call attempt Initiating a call which may or may not be completed. If a large number of call attempts are not completed, diagnostic and troubleshooting steps should be taken. Solutions may include training, additional lines, staff changes, or equipment changes or repairs. See call abandons, abandoned call cost.

call barring Prevention or elimination of all calls, or specific calls, associated with a specific phone. Usually implemented to prevent unauthorized use, or abuse. See call blocking.

Call Before Digging A safety sign to warn area workers that they must call for information on underground cables or hazards before digging.

call block A restriction put on a phone line to prevent connection of certain calls. More recently it has come to mean retaining anonymity from Caller ID by blocking the caller's identity from the receiver if the caller has keyed in a blocking code. Call blocking in this sense is free, whereas Caller ID costs money. This may seem backwards, like allowing a stranger through the door unseen, while the person opening the door has to pay money to see who it is, and still may not get the information because that person is disguised (blocked). In terms of personal safety and security, it should have been set up the other way around, with the person answering the phone being provided the identity of the caller for free, and the caller having to pay to hide his or her identity. However, the system was probably set up this way because it generates more revenue for the phone company.

call card A manual call management and billing system in which the information about the caller, callee, distance, and duration of the call is recorded on a call by the operator. See call accounting system.

call center Any centralized telephone call facility which handles a large number of incoming and/or outgoing calls. Call centers are often specialized for handling many incoming calls, such as those resulting from television marketing through toll free numbers, or many outgoing calls, as those originating from telemarketing or teleresearch firms. In these specialized environments, automatic call distributors, head sets, computerized dialing, and automated answering are commonly used.

call clearing The process by which a call connection is released and the call resources made available to other users. This is particularly important in end-to-end transmissions in which the line must be freed up before it can be used again.

call control The entire process of detecting the request for a call, setting up the physical and logical connections, rerouting to available trunks if necessary, facilitating the transmission, shutting down the call and freeing up the resources for other callers. Most of these functions are now computerized, although occasional operator assistance, directory assistance, or services for special needs users are still handled by human operators. Call control may go through more than one system, as when a call goes into or out of a private branch exchange and through a public exchange, or where wireless and wireline services from different providers are used together to complete a call.

call control signal Any signal used in automatic connection and switching systems which controls the call sequence. In older systems, the control signals were transmitted by means of tones on the same line that was used for the voice transmissions. In newer Signaling System 7 (SS7) systems, the control signals, and the voice transmissions are handled over separate channels. See Signaling System 7.

call data The statistical information associated with a call. This is used for monitoring, accounting, management, and planning, and these days is usually stored in a computer database, and sometimes organized and analyzed by computer software.

Call Detail Record CDR. A telephone record-keeping system, usually used for accounting and administrative purposes, which tracks and records details about incoming and outgoing calls such as the call duration, caller and/or callee, time of day, etc.

call diverter A subscriber surcharge service, or phone peripheral device which intercepts an incoming call and forwards it to a phone operator or phone message, or to another number, as in Call Forwarding. Depending upon the service or device, the caller may or may not be aware that the call has been diverted.

call duration The period of time from actual connection of the call, until its termination. On phone lines and data networks, call duration information is used for statistical purposes for tuning the system, and determining peak hours, and for billing purposes. It may also be used to detect fault conditions.

call establishment The process of routing and connecting a phone call or data transmission path.

Call for Votes CFV. A formal process used as part of the sequence of events necessary to create a new public newsgroup on USENET. The various steps in the process involve writing a charter for a group, assigning a name, announcing the intention to begin a group, alerting people at specified intervals of the progress of the request, and calling for votes to support the creation of the group. Designated trusted volunteers monitor the process and tally and report the results of the voting. Certain minimum numbers and percentages of votes are needed before the group can be created. The requirements have gradually become more stringent as the volume on USENET increases. See newsgroup, USENET.

Call Forward A surcharge phone service which permits the subscriber to automatically redirect a call, intended for one number, to another number. This is useful in cases where the callee is temporarily at another location, or where the callee wishes someone else to handle calls (like an answering service). On consumer systems, the call forwarding is usually enabled by using a touchtone phone to dial a code (72# in N.A.) followed by the number to which the calls are being forwarded, and disabled by dialing a code (73# in N.A.). Some newer phone systems have an indicator light to show that the calls are being forwarded to prevent the subscriber from forgetting to deactivate Call Forward after returning to the original location. See Call Forward Busy, Call Forward No Answer.

Call Forward Busy Similar to Call Forward, except that calls are rerouted to a predetermined number only if the called number is busy, and otherwise rings through to the original number. See Call Forward.

Call Forward No Answer Similar to Call Forward Busy, except that calls are rerouted to a predetermined number only if not answered after a specified number of rings. See Call Forward Busy, Call Forward.

Call Forward Variable A combination of Call Forward Busy and Call Forward No Answer in which the call is rerouted to a predetermined number if a busy signal is encountered, or if there is no answer after a specified number of rings. See Call Forward No Answer, Call Forward Busy, Call Forward.

Call Girls One of the many colloquial names given to the early female telephone operators. Others include *Hello Girls*, *Central*, and *Voice with the Smile*. See telephone history.

call handoff In mobile phone systems based on passing the transmission on to another transceiver while the call is taking place, as in cellular communications, the handoff is the point at which the call is transferred during the conversation. Mobile providers strive to create systems where the handoff is seamless and does not create delays, noise, or significant volume changes.

Call Hold A surcharge phone service or multiline subscriber feature in which the subscriber can put a current call on hold, accept or place a second call, and then return to the original call. This service is similar to a hold button on a multiline phone, and the person on hold is not able to hear the second conversation.

call horn alert A mobile system set to beep a car horn to signal an incoming call, when the driver is away from the mobile handset or receiver.

call mix Telephone calls are of many kinds, as are logons on a computer. In a telephone system, the calls may be long or short; busy, abandoned, or completed; local or long-distance. On a computer system, the logons may result in downloads, modem access, running of applications, file maintenance, etc. The call mix is a statistical look at the types of usage that occur on a network.

Call Park A subscriber service or console feature that allows a user to set the call so it can be answered on any other phone on the system. Call Park is useful in situations where the callees are moving around, and where they may be alerted to the presence of the call through a paging system. The parked call can then be retrieved from another line by dialing in a call unpark code.

Call Pickup A surcharge phone service or multiline subscriber feature which permits a subscriber to intercept a call to another prearranged number by typing in a code and then answering the other call. Suppose you and your housemate have separate lines, and your housemate has asked you to answer his or her calls, you can do so from your own phone. See Call Pickup Group.

Call Pickup Group CPUG. All the phones in a system through which Call Pickup is activated and which can intercept the calls of the others. See Call Pickup.

call processing A combination of computer and human operations in which the call is often set up and connected electronically, and then 'handed off' to a credit collector, researcher, telemarketer, technical supporter, or other agent, once the connection has been established. See call center.

call progress signal A telephone switching signal which indicates whether the call is generating a busy tone, a ringback tone, or an error. See ringback.

Call Record A data record of call details which includes information such as date and time, call duration, call routing, stations used, time on hold, etc. This information may be used for billing and administration.

call release time The duration during which a call is shut down and the line released for the next call.

call reorigination A handy feature in which calls can be initiated one time during a multicall session with a debit card, charge card, credit card, or calling card account. In other words, a series of calls can be made at the same time without having to re-enter codes, or having to re-insert the card to make the subsequent calls. Between calls, a code is usually pressed, and the caller receives a signal to continue with the next call. This is particularly useful when having to make several calls at an airport to let people know your flight plans have been changed and you are catching a plane at a different departure gate.

call restrictor A physical or virtual call blocking mechanism which controls the type of outgoing calls that can be made on a line. Examples include blocking long distance calls from a phone near a public area, or blocking 900 calls from phones used by teenagers.

Call Return A subscriber surcharge option that allows the last caller, whether the call was answered or not, to be dialed back automatically. This can be handy for Crisis centers and other emergency services.

call routing tree A diagrammatic representation of call routing configuration and logic. See call tree.

call screening The most familiar call screening is a receptionist who says the boss is in a meeting and can't be reached at the moment when the boss is actually watching the World Series with his or her feet up on the desk. More legitimate uses of call screening involve getting enough information from the caller to direct the call to the best person equipped to handle it. In automated systems, call screening is a setup which uses Caller ID, or some other identification tool, to monitor the origin of the call, and to patch it through accordingly, or which uses a speech recognition system to direct the call.

call sequencer An automated system for evaluating incoming calls, queuing them, if necessary, and assigning them to agents depending upon priority, availability, or caller characteristics.

call setup time In a circuit-switched network, as most phone networks, the amount of time it takes to patch through the route from the caller to the destination in order to set up an end-to-end path for the communications. During the course of a call, the resources are dedicated to that communication and cannot be used by others. For a phone call, the call setup time includes the time it takes to dial, for the call to be switched through the system and the appropriate trunks, to the destination. This is usually not billed for land lines (wireless may be billed for air time) since it is not known during setup whether the call will be answered, and the length of the call duration.

call shedding A situation in which a call made with an automated dialer connects with the callee even if no agent is available to take the call. This practice results in the callee picking up the phone and either being put on hold, or getting an automated message, or finding the call has been hung up (shed). The callee does not get the opportunity to speak with the human agents who are originating the calls through the dialers. This practice is very annoying to callees and is illegal in many areas.

call sign See callsign.

call splitting A subscriber surcharge or private branch service in which a conference call participant can speak to any one of the other members of the conference privately, that is in non-conference mode. When a phone attendant is involved in the call, the attendant may relay the information privately to one of the called parties.

Call Stalker An AT&T commercial software product which provides 911 emergency service agents with information about the caller, such as address and calling phone number.

call supervision A process whereby it is determined whether a telephone communication was actually answered, such that billing is not activated unless a connection was made.

Call Trace A surcharge phone service or emergency service in which the tracing of the origin of the last call is provided and recorded in the event that it may be needed later for legal reasons. The results of the trace are not provided to the customer under privacy laws, but may be revealed later in the course of investigations, through proper legal channels.

Call Transfer A surcharge phone service, or capability of a multiline phone system, which allows a call to be transferred to any other phone on the system. Transfers are accomplished by typing in codes and the transfer number, or by using a transfer button followed by the callee's line. Call transfer is commonly used in business, and the console often staffed by a full-time operator or receptionist. Callers are not tolerant of calls that are incorrectly transferred, or accidentally terminated, and it's important that personnel responsible for transferring calls are well trained on the equipment, and in business etiquette.

call tree A diagrammatic representation of call sequence information (usage) used for statistical analysis and planning. See call routing tree.

Call Waiting A surcharge phone service which becomes active if a call comes in while the callee is already engaged in a call. Call Waiting signals the callee that there is another party trying to call, either by an audio signal or blinking light, and provides the callee the option of ignoring the incoming call, or terminating (or putting on hold—if hold is available) the current call and then answering the second incoming call. This is useful for emergency calls, or for terminating a casual conversation to carry on with other calls. Call Waiting can interfere with transmission, or even cut you off, if you are connected through the line via modem. Call Waiting can usually be temporarily disabled in order to avoid this problem, or the modem can be reconfigured to ignore this type of interruption. The first option is easier, and preferable. Information on how to disable Call Waiting is listed at the front of most local phone directories.

Callan, Nicholas J. An Irish priest and educator who devised an early induction coil in 1836. See induction coil.

callback facsimile A system in which you dial a phone number, key in your callback facsimile number and documents that are of interest, usually from a numerical list given by a voicemail system over the phone, hangup, and then wait for a facsimile callback service fax machine to call your fax machine and deliver the documents requested. A lot of computer industry technical support and product information is now delivered this way.

callback modem A modem which is set to receive a phone call through a network that acts as a callback request. A password may be required, and then a phone number to be dialed is provided to the system. The modem then is set by the computer to dial the number provided. Why do this instead of dialing directly? This system provides better security, so there is a record of what numbers have been connected to the network and accessing the data, and it sometimes reduces toll charges, so that the toll is billed to the network number and handled by the business accounting office, rather than being billed to an employee.

Callender Rapid Phone Company One of the earliest automatic switching phone services, established in England in 1896 by musician and inventor Romaine Callender.

Callender, Romaine A Canadian music teacher and instrument maker. Callender was an associate of Alexander Graham Bell and founded the Callender Telephone Exchange Company in Ontario, Canada. Between 1892 and 1896, he submitted three series of patents for telephone switching inventions, but failed in implementing them in Ontario and subsequently traveled to New York to seek financing and open another firm. Traveling with him were two brothers, George William Lorimer and James Hoyt Lorimer, who assisted him in his further experiments. They finally succeeded in developing an automatic switching system in 1895. They later returned to Brantford, Ontario, and Callender sailed to England in 1896, where he formed the Callender Rapid Telephone Company. See Lorimer, George and James.

Callender switch A very rudimentary, early telephone switching system developed by Romaine Callender and the Lorimer brothers in the late 1800s. See Callender, Romaine; Lorimer switch.

Caller ID, Call Display A phone carrier 'added value' pay service which provides the call recipient with the phone number identity of the calling party. You may have to pay local and long-distance Caller ID charges separately. In North America, the Caller ID information is usually passed to the receiving phone between the first and second ring.

You need two things to take advantage of Caller ID: a subscription through the phone carrier to the Caller ID service, and a phone or separate device with a Caller ID display.

A Caller ID display device can be leased or purchased from the carrier, or other commercial vendors. Many phone companies are now giving them away free with the service, to encourage subscribers to pay the monthly charge. Add-on phone peripherals, and newer phones, make it possible to associate a human name or code with the called name, and read it on a small LCD screen. The recipient can then judge whether to answer the call, let the answering machine take it, or ignore it. This may afford some protection from telephone solicitors, and nuisance callers. Unfortunately, it's not useful if the calling party is using *call blocking* to obscure their identity.

Business software now exists which allows the calling number to be entered directly from the phone system into a database through a computer interface, in order to display any files associated with the caller (previous sales, problems, VIP status, etc.) in order to transact business in a faster and more personal manner. Some callers are concerned about the amount of personal information that is available to a business without the caller's knowledge, especially when this is stored in a file, or being accessed through the net. See call blocking, Class, ANI.

Caller Independent Voice Recognition An automated voice recognition system which can interpret voice input without being specifically 'tuned' to a particular caller's voice. It is useful in phone applications which can accept spoken numbers or commands for processing a call, and in word processing applications voice recognition.

Caller Name A phone carrier 'added value' pay service which takes an incoming Caller ID number (assuming the call is not blocked), looks it up in a directory listing database, and transmits the Caller ID number and its associated listing, if it exists, to the recipient's add-on Caller Name display, or a phone providing Caller Name display. This is not as flexible as a user-configured system where you can associate any name or code you wish with a specific incoming number, but it is very useful for identifying a first-time caller or stranger (and it may be possible to use them together if you have compatible peripherals). See Caller ID, call blocking.

calling The act or process of contacting another party or entity across some type of communications medium, or at some distance, with the intention of establishing a link between humans and/or devices.

calling card A remote or off-premises phone service provided by common carriers to allow local and long-distance calls to be charged back to the subscriber's local phone number or other authorized billing number. There may or may not be surcharges associated with such a call. The name derives from a wallet card typically issued to the subscriber with instructions and digits to be dialed to gain access to the service. In many cases, you don't need the physical card to make the call, but automated phones are becoming prevalent in which the card is physically inserted in a slot, or swiped through the phone to expedite the processing of the call. Bell Canada claims a trademark over the Calling Card name, but the term is so widely used it would be difficult to enforce the trademark.

calling jack In manual switchboard systems, the jack that is used by the operator to connect the call that came in through the answering jack, to the circuit for the *subscriber* who will be receiving the call.

calling number display See Caller Name, Caller ID.

calling party, calling station A person or entity originating a call. See call.

Calling Party Number CPN. In telephony, a common channel signaling (CCS) parameter in the initial address message which identifies the calling number and is sent to the destination carrier.

calling sequence The sequence of numbers, letters, steps, and other information needed to connect a call through a traditional phone line or digital computer phone system. When calling through a modem, the calling sequence includes not just the number being dialed, but also the parameters for the line, the baud rate, whether it is pulse or tone, the speaker level, pauses, wait for tone to continue with extension numbers, etc. In computer software, the calling sequence may include linking in to an address book or other database, and saving statistic information gathered on the call.

C

callsign, call sign In radio communications, a series of identification characters which is assigned by local regulating authorities to every licensed radio operator or station. The callsign identifies the country, and sometimes also the region of the country. One of the most famous callsigns in radio history is 8XK which Frank Conrad used from his Pennsylvania garage, and which was later licensed as the history-making KDKA radio broadcast station. See KDKA.

CALS Continuous Acquisition and Life-Cycle Support (formerly Computer-aided Acquisition and Logistics Support). A Department of Defense (DoD) strategy for the creation, use, and exchange of weapons-related digital data.

CAM 1. carrier module 2. Call Accounting Manager 3. Call Applications Manager. A Tandem telephony software interface for linking computers with telephone switches. 4. See computer-aided manufacturing. 5. computer-assisted makeup, composition and makeup. A WYSIWIG terminal for previewing type composition and page layout.

Camcorders have become smaller and more powerful. This Canon ES camcorder offers high quality Hi-8 picture resolution, stereo sound, calibrated eye-sensing focus, and many editing features formerly found only on full-sized editing decks.

CAMA Centralized Automatic Message Accounting. A billing and statistical system for recording calls on tape. It is sometimes also used to trace fraudulent use of phone services.

camcorder A combination video recorder and camera unit. Increasingly, consumer camcorders include playback, editing, and special effects capabilities. Newer digital camcorders can be used as both digital still-frame and motion recorders, and can be interfaced directly with software for scanning, image processing, and Web applications. Camcorders may eventually supersede still film cameras, since no film processing is required, and consumers frequently favor convenience over image quality (film is about 16 times higher resolution than current consumer digital systems).

Cameo Personal Video System A Macintosh-based commercial videoconferencing product from Compression Laboratories Inc. which supports audio, video, and file transfers. It works over Switched 56, ISDN, and Ethernet networks. Cameo uses a proprietary CLI PV2 compression scheme. See Connect 918, CU-SeeMe, MacMICA, IRIS, ShareView 3000, VISIT Video.

campus A physical and geographic environment (primarily the grounds) associated with learning and/or research facilities, such as universities, hospitals, and some businesses.

campus backbone The primary network of wires/cables which interconnect a campus. See backbone circuit.

Campus Wide Information System CWIS. A system of interactive kiosks and public information sources which provides directories, product or course offerings, maps, calendars, and other general public services of interest to educational institutions, businesses, expositions, and shopping complexes.

CAN Control Area Network.

Canada Machine Telephone One of the earliest phone companies to use automatic switching, technology that was developed jointly by the Lorimer brothers and Romaine Callender. The Lorimer brothers established CMT in Peterborough, Ontario, Canada in 1897, and there produced the first commercial Callender Exchange. The Lorimers continued to improve on the technology until it bore little resemblance to the original Callender switching system. The company lost its technical expertise when James Hoyt Lorimer died, but his brothers continued to market in North America and Europe. Unfortunately, due to lack of reliability and long installation times, the company didn't thrive and was acquired by Bell in 1925.

Canadarm A remote manipulator system designed and made in Canada for the U.S. space shuttle program. The National Museum of Science & Technology has constructed a full-size replica and accompanying video available as a traveling exhibit.

Canadian Amateur Radio Advisory Board CARAB. A nonprofit consulting group comprised of members of the Radio Amateurs of Canada (RAC), and the Radio Regulatory Branch of Industry Canada (IC). CARAB works as a communications liaison between RAC and IC. http://www.rac.ca/carab.htm

Canadian Association of Broadcasters, L'Association canadienne des radiodiffuseurs CAB/ACR. A trade organization founded in 1926 by 13 broadcast pioneers. The CAB supports over 500 radio, television and specialty broadcast providers in Canada.

Canadian Broadcast Standards Council CBSC/CCNR. An organization incorporated in 1990 to encourage high standards of broadcasting and professional conduct by private radio and television broadcasters. The CBSC keeps broadcasters informed about societal issues, administers codes of industry standards referred by the Canadian Association of Broadcasters (CAB), and provides information resources to the public. http://www.cbsc.ca/

Canadian Broadcasting Corporation CBC. The primary broadcasting organization of Canada, CBC is a public broadcasting service providing television and radio programming in both English and French. The CBC was initially established in 1936 to ensure Canadian content in broadcasting. CBC's first television broadcast took place in 1952, in Montreal. In 1966, it began color broadcasting, the first in Canada to do so. See ANIK, CKAC. http://www.cbc.ca/

Canadian Business Telecommunications Alliance CBTA. A national, nonprofit organization representing over 400 business and telecommunications users in Canada. The CBTA serves to support members and to facilitate Canada's competitive participation in telecommunications markets through quality and innovation.

Canadian Datapac The world's first public data network which began operating in 1976.

Canadian Information Processing Society CIPS. Founded in 1958, CIPS defines and promotes information processing in Canada.

Canadian Journal of Communication CJC.

A scholarly professional journal which deals with many historical and sociopolitical aspects of communications in Canada and abroad.

Canadian National Museum of Science & Technology, Musée National Sciences & Technologie Canada's largest technological museum, located in southeast Ottawa, featuring permanent and special exhibits, traveling exhibits available for loan, school programs, workshops, lectures, publications, and more. http://www.nmstc.ca/

Canadian Radio Television and Telecommunications Commission CRTC. The Canadian regulatory commission, based in Ottawa, Ontario. This important organization is similar to the Federal Communications Commission (FCC) in the United States in that it allocates frequency spectrums and carries out other commercial and amateur radio and television broadcasting administrative functions.

Canadian Satellite Users Association CSUA. A trade association of broadcasters using Telesat facilities, and suppliers of goods and services to CSUA voting members. THE CSUA sponsors an annual trade convention. See ANIK, Canadian Space Agency.

Canadian Space Agency CSA. One of the more ambitious of the CSA's various projects was the Communications Technology satellite (HERMES) project which was undertaken jointly with the U.S. Canada was to supply the satellite, and the U.S. the traveling wave tube amplifier. This high power, high frequency, communications satellite project got underway in 1971 and was intended to test direct-to-home broadcasting technology. HERMES was successfully launched in 1976 aboard a three-stage rocket. The satellite operated for almost twice its expected lifetime, almost four years.

Canada competes at the international level in spacecraft assembly, integration, and testing through its David Florida Laboratory (DFL), west of Ottawa, Ontario, established in 1972. Besides the HERMES satellite, the CANADARM and various ANIK satellites have been developed and manufactured at the DFL. See ANIK.

Canadian Standards Association CSA. A Canadian independent, not-for-profit standards-setting body established in 1919. The CSA is a strong participant in international standards discussions and directions. It engages in a consensus approach to standards adoption, and provides educational services, including publications, conferences, and seminars. The CSA operates a Certification & Testing Division and indicates that products or systems have passed a formal evaluation process at stated levels.

The CSA is recognized by the U.S. as a Nationally Recognized Testing laboratory (NRTL) in order to eliminate the need for duplicate testing for products marketed in both Canada and the U.S., and provides assistance to manufacturers marketing to the European Union.

CSA has an official mark which is recognized as indicating that the product or system with which it is associated, meets certain industry standards. See Standards Council of Canada.

Canadian Telecommunications Consultants Association CTCA. A Canadian association of independent telecommunications consulting professionals. http://www.ctca.ca/

Canadian Wireless Telecommunications Association CWTA. A trade association representing the Canadian wireless telecommunications industry, including satellite, cellular, and other mobile communications services.

cancel Stop a process, function, or action. On a copying machine, to abort the current copy if it has not already gone through the machine, and any additional copies that may have been originally requested.

In a computer application, to stop or abort the current operation or process. Control-C (two keys held down together), sometimes designated in print as ^*C* or *Ctrl-C* is a very common key code combination for aborting a process. It should be used with care as it may abort the user right out of the program. In many applications, a Cancel button is provided to close a dialog or window without carrying out any actions (when you change your mind), or to stop a process in progress. In some older systems, ^Y works in a manner similar to ^C. ^Z is somewhat related, and usually less dangerous, it may suspend the current process (rather than closing it down), and allow you to carry out other activities, so you can later return to the original process. With Unix system shell commands, a process can be resumed with *fg* (foreground) after having been suspended with ^Z.

On phone systems, many services are enabled and disabled, or canceled, by typing in two or three digit codes, sometimes followed by a # or * symbol. This applies to services such as Call Forwarding, Call Waiting, etc. It is advisable to cancel, or disable, Call Waiting before using a modem on a phone line in order not to be interrupted during a big data transfer. The codes for the subscriber's region for disabling these various services are usually listed at the beginning of local phone directories.

cannibalize To strip parts from a system, usually for building another, or repairing a similar system. This often happens in field work and basement labs, where the technicians are making do with what is available.

CAP 1. See carrierless amplitude and phase modulation. 2. Cellular Array Processor. 3. See Competitive Access Provider.

Capabilities Exchange In Data Link Switching (DLSw), a Switch-to-Switch (SSP) control message which describes the characteristics of a sending Data Link Switching (DLSw) router to allow inter-router information exchange, and to provide greater compatibility among different implementations. See Data Link Switching.

capacitance (Symbol - C) The ratio between an electric charge and the resulting change in potential, or the time integral of the rate of flow of electric charge, divided by the related electric potential. Capacitance is measured in farads. See capacitor, capacity, Leyden jar.

capacitor A device arranged of conductors separated by dielectrics, which may be fixed or variable. A capacitor designed to store electrical energy and used in a wide variety of electronic devices. See capacitance, capacity, condenser, Leyden jar.

This historic capacitor from the Bellingham Antique Radio Museum shows how much capacitors have changed. From Leyden jars to tiny solid state components, various means of storing electrical energy have been devised.

capacity 1. The maximum number of objects or occupants which can be contained on or in a system or environment under normal operating conditions (e.g., theater or bridge capacity). 2. The maximum information-carrying capability of a communications system. The unit of capacity varies from system to system; on a network, it might be described generally in terms of number of users, or more specifically in terms of a calculation based on speed or access or load on a CPU, or it may be based on transfer rates for cells or frames.

Capasso, Frederico (1940s-) An Italian-born Bell Laboratories scientist who has made numerous contributions to electronics, particularly photonics. Capasso has contributed to bandgap engineering innovations in optoelectronics, semiconductor, and solid state electronics, and in 1994 co-invented the quantum cascade (QC laser). Capasso has developed components which function in ways not previously observed in nature and which are based on relative thickness and proximity, rather than chemical composition. See quantum cascade laser; Townes, Charles H.

Cap'n Crunch One of the more infamous of the phone hackers (phreakers) from the 1970s and 1980s, John Draper adopted this handle (techie nickname) and served jail time for illegal (albeit creative) tampering with the phone system, particularly in using technology and tones to make unpaid-for long-distance calls. His adventures and discoveries resulted in the phone company making some significant changes to their technology, and plugging a number of security loopholes. Some of his exploits are described in Stephen Levy's book Hackers, and in a '71 article in Esquire Magazine entitled "Secrets of the Little Blue Box."

Legend has it that John Draper's monicker stems from a whistle he acquired from a cereal box of the same name, one which produced a 2600 Hz tone which could be processed by the phone system as a hangup signal when blown into the telephone mouthpiece. The line would stay connected, but the call would not be billed. This type of caller signaling is not possible on newer phone systems which use 'out of band' signaling, because the voice conversation and the phone control signals are on different circuits.

Between jail sentences, in 1985, Draper found time to write a series of Amiga computer technical tutorials, which he distributed free over the net, at a time when the Amiga was an underappreciated new entrant to the field of multimedia microcomputing. See blue boxing.

Capstone chip A hardware security device which uses the same SKIPJACK cryptographic algorithm as the Clipper chip. It incorporates a Digital Signature Algorithm (DSA), a Secure Hashing Algorithm (SHA), a public key exchange, and various associated mathematical algorithms. It's a complex, powerful system, requiring almost 1 Gigabyte on an automated design system to set up the chip. The chips are being installed in various electronic devices for the U.S. Defense Messaging System. See Clipper chip, Pretty Good Privacy.

Capture Division Packet Access CDPA. A packet-oriented cellular communications network architecture designed to handle constant bit rate (CBR) and variable bandwidth multimedia telephony applications such as videoconferencing. Unlike some other protocols, CDPA is bandwidth-adaptable, that is, it can support increased channel access for individual users for brief periods.

capture effect, captive effect In radio communications, signals often compete with one another if the frequencies are very similar, or if two stations are coming in with approximately similar strength. In amplitude-modulated transmissions (AM), the two sound sources will be heard overlapping one another and it's hard to make out what is being heard. In frequency-modulated transmissions (FM), the receiver will filter out the weaker signals, resulting in the 'capture' of the weaker signal and the exclusive broadcasting of the stronger one. If the signals are equal in strength, the receiver may switch back and forth between the two, but it won't play them both simultaneously as in AM.

capture ratio The ability of a tuner to reject unwanted transmissions (other stations, interference) that are on the same frequency as those desired. Measured in decibels, with a lower figure indicating better performance.

CAR computer-assisted retrieval.

car phone A cellular communications unit installed in a vehicle. While handheld, battery-operated systems are often called *car phones,* the phrase more properly distinguishes larger units that use power from the car's battery and connect to an antenna physically attached to the car (the center of the roof, or elsewhere). Generally they consist of two parts, a trunk or below-seat unit, and a handset. Car phones generally have higher power and better transmission than handheld cellular phones, although they lack the convenience of portability. See cellular phone, mobile phone, AMPS.

CARAB See Canadian Amateur Radio Advisory Board.

carbon-based life forms A phrase jokingly used to refer to human beings (and other earth animals) from the perspective of sentient alien life forms.

carborundum A substance with rectifying properties which was used in early radio wave crystal detectors. Unlike the popular galena which required very delicate contact and tuning, carborundum could be clamped tight and sealed firmly within the detector unit, making it suitable for field work and rough handling. Much of the pioneer work on carborundum detectors was done by H. Dunwoody of the U.S. Army, who received a patent in 1906.

carcinotron An electron tube-based backward oscillator designed to generate extremely high frequency signals. See magnetron.

card An electronic printed circuit board, especially one which is easily dropped into a *card slot* by a dealer or consumer. See printed circuit card.

card column In punch cards, a vertically aligned slot in which holes may be punched.

card hopper In mechanisms that hold and feed punch cards, the holder in which the cards are stacked next to the feed mechanism for processing. See card stacker.

Card Issuer Identifier Code CIID. A calling card identification scheme. There are restrictions on which carriers can use CIID cards.

card slot A slot-shaped data connector within an electronic system designed for the insertion of printed circuit board peripherals. Cards frequently consist of graphics controllers, drive controllers, serial and parallel ports, network connectors, and others. See edge connector, printed circuit board.

card stacker In mechanisms that hold and feed punch cards, the exit tray in which the cards are stacked after processing. There may be several of these, with a card sorter determining the destination stacker from holes in the cards. See card hopper.

Cardan, Jerome An Italian mathematician and physician who authored *De subtilitate* (On subtlety) in 1550 which described the accumulated knowledge about amber, and made a definite statement that the properties of lodestone and amber differed in important ways. He further described these differences. See amber, lodestone.

CardBus A 32-bit computer data bus designed for use with PCMCIA cards. The CardBus was designed to succeed the PC Card standard. See Personal Computer Memory Card Interface Association.

cardiode pattern A diagrammatic representation of the directional response of various transmitting and receiving devices: antennas, speakers, etc. It derives its name from the symmetrical, heart-shaped pattern that is typical. See antenna lobe.

caret (Symbol - ^) A character located above the number six on most English language computer keyboards. It is often used in computer-related communications to represent the control (Ctrl) key, and is sometimes used to represent the mathematical 'power of' notation, as in 10^2 (10^2 - ten squared) in character sets that don't support superscripted letters.

careware A variation on the computer software *shareware* concept, except that the license agreement accompanying the software directs users to donate to a specified charity, or charities of their choice, rather than forwarding payment to the author. See shareware.

Carnegie Mellon University This U.S. educational institution is known for many contributions to telecommunications, one of the more familiar is the Andrew File System (AFS). More recently, they have developed a working, campus-wide wireless data communications system which is serving as a model for similar installations elsewhere, and AFS is evolving into a powerful, distributed network protocol with a new name and some interesting new capabilities.

Carpal Tunnel Syndrome CTS. A medical condition, usually of the arms, resulting in inflamed tendons, and abnormal pressure on nerves resulting in pain, tingling, and numbness. Contributing factors may include poor equipment design, incorrect posture, and/or excessive repetitive movement. CTS can result from typing incorrectly or for over-extended periods. Prevention includes frequent rests, good posture (wrist aids are helpful for some people, keyboards set at the right height, and good chairs are essential), cycling through a variety of activities, and of course, good diet and exercise. Consult a medical professional if you suspect this condition.

carriage return Descended from typewriter days where, when the end of a line was reached, the typist would have to fling the roller (carriage) back to the starting point of the line, simultaneously advancing it down the page a designated distance. Thank goodness those days are over.

Carriage return now generally refers to the action of moving the position of a pointer, cursor, or head of a printer back to the beginning of a line, and advancing down the page, although technically, that's a carriage return and a line feed. (Various systems and software programs do it differently.) In ASCII, the carriage return is character 13. In text editors that show control characters, the carriage return is sometimes designated as Ctrl-M. Because of the lack of uniformity among file formats that include carriage returns, especially when transferring a file from one kind of computer to another, there are many utilities to strip or insert the return character, or line feed character, whichever didn't work correctly.

carriage return key A key located on most computer keyboards approximately to the far right of the middle line of letters. It is generally labeled [Return] [Ret] [Enter], or designated with a [Right-angle Arrow] symbol. See carriage return.

carrier 1. A wave of constant amplitude, frequency, and phase, which can be modulated by changing one of these characteristics. See carrier wave, carrier frequency, T1. 2. An entity which can carry an electrical charge through a solid. 3. An information-providing radiant energy from space. The four known categories of carriers are electromagnetic radiation, solid bodies, elementary cosmic rays, and gravitational waves.

carrier, communications A provider of communications circuits. *Common* (usually the local phone company) and *private* carriers are distinguished by degree of regulation, and right to access of service by the public. The term was intended to mean companies with their own transmission facilities, as opposed to companies that lease or buy equipment or services for resale, but the public often uses it more loosely to include all long-distance companies.

carrier, GPS A radio wave with at least one characteristic (such as frequency, phase, amplitude) which, by modulation, can be varied from a known reference value. See Global Positioning Service.

Carrier Access Code CAC. See Access Code.

carrier band A range of frequencies that can be modulated to carry information, such as radio broadcast waves (which, without a carrier wave, would not be able to be transmitted without signal overlap and disruption). See band, carrier, carrier wave, modulation.

carrier bypass A phone service provider direct-connect link to the customer's lines, bypassing the local phone carrier. Some long-distance companies provide services through a carrier bypass in order to provide faster service, or less expensive service by avoiding Carrier Common Line Charges. See Access Charge.

Carrier Common Line Charge CCLC. A charge paid by phone services providers to a primary carrier for using their switched network lines. Typically paid by long-distance providers. See Access Charge, carrier bypass.

carrier detect CD. A signal generated by a modem which operates over phone lines to indicate whether the phone carrier is present, and the line can be dialed. Many modems have an LED to indicate the presence of the carrier signal. The command to the modem for carrier detect is typically `&C1`.

carrier frequency 1. The frequency of a carrier wave intended to be modulated by the wave containing the information. See carrier wave. 2. In Global Positioning Service (GPS), the frequency of the unmodulated fundamental output of a radio transistor. 3. The reciprocal of the period of a periodic carrier. See center frequency.

Carrier Identification Code CIC. A short code to uniquely identify a secondary phone service carrier for routing and billing. It was formerly three digits, but Bellcore informed the Chief of the Common Carrier Bureau in 1989 that four digits were needed to meet increasing demand. By the mid-1990s, Bellcore began assigning four-digit Feature Group D CICs. In 1998, this was further changed to a prefix (e.g., "10+10*xxx*") followed by the number, thus bypassing the subscription carrier (which would use the prefix code "1"). CIC is part of the North American Numbering Plan (NANP). See Access Code, NANP.

carrier select keys Buttons included on a phone (usually a payphone) to provide the caller with a quick way to select a long-distance provider, thus not having to key in extra digits for access codes.

carrier selection Selection by a phone customer of a long-distance provider, usually done at the time of ordering the service, but can be changed at any time. If you select a primary long-distance carrier, you will be able to access the service by dialing "1" plus the number. For alternate long-distance companies, you have to enter additional digits, or Access Codes in order to complete a call. There are many long-distance companies, each offering better features, and lower prices than the next. Evaluate these carefully before you switch services, as there may be inconveniences, hidden charges, or limitations which are not apparent from the advertising literature, and may result in service that is limited, and not necessarily cheaper in the long run. See Access Code, carrier bypass.

carrier sense The capability of a station to continuously monitor other stations to see if they are transmitting. See Carrier Sense Multiple Access.

Carrier Sense Multiple Access CSMA. A listen-and-send protocol used on local area networks (LANs). A system readying to transmit first probes the network to see if the line is clear, in other words, it ensures that another workstation is not transmitting. If the coast is clear, it sends the transmission. This does not guarantee that collisions don't occur; it simply reduces the likelihood of an immediate collision. Various versions of CSMA exist to enhance its efficiency and provide greater collision detection and avoidance.

Carrier Sense Multiple Access with Collision Avoidance CSMA/CA. A version of Carrier Sense Multiple Access which is used in Ethernet systems in association with Media Access Control (MAC) protocols to integrate collision detection with time-division multiplexing (TDM). This aids in improving efficiency in CSMA systems. See Carrier Sense Multiple Access.

Carrier Sense Multiple Access with Collision Detection CSMA/CD. A version of CSMA with added traffic flow control capabilities to detect collisions, in order to increase the efficiency of flow of information on a local area network (LAN). CSMA/CD is not ideal for all implementations. In satellite communications, for example, the transmitting Earth stations cannot engage carrier sensing on the uplink due to its point-to-point nature. See Carrier Sense Multiple Access.

carrier shifting A technique of moving an entire modulated wave sequence positive or negative with respect to its midpoint, without changing the overall shape of the envelope. Carrier shifting is often carried out to mathematically manipulate a wave, or to recreate a wave based on only partial information (e.g., sideband).

carrier shifting fault An undesirable condition in transmitting a modulated carrier wave, in which the envelope, the range of amplitude-modulated signals above and below the midpoint of the waves, is unbalanced.

carrier signal A continuous radiant wave which can be modulated to add information to the wave. Carrier signals are modulated in a variety of ways, the two most familiar being amplitude modulation (AM) and frequency modulation (FM). One of the first researchers to search for a way to add information to a carrier wave was R. Fessenden, an American inventor who devised the hot-wire barretter, and a high-frequency wave generator in 1901. Later, J. Carson studied the mathematical properties of carrier signals and proposed ways of carrying information by manipulating and recreating the signal at the receiving end, thus saving transmissions bandwidth. See Carson, John R.; Fessenden, Reginald Aubrey; modulation; single sideband.

carrier synchronization In radio broadcasting, a carrier wave is used to carry a signal through a process called modulation, wherein information is added to the carrier wave. Various means of sending the modulated wave have been developed, some of which send only the sidebands, some of which send one side of the signal and recreate the other, etc. Consequently, at the receiving end, the receiver has to be designed so it can properly process the type of wave that is being received. In some cases, this involves the creation of a reference carrier which is synchronized with the received signal. See Carson, John R; single sideband.

carrier wave A single-frequency wave which 'carries' the transmission by being modulated by another wave containing the information. A carrier wave provides multiple channels, and a means to reduce signal overlap through multiplexed broadcast waves. See carrier.

carrierless amplitude and phase modulation CAP modulation. A coding technique, based on quadrature amplitude modulation, used in Digital Subscriber Line (DSL) transmissions. See discrete multitone, modulation, pulse amplitude modulation.

CARS See Cable Television Relay Service Station.

Carson, John R. A mathematician and researcher at Bell Laboratories who, in 1915, contributed mathematics fundamentals related to modulation of communications waves which provided a way to recover the whole band from sideband transmissions. In 1915, he and Arnold demonstrated that separate channels could be carried on each of the sidebands of a modulated carrier wave.

Carson's rule A method for calculating the minimum bandwidth of a frequency modulated signal that is needed to transmit the desired communication. A larger number of subcarriers will necessitate a wider Carson's bandwidth.

Carterfone A commercial device developed in the 1960s for acoustically connecting two-way mobile radio communications to a telephone network system. Developed by Thomas Carter, who battled for the right to connect into the public phone network, the system became known through an important judgment by the Federal Communications Commission (FCC). Carter didn't sell a lot of the devices, but AT&T saw the precedence as threatening enough, in terms of its implications for other vendors, to obstruct its use. See Carterfone Decision.

Carterfone Decision A 1968 landmark judgment by the Federal Communications Commission (FCC) in which existing interstate telephone tariffs that prohibited subscribers from attaching their own phone equipment to existing phone lines was struck down. Carter Electronics had sought since 1966 to acoustically interconnect their private mobile radio systems to the national exchange network through a voice-activated system that started the radio transmitter. In pursuing their own right of access, they were paving the way for other companies as well, and in a sense, foreshadowing the divestiture of AT&T.

As a result of the Carterphone Decision, manufacturers other than Western Electric, which had had exclusive arrangements with AT&T, no longer were prevented from using the resources, and the interconnect industry was born. See Carterfone, Hush-a-Phone decision.

Computer data storage cartridges have become popular because they are fast, inexpensive and portable. This Syquest cartridge holds 230 MBytes. They are especially popular for still and animated graphics, sound, large page layout documents, and data archiving.

cartridge A common type of removable data storage medium which works like a floppy diskette, but is a little larger, much faster, and holds substantially larger amounts of data. Cartridges hold between 200 Mbytes and 1 Gbyte of information uncompressed. With the introduction of super-capacity disks, 3.5" floppies that can store more than 100 megabytes of data, the distinction between the storage capacity of cartridges and floppies is less.

cartridge, phonograph In a phonograph

pickup, a small enclosed mechanism into which the stylus is inserted to translate the vibrations from the phonograph record into electrical impulses.

Carty, John J. Chief engineer of AT&T in the early 1900s after serving as the head of Western Electric's cable department. Carty developed the first two-wire telephone circuit and the phantom circuit, through which three conversations could be transmitted at one time over two pairs of wires. Western Electric purchased the rights to Lee de Forest's *Audion* a three-electrode tube in time for Carty to fulfill a promise he made in 1909 to provide a transcontinental telephone service to the west coast by 1914. See AT&T, phantom phone.

CAS 1. Centralized Attendant Service. A centralized group of operators servicing systems which may have a number of branches within a region. 2. See channel associated signaling. 3. See Communications Applications Specification.

cascade *v.* To arrange or pattern into a series or succession of steps or stages, each dependent upon, or derived from, the preceding, often in a falling, or downward hierarchy. Computer menus, file systems, applications windows, and other graphical and logical structures are often developed with a cascade structure. Text editing applications sometimes have telescoping and cascading outline capabilities. Cascading principles and properties are now being studied with relation to quantum effects, and harnessed for commercial applications. See quantum cascade laser.

CASE computer-aided software engineering, computer-assisted software engineering.

case sensitive 1. Computer software data or process in which the case (lower or upper) is considered significant to the meaning of the text. For example, file names on Unix systems are case sensitive, that is, "MyFile.txt" is different from "myfile.txt". Case sensitive file names result in a greater range of descriptive naming possibilities. MS-DOS systems are case insensitive. AmigaOS is partly case-sensitive, forgiving about case when traversing directories in the shell, but allows file names which can be distinguished from one another by case, Unix is case sensitive. 2. In word processing, search and replace routines can usually be configured to be case sensitive or case insensitive, depending upon your needs. 3. On the Web, URLs are case sensitive, but for the convenience of users, many browsers will resolve the cases in a forgiving manner, to load a Web page even if the case is mis-specified. This is not true of all browsers, so it's usually better to type in the case correctly.

case sensitive password For access to secure computer systems, most password fields require an exact match, and thus are case sensitive, in order to provide a greater variety of possible passwords, and thus increased security. This also increases the total number of possible passwords, which is important if it is a multiuser environment with a limit on the number of characters in the password (e.g., eight).

Cassegrain antenna A parabolic antenna arrangement in which the feed is located near the vertex of a concave surface of the main reflector, and a secondary reflector is located near the focal point and aligned to be within the focus of the main reflector. A beam is thus redirected from the feed unit through the secondary reflector to the main reflector to radiate a beam that is parallel to the axis of the main reflector. A Cassegrain feed is one type of arrangement, horn feed is another. Cassegrain antennas require more careful alignment and more parts than a horn feed antenna and thus tend to be used in higher end, more expensive applications. Due to the redirection of the reflection, they stay cooler than horn feed arrangements, and are thus suitable for hotter climates. See antenna, horn feed, parabolic antenna.

cassette tape A portable recording and playing medium consisting of a long narrow magnetic tape wound onto reels protected by a roughly rectangular plastic case. Cassette tapes come in a variety of tape widths and are used for both sound and video.

Some audio tapes are wound onto the reel in a loop, to enable continuous playing but most are manually turned over, or mechanically rewound. Cassettes are commonly used for consumer audio and backup and archiving of computer data. Very small cassette tapes are used in answering machines and on small tape recorders designed for maximum portability (and, in some cases, minimum visibility). Cassette tapes for consumer audio have almost completely replaced reel-to-reel tapes, and are gradually being supplanted by audio CDs and other digital audio technologies. See CD, DAT, leader, reel-to-reel.

Cassini A spacecraft designed to travel out into the solar system and send back information to be evaluated by scientists to teach us more about our planetary environment and the universe. The Cassini spacecraft has many tasks to perform on its way to the planet Saturn. Flyby targets include Venus, Earth, Jupiter, and various asteroids. Cassini's launch window was very small. It had to be launched between 6 Oct. 1997 and 4 Nov. 1997 to execute the various maneuvers and flybys needed to take it on its seven year journey to Saturn. The telecommunications and guidance systems associated with the Cassini mission are some of the most sophisticated to date, and will teach us much about how far and how well we can transmit information to and from the corners of the Galaxy.

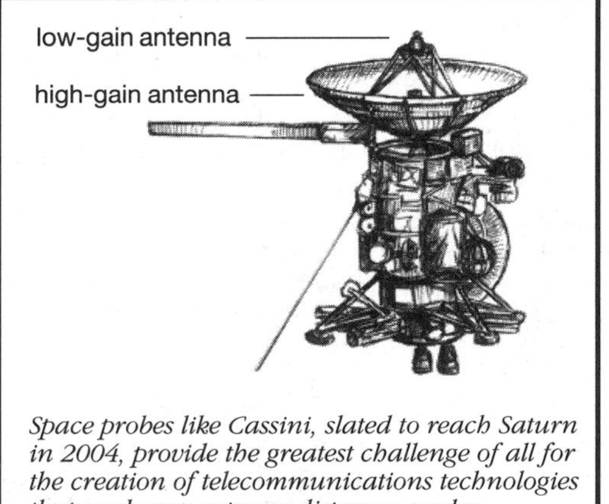

Space probes like Cassini, slated to reach Saturn in 2004, provide the greatest challenge of all for the creation of telecommunications technologies that work over extreme distances under unfamiliar and unpredictable circumstances.

CAST computer-aided software testing, computer-assisted software testing.

castellation An indented pattern or surface of a regular, repeated nature. For example, the battlements on castles are castellated. Castellated protruberances, or thin pads of conductive materials, are often incorporated into the edges of electronic circuit boards to provide contact points for electrical connections.

cat, Cat *abbrev.* category. In cabling, *Category*, or *Cat* followed by a number denotes industry-specific cabling standards. See category of performance.

CAT 1. See computer-aided teaching, computer assisted instruction. 2. Call Accounting Terminal. An AT&T term for a microprocessor-equipped device that records call activity in order to provide automated accounting information.

category, wiring See category of performance.

category of performance Cabling and component standards have been defined to promote and facilitate intercompatibility of products from different vendors. These are widely used in the phone and computer network industries, particularly Cat 5.

The categories of performance focus on the throughput of the transmissions rather than on the specific materials that are used to construct individual cables. They are 'self certifying' in the sense that the vendor is responsible for testing and maintaining quality and manufacturing standards to provide the performance categories detailed in the preceding chart.

Categories of Performance

Category	Transmission rate	Notes
Cat 1	not used	
Cat 2	not used	
Cat 3	up to 16 megahertz	24-gauge wire. Typically used in voice communications and lower end data communications, such as Token-Ring and 10-Mbps Ethernet.
Cat 4	up to 20 megahertz	Digital voice communications and data networks, e.g., Token-Ring.
Cat 5	up to 100 megahertz	24-gauge wire with more stringent fabrication requirements than Cat 3 (e.g., better shielding). Typically used in higher end data communications and high-grade or digital voice applications, particularly high-bandwidth ones like videoconferencing. Examples include FDDI, 100Base-T, 100-Mbps Token-Ring or Ethernet.

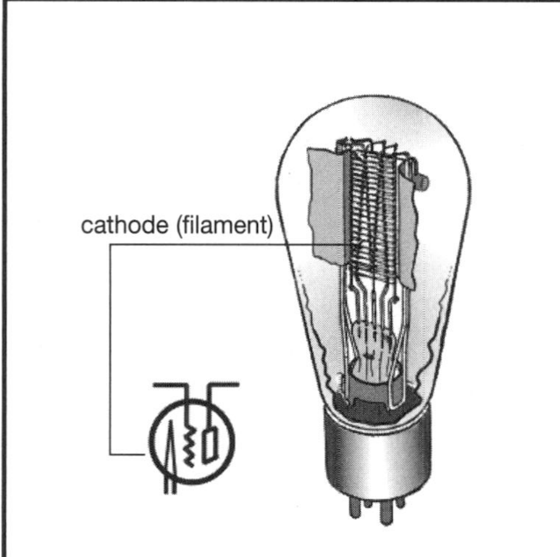

The cathode is one of the three essential elements of an electron tube, emitting the electrons that are attracted to the anode. In this tube, the filament acts as the cathode.

cathode 1. The negative terminal of an electrolytic cell. 2. The positive terminal of current-supplying primary cell. 3. In a moving electron system such as an electron tube, the electron-emitting portion, directed toward a cathode, often a thin metal plate, sometimes passing through a controlling grid. The electron beam-emitting end of a cathode ray tube (CRT). See cathode ray tube.

cathode ray An ionized region, composed of a stream of electrons influenced by an electric field, emanating from a cathode. See cathode ray tube.

cathode ray tube CRT. A display device consisting of a closed tube of glass with the air removed which contains an electron gun at one end and a coated surface at the other. The cathode ray electron beam emanates from the cathode and passes through a magnetic field that controls the beam. By sweeping across the coated inside surface of the glass, a *frame* is formed on a *raster* display and a *vector* is formed on a *vector* display, either of which can be seen through the glass from the outside.

The movement of the electron beam across the display surface excites the phosphors so that they selectively light up (fluoresce) and remain visible for a few moments. The sweep of the beam is very fast so that perceptually humans will 'see' the entire frame as one image rather than as a series of constantly refreshed lines.

Sometimes a grating called a *shadow mask* is inserted between the beam and the coating to further control and focus the beam to provide a crisper display.

Cathode ray tubes can be *short persistence* or *long persistence*. Long persistence means the phosphors remain lit for a longer period of time, and the screen may not have to be refreshed as often to keep an image visible.

The refresh rate on most current monitors is 60 frames per second, a rate that is fast enough to appear stable, and not flickering to the human eye. Color CRTs typically have three beams, red, green, and blue (RGB). The cathode ray tube is fragile and large (regrettably), and is typically encased in a protective console. It is not advisable for laypersons to open up the back of a CRT device, as there is a danger of electric shock. CRTs were being used as computer display devices by the early 1950s but did not become regularly associated with microcomputers until 1976. CRTs are commonly used for monitors (computers, video editing, scopes) and television screens. See flat panel CRT; frame; interlace; screen saver; shadow mask; Zworykin, Vladimir Kosma.

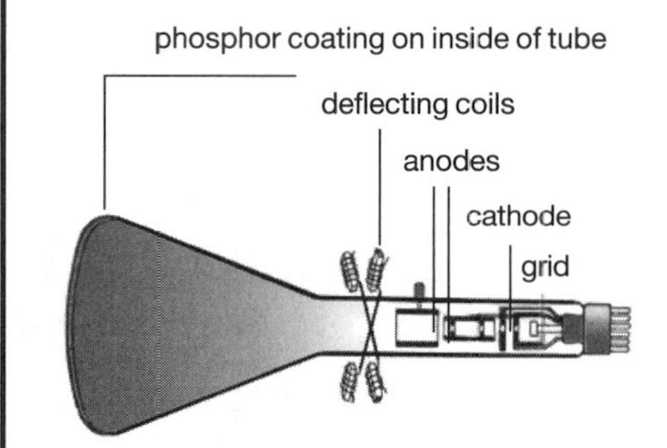

Cathode ray tubes are essential components in many aspects of electronics, particularly as display devices for televisions, radar scopes, oscilloscopes, computer monitors, etc. This very basic diagram shows elements common to an electromagnetic-deflection cathode ray tube.

cathodic protection In many wiring installations, bare wire is used, so corrosion is a significant concern. One of the ways to prevent corrosion and buildup is by running a negative charge through the wire to repel negative ion materials such as chlorine.

CATNIP See Common Architecture for Next Generation Internet Protocol.

Caton, Richard A researcher who described the human brain as an electrical device in 1875. See neural networks.

CATS See Consortium for Audiographics Teleconferencing Standards.

CATV 1. Cable Television. A system which delivers frequency-segmented television programming channels to subscribers through physical cables, usually 75-ohm coaxial cables. The full bandwidth of cable is not typically used, partly due to the extent of the subscriber's service and partly due to the insertion of non-program-carrying 'guard' bands which act as separators to keep individual channel transmissions from interfering with one another. 2. Community Antenna Television.

catwhisker A fine metal thread resembling the arched shape of a cat's whisker, used in early radio wave detecting crystal sets. The catwhisker contacted the crystal on one end and was secured to a metal support on the other end. The contact pressure was often adjustable to accommodate different types of crystals. Some enclosed sets incorporated a catwhisker that was fixed in place at the factory. Greenleaf W. Picard filed a patent for a catwhisker radio detector in 1911. Many variations of the catwhisker came about, with various twists and spirals incorporated into the shape to increase its effectiveness. Prior to the commercialization of crystal sets, the filament was known as a *feeler*. See crystal detector.

CAU controlled access unit.

CAV See Constant Angular Velocity.

cavity A depression, hole, indentation, or pit, which may be of any size. In various media, cavities of precise characteristics are created in the surface so they can later be used to deflect radiant waves or carefully focused laser light beams. The deflections pass into some kind of pickup mechanism (read mechanism) so the encoded information can be recreated and presented.

cavity magnetron An early British innovation in radar systems development during World War II which permitted the use of extremely short waves (microwaves) and improved the quality of the information and images possible through radar systems. See magnetron.

CB radio See citizen's band radio.

CBC See Canadian Broadcasting Corporation.

CBDS Connectionless Broadband Data Service. In ATM, a connectionless service similar to the SMDS (Bellcore) which is defined by the European Telecommunications Standards Institute (ETSI).

CBM See Amiga computer, Commodore Business Machines.

CBR An ATM traffic flow control concept. See constant bit rate.

CBS See Columbia Broadcasting System.

CBSC See Canadian Broadcast Standards Council.

CBT 1. See Canadian Business Telecommunications Alliance. 2. Computer-Based Training. See computer-assisted instruction.

CBTA See Canadian Business Telecommunications Alliance.

CBX Computerized Branch Exchange. A commercial private telecommunications system trademarked by the ROLM Corporation which is their version of a private branch telephone exchange (PBX).

CC 1. Abbreviation for carbon copy. A copy of a document created by placing a sheet of carbon paper and another piece of paper under the original document, and exerting pressure so that the impression on the original document selectively presses the carbon through to the copy. These were common on typewritten copies before photocopy machines and word processors became prevalent. The term is loosely applied to exact duplicates even if they do not use carbon.

Benjamin Franklin developed a printing technique to create copies of documents to cope with the great volume of correspondence he handled on behalf of America in France during the American Revolution.

Carbon copies, or 'carbonless' copies are incorporated into multipart forms even now, especially those used on dot matrix impact printers. The letters CC are transcribed on letters and memos for legal reasons and as a courtesy, to indicate that someone else has received a copy of the document or transmission.

It is recommended that the sender use *EC* for electronic copies instead of CC, because there are instances where the receiver needs to know if the 'carbon' copy is electronic or paper, given the volatile nature and mass distribution properties of electronic copies. 2. See Closed Captioning. 3. Call Control. In wireless communications, circuit communications administration and management. 4. Country Code. One to three digits used to specify an international phone number. 5. connection confirm. A signal in packet-switched networks, a transport protocol data unit.

CCB See Common Carrier Bureau.

CCC 1. clear channel capability. That portion of a data transmissions capacity which is available to users, above and beyond the various control and signaling transmissions. 2. Communications Competition Coalition. A Canadian support and lobbying organization established to encourage Canadian telecommunications competition.

CCD 1. See charge coupled device.

CCIA See Computer and Communications Industry Association.

CCIR See Comité Consultatif International des Radiocommunications.

CCIRN Coordinating Committee for Intercontinental Research Networks. Established by the U.S. Federal Networking Council (FNC) and the European Réseaux Associées pour la Récherche Européenne (RARE), CCIRN promotes international cooperation and sponsors a number of working groups which meet in different parts of the world.

CCIS See Common Channel Interoffice Signaling.

CCITT Comité Consultatif Internationale de Télégraphique et Téléphonique. International Telegraph and Telephone Consultative Committee. An influential United Nations-sponsored international telecommunications standards committee based in Geneva, Switzerland. It changed its name to ITU in 1990. See International Telecommunication Union.

CCITT Study Groups These subgroups, operating under the CCITT (now the ITU) study and make recommendations for specialized areas of telecommunications. See International Telecommunication Union.

CCP Compression Control Protocol.

CCR See current cell rate.

CCS See Common Channel Signaling.

CCS/SS7 Common Channel Signal/Signaling System 7. See Signaling System 7.

CCT 1. Calling Card Table. See unmatched call. 2. See Consultative Committee Telecommunications.

CCTA See Central Computer and Telecommunications Agency.

CCTV Closed Circuit TV. See closed circuit broadcast.

CCU 1. camera control unit. 2. communications control unit.

CD 1. See carrier detect. 2. See compact disc. 3. count down. A concept in broadcasting related to the signaling of the beginning of taping, editing, or live broadcasting.

CD-Audio compact disc audio. A digital sound representation standard which is incorporated into CD-ROMs. Conversion from digital to analog for listening occurs in the computing hardware. Also known as *Redbook Audio*.

CD-I compact disc interactive. An interactive multimedia standard developed by Philips and Sony. CD-I players are designed to accept and play a variety of CD-encoded data and can typically be interconnected with a computer or TV playback system.

CD-Plus compact disc plus. A standard developed by Philips and Sony which enables audio CD players to play multimedia (graphics and sound) discs by skipping over the non-audio segment that is stored on the first track.

CD-R, CD-ROM R compact disk recordable. A format for read/write CD-ROM systems. See compact disc for a fuller description.

CD-ROM compact disc read only memory. A standardized, widely used format for storing digital information on small flat optical platters which are read with laser technology and played on CD players. See compact disc for a fuller description.

CD-ROM drive A computer peripheral which reads, and sometimes writes, digital information to a compact disc. Most consumer CD-ROM drives are read only, although read/write drives are now under $500 and may soon be a consumer item. A CD-ROM drive can be used to run applications, read text files, images (PhotoCD), and audio. Many CD-ROM drives come with software to play audio CDs through a speaker.

CD-ROM XA compact disc read only memory extended architecture. A format developed by Microsoft that allows the interleaving of audio and video, rather than recording them on separate tracks. It requires a player that can understand the format. If played on a regular player, the audio will not be detected and played.

CD-UDF A standardized format for CD-recordable (CD-R) media which enables a variable packet-writing scheme to be used as an incremental approach to the recording of compact discs (CDs). This provides a means for easily recording files on a CD in much the same manner as on a floppy disk.

CD-V compact disc video. A standard for storing video images on compact discs that hasn't really caught on. It is being superseded by

CD-XA which allows the interleaving of video and sound.

CD-WO compact disc write once. A format designed for mastering a CD, which is then used in-house, or in limited quantities, or which is sent to a duplication factory for mass production. Drives that are able to write a master CD were once out of the price range of small companies and consumers, but they have dropped to below $500 and can now be used by software developers, composers, and small record companies to produce masters or small production runs of specialized recordings.

CDA See Communications Decency Act.

CDCS Continuous Dynamic Channel Selection.

CDDI See Copper Distributed Data Interface.

CDE See Common Desktop Environment.

CDF See cutoff decrease factor.

CDLC See Cellular Data Link Control.

CDMA See Code Division Multiple Access.

CDMP Cellular Digital Messaging Protocol.

CDO See community dial office.

CDP 1. Cisco Discovery Protocol. 2. Customized Dial Plan.

CDPD See Cellular Digital Packet Data.

CDPD Forum, Inc. A not-for-profit organization established in 1994 to promote the development and acceptance of Cellular Digital Packet Data (CDPD). It supports vendors which develop and distribute CDPD products and services. http://www.cdpd.org/

CDR See Call Detail Record.

CDT credit allocation. In packet-switched networks, such as OSI, it applies to flow control.

CDV 1. cell delay variation. 2. See Compressed Digital Video.

CDVT See cell delay variation tolerance.

CE Connection endpoint. In ATM, a terminator at one end of a layer connection within a SAP. 2. circuit emulation.

CE Mark A sign that an object has been certified through the overseeing European regulatory body, the European Telecom Directive, and does not require further testing or approval within the individual participating countries. This provides identification of products that conform to certain specified safety, electromagnetic, and interoperability requirements. CE certification is required for all telecommunications terminal equipment (TTE) sold in the European Union. See Underwriters Laboratory, Inc.

CeBIT, Hannover Fair A huge office automation/computer trade show operating since the early 1980s, attracting more than half a million visitors to Germany in the spring of every year. Trade shows in Europe are remarkable. The author has visited professional computer trade shows in which attendees stood 100 people wide for many city blocks waiting to get in the front door. Inside the wide aisles, people will often be shoulder-to-shoulder throughout the concourse, with no open space between facing rows of booths.

CEBus Consumer Electronics Bus. A home automation standard managed by the CEBus Industry Council, and accepted by the Electronics Industry Association (EIA). CEBus specifies a common format for connectionless peer-to-peer communications over standard electrical wiring. CEBus is a two-channel specification, with one channel assigned to realtime control functions, and the other for informational data. It uses a CSMA/CD protocol that includes various error detection and retry functions, end-to-end acknowledgment, and authentication.

CEDAR The Center of Excellence for Document Analysis and Recognition. An organization at the State University of New York at Buffalo which provides a number of interesting services including informational CD-ROMs.

CEI In ATM networking, a connection endpoint identifier.

Celestri A downsized version of the original M-star project, Celestri is a low Earth orbit (LEO), geostationary hybrid satellite system from Motorola. In May 1998, the Celestri expertise and technology was rolled into the Teledesic project, when Motorola Inc. bought in as a major partner. See Teledesic.

cell, ATM In asynchronous transfer mode networking, a unit of transmission consisting of a fixed-size frame comprising a header and a payload. See asynchronous transfer method, cell rate.

cell, battery Minimally, a receptacle containing an electrolyte and two electrodes arranged so the electricity can be generated from the cell by chemical actions. Development of modern cells stems from the experiments of C. A. Volta. Two or more cells can be combined to form a *battery*. See battery, storage cell.

cell, mobile phone In mobile communications, the basic geographic unit of a distributed broadcast system, within which a low-power transmitting station is located. Roughly hexagonal in shape, depending upon terrain. Its size varies with available channels generally increasing as the radius of each cell decreases. Cells are further grouped into clusters. See cluster, cellular phone, mobile phone.

cell delay variation CDV. In ATM networking, a traffic flow buffering and scheduling concept. CDV parameters are associated with constant bit rate (CBR) and variable bit rate (VBR) services and relate quality of service (QoS) information by indicating the probability that a cell may arrive late. See cell rate.

cell delay variation tolerance CDVT. In ATM networks, a traffic flow control mechanism which allows cells to be queued during multiplexing to allow others that are being moved onto the same communications path to be inserted. Cells may also be queued to allow time for the system to insert control cells of one sort or another. See cell rate.

cell error ratio In ATM networks, the ratio of cells in a transmission which are errored, to the total cells in the transmission, over a specified time interval. Preferably as measured on an in-service circuit. See cell loss ratio, cell rate.

cell interarrival variation CIV. In ATM networks, a description of changes in arrival times of cells nearing the receiver. If the cells are carrying information in which the arrival of the cells at the same time is important to the synchronization of the final output, as in constant bit rate (CBR) traffic, then latency and other delays that cause interarrival variation can interfere with the output. For example, in video conferencing, synchronization of images and sound might be affected by cell delays. See cell rate, cell delay variation tolerance, jitter.

cell loss priority field CLP. In ATM networks, a bit field contained in the header cell which indicates the cell discard eligibility of the cell. In congested situations, this cell may be expendable.

cell loss ratio CLR. In ATM networks, cell traffic is handled in many ways in order to maximize throughput, to synchronize arrival times where appropriate, and to minimize delays, latency, jitter, or loss. The cell loss ratio is a negotiated quality of service (QoS) parameter that depends upon the network traffic flow control setups. It is computed as a ratio of lost cells to the number of total cells transmitted, expressed as an order of magnitude. See cell error ratio, cell loss priority field, cell rate, leaky bucket.

cell misinsertion rate CMR. In ATM networking, a traffic flow evaluation parameter which indicates the ratio of cells that are received at the endpoint which were not originally transmitted by the source compared to the total number of cells correctly transmitted.

cell phone See cellular phone.

cell rate In ATM networks, a concept that expresses the flow of basic units of transport used to convey data, signals, and priorities. See the accompanying chart for cell rate concepts. See leaky bucket, cell rate margin.

cell rate margin CRM. In ATM networks, an expression of the difference between the effective bandwidth allocation for the transmission, and the sustainable cell rate allocation in cells per second.

cell relay A type of fast packet switching network architecture using small fixed length packets that can be used for a variety of data types. The cell format is typically 53 octets comprised of 5 bytes of address information, and 48 bytes of informational data. Cell relays can also provide quality of service (QoS) guarantees to a variety of services. See frame relay.

cell relay function In ATM networking, a basic service provided to ATM endstations. See cell relay.

cell relay service CRS. An ATM carrier service.

cell reversal In a battery, a reversal of the polarity of the terminal cells resulting from discharge.

cell site In cellular wireless communications systems, an individual transceiving unit, of which there are many, used to provide roaming capabilities. The cell site serves the local cell and slightly overlaps with adjoining cells to avoid dead spaces between transmissions when a subscriber passes from one to another.

cell site controller Cellular radio operates with numerous cells, each associated with a transceiver. The cell site controller manages the various radio channels within that cell, allocating them when a user moves into range of the cell, and deallocating and reusing available frequencies as the user moves out of range again, or terminates the connection.

cell splitting A means of increasing the call capacity of a cellular system by splitting cells into smaller subunits.

cell switch router CSR. A network routing device that incorporates ATM cell switching in addition to conventional IP datagram forwarding, in order to provide improved service over traditional hop-by-hop datagram forwarding, especially with transmissions that pass through subnetwork boundaries. See RFC 2098.

cell switching In cellular mobile phone systems, this is the overall process of handling calls, monitoring signals as users move in and out of range of the transceivers in the various cells, and allocating and deallocating frequencies as needed to provide seamless service through a series of cells. Cells are designed to overlap somewhat so that there is no gap when switching from one to another, and to com-

Cell Rate Concepts

Abbrev.	Name	Notes
ACR	allowed cell rate	A traffic management parameter dynamically managed by congestion control mechanisms. ACR varies between the minimum cell rate (MCR) and the peak cell rate (PCR).
CCR	current cell rate	Aids in the calculation of ER, and may not be changed by the network elements (NEs). CCR is set by the source to the available cell rate (ACR) when generating a forward RM-cell.
CDF	cutoff decrease factor	Controls the decrease in the allowed cell rate (ACR) associated with the cell rate margin (CRM).
CIV	cell interarrival variation	Changes in arrival times of cells nearing the receiver. If the cells are carrying information which must be synchronized, as in constant bit rate (CBR) traffic, then latency and other delays that cause interarrival variation can interfere with the output.
GCRA	generic cell rate algorithm	A conformance enforcing algorithm which evaluates arriving cells. See leaky bucket.
ICR	initial cell rate	A traffic flow available bit rate (ABR) service parameter. The ICR is the rate at which the source should be sending the data.
MCR	minimum cell rate	Available bit rate (ABR) service traffic descriptor. The MCR is the transmission rate in cells per second at which the source may always send.
PCR	peak cell rate	The PCR is the transmission rate in cells per second which may never be exceeded, which characterizes the constant bit rate (CBR).
RDF	rate decrease factor	An available bit rate (ABR) flow control service parameter which controls the decrease in the transmission rate of cells when it is needed. See cell rate.
SCR	sustainable cell rate	The upper measure of a computed average rate of cell transmission over time.
UBR	unspecified bit rate	An unguaranteed service type in which the network makes a best efforts attempt to meet bandwidth requirements.
VBR	variable bit rate	The type of irregular traffic generated by most non-voice media. Guaranteed sufficient bandwidth and QoS.

pensate for the fact that the signal is weakest on the periphery of the transmitting area. The sophisticated moment-by-moment monitoring and orchestrating of this process is handled by monitoring systems and cell-switching software.

cell transfer In cellular mobile phone systems, the logistics of keeping an ongoing connection at acceptable volume and quality levels when switching the user from the transceiver in one cell to the transceiver of the cell that is being entered. This involves allocating a frequency channel in the entered cell, and deallocating and reassigning, if needed, the frequency channel of the excited cell.

cell transfer delay CTD. In ATM networking, the time elapsed between a cell exit event at the first point of measurement, and the corresponding cell entry event at the second point of measurement for a particular connection. The cell transfer delay between the two points of measurement is the sum of the total inter-ATM node transmission delay and the total ATM node processing delay. See cell rate.

Cello A graphical Web browser created at the Cornell Legal Information Institute.

cellphone See cellular phone.

cells in flight CIF. In ATM networking, a descriptive phrase for a traffic service parameter, the available bit rate (ABR). CIF is a cell number limit negotiated between the receiving network and the source of the cells during the idle startup period, prior to the first RM-cell returns. See cell rate.

Cells in Frames CIF. The name given to a number of mechanisms for carrying ATM network traffic across a media segment and network interface card. CIF was developed by the Cells in Frames Alliance, a diverse group of professionals and commercial vendors. This group released the first CIF specification in 1996 for carrying ATM over Ethernet, Token-Ring, and 802.3 networks. As ATM is not tied to a particular physical layer, CIF has been defined as a pseudo-physical layer for carrying ATM traffic.

CIF provides a frame-oriented means of using ATM layer protocols transparently over a variety of existing local area network (LAN) framing protocols.

The Cells in Frames project has been funded in part by the National Science Foundation (NSF). More information is available through Cornell University. http://cif.cornell.edu/

Cells in Frames Alliance An open membership organization which promotes the affordable deployment of ATM technologies, and better use of existing local area network (LAN) infrastructures, while providing applications developers direct control over Quality of Service (QoS) networking issues.

Cellular Data Link Control CDLC. An open data communications protocol suitable for wireless communications. In cellular phone systems, it provides a means to interconnect a data terminal and a cellular phone. The protocol includes various error detecting, correcting, and data interleaving features that make it suitable for wireless transmissions.

Cellular Digital Packet Data CDPD. An open standard originally developed and released by vendors in 1993, which was further defined by the CDPD Forum. Then, it was passed on for maintenance and enhancement to the TIA in 1996. CDPD is suitable for packet data services for mobile communications, conceived as an extension to landline services for mobile users. The originators of CDPD wanted to develop a means to use existing cellular networks for wireless data, in other words, to overlay newer services on the existing infrastructure.

CDPD is a packet-based system, defined to operate over AMPS analog voice systems. Transmission speeds of 19.2 Kbps are possible over the traditional infrastructure. Internet Protocol (IP) is typically used with TCP. Packets are routed into and out of the CDPD network through an Intermediate System (IS), which acts as a relay. Specific routing functions and monitoring of mobility are handled by the Mobile Data Intermediate System (MD-IS). The MD-IS keeps track of locations through a Home Domain Directory (HDD) database.

CDPD is not intended to specify the various types of service which can be carried over the system, but rather describes the architectural structure of the service, and its integration with the existing infrastructure. Specific value-added services are up to individual vendors. See A interface, E interface, I interface.

Cellular Digital Packet Data Forum CDPDF. An association formed in the early 1990s to develop and promote a standard for cellular digital packet communications. This is now called the Wireless Data Forum. See Cellular Digital Packet Data, Wireless Data Forum.

cellular modem A modem integrated with cellular phone technology to provide ease of access to telecommunications services through mobile, wireless transmission.

Cellular modems are frequently used with laptop computers, often available in the form of PCMCIA cards, which are small, slender peripheral components that fit easily inside a portable device. Dialing and data transfer is controlled by the software. The communication may originate with a wireless system and hook into a landline system. Cellular modems are favored by traveling professionals, such as journalists and scientific researchers, who may be relaying information to a central facility on a regular basis. See cellular phone.

Lightweight, handheld cellular phones are now a popular way to communicate without wires. Cellular phones like this one from Motorola, typically work with battery packs that are attached to the back of the device. Some makers have optional adapters to plug battery chargers into a 12-volt car lighter power source.

cellular phone, cell phone An analog, digital, or hybrid mobile communications system, employing hardware interfaces resembling traditional phone handsets, linked through a gridlike network of low-power wireless transmitters each servicing a geographic area, with a small amount of overlap with adjacent cells. As the user travels through these areas, or *cells*, the transmission is *handed off* to provide continuous service, and frees up previous channels no longer required. Carrying on communications in various cells while on the move is called 'roaming.' The power levels in each cell are optimized according to demand and subscriber density.

The cell concept was introduced in the late 1970s to improve on older single-transmitter mobile systems. The commercial cellular phone system in the U.S. was available by the early 1980s. More recently, digital cellular systems have been devised to increase capacity and call security (through encryption). It is estimated that there are now more than 150 million cellular phone users worldwide, and ambitious programs for launching communications satellites to handle seamless global cellular communications are in progress. See cell, cluster, cellular modem, mobile phone, PCS, AMPS, TDMA.

cellular phone security Cellular phone technology is, for the most part, not transmitted in a secure manner. In analog systems, the signal goes over airwaves which can be tapped by a radio scanner operating in the same frequencies. Although encryption is starting to be incorporated into digital cellular communications, it is not yet universal. Since cell phones are used for many sensitive business and law enforcement communications, this is of some concern to cell phone users. A number of security systems have been developed by cell phone providers, including cell phone attachment peripherals that provide fully digital voice encryption capabilities.

cellular radio Very similar to cellular phone, and in fact, a precursor to cellular phone, in which a region is organized into cells, each with a transceiving unit which overlaps with the coverage of adjacent transceiving units. Cellular radio provides for reuse of frequencies and greater capacity than non-cellular mobile radio services, and allows users to purchase cheaper equipment, since the power requirements are not as high. See cellular phone.

cellular security devices CSD. Add-on peripherals, or all-in-one cell phone sets, which incorporate various security means, primarily digital encryption, that will sound like random noise to anyone attempting to monitor the communication. See cellular phone security.

Cellular Telecommunications Industry Association CTIA. http://www.wow-com.com/

celluloid A durable, though flammable, plastic material composed mainly of cellulose nitrate and camphor. In 1885, Hannibal Goodwin developed celluloid film which came to be so widely used in the motion picture industry, that the films were known for many years as *celluloids*. In spite of the greater resolution and durability of images captured on film, digital imagery, due to its flexibility, is beginning to supersede the use of film in consumer markets. Estimates vary as to how high the resolution of a digital image needs to be to match or exceed the film quality, but film-makers generally accept that resolutions of at least about 4,000 dpi or 4,0000 line screens for each frame are needed when displayed at cinematic quality. While not all viewers notice the difference between 4,000 dpi and 2,000 dpi images, to artists and film professionals the difference is evident and dramatic. See digital video.

CELP code-excited linear predictive. An analog-to-digital voice encoding scheme which can be used to send voice conversations through digital networks like local area networks (LANs) and the Internet. This makes long-distance carriers very nervous, as existing phone lines are typically used for portions of the transmission, without the requirement of paying existing long-distance phone charges.

Celsius scale A scale developed in the 1700s to describe temperature with the boiling point of water referenced as zero and freezing point as 100 degrees. A year later, the reference points were reversed by Christin, and the scale today continues to use zero as the freezing point of water and 100 degrees as the boiling point. Also called centigrade scale.

CEN/CENELEC Comité Européen de Normalisation (European Committee for Standardization)/Comité Européen e Normalisation Electrotechnique (European Committee for Electronic Standardization) CEN is one of three organizations responsible for overseeing voluntary compliance with standards in the European Union, while CENELEC develops technology and standards. CEN cooperates with ISO. See European Telecommunications Standards Institute.

centigrade scale See Celsius scale.

Central One of the many colloquial names given to the early female telephone operators. Others include Hello Girls, Voice with the Smile, and Call Girls. See telephone history.

Central Computer and Telecommuncations Agency CCTA. A United Kingdom government agency promoting good practices in information technology and telecommunications in the public sector.

central office CO. 1. Headquarters or main service- or administration-providing center. 2. In telephony, the switching station from which subscriber loops are established. In Europe, *public exchange* refers to the switching center. The purpose of the office or central exchange is to provide a connecting point through which a subscriber can establish a connection to any other public subscriber on the circuit, or to a trunk line leading to other central offices. The office secondarily provides power requirements, signaling and control devices, and subscriber line services. The lines are further equipped with protective devices, fuses, coils, etc., to guard against unusually high voltages.

C

central office battery The power source that provides direct current to the connected lines for phone conversations. Historically, this was accomplished with a 24 or 48 volt *talking battery*. Later, the 48 volts was transformed from an alternating current at the central office. See talking battery.

central processing unit CPU. A circuit on a single chip which provides the basic, essential logic for performing general purpose computing computations and decisions. The most celebrated early 'computer-on-a-chip' was the Intel 8008, released in the early 1970s, which became the basis for a line of early microcomputers and the inspiration for a whole new industry. CPUs may be set up in parallel, or may serve as the central processor for an individual system. Desktop computers are typically based around one CPU, while some workstations, especially those for scientific computations, or high-end graphics, have multiple CPUs.

Just as computers have different ways to organize the various circuits and processors, the circuitry within a chip can be organized in many different ways. The architecture of the chip affects other aspects of the system, such as memory management, bus addressing, and programming procedures, particularly machine language and assembly programming.

The term central processing unit was coined somewhere around the late 1960s, but did not become prevalent until the late 1970s. During the 1970s, the CPU was often called a main processing unit (MPU). See complex instruction set computing, reduced instruction set computing.

Centre National d'Etudes de Télécommunication CNET. The French organization that approves telecommunications products for the French market.

Centrex (from *Cent*ral *ex*change) A commercial telephone service provided by local telephone exchanges in which the subscriber-specific switch is physically located either: on the premises of the phone exchange (CO); or at the customer's premises (CU). Used primarily by businesses, as a lower cost alternative to a private branch exchange (PBX), Centrex systems have a number of extra calling features (Caller ID, Call Conferencing, etc.), with a wider selection of options than are available to residential subscribers.

An on-premises Centrex system is similar to a PBX system, in that it is located on the customer's premises, but a PBX is owned by the customer, whereas the Centrex system is leased from the phone company. Each choice has pros and cons in terms of upgrades and maintenance. A Centrex central office system can also be combined with a PBX system, but care should be taken not to order redundant options for the service, since many can be provided by either. A Centrex system (or PBX) can be combined with Automatic Call Distribution (ACD), to provide self-contained, sophisticated automated business telephone services, like those used by many mail order retailers.

Centronics A printer manufacturer that was well known for establishing a parallel data transmission standard for computer printers, especially dot matrix printers that were popular in the 1980s.

Centronics parallel data standard A data transmission standard established by Centronics and accepted de facto by much of the printing industry in the 1980s, particularly for cabling dot matrix and some daisy wheel printers to parallel ports on desktop computers. The cable is usually a flat or pinned D connector. This parallel standard is generally faster than similar serial cable attachments because data can be carried over eight wires at once rather than just one. Attachments vary, but most systems employ a pin connector on the computer side and a flat connector on the printer side.

CEO Chief Executive Officer. Typically the highest member of the corporate hierarchy, in charge of overall business direction, goals, and strategies.

CEPT See Conférence Européenne des Administrations des Postes et Télécommunications.

CEPT1, CEPT2, CEPT3 See E1, E2, E3.

CER See cell error ratio.

CERB Centralized Emergency Reporting Bureau. A Canadian reporting organization to safeguard the public.

Cerf, Vinton Vinton Cerf is credited with some of the early ideas for network gateway architecture. He has held various engineering, programming, and teaching positions in businesses and educational institutions. In the early 1970s, he researched networking and developed TCP/IP protocols under a DARPA research grant, and in 1974, co-authored "A Protocol for Packet Network Internetworking" which describes Transmission Control Protocol (TCP). During this same period, he became a founding chairman of the International Network Working Group (INWG). In 1977, Cerf co-demonstrated a gateway system that could interconnect packet radio with the ARPANET. In 1978, he co-developed a plan to separate TCP's routing functions into a separate protocol called the Internet Protocol (IP). See Kahn, Bob.

CERFnet See California Education and Research Federation Network.

CERN Centre European des Recherche Nucleaire. The European Laboratory for Particle Physics Research in Geneva, Switzerland. http://www.cern.ch/

CERT See Computer Emergency Response Team.

certificates Authentication entities used in a variety of cryptography transmissions designed to safeguard privacy and authenticity in electronic messaging and transactions, particularly contracts, payments, etc.

CEST Centre for the Exploitation of Science and Technology. UK industry-funded organization formed in 1988.

CEV controlled environmental vault. See cable vault.

CFB Call Forward Busy.

CFDA Call Forward Don't Answer.

CFF See critical fusion frequency.

CFGDA Call Forward Group Don't Answer.

CFP Channel Frame Processor.

CFR Confirmation to Receive. A notification in networks that a frame can be forwarded.

CFUC Call Forwarding UnConditional.

CFV See Call for Votes.

CFW Call Forward.

CGA 1. Carrier Group Alarm. An out-of-frame alarm signal generated by a channel bank, which may be followed by trunk rerouting and error control. 2. See Color Graphics Adapter.

CGI See Common Gateway Interface.

CGM Computer Graphics Metafile. A standardized graphics interchange format.

CGSA Cellular Geographic Service Area. The area within which a cellular company provides services. CGSAs may include multiple counties and may even cross state lines.

CGSA Restriction A subscriber option to restrict calls outside a local Cellular Geographic Service Area.

chad The small punched out pieces from encoded punch cards or paper tape. Chad was collected in *chad boxes* which had to be periodically emptied. Originally discarded, some bright marketing person started packaging them as *confetti* for parties and celebrations.

chad tape Paper tape designed for the encoding of messages or algorithms. Originally designed for devices such as telegraphs and stock ticker machines, the tapes were narrow and very long. The punched out pieces are called *chad*.

chadless tape A tape designed for encoded messages or algorithms which is only partially perforated in order to avoid large quantities of waste in the form of punched out *chad*. By only 3/4-perforating the tape, there is sufficient contrast in texture for the holes to be mechanically read, while avoiding the chad waste buildup. Because the holes are not consistently open and empty, optical readers are not suitable for reading messages from chadless tape.

Chadwick, James (1891-1974) An English physicist who is credited with discovering the neutron in 1932 at Cambridge, England. Chadwick actively engaged in radiation experiments, some of which stemmed from the work of Ernest Rutherford.

chaff, window Materials such as metal strips or fine wires which are highly reflective to radar waves. Chaff may be strung or shot into the air, for the purpose of scattering, deflecting radar waves.

chaining A common modular software execution technique in which a process is handed off to another program, or launches other program modules on an as-needed basis. By having only the necessary components memory-resident, and by off-loading modules which are no longer needed, complex programs can be managed with limited resources.

Challenge-Handshake Authentication Protocol CHAP. An Internet standards-track protocol descended from a variety of semi-secure network implementations in the mid-1970s and 1980s, which provides a method for key authentication using Point-to-Point Protocol (PPP). It employs a three-way handshake upon link establishment to verify the identity of the peer and may repeat it at intervals. After link establishment, the authenticator *challenges* the peer, which responds with a hash value. If there is a match, authentication is acknowledged; if not, the connection should be terminated. The key is known only to the peer and the authenticator, and is not transmitted.

This protocol is suitable for small or medium connections with an established trust relationship. Large tables of keys are not practical, and the information must be available in plaintext form, with a secure central repository to store it. See RFC 1994.

channel 1. In its most general sense, a path along which signals can be transmitted. 2. In radio broadcast, the electromagnetic frequency spectrum extending roughly from VLF to UHF (further above that is the microwave frequency). 3. A portion of the spectrum assigned for the use of a specific carrier, e.g., the FM band is divided into channels of a specified kilohertz range. 4. In a GPS receiver, the radio frequency, circuitry, and software needed to tune the signal from a satellite. 5. In audio, a sound path (e.g., stereo requires at least two sound paths). 6. In telephony, voice-grade transmission within specified frequencies and bandwidth. See circuit. 7. In a frame relay, the user access channel across which the data travels. If a physical line is *unchannelized*, then the entire line is considered a channel. If the line is *channelized*, the channel is any one of a number of time slots. If the line is *fractional*, the channel is a grouping of consecutively or nonconsecutively assigned time slots.

channel aggregation Inverse multiplexing. Bonding multiple lines (e.g., phone lines) in such a way that transmission speeds are faster. This technology is sometimes incorporated into modems to speed upload and download times.

channel associated signaling CAS. In ATM, a form of circuit state signaling, in which the state for that specific circuit is indicated by one or more bits of signaling status, which are repetitively sent. In T1 voice applications, inband signaling information, which is carried in the data stream with each channel, rather than as a separate out-of-band transmission.

channel bank A network interface device which provides multiplexing and flow control functions on multiple transmission channels, such as voice and data, which are being brought into a single electrical data stream. A channel bank is not intended to provide switching functions. A channel bank is commonly used as an interface between multiple lines coming out of an analog systems, such as a private branch exchange, which are brought together and multiplexed onto a larger bandwidth digital transmissions medium such as a DS-1 stream associated with a T1 line. See multiplexing.

channel bonding A technique for increasing transmission speed by means of aggregating multiple lines. It is a way of getting a little more performance from existing lines without investing in higher end technologies, like a T1 line. Channel bonding for modem communications is usually only cost effective for two or three lines; above that number, ISDN may be a better option.

channel hopping 1. Changing channels on a radio phone, usually to find one with clearer reception. 2. Changing channels constantly on TV, usually with a remote control, and often during commercials, to the consternation of other viewers who don't have control of the remote. 3. On IRC, switching chat channels and barging in, or listening in, on many conversations in a short period of time.

channel separation In broadcasting, the space that is left between designated adjacent channels. Since signals may have a tendency to interfere with one another, leaving a gap increases the likelihood of clean signals on either side of the gap. Commonly used in radio transmissions, especially FM broadcasts where the width of the separator may be greater than the width of the channel through which the FM transmission is passing.

Channel Service Unit CSU. Typically integrated with a Data Service Unit as a CSU/DSU. Often the first device within a digitally networked facility which routes information, and may also protect equipment within the facility from electrical interference and damage. It may further regenerate a signal to maintain signal strength and integrity, or convert the signal from one format to another. Commonly used between incoming T1 or E1 lines and internal channel banks, at a central office offering end-to-end digital line services to subscribers, usually through leased lines. See Data Service Unit.

channel surfing Flipping through stations on a television looking for something worth watching, or avoiding the commercials. On a TV, it is usually done with a remote, from a comfortable position on the couch. Web surfing on the Internet is similar, and may eventually evolve into channel surfing, as more and more companies are starting to broadcast multimedia on their sites.

channelize To increase the capacity of a wide- or broadband transmission medium by subdividing it into smaller channels, often with a small 'gap' between channels to reduce interference.

CHAP See Challenge Handshake Authentication Protocol.

Chappé, Claude (1763-1805) A French scientist who, in conjunction with early experiments with his two brothers, advanced communications in the late 1792 by adapting semaphores (visually encoded messages) to a pair of arms mounted on a system of towers. This allowed a message to be transmitted from Paris to Rhine, a distance of about 150 miles, in minutes. There were disadvantages to this system including visibility, the need for many towers, and the 24-hour monitoring that was necessary in order to change the arms when a new message came through. Nonetheless, it was an improvement over previous systems, and grew in France to over 500 signaling stations. The first formal *telegramme* is said to have been sent on 15 August 1794.

Chapuis, Robert J. A researcher and author who has enriched our understanding of telephone history. Chapuis has brought to light the inventions of individuals who might otherwise remain uncredited because they were not directly associated with the larger more successful phone service companies. See Lorimer telephone.

character code The specific coding system, or protocol, used to formulate characters for transmission. Examples of character codes include Baudot code, Morse code, ASCII, UNICODE, and EBCDIC.

character generator A process, device, or chip for storing and rendering a character on a printed page, or display output device such as a computer monitor. Character generators can be very simple, from those on 9-pin dot matrix printers, which use a simple internal system, to Display PostScript systems which generate sophisticated character shapes for a wide variety of fonts and display them as raster graphics at many resolutions.

character generator history The evolution of character generators on microcomputers is quite interesting. For example, on a 1978 Radio Shack TRS-80 Model I computer, the character set was limited to the shapes programmed into a chip that was imbedded on the motherboard in the keyboard. The first chip contained only upper case letters, and they were quite unattractive. Third-party vendors developed improved character sets with upper and lower case, so a hobbyist could open the keyboard and swap the chips.

On IBM and IBM-compatible Microsoft DOS-based systems in the early 1980s, the character set was rendered as text, which was fast, but did not give much flexibility in terms of fonts or foreign characters. The keymapping was originally limited to all the upper bank or all the lower bank, without individual variations.

On the Apple Macintosh computers, the characters could be rendered as raster fonts on the screen, with the character shapes optimized for small screen sizes, but mapped to PostScript vector fonts when printed on a laser printer with PostScript capabilities. This made WYSIWYG desktop publishing possible.

On the 1985 Amiga, characters were displayed as *graphics characters* providing much more flexibility, with only a very slight loss of speed of the display. This provided keymapping options, and made it easy to support a wide variety of foreign character sets including Cyrillic.

Later still, the 1988 NeXTStep operating system incorporated Display PostScript, providing almost unlimited flexibility in combining text and images, and very high quality screen characters chosen from professional quality PostScript fonts. This means of rendering characters also made it possible to print high quality images on an inexpensive printer, because the rasterizing process was done on the computer, rather than in the printer.

About this time, TrueType fonts were being developed to provide better quality fonts for users of Microsoft Windows-based systems.

There is a distinction between computer fonts which are designed to be typefaces, and PostScript fonts, because a PostScript font is only one possible manifestation of shapes in the PostScript page description language, which is a graphics programming language. Thus, it is theoretically possible to animate, or individually vary the letters within a PostScript font, and to spiral, shear, and otherwise manipulate the characters within the context of the language in a way that is not possible with character sets that are self-contained and limited in their purpose. The many capabilities of PostScript-generated characters have not yet been fully appreciated and exploited.

To summarize, early computer fonts were hardware generated by a character-generating chip, and many systems only supported upper case letters, unless a third-party enhancement chip was installed. In the early 1980s, some of the character sets began to be created in software, and were displayed on the screen in graphics mode. This system was far more flexible, as it supported multiple character sets in many sizes. By the early 1990s, almost all computer displays supported graphics text rather than hardware-dependent text.

character printer A printer which images and impresses one character at a time, as a typewriter, or consumer level impact printer. There are also *line printers*, which stamp an entire line of dots or text on the page at one time, and *page printers*, which image the entire page and print it at remarkably high speeds (and usually with a very high noise level requiring earplugs).

In a sense a laser printer is a type of *page printer*, in that the entire page is imaged before the page ejects, but since some laser printers image part of the page while another part is being fused with toner, they are not all page printers in the strictest sense of the phrase.

character set 1. All the symbolic character representations (letters, numbers, punctuation, diacritical marks, etc.) available to a system, such as a computer, printer, or typewriter. Character sets vary widely from device to device and culture to culture, and interpretation between different character representations is one of the big challenges in global computing. 2. In network programming, *character set* is associated with mapping tables which are used to convert octets to characters, related to *character encoding*. These are often identified by tokens. See ASCII, RFC 1345, RFC 1521.

characters per second In a telegraph key transmission in Morse code, the number of characters a human operator can send, or read, in a second. In a printer, the number of characters that are printed in a second. In a data transmission, the number of characters that are sent through a telephone line, through a modem, or through a network. These are sometimes expressed instead in terms of bits per second, since a character may take one byte in some systems, and two bytes in others.

charge coupled device CCD. A sensing system that uses lines or arrays of light-sensitive photo diode elements. As light comes in contact with various CCDs, the intensity of the light is registered and translated into digital information. Resolution is related to the type, placement, and quantity of these elements. CCDs are used in many photographic and scanning devices, digital cameras, camcorders, computer scanners, robot vision systems, etc.

chase In mechanical printing, an open steel or wood frame used to hold the various blocks of type and engravings that make up a page layout. *Furniture* (nonprinting spacer blocks), and *quoins* (nonprinting pressure blocks) employ friction and pressure to hold the various printing elements in place. The contents extend slightly above the surface of the chase, and are inked for printing on a traditional press.

chase trigger In digital recording techniques, there are some legacy analog time code technologies that cause problems when applied to a digital environment. Digital audio, for example, is recorded according to a system clock. If the frequencies vary, there may be audio artifacts that cause unwanted noise in the recording. A chase trigger is a means of synchronizing time code by starting the sound segment when a particular time trigger occurs. Once it is triggered, it follows its own clock speed irrespective of whether the underlying recording that initiated the trigger has remained steady. If the underlying signal is stable, it's not a problem, but in some recording environments, stability is hard to guarantee. See house sync, reference clock, time code.

chat In online computer telecommunications, chat refers to private or public message areas in which participants type messages to one another in a somewhat real-time manner. Internet Relay Chat (IRC) is the largest chat forum on the Internet, although some of the large service providers have their own subscriber chat channels. Chat lines can be set up to be keyword protected, to offer private or group conference conversations. Chats with celebrities are sometimes moderated to keep the conversation to a level in which the comments are not too rapid or overwhelming. Anyone can open a public chat channel on IRC. To create a new chat, or join a current chat, you enter the IRC server and type `#join mychatchannel` (e.g., `#join gardening`), or select join from the menu, if using a menu-based IRC software client. It is not acceptable on IRC to make off-topic comments, or to denigrate other participants or their viewpoints. See Internet Relay Chat, Netiquette, Netizen.

chatter 1. In circuits, a repetitive, undesirable, fast clicking or opening and closing of a circuit. Power fluctuations can sometimes cause chatter. Unchecked, it can lead to damage of equipment, and interference with communications. 2. In servos, styluses, and other moving control mechanisms, quick, short oscillations in a direction other than the desired direction (often perpendicular to the desired direction) cause by friction, or power fluctuations, or improper calibration, or improper mounting (too loose or too tight).

cheapernet *jargon* Cheaper, maybe-not-as-fast, affordable networks, such as Ethernet running over thin coaxial cable.

check bit A bit, or a group of bits, used for a variety of error housekeeping functions. A single bit is often used for parity checking, whereas 7 or 16 check bits may be used for various cache functions. See checksum, parity.

checksum A computed value commonly used for assessing data integrity and detecting errors or anomalies. Checksums are used in file systems, encryption systems, and packet transmission protocols. In networks, checksums can help to determine, with a reasonable degree of confidence, whether a packet has arrived at its destination unchanged. See check bit.

cherry picker *colloq.* An industrial crane arranged with a one or two person 'bucket' to raise workers to levels that cannot easily be reached by other means. These are used to access fruit trees, windows, utility poles, and other high places. See lineman.

cherry picking *colloq.* Selecting only the calls most likely to 'bear fruit' and assigning them appropriately. In other words, when a call comes in, over a phone, or over a modem, those callers which are in some way identified or prescreened to be the most likely to benefit the callee, usually by purchasing products or services, or by investing in the company, are processed as a priority.

Cherry picking is also done with reader service inquiries. When magazine readers send in reader service cards, these inquiries are forwarded to the appropriate vendors. The vendor sheets sometimes include statistics gathered by the magazine, such as how many boxes the inquirer checked, and their job titles. Thus, the vendor can select and respond first to those most likely to produce revenue for the company.

Chicago The development code name for Microsoft's Windows 95. It is common for developers to assign code names to products for many reasons. Sometimes it helps protect information about the nature of the product, it assists developers in communicating about the product, and it serves as a holdover until marketing personnel come up with a good name and take care of any preliminary trademark searches, and legal registrations that may be associated with the name.

Chiclet A tradename for a type of gum. The name is derived from 'chicle' which is a latex gum substance from a sapodilla plant. In the early days of microcomputers, International Business Machines (IBM) produced a flat keyboard that came to be known as a 'chiclet' keyboard. It didn't last long for desktop computers. Although it was practical in terms of being compact and inexpensive to manufacture, the flat surface was not easy to use, the response was not comfortable, and users complained. While IBM doesn't produce them for desktop systems any more, similar keyboards still have specialized applications, in kiosks, for example.

Chief Information Officer CIO. Usually a person connected with Management Information Services (MIS) who assigns, evaluates, specifies, recommends, or otherwise manages information resources within a company, which broadly includes computer and other telecommunications systems.

child process In computer programming, a process which is spawned, or spun off from a parent process, that is, the child process is subordinate to the parent process in terms of the sequence of its creation, but which may be designed in such a way as to outlive or outpower the parent process.

chime A tone on a telephone, computer, or alarm system which sounds like a pleasant musical tone rather than a buzz or a beep. It's straightforward to generate many types of sounds and chimes on a computer system through sampling and storage as sound files.

chimney effect One type of cooling system which is based on the natural tendency of heat to rise, it consists of ventilation slots or holes in the top and bottom surfaces of a cabinet or component in order to facilitate natural air circulation.

chip 1. In computers, most commonly a chip is a component integrated circuit which is attached to the circuitry by conducting legs that can be inserted into a PC board, or onto another chip. Chips are generally not soldered. Chips come in many shapes and sizes, but most of those used in consumer appliances and desktop computers range up to about 2" or 3" in length. Sometimes more than one chip is closely coupled or laminated together. Larger, heat-producing chips may be coupled with a fan, or require that a fan or heat sink be installed nearby with adequate air circulation. Chips can be general purpose, as with central processing units, or very specialized, as with device driving chips optimized for a particular device. See semiconductor, silicon, large scale integration. 2. The transmission duration for a bit or single symbol of a PN code. 3. In spread spectrum wireless communications, a chip is a redundant data bit inserted at the origin of a transmission, which subsequently is removed when the transmission is deciphered at the receiving end.

C

Chireix antenna, Chireix-Mesny antenna A bidirectionally resonant, cascading array, wire or tubing antenna consisting of series-fed square loops with sides the length of a half-wave or quarter-wave. Used for very high frequencies, typically (VHF) and higher.

choking In printing, a technique for assuring that adjacent inks meet, or slightly overlap so as not to cause an undesirable, paper-colored gap between the inks if the registration of the press is slightly off. When desktop publishing page layouts that are intended for printing at high resolutions, there are usually settings provided in the print dialogs to set choking. Correct use of choking and trapping can dramatically effect the quality of the final product. Typically, large, or outer lighter edges are choked inward (slightly overlapped) around darker print elements. See trapping.

Chooser A Macintosh desk accessory which brings up an icon menu of various available devices, file servers, and printer drivers. The user can select the desired device, and may be prompted for additional options. For example, if the AppleTalk file server is used to provide access to another machine, the user may be prompted to specify which drives are to be mounted.

chopper A device for quickly opening and closing an electrical circuit or beam of light at regular intervals. A chopper may be used in this way to interrupt a signal to permit amplification, or to demodulate a circuit; or to interrupt a continuous stream of particles. The process is called chopping.

Christie, Samuel Hunter An English mathematician and educator who investigated the electrical properties of various metals, and created a balanced bridge resistance-measuring circuit in 1833 which was used and described by Charles Wheatstone, and became known as the *Wheatstone bridge*.

chroma A color as characterized by its hue and saturation, without reference to the relative brightness of the color. Black, white, and shades of gray are not considered to be chroma.

chroma key In cinematography, a technique for eliminating a background such that the foreground can be superimposed on a different background. Everyone who has watched television shows, especially action shows, has seen chroma key effects. High speed chases on motorcycles and cars are often photographed from the front with the bike or car mounted on a frame that sways back and forth before a blue or green (the chroma *key*) curtain. In other words, it doesn't actually drive anywhere. This way the actors can be safely filmed, and their conversations recorded without extraneous noise.

Later, in the editing room, the foreground is superimposed on a background of a road, or desert, or foreign planet, and everything that was blue or green becomes the background. Chroma key effects are often cheesy looking, with a halo around the actors giving away the illusion, but with digital photography, it will become increasingly difficult to distinguish what is 'real' and what isn't.

chrominance The color-carrying portion of a video signal, which for various reasons related to compatibility with black and white television sets, is carried separately from luminance. The chrominance comprises hue (color) and saturation (the 'intensity' of the color).

chronograph 1. An instrument which graphically displays the time. 2. A diagrammatic representation of time intervals where some quantity or variable is expressed as a function of time.

chronoscope An instrument for measuring time in very precise increments. Used in scientific applications.

CHU A specialized AM band radio station which broadcasts the time with brief tones each second, and the spoken time each minute in English and French. The time broadcast is Eastern Standard Time (EST), which is five hours after Coordinated Universal Time (UTC).

churn A term to describe customer subscription turnover in the cellular telephone and cable TV industries. Churn is of concern because the setup fees are often spread out over future monthly revenues, and are not recouped unless the customer retains the service for some time. Churn is usually computed as some percentage formula based on connects and disconnects.

CI 1. See congestion indicator. 2. Certified Integrator.

CIAJ Communications Industry Association of Japan. An organization committed to promoting the development of manufacturing of telecommunications equipment and related business activities.

CIC See Carrier Identification Code.

CIDR See Classless Inter-Domain Routing.

CIE The Commission Internationale de L'Eclairage (Internal Commission of Illumination) established in 1931, developed an international color model for light, with the aid of technology that enabled more precise measurement of wavelengths, especially those within the visible light spectrum. Maxwell's triangle was taken for the basic model and three primary colors selected and assigned to the system. The CIE chromaticity model is two dimensional, but can be extended to three dimensions by reducing the value, or amount of light, thus diminishing the brightness until it reaches black, the absence of light. See Maxwell's triangle.

CIF 1. See cells in flight. 2. See Cells in Frames. 3. See Common Intermediate Format. 4. cost, insurance, freight.

CIID See Card Issuer Identifier Code.

CIIG Canadian ISDN Interest Group.

cinching A tightening, or increase in pressure in which something becomes more difficult to undo, or becomes locked up. This effect is sometimes seen in reel-to-reel systems, where the pull on a reel is greater than the speed at which it unwinds so the remaining tape, or other material, slips and becomes very tightly packed.

CIP 1. Carrier Identification Parameter. In ATM, CIP is a 3- or 4-digit carrier ID carried in the initial address message, used in establishing a connection. 2. Channel Interface Processor. A Cisco Systems channel attachment interface for their 7000 series routers which connects a host mainframe to a control unit.

CIR See Committed Information Rate.

circuit A physical or virtual collection of pathways, channels, or conductors interlinking given points or nodes in an orderly fashion to create a means for communications or electrical links. Computer circuits include traces, wires, chips, resistors, capacitors, etc. A circuit can be open or closed.

circuit line The physical line carrying network traffic between user equipment and an

IPX or IGX node. Most network services (voice, data, ATM, etc.) require that the circuit line be configured and 'activated' on the system to which it is attached before they can be used.

circuit switching A type of end-to-end transmission system commonly used for phone connections. In the process of setting up the connection, a number of resources are allocated to that specific call, most of which cannot be used by others until the call is completed and the connection terminated. One advantage of this system is that it can guarantee a certain level of performance for the duration of the call. A disadvantage is that the resources are tied up whether or not much communication is taking place. See message switching, packet switching.

circular antenna A horizontally polarized, half-wave dipole antenna formed into the shape of a circle except that the terminating ends do not touch to make a continuous loop.

circular magnetic wave A magnetic wave in which the lines of force describe a circular pattern.

circular polarization An electromagnetic wave whose lines of flux are oriented in a plane, usually horizontal or vertical, or as in this case, where the 'edge' of the field describes a circular shape. Circular polarization is used in antennas, where electricity is used to uniformly rotate the electromagnetic field through the antenna. It is possible to use one circularly polarized wave to communicate with another, or the circularly polarized wave can be manipulated to yield linearly polarized waves perpendicular to one another.

circular scanner Scanning in which the sweep describes a full 180 degree arc, which can be pictured as a cone-shape spreading out toward the direction of that being scanned.

circulator 1. A process or device that moves something from hand to hand, or device to device. 2. In microwave transmissions, a multiterminal coupling device in which the transmission is passed down through adjacent terminals. 3. In radar transmissions, a device that alternates the signal between the transmitter and the receiver. 4. In data communications, a means to allocate, or transfer information or control among ports.

CISC See Complex Instruction Set Computing.

CISCC Collocation Interconnection Service Cross Connection.

Cisco IOS Cisco Internetwork Operating System. An OS incorporated as part of the CiscoFusion architecture designed to provide centralized, integrated, automated installation and management of internetworks.

Cisco Systems Inc. A significant vendor of routers, switchers, and related hardware and software for network systems. The author gained a greater understanding of the function and implementation of network routing systems through Cisco seminars.

CiscoFusion A Cisco Systems internetworking architecture which integrates scalable, stable, secure technologies with ATM, local area networks (LAN), and virtual local area networks (VLANs).

CiscoView A graphical device-management application which dynamically provides administrative, monitoring, and configuration information for Cisco internetwork devices.

CISE See Computer and Information Science and Engineering.

CITEL Inter-American Telecommunications Commission.

citizens band radio, citizens radio service CB radio. Radio frequencies set aside for the use of relatively low power consumer radios and radio controllers (for model cars and planes). These have a limited range (up to about 10 or 15 miles for mobile units), although sunspot activity and local weather can sometimes provide some surprisingly long connections when broadcasting conditions are optimal.

In the United States, CB radios are commonly used by truckers, travellers, and radio hobbyists. Communications over 150 miles are prohibited by the Federal Communications Commission (FCC). The frequencies originally allocated by the FCC were around 27 Mhz, but this has been changed to around 463 to 470 MHz. Before computer bulletin board systems, and the Internet, CB radio was a popular means of community interaction. Not all countries are free, and civilian use of radios is not permitted in some regions of the world. See OSCAR, AMSAT.

CITR Canadian Institute for Telecommunications Research

City and Suburban Telegraph Company The first company in Cincinnati to provide direct communication between homes and businesses, incorporated in 1873. In 1878, they contracted with the Bell Telephone Company of Boston to provide Bell services in the Queen City area, and in 1882, contracted with American Bell to provide long-distance services. Their first payphone was installed in 1904 and mobile phones were introduced in 1946. In 1952, they became the first Bell company to provide 100% dial service. The company became Cincinnati Bell Telephone in 1971.

CIV See cell interarrival variation.

CIVDL See Collaboration for Interactive Visual Distance Learning.

CIX See Commercial Internet Exchange.

CJC See Canadian Journal of Communication.

CKAC The first Canadian television broadcasting station, which began experimenting with mechanical television transmitted over wires in 1926. A Baird disc camera and Jenkins scanning disc television receiver were some of the early inventions which were tried around this time. Sound and images were transmitted separately so the sound could be played on a radio receiver. Shortwave bands were used for the images. Alphonse Ouimet, who later became the president of the CBC, was a technician for the first historic CKAC broadcast in 1931, a musical performance that was sent out to 20 viewers.

cladding 1. A coating, something which overlays, a protective covering, sheath. 2. A substance, as metal, bonded to another to cover it by various means, such as pressure rolling or extruding. A process sometimes used in producing transmissions cables.

cladding diameter In a cable which includes a cladding layer, as a metal wire with a bonded coating, or a two-glass cladded fiber cable, the diameter that includes the cladding layer.

cladding glass A type of glass or other transparent material used in fiber optic cables which has a lower refractive index than the glass used in the core.

cladding mode In a transmission through a cladded conductor, a signal conducted through the outer cladding in addition to any signals which may be transmitted through the cladded core. See cladding ream.

cladding ream In fiber optics transmission, a beam that transmits within the core and cladding layers by being reflected off the edge surface of the cladding glass. See cladding mode.

CLAMN Called Line Address Modification Notification.

clamping 1. Holding within an established operating range, or baseline or midline range in a circuit, in order to maintain various processes or electrical charges at a stable or safe level. 2. In a cathode ray tube (CRT), a process which establishes a level for the picture display at the beginning of each scan line within a frame.

clamping voltage An established level of voltage around, under, or over which an electrical device is permitted to operate. For example, clamping voltages can be used to establish a range within a device operates, by setting them so that any fluctuations above or below that voltage will cause a system shutdown or other protective reactions.

Clark cell A type of early low-volt energy cell using mercury and zinc amalgam in the cathodes and anodes.

Clark, David David Clark has been a chairman and active participant in various Internet associations, including the IRTF and IAB. He has participated in numerous research efforts in high speed, very large networks, and network video applications, and various development efforts including the Swift operating system, Multics and Token-Ring local area networks (LANs).

Clark, Jim Formerly of SGI, in 1994, Clark co-founded Mosaic Communications Corporation with Mark Andreessen, which later become Netscape Communications, distributors of the most broadly used browser software on the World Wide Web. See World Wide Web.

Clarke, Arthur C. (1917-) An English-born scientist and writer. With remarkable prescience, Arthur C. Clarke anticipated the 'age of satellites' and long-distance communications. He was talking about it as early as 1942, while still in his twenties, and published an article about it called "Extra-Terrestrial Relays" in *Wireless World* in 1945. Clarke further wrote detailed descriptions of geostationary satellite orbits, and satellite transmitting and receiving stations in the 1950s, years before the first Sputnik was launched. In the 1960s, he collaborated with Stanley Kubrick in the making of the movie 2001: A Space Odyssey (1968), which has since become a classic. See satellite, Sputnik I.

class In object-oriented programming, a means of describing the content of a specific object and its related programming interfaces. See object-oriented programming.

CLASS Custom Local Area Signaling Services. Telephone subscriber calling options including, but not limited to, Automatic Callback, Call Trace, Caller ID, Selective Call Rejection. In the past, when demand for these services was less, they were billed individually, depending upon which ones were selected. More recently, phone companies have been offering monthly flat rate 'bundles' on a variety of these caller options.

Class, facsimile For information on Class 1 and Class 2 facsimile standards and related concepts, see facsimile, formats.

Class, IP Internet Protocol addresses have been organized into five general classes, and some special cases also exist. Classes A, B, and C were designated for unicast addresses, Class D were designated for multiclass addresses, and Class E was set aside for future use. Classless addressing also exists.

class of service CoS. A general designation for an agreed, or specified level of functioning, or security which varies from industry to industry. In telecommunications, network configuration and tuning, and sometimes billing levels are established according to class of service parameters. See quality of service.

Classical IP A set of specifications for an asynchronous transfer mode (ATM) implementation model described in the early 1990s by the Internet Engineering Task Force (IETF) for local area internetworking. In Classical implementations, IP headers are processed at each router, creating latency and limiting throughput. Due to the increase in demand for multimedia capabilities, Classical IP is showing its age. One of the limitations of Classical IP is that direct ATM connectivity exists only between nodes with the same IP address prefix. See ATM models for a chart of some historic and new ATM models. See RFC 1577.

Classmark An electronic designation which identifies privileges and restrictions associated with a particular line or trunk. See class of service.

CLC 1. Carrier Liaison Committee. 2. Competitive Local Carrier.

clear 1. In computer monitor displays, to blank a screen, applications window, or terminal window. The *clear* command provides a clean slate, a visual working space without clutter, obsolete information, or distractions. 2. In programming, to set a storage location (a buffer, address, etc.) to a zero state, or a blank state (as with space characters), or a previous

Amplifier Categories	
Class A amplifier	A single-ended circuit in which output current flows during the input cycle, as related to the grid bias and grid voltage. Provides good fidelity at low receiving levels.
Class AB amplifier	A circuit in which output current flows for more than half, but less than the full duration of the input cycle. Better efficiency than a Class A amplifier, but also has higher power requirements.
Class B amplifier	A circuit in which output current flows for half of the input cycle. More efficient than Class A or Class AB, but has higher power requirements, and can't be configured as a single-ended circuit.
Class C amplifier	A circuit in which output current flows for somewhat less than half of the input cycle. This provides high efficiency, but also has higher power requirements.

Emissions Categories	
Class A0 emission	Incidental radiation emanating from an unmodulated carrier wave transmission.
Class A1 emission	A low-speed carrier wave (as those used for early telegraphy) unmodulated by an audio signal.
Class A2 emission	An amplitude-modulated carrier wave modulated by low audio signals to transmit simple tones or Morse code.
Class A3 emission	An amplitude-modulated carrier wave modulated by audio signals so intelligible conversation can be transmitted.

or default state. 3. In communications, a clear signal is one without noise or interference, and which is of sufficient volume or intensity to be heard or seen distinctly.

clear channel 1. In telephone communications a transmission line which is used entirely for communication, and no control or other signaling bits are being transmitted. In other words, all the resources are available for the informational communication. 2. In radio communications, a station which is permitted to dominate a frequency and broadcast at a certain power level, or up to a certain distance (e.g., 750 miles) during a specified time of day. A type of exclusive frequency arrangement.

clear to send CTS. A signal provided when the communication has been set up over a serial link, and the called modem is ready to receive information. See RS-232.

clearance In electrical installation, the shortest distance between separated live conductors, or between live conductors separated from physical structures, or between live conductors separated from associated grounds. See gap.

cleaving The process of cutting or breaking a wire or component so the broken surface meets certain needs, as in junctions, solder joints, or connections. With simple wires, the type of break in a line is not usually critical, as wire can be readily wrapped or soldered in place. However, with layered wires or certain electronic components, where a smooth or straight surface might be important to the electrical contact, or in fiber optic cable where rough edges can effect the light-carrying properties of the fiber, cleaving can be quite critical.

CLEC See Competitive Local Exchange Carrier.

Clerk-Maxwell, James See Maxwell, James Clerk-.

click tones A signaling system, common on phone systems, especially wireless phones, that alerts the user that the call is being processed.

clickstream *slang* A description of the flow of events and sites visited when a user navigates the Internet, particularly the Web, which is connected through clickable hyperlinks. Product vendors have an intense commercial interest in monitoring and maneuvering users to their sites.

client A system or application which serves the user, but which may seek or require information, or operating parameters through a host with a higher priority, or greater capabilities. In the past, host and client have had almost opposite meanings for some computer administrators, but for consistency in this dictionary, and because the trend is in this direction, client is defined as the adjunct or subservient system or application. See host.

client application In a client/server computer software application, the client is typically the application used by the user to communicate to a source or destination through a related higher priority or more powerful (or just different) server program. The Netscape Communications Web browser is an example of a common client application which communicates to Web sites through a Web server which handles the traffic and provides some measure of security.

client operating system On a network, the operating systems run on client machines, user terminals, and subsidiary machines. These do not have to be the same as the server operating system. A good server can handle a variety of client OSs and network between them seamlessly, using standard network protocols. For example, you may have a network which is configured with a Sun workstation and Sun operating system (SunOS, Solaris) as the main server, with a number of different client platforms connected to it, running different client operating systems and operating environments, such as Linux, Apple Computer's Rhapsody, Be Inc.'s BeOS, or Microsoft Windows.

client/server model A computer processing method of improving efficiency, and sometimes security, by selectively distributing activities. In human enterprises, there is often a manager with an overall knowledge of the work to be done, security clearance, and the authority to designate tasks and respond to requests. In conjunction with the manager are workers with knowledge of specific tasks and needs, lower security clearance, and instructions to report their findings, and to direct their questions and requests for resources, to the manager. A client/server model on a computer system is similar to this. An ISP's Web server has the logic and security clearance to accept requests from many Web browsers, and to fetch the information and serve it back to the browsers which then format and display the information for the user. Most networks work on client/server models, where the server handles administrative details, file management, and security, and the client machines, usually terminals or desktop computers, handle input and output, local processing and display.

clipboard In most operating systems, and in some software programs, an area of memory or a file on a hard drive designated to hold information (usually images or text, but may also be sound or video clips) that has been *cut* by the user for later retrieval. Most clipboards store only one clip at a time, with subsequent clips over-writing previous ones, so that only the most recent can be retrieved. Some clipboards can handle multiple clips, and some store the information on disk for later retrieval. On the Macintosh, for example, a user can save clips by copying or cutting them, and storing them in the Scrapbook, and retrieve them later by paging through the clips, selecting those desired, and copying and pasting them back into an application.

Clipper chip A microprocessor chip which provides security encryption features that can be incorporated into electronic devices. The Clipper chip has become the focal point for broad and heated debates over the privacy of global communications. The U.S. Federal governing bodies had initiated plans to include this in consumer telecommunications products, to 'secure' conversations from anyone but the government. This was initially announced in 1993 through the White House Escrowed Encryption Initiative. The system was designed by the National Security Agency (NSA). Three versions of the proposal, Clipper I, II, and III were promoted between 1993 and 1996.

Most people agree that there is a need for voluntary, widely available encryption options for government and private use, and vendors agree that standards provide a means for them to distribute products with intercompatibility. But on this issue, concern has been expressed about how the government is intending to implement and enforce encryption policies, and on their assertion that the system will only work if made mandatory. There has been a considerable outcry from vendors and the public, who questioned the robustness of the technology, and who are gravely concerned about too much power being in the hands of too few people.

In spite of the negative feedback, the Department of Commerce approved the Escrowed Encryption Standard (EES) as a voluntary Federal Information Processing Standard (FIPS) in 1994. One of the requirements would be that every Clipper chip would have its unique key registered with the Federal government, and held in split form by two Federal escrow agents (NIST and the Treasury Department), creating accessibility for the Federal government to wiretap 'secure' communications. The debate over the chip, privacy, and law enforcement led in the fall of 1994 to the Encryption Standards and Procedures Act, which described Federal policy governing the development and use of encryption technology for unclassified information. Back references were made to The Computer Security Act of 1987.

The public responded in many ways to the various proposals regarding the Clipper chip. Some sought to point out flaws in the process and design, others created free user-encryption programs which would defeat the Clipper system. One of the more significant challenges to the system was the X9 Accredited Standards Committee (ASC) announcement that it would develop a competing data security standard based on triple-DES. The ASC sets data security standards for the U.S. financial industry.

The Clipper chip uses a nonpublic encryption algorithm called SKIPJACK which cannot be read off the chip, designed so that it cannot be reverse engineered. According to the EES, when two devices negotiate a communication, they must both have security devices with Clipper chips, and must agree on a *session key*, which may be a public key such as RSA or Diffie-Hellman. The message is then encrypted and sent with a law enforcement access field (LEAF), a serial number, authentication string, and a family key. When received, the LEAF is decrypted, the authentication string verified, and the message decrypted with the key. See Capstone chip, LEAF, Pretty Good Privacy.

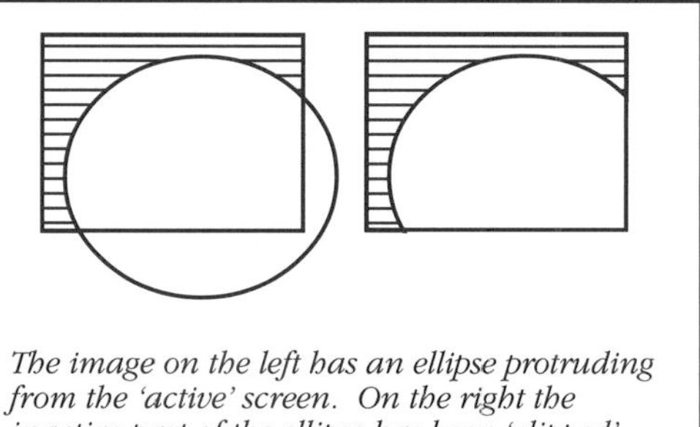

The image on the left has an ellipse protruding from the 'active' screen. On the right the inactive part of the ellipse has been 'clipped', that is removed outside the desired boundaries, or in some cases, removed from view, although the data may still be retained in memory.

clipping 1. In software applications, cutting out information, such as graphics, text, sound or video, usually for later retrieval or insertion elsewhere. See clipboard. 2. In graphics programs, the process of removing parts of an image, or of the display outside some designated boundary, usually the outer margin of a picture, or of an application's window. Information which is *clipped* may or may not be retained in memory. Often a program will retain the information, even if the user can't see it, so the user can quickly restore the information, or scroll quickly through the image without recreating it or having to wait for the computer to reread it from disk. 3. In audio, a brief loss of sound, especially at the beginning or end of a transmission due to limitations of the technology (limited frequency range, direction flipping, ramp-up time). 4. In audio communications, especially phone calls over satellite links, the equipment may be operating part of the time in half duplex mode, transmitting in only one direction at a time, so gaps in the conversation may cause a switch in the other direction and clip part of the conversation.

CLLI See Common Location Language Identifier.

clock 1. A time-keeping and reporting device that uses various gravity (sand, weights) or oscillating mechanisms (radioactive decay, emissions, crystal vibrations) to track time. Quartz crystals have extremely consistent vibrations which are sometimes used to make very accurate clocks. 2. A device that provides regular signals for use as a timing reference. On a computer, instruction speeds are expressed in *clock cycles*.

clock bias The discrepancy between the time indicated on a clock and True Universal Time. See Coordinated Universal Time.

clock doubling A means of getting a little more performance out of a computer instead of having to replace the system. With constant demands for faster systems, balanced by the high cost of replacing a system which may only be a year old or less, some manufacturers have provided versions of the CPU chip, or accelerator accessories which effectively 'double' the speed of the CPU. This does NOT mean performance is doubled. The CPU is only one part of a system, and the bus rates, coprocessing chips, software design, operating system parameters and other factors will affect the actual performance increments to a great extent. In other words, the speedup is usually more on the order of 20% or so, but for graphics computations, or using resource-hungry software, might be an important 20%. It's sometimes worth it; it depends upon the cost of the doubler.

clock speed In computer systems, an expression of the speed of a central processing unit (CPU) or other processing chip, usually expressed in megahertz. Microcomputers in the 1970s ran at clock speeds ranging from about 1 to 4 MHz. Current microcomputers in consumer price ranges, run at about 200 to 300+ MHz.

Clock speed is *not* equivalent to system speed. Doubling the clock speed doesn't mean doubling the computing speed; sometimes the efficiency is just slightly more, and sometimes it is three or four times more. Determining the overall speed of a computing system is complex, and requires evaluation of the general architecture of the system, the efficiency of the operating software, the amount of memory, the inclusion of co-processing chips, and the type of application being run.

For example, the author's first 8-kilobyte RAM (yes, kilobyte, not megabyte), 1.8-MHz system ran telecommunications software and word processors very effectively at typing speeds of over 80 wmp. A feature-rich, well-written graphical word processor can run very efficiently on an 8-MByte RAM, 10-MHz system. The same software running on a 16-MByte 233-MHz system often is not *perceptually* faster because text entry, at its basic level, is not a computing intensive application.

In contrast with basic word processing, computing-intensive applications, however, can be dramatically effected by clock speed. A stock 1.8-MHz system is essentially incapable of doing 3D ray-tracing in a reasonable amount of time, whereas an older Amiga computer with a clock speed of only 7.16 MHz can render a complex 3D scene in 3 or 4 days, faster than many 25-MHz computers with different architectures. Amigas with 40-MHz accelerator cards can render the same scene in 3 or 4 hours, and dedicated graphics systems, running on parallel processing systems, or current Silicon Graphics Machines, for example, can accomplish the same feat in minutes or seconds.

Since computing speed is important to computer electronics designers, a number of measures have been established to provide information for comparing chips, systems, or architectures. These *benchmark* tests are not absolute measures of clock speed, but they provide some information that is helpful and they generate some pretty entertaining controversy. See benchmark, clock doubler, Dhrystone, Whetstone.

clone *n.* 1. Duplicate, exact copy, genetically identical individual. 2. A software program or device configured to masquerade as another device, either for diagnostic purposes, interim use, or fraud.

clone fraud A method of gaining entry to a system, or using a device, by simulating a user, serial number, or access code. Cellular phones are particularly susceptible to clone fraud, as it is not difficult to program a legitimate serial number into another cellular unit. See tumbling.

closed architecture A proprietary design which is supported and enhanced by peripherals which conform to its particular specifications, and which may not be manufactured by third party vendors, except perhaps by obtaining special permissions or paying royalties. Contrast with open architecture.

closed captioning CC. A broadcast technique for transmitting text, usually to be superimposed over a corresponding television image. Provided mostly as an aid for the hearing impaired, although it may in some cases also be used to provide subtitle translations. It is typically sent on the vertical blanking interval of the transmission, and a decoder may be required to interpret the signals.

closed circuit A broadcast circuit in which the sending and/or receiving components are limited to a certain frequency range. Thus, a closed circuit radio system within a complex may be set to send and receive FM signals at 89 hertz. A radio station may have permission to broadcast at only 91.7 hertz on frequency modulated (FM) signals. In contrast, an open circuit is one which is not restricted to a narrow frequency range, as a CB radio, for example, which may be set to pick up signals broadcast over a variety of channels.

closed circuit broadcast, closed circuit TV A radio or television transmission that is broadcast to a small or restricted audience, often within a specific building complex or campus. Low power frequency modulated (FM) ranges are often used for this type of transmission because they are not as strictly regulated as higher power transmissions.

closet A room, cabinet, or case used for terminating blocks or patch panels for wiring configurations. The closet serves a variety of aesthetic, safety, organization, and security purposes.

cloud network Frame relay network connections are now offered as a lower cost alternative for small businesses and educational institutions, and a cloud relay is one connectionless option of this type in which resources are shared, usually among four or five small subscribing organizations.

CLP Cell Loss Priority. A one-bit ATM networking cell header toggle indicating the relative importance of the cell. This is important as there are various mechanisms in ATM for prioritizing cell traffic, or discarding cells in congested situations. See cell rate.

CLR See Cell Loss Ratio.

CLTP Connectionless Transport Protocol. A protocol which provides a means to send to a recipient which may or may not be connected to the network at the time the data is transported. CLP allows end-to-end transmission addressing and error control, but does not guarantee delivery.

CLTS Connectionless Transport Service.

cluster 1. In cellular communications, a unit consisting of a group of adjacent *cells* within which channels are not reused. See cell, cellular phone, mobile phone. 2. A set of workstations or terminals in the same general physical or virtual networked grouping. These may share more than physical connectivity, they may also have shared devices that manage processing input and output, or specialized requests of the cluster. See cluster controller. 3. A combined unit of disk storage allocation, usually consisting of four or more sectors.

cluster controller A device that controls communications input and output for multiple interconnected devices.

clutter Wave reflections from obstructions such as terrain and buildings, which may show up as echoes or unidentifiable blips on a radar screen, thus interfering with scanning.

CMC See connection management controller.

CMDS See Centralized Message Distribution System.

CMOS Complementary Metal Oxide Semiconductor. A semiconductor chip that combines p-channel and n-channel MOS in a single substrate with push-pull circuits. Slow, but noise resistant, and good for battery-operated devices. CMOS RAM needs a small stream of constant power to preserve information stored in its memory, and this is typically supplied by a lithium battery (available in photography and electronics stores). Default settings and sometimes video card and other peripheral parameters may be stored in CMOS RAM linked with a lithium battery on a computer's motherboard. See PRAM.

CMR See cell misinsertion rate.

CMRS/PMRS Commercial and Private Mobile Radio Service.

CMTS Cellular Mobile Telephone System.

CMY A color model widely used in the printing industry. The initials signify cyan, magenta, and yellow, the three colors that are combined as tiny dots in *process color printing* jobs to simulate all hues. This system is more economical than CMYK, but does not create the strong, dense blacks that are possible with CMYK, since dots of C, M, and Y do not create a completely consistent black. However, if there is no black in the image, or its use is sparing, this is not a problem.

Metallic colors cannot be produced within this color model, and extra runs through the press, or spot application of metallics on a multicolor printer are necessary to create metallic effects. See CMYK.

CMYK A color model widely used in the printing industry. The initials signify cyan, magenta, yellow, and black, the four colors that are combined as tiny dots in *process color printing* jobs to simulate all hues and black. Black is included because the combination of the first three does not give a dark, rich black pigment. Metallic colors cannot be produced within this color model, and extra runs through the press, or spot application of metallics on a multicolor printer are necessary to accommodate metallic effects.

Sometimes spot colors are used in conjunction with color process printing to create fresh, clear colors for certain areas, especially a large area of a single color, or overlaid text or embellishments. These are handled in a separate run through the press, or as a separate application of ink on a multicolor press.

CN Complementary network.

CNA 1. Centralized Network Administration. A means of consolidating network-related connections in a single location, usually a wiring closet or panel, rather than distributing them in various parts of the premises. 2. Cooperative Network Architecture.

CND 1. Calling Number Delivery. 2. Calling Number Display.

CNET Centre National d'Études de Télécom–munication.

CNG A calling tone emitted by facsimile machines which lasts about half a second and repeats as many times as the software dictates, to signal its presence and try to establish a negotiation with a receiving fax machine. Most machines default to about 45 seconds of tone sequence before they disconnect if there is no successful connection. This may not be enough for some systems, or if it is a long-distance connection, and some fax machines and fax modems have an option for extending this.

Most fax machines now automatically dial and emit the CNG. However, some of the older fax machines, or bargain basement varieties, still require a human operator to dial the number. The operator must then wait to hear a fax response, and start the fax machine CNG by pressing a button. This is a problem if the system that has been dialed has a sensing device to route incoming calls to a phone, modem, or fax machine depending upon the tone. If a human dials the line as a voice call, the switcher will route it to a phone, and then starting the calling fax's CNG does no good, as the phone has no way of routing the call back through the switcher to the fax machine. However, with increasing automation and decreasing cost of better fax machines, this problem is becoming less prevalent.

CNIS Calling Number Identification Services.

CNR 1. See Complex Node Representation. 2. customer not ready.

CNRI Corporation for National Research Initiatives.

Co Symbol for cobalt.

CO 1. cash order. 2. See central office. 3. commanding officer.

COAM Customer Owned And Maintained (equipment).

coaxial cable A transmission cable consisting essentially of an inner conducting core surrounded by a conducting tube, each insulated and all wrapped together in an outer protective sheath. The inner core is a metallic conductor surrounded by a metal shield, which acts as a *Faraday cage,* with a dielectric material interposed between them. Typically, the signals are propagated in one direction along the conducting core.

Coaxial cable was an important development for the transmission of telegraph, telephone, and television signals as it conducts radio frequency (RF) signals well. By the late 1940s, much of the eastern United States was interconnected with coaxial cable. In computer networking, 75-ohm coax is commonly used for unbalanced E1 connections, and 100- to 120-ohm twisted pair for balanced E1 connections, and subrate cabling in trunk/circuit lines.

COB Close of Business.

COBOL Common Business-Oriented Language. A verbose, high-level programming language once widely used for business applications, and still taught in business schools, but which is slowly being replaced by other languages.

COBRA A frequent misspelling of CORBA, Common Object Request Broker Architecture. See CORBA, Object Request Broker.

COBRAS Cosmic Background Radiation Anisotropy Satellite.

COCOT customer-owned coin-operated telephone. See payphone, private.

COD connection-oriented data.

code 1. A system of symbols, cyphers, characters, images, movements, sounds, or other meaningful marks or actions which serve to represent ideas and language in a way that is not commonly understood or recognized. Not all symbolic forms of communication are considered to be codes. For example, American Sign Language is not understood by many, but is not considered a code in the sense that information on learning it is readily available in schools and libraries.

Social changes can alter the perception of whether something is a code. Before the development of the printing press and public education, text and reading were mainly restricted to the elite political leaders, and common people probably considered it as a sort of code. The use of coded information is common in wartime, or with politically or economically sensitive information. Some codes are exceedingly sophisticated and difficult to break. Until recently, most analog communications have not been coded to protect privacy, due to the difficulty of doing so. With recent digital technology, it becomes much easier to code communications, and many software developers and equipment makers are adding encoding to their products. Many satellite communications, cell phone messages, and computer data communications are now encoded.

2. An abbreviated means of representing information in order to save time in its transcription or transmission, or to send it over limited transmissions devices, and sometimes also to shield it somewhat from prying eyes. Shorthand is a type of code intended to save time in taking oral dictation. Drumbeats or smoke signals are two types of codes designed to abbreviate information so that it is practical to transmit it through these basic means.

Basic telecommunications codes have been in development since the 1600s. Schilling developed a needle telegraph code in 1832. Morse (Vail) code is a widely used alphabet coding system developed in the early 1830s. It is still often used in telegraph and radio communications, particularly in countries with limited access to computer equipment. See semaphore, Baudot code, Hollerith, Morse code.

3. Computer programming code is a system of linguistic and symbolic characters and syntaxes which serve to represent computer instructions so they can be run directly by the machine, or compiled into machine-readable form.

Code designations in packet networking See Link Control Protocol codes.

code division multiple access CDMA. A digital, wireless communications service based on spread-spectrum technology, which claims to provide about 10 times the capacity of analog. Access to the local exchange is wireless.

This technology was originally used in military satellites for its security features and resistance to jamming. Now more widely used in commercial applications, it provides access to many users at a time without the multiple user interference associated with other modulation techniques. The same frequencies in adjacent beams can be reused by assigning varying spreading codes to users. The method offers authentication of the source transmitter and is very secure against eavesdropping.

Frequency reuse logistics in AMPS and DAMPS systems are eliminated in CDMA by assigning codes to users so they can share carrier frequencies. The system capacity is not fixed, but is influenced by the accumulated noise and interference associated with power levels and simultaneous users. There are two common CDMA techniques:

Technique	Notes
DS-CDMA	Spread spectrum technology in which codes are used to modulate information bits such that each code is assigned to prevent the overlap of signals from user to user. The receiver regenerates the code and uses the information to demodulate the transmission.
FH-CDMA	A group of changing frequencies are modulated by the information bits in a two step process. First, the carrier frequency is modulated, and these modulated frequencies further modulate frequencies while still keeping them independent.

CDMA, supported by companies like Sprint and PrimeCO, is somewhat similar to TDMA, with somewhat less built-in support for private branch applications. B-CDMA is also in development as a proprietary technology by a group of vendors supporting InterDigital Communications. See B-CDMA, spread spectrum.

code rejects In packet networking, codes which are not used or not recognized are processed as code rejects.

codec en*code*/*dec*ode, *co*der/*dec*oder. A system to convert analog signals, such as video and voice, to digital for transmission, then back to analog at the destination. Contrast with modem.

cohere To come together firmly, to be cohesive, to coalesce, to hold together, join, unite, merge, especially small, discrete parts or granules.

coherent light Light in which the wave lengths are aligned or *in phase* to create a very straight, narrow beam, in contrast to light from lamps and flashlights that spreads out and quickly diminishes in intensity. Coherent light can be generated by lasers, and by some light-emitting diodes (LEDs). Both lasers and LEDs are used as light sources for fiber optic cables.

This historic Marconi coherer is only a couple of inches long, the delicate glass tubing supported by an ivory base. It is part of the Bellingham Antique Radio Museum collection.

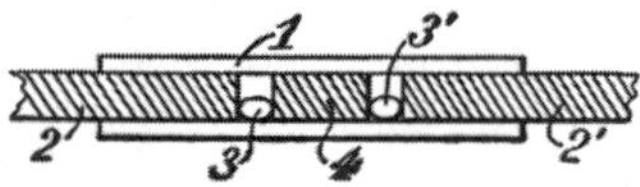

This diagram of a Castelli coherer shows a tube (1) within which are conductor plugs (2,2') separated by an iron plug (4) and two mercury pockets (3,3'). This coherer was used by Guglielmo Marconi in transatlantic experiments. Scientific American, Oct. 4, 1902.

coherer A device which causes particles to join, lump, or clump together when exposed to a nearby discharge of electricity, or to a current running to the particles through a wire. As the particles are stimulated to arrange themselves in a more 'coherent' fashion, that is to align themselves so that resistance is lowered, they together provide a better conducting surface. Many early coherers consisted of a glass tube containing filings which was corked at each end. This was connected in series with a battery-driven electrical circuit.

Early experiments by O. Lodge in 1894, D. Hughes in 1878, and É. Branly in 1890 resulted in cohering apparatus which could behave as an on/off switch by serving as a nonconductor, unless stimulated by a spark, and returning quickly to nonconducting status once the spark and the current had passed through. This useful device was adapted by Marconi for improvements in radio devices. See Branly detector.

coil In its simplest sense, a loop or number of continuous turns of wire or other material. The coil may have successive windings which are touching, or which may be spaced and stretched out like a spring. Coils are often used in wireless communications technologies where a long length of wire must be fit in a small amount of space, where a broader conductive surface area is needed, or where the proximity of the wire changes its conductive properties.

In antennas, receptivity to electromagnetic waves is based in part on matching the length of a long wave, consequently, very long wires are needed for some applications.

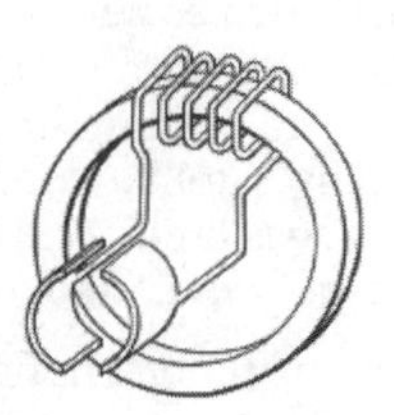

Two simple types of armature coils are shown here. On the left is a single coil, on the right, a double coil wound in parallel. Armature coils can be quite large and intricate, and are the basis of electric 'dynamos,' now more often called generators.

There are many ways to wind and use coils. Tables are published in electronics guides which describe the length and diameter of cores, and the gauge and number of windings needed for the wire. Open coils with few turns are used as *load coils* in voice grade telephone wire installations. Wound coils, wrapped around a metal core, can be used to create an *armature*. Sending/receiving coils can be created with many windings over a core or a frame, utilizing the thickness of the wire, the shape of the coil, and other characteristics to control which frequencies are transmitted or received. Sometimes dual windings are used, that is, a smaller coil inside a larger one, with an insulating layer in between. A spark coil for a basic wireless transmitter can be constructed with an inner primary winding coil and an outer secondary winding coil encasing a soft iron core. Commercial induction coils, based on the same structure as the simple spark coil, were used for decades to generate intermittent high voltage.

One unsettling historical fact is that X-ray coils were used in the early part of the century for sending wireless communication signals.

Load coils are commonly used on copper telephone wire installations to improve signals at voice grade levels, but they cause problems when data is sent at high speed through the wires, as in digital subscriber line (DSL) services; DSL transmissions are highly sensitive to noise and distance. See antenna, armature, basket winding, induction coil, load coil, winding, winding machine.

coin telephone See payphone.

cold docking Hooking components into a base or desktop unit while one or, preferably, both units are powered off. This is done to prevent danger of electrical shock or damage to sensitive electronic components. See docking.

cold start Starting a system from a power off condition. In a computerized system, it also means there is no software online. From a cold start, many systems will run through physical and logical self-test sequences, and bootstrap sequences to load device drivers or other software which may be needed to recognize and bring online the rest of the system, and eventually the whole operating system.

Collaboration for Interactive Visual Distance Learning CIVDL. Videoconferencing technologies applied to distance education for engineering programs. The CIVDL is a member of the PUG Alliance.

collapsed backbone A backbone is a main 'artery' or trunk in a network system. A collapsed backbone is one in which the physical connections are incorporated into a centralized intelligent hub or *network center*, providing easier access and administration.

collate To assemble in the desired order. Many printing programs, word processors, desktop publishing programs, and photocopy machines now have settings that allow you to choose, for multiple printouts of a multipage document, whether it is to be printed sequentially, or in page groups. This is the electronic substitute for lining up three card tables in a row with a pile of each page of a twenty-page document lying side-by-side, and having friends and coworkers walk down the line picking up one of each page. I'm sure most readers have done this at least once in their lifetimes. Collating settings and devices are great time savers.

collect call A call, usually on the telephone, in which the receiver pays for the call once it has been initiated. Most collect call systems require the prior approval of the person receiving the call before the call is permitted to continue. Person-to-person calls are generally more expensive than station-to-station calls. It is more difficult now to connect collect calls, as many people have answering machines to screen calls, and may not hear the operator requesting authorization.

collimate *v.* To make parallel, to cause to follow parallel trajectories.

collimation 1. The process of making something travel parallel, with a minimum of divergence or convergence. 2. The process of making light waves travel parallel without diverging or converging. This is useful in the testing and alignment of optical instruments.

collision In data networks, there are commonly many devices trying to send signals at the same time. If this happens at exactly the same time, collisions may occur. To manage the collision-detection and traffic flow, there are a number of mechanisms, including jam signals, to cause the devices to stop sending at the same time. Typically, the jam signals will cause them to back off and wait for a random period of time before trying again. The introduction of the random time factor reduces the chance of the same devices starting the transmission again at exactly the same time. Care must also be taken to ensure that not too many collisions occur. If there are many collisions, and devices are constantly backing off and trying again, then throughput may be compromised. Excessive collisions may mean that an additional router or bridge needs to be added to the system, or that some devices need to be disengaged.

collision detection On data networks, the means by which the system detects that more than one device is attempting to transmit data at the same time. This may be done in a number of ways, with acknowledgments being one means of signaling a system that data has made it successfully through. If it hasn't, and no acknowledgment is received in a reasonable amount of time, then there may have been a collision, and the system reacts accordingly. One type of mechanism triggered by collision detection is a jam signal, which alerts devices to back off until the jam is cleared. See collision, jam, jam signal.

collocation 1. Adjacent placement. 2. Physical placement of customer transport and/or multiplexing equipment within the carrier's premises.

collodion A viscous solution which was introduced into the processing of photographic prints in the 1850s.

colophon A description, usually in a book, of the methods and materials which were used to produce the publication. A colophon will often include detailed information about the typefaces, software programs, and printing processes used.

color bars A broadcast transmission and monitor display pattern that provides a reference for adjusting chrominance. Typically, it consists of vertical bars, from left to right, of yellow, cyan, green, magenta, red, and blue.

color burst. See burst.

color carrier reference A continuous signal, related to a color burst signal, used for modulation and demodulation.

color code An identification system based on colors, or specified widths or patterns of color. Many industries color code their dials, wires, and components for quick recognition and selection. Electronic components like resistors are often labeled as to their values with bars of colors in particular sequences.

Color Computer See Tandy Color Computer.

Color Graphics Adapter CGA. A color graphics standard introduced by International Business Machines (IBM) in 1983 as their first color graphics controller card. Until then, IBM computers with native controllers displayed only in monochrome. CGA supported a display resolution of 320 *x* 200. It has since been superseded, first by EGA, and then by VGA, and now, almost entirely by SVGA.

color model A conceptual description of how colors are detected, perceived (usually by humans), or reproduced. Human color perception is an exquisitely sophisticated phenomenon, as is described insightfully and anecdotally in Oliver Sacks' "An Anthropologist on Mars" and many, many color models exist, none of which is complete or generalizable to every situation. See CMYK, color space, Maxwell's triangle, Munsell's color model, RGB.

color monitor A monitor that is coated on the inside front of the tube with phosphors which, when excited, glow in particular colors (usually red, green, and blue), which combined, can appear as any of millions of colors. Red, green, and blue are considered primary colors in light, because their combination in different intensities produces 'any' color. (Pigment systems define red, yellow, and blue as the primary colors.) Thus, most color systems in cathode ray tubes employ three electron guns, and are commonly known as RGB systems.

color space A model or scheme for objectifying the representation of color. Many color spaces exist, most of them devised to work with specific technologies. Color spaces for printing pigments assign numeric values to particular hues which are further coded so that the printer can mix the correct inks for use on the press.

color subcarrier A monochrome broadcast signal which is modulated with sideband information in order to convey color.

color television standards Different parts of the world have standardized on different formats, and even different subformats, many of which are not intercompatible. The common ones for color television are NTSC, PAL, and SECAM.

color wheel An ordered physical representation of a color space which resembles a wheel. See Maxwell's triangle.

colorimeter An optical instrument for measuring and comparing colors from different sources, often used to match or calibrate colors according to a color model or sample.

colorimetry A quantitative method of specifying colors through attributes such as wavelength (color), excitation purity (saturation), and luminance (intensity).

Colossus Mark I A code-breaking machine developed by Alan Turing and others, put into service in 1944 in Bletchley Park, England, to help decrypt messages from other nations, particularly Germany, being transmitted during World War II. It was delivered under the leadership of Tom Flowers, representing the Telephone Research Establishment; Max Newman and Harry Hinsley played prominent roles. The existence of this machine was not publicly known until almost three decades later. See Manchester Mark I; Turing, Alan.

Columbia Broadcasting System CBS. This major U.S. network was granted its first commercial broadcast license in 1941 and not long after began to develop a color television system.

column The grouping, between top and bottom edges, of more-or-less vertically aligned elements arranged within a grid or tabular format. Commonly used to reference screen locations, or positions within a spreadsheet. See row.

COM 1. See Component Object Model. 2. See continuation of message.

combination antenna An antenna designed to cover a range of frequencies, usually UHF, VHF, and FM, in a single unit. Combination antennas have a variety of elements including reflectors, Yagi-Uda arrays, and log-periodic components to accommodate a variety of signals with good gain. Since several signals are being received, the down-lead will usually require a splitter to feed the individual signals into the appropriate components, or in a combination component, into the appropriate input receptacles. See antenna, UHF antenna, VHF antenna.

combiner circuit A circuit which combines the separate luminance channel, chroma channel, and sync signals generated by color cameras.

COMETT Community Action Programme in Education and Training for Technology. An initiative of the European Union.

Comité Consultatif International Télé–graphique et Téléphonique CCITT. This important standards body is now known as the ITU. See ANSI, International Telecommunication Union, CCITT.

Comité Européen des Postes et Télécommunications CEPT. See E1.

command buffer A portion of memory which stores recently executed commands, or frequently executed commands, so that the command can quickly be fetched and re-executed if needed. This is a type of simple memory cache used to speed up the overall performance of a system. See cache.

command interpreter In computer operating systems that include a text or *shell* interface, the commands typed in at the *command line* are passed to a *command interpreter* for immediate execution when the [Return] or [Enter] key is pressed to signal the command. The interpreter then converts the request into machine instructions which the operating system executes, or relays back to the user as an error message if the command is not understood, or cannot be carried out.

Examples of common interpreted commands are *dir*, *ls*, *copy*, *cp*, *mkdir*, etc. On systems that use a graphical user interface, commands are usually sent to the interpreter through double-clicking an icon, or making a graphical or text selection in a graphical window or dialog box. The end result is the same, only the method of user interaction with the interpreter changes.

command line A input/output interface in a text environment such as a shell window. The command line allows the user to type commands to the operating system or a scripting language or command interpreter, which are then usually executed directly, returning a text response, directory, answer, beep, or error message to the user. The presence of a cursor indicates readiness for input, and asterisks or dots are sometimes displayed during long processing times to inform the user that the process is still executing and is not dormant, or interrupted. See command line interface.

command line interface, command line interpreter CLI. The software interpreter which accepts text commands input by the user, attempts to fulfill the request by interpreting them into machine language, then responds with an answer, information, or error message. Most operating systems come standard with a command line interpreter, the Macintosh being a notable exception. On many computers, like Amiga and Unix systems, new commands can be readily added to a *bin* directory and henceforth executed in the same manner as the default command set. See command line.

command path A location designator for directories on a system which hold system commands, or commands which are to be activated from anywhere on the system without having to type the full path from the current directory. Most systems have a configuration file which allows common path names to be established at start-up time, and these generally stay active while the system is powered up. If path names are changed, it will be necessary to re-read the path file to establish the new paths, and on some systems, you may have to actually reboot the machine (very inconvenient).

command prompt In text-based command interfaces to a command interpreter, the command prompt is a means to let the user know the computer is ready for input. In fact, in old microcomputers that booted up into a programming language like BASIC, or into simple text-based operating systems, the words "READY" and "OK" were used as the command prompt. This gradually gave way to the practice of using a symbol such as a dollar sign or question mark, or a designation of the current directory followed by a symbol. On many systems, the command prompt symbol can be customized (the author used "*path/* M'Lady?" on VMS systems through graduate school) and a partial or full pathname can be specified so the user can determine the current working directory at a glance.

comment out In text editing or software coding, to delimit a section with comment markers so that the section is separated from the rest of the text. In text editing, this may be done in an article in which the length of an article has to be reduced, but the final decision on where the cuts will be has not yet been made. In the case of software code, text which is *commented out* is not executed, and is there as a placeholder, explanatory description, or as code which is being temporarily removed from execution, as for debugging purposes, a very common practice in software development.

Commercial Cable Company A historic communications cable company founded in 1883 by John W. Mackay and James Gordon Bennett, Jr. The company laid some of the earliest cables between Ireland and the west coast of North America, and later to continental Europe as well. The company was hotly competitive with Western Union, but needed land systems to be completely independent of Western Union. As a consequence, Mackay purchased a controlling share of Postal Telegraph Company.

Commercial Internet Exchange CIE. An alliance of CERFnet, Uunet, and PSI in 1991. Since that time, other services have formed agreements with CIX to allow unrestricted flow of traffic across networks in the CIX backbone. For a fee, service providers may access and send traffic across the network.

Commercial Internet Exchange Association CIX. A nonprofit trade association established to promote and support the use of the Internet for commercial activities. Its members consist of public data internetwork service providers supporting public data communications. CIX provides a forum for the exchange of ideas and information, and encourages technical research and development. Membership is open to organizations offering TCP/IP or Open Systems Interconnect (OSI) public data internetworking services to the general public. http://www.cix.org/

Commercial Space Launch Act of 1984 A U.S. act of Congress which provided support for private satellite communications systems launching and operation. The regulation at present is light, mostly related to Federal Communications Commission (FCC) frequency assignments, and the positions of satellite orbits, but this may change in the decades ahead as more and more satellites vie for space in Earth orbit. See Telecommunications Act of 1996.

committed burst size Bc. In frame relay networking, the maximum amount of data that the network will agree to transfer during a time interval Tc. See excess burst size.

committed information range CIR. In ATM networking over frame relay, the CIR is a transmission rate that has been committed for use under typical conditions. Since rates may vary, it is a computed average over a specific period of time. See cell rate.

committed information rate CIR. A service rate and traffic flow commitment level established for service in a frame relay network. That is, the CIR is a level that is agreed upon for data transmission rates. If the user wishes to use higher transmission rates, this may be done, but the excess data will be marked as discard eligible (DE) in the case of network congestion.

committed rate measurement interval Tc. In networking, the nonperiodic time interval used to measure incoming data, during which the user can only send *committed burst size* committed amount of data and *excess burst size* excess amount of data. Generally, the duration of this *measurement interval* is proportional to traffic burstiness. See committed information rate, committed burst size.

Committee T1 An ANSI-accredited organization established in 1984 which develops and publishes U.S. network reliability standards, technical information, of interest to network equipment developers, installation and maintenance personnel, and system administrators. The organization contributed to the ITU-T I-series recommendations for B-ISDN among others.

Documents related to safety, power, ISDN, SONET, SS7, and wireless communications are available through Committee T1's sponsor, the Alliance for Telecommunications Industry Solutions (ATIS) in Washington, DC. Committee T1 works in cooperation with organizations such as the Network Reliability Council. See Alliance for Telecommunications Industry Solutions. http://www.t1.org/

Committee T1 Technical Reports The Committee provides a series of telecommunications technical documents available for a fee, and some which can be freely downloaded off the Internet in Adobe PostScript or Adobe .pdf format (which can be read with one of the many freely distributed Adobe PDF readers). Abstracts for *Approved ANSI T1 Standards* are also available. Since many of these are of direct interest to people developing, installing, and maintaining communications networks, a few are listed in the accompanying chart.

Commodore 64 computer C64. A low-cost 8-bit computer introduced by Commodore Business Machines in the early 1980s, aimed at the home and school markets. Listed at under $600 U.S., the C64 included a 6510 CPU with 64K RAM, a built-in sound generator, the Digital Research CP/M operating system, and game controllers and cartridge slot. It featured 320 *x* 200 pixel color graphics, and was competitive with the Apple IIe (48K) and the Atari 800 (16K) and continued to be popular for a couple of years after the Amiga was introduced by Commodore in 1985. The C128 was an expanded version of the C64.

Commodore Amiga See Amiga computer.

Commodore Business Machines CBM. Formerly an office equipment company selling calculators, and later, the Commodore PET (Personal Electronic Transactor) computer. In the mid-1980s, at a time when Radio Shack had lost their enormous market share to IBM computers, Commodore acquired a computer named the Lorraine, and launched it in the fall of 1985 as the Amiga, despite protestations from the developers that the operating system (OS) wasn't finished, and that the hardware should have slots and more memory. Despite these concerns from the designers, the Amiga was called revolutionary when Commodore released the Amiga 1000, which was hotly competitive with the Atari Computer, and somewhat with the Apple IIGS. See Amiga computer, Apple Computing; Miner, Jay.

Commodore PET Personal Electronic Transactor. One of the earliest commercially successful microcomputers, the PET was introduced early in 1977 by Commodore Business Machines. It was a 6502-based black and white system, with 14 kilobytes of ROM and 4 kilobytes of RAM. It was competing with the Tandy Radio Shack TRS-80 which initially was also black and white with 4 kilobytes of RAM.

Common Applications Environment CAE. A set of standards intended to provide a framework for integrated systems, developed by the X/Open Company. See Single UNIX Specification.

Committee T1 Technical Reports

Number	Date	Title
TR-7	June 1986	3-DS0 Transport of ISDN Basic Access on a DS1 Facility
TR-13	Dec. 1991	A Methodology for Specifying Telecommunications Management Network Interface
TR-15	March 1992	Private ISDN Networking
TR-21	Sept. 1993	System and Service Objectives for Low-Power Wireless Access to Personal Communications
TR-36	May 1994	A Comparison of SONET and SDH
TR-45	Dec. 1995	Speech Packetization
TR-47	June 1996	Digital Subscriber Signaling System Number 1–Codepoints for Integrated Services Digital Supplementary Services
TR-53	June 1997	Transmission Performance Guidelines for ATM Technology Intended for Integration into Networks Supporting Voiceband Services

Common Architecture for Next Generation Internet Protocol CATNIP. When IPv6, the successor to IPv4 for the Internet, was in the design stages, a number of proposed formats were submitted. CATNIP is one of three formats which were incorporated in the IPv6 specification by the Internet Engineering Task Force (IETF). See IPv6.

common battery In early telephone central offices, a 24- or 48-volt battery called a *talking battery* was used for supplying the power for a phone conversation. Later, starting around 1893, these were replaced by 48-volt *common batteries* at the central office which supplied 'talking battery' to each subscriber through the wireline, rather than each subscriber individually providing battery power. This made it possible for smaller home phones to be designed. See battery.

common bell A bell that rings when any of the designated lines on a phone system ring. Often used on main consoles, to allow an operator to intercept calls if someone is away from his or her desk, or on night systems, so a single person can answer calls on several lines that would normally be answered individually.

common carrier A public communications service carrier, usually regulated and licensed by a government agency. A common carrier cannot withhold service, or discriminate against any public purchaser of the services.

Common Carrier CCB. A large bureau of the U.S. Federal Communications Commission (FCC) which recommends and implements regulatory policies for interstate telecommunications through enforcement, pricing, accounting, and program planning of network services, and wireline services.

Common Channel Interoffice Signaling CCIS. An out-of-band telecommunications signaling system that encodes information and sends the signaling data over separate channels than voice, using digital time-division multiplexing. This system is more efficient as full voice-grade paths are not needed for sending signaling information, and more secure than older signaling systems which employed 2600 Hz and 3700 Hz tones as supervisory signals. See Signaling System 7.

Common Channel Signaling CCS. In ATM networks, a type of signaling architecture in which circuits share signaling channels. See Signaling System 7.

Common Desktop Environment CDE. An integrated graphical user interface for open systems featuring a standard interface for management of data and applications. An IETF platform Human Computer Interface (HCI) standard. See X Window System.

Common Gateway Interface CGI. A means of communicating instructions to a Web server through scripts or code, in order to enhance the utility of Web pages. HTML, a markup language used on the Web, is designed for formatting, not processing of data interactions. To overcome this limitation, CGI can be used in conjunction with input to Web pages to process forms, messages, chat rooms, database records, searches and more.

Common Intermediate Format CIF. A subsection of the ITU-T H.261 standard which specifies various broadcast format parameters for ISDN videoconferencing. See chart below.

Common Intermediate Format (CIF) Types

Format	Lines *x* Pixels	Defined within standard	Notes
General		H.320	An umbrella which encompasses the following standards. Now more commonly known as p*64.
CIF	352 *x* 288 color	H.261.	Suitable for large format video-conferencing. Requires two B channels to support both audio and video.
		H.221, H.230, H.242	Communications, control, and indication.
		H.711, G.722, G.728	Audio signals.
FCIF	352 *x* 288		
QCIF	176 *x* 144	H.276.	Requires less bandwidth than CIF, but also provides less resolution.

Common Location Language Identifier CLLI. A unique identifier system, developed by Bellcore, for certain regions and equipment. Thus, various exchanges, buildings, and facilities could be coded. A CLLI consists of four characters for the location, followed by two characters for the region, and five characters for the item.

Common Part Convergence Sublayer CPCS. In ATM networking, a portion of the convergence sublayer of an ATM adaptation layer (AAL) that remains common to different types of traffic.

common part indicator CPI. In ATM networking, a 1-byte field which is used to interpret the remaining fields in the header and trailer.

Common Object Request Broker Architecture See CORBA.

Communications Act of 1934 A U.S. Federal regulations act established to organize and promote competitive communications technologies and services. This act established and described the responsibilities and jurisdiction of the Federal Communications Commission (FCC) which was descended from the Federal Radio Commission (FRC) formed from the Radio Act of 1927. The Communications Act of 1934 was amended by the Omnibus Budget Reconciliation Act (OBRA) to preempt state jurisdiction in such a way that individual states were no longer regulating rates and entry by companies offering wireless services. It further organized wireless into two categories: commercial mobile radio services (CMRS) including cellular radio services and personal communications services (PCS); and private mobile radio services (PMRS), including public safety and government services.

Communications Act of 1996 See Telecommunications Act of 1996.

Communications Applications Specification CAS. A communications protocol developed in the late 1980s by Intel and DCA for use with computer peripherals in order to enable software applications to communicate with fax/modem interfaces.

Communications Decency Act of 1996 A provision of the Telecommunications Reform Act which aroused extreme controversy and opposition by the Internet community as it made it a federal crime to send certain 'lewd' or 'indecent' or other objectionable communications across networks. The Internet community rallied against it, and, in a June 1997 decision in the case of *Reno versus ACLU*, it was declared unconstitutional, as it violated individual rights to freedom of speech. See Telecommunications Act of 1996.

Communicator III An IBM-licensed/Intel-based PC videoconferencing product with audio, video, whiteboard, and file transfer capabilities from EyeTel Communications, Inc. Communicator III works over Switched 56, ISDN, T1, Ethernet, and Token-Ring networks. It uses ITU-T H Series Recommendation standards and encoding.

Communique! A Sun SPARC-based videoconferencing program from InSoft which works over ISDN, FDDI, SMDS, Ethernet, ATM, and frame relay networks. It supports audio, visual, whiteboarding, file transfers, and a number of applications. CellB, JPEG, and Indeo standards and encoding are supported.

community dial office CDO. A type of central telephone switching office that is most often found in small rural communities. It is an unattended switching center that is serviced only as needed, and maintained on an occasional basis by a traveling maintenance technician.

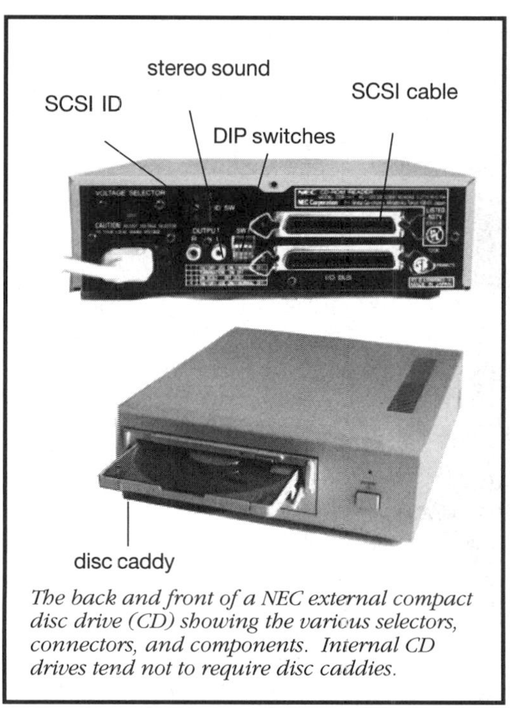

The back and front of a NEC external compact disc drive (CD) showing the various selectors, connectors, and components. Internal CD drives tend not to require disc caddies.

compact disc A small, flat, circular, optical, digital random-access storage and retrieval medium. CDs are written and read with laser devices. CDs are used for audio recordings, audio/visual sound and graphics, and computer data and multimedia applications.

The CD format has been standardized to 120 mm (4.75") diameter. It consists of a thin layer of metallic film, etched with microscopic indentations called *pits* spiraling literally for miles around the recording surface. This structure is coated with a smooth plastic surface. The data is stored in a format that was developed by Sony and Philips, and agreed upon by electronics vendors in 1981.

CD players first began to be marketed in Japan and Europe, and to a limited extent, in Canada, in 1982. They did not begin to be distributed widely in the United States until 1983. By 1986, consumer players were inexpensive enough to promote an explosion of interest in audio CDs. See SPARS code.

compact disc interactive CD-I. A more recent version of CD formats with read-only players based around Motorola 68000 technology. It was developed by Sony and Philips, and released in 1988. CD-I allows interactive multimedia use of compact discs. The CDs can be recorded with information in various forms, including computer data files; video images and still frames at more than one resolution; and audio in three formats.

compact disc types and uses The two most common types of CDs are music CDs and multimedia computer application CDs. Music CDs are supplanting music on cassette tapes and vinyl records due to the greater clarity of the sound (no scratches or hiss) and greater stability of the medium (magnetic data, and the thin tapes themselves are somewhat fragile). CD-ROM discs hold about 680 MBytes of data, although actual informational content may be greater, if the data has been compressed.

Typically, CDs are written once and read many times, although the data on PhotoCD discs may be extended in several sessions, with the new data being written to an unused section of the disc. A *multisession* CD player is needed to read discs that have been recorded in more than one session. See bar code, compact disc; digital video disc, laserdisc, PhotoCD.

compact disc video CD-Video. A variation on compact disc technology, announced in 1987, which delivered audio and video on one disc. The inner portion of the disc is the recorded music, and the outer portion contains up to about five minutes of analog video and sound, similar to a small laserdisc. CD players that support the video portion, spin the disc faster than when playing the standard audio track on the inner portion of the disc.

compander A device, or more commonly two devices at each end of a transmission, which work together to *comp*ress and then ex*pand* a signal, usually to save transmissions time. Modems which use compression techniques on-the-fly are types of companding devices.

companding A combination and telescoped word derived from *compressing* and *expanding*. Companding is a process of compressing and expanding a signal which is used for a variety of purposes, including noise reduction, security, and increased transmission speed.

Compaq Computer Corporation A successful computer company established in 1982, which shipped its first product a few months later, in January 1983, and made phenomenal first-year sales. Compaq made the Fortune 500 list in 1986 and is still active in 1998, having bought out Digital Equipment Corporation (DEC), one of the long-time, well-known companies in the computer industry.

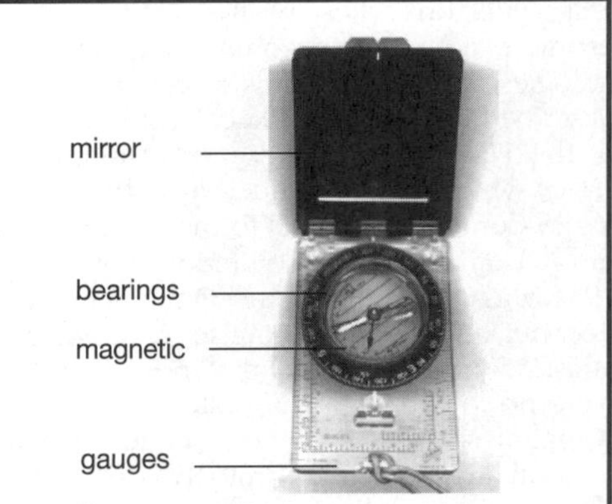

Compasses now come with a number of gauges and measures in addition to the directional markings circling the magnetic needle. The top surface of this Silva compass, on the inside of the case, is mirrored.

Compasses contributed to much of our understanding of electromagnetism, and are indispensable tools for mariners, hikers, and other travellers.

compass An instrument for determining direction, commonly through a freely turning magnetic needle on a pivot point which is oriented by the user with respect to earth's magnetic north, which is near the geographic North Pole. Primitive compasses have been described in Asia as early as 4500 years ago, and have been used for thousands of years for navigational purposes. Simple forms of magnetic compasses as we know them existed around the first century AD in the form of a sliver of suspended lodestone. See magnet.

compatibility Interconnectivity of physical parts and interoperability of transmissions protocols, electrical requirements, and data formats. Compatibility is one of the biggest issues for consumers and vendors in the computer industry. Vendors prefer in general to establish industry standards, so their equipment will work with that of other vendors, and be accepted by the marketplace, but large, powerful vendors prefer proprietary products because it gives them an edge in the marketplace, control over other vendors, and is attractive to investors. Thus, the industry seesaws back and forth constantly, with standards gradually getting established (this is usually good for consumers), and proprietary standards then being controlled or promoted by one or two top companies, thus edging out small companies and reducing competition. That is not to say that widely accepted standards are always in the consumers' best interests. If a standard is too entrenched, innovation becomes difficult or impossible. The low resolution North American television standard, NTSC, is very outdated, and everyone in the industry knows it. Many organizations for the last 15-20 years have worked hard to try to update this standard, but the proliferation of industries supporting the older NTSC have too much at stake to change, and the harangue about *which* of the new proposed standards would be best to supersede NTSC continues to be somewhat unresolved.

Competitive Access Provider CAP. A competitive local carrier which is permitted to compete with Local Exchange Carriers (LECs) and Inter Exchange Carriers (IXCs) to provide voice or data services. See Competitive Local Exchange Carrier.

Competitive Local Exchange Carrier CLEC. A competitive carrier that is permitted to compete with established local voice and data service providers, as a result of the deregulation in the Telecommunications Act of 1996. CLECs may build their own wirelines, or lease existing lines for resale of services. CLECs include CAPs, IXCs, CATV service providers, and others. See Incumbent Local Exchange Carrier.

Competitive Telecommunications Association CAT. A Canadian-based association representing new entrants in the telecommunications service business, including interexchange carriers (IECs), competitive access providers (CAPs), and resellers.

compiler A software program which translates code written in higher level computer languages into machine language in order to speed execution, and take advantages of any special features which may be available on a particular platform. Once all the code is compiled and linked, it can be run directly many times, without the need of interpreting each command as it is encountered, which is slower. Compilers typically include diagnostic tools that provide messages to the programmer if compilation errors are encountered, in order that these errors can be fixed more readily. The entire compilation need not be done in one pass. Compilers may check syntax and structural integrity in the first pass, and perform translation on the second pass. From a programming point of view, this saves time in catching and correcting errors. Compilers are machine and language specific. See linker. Contrast with interpreter.

complete document recognition CDR. A process which goes beyond object character recognition (OCR), in that it recognizes not only text and individual blocks or elements on a page, but the general layout and types of data. CDR software is quite sophisticated and can fairly reliably distinguish the difference between text and images, headlines and regular text, and columns and sidebars.

completed call In the telephone industry, this has a fairly specific meaning, as it describes a call that has reached and been answered by the callee, but does not include the time that the callee actually spends on the conversation. In other words, the meaning of *completed call* concerns the establishment of the connection with the person being called, and not the actual length of the communication.

complex instruction set computing CISC. A microprocessor architecture which accommodates complex machine language instructions in which a single operation may be comprised of many small instructions of different sizes, and which may take longer to execute than the same operation carried out on a *reduced instruction set computing* (RISC) chip.

CISC chips are more common on older architectures. A CISC processor command is translated into microcode, a series of smaller instructions, which are in turn queued and processed one at a time by a nanoprocessor. See reduced instruction set computing.

Complex Node Representation CNR. In ATM networks, a collection of state parameters that provide information about a logical node.

Complex Text Layout CTL. An IETF Human Computer Interface (HCI) platform standard.

Component Object Model COM. Microsoft's approach to object-oriented programming. The COM is a means for creating components which are resuable across a variety of applications, thus reducing programming time, and increasing interoperability across applications. Microsoft's Object Linking and Embedding (OLE) provided a subset of the functionality now associated with COM. See Object Management Group. For a more complete discussion of the basic concepts associated with programming objects, see object-oriented programming.

Component Software Microsoft's description for object-oriented programming components associated with their Component Object Model. See Component Object Model, object-oriented programming.

composite Combined, bundled, aggregated, interleaved, entwined, mixed.

composite video A color composite video signal is one in which the luminance (brightness) and chrominance (color) are combined, with the chrominance modulated onto the luminance as a subcarrier, and may have to be separated out by the receiver, depending upon the system. Videogame systems which plug into a TV set send out a composite signal, as opposed to an RGB signal that might be sent to a computer monitor.

compound modulation A successive modulation technique in which the modulated wave from one step becomes the modulating wave in the next step.

compress Condense, contract, shrink; reduce in size, transmission time, or byte count.

compression The act of reducing, shrinking, or shortening items or data in order to more easily store or transmit the objects or information. Data compression is based on the premise that most files or transmissions include white spaces, noninformational sections, or redundancies that can be removed without affecting or significantly degrading the meaning or quality of the information when it is decompressed. Compression is sometimes also based on human perceptual characteristics, or multiple means of representing the same data, some of which may be more space-conserving than others. See data compression, decompression, lossless compression, lossy compression, run length encoding.

compression algorithm The computer logic and code designed to automate the process of saving or transmitting data in less space or less time than if the data were stored or transmitted *raw* (unaltered). Compression algorithms are used on many types of data: video, still images, sound, text, etc. and the degree of compression is often tied to the type of data, and even the specific character of the particular data being compressed. A 'compressed' file is not always smaller than the original. Compression algorithms may be *lossless*, that is, the information can be reconstructed to be the same, or to appear the same, as the original, or *lossy*, in which the information is reconstructed to be *essentially* the same as the original, or perceptually similar, but not identical.

CompTel Competitive Telecommunications Association. An association that includes WorldCom, and a number of medium-sized communications carriers.

CompuServe A large, commercial dialup Internet Services Provider (ISP), which also provides Compuserve-specific services available only to its members, including airline reservations, stock listings, chat services, etc.

computer A logic-processing device, which usually includes temporary or long-term storage, and input and/or output devices for interaction with the user. It may or may not be programmable, and may or may not be constructed with binary architecture (binary computers are prevalent). A computer doesn't have to be strictly electronic, and researchers have explored ways of incorporating biological parts or processes into computing devices. Quantum computers have been proposed, with science fiction possibilities, but none has yet been devised. However, individual quantum processes have been developed successfully, and their incorporation into computing devices in the future is not unlikely.

The most common configuration for digital desktop computers consists of a central processing unit (CPU) for performing mathematical logical instructions, sometimes cooperating with coprocessing chips for graphics and sound; volatile storage, usually in the form of RAM; read/write semi-permanent storage, usually on magnetic, or magneto-optical media; user-interaction input/output devices such as monitors, keyboards, mice, microphones, cameras, speakers, and joysticks; and program instructions in the form of operating systems and applications programs.

To enhance the usefulness of basic computers, printers, scanners, modems, and network interfaces have been developed which communicate with the computer through a printed circuit board, or various SCSI, IDE, serial, parallel, and networking ports. Many people make the mistake of assuming the software that runs the system, the operating system (OS), is the computer itself. While it is true that a particular OS is usually optimized for a particular platform, operating systems can be adapted to run on many systems. Early computers ran several operating systems (see TRS-80), and the trend is moving back in that direction. Linux, Be, Inc.'s BeOS, and Apple Computer's Rhapsody, as examples, are designed to run on a number of hardware platforms, providing the user freedom to choose his or her hardware/software combination.

Computer and Communications Industry Association CCIA. A trade organization based in Virginia which represents data processing companies and common carrier service companies. The CCIA provides education and lobbying support to its members.

Computer and Information Science and Engineering CISE. A U.S. National Science Foundation Directorate which promotes basic research and education in the fields of computer information sciences and engineering. http://www.cise.nsf.gov/

Computer Emergency Response Team CERT. Established in the late 1980s by the Advanced Research Projects Agency (ARPA), based at Carnegie Mellon University. The service provides assistance to computer operators wrestling with various network security and operations issues.

computer fraud Misrepresentation or theft accomplished on, with, or with regard to computers. The computer data may itself be the target of the fraudulent activities, or a computer may be used as a tool to aid in non-computer-related fraud (as for record-keeping, spying, or unauthorized access). Unsecured computer data, in the form of accounts, confidential business or investment information, personnel files, etc. is especially subject to tampering.

computer history See the timeline in the appendix for information about early computers. Also see listings under specific computers: Altair, Atanasoff-Berry Computer, ENIAC, EDVAC, ILLIAC, Kendak-1.

Computer Support Telephony See CST.

Computer Supported Telecommunications Applications CSTA. A standard published by the European Computer Manufacturers Association (ECMA) in 1992. CSTA was developed for integrating computers and telephone technology into a unified system.

computer-aided design/drafting CAD. A CAD system is one in which drafting and design are facilitated with drawing tools, primitives, blocks, rendering, 3D projections, printing routines, scripting languages, and other aids to creating architectural renderings, engineering drawings, schematics, proposal plans, flow charts, etc. CAD programs are very popular and very useful. Consumer CAD programs allow people to plan room designs, houses, and gardens. Professional CAD programs allow engineers and architects to design and specify bridges, commercial buildings, houses, highway systems, electrical distribution systems, telephone switching stations, and more. When combined with a parametric design program, which automates the creation of drawings of similar types of items (bolts, boxes, keys, etc.), CAD can be used directly in the manufacturing process, particularly if it is interfaced with NCR and other production machines. See parametric design.

computer-aided dispatch CAD. A system in which the administration of services is done with the aid of a computer. For example, emergency or law enforcement systems may track the location of vehicles, and their direction of travel in order to dispatch calls in an efficient manner. Similarly, taxi and limousine services can be managed with the aid of a computer. Billing and mileage factors may also be stored by the system, and statistical measures tracked in order to allow a company to better manage its resources.

computer-aided learning, computer-assisted learning See computer-assisted instruction.

computer-aided manufacturing The process of using a computer to directly control manufacturing equipment, such as drilling machines, production lines, bottle cappers, saws, chisels, and any of the fabrications equipment which normally may have been driven by manual, electrical, or mechanical machines without the benefit of logic programming. One of the best applications of computer-aided manufacturing is integrating the production machines with parametric design software (usually used with CAD) so that objects that are generally the same, but perhaps different in specifics, can be manufactured with the same equipment, under the control of the computer. See parametric design.

computer-assisted instruction CAI. Instructional media and techniques offered in conjunction with computer software, or entirely by computer software, in the form of tutorials, demonstrations, white papers, online mentors and instructors, educational software, tests, multimedia presentations, etc. CAI has been around for decades as a number of instructors were quick to grasp the significance of and opportunities provided by CAI, particularly for individual learning and distance learning. However, the resources and time to provide good educational programming, and the prohibitive cost of systems up until recently greatly limited the practical application of these ideas.

Computer+Science Network CSNET. CSNET merged with BITNET in 1989 to form the Corporation for Research and Educational Networking (CREN); CSNET was discontinued in 1991.

computerized braille A method of transferring printed text to braille developed by Roland Galarneu, originally using telephone relays and a teletypewriter that created a perforated tape. After six years of development, it was ready for production in 1972.

Computer Systems Policy Project A lobbying organization formed in 1991 which represented interests that felt national networks were too oriented toward research and science and not enough toward everyday users. In actual fact, statistics show that a great majority of Internet use *is* devoted to conventional everyday user traffic, predominantly business and personal electronic mail and file transfers unrelated to research and science.

computer telephony integration CTI. Integration of computer database, dialing, and other features, with voice communications through a headset, handset, or other computer peripheral voice transmitting and receiving device. See computer telephony.

computerTV See telecomputer.

Computists International CI. A professional association for information science, artificial intelligence, and computer science researchers. CI provides information on industry trends, leading edge technologies, research, and job opportunities. CI publishes a weekly Computists' Communique reporting on artificial intelligence, neural networks, genetic algorithms, machine learning, natural language processing, fuzzy logic, and computational linguistics. See artificial intelligence.

COMSAT Corporation Originally created by the U.S. Congress, COMSAT merged with Continental Telephone to form COMSAT Corporation, an international provider of satellite communications and networking services. COMSAT operates through the INTELSAT and Inmarsat systems, and is currently the largest user of both systems. COMSAT operates the COMSAT Laboratories for research and development, and technical consultation in pioneering satellite communications technologies. http://www.comsat.com/

concentrator A point or device at which a number of elements are brought together either for simplicity of cabling and management, or to more efficiently provide a means to allocate shared resources. A start topology network is a type of concentrated configuration. A printer room with several kinds of printers, available to the general office, is another type of concentration point. The concentrator itself may be a smart device, like a router, which processes the incoming information and sends the task or communication to the best destination.

concurrent Functioning, processing, or operating at the same time; parallel, in conjunction with; coexistent, simultaneous, synchronous.

concurrent programming Techniques and associated notation systems for parallel processing implementation. Distribution, synchronization, prioritizing, and signaling are important aspects of concurrent programming. For example, computer graphics special effects rendering is computing intensive, and 'farming out' various objects to various processors or workstations, and then combining them in one frame when each is rendered, can greatly decrease the time it takes to create each image (these setups are known as 'render farms' in the visual media industry). Not all types of operations benefit from concurrent programming. The overhead involved in setting up the distribution and coordination of the data must be smaller, in proportion to the effective processing which occurs, or it may not be worth the resources.

concurrent site license In the software industry, there are a number of common schemes for assigning software use rights. Exclusive operation on one machine at a time only is the most common, but it is also possible to get concurrent licenses that permit up to a specified number of users to access the software at any one time from a networked server, or which permit up to a specified number of users (five is common) to install the software on their individual workstations.

condenser An apparatus which concentrates, or condenses a beam, ray, wave, or collection of particles. In the process of concentrating a substance, wave, or particles, the condenser may also secondarily store them, as in electrical energy. A device which focuses radiant energy, as a lens for concentrating light, or a parabolic antenna for concentrating satellite waves, can be considered a basic type of condenser. See condenser, electrical.

condenser, electrical Condensers range widely in complexity and construction. They are used with a variety of types of electrical apparatus, for example, spark coils. Condensers employ a dielectric, that is a material that doesn't readily conduct direct current (DC). Dielectrics vary from paper to ceramic or glass, with the better insulators being used in higher voltage applications. A Leyden jar is one of the earliest condensers used for concentrating and storing electrical energy. In the Leyden jar, the glass acts as the dielectric. A variation on the same idea, using glass plates in a rack rather than a jar, were used for early wireless condensers. See capacitor, Leyden jar.

conditioning The processing of current to make it suitable for specific tasks. Some electrical appliances can tolerate variations in current or noise, and others are very sensitive to variations and noise, particularly small electronic components, requiring that the raw current that may come from a wall socket or other source, first be conditioned to meet the needs of the device.

conductance The ability, readiness, inclination, or disposition of a material or system to carry an electrical current, expressed in the practical unit mho (ohm spelled backward). The reciprocal of electrical resistance. See conductor.

conductivity method A pioneer experimental method of sending wireless communications using the ground as the conductor. This was demonstrated successfully up to distances of over five miles by Preece, before Marconi demonstrated practical applications of wireless communications. Terminals of strong batteries were set up in series from a sending key, grounded at a distance of about fifty feet apart. A symmetric arrangement was set up at the receiving end, except that it used a telephone receiver or galvanometer. Other researchers experimented with this method, but little documentation of their efforts is available. See Preece, Steinheil.

conductor A material which readily carries an electrical current or heat. Some metals make especially good conductors (e.g., silver, copper, gold, aluminum) and are widely used in the manufacture of wire. Less conductive materials, such as rubber, are sometimes impregnated with metal to increase their conductivity, while still retaining attributes which are difficult to achieve with metal alone, used in specialized parts such as gaskets and seals. The term *conductor* originates from Desgauliers in the 1730s. Contrast with insulator.

conduit 1. In its most basic sense, a channel for directing physical objects or virtual data along its path. 2. A liquid conduit is a pathway which is often used for temperature regulation, dispersion of lubricants, or channeling of fluids from one area to another. See duct.

C

conduit, wiring 1. A tubular, hollow, physical pathway providing a channel for materials which are installed inside, or directed through, its core. Plastic, metal and ceramic are common conduit materials. 2. A pipe which provides a protected pathway for wire, cable, or other conductive materials. Conduit is commonly used to run wires in a building, and may also include insulation, color coding, and other attributes to protect or identify its contents. Conduit can be a good way to hedge against obsolescence, since it can be rethreaded more easily than cables which have been attached directly to the structure of a building inside its walls.

cone of silence See zone of silence.

Conférence Européenne des Administrations des Postes et Télécommunications (European Conference of Postal and Telecommunications Administrations) CEPT. An international standards body which represents telecommunications providers in most nations other than Japan, Canada, U.S., and Mexico. It cooperates with CEN/CENELEC.

configuration Setup, organizational structure, architecture, topology, assemblage, physical and logical parts and interactions taken as a whole.

congestion indicator In ATM networking, a traffic flow control signal to reduce the allowed cell rate in order to reduce the likelihood of increasing congestion. The information is contained in the RM cell. See cell rate, leaky bucket.

conical array antenna An antenna which can receive a range of VHF signals through a central rod with reflector elements extending out at right angles to the support rod and, at the other end, forward-oriented driven elements fanned out more or less from a single connection point on the rod.

conical monopole antenna A vertically polarized, broadband antenna shaped like a cone with the narrow end oriented towards the top. The frequency response is related to the size and the angle of the cone.

conical scan In radar antennas, a circular scanning motion, often used on aircraft, that provides more complete information on the location and characteristics of the object of the scan. The conical scan can provide angular information.

Connect 918 A Macintosh-, and IBM-licensed PC-based videoconferencing product from Nuts Technologies, which supports video, audio, whiteboarding and screen sharing over analog, Switched 56, ISDN, and Ethernet networks. It uses ITU-T H Series and G Series Recommendations standards and encoding. See Cameo Personal Video System, CU-SeeMe, IRIS, MacMICA, ShareView 3000, VISIT Video.

connection A system or circumstance of physically or logically joined entities, objects, or processes. An electromagnetic connection is one in which current from one system can pass onto another, either through a splice, jack, plug, or other connector, or by a spark, or induction system.

connection management controller CMC. Works in conjunction with a broadband integrated gateway (BIG) to take data from incoming ATM cells for processing and routing. See broadband integrated gateway, HFC.

connection protocol The software protocol which negotiates a pathway for a transmissions connection session.

connection-oriented A type of communication in which the sender/receiver connection is established prior to transmission, as in a phone call. This may sound like a logical way to do things, but a substantial amount of network traffic does not follow this model. In sending an email message, for example, the message will be sent irrespective of whether the receiver is online at the time the message is sent. Then, if too much time elapses, or a certain number of attempts to deliver the message have failed, it will be returned to sender. Modem communications are connection-oriented. If there is no answering handshake at the other end of the transmission, no transfer of data takes place. Contrast with connectionless.

connection-related function In ATM networking, a traffic management and policing function related to a network element (NE) where connection-specific functions are carried out.

connectionless A type of network transmissions architecture in which the data is sent without first establishing that the receiver is connected and available to receive transmissions. Large distributed computing environments frequently employ connectionless services. In contrast, phone calls and modem data transfers are connection-oriented end-to-end communications. Contrast with connection-oriented.

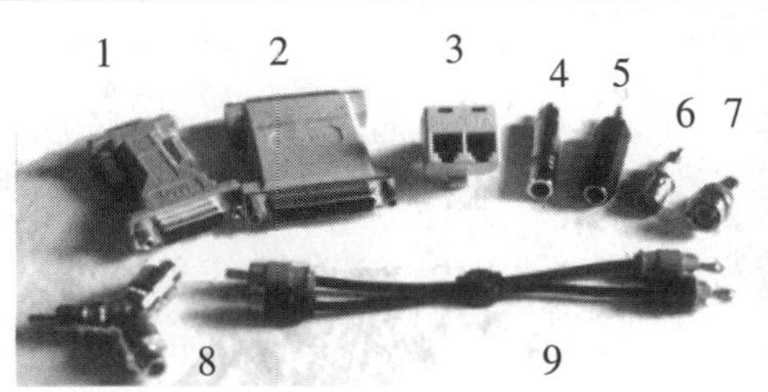

A variety of common, standardized computer, phone and video connectors.

1. *9 to 15-pin D computer data adapter*
2. *25-pin D null modem data adapter*
3. *RJ-11 phone line splitter*
4. *and 5. stereo sound adapters*
6. *and 7. coax F and BNC video adaptors*
8. *RCA video/audio splitter*
9. *RCA video/audio cross connector*

connectivity A property of mechanical and electronic systems that allows them to interconnect with other devices or systems. While connectivity generally refers to physical connectivity, the increasing importance of data in communications and hardware configuration has extended the term to software as well. When systems or devices can be interconnected in terms of hardware and software, they are said to be *compatible.*

connector 1. A device to join or combine two or more objects or circuits in its simplest sense, which assumes that the two systems or objects connected communicate in the same way. The connector may be incorporated into the device being connected, or may be a separate item. See gender changer, jack. 2. In a flow chart, a connector is a symbol which can be used to indicate a break in a flow, or the divergence of the flow into additional paths.

Connon, John A Canadian inventor and historian who devised various mechanisms, including a type of dynamo, and who was granted the first patent for a cinematic camera in 1888. The innovative camera could photograph a continuous image, without seams, while rotating 360 degrees. See panoramic camera.

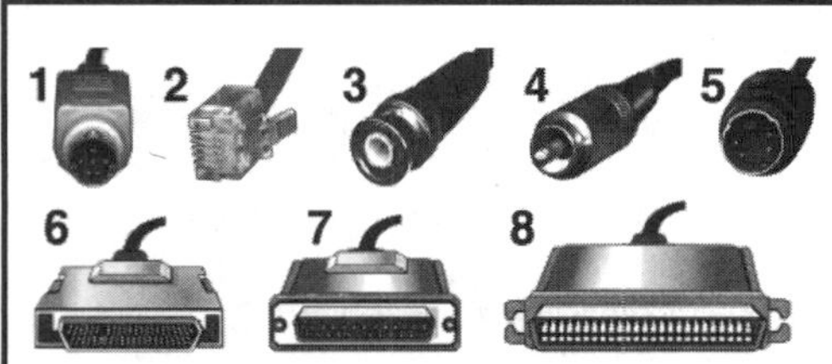

1. 6-pin DIN computer connector
2. RJ-45 10Base-T computer network connector
3. Video BNC coaxial cable connector
4. Audio/video RCA connector
5. Super-VHS video connector
6. SCSI-2 50-pin computer data connector
7. 25-pin D computer data connector
8. 50-pin flat SCSI data connector

Conrad, Frank (1874-1941) An American broadcaster who began as callsign 8XK in his Pennsylvania garage, which was later licensed as the history-making KDKA radio broadcast station. See KDKA.

consecutive Continued presentation of objects, data, or actions one after the other; successive, sequential, following. A distinction may be made between sequential and consecutive in that sequential implies that there are no gaps between each step, whereas consecutive may have gaps or delays in the successive presentations, depending upon the nature of the information or actions. For example, it would not be unusual to say that a person was working on a gardening project on consecutive weekends, but the phrase 'sequential weekends' would not be used, as there is a weekday gap between each weekend. Sequential events are always consecutive, but consecutive events are not always sequential. If this is confusing, think of the fact that a square is always a rectangle, but a rectangle is not always a square. See concurrent, parallel, sequential, serial.

Consent Decree of 1956 An agreement between the Justice Department and American Telephone and Telegraph (AT&T) to separate Northern Electric (later Northern Telecom) from Western Electric.

Consent Decree of 1982 The divestiture of AT&T that took place in the mid-1980s under the direction of Judge Greene. It is now more commonly known as the Modified Final Judgment (MFJ), since it was a modification of the Consent Decree of 1956. See Modified Final Judgment.

console 1. Floor-standing cabinet, typically holding consumer broadcast receivers (radio, TV). 2. A primary operations physical unit which holds main electronic controls and monitors (such as lab equipment, medical monitors, industrial plant operations equipment, etc.).

console, computer operations A computer terminal typically used to monitor and control computer operations, printers, etc. On a secure network, the operating console is often password-protected to control access, and may even be locked away in a separate room to prevent access or physical theft. The main server sometimes serves also as the console, although on larger systems, the server and the console may be separate physical or virtual systems.

console, telephone A primary multiline telephone unit used by an operator to answer and route calls (a replacement for the old physical cord-and-stereo-jack-style switchboards). These come in a wide variety of configurations. Some are programmable, by entering letters, features, and numbers through the keypad, which may further be displayed on a small character display. See PBX.

consoleless operation Automated operations or routing, an option for companies whose needs are simple enough that they can function without a central unit, or without the expertise of an operator. See PBX.

Consortium for Audiographics Teleconferencing Standards CATS. A nonprofit organization based in California, to promote acceptance and development of audiographics teleconferencing standards. Audiographics teleconferencing in its ideal form, is the simultaneous realtime use of images and sound, in a cooperative environment, by participants in different locations.

constant angular velocity CAV. A playback mode for magnetic and optical discs in which the disc rotates at a constant speed. A CAV disc generally requires more space on the disc to hold the same information as can be stored on a constant linear velocity (CLV) disc, but CAV format has the advantage of providing frames that can be viewed individually in 'freeze frame' mode as still images. See constant linear velocity.

constant bit rate CBR. In ATM networks, a cell rate traffic flow Class of Service (CoS) category which supports a constant or guaranteed rate of transport, and circuit emulation. Constant bit rates are important for types of communications that require synchronization of signals at the receiving end. For example, synchronization of sound and audio in a videoconferencing application is important, and unacceptable delays might occur if related cells are not arriving at the same time. See cell rate.

constellation 1. A group related by proximity and physical or conceptual connectivity (such as workstations, celestial bodies). 2. In GPS, the set of satellites used in a position calculation, or, all the satellites within communications range of a GPS receiver at a specific time. See Global Positioning Service.

Constellation Communications, Inc. A U.S.-based commercial provider of satellite communications services. Constellation is developing a low Earth orbit (LEO) system comprising 46 satellites called the ARIES satellite system. Eleven ARIES satellites will be placed in circular equatorial orbits at 2,000 kilometers, and 35 will be divided into seven circular inclined orbits at the same altitude.

consult To seek advice, opinion, or information from reference materials, or from another person, presumably with expertise in the area of inquiry.

consultant Professional, or other expert offering advice or information services, usually specialized.

Consultation Hold A surcharge phone service, or multiline subscriber service, which allows the operator to put an incoming call on hold while engaged in another call.

Consultative Committee Telecommunications CCT. A three-nation industry trade association which promotes trade expansion, and the evolution of telecommunications equipment and services within NAFTA and North and South America. The CCT represents more than 50 industry telecommunications equipment and services suppliers, as well as regulatory and certification agencies. CCT liaises with CITEL, and serves as industry advisor to the NAFTA Telecommunications Standards Subcommittee.

contact A point in a circuit, usually at a junction, binding post, or terminal, where other parts of the circuit interconnect or are attached.

continuation of message COM. In ATM networks, a status indicator used in the asynchronous transfer mode (ATM) adaptation layer (ATM AL or AAL) to indicate that the cell is a continuation of a communication which has been segmented, that is, broken up and sent in different sections, sometimes over different pathways. See asynchronous transfer mode.

continuous wave A wave which is constant or unvarying in its major characteristics, such as amplitude.

continuous wave transmissions Any transmissions technology which employs continuous signals rather than pulses. Since most communications media rely on pulses or *modulation* to send meaningful information, continuous wave transmission is more specialized, and generally used in signaling situations, such as security systems, in which an interruption of the continuous wave serves as an alert or system startup signal.

contract A binding agreement between two or more persons or entities. In many regions, a verbal contract is legally binding, and may be taken up in the justice system, where at least one of the parties must establish that a verbal agreement took place. Many software agreements go into effect at the moment the product is opened, provided the license is displayed on the outside of the package, where it can be seen before the package is opened. Other software agreements go into effect when the product is installed, or when it is first used. See license agreement.

control character. A character in the ASCII standard which performs some action other than printing an alphanumeric character. Sometimes control characters show up as symbols, but often they are considered to be *unprintable characters*. Examples of control characters are line feed, carriage return, bell, etc.

The control key is a special function key common on computer keyboards which is held down simultaneously with other keys to modify them, usually for system interrupts and similar tasks. The control key is sometimes used to access control characters in the ASCII set, but they are not directly related. The control key is used for other functions as well. For example, *Ctrl-C* may be used on most systems to interrupt the current process. See ASCII.

control field In many types of data communications that employ 'fields' as information units, a control field is assigned to contain information about how to process related data.

control panel A console used to control operations of a system, vehicle, aircraft, or network. The control panel may consist of physical switches or dials, or may be a text or graphical interface on a computer screen (often simulating switches and dials).

Control Panel On the Macintosh computer, a collection of utilities, accessible through the

Apple Menu, which provides access to many basic operating parameters, including sound, memory, monitor settings, network configuration settings, etc.

control segment In the Global Positioning System (GPS), the control segment is a global network of control and monitoring stations which ensure the accuracy of satellite positions and their clocks. This coordination is an important part of GPS, as the data is derived in part from the relationship of the satellites to one another and their signals. See atomic clock.

control terminal A workstation or personal computer configured to provide control of a network from any routing node. It acts as a command input and display console. Through remote access commands, it is possible to control a network from a node other than the one which is physically connected to the terminal (virtual terminal).

controller A software-supported computer hardware device, that works in conjunction with the operating system, through the various system interfaces, handling input and output and control of that device. Thus, a disk controller provides functions to handle a hard drive. The most common desktop computer disk controllers follow SCSI and IDE standards. SCSI controllers are also commonly used for scanners, CD-ROM drives, and many types of cartridge drives. A serial controller handles serial communications into and out of the computer, usually to a printer or modem. RS-232 and RS-423 are two of the most common desktop serial interface standards.

When a computer first boots up, one of the processes that occurs is bringing the controller hardware and software online. The computer needs to locate the various devices, and will often load in a variety of software device drivers that support the hardware functions. See controller card.

controller card A computer peripheral card which fits into a slot, or piggybacks on the motherboard, to provide an electrical and logical connection between a device and the main circuitry of the computer. There are a limited number of slots available for controller cards, usually more in a tower model, and extra power consumption may be needed to handle the extra load. There may be jumpers or dip switches on the controller card to fine tune the settings, as on a graphics card, or hard drive controller. The communications standard used by the controller card must fit that of the slot into which it is inserted. An EISA card cannot be put in a PCI slot, and vice versa. Software may need to be loaded onto the computer for the operating system to recognize the controller card functions. See controller.

CONUS Contiguous United States. The continental, contiguous U.S. consisting of 48 states. Used when referring to U.S. travel, transmission, or broadcast regions.

converter A very broad term for anything which changes the incoming signal to a different outgoing signal, or which converts one type of connection to another. Converters are frequently used to convert between alternating current (AC) and direct current (DC), or to convert one cable type to another. A converter may be used to convert a frequency spectrum from one range to another, but this is pushing the boundaries of the term, as 'converter' usually implies fairly simple conversions. More complex ones tend to have their own descriptive terms. When a converter is used simply to change one type of plug or jack into another, without any electrical changes, it is more commonly called an *adapter*. When it simply links two components, with no changes in the signals, it is called a *connector*.

Conway's law This saying has been variously restated (and probably improved). The idea is that there is congruency between the composition of the software team, and the final design of the software (and in this version, an implied dig that a single programmer wouldn't ever finish the project), stated as, "If you assign *n* persons to write a compiler, you'll get an *n-1* pass compiler." Another version of this, not quite as apropos to computereze as the one just stated, but perhaps closer to the original, is "If you assign four groups to working on a compiler, you'll get a 4-pass compiler." is attributed to Melvin Conway, an early Burroughs computer programmer.

Another Conway's law has been stated in Dilbertian fashion as follows, "There is always one person who knows what is going on. That person must be fired."

Cook, Gordon Author of the Cook Report, a newsletter devoted to issues concerning the commercialization and privatization of the Internet. Cook was formerly Science Editor at the John von Neumann National Supercomputer Center. Later he served for 18 months as a director for the U.S. Congress Office of Technology Assessment, assessing the National Research and Education Network (NREN)

Cooke, William Fothergill A British researcher who collaborated with Charles Wheatstone in developing the telegraph. See telegraph history; Wheatstone, Charles.

cookie A token or other transaction acknowledgment or ID passed between transacting processes or programs to keep a record of an access or action. Cookies may be passed transparently between systems as part of normal operational protocols. A hat check tab is an example of a cookie, whether or not the hat is retrieved later.

Cookies are an integral part of Internet commerce, especially in the form of identifiers in Web link referrals and shopping cart purchases made online. Cookies can be passed from the browser to the shopping cart site to identify the visitor for later purchasing or statistical purposes, and may or may not be acknowledged with information from the vendor being deposited on the buyer's system. The cookie may also be used to track a customer, as they browse other sites (common on the Internet), and then return to finish their online shopping. Some people object to these specific types of cookies, which are automatically offered to the visited site by the browser, and will disable this capability, and many people have objected to the *reverse cookie*, the one which is deposited back to the visiting browser's system, often without the knowledge of the user, as this opens up a porthole for viruses, vandalism, and unfair trade practices. See Caller ID.

cookie monster A software virus program, widely distributed in the mid-1980s, named for the popular children's television program character. The program would prompt the user with "Give me cookie ..." at increasingly shorter intervals, gradually taking up more and more CPU time, and if the user didn't type in the word "cookie" it would eventually print so frequently, and steal so much CPU time, that it would make the terminal unusable. The author encountered this virus on the west coast on a VMS system in 1984; the original was rumored to have come from MIT. See virus.

Coordinated Universal Time, Temps Universal Coordonné UTC. An international astronomical time reference devised in 1970 by the ITU. UTC is related to the Greenwich meridian, that is 0 degrees longitude on the earth's surface.

UTC uses a 24-hour clock, thus, 2:00 pm is designated as 1400 hours. Since UTC cannot exactly match Earth's slightly varying rotation, UTC was set to a UT1 reference with the Earth's position as of 0000 hours on 1 January 1958. Deviations are adjusted with leap seconds.

Coordinated Universal Time is based on the average period of the rotation around the sun. UTC receives its frequency and time information from over 50 centers around the world and broadcasts it over a number of radio frequencies, usually with tones to indicate seconds, and spoken words to signal upcoming minutes.

copper A malleable, metallic chemical element with high conductivity which makes it invaluable in the manufacture of electrical wire and heating implements. It is the most widely used conductor in electrical work due to its properties, availability and price. Gold and silver are also good conductors, but not economically practical for most electrical work.

Copper Distributed Data Interface CDDI. An ANSI standard version of Fiber Distributed Data Interface (FDDI) which runs on twisted-pair copper wiring rather than on optical fiber.

copper twisted pair Copper is very commonly used for electrical wires, and twisting two wires together is a very common means of manipulating copper wire to improve its transmission properties. A pair of wires is intertwined in a helical pattern over distance in order to reduce capacitance. Over longer distances, the distributed capacitance can build up, and *load coils* are introduced, at intervals, to help balance capacitance and inductance. The use of load coils for telephone voice connections is common, however, they can cause problems when the same wires are used for data transmission. Copper pairs used for data transmission may be constructed differently from voice lines, with the insertion of a metal screen to differentiate the transmit and receive.

Compared to fiber optic cable, it is very easy to work with twisted pair. Cutting and splicing is relatively straightforward, whereas the cutting and splicing of fiber optic must be done with great care so as not to alter the alignment properties of the optic waveguides.

Sometimes the twisted pairs are further aggregated into 'binders,' a group of 25 twisted pairs. This simplifies installation in multiconnection installations. Color coding is often used to keep track of the connections and binder bundles. See copper wire, load coils.

copper wire The most commonly used transmissions medium for telephone calls and related telecommunications. Copper wires were first widely installed in the 1880s, superseding some of the earliest galvanized wires used for telegraph signals. Copper wire with an iron core was developed by Bell's Thomas B. Doolittle in 1883, and it became popular due to its combination of conductivity and durability.

Copper wire for phone communications was most commonly installed as a single wire, strung on utility poles, or as twisted pair. More recently, gel-filled multicables included up to almost 5,000 twisted pairs has been used where many connections are required.

Many bare conducting wires have been strung without insulation, but insulation is often used to protect the wire from damage, interference, and corrosion. Rubber, gutta-percha, latex, plastic, and wound paper have all been used as wire insulators. Air and jelly inside an outer core have also been used. See coaxial cable, copper twisted pair, Copperweld, fiber optic.

Copperweld A trademark name for an early combination of copper wire with an iron core. This combined the flexibility and conductivity of copper with the durability of iron, and increased the longevity of the wires.

coprocessor A computer processor which is not considered the main or central processing unit (CPU), but which assists the CPU in handling heavier processing loads, or specialized processing loads. CPUs are designed as general-purpose chips, and are not intended specifically for any one type of task. The Amiga computer, released in 1985, is an example of one of the first desktop computers to make extensive use of coprocessing chips to handle resource-hungry graphics and video operations in order to prevent these computations from slowing down the CPU.

The interaction of a CPU with support circuitry such as coprocessors is one of the reasons the raw speed of the CPU is not a perfect indicator of the performance of a system. Computers with coprocessing chips and average speed CPUs have often been shown to outperform faster CPUs if they don't have coprocessing support. Coprocessor chips are gradually becoming more common in desktop systems, with math coprocessors becoming prevalent in the 1980s, and graphics and sound coprocessors beginning to show up in many consumer systems in the 1990s.

COPT Coin Operated Pay Telephone.

copy A text message on a telegraph printer or typewriter, or which was hand-written by an operator as the message was received. The term is now broadly used to refer to almost any text document, especially advertising copy, in which the text is often received separately from the images, and combined in the page layout process later on.

Hard copy is a somewhat redundant term, but may now be a relevant distinction, since many messages are received in data format (as e-mail) and may never exist on paper. *Solid copy* indicates a message that has come through cleanly and clearly.

In radio communications, the quality of the message transmission, the copy, can be described in terms of a code number from 1 to 5, with the number increasing as transmission quality increases.

copyright Certain legal safeguards conferred by government agencies. Copyright protections are granted to original works for a specified period of time, depending upon the type of work. Original drawings, musical compositions, software programs, and stories are copyrightable. Inventions are sometimes copyrightable, sometimes patentable, and sometimes both. It is important to include a copyright symbol © and a date, with the name of the copyright holder, or the word *copyright* and the date on each presentation of the original work, or reproductions thereof. The C with the circle is recognized internationally by those countries cooperating in international copyright treaties, such as the Berne Convention.

A fee-based formal copyright process is available in most countries to provide a record of the type and date of the copyright materials. This is a good source of evidence in a legal dispute, but the copyright registrar does not police the copyright; that is the responsibility of the copyright holder.

Many researchers, academics, and business employees are mistakenly under the impression that they automatically own the copyright for something they create. This is often not true, although the laws have swung slightly more in the direction of the creator in recent years. However, if it is work for hire, or work paid for in the normal course of an employee's duties, the educational institution, or the corporation *usually* owns the copyright to the work, unless there is a specific written agreement stating otherwise. This has caused many surprises and unfortunate situations in the past. If the employee, on the other hand, creates some original work on evenings and weekends, and can *prove* it wasn't done under the direction of the employer or during working hours, the employee may have a case for copyright ownership.

The most well-known Internet browser, Netscape Navigator, was originally known as Mosaic, as it was descended from a software program written at a university. When Mosaic's developer left the university to found Netscape Communications, he had to rewrite the software completely, in order to not infringe on the university's copyright, and also ended up changing the name of the commercial product from Mosaic to Netscape Navigator. Even a software rewrite will sometimes be disputed by the original copyright owner, if the owner can demonstrate that the software rewrite is copied from and not essentially different from the original.

Copyright does not protect the owner if someone independently comes up with the same idea, and has not copied the original idea. However, it can be difficult to prove the idea was conceived independently, if it is very similar to another, and that may be necessary in court, if a legal proceeding is initiated. Most public libraries have excellent references on copyright requirements and registration guidelines, as do all U.S. government documents repositories. See intellectual property, patent, trademark.

CORBA Common Object Request Broker Architecture. In the current state of computing, there are many different vendors, many different computer platforms, and many different software applications, resulting in a lot of duplication and incompatibility. Now that we have the Internet as a common ground for sharing development strategies, applications, and application-development tools, it is not necessary for these incompatibilities to exist, and neither is it necessary for a consumer to be forced to use any one particular computer platform.

CORBA is a strategy, and a set of tools which enables reusable programming objects to be used by many applications in a platform-independent manner. It is the combined effort of more than 500 vendors, engineers, and end users, organized as the Object Management Group (OMG). CORBA is a set of specifications for platform-independent, interoperable, distributed object-oriented applications. By using CORBA specifications, software vendors can create truly global software that can be distributed over the Internet, and run on a multitude of systems.

CORBA is an infrastructure that provides general services, and request and response capabilities at a low level. The distribution of objects written in a variety of programming languages is supported. CORBA does not define the upper level architecture; this is left up to individual developers. See Object Request Broker.

cord lamp In manual switchboards, or any type of cord panel where indicators are used, the lamp is a small bulb associated with a physical socket connection for a cord jack, which shows whether that specific circuit is active. Cord lamps were used on old telephone switchboards as indicators of active switches.

cordboard A very early type of switching panel, human-operated with long cloth-wound patch cords which plugged into jack receptacles on the desk level and interconnected, as needed, with jack receptacles on the wall corresponding to the 'local' phone desk. See switchboard.

cordless switchboard Unlike cordless phones, which are used in wireless systems, a *cordless switchboard* is not a new wireless technology, but rather an older switching technology in which human operators used keys instead of patchcords to connect call circuits. While this was a great improvement over patchcords, and was still used on many long-distance circuits until the 1970s, it was slow and expensive compared to all-electronic automatic switching systems.

cordless telephone A battery-powered telephone handset with a short antenna, and a separate charging element and AC adaptor, used for wireless communications, usually through very short range radio signals. While a cellular phone is a type of cordless phone, the phrase *cordless telephone* is usually used to refer to very short-range phones used within buildings or circumscribed areas. Many of the better cordless phones operate in the 900 MHz frequency range, and channel switching is provided in the event that interference occurs on the current channel.

The charging element and AC adaptor that comes with a typical cordless phone is used for cradling and recharging the phone when not in use. It sometimes also functions as an intercom station and speakerphone, depending upon the brand.

Cordless phones are predominantly analog, but more digital phones are being produced, resulting in more options for interfacing with a computer or providing secure or semi-secure communications.

core 1. Center, inner, inmost. 2. A central strand or wire around which other conductive or protective layers, strands, or insulating materials may be wound. Usually the main conducting portion of a wire assembly. In most electrical installations, copper wire is used as a conductive core. In fiber optic cable, the core glass is usually surrounded by a layer of lower refractive cladding glass. See cladding glass. 3. A central bar, often of iron, around which a coil is wound to create an electromagnetic part. See armature, coil, electromagnet. 4. A small doughnut-shaped magnetic component used for computer storage, with polarity representing binary states. See core dump. 5. A central, removable strand around which other materials may be wound or braided in order to provide a brace for molding their shape.

Core, the 3D Core Graphics System. A baseline specification developed in the mid-1970s to encourage standards for device-independent graphics. This led to development of the Graphical Kernel System (GKS), an official standard for 2D graphics. For 3D graphics, GKS-3D and the Programmer's Hierarchical Interactive Graphics System (PHIGS) became official standards in the late 1980s.

core diameter A description of the thickness of an inner, usually conducting material, as of copper wire or the inner layer in a fiber optic cable. See American Wire Gauge, Birmington Wire Gauge.

core dump A copy of the contents of core memory from a process error condition, usually consisting of undecipherable symbols and unprintables that can make a terminal or printer go crazy. On large systems, the output can be voluminous. Irate receivers of "spam," unsolicited commercial email, have been known to retaliate by sending back large core dumps.

Ezra Cornell was involved with many early telegraph line inventions and installations and founded Cornell University.

Cornell, Ezra (1807-1874) Cornell was talented in both business and mechanics and was associated closely with Samuel Morse, and Hiram Sibley, founder of the company which became Western Union. Cornell contributed to early telegraph installations and helped Morse construct the historic Washington, D.C. to Baltimore line that was funded by the U.S. Congress. Cornell designed early insulators from glass plates. Cornell remained a lifetime director of Western Union and was the chief founder of Cornell University.

corner reflector In its simplest form, two intersecting, or joining flat reflective surfaces with enough of an angle between them (usually 20 to 160 degrees, and often 90 degrees on the reflective side) to allow reflectance of a beam. The corner may also be in three planes, shaped like the corner of a room where the walls join the floor or ceiling. Corner reflectors are common in radar applications. The materials vary, but mirrored glass and metals are often used.

corner reflector antenna A type of antenna which combines a primary radiating element in relation to two angled metallic surfaces, or rods arranged in a plane. Various styles of corner reflectors are used in UHF television reception and radar applications.

corona A halo, glow, or other luminous surrounding from various causes including refraction, particle movement, ionization, radiation, reflection. St. Vitus' Fire, as reported by sailors, is likely a kind of corona effect. Voltages around power lines and antennas can sometimes ionize the surrounding air, resulting in a whitish-blue corona effect.

Corporation For Open Systems International COS. A nonprofit vendor-supported organization created in 1986. It was established to further acceptance and use of data processing and data communications equipment, and to encourage multivendor product compatibility in these areas. COS is involved in various standards efforts, particularly those involved with test methods and certification requirements.

Corporation for Research and Educational Networking CREN. Formed in 1989 when BITNET merged with the Computer+Science Network (CSNET). http://www.cren.net/

corresponding entities In ATM networking, peer entities with a lower layer connection between them, coordinated through protocol control information.

corrosion A wearing away, or alteration by chemical action, often leaving a residue such as rust or film as a byproduct of the corrosion. Many electrical wires and components are coated or bonded in order to prevent corrosion. Some elements resist corrosion, making them useful for applications in corrosive environments. See oxidation.

C

CoS See class of service.

COS 1. compatible for open systems. 2. See Corporation for Open Systems International.

COSETI Columbus Optical SETI. See SETI.

COSINE Cooperation for Open Systems Interconnection Networking in Europe. A program established by the European Commission to utilize Open Systems Interconnection (OSI) to interconnect various European research networks.

COSPAS/SARSAT A cooperative effort begun by the United States, the USSR, France, and Canada in the later 1970s. It supports satellite communications-related search and rescue operations which enables information such as the location of distressed aircraft or marine vessels to be communicated to rescue systems. COSPAS/SARSAT operates in conjunction with the emergency position-indicating radiobeacon (EPIRB) to support the Global Maritime Distress and Safety System (GMDSS).

COTS 1. commercial off the shelf. 2. Connection Transport Service.

Cotton-Mouton effect When light passes through a pure liquid in a direction perpendicular (normal) to an applied magnetic field, the liquid becomes doubly refracting. See Kerr effect.

couch potato *colloq.* An individual who spends an excessive amount of time lounging on a couch watching television, playing video games, or doing nothing in particular.

coulomb A unit of electrical quantity named after Charles A. de Coulomb. A unit for the amount of electrical charge in the meter-kilogram-second (MKS) scale that passes through a circuit in one second at one ampere (unvarying) current. The international coulomb is the quantity of electricity which will deposit 0.0011180 grams of silver when passed through a neutral solution of silver nitrate in water. A coulomb can also be expressed as the quantity of electricity on the positive plate of a condenser of a capacity of one farad when the electromotive force is one volt. See ampere.

Coulomb, Charles Augustin; de Coulomb, Charles Augustin (1736-1806) A French physicist and engineer who experimented with applied mechanics and electromagnetism. In 1785, he demonstrated the laws of electromagnetic force between elements using artificial magnets with well-defined poles in which the associated phenomena could be more clearly observed. The Crater Coulomb on the moon, and the coulomb unit of electric charge are named after him.

Coulomb's law A description of the magnitude of an electromagnetic charge. Two electromagnetic point charges will attract or repel one another with a force directly proportional to the product of their charges, and inversely proportional to the square of the distance between the two point charges. This phenomenon is more easily observed than many in the field of physics. Named after Charles Coulomb.

Charles Augustin de Coulomb experimented with electromagnetism.

counter-rotating ring In ring network topologies, data typically travels in one direction along each path. In a counter-rotating ring, there are two signal paths, each one traveling in the direction opposite to the other. See Fiber Distributed Data Interchange, Token-Ring.

counterpoise 1. A state of balance, counterbalance, equilibrium. 2. A power or force acting in opposition such that the opposing forces are equivalent, or balanced. 3. A structure designed to balance the transmitting or conductive properties of a circuit.

cover page The first page of a printout or facsimile that identifies information about the nature and date of the printout, and often the sender and intended recipient. On facsimiles, it is wise to note how many pages are following, so you can be sure the entire transmission was received. See banner.

coverage area In news and entertainment broadcasting and cellular communications, the geographic range of users/subscribers. Outside of the coverage area, signals will be weak or absent. This is not a problem for broadcast subscribers, who usually know they are either inside or outside the range of a certain channel, but for mobile communications users, driving out of range in the middle of a conversation can be a problem. For this reason, some mobile communications will provide a signal that indicates that the limits of the range are nearby, and some handsets will have a light or message which indicates the user is outside the service range.

CP/M Control Program/Monitor, Control Program for Microcomputers. A widespread operating system in the late 1970s and early 1980s written by Gary Kildall. It ran on Intel 8080 and Z80 microprocessor families. Kildall formed Inter-galactic Digital Research, later Digital Research. The initial version of MS-DOS was derived from CP/M and at a casual glance, exactly resembled it in syntax and function. Digital Research continued to develop CP/M into CP/M86, and later DR-DOS. Kildall also created a multitasking version of the operating system. See Digital Research.

CPAS Cellular Priority Access Service.

CPCS See Common Part Convergence Sublayer.

CPD See Call Processing Data.

CPE See Customer Premises Equipment, Customer Provided Equipment.

CPI See common part indicator.

CPL commercial private line.

CPN See Calling Party Number.

cps See characters per second.

CPU See central processing unit.

CQ In early radio transmissions, CQ was often used as a call to operators (all stations), a way of getting general attention. There are some historians who believe this may also have been used as a distress call and some have interpreted it as meaning 'Come Quick' although this may have been attributed after the abbreviation had been around for a while. Baarslaq has written that the Marconi Company requested *CQD* to be established as a distress call (presumably *CQ Distress*) to distinguish it from a general CQ call in early 1904.

CQD A radio distress call which predated SOS. See CQ.

CR 1. See carriage return. 2. call reference. 3. connection request. Open Systems Interconnect (OSI) transport protocol data unit. 4. customer record.

cracker A specific subset or connotation of 'hacker' for someone who gains unauthorized access to systems. Since breaking into systems can sometimes be done through sheer persistence and brute force techniques, not all cracking is hacking in the sense of the word that implies the application of brilliant and elegant solutions. The term was derived from *safe cracker*. See hacker.

C

CRAFT Cooperative Research Action For Technology.

crankback In ATM networks, a mechanism that allows the release of a connection setup that has encountered an error condition (as in a failed Call Admission Control) to permit rerouting of the connection.

crash System failure, lockup. On a computer, a crash usually implies a complete lockup of the operating system (this shouldn't happen, and very rarely happens on good operating systems), and usually requires a reboot.

Cray, Seymour Seymour Cray founded Cray Research in 1972, a company that was well-known for supercomputers for over 20 years. The Cray 1 supercomputer was announced in 1975 and the Cray 2 in 1983.

Supercomputers attract less excitement now than they did a decade ago because the general processing speed and level of functionality of desktop computers have both increased so dramatically that the distinction between the high end and the low end seems less dramatic. Many supercomputing operations are now run on desktop computers internetworked with the Linux operating system. This is very cost-effective for research labs, educational institutions, and government agencies.

CRC See cyclic redundancy check.

Creative Computing A pioneer popular magazine for computer consumers, established in 1974.

credentials Documents or expressions of trust and confidence, particularly those which indicate the capability to perform a function or task. The Pretty Good Privacy encryption system has an interesting aspect in which parties can vouch for the good name and veracity of public key holders. See Pretty Good Privacy.

credit card phone A pay telephone equipped with a card slot instead of or in addition to a coin slot to read magnetized credit cards and calling cards. These are becoming increasingly common, and are handier than phones where long credit card numbers and passwords have to be manually keyed.

Creed Telegraph System A system designed

by a Canadian inventor Frederick Creed. Creed devised a typewriter-style tape-punching machine that improved on the speed and utility of existing manual punchers, and which began commercial distribution in 1908. His automated perforators could operate up to 150 words per minute. To this he added a translating and printing system which eventually became a teleprinting transmitter/receiver system sold in England by 1927.

Creed, Frederick George (1871-1957) A Canadian inventor and telegraph operator who developed the Creed Telegraph System, an automated teleprinting transmitting/receiving system, in 1889. He then traveled to England to manufacture the Creed Printer. By 1898 he had demonstrated that he could send telegraph messages at sixty words per minute, and his inventions were put into commercial use by 1913. Creed was a member of the Board of Directors for the International Telegraph & Telephone Company (ITT). See Creed Telegraph System, telegraph history.

Creighton, Edward (1820-1874) Creighton was experienced in building communications lines and roads, and thus became the organizer of the western expansion of the telegraph system for Western Union, under Sibley and Cornell. Creighton surveyed the first transcontinental route across the wilderness in 1860, which followed the trails of the recently established Pony Express. Creighton worked in conjunction with the Overland Telegraph Company to carry out construction on the section west of Omaha.

In those days, building a line involved more than muscle and materials, it involved traversing roadless, supplyless, lonely distances without comforts of any kind. It also involved negotiating with native inhabitants and working out problems associated with roaming herds of buffalo who thought the telegraph poles were installed for their convenience as backscratchers. The line construction was originally estimated to take two years at a cost of over a million dollars. Under Creighton's supervision, the job was done in four months at a fraction of the original cost estimate. Fortunately, Creighton owned stock in Western Union and was able to benefit from this astonishing engineering feat. He used his gains to found Creighton University, and to contribute to many civic projects.

CREN See Corporation for Research and Educational Networking.

CRF 1. See cell relay function. 2. See connection-related function.

crimp tool A common handheld wiring installation tool which resembles a fat, snub-nosed set of pliers. It is designed to facilitate the cutting, stripping, and crimping of wires.

CRIS Customer Record Information System.

critical angle The angle at which a beam is reflected within a transmissions medium such as optical fiber. This can be very important to the strength of the signal over distance, and the total distance the signal can travel. It may be modified by the angle of the beam, the thickness of the fiber, and various impurities (doped elements) that may have been introduced.

critical charge The amount of charge needed to initiate a process or to change the state of or value of data being stored or processed.

critical fusion frequency CFF. In a display device, the refresh rate frequency above which the individual scans are fused by human perception into a single frame or image. This is not a set number, as, for example, on a cathode ray tube (CRT) display, it is related to the rate of persistence of the visible light from the excitation of the phosphors. As a rule of thumb, though, most systems show a nonflickering image at about 60 frames per second, or at about 85 Hz.

Critical Technologies Institute CTI. Organization established within RAND by an Act of U.S. Congress in 1992, and primarily sponsored by the White House Office of Science and Technology Policy. CTI works with the Federal Coordinating Council for Science, Engineering, and Technology to assure that technologies critical to national interests are identified and supported.

CRM See cell rate margin.

CRMA See Cyclic Reservation Multiple Access Protocol.

Crookes Dark Space In a cathode ray tube (CRT), as the gas pressure in the tube is gradually diminished, the glow surrounding the cathode detaches, leaving an area that is dark around the electrode, an area that may become quite large at low pressure levels. In a tube that has some air in it, this region can be more easily distinguished as being between the cathode glow on the inside, and the negative discharge glow on the outside. Outside the negative glow is the Faraday Dark Space. See Faraday Dark Space.

Crookes, William (1832-1919) An English physicist and chemist who investigated the effects obtained by passing electrical charges through various gases. Crookes tubes were subsequently used by Röntgen in his discoveries of X-rays.

crop marks In printing and desktop publishing, small marks, usually resembling crosses, which are placed at the corners of an image to be printed on paper larger than the final *trim size*. Note that there will be no room for crop marks if the extents of an image reach the edge of the paper on which they are laid out. If you are desktop publishing an 8.5 *x* 11 image with *bleed*, or with crop marks, you will have to select a paper size of *larger* than 8.5 *x* 11 (*letter extra* is a common setting) for there to be room for crop marks and other printers' marks. See registration marks.

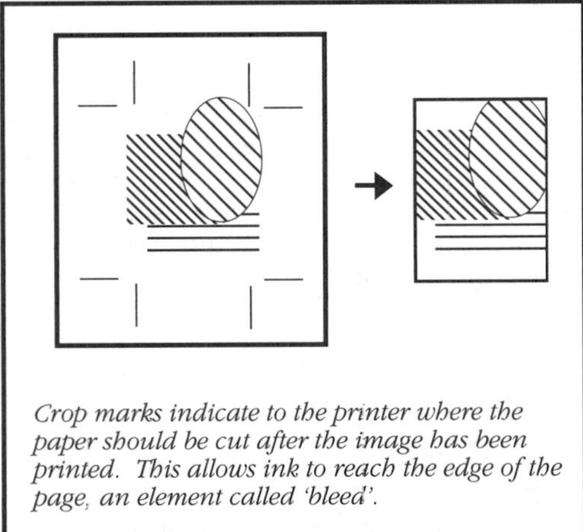

Crop marks indicate to the printer where the paper should be cut after the image has been printed. This allows ink to reach the edge of the page, an element called 'bleed'.

cropping The process of truncating information, usually an image. Cropping refers to the 'chopping off' of unneeded elements or data. For example, excess blue sky may be *cropped* from a photo, during the development process, or after the photo is printed. Cropping does not imply any change in size or proportions to the part of the image which remains. In computer imagery, the cropping may involved removing the data only from the display area, the data regarding the 'hidden' or cropped information may still be in the computer's memory, so it can be quickly restored, if needed. Cropping or *trimming* is often done by printers when cutting down print jobs with *bleed* to their finished size. See scaling.

cross assembler An assembly language programming and translation tool which allows assembly language symbolic coding to be written on one system, which is intended to run on another system. This is convenient because development machines often require more resources (more speed, memory, etc.) than the system on which the software will eventually run. There may not even be a machine available on which to run the software until the software is partly, or mostly written, due to production schedules, or cost. A cross assembler allows software development to proceed without hindrance from the hardware schedule.

cross connect A point in a circuit where a new or temporary connection is created by wiring between existing circuits or between facilities. Used variously for diagnosing problems, rerouting, or adding circuits.

Cross-Industry Working Team XIWT. An organization established to promote the understanding, development, and application of cross-industry National Information Infrastructure (NII) visions into practical technologies and applications, to facilitate communication between stakeholders in the public and private sectors. The XIWT Web site provides links to information, many reports, and white papers. http://www.xiwt.org/

crossbar switch In older mechanical telephone switching systems, a crossbar switch was similar to a relay, except that it was controlled by two external circuits, and was used in more complex switching arrangements, such as those needed for long-distance connections. It was devised in the late 1930s, and AT&T developed a version based on pioneer work by Swedish engineer Gotthilf Ansgarius Betulander. In 1938, the first crossbar central office went into service in Brooklyn, the same year the infamous "War of the Worlds" broadcast frightened credible residents of the area (who tied up phonelines in their panic).

The crossbar switch eventually succeeded the widely used, but troublesome panel switch in the 1950s, and step-by-step switches in the mid-1970s. See Callender switch, Lorimer switch, panel switch, step-by-step switch.

crossover cable A cable in which a pair of wires are reversed at one end of the connection. This is commonly done to convert a serial communications cable to a 'null modem' cable. In this case the transmit and receive wires are crossed, or switched over. In RS-232 specification cables these are lines two and three.

crosspost *v.* A USENET term for a message which is simultaneously submitted to two or more (sometimes hundreds) of newsgroups at one time. Crossposting is discouraged and looked upon as bad Netiquette except in very specific circumstances. It is important to look at the TO: field of a crosspost, if you reply to it, to make sure you remove all unnecessary newsgroups and don't perpetuate the extraneous posts.

crosstalk A term for undesirable electrical interference, usually from nearby lines. The crosstalk may be so excessive that a telephone conversation from another line can actually be heard. Crosstalk usually occurs when inadequate spacing or shielding has been provided.

CRS See cell relay service.

CRT See Cathode Ray Tube.

CRTC See Canadian Radio Television and Telecommunications Commission.

cryptanalysis The research and analysis of cryptography, that is, message or data encryption. While cryptanalysis is generally the art and science of a broad range of cypher-related concepts, it is also more narrowly understood as the actual analysis and breaking of a cyphered message without foreknowledge of its content, structure, or any keys that might be needed to discern its contents.

cryptochannel A communications channel which is encrypted in some way to provide privacy and security to the conversants. When carried out through a computer network, or over a digitally encrypted mobile communications line, or digital telephone line, many means can be used to hide the signal, or the contents of the signal. These include key encryption, scrambling, frequency hopping, and others. Cryptochannels were not generally available to the public before digitally encrypted data communications, introduced to consumers in the late 1980s and 1990s; they were mainly used in government communications, particularly in the military. Now that encryption and secure channels are becoming available to almost everyone, this may change the way society communicates.

cryptography The process and study of concealing the contents of a message, or transmission from all except the intended recipient. It is the primary means of security in telecommunications. The development of digital communications (ISDN, digital cellular, etc.) makes it easier to provide security, as typical unscrambled raw data or broadcast signals can be intercepted by unauthorized viewers. See Clipper Chip, cryptochannel, PGP.

Cryptolope A type of electronic cryptographic container for the secure packaging of digital information, introduced in 1996 by International Business Machines (IBM). A Cryptolope is a public/private key encryption specification which provides a means to package and distribute control information and content in one package. The control information includes pricing, licensing, and conditions of usage. Also included are network addresses and usage data distribution instructions. Cryptolopes are implemented through Web browser plugins.

The Cryptolope package is organized in data layers, including a bill of materials (BOM) describing container contents; a clear text abstract of the contents, author, etc.; the encrypted contents; intellectual property rights and related copyrights and usage rights, etc.

crystal 1. A substance which is characterized by a repeating internal structure occurring during the solidification of an element or mixture. The characteristic repeating structure is often manifested in the outward appearance. Many crystalline forms are transparent, or nearly so. See piezoelectricity. 2. A piece of transparent, or semi-transparent quartz, usually colorless. See quartz. 3. A crystalline material used in electronics for various purposes such as timing, rectification, and frequency evaluation. See crystal detector. 4. A wave-sensitive semiconductor used in electronics for applications such as radar detection.

crystal detector An elegantly simple, early form of radio device which superseded the coherer. A crystal detector took advantage of the rectifying properties of various natural and synthetic substances, commonly galena and carborundum. These materials have a property of allowing electrical alternating current impulses to pass through in one direction only. Thus, they can be used to convert AC frequencies to a direct current (DC) *half-wave*. AM radio signals are converted from radio frequencies to audio frequencies which are audible through headphones or speakers.

Crystal detectors could be built on a very small scale, and could be used without power sources or amplification, when carefully tuned and connected with high impedance headphones. In essence, they were the first portable radios, and were popular for field and hobbyist uses.

The earliest sets used natural crystals, but later a number of synthetic crystals were developed, with various properties and degrees of sensitivity. Portability could be increased with sets that used crystals that could be tightly coupled with the catwhisker. Some of the more elaborate sets included tuning coils. Eventually crystal sets were superseded by vacuum tube radios, which provided amplification and a much higher degree of electronic manipulation and control

Crystal radio sets are still sold as hobby kits from electronics suppliers, many of whom are on the Web. See catwhisker; Pickard, Greenleaf Whittier; piezoelectric.

This crystal detector, with its tuning coil wound around a hollow core, is from the Bellingham Antique Radio Museum collection.

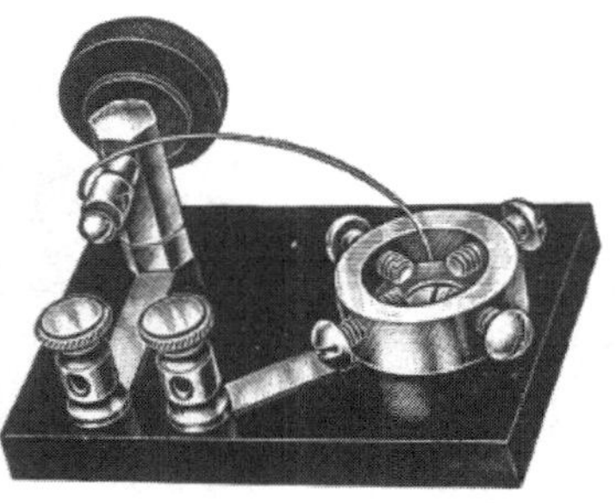

This diagram clearly shows the catwhisker and mounting base for the crystal, with screws to hold the crystal in place. The mounting posts on the left are for connecting the wire for the headphones. The screw provides fine adjustment (tuning) for the catwhisker.

crystal microphone An early type of microphone employing a piezoelectric crystal.

crystal pickup A particular type of stylus on an instrument such as a phonograph, which is created from a piezoelectric crystalline material that changes in shape and consequently generates an electrical impulse which is then interpreted by the electronics into sound.

crystal shutter A type of safety mechanism which is used in conjunction with crystal detectors to block excess radio frequency (RF) energy from reaching and possibly damaging the components.

CS communications satellite.

CSA 1. Callpath Services Architecture. 2. See Canadian Standards Association.

CSC customer service center.

CSMA See Carrier Sense Multiple Access.

CSNET See Computer+Science Network.

CSPP See Computer Systems Policy Project.

CSR 1. See cell switch router. 2. customer service record.

CST See Computer Supported Telephony.

CSTA See Computer Supported Telephony Application.

CSU See Channel Service Unit.

CSU/DSU Channel Service Unit/Data Service Unit.

CSUA See Canadian Satellite Users Association.

CT 1. Call Type. 2. Cordless Telephone. 3. Conformance Test. A test intended to determine whether an implementation complies with the specifications of, and exhibits behaviors mandated by, a particular standard.

CT3IP Channelized T3 Interface Processor. A Cisco Systems commercial fixed-configuration interface processor used with Cisco 7*xxx* series routers. The CT3IP provides 28 T1 channels for serial transmission of data, each with *n x* 56 kbps or *n x* 64 kbps bandwidth. Unused bandwidth is filled with idle channel data. The CT3IP does not support multiple T1 channel aggregation (bonding).

CTCA See Canadian Telecommunications Consultants Association.

CTD 1. See cell transfer delay. 2. Continuity Tone Detector.

CTI 1. Call Technologies, Inc. 2. See Computer Telephony Integration. 3. See Critical Technologies Institute.

CTIA 1. See Cellular Telecommunications Industry Association. 2. Computer Technology Industry Association.

CTL See Complex Text Layout.

ctrl, CTRL, Ctrl *abbrev.* Control. The control key is a special function key common on computer keyboards which is held down simultaneously with other keys to modify them, usually for system interrupts and similar tasks. For example, *Ctrl-C* indicates that the Control key and the C key are held down at the same time. On most systems, Ctrl-C is standardized to interrupt the current process. Control characters described in print, or indicated on a screen, are commonly designated with a caret

(^) for a prefix, e.g., Control-D would be written *^D*.

CTS 1. See clear to send. See RS-232. 2. Communication Transport System. 3. Conformance Testing Services.

CTSS Compatible Time-Sharing System. A developmental time-sharing system from the early 1960s.

CTTC coax to the curb. Coaxial cable installed into residential areas. See fiber to the home.

CTX See Centrex.

CU-SeeMe A Macintosh- and IBM-licensed PC-based videoconferencing program from Cornell University. It supports video, audio, and other utilities over Internet, with plans to make it Mbone-compatible. The encoding is proprietary. See Cameo Personal Video System, Connect 918, MacMICA, IRIS, ShareView 3000, VISIT Video.

cube *colloq.* An early model of NeXT computer that was shaped like a black cube, essentially a 'tower' model which could be easily accessed and upgraded. The shape of the NeXT later went to a more conventional thin 'slab' which could sit under the monitor.

cure *v.t.* To process so that the essential properties of a substance are changed, usually to improve them, as in curing a metal to give it strength or resilience, or curing a hide to preserve it.

curie A unit used for describing the strength of radioactivity, which is equal to 3.7×10^{10} disintegrations per second. It is named after Nobel scientist Marie Curie who did substantial pioneer work in radiation.

Curie point, Curie temperature A temperature at which peak levels of a dielectric constant occur in ferroelectric materials.

current (symbol - I) Movement of electrons through a conducting medium, usually expressed in amperes. Electricity moving through a wire or cable is current. See ampere.

current amplifier Any natural, mechanical, or electronic device which provides greater output of an electrical signal than the input signal. A public address system (PA) is a type of current amplifier, as are other microphone and speaker combinations.

current cell rate CCR. In ATM networking, a traffic flow control concept that aids in the calculation of ER, and which may not be changed by the network elements (NEs). CCR is set by the source to the available cell rate (ACR) when generating a forward RM-cell. See cell rate for a chart of related concepts.

cursor A solid, blinking, underlined, colored, highlighted, or otherwise visible means of indicating to a computer user that the computer is ready for input, and where that input can be received.

On a command line, the cursor is usually positioned on the far left, after the last line executed, or last output, next to a *command prompt*. Pipe symbol or *I-beam* cursors are favored for text entry, as they help the user to locate the baseline of the text font, or to accurately position the cursor between characters. Pointers, grippers, and hands are favored for graphical pointing, selecting, or dragging of various screen display elements.

customer access line charge CALC. The charge for connecting a private branch phone exchange (PBX) to the central office exchange (Centrex).

cutoff decrease factor CDF. In ATM networking, CDF controls the decrease in the allowed cell rate (ACR) associated with the cell rate margin (CRM).

cutover That moment when a system is switched from one to another, as from an old system to a new one, which has often been installed redundantly until it is fully tested and operational. It is the goal of most people executing a cutover, that it happen as quickly and uneventfully as possible, preferably so users on the system don't even notice the change, or are only momentarily inconvenienced. See half tap.

cutter A mechanism for inscribing grooves in a recording medium such as a phonographic record. The mechanism is used to translate electrical impulses into physical patterns that can later be read and converted back into electrical pulses, usually auditory.

CWIS See Campus Wide Information System.

CWSI CiscoWorks for Switched Internetworks. Integrated management control technology (for topology, device configuration, traffic reporting, VLAN, ATM, and policy-based management) from Cisco Systems, Inc.

cyber- A prefix widely used with almost anything these days to indicate an electronic version of something. William Gibson is credited with popularizing the word 'cyberspace' to indicate an interconnected science fiction environment, in Neuromancer, in 1984. Cyber- and sometimes just Cyb- has since been used in many contexts from computers to music, as in cybrarian, cyberceleb, cyberphile, cyberspace, cybercast, cyberphant, cyberphobe, etc.

cybernetics A term introduced by American prodigy, logician and mathematician, Norbert Wiener, who collaborated with Arturo Rosenblueth and a group of scientists from various disciplines in developing many fundamental concepts of artificial intelligence. He authored "Cybernetics: or, Control and Communication in the Animal and the Machine" in 1948 to discuss ideas about self-reproducing machines and self-organizing systems. Cybernetics refers generally to the field of control and communications theory, encompassing both human and nonhuman systems. Wiener further described feedback theory in mathematical terms, and studied the flow of information from a statistical point of view. These disciplines have many practical applications in robotics.

In "Cybernetics," Wiener poses some provocative (and revolutionary at the time) parallels between neuron states and the electrical states of a binary computing device.

cyberspace 1. A term popularized by William Gibson in his popular science fiction/fantasy novel "Neuromancer" to describe a computer society. 2. A conceptualization of the computing machinery and culture which sees it as existing beyond the role of tool and communications device, as a meta-environment in which we can interact as part of a larger, perhaps not fully knowable, dynamic organism. See Dyson, George. 3. A content-rich virtual reality environment in which participants interact through a variety of sensory data input devices.

CyberStar A commercial global satellite communications system designed to provide broadband, interactive multimedia data transmissions. CyberStar is a venture of Loral Space & Communications. CyberStar operates on leased Ku-band transponders on the Telstar system.

cybrarian A compound word derived from *cyber[space]* and *librarian*. A cybrarian is a librarian or other research professional conducting research and information retrieval online, especially on the Internet. Given the astounding volume of free information on the Net, and the difficulty of narrowing the search and finding relevant information, in essence, locating the needle in the haystack, cybrarians provide valuable filtering and organization services. You can find almost any type of information on the Web, from people's names and addresses, to scientific abstracts, and more, and this could easily be a full-time occupation. See data mining.

cyclic memory A type of memory which can be accessed only when the process of memory access passes through that portion in its cycle that contains the information desired.

cyclic redundancy check CRC. A file integrity mechanism, and a file transmission error checking mechanism widely used in many areas of computing. The CRC is a calculate-and-compare mechanism. A block of data, the total data content of a file, or a group of transferred data can be scanned in order to create a numeric sum total intended to provide a simple representation of its contents. This total is then compared with one computed the same way after file compression, manipulation, or transfer and the values matched. If they do match, an error *may* have occurred, but with a low probability of likelihood. If they don't match, it is unlikely (although not impossible) that an error occurred, and the process can be repeated.

File transfer programs such as ZModem often included CRC methods to monitor data transfer. Some file compression formats, such as PNG, are divided up into logical data chunks, with each chunk incorporating a CRC to provide a reference for data integrity so the file can be checked without opening the image in a viewer. See checksum, magic signature.

cyclic shift A system in which anything that comes out one end of the production line, register, stack, or other physical or logical conveyance, goes back in at the beginning of the system in a continuous loop. Programmers sometimes use cyclic shifts to rearrange data.

cyclotron A device for devising nuclear manipulations which follow a helical path. There are many makes and models of cyclotrons, ranging also greatly in size. A common, basic configuration is an evacuated contained space in which charge atomic particles are guided and accelerated through a spiral path by various magnetic means. The centripetal path of the particles can be used to effect radiant emissions.

cypherpunk An individual advocating the prevention of tyranny through public access and widespread dissemination of electronic cyphers, encryption methods, and other digital security technologies, in order to ensure that their power and accessibility are not concentrated in the hands of only a few people or organizations. See Pretty Good Privacy; Zimmermann, Philip.

Czochralski technique A means of creating crystalline structures which are useful in semiconductor technologies. By passing the materials through a molten state, large single crystals can be grown. When crystals are drawn from a melt, it is known as "crystal pulling."

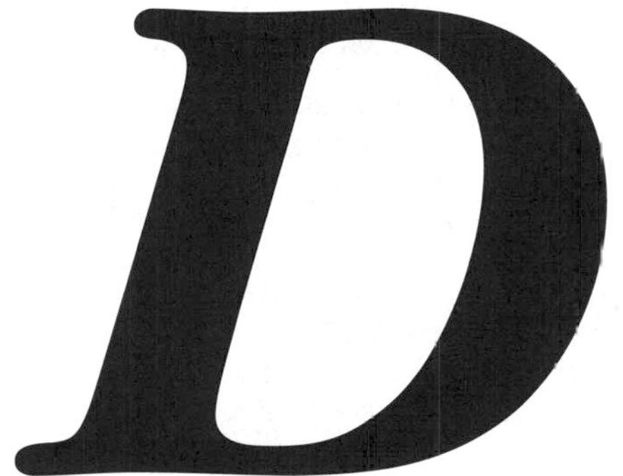

d Symbol for deci-, a prefix used to denote one tenth, or 10^{-1}, as in *deci*meter, *deci*bel.

D Symbol for electrostatic flux density. See flux.

D Block A Federal Communications Commission (FCC) designation for a Personal Communications Services (PCS) license granted to a telephone company serving a Major Trading Area (MTA). This grants permission to operate at certain FCC-specified frequencies. See A Block for a chart of frequencies.

D channel delta channel, data channel. In ISDN, a full-duplex signaling channel that carries control and customer call data information. The D channel handles various call setup and teardown functions. Depending on the type of service, the D channel controls associated B channels at 16,000 bps for Basic Rate Interface (BRI), or 64,000 bps for Primary Rate Interface (PRI) as shown in the following chart.

D connector A cable connecting standard housed in a shell that resembles the letter D. D connectors, especially DB-9, DB-15, and DB-25 are commonly used to interconnect computers and peripheral devices.

D region A region (as opposed to a layer) of the Earth's ionosphere which exists only in the daytime, starting from around 70 or 80 kilometers above the Earth's surface, and extending up to, and overlapping the E region, which is more clearly defined. The D region can have a significant impact on the propagation of radio waves, causing greater dissipation and attenuation when the region is active in the daytime. See ionospheric sublayers for a chart.

D-scope A type of C-scope radar display in which the target blips extend vertically to provide an estimate of distance.

DAMPS Digital Advanced Mobile Phone Service. Originally, AMPS was used as a 900 MHz frequency modulation (FM) transmission technology with bandwidth allocated according to frequency division multiple access (FDMA) schemes. To increase capacity and security, digital techniques for cellular were introduced and systems are being converted from AMPS to DAMPS. The two most prevalent means of dividing frequencies in DAMPS are time division multiple access (TDMA), and code division multiple access (CDMA). These two formats are not directly compatible. See AMPS, NAMPS.

D'Arsonval, Jacques Arsène (1851-1940) A French physicist who proposed utilizing the thermal energy of the ocean, but was unable to achieve a net gain in power generation. Succeeding generations of researchers contin-

ISDN Channel Functions

Abbrev.	Service	Notes
BRI	Basic Rate Interface	16 kilobytes/second using DSSI to control the two B channels and/or the X.25 format user data.
PRI	Primary Rate Interface	64 kilobytes/second using DSSI to control all the B channels. In conjunction with NFAS, the D channel can also control B channels on multiple PRIs.

ued to pursue this idea with better success due to experimentation and newer methods.

D'Arsonval introduced the first reflecting, moving coil galvanometer in 1882, an improvement on previous arrangements. By means of a small, concave mirror mounted on the coil, the instrument could reflect a beam of light to a calibrated scale. It could measure the current and voltage of direct currents, and was widely distributed in many forms. A number of electrical concepts and inventions are named after him. See D'Arsonval galvanometer.

D'Arsonval current A high-frequency, somewhat high-amperage, low-voltage current.

D'Arsonval galvanometer An early, simple type of galvanometer, consisting of a narrow, rectangular coil, suspended so that it could move between the poles of a permanent magnet to register a reading of direct current. Later adaptations of thermocouples or rectifiers to the D'Arsonval galvanometer permitted the conversion of alternating current (AC) to direct current (DC) in order to measure alternating current. See astatic galvanometer, galvanometer.

D'Arsonval movement A description used in contexts where a pointer associated with a dial moves to show a reading when stimulated by direct current.

D/A digital to analog. See D/A converter.

D/A converter digital to analog converter. A device which takes discrete digital information and converts it to a continuous form that can be carried over analog circuits, usually through one or more modulation processes. Thus, information from a computer can be converted by a computer modem to analog audio signals that are sent through a phone line, or digital signals from an Internet phone can be converted to analog pulses that can be heard over an analog headset. See modem.

D4 In T1 digital transmission lines, D4 is a type of channel bank. Channel banks carry out a variety of interface tasks including time slot framing, and detecting and transmitting signalling information. See SuperFrame.

DA 1. See desk accessory. 2. See destination address. 3. See Directory Agent. 4. See Directory Assistance.

DAA Data Access Arrangement.

DAB 1. See digital audio broadcasting. 2. dynamically allocatable bandwidth.

DACS Digital Access and Cross-connect System. A technology for reconfiguring a circuit, without manually changing the interconnections. It is similar to a multiplexer, except that changes can be made with software, rather than physical rewiring.

daemon A computer process which 'lurks' in the background to do work that is low priority, or that is requested intermittently, especially in Unix environments. Daemons often perform low-level operating tasks, to automate some aspect of a system administrator's responsibility, and are transparent to most users. A daemon may be a continuous background process, or may be intermittently generated as needed. Daemons are useful as print spoolers, mail message managers, and general resource allocators, especially for client/server requests that are invoked irregularly.

DAF Destination Address Field.

Daguerre, Louis Jacques Mandé (1789-1851) A French artist and inventor who made significant improvements in photographic imagery in 1839. His early photos, called *daguerreotypes*, were impressed in silver plated onto copper. They have a very soft, low contrast quality to them and the clarity of the image is affected by the angle at which the plate is held when viewed, due to the reflectivity of the metallic medium. They are fade-resistant, and many of the original daguerreotype images that still survive retain their images. See Talbot, Fox.

daisy chain *v.* To connect items individually one to another in a series. Communication through a daisy chain of electronic units may be unidirectional or bidirectional. SCSI devices such as hard drives, scanners, cartridge drives, and CD-ROM drives are frequently daisy-chained to one another and to one controller on the logic board. When chaining SCSI devices, care must be taken to assure that each device has a unique ID number (usually from 0 to 7), and that the last member of the series (the one farthest from the SCSI controller), or *chain* is terminated, either with a physical connector attached to the outside, or by setting external or internal switches accordingly.

daisy wheel printer A printer that uses a rotating ball with characters embossed on the surface of the ball. When the ball rotates and strikes the paper through a ribbon, the character is imprinted on the paper. This technology was once common on electric typewriters. Because of the mechanical encumbrance of the rotating ball (which usually also moves up and down), daisy wheel printers are somewhat slow and noisy. They are also limited in that you cannot change type styles without physically changing the printing ball. Nevertheless, in the 1980s, when many dot matrix printers had only low-quality nine-pin printing capabilities, the daisy wheel printer provided a higher print quality option. Daisy wheel printers have been superseded, for the most part, by faster, quieter, more flexible laser and ink-jet printers.

DAL See Dedicated Access Line.

Dalton, Orv An amateur radio enthusiast who contributed substantially to the design and construction of the first three OSCAR satellites. See OSCAR.

DAMA See Demand Assigned Multiple Access.

damped wave Radiant wave oscillations which gradually diminish in amplitude or which are being deliberately suppressed so that the amplitude diminishes.

damping *v.t.* The process of decreasing the amplitude of wave oscillations. The term is often used in reference to progressively suppressing sound waves (*sound damping*), though it can generally be used to indicate the suppression of a variety of types of energy, as electrical oscillations in a circuit.

Daniell battery An early, fairly simple chemical battery, providing approximately 1.1 volts per cell which was suitable for providing current for early telegraphic systems, like the Morse system, as it tended to last longer than other types in a closed circuit. It was comprised of a copper electrode in a copper-sulfate solution on one side of a porous separator, and a zinc electrode in a diluted sulfuric acid or zinc-sulfate solution on the other side. It is named after J. F. Daniell. See cell.

Daniell, John Frederic An English chemist who invented a chemical battery known as the Daniell cell which was used in early telegraph systems. See Daniell battery.

DAP See Directory Access Protocol.

DAQ Delivered Audio Quality.

DAR See digital audio radio.

dark conduction The property of a substance, such as a photosensitive material, to retain electrical conductance in darkness. This is usually a residual effect, and tends to diminish over time until restimulated by light.

dark fiber, dry fiber Plain, unconnected fiber optic cable, not currently carrying a signal. Since fiber is often sold as the hardware portion of a subscriber service, this phrase was coined to indicate fiber which is sold just as fiber, with the purchaser doing the wiring of the components and transmitters. See dim fiber.

DARPA See Defense Advanced Research Projects Agency.

DARPANET A network of the U.S. Defense Advanced Research Projects Agency, originating in 1969, from a desire on the part of the U.S. military to exchange information between different sites, and to provide redundancy in the event of an attack. This project grew to become ARPANET by 1972. In 1983, ARPANET had grown so large that it was split into MILNET specifically for U.S. military use, and NSFNET (National Science Foundation Network), which opened it up to researchers and scientists. See ARPANET, Internet, RFC 791.

DARS Digital Audio Radio Service.

DASD See direct access storage device.

DAT See digital audio tape.

data 1. Constituent basic elements of information which can be formally organized and combined to provide communication, most commonly through written means, though the term is not restricted to written communications. 2. Building blocks that can be manipulated and presented by electronic means, or which are used, interpreted, and organized by human perception and thinking.

data base See database.

data bus See bus.

data circuit A circuit which uses transmission wires and components suitable for the fast transmission of digital information.

Data Communications Channel DCC. In SONET, a channel related to the OAM&P which includes security and performance information associated with facility and network elements (NEs). Both generic and vendor-specific information can be included. The DCC is incorporated into both the section and line overhead.

data communications equipment DCE. A category of devices specified by the Electronic Industries Association (EIA) which typically includes common serial communications peripherals, such as modems and printers. These in turn interface with data terminal equipment (DTE). In frame relay networking, DCE has a more specific meaning, as switching equipment that is separate from the various peripheral devices that are connected to a network or workstation. See data terminal equipment.

data compression The process of encoding data to store it in a smaller amount of space. Data compression is typically achieved with specialized software tools, or with software built into data transmissions hardware. Data compression may be done in advance, if files are to be stored or transmitted later, or it may be done dynamically at the time it is needed, sometimes called *realtime* or *on-the-fly*, a capability that is built into some modems. There are many different general-purpose and specialized means of compressing data. Some data compression algorithms are paired with data decompression algorithms, for archiving search and retrieval, and audio/video recording and playback.

Data 'compression' tools do not always make the data smaller. For example, a very tiny icon file, when encoded with an image or general purpose data compression program, may actually be larger than the original by the time the header, decompression, or statistical information about the file is inserted by the compression program. Yet the same tool may quite effectively achieve as much as 60% compression on large images, so the selection of data compression technologies depends on finding the right tool for the job. To overcome this problem, three developers at Western DataCom have developed Ardire-Stratigakis-Hayduk (ASH), a compression scheme which incorporates some of the pattern-matching, and predictive concepts associated with artificial intelligence programming. This scheme attempts to broaden the scope of compression to handle many different types of data in the increasingly media-rich communications that are evolving. See Ardire-Stratigakis-Hayduk, Lempel-Ziv.

data compression approaches There are many practical approaches to data compression. The simplest means is to remove redundant data, such as gaps, spaces, or repetitions. This approach is used in encoding voice conversations, which typically have a lot of pauses; graphics which have large areas of similar colors; and text documents with repetition and blank spaces. Another means for compressing information is expressing it in a different way. For example, a bitmap image of a large letter *O* may require 15 kilobytes, whereas the mathematical definition for an ellipse that can define the letter *O* may require only 5 kilobytes. A third means to compress data is to try to match the human perceptual recognition of the presented data rather than the structural data characteristics of the original presentation. In other words, there are ways to display graphic images, or to play sound files so that they 'look the same' or 'sound the same' to the general viewer/listener, even though the construction and 'dynamic range' of the information may have been altered. Humans have a remarkable ability to conceptually 'add' information or 'construct' a view from a few clues. If you've ever watched a black and white TV show, and 'could have sworn' you had seen it in color, you've experienced one aspect of this phenomenon.

Data compression can be *lossy* or *lossless*, that is, it can retain most of the information in a file, or all of the information in a file. A commonly used lossy image format in which most of the information is retained is JPEG, often for displaying Web graphics and videoconferencing images.

With the ever-growing volume of data being unleashed by the capabilities of computer technology, and greater demands for perceptually-rich multiple media, the demands for data compression to reduce file space, speed transmission, and lower costs is very high. Some of the most promising recent data compression programs incorporate fractal and wavelet theories into their encoding techniques. See Ardire-Stratigakis-Hayduk, JPEG, Lempel-Ziv, PNG, wavelet theory.

data conversion The process of converting computer data stored in one format to another. The three most common reasons for converting data are: achieving compatibility (upward, downward, and inter-application); saving space and/or saving time (compression/decompression); and needing to convert between digital and analog forms of information.

When software applications are upgraded, they often incorporate new features that are not available in the older versions. Data conversion may be necessary to store information in the older or newer file format, and some of the information may be lost in the conversion process.

Computer data conversions tend to happen within 'families' of data. That is graphics formats are frequently interchanged, audio formats are frequently interchanged, but there isn't much need to convert audio data into visual data, except for experimental applications. That is not to say computer data has to be rigidly defined; it doesn't. For example, the Interchange File Format (IFF) developed jointly in the mid-1980s by Electronic Arts and Commodore Business Machines is a broad specification for data definition that can be generically applied to text, sound, and graphics. Similarly, Adobe PostScript fonts, while following specific guidelines, are not just fonts, but rather are shapes that fit in the context of a larger picture, that of a page description language which is capable of describing many types of graphical elements besides fonts.

Many shareware and commercial data conversion utilities are available, especially for converting among the myriad graphics formats such as PNG, JPEG, Compuserve GIF (which now uses the PNG specification), BMP, ILBM, and TIFF. PNG, JPEG, and GIF are the most commonly used raster graphics formats on the Web, and TIFF is the most widely used graphics format in the publishing and document industry (faxes are also defined within the TIFF specification). ASCII is very widely used in text conversions, and Microsoft's Interchange Format (Rich Text Format or RTF) can be used for text conversions that retain formatting such as bold, indents, fonts, etc. For database information, dBASE formats are often used for converting between one program and another. See data compression, digital to analog conversion.

Data Country Code DCC. In networking, the DCC is a numeric code which specifies the country in which an address is registered for a public network. Each data country code is in Binary Coded Decimal (BCD) format, contained in two octets, in ISO 3166 format. See Data Network Identification Code.

data description In a data dictionary, a unit or group of information which may comprise one or more of: a definition of meaning and usage, attributes or characteristics, and category or classification information.

data dictionary 1. A reference set of data descriptions which can be machine-processed, and shared by a variety of applications. 2. In database management, a lookup reference of data descriptions with a format or relationship such that the database engine can efficiently save, extract, or scan information to/from the dictionary according to the needs of the database.

data element A basic unit of information defined generically, or for a specific application. For example, a data element in an employee database might consist of a name or job category. A data element may be further defined as including data items components or subcategories.

Data Encryption Standard DES. A cryptographic system consisting of an algorithm and a key comprised of a long series of numbers which are used together to transform data into information which appears unintelligible, and back into data by the person for whom the information is intended. DES was developed by the National Bureau of Standards (now the National Institute of Standards and Technology (NIST)), and is intended for public use, and for government protection of certain federal unclassified data. See Clipper Chip, Pretty Good Privacy.

data entry The act of using a hardware interface to input data to a computing device. Data entry is commonly accomplished through a keyboard and mouse, but voice recognition systems, touch screens, and pen computers are broadening the choice of input devices. Typically, data entry is used to describe repetitive, discrete types of data, like database entries (names, addresses, order numbers, etc.), spreadsheet entries, etc. When the data is more fluid and conceptual and less repetitive, it is still, in its broadest sense, data entry, but is more likely to be described in terms of the type of application being used, such as word processing.

data exchange interface DXI. A layer 2, frame-based interface installed between a packet-based router, and an SMDS or ATM CSU/DSU. The DXI performs assembly and reassembly tasks on behalf of a router which may not have these capabilities. Since most routers now can handle these tasks, use of this particular type of interface is diminishing.

Data General DG. One of the better known computer companies in the 1970s, Data General was founded in 1968 to develop minicomputers, and grew to make the Fortune 500 list a decade later. With almost half a million systems installed worldwide, Data General targets high-performance computing environments, including scientific, technical, and industrial sites.

data grade circuit A distinction made to indicate the more stringent needs of computer data transmission, as compared to voice grade transmissions. Data is transmitted at different frequencies, and is more precise and easily interrupted than a phone conversation. Phone conversations use a narrow frequency range, and have a great tolerance for pauses, spaces, and extraneous noise, particularly since part of the processing equipment in a voice conversation is the human brain, which understands context and innuendo, as well as just the words. Data on the other hand, requires a cleaner line, less interference, a greater frequency range, and has low tolerance for pauses and spaces if they affect the integrity of the information that is being transmitted.

Voice grade circuits over phone lines are improved by load coils, a system of looping the wires that are strung along utility poles. Data grade circuits built in the same basic way are hindered by load coils, as they introduce noise at the higher frequencies used.

Data Link Connection Identifier DLCI. In frame relay networks, a unique 10-bit identification number assigned to a virtual circuit (VC) endpoint that identifies the endpoint within a local access channel. This is used for switching and multiplexing. In ISDN, a unique 13-bit identifier.

Data Link Control DLC. A layer in the Open Systems Interconnection (OSI) model. DLC is responsible for a number of administrative and error-checking functions. In satellite communications, some special adaptations are needed at this level to accommodate the high bandwidth/delay characteristics of these transmissions.

data link layer DLL. In the Open Systems Interconnection (OSI) reference model, the layer that ensures transmission of data between adjacent network nodes. Bridges work at the data link layer. See Open Systems Interconnection.

Data Link Switching DLS, DLSw. Originally developed by International Business Machines (IBM), and in 1993 submitted to the IETF as an informational Request for Comments (RFC). DLS defines a reliable means of transmitting SNA and NetBIOS TCP/IP traffic using IP encapsulation through multiprotocol router networks. See RFC 1795.

Data Link Switching Workgroup DLSW. A group which worked on the development of a new switching standard for integrating networks over TCP/IP. See Data Link Switching.

Data Link Switching Special Interest Group DLSw SIG. A vendor implementation group created in 1993 to address some of the issues raised in regard to RFC 1434 in which International Business Machines (IBM) provided preliminary information on Data Link Switching. This resulted in a new RFC being submitted to the IETF as RFC 1795 which obsoleted RFC 1434. See Data Link Switching.

data mining The process of seeking out relevant information from a large storehouse of electronic information that may be on many different systems, in many different formats. Data mining involves using intelligent algorithms to search out relevant materials based on various parameters: previous search history, data patterns, information correlations, preferences of individual users, keywords, and other triggers implemented in a manner that maximizes the relevance of the information that is retrieved or flagged.

Data mining has become a topic of substantial interest and development due to the vast amount of information that is flowing onto the Internet. Anyone who has used a search engine and received 300,000 *hits*, after narrowing a search a couple of times, can see the value in data mining algorithms that can carry out some or most of the work in advance. See cybrarian.

data multiplexing See multiplexing.

Data Network Identification Code DNIC. An ITU-T internationally specified system of network host identification which permits individual local networks, tied to public networks, to be located and recognized for internetwork communication, much as a country code and local phone number identifies a phone line. This data network identification scheme, somewhat analogous to a phone number, is used to locate hosts on interconnected public networks by means of X.75. The DNIC is the first four digits of a longer 14-digit international code. The first three digits are assigned by the ITU-T to specify a data country code (DCC) and the fourth digit is assigned by the national Administration to specify the public data network within that country. Network Terminal Numbers (NTNs) are the responsibility of the administrators of the public network. See X Series Recommendations.

Data Numbering Plan Area DNPA. An ITU-T X.25-specified system of endpoint terminal identification implemented in the U.S. using the first three digits of a 10-digit network terminal number (NTN).

data over voice A means of including data on a transmissions line carrying voice signals by using frequency division multiplexing (FDM) technology to secure and utilize the remaining available bandwidth for the data signals.

data processing A broad category of activities encompassing any means of manipulating data. When computer applications such as word processors, spreadsheets, paint programs, etc., are being used, data is being processed.

Within database management systems, data

processing has a more specific meaning, which involves the creation, access, retrieval, manipulation, and analysis of textual and financial data. Data processing is commonly used in payroll accounting, statistical analysis and reporting, customer profiling, and many other common tasks related to commerce and business management.

data protection A broad category of actions and systems that are designed to protect data. There are two general categories of data protection: keeping the data available and in its desired form (uncorrupted), and keeping the data safe from unauthorized use.

In the first category, data backups, archiving, mirroring and other means are used to protect data from being lost or corrupted. This can occur at the local applications level, in the hardware, and at the overall systems level. Many file system directories are duplicated to provide access if corruption occurs in one. Some operating systems allow multiple versions of a file to be saved automatically, so that there is always a history of recent changes, and a previous version that can be used if needed. Backup hardware in the form of tapes, cartridges, optical media, and redundant drives are used by many for scheduled or dynamic backups.

In the second category, passwords, digital encoding/encryption, secure channels, proprietary formats, vaults, safe-deposit boxes, data certificates, digital signatures, etc., are all used to protect the data from unauthorized access, use, or abuse. See backup, backup file, encryption, mirroring, Pretty Good Privacy, RAID.

data rate A quantification of the input or transmission of computer data. Data rates are very situational. For example, in data entry jobs, the data rate may be the number of fields filled per minute, or the number of customer orders entered per hour. In network communications, it may be the number of bits or packets transmitted per second. See baud rate.

data service unit DSU. A device used in ISDN systems to permit computers to interconnect with digital phone services for end-to-end digital communications. It is similar to a modem in the sense that it fits between the computer and phone line service, but differs in that it does not perform analog to digital, and digital to analog conversions. The DSU is installed in the customer's premises, and connects the synchronous communications system through a four-wire line (usually a leased line), to the local central office. The DSU is used in conjunction with a Channel Service Unit (CSU) which is installed at the central office.

data set ready DSR. A control signal commonly used in serial communications, and included in the pinout specifications for the ubiquitous RS-232 electrical connector. The DSR indicates whether the communications device is connected and ready to begin handshaking. For example, assume the user has dialed a BBS or Internet Access Provider (IAP), and the called modem has just picked up the line. The DSR senses the connection and provides a signal that lets the hardware/software know that it can continue to the next step of negotiating a connect speed and beginning the communications. See data terminal ready, RS-232.

data striping A means of distributing data across drives in an array. A fault tolerant means of providing data security that is incorporated into *redundant array of inexpensive disks* (RAID) systems.

data terminal equipment DTE. A communications data terminal hardware specification. See data communications equipment.

data terminal ready DTR. A control signal commonly used in serial communications, and included in the pinout specifications for the ubiquitous RS-232 electrical connector. The DTR signals whether the communications device is connected and ready after it has successfully begun handshaking. For example, assume the user has connected with a BBS. The DSR verifies the connection, a connect speed is negotiated, handshaking begins, and the terminal is ready to continue communicating. The DTR signals this state of readiness. See data set ready, RS-232.

data typing The process of specifying or determining the format of a variable, file or block of data.

At the applications level, the trend is for data typing to be incorporated into applications so the user doesn't have to verify the data type, or convert it before using it. For example, the user may have a word processor that can read a variety of document file types, including various proprietary and open formats. The user selects the LOAD option, the software checks a list, extension, or file header, to try to determine the data type, and loads it with an appropriate filter to convert the data into a format that can be displayed by the program. The list of data types may be internal to the application, supplied semi-externally by a system of *plugins*, or may reside in a predetermined directory on the operating system.

Automatic data typing is very convenient. For example, a word processor or mail program may be able to automatically recognize and load various types of sound, image, and text files in many different formats, without intervention from the user.

data typing, programming At programming levels, data typing is more specific. Particular variables are required to be identified as to whether they are strings, integers, floating point numbers, etc. Some languages, like Pascal, are strongly typed, and others, like ARexx are untyped. Programmers constantly discuss the merits and demerits of each. Strong typing is intended to promote uniformity and reduce errors; liberal typing is intended to let the programmer control what is happening in the software, and take responsibility for finding and fixing errors.

data warehousing A phrase that was used primarily by large corporations with very large databases until the Web made huge databases easily accessible to almost anyone through the Internet. Data warehousing is a system of storing and extracting information from vast databases which are comprised of smaller storehouses of databases. Large databases present unique logistical and programming challenges. Many database storage and retrieval systems are limited in the number of records or files they can handle, in other words, many of them are not *scalable*. Consequently, various new strategies for data warehousing are being developed.

The Internet has made accessibility, through a local Internet Access Provider (IAP), so easy and inexpensive that companies are demanding increased access to databases not only online, but at branch offices in other states and countries. Thus, data warehousing through Web browsers is developing, even though the suitability of HTML to this task is somewhat limited. With one of the Web-friendly programming environments, such as Sun's Java, the job becomes easier, but the logistical demands of taking geographically divergent databases that may be in a variety of formats to meet local needs, and accessing them all as a conceptual unit over the Net, is an ongoing programming challenge that will probably continue to evolve for some time. Efforts to promote open systems, and object-oriented programming strategies may contribute to streamlining the process of data warehousing. See CORBA, Open Systems Interconnect.

database Any collection of data organized in some form for storage, or for storage and retrieval. A database can be as simple as a list of names, or as complex as a relational, distributed, multi-site archive of integrated images, ideas, text, facilities, actions, and processes.

A file system hierarchy on a computer storage device is a type of database, as is an employee file that includes pictures, birthdates, addresses, and social security numbers.

Database creation and management programs typically have either text-based interfaces, graphics interfaces, or both. With text-based interfaces, information is organized into lines and fields, and usually listed sequentially from top to bottom. In more flexible graphical databases, a *screen mask* or input template can be created almost as though using a paint program to draw the input screen. Lines, boxes, colors, and other visual elements can be used to make the database appealing and its functions and input actions apparent to the user. Fields are then assigned to the graphical elements, and the order of input defined. More sophisticated databases include scripting or symbolic programming languages to allow automation of the database so error messages, prompts, help windows, and other applications elements can be presented when appropriate.

There are many ways to store data in a database: compressed or uncompressed, encrypted, encoded, or plain ASCII. The format of the data isn't usually what creates compatibility problems. More often the organization of the data, which can vary widely, is the hurdle that must be overcome when interchanging data among applications or systems.

See data warehousing, expert system.

database reports Charts, graphs, lists, and other statistical reports that can be selected, or computed from information in a database, or more than one database. Many database programs have limited reporting capabilities, so a number of separate applications have been developed which specifically concentrate on providing good selection and formatting of the desired information, usually so it can be printed, or transmitted to other sites electronically. Reports are widely used for financial statements, business plans, demographics, research, etc.

database server A system, computer, or application specifically designed to provide database capabilities, security, and file access to multiple users on a system. There are two aspects to a database usually incorporated into a client/server model: the application that generates, searches, and retrieves the data; and the data itself. Sometimes the data is on the server, and the application is on the individual user's machine. Sometimes it's the other way around, and sometimes all aspects of the database system are handled by the server. It depends on the sophistication of the system, and the needs of the users. In high security situations, the server usually handles everything. In smaller networks, where security is less of an issue, the applications may be on individual machines, so they will run faster, and have the data on the server, with password access, number of user restrictions, etc., enforced centrally by the server software.

datagram This term is used in a general sense to mean a unit of information in a packet-switched network without regard to previous or following packets, and in more specific senses, depending on the type of network and its architecture. In layered architectures, the datagram may be associated with a specific layer or layers. A datagram may be encapsulated and subsequently decapsulated at the receiving end, for example, when tunnelling through different systems. Internet Protocol (IP) datagram transmission over connectionless X.25-based public networks has been defined by a variety of organizations. Most specify that source and destination information be associated with the datagram. See Point-to-Point Protocol, RFC 877, RFC 998.

date and time stamp A common function of computer applications and operating systems (OSs) which records when some event occurred. For example, files are usually date- and time-stamped as to the time of their creation, or the time they were last updated (or both). Entries to databases are frequently date- and time-stamped, as are computerized physical premises access systems, electronic timecards, and many more. The only problem with date and time systems on computers is that not all computers take the time or date from a reliable source. Some have lithium-battery powered realtime clocks, but many do not. It may be up to the user to set the date and time manually, and a power outage can change the settings. Date and time stamps aren't very useful if they are incorrect, and may cause serious problems for backup systems.

daughterboard, daughtercard *jargon* A printed circuit board which piggybacks onto a motherboard (which contains the main processing circuitry) in an electronic system. The daughterboard is frequently, though not necessarily, smaller than the motherboard, and usually adds some specific type of functionality: more memory, acceleration to the main CPU, a device interface, etc. A fatherboard has further been described as a connection to a motherboard which provides a series of connectors, into which several daughterboards can be connected.

DAV See digital audio video.

DAVIC See Digital Audio-Video Council.

Davisson, Clinton J. An American Bell Laboratories researcher, and winner of the Nobel Prize in physics in 1937.

Davy, Edward An English inventor who created one of the early telegraph systems.

Davy, Humphry An English scientist and educator who passed a current through potash in 1807, decomposing it and discovering a new element (potassium). Davy subsequently discovered more elements, and clarified that some substances, considered elements, actually were not, and proposed a theory of electrolysis. In the early 1800s, he observed the properties of carbon when connected to an electrical source, and in 1808, he invented the arc lamp by connecting the terminals of a voltaic cell to a piece of charcoal, resulting in a brilliant light now known as an arc light, or electric arc. Michael Faraday became his laboratory assistant in 1813.

dB *abbrev.* See decibel.

DB 1. data bus. 2. See database.

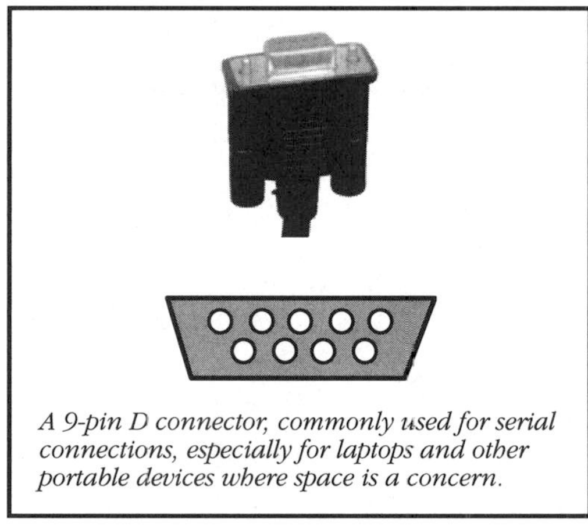

A 9-pin D connector, commonly used for serial connections, especially for laptops and other portable devices where space is a concern.

DB-9 A common designation for a 9-pin D-shaped computer connector, used on many laptops and desktop computers, especially for serial connections through an RS-232 cable. DB-9 simply describes the physical connecting portion and does not define the electrical relationships of the pins to the wires in the cable to which the connector attaches. RS-232, on the other hand, defines specific pinouts and pathways for various types of signal and information data.

DB-15 A common designation for a 15-pin D-shaped computer connector most often used for monitor cables and Ethernet transceivers. DB-15 simply describes the physical connecting portion and does not define the electrical relationships of the pins to the wires in the cable to which the connector attaches.

DB-25 A common designation for a 25-pin D connecter very widely used for computer data transfer, especially serial cables, and one end of many parallel and SCSI cables. DB-25 simply describes the physical connecting portion and does not define the electrical relationships of the pins to the wires in the cable to which the connector attaches. Many of the common, inexpensive A/B switchboxes are installed with DB-25 female connectors.

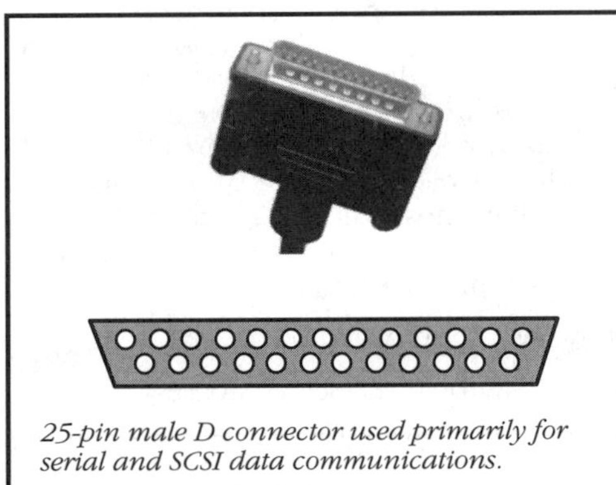

25-pin male D connector used primarily for serial and SCSI data communications.

DBMS See Database Management System.

DBS See direct broadcast satellite.

DBT Deutsche Bundespost Telecom.

DC 1. In telephone communications, Delayed Call. 2. See direct current. 3. disconnect conform. In Open Systems Interconnection (OSI), a transport protocol data unit.

DCA 1. Defense Communications Agency. A U.S. government agency involved in military standards development. 2. Document Content Architecture. An International Business Machines (IBM) system of specifying a document series from draft to final document. 3. Dynamic Channels Allocation. A wireless concept which is used in DECT PCS services.

DCC 1. See data communications channel. 2. See Data Country Code.

DCD 1. Data Carrier Detect. Signal from the DCE to the DTE, indicating a valid signal between the DTE and DCE devices. Typically used to set port status for a connection and to generate a signal indicating the loss of a connection. The DCE is commonly a modem or serially connected printer, and the DTE, the terminal or computer.

DCE See data communications equipment.

DCM See dynamically controllable magnetic.

DCP See Digital Communications Protocol.

DCS 1. digital communications system. 2. digital cross-connect system. 3. distributed computing system.

DCT See discrete cosine transform.

DCTI desktop computer telephony integration.

DDB digital databank. See data warehousing.

DDCMP Digital Data Communications Message Protocol. A station-to-station, byte-oriented, link-layer protocol developed by Digital Equipment Corporation (DEC).

DDD See Direct Distance Dialing.

DDE dynamic data exchange. Any process in which data is transferred between systems, or between applications without intermediary steps, such as saving out the information and transmitting with a different application. See drag and drop, Object Linking and Embedding.

DDN See Defense Data Network.

DDP distributed data processing. The process of 'farming out' a task to more than one processor to speed up the data handling.

DDS 1. digital data service. 2. digital data storage. 3. distributed data system.

DE See Discard Eligibility.

de Coulomb, Charles A. See Coulomb, Charles A.

de facto standard A format, specification, or design, usually from a self-interested commercial source, which has become widespread. Secondarily, this often confers a large degree of industry control to the major stakeholders.

Occasionally de facto standards are good, if a public-service body hasn't provided a standard, and if the standard brings down the cost of goods to make them more widely available to the general public. Sometimes de facto standards are bad, since the specification or products themselves may be of poor quality, and may only have become widespread through aggressive advertising, or consumer trust or lack of understanding of the technology. By the time the consumer learning curve catches up with the technology, the standard may be too entrenched to change.

Very frequently, the cheapest product becomes the de facto standard, not because it is a good product, based on good design principles, but simply because it's more affordable. The VHS video format sold more than Beta because it was cheaper, even though people openly acknowledged that Beta had a better picture quality. The IDE specification is limited compared to SCSI, but IDE drives are less expensive, and hence outsell SCSI drives on low-end desktop systems (SCSI is still preferred for workstations and redundant array systems).

Sometimes the most convenient product becomes the de facto standard. If a product is easier to use, or more portable, it may outsell more flexible or powerful designs.

Sometimes the first product to hit the market becomes the de facto standard, and manufacturers are sometimes tripping over one another in a race to be the first. This scenario has occurred many times in the modem industry. The first to get the new, faster modem on the shelves often had a big say in specifying the format for communications, and a short-term monopoly.

de Ferranti, Sebastian (1864-1930) An inventor who collaborated with Elihu Thomson and William Stanley in the development of the transformer. He was also responsible for developing the first high voltage alternating current (AC) distribution system at a time when direct current (DC) distribution systems were prevalent.

de Forest, Lee (1873-1961) A highly ambitious American inventor who harnessed the power of electrons by inventing the Audion vacuum tube, which was granted a patent in January 1907. This significant technology has been used in electronics in many industries for decades, though eventually transistors superseded vacuum tubes, except for some specialized high frequency applications. Although he was loathe to acknowledge his predecessors, de Forest's invention stemmed from the work of T. Edison and J. A. Fleming. However, he is to be credited with the introduction of the electron tube *grid* unit, creating a *triode*, which was a significant advance over the design of the Fleming tube. Lee de Forest's three-electrode vacuum tube made transcontinental communication possible, and the proliferation of vacuum tubes for radio wave detection caused the decline of crystal detector radio sets. In the 1920s, de Forest contested E. H. Armstrong for the invention of regeneration and won. (Lee was apparently born "*De Forest*", but later in life is said to have preferred "*de Forest*". His wireless company was spelled "DeForest".) See Audion, DeForest Wireless Telegraph Company, Edison effect, transistor.

Like Bell, Edison, and Marconi, Lee de Forest created a number of commercial products from his inventions, including this transportable de Forest telegraph receiving unit.

de-encapsulation An important aspect of packet-switched networking in which data from the user data field of an encapsulated packet is extracted on receipt.

dead band In guidance systems, a means of introducing hysteresis by preventing errors from being corrected until they have exceeded a certain specified magnitude. This is a safety precaution against the guided object reacting prematurely to interference or spurious signals.

dead spot A phrase often used in broadcast communications (radio, TV, cellular) to describe a region in which there are no signals, due to terrain, or other obstructions. Cellular customers in particular are susceptible to dead spots, because they may be constantly moving between buildings, boulders, mountains, etc.

debit card A monetary transaction card that

resembles a credit card, which allows remote transfer of existing funds. A debit card assumes the money is available somewhere, and can be immediately or quickly transacted and received, as opposed to a credit or charge card in which the user is billed and can pay later. Prepaid phone cards and checking or savings account debit cards are common. Some debit cards resemble credit cards, in that they are labeled for a particular type of credit card, and can be used in locations that accept that type of credit card, but instead of it being a credit transaction, the money is taken from a checking account. This is handy when travelling out of town and purchasing from a vendor that doesn't take out of town checks.

debug *v.* To systematically rid a system of problems or *bugs*, especially in computer software. In software, bugs include syntax errors, looping errors, logical errors, and user interface design ergonomics problems. This process has become easier, with the availability of debugging software and higher level programming languages, but is still an arduous, exacting, painstaking activity. In the programming community, not all good programmers are good debuggers, and code is sometimes passed on to a programmer who has a particular mindset and talent for this exacting, detail-oriented work. Software databases for tracking and reporting bugs are becoming popular, and programs which do automated tests of software, in order to find problems and report bugs, may be included in the debugging arsenal.

declination 1. The angular distance from the celestial equator, north or south. 2. When using a compass, the deviation angle from geographic north, of the magnetic north-seeking direction of the compass needle. Most compasses provide a way to set the declination for a particular geographic area to compensate for the fact that magnetic north is not coincident with geographic north.

DECCO Defense Commercial Communications Office.

decibel dB. One tenth of a bel. A dimensionless ratio of two powers, which, in electricity related to telecommunications circuits, may be referenced to milliwatts. More familiarly, in acoustics, a decibel may be expressed as a calculation of the sound pressure ratio to a reference pressure expressed in pascals. Since that's a little difficult to understand without knowing the formulas and individual frames of references, it may be easier to understand decibels in terms of examples in acoustics. Sound volume is typically expressed in decibels, on a logarithmic scale (the bigger it gets, the proportionally louder it sounds) with lower numbers representing lower volumes. At the high end of the scale, around 150 - 200 dB, permanent hearing damage occurs. Even at sustained levels of 100 to 150 dB (and sometimes lower), loss may occur. Sudden contrasts from low to high volume sounds can be especially harmful to human hearing. Environmental noises tend to range from about 5 to 100 decibels, with low volume sounds emanating from appliances and traffic, and high volume sounds coming from horns, explosions, collisions, etc.

decimal system Based on a base 10 numbering system, the most common one in human culture, probably due to the fact that we have 10 digits each on our hands and feet. Computers, on the other hand, are commonly based on binary systems, base 2, due to the fact that electrically powered computers are easy to design around systems that use two states: on or off, high or low, etc.

DECNet Digital Equipment Corporation's proprietary Ethernet-based local area network (LAN). See Digital Equipment Corporation.

decompression The process of decoding, expanding, and otherwise reverse-engineering the information in a compressed file. Decompression is usually done to restore a file to its original state, or to restore the information to a form that is comprehensible to a viewer/listener/reader, but which may approximate, rather than duplicate the original state. Decompression may happen in advance of using the information, or may be carried out as the data is being read, as in MPEG animation playback systems. See compression, data compression.

DECT See Digital European Cordless Telecommunications.

Dedicated Access Line, Dedicated Line DAL, DL. A private network connection between a business or individual, and a phone carrier or network service provider. Calls through a dedicated line are automatically routed to the phone carrier or other service provider. Dedicated lines can be installed on various types of circuits, ranging from copper phone lines to fiber optic.

dedicated array processor DAP. A processor in a redundant array of inexpensive disks (RAID) system which specifies various array-specific tasks related to management of the multiple disks, and the information and organization of the information on the disks. This is particularly important if a problem has occurred with the data, and the disk array needs to be adjusted or rebuilt.

default The initial setting, factory setting, reset setting, parameters or configuration. The state or parameters that are in effect at the start of a program, or available in a dialog box, or stored in a file. Sometimes the user can modify the default settings, and sometimes not.

default carrier In telephone communications, the long-distance carrier which is assigned to handle calls for customers who have not specified an alternate carrier.

Defense Advanced Research Projects Agency DARPA. Formerly ARPA, an agency of the U.S. Department of Defense (DoD) which handles research and development in basic and applied research, and which takes on imaginative, high-risk/high-payoff projects which may result in dramatic technological advances on behalf of the DoD.

DARPA was established in 1958 to assure U.S. global technological advancement in coordination with, but independent of, the U.S. military research and development establishment. DARPA has, as part of its mandate, a design which is a deliberate counterpoint to traditional research ideas and approaches. It manages a budget of about $2 billion.

DARPA's technical staff is rotated every few years to encourage new perspectives, and is drawn from industry, academia, and government laboratories.

Defense Data Network DDN. The packet switching network of the U.S. Department of Defense, established in 1982, which later became MILNET. DDN was separated from ARPANET (which evolved into the Internet), and has a number of classified and unclassified sections (DISNET*x* and MILNET).

defense information infrastructure DII. The facilities, networks, software and information that comprise both peacetime and wartime resources of the U.S. Department of Defense.

Defense Satellite Communications System DSCS. A global military communications satellite network operating in super high frequency bands. DSCS is administrated by the DISA.

deflection modulation On a scanning cathode ray tube (CRT), a system of deflecting the scan vertically from a base horizontal path. This makes the signal appear like a series of peaks above the baseline.

DeForest Wireless Telegraph Company A company formed in 1902 by Lee de Forest to commercially distribute the results of his inventions, many of which related to early electronics and telecommunications. See de Forest, Lee.

degauss To demagnetize; to bulk erase; to provide electrical current-carrying coils to neutralize magnetism. Cathode ray tube (CRT) monitors are equipped with magnets, to influence the direction of the electron beams. Some monitors will have a degaussing button to reset the magnetic environment in the monitor. Named after Karl F. Gauss.

deinstall To remove. In computer software, to deinstall an application can sometimes be a hit-and-miss process, as some programs will install bits and pieces of files in many different directories, sometimes with cryptic filenames, and don't always provide documentation on what those bits and pieces are, or where they are. To facilitate deinstallation, many developers now include deinstall software with a program, which seeks out the parts in the various locations and removes them cleanly, and hopefully without interfering with anything else on the system. In computer hardware, deinstalling a component usually involves turning off the computer, and removing a peripheral card, or cable. It is not wise to do this with the computer running, unless it is designed as a *hot swap* backup device, such as a redundant array of inexpensive disks (RAID) system.

Delaney lamp detector A novel solution to creating a radio wave detector is the Delaney lamp detector, devised by U.S. naval electrician Delaney. It was an electrolytic detector which consisted of an incandescent lamp bulb with the top broken off, the filament removed, and a 20% solution of nitric acid poured into the globe. This detector was found to respond well to signals originating nearby, without burning out from the oscillations of a strong signal coming from a nearby wireless station, a problem that was common to other types of electrolytic detectors. See detector.

delay Lag, hysteresis, gap, retardation, lapse of time. Delay is inherent in many types of telecommunications technologies, and systems design and utilization are affected by the types and degrees of delay that might occur. Transatlantic phone callers used to experience appreciable delays between the time one side of the conversation was heard and understood, and the other side responded. This type of delay is now rare, or within acceptable levels.

Delay can be a serious problem in systems that use more than one line to transmit, as in many videoconferencing systems, or in packet switched networks in which packets are split up and sent through different routes, and reassembled at the receiving end. If portions of the transmission are delayed, it may affect the entire message.

Delay is sometimes deliberately introduced in data transmission systems to prevent overly quick reactions to situations that might be nonmeaningful (as in guidance systems, or brief power brownouts). Delay is inherent in store-and-forward systems that may collect email messages, or other types of network traffic, and dispatch them in batches, or when CPU time is at a low ebb. See cell rate, hysteresis, jitter.

delimiter Any symbol, code, character, or other data used to signal a gap, break, boundary, or stopping/starting point. Delimiters are very important in both human and computer communications. People use spaces, commas, paragraphs, and other punctuation and symbology to serve as delimiters in written text. This greatly enhances the ease with which the information can be understood. Computer programs use symbols, punctuation, and spaces as common delimiters for arrays, lists, paths, and other types of information. Radio broadcast systems use spaces as delimiters between songs, or tones as delimiters before and after emergency announcements.

Dellinger fade-out In radio communications, sun spot activity may be associated with highly absorbing areas in the ionosphere which can impair short wave transmissions.

delta channel See D channel.

delta matched antenna See Y antenna.

delta modulation A common method for converting analog signals, such as voice, to digital signals, usually for transmission over data networks. Delta modulation involves comparing a scanned value to the previously scanned value, to see if it is greater or less than the previous value, sending a one if it is greater, and zero if it is less. This simple scheme results in very fast encoding. There are many different versions of delta modulation, and many are not compatible. Pulse code modulation (PCM) is the other common method for converting from analog to digital. Because more information is encoded, it's not as fast as delta modulation, but it can be more readily converted between different versions of PCM. Unlike delta modulation, PCM doesn't require an intermediary analog stage to convert between different versions, and thus is more popular than delta modulation. See modulation, pulse code modulation.

Delta Project A large-scale European initiative, comprised of universities and interested businesses, which applies technology to distance education and training. The Delta Project includes more than 300 subprojects in multimedia and networking. See distance education.

Demand Assigned Multiple Access DAMA. On demand channel sharing accomplished by assigning a call to a channel which is currently idle, or to an unused time slot.

demand circuits Data network segments whose costs are related to usage.

demand publishing, document on demand A marketing phrase for very fast turn-around in the publishing industry. There are many demand service bureaus now which can take a computer file from a diskette or FTP site, feed it directly into a dry toner copy system, without printing out a master first, and produce 1,000 copies of a 100-page cut and bound document in a few hours. With overnight shipping, the customer gets the finished publication back the next day. Demand publishing enables businesses and individuals to produce short-run publications on a short schedule.

Human society's ability to read all this new printed material, and the ecosystem's ability to regrow the raw resources needed for the paper, have been outstripped by the speed with which the new information can be produced. A wise person once said, "There is more to life than increasing its pace." Eventually, when computer monitors can be designed so they look and feel like a piece of heavy paper or cloth, and can be carried around rolled up in a pocket, the trend will shift from paper printing to electronic printing. The research at the Massachusetts Institute of Technology (MIT) has already resulted in experimental technologies that show promise. In the meantime, demand resources allow publishers to create new paper-based products very quickly.

demarcation point The point at which a telecommunications carrier's equipment, or responsibility ends, and the subscriber's equipment or responsibility begins. This may be at a junction box on the side of a building, or a patch panel within a building. Residential lines tend to be demarcated outside, while multiline phone systems in businesses tend to be demarcated at a patch panel, or terminal block inside the premises. The block itself is sometimes called a demarcation strip.

DEMKO Danmark Elektriske Materielkontrol. Denmark's electrical testing institute.

demodulation The process of taking a signal which has been manipulated to carry information, and extracting that information from a carrier wave, or mathematically recreating that information from sidebands, or other parts of the original transmission.

A crystal detector uses a crystal as a type of one-way 'valve' to detect and demodulate radio waves. A modem is a device that demodulates an analog phone signal to convert it to digital information that can be understood by a computer. A broadcast receiving station takes modulated airwaves and demodulates them for local transmission through cable or airwaves.

demon dialer, war dialer A function of a program, usually for telecommunications or telemarketing, that calls a phone number repeatedly, or calls down through a list of phone numbers, cycling back to the top until one answers.

DEMS Digital Electronic Message Service.

demultiplexing 1. In a multiplexed signal, a process for recovering signals combined within it, usually to restore distinct channels contained within the transmission. See carrier wave, band, channel. 2. In ATM, a function performed by a layer entity which identifies and separates SDUs into the individual connections of which it is comprised.

DENet Denmark's Ethernet network which interconnects academic institutions.

densitometer A photoelectric instrument for measuring the opacity or relative degree of light absorption of a material. Darker materials absorb more light, and thus have a higher optical density. There are different types of densitometers, some which measure transmission, and some which measure reflection. They are commonly used in the printing industry to monitor consistency and quality. See hygroscope.

density A general descriptive or comparative term describing the proximity of individual elements, and hence the total number that can be fit within a specified area. Density is used to describe optical and magnetic storage capacities, screen and printer resolutions, compression efficiencies, etc. As a general rule of thumb, especially for storage media, the denser the information capacity, the higher the cost.

Denver Telephone Dispatch Company One of the earliest phone companies in the United States, established by Frederick O. Vaille in early 1879.

depletion layer A barrier region in a semiconductor in which the mobile carrier charge density is not sufficient to neutralize the charge density of donors and acceptors. The donors contain impurities to facilitate electron activity, and the acceptors, include trivalent impurities, forming 'holes' to accept the electrons.

depolarization In many transmission systems, the radiant wave signal may be horizontally or vertically polarized and may gradually lose that polarization as it passes through various media (moisture, particles, terrain obstructions, etc.). The wave may also be deliberately depolarized at the reception point.

deregistration The process of dissociating two entities, which may be names, processes, objects, etc. Deregistration is a process that happens often in computer applications, particularly those associated with networks. The process of creating virtual links between applications, icons, and processes allows for keeping track or creating shortcuts. Often these virtual links are removed or reorganized, particularly in a dynamic network environment.

In public or pay systems, deregistration may be associated with billing or security, in that a subscriber, cellular handset, computer, or other entity is registered in order to assure service to the subscriber, or to provide privacy and authentication. In these systems, registration is often with a unique identifier which must be deregistered before services can be discontinued, or before the identifier can be assigned to another.

deregulation A process of removing authority from a governing body in general, or of removing authority of the government body over specific jurisdictions. As examples, control over banking, telecommunications, or other large public or commercial systems have at various times been removed or reduced in order to stimulate competition and innovation. Sometimes deregulation has the intended positive effect. Sometimes deregulation causes a change in the system that attracts unscrupulous members of society, who seek new loopholes that can be used to take advantage of the system to the detriment of others.

DES 1. See Data Encryption Standard. 2. destination end station.

Desbarats, George Edouard A Canadian newspaper publisher who printed the first halftone photograph in The Canadian Illustrated News in 1869. Until this time, printed illustrations were limited to line drawings. Half-tones are still widely used in the paper publishing industry.

descender In typography, the portion of a character that extends below the main body of a typestyle. That is, the lower stem in the letters *g* or *y* are called descenders. See ascender.

descramble Scrambling is 'mixing up' a signal or digital data file so that it cannot be used without first being processed by a descrambling device or algorithm. Scrambling is a means for providing security, protection from unauthorized viewing or copying. Scrambling is a common way to commercially protect cable or satellite TV programs from being viewed by those who haven't paid for the service. In most regions, unauthorized descrambling devices are illegal.

Design System Language DSL. A predecessor to the PostScript page definition language, DSL originated in the mid-1970s at Evans & Sutherland Computer Corporation. Evans & Sutherland (E&S) became well-known for their pioneering work in flight simulation and other 3D graphics software, and the Design System was one of the outcomes of a research project using an interpretive language to build complex graphics databases. The Design System Language was later put to use in CAD applications. See PostScript.

Designated Transit List DTL. In ATM, network routing policy data consisting of a list of nodes and any associated link identifiers, which specifies the path across a PNNI peer group.

desk accessory DA. A designation for a number of common computer software applications that are placed in a higher level directory for easy access. For example, on a Macintosh computer, the various desk accessories include a calculator, calendar, clock, etc. These types of accessories are so handy, they are commonly included with most computer operating systems.

desktop A broad term in the computer industry describing any common workspace found in a home or office (or garage) where the various user tools and appliances are contained. Desktop computers are those which can fit comfortably on a desk (as opposed to dishwasher-sized minicomputers, or room-sized supercomputers). See desktop metaphor.

Desktop Management Interface DMI. A specification developed by the Desktop Management Task Force (DMTF) to allow easier, more consistent access to management database information and applications through Management Interfaces (MIs). The DMTF is a consortium of hardware and software vendors.

desktop metaphor A phrase to describe the conceptual rationale for the design and display of information on a computer output device, usually a monitor. The idea is to take objects and actions which are familiar to workers and home users, and represent them in an easily recognizable form on the computer. The desktop metaphor strives to be fairly universal, though there will likely always be cultural and individual differences in how symbols are perceived. Prevalent implementations of the desktop metaphor are file folders to represent directories, or collections of information; trash cans for discarded files; clocks for time and date functions; magnifying glasses for zoom functions; etc.

desktop pattern A pattern that forms a background on a computer display in front of which text and various graphical elements remain visible. The ability to enable, or create and add, desktop patterns has been common on workstation and microcomputer systems since the mid-1980s. On graphical user interfaces on some systems, separate patterns can be set within individual windows, and on the background. Desktop patterns are more than just a decorative addition, they can be used to reduce contrast and eye fatigue, to help distinguish various windows, and to customize multiuser systems so the person logged on can be identified at a glance.

desktop publishing DTP. Many of the pioneer tools for the evolution of desktop publishing were developed at Xerox PARC, subsequently implemented and popularized on the Macintosh computer in the mid-1980s, and enhanced through the development of laser printers. DTP has had a resounding impact on the publishing industry, and has put publishing into the hands of tiny companies, and individual publishers.

Quark XPress, Adobe PageMaker, Adobe FrameMaker, and TeX are DTP software applications widely supported by users, service bureaus, and publishers. Ventura Publisher was popular in the 1980s. Quark is favored for ad layout, posters, and other complex color projects and, to a lesser extent for books, mainly those which have many illustrations, and those that don't require a lot of extensive indexing and cross-referencing. Design professionals tend to like Quark. PageMaker is favored for general purpose books and document production. Users like its ease of use and general support of many types of documentation. FrameMaker is favored for more technical documentation especially that requiring general support for mathematical symbols, and for general purpose publishing. As yet, FrameMaker is not as well-supported at service bureaus as Quark and PageMaker, but now that it is owned by Adobe, its compatibility may increase, and many people like the product. FrameMaker is well-supported on workstation platforms.

TeX is preferred for sophisticated technical symbols as are found in mathematics and physics treatises and textbooks, and while its interface is less intuitive, it has capabilities in symbol representation and formatting beyond most general purpose desktop publishing programs. Corel has been promoting its products as desktop publishing tools, and for illustrations, it can work well, but many service bureaus and professionals prefer some of the other products, at least at the present time. However, for creating and inserting PostScript illustrations into other desktop publishing programs, many people swear by Corel Draw! for the creation of *eps* (encapsulated PostScript) images. There are also a number of other publishing programs available, and word processors that are sometimes stretched to perform basic desktop layout tasks.

desktop video Just as the Macintosh computer facilitated the development of a new mass market field called *desktop publishing*, the Amiga computer facilitated the development of a new mass market field called *desktop video*. Within two years of its debut in 1985, many video genlock and broadcast video products were introduced for the Amiga, the most significant being the NewTek Video Toaster, which is still used today by many local cable companies and some of the large broadcast networks. A desktop video-equipped computer enables the development, editing, and merging of computer images, video taped footage, and live broadcasts at a fraction of the cost of using older video equipment.

desktop videoconferencing Video phone capabilities provided on a small desktop dedicated system, or microcomputer equipped with a microphone, video camera, and fast network or phone access. Businesses are considering videoconferencing as an option to expensive travels, especially for meetings involving participants who are widely distributed geographically. Videoconferencing systems are almost affordable now, but the speed of the line greatly affects the refresh rate of the image. If the transmission rate is slow, the image will be blurry and slow to update, and will look more like a series of still shots than natural movement.

despun antenna A type of antenna which rotates on a platform, and is used on satellites to orient a cone-shaped beam in a specified direction, usually a region on the Earth. There were early electronically despun antenna experiments on U.S. military satellites in the late 1960s which met only limited success. Later, mechanically despun antennas were also tried.

destination address DA. A flag, field, or other indicator commonly used to designate the receiving point for a data transmission, call, or physical correspondence. In Token-Ring, Ethernet, and Fiber Distributed Data Interface (FDDI) networks, the DA is a data field that is sent in the direction of the recipient which describes a unique Media Access Control (MAC) address. See address, domain name, Media Access Control.

destination end station DES. In ATM networking, the end or termination point of a transmission. It is used as a reference point for available bit rate (ABR) services. Since ATM has a number of traffic flow control mechanisms, which often depend on cells arriving at the destination at a certain time or in a certain manner, ABR and constant bit rate (CBR) distinctions are important with relation to the defined end station. See cell rate.

Destriau effect When exposed to an alternating electric field, certain phosphorescent inorganic materials suspended in a dielectric medium will luminesce, that is, emit *electroluminescent light* rather than incandescent light. Certain semiconductors can be designed and excited to emit *carrier injection luminescence*.

destuffing Sometimes data bits are 'stuffed,' that is inserted in a transmission, for one reason or another. Destuffing is the process of selectively removing the stuffed bits to recreate the original information prior to stuffing.

detector 1. In radio electronics, a device to apprehend, or detect, radio waves. In early schemes, many materials were tried, including barretters, natural rectifying crystals, synthetic crystals, and electrolytic cups. The challenge with most detectors was solving the problem of amplifying the signal, or shifting the range of frequencies into the audible range. Current electronics have sophisticated and effective ways of achieving it, but in the late 1800s and early 1900s, ways of capturing and amplifying the waves were just being developed. 2. A substance or circuit which reacts when exposed to electromagnetic waves such as radio waves. The detector may also convert or 'rectify' the electromagnetic oscillations so they can be incorporated into a circuit with practical applications. A crystal detector works as a type of one-way valve to rectify radio waves, without outside power sources. Later, vacuum tubes were designed that could accomplish this task in ways that allowed great selectivity over the frequencies detected (tuner circuits), and which could amplify the signal so the broadcast could be played through a speaker rather than through small headphones. See crystal detector, Delaney lamp detector, electrolytic detector, electron tube, Lodge-Muirhead detector, magnetic detector, Massie Oscillaphone, optical detector, Shoemaker detector, silicon detector, vacuum tube.

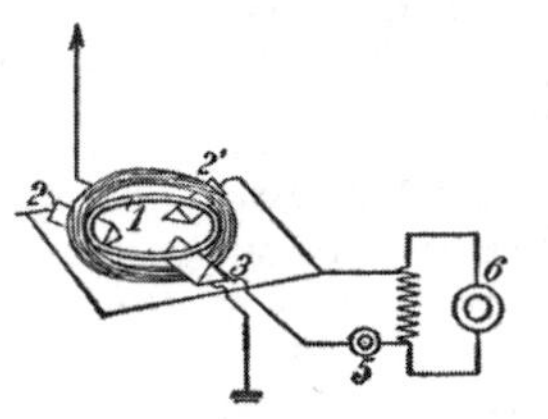

One type of magnetic wireless telegraphy detector which uses magnetic permeability properties, which is connected in series to a telephone receiver. Current can be used to vary the resistance of the carbon wedges, registering the changes in the telephone receiver.

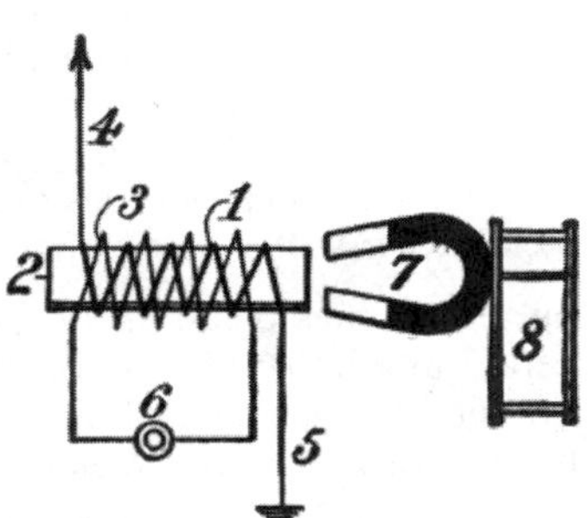

A Marconi magnetic detector which utilizes a varying magnetic field produced by high-frequency oscillations using magnetic hysteresis. The dual coil surrounding the core is of thin copper and iron wires. Scientific American, Oct. 4, 1902.

Deutsch, Peter Along with Alan Emtage, one of the developers of the Archie system created originally at McGill University, and a cofounder of "the Archie group," a group of volunteers dedicated to supporting and enhancing the Archie project. Archie is very widely used on the Internet.

Deutsche Telekom The German telephone services authority. DT includes a Technology Centre, which does research, development, and demonstration of innovative telecommunications systems.

device configuration management In networks, the configuration, tracking, and management of various devices, ports, and interface cards. These devices are increasingly software configurable, and the software used to manage them will often show actual images of the physical switches through a graphical user interface.

device driver Computer software that controls, or provides an interface to, one of a variety of devices such as hard drives, printers, video cards, CD-ROM players, etc. A device driver is a software interface through which the operating system interacts with various peripherals attached to the system. On larger networks, whole databases, and device servers may be installed to manage various devices on the system. See display driver.

DEW line Distant Early Warning line. A line of radar stations stretching across the northern frontier of North America to provide advance warning to the United States government of approaching aircraft or missiles.

DFA doped fiber amplifier. A fiber optic cable which has been impregnated or 'doped' with substances, usually rare earths, which alter its transmission properties. See doping.

DGPS Differential Global Positioning System. See Differential GPS, Global Positioning System.

DGPT Department General of Posts and Telecommunications, Viet Nam.

DGT Dirección General de Telecomunicaciones. Spanish telecommunications authority.

DHCP See Dynamic Host Configuration Protocol.

Dhrystone A relative performance test intended to evaluate system program execution other than floating point and input/output operations. As with many benchmarks, the information is only useful within a narrow viewpoint, as in controlled experiments. The Dhrystone was developed in Ada by R. P. Weicker of Siemens, and is described in overview with other benchmarks in IEEE Computer, Dec. 1990.

See benchmark, Rhealstone, Whetstone.

diagnostic programs Applications programs used to run hardware test suites, or to evaluate processor functioning, or both. Most computers will run a systems check of the hardware before booting up the higher level operating functions. Network servers now often have diagnostic tools with graphical user interfaces to show virtual and physical connections, traffic flow, routers, switches and various configuration settings. Diagnostic software is available for many computers to self-check and indicate possible problems in processing rates or connections. Many diagnostic routines are intended to run on a regular schedule, to locate potential problems before they become serious. *Engine diagnostics* used to mean getting grandpa to stick his head under the hood of the car to listen to the engine. Now it means taking the car into a shop and getting it hooked up to a computerized engine evaluation system that displays various system parameters on a computer monitor with charts and graphs.

diagnostic techniques Various means of ascertaining and measuring the properties of a system or circuit in order to understand its characteristics, monitor its behavior, or detect any problems or anomalies. Diagnostic techniques form a part of troubleshooting a system, and are often part of, or combined with installation and testing. Some systems incorporate automated diagnostics which will check systems on startup, at random intervals, or at scheduled intervals. See ammeter, tap, voltmeter, Wheatstone bridge.

dial *n.* A circular, movable mechanism for entering a number or other code. While dials are used in many electromechanical devices, they are most commonly associated with rotary telephone sets. In a telephone system, turning the dial causes a pulse of a predetermined length to be sent along the line to indicate the desired destination address. This allows the system to set up an end-to-end connection for a conversation or data communication. The dial was also known as a *finger wheel*.

The dial superseded the hook on old telephones, and effectively obsoleted human operators for local calls within a few years. Dials were commonly used until the 1970s, when touchtone phones began to become prevalent in North America (many other countries still use dial phones). See Strowger switch.

dial tone An audible signal that indicates that a phone line is active and ready to be dialed. The dial may be provided by the local phone carrier, or by a private branch system. A different type of dial tone, called a *stutter dial tone* has recently been introduced by phone companies offering voice messaging services to their subscribers, to indicate that messages are available to be retrieved.

dial-up A phone-in connection to a computer network service.

D

Dialed Number Identification Service DNIS. This service allows a user to know which number the caller dialed on an incoming phone call. This is useful for regional numbers that are rerouted to a central administration or sales system. For example, an economic region may have several area codes, but there may be one sales office handling all the calls. The DNIS allows the person answering the call to see whether the caller dialed the local number, the toll-free number, or a long-distance number, giving some feedback as to which lines are being used, and where callers are calling from.

This service can also be used for 'split run advertising' analysis. Suppose a company runs two different ads for the same product and inserts different phone numbers into each ad. The two different ads for the same product are run in a number of publications. Let's say a company was selling a new electric car, and one ad focused on college students on a budget, the other on grandparents on a pension. The incoming calls give an idea of which ad results in the most calls, and the DNIS number keyed to the ad provides this response information. This feedback allows further targeting of future advertising slant and funds. DNIS can also help distinguish an employee from a customer, by indicating which line was called, but saving resources by routing both through the receptionist. DNIS does not provide the number of the calling party; a service like Caller ID is needed to provide this additional information.

dialing parity This is defined in the Telecommunications Act of 1996, and published by the Federal Communications Commission (FCC), as

"... a person that is not an affiliate of a local exchange carrier is able to provide telecommunications services in such a manner that customers have the ability to route automatically, without the use of any access code, their telecommunications to the telecommunications services provider of the customer's designation from among 2 or more telecommunications services providers (including such local exchange carrier)."

This measure was designed to 'level the playing field' and encourage fair practices. See Federal Communications Commission, Telecommunications Act of 1996.

dialog box A computer user interface device resembling a window or box, in which there are prompts alerting the user of a situation that requires a response, and gadgets (usually buttons) that provide options or some other means for providing that response through an input device such as a mouse, keyboard or touchscreen. In most cases, a dialog box is displayed when a response is immediately required, and the program will not continue until that response is received. However, there are some situations, in multitasking systems, depending on the circumstances of the dialog box, where other actions can be taken before responding to the dialog prompt.

Dialogic A New Jersey-based commercial provider of voice processing applications for desktop computers.

diaphragm A thin, flexible sheet or membrane, usually with a curved surface. In electronics, a diaphragm usually is designed to respond to vibrations, and has an interface to a device which can interpret those vibrations into electrical signals. Or conversely, it may take electrical pulses and interpret them into vibrations. Diaphragms are common in microphones and speakers.

DIB Dual Independent Bus. The communications bus between an Intel Pentium II processor and its Level 2 cache. See Intel Pentium II.

DID See Direct Inward Dialing.

dielectric Nonconducting material that provides an insulating layer by impeding or resisting the passage of current. Dielectric materials are often used as cable shieldings, or applied in layers between sheets of conducting materials in condensers. Common dielectrics used over the decades include paper, cloth, air, Bakelite, glass, ceramic, and certain synthetics. Glass and ceramic are the ones used in Leyden jars and utility pole insulators. See insulator.

dielectric feed In communications, a dielectric feed is a type of microwave lens that fits into the mouth of an antenna waveguide. It provides a wideband alternative to scalar feeds.

differential GPS DGPS. An implementation of the Global Positioning Service designed to improve local accuracy of the data. One or more high-end GPS receivers are placed at known locations where they receive GPS signals. These become reference stations which estimate the variations of the satellite range measurements, forming corrections for GPS satellites within current view, broadcasting the correction information to local users. See local differential GPS.

differential modulation A means of relative modulation based on the detected state of the previous instant, rather than on an absolute predefined parameter. See delta modulation for an example of a commonly used, simple type of differential modulation.

differential phase shift keying DPSK. A means of relative modulation in which the previous state of the carrier signal phase is detected, and the subsequent state based on the previous, rather than on an absolute predefined parameter. See phase shift keying.

differential polar relay A telegraphic transmissions relay in which the armature is polarized by contact with a permanent magnet, and is operated by the difference in the strength of the currents. The direction of the currents constantly changes, and can be controlled with a pole changer.

differential pulse code modulation DPCM. A means of sampling a signal, subdividing it, and assigning values to the individual parts (quantization) in order to add this information to a carrier signal. This can be done in a number of ways, and not all PCM transmissions are compatible. PCM is a very common means of converting analog to digital signals, and is widely used in telecommunications. In differential PCM, a transmitted digital signal is used to represent the difference between consecutive analog signals. These differences are obtained by using a fixed quantization step size. See quantization, pulse code modulation.

differential quadrature phase shift keying DQPSK. In general, quadrature phase shift keying (QPSK) is a modulation scheme in which four signals are used, each shifted by 90 degrees, with each phase representing two data bits per symbol, in order to carry twice as much information as binary phase shift keying. DQPSK is a subclass in which the difference between the current value of the phase, and the previous value of the phase, are used instead of the absolute value of the phase. See modulation, quadrature phase shift keying.

Diffie-Hellman A fairly fast public key encryption system described by Whitfield Diffie and Martin Hellman in a 1976 IEEE issue of Transactions on Information Theory entitled

"New Directions in Cryptography." This concept has since been incorporated into many encryption schemes, including some Cellular Digital Packet Data (CDPD) systems, and the well-known Pretty Good Privacy (PGP) program developed by Philip Zimmermann. While the inventors patented the system, it came under dispute due to its public disclosure prior to the patent application. See Hellman-Merkle, Pretty Good Privacy.

dig A Unix command which provides information about domain names.

digicash *slang* A term for 'soft transfer' electronic transactions carried out in a manner which is similar to purchasing with money, and which uses the same monetary equivalents, but which does not involve the actual transfer of paper from the purchaser. Digicash eventually involves a 'hard transfer' of actual funds, sometimes directly from the bank to the vendor, sometimes through a middle agent which may be human, corporate, or electronic.

digit punch A section of a Hollerith punch card which can be individually punched to represent a number, or punched in conjunction with a zone to represent a letter. See punch card, zone punch.

On the Equity alarm clock on the left, a digital readout shows the time incremented in minutes. On the right, an analog display with a 'sweep hand' travels through its arc in a continuous movement.

digital A means of representing information in discrete units, rather than as a continuous stream. In communications technologies, information is typically represented in terms of binary units. There are many ways to represent information in a binary system: on/off, high/low, large/small, changed from previous state, loud/soft, fast/slow, lit/unlit, up/down, present/absent, etc. In digital computing, the binary units are usually ones and zeros. In electronic circuits, the units are often represented electrically by on/off, high/low, or change from previous state. Despite the simplicity of a binary system, it is powerful and flexible, and extremely sophisticated processes and information can be achieved by combining, organizing, and variously encoding this digital information. Most communications systems prior to the 1970s were analog, but the trend is strongly towards converting analog signals into digital signals. This allows a far greater degree of control, security, compression, noise control, and modifiability over analog systems.

An analog watch has a hand that sweeps around in a 360° arc, showing hours and minutes, and the positions in between. A digital watch has a readout that displays the time incrementally, usually in one second increments.

An analog dial on an AM radio allows the tuner to be gradually adjusted through adjacent stations. As the dial moves, the signal can be heard to increase and decrease, and there may even be periods where multiple radio signals overlap. This type of radio dial does not allow the listener to jump directly from a low frequency station to a high frequency station. With digital pushbuttons on AM and FM radios, the tuner can be set to 'jump' to a specific frequency, and stations can be selected this way in any order, even if they are not in adjacent frequency ranges.

In an analog phone conversation, the phone equipment converts sound waves into electrical signals that are sent through the lines to the person being called, but it is also possible to encode the conversation as a digital signal and send it through a computer network, or computerized phone system. Digital encoding allows the information to be compressed, modified, stored for later retrieval, or sent in conjunction with other data signals (such as a computer data file transfer) at the same time. See analog, ISDN, quantization, voice over ATM.

Digital Advanced Mobile Phone Service See DAMPS.

digital audio broadcasting DAB. A transmission modulation technique which sends digital rather than analog audio signals.

digital audio radio DAR. A new audio broadcast technology which provides high quality sound over the airwaves, and which provides a wider selection of regional programming. It also integrates with various news, paging, and email services. There has been talk of putting DAR in the S-band, but a number of technical characteristics of DAR indicate this may not be the best choice. In 1995, the Federal Communications Commission (FCC) assigned a frequency spectrum for DAR use.

digital audio tape DAT. A high-quality, high-capacity digital audio recording format. DAT is most commonly used for high quality digital audio recordings and computer data storage. For audio recordings, the sound is sampled, quantized, and converted to a specified encoded format. The encoding includes error checking mechanisms, and tracking information to facilitate searching for a particular location on the tape. DAT became popular in Europe in the early 1990s, but American vendors were so concerned about audio piracy on DATs, that they effectively blocked the spread of the technology in the United States. DAT is now used to some extent for computer tape backup systems.

digital audio-visual DAV. Digitized audio/video data that typically bypasses a computer's main bus.

Digital Audio-Video Council DAVIC. A non-profit association established in 1994 in Geneva, Switzerland, to promote global open interfaces and protocol specifications (DAVIC specifications) in audio-visual applications and services. There are over 200 member companies from more than 25 countries worldwide, representing manufacturing, service, research, and government agencies.

DAVIC concerns itself with the specification and development of tools rather than systems, with a focus on identifying and specifying components which are relocatable on a specific platform, and which are also cross-platform.

The DAVIC 1995 specification recommends SDH/SONET as the core network physical layer to which ATM cells, as standardized by various international bodies, can be mapped. Timing involves the use of a transmit clock derived from the network. Jitter is also managed with the network clock as the reference clock. There are five main entities within the specification, as shown in the chart.

Entity	Abbrev.	Notes
Content Provider System	CPS	
Service Provider System	SPS	
Service Consumer System	SCS	
CPS-SPS Delivery System	--	Connects CPS to SPS
SPS-SCS Delivery System	--	Connects SPS to SCS

Some technologies incorporated into DAVIC are the intellectual property of the contributors; they have agreed to make the technology available for free or for reasonable royalty fees available to anyone. http://www.davic.org/

digital camcorder A digital camera which is capable of capturing and storing information at a rate that is fast enough to create a series of digital frames which, when played back, show full motion video. See camcorder, digital video, dry camera.

digital cellular service A digital version of mobile cellular telephone communications in which the voice conversations are sampled, quantized, and encoded for transmission. This permits increased security, privacy, capacity, and better handling of noise interference, and corrective processes during roaming across cells.

Digital Equipment Corporation DEC. A well-known computer hardware/software/services company which was established in the 1950s by Kenneth H. Olsen. DEC is perhaps best known for its PDP minicomputer series, the subsequent VAX series (VMS and UNIX operating systems), and the DEC Alpha. Many universities are equipped with VAX machines. In 1998, DEC was bought out by Compaq, one of the leading makers of desktop computers. See Compaq.

Digital European Cordless Telecommunications DECT. Now called Digital Enhanced Cordless Telecommunications. An organization and wireless standard developed in Europe and adopted by the European Telecommunication Standard Institute ETSI in 1992. It was originally proposed as a unifying digital radio standard for European cordless phones. It has since been adopted by other countries, including Britain and some Asian countries. The DECT standard improves on previous technologies by supporting two-way calling, in addition to better mobility.

Open Systems Interconnection (OSI) principles have been incorporated into DECT in the sense that it consists of a physical layer, a data link layer, and a network layer.

DECT is implemented with transceiving base stations and mobile handsets. As it is optimized for capabilities different from those developed for cellular, it requires more cells to be used in a manner similar to cellular, due to the low power signals of DECT, but higher densities are then also possible.

DECT incorporates handover capabilities, and

Dynamic Channels Allocation (DCA) instead of fixed channels, with the hand unit scanning for the best signals.

Digital Loop Carrier DLC. Similar to a Local Loop Carrier, which provides a physical connection between subscribers and a main distribution switching frame, except that the DLC is committed to *digital* services over twisted pair copper phone wires. The DLC is a system of switches and multiplexers which concentrates low-speed services prior to distribution through a local central switching office or controlled environment vault (CEV). By multiplexing signals up to a local terminal where it then splits to provide service to subscriber pairs, the cost of wiring can be reduced. DLC systems developed in the early 1970s. See Next Generation Digital Loop Carrier.

digital multiplexer A system for aggregating or interleaving two or more digital signals, so they can be carried over fewer transmission lines, and sometimes also to aid in synchronization of multimedia applications that may require more than one signal (e.g., audio and video for videoconferencing). The signal is frequently demultiplexed at the receiving end in order to separately handle the various component signals.

Digital Network Architecture DNA. 1. An architecture which incorporates many aspects of the Open Systems Interconnection (OSI) model used by Digital Equipment Corporation (DEC) to develop applications. 2. A commercial network system from Network Development Corporation.

Digital Research Inc. DR. Originally called Inter-Galactic Digital Research, Digital Research was founded by Gary Kildall and his wife at the time, Dorothy McEwen. Gary was the developer of CP/M (Control Program for Microcomputers), a popular text-based operating system for microcomputers. It produced a line of good quality products starting with CP/M-80. GEM, the Digital Research graphical operating system predated working versions of Microsoft Windows by several years, and DR-DOS was often described by reviewers and users as superior to MS-DOS.

DR's efforts were not limited to software. In 1984, the company released an expansion board for Intel 8088-based personal computers which allowed four terminals to be networked to a PC using standard RS-232. With Concurrent PC-DOS, it provided the user the ability to run up to four MS-DOS or CP/M-86 applications concurrently, along with the program running on each individual terminal.

Over the years, Digital Research introduced many basic desktop computing and networking tools that have become intrinsic to the industry. The company was bought out in the 1990s by the Novell Corporation, who subsequently transferred DR-DOS to Caldera who released it as OpenDOS. Unfortunately, Kildall, who pioneered so many fundamental contributions to the microcomputer industry, was found dead at the age of 52. See CP/M; Graphics Environment Manager; Kildall, Gary.

digital signal hierarchy DS-. A North American time division multiplex (TDM) signal hierarchy, which is used in connection with data communications protocols. See DS-0 through DS-4, T1.

digital signal processor DSP. A specialized computer processor designed to work with digitized waveforms, often audio and video samples, in order to speed execution and provide more complex operations. Their computing power and flexibility allow them to be used for a wide variety of applications, such as the compression of voice and video signals, multimedia applications, medical imagery, combination phone/fax/modem devices, etc.

digital signature A type of digital identification which is sufficiently unique, secure, and resistant to forgery that it can be used for confidential and commerce-related online messages and transactions. A number of digital signature schemes are already in use for stock-related transactions, contracts, and general messaging. Digital signatures typically employ key encryption methods. See encryption, Pretty Good Privacy.

Digital Signature Standard DSS. A draft standard to permit the creation and transmission of a secure digital 'signature' through a Digital Signature Algorithm (DSA) to provide authentication of documents and transactions. Web commerce is eagerly seeking means by which documents can be electronically secured in order to use it for trade, banking, stock transactions, contract negotiations, etc., and will probably quickly adopt this or another scheme when sufficient confidence in its efficacy is attained. See Electronic Certification.

Digital Subscriber Line DSL, *x*DSL. Imagine turning on your computer, connecting to the Internet, and finding something interesting that you want to show a business colleague or friend. If you have only one phone line, and you're using a modem to change the computer's digital signals into analog signals that can be sent over the phone line, you will have to hang up the modem, wait for a dial tone, and *then* call your colleague or friend. Digital Subscriber line is a family of two-way communications services which makes it possible for you to talk to your friend *without* hanging up the computer connection first. You can do both at the same time, which means you can talk through the phone while you navigate the Net together, discussing the things that you both can see. This is how it is done.

Phone services historically have been analog systems, and there are millions of miles of copper wires installed around the world to provide these services. With the development of computers, phone switching centers began, in the late 1980s, to convert to digital equipment and software. This allows voice and data to be carried on one line at the same time, and instead of using a modem to change the computer signal to analog, and leaving the voice as an analog signal, it can be done the other way around. In other words, now the *voice* call is changed to digital and the computer signal *remains* digital. This opens up a world of possibilities for faster transmission, better compression, and simultaneous data/voice communications, without having to replace those millions of miles of copper wires.

That sounds very practical, yet relatively few people have switched to DSL services. One of the reasons is distance. While most subscribers are within the 12,000 feet or so in which DSL services can operate at their best speeds, about 20% of the population is not. Crosstalk and other types of interference are problematic, and are still being resolved. Traditional phone lines have loading coils installed at intervals, to extend the signals on voice grade communications. Unfortunately, at higher digital data rates, these coils cause interference.

Perhaps more important is the way in which DSL services were deployed. Originally, subscribing to DSL involved having the phone company install a special voice/data splitter on the subscriber premises, and further, installing a special peripheral device in the subscriber's computer. This was costly and not very practical, and most consumers are resistant to having proprietary peripheral cards installed in their computers. Most prefer the option of choosing a vendor and interface, and also of installing the hardware external to the computer, so things can be changed around as needed. For this reason, a number of commercial vendors have proposed several variations of DSL services, such as DSL Lite.

DSL was first developed by Bell Communications Research Inc. in 1987 to provide a means to deliver interactive TV and video-on-demand,

Varieties of Digital Subscriber Line Services

Type	Abbrev.	Speed	Notes
asymmetric DSL	ADSL	6 Mbps +	Twisted pair copper phone wires. The possible maximum rate of transmissions is inversely proportional to distance. Typically uses discrete multitone (DMT) line coding for data; frequency division multiplexing (FDM) or echo cancellation is used to subdivide the bandwidth.
high bit-rate DSL	HDSL	T1/E1 speeds	Symmetric. Longer distances can be supported through the use of repeaters. See high bit-rate Digital Subscriber Line.
single line DSL	SDSL		Still in development, can be used over a single wire pair.
rate adaptive DSL	RADSL	up to 8.7 Mbps	Bandwidth can be tuned to subscriber needs. It works over longer transmission lines. Rate and speed adjusts to the line length and quality.
very high rate DSL	VDSL	13-60 Mbps	Used in conjunction with FTTC or FTTB. Different downstream and upstream speeds. (Upstream speed is 1.5 to 2.3 Mbps.) Shorter maximum distance. Still in development.

over copper wires. The name is somewhat confusing, since it is not the *line* which is being installed, but rather the *interfaces* at each end of the line. The point of DSL was to create technology which would make use of existing lines. In fact, a DSL 'line' typically consists of two telephone lines. Since the introduction of DSL, further variations have been adapted, as shown in the Digital Subscriber Services chart.

Digital Subscriber Line coding and variations Since DSL is a multichannel service, it is necessary to split the available bandwidth. This is typically done with echo cancellation (EC) or frequency division multiplexing (FDM).

There are two predominant schemes for subdividing the available bandwidth into smaller units to individually evaluate their suitability for transmission. This is useful over twisted pair copper lines, which can vary widely in their characteristics. The two most common are discrete multitone (DMT) and carrierless amplitude and phase modulation (CAP), and others are being developed. Some of these modulation techniques have descended from the telegraph and radio broadcast industries, and some, such as wavelet encoding, are relatively new and still being explored. Each of these has various tradeoffs in terms of availability, cost, speed, and susceptibility to interference, as shown in the Common Modulation Schemes chart.

digital to analog conversion The conversion of data stored as discrete units, usually in ones and zeros on computer systems, to modulated analog wave patterns. A modem is a common device which performs digital to analog conversion when changing computer signals to modulated analog signals that can be carried electrically through a phone line connection. The process is reversed at the receiving end. See modem.

Digital Versatile Video DVD. A high-density compact disc standard for encoding large amounts of digital data on small discs. DVD designates MPEG-2 as the digital compression standard for video recorded on DVDs. MPEG-2 is a fast digital motion recording and playback specification. A DVD can be recorded on both sides for up to a total of about 18 Gbytes of data. This is very attractive to developers, as it means a two-hour MPEG-2 encoded movie can fit on one side. Sound is encoded in either MPEG audio or Dolby AC-3.

DVD specifies more than the compression and playback format, it also provides functionality to build interactivity into the medium through menus, multiple languages, and other features. This makes it attractive for educational software, and games programming.

DVD can be played on a stand-alone system similar to a combination CD/laserdisc system, that is, it works like a laserdisc player, but is small like a CD player. It can also be played on a computer through a DVD computer peripheral player. Some vendors are offering combination CD/DVD players in order to entice users into accepting the new technology. It is possible that DVD or something like it could eventually supersede traditional video cassette tape movies. The DVD discs are smaller, lighter, and more permanent than tapes. See MPEG.

Common Modulation and Signal Subdivision Schemes

Modulation or subdivision scheme	Abbrev.	Notes
carrierless amplitude/phase modulation	CAP	Carrier signal is suppressed, and reassembled at the receiving end. Single channel makes it more susceptible to interference.
quadrature amplitude modulation	QAM	Variations in signal amplitude are used to represent data.
discrete multitone	DMT	Frequencies are divided into discrete subchannels in order to optimize the throughput of each channel; faster, less susceptible to interference.
discrete wavelet multitone	DWMT	Provides some interesting means of implementing better performance, and low-loss compression, with less susceptibility to interference.

digital video DV. Technologies which permit the recording and playback of digitally encoded moving image information and sound fall into the category of digital video. One of the big barriers to inexpensive digital video has been the large amount of data that is required to record even small segments of video. When still images from film frames using cell animation techniques are individually digitized and stored, each frame may require up to 24 Mbytes, if it is to closely approximate the image quality of 35mm film. Since each second of animation requires between 24 and 30 individual still frames, as much as 720 Mbytes may be needed to store a second of video. A full-length movie is usually about 7000 seconds or more, which would require more than 5,000,000 Mbytes of storage. That's a lot, and that's not including sound, or data that might be added to provide search and retrieval markers interspersed with the images.

In order to make digital video technology possible, a number of innovations and trade-offs have been implemented. Data compression and decompression techniques are used to store images in less space, but the picture quality does not equal film, and fast processors and frame buffers are required to handle playback in realtime.

In spite of its limitations and technical requirements, developers are forging ahead with digital video products partly because digital video can be edited and manipulated in remarkable ways. Special effects that are impossible or difficult to achieve with analog film, are possible with digital video. DV also has greatly increased possibilities for interactivity and access through the Internet. Of further importance is the fact that it doesn't have to go through a chemical photofinishing process before it can be used. See animate, celluloid, Digital Versatile Video, interactive video, MPEG, video-on-demand.

Digital Video Broadcasting Group DVBG. A European organization that provides support and specifications for traditional and emerging broadcast technologies like cable and satellites, and schemes for the protection of commercial programming.

Digital Video Interactive DVI. A digital recording and playback chipset technology developed at the David Sarnoff Research Center. The technology was acquired by Intel Corporation which subsequently developed it into Indeo 2 and Indeo 3. It is now known as Intel Video Interactive (IVI). See Intel Video Interactive.

digital videodisc, Digital Versatile Disc DVD. A vendor consortium-developed ISO-9660-supporting optical disk format specification, similar to compact disc, except that it is designed to store a much larger quantity of data. DVD physical discs are the same diameter as the common compact discs popularly used for music, but very slightly thicker, bringing the recorded surface of the disc a little closer to the laser pickup, permitting a higher 'resolution' or density. That is, smaller, more precise pits can be used to store the information, up to 17 Gbytes, depending on whether the disc is recorded single- or double-layered, and single- or double-sided. Further flexibility is possible through the use of dual lens apertures in the laser pickup to provide a dual CD/ DVD player.

Vendors are particularly interested in offering movies on DVD rather than on cassette tapes or the larger videodisc formats. The DVD medium is more robust and convenient than tapes or large discs, provides better sound capabilities than tape, and can store motion pictures of up to 133 minutes in length, including subtitles. While this isn't sufficient to hold every type of movie, it seems reasonable that some of the longer films, many of which are popular classics, may be offered on two discs, as some laserdisc movies are now.

digital voice encoding A process of sampling and quantizing voice signals and storing them as digital data. Since voice encoding can require a lot of memory, the information is usually compressed for storage, and decompressed when replayed. Fractal and wavelet compression techniques are becoming popular choices. Much of the technology for digital voice encoding has come from the music industry. Research into the sampling and playback of synthesized music can be applied effectively in voice encoding and playback applications.

Digital voice encoding is used in speech and voice recognition systems, and is used to send conversations over digital voice telephone channels, such as Internet telephone applications, ISDN, or digital PCS. To save memory and cut down on transmission time, interesting algorithms for removing pauses and spaces are used in conjunction with digital voice. Encryption to ensure privacy is also possible with digital voice communications.

Digital voice encoding is a means of creating a data library of sounds, phonemes, words, or other sound units which can be used in digital applications such as automated voice menu systems, speaking computer applications, digital talking books, digital answering machines, voice mail systems, and more. Digitally encoded voice tends to be more pleasant and natural sounding than mathematically generated voice, and hence is favored for applications where instructions or responses are given to a human listener. See quantize, silence suppression, speech recognition, voice recognition.

digitize *v.* To convert from analog to digital, that is, to take an analog signal and convert it into data that consists of discrete units, such as ones and zeros, usually by sampling the analog signal at discrete points and assigning a value to the data at that point. Since most broadcast media are transmitted as waves, they are analog systems of communication. However, in order to process the information, or interface it with networks, it must be converted to digital format at some point in the transmission process. Thus, digitizing a phone signal makes it possible to add features such as compression, encryption, and voice recognition. Digitizing a video signal allows processing of the image: palette changes, overlays, image composites, split screen viewing, videoconferencing and more. A desktop scanner is a type of digitizer, as is a digital camera. See analog, digital, digitizer, pulse code modulation.

digitizer A device which converts a signal from analog to digital, usually by sampling the analog signal at discrete points over time, and assigning a value to the measurement. In most cases, the more frequent the sampling, the more true the digital representation is to the original communication. A digitizer is commonly used in video and audio applications. The *sound patches* used in electronic music are digitized from analog sound samples. TV weather forecast images are created by overlaying digital signals or digital and analog signals, usually from a combination of digital and analog sources.

Digitizers need not be restricted to 2D representations. One type of digitizer consists of a pen or robot arm, which traces contours on an object and converts the spatial information into a 3D coordinate system for use with rendering, ray tracing, or CAD programs. One of the first widely distributed 2D digitizers for microcomputers (1986) was NewTek's DigiView for the Amiga. It is historically significant because it was the precursor to the Video Toaster, a product which initiated the desktop video industry.

DII See defense information infrastructure.

Dijkstra's Algorithm In ATM networking, an algorithm sometimes used in conjunction with link and nodal state topology information to calculate routes.

Dilbert The name of a syndicated comic strip and its main character, written and illustrated by Scott Adams. Dilbert engages in satirical adventures as an engineer in a red-tape high-tech company. The close-to-home quality of his absurd interactions and adventures have made it extremely popular among software development and corporate cultures.

DIM See document image management.

dim fiber Fiber optic cable in which the carrier provides the means to carry the signal through the fiber, but does not originate the signals at either end of the circuit. See dark fiber.

dimmed See ghosted.

DIMS Document Image Management System.

DIN 1. Deutsches Institute för Normung. German institute for standardization, one of the major standards bodies in Europe. DIN is located in Berlin. The DIN specification is a widely used standard for computer connectors. 2. dual inline.

DINA Distributed Intelligence Network Architecture.

dinosaur *slang* A large, obsolete, or aging system that requires excessive resources to keep in operation. Given the speed of technological obsolescence in computer technology, systems quickly become dinosaurs, but firms often hang on to them because the cost of reinstalling and relearning new technology may be higher than continuing to use old, slow, but tried-and-true systems. The term dinosaur applies especially to old room-sized supercomputers that are now less powerful than many desktop systems.

diode In older electronics systems, an electron tube with only two electrodes, the cathode (electron-emitting) and the anode (electron-attracting), the electrons flowed freely and uncontrolled. This type of tube wasn't very useful. It wasn't until L. de Forest developed a triode, a three-electron tube that control over the electron flow became practical. In terms of modern transistor electronics, a piece of semiconductor material, positive on one side and negative on the other, with a terminal at each end constitutes a simple diode. Where the electrical positive and negative regions come together in the semiconductor, it is called a p-n (positive-negative) junction. Like a two-element electron tube, the electrons normally flow in one direction, serving as a rectifier. This allows the conversion of current. Conversion of energy from one form to another is an essential aspect of communications technology. An old crystal radio rectifier can change radio waves into audio waves that can be heard through earphones. A semiconductor rectifier can change alternating current (AC) to direct current (DC) that can be used to power electronic equipment.

Forcing the flow of electrons to go opposite to the natural direction can be accomplished in some circumstances with sufficient voltage, resulting in reverse bias. This is sometimes useful in semiconductor technology for altering the information stored in a chip. See avalanche breakdown, erasable programmable read-only memory, Zener diode.

diode transistor logic DTL. A logic circuit employing diodes to perform logic functions that activate the circuit transistor output. See transistor-transistor logic.

DIP 1. document image processing. 2. dual in-line package. See DIP switch.

DIP switch dual in-line package switch. A very small switch that is meant to be toggled to one side or the other (hence dual). It can be changed with a pencil or other pointed object. A few DIP switches are large enough to be toggled with the fingers. The early desktop microcomputers like the Kendak-1 and the Altair were programmed with DIP switches.

The smaller DIP switches are often found on SCSI devices, for setting SCSI ID numbers, or for adjusting other settings. Graphics controller cards sometimes have DIP switches to adjust resolution or scan rate settings.

dipole antenna Simply implemented, an antenna with two poles mounted horizontally to produce one long rod which is a subdivision of the length of the wavelength it's designed to receive. Ideally the impedance (expressed in ohms) of the lead connecting the dipole antenna to the receiving equipment (radio, television, etc.) should match the impedance of the antenna at its strongest point (usually the center in a symmetrical antenna). The frequency response of a simple dipole antenna can be increased by placing the two rods parallel to one another, close together, and connecting them at either end. This is known as a folded dipole antenna.

DirecPC A commercial service from Hughes Network Systems which allows data communications through Very Small Aperture Terminal (VSAT) satellite systems by means of a personal computer. The satellite signal is received by a two-foot parabolic dish antenna, which feeds the transmission to a peripheral card in a personal computer where the signal is demodulated, demultiplexed, and decoded, and then sent to the software interface. In turn, the software permits orientation of the satellite dish.

An innovative configuration of this system for Internet use combines the satellite receiving system, and a normal analog connection to an Internet Services Provider (ISP) as an upload/download hybrid system. This service allows the user to connect to an ISP through the normal phone line with a conventional modem, to interact through a Web browser. When files are requested, rather than downloading through the ISP phone connection, the files are transferred from the DirecPC network operations to the VSAT satellite, and to the user's DirecPC dish, thus providing downloads at 400 kbps compared to about 39.6 kbps.

direct access storage device DASD. Quick access computer storage devices such as hard drives and memory chips.

direct broadcast satellite DBS. Originally this phrase was intended to describe a particular service transmitting in the 12.2 to 12.7 GHz range, a frequency approved by the World Administrative Radio Conference (WARC). However, since the availability of smaller, more convenient receiving dishes, the term is also used in a broader context, to describe any satellite that transmits a signal that can be picked up by individual home and business subscribers, without going through an intermediary station.

Since direct broadcasts are transmitted to small consumer dishes, and the size of the dish is related to the length of the radiant waves being received, the higher wavelengths are used, such as Ku-band frequencies. This permits a dish as small as two or four feet in diameter to be effective. Further, as DBS systems are transmitted to the small consumer dishes rather than to a station with a powerful receiving antenna, it was necessary for them to use high power transmissions, usually around 40 watts to 160 watts, far more than was being used for C-band communications in the 1980s when DBS began to develop. (Today satellites can deliver almost ten times that power.)

DBS to the home presents a number of the

same moral challenges as widespread access to the Internet. DBS providers are concerned about illegal consumer copying and redistribution of programming, and consumers are concerned about program content. These issues are still being studied and resolved.

DBS system guidelines for Europe were originally established by segmenting the 11.7 to 12.5 GHz frequency spectrum into 40 channels, to be shared among the various European member nations.

U.S. DBS systems fall under the jurisdiction of the Federal Communications Commission, through guidelines established in the various Telecommunications Acts, and must conform to internationally established technical standards. See C-band, Ku-band, microwave.

direct connect modem A type of modem which began becoming popular on desktop systems in the early 1980s, gradually superseding acoustic modems. The acoustic modem connected to phone lines by setting the handset of a telephone into two suction-cup style holders. This was impeded by noise and didn't always provide a good connection. The direct connect modem, in contrast, was cabled electrically between the computer serial port and the modem, if it was an external modem, or between the computer and the internal modem card to the telephone line, through the telephone jack that normally connected to a phone. This system was not only more convenient, and more effective in terms of reducing noise and ensuring a good connection, it also made it possible to use higher baud rates. See acoustic modem.

direct current DC, dc. A more-or-less constant electrical current flowing in one direction. DC current is the type supplied to many small appliances by batteries.

Much of early communications history was based on direct current (DC) as a power source. Telephones had 'talking batteries' and 'common batteries.' The batteries tended to be large, leaky wet cells, that caused inconveniences if moved or subjected to temperature fluctuations. More than fifty years after the invention of the telegraph, AC power for telegraph systems was still considered a novel idea, but the shortage of batteries, and their cost, provoked French and Swiss engineers to experiment with AC generators, as described in the *Annales des Posts, Télégraphes et Téléphones* in September 1919. Eventually the advantages of AC power were better understood, and its use became common, and DC took its place as a current source for small, portable electric conveniences such as calculators, radios, wristwatches, laptops, cameras, etc. See alternating current; resistance; Tesla, Nicola.

Direct Distance Dialing DDD. A commercial name to indicate the capability of a network to connect a long-distance call without operator intervention.

Direct Inward Dialing DID. In the past, calls going through a central office (CO) to a private branch extension, had to go through an attendant. With increasing automation, this is now rarely necessary. With DID, the called digits are passed through central office DID lines directly into the private branch exchange. DID lines do not offer a dial tone and hence cannot be used for direct outgoing calls.

Direct Inward System Access DISA. A telephone system setup in which outside callers can dial into a telephone system, usually a private branch exchange, and have the use of the system's capabilities as though they were on the premises using the system from the inside.

direct memory access DMA. A means to bypass the central processing unit (CPU) in a computer, and interact directly with memory. This is used to reduce processing time and increase speed.

direct outward dialing DOD. The capability of a private branch exchange to dial calls outside the exchange without first dialing an access code (typically '9'), or going through an operator.

Direct Public Offering DPO. A mechanism for raising investment capital through the sale of shares. A DPO is a state-regulated type of Initial Public Offering. For a fuller explanation see Initial Public Offering.

direct sequencing A spread spectrum frequency changing broadcast technology. Spread spectrum broadcasts spread a transmission over a broader range of frequencies than is typically needed to contain the broadcast. A pseudorandom digital sequence directs a phase modulator to distribute the original RF transmission over a bandwidth which is proportional to the clock frequency of that sequence. The receiver must then be synchronized to the same pattern as the broadcast generator in order to remove the phase modulation and recreate the original signal.

Although it's not commonly done, it is possible to combine direct sequencing with frequency hopping. This would likely only be used in very high security transmissions, as the synchronization and receiving techniques are not trivial. See frequency hopping, spread spectrum.

direct set In ATM networking, a set of host interfaces which can establish direct communications at layer two, for unicast.

direct view storage tube DVST. A cathode ray display device introduced in the late 1960s to overcome the slow refresh and large storage buffer needs of early vector display monitors. By employing a slow-moving beam and a storage mesh, the DVST design substantially decreased the cost of display devices. See vector display.

directional antenna A radio antenna which is designed to concentrate its signal transmission or receiving strength, resulting in a stronger signal, but one which is not of equal magnitude in all directions. Commonly found in AM radio transmissions.

directory 1. A list, usually of names and associated information, often sorted, or organized and displayed to enhance visual clarity. These may be the names of people, companies, institutions, files, or subdirectories, etc. 2. A table of organizational identifiers that provides addresses to individual items within the organizational path. This path is frequently hierarchical in structure.

directory, file On a computer file system, an organizational structure, comprising a file storage area, under which there can be further files or directories. A directory listing typically includes other information about the directory and its associated files, or subdirectories, such as creation date, permissions, file type, and byte size. The display may be in text or graphical mode, and an icon resembling a file folder is often used to symbolize the directory.

Directory Agent DA. A network agent is a database service which accepts advertisements for SLP devices from Service Agents, and answers queries. See Service Location Protocol, Service Agent, User Agent.

Directory Assistance DA. A dialup telephone service in which the caller dials a number set aside for directory assistance service in order to get the number of a person outside the local calling area, or within the local calling area if the number is not listed. The number may not be listed because the individual just moved in, just changed a phone number, or installed the phone just after the directory was published. Subscriber unlisted numbers may not be given out through Directory Assistance. In most areas, there is now about a $.75 charge associated with a Directory Assistance request, and there is usually a two-number limit on each request.

Directory Assistance used to be handled entirely by human operators, but there are now automated systems which will request the city and name of the person whose number is being sought and dispense the number, or hand over the call to the operator to complete the transaction, or to clarify the information provided by the caller. These automated systems combine speech recognition and speech synthesis to carry out their tasks.

directory caching A time-saving function of many operating systems which stores the local directory listing in memory so that each time the user accesses the directory information (for listing parts or all of the files there), it will be displayed very quickly. For small directories, there isn't much difference, but for very long directory listings, it can be handy, especially if the information is being output to a shell or MS-DOS text window. Further, the information can be used to more quickly access the files in that directory. Since the directory information is accessed from memory rather than from disk, the transfer time for the information is faster.

Directory Number Call Forwarding DNCF. An interim Service Provider Number Portability (SPNP) which is provided through existing available telephone services, such as call routing and call forwarding. The DNCF is set up so that calls are forwarded to a new number. While it is called Directory Number Call Forwarding, unlisted numbers can also be set up with the service.

Directory System Agent DSA. A directory application process in the Open Systems Interconnection (OSI) system. OSI utilizes a system of 'agents,' that is, helper applications which can query a local database, communicate with other agents, or hand off requests to other agents when appropriate. The DSA specifically provides associated Directory User Agents with access to the directory information base (DIB).

directory tree The name given to a hierarchical computer file structure. The 'tree' includes the current directory and subdirectories and files associated with the current directory both above it and below it in the hierarchy. Directory tree commands are important because the user may wish to use file creation, deletion, renaming, protection, or searching commands on a series of files or all the files located in the tree.

Directory User Agent DUA. A directory application process in the Open Systems Interconnection (OSI) system. OSI utilizes a system of 'agents', that is, helper applications which can query a local database, communicate with other agents, or hand off requests to other agents when appropriate. The DSU specifically assists and represents the user in accessing a database through a Directory System Agent (DSA). See Directory System Agent.

dirty power Electrical power that is spiky, bursty, noisy, or otherwise unreliable. Dirty power can be dangerous to delicate electronic components. Laptops plugged in on ferries and trains should be used with power conditioners and good quality surge suppressors.

DISA 1. Data Interchange Standards Association. 2. Defense Information Systems Agency. 3. Direct Inward System Access. A subscriber option for external access to a private branch exchange (PBX), usually through a security code.

disable 1. *v.* To prevent from functioning. There are many ways to disable a process or mechanism, but the most common are interrupting the power, or the transmission. 2. *n.* A signal, tone, or command sent through a circuit to disable a device at the end of the transmission. Telephone companies have the capability of electronically enabling and disabling a number of phone services.

disaster recovery The procedures and resources necessary to bring a system back into functioning after a disaster. Disaster recovery applies to everything from 1. repairing or replacing facilities and equipment damaged by floods, hurricanes, or bombing; to 2. recovering data from a hard drive that was damaged from a lightening strike, or a toddler poking paper clips into the mechanism. Data recovery is a fact of life. Whether or not you archive your data, your drive *will* fail at some point, or you *will* inadvertently overwrite an important file. Prevention is usually better, but if recovery of data becomes necessary, there are many shareware and commercial tools to assist with the task.

disc A round, flat data storage and retrieval medium. Usually these are *write once/read many* (WORM) devices, since optical media are not as easy to write, and rewrite, as floppy disks. Although in English disc and disk are used somewhat interchangeably, in computer parlance, disc tends to be used more often to designate optical storage media such as audio CDs, laserdiscs, etc., whereas disk is used more often as a contraction of *diskette* to designate an encased floppy diskette (flexible magnetic medium). See disk.

discard-eligible DE. In data-segmented networking systems, there are often packets or *frames* that are discarded for various reasons, including redundancy, congestion, timing out, low priority status, etc. Sometimes the decision to discard the packet is made on the part of the network's lower level functions, and sometimes the discard status is explicitly represented by a signal bit. In frame relay, there is a bit in the HDLC frame address field which can be set to mark the frame as expendable in case of congestion.

disclaimer Because of the public nature of many network communications, especially on USENET newsgroups or Internet Relay Chat (IRC), and the fact that many people post from work accounts, people's remarks or postings are frequently accompanied by disclaimers specifying limitations of liability to themselves or their employers. Software license agreements often have disclaimers of liability for any damage to computers on which the software is run. It's pretty unusual for software to damage hardware (yes, it is possible), but it doesn't happen under normal operating conditions. Nevertheless, attorneys prefer to have all bases covered.

disco tech *slang* A rather good pun for *disconnect technician*.

disconnect 1. To separate two discrete units or devices from one another, usually with the implication that the separation broke some type of electrical, inductive, or communications link between the two. 2. To terminate a communication, as in a phone call. 3. To break a circuit.

discrete Individual, separate, distinct.

discrete cosine transform DCT. A mathematical means of manipulating information by 'overlaying' cosines in order to analyze or use it from another point of view. DCT techniques are used in a number of digital compression schemes, including a lossy compression technique that provides a practical balance between a high degree of compression and relatively good perceptual clarity in the decompressed image (although very high compression ratios may create a blocky effect). An adaptive version of this technique is used in JPEG image compression. JPEG is one of the primary image representation formats used to display graphics on the World Wide Web. One of the disadvantages of DCT is a visual artifact called Gibb's effect which manifests as ghostly ripples running along distinct edges. See discrete wavelet multitone, Fourier transform, fractal transform, JPEG, MPEG, lossy compression, wavelet.

discrete multitone DMT. A multicarrier technology used for transmitting multiple media over existing copper wires. This lends itself to a variety of data delivery services, including Digital Subscriber Line (DSL) and Hybrid Fiber-Coax (HFC). DMT uses discrete Fourier transforms for creating harmonics along the main lobes, and demodulates at the receiving end. By dividing available bandwidth into smaller units, portions of the available bandwidth can be individually tested to evaluate speed, availability and suitability for transmission. This allows the optimization of transmissions over existing twisted pair installations, which can vary widely in their characteristics from region to region. This scheme is currently used in ADSL installations, and is being evaluated as a potential standard for VDSL. See carrierless amplitude and phase modulation, Digital Subscriber Services, discrete wavelet multitone.

discrete wavelet multitone DWMT. A multicarrier technology developed by Aware, Inc. for transmitting multiple media over existing copper wires, based on the same general principles as discrete multitone. DWMT is being promoted for use with Digital Subscriber Line (DSL) and Hybrid Fiber-Coax (HFC). DWMT divides the available bandwidth into smaller units to utilize suitable portions. It differs from DMT, which uses Fourier transforms, by using wavelet transforms for encoding subchannel bits. DWMT produces lower energy harmonics than DMT, making it easier to demodulate the encoded signal at the receiving end, where a forward fast wavelet transform (FWT) is used. It is less susceptible than other schemes to channel distortion, and requires less overhead. See discrete multitone, wavelet.

dish *colloq.* Common terminology for a parabolic satellite or terrestrial receiving or transmitting dish.

dish aperture The diameter of a parabolic communications receiving dish. In general, the larger the aperture, the broader the scope, or 'more' signal is apprehended (there are exceptions depending on the nature of the signal, and the shape and materials of the dish). The amount of aperture is also partly dependent on the size and position of the feed apparatus, as it will block some of the signals.

dish focal point The distance, in a parabolic antenna, from the reflective surface to the focusing point of the signal. It is important to know this point in order to position the feed mechanisms efficiently. The focal point is dependant on the breadth and curvature of the dish, with flatter dishes generally having a focal point that is farther away from the reflective surface than more concave dishes of the same diameter.

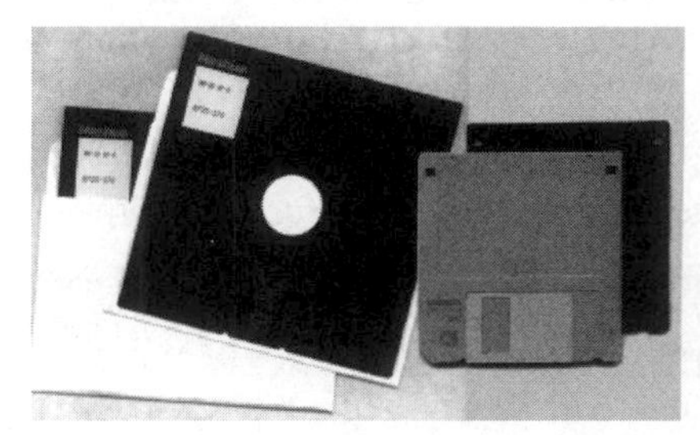

In the older 5 1/4" soft-sided diskettes on the left, the read/write area is exposed, and vulnerable to fingerprints and other damage. The 3 1/2" diskettes on the right have a hard, protective shell, and a sliding media protector which is attached to a spring. When inserted in a drive, the sliding shutter opens to allow access to the rotating floppy disc media inside.

disk, diskette A round, flat, flexible, encased medium, typically used for data storage and retrieval. The flexible medium inside the case is coated with magnetic particles that can be rearranged to provide read/write capabilities.

In the late 1970s and early 1980s, desktop computers were equipped with 8" disk drives, and the accompanying diskettes were flexible and highly subject to damage. By the mid-1980s, 3.5" disk drives were becoming popular, even though the diskettes were expensive at $6.00 each, because they had a hard shelled protective covering. Prices dropped until diskettes of double the capacity were only $.30 each. By the mid-1990s, superdisks were starting to be developed, but it was not until 1998 that they were widely announced. The superdisks are downwardly compatible with regular floppy disks, but can hold over 100 Mbytes each. Very convenient, compared to 1.4 Mbytes, and less expensive than most cartridge drives.

Although in English disc and disk are used somewhat interchangeably, in computer parlance, disk is used more often as a contraction of *diskette* to designate an encased floppy diskette (flexible magnetic medium) rather than an optical medium. Disk also is short for *disk drive,* a high capacity hard storage medium. See disc, diskette, hard drive.

disk controller A hardware peripheral circuit which provides an interface between a computer's main circuitry and a disk drive. The most common formats for disk controllers are SCSI and IDE. SCSI is predominantly used on Sun, Amiga, Macintosh, DEC, HP, NeXT, SGI, and server-level IBM-licensed desktop computers. IDE is predominantly used on IBM-licensed consumer-model desktops, and some recent Macintosh computers. SCSI allows up to six devices to be chained to each controller port (the controller is considered the seventh device). IDE allows up to two devices, designated master and slave, to be chained to each controller port. SCSI disk controllers can be used to interface with hard drives, scanners, cartridge drives, CD-ROM drives, and redundant array of inexpensive disk (RAID) systems. IDE can interface with hard drives, cartridge drives, some CD-ROM drives, and some scanners. To enhance the limited capabilities of IDE, an Enhanced IDE specification has been developed.

disk mirroring See mirror, RAID.

disk operating system DOS. The low level operating functions of a computer that are used to read or write a floppy disk or hard disk, and engaged when the system 'boots up.' The phrase has since taken on a more general meaning to encompass *all* the low level operating functions of a computer. In the late 1970s, there were several disk operating systems made available for the various microcomputers. For example, the Radio Shack TRS-80 Model I could be run with CP/M, TRS-DOS, LDOS and others, and you could select the one you wanted to run by reading it from tape (technically, a tape operating system), or from a floppy disk (and later from a hard drive). Gradually, however, as microcomputers came into the mainstream, people began to associate the operating software with the computer on which it came, and many do not realize that each computer hardware platform is capable of running a variety of operating systems.

For modern computers, there are many choices including Rhapsody, BeOS, Windows, MacOS, OpenStep, Linux with The X Windows System, etc. The user doesn't have to be tied to the operating system that comes bundled with the computer. In the future, if more developers adapt the various emerging international standards for open systems, particularly object-oriented technologies, applications may finally become operating system independent, and then users will be able to choose different computers the way they choose different cars, with the common object format being the 'gas' that is standardized to power them all.

disk server A system dedicated to file storage, retrieval, and handling on a network with shared disk resources. It may also implement user access passwords, file locking, license restrictions, and other administrative tasks associated with dynamic file sharing on a multiuser network. On large computer systems, the disk server may manage many dozens or hundreds of disks, and may require a room of its own with special fast cabling such as Fibre Channel cabling. See RAID.

diskette See disk.

diskless computer A computer with no user-accessible disk drive. Typically, diskless PCs are used in areas where vandalism or security are an issue, or where the interface is simple in order to make it accessible to inexperienced computer users. The absence of a way to access the PC storage prevents data theft, and the danger of introduced viruses. Diskless PCs are generally attached to a console, in which the storage is housed, or use flash memory for data storage, or are networked to a remote server. They are often found in kiosks, copying centers that offer access to computers by the hour, public shopping centers, libraries, amusement parks, and science and technology museum exhibits.

DISN See Defense Information Systems Network.

dispatch 1. *v.* To send out a verbal or written communication, or package. 2. *n.* A missive, telegram, or other written or electronic communication usually intended for someone out of normal hearing range, or intended to provide a record of the communication. 3. *n.* A communication intended for a group of recipients, such as field workers, taxi drivers, law enforcement officials, etc. Often sent verbally as a radio communication.

dispersion The spreading and gradual loss of signal strength that occur in electromagnetic or optical data transmissions over distance. Different media have different dispersion characteristics. Dispersion is usually, though not always, undesirable, and steps are taken to minimize dispersion in most transmission technologies. In fiber optics, multimode cables are more subject to dispersion than single mode cables, due to refraction, and hence have a shorter effective distance. In airwave broadcasts, undesirable dispersion, called scattering, can occur as a result of moisture, particles, and terrain.

dispersion, wave The separation of an electromagnetic wave into its component frequencies. For example, light rays passing through a prism will be dispersed into individual color frequencies.

display *n.* A presentation of visual information, usually on a display medium such as a screen, wall, or computer display device such as a cathode ray tube (CRT), plasma display, liquid crystal diode (LCD), or light-emitting diode (LED) surface. See cathode ray tube, light-emitting diode, liquid crystal diode, multisync.

display driver The software which translates instructions from a computer processor to the correct operating instructions for the attached display device. A display driver may be written to control one specific type of display, a family of devices, or a wide variety of devices. Most computers have a set of generic display drivers incorporated into the operating system, or shipped with the system to work with the operating system. However, as there are many different types of vendor products, and new ones all the time, many vendors will provide a display driver software utility with their product. Sometimes the display driver will be incorporated into the peripheral card that interfaces with the display device. Sometimes the display driver is a two-way communications channel. Some display devices will send information to the driver about their capabilities and configuration parameters so that the driver can configure itself to optimally take advantage of the installed device.

display hack A clever, imaginative program designed to show off some entertaining characteristic of computer display operations, or to simulate a humorous problem or fool the viewer. Some display hacks include windows unexpectedly melting, bouncing, or eating one another up. Because of its good graphics capabilities, a profusion of display hacks were circulated on Amiga computers from 1985 to 1987, and so many favorite ones were created by a programmer named Leo Schwab, that they became popularly known as Schwabbies.

display interface card A peripheral card which is interfaced between the computer processor and the physical display device in order for electrical signals and control data from the device driver to be transmitted between the computer and the display device. Sometimes the peripheral card itself comes installed with display driver software.

Some display interface cards have hardware settings to indicate the parameters of the installed display device. Thus, there may be a dial, jumpers, or dip switches that have to be set to configure the screen resolution, number of colors, refresh rate, and other parameters.

distance learning Receiving education through telecommunications media, including mail correspondence, email, Web sites, audio and video tapes, video conferencing, etc. A number of educational institutions are providing course content, references, etc., on the Internet, especially on the Web, and, with digital signatures, they may also eventually provide testing, assignments and critiques through electronic means. See audiographics, videoconferencing, whiteboarding.

distant learning, distance learning A concept that has been around since early correspondence school days, distant learning involves interacting with learning materials and instructors over distance. The Internet has increased opportunities for distant learning, and there are many educational institutions which now publish course materials, and even take-home tests, on their Web sites. Distant learning is also being furthered through videoconferencing classrooms where students can see lectures and labs through video phones or personal computers. Distant learning is beginning to fulfill some of the promise hoped for by its early pioneers. See audiographics, distance learning, videoconferencing.

distinctive dial tones This is a method by which dial tones are distinguished from one another by properties such as pitch, so that the caller can tell what type of call is being connected. This is particularly useful on private branch systems where internal and external calls require different sequences of numbers. For example, it is common to dial '9' to get an outside line on private systems, and the change in tone helps the caller know the call has been given outside access and will now accept an outside number.

distinctive ringing A subscriber option, or feature of some phones which uses a different tone or ringing sequence to identify incoming calls as coming from inside or outside a branch system, or when calling internally to another extension. Different phone carriers offer this option under a variety of names, such as Feature Ring, and Ident-a-Call.

While distinctive ringing is common in businesses with private exchanges, there are now phones and peripherals for residential and small business users which will distinguish calls on separate lines, or provide a means to call through to another callee with codes that make a different ringing sound or sequence. It's handy if you get a lot of calls for teenagers that you want them to answer themselves.

distort To deform, contort, warp or pervert out of a normal sound, shape or condition. Distortion is sometimes intentional, as in distorting the sound from an electric guitar or synthesizer to create some special effect. In most cases, however, distortion of a transmission is an undesirable fault condition.

distortion An undesirable change in the basic characteristics of a wave or data transmission sufficient to interfere with the information, or its perception. Extraneous noise is not technically considered to be distortion. In visual images, distortion usually involves undesirable aberrations in the basic characteristics of the image, such as color, shapes, or lines which misform or obscure the original features, or of the entire image, in which case the outer contours may become squeezed or twisted. In sound distortion, the pitch, speed, or timbre of the sound may be altered, to make it difficult to discern the content or source of the sound.

distribute To apportion, spread out, scatter, dole, deal out, or dispense; to give out, broadcast, or deliver to members in a group.

distributed backbone A backbone is a central 'artery' or trunk in a network. A distributed backbone is one in which network segments are interconnected through hubs joined with backbone cables. Thus, there may be multiple segments or rings, joined to one another through a backbone segment. See backbone, collapsed backbone.

distributed computing A system in which the computing processes are divided up, parcelled out, or otherwise handled simultaneously, or in which computing services are apportioned, broadcast, or delivered to users. A local area network (LAN) or wide area network (WAN) are examples of distributed computing environments. Users of individual workstations can work independently of one another on tasks that are frequently carried out, yet can share common files, applications, and devices distributed around the system. This permits more open and efficient use of resources.

A render farm is another example of distributed computing in which a centralized system parcels out individual tasks related to constructing 3D models and images. These are assigned to various machines, and the completed processes or objects reintegrated to create a completed rendering, thus potentially speeding up processing time. The Internet represents an example of distributed computing in that routing and transmitting of messages occurs through a cooperative network of many interconnected systems; and many common resources (Web sites, archives, applications, search utilities, online chats) are shared among users.

Distributed Computing Environment DCE. A commercially developed set of services from Digital Equipment Corporation (DEC) that supports the development, maintenance, and use of distributed computer applications. It is based on the Open Software Foundation's DCE. (The Open Software Foundation is now The Open Group.)

distributed file system A file system which is distributed across more than one partition, more than one device, more than one workstation, or more than one system. Some operating systems can set up a distributed file system across partitions, or across disks, in such a way that a directory or set of files appears to the user as though it were all in one place, thus making it transparent, appearing as a single logical unit, even though the physical underpinnings are different devices. See redundant array of inexpensive disks (RAID).

distributed network administration DNA. 1. A wiring and services distribution scheme which divides up the administration and actual physical connections into smaller units, which may be one per department, one per floor, or some other arrangement. This is convenient for systems where there are significant physical limits on the length of individual transmissions cables (as in some fiber optic installations), or where operations or maintenance personnel are assigned to different sections of the building, or where different types of wiring are used for different departments or areas of the premises. See distribution, distribution frame. 2. A wireless services system in which administration is subdivided into particular regions or sections of the service (a local area, city, or state). This is a common administrative system for nationwide wireless systems that cover a large area of territory, but through which many users roam from region to region.

Distributed Queue Dual Bus DQDB. A connectionless packet-switched network protocol with a 53-octet cell and header/information structure somewhat like ATM, used for telecommunications services in Metropolitan Area Networks (MANs). DQDB is described in IEEE 802.6, and descended from Queued Packet and Synchronous Exchange (QPSX). DQDB supports isochronous and nonisochronous communications.

DQDB employs the *slot* as the basic unit of data transfer. A slot is further subdivided into a one-octet Access Control Field (ACF) and a 52-octet segment.

distribution 1. A cabling term referring to the delivery of services through cables and/or wires. A distribution system is the combined media, connections, and topology which provide services through wires or cables, usually consisting of electricity, voice, or data network services. Many different types of distribution arrangements exist and are described throughout this reference. See distribution frame, horizontal distribution. 2. Apportioning, assigning, sending out, delivering of products, services, or computing processes.

distribution frame A centralized circuit management structure for creating, troubleshooting, and accessing a variety of incoming and/or outgoing lines, sometimes in the thousands, if it is a commercial switching system. A distribution frame may be for supplying electrical power, or may organize lines and connections for data communications. Frames are often built into closets, floors or ceilings, depending on available space, and whether or not frequent access is required. The frame may include blocks, conduits, and other physical structures for facilitating the cabling.

distribution panel A grid-like frame, usually of metal, with rows and columns of punched out holes through which cables can be threaded and mounted. These are often designed to fit standard 19" rack mounts.

dithering A term that the imaging field has adapted from physical appliances in scientific labs where dithering devices are used to shake up items to minimize clogging, settling, or the effects of friction. In computer imaging, dithering is the blending and combining of pixels to simulate the effect of a shade of gray, or a color that may not be available on the system. For example, in a 16-color palette, dithering yellow and red pixels together is a way of simulating various shades of orange, depending on the proportion of yellow to red elements. Since the user perceives the adjacent colors as blended, when viewed from a distance, it is a means to extend the palette. Similarly, on a monochrome screen, shades of gray can be simulated by dithering together white pixels with black in ordered or random sequences. Automatic dithering is done on many operating systems with limited palettes when for example, displaying 24-bit images (over 16 million colors) on 256-color display systems. There are many different dithering algorithms, often named after their inventors. See Floyd-Steinberg.

divest To dispossess of authority, property, or jurisdiction; to take away, to deprive, to disinherit; to dismantle.

divestiture The act of breaking up, dispossessing, or otherwise dismantling an institution. In the U.S. legal system, divestiture is a process whereby the U.S. Justice Department oversees the alteration of a company's organizational and asset structure in order to deconcentrate power or enforce fair competitive restrictions. This is not a common procedure, but it occurs when a company's activities are monopolistic in nature, or where a company has engaged in unethical or illegal practices that provide the company with an advantage. The process of divestiture thus is intended to re-establish safeguards to competitive access by new or smaller contenders. In the telecommunications industries, the three most high profile Justice Department actions against major vendors include the AT&T divestiture in the mid-1980s, and the investigations into the competitive policies and practices of Microsoft Incorporated and Intel Corporation. At the time of writing, no conclusions have been drawn, and no divestiture proceedings have taken place, but this is one of the possible steps the Justice Department may follow if any of the allegations of unfair practices are found to be true. For information on the AT&T divestiture, see Consent Decree of 1982, Judge Green, Kingsbury Commitment, Modified Final Judgment.

DL 1. See distant learning. 2. distribution list. A list to whom correspondence, items, or processes are distributed. Email is often handled through distribution lists, as are postings to online newsgroups.

DLC 1. See Data Link Control. 2. See Digital Loop Carrier.

DLCI See Data Link Connection Identifier.

DLL 1. See data link layer. 2. See Dynamic Link Library.

DLR Design Layout Record.

DLS, DLSw See Data Link Switching.

DLSW See DataLink Switching Workgroup.

DM See delta modulation.

DMA 1. See direct memory access. 2. Document Management Alliance.

DMD differential mode delay.

DMI 1. See Desktop Management Interface. 2. Digital Multiplexed Interface. An AT&T interface that interconnects and multiplexes transmissions between T1 trunks and private branch exchanges.

DMS Digital Multiplex System. A series of programmable communications switches. Northern Telecom (NTI) provides one of the switches in common use in telecommunications, called the DMS-250 (there are other models as well).

DMSP Defense Meteorological Satellite Program.

DMT See discrete multitone.

DMTF Desktop Management Task Force.

DN See Directory Number.

DNA 1. See Digital Network Architecture. 2. See distributed network administration.

DNC 1. dynamic network controller. 2. distributed networking computing. See distributed computing.

DNCF See Directory Number Call Forwarding.

DNIC See Data Network Identification Code.

DNIS See Dialed Number Identification Service.

DNR Dynamic Network Reconfiguration. A feature on an International Business Machines (IBM) network which allows network address designations to be reorganized without closing down the system.

DNS See Domain Name System.

dock *n.* A computer interface console or device that allows a laptop or other similar portable unit to be hooked into a more comfortable-to-use desktop system. Docks are often available to hook laptops into larger desktop monitors and keyboards while being used at the desktop. The dock sometimes also provides an interface to a desktop network system (through which additional peripherals such as CD-ROM drives and printers may be accessible). To go on the road, the user disconnects from the dock, and uses the laptop's small built-in LCD screen and keyboard. Shipping agents use docks to interface handheld bar code readers to computer systems. Some radio phones can use a console for a dock. The phone can be used away from the console as a cordless, but may have more features available (such as a digital display) when linked into the wall or desk console. A hot dock is one in which a portable device can be linked into the dock when the power is on. These are rare. In most cases it is recommended, or required, that the power be turned off (cold docking) before interfacing the sensitive electronic components.

docking station See dock.

document *colloq.* A text file stored in a format specific to the word processor which created the file. The *.doc* file name extension is often used to identify this type of file. Document files in this sense of the word are not the same as standardized ASCII text files. Doc files provide formatting, font, and color information specific to the applications program that created the file. One of the best formats for exchanging files between applications which are not intercompatible is the Interchange format, also known as Rich Text Format (RTF). It holds more information than ASCII text files (font names and sizes, and text attributes such as bold and underline are retained), and it is widely supported on many platforms, and by every major word processor and desktop publishing program. In fact, it would be great if more people would use RTF files on the Web, for document downloads, instead of using word processor-specific document files that can't be loaded by many users.

document camera A camera specially adapted for photographing documents, usually for historical preservation, archiving, or replication. This type of camera is optimized for a suitable focal length and usually for artificial lighting conditions. It may be combined with a bank of lights to illuminate the documents evenly. It is typically mounted on a stand that can be readily raised or lowered like an enlarger.

More recently, document cameras have been equipped with electronics that allow the signal from the camera to be fed into a computer system. This may be a digital still camera, digital camcorder, or analog camcorder. In this way, documents can be digitized, stored directly on networks, or used in videoconferencing sessions.

document database A digital repository of document information, usually consisting of document images which show documents as individual pages as they might be viewed in print form, or which individually stores the elements of the documents (pictures, text, mathematical formulas, etc.), in such a way that the information can be manipulated, stored, and retrieved with word processors, drawing programs, etc. The design and programming of document databases is a great challenge, as it is not always known how the information might be used in the future. It is also becoming imperative to design systems which can handle vast amounts of information, yet can still search and retrieve the information with reasonable speed.

document image management DIM. Electronic storage, access, processing, and retrieval of documents stored in image format. Image format is a very common way to archive information which would otherwise require massive space. It means simply that text, graphics and everything else is stored all together as a picture, with no explicit differentiation of textual content. Microfiche is one of the systems designed to store image directory lists and image 'snapshots' of newspapers, journals, documents, certificates, etc., in order to provide reasonably quick lookup, and to save storage space.

With computer database and network capabilities, image management becomes both simpler and more complex. Lookup of information is simpler, display of images on a computer screen, while somewhat low resolution, is still more comfortable and convenient than peering at microfiche readers and manually moving around thin sheets of microfiche film. At the same time, the process becomes more complex because the greater capabilities of computers creates a greater demand for document processing and archiving, and higher expectations for the technology. In other words, the amount and type of information that can be processed increase, and correspondingly the amount and type of information that are desired and required to be stored increase. This requires more sophisticated compression algorithms, vast storage capacities, and some type of mechanism to keep the information from becoming obsolete due to unknowable, or unanticipated changes in computer technologies, particularly file formats.

People who have a lot of old 8mm film home movies appreciate this dilemma. If the 8mm projector breaks down, the user has no way to view the old films. 8mm projectors are no longer a local department store item, so it's not possible to just drive down the street and buy a new one. Converting to video is a compromise option, since a lot of information and quality is lost in the conversion, which is usually copied manually onto low resolution VHS tape. In the future it might be possible to digitally convert the film in order to retain the usefulness of the records, but that day is still in the future. Given this relatively simple example, and that computer data storage formats are more complex and more volatile, similar scenarios on a much larger scale will continue to occur in the never-ending document image management process. See optical character recognition, optical document recognition.

document image processing DIP. The process of taking print-related documents and converting them to visual form to be manipulated and/or viewed by electronic means. A transitional technology to computerized document processing is microfiche processing. By photographing documents onto film, and reducing them in size so that a special viewer was needed to make use of the information, the documents were being processed so that they would take up less storage space. Now, with computer technology, the information is not just reduced in size and redisplayed, it can be combined, manipulated, evaluated, and more.

In the past, converting text information to image format somewhat limited what could be done with the images. It wasn't possible to apply search/retrieval, and editing to the text itself, because it was not in a form amenable to database software. That is no longer true. While work still needs to be done on document recognition systems, there are now programs which can take information stored in image format and process the characters into text; some can even recognize which parts of the page are images, and which parts are text, in order to intelligently handle the content of the page. See document image management; zone, optical recognition.

document recognition The capability of an application to capture and analyze information on a page beyond that of performing straight character recognition. Document recognition may evaluate whether an area is an image or text, may recognize the structure of columns, and be able to reproduce them, and may even recognize the type of document (memo, letter, policy station, etc.) and preprocess, file, or forward the document as defined. See zone, optical recognition.

documentation Text and/or images which, taken together, describe how a system works, and how to use it. Good documentation goes beyond simply describing system features and functions, it describes them in such a way as to provide a context as to *when* and *why* they should be used. Documentation is often subdivided into tutorial and reference sections. This is generally a good idea, as tutorials are useful when you are first learning a system, and references are useful once you have a grasped how to use a system. The worst computer documentation is a manual which simply repeats and describes the contents of the menus. The better ones tell you *what* the program does, *how* you should do it, *why* you should do it, and *when* you should do it.

DOD 1. Department of Defense. 2. See direct outward dialing.

Dolbear, Amos E. A Tufts College professor who was awarded a patent for an induction-based wireless telegraph in 1886. In the 1890s, he was researching and writing about discoveries related to the effects of temperature on the properties of metals and their relationship to conductivity and stated "... that at absolute zero their electrical conductivity becomes infinite, or, as it is more generally stated, the electrical resistance of metals becomes zero.... so it seems altogether probable that the qualities and states of matter so familiar to us as solids, liquids and gases depend absolutely upon temperature and that at absolute zero there would be neither solid, nor liquid, nor gas, and that electrical and magnetic qualities would be at a maximum."

Dolby-NR Dolby Noise Reduction. A system developed by Ray Dolby to improve sound quality by reducing noise. Quality levels are designated with letters (Dolby SR is professional level quality). Dolby-NR is widely applied in the sound portion of the motion picture industry. See THX.

domain Sphere of influence or activity, province, dominion; the set of processes, items, or actions which constitute a sphere of influence or activity. There are two aspects to the concept of domain in telecommunications. The first is recognition and access to and from the domain. How does the domain provide a presence to other systems, and how do they recognize and acknowledge its presence? The second is what constitutes a domain, what is included within the organizational grouping that is called a domain? See domain identification and domain organization.

domain identification In most data networking architectures, a domain is perceived from outside a system as a unique address or identifier describing the logical and sometimes physical access points to a system, since systems external to the domain frequently know very little about the size and composition of elements within the domain, or their functioning.

Since computers work with binary digits, the domain on a computer network, that allows it to be identified and signaled by other domains, is usually an *address* that can be expressed as a numeral, or alphanumeric series. This address will often have a name or symbol associated with it, to make it easier for humans to recognize and remember it. Information about the nature of the domain, such as its basic function or geography, are sometimes expressed in the name. For example, on the Internet the alphanumeric expression of a domain name typically indicates the type of domain (commercial, not-for-profit, educational, etc.), and the location (*.ca* (Canada), *.au* (Australia), etc.). This naming scheme is not completely consistent, but it is helpful in most circumstances. It also is not a guarantee of location, as a system may remotely dial into a domain in another country. See domain name for further details on Internet naming, and domain name server, Domain Name System, firewall, host, Internet, server.

domain name A unique identifier, usually expressed in alphanumeric characters, to a computer domain. Local networks which are not connected to outside entities can set this up in any way they choose, or according to the domain naming parameters of the network software that is running on the system. On the Internet, however, which is a global distributed network, it is necessary to maintain an extensive database of domains in order that each one is uniquely identified. Thus, a domain name on the Internet must be registered with one of the assigned registration entities, the oldest of which is InterNIC. See domain name, Internet.

domain name, Internet A globally unique identifier for a domain which is continually, or occasionally online on the Internet through an Internet Access Provider (IAP), which may provide additional services as an Internet Services Provider (ISP). This identifier is used by the system to locate a domain in order to send and receive files, email, messages, routing information, and other network traffic. A certain flexibility is inherent in this scheme in that levels below the IAP can be rearranged according to the needs of the local network.

The domain name is actually a name resolved to an Internet Protocol (IP) address which is composed of numbers, but the name is what is familiar to most users, and automatically converted by the system. Once an Internet domain name has been assigned, let's assume it's *ourdomain.org*, subdomains can be locally assigned, such as:

```
accounting.ourdomain.org
administration.ourdomain.org
sales.ourdomain.org
```

Similarly usernames associated with email addresses can be expressed in a variety of ways:

```
max@sales.ourdomain.org
bighoncho@ourdomain.org
```

An email server at the local domain handles the processing of mail once it is received from the ISP, and can route it according to the needs of the local system.

The domain name is expressed alphanumerically with dots between each of the levels or portions of the domain, with the assigned domain as the last two parts of the name. Within the U.S., domain names are subdivided into various categories. Outside the U.S., the extension is usually a designator for the country. The North American Domain Name Extensions chart shows a sample of familiar domain name extensions. The appendix includes a more complete list of over 200 Internet domain name extensions from around the world.

domain name server A computer which communicates with the Internet after having registered its own unique ID, which provides a corresponding Internet Protocol (IP) address for a domain name which is not fully qualified (does not end in a dot). For individual users, this is usually handled on one of the machines used by a local Internet Services Provider (ISP). For campus/commercial users, it may be handled by a local campus backbone server. The local *Start of Authority* is delegated to providing Domain Name Server services for the assigned domain. This information is essential to providing connections between local networks and the Internet at large. See Domain Name System.

domain name registration A necessary administrative step for individuals and organizations who wish to have a unique domain name

North American Domain Name Extensions

Primarily U.S.A.		
.us	United States	U.S., not commonly used.
.um	United States	Outlying islands
.com	commercial	Typically business.
.net	network	Net-related.
.org	organization	Nonprofit, not-for-profit, charitable.
.edu	education	Schools, colleges, universities, other educational.
.gov	U.S. government	Local, state, and federal government agencies.
.mil	U.S.A. military	Military agencies, bases.
.arpa	ARPANET	Advanced Projects Research Agency
Countries other than the U.S.A.		
.int	International	
.nt	Neutral Zone	
.ca	Canada	
.mx	United Mexican States	

on the Internet. Unlike state registrations of company names, or companies in different industries which share a common name, there can only be *one* of a particular name on the global Internet. This has made domain names a hot commodity, with companies buying and selling domain names, the way logos are sometimes bought and sold. To register a domain name, you must have a site ready to come on line, and will usually go through your ISP to establish the name for the site. In 1995 registration changed to a fee system, and at the time of writing, it is about $100 for initial registration, and $50 per year to retain the name. There is dispute over the whether new name extensions should be added to the currently established *.com, *.net, *.edu, and *.org domain extensions, with many schemes for additional names being proposed. See Domain Name System, InterNIC, IANA.

Domain Name System, Domain Name Service DNS. A domain name distributed database established in the early 1980s at the University of Wisconsin. DNS provides mapping between host names and Internet Protocol (IP) addresses. DNS evolved out of a need for a distributed system to handle a very large number of domain names. In older systems, host files were regularly distributed, until they became too large to be managed on most systems.

To become a node on the Internet, it is necessary to formally register a unique domain name. The extensions familiar to Internet users as *.com*, *.edu*, and *.net* are part of the domain naming scheme called *zones*. Additional zones have been added to provide more options. Each domain name is then stored in a central repository on the Net, and addresses are resolved through this database.

The demand for domain names on the Internet has risen from a trickle to a flood. Businesses are realizing that the unique name requirement is different from traditional business naming schemes. For example, it is possible for two businesses in the same state to have the same business name, providing they are in different lines of business. For example, the business name "Great Expectations" might be used by a prenatal care center and by a bookseller selling old classics in the same state. It might even be used by a different prenatal care center in another state. In fact, the business name "Great Expectations" might be in use for several thousand different types of businesses in North America. On the Internet, there can only be one *greatexpectations.com* in the world, on a first-come, first-serve basis. See domain, domain name server, InterNIC, RFC 830.

domain organization In its general sense, the organizational structure of a digital network domain includes the operations, devices, and other elements under the general control of a processor, system or network. The overall controlling and administrating entity may be called a *host* and may function as a server, or have jurisdiction over a number of servers, and maintain some type of access and management mechanism to a database or other record of other computers and devices on the system. This 'entity' may be a single computer or software program, or a logical amalgamation of several computers and/or software programs. Similarly, security mechanisms are generally orchestrated by the host or other controlling member of the domain. The organization of the elements associated with the host, and contained within the domain can vary substantially with the type of network and various devices that are included.

domestic arc A portion of an orbiting satellite's path or range which provides transmissions between the satellite and the country it is serving. There are a number of domestic satellites in use which specifically are launched to cover a particular country or territory, as in Alaska and India. Others serve domestic needs during a particular portion of their orbit. See satellite.

dominant carrier A designation for a long-distance telecommunications provider that dominates a particular region or market. In most cases, a dominant carrier is more stringently regulated in order to balance a monopolistic advantage with opportunities for competition.

DOMSAT Domestic Communications Satellite.

done deal A phrase common in business transactions, especially verbal agreements, and those involving written contracts, that indicates that the interested parties have mutually accepted the terms of the agreement.

dongle A small hardware device or *security key* used for software or system security. Dongles were widely introduced on microcomputers to protect commercial products in the 1980s, and were raucously opposed by hackers, technical users and enough general users that their use has been almost abandoned in North America. The argument was that a dongle sometimes interfered with the basic workings of the system's port to which it was attached, especially if multiple dongles had to be attached to the same port. A dongle functions by checking for a signal, password, or serial number to verify the user or system to which the program is licensed before proceeding. Thus, if someone makes a pirate copy of the software and tries to run it on another system and the dongle is not encountered, the software will not run.

dongle disk A diskette that includes special codes or routines that must be present before a piece of software will run. A number of programs have been distributed with dongle disks so the user can copy the software to any computer hard drive, and run it faster from the drive, but only the one that has the dongle disk in the floppy drive will actually run at any one time. This unfortunately ties up a floppy drive, but still provides a compromise for the user to run the software off of a hard drive rather than a diskette, and provides a measure of security for the developer. Since most people don't use other software while they are playing computer games, games are the one commercial area in which dongle disks have been relatively successful.

This dongle from Onyx Computing attaches to the serial port of a Macintosh computer, to allow use of "Tree", a graphics application program. Most dongles provide 'pass-through' capabilities so the port can still be used for other devices.

dongle, hardware A small computer peripheral that pokes out as an attachment or plug interface. The most common use for hardware dongles is as copy protection devices for computer software. See dongle.

donor An n-type dopant, such as phosphorus, which is used in solar photovoltaic devices. The dopant puts an additional electron into an energy level near the conduction band to increase electrical conductivity. Doping is a common way of manipulating the properties of electromagnetic materials. See doping.

door A term often used on computer bulletin board systems (BBSs) to indicate a category of user access activity which is external to the BBS software itself. Thus, a separate software program like a game or quiz, which is external to the BBS software, is launched when the user selects a specific door.

dopant An industrial chemical which is added in minute amounts to pure semiconductor materials, usually to improve the conducting properties of the materials. See donor, doping.

doping A means of adding small amounts of materials with particular properties to another in order to enhance it for the purpose for which it is being used. For example, in semiconductor manufacture, materials are doped to enhance or inhibit particular tendencies to give up electrons or form 'holes.' Optical fibers are often doped with rare earth elements to alter their transmission characteristics.

Doppler shift A perceived or measured shift in frequency when the source of radiant energy moves relative to the position of the observer or receiver.

DOS See disk operating system.

dot addressable graphics Not all computer display graphics systems can address, that is control or sense, a particular point on the screen. Vector graphics are created with lines rather than with dots, and some raster systems which use dots can only address every other pixel on the screen. Dot addressable graphics imply those in which any particular dot position, or picture element (pixel), can be addressed ('queried' or 'sensed') as to its colors or other properties (e.g., brightness), and/or can be written (controlled, lit up, colored, etc.). Most current raster graphics systems are dot addressable.

dot gain A term in the printing industry to indicate the *spread* of the printing medium (ink, dye, etc.) beyond the area in which it is aimed, or makes contact with the printing image or block. In other words, if a dot of ink that hits the paper spreads by 16% because it's a warm day and the ink is more viscous, this is dot gain. In process color printing, where four colored dots are used to simulate millions of colors by their proximity and orientation, dot gain can have a substantial impact on the look of the final product. Dot gain is influenced by many factors, including the temperature, the chemical composition of the ink or dye, the porosity of the paper, and the tightness of the press's ink roller. Professional desktop publishing programs have settings for stipulating dot gain. Call your local printer to find out the suggested setting for the type of paper and printing process you are planning to use. If you are printing to a small printer attached to a computer, and the image seems 'thin' or, at the other extreme, blurry or smudgy, you may be able to improve the output by adjusting the dot gain.

dot matrix printer A type of one-color impact printer which became widespread in the 1980s. The early consumer dot matrix printers released in the 1970s had nine pins, very limited capabilities, and cost about $2,000. Current dot matrix printers usually have 24 pins, scalable fonts, buffers, special feed mechanisms and sometimes graphics programmability built in for about $300.

A dot matrix printer has a small 'bundle' of very fine short wires usually arranged in one or two lines. When a series of character codes is sent to the printer, the letters are imaged from top to bottom, across the length of the line, as the paper is fed through the roller in tiny increments. The combination of several passes across the horizontal length of the page recreates the characters on the printing medium. The pins are 'fired' individually as appropriate against a carbon or colored ribbon, essentially the same as a typewriter ribbon, to deposit a dot of ink on the page. Once the ribbon has been through three or four cycles, the type will become faint. New life can be restored to a ribbon to a limited extent by exposing it to alcohol vapor, if it has become dry, or by reinking it, if the ink has been used up.

Special type attributes like oblique, extended, etc., are available by offsetting the dots, or increasing the space between them according to a formula. Features such as bold are created by double-striking the dots, or restriking them just slightly to the side of the previous set of dots.

As ink-jet printers and laser printers become less expensive, dot matrix printers, which are slow and noisy, are less commonly used in homes. They are still the best solution, however, to printing business-related continuous-feed multi-copy documents such as computer checks, invoices, purchase orders, etc., as the impression will easily go through two to four pieces of carbon or carbonless paper. See dye sublimation printer, ink-jet printer, thermal wax printer.

dot pitch A quantitative description of the resolution of a device displaying images in point or pixel form. Dot pitch is commonly used to describe the distance between electron beam positions on cathode ray tube computer monitors. A resolution of about .28 dot pitch is considered minimal for a reasonably clear screen (larger numbers appear grainy), and .26 is quite fine and crisp, and usually more expensive.

double buffering A technique used in computer programming in which a section of memory, a buffer, outside of the sight or awareness of the user is set aside to build the next screen, or next block of information, so that when the user next perceives or requests a screen, the information can be presented instantly, without having to build it while the user is waiting. For example, in animations, single frames have to be displayed very quickly, and the access of the information from a hard drive may result in a delay which ruins the effect. Double buffering the sequence of still images, that is building the next image frame in a buffer while the current image frame is being displayed, allows the new image to be quickly 'swapped in' without waiting for disk access, thus giving the impression of a smooth transition between frames, and the illusion of motion.

The same technique can be applied to mathematical or textual data in a spreadsheet or database. The software can search and double-buffer the results of the search, and display them as needed without delays that might otherwise be annoying to the user.

double click A common graphical user interface convention in which two clicks from a pointing device such as a mouse, executed very close together, are interpreted by the software as a *select and continue* event. Thus, one click might indicate a selection, whereas two clicks may indicate *select and launch*. Double clicks are usually executed with the left mouse button. On some systems, the same operation may use the left mouse button to select, followed by the right mouse button, to launch the process or program selected.

double density A term applied to the physical configuration of magnetic particles on a floppy diskette, and hence the maximum amount of digital data it can hold. Three and one-half inch, double density diskettes hold approximately 720 - 880 kilobytes of data, depending on the platform. The diskettes themselves are interchangeable between systems, provided that they are formatted for the operating system on which they are being used, or used with systems (e.g., OpenStep) which automatically recognize, read and convert a variety of formats. Double density diskettes have given way to high density diskettes (1.4

Mbytes), SuperDisks (100+ Mbytes) and higher capacity cartridge formats.

DOV data over voice. Technology that allows data transmissions to be carried over traditional phone connections, usually copper twisted pair. See ISDN.

down converter A technique and device used in communications in which the incoming frequencies are 'shifted.' There are two common reasons for doing this. By shifting incoming frequencies so they are different from the outgoing frequencies, it is possible to reduce interference between the two sets of signals. Some frequencies are shifted up when they are transmitted, in order to put them into a particular broadcast band or slot. A down converter is then needed at the receiving end to downshift the frequencies back to levels that can be used by the playing or viewing equipment that provides the information to the user.

downlink The common name for the *satellite to Earth* portion of a transmission, with the uplink being the Earth to satellite portion. The downlink is often frequency-shifted from the uplink in order to reduce interference between the two sets of signals. Uplink and downlink services may be carried by different providers, and may be subject to different usage restrictions or billing arrangements

download To receive computer data from a source on another system, usually through a network or modem connection. Downloading is typically done with a communications program file transfer protocol (ZModem is commonly used), FTP client, or a Web browser. Common types of downloads include files from BBSs, or Internet file archives, and World Wide Web images and text.

When using browsers on the Web, the information that you view is typically downloaded or *cached* on your machine. (This is because loading the source code of the Web page on your local computer allows the browser to more quickly redisplay the previously-viewed pages.) This is usually convenient, but it has disadvantages as well, and constitutes a security hole on your system. It may also fill up your hard drive. Make sure your cache is flushed (erased) when you are finished with the Web files. You can always save ones of interest in appropriate directories for later viewing while offline.

While downloading files from FTP sites or Bulletin Board Systems, make sure you don't accidentally overwrite an existing file of the same name. Not all download software will inform you of a duplicate filename. It is a good policy, at any time, to download into a separate directory, or even a separate partition until you have run a virus checker on the downloaded files. All foreign files should be suspected of possible viruses until you have determined that they are problem- and virus-free. See FTP, upload.

downsizing An economic strategy, often employed by businesses to cut costs. Downsizing may mean closing out departments, laying off staff, reducing the quantity or types of equipment and facilities, or benefits or services to employees. With regard to computers, many companies have 'downsized' to microcomputers rather than higher end systems to reduce initial cost and maintenance. Businesses should be careful to avoid 'false economy' in downsizing to microcomputers. The initial cost may appear to be lower on some systems, until you factor in upgrades, networking, and the enormous cost of loss of productivity due to downtime on unreliable or incompatible systems. Paying more for a workstation level computer may be less expensive over the long run, or even in the short run. In fact, if a network consists of more than a dozen computers, the payback in the greater initial cost of a workstation level server rather than a microcomputer level server can often be recouped in as little as six months to one year. Shop around. See microcomputer, server, Unix, workstation.

downstream A designation for any of the systems, nodes, legs, or hops in a transmissions pathway that are subsequent to the current one. Thus, a printer is generally downstream from a computer, a radio listener is downstream from the radio broadcaster, various workstations may be downstream from a server. See download. Contrast with upstream.

downtime The block of time during which a system is nonfunctional. Downtime on a computer may be caused by software or hard drive crashes, broken network connections, etc. Downtime on phone systems may be caused by power outages, overloaded lines, or breaks in the lines.

downwardly compatible Software or hardware which is designed to work in some way with older software or hardware, sometimes called *legacy* applications or equipment. Often the downward compatibility is only partial. For example, a software program may be able to export a file in the older format, but it may not include all the characteristics of the file when loaded into the earlier application. Similarly, a new computer may work with an older monitor, but that doesn't mean the monitor can support all the graphics modes that might be built into the graphics controller on the new computer. Downward compatibility is a way of safeguarding a financial investment, and of maintaining a minimum level of continued data access, and use of existing software. Contrast with upwardly compatible.

DP 1. See data processing. 2. See demarcation point. 3. Dial Pulse. A standard Hayes modem command with "P" used to designate a pulse dial setting for subsequent dialouts. See ATD.

DPA 1. Demand Protocol Architecture. 2. digital port adapter.

DPBX digital private branch exchange (PBX). Most private branches in North America are becoming digital, so the D is now commonly assumed when using PBX.

DPCM See differential pulse code modulation.

DPE distributed processing environment. See distributed computing.

DPLB Digital Private Line Billing.

DPNSS Digital Private Network Signaling System. A standard for integrating private branch systems with E1 lines.

DPO Direct Public Offering. See Initial Public Offering for description of this specific state-regulated subcategory for securities offerings.

DPP Distributed Processing Peripheral.

DPX DataPath loop extension.

DQDB See Distributed Queue Dual Bus.

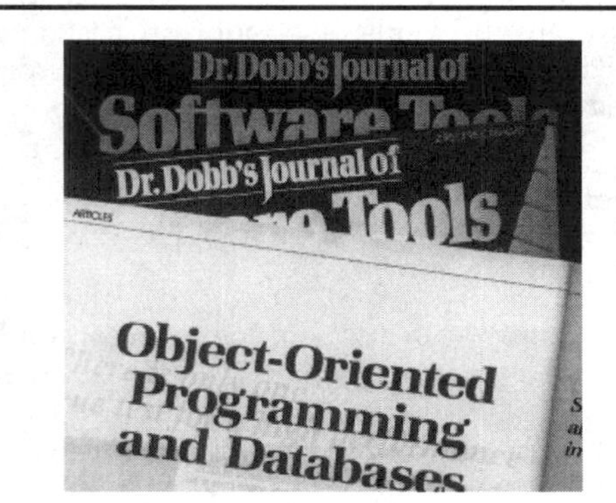

Dr. Dobb's journal has been a favorite of amateur and professional programmers since the early days of microcomputers, inspired by the Altair and the potential of desktop computers.

Dr. Dobb's Journal of Software Tools This journal has been a perennial favorite with programmers, providing technical information on a wide variety of platforms and programming languages since 1976. It originated as a newlsetter in 1975 documenting Tiny BASIC. The name is somewhat a collapse of the originators' first names, Dennis and Bob.

Draft RFC A formal stage in the Request for Comments standards and information distribution process in which the proposal is submitted for evaluation and comment. On the Internet, this process is widely used to encourage open standards and professional and public participation. See Request for Comments.

drag *v.* A computer user interface term which refers to a sliding motion of a pointing tool (mouse, trackball, etc.) that visually drags or draws some object or element of the screen from one point to another. The drag is generally considered to be the time during which the pointer is activated by a button press or some other mechanism and subsequently moved. Dragging is used everywhere on graphical user interfaces. Windows can be made larger or smaller, objects can be sized, menu bars can be viewed or selected by dragging. Since reaching out and drawing something back is one of the most natural things humans do, this screen metaphor has been successfully adapted to many software interfaces. See drag and drop.

drag and drop A computer user interface facility that enables the user to select an area, or object from one window or application and slide and insert it, with the pointing tool, to another window or application without having to do an interim save to a clipboard, scrapbook, or other temporary buffer on the hard drive. Drag and drop is a great convenience and time saver. Drag and drop does more than just move the information, it often converts it to a format which is compatible with the destination application and window, which may be quite different from the original format, or it may maintain a list of links to the original source for the dragged information. Before drag and drop, the same functions were usually performed by copying a region, pasting it to a buffer, clipboard, or scrapbook, or saving it to a file, converting it, if necessary, and then pasting it into the receiving region, thus involving many more steps.

drag line A wire, rope, or other line used to facilitate the threading of wire and cable through narrow channels (pipes, conduits, walls, etc.). The drag line may be preinserted during building construction and left for later use. See birdie.

dragon A program running low-level secondary systems tasks, especially on Unix systems, which are generally transparent to the user. Monitors and statistical programs are often run in the background as dragons, and the results of their activities may be viewable by the system administrator, or those with sufficient security clearance. See daemon.

DRAM 1. digital recorder, announce mode. 2. See dynamic RAM.

Draper, John See Cap'n Crunch.

Draper, John W. An American educator and experimenter who developed some of the early photographic processes and is credited with creating the first American portrait in 1840. Draper's ten-minute exposure was a big improvement over previous techniques that required many hours of exposure.

drift 1. Variation from a desired signal or current over time from factors other than line, load, environment, or warmup period. See calibration. 2. In radio technology, signal drift is not uncommon. For example, if you set an analog radio to a favorite station, and gradually lose the setting as the tuning changes, this is drift.

drift, travel 1. Aircraft drift occurs when wind factors cause displacement from the desired course. This type of drift is usually reported in degrees. 2. Highway drift occurs when you're too tired to drive, and you stray outside your lane markers. When this happens, you really should pull over and rest.

drive type The make, model, and configuration of a hard drive. A drive can conform to one of the common desktop standards, SCSI or IDE, for example, and once it is formatted, it is important for the system to have a record of the drive type in order to keep track of sectors, blocks, partitions, and files, the various data configurations that can be set up dynamically on the drive during use. The 'drive type' is really a combination of the cabling, the data bus characteristics, the physical properties of the drive, and the magnetic data configuration which is superimposed on the magnetic recording surface.

Some types of drives can be used in combination with one another, and others cannot. For example, SCSI and IDE drives are not mixed on one data bus. A different controlling mechanism is used for each type. SCSI devices can chain up to six devices (the controller counts as the seventh device), while IDE drives can chain up to two, with one designated as a *master* and the other as a *slave*.

When a formatting software program is run with a new drive, or one which is being formatted for another computer system, it may query the drive for information about its characteristics, and display that information on the screen. For example, it may show the brand, model, and size of the drive, and whether there are any existing partitions. Many drives now come preformatted, but it may still be necessary to set up partitions, if desired.

driver 1. In software, a program which includes code that can translate commands into instructions recognizable by a specific device, such as a facsimile modem, printer, scanner, hard drive, etc. Desktop publishing programs typically include a directory full of drivers for various printers which translates the print instructions from the software into the closest approximation possible by the printer through the print driver software. 2. In software event-processing, code that receives commands and distributes them appropriately for execution.

droid *colloq. abbrev.* android. 1. An anthropomorphic robot, generally more human than machine which may be a combination of biological and mechanical/electronic parts. 2. A company drone, someone who unquestioningly follows instructions and mechanically goes about the business of work (or living) without much enthusiasm or introspection, possibly due to apathy, or unquestioned acceptance of authority. 3. A person hired as a 'human robot' that is, to do a mindless, repetitious, production-line job that offers few opportunities for variety or creative interaction. These are the kinds of positions that should be handled by machines, in order for people to have more leisure and creative time.

drop 1. A short cable connection, often between a utility pole and a building, or between one panel and another, or a panel and other distribution entity.

drop, interapplication In computer software, to 'drop' is to move one piece of information into another application or window and have the system take care of the details of links, or format conversion and recognition and incorporation of the new information. See drag and drop.

drop frame In television video broadcast recording and playback, North American television was designed to play at 30 frames per second. Then when color signals were introduced, the differences between black and white and color technology resulted in a compensatory adjustment of the frame rate down to ~29.97 frames per second, a situation which altered the time code and reduced its usefulness for station timing.

Drop frame mode, also known as compensated mode, is a technique in which the system skips ahead a very small amount at specified intervals, skipping over the first two 'bits' in each minute. However, this is minutely too much of an adjustment, so each ten minutes, only a single 'bit' is skipped. It's similar to the way in which we adjust our calendar to celestial events by introducing leap years (except that a day is added rather than skipped in a leap year) where needed, to better synchronize them.

Drop frame modes are important to broadcasters because programming is interspersed with commercial announcements, shorts, special programming, and other timing-related items, that need to be carried out in a smooth, seamless way. See SMPTE time code.

drop loop In telephone wiring between the switching office and local subscribers, the circuit is called a 'local loop' and the specific section of the circuit from the utility pole or other nearby junction point and the subscriber's home or office is the drop loop.

dropout, drop-out An undesirable low-level, irregular loss of information when transferring from one system, medium, or format to another. This happens, for example, in video editing, when copying or editing tapes, especially with less robust formats and inexpensive equipment. Dropout can sometimes be seen as white dots appearing somewhat randomly on the screen. They are especially noticeable if the screen is mostly areas of dark or solid color.

dropout, transmission A short interruption in a transmission, usually caused by a problem in the transmitting or receiving equipment. Different industries have different objective measures for the length of interruption that constitutes a dropout.

dropped call A call terminated without the express desire of the parties engaged in the call. In radio phone communications, dropped calls are not uncommon, as the signal can easily be interrupted by terrain, weather, a stronger signal from another source or distance.

drum *n.* A generic term to describe a wide variety of rotating mechanisms in electronic devices. Printers, high-end scanners, photocopiers, facsimile machines and many recording mechanisms include rotating drums. Because they are subjected to wear and tear, pressure and movement, drums require regular maintenance and replacement. Technicians warn laser printer owners against using refilled toner cartridges in their laser printers, because part of the drum mechanism is incorporated into the cartridge, and has a limited lifespan. An old drum may adversely affect the printer even if the toner is fresh.

drum speed The rotational speed of a drum, usually described in revolutions per second or revolutions per minute.

dry cable, raw cable, dark cable Conductive cable or wire with no added electronics, and no signal passing through. The cable you find on spools in the hardware store is dry cable. Raw fiber optic cable is called dark cable.

dry camera A camera which allows the image to be viewed or transferred without going through a 'wet' chemical process (such as film development). Digital cameras that hook into computers or scanners are examples of dry cameras. Currently most dry camera images do not match the quality of film. Some of the highest end digital cameras are beginning to be able to provide a reasonably good high resolution image, however. Digital 'dry' images are very flexible and can be image processed to change colors, add titles, etc. Dry photography will eventually have a profound effect on the film and photo finishing industries. See digital camcorder, digital camera.

dry cell A common, compact type of battery descended from the wet cell, but differing in that it employs non-liquid electrolytes in the form of paste or gel. Dry cells were invented by Gassner in 1888 and manufactured in the early 1900s. Since they do not use liquid acids, they are easier to handle, and more portable than wet cells, and can be used in any orientation. They are commonly used in flashlights, small appliances, and many hand-held devices. Many dry cells include toxic chemicals and heavy metals, and should be recycled through local centers, not thrown in the trash. See Gassner, wet cell.

DS 1. Dansk Standardiseringsrad. The Danish Standards Institute, located in Hellerup. 2. digital system. 3. See Distributed Single Layer Test Method.

DS- A series of signal speeds for transmitting digital data through a variety of modulation and multiplexing schemes, designated DS-1 through DS-4, with higher numbers representing faster possible transmission speeds. This system is primarily used in North America and Japan. A similar system, which differs in data rates, encoding, and numbers of channels, the *E*-system, is used in Europe. The *DS*-system first began to be used by telephone carriers, for connecting main switching centers. Gradually, as the technology became less expensive, it began to be used in the backbones of larger private branches and institutions, and now is used in telephone feeder plants, and local area network backbones as well. See E-.

DS Facility A categorization system for describing digital transmission capacity. See DS-, Digital Transmission Speed Categories chart following.

DSA 1. data service adapter. 2. Digital Signature Algorithm. See Digital Signature Standard, Electronic Certification. 3. See Directory System Agent.

DSAT Digital Supervisory Audio Tones.

Digital Transmission Speed Categories

DS-0	Digital Signal Level 0	A 64,000 bps standard for transmitting digital data through pulse code modulation (PCM). A sampled signal is quantized and transmitted with bits that represent quantization levels being transmitted separately. A standard used in telephone systems.
DS-1	Digital Signal Level 1	A frame format standard for transmitting data at 1.544 Mbps, developed in 1962. Used on T1 systems. It incorporates time division multiplexing (TDM) to combine 24 DS-0 signals, and adds a single framing bit. Signals are transmitted with bipolar (B8ZS) pulses or alternate mark inversion (AMI). In 1969, the standard was extended to SuperFrame to increase the signal-to-noise ratio, and later was further modified to create Extended SuperFrame which is more robust. Europe uses a 32-channel 2.048 Mbps system which is somewhat similar, but which incorporates different synchronization and signaling formats. See Extended SuperFrame, SuperFrame.
DS-1C	Digital Signal Level 1C	This signal system was designated 1C because it fits somewhere between DS-1 and DS-2 in terms of its 3.152 Mbps signaling rates. Used on T1C systems. It was introduced by AT&T in 1975. DS-1 signaling bits are bit-interleaved into the information bits.
DS-2	Digital Signal Level 2	A frame format developed for longer transmission lines, and to accommodate AT&T's Picturephone technology (which was developed many years before the technology to use and support it became sufficiently widespread). Used on T2 systems. It combines four DS-1 signals or 96 DS-0 signals, employs two framing stages, and transmits at 6.312 Mbps. Europe uses a different ITU-defined system that operates at 8.448 Mbps (2.048 Mbps primary rate).
DS-3	Digital Signal Level 3	A frame format developed for signaling over broad bandwidth signaling systems. Used on T3 systems. DS-3 uses Bipolar with Six Zero Substitution (B3ZS). The DS-3 signal combines 7 DS-2 or 672 DS-0 signals, is framed in two stages, and transmits at 44.736 Mbps. Through multiplexing, the asynchronous signals are transmitted over synchronous links. Europe uses a different ITU-defined system that operates at 34.368 Mbps (2.048 Mbps primary rate).

DSC Digital Selective Calling.

DSCS See Defense Satellite Communications System.

DSE Distributed Single Layer Embedded (Test Method). In ATM networking, an abstract method for testing a protocol layer, or sublayer, which is part of a multi-protocol Implementation Under Test (IUT).

DSH double-superheterodyne. See superheterodyne.

DSI See digital speed interpolation.

DSL See Digital Subscriber Line.

DSLAM DSL access multiplexer.

DSP 1. see digital signal processor. 2. Display System Protocol.

DSR See data set ready.

DSRC Dedicated Short Range Communications.

DSRR digital short range radio.

DSS 1. See Digital Signature Standard. 2. Direct Station Select. 3. direct satellite system. See direct broadcast satellite.

DSTO Defence Science and Technology Organisation. Australian developers of a trial high frequency (HF) radio data network.

DSU See Digital Service Unit

DSU-CSU Digital connecting device, usually connecting a transmission line and a router. Some units combine a digital modem, router, and terminal server, a combination which is popular with Internet Services Providers. See digital service unit, channel service unit.

DSX panel A type of electrical cross-connect bay which facilitates the interconnection of digital telecommunications facilities and equipment.

DT Deutsche Telekom.

DTE See Data Terminal Equipment, End Device.

DTL 1. See Designated Transit List. 2. diode transistor logic.

DTMF See dual tone multifrequency.

DTP See desktop publishing.

DTR Data Terminal Ready. A communications control signal, commonly used in serial flow control, which is an output for DTE devices and an input for DCE devices. See DSR, DCD and RS-232.

DTRS Digital Trunked Radio System.

DTSR Dial Tone Speed Recording.

DTT Digital Trunk Testing.

DTU Digital Test Unit.

Du Fay, Charles François de Cisternay (1698-1739) A French soldier and scientist who discovered that electricity had two basic attracting and repelling properties, which could be demonstrated, for example, by rubbing amber with wool and rubbing glass with silk. He called these *resinous electricity* and *vitreous electricity*, making the distinction in a context that had eluded previous researchers.

Du Fay developed some of the ideas first investigated and described by S. Gray in England, and made some important observations about the composition of the materials of the conducting medium. Perhaps most important of Du Fay's observations is that

"... an electrified body attracts all those that are not themselves electrified, and repels them as soon as they become electrified by ... the electrified body."

DUA See Directory User Agent.

dual attachment concentrator DAC. A connecting device used in double ring Fiber Distributed Data Interface networks which employ a token-passing scheme over a redundant ring network. Dual ports and attachment points are used in connection with the concentrator to reroute data if a problem arises. See Fiber Distributed Data Interface for more detailed information.

Dual Attachment Station DAS. A configuration of a Fiber Distributed Data Interface token-passing, dual attachment network. The dual attachments are incorporated to provide fault tolerance to the system. The dual attachment consists of a primary ring and a secondary ring, the first of which is usually used for data transmissions, and the second as a backup in case of problems with the first. A Class A, dual attachment station (DAS) connects to both rings and a concentrator, which in turn ensures that the ring is not interrupted. Failure in a ring causes a series of adaptations which result in the ring wrapping back on itself and temporary elimination of the failed station. See A port, B port.

dual cable A two-cable configuration, usually in a local area network (LAN), often implemented to provide redundancy and fault tolerance, as in Fiber Distributed Data Interface (FDDI) systems. See Fiber Distributed Data Interface.

dual homing 1. A means of providing backup and fault tolerance on a network system. This is particularly characteristic of Fiber Distributed Data Interface (FDDI) networks. FDDI networks utilize stations, which can be eliminated through rerouting if a problem is found, without interrupting transmissions. Dual homing utilizes two concentrators, a primary and a secondary, with the secondary used as backup that is automatically activated if a problem occurs. See Dual Attachment Station, optical bypass. 2. In a frame relay network, a means of providing fault tolerance by using dual port connections in different locations.

dual mode There are many dual mode devices in telecommunications. Many modems are dual mode in order to support both vendor proprietary protocols, and industry standard protocols. Dual mode monitors will sometimes support both NTSC and RGB signals. Some phones have dual pulse and tone dialing capabilities. Many devices, such as cellular phones, are now being designed to support traditional analog signals, and emerging technologies that use digital signals. Dual mode devices tend to come about when there are competing standards, or when technology is transitioning from one stage to the next. Some are autosensing, switching to the correct setting unaided, and some have to be explicitly set with a switch or software.

dual tone multifrequency DTMF. Touchtone signaling on a phone system. The tones are actually a combination of two frequencies that are variously combined to provide unique codes for each key on the telephone, when pressed. This signal is sent through the line to indicate the desired number to be dialed, or the desired selection from an automated phone menu system. The tones were chosen for frequencies that carry well on voice-grade lines.

Phone phreakers used to exploit these tones for dialing unauthorized long-distance numbers with small devices called blue boxes. With the increased use of 'out-of-band' signaling systems, such as Signaling System 7, which send the signals separately from the conversational information, this is becoming less prevalent and will eventually be impossible. See touchtone phone.

duct Protective pipe or tube through which lines or fluids are run. See conduit.

ductile A property of being malleable, a material that can be shaped, drawn out, flattened, or otherwise bent or manipulated without significant stress or breakage. Pliable.

Duddell, William An early English experimenter and engineer who discovered that electric arcs created in a circuit with coils and condensers could generate very high-frequency audible tones in the low radio wave frequencies.

dumb switch A switching device which can channel signals through a desired pathway as needed. For example, a dumb A/B switch can be used for two computers to share one printer or one modem. Dumb switches are very common, particularly as A/B or A/B/C switches, because they are inexpensive and easy to set up and use. They make no logical or electrical evaluations or decisions about the incoming or outgoing data, they simply route it mechanically.

In its simplest form, a dumb switch does not alter or boost the electrical connection, though some may be equipped to amplify or condition a signal, without changing its informational content. Thus, switches may be electrically passive or active. In video, where a great deal of switching occurs, banks of both passive and active switches are used to channel the desired video or audio feeds into the broadcast or recording channel.

A passive dumb switchbox with a simple mechanical dial is the type most commonly used in small local computer installations to connect printers, modems, and scanners to one or more computers. The 'dumb' switches which are used in network connections are typically smarter than simple mechanical switches, and may have the capability of taking instructions from the computer operating system and switching the system electronically, without the user physically turning a dial, or changing the connectors. On a network, 'smart' switches may evaluate the incoming data and perform some rudimentary routing. Some switches are so smart, in fact, that the distinction between switches and routers at the high end is becoming blurred. On layered networks, switches typically operate at the second layer.

Other than the inconvenience of threading cables behind desks and through walls, installing a dumb device switch is pretty straightforward. Ensure that the interconnected devices are compatible, and use gender benders and converters that are the right sizes and numbers of pins to hook everything together. Most dumb switches for computer applications have

25-pin female D connectors. Make sure all the systems are powered off before making any connections, and test new connections individually rather than all at once, so that a problem can be isolated and corrected right away. See A/B switchbox.

dumb terminal A minimally configured computer terminal with no direct processing capabilities, which isn't very useful unless networked to a central processing system. Universities and libraries often use dumb terminals to provide cost-effective user access to the main system. The advantage is that the dumb terminals are inexpensive, easy to maintain, and not highly subject to abuse or vandalism. In their simplest form, they consist only of a touchscreen monitor, or teletype interface, one step up is a keyboard or mouse, and a monitor. Some not-quite-as-dumb terminals will include both keyboard and mouse, monitor, and sometimes a floppy drive or CD-ROM drive, but still rely on a remote system for actual processing of data and commands. Hybrid terminals may include a processor for simpler tasks, but still rely on the remote system for most of their computing power.

dumpster diving Searching for mechanical parts, discarded electronics, trade secrets, software and hardware manuals, access codes, login names, and passwords in large outside trash cans. Shredders are employed by many companies to protect sensitive documents from prying eyes.

Dunwoody, Henry H. C. Dunwoody patented the use of carborundum in early radio wave crystal detectors in 1906. Carborundum was robust in the sense that it could be clamped down, and thus used in portable units.

duopoly A market situation in which two major sellers greatly influence, though they may not necessarily control, the market. In industries that require licenses, a situation where exclusive operating licenses are issued to two businesses rather than one. Duopolies are very prevalent in competitive markets. As markets mature, they tend to 'shake down' into a couple of major players, especially as there are legislative restrictions against monopolies. It is interesting to note that thousands of independent telephone switching stations opened when the Bell patents expired after their 17-year term of protection, and yet it was only a few years before AT&T again began to dominate the market. This pattern is repeated throughout telecommunications history, with buyouts of small companies by larger ones resulting in large influencing or controlling companies, and small competitors finding it impossible to break into the existing market, succeeding better in gaining footholds in new markets.

duotone In printing, a two-color halftone. Duotones are used to achieve artistic effects, or to simulate the feeling of tonal, colored images while using only two colors of ink, toner, or dye.

duplex Double, bi-directional.

duplex connection, duplex transmission Data transmissions in which a message can be sent in both directions along the same transmissions line or path. In many duplex systems, the messages can be alternately sent in one direction or the other, but not in both directions simultaneously, whereas in full duplex, the messages can be sent in both directions at the same time. Serial communications software often has half- and full-duplex settings.

duplex printing Printing on both sides of a print medium. Duplex printing has traditionally not been done on both sides simultaneously (since ink needs time to dry), but with the increased use of dry transfer processes, duplex printing can be done by printing side two after side one, or less often, by printing both sides simultaneously. Not all laser printers can print duplex by running the page through a second time. It may result in toner being fused to the drum or page feed mechanism, resulting in undesirable ghost images on subsequent pages.

duplex telegraphy An early innovation in telegraphic communications in which two messages were sent, in opposite directions, at the same time over the same wire.

duplexer A switching device that provides alternating transmitting and receiving through the same transmissions system (data line or antenna).

DVBG Digital Video Broadcasting Group.

DVD See digital videodisc.

Dvorãk keyboard A type of keyboard layout designed by August Dvorak and William Dealey after they studied the natural movement of fingers, and of the hand over typewriter keys and researched ways in which to conform the key positions to the comfortable hand use, rather than conforming the hand to unnatural keyboard lettering layouts.

DVST See direct view storage tube.

DXI See data exchange interface.

dye sublimation printing A technology in which dyes are transferred to paper through a heat process that changes the solid inks to a gas (without going through a liquid stage) before they reharden on a special coated paper. The result is a printed image that has the rich colors and smooth texture of a photo. Since adjacent dyes will blend to provide a tonal spread between their neighboring dyes, a smooth, continuous-tone quality results, unlike the stippled results from drier processes like dot matrix printing.

Dye sublimation printers are more expensive than ink-jet, ribbon cartridge, or bubble-jet, but they're now in the *prosumer* (low-end *pro*fessional or high-end con*sumer*) price range. Dye sublimation prints are not generally suitable as match prints for color process printing, because printing inks do not melt together the way dye sub inks melt together. Printing inks for process printing are distributed as discrete dots that can be seen with a magnifying glass, and as such, do not create intermediate color tones.

For t-shirt transfers, thermal wax printouts are generally preferred over dye sublimation printouts, because the thermal wax can be reheated and thus, transferred to fabric. Dye sub printers will probably become widespread for printing small consumer photographs taken with digital cameras, a combination that will challenge the future of photofinishing services. See dot matrix printer, ink-jet printing, thermal wax printing.

dynamic bandwidth allocation The process of assigning bandwidth 'on demand' or according to algorithms to maximize the efficiency of the system, rather than transmitting on particular frequencies or at particular times.

dynamic beam focusing In cathode ray tubes (CRTs), the sweep of the beam from an electron gun across the inside surface of the screen that displays the image. This beam forms a series of motions, which if kept equal, are curved. To keep the distance equal across the sweep of the beams, earlier television screens and computer monitors were curved to match the length of the beams. Early flat screen monitors were rare and expensive. With more sophisticated hardware and software algorithms, manufacturers have devised ways of compensating the travel distance of the beam to adjust to the characteristics of a flat surface. One of these techniques is dynamic beam focusing, adjusting the beam focus as needed, depending on which part of the screen is illuminated, and its distance from the gun.

Dynamic Host Configuration Protocol DHCP. An expanded client/server configuration protocol descended from, and downwardly compatible with, Bootstrap Protocol (BOOTP). DHCP provides manual, automatic, and dynamic allocation of IP addresses, and a complete set of TCP/IP configuration values. It utilizes ports 67 and 68 and retains BootP's *bootrequest* and *bootreply* packet formats. See RFC 1533, RFC 1541.

dynamic IP addressing When logging onto the Internet, or any system using the Internet Protocol (IP), it is necessary for a unique number to be assigned to the session to handle the flow of data to and from the user. Dynamic IP addressing is a scheme for automating the process of assigning an address when a user connects to an Internet Access Provider (IAP) or other network access point. As part of the user login, a unique number assigned for that session. The number is typically freed up when the user logs off, so the IAP can reassign it to the next user, if needed. Freeing up the address is an important part of the process on large distributed networks like the Internet, where there may be millions of users online, with some of the large IAPs handling tens of thousands of simultaneous users.

Dynamic Link Library DLL. In software programming, a Microsoft product format for consolidating a number of frequently used routines, or routines which may not be available by default in an application such as Visual BASIC. The DLL is an organizational programming tool which allows a 'library' of routines to be written once, and bundled together, and thereafter linked into a program and called by the application program as needed.

dynamic RAM DRAM. Random access memory which requires a supply of current through the chip at all times in order to retain and refresh the stored information. When you turn a computer off, the data currently in RAM is lost. RAM is one of the most prevalent types of dynamic fast storage used in computers. Most systems these days require about 16 or 32 Mbytes of RAM for basic functioning. This is in stark contrast to desktop computers in the 1970s which could run telecommunications programs, word processors, and spreadsheets in less that 8 kilobytes (not megabytes) of RAM, and systems in the mid-1980s which could run music and graphics simultaneously in a fully multitasking environment in only 4 megabytes of RAM. See static RAM.

Dynamic Random Access Memory See dynamic RAM.

dynamic range A range of intensities, between the minimum and maximum extremes. It's a phrase that is often applied to concepts of light or sound. In imagery, the dynamic range of a scanner, for example, is the range of light levels, from the brightest highlight to the darkest shadow that can be picked up and transmitted. In music, the dynamic range of a recorded symphony performance is the range from the softest note to the loudest, expressed in terms of decibels. Dynamic range is sometimes described more objectively in terms of the maximum and minimum levels of a parameter as measured by an instrument designed for that use. See gamut.

dynamic resource allocation In various types of communications, the administration, allocation, and dynamic reallocation of resources, such as frequencies, channels, processes, programs, and access to shared peripheral devices. Dynamic resource allocation usually entails intelligent algorithms for determining authorizations, priorities, and needs, and often includes sophisticated queuing, routing and multitasking capabilities.

dynamic routing In general, the creation and adjustment of communications paths on an as-needed, or as-optimized basis so paths will change to fit the needs of a situation as specified. In data networks, dynamic routing allows the system as a whole to stay online even if individual systems or routes change or are unavailable. This is accomplished through routers which can communicate with other routers, usually those topologically nearby, and which may increase, or modify routing tables as needed.

Dynamic routing works well on large, changeable, packet-switched systems like the Internet. Routers can relay data around distressed or suddenly unavailable systems or trunks. On small systems, the overhead of dynamic routing may not be worth the loss of speed that the processing takes. Static routing may be used quite effectively on small systems with known, stable characteristics. See router, Routing Information Protocol.

dynamic sector repair A fault correction and prevention system built into hard drive systems, particularly multiple disk arrays such as RAID, which seeks out faulty sectors on a disk, repairs the data, if possible, and records bad sectors to prevent the system from trying to write to those sections in the future. See redundant array of inexpensive disks, SMART.

dynamic storage In computing, the allocation of temporary or permanent storage space in an intelligent manner, so unused space can be optimally used, and unneeded data is removed to allow the reuse of storage for other applications. It may also involve occasional reorganization of information, if extra processing cycles are available. See garbage collection.

dynamically controllable magnetic DCM. Magnetic materials that can change permeability in realtime when stimulated by a magnetic field. DCM materials that have this property have been found in the VHF to microwave frequency ranges. There are some communications antennas which need to be transparent at some frequencies and reflective at others, and DCM materials are being tested for their effectiveness for this use.

dynamo A generator, a machine that converts mechanical energy into electrical energy (direct current). See alternator; generator; Siemens, Werner. 2. An energetic, dynamic individual.

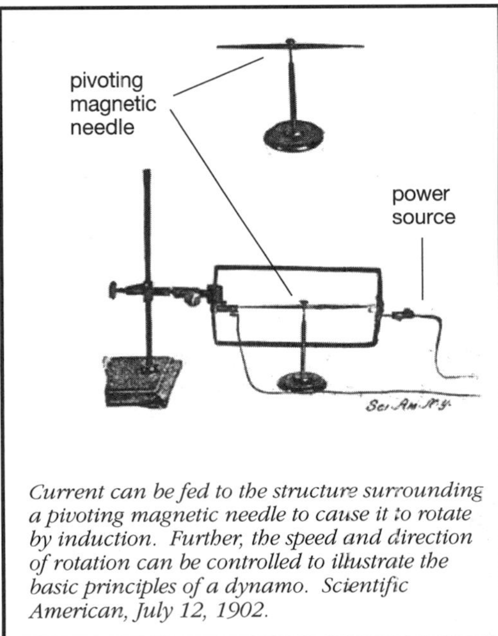

Current can be fed to the structure surrounding a pivoting magnetic needle to cause it to rotate by induction. Further, the speed and direction of rotation can be controlled to illustrate the basic principles of a dynamo. Scientific American, July 12, 1902.

dynamometer, electrodynamometer A sensitive current, voltage, and power detecting instrument similar to a D'Arsonval meter except that it employs a field coil, or coils, rather than a permanent magnet.

A dynamometer functions through a rotating coil controlled by the interaction between the magnetic fields of a moving coil and field coil(s). It can be used in conjunction with both direct current (DC) and alternating current (AC). See D'Arsonval galvanometer.

Dyson, George (1953-) Author of "Darwin Among the Machines: The Evolution of Global Intelligence," a provocative book about the origins of computers and networking, and philosophical speculations about intercommunication and emerging digital intelligences.

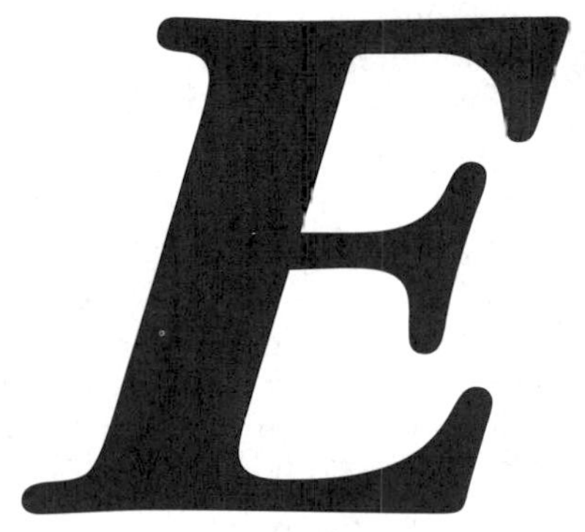

e 1. Symbol for a basic unit of charge. Proton-associated charges are designated *+e* and electron-associated charges are designated *-e*. See coulomb. 2. Symbol for voltage.

E & I engineering and installation.

E & M signaling A signaling method which is communicated over two leads or wires which are separated from the signal path for the analog information and used for signaling and supervisory information. The two leads are labelled E and M.

E Block A Federal Communications Commission (FCC) designation for a Personal Communications Services (PCS) license granted to a telephone company serving a Major Trading Area (MTA). This grants permission to operate at certain FCC-specified frequencies. See A Block for chart of frequencies.

E interface, external interface A general interface so that Cellular Digital Packet Data (CDPD) deployed over AMPS can interface with external networks. See A interface, I interface.

E region A portion of the Earth's ionosphere, above the troposphere, which ranges from about 100 kilometers to about 130 kilometers above the surface of the Earth. This is also known as the Heaviside layer, or the Kennelly-Heaviside layer, and is used for deflection of short wave radio waves. See Kennelly-Heaviside layer. See ionospheric sublayers for a chart.

E1, E-1, CEPT1 The European ITU-T-specified analog to the T1 high-speed communications system used in North America. They are the same in many general aspects, but differ as to details, such as speed of transmissions, number of channels, and the lack of bit-robbing. E1 transmits at 2.048 Mbps (compared to T1 1.544 Mbps) using time division multiplexing (TDM), and pulse code modulation (PCM) simultaneously on up to 30 64-Kbps digital channels. Two additional channels are used for signaling and framing.

E-911 service Enhanced 911 service. A 911 emergency service with extra features like automatic identification of the caller's phone number and calling address.

E-commerce electronic commerce. Many forms of electronic commerce have been around for the last twenty years. Banks have been using computers longer than most businesses and ATMs have been common since the early 1980s, but E-commerce has extended to consumers over the Internet and now the term has a broader meaning for any type of electronically facilitated or direct electronic transaction carried out over a private or public network, in addition to traditional electronic transactions.

E-IDE Enhanced Integrated Drive Electronics. An enhanced computer peripheral connection format, descended from IDE. IDE hard drive controllers are very common on Intel-based desktop computers. Enhanced IDE has a greater storage capacity than IDE and faster data transfer rates. See IDE.

E-TDMA See extended time division multiple access. See time division multiple access.

e-zine electronic magazine. An electronic publication, usually provided over public networks, which retains many of the format and editorial features and characteristics of a print magazine. E-zines may be freely accessible or may require the payment of a fee and password access. Some magazines provide back issues for free, but charge for access to recent issues. Originally e-zine was a term used for publications from smaller presses, or individual specialized publications, but the term has broadened to include most online popular journals and trade magazines as well. An e-zine distributed over the Web is sometimes distinguished as a Webzine.

EA Equal Access. A moral and regulatory stipulation that all persons have equal access to telecommunications services.

EACEM European Association of Consumer Electronics Manufacturers. A trade organization established to promote and support production and distribution of consumer electronics products and services.

EAGLE Extended Area Global Positioning System (GPS) Location Enhancement. EAGLE is a commercial GPS system implemented by Differential Corrections Inc. (DCI) to provide services in North America. The EAGLE system employs a network of reference nodes or stations and a central processing hub, to provide more precise GPS location information than is provided by an unenhanced GPS system functioning on civilian frequencies. EAGLE employs separate error estimates to generate local area corrections, and frequency modulated (FM) subcarrier broadcasts to provide correction information to users.

Current EAGLE nodes are widely installed across North America, from Seattle and San Diego, to Halifax and Miami, and tie in with frame relay networks.

The frame relay system is used for the transmission of data to two network hubs which create separate estimates for the position and clock errors of each of the satellites being used. A grid of ionospheric corrections is also used. The composite corrections are then transmitted to all the FM stations in the network through a geostationary satellite sending to small dish antennas at the receiving facilities. The vector corrections are converted into local corrections to produce scalar corrections for each satellite. The resulting data stream is broadcast to the mobile user through a frequency modulated (FM) subcarrier.

Early Bird INTELSAT's commercial communications satellite, which is claimed to be the first commercial satellite in orbit, launched in 1965 over the Atlantic Ocean.

Early Packet Discard EPD. In ATM networking, a traffic flow control service guarantee technique used in situations where congestion occurs on ATM networks, usually in unspecified bit rate (UBR) services. Cells early in the packet set are discarded, perhaps right down to the final cell, which is not discarded, as it is needed as a signal for the receiving station that it is the end of the packet set. See cell rate.

early token release In a token-passing network, as a Token-Ring network, a means by which a station sends out a token without first checking to see if the receiving system has acknowledged the transmission. This can increase the efficiency of transmissions around the ring in some situations, as normally the system only sends one token at a time in one direction.

EARN See European Academic Research Network.

EARP See Ethernet Address Resolution Protocol.

Earth grounding Grounding an electrical circuit by placing a lead into the Earth. It works best when it is inserted a few feet into damp soil. Earth grounding was often placed near outhouses, where the soil was damp and softer, in early telephone installations before inside plumbing became prevalent. See ground.

Earth Observing System, Earth Observation Satellite EOS. A central project of NASA's Earth Science Enterprise (ESE) consisting of scientific research and data supporting a series of coordinated polar-orbiting and low-inclination satellites designed for long-term global observations and experimentation. See ASTER. http://eospso.gsfc.nasa.gov/

Earth Resources Technology Satellite ERTS-1. The historic first Earth remote sensing satellite launched in 1968 from an Air Force Base located in California. It was equipped with the controversial, but ultimately successful, Hughes Aircraft scanner. This program developed into the Landsat series in 1975 and two very similar satellites were launched in 1975 and 1978. See Landsat, scanner.

Earth station The portion of structures and transmission equipment associated with a satellite which are stationed on the Earth. They may include facilities, antennas, orientation systems, transceivers, etc., in a building or on a mobile unit.

Earth-Moon-Earth bounce EME. See Moonbounce.

earthing See grounding.

EAS 1. See Emergency Alert System. 2. See Extended Area Service.

easter egg A surprise, usually in the form of a message, sound, or image, hidden in software. Many software applications have unlikely, undocumented key combinations that will pop up information about the authors, or, in some cases, not very flattering critiques of their employers. Earlier versions of the Commodore Amiga operating system included an uncomplimentary message from the developers to Commodore that could only be seen after entering a complicated combined-key sequence. Easter eggs may or may not be readily detectable when browsing the code, or trying to reverse-engineer the compiled code. Sometimes the easter egg can only be seen by looking at the code, and will never be known to a user who only runs the software.

Eastern & Associated Telegraph Company An early submarine telegraph company, which had become the largest by 1900.

EBCDIC (*pron.* eb-si-dik) Extended Binary-Coded Decimal Interchange Code. A family of 8-bit (256 character) encodings adapted by International Business Machines (IBM) from punch card codes, in preference to ASCII which is more widely used by the rest of the computing community. See American Standard Code for Information Interchange.

Ebone The European network backbone.

EBS See Emergency Broadcast System.

EBU European Broadcasting Union.

EC See exchange carrier. 2. European Community, European Common Market, European Union (EU). An organization of member European nations who have been developing, over a number of decades, a common currency, common passports, common network resources, and intercountry work, commerce, and decision-making alliances in order to promote trade, both within the EC and between the EC and other nations.

ECC 1. elliptic curve cryptography. 2. Emergency Communications Center.

eccentric circle In a phonograph record, a blank, nonconcentric groove cut into the inner part of the platter to trip the automatic stylus pickup mechanism when the record has finished playing.

eccentricity 1. Deviation from normal or expected. 2. Deviation from a regular or expected path, as a straight line or a circle. 3. In orbits, the deviation from a circular path. 4. In conductive materials, as wires, the deviation at a particular point, of the diameter of the conductor with the insulation when measured in cross-section.

ECCO A system of 11 satellites plus one spare in low Earth orbit (LEO). ECCO is a joint venture of Constellation Communications, Inc., and Telebras. Commercial services include mobile and fixed-site voice, data facsimile, and positioning are expected to start coming online by 1999. The 12 satellites will share a ring orbit around the equator at 2,000 kilometers using CDMA in the L- and S-band frequencies for transmissions, and C- or Ku-band for feeders. The service area is targeted at the equatorial belt between 23 degrees north and south latitudes, comprising over 100 countries.

echo 1. Repetition of a sound, sometimes repeatedly, with the sound gradually dying away through attenuation. Undesirable echo is sometimes experienced on phone lines and radio links where there is a delay or other technical problem. 2. Output to a command line or output window on a computer. Echo is a command used by many batch and other scripting languages to 'echo' or 'print' to a terminal, whether that terminal is a window on a computer screen, a teletype, or a printer.

ECHO European Commission Host Organization.

ECHO 1 A telecommunications and geodesy satellite launched by the United States on 12 August 1960, historically important because it provided the first government satellite telephone links and television broadcasts on 24 February 1962. This satellite's orbit decayed in 1968. See ANIK for information on the first commercial television broadcast.

echo cancellation A technique for isolating and filtering unwanted echo signals which may accompany and interfere with the main analog transmission. Echo cancellation is often used on voice circuits, especially satellite transmissions, and may also be used in frame relay systems. In general, echo cancellation attempts to maintain a full-duplex circuit, although there are exceptions as in clear channel or ISDN calls. See echo suppressor, interference, noise.

echo check A diagnostic technique in which data is transmitted and then echoed back from the receiving end to the sender to check the completeness and integrity of the data.

ECHO satellites A series of satellites launched by the United States, beginning in 1960. The early ECHO Project launched large highly reflective 'balloons' capable of bouncing back radio signals. Without active relays, the communications signals were weak, but much was learned from these early experiments. The series included ECHO A-10 (never achieved orbit), ECHO 1, ECHO 2 (delayed from 1962 to 1964). Improvements in active relays superseded the ECHO Project. See ECHO 1, West Ford satellites.

echo suppression A means of reducing undesirable echoes, especially in satellite voice communications. Echo suppression differs from echo cancellation, in that echo suppression disables the reverse transmission while the person continues to talk, thus functioning more like a half-duplex line. See echo cancellation, interference, noise.

echo suppression disabler A means to coordinate *echo suppression*, the removal of undesirable echoes, especially on satellite voice lines. Since echo suppressors limit the capability of the system to half-duplex transmission by suppressing the signal in the direction opposite to the sending signal, it is important to be able to disable the echo suppression to restore full-duplex operation. This is typically done by sending a high-pitched signaling tone from an answering modem.

ECI emitter coupled logic.

Eckert, John Presper (1919-1995) An electronics inventor and collaborator with J. Mauchly on the historic ENIAC computer project. With Mauchly, he formed the Eckert-Mauchly Computer Corporation in 1946, which was acquired by Remington Rand Corporation in 1950. When Remington Rand merged with Sperry Corporation in 1955 to form Sperry Rand, Eckert became an executive with the company. This eventually merged with Burroughs Corporation to become Unisys. See BINAC, ENIAC.

ECMA European Computer Manufacturers Association.

ECP 1. See Encryption Control Protocol. 2. Enhanced Call Processing.

ECPA Electronic Communications Privacy Act.

ECSA See Exchange Carriers Standards Association.

ECTF 1. See Enterprise Computer Telephony Forum. 2. See European Community Telework Forum.

ECTUA European Council of Telecommunications Users Association.

ED See Electronic Directory.

EDA electronic design automation.

EDAC error detection and correction.

EDACS Enhanced Digital Access Communications System.

EDDA European Digital Dealers Association. A vendor organization comprised of resellers, service providers, and consultants of Digital Equipment Corporation (DEC) products, and related third-party products.

eddy current An electrical current induced by an alternating magnetic field which can be found in good conductors such as iron, and which may contribute to signal loss in electrical circuits.

EDF erbium-doped fiber. Erbium, a rare earth element, is used in fiber to manipulate its transmission properties. See doping.

EDFA erbium-doped fiber amplifier. An amplification technique used in fiber optic cables to increase the distance of the transmission by using a laser pump diode. See doping.

EDGAR Electronic Data Gathering Archiving and Retrieval. A database maintained by the U.S. Securities and Exchange Commission (SEC). It provides a means of electronic filing of securities-related documents through File Transfer Protocol (FTP), replacing the earlier IBM BISYNC communications protocol. Securities regulations for reporting are stringent, and electronic means to do so have been provided for some time through EDGAR. See Initial Public Offering.

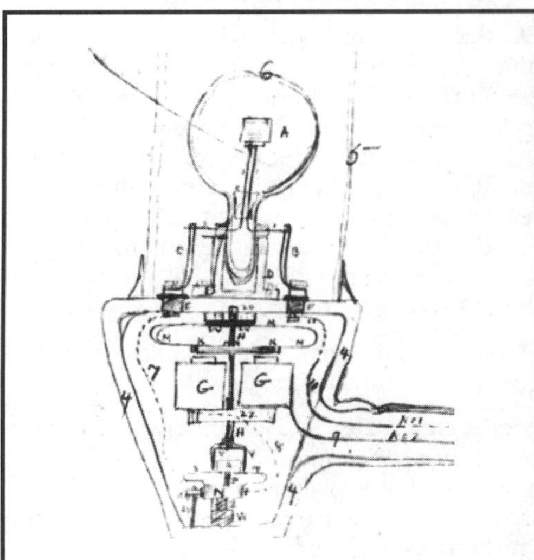

A diagram of an idea for a lamp with an elaborate pneumatic regulator, from Thomas Edison's notes.

Edison's quest for a replacement for lamps powered with gas or oil was inspired by bright arc lamps he saw in a workshop in Connecticut. He felt there was an opportunity to create a lamp which could provide power for mundane tasks and he set about finding a way to accomplish this, creating the light bulb.

edge connector A common type of thin, printed circuit board foil-imprinted extension used for making an electrical connection to a slot, usually inside a computer or in a peripheral card bay.

EDH electronic document handling.

EDI See Electronic Data Interchange.

EDIFACT Electronic Data Interchange for Administration, Commerce and Transport. See Electronic Data Interchange.

EDIS Emergency Digital Information System.

Edison base The standard screw-in lightbulb base common in North America.

Edison cell A type of variable storage nickel hydrate (positive) and iron oxide (negative) cell with an electromotive force lower than that of a lead cell. An older battery that was used in automobiles due to its ruggedness. See battery for chart of other types of cells.

Edison effect A phenomenon which Thomas Edison observed in 1883, and patented in 1884. While working with electrical illumination, he sealed a metal wire into a bulb near the filament, and noticed that electricity flowed across the gap between the hot filament and the metal wire, a discovery that became important to later electronic researchers in the development of broadcast technologies. See Audion.

Edison General Electric A company formed by Thomas Edison in 1889. Throughout his prolific career, Edison sought practical, commercially viable applications of his ideas, and actively marketed many of the products that were invented in his laboratory.

Edison, Thomas Alva (1847-1931) An American inventor who, at the age of 15, learned telegraphy and soon became a very fast and competent telegrapher. In 1868, he invented a device to record votes, and later, in New York, invented a stock ticker. In 1876, he set up a pioneer industrial research lab in Menlo Park, New Jersey, where he turned out hundreds of inventions, including the phonograph, electric typewriter, and fluoroscope. He developed the ideas of others, as well, making practical improvements to Bell's early telephone devices.

Edison's inventive output was prodigious and he pursued his interests almost to the complete exclusion of his business affairs, practical matters, and family. In all, he received more than 1,000 patents.

Edison is probably best remembered for developing the incandescent lamp in 1879, following the invention of the arc lamp by Humphry Davy early in the century, and a short-life incandescent bulb by Joseph Wilson Swan. Almost immediately following his invention, Edison began an electrical utility company in New York city in 1882 called the Pearl Street central station. This and other early utility companies provided direct current (DC) which has since been almost universally been superseded by alternating current (AC), an approach promoted by Nikola Tesla, and denounced by Edison. See incandescent lamp. See Tesla, Nikola.

Thomas Alva Edison from a portrait in the U.S. National Archives collection.

edit To change, alter, mend, or improve so as to meet certain standards or requirements of correctness, consistency, or professionality.

editing In computer applications, altering the content or presentation of information. Editing can be done with text editors, word processors, raster paint programs, vector drawing programs, audio sequencers, desktop publishing programs, and many other tools that allow the user to manipulate the information. Text editing is one of the most common activities carried out with computers.

editor A software program for manipulating information, particularly textual information. While graphics can be edited online, the tools to do so are not usually called editors, but rather paint or drawing programs, or image processors. The term editing is used more in the context of line-oriented information carried out with word processors, desktop publishing programs, and text editors.

Programming editors are often optimized for the special formatting needs of programmers. For example, they may monitor parenthetical statements, and alert the programmer if the parentheses, or brackets, are unbalanced, which would result in a syntax error in the program. They may provide different colors for different kinds of information, as for comments, variable names, or procedural labels. They may also provide the capability of telescoping or expanding the text (a feature also found in some word processors).

Scriptwriting editors are often set up with templates that indicate the correct margins and line spacing required by the theater and motion picture industries.

Word processors are optimized for document creation and basic formatting of text. Extensive formatting of text and graphics shouldn't be done with a word processor. Yes, it's possible, and yes, individuals have created some great documents with word processors, but it's also possible to hammer with the backside of a hatchet. It just isn't very efficient or comfortable. Complex page layout should be done with a page layout program. In a business environment, the extra cost of the software is incidental compared to the extra hours, weeks or months that have to be paid to someone to use a word processor for an unintended purpose.

GNU-Emacs is one of the most powerful text editors in existence, developed by Richard Stallman in the mid-1980s. It is configurable, scriptable, full-featured, and freely distributable.

EDLIN A Microsoft MS-DOS line editor which is somewhat limited in capabilities, but since it comes with the system is very widely used on Intel-based desktop computers.

EDM electronic document management.

EDO RAM extended data-out random access memory. A type of faster random access memory (RAM) that began to become prevalent around 1997. Many Intel-based desktop architectures now have expansion slots for EDO RAM.

EDP See electronic data processing.

EDSAC Electronic Delay Storage Automatic Computer. A significant large-scale, stored-program, electronic, digital computing machine developed in 1949 at Cambridge University, England, under the leadership of Maurice Wilkes. See ENIAC.

EDTV See Enhanced-definition TV.

EDUCOM An association of colleges and universities dedicated to support and evolution of educational computer network technologies.

edutainment A compound word that combines *edu*cation and enter*tainment*. Ideally, computer applications have the potential to provide education through highly appealing, entertaining, interactive means. Unfortunately, just as there are many products that are labelled 'screen savers' which do *not* save the phosphors in a monitor, there are many commercial titles labelled 'edutainment' which are not educational, and sometimes not even entertaining. Nevertheless, there are great possibilities for enlightening multimedia educational titles for all ages, which integrate relevant content, sound, visuals, and scenarios with computer interactivity. See multimedia.

EDVAC Electronic Discrete Variable Automatic Computer. A historically significant room-sized binary computer developed by von Neumann, Mauchly, and Eckert before ENIAC, its precursor, was fully operational. This high maintenance, vacuum-tube-based computer incorporated some of von Neumann's concepts of storing the program information in the machine itself.

Running the room-sized EDVAC involved the maintenance of hundreds of vacuum tubes and thousands of feet of wires. Modern handheld desktop calculators are now faster and more powerful. U.S. Army Photo.

EEC European Economic Community. Now the European Union (EU). A European common market in which, gradually, European currency, European passports, inter-country networks, and greatly reduced border restrictions are being phased in.

EEI external environment interface.

EEMA European Electronic Messaging Association. http://www.eema.org/

EEPROM electronically erasable programmable read-only memory. See erasable programmable read only memory.

EF entrance facility.

EF&I engineer, furnish and install.

EFCI See explicit forward congestion indicator.

EFF See Electronic Frontier Foundation.

effective competition A market regulation status. Broadcast cable providers must meet certain criteria to claim *effective competition* status. See cable access.

effective radiated power ERP. The transmitting power of a broadcasting antenna. This is sometimes directed by means of a directional antenna.

EFI&T engineer, furnish, install and test.

EFS See error free seconds.

EFT See Electronic Funds Transfer.

EFTA See European Free Trade Association.

EFTPOS Electronic Funds Transfer Point of Sale.

EGA See Enhanced Graphics Adapter.

Eggebrecht, Lew Chief of Commodore Engineering about the time the company was sold to Amiga Technologies in Germany.

EGNOS European Geostationary Navigation Overlay Service. A European land and marine communications service which augments the U.S. Global Positioning System (GPS), and, when it comes online, the Russian GLONASS system (similar to U.S. GPS). See Wide-Area Augmentation Service.

ego surfing *slang* Searching for your own name on the Net, in the media, or in databases. There are many legitimate reasons for 'ego surfing' and everyone should probably do it once in a while to make sure names or net addresses are not fraudulently distributed, or used by imposters to post to newsgroups. This could result in misunderstandings and embarrassment (and sometimes even litigation) directed at the legitimate owner of the name or address.

There are search engines on the Internet which allow users to check newsgroups for postings under a particular name. Many Web directory services on the Web give out phone, address and sometimes even personal information on individuals, drawn from phone books and less legitimate sources. You must alert these services if your personal statistics are being used indiscreetly or illegally, and request that they stop. (There is probably information about you on the Net, whether or not you even use a computer.)

Ego surfing is routinely engaged in by those who enjoy studying family histories, and are on the lookout for more information to add to their genealogy databases. Ego surfing also can help those who are publishing scientific, political, or other information with wider social implications to follow the dissemination, and sometimes the impact, of their communications, in order to engage in global dialog, or to correct misrepresentations or misunderstandings.

EGP See Exterior Gateway Protocol.

egress 1. Exit, way out. 2. In frame relay networks, frames which are exiting *away* from the frame relay towards the destination. The opposite of ingress.

EIA See Electronic Industries Alliance.

EIA Interface Standards A collection of standards describing configurations, signals, and other communications parameters for various electronic connecting interfaces. These are often used in conjunction with ITU-T specifications for protocols and functions.

Probably most familiar of the Interface Standards are EIA-232-D and EIA/TIA-232-E which are 1987 and 1991 updates to the decades-old RS-232 specification for serial transmissions between data terminal equipment (DTE) and data communications equipment (DCE). This standard has been widely implemented in desktop computers and other devices, and is commonly used for communicating with modems, remote terminals, and printers.

Most systems support EIA-232-D and EIA/TIA-232-E through 25-pin D connectors, though minimally nine pins are needed to implement the specification, and 9-pin D connectors (EIA-574) are sometimes used. The EIA has also defined faster standards for serial communications, including EIA-422 (balanced signals), EIA-423 (unbalanced), EIA-485 (multipoint), and EIA-530 (EIA-422 with 25-pin D connector).

EIA standards additionally encompass wiring connectors and topology, including building wiring and network backbones.

EIA/TIA categories Standardized specifications for cable transmission speeds. See the EIA/TIA Transmissions chart for categories.

Eiffel, Alexandre Gustave (1832-1923) The French designer and engineer who studied aerodynamics, and designed the Eiffel Tower, the western train station in Budapest, Hungary, and the Nice, France domed observatory. More familiar to Americans, he created the great structural support for the 151-foot high Statue of Liberty, over which Bartholdi overlaid the majestic sculpture. It was intended to commemorate the U.S. centennial, and the concept of liberty, though it wasn't possible to finish it and deliver the immense undertaking until 1886, a decade after the centennial celebrations.

Eiffel Tower, Tour Eiffel The Eiffel Tower, in Paris, France, is a remarkable historical monument and engineering milestone. It was constructed in 1889 of structural steel, unusual at the time, as most large structures were built of stone, setting a precedent for future highrises. A. Gustave Eiffel designed the monument, which became the largest radio tower in the world, for the 1890 World's Fair. A Bell Telephone engineer and Lee de Forest, inventor of the Audion, were two of the first people to carry out transatlantic broadcasts between the Eiffel Tower Paris and the United States.

The almost 1000 ft. Eiffel Tower and surrounding environs of Paris, France at the time it was constructed for the upcoming 1890 World's Fair.

EIG See Electronic Information Group.

eight hundred service See 800 service.

Einstein, Albert (1879-1955) A German scientist who is considered one of the greatest in the history of physics, best known for his special theory of relativity. Einstein moved first to Switzerland, where he was educated as a teacher of mathematics and physics, and then later to the United States. Unable to find a teaching or research position, he worked in the Bern patent office for seven years. Within two years of starting as a junior clerk, Einstein had become a technical expert in the patent office. While there, he made good use of his free time in writing a large volume of theoretical physics articles.

By 1905, Einstein was writing about his ideas in electromagnetic energy, and the photoelectric effect, building on and extending the work of M. Planck. In the same year, he wrote his paper on the special theory of relativity, reinterpreting classical physics, and supported Maxwell in hypothesizing that the speed of light was constant in all frames of reference. His concepts about the equivalence of mass and energy were tied in with the other writings at that time. This creative intellectual output was astonishing, and the later recognition and corroboration of his ideas catapulted him into the history books and out of the patent office into a series of posts at universities in Eastern Europe, Europe, and America. In 1940, he became a citizen of the United States.

In 1915, he published the definitive culmination of his writings on relativity, and in 1917, he proposed electromagnetic emission principles which lead to the development of lasers. In 1924, he made further discoveries concerning the relationships between waves and matter.

In 1921 Einstein was awarded the Nobel Prize in physics for his studies of the photoelectric

EIA/TIA Cable Transmission Categories

Categ.	Transmission Speed Rating	Typical installations
Cat 1	No performance criteria specified.	
Cat 2	Rated to 1 MHz	Telephone wiring installations
Cat 3	Rated to 16 MHz	LAN, Ethernet, 10Base-T, UTP Token-Ring
Cat 4	Rated to 20 MHz	LAN, Token-Ring, 10Base-T, IEEE 802.5
Cat 5	Rated to 100 MHz	WAN, Fast Ethernet, 100Base-T, 10Base-T

effect, and a crater on the moon has been named after him.

Albert Einstein was a patent clerk in Switzerland for seven years before he was able to find the type of position he desired in an academic research environment.

EIR Equipment Identity Register. A mobile services security database.

EIRPAC Eire packet network. An X.75 packet-switched network in the Irish Republic.

EIS Expanded Interconnection Service.

EISA Extended Industry Standard Architecture.

EIU Ethernet Interface Unit

EKE electronic key exchange.

EKTS Electronic Key Telephone Service. An ISDN-1 standard for supplementary ISDN services.

electric eye *colloq.* 1. A type of simple motion detector which consists essentially of a beam shining across the space to be monitored, hitting a sensor on the other side. When the beam is interrupted, the sensor triggers associated mechanisms such as alarms, locks, or surveillance cameras. Electric eyes are common in automatic doors, and on-demand elevators, security systems, and traffic evaluation systems, and can be coupled with computer programs that produce records of entry and exit. An electric eye could also be used to automate Internet cams, to trigger a recorder to begin when someone sits down at the system or enters a room. A number of creative people on the Net have aimed video cameras on their pets who may not always be hamming for the camera. An electric eye could be used to shoot some frames when the pet is more active, and more apt to add interest to the images. 2. A tuning indicator, consisting of a fluorescent cathode ray tube (CRT) screen with a dark region that varies in proportion to an incoming radio frequency (RF) signal strength.

Electric Telegraph to the Pacific Act A U.S. Congress act to solicit bids for construction of a government line connecting San Francisco to major centers. The line was to be open to the use of all U.S. citizens on payment of the appropriate charges.

electrical wire Any conductive filament, usually metallic, which readily transmits current. Copper is very commonly used for telecommunications wires.

One of the earliest documented 'long' wires was a thread that conducted 'electric virtues' over 600 feet in 1729. The thread was strung during electrical experiments conducted by S. Gray and G. Wheler.

electricity A fundamental constituent of nature that is observable as positive and negative charges and currents through materials according to their conductivity. The discovery and harnessing of electricity is the basis of our current technological society. Electricity arises from various sources, including friction (static electricity), light (photoelectricity), heat, chemical activity, piezoelectricity (pressure, especially in crystalline substances), and mechanical energy (as from a generator). Ancient observers were aware of various electrical phenomena, but it was not until the last 200 years that they came to understand their various properties, and the fundamental principles that underlay them.

electrode 1. An essential component of an electron tube. Any of the basic components, the electron-emitting cathode, controlling grid, or electron-attracting anode, are considered electrodes. 2. A plate in a battery.

electroluminescence The direct conversion of electrical energy into light. This process is used in display technologies, with electroluminescent materials such as zinc sulfide doped with manganese. See electroluminescent display.

electroluminescent display EL display. A gridded display technology which incorporates an electroluminescent material sandwiched between outer panels. When exposed to high electrical fields, the inner material emits light. Like plasma panels, individual points are selectively lit through matrix addressing. EL displays are brighter then passive LCD displays, but also require more power. See liquid crystal display, plasma display.

electrolysis The production of chemical changes by passing an electrical current through an electrolytic material. See electrolyte.

electrolyte A nonmetallic substance which, when chemically or electronically stimulated, becomes an ionized conductor. Electrolytic properties are widely employed in electronics. Electrolytes are used electrical cells, rectifiers, etc. Various acids were commonly used as electrolytes in early inventions.

electrolytic cell A power-conducting (as opposed to power-generating) cell comprised of a conducting liquid (the electrolyte) and two identical electrodes. Electrolytic cells are used in refining, reduction, and electroplating processes.

electrolytic detector, liquid detector A radio wave detecting device patented by R. Fessenden in 1903. It was discovered accidentally when Fessenden was seeking ways to improve on his hot-wire barretter. A broken filament led to the discovery that signals could be received better through the separate pieces of filament in an electrolytic solution, than with a single piece of filament. From this, Fessenden combined nitric acid and a platinum wire into a rectifier which could detect both continuous and damped waves.

The electrolytic detector was an important milestone in radio history, as it provided a means to create a much more sensitive receiving instrument. Most electrolytic detectors required an outside power source, though some were manufactured with a built-in battery integral to the design. See barretter, Shoemaker detector.

electrolytic paper tape A type of paper tape used on some of the old telegraph systems in which a stylus passed an electrical signal onto the coated tape to produce an image of the message being transmitted. The image on early systems was often blue, though the amount of current on an electrolytic system can influence the color of the image.

electromagnet A device which has a significant magnetic field only when current is flowing through it. The strength of the wire is dependent on the size and type of materials used, the amount of current, and the number of coils. Electromagnets are used extensively in appliances, industrial hoists, telephones, public address speakers, and bulk erasers. You can make a simple electromagnet by wrapping a conductive wire around an iron nail and passing current through it (taking care not to touch any of the live wires). This in turn can be used to magnetize the end of a screwdriver by stroking the nail in one direction over the screwdriver. Quite handy for holding screws in place, but be wary of using magnetized screwdrivers near electronic components. See bulk eraser, solenoid.

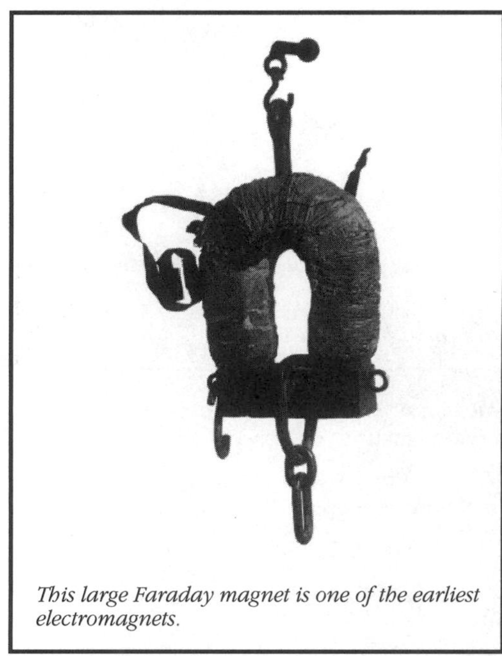

This large Faraday magnet is one of the earliest electromagnets.

electromagnetic Embodying electric and magnetic properties.

electromagnetic communications Communications which employ the propagation of transmission waves through space. Meaningful signals are sent in many ways, with light, radio waves, microwaves, etc., usually by modulating the transmission of the radiant energy in some way.

electromagnetic deflection The directing of the path of an electron beam by means of a magnetic field (often in the form of a coil).

electromagnetic field 1. A field of magnetic influence around a conductor produced by a current flowing through the conductor. See electromagnet. 2. Together, an electric field and its associated magnetic field. The magnetic field is perpendicular to the lines and direction of force. See right-hand rule.

electromagnetic interference EMI. Undesirable noise, degradation, overlap, or echo in an electromagnetic transmission.

electromagnetic pulse EMP. A large or fast-moving electromagnetic transmission which is quicker, or burstier than the immediately preceding and succeeding transmission. Sometimes a pulse is a natural phenomenon, such as lightning, or it can be a deliberate means of creating a signal, or carrying information.

electromagnetic spectrum The range, or diagrammatic relationships, of the known types of electromagnetic radiation, organized by wavelength.

electromagnetic wave The radiant energy produced by an oscillating electric charge. The infrared, ultraviolet, gamma, visible wavelengths, and cosmic rays are some examples.

electromotive force Descriptive of the 'pressure' of the movement of electrons through a circuit, somewhat analogous to the movement of liquid through a closed piping system. Indeed, for many decades the two kinds of movement (electrical and liquid) were assumed to be the same, and early scientists spoke of the 'electrical fluid.' See volt.

electron A minute, elementary particle of matter, carrying a negative electrical charge. Electrons are normally found surrounding a positively charged nucleus. The term is derived from the Greek word elektron (amber), but was first used with this specific meaning by G. J. Storey in 1891. See positron.

electron beam A stream of electrons travelling together in the same trajectory, directed by a magnetic field. An important constituent of cathode ray tube technology which is widely adapted to TV and computer monitors.

electron microscope An optical-electronic instrument which provides magnification of minute structures by means of recording the movement of a focused beam of electrons. The results were originally displayed on fluorescent screens or photographic plates, and now computer monitors can also be used. Early electron microscopes could enlarge images 100 times more than the finest optical microscopes of the time, but the images were limited to black and white still objects. New techniques were continually being sought to increase the range of objects which can be imaged, and the ways in which they can be represented. Computer enhancement and interpretation has opened up a wide range of possibilities.

electron tube A device in which the movement of electrons is conducted within a sealed glass or metal container. While electron tubes were made of glass for many decades, some all-metal tubes came into use in the mid-1930s. The most common implementation of the electron tube is the *vacuum tube*, since the life of the electron-emitting materials could be extended by removing the air, or encasing a controlled mixture of gases.

The most important evolutionary development in the history of the electron tube is the *Audion*, a *triode* that is a three-element tube with a control grid invented by Lee de Forest. One of the most important adaptations of the vacuum tube is the *cathode ray tube* still widely used in television sets and computer monitors. See Audion, cathode ray tube, vacuum tube.

The most common types of electron tubes used in electronics for many decades were three (or more) element vacuum tubes. Experimentation led to the development of many different types of tubes for different purposes, and numbering systems were set up to keep track of parts so consumers could replace broken or burnt out tubes. For the most part, semiconductor components have replaced electron tubes, except for some high frequency applications. This interesting assortment is from an exhibit at the Bellingham Antique Radio Museum.

electronic blackboard See whiteboarding.

electronic bulletin board See bulletin board system.

Electronic Certification An electronic signature which serves the same purpose as a written signature on a physical document (usually a letter, contract, or administrative approval). Electronic Certification is accomplished through cryptography, typically key cryptography. See DSS.

Electronic Classroom A commercial Macintosh-based distance-learning 'audiographics' multimedia videoconferencing tool written by Robert Crago, an Australian developer. Electronic Classroom provides images, QuickTime compressed video, and voice over public switched telephone networks (PSTN).

electronic commerce Financial and barter transactions conducted across data networks using electronic means of communications and agreements, including the exchange of documents, signatures, virtual money, etc. See Electronic Data Interchange, electronic mall, Pretty Good Privacy.

Electronic Commerce Service ECS. ECS is a set of electronic mail and verification services developed by the U.S. Postal Service to offer secure electronic mail so that it becomes an electronic extension of the U.S. Postal physical mail system. USPS is cooperating with private firms to develop this technology.

Various aspects of the USPS plans include personal and professional certificate services through a Certification Authority (CA), time and date stamping, in essence an electronic postmark, certified email, return receipt, verification, and archiving.

It is important to consider that USPS email differs from personally forwarded email in some of its legal safeguards, and has a history of statutes and precedences which may make it attractive to business users.

As with all major milestones in U.S. Postal history, the USPS released a commemorative stamp, in early 1996, to launch their electronic venture. The computer whose birthdate was commemorated was the ENIAC, which certainly deserves credit for its historical importance, but the system that should probably have been honored as the first large-scale computer is the Atanasoff-Berry Computer, which preceded the ENIAC.

Electronic Data Interchange EDI. A series of standards developed primarily for business communications, EDI is a process for network interchange of electronic messages and documents, often between different companies or government agencies. EDI software works in conjunction with applications software. Files are extracted from an application, converted into a standard EDI format, and passed on to the communications software for transmission. EDI is not secure in and of itself, and may be combined with authentication and encryption schemes. Practical applications include the interchange of invoices, purchase orders, policy documents, RFQs, waybills, cost estimates, etc.

The primary international standard for formatting EDI messages is Electronic Data Interchange for Administration, Commerce, and Transport (EDIFACT). EDIFACT messages are included within an EDI 'envelope.' The interchange IDs of sender and receiver must be agreed upon. Their addresses are described in the X.400 standard (ITU-T X.400/X.435). See X Series Recommendations.

electronic data processing EDP. A system for taking in a lot of data, manipulating it, and storing, or displaying it in another format. For example, numbers may go in, and paychecks, employee statistical information, or sales demographics may come out.

Electronic Directory ED. An informational database based on a directory standard, such as X.500 or LDAP, intended to help integrate various directories on a network. Thus, access is improved over that of searching and querying various directories, and different formats and protocols are made transparent to the user. EDs are of interest especially to corporations, educational institutions and libraries. See X.500.

Electronic Frontier Foundation EFF. A well-known, influential civil liberties association cofounded in 1990 by Mitch Kapor and John Perry Barlow, which acts as a forum and advocate for social and legal issues related to the information and electronic revolution. http://www.eff.org/

electronic funds transfer EFT. A system through which financial transactions are enacted electronically over a network without the actual exchange of physical currency. Stock transactions have typically been carried out this way, especially in the last decade, and the exchange of actual stock certificates is not required.

Electronic Industries Alliance EIA. A national trade association representing the United States high technology industry. The related Government Electronics and Information Technology Association (GEIA) includes companies which manufacture or engage in research, development or systems integration to meet the needs of United States government agencies. http://www.eia.org/default.htm

Electronic Industries Foundation EIF. The philanthropic arm of Electronic Industries Alliance, established in 1976. EIF addresses social/educational issues in the electronics industry.

Electronic Information Group EIG. The EIG is associated with the Electronics Industries Alliance. http://www.eia.org/

electronic key exchange The process of exchanging public keys in a public key encryption scheme, so that messages can be forwarded to the owner of the public key and decrypted with a corresponding private key. This is used for authentication of electronic communications. Public keys are sometimes uploaded to public databases.

electronic leash Any mobile communications unit that ties the user to an employer or other entity to which the user has a responsibility or obligation to respond. It particularly applies to those devices which clip to a belt or pocket. Beepers and pagers are good examples, though short-range radios and cellular phones may be included. Electronic leashes are frequently used by professionals in the field, or 'on call.' Short-range radios are now also being marketed to families as child-monitors for family outings at parks or malls. It's considered bad manners to keep a beeper or pager active while watching a live stage performance.

electronic lock Any locking mechanism which is activated through software. Electronic locks are being incorporated into laptops and mobile phones to deter unauthorized use. They are also incorporated into some door entry systems. Using an electronic lock is similar to establishing and using a password.

electronic mail, email The invention of email is attributed to Ray Tomlinson of BBN in the early 1970s.

Email is the exchange of correspondence through computer networks. In its broadest sense, it includes text files, sound files, graphics, and data files. The exchange of electronic mail is one of the heaviest used capabilities of the Internet, where messages can be sent to almost anyone in the world with an Internet connection, often without any extra charges (unless the messages are unusually large), depending on the Internet Services Provider.

electronic mail gateway In a network where there are links to other networks which may be using a variety of network mail protocols, a gateway system handles receipt, translation (if necessary), and delivery of the messages to the local system, the external system, or both. Mail filters and distribution agents may be incorporated at the gateway level to selectively process and deliver email. See electronic mail, Post Office Protocol.

electronic mall An online storefront, usually Web-based, which allows consumers to view, review, or request information about products and services. Although Web page layouts differ widely, electronic malls are nevertheless beginning to use some common conventions like 'shopping carts' which are lists of products you have selected by pressing buttons on the Web page to queue for the final order entry and payment section. As no consistent electronic commerce money transfer schemes have emerged as yet, most of the electronic malls which are successfully getting orders are using credit card ordering.

E

Electronic Messaging Association EMA. An organization established to support the development and use of secure global electronic commerce services, including electronic mail, faxing, paging, and computing services. EMA contributes to education, public policy discussions, and connectivity and interoperability issues. http://www.ema.org/

electronic news gathering ENG. The process of seeking out new information on selected topics, or general topics from among the many sources available online. This is beginning to become a monumental task. Our ability to create information as a society far exceeds the ability of any individual or organization to use that information. See data mining.

Electronic Payments Forum EPF. An alliance of commercial, non-profit, academic, government, and standards bodies, committed to furthering the development and implementation of electronic payment systems to promote global electronic commerce. The EPF is organized under CommerceNet, the Financial Services Technology Consortium (FSTC), and the Cross-Industry Working Team (XIWT). http://www.epf.net/

electronic phone Newer phone technology in which many of the mechanical parts common to traditional phones are replaced with circuit board logic design. This results in components, such as cellular phones, which are smaller, lighter, and more portable.

electronic receptionist A type of Auto Attendant system with extra logic and capabilities that allow the personalized notification of messages through a network, sometimes even with an animation of a human receptionist delivering the message. It is likely that electronic receptionist virtual reality scenarios and videoconferencing will create some remarkable systems in the future. See Auto Attendant.

electronic ringing A digitally encoded sound file that is played instead of the conventional mechanical telephone ringer. Since it's digital, it can be made programmable, and thus any sound patch could technically be used as a ringing sound. Just as the error beep on most computer systems can be selected or recorded by the user, electronic ringers will probably be selectable in the future. Given the creative bent of many individuals, some likely ringers would include car horns, buzzers, voices saying "You've got a call.", dogs barking, etc. Now, if you consider the availability of Caller ID, it won't be long before someone invents a ringer that not only alerts the person about an incoming call, but also speaks out a message that says, "It's your girlfriend, better answer before she changes her mind!" Phones integrated into personal computers already exist, so it's just a matter of time before these are integrated with electronic voice mail and customized ringing sounds.

electronic serial number ESN. In general, an electronic serial number is an encoded identifier created for locating a device, or authenticating a device. The degree of security of these numbers varies widely. In the cellular phone industry, the ESN is a 32-bit code which uniquely identifies a cell phone unit. It is standardized to include information about the manufacturer and a unit number. In addition, there is reserved space for future or specialized use. In some cell phones, the ESN cannot be changed. In some, it can. Those who are handy with electronics are capable of changing electronic serial numbers with a custom reprogramming cable and computer connection.

When using a cellular phone to make a call, the cell switching center uses the information to check the validity of the caller, and sometimes checks a database of cell phones that are reported as stolen. If the number is valid and there are no problems with the ownership of the unit, or the currentness of the account, the call is put through.

The ESN is also used as an access code on the Web. Different cell phone manufacturers have various online services, including customer assistance, warranty information, etc., which can be accessed by users with legitimate ESNs.

electronic switching system In old phone systems, call connections were established by human operators plugging phone jacks into jack panels. Later, mechanical switches, and electromechanical devices began the automation process. Still later, logically programmed solid-state components took over the switching functions in conjunction with the telephone switching equipment to create electronic switching systems.

electronics The art and science of investigating and harnessing electrons. Some lexicons make a distinction between electron movement through the air or through a vacuum (electronics) as distinct from electron movement through a wire or other similar conductor (electricity). While this distinction may be important in a theoretical sense, for the purpose of this general purpose telecommunications dictionary, the broadest sense of the term *electronics* has been used. The harnessing of electrons is fundamental to telecommunications technology. The term came into general use in the 1940s. See magnet, vacuum tube.

electrophoretic display A display technology which incorporates positively charged, suspended color particles sandwiched between two outer plates. The suspension medium consists of a contrasting color such that the suspended particles are visible or not visible depending on their position within the suspension medium. The particles are moved to the front (visible) or back of the display (not visible) by applying negative and positive voltages through matrix addressing.

electrophotography An image transfer printing process commonly used in photocopiers, which images by means of electrostatic charges on a photoconductive medium. Selenium and zinc oxide are two types of coatings commonly used.

electroplating The process of using electrodeposition to cause a usually metallic substance to adhere strongly to another. Metals are often electroplated to prevent corrosion, to improve conductive properties, or to increase their value (as with gold or silver electroplating on jewelry or cutlery).

electroscope An instrument for detecting the presence of an electric charge, and whether it is positive or negative. It may also be designed to indicate the intensity of that charge.

electrostatic An electrical charge at rest, familiar to many people as *static* electricity. Static electricity associated with rubbing amber (*amber* in Greek is *elektron*) was known to the Greeks by at least 600 BC when Thales recorded that a 'fossilized vegetable rosin' (amber), when rubbed with silk, acquired the property to attract very light objects to itself.

electrostatic charge An electrostatic charge

which is stored, as in an insulator, or capacitor. Subsequent discharge must be carefully controlled or prevented in electronic components, as they have the potential to cause harm. See condenser, Leyden jar.

electrostatic deflection The process of controlling the path of an electron beam by passing it between charged plates. This system is widely used in cathode ray tubes (CRTs). See cathode ray tube.

electrostatic focusing The process of using an electrical field to focus an electron beam.

electrostatic plate A type of printing plate used for high speed laser printing which utilizes various metals, chemicals and photoconductors.

electrostatic printing A printing process in which the imaging is primarily done with electrical rather than chemical means; it is a 'dry' replication process very common in photocopiers and laser printers. Typically, laser beams trace an electrostatic pattern that attracts very fine particles of powder, usually toner, which are subsequently fused onto the printing medium with heat.

electrostatic voltmeter A voltage-driven voltage measuring instrument which requires almost no power for its operation. The electrostatic voltmeter functions by using the attracting and repelling properties of magnetic bodies. It can be used on both alternating current (AC) and direct current (DC).

ELIU electrical line interface unit.

ELIZA An early human conversation modelling program developed by Joseph Weizenbaum. It became quite popular and generated a lot of discussion and controversy about the ability of machines to think and converse in such a way as to be able to interact with humans, and potentially fool them into thinking they were interacting with another human rather than a computer program. Eliza used pattern recognition and substitution, and interacted like a Rogerian psychotherapist, returning many of the statements which were posed to it as questions to the user. See artificial intelligence, Turing test.

Ellipso A commercial satellite communications system based on groupings of low Earth orbit (LEO) *bent pipe* transponders, operating in a unique elliptical obit through CDMA technology.

The Ellipso project is technically interesting in that its orbits are adapted to the distribution of land masses and populations on the earth. Two complementary and coordinated constellations of satellites comprise the Ellipso constellation. The Ellipso-Borealis Subconstellation covers northern latitudes through 10 satellites in elliptical orbits in two planes with apogees of 7,846 kilometers operating in a three-hour orbital period. The Ellipso-Concordia Constellation provides coverage of southern latitudes through six quasi-circular equatorial orbits at 8,040 kilometers, and four complementary elliptical orbits to increase daytime capacity.

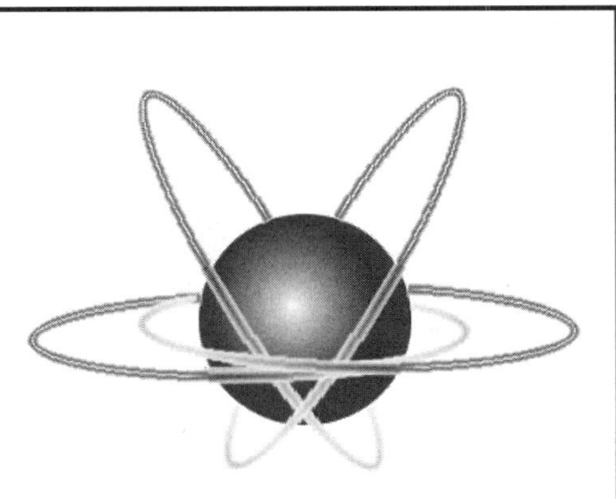

The Ellipso project employs some unusual orbits, different from the common geostationary or regularly elliptic orbits used by other satellite service providers.

Ellipso is a service of U.S.-based Mobile Communications Holdings, Inc. Mobile's communications services are designed to enhance rather than displace existing phone and data services. Mobile Communications is targeting low-cost services especially to isolated and rural areas.

elliptic curve A geometric structure of interest to cryptographers because it can be incorporated into algorithms with calculations that can be performed with relative ease in one direction, but which are difficult or impractical to reverse.

ELOT Hellenic Organization for Standardization, located in Athens, Greece. ELOT handles standardization, certification, and quality control in Greece.

ELSU Ethernet LAN Service Unit.

em In typography, a unit of measure related to type size indicating a distance equal to the point size height of a font. Thus, an em in a 12-point font is 12 points wide. See en, point size, thin.

EMA See Electronic Messaging Association.

EMACS, GNU EMACS A powerful, flexible, configurable, network-savvy text editor written originally by Richard Stallman, and promoted and supported by a large community of programmers. Emacs works on a wide variety of platforms, with GNU EMACS running principally on Unix systems. MicroEMACS is a popular derivative, and CygnusEd is a fast commercial editor with many EMACS-like characteristics. While EMACS has a bit of a learning curve, it has many useful capabilities not found in other editors. It can work directly as a mail client, can read and write files directly over an FTP link, and can be modified and extended with LISP. See text editor.

email See electronic mail.

EMBARC Electronic Mail Broadcast to a Roaming Computer. A North American paging service from Motorola based on the X.400 standard. EMBARC transmits 8-bit text and binary files containing up to 1500 characters. See ERMES, SkyTel, X Series Recommendations.

embossing 1. A printing process in which the printed medium is passed across a dye which raises the embossed portions of the medium above the surface of the medium, usually by about 1/16th of an inch. Embossing ink is a special ink which is designed to rise up from the surface on which it is printed, usually after being exposed to heat (thermography). 2. A labelling, or marking system in which raised symbols or lettering are imprinted on a surface. Used on wiring/cabling sheaths, circuit boards, etc.

EME Earth-Moon-Earth. See Moonbounce.

emergency access An alert system incorporated into some private branch systems which rings selected phones, or all phones, in the event of an emergency.

Emergency Alert System EAS. A new emergency system designed to replace the Emergency Broadcast System (EBS). The EAS is based on digital technology, and all broadcast stations are required to have EAS decoders subsequent to 1 January 1997. The EAS provided for a one-year transition period during which old equipment had to be retained to ensure that new equipment could take over the functions effectively. Further, television stations must be able to automatically send a visual message. Specific equipment for delivering the emergency messages has not been defined on a physical basis. Instead, the type of message functions which must be supported are described. The equipment allocated to these functions must be certified. Live monitoring, logging, and documentation requirements form part of the EAS. See Emergency Broadcast System, Global Maritime Distress and Safety System.

Emergency Alert System (EAS) activation signal This is a mandatory protocol of the EAS, which consists of a digital header, an attention signal, a message, and an end of message (EOM) code. The header contains information about the origin of the message, its nature, location and time during which it is valid.

Emergency Broadcast System EBS. A warning system which has been replaced by the new digitally-based Emergency Alert System (EAS). See Emergency Alert System.

emergency dialing A type of speed dialing in which a shortened sequence of numbers, or sometimes even one button, facilitates a fast connection with emergency telephone services such as the police, ambulance, or fire department.

emergency hold A capability in which an emergency professional (fire, police, crisis center, etc.) can keep a connected call live even if the caller hangs up or presses the switch hook. Since it is not unusual for people in emergency situations to behave erratically, or to hang up inadvertently, and then try to reconnect, it gives the emergency personnel the opportunity to retain or regain contact with the caller.

emergency telephone A phone designed to provide service in spite of unusual circumstances, such as power outages. It may have backup batteries, or access to emergency power in order to maintain operability.

EMF, emf electromotive force.

EMI See electromagnetic interference.

emission 1. Outward radiation or conduction of waves. 2. Ejection of electrons, which could arise from a variety of causes such as impact or heat.

emission current A current produced in the plate of a cathode ray tube (CRT), when the electrons beamed from the cathode pass into the plate.

emission velocity The initial velocity of electrons beamed from the cathode in a cathode ray tube (CRT).

emissivity A measure of the ratio of flux emitted by a radiation source to the flux radiated by a black body with the same area and temperature.

emitter 1. A device which provides periodic pulses at regular intervals as is used, for example, on punch card machines. 2. An electrode in a transistor (minority carrier).

emoticon An apt name for an online grassroots evolution of tiny ASCII icons to represent moods, emotions, and expressions. Since networks before the Web were traditionally text-based, and since communications online can often be misunderstood without facial and voice expressions to express subtle distinctions, many users have adapted an iconography into text messages to provide a wider range of expression. In some cases, emoticon faces (and bodies) are used as symbolic representations of a person's physical appearance. Because of the difficulty of creating vertical drawings in a text-based message, most emoticons are designed to be read sideways, that is, with your head turned 90 degrees, usually to the left. They are commonly used in public forums and email, and many are known as 'smileys.' The list includes a few examples from the hundreds extant.

Emoticons (Smileys)

<table>
<tr><td>:-)   :)</td><td>two variants of the basic smiley, the most common emoticon</td></tr>
<tr><td>;-)   ;)</td><td>a smile with a wink</td></tr>
<tr><td>:-D   :D</td><td>laughter, a big talker, delight</td></tr>
<tr><td>:-O   :o</td><td>surprise, astonishment, awe</td></tr>
<tr><td>:-/   :-(</td><td>a variety of smirks</td></tr>
<tr><td>:-\   :-|</td><td>and frowns</td></tr>
<tr><td>:~P</td><td>sticking a tongue out</td></tr>
<tr><td>:^b   ;^3</td><td>caricatures</td></tr>
<tr><td>:-#</td><td>a zipped lip, lips are sealed, embarrassment</td></tr>
</table>

EMP See electromagnetic pulse.

empty slot ring In a ring-based local area network (LAN) topology, such as Token-Ring, a free packet, that travels around the ring through each node, and is checked at each workstation for messages.

EMR See Exchange Message Record.

EMS See Expanded Memory Specification.

EMT electrical metal tubing.

EMU European Monetary Unit. A common market currency in effect as of 1 January 1999.

emulate *v.* To copy, simulate, provide the working functions of.

emulation, software To recreate the functioning of an application or operating system. Emulators abound in the computer world. There are Macintosh OS emulators for Amigas and NeXT computers, Windows emulators for Macintosh and Amiga computers, etc. It is not strictly true that an emulator *always* runs more slowly than the original system. With fast hardware, one operating system can sometimes run an emulator faster than it may run on the machine for which it was originally designed.

en In typography, a unit of measure related to type size indicating a distance equal to half the point size height of a font. Thus, an en in a 12-point font is 6 points wide. See em, point size, thin.

encapsulated PostScript EPS. A PostScript file format which includes *bounding box* information, that is, explicit statements about the lower left and upper right extents of a rectangular area which includes the entire image contained in the file. The EPS bounding box information is commonly found in the header at the beginning of the program. Many programs cannot print a PostScript page without the EPS information, if it is an imported file, since the size and boundaries of the image may not be explicitly stated in the page description code. (It's like trying to find a frame for a picture, without first knowing the size of the picture.)

EPS is a widely supported vector format, and is useful for desktop published documents, posters, typography, and other applications in which one wants to take advantage of the highest possible quality available on a particular output device. See PostScript.

encapsulation 1. To encase or completely surround by a protective covering. When something is encapsulated, there's not usually much room between the covering and that which is being covered. Encapsulation is used for insulating, sealing, or separating, and may or may not make the contents water- or airtight. 2. In frame relay, a process by which an interface device inserts the protocol-specific end device frames into the frame relay frame. 3. A means of 'wrapping' a bundle of network information in order to 'tunnel' it through a system, that is, to pass it through a system which may use different protocols, without altering the wrapped data. It is subsequently de-encapsulated to retrieve the wrapped information. See Point-to-Point Tunneling Protocol, tunneling.

encoding The conversion of data from one form to another, sometimes to obscure the contents, sometimes to reduce the amount of space necessary to hold the information, sometimes to change it to a compatible format. Compression and encryption are two means of data encoding.

encryption The process of encoding data in a way that makes it difficult or impossible to decode. Encryption is widely used in times of conflict to hide the contents of sensitive communications. It is also used for encoding private messages, or business transactions and documents. Encryption is often combined with authentication. Key encryption is a type of encoding which uses public and private keys to encode and retrieve messages. See Clipper Chip, key encryption, Pretty Good Privacy.

Encryption Control Protocol ECP. A means for configuring and enabling data encryption algorithms on both ends of a point-to-point (PPP) link. ECP uses the same basic packet exchange mechanism as the Link Control Protocol (LCP). See RFC 1968.

end delimiter An end of file, (EOF), end of block (EOB), end of message (EOM), end of transmission (EOT), or other signal that indicates the final byte or section of a data block has been reached.

end device 1. The terminal point, whether it be the source or the destination, of a data transmission. 2. A hardware or software terminal device, which may be a file server or host. 3. In a frame relay, data is sent from the source to an interface in which it is encapsulated into a frame relay frame. At the destination end device, the frame relay frame is stripped off again (de-encapsulated). See Data Communications Equipment, Data Terminal Equipment.

End Office switches Switches from which telecommunications subscriber Exchange Services are connected.

end of file EOF. A control character or other signal which indicates the end of a file record has been reached. EOF markers are important to programmers. Many programs have to be explicitly programmed to react accordingly (quit, stop reading, reverse direction, return, etc.) to an EOF marker.

end to end digital path Communications link between two hosts capable of sending wholly digital data (without analog conversion along the way) over an arbitrary number of routers and subnets.

end to end path Communications link between two hosts capable of sending data over an arbitrary number of routers and subnets.

end to end service A network service that links one subscriber to another subscriber, or another branch of the first subscriber. It typically consists of the local loops at each end of the transmission connected through an InterExchange Carrier (IXC).

end user 1. The service or product user at the end of the retail or service chain, in other words, it's not a user who is a wholesaler, broker, or agent. Some people use this phrase interchangeably with *user*, but it might be useful to maintain a distinction. 2. In telephone services, an end user is distinguished as a subscriber using loop start or ISDN. 3. In network communications, the computer users who are not the technical administrators of the system, e.g., students and faculty on a college campus.

Energy Sciences network ESnet. The network of the U.S. Department of Energy.

ENG electronic news gathering.

engine The internal nuts and bolts of an application, the essential working features of a program exclusive of the interface. For example, a *database engine* includes the search and retrieval capabilities upon which the software application is built. Database engines can be purchased or licensed, then configured with an interface (front-end) and sold for specific vertical markets, as security databases, CD databases, equipment databases, etc. A graphics engine supplies the floating point routines and graphics primitives on which a paint, drawing, or rendering program can be built.

Englebart, Doug An innovator at the NLS laboratory at the Stanford Research Institute in the 1960s, and at Xerox PARC in the 1970s, Englebart is credited with inventing the three-button mouse in the late 1960s, and an accompanying 'keyset,' and many of the graphical user interface ideas that are now prevalent on desktop computers. His concept of 'windows' as a means of expressing a work area on a computer monitor was further enhanced by Alan Kay, the originator of Smalltalk, and others in the lab at the time. See Kay, Alan.

Enhanced Graphics Adapter EGA. A 16-color graphics display standard developed to supersede the Color Graphics Adapter (CGA) standard that was prevalent on Intel-based desktop computers in the mid-1980s. EGA could display up to 640 *x* 350 pixels. See Color Graphics Adapter, VGA.

Enhanced IDE An enhanced computer peripheral device controller specification which provides faster transfer speeds and larger addressing space than the original IDE specification.

Enhanced Parallel Port EPP. A parallel data transmissions specification intended to provide faster parallel speeds over common desktop computer parallel ports, without changing the basic hardware (cables, connectors, etc.) used to implement the connection. Parallel ports are commonly used with impact printers, and some ink-jet and bubble-jet printers.

Enhanced Serial Interface ESI. A freely distributed interface specification developed by Hayes Microcomputer Products. It is an extension to the COM Card specification for desktop computers.

Enhanced Serial Port ESP. A Hayes serial port specification introduced in the early 1990s which is intended to extend and replace the COM1/COM2 ports that are familiar to most users of Intel-based desktop computers. It provides two independent operations modes, to support existing standards and the ESP format. Depending on the version, defaults are set either with the Hayes Programmable Option Selection (POS), or with DIP switches.

Enhanced Specialized Mobile Radio ESMR. A traditional nationwide two-way radio dispatch service which has been expanded to offer services somewhat similar to cellular, but with more limited allocation of bandwidth.

Enhanced Trivial FTP Enhanced Trivial File Transfer Protocol. An experimental implementation of Network Block Transfer Protocol (NETBLT) which uses the User Datagram Protocol (UDP) for its transport layer. It was designed as a means to improve data transfer throughput for the specialized needs of half duplex radio networks using the Internet Protocol (IP). Through ETFTP, transmission parameters can be customized for low-speed, long-delay radio links. See File Transfer Protocol, Simple File Transfer Protocol, Trivial File Transfer Protocol, RFC 998, RFC 1350, RFC 1986.

ENIAC Electrical Numerical Integrator and

This photo shows two staff members changing the programming on ENIAC. In the early days of computers, this was done by reconfiguring the many wire connections. Grace Hopper was one of the first to advocate 'reusable code,' that is subroutines and other programming structures that could be run more than once. Unfortunately, at the time, many of her ideas were met with scepticism. Most engineers conceived of 'programming' as rearranging wires and Hopper's ideas were not understood by those who lacked her foresight. U.S. Army Photo.

Calculator/Computer. A historically significant, post World War II, room-sized, vacuum tube, punch card computer dedicated in 1943, following the success of the Harvard Mark I. ENIAC was developed by John W. Mauchly and J. Presper Eckert at the University of Pennsylvania, under the guidance of John Brainerd. It was unveiled in Philadelphia in 1946. Hand wiring was necessary to configure ENIAC to handle different problems.

The ENIAC was derived in part from the Atanasoff-Berry Computer (ABC). At the very least, the ABC provided inspiration for the ENIAC, and it is possible that it also provided some design ideas. In the 1930s, Mauchly is reported to have visited with Atanasoff and left with notes on the ABC. See Atanasoff, John V.; Atanasoff-Berry Computer.

ENN Emergency News Network.

ENOS Enterprise Network Operating System.

ENS Emergency Number Services.

Enterprise Computer Telephony Forum ECTF. A nonprofit association established in 1995 to promote interoperability among computer telecommunications products, and to provide a framework for Computer Telephony (CT) interoperability, including standards and education. http://www.ectf.org/

enterprise network, corporate network A network which broadly serves the needs of an enterprise (a larger business), and reaches most of the network-using members of the organization. Since larger networks often have different types of users, they may integrate a variety of technologies, such as a private branch exchange, data/voice/videoconferencing, etc.

entrapment 1. To ensnare, to manipulate a situation so an individual is encouraged to reveal actions or attitudes. 2. To deliberately provide a security loophole for the purpose of detecting unauthorized access or attempts at access.

envelope 1. Boundary, encasement, encapsulated entity, bounding box, extent. 2. The globe around a vacuum tube or bulb. 3. In amplitude modulation, the extents of the frequencies of the modulated wave. 4. A means of characterizing the content and characteristics of a message.

EO 1. end office. 2. erasable optical.

EOB end of block

EOF end of file.

EOM end of message.

EOS See Earth Observing System.

EOT 1. end of transmission. 2. end of tape.

EOTC European Organization for Testing and Certification.

EPD See Early Packet Discard.

EPF See Electronic Payments Forum.

Ephemeris A tabular prediction of the position of a celestial body, or orbiting satellite. This may be calculated from Earth information, or be supplemented with information transmitted by the satellite itself, as in Global Positioning System (GPS).

EPLRS Enhanced Position Location Reporting Systems. A tactical radio system. See Enhanced Trival FTP.

epoch A moment in time, selected as a reference point. In computer timing and timestamping, the epoch is usually 00:00:00 Greenwich Mean Time of a specified date, with system time measured from this point on.

epoch, historic A moment in time marked by a significant event after which history takes a different course, or a period in time marked by significant events (or the lack of significant events).

epoch, geological In geology, a segment of time which is greater than an *age* and less than a *period*.

EPP See Enhanced Parallel Port.

EPROM See Erasable Programmable Read Only Memory.

EPS See Encapsulated PostScript.

EPSCS Enhanced Private Switched Communications Service. A commercial switch-renting service from AT&T serving businesses.

Equal Access A requirement resulting from the AT&T divestiture Modified Final Judgment (MFJ) in the 1980s, that holds that each Bell Operating Company must provide network services access to competitive companies equal in type and quality to that used by the Bell Operating Companies themselves. See Feature Groups.

Equal Charge Rule A stipulation of the 1980s AT&T divestiture Modified Final Judgment requiring Bell Operating Companies to charge rates such that they don't vary as the volume of traffic varies.

erasable programmable read-only memory EPROM. A read-only computer memory chip which can be electronically erased and reprogrammed. An EPROM is like an Etch-a-Sketch, it's a general purpose shape and size, and you can imprint information on it, but change it later if desired. The EPROM can be programmed for specific needs, which is especially handy for hardware design of peripheral devices, and doesn't need a constant current to keep the memory refreshed. In other words, it can keep the information if the system is turned off. EPROMs are handy for technology that is frequently upgraded. The information most likely to change can be put in an EPROM, and the chip can be swapped out for an updated EPROM, leaving the rest of the circuitry intact.

erasable storage Any storage medium on which the information can be readily removed after it has been recorded. Most erasable storage media are magnetic in nature, as magnetic particles are amenable to being rearranged to remove the data encoding, or to overwrite new information. Semipermanent magnetic storage media include hard disks, floppy diskettes, data cartridges, and audio/video tapes.

Many types of memory chips are also erasable, and are commonly used as temporary storage on computers. The most commonly used memory chips will lose the information when the power is discontinued.

A few optical storage media are also erasable, as some have been developed such that the layer in which the encoded 'pits' are stored can be subsequently altered, but this is a property not common to most optical media. See erasable, programmable read-only memory, superparamagnetic.

erase head A head specifically designed to remove the data from magnetic media. On some systems, erasing the information that exists on the tape or disk results in 'cleaner' subsequent recordings (this is particularly true of video tapes) than just recording over the information that is there. For this reason, an extra erase mechanism, such as a 'flying erase head' on a video camcorder, rearranges the magnetic particles before recording, particularly for insert dubs and other types of edits. Erase heads are not always included on less expensive consumer models.

erbium A rare earth element commonly used in the doping of fiber optic cable in order to manipulate its transmission propagation characteristics, especially to enable the cable to carry signals for longer distances.

erbium doped fiber amplifier EDFA. Erbium-doped amplifiers have become an important means of reducing signal loss on long fiber optic transmissions. These amplifiers came into use in the late 1980s and are now used to extend the range of fiber optic transmissions.

erbium doping A technique of using erbium, a rare earth, to impregnate another material in order to alter its transmission characteristics. Erbium doping is a technique used in the manufacture of fiber optic communications cables amplifiers to minimize signal loss over distance. See doping.

ergonomics The study and application of human engineering, that is the design of systems and products which adapt those systems and products to the needs and comforts of their users, rather than the other way around. Ergonomic applications require a knowledge of human anatomy, movement and orientation, as well as human perception, and human preferences—psychological and sociological. This information is then incorporated into design and manufacturing with a result that, more often than not, is an economic and social compromise. Nevertheless, ergonomic designs are to be encouraged. A number of interesting ergonomic adaptations can be seen in the design of chairs, computer keyboards, and phone head and hand sets.

Ericsson Telecommunications A major equipment supplier and research and development organization serving the Canadian communications market.

erlang A unit of measure of telephone traffic. This has been variously interpreted in the telephone industry as equal to a full traffic path, or equal to a specified number of calling seconds, or equal to a ratio of full traffic to no traffic. The term is based on mathematical analysis of the characteristics of telephone transmissions by the Danish telephone engineer A. K. Erlang. Erlang was analyzing traffic flow and congestion in the Copenhagen Telephone Company beginning in 1908, which led to changes in the design of telephone switches.

Erlang's theories have had practical applications in phone system design for many decades, but now have to be re-evaluated in light of changing characteristics of phone calls, since the rise of the Internet. Two-hour voice calls are rare, two-minute voice calls are common.

But when computer users log onto the Internet, two-hour connect times are common, as are four-hour connect times. The theories used to develop trunk use and capacity algorithms may have to be reapplied to the new types of usage patterns. See queuing theory.

Erlang, Agner Krarup (1878-1929) A Danish mathematician, educator, and telephone engineer who studied the mathematical characteristics of telephone transmissions in the early 1900s and described his findings in a number of publications, including "*The Theory of Probabilities and Telephone Conversations*" in 1909. He described how random calls follow a Poisson pattern of distribution. This not only led to some practical design changes in telephone switching systems, it also was the beginning of the study of queuing theory, an area of research that has many implications for current research and applications in data network traffic.

ERMA Electronic Recording Method/Machine, Accounting. A historic banking system, first demonstrated in 1955. By a year later, the system had been enhanced with solid-state components and released as ERMA Mark II. In 1959, General Electric began delivering the system, and one was installed in a Bank of America location in California, considered to be the world's first electronic banking system.

ERMES European Radio Messaging System. A European wireless mobile communications paging protocol specified by the European Telecommunications Standards Institute (ETSI) in 1986. ERMES operates at 169.6 to 169.8 MHz at 6.25 Mbps. See EMBARC, SkyTel.

ERP effective radiated power.

error control In computing, there are many schemes, philosophies, and protocols for safeguarding the integrity of data. Error control encompasses several aspects of data handling: error detection and error correction, if appropriate or possible, which may be part of more extensive data recovery, if appropriate or possible. One of the most rudimentary types of error control is detect-and-drop-if-bad. In other words, if cyclic redundancy checking (CRC), noise sensing, or some other error detection mechanism detects a problem, drop the transmission. While this sounds harsh, it actually was the predominant strategy for file transfer protocols for many years.

Error control is related to every aspect of computing, not just file transfers. It involves user interaction with applications programs, file loads and saves, data protection while files are open, and information protection for cached data. The most common implementations of error control, however, are in network transmissions, and dialup data transfers through modems.

The arsenal of error control mechanisms is growing, and more and more, error control schemes are a mix of software and hardware functions. Error control protocols now sometimes include sophisticated check, compare, and evaluate algorithms, some incorporating artificial intelligence concepts.

There are a number of error-correcting protocols now widely used in data modems, including MNP4, HST, and V.42 (which includes MNP4 and Link Access Procedure (LAP-M)). See checksum, cyclic redundancy checking, Microcom Networking Protocols (for a chart), XModem, YModem, ZModem.

error free seconds EFS. A unit of measure of the quality of a transmitted signal expressed as a percentage of bit errors over a specified period of time. EFS is defined in the ITU-T O Series Recommendations (0.151). See bit error rate.

error trapping In software programming, various means of catching, locating, or monitoring error conditions, or suspected error conditions. Error types vary, and may include syntax errors, mathematical errors, stack errors, I/O errors, collisions, incorrect memory management, and much more. A debugger is a software development and diagnostic tool which aids in setting error traps, and in detecting and diagnosing errors. It usually provides more control and feedback than trying to trap the error conditions individually in the code itself.

ERS European Remote Sensing Satellite.

ERTS-1 See Earth Resources Technology Satellite.

ESA 1. emergency stand-alone. 2. European Space Agency.

ESC, Esc Abbreviations for escape. Most computer keyboards include an ESC key in the top left corner. This is implemented into many applications programs as a 'friendly break', that is, a break or abort in the current function that doesn't interrupt the application itself. In many programs, the ESC will allow a long, slow process to be exited without the user having to wait for a dialog box, menu, or cursor to return to active status.

The ESC key sometimes is used to introduce a

sequence of control characters. A number of popular text editors use the ESC as a symbol that the next character is to be processed as a control code rather than as a text character insertion in the document. In some cases, when many control codes are available, the sequence may even require a double-ESC (ESC typed two times in a row) followed by a key. With the increasing prevalence of graphical interfaces, long escape or control code sequences are less common.

ESCA European Speech Communication Association.

ESD See electrostatic discharge.

ESF See Extended SuperFrame.

ESI 1. Enhanced Serial Interface. 2. End System Identifier. In ATM, an identifier which distinguishes multiple nodes at the same level, in case the lower level peer group is partitioned.

ESMR See Enhanced Specialized Mobile Radio.

ESMTP See Extended Simple Mail Transport Protocol.

ESN 1. See electronic serial number. 2. electronic switched network. 3. See emergency services number.

ESnet See Energy Sciences Network.

ESP 1. Encapsulating Security Payload. In secure network transmissions, the ESP can be used in conjunction with an authentication header mechanism. See link encryption. 2. See Enhanced Serial Port

ESPA European Selective Paging Association.

ESPAN Enhanced Switch Port Analyzer. An external network diagnostic and analysis tool which captures information that has been copied to a switched interface.

ESS See electronic switching system.

essential service A regulatory distinction that provides special access to some types of telecommunications equipment or services, and which provides more relaxed regulations in some aspects, and more stringent regulations in others. Essential services are classified differently in different nations, but tend to include some types of medical personnel, firefighting services, transportation administration, emergency broadcast channels or stations, etc.

ESTO European Science and Technology Observatory. The monitoring arm of the Institute for Prospective Technological Studies (PROMPT).

ETACS Extended TACS. The wireless transmission technology used in the United Kingdom and northern Europe, derived from U.S. AMPS systems.

ETB A terminating marker, indicating the end of a transmission block.

ETF European Teleconferencing Federation. An industry trade association for video- and teleconferencing. See audiographics, videoconferencing.

ETFTP Enhanced Trivial File Transfer Protocol. See Enhanced Trivial FTP.

ether, aether A ubiquitous, invisible medium throughout space and the Earth's atmosphere postulated by early philosophers and scientists to explain various concepts such as gravity, gravitation, transmission of light and radio waves, and much more. It was further conjectured that the ether was divided into three kinds: heat, light, and actinic rays. That is not to say such a thing actually exists, or that an 'ether' is necessary for these phenomena to occur.

Our current knowledge is that there are layered envelopes around our planet, and other planets, but there isn't any ether out in space, or an all-pervasive 'ether' that can be considered the basis for the behavior of electromagnetic waves, though neutrinos have interesting properties.

We now assume that particles, particle-like transmissions, and radio waves can beam through space without a transmissions medium. Nevertheless, when speculations about the ether were occurring around the time of the discovery of radio waves and transmissions which do require a medium, the ether concept was a useful stepping stone. It aided researchers in beginning to understand wave propagation and helped them explore and recognize the characteristics of the ionosphere. Many otherwise accurate, and well-respected texts up until the early 1900s used the ether as an accepted phenomenon to describe the properties and behavior of radiant energy. As late as 1994, there were still acknowledged scientists supporting the possibility of an etheric force, so the idea lives on, at least as an occasional interim model, and as speculation about a 'unifying theory' of the universe continues. In general, however, conventional science discounts the existence of an ether, and no one has found compelling evidence to support the theory.

Ethernet An important, widely implemented local area network (LAN) and metropolitan area network (MAN) network transmissions standard developed in 1973 by Dr. Robert M. Metcalfe and David Boggs and patented in 1975. Tat Lam designed the first transceivers for Ethernet, and Ron Crane provided hardware expertise for the eventual IEEE 802.3 standards. Crane and Metcalfe founded 3Com Corporation in 1979.

The early Ethernet ran at approximately 3 Mbps. Much of the early work was done by the Xerox research lab (PARC), and further development was undertaken by a multi-vendor consortium. Ethernet was formally specified as a production-quality standard called the DEC-Intel-Xerox (DIX) or "Blue Book standard," transmitting at speeds up to 10 Mbps. It was subsequently adopted for standardization for a wide variety of media by the Institute of Electrical and Electronics Engineers and designated IEEE 802.3 CSMA/CD, in 1985.

Each Ethernet interface card requires an Organizationally Unique Identifier (OUI) which is assigned as a three-octet number for the IEEE. The organization further subdivides this locally into unique six-octet numbers known as a Media Access Control (MAC) address or Ethernet address. The IEEE organization handles identifier allocation by online registration forms, or by phone at the IEEE Registration Authority.

Current Ethernet protocols can run over thick and thin coaxial cable, multimode fiber, and unshielded twisted-pair. Physical standards for running Ethernet include 10Base-5, 10Base-T, 10Base-2 and others. It's easier to understand these physical standard designations if the three component parts are analyzed as follows: the '10' in each of these indicates a signaling speed of 10 Mbps, while the 'base' stands for *baseband*, and the suffix describes the maximum run of an unrepeated cable segment (in hundreds of meters), if it is a number, or refers to fiber (F) or twisted-pair (T).

Ethernet is now a worldwide networking standard, having been adopted by the International Organization for Standardization (ISO) as ISO/IEC ANSI/IEEE Std. 802.3 in 1992.

Ethernet Address Resolution Protocol EARP. The addresses of hosts within a particular protocol may not be compatible with the corresponding Ethernet address. That is, the lengths or values may differ. EARP deals with an incompatibility by allowing dynamic distribution of the information needed to build tables to translate an address from the foreign protocol's address space into a 48-bit Ethernet address. This can also be generalized to non-10Mbps Ethernet systems such as packet radio networks. See RFC 826.

Ethernet Digital Subscriber Line EDSL. A Digital Subscriber Service which uses copper wires running between subscribers and the central office as a shared communications medium. Crosstalk is a limitation, but, like Ethernet, and unlike earlier xDSL technologies, EDSL has some ability to adapt to traffic interference. See Digital Subscriber Line. Contrast with Asymmetrical Digital Subscriber Line.

EtherTalk An IEEE 802.3 standard Ethernet protocol implemented for local area networks (LANs) on Macintosh and G3 computers by Apple Computer. See AppleTalk.

ETNO European Public Telecommunications Network Operations Association.

ETS European Telecommunication Standard.

ETSI See European Telecommunications Standards Institute.

Eudora Light, Eudora Pro A widely distributed commercial personal computer electronic messaging applications program compatible with the Macintosh operating system (OS) and Windows. Eudora Pro is a pay version, and Eudora Light is freely distributed by Qualcomm Enterprises.

Euro ISDN A version of Integrated Services Digital Network (ISDN) implemented for the European networking system, which differs in a number of respects from North American systems. While the signals are compatible (it is possible to place transatlantic ISDN calls), the equipment is not.

European Academic and Research Network EARN. A European networking system which joined with the BITNET system in 1982, making BITNET an international network.

European Community Telework Forum ECTF. An organization formed in 1992 to further and coordinate European developments in telework, and to provide an open forum for discussion of related issues. It is funded through a non-profit European Economic Interest Group. See ADVANCE Project, TelePrompt Project, telework.

European Space Agency The research, development, and administrative organization for space exploration for Europe. It is roughly equivalent to the U.S. National Aeronautics and Space Administration (NASA).

European Telecommunications Standards Institute ETSI. The standards body in Europe which corresponds to the American National Standards Institute (ANSI) in the US. ETSI's role is to promote and support communication, cooperation, and integration of technologies in the European community in support of the European Union, Europe's common market. It incorporates the cooperation of a broad range of network administrators, service providers, manufacturers, researchers, and users. ETSI was founded in 1988, and is currently located in France.

ETSI has a number of subgroups, including the Radio Equipment and Systems (RES) 10 group, which is responsible for high speed wireless data standards. This is in the process of being reorganized into another body. See CEPT. http://www.etsi.fr/

europium A soft, ductile, relatively expensive rare earth metal discovered in the late 1800s and separated in fairly pure form by Demarcay in 1901. Europium is used in conjunction with yttrium oxide to create the red phosphors in color cathode ray tubes (CRTs), and is used as a doping material for plastics used in lasers. See doping, erbium, gadolinium, yttrium.

EUROTELDEV European (Regional) Telecommunication Development

EUTELSAT *Eu*ropean *Tel*ecommunications *Sat*ellite organization. The largest satellite operator in Europe, founded in 1977 and formally established in 1985, with 11 satellites in orbit, and a further six under construction. The EUTELSAT orbital test satellite (OTS) project, designed as a direct broadcast satellite test system, preceded the EUTELSAT F*x* series Ku-band systems.

EUTELSAT provides hundreds of television and radio stations to subscribers in more than 40 European member countries equipped with DTH or cable television reception services. http://www.eutelsat.de/

EUV extreme ultraviolet.

eV electron volt.

event In computer processing, a signal or other indicator that a device or process requires attention, is relinquishing resources, or otherwise needs to communicate its activities to a central processor or other control unit.

event driven An event-driven hardware device, process, application, or network communication topology is one which proceeds when triggered by an external event, such as a token, trigger, tickler, interrupt, or alert. Many telecommunications devices, appliances, and input/output peripherals are event-driven, or event initiators.

EW Electronic Warfare. The use of electronics in guidance systems, control systems, simulation and modeling systems, and other activities related to conflict resolution and defense.

EWOS European Workshop for Open Systems. See Open Systems Interconnection.

EWP electronic white pages. An electronic database of personal, and sometimes business, phone and address listings. There are many EWP lookup services on the Web.

ExCa Exchangeable Card Architecture. An open socket architecture extension to PCMCIA 2.0 for use on Intel x86-based computers, introduced by Intel in the early 1990s. The software specification provides standardized socket, card, and client services. ExCA allows interfacing of PCMCIA devices with computers, particularly mobile computers, which are more likely to have PCMCIA slots. See Personal Computer Memory Card International Association.

exception 1. error or unusual occurrence, such as an abnormal signal, data falling outside a certain specified range, or deviation from normal program execution. Common exception conditions in programming include stack overflow and divide-by-zero errors. In software development, exception handlers can be included in the code to detect and manage error conditions and resume program execution. 2. In ATM, a connectivity advertisement in a PNNI complex node representation that represents something other than the default setting of the node representation.

excess burst size Be. In frame relay, a cell traffic descriptor for the maximum amount of uncommitted data in excess of Committed Burst Size that the network can attempt to deliver within a time interval (Committed Rate Time Interval). This type of data is eligible for discard, if necessary, to provide quality of service (QoS). Bursty traffic and congestion are two of the common traffic management challenges that are handled with a variety of procedures. See cell rate, committed burst size.

excess noise, current noise Undesirable noise that results from current passing through semiconductor components.

exchange A central location for making connections, directing traffic, and redirecting traffic. A public telephone switching office or regional system is often called a telephone exchange.

exchange access This is defined in the Telecommunications Act of 1996, and published by the Federal Communications Commission (FCC), as

"... the offering of access to telephone exchange services or facilities for the purpose of the origination or termination of telephone toll services."

See Federal Communications Commission, Telecommunications Act of 1996.

Exchange Access SMDS XA-SMDS, Exchange Access Switched Multimegabit Data Service. A connectionless, cell-switched, security-enabled data transport service for extending network features through standard interconnections with interexchange carriers (IXC). XA-SMDS is similar in structure to ATM, and is designed so that migration to ATM may be possible as ATM becomes more widely implemented. Multiple node local area networks (LANs) and wide area networks (WANs) can be interconnected without installing a dedicated path, at speeds ranging from 1.17 to 34 Mbps. XA-SMDS is a public level service, with a universal addressing plan, so various XA-SMDS networks can intercommunicate as desired.

exchange carrier EC. A telecommunications provider operating under specified territorial and operating parameters designated within the industry.

Exchange Carriers Association ECA. An organization established to support the interests and accounting administrative concerns of long distance telephone companies.

Exchange Carriers Standards Association ECSA. More familiarly known as the Alliance for Telecommunications Industry Solutions (ATIS) since 1993, the Washington, DC-based ECSA was established in 1983 to develop and promote standards related to the needs of various telecommunications carriers. The ECSA works in conjunction with a number of committees, including the Carrier Liaison Committee (CLC), the Information Industry Liaison Committee (IILC), and the Telecommunications Industry Forum (TCIF). See Alliance for Telecommunications Industry Solutions for more information.

exchange line The connection between a telephone subscriber and the local telephone switching exchange. See local loop.

Exchange Message Record EMR. An industry standard for the exchange of sample, study, and billable data messages among local exchange carriers (LECs).

Exchange Service ES. Basic subscriber phone service with a unique local telephone number, and access to the public switched telecommunications network. Includes residence and business services, and private branch trunk line services. Private lines, and Special Access services are not considered to be Exchange Services.

excitation The application of an external stimulus to a system resulting in a reaction or response. The application of a charge, potential, or electromagnetic influence.

excitation voltage The minimum or sufficient voltage required for a circuit to be functional.

exciton An excited state in a crystal substance with the characteristic of moving and recombining holes and electrons. See p-n junction, quantum.

execution In a software process, the carrying out of preprogrammed, realtime, or heuristic steps in order for the program to 'run' through its instructions or logical structure. This may or may not be an interactive process.

execution time A measure of the time in steps, minutes, or machine cycles that a process, or a particular computer instruction, takes to be carried out.

exosphere A region beyond the Earth's surface which is at the edge of the atmospheric 'envelope' surrounding the planet. See ionosphere.

Expanded Interconnection Service EIS. A collocation arrangement, in which the switch services for a private branch are located within the offices of the local carrier.

expanded memory EM. An addressable region of random access memory beyond what was previously a 640-Kbytes barrier on Intel-based microcomputers running MS-DOS. See Expanded Memory Specification.

Expanded Memory Specification EMS, LIM-EMS. A vendor-developed standard (Lotus, Intel, and Microsoft) for accessing memory in MS-DOS running on Intel-based microcomputers. Expanded memory is the random access storage beyond 640 Kbytes. The EMS describes a means to implement 'paging' through use of a portion of reserved memory between 640

Kbytes and 1 MByte to reference further memory storage areas beyond 1 MByte. Paging is a way of swapping blocks of memory in and out as needed.

expansion slots These are peripheral slots in an expansion bay or a computer intended for the placement of controllers, cards, and other device interfaces, usually comprised of printed circuit boards, which are used to extend a system. VESA, EISA, ISA, MCA and PCI are various common standards for the electrical and transmissions protocols used with slot peripherals for personal computers.

expert system An expert system is a type of information-handling approach which grew out of artificial intelligence research. Various types of expert systems exist for information creation, storage, and manipulation. An expert system is one which involves the manipulation and creation of information in a way that is rule-based and evaluative, rather than search-and-query. The traditional means of providing information to computer users is through a database, which usually involves storing and retrieving the data on a keyword basis, but an expert system can take in a richer mix of inputs, or nontraditional inputs, including natural language queries, visual queries, or other contextual input. An expert system also incorporates the combined knowledge of many 'experts' in that it is not just a collection of facts, but may further include data relationships, means of analyzing and evaluating the data, and other pertinent evaluative characteristics. Expert systems grew out of efforts to mimic the ease and naturalness of human communications through machine interfaces, in order to enhance the usefulness of computers.

Because expert systems often handle different types of data, or different types of input, they sometimes work with different programming languages than those commonly used for commercial applications. Cobol, Fortran, C, and BASIC are used for many programs used in business and educational settings. However, because expert systems often require a different programming approach, good text parsing languages like Perl, and good information parsing and rule-based languages like LISP and Prolog may be used.

Of the various types of products which have evolved from artificial intelligence research, expert systems are some of the most commercially successful results.

explicit forward congestion indicator EFCI. In ATM networking, a traffic flow control congestion, or impending congestion, indicator contained in the ATM cell header. The congestion signal is sent to the end destination to adjust accordingly. See cell rate, leaky bucket.

Explicit Rate ER. A network congestion feedback mode provided in available bit rate (ABR) service. Network rates which can be received are indicated within Resource Management cells. See cell rate.

E

Explorer See Microsoft Explorer.

Explorer I America's first successful satellite launched on 31 January 1958. Its mission was scientific, and it included instruments to measure radiation in space. At first, it was thought that the instruments might be defective, as the readings were much higher than expected, but the measurements were later verified.

Explorer 8 The first NASA satellite launched by the United States. The Explorer 8 was launched on 3 November 1960, to study the ionosphere.

export To save out information in a format which is not the native format of the application doing the saving. For example, a word processed document may be saved in ASCII to facilitate transfer over a 7-bit network. This is often done to create a version of a file which is compatible with other applications or transport mechanisms. Exporting is usually done through a conversion filter, and there may also be filters for importing.

exposure Contact with radiant energy, bacterial or viral toxins, or chemicals. Sun exposure can cause fading, burning, melting, or other chemical reactions. Exposure to radiation can cause burns, deep cellular damage, chemical changes, or death to biological organisms at high doses. Exposure is a concern in industrial environments for both equipment and humans. It is also a concern in medical environments, where exposure to viruses, bacteria, X-radiation, chemicals, and other contaminants or hazards may cause harm.

express circuit An interurban phone carrier circuit connected without multiplexing equipment.

extended ASCII A colloquial designation for a variety of non-compatible 8-bit character code designations in which the first 128 characters conform to the ASCII standard, but the subsequent 128 characters are variously assigned by different developers.

Extended-definition Television See Enhanced-definition TV.

Extended Digital Subscriber Line EDSL. A version of Digital Subscriber Line (DSL) services which supports 23 B channels and one 64-Kbps D channel transmitted over a single line. See Digital Subscriber Line, Primary Rate Interface.

extended graphics adapter EGA. A color graphics standard introduced by International Business Machines (IBM) in 1984, considered the successor to color graphics adapter (CGA). EGA was widely implemented by third party developers on Intel-based personal computers. EGA could display up to 640 x 350 in 16 colors. Actually, this is stretching it a bit, because in fact there were eight colors, plus eight half-intensity versions of those same colors, rather than 16 colors selected for their usefulness to a limited palette, so the second eight were derivative rather than individually selected colors, accounting for the lack of appeal and variety in the colors. Not long after, IBM introduced PGA, which had slightly better vertical resolution than EGA (640 x 400), but was otherwise not a significant evolution. See color graphics adapter.

extended graphics array XGA. A 1,024 x 768 color graphics format used in liquid crystal display (LCD) data projectors.

Extended Industry Standard Architecture EISA. A 188-pin bus interface specification to succeed Industry Standard Architecture (ISA), which in turn succeeded the IBM PC/AT bus specifications. EISA supports 32-bit memory addressing, and 16- or 32-bit data transfers. EISA was designed to support 32-bit Intel 80386 and 80486 processors. The specification works with various system resources, including input/output ports, memory, and DMA channels.

On EISA boards, configuration is done with EISA Configuration Utility (ECU) software, rather than through hardware, using a CFG file supplied with the board. EISA boards, while faster, are somewhat physically compatible with legacy boards, preserving the old AT pin specifications on the upper 98 pins. The rest are used for the EISA bus signals. The slot into which an EISA card is inserted is assigned a unique address so that the system can recognize and initialize the interface.

EISA is widely supported by many manufacturers, but is being gradually superseded by newer formats.

Extended Memory A designation, on 80286/80386/80486/Pentium-based personal computers, for expansion memory, beyond that which is directly addressed by the operating system. When running in a special mode called *protected mode* or *virtual real mode*, it is possible to use this memory in a limited way, for specific types of uses. A High Memory Area (HMA) designation, within Extended Memory, further subdivides the memory, and provides space in which to relocate program data and device drivers, according to the management of the Extended Memory Specification (XMS).

Extended Memory Specification EMS. A memory management specification designed to handle some of the eccentricities of memory addressing on various common 80286/80386/80486/Pentium-based personal computers. Extended memory is a means of providing memory above the 640K of base memory which can be accessed directly by the operating system (in this case, usually MS-DOS). The High Memory Area (HMA) portion of extended memory (up to 64 kilobytes) manages Extended Memory, and can be used to relocate device drivers and certain data buffers within several restrictions. Management of HMA is done through the EMS. See cache memory.

extended play A designation for a technology which plays beyond that which is generally expected to be the maximum limits. Historic phonograph cylinders played for two or three minutes, but some companies found a way to make them play for four minutes on standard equipment, thus creating extended play albums.

Extended SuperFrame ESF. A frame format for 1.544 Mbps communications (2.048 in Europe with 30 channels) evolved from DS-1 in 1962 and SuperFrame in 1969, and widely used in T1 systems. ESF provides improved error correction and can be serviced without taking down the entire system. Twenty-four frames are combined to create one Extended SuperFrame. Six frames are used for frame synchronization, six for error tracking, and twelve for Facility Data Link (FDL). Signaling is accomplished through robbed bits in frames six, 12, 18 and 24, except in transparent mode, in which the 24th channel is used in order to provide Clear Channel Signaling (CCS). Facility Link Data (FDL) is used to transmit to telephone monitoring stations.

extended text A rather subjective phrase for text display on computer terminals that is wider (provides more columns, and hence more characters) than the accepted average at the time. In other words, when computer terminals only displayed 40 or 64 columns, an 80-column screen was considered 'extended.' When display resolutions increased so that 80 was typical, extended text came to mean 132 columns. The same designation is often applied to output widths on printers, with 132 typically being considered extended.

extended time division multiple access E-TDMA. A type of digital transmission scheme favored by cellular providers over older analog-based systems. See time division multiple access.

Extensible Markup Language XML. XML is a markup meta language which allows more flexibility and complexity of presentation than HyperText Markup Language (HTML), and is not limited to Web publishing. Like HTML, it is based on the Standard Generalized Markup Language (SGML). Some have promoted it as the successor to HTML, but the industry has not yet formed a consensus on this possibility. See HTML, SGML, World Wide Web Consortium.

extension, file name A suffix, often preceded by a period as a subsection delimiter on systems that require it, and as a visual locator on systems that don't, which helps to indicate the type of file format it might be, such as *.txt, .bmp, .tiff, .ilbm, .frame, .wrd,* etc. File extensions need not be restricted to three characters except on some systems with older types of file structures. In the mid-1980s most types of computer platforms (Atari, Macintosh, Amiga, Sun, Apollo, NeXT, SGI, etc.) did away with the mandatory period and three-character limitation, as did Intel-based machines running OS/2 and Microsoft Windows NT. Surprisingly, many consumer Intel-based desktop computers running Microsoft Windows retained the limitation until 1996 (and later on, older machines still in use).

extension, phone An extra phone line that uses the same phone number as the originally installed phone with which it is associated.

Exterior Gateway Protocol An Internet protocol developed in the early 1980s for the exchange of routing information between autonomous systems. See RFC 827.

External Data Representation XDR. A standard representation for platform-independent data structures for remote procedure call systems, developed by Sun Microsystems. See RFC 1014.

external memory Any memory outside the direct access memory peripherals or chips in a system. Thus, information stored on punch cards or paper tape would be considered external memory, as would removable cartridges or tapes. Chip memory and internal hard drives would be considered internal memory.

extinction potential The lowest voltage level at which plat current in a plat will flow in a gas-filled electron tube.

EXTN Extension. An industry abbreviation designating the last four digits of a phone number. A 10-digit number is expressed with symbolic characters as: NPA-NXX-EXTN.

extraneous emission Any emission in addition to, or external to, the desired emission. Thus, emissions outside the case of a computer system or outside the sheath of an insulated wire are considered extraneous. Since these emissions can interfere with radio transmissions (radio broadcasts, intercoms, cordless phones, etc.), they are strictly regulated by the Federal Communications Commission (FCC), and electronic components must conform to stated emission requirements.

extremely high frequency EHF. The frequency spectrum designated as 30 GHz to 300 GHz, typically used for satellite communications. These very short wavelengths can be apprehended with small antenna assemblies. See band allocations for a chart.

extremely low frequency ELF. The frequency spectrum designated as 30 Hz to 300 Hz. Waves in this range are extremely long and not of much practical use for communications with our present technologies (the same was formerly said of microwave frequencies, and they are now widely used). See band allocations for chart.

extrinsic semiconductor A type of semiconductor which includes impurities which contribute to its electromagnetic properties, usually to enhance them. See doping.

extrusion 1. Forming by forcing through an opening, which may contribute to the shape of the extruded material. Extrusion is often accompanied by a heating or cooling process in order for the extruded material to retain the desired shape. 2. A means to produce or apply insulating materials to wire or cable by forcing plastic or other materials through an opening.

eye candy *slang* A visually appealing screen display designed for aesthetic enjoyment, or to provide a pleasant visual distraction while software is initializing, or performing background tasks. Commercial developers often include a title page, sometimes called a 'splash page,' as company image eye candy while a software application is launching, and many so-called screen savers are interesting eye candy.

eyelet A small, washer-like flat cylinder or short tube for threading or supporting various wires, cables, or other narrow parts.

EYP electronic yellow pages. An electronic database of business address listings. There are many EYP lookup services on the Web which allow the user to query for a business name, or business-related keyword in order to locate a company. The company does not have to be connected to the Web to be listed, as many electronic yellow pages services get their listings from local communities or printed directories. See yellow pages, Yellow Pages.

f 1. Abbreviation for farad. See farad. 2. Symbol for femto-. See femto-. 3. Symbol for focal length (usually italicized). See focal length. 4. Symbol for frequency. See frequency.

F 1. Abbreviation for Fahrenheit. See Fahrenheit. 2. Symbol for filament. 3. The symbol for 15 in the hexadecimal number system. The symbols used are 0 1 2 3 4 5 6 7 8 9 A B C D E F. Thus, 'A' in hexadecimal represents '10' in the familiar decimal system, and 'F3' represents '18' in decimal. Hexadecimal numerals are sometimes preceded with 'X' or '0X' to indicate that the subsequent digits are represented in the hexadecimal system. For example, '15' in decimal may be represented as '0X0F.' See the chart under ASCII for a list of decimal, hexadecimal, and octal equivalents up to 127 decimal.

F Block Federal Communications Commission (FCC) designation for a Personal Communications Services (PCS) license granted to a telephone company serving a Major Trading Area (MTA). This grants permission to operate at certain FCC-specified frequencies. See A Block for a chart of frequencies.

F connector A small coupling connector used at the end of coaxial cable which is common in video editing, broadcast components, and local area network (LAN) cables. See connector for a diagram.

F-display, F-scope A type of rectangular radar viewing screen. The blip which indicates the object being scanned is indicated in relation to its horizontal and vertical position on the screen.

F region A region of the Earth's ionosphere in which F1 and F2 regions tend to form. The F1 region is active in daytime. The F2 region is commonly used for the propagation of radio waves, due to its high ionization levels. See ionospheric sublayers for a chart.

***f*-stop** A specified setting for the opening of an aperture. The term is commonly used in conjunction with lenses to indicate the amount of light coming through a shutter. Lower numbered *f*-stops indicate larger openings, and hence more room for light to enter. Typical camera *f*-stop settings range from 4 to 22. The *f*-stop is often balanced with the aperture opening speed to further control the entry of light. See focal length.

F2F An abbreviation for "face to face" used in email and online public forums to indicate a personal encounter.

fab, fab plant A fabrication facility, a phrase often applied to a plant that produces computer chips.

face model A graphically modeled image of facial features, particularly eyes, lips, nose, and the general contours of the face. These images can be used in conjunction with computer applications for videophone, graphical answering services, simulations, police identification, artistic works, educational applications, and computer animations. Just as we now have synthesized speech for responding to user inquiries, we may someday have computer generated facial images responding to videophones and other interactive electronic devices, in effect, electronic receptionists.

faceplate A plate, usually of plastic or metal, that fits over the front surface of a console or other device, which may have openings to accommodate various knobs or dials. The faceplate may be engraved, painted, or otherwise labelled to indicate settings. See bezel.

facilities An installation designed for a particular purpose. The term typically encompasses related buildings, equipment, and operations, though some people use it loosely to include the personnel as well.

facilities-based carrier FBC. Phone carriers which use their own facilities and switching equipment to provide phone service, often long-distance service. Contrast this with those who lease or resell services from established carriers, although even facilities-based carriers enlist other carriers as needed.

facilities management The administration and day-to-day operating of a facility which may be carried out by members of the company which owns the facility or may be outsourced to specialists. Facilities management outsourcing falls somewhere between in-house operations, where equipment and personnel are company owned and managed, and service bureaus, where equipment and personnel are externally owned and managed.

facom A distance radio navigation system or measuring system. Facom is a means of analyzing local signals, and received signals using the low frequency band for distances up to several thousand miles to determine distances. See band allocations.

FACOM fully automatic computer. A series of commercial large-scale computing systems first introduced in the mid-1970s by Fujitsu Limited, at about the same time the first microcomputers were being developed for commercial distribution. The FACOM M-190 is significant for being based on the large-scale integration (LSI) semiconductor circuitry that was new at the time.

facsimile device, fax device A device for sending an image transmission through wireless or wireline phone transmissions. The word 'facsimile' implies an exact copy, though on some of the cheaper fax machines, that's wishful thinking.

The two most common ways to transmit and receive facsimiles (faxes) are through dedicated fax machines and through fax modems. Often a fax generated on one type of system will be received on the other type.

The most common form of fax machine is a dedicated system which resembles a small printer that connects to a phone line. It sends faxes by scanning a piece of paper which is fed through the machine. Received faxes are printed in much the same way as they would be on a computer printer. Some faxes use continuous feed thermal paper, though more commonly now fax machines can work with sheet-fed plain paper. Many faxes optionally function as photocopiers and printers. See facsimile modem.

facsimile formats In order for facsimile (fax) machines to interchange data, a number of international standards have been defined. In addition to this, dialup modem transmission protocols with compression and error correction for fax/modems, called the V Series Recommendations, are used for data transfer in conjunction with fax formats. See facsimile mode.

Basic Facsimile Formats	
Class 1	EIA/TIA (EIA-578) standard for basic computer fax/modem interface.
Class 2	EIA/TIA standard for extended computer fax/modem interface which includes AT commands.
Group 1	Single page transmission in six minutes. Common in the 1970s.
Group 2	Single page transmission in three minutes. Common in the late 1970s.
Group 3	(There is also a Group 3 bis, or Group 3 enhanced format.) 14,400 bps facsimile protocol. Two resolution modes include 103 *x* 98 dpi, and 203 *x* 196 dpi (fine). Compression is supported. This is the most common protocol used with fax machines and fax modems, either Class 1 or Class 2. See V Series Recommendations V.27. Single page transmission in under 30 seconds.
Group 4	ISDN B facsimile protocol adopted in 1987.

facsimile history The technology to send images over distance has been around since the early days of telegraphs and radios. Alexander Bain, in 1842, proposed a telegraph facsimile based on the earlier photoelectric discoveries of Edmond Becquerel, a French physicist.

In Great Britain, F. Bakewell patented a chemical telegraph in 1847. It used a system of synchronized rotating cylinders instead of the scanning pendulums incorporated into Alexander Bain's machine. It was further improved in 1861 by Italian inventor Abbe Caselli, who wrapped tin foil around the cylinders. These various inventions lead to the development of scanners, fax machines, and television.

Surprisingly, considering some of the reasonably good success of early fax systems, faxing did not become widespread until the late 1980s, 140 years after the initial experiments. See facsimile history inset.

facsimile mode Facsimile machines have a number of operating modes, including various regular and fine resolutions, and can be manipulated to send in monochrome or grayscale, depending on the capabilities of the sender and receiver, and the software or hardware. Most fax machines and fax/modems send in Group III standard and fine modes, that is 203 *x* 98 pixels and 203 *x* 196 pixels. Other modes have been defined (Group III 203 *x* 391 - superfine; Group IV 400 *x* 400 - standard), but they are not widely supported in consumer-priced products. A fax can only be as good as the weakest link in the transmission. If the sending fax sends in the lowest resolution, a higher resolution receiving fax doesn't improve the image. Conversely, if the sending fax uses fine resolution, but the receiving fax can only support standard resolution, the details will be still lost. Since the orientation of most faxes is *portrait*, and the orientation of most computer monitors is *landscape*, fax/modem software usually has zoom, pan, and rotate features to aid in viewing documents.

facsimile modem, fax modem A fax modem system consists of a fax-enabled data modem (one which works in two modes) hooked to a computer, sometimes combined with a scanner. Instead of creating a document, printing it, feeding it through the fax machine, receiving a printed page at another destination fax machine, and then perhaps even typing or rescanning (and OCR-ing) the printout back into a computer at the other end, the fax modem system sends a digital fax directly from the software application that created the document, to the receiving end. Or, if a scanner is used, the system sends the scanned file as the fax. If the receiving device is a fax modem system, rather than a fax machine, then the fax goes directly to the computer hard drive, and no paper is used in the transaction.

Optionally, a fax machine may connect to a local network so that users can be notified if a fax has been received, or even select an option to view the fax on a computer monitor (thus providing hybrid fax machine/fax modem capabilities).

In business environments some people erroneously use fax machines when they should be doing direct data transfers. This is a common scenario: the main office of a corporation creates a new 80-page policy manual and wants to distribute it to all ten branch offices. The branch offices would like an electronic copy in order to customize it for their needs, or to easily make corrections as directed by the main office, etc. The typist types a copy, faxes 80 pages to each of the ten branch offices. 880 pages are generated in all, the original, and the 10 branch copies. Now the typists at each branch office retype the document into their word processors, thus duplicating the work already done.

Rather than always using a fax machine, there are better ways to distribute some types of documents. The first is a slight improvement. By using a fax/modem software program to send the document directly to each branch to another fax/modem program, no paper is printed until the documents are complete to each branch's satisfaction, and the completed customized documents can be OCR-scanned back into a word processing text file at the destination.

A better solution is to send the original file, in document format, through a modem or through the Internet, to each branch office, where the secretaries can load the received file directly into the word processor. This can be accomplished by putting the file on an FTP site, and notifying the branches that they can access the site and download the latest version of the file. If different word processors are being used, the original can be saved in Microsoft Interchange format (also known as RTF or Rich Text Format), a widely supported format that can be read and saved by all major word processing programs.

The best solution to document exchange may be to have a secure centralized online document repository which can be accessed and modified dynamically by all branches through an Internet or private network connection. Fax machines are a great resource for sending short documents, but they are not the best solutions for all types of document transfers, and the Internet is providing distant branches a means to dynamically interact in the production and maintenance of documents without incurring long-distance charges. See facsimile device, facsimile format, facsimile history, facsimile modes.

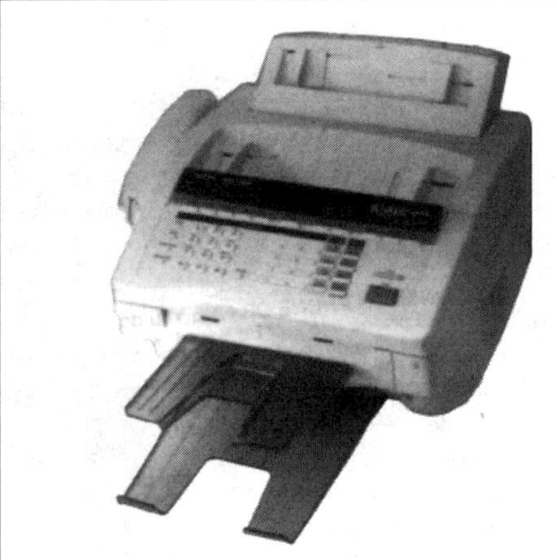

The trend is for facsimile capabilities to be combined with other similar functions, as in this Brother MFC 600 dpi printer/scanner/ facsimile/phone combination. A serial port allows it to be used as a peripheral for a desktop computer, or integrated into a computer network as a shared resource.

facsimile switch An external switching device which allows a single phone line to be used for more than one phone-related piece of equipment. Fax switches often can also handle telephone answering machines and computer modems. The fax switch is attached between the phone line plug and the various phone devices. When a call comes through, the device evaluates the tones and decides whether it's a voice call, a modem call, or a fax call, and routes the call to the appropriate device. Unfortunately, a fax switch can't automatically detect when a manual fax machine is going to send a fax. If the person dials the phone as a voice call, and then wants to switch over to a fax call when the connection is established, it won't work on most fax switches, as it has already detected the call as a voice call and can't switch back. In spite of that limitation, it's a great tool for homes, home offices, and small businesses that can't afford extra phone lines.

fade To diminish in strength, loudness, or visibility. In video or audio editing, fading is deliberately used to provide transitions that are perceptually pleasing. In data or broadcast transmissions, fade is usually an undesirable effect due to various factors such as distance, loss of signal, obstructions, interference, etc. Undesired fade can sometimes be reduced or eliminated by amplifiers, repeaters, robust wiring mediums, and good insulation.

fade margin Signal losses in satellite systems can occur from scattering, absorption, and various subtle types of interference. Consequently various fade margins are incorporated into the design of the systems, and they will vary depending on the degree of fade expected from various sources and on the length of the broadcast waves, with shorter waves generally being more subject to fade.

Fahnestock connector An innovative electrical binding post patented in 1915, and produced in the early 1920s. It was easy to mount and made a good electrical contact.

Fahrenheit scale A temperature scale which designates 32 degrees for the freezing point of water at normal pressure and 212 degrees for the boiling point of water at normal pressure, and other points relative to these. See centigrade scale, Celsius scale.

Fahrenheit, Daniel Gabriel (1686-1736) A Polish-born German scientist who established the widely-used Fahrenheit scale. Zero degrees was designated as the temperature of a mix of ice, water and salt, and ninety degrees was considered to be the temperature of the human body (in fact it's closer to 98.6°). See Rümer, Ole Christensen.

failsafe A designation that indicates that failure of a system is unlikely or impossible, or that backups are available if needed. In networking, few, if any, systems are completely failsafe, but there are steps that can be taken to prevent problems, such as the use of surge suppressors, backup power systems, redundant data storage or broadcast signals, etc. See fault tolerant, redundant array of inexpensive disks.

fair condition A product which has been used, may not be in perfect working order, and may have cosmetic abrasions, dents, and scratches. It may have had a checkup and some basic servicing. It is generally considered only one step above 'as is' and usually sold without warrantees of any kind.

Fair Market Value FMV. A proposed or theoretical designation of the value of an item if it were sold on the open market in a fair system. Fair Market Value is used to evaluate items for sale, barter, lease, tax, property distribution (inheritance, divorce), or bankruptcy filings.

fake code See pseudocode.

fallback A contingency mode, plan, or operation. In communications, a designation for

Historic Facsimile Technology

The idea of sending facsimiles over distances has been around almost since the dawn of telegraph technology. There were many names for early facsimiles, the facsimile telegraph shown here, called an *electrogram*, was developed in 1901 by Herbert R. Palmer and Thomas Mills. It was one of the first to print on plain paper, that is, it didn't require thermal or other chemically treated paper.

The example shows a picture of President McKinley next to the cylinder on which the image was inscribed by a pen mounted on a carriage. As the cylinder rotated, the electrical transmission was alternately connected and broken to create the marks on the page. A motor and drive belt, not unlike early sewing machine mechanisms, served to rotate the cylinder as needed, as shown in the illustration.

When the transmission was complete, the paper was removed from the cylinder, spread out to reveal the image, and a new sheet of paper wrapped in its place.

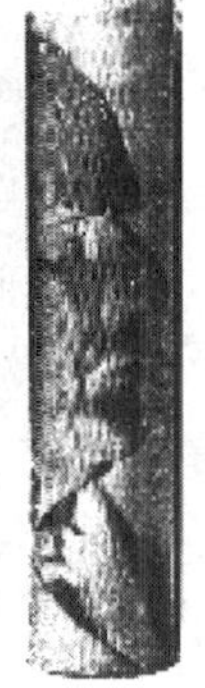

Early facsimile electrograph of President McKinley shown stretched out on the left, and wrapped around the receiving cylinder on the right.

1902

Images: Scientific American, Nov. 15, 1902.

another speed or mode of operations if the current mode is not functioning as well as might be desired. Many modems may fall back (two words) to a slower speed if the connect negotiation doesn't work at higher speeds. Many communications programs may fall back to smaller packet sizes if there is a lot of noise, or other impingements on a data file transfer. In software, a fallback (one word) may be one in which the application or operating system goes to another mode or another program, if some error condition or slow-down is detected. A network may go to a fallback route if the usual one is not available, or not responding as expected.

falsing Spurious signals which accidentally are interpreted by a system as commands, or which are deliberately introduced to fool a system, usually for unauthorized purposes. In telephone systems, certain situations can be simulated by playing particular tone sequences, so the system is fooled into switching, transferring, or connecting long-distance or other types of calls. In transmissions control for satellites and other radio-controlled devices, environmental noise, falsely interpreted signals, etc. can have major consequences if the system thinks it's a command and acts on it.

FAM fast access memory.

fan 1. Fan of science fiction. Since there are a large number of software developers who are science fiction fans, they have co-opted this term into many computer-related situations, video games, simulations, and virtual reality environments. 2. An active cooling device (as opposed to passive devices such as heat sinks) often used to cool computers so that chips and other components are kept at optimum operating temperatures.

fan dipole antenna A type of antenna used for ultra high frequency (UHF) television broadcast signals which is in some ways similar to a basic dipole rod antenna. It differs from the dipole in that it uses flat triangular sheets of metal, rather than rods, in a symmetrical triangular arrangement resembling a butterfly or bowtie. The fan dipole has a broad range, covering the entire UHF spectrum. See antenna, UHF antennas.

fanfold See z-fold.

FANP See Flow Attribute Notification Protocol.

FAQ See Frequently Asked Question.

far end crosstalk FEXT. When wires are packed together tightly, and signals are travelling through most or all the wires, the potential for interference from crosstalk increases. Far end crosstalk is a type of interference originating from multiple signals travelling in the same direction, typically through wire pairs, as in common copper twisted pair installations. FEXT directly effects bit error rates (BERs), as it cannot be cancelled as easily or as effectively as near end crosstalk (NEXT). See near end crosstalk.

farad A unit of capacitance which is equal to one coulomb (of electricity) divided by (a potential of) one volt. Named after Michael Faraday.

faradaic Relating to an asymmetric alternating current (AC) produced by an induction coil.

faraday A measure of electrical charge transferred in the process of electrolysis per weight of an ion, or element, that is equal to about 96,500 international coulombs (or 96,490 absolute coulombs). Named after Michael Faraday.

Faraday cage A structure, usually mesh- and cage-like, to isolate a person, device, or electronic system from damage or interference from outside electrical sources. These may sometimes be seen in science museums where electrical devices, especially large Van de Graaff generators, are demonstrated. Named after Michael Faraday.

Faraday Dark Space In a cathode discharge tube, a region between the positive column and the negative glow which appears dark. Regions in the tube become easier to distinguish if the pressure is lowered in a tube that has some air in it (normally air is removed to extend tube life and effectiveness). Then it becomes possible to distinguish the Faraday Dark Space as a region just outside a pale negative discharge glow, which in turn terminates in Crookes Dark Space, which borders the outside glow of the cathode. Named after Michael Faraday. See Crookes Dark Space.

Faraday dynamo A historic electrical generator developed in 1832 by Michael Faraday.

Faraday effect A plane of polarization of light in a magnetic field, travelling parallel to the lines of magnetic force, can be rotated to another plane by a transparent isotropic medium. In satellite communications, the plane of polarization of radio waves travelling through the ionosphere rotates about the direction of propagation, particularly at lower frequencies. Named after Michael Faraday.

Faraday, Michael (1791-1867) An English physicist and chemist, who was apprenticed to a bookbinder at the age of 13. He took time to read the books, and to listen to local lectures by Humphry Davy, becoming his laboratory assistant in 1813. Faraday went on to conduct extensive experiments in electricity and magnetism. He passed electrical currents through solutions, and observed their effects, adding new knowledge to the discoveries of A. Volta. Faraday demonstrated that the amount of an element deposited at an electrode is proportional to the current flowing through the solution. In 1831, he demonstrated that an electrical current can induce a current in a different circuit and made a historic entry in his journal linking electricity and magnetism. The following year he constructed a basic generator, calling it a *dynamo*.

Faraday also studied the properties of metals and glass, and developed new types of optics. He coined the terms *electrolyte*, *electrode*, and *ion*. Further important investigations of inductance in electrical circuits by other scientists grew out of Faraday's work. Many electrical effects have been named after him. See Davy, Humphry.

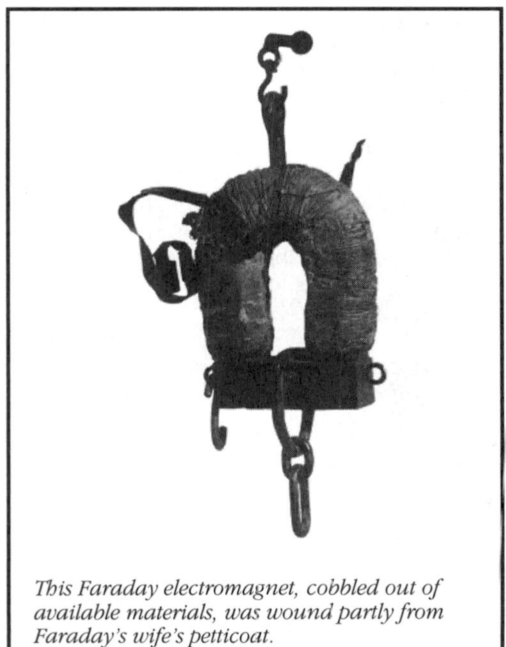

This Faraday electromagnet, cobbled out of available materials, was wound partly from Faraday's wife's petticoat.

Faraday's laws Michael Faraday investigated the phenomena related to decomposition by galvanic current and made some important discoveries that have been investigated and variously stated by succeeding scientists. Generally, 1. in electrolytic decomposition, the number of ions charged or discharged at an electrode is proportional to the current passed; 2. the amounts of different substances deposited or dissolved by the same quantity of electricity are proportional to their equivalent weights; 3. when passing a constant quantity of electricity through different electrolytes, the masses of the ions set free at the electrodes are directly proportional to the atomic weights of the ions divided by their valence. Faraday called his discovery the "law of definite electrolytic action." It was opposed by Berzelius and those who adhered to Volta's theory of galvanism. Through subsequent experiments, Faraday's concepts have been refined and confirmed, and his discoveries are known as Faraday's laws.

Farber, David Originally a computer consultant to the Rand Corporation in the late 1970s, Farber later became a co-founder of CSnet (Computer Science Network). He is known for his online discussion list "Interesting People."

FARNet See Federation of American Research Networks.

Farnsworth, Philo T. A precocious American inventor who described a television system to friends, and is reported to have shown a drawing of the idea to a teacher in 1922 when he was only 15 years old. He kept working on the idea and five years, later transmitted his first TV image. See television history.

fascicle In printing, one segment or division of a book that has been published in parts. See imposition, signature.

fast busy A telephone busy signal that is distinctive in that it repeats at twice the rate of a regular busy. A regular busy signal indicates the caller's phone is unavailable (it's off-hook or in use), whereas a fast busy indicates that all trunk lines are busy, and the call cannot currently be routed to the destination.

Fast Ethernet A version of Ethernet enhanced to increase its 10 Mbps capacity up to 100 Mbps over copper or fiber, which brings it into the high speed networking range along with asynchronous transfer mode (ATM) and FDDI. This enhanced capability requires the upgrade of other devices such as hubs and network cards, partly because Ethernet hubs can be cascaded, whereas Fast Ethernet hubs are stacked. Fast Ethernet is an international open IEEE standard (802.3u - 1995) which is used in medium-scale networks such as campus backbones. See Gigabit Ethernet

Fast Ethernet Alliance A trade association

of vendors established to develop and promote Fast Ethernet technologies for existing voice-grade traditional copper twisted pair lines.

fast SCSI A means of configuring SCSI to provide faster transmission speeds.

FastIP, Fast Internet Protocol A 3Com commercial in which just the first datagrams of the IP traffic are passed through the router, and, if a direct path is found, subsequent ones may bypass the router using Next Hop Resolution Protocol (NHRP). It is embedded in local area network (LAN) adaptors and implemented in LAN switches rather than in Internet Protocol (IP) routers. See IP switching.

FAT 1. File Allocation Table. See FAT format. 2. final acceptance testing.

FAT format File Allocation Table. The FAT, a file management scheme which is an integral part of MS-DOS, was developed and coded in the 1970s by Marc McDonald, in consultation with Bill Gates. In its original form, it consisted of a linked list consolidated in one location, with chained references to the storage locations of files on the disk.

The FAT is a map to the information that is saved on various locations on a disk. In fact, any individual file, especially a large one, may be broken up in portions, and the file structure has to provide a means to find all parts of a file. In addition to this, files on read/write data are constantly being written and erased, in no predictable order; the file structure helps the operating system keep track of which portions of the disk are available, and which are in use. The FAT serves as a liaison between the disk space and the operating system, indicating which clusters are allocated to stored files, and in what order.

The FAT is located on the disk after the boot sector, and space is reserved for the information which will be written there, as files are added to the disk. When a file is saved, a directory entry is made under an assigned FAT entry number. If the file takes up more than one cluster, it is said to be 'chained'; the first FAT entry indicates the entry number of the second cluster, or a last-cluster number if there is only one. Thus, a six-cluster file will have six FAT entries and 5 links. In order to provide extra protection, in case of damage to the information in the FAT, most disks store two FATs, one after the other.

Fat MAC A 512 kilobyte version of Apple Computing's Macintosh personal computer line, an enhancement to the original 256K Mac, released in early 1985.

FAU See fixed access unit.

fault isolation In circuitry, or software development, a troubleshooting strategy for isolating the location of a problem. In circuitry, this may involve shutting down parts of the system, wiring in shunts or bridges, or selectively stimulating particular areas. In software, it may involve setting breakpoints, printing out debug messages, or tracing particular variables. See bridge, shunt, trace.

fault tolerant A fault tolerant system is one which is designed so that if a problem occurs, the entire system or important parts of the system will continue to function until the problem is corrected. Thus, system redundancy, backups, secondary routines or hardware paths, etc. can be incorporated to increase fault tolerance. Good computer operating systems are designed so that individual applications don't crash the system. The application itself may crash, or need to be 'killed' (by killing the individual processes associated with the program), but the system can handle the crash without affecting other programs, the general operations, and will clean up stray files, memory, etc. See failsafe.

fax *colloq.* facsimile. See facsimile machine.

fax mode See facsimile mode.

fax on demand, faxback A combined facsimile and phone, or network application which allows a user to query a document archive and request, by means of software or phone menu tones, desired documents. The fax on demand system stores the requests, and prompts the user to leave the number of his or her facsimile machine. The fax on demand server then automatically dials the user's facsimile machine and dispatches the requested faxes. There may or may not be a charge for the service. Product information and technical support documentation are often dispensed this way.

faxback See fax on demand.

FB Framing bit.

FBT Fused Biconic Tape.

FBus Frame Transport Bus.

FC 1. feedback control. 2. frame control.

FCA See Fibre Channel Association.

FCC See Federal Communications Commission.

FCLC See Fibre Channel Loop Community.

FCS 1. Federation of Communications Services. 2. See Frame Check Sequence. 3. Fraud Control System.

FCSI Fibre Channel Systems Initiative.

FDD floppy disk drive.

FDDI See Fiber Distributed Data Interface.

FDM See frequency division multiplexing.

FDMA See frequency division multiple access.

feather In computer imagery, a very handy software tool that smooths out a specified edge to form a transition between it and the area surrounding it, a form of antialiasing. This works well for objects that have been pasted into a new background, to make a natural-looking transition to that background. In some programs, feather is the same as, or similar to, *smooth* or *defringe*. See antialias.

feature code A number or character sequence used to activate a feature on a phone system, such as speed dialing, last number redial, etc. These are more common on multiline business phones than on residential phones.

feature connector A means of connecting a peripheral card or device to another peripheral card, such as a video graphics adapter, so the second card can perform direct memory access (DMA) through the card's bus, without having to load the system bus. The feature connector is commonly used on VESA-compatible systems.

feature creep, creeping featurism A phrase to describe the sometimes insidious addition of numerous features to a system *after* it is found to be functional and to have most of the capabilities most users desire. In other words, cool bells and whistles that might not really be necessary, and might even interfere with the quick and easy use of a product. This happens in many aspects of programming and technology.

Feature Groups Types of long-distance carrier switching arrangements that are part of the Bell Operating Companies (BOC) system.

Feature Group Switching Arrangements	
Feature Group A	A subscriber line connection rather than a trunk connection to a local exchange carrier's network.
Feature Group B	A trunk connection which uses an authorization code for billing. Used in areas where it is not practical to offer Feature Group D (Equal Access services), such as some older switching systems, and independent services.
Feature Group C	The older long-distance services offered by local exchange carriers to AT&T before divestiture. Mutually exclusive with Feature Group D.
Feature Group D	Equal Access services, facilities and signaling specifications, established since divestiture, and implemented in the mid-1980s. Mutually exclusive with Feature Group C.

feature phone A phrase for phones that have extra features, sometimes the features improve functionality (redial, speakerphone, etc.), but sometimes they are just there to entice the consumer and may not be especially useful.

FEC See forward error correction.

FECN See Forward Explicit Congestion Notification.

Federal Communications Commission FCC. A U.S. federal regulatory organization

EIA/TIA cable transmission categories

Categ.	Transmission Speed Rating	Typical installations
Cat 1	No performance criteria specified.	
Cat 2	Rated to 1 MHz	Telephone wiring installations
Cat 3	Rated to 16 MHz	LAN, Ethernet, 10Base-T, UTP Token-Ring
Cat 4	Rated to 20 MHz	LAN, Token-Ring, 10Base-T, IEEE 802.5
Cat 5	Rated to 100 MHz	WAN, Fast Ethernet, 100Base-T, 10Base-T

which was created through the Communications Act of 1934, subsequent to the formation of the Federal Radio Commission (FRC) in the Radio Act of 1927. Its original mandate was to regulate the broadcast industry in the United States by the granting and administration of radio licenses. As such, the FCC administrated the allotment of frequencies, time slots, and callsigns. Since then, its jurisdiction has increased. The Commission is directly responsible to the U.S. Congress.

The FCC has a powerful role to play in the fair and equitable enactment and distribution of telecommunications resources in accordance with the Telecommunications Act of 1996. It is the responsibility of the FCC to see that the Act meets its goals of opening up the telecommunications business to anyone, and of promoting fair competition in the industry.

The FCC now also oversees product emissions, ensuring that computing devices do not emit harmful radiation or unharmful radiation at levels which may nevertheless interfere with other radiant technologies such as radio waves.

The FCC overall organization consists of a number of commissioners, about nine offices (public affairs, plans and policy, general counsel, etc.) and six bureaus. See Primary Divisions chart. See Communications Act of 1934. http://www.fcc.gov/

Federal Communications Commission classes A series of designations or ratings applied by the FCC to electronics devices. These are primarily intended to help prevent interference from devices like computers which may affect electromagnetic broadcast waves such as radio and television signals. If you have tried to use a cordless phone near a computer, and experienced interference, you are familiar with the type of problem excess emissions can create.

Category	Notes
Class A	Computing devices rated for office use which may not be used in the home.
Class B	Computing devices rated for home use.

Federal Networking Council FNC. The FNC reports to the Federal Coordinating Committee on Science Engineering and Technology and was chartered by the National Science and Technology Council's Committee on Computing, Information and Communications (CCIC). It provides a forum for networking collaboration among U.S. federal agencies with regard

Primary Divisions of the Federal Communications Commission

FCC Bureau	Responsibilities
Common Carrier (CCB)	Enforcement, pricing, accounting, program planning, network services, and wireline services.
Wireless Telecommunications (WTB)	Domestic wireless communications, including paging, cell phone, PCS, and radio, excepting satellite communications. This bureau is further subdivided into Commercial Radio, Enforcement, Policy, Private Radio, Licensing, Customer Services, and Auctions divisions.
Mass Media (MMB)	Audio service, enforcement, policy and rules, video services, administration, and inspections.
Compliance & Information (CIB)	A national call center, and information resources, management, compliance, technology, and regional offices.
International	International planning and negotiations, satellite and radio communications, and general administration.
Cable Services (CSB)	Consumer protection and competition, engineering and technical services, policy and rules, public outreach, management.

to education, research, intercommunications, and network operations. Since 1997, the various activities of the FNC have been carried out through the Large Scale Networking (LSN) group. http://www.fnc.gov/

Federal Telecommunications Standards Committee FTSC. A U.S. government agency which promotes the standardization of communications interfaces, including computer networks.

Federal Telecommunications System FTS. The intercommunications network used primarily by U.S. government civilian agencies. It includes interconnections to other agencies and to the public switched telephone network (PSTN).

Federation of American Research Networks FARNet. An organization comprised of commercial providers, some of the telephone providers, and mid-level NSFNet networks that meet to discuss commercial and other issues related to these businesses and the Internet.

feed holes A series of holes punched in a medium to provide an alignment guide, rather than to convey information. Used in various applications, including industrial production lines, computer programming paper tapes (for alignment, other holes convey information), and continuous tractor feed printing paper.

feed horn, feedhorn A basic signal-capturing component in satellite receiving antennas that is mounted at the focal point. It must either be rotated to correspond to the polarity of the incoming signal (horizontal or vertical), or be attached to a dual coupler. The focal length of the feed horn is dependent on the depth and diameter of the parabolic dish in which it is mounted. The feed horn is attached to a signal amplifier. See antenna, low noise amplifier, microwave antenna, parabolic antenna.

feedback *n.* 1. Returned information about something which has been received or perceived. 2. An opinion offered in response to some preceding event or information. 3. Returned information about data that has been received or passed through. In networks, there are many feedback mechanisms providing information data rates, congestion, traffic in the opposite direction, and the progress or success of a transmission.

feedback signal 1. A signal which loops back around to its source. An undesirable audio or visual artifact can occur when the same signal that is being transmitted, travels back through the original transmissions media. In sound systems, this commonly manifests as a piercing, shrieking sound, as when a microphone is located too near a speaker carrying signals from that microphone. If carefully controlled, audio feedback can sometimes be used to boost a weak signal. In visual systems, feedback often manifests as ghost images, or wiggly distortions. 2. An intentional diagnostic looped back signal. In diagnostic systems, when a signal is transmitted, and then compared with a reference when it returns (the returning signal is the feedback signal), it is possible to evaluate the correspondence, or differences between the two signals, or the information carried on those signals.

feeder cable 1. A primary cable, extending from a service provider, or central switching location, to a distribution panel or end-user. In large installations, there may be a main feeder cable, and branch feeder cables. 2. The cable that connects a primary distribution frame with intermediate distribution frames. 3. A main network backbone cable, which may have branch feeder cables leading to the main host computers. 4. A heavy duty, primary, or high bandwidth wire or cable intended to carry the main part of traffic from the transmission source to its primary dropoff points or hosts. Thus, fiber optic cables and 25-pair cables are common feeder cables.

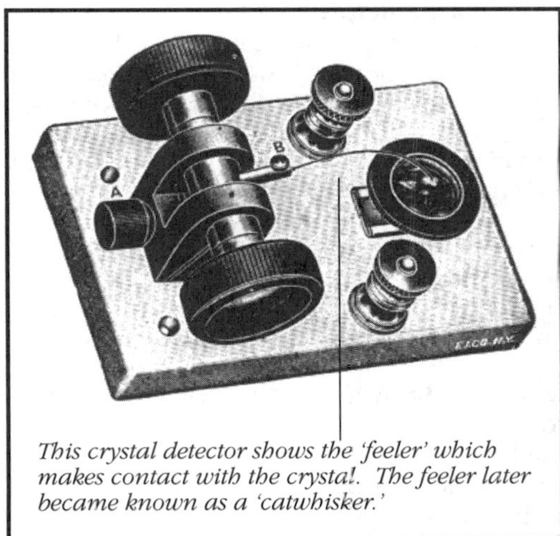

This crystal detector shows the 'feeler' which makes contact with the crystal. The feeler later became known as a 'catwhisker.'

feeler In old crystal detecting radio sets, a long, fine conducting filament that was placed in contact with a natural or synthetic oscillating crystal to 'detect' radio waves. It was discovered by researchers like H. Hertz that the length of the detector was an important factor in its ability to respond to waves produced by a resonator, thus leading to the discovery that radiant waves were of specific lengths. In commercial detector sets, the feeler came to be known as a catwhisker, which is poetically descriptive of its size and shape. See catwhisker.

FEFO first ended, first out. A priority queuing arrangement in which the first item processed, or the first process completed, is the first to be passed on, or further processed. Thus, tasks that are finished are taken out of the queue in order to leave space or processing time for others. See FIFO, FILO, LIFO, LILO.

femto- (Symbol - f) An SI unit prefix for 10^{-15}, a very, very small amount. In decimal, femto- is expressed as 0.000 000 000 000 001. See atto-.

FEP See Front End Processor.

FER Frame Error Rate.

Fermat, Pierre de (1601-1665) A French lawyer, linguist, and mathematician who made many contributions to our understanding of mathematics and optics, in spite of his recreational approach to mathematics, which meant that many of his discoveries initially went unpublished. Fermat's principle is named after him.

Fermat's principle When electromagnetic radiation travels by reflection off a surface from one point to another, it will take the path that can be traversed in the least amount of time.

Fermi level A value designated for electron energy at half the Fermi distribution function.

Fermi, Enrico An Italian physicist who investigated atomic physics by systematically irradiating the elements, work derived in part from the investigations of James Chadwick.

ferric oxide A metallic compound which is commonly used to coat thin tapes or platters used in magnetic storage media. The ferric oxide molecules can be selectively rearranged by magnetic impulses in order to encode the desired information on the medium. There are other types of coatings available for applications such as sound or video recording; the differences in various coatings can effect the quality of the recording.

ferromagnetic Having the property of being very easily magnetized with high hysteresis, that is, magnetism that changes readily with changes in the magnetizing force. See electromagnet.

ferrule 1. A snug ring or cap encircling a tool, pipe, or wire which helps to strengthen or secure it. Commercial ferrules are frequently made of metal. 2. A short length of tubing or bushing (insulating liner) for providing a snug connection between wires or pipes. Ferrules are used to secure and align wires and cables.

FES Fixed End System.

Fessenden, Reginald Aubrey (1866-1932) A prolific, Canadian-born, American inventor and radio pioneer who was one of the first to try to devise ways to carry information on top of a carrier wave. In the process of trying to achieve this, he developed a high-frequency generator in 1901 that could create radio waves, and a hot-wire barretter, which was developed into an electrolytic detector, for detecting radio waves. On Christmas Eve 1906, to the astonishment of those who heard the broadcast, Fessenden succeeded in transmitting voice and music, using an Alexanderson alternator, over public radio waves to the U.S. east coast. See barretter, carrier wave, electrolytic detector, radio history.

Fessenden Station A radio station installed in Brant Rock, Massachusetts in the early 1900s which was famous for having one of the highest masts in the world at the time, over 400 feet tall. The mast consisted of cylindrical steel sections up to three feet in diameter, supported by a base made of insulating material. Fine guy wires held the mast in place.

FEXT See far end crosstalk.

Feynman, Richard Phillips (1918-1988) A charismatic, individualistic American physicist who contributed greatly to our understanding of physics, especially in quantum electrodynamics (QED quod erat demonstrandum - that which has to be demonstrated), who developed Feynman diagrams, and provided insights into the theory of computing.

FF See form feed.

FGDC Federal Geographic Data Committee.

fiber 1. A strand or filament, or other structure with thread-like qualities. 2. Colloquial for fiber optic (or optical fiber). See fiber optic.

Fiber Channel See Fibre Channel Standard.

Fiber Distributed Data Interface FDDI. An American National Standards Institute (ANSI X3T12 - formerly X3T9.5) standard high-bandwidth 100 Mbps packet-switched protocol developed by the X3T9.5 committee.

FDDI is packet-switched, based on token-passing technology, with multiple frames travelling the ring at the same time, that is, a dual-ring network with a primary and a secondary ring. On most configurations, the primary ring is used for data communications, and the secondary ring is used as a backup. A *concentrator* is typically used as an attachment point, with dual attachment stations (DASs) attached to both of the rings.

FDDI is typically used on local area networks (LANs) with fiber optic lines such as campus backbones. Two types of fiber optic cable are typically used: single-mode and multimode. FDDI supports synchronous and asynchronous transmissions, and transmits over fiber optic cables to distances of about 200 kilometers, and can service about 1000 individual workstations, depending on the topology. The specified wavelength is 1300 nanometers for data transmission. Short hops within the network may be handled by twisted pair copper wires.

FDDI has a theoretical maximum speed of about one half million packets per second; however, with padding, and various other factors, achieved speed in commercial implementations is about one third of this.

There are four documents associated with the FDDI architecture standard, as shown in the FDDI Standard Documents chart. See A port, Copper Distributed Data Interface, Dual Attachment Station, dual homing, optical bypass.

Fiber Distributed Data Interface II FDDI-II. Based on FDDI for high bandwidth rates of up to 100 Mbps, but enhanced to handle circuit-switched (in addition to packet-switched) PCM data for ISDN or voice, in addition to data. FDDI-II supports both basic mode and hybrid mode transmissions, and in hybrid mode, adds isochronous support to the asynchronous and synchronous services pro-

Fiber Distributed Data Interface Data and Token Frame Formats

```
bits  64      8     8    (16 or 48) (16 or 48)   0      32     4     1
-------------------------/-----------/--------/--------------------
|  PA      | SD | FC |    DA     |     SA    |  Info  | FCS | ED | FS |
-------------------------/-----------/--------/--------------------
```

PA	Preamble: frame synchronization with each station clock
SD	Starting delimiter
FC	Frame control
DA	Destination address
SA	Source address
Info	Information field: 0 to 4478 bytes
FCS	Frame check sequence
ED	Ending delimiter: end of frame
FS	Frame status

The above diagram shows the data frame format, the basic token frame format is as follows.

```
bits   16+     8    8    8
---------------------------
| Preamble | SD | FC | ED |
---------------------------
```

PA	Preamble: frame synchronization with each station clock
SD	Starting delimiter
FC	Frame control
ED	Ending delimiter: end of frame

vided in basic FDDI. See Fiber Distributed Data Interface.

Fiber Distributed Data Interface frame format The frame format for FDDI frames includes a data frame and a token frame. The data frame contains packets from higher-level protocols enroute to another node. The token frame contains three bytes and a preamble for setting the signal clock. The basic FDDI frame format is shown in the Frame Format chart.

Fiber Distributed Data Interface physical layer The FDDI physical layer implementation differs somewhat from other IEEE standards. It consists of a medium-independent portion (PHY) and a medium-dependent or medium-specific portion (PMD).

Fiber Distributed Data Interface ports There are four types of ports defined for an FDDI network as shown in the FDDI Ports chart.

fiber loss Signal deterioration (attenuation, fade) in a fiber optic transmission. See fiber optic amplifier.

fiber optic, fiber optics, fiber optic cable, optical fiber These are cables constructed of parallel, bundled, slender, transparent fibers of glass or plastic, encased in a lesser refractive material, which carry transmitted light through the length of the cable through internal reflection. The low loss of signal (light) over distance, and the reflective nature of the transmission which make it able to carry the signal around curves, make this an excellent medium for data transmission and networking. Fiber optic cables are generally categorized as single mode or multimode. Multimode cables provide an optical pathway that is wide enough that the signals refract at various angles as they travel along the cable. Eventually, however, they tend to run together, thus limiting cable lengths for this type of transmission to distances of less than one or two kilometers. Since dispersion in single mode cables is not of the same nature as that of multimode cables, transmission distances are longer.

There are different ways to generate the light that travels through the cable. Single mode cables typically use lasers, whereas multimode cables typically use light-emitting diodes (LEDs). The wavelengths of light which are used vary with the installation.

The unenhanced capacity of fiber optic is about 2.5 Gbps, but technologies are being developed and implemented for boosting this. Fiber optic cables have far greater bandwidth, that is information-carrying capacity, than traditional copper phone wires, and are suitable for transmitting high information content signals including full-motion video. See blown fiber, coaxial cable, copper wire, multimode optical fiber, single mode optical fiber.

fiber optic amplifier A fiber optic amplifier can improve the signal to noise (S/N) ratio in a fiber transmissions medium, and extend the distance of a transmission. Fiber optic cables most commonly use light from lasers, or from light-emitting diodes (LEDs). The light source can be modulated by directly manipulating it on and off, or can be indirectly manipulated by an outside controlling device. A lithium niobate modulator is an example of an external modulator which is used in the cable TV and digital network industries to extend a signal. Because it allows a stronger laser to be used, the signal can be sent farther. Pump lasers are a means to use semiconductor technology to amplify a laser signal. Doping with rare earth elements such as erbium, alters the transmission properties of a cable, and the technique is commonly used to amplify signals. See doping, modulation.

FDDI Basic Port Types

Port	Type	Characteristics
M port	Master port	Connects two concentrators and can communicate with DASs and SASs.
S port	Slave port	Connects single-attachment devices for interconnecting stations, or for connecting a station to a concentrator.
A port	Dual-attachment	Connected to the incoming primary ring, and outgoing secondary ring. See A port dictionary entry.
B port	Dual-attachment	Connected to the incoming secondary ring and the outgoing primary ring. See B port dictionary entry.

Fiber Optic Association, Inc. An international non-profit professional association representing the fiber optic industry. It provides education, seminars, publications, member support, and input into certification programs. http://www.std.com/fotec/foa.htm

fiber optics probe A probe used in medical or industrial applications that employs a bundle of fine, aligned optical fibers with light to transmit an image to a monitor, computer, or video recording device. Fiber optics probes are usually designed to be flexible so they can be threaded through small, curved spaces. This makes them valuable for medical examination of living bodies, and industrial inspection of pipes and internal structures. A fiberscope is a similar tool, with a lens on one end, and a viewing eyepiece, or connection to a display monitor on the other end.

fiber to the home The process of bringing fiber optic cables and telecommunications services to residential subscribers. It's an ambitious undertaking, but BellSouth and the Nippon Telegraph and Telephone Corporation (NTT) announced in June 1998 that they were embarking on a joint project to make this a reality. The specifications being developed by the two telecommunications corporations will be disseminated through the international Full Service Access Network (FSAN).

fibre Alternate spelling of fiber, common in most English-speaking countries except America. See fiber.

Fibre Channel See Fibre Channel Standard.

Fibre Channel Association FCA. An organization supporting Fibre Channel technology. http://www.fibrechannel.com/

Fibre Channel Consortium FCC. An association in the UNH InterOperability Lab which provides support for product interoperability and educational demonstrations of Fibre Channel technology.

Fibre Channel Loop Community FCLC. An organization supporting Fibre Channel technology.

Fibre Channel Standard FCS. A high-speed, fully bidirectional data transfer interface for interconnecting workstations, mainframes, display peripherals, and storage devices. FCS has been standardized by the ANSI X3T11 committee.

The FCS works with both wire and fiber optic systems, from 133 Mbps to 1,062 Mbps at distances up to 10 kilometers (contrast this with a standard SCSI electrical cable transmission, for example, which has a distance of only a few feet). This provides excellent opportunities for aggregating peripherals under desks, or in server rooms and secure areas, for standardizing a wide variety of computer peripherals, and for increasing architectural flexibility in the placement of equipment.

Fibre Channel Systems Initiative FCSI. A group organized in 1993 to promote use and distribution of Fibre Channel technologies.

FDDI Architecture Standard Documents

Abbrev.	Item	Notes
MAC	Media Access Control	A network control mechanism for defining formats and methods. Like the PHY layer, the MAC layer is directly implemented in FDDI chips. The higher LLC sublayer provides data to the MAC.
PHY	Physical layer	An electronic signal encoding/decoding layer which mediates between the higher MAC layer and the lower PMD layer.
PMD	Physical Medium Dependent	The lowest sublayer, which specifies various physical media such as interface connectors, cables, power sources, photodetectors, etc.
SMT	Station Management	A node manager and bandwidth allocator. The SMT is further subdivided into connection management (CMT) which controls access, ring management (RMT) which provides diagnostic capabilities, and frame services.

Fibreoptic Industry Association FIA. A professional organization representing suppliers, educators, and installers in the fiber industry.

fiche See microfiche.

FID Field Identifier, part of an ISDN Service Profile Identifier. See SPID.

FidoNet Established in 1984 by Tom Jennings with the second node belonging to John Madill, this networked bulletin board system (BBS) became a major communications tool for techie discussions, email, and file transfers as bulletin board operators all over the country started to establish Fido boards for their local users. The initial concept was simple enough, the two originators wanted a way to automatically transfer source code files inexpensively across the continent (during off-peak hours), and then to exchange electronic correspondence as well. Once they had a system working, it was incorporated into the Fido BBS software. Within two years, NetMail had been extended by a Fido sysop to permit BBS special interest groups (SIGs) to run public discussions like those now familiar on the USENET newsgroups, establishing EchoMail. FidoNet grew to more than 11,000 nodes worldwide, with many gated to the Internet.

field In a scanning video broadcast display, a field is every other line of the full picture 'frame.' Thus, it is all the odd numbered lines taken together, or all the even numbered lines taken together, in an interlaced image.

field, data A record-holding or record-entering entity in a database. The definition of field types facilitates program setup, management, and data manipulation by alerting the software as to the nature of the information being entered into a field. That is, a field may be given a data type (number, string, date, etc.), or it may be untyped, but either way, this tells the system something about the data.

field mode In video image capture, a mode that captures only half of the scan lines in order to save an image in lower resolution, thus taking less storage space. See field, frame mode.

field winding A mechanism for energizing electromagnets in a generator. See winding.

FIF See Fractal Image Format.

FIFO first in, first out. In programming, a means of processing data so the first item to be stored, or placed on a stack, is the first to be fetched, moved, or discarded. Imagine a narrow vertical tube for gerbils (or hamsters, if you prefer fluffy rodents); the first gerbil to squeeze in through the top is the first to slide out the bottom. In general terms of telecommunications, in a FIFO system, the first person who calls, is the first to be referred to an agent.

filament A fine metal conducting wire commonly used in tubes and bulbs. By passing a current through a filament in a specialized, enclosed environment, it becomes incandescent, giving off light. See cathode.

file A set of data stored so that a pointer to the information identifies and encompasses the contents of that file as an accessible, readable unit. This is one of the most common basic units of storage in a computer system, and file hierarchies, file folders, file types, and file management are all computer structures and processes that are constructed to manage files.

File Allocation Table FAT. A means of storing and tracking file locations on a disk in the MS-DOS file system. The size and location information for a file may or may not be stored in contiguous clusters, so a map of it is stored in a table for locating the information as needed, or the space freed up and reused, if there is a request to the system to delete the file. See FAT for a more complete explanation.

file attachment Most email systems are text-oriented 7-bit messaging media. So how do you send someone an 8-bit binary file? To meet this need, many text email systems have the capability of sending binary files as file attachments to a message. Since binary files include symbols and characters which cannot be displayed in a plain text window, and since the symbols are not meaningful to humans, it is more practical to send the file (which may be sound, graphics, or a computer application) as an attachment, rather than as a postscript to the email text message. In most cases, all that is necessary is to specify the name of the file in the `Attachment:` text box or e-mail message header, and the system will take care of the transfer of the information.

file cache An area of memory allocated by an operating system or computer applications program, to temporarily store a file which may need to be accessed or modified frequently. Many database and spreadsheet programs use file caches to allow quick updates and redisplays of information, and the data may also be periodically stored on disk as a background task so as not to lose information and updates in the case of a software crash or power outage.

file extension A syntactic convention which aids in the identification of computer data file types. There are many categories of computer files: text files, graphics files, sound files, and within these basic categories are many subcategories, such as JPEG, TIFF, etc. A convention of adding a period and a short suffix to identify the type of file, so it can be found at a glance, has become widespread, and some applications and systems will even enforce certain file extensions.

Since the mid-1980s, every significant microcomputer and workstation level operating system except MS-DOS has allowed file extensions of any reasonable length (up to 16, 32, 64, or 256 characters for the whole file name, depending on the system). MS-DOS restricted its users to only three characters, and enforced the use of the period (dot) as the file extension symbol. Since there were so many DOS-based machines, users of other systems had to truncate their file extensions (and in fact, the rest of the file name) when transferring files to other systems. This impractical three-character extension limit is still prevalent, in spite of the fact that most Windows-based systems now support longer filenames (even now it is common to see HTML file extensions on the Web listed as ".htm" instead of the ".html").

On most other systems, the dot (.) is not mandatory for specifying the extension. The user can save a file with no dots, or with a dozen dots. However, since several early systems in the 1970s required a dot, users have become used to this naming convention.

On many systems, a file extension is used to indicate a version number, so backups and a revision history can be maintained at all times. For example, the following versions may be automatically stored by the system:

```
mygreatimage.tiff.1 or mygreatimage.tiff;1
mygreatimage.tiff.2 or mygreatimage.tiff;2
mygreatimage.tiff.3 or mygreatimage.tiff;3
```

If the default for the revision level is 'three,' then the next file to be saved under the same name will supersede the oldest file, in a first in, first out (FIFO) sequence, so that no more than three files with the same name are stored at any one time. This version number extension/revision system is very handy when something is saved accidentally, and the previous version needs to be retrieved.

file gap A blank inserted to indicate a stopping point, or a division between sets of information, especially on a sequential file recording system. On an audio tape or digital data tape, a file gap indicates the beginning or end of a song or file.

file server Generally, a system on a network which administrates the storage of and access to files, often through a client/server model, in which multiple users make requests to the file server through the client software. This system reduces redundant storage of files on individual systems, and makes it easier and faster to update individual files. The server also handles file locks so that data files cannot be simultaneously updated and saved by multiple users. Usually, a dedicated file server is equipped with high storage capacity, and may manage security levels for access to the files. Network File System (NFS) is a commonly used Unix file server system from Sun Microsystems that is implemented on many platforms.

file server, frame relay In a frame relay network, the file server is a device which provides connections with terminals, controls transmission flow, and provides end-to-end acknowledgment, and error recovery.

File Service Protocol FSP. A file transfer protocol somewhat similar to File Transfer Protocol (FTP).

file sharing Access by more than one user, sometimes at the same time, depending on the nature of the data, to files that may be on one system on a network, or spread out over several workstations that are interaccessible. On the Apple Macintosh, file sharing is easily set up via utilities in the Control Panels so that passwords can be assigned, and files shared with designated users on the system. On a larger network, a particular machine or set of machines, usually with large hard disk storage capacity, may be dedicated to file serving and sharing activities. See file server.

File Transfer Protocol FTP. A user-level file sharing protocol established by the early 1970s on the ARPANET, and now widely implemented on the Internet in the form of FTP archive sites. The concept of FTP sites was to provide a simple, consistent means of presenting and accessing file information on a variety of types of file archive sites, so the user could easily navigate the site and upload or download files unassisted. In other words, FTP sites have a consistent look and feel; once you've learned a few easy commands, you can log in, look around, and get what you need without having to worry about the individual characteristics of the system on which the files are stored.

Many FTP sites provide public access through a user login in which you type "anonymous" as the username, and your full email address as the password. If you have a Unix shell

Sample Anonymous FTP Login Session

```
abiogen@frodo: /1.2GB/users/abiogen $ ftp ftp.peanuts.org
Connected to ftp.peanuts.org.
220 arcadia FTP server (Version wu-2.1c(4) Sun Jan 23 18:16:17 MEZ 1994)
ready.
Name (ftp.peanuts.org:abiogen): anonymous
331 Guest login ok, send your complete e-mail address as password.
Password:

230-         ==================================================
230-         ==  The University of Munich Software Archive   ==
230-         ==  Peanuts NEXTSTEP/OPENSTEP/Rhapsody Archive ==
230-         ==================================================
230-
230-             This server is in Munich, Germany, Europe.
230-                                  =-=-=-
230-
230-  All transfers are logged with your host name and email address.
230-            If you don't like this, disconnect now.
230-
230-  ________________________________
230-  The following FTP sites maintain a mirror of the Peanuts archive:
230-          ftp.evolution.com           /pub/next     USA
230-          ftp.mb3.tu-chemnitz.de      /pub/NeXT     Germany
230-  ________________________________
230-
230- For many directories an alias exists. Try a 'cd next' in any directory.
230-
230- The NeXT section of this server in now also available on CD-ROM!
230- Please see the file 00INFO/Peanuts_CD-ROMs in the NeXT part of our
230- archive for further details about the Peanuts Archive Discs (PAD).
230-
230- You are user number 15. The maximum number of users is 90.
230-
230 Guest login ok, access restrictions apply.
ftp> cd pub
250 CWD command successful.
ftp> dir
200 PORT command successful.
150 Opening ASCII mode data connection for /bin/ls.
total 7468
drwxr-xr-x   5 ftpadm   ftpadm       1024 Jun 30 19:17 .
drwxr-xr-x  11 ftpadm   ftpadm       1024 Jun 30 19:17 ..
-rw-r—r—   1 ftpadm   ftpadm       1182 Sep  8  1993 00COPYING_POLICY
-rw-r—r—   1 ftpadm   ftpadm    7621825 Jun 30 19:17 00INDEX.gz
drwxrwsr-x  23 13222    10676        1024 Jul  1 01:46 FreeBSD
drwxr-xr-x  14 ftpadm   ftpadm       1024 Feb  6 14:07 comp
drwxr-sr-x  10 ftpadm   ftpadm       1024 Jun 16  1995 local
drwxrwsr-x  24 ftpadm   ftpadm       1024 May 27 17:15 next
drwxr-xr-x   4 root     sys          1024 Feb  6 14:06 rec
drwxr-xr-x   4 ftpadm   ftpadm       1024 Feb  6 14:06 science
226 Transfer complete.
641 bytes received in 0.92 seconds (0.68 Kbytes/s)
ftp> bye
221 Goodbye.
abiogen@frodo: /1.2GB/users/abiogen $
```

account with an FTP client, you simply type "ftp" (in all lowercase), followed by "help" when it activates, to learn its basic commands and capabilities. The inset shows an example of a simple anonymous FTP login.

In the example session shown, the user logs in as anonymous, supplies a legitimate email address as the password, and is dropped into a limited environment where basic directory traversing commands and file download commands can be used. The message "Guest login ok, access restrictions apply." is displayed. This session is very typical in that the user is prompted to disconnect if logging of his or her activities is objectionable, and is notified that there are restrictions. The user's logon number is shown, in addition to the total number of people on the system. If the system is at capacity, the user may be asked to try again, or may be provided with a message giving the addresses of mirror sites (sites with the same files in other locations).

For file transfers, the 'get' command, followed by a filename, will initiate a file download. The commands 'bye' or 'quit' will end a session.

On Unix systems, you can type "man ftp" at a shell prompt to read the manual pages for FTP, which includes a list of common commands. FTP file download capability is built into most Web browsers, and works transparently with many Web file archives. Several variations of the File Transfer Protocol exist, and cut-down, easier-to-implement versions have also been developed, which are described on the Internet in various RFCs. See File Service Protocol, Simple File Transfer Protocol, Trivial File Transfer Protocol, RFC 171, RFC 172, RFC 959.

file transfer protocol In its general sense, any program that facilitates the movement of files from one system to another, particularly through phone, null modem, or other serial data links, or through the Internet. There are many file transfer protocols, but two of the most popular implementations are ZModem, for telephone line transfers, and File Transfer Protocol (FTP) for Internet transfers. Other popular programs include Kermit, XModem, and YModem. More detailed information is included in this dictionary under individual listings for the various protocols.

fill *v.* In image processing, a function of a software program which allows a bounded object, or unbounded expanse (such as a scalable background), to be filled or *flooded* with a color, a gradation of color, a texture, or a pattern. Programs typically permit the user to create automatically filled geometric patterns, or irregularly shaped patterns that can subsequently be filled. The fill may or may not include the actual boundary, or extents, of the shape, depending on the software (*outline* usually applies to the outer extents). Fills are typically associated with 2D paint programs; in 3D modeling programs, it's more common to say that you *texture map*, *skin*, or *wrap* the surface of an object rather than fill it.

film *n.* 1. A thin membrane, skin, or coating. 2. A thin, light and/or chemical sensitive material commonly used in the photographic industry. 3. The collective name for a sequential, related set of still frames which, taken together, form a story or cohesive idea that is viewed by playing the frames through a projector. Also called movie.

F

filter *n.* 1. A porous material through which mixtures are screened in order to selectively prevent larger bits of the mixture from passing through. 2. A device or material through which particular waves, frequencies, or particles do not pass. A filter may be used in combination with another device, such as an amplifier, in order to filter out noise, while propagating the desired portion of a signal. Electrical and audio filters are common.

filter, file File filters are not necessarily exclusionary tools, as in some senses of the word 'filter,' but rather may be conversion utilities available in many application programs to input or output files in a format that is not native to the application. Thus, a TIFF file might be imported into a paint program with a proprietary format, through a filter, and may be exported through another filter to a JPEG file, for example, for use on the Web.

A filter may also be coordinated with a database to selectively provide access to higher priority messages or processes, while filtering out, or queuing those of lower priorities. Email filters are especially useful to those who get hundreds of messages a day, as often happens on email mailing lists. A good email client will let you set up filters that file the messages in separate folders to be selectively read later, so the user can more easily determine which messages to check first. Exclusionary file filters also exist. For example, an email file filter may exclude all messages received from `username@hotmail.com`, or relegate them unread to the bit bucket.

filter, network In network transmissions, there are physical filters and logical filters. Logical filters function on every level of the system from low-end operating functions, to high-end user applications. Logical filters employ algorithms to selectively block the continuation of certain information, such as extraneous packets, unrecognized characters, extra information not supported by the receiving protocol, unwanted email, messages from sites operating unlawfully, etc.

filtering Using physical or logical means to selectively permit access of only the desired information. Thus, unwanted information can be screened out, or a lower capacity system can be used to view or use part of the information according to its capabilities. For example, filtering out parts of a transmitted image makes it possible to display it on a system with low resolution, or a slow image display, a solution that may be preferable to no image at all. See compression, MPEG.

filtering agent, filtering client A software program that can be configured to selectively reject or keep information according to a set of parameters or keys. With the excess of information available through the Internet, filtering agents are increasing in importance. See data mining.

filtering traffic On a network, the selective acceptance or rejection of certain packets, messages, or processes according to a set of priorities and parameters. High and low usage times may also be factors in setting up filtering instructions. Traffic filtering is usually accomplished by combining a database with a list of priorities. See firewall.

fin waveguide A structure which can be used in conjunction with circular waveguides to increase the range of wavelengths that can be transmitted by attaching a longitudinal metal fin.

finder A name used on several computer systems for applications which aid in locating information on a system, whether it be files, directories, or the specific content of files.

Finder On the Apple Macintosh, the graphical user interface and operating system processes through which the user interacts with the system. Multifinder allows more than one program to be executed at a time and is available on the more recent versions of MacOS. It is also a generic name for a file finding tool that comes with the operating system.

Finger The name of an online information utility, based on the Finger Protocol, which allows the user to retrieve and display information about users of a system, or the owner of a particular account on the network, provided no firewalls exist to block the *finger* command (as a command it is spelled all in lowercase). Login and logout times may be displayed, or the length of time since the last login. If the user queried has particular 'dot' files configured, such as *.plan (dot plan)*, additional information from this file will be displayed. Users often use the *.plan* file to list philosophies, home addresses, office hours, interests, or professional credentials. See firewall.

Finger Protocol A network information protocol which is an elective proposed Draft Standard of the IETF. See finger, RFC 1288.

Finkel, Raphael First disseminator of the infamous Jargon File, distributed from Stanford University in 1975. See Jargon File, The.

FIPS Federal Information Processing Standard. A set of standards for document processing, search, and retrieval.

firefighting Trying to fix something after the fact. Often used in a derogatory sense to indicate the frustration of trying to rescue a situation that would not have occurred if proper steps, or prevention methods had been used in the first place. The term is used to describe distressing, expensive catch-up or fix-up situations resulting from bad management decisions. For example, shipping a software product before it is fully tested and debugged can result in a great loss of confidence on the part of customers, and enormous extra firefighting expense to the company in terms of subsequent upgrades and tech support that would not have been required if the product had been properly completed before shipment.

firewall A physical screen created to prevent the spread of fire. This may be a wall of heavy, fire-resistance materials. See wiring vault.

firewall, network A computer network security configuration designed to completely or selectively limit access to a system. At one time, firewalls were usually implemented on a specific *gateway* machine, but hardware and software firewalls now are set up in a number of ways, using filters, proxies, and gateways at the circuit level. A network traffic firewall examines incoming packets and selectively lets them pass through, and may also edit outgoing traffic in order to protect the identities of the senders, as in some government networks. Many local area network (LAN) firewalls are one-way, with unlimited access out of the LAN, and selective access into the LAN. Systems with firewalls frequently log all activities through the point or points of entry, with or without notification. See packet filtering, proxy server.

FIRMR Federal Information Resources Management Regulation.

firmware Programmed circuitry which is semi-permanent. Software on a disk is easily changed and rewritten. Software on the circuitry of a microchip is not easily changed and rewritten. In between these are EPROMs, erasable, reprogrammable chips which can be changed with the right equipment, and which retain the information during a power-off.

first call date A record of the first time a subscriber line is used, sometimes used in billing, or in settling disputes.

first in, first out See FIFO.

fish job *slang* Phrase to describe a difficult wiring installation in which the wiring has to be pulled and threaded through constricted or hard-to-reach spaces.

fish tape *slang* A smooth surfaced, non-conductive tape that is threaded through tight areas, such as conduit, in order to attach it to a wire and pull it more easily back through the wiring path. See pulling eye.

fishbone antenna An antenna named for its resemblance to the ribs of a fish because it includes a series of coplanar antenna elements arranged in pairs. The fishbone antenna is used in conjunction with a balanced transmission line.

Fisher, Yuval Author of "Fractal Image Compression," which describes the current knowledge of fractal compression in down-to-Earth terms with C source code examples. See fractal transform.

five-level code In telegraph communications, a code that utilizes five impulses to describe each character. See Baudot code.

FIX Federal Internet Exchange.

fixed access unit FAU. A wireless telephony designation for a wireless phone unit which is not intended to be carried around, but rather to provide wireless communications within a limited region. Thus, local wireless phone service can be installed without going through a local phone provider, much like a fancy intercom unit, or can be subscribed through an alternate vendor as a limited cellular or PCS service.

Fixed End System F-ES. A non-mobile data communications system through which a mobile subscriber accesses landline network services. F-ESs typically comprise modems installed into desktop computers. See Cellular Digital Packet Data, Mobile End System.

fixed satellite service FSS. One of two divisions into which Ku-band satellite broadcast services have been split. FSS operates in the 11.7 to 12.2 GHz range. The other is *broadcasting satellite service* (BSS). See Ku-band.

Fixed Wireless Access FWA. In regions where the cost of installing wireline may be prohibitive due to rough terrain or sparse population, or where regional growth outstrips wireline installation capacity, FWA provides a long-term or temporary alternative. It combines radio-based phone service, in the place of the local wireline loop, with common carrier phone service. See time division multiple access, code division multiple access.

fixed-width font See monospaced font.

FK foreign key. A designation in a key cryptography scheme.

flag *n.* 1. A device or signal used to attract attention, or to indicate the state of a situation. In software programming and operation, flags are frequently used to indicate the state of processes or variables, often under changing conditions.

FLAG Ltd. Fiberoptic Link Around the Globe. A commercial fiber services carrier with installations of more than 18,000 miles of fiber optic cable installed worldwide.

flame *n., slang* Derived from "in*flam*matory remark." A highly critical posting, sometimes negative or abusive, in email, or on an online public forum. Public flames are discouraged on the Internet. Sometimes words written all in capitals are used to indicate shouting, so don't use all caps unless you intend it that way. Flames on people's spelling, or grammar, except in forums where that is the appropriate topic of conversation (as in English usage newsgroups) are considered very bad etiquette, particularly as there are many users with physical handicaps, learning disabilities, or for whom the language being used (usually English) is not the native tongue. See emoticons, flame fest, Netiquette, newsgroups.

flame bait A communication intended to incite strong opposition or excessive negative responses. Devil's advocate postings sometimes *unintentionally* incite more flames than the original poster expected. The individual may have been offering a thoughtful opposing viewpoint. Typically, the phrase flame bait refers to *intentional* inflammatory messages.

flame detector An early experimental type of very sensitive and delicate radio wave detector devised by Ernest Ruhmer, which uses a receiving circuit with two platinum contacts which are inserted into the coldest and hottest part of a gas flame. Vaporizing salts are placed in a trough on the lower contact. The battery power for the circuit flows through the flame to the telephone receiver. When a radio wave comes in contact with the flame, the properties of the flame are altered, and the current flowing through translates these changes into corresponding signals in the receiving unit. Care must be taken to prevent wind or other non-radiowave interference from altering the flame. The flame detector is an important evolutionary step in radio technology because it represents a stage in development towards the Audion, a historic invention by Lee de Forest. See Audion, detector.

flame fest, flame war *slang* A descriptive phrase for a barrage of critical (inflammatory) messages in email, or on a public forum online. Although there may be occasional situations where a strong criticism is appropriate, in general, a flame war is a situation where negative opinions, or excessive personal criticisms, are unkind and uncalled-for, or are taking up too much bandwidth, and crowding out normal, productive discussions. It is not advisable to encourage or extend flame wars. If you have a genuine beef, say it once, succinctly, then let it go. If you desire to continue it through constructive criticism, do so in private email so that other members of the public forum are not dragged into the flames. See flame.

flame mail An email communication of a highly critical, negative, or abusive nature. See flame.

flame resistant, flame retardant A medium which is inherently resistant to catching fire or spreading flames, or which is treated or manufactured to increase these properties. Flame resistant and retardant materials are used in many industries including construction, electrical installation, and clothing manufacture.

flammable A property of easily catching fire, or continuing to burn readily.

flange A rim or rib on an object to add strength or aid in alignment.

flash *v.* On a phone or intercom system, to send a signal through the line by pressing the switch button on the handset holder, or a button designated as a *flash button*. The flash button is used on some local multiline systems to transfer a call, and may be followed by the keying in of the number of the extension.

flash button A button designated on a phone or intercom system to send a signal which is the same as pressing the switch button on the handset holder. See flash, flash hook.

flash cut See hot cut.

flash hook See switch hook.

flash interference In television transmission and display, a 'flash' is a very brief interference, sufficient to distort the picture information.

flash memory A type of nonvolatile, rewritable computer memory technology, developed by Intel, which provides an alternative to large storage devices. Since flash memory is physically compact, and doesn't lose its data when the device is not in use, it has been incorporated into PCMCIA cards for portable computing applications. Flash memory is also starting to show up in portable telephone devices and digital cameras. See memory, PCMCIA.

flash tube A bulb or tube used to create a bright, momentary burst of illumination through application of a high-voltage pulse. One-time flash bulbs were used in older cameras; electronically activated, reusable bulbs are now common.

flat panel Any of a number of types of display systems which are narrower and flatter than traditional CRT displays. These may be special flat panel CRTs, gas plasma displays, liquid crystal displays (LCDs), or light emitting diode displays (LEDs). Flat panel displays are especially favored on mobile systems like computer laptops.

flat panel CRT A type of cathode ray tube (CRT) color display technology in which the electron beams are aimed parallel to the front of the display device, then deflected 90 degrees onto the viewing surface. This permits the construction of a much flatter, smaller, more convenient display device. While this technology is still relatively new and expensive, the bulkiness of traditional CRTs makes the flat panel CRT commercially attractive.

flat plate antenna A commercial/industrial/military satellite communications focusing antenna based on microcircuit design. It is similar to a common parabolic antenna, except that it incorporates a series of concentric rings laid over a transparent sheet to create a lens that can be used to redirect signals.

flat rate service A very common subscriber billing technique. Flat rates usually arise in services where the overhead of keeping track of many different types and quantities of usage would cut into profits. Flat rate services are also attractive to many subscribers, as they know in advance what it will cost, and don't have to watch the clock or keep track of usage. In computer network access and telephone services, flat rate billing is very common. Since users of these services vary dramatically in time of access, connect times, and types of services used while connected, it probably is more economical in the long run to assign average usage fees, than to try to track and bill widely varying usage. Flat rates for businesses tend to exceed those for residential use by roughly a factor of three, depending on the type of service. Local phone calls in many areas in North America are billed on a flat rate. In Europe and some parts of North America, per call charges are levied instead, and in most cases, long-distance services are billed on a per call basis. Cellular phones are typically billed as a flat rate for the basic service, with per call charges for each call. The newer digital cellular technologies sometimes have a flat rate billing option.

flat top 1. Something with a flat surface on top, as a flat-roofed building, or aircraft carrier. 2. The portion of an antenna that lies horizontal.

flat top antenna An antenna which has two or more parallel, horizontally strung wires.

flatbed scanner A type of desktop scanner which allows the object to be scanned to be placed directly on the scanning surface, that is, it lies flat, and doesn't have to roll through a drum or other moving mechanism. This type of scanner is preferred for scanning books, and other large or three-dimensional objects.

flavor A slang term for type or model. Programmers frequently refer to different 'flavors' of software languages or operating systems to indicate that they are essentially alike, but differ enough to create compatibility issues. The distinction is somewhat like a 'dialect,' in languages, or a 'model' in a particular type of car.

FLC ferroelectric liquid crystals. Crystals which are incorporated into spatial light modulators (SLMs) in optical computing technologies.

FLCD ferroelectric liquid crystal display (LCD).

FLEA memory flux logic element array memory. The whimsical acronym for a type of computer memory developed by RCA in the early 1960s. The FLEA was created photographically, and was capable of storing 128 bits of information. Its processing speed was 100,000 items per second.

Fleming, John Ambrose (1849-1945) An English electrical engineer who investigated the Edison effect, and experimented with improvements to wireless receivers in 1904. By modifying an electron bulb so that it incorporated two electrodes, and attaching it to a radio receiving system, the radio waves could be converted to direct current (DC). Unfortunately, this new *diode* did not constitute a significant improvement over previous electron tubes, but it was important in the evolution towards more sophisticated tubes that came later. The most important of these was the *triode* in which L. de Forest took the two-element Fleming tube as the basis for the invention of the *Audion*, which included a controlling grid as a third element.

Fleming oscillation valve An electron tube developed by J. A. Fleming, based on Edison's work with electric light bulbs. This diode tube was in essence a two element rectifier. While it did not achieve the practical utility of later tubes, it led to the development of the *triode* by Lee de Forest.

Fleming's rule See right-hand rule.

flicker A characteristic of display devices, such as cathode ray tubes (CRTs), in which the scanning of the screen is visible to the human eye as a light-dark flashing flicker. Flicker can result from a number of causes, including the quality of the monitor, the mode of display (interlace or noninterlace), or the speed of the screen refresh as the electron beam 'sweeps' the screen. Generally, slower sweeps will appear to flicker more, as do interlace screen modes.

Apparent flicker is eliminated on better multiscan monitors. Most individuals can comfortably watch displays which are refreshed at about 70 Hz to 80 Hz; above that level, the trade-off in cost and computing is not sufficient to justify the insignificant or nonexistent improvement.

While flickering on screens may be uncomfortable to watch, sometimes an interlaced mode has a practical purpose, as when an NTSC-compatible signal is being generated to output to video. See cathode ray tube, frame, interlace, multiscan.

flip-flop 1. Quick reversal of direction or

opinion. 2. A circuit or logic state which can assume one or the other of two stable states (on/off, high/low, etc.). A trigger circuit, or toggle.

flippies A term coined by the author in the 1970s to describe those subscription and advertising cards deliberately placed so they flip out of a magazine when you try to leaf through it. The term has no doubt been independently coined by others.

floating point A mathematical representation system in which a number is expressed as a product of a bounded number (mantissa) and a *power of* scale factor (exponent) within a number base (e.g., base 10); hence, 123.45 can be expressed as $.12345 \times 10^3$. *Floating point* refers to the flexibility inherent in placing the decimal point by adjusting the exponent.

floating point unit FPU. Math coprocessors often paired up in computers with central processing units (CPUs) to take the load of the math calculations, which are usually processor-intensive; this frees up the CPU for other tasks.

floating selection In graphical user interfaces, a selected text or image area which can be manipulated and moved separately from its background, and thus appears to *float* over the other elements on the screen. This is useful for cut and paste, drag and drop, and image processing applications.

flood *v.* 1. To inundate, overflow, or cover a broad area all at once. 2. In scanning and printing technologies, flood lamps are often used to process plates and provide illumination for the recording of images. 3. To inundate with data, often unintelligible, as an incendiary or retaliatory action. 4. The outpour of vast quantities of digitally generated or stored information. See data mining.

flooding 1. Overflowing, inundating. 2. In networks, a technique of sending many identical packets through various routes so redundancy increases the chances of the data reaching its destination. 3. In networks, a deliberate act of vandalism in which data is directed toward a system, or an email address, to fill up the hard drive space, or tie up the processor, to render the system useless. Users caught flooding are usually denied further access to a system. See mail bombing.

FLOP Floating Point Operation. Mathematical manipulation of a floating point number. FLOPS (Floating Point Operations per Second) is often used to describe and compare microprocessor speeds. See MFLOP.

floppy diskette, floppy disk A thin, compact, portable, flexible, read/write, random-access data storage medium originally encased in a soft protective case, or a hard protective case. Data is stored and modified by rearranging magnetic particles, and as such should be kept away from magnetic surfaces to reduce risk of loss. Generally, magnetic media are not reliable for long-term storage (see superparamagnetic).

Floppy diskettes were originally developed at IBM, with additional contributions to the technology by Shugart. Floppy disks were originally nonrigid, rotating plastic disks in flexible coverings with open windows for the read head to access the data. The older 8" floppy disks came into general use in the early 1980s by gradually superseding tapes, which were slow, and which only provided sequential read/write capability. 5 1/4" floppies, developed at Shugart, were common in the mid-1980s, but were superseded in the late 1980s by 3 1/2" high density (approx. 1.4 megabytes) floppy diskettes encased in a hard shell. The 3 1/2" floppies were developed by Sony, not long before the Macintosh was introduced. More recently, floppy diskettes are starting to give way to cartridges and superdisks that hold 10 to 20 times more data for a similar price.

Floppies have become progressively smaller, denser, and more robust. They have decreased in size from 8" to 5 1/4" to the current 3 1/2," and have increased in capacity from a few kilobytes to over 100 Mbytes. The 100+ Mbyte floppy drives are downward compatible with standard 1.4 Mbyte floppies.

FLOPS Floating Point Operations Per Second. See FLOP.

Flow Attribute Notification Protocol FANP. In packet-switched networks, a protocol for management of cut-through packet forwarding functions between neighbor nodes. FANP indicates mapping between a datalink connection and a packet flow to the neighbor node, and helps nodes manage the mapping information. This allows the bypass of the usual Internet Protocol (IP) packet processing by allowing routers to forward incoming packets. See RFC 2129.

flow chart A somewhat standardized diagrammatic representation of processes, procedures, conditions, and directions of traffic or information flow. Flow charts employ geometric shapes, symbols and connecting lines to indicate the relative importance and relationships of the concepts being illustrated. Programmers are often required by managers to provide flow charts of their software designs, but many programmers argue that outlines and pseudocode are more useful in representing the relationships and flow within a software program than conventional flow charts.

flower key, cloverleaf key A control key next to the spacebar on Macintosh keyboards which allows modification of other keys when they are held down together. The origin of the symbol is attributed to old Scandinavian where it is now seen on roadside historical markers in Sweden.

Floyd-Steinberg A dithering algorithm, that is, a means of creating a perceptual tone, or range of tones by intermixing colors related to those tones. This is a way of 'stretching' a limited palette. In the Floyd-Steinberg error diffusion algorithm, the error between the approximate output value of a pixel, and the actual value of a pixel, is sequentially diffused to its near neighbors. See dithering.

fluorescent lamp Manufactured since the late 1930s, the fluorescent lamp doesn't use a filament, and provides more light than an incandescent lamp for the same amount of current. Since less current is required, the bulb emits less heat. A fluorescent bulb typically consists of a long glass tube equipped with an electrode at each end, and specialized vapor and gases inside the tube. When electricity passes through the tube, light waves are emitted, causing phosphors coated on the inside of the tube to glow.

flutter 1. A rapid, repetitive, agitated back-and-forth movement; any erratic vibration or oscillation. In most systems, flutter is an undesirable characteristic that interferes with the main signal. See drift, wow. 2. Undesirable phase distortion variations that may result from more than one frequency transmitting at the same time. 3. In radio terminology also loosely called drift and wow.

flutter bridge A device to measure flutter (undesirable variations from a constant oscillation, movement, or signal). This is used for testing and diagnostic purposes for various playback devices which should be playing at a constant speed, such as phonographs, tape recorders, film projectors, or disc players.

flutter rate The speed at which an oscillating body moves back and forth, which may be expressed in units per second.

flux 1. Stream, continued flow. 2. An expression of the rate of transfer across or through a unit area of a given surface, per unit of time. See watt. 3. A substance used to facilitate the fusing of materials, as the use of rosin in soldering or welding. 4. Magnetic lines of force in a magnetic field taken as a group (symbol - B). When expressed in terms of density per unit area, it is called *flux density* (symbol - D).

fly-by *n.* A representation of movement from a point of view above the ground, commonly used in animations, especially video games with flight simulation. Fly-by animations give a wonderful sense of being inside the scene that is being imaged. NASA has produced some wonderful fly-by animations of the surfaces of other planets like Mars. Satellite geophysical data make it possible to create fly-bys of Earth's surface right down to individual buildings and streets. Virtual reality fly-by simulations are startlingly real, with participants ducking moving images so as not to be 'hit.' See virtual reality.

fly-page See banner.

flyback retrace. In a cathode ray tube (CRT), the movement of the electron beam tracing the image on the screen from the end of the trace to the beginning where it starts over on the next line. This is usually associated with a 'blanking' interval in which the beam is turned off so as not to interfere with the image already displayed. There is more than one type of flyback on a monitor. The flyback associated with scanning each line is similar to the line feed and carriage return on a typewriter, in that the scan finishes at the end of one line, and flies back to the next line down (or two lines down in interlaced screens) and the beginning of the subsequent line, in a zigzag (sawtooth) pattern. The other type of flyback is when the full video frame is finished, the beam is at the bottom or last line of the screen, and then flies back to the top, or first line of the screen. (This example assumes a typical CRT in which the scanning is left to right and top to bottom.) See blanking, frame.

flying erase head A mechanism on prosumer and industrial level VCRs and camcorders which erases previously recorded video traces that might otherwise interfere with new information being recorded on top of the same section. This head is typically found on systems that support insert editing. Rainbows and other undesirable artifacts are thus avoided.

flytrap A firewall or other security system that logs unauthorized attempts at access to provide information that can help identify or apprehend the intruder.

flywheel 1. A wheel that works in conjunction with other mechanisms to smooth out and reduce inconsistencies in the rotational speed of the equipment. 2. A wheel that is used with other mechanisms, whose purpose is to store kinetic energy. Often used in conjunction with power generators, to continue motion during times when the mechanism slows down or is idle.

flywheel effect In a transmission that experiences fluctuations, the maintenance of a steadier, more consistent level of current, information, or oscillation by physical or logical means. Analogous to the function of a flywheel.

FM 1. fault management. 2. See frequency modulation.

FM broadcasting Transmission through frequency modulation technologies on approved FM frequencies with the appropriate FM broadcasting license. In the United States, FM stations are spaced at 0.2 kHz intervals, ranging from 88.1 kHz to 107.9 kHz. Low power FM broadcast signals are used for mobile intercoms, indoor intercoms, monitors, and cordless phones. See broadcasting, FM broadcasting, frequency modulation.

FM transmitter A device which, in its simplest form, includes a microphone, a circuit, and an FM transmitter, and which in more sophisticated forms includes the various commercial/industrial transmitters costing thousands of dollars which are used to broadcast from licensed radio news and entertainment and other FM communications stations.

Building simple FM transmitters in the 88 to 108 MHz frequency range is a very popular hobbyist introduction to electronics. With current technology, it is possible to create very compact, working FM transmitters for under $30, that will broadcast to a few hundred feet, or even up to two miles under good conditions. Before conducting hobbyist experiments with low power FM transmitters, it is important to learn the various Federal Communications Commission (FCC) restrictions on broadcasting, and to honor laws protecting the safety and privacy of individuals.

FMAS Facility Maintenance and Administration System.

FMV See Fair Market Value.

FNC See Federal Networking Council.

FNEWS A fast full screen news reader for UNIX, ALPHA-VMS, and VAX/VMS systems, similar to NEWSRDR and ANU-NEWS. News articles for groups which are read are cached, and dynamically loaded.

Fnorb A CORBA 2.0 Object Request Broker (ORB) written in Python and a tiny bit of C by the Hector project participants at the CRC for Distributed Systems Technology at the University of Queensland, Australia. Fnorb supports CORBA datatypes and full implementation of IIOP. It is freely distributable for noncommercial use. See CORBA, ORB.

FNR fixed network reconfiguration.

FNS Fiber Network Systems.

FOA 1. fiber optic amplifier. 2. See Fiber Optic Association, Inc. 3. First Office Application. Testing of systems within an office application once in-house testing is complete or nearly complete. Most of the problems in the system have been worked out and what is now needed is feedback from a real-world installation, a type of beta testing.

FOC Firm Order Confirmation. A product or service agreement confirmation document.

focal length (Symbol - *f*) In a viewing or recording mechanism, the distance from the focal point on the surface being viewed or recorded, to the center of a lens, or surface of a mirror, as on a camera.

focus *n.* 1. In an optical viewing or recording mechanism, the point at which rays diverging or converging from a surface intersect in the mechanism (through a lens or on a mirror) to produce a clear, unblurred image of the surface. 2. In a projected image, the point on the projection surface in which the rays converge to produce a clear, unblurred image. 3. In a color cathode ray tube (CRT), convergence of the electron beams on a precise point on the coated inner surface of the glass to provide a clear image on the front surface of the tube. 4. In human vision, the point at which the distance of the object being viewed, and the angle of the individual parts of the eye, and the angle of two eyes are correlated so that the image appears clear and unblurred. 4. Center of attention or activity.

focus group A group organized to concentrate on or discuss a specific issue.

FOD See fax-on-demand.

foil See overhead transparency.

folder A term in software directory listings, and graphical user interfaces (GUIs), to stipulate a group of files, as in a directory, or subdirectory. In graphical user interfaces (GUIs), it is often represented by an icon that looks like a paper file folder.

font Collectively, a set of characters and symbols with an overall aesthetic relationship in one style and size of typeface. For example, Helvetica and Times are common typefaces, and 12 point Helvetica Bold or 10 point Times Italic would each be an individual font within its respective typeface. Fonts are commonly used in printing, and computer printers are capable of rendering a wide variety of fonts. Here are four examples of font variations of the Times typestyle:

Times italic

ABCDEFGabcdefg1234567!@#$%^&

Times bold

ABCDEFGabcdefg1234567!@#$%^&

Times extended bold

ABCDEFGabcdefg1234567!@#$%^&

Times italic bold

ABCDEFGabcdefg1234567!@#$%^&

See PostScript, Truetype, typeface.

font size A designation of the overall height of a font, usually described in *points*, as in "14-point Palatino." Fonts of the same size may appear very different from one another, depending on the ratio of the baseline, descenders, and ascenders to the total height of the font. Thus, a 12-point script font may appear much smaller than, 12-point Helvetica, a plain nonserif font, even though their height from top to bottom may be the same. The actual font size is measured approximately from the top of the highest ascender to the bottom of the lowest descender, with perhaps a small amount of extra space above and below (usually below) these extents. The extra space comes from the fact that traditional physical printing blocks sometimes extended slightly beyond the top and bottom of the ascenders and descenders. Further distance between lines of type is accomplished with *leading*. See Font Sizes chart. See leading.

foo 1. *slang* Term of frustration. There are many stories as to the origin of the term, but the most common explanation is that it is a shortened derivation from a World War II acronym FUBAR (Fowled Up Beyond All Recognition or Repair), with the variant phonetic spelling of *foobar*. Forward Observation Officer (FOO) has also been offered as a possible origin in some dictionaries. It has further been suggested that foo may be a contraction of *fooey*, a common expression of frustration or disappointment. 2. A term adapted by many programmers as a general-purpose or temporary variable name, or as a counter variable. 3. An online general-purpose placeholder.

footprint 1. An area or impression on a surface which comprises a more or less contiguous region of contact with the 'bottom' of some object or signal. 2. The desk space or floor space taken up by a piece of furniture or equipment, usually considered the area of actual contact, or the area of contact plus everything within its boundaries, and the small area surrounding it, which may be taken up by connectors or protruding knobs. 3. The terrain or surface of the Earth over which a transmission signal can be received. A transmission footprint is a little less defined than a physical footprint, as a transmission tends to gradually decrease in intensity (this may be

Font Sizes Shown in Points

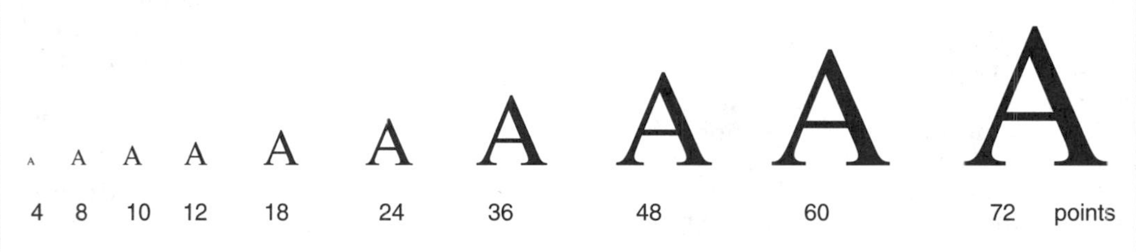

A	A	A	A	A	A	A	A	A	A	
4	8	10	12	18	24	36	48	60	72	points

shown by contour lines on a map or chart), and there is often no definite cutoff point, unless specified as signals below a certain level. 4. An audit trail, or traces left by a transaction or process which has concluded, or aborted. 5. The resource requirements of a system. For example, the Amiga is said to have a *small system footprint* because it can adroitly handle preemptive multitasking, sound and simultaneous animated graphics in a Megabyte of memory on a 25 or 40 MHz processor.

forecasting Predicting future events, usually based on an analysis and evaluation of past events. Forecasting is needed in all areas of telecommunications to choose technologies that are powerful, economical, and that won't be quickly outdated. It is also used by system administrators to configure and tune systems to handle predicted needs and traffic loads. It is utilized by businesses to select local area network topologies and workstations, and by managers to organize employee loads and working schedules. See erlang, queuing theory, traffic management.

Foreign Agent A service which enables nodes to register at a remote location, providing a 'forwarding address' to a home network in order for forwarded packets to be retransmitted to the remote location. Foreign agents are an important aspect of Mobile IP systems.

Foreign Exchange Service FX, FEX, FXS. A service which connects a subscriber's telephone to a remote exchange as though it were a local exchange. Commercial vendors provide a variety of multiplexing interface cards to telecommunications carriers to facilitate provision of subscriber Foreign Exchange Services.

form feed FF. A formatting instruction in electronic and print publishing that directs the cursor or print head to position itself at the top of the next form or page. In electronic documents, and continuous feed paper (as on unperforated rolls, sometimes used in facsimile machines), the form feed may be arbitrary, situational, or user-configurable. Commonly, a form feed defaults to 60 lines, or the top of a letter- or legal-sized piece of paper. A form feed is sometimes symbolized in documents with CTRL-L.

Forrester, Jay Wright (1918-) A computer pioneer who investigated memory devices for computers in the 1940s, and also in the 1950s while working on the construction of the Whirlwind computer at the Massachusetts Institute of Technology (MIT). Forrester was at the forefront of transition technology from analog to digital systems, and invented core memory with assistance from William N. Papian in 1951.

FORTH *Fourth* Generation Language. An extensible, high-level programming language typically used in calculators, robotics, and video gaming devices.

FORTRAN *For*mula *Tran*slation. A high-level computer programming language that was commonly used in the 1980s for math-oriented applications, and from which BASIC has derived many of its syntactical characteristics. It grew partly out of conceptual ideas and examples of reusable code promoted by Grace Hopper, and further from the encouragement of John Backus that a language be developed that could express and solve problems in terms of mathematical formulae. With the advent of BASIC, C, C++, Perl, and Java, the use of FORTRAN is declining.

fortune, fortune cookie A witty, profound, pithy, satirical, or quietly thoughtful quote, proverb, maxim or remark often displayed immediately after login on a networked computer or terminal.

forum Discussion group, private or public meeting, judicial assembly.

forum, online A network virtual environment for discussions. Internet Relay Chat (IRC) channels, USENET newsgroups, discussion lists, and various meeting places on Web sites are examples of global forums where topics are ardently and enthusiastically debated. When forums include celebrities, they are usually moderated to keep the questions and comments to a reasonable level.

forward error correction FEC. A means of ensuring a transmission in advance by duplicating information, or otherwise improving the chances of it being received the first time. For example, characters, or groups of characters, may be sent two or more times (called *mode B* in amateur radio transmissions) according to a predetermined arrangement. Repeating characters, or groups of characters in data transmission gives a receiver an opportunity to compare the groups, and, if any of the information doesn't match, then ask for a retransmission. The basic idea is to minimize the back-and-forth nature of handshaking in order to speed up a transmission while still giving information which may be used to check the integrity of the information being received.

Forward Explicit Congestion Notification FECN. In a frame relay, a bit which notifies an interface device to initiate congestion-avoidance procedures in the direction of the received frame. See Backward Explicit Congestion Notification.

FOT Fiber Optic Terminal. A point or device at which a fiber optic circuit connects to copper wire circuit.

FOTS fiber optic transmission system.

Foucault, Jean (1819-1868) A French physicist best known for his studies of light and the rotation of the Earth through the use of pendulums.

Fourier, Jean Baptiste Joseph (1768-1830) Fourier, a French mathematician and lecturer, discovered in the early 1800s that the superposition of sines and cosines on time-varying periodic functions could be used to represent other functions. He made practical use of these techniques in the study of heat conduction, work that was developed further by G. S. Ohm in the 1820s in his mathematical descriptions of conduction in circuits. Work on linear transformation mathematics that predated Fourier's publications was carried out by Karl. F. Gauss, but went unpublished until after Fourier's descriptions. See Fourier transform.

Fourier's mathematical writings inspired many subsequent mathematicians and physicists to apply various mathematical manipulations to various types of data. These techniques are still used in mathematical modeling, analysis, and computer data manipulation techniques.

Fourier analysis A means of representing physical or mathematical data by means of Fourier series or Fourier integrals.

Fourier transform A linear mathematical data manipulation and problem-solving tool widely used in optics, transmissions media (antennas), and more. The superposition of sines and cosines on time-varying functions can be used to represent other functions, in other words, to represent the data from another point of view. The result of such a transformation is to decompose a waveform into subsets of different frequencies, which together sum up to the original waveform. In this way, the frequency and amplitude characteristics can be separately and more easily studied.

A rudimentary application of Fourier series calculations was applied to astronomical attempts to understand planet orbits in Greek times. Their development was in part hampered by the Greeks' mistaken assumption that the Earth was the center of the universe.

Fourier transforms differ from wavelet transforms in that they are not localized in space, however they also share many common characteristics. Named after J. B. J. Fourier. See discrete cosine transform, wavelet transform.

Fourier transform, fast FFT. This optimized version of a Fourier transform was developed in 1965 by Tukey and Cooley. It substantially reduces the number of computations needed to do a transform, hence the name.

fox message A test sentence that includes all the letters of the English alphabet, which is commonly used to verify if all letters of the English alphabet in a device or coding system are present and/or working correctly. Familiar to most as "THE QUICK BROWN FOX JUMPS OVER THE LAZY DOG" (which is then repeated as all lowercase, as needed).

Fox, Talbot An early experimenter who developed a salt print process which was an important step in efforts to create photographs on paper. The early photos up to this time had been printed on ceramic and silver-plated copper. See Daguerre, L.J.M.

FPLMTS See Future Public Land Mobile Telecommunication System.

fps See frames per second.

FPU See floating point unit.

FRA fixed radio access.

fractal A term popularized by Benoit Mandelbrot to describe his geometric discoveries and descriptions of structures which can be described and reproduced as mathematical formulas, and which have the characteristics of self-similarity in increasingly fine degrees of detail.

Fractal concepts have since permeated almost every aspect of computing, especially computer imagery, as the geometry provides the means to model surprisingly complex and natural looking structures with simple mathematical formulas. Fractal geometry has also been applied to a number of compression and generation techniques in a variety of fields. See the "Fractal Geometry of Nature" by Benoit Mandelbrot.

Fractal Image Format FIF. A proprietary image compression format developed by Michael Barnsley and Alan Sloan, who together founded Iterated Systems, Inc. to exploit the technology. Very high rates of compression are possible. The technology is asymmetric, that is, it takes a while to compress the information, but it decompresses relatively quickly. See fractal transform.

fractal transform, fractal compression A resolution-independent, lossy image compression technique which provides a high degree of perceptual similarity with excellent compression results. Fractal compression works by storing image components in terms of mathematical algorithms, rather than as individual pixels of a particular location and color. The organization of the image is evaluated for its intrinsic characteristics of self-similarity, and those characteristics are coded so they can be reproduced by repetitions in increasingly fine detail, up to the resolution of the output device.

With their excellent image fidelity and high compression ratios, the trade-off in fractal compression is the time it takes to encode or decode and display the decompressed image. With faster processors, this is becoming less of a limitation. See lossy compression; discrete cosine transform; Fisher, Yuval; JPEG; Mandelbrot, Benoit; wavelet transform.

fractals, fractal images A term borrowed from fractal geometry to describe visual images which have recognizable visual and mathematical characteristics of self-similar, repeating branches and curves resulting from the rendering of fractal formulas. Colored fractals can be beautiful, and adorn many calendars, posters, and t-shirts. Many familiar fractal formulas have been given names, such as Julia Set, Mandelbrot Set, etc. See Mandelbrot, Benoit.

FRAD See frame relay access device.

fragmentation 1. Breaking up, separating into units or groupings, losing connections or cohesiveness, becoming physically or logically separated over time. 2. In hard drive storage, fragmentation is a gradual process of the available or used areas of a drive becoming smaller, and more widely dispersed. When information is stored on a hard disk or other similar directory-based system, files are placed where there is room on the drive, and sometimes spread over a number of areas on the drive. When a file is deleted, its directory entry is removed, and the space it occupied becomes free for other files. However, over time, especially if there is a lot of disk activity, the free areas get smaller and farther apart, and files stored on the drive need an increasing number of sections and links to keep track. This slows down the system. It is sometimes advisable to defragment or 'defrag' a drive to clean up the tables and file data locations. Some operating systems have built-in utilities for rebuilding a drive or system. It is important that sufficient memory and swap space are available on a system before defragmenting a drive, and it is highly advisable to back up the drive before proceeding.

frame 1. A visual or logical unit, or block of related information, sometimes delimited with visual or binary flags or markers. A frame is sometimes a natural unit, as in a cyclic event in which the information repeats in some general sense (though the content may vary), and sometimes is an arbitrary unit, chosen for convenience, or by convention. 2. A physical unit, border, containment area, skeleton (framework), or inclusive extent. 3. A full-screen perceivable image on a monitor or TV screen consisting of the sum of all the sweeps of an electron gun during a full cycle of oscillations across the screen. 4. A unit of information in data networks such as frame relay systems. 5. A contained group of information on an HTML layout, such as a Web page. 6. A housing or support structure for components or wiring. See distribution frame, rack.

frame, data In most networking architectures, a frame is group of data bits of a fixed or variable size, often in a specified format. It is common for frames to be organized into two general types: those which carry signaling, addressing, or error detection/correction information, and those which carry the contents of the communication itself (sometimes called payload), although even these are sometimes combined. The format and organization of the frames is defined by a data 'protocol,' and there are many general-purpose and specialized protocols in use, most not directly compatible with one another. Interprotocol frame traffic can be 'carried' or 'tunneled' through other protocols, or can indirectly communicate through conversion agents or filters.

Frames are organized into larger units that

comprise a communication, and then may be sent all together to the destination, or disassembled, sent along different paths, and reassembled at the destination. Frames may also be 'encapsulated,' that is wrapped in an outer envelope, to carry them through a system that requires another format, or to tunnel through a system without having the contents of the encapsulated package changed in any way. It is then de-encapsulated at the exit point, or at the destination.

When frames carry different types of information, such as graphics in one and sound in another, they are sometimes sent simultaneously through separate wires or data paths, and reassociated at the receiving end, as in a videoconferencing application. In these situations, synchronization or alignment of information is important and information for achieving this may be included in the frames. See frame relay, frame relay frame format, protocol.

frame, distribution A wiring connection supporting structure. See distribution frame.

frame, video In video displays which cyclically sweep the full screen to create an image, a frame is the extent of the sweep that is required to cover the full screen. In the NTSC system prevalent in North America, the sweep is ~29.97 frames per second, and, on an interlaced screen, is further subdivided into two sets of 'fields' (all odd lines or all even lines). The formats that are common in Europe (PAL, SECAM), display at 24 or 25 frames per second.

It is important to time the frame presentations at a broadcast station, so that news briefs, commercials, and regularly scheduled programming can be organized into precise time slots. NTSC displays are generally 525 scanlines, though not all the bottom scanlines may be visible on the screen. European standards are 625 lines. A frame is an important unit in video display not only for physical synchronization of the signals, but also because the rapid sequential presentation of 'still frames' creates the illusion of movement, and the properties of this illusion must be taken into consideration if creating 'still-frame' animation sequences. See station clock. See television signal for a chart of common formats.

frame buffer A storage area which is used for preconstructing digital images in order to facilitate the quick display of those images, especially if they are to be displayed one after another, as in animation. The image in the frame buffer is not necessarily displayed all at once. For example, in video games, it is very common to store a wide narrow *panoramic* landscape in a frame buffer and to display only a portion of the scene at any one time. Then, as the characters in the game move along the landscape, the display scrolls smoothly to right or left, without display artifacts such as flicker or jumping. Frame buffers are commonly used for computer animations, arcade games, and video walls. See frame store.

Frame Check Sequence FCS. A mathematical algorithm which derives a value from a transmitted block of information and uses the value at the receiving end of the transmission to determine whether any transmission errors have occurred.

FCS is used in bit-oriented protocols such as SNA SDLC to determine if sent and received messages are the same. For example, in SDLC the two-byte (16-bit) FCS field includes a cyclic redundancy check (CRC) value which is used to assess the validity of the received bits.

frame grabber A computer hardware/software peripheral device designed to capture and digitize a frame, or series of frames, from a continuous signal, usually from an NTSC source. The signal generally comes from live video, laserdisc, or prerecorded tape. It is sometimes called a video capture board or video digitizer. See frame buffer.

frame merge Over frame-based media, a stream merge.

frame mode In video image capture, a mode that captures a full frame of scan lines. Full frame images preserve image integrity, but also take more storage space than some modes. See field mode.

frame rate, video The speed at which a series of images is presented or a screen of visual information is drawn, usually expressed in terms of seconds. Due to *persistence of vision* in human perceptual systems, individual still images presented at about 20 frames per second or faster give the illusion of motion. At speeds of over 30 frames per second, no substantial improvement in the animation quality is perceived by most people.

Motion picture film is usually displayed on 35 mm projectors at 24 frames per second. Home 8mm and Super-8mm projects are somewhat variable around 20 to 24 frames per second, since most of them have dials to speed up or slow down the film.

North America TV is broadcast at about 30 frames per second. (In actual fact, due to differences between black and white and color technology, the rate is closer to 29.97 frames per second.) On various European systems, such as PAL and SECAM, broadcast frame rates are 24 or 25 frames per second.

On computer systems, frame rates vary with the software that is creating the frames or with the software playing the frames. Smaller video 'windows' can be played back faster than large ones, as they take less time to compose and require less computing power to display. Displays of 256 colors also refresh faster than 24-bit (~16.9 million colors) displays, although this will vary with the system speed and type of graphics card used. On the Amiga 4000 computer, one of the better-adapted systems for video, combined graphics and sound can be played back simultaneously at speeds from 30 to 60 frames per second without degradation of the audio or visual. On systems less well adapted to video, the rates may vary from 20 to 30 frames per second.

Videoconferencing systems running over analog phone lines may only refresh at frame rates of 5 or ten times per minute, as the voice-grade lines and modem create a bottleneck. On ISDN and other digital lines that run at faster rates, 20 or more frames per second may be possible, depending upon the type of system and the size of the image window. See drop frame, MIDI time code, SMPTE time code.

frame relay Frame relay is a networking connection option often selected by small businesses as a cost-effective means of setting up a reasonably fast and powerful local area network (LAN) that can connect with public networks. It is a connection-oriented packet-switching network protocol designed to provide virtual circuits for connections attached to the same frame relay network. It evolved from, and is somewhat simplified over X.25, for example, in that it concentrates on packet delivery without sequence and flow control, resulting in faster throughput and sometimes lower cost. Frame relay has been shown to work in practical situations up to about 50 Mbps. Frame relay can be used to interconnect multiple LANs at lower rates than the cost of leased lines. It has been commercially implemented since the early 1990s.

Control signals must be provided to indicate the connection status of the link. In a Point-to-Point (PPP) system, frame relay framing is treated as a dedicated or switched bit-synchronous link. Frame relay services are available from commercial vendors in most urban areas in the U.S.. See cell relay.

frame relay access device FRAD. This is another name for the switch, router, or other network device that assembles and disassembles frame relay frames as they are transported through a system. When data frames

Frame Relay Frame Format

```
Frame Format

0         1         2         3         4         5         6         7         8
01234567890123456789012345678901234567890123456789012345678901234567890123456789 0
+-------------------/----------------/-------------------------------------+
| flag    |       address      |  information     |        FCS        |  flag  |
+-------------------/----------------/-------------------------------------+
    1 octet    2 to 4 octets        variable            2 octets         1 octet

Header Structure

+--------------------------------------------------------------------+
|         DLCI upper                                |   C/R   |   0   |
+--------------------------------------------------------------------+
|         DLCI lower             |  FECN   |  BECN  |   DE    |   1   |
+--------------------------------------------------------------------+


   DLCI    data link connection identifier
   C/R     command/response
   FECN    forward explicit congestion notification
   BECN    backward explicit congestion notification
   DE      discard eligibility
```

are sent over a frame relay network, they are 'packaged' with various types of information, often at the beginning and end of the block of frames, and 'unpackaged' again, often at the access point to the destination system to recover the structure and contents of the original communication.

frame relay flow control Flow control, the management of movement of frames within and between networks, is not explicitly defined in the frame relay specification, and the ITU has defined general concepts and standards for handling flow and congestion. In practice, congestion can be prevented in frame relay networks by establishing committed information rates (CIRs) to each user, or denying the connection if insufficient bandwidth is available, and by discarding frames above the CIR. Existing congestion can be signaled to the user in the form of backward explicit congestion notification (BECN) and forward explicit congestion notification (FECN).

Frame Relay Forum FRF. An international professional association of corporations, vendors, carriers, and consultants promoting the frame relay networking technology, and supplying commercial frame relay products and services. The FRF was established in 1991. http://www.frforum.com/

frame relay frame format The format for a frame is based on link access protocol D (LAP-D) for ISDN. Frames are also known as protocol data units (PDUs). Frame relay frames are similar to DXI and FUNI.

The format specified for a frame in a frame relay system includes a one byte flag, followed by two- to four-header address bytes, followed by a variable number of information bytes, followed by a two-byte CRC code, followed by a one byte flag. There are a number of possible configurations of the address field, as it may be two, three, or four bytes in length, as determined by the extended address (E/A) bit. See the Frame Relay Format chart.

frame relay installation Frame relay communications connection services are generally available for a monthly subscriber fee or per-data rate from a local commercial provider, depending on the speed of transmission, along with a one-time connect charge each for installation and port installation. Transmission speeds up to 56 Kbps or 64 Kbps are typical, although most vendors offer higher speeds for more money.

frame relay physical layer interface The specification for frame relay does not stipulate particular physical connectors or cables. In practice, however, unshielded twisted pair (UTP) is commonly used in ISDN implementations of frame relay.

frame relay service Frame relay service consists of a combination of hardware, software, and transmission services. It provides multiple independent multiplexed data links to another destination or to several destinations through a process which is at least as transparent as a leased line and less expensive. See frame relay installation.

frame relay, voice over VoFR. Frame relay technology provides an opportunity to combine data and voice communications services over the same network. Analysis of typical voice communications indicates that much of it is unnecessary (background sounds, pauses, etc.) and can be screened out before transmission over data networks. This offers possibilities for processing and compression to provide for efficient transfer of digitally encoded voice conversations. Initially, there was no uniform standard for carrying voice over frame relay and various schemes for its implementation had been developed. However, work on interoperability and transport standards are ongoing.

frame relay capable interface device FRCID. A device which performs frame encapsulation within a frame relay. See bridge, encapsulation, router.

Frame Relay Implementors Forum An association of vendors that supports standards of interoperability for frame relay implementations. A common specification was first introduced in 1990 based on the standard proposed by the American National Standards Institute (ANSI). See Frame Relay Forum.

frame store A high-capacity digital video storage buffer. Frame stores are most commonly used in two categories of applications: 1. those which require image buffering to provide sufficient speed for continuous display (see frames per second), such as computer editing or display systems, and 2. those which require image buffering in order to create complex, composite, or multiple display systems (such as video walls).

In the first instance, the device from which the frames are being displayed or the display software may not be fast enough to read and display at 30 or so fps. By using a frame store, sufficient frames can be 'buffered' in fast access memory (or on a very fast drive) to provide quick display and the illusion of continuous motion. If the display software creates unwanted effects on the screen when loading the next frame, the transition can sometimes be smoothed out with double-buffering or grabbing the next frame from the frame store rather than from a hard drive. In other words, the new image is preconstructed in memory while the current image is being displayed, and the buffered image can then be displayed instantly over the previous frame, rather than reading in and decompressing the frame and then displaying it line by line over the previous frame.

In the second type of application, a frame store can help compose a complex image, such as computer graphics effects for a movie, which may have been raytraced one frame at a time, but which, when combined with footage of the actors, needs to match the speed of the action. A frame store can also be used as a component of a video wall, say 20 monitors in a four by five grid, which shows 1/20th of the actual image on each monitor. Since this takes some computing power to split up an image into 20 separate subimages, the image grid could be segmented and prestored, so all the monitors display the correct parts of the grid at the same time. See buffer, frame buffer, desktop video.

frames per second A phrase which describes display speed for TV broadcasts, video, and film animations. The two most important aspects that determine this speed are human perception and display technology. Through persistence of vision and expectation, humans perceive still frames displayed quickly one after the other as motion. It requires only about 15 to 30 frames per second (depending upon the amount of detail, and speed of the action) for these images to appear to be continuous motion. Most animations are created with 24 to 30 frames per second. Since motion media can be displayed only at the fastest speed of the display medium (usually a cathode ray tube), the technology also determines the display rate, with speeds of 15 to 60 fps being implemented, and about 20 to 30 fps most often used. See frame; frame rate, video; refresh.

franchise A government granted right to offer community public 'right-of-way' for exclusive commercial communications services, such as phone services or cable broadcast services. The franchise fees, or a portion of them, may be used by local government agencies, a portion of which may be allocated to local Designated Access Providers (DAPs) for facilities funding. Some of the earliest local phone companies may be partially exempt if they gained their exclusivity prior to regulation (grandfathering).

Franklin, Benjamin (1706-1790) An American businessman (printer), statesman, scientist, and philosopher who did numerous experiments in electricity and printing. He shared his discoveries openly, and coined many of the terms used to describe mechanics and electricity today. He called vitreous electricity, demonstrated by rubbing glass with silk, *positive* electricity, and resinous electricity, demonstrated by rubbing amber with wool, *negative* electricity. He did a number of experiments with lightning and stored up electrical charges in a device called a Leyden jar, and established that man-made electricity and atmospheric electricity had the same properties. These experiments were enthusiastically received and replicated throughout Europe, spurring interest and development in the field of electricity.

Ben Franklin also developed some early document duplication techniques which he used on his own printing press to help him manage his voluminous records and correspondence. Ben Franklin always hoped to retire early to devote the rest of his life to scientific inquiry and his various hobbies, but the American Revolution and the overwhelming public demand for his diplomatic skills kept him occupied for long hours right up to the time of his death in his mid-80s. See electrostatic, Leyden jar.

Franklin Institute A significant organizer and promoter of activities related to electrical education, professional development, and technological development in electronics. It organized many key American and international exhibitions starting in the 1800s.

fraud *n.* Deceit, trickery; unauthorized access or use, especially under an assumed identify such as a false username or through the use of unauthorized equipment.

Fraunhofer region In an antenna, a region of the field from which the energy flow proceeds as though emanating from a point source near the antenna. See Fresnel region.

Fraunhofer spectrum The portion of the solar spectrum visible to humans.

Free Software Foundation FSF. A Massachusetts-based association committed to the development, acceptance, and promotion of open, free software standards and applications to benefit the world at large. The freedom to copy and distribute software, and the freedom to modify, enhance and improve software are encouraged by the FSF. Thus, the programming and user communities benefit by the availability of constantly improving software and standards, and programmers have a broad, ready base of software from which to learn and to improve their skills.

The FSF has developed the integrated GNU software system, which includes assemblers, compilers, and more. Donations to the FSF are tax deductible. http://www.fsf.org/

FreeBSD A Unix computer operating system descended from Berkeley Standard Distribution (BSD), which flourished with the development of the ARPANET, the forerunner to the Internet. The Computer Systems Research Group (CSRG) enhanced BSD with 32-bit addressing, virtual memory, and a fast file system supporting long filenames. They further introduced BSD Lite that was BSD without the licensed AT&T code, which could be freely distributed. The CSRG disbanded in 1992, and the community at large adopted BSD and developed FreeBSD. In 1994 some of the CSRG briefly came together and further developed BSD 4.4 Lite. See Berkeley Standard Distribution, Unix, UNIX.

freeware A category of product, usually software, which may be distributed and acquired without cost. Freeware does not mean 'copyright-free.' A developer has the right to retain the copyright to an original work and still distribute that work or product free of charge, keeping the right to revoke freeware privileges. Freeware is *not* the same as public domain software, in which the owner has given up the copyright, and is *not* the same as shareware, in which there is a moral obligation to pay the stipulated fee. Freeware (public domain) and shareware are two common types of freely distributable software. See public domain, shareware.

freeze frame A mode of visual display in which only one screen-full, or cell, of an animated sequence is shown. On digital systems, it's easier to show a single frame of information; displaying a single frame on film or on a CAV laserdisc is pretty straightforward. In analog systems, or those which are recorded with overlap of information or no firm transition from one 'cell' to the next, it is more difficult, as VCR tapes and the freeze fame mode may be of limited duration and quality. Some analog/digital systems will take a digital 'shapshot' of the frame to be displayed and display it as a digital image, usually with better results than trying to display the analog image. Many video conferencing systems don't show actual realtime motion, but rather snapshot a digitized freeze frame every few seconds, to provide the illusion of seeing what is going on at the other end without seeing actual movement. These are sometimes distinguished as 'audiographics' systems. As transmission media become faster, full-motion video will become standard.

F

frequency (Symbol - f) The number of periodic occurrences in a specified unit of time. In electricity, frequency is the number of times a current alternates in hertz (named after Heinrich Hertz). Radio signals are usually measured in kHz, or in MHz at high frequencies (above 30,000 kHz). Many older radio and electronics manuals will describe the frequency in terms of cycles per second instead of hertz (cycles per second), as the naming of the unit did not begin to be widely used until the 1960s.

frequency bias A constant adjustment to a signal frequency which prevents it from reaching zero.

frequency departure The degree of variation of a carrier frequency, or reference frequency from an expected, assigned value.

frequency division multiple access FDMA. One of the simplest techniques for increasing capacity over communications channels, since the radio frequency spectrum is not unlimited. FMDA is a way of dividing up the available spectrum according to frequencies. The communications station typically assigns a unique frequency or frequency sequence to each user currently engaged in communication, and tracks these as needed to provide many simultaneous links. This technique is used in cellular phone and satellite transponder systems. See code division multiple access, complex scheme multiple access, multiple access, time division multiple access.

frequency division multiplexing FDM. A technique used to more efficiently utilize a fixed or limited amount of bandwidth by subdividing it into narrower channels. Typically *guard bands* are inserted between communications bands to reduce interference. Multiplexing can be used to increase the number or types of transmissions within a fixed medium. For example, it may be used to simultaneously transmit voice and data.

One of the earlier contributors to FDM technology was George Ashley Campbell, who invented the electric-wave filter in 1915, a device which is used in FDM. FDM is a widely used transmissions technique which is only just now being superseded by other methods, such as time division multiplexing (TDM), which is prevalent in fiber optic communications systems. See single sideband, time division multiplexing.

frequency frogging See frogging.

frequency hopping In mobile communications systems, a spread spectrum technique in which frequencies are jumped during the course of a transmission. This may be done for many reasons: to try to find a cleaner or more stable signal or to try to avoid detection (sometimes used in military zones).

Frequency hopping was invented by Hedy Lamarr (born Hedwig Eva Maria Kiesler) while trying to develop a secure guidance system for a torpedo, using radio signals that would not be detected and subsequently jammed. Her collaborator, George Antheil, suggested a way to synchronize the varying frequencies with paper tape, but the synchronization system was somewhat cumbersome, and it was not until the development of computer electronics that Lamarr's idea was fully implemented. Lamarr and Antheil received a patent for the technology in 1942 and it has since been extensively used in military and civilian communications systems. Unfortunately, neither Lamarr, better known for her film career, nor Antheil, received any of the compensation or credit which was due for the invention. See direct sequencing, multiple access, spread spectrum.

frequency modulation FM. A sine-wave modulation technique widely used in broadcasting that works by varying the frequency of a constant amplitude carrier signal with an information signal.

FM radio broadcast signals typically require about 200 kilohertz of bandwidth and are not as subject to noise and interference as amplitude modulation (AM).

Many scientists insisted that frequency modulation was not possible. Edwin Armstrong thought it was and devoted a decade of intense research to the problem, ultimately proving successful. FM radio stations began broadcasting in the early 1940s. In the United States, the Federal Communications Commission (FCC) approved FM stereo broadcasting in 1961. It has approved the range from 88 to 108 MHz for FM broadcasting.

In one type of telephony, a frequency-modulated carrier signal can be transmitted over wires. Frequency modulation can be used when digital data is routed through an analog system for part of the transmission.

FM is also commonly used for very short range communications for cordless phones, home and business intercoms, baby monitors, short-range television security systems, and burglar alarms. See amplitude modulation; Armstrong, Edwin; carrier; channel; modulation; Moonbounce.

frequency shift keying FSK. A modulation technique used in data transmissions such as wireless communications in which binary "1" and binary "0" (zero) are coded on separate frequencies. This scheme can also be adapted to regular phone lines by assigning binary "1" to a tone and binary "0" (zero) to a different tone. There are other keying schemes for carrying information such as *on/off keying* and *phase shift keying*. See frequency modulation, phase shift keying.

frequency swing In frequency modulation, the difference between the maximum and minimum values at a given frequency. In other words, the limits within which the oscillations range.

frequently asked question(s) FAQ. A query, or question and answer list, of questions which have been asked and answered many, many times, so often, in fact, that someone has taken the time to write up the question/answer and post it, usually in a public forum on the Internet. FAQs comprise an important part of the information base of the Internet, on private and public forums, chats, special interest groups (SIGs), USENET newsgroups, and Web data sites. All Internet users are strongly advised to read the FAQ *before* posting on any online forum or risk being soundly scolded or flamed by other users. See Netiquette, RTFM.

Fresnel region A region around an antenna designated as that between the antenna itself and the Fraunhofer region. The transition between the Fresnel and Fraunhofer regions can be mathematically calculated if the length of the antenna and the wavelength are known. See Fraunhofer region.

friction feed A feed mechanism in a machine that relies on friction or pressure to feed the sheets (paper, card stock, thin metal plates, etc.) as in a printer, press, or photocopying machine. Friction feed devices often use materials such as rubber to help adhere the medium to the feeder. See tractor feed.

friendly name A name which is easy to recognize and remember, used in place of cryptic or long names or codes. For example, a printer with a computer designation of LSL2345-b may be assigned a friendly name of *AdminLaser* in lists of available output devices. Domain names on the Internet have been given friendly names. The computer system doesn't require a familiar name like "coolsite.com" to locate a site; a binary address is more direct. But humans prefer language to numerals or binary addresses, and so domain names have been associated with data addresses to make it easier to use 'FTP' or 'Telnet' commands, or to access a site through the Web. See alias.

fringe area A region just outside the major transmission area of a broadcast signal where the signal is degraded and inconsistent. Sometimes those in fringe areas can improve the quality of service, up to a point, with better antennas.

fringing 1. An undesirable visual artifact, especially on cathode ray tube (CRT) color displays, in which the electron beams are converging incorrectly so as to appear unfocused, with a fringe or edge of color slightly offset. 2. A visual artifact on an object displayed in a computer paint program, in which the color of the previous background of the object (e.g., a blue sky) shows up distinctly as a halo around the edges when the object is placed on another color (e.g., a red brick wall). A defringing option to blend the edge colors is available in many paint programs, to smooth out the transition in a process called 'antialiasing.'

FRM focus-rotation mount. A pivoting antenna focusing structure.

FRND frame relay network device.

frogging 1. An equalizing technique in which incoming high or low frequencies are inverted to become outgoing low or high frequencies. 2. Corruption of transmissions data in which incorrect data is inserted into, or overwrites some of the expected data in a nonrandom way.

front end The portion of an application or device that interacts with, or is accessible to, the end user. On computers, shell command lines and graphical user interfaces are the most common types of front end. In audio/video equipment, the front end consists of the knobs and dials that are within easy reach of the user. The Web has become an interactive front end to the Internet. In commercial facilities, the front end is usually a store front or reception area that provides customer services, as opposed to storage or personnel-only work areas. In broadcast circuitry, the front end consists of the knobs and components which tune in the desired frequency.

front end system A system that acts as an intermediary gateway, filter, or console for a more powerful, but less user-friendly or accessible system. A desktop computer-based telecommunications server can serve as a front-end to a mainframe, sparing it for more computing-intensive tasks. An information kiosk with a simple touchscreen or touchpad input system that hooks into a more powerful network system is a type of front end for the general public. An automated teller machine (ATM) is a banking system front-end for the public.

frownie :-(:-/ 8-(A form of ASCII-based visual communication of an emotion, used on communications networks. You can look at these little symbols with your head turned to the left to see the images of the frowning faces. See emoticon for additional examples.

FRSE frame relay switching equipment.

FRTE frame relay terminal equipment.

frustum The surface of a solid cone or pyramid that would be created if the top of the cone or pyramid were cut off parallel to its base. A concept of interest to programmers and users of 3D modeling software.

FSAN See Full Services Access Network.

FSF A nonprofit educational association supporting GNU. See Free Software Foundation.

FSK See frequency shift keying.

FSO Foreign Service Office.

FSP See File Service Protocol.

FSS See fixed satellite service.

FSTC Financial Services Technology Consortium.

FTA See Federal Telecommunications Act.

FTIP Fiber Transport Inside Plant.

FTNS Fixed Telecommunications Network Service.

ftp The command typed at an FTP site to access an archive based on File Transfer Protocol (FTP). It is important to type the command all in lower case, however, the protocol itself is usually written in upper case.

FTP See File Transfer Protocol.

FTP mail server A mail server that facilitates the retrieval of files from FTP archives by sending them to the user's email address. Since files on FTP sites can be text or binary, and some email addresses cannot directly accept binary files, the retrieved files may be sent as a binary *file attachment*. See file attachment, ftp, FTP.

FTS 1. file transfer support. 2. Federal Telecommunications System. A government private telephone network. See FTS2000.

FTS2000 Intercity telecommunications services provided to federal agencies by the U.S. General Services Administration (GSA) through two networks (A & B) transmitting through fiber optic cable.

FTTC fiber to the curb.

FTTH See fiber to the home.

FTTL fiber in the loop.

FUBAR fouled up beyond all recognition. A phrase purportedly originating in military speak in World War II. Less polite versions of it fit the acronym as well. See foo.

FUD fear uncertainty doubt. A sales strategy attributed to Gene Amdahl as "FUD is the fear, uncertainty, and doubt that IBM sales people instill in the minds of potential customers who might be considering [a competitor's] products." The marketing spin is that customers are safer with International Business Machines (IBM) products.

fudge *v.* To hedge, approximate, overstate, or talk around a subject so as to appear to know what you are talking about, to use 'bafflegab'; to cobble together so it appears as though it might work, or so that it approximately works but may not be complete or robust.

fudge factor Tolerance factor, buffer, safety net. See fudge.

fugitive glue A type of glue designed to hold things in place 'just long enough,' as in magazines, where it is strong enough to hold *flippies* in place until they reach the newsstand or your mailbox, but drops them at your feet as soon as you leaf through the publication.

Fujitsu Limited A large Japanese commercial conglomerate originating in the 1920s. Fujitsu is known for a number of large-scale computing products, is a world leader in industrial robotics, and manufactures many consumer computer-related accessories. See FACOM.

Fujitsu Network Communications, Inc. A designer and manufacturer of fiberoptic and broadband switching systems, and provider of telephone network management software which is marketed to ILCs, CLECs, VPNs, and cable TV companies.

full duplex A system that supports simultaneous transmission and receipt at both ends of the circuit. Full duplex operation requires a balance of hardware and software protocols. In some systems, full duplex operation creates a digital 'echo' in which each unit of textual information is repeated. Some systems can technically support full duplex operation, but are selectively operated in half duplex mode to improve the quality of the communication, as in some satellite voice systems and speakerphones. These systems have tones and sensors that coordinate the back-and-forth nature of the conversation so the transmission favors the direction in which the current user is transmitting. Systems with bandwidth limitations may work in full duplex mode for some operations (e.g., voice conversations) and then may switch to half duplex for more bandwidth intensive operations (e.g., videoconferencing). See half duplex.

full scale The full functional range over which an instrument or device operates.

Full Services Access Network FSAN. A group of cooperating international telecommunications companies, including Bell Canada, BellSouth, BT, Deutsche Telecom, Dutch PTT, France Telecom, GTE, Korea Telecom, NTT, SBC, Swisscom, Telefonica, Telstra, and Telecom Italia. FSAN shares its documentation with relevant standards bodies. One of the groups associated with FSAN is the Optical Access Network (OAN). Nippon Telephone and Telegraph (NTT) and BellSouth are developing fully FSAN-compliant ATM-PON systems for 1999. See fiber to the home.

function key A configurable, or special-purpose keyboard button. Function keys are often programmed as 'shortcuts' to produce the same effect as typing several keys, or selecting an operation several menu items deep. Many computer keyboards have 10 or 12 function keys with a variety of uses, depending upon the currently active software. They may be located in a vertical line above the other keys in the keyboard, or they may be organized in two rows to the right or left of the keyboard. For touchtyping, the double row to the right or left of the keyboard is more practical and easier to use.

Function keys are programmed to save the user time. For CAD programs, for example, the function keys may be shortcuts for setting line types or thicknesses or inserting combinations of blocks. For drawing programs, the function keys often activate palette selectors or different graphics modes. Many telephones have prelabeled or configurable keys for redial, speed dialing, and other optional functions.

fundamental frequency 1. The lowest natural frequency in an oscillating system. 2. The reciprocal of the period of a wave.

fundamental group A group of trunks in which each local switching center is interconnected to one of a higher order.

fuse 1. *n.* A protective mechanism which reacts to break an electrical circuit when the current through the circuit exceeds a specific value. The mechanism may consist of a wire or chemical junction mounted in serial, which melts or breaks at a specified value, usually indicated with a number. Fuses are designed to break first, if there is a problem, in order to protect more expensive electrical components from harm. The fuse was first patented by Thomas Edison in the early 1880s. Circuit breakers have replaced fuses in most new home electrical installations, as they are more convenient than replacing a fuse. See circuit breaker. 2. *v.* To join or blend together, usually by melting. To *fuse* implies a stronger or more consistent bond than to *bond* (as with an adhesive), since there may be momentary heat or chemical alteration to create the bond.

fuse alarm A type of fuse which is connected with an audible device or a flashing light (or both) to indicate that the fuse is blown and must be replaced.

fuse block An insulated mounting structure for a fuse or bank of fuses. In smaller electronic devices, the block may secure a small clip that holds the fuse in place. In larger wiring installations, as in houses and offices, the fuse block may be a large metal electrical cabinet with several rows of fuse mountings. When this is further enclosed, it may be called a fuse panel or fuse box.

fuse wire The wire inside the fuse housing which breaks when subjected to excessive current loads. This breaks the circuit with the intention of protecting sensitive electronic parts. In the illustration of fuses, three different types of fuse wires can be seen, with straight wires on the three right fuses, a corrugated wire third from the left, and a flat 'zig-zag wire in the two left fuses.

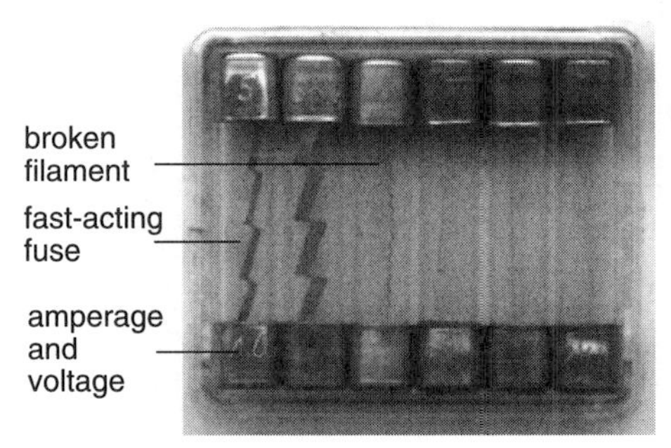

The above diagram shows three different types of fuses commonly used with electronic devices to protect sensitive components. The two on the left are fast-acting fuses. The two in the middle can no longer be used, as the filaments are broken.

It is important to select fuses with voltages that equal or exceed the voltages of the circuits they are intended to protect. The amperage and voltage are indicated on the package and sometimes are also engraved into the metal cap of the fuse.

fused quartz, fused silica A glassy substance made from quartz crystals which is highly resistant to chemicals and heat. Quartz has remarkable noncorrosive and vibrational qualities that makes it a valuable industrial material. See quartz.

fused semiconductor In semiconductor fabrication, the materials can be subjected to heat in such a way that they cool and recrystallize on a base crystal to form an electronic junction. See p-n junction, semiconductor.

fusion 1. A heat-induced liquid state. 2. Union of parts by applying heat or chemicals to liquify one or both of the parts (or a binding substance such as solder or weld) to form a permanent bond between them. 3. The union of atomic nuclei to form heavier nuclei, a process which generally requires enormous amounts of heat or pressure, and can result in the release of a great amount of energy.

Future Public Land Mobile Telecommunication System FPLMTS. A standards effort initiated in the early 1990s to create a global mobile communications system that encompasses cordless and cellular technologies. It is intended to form a basis for integrated voice, paging, data services at rates up to 9600 bps, and perhaps up to 20 Mbps, in both connection and connectionless modes, and for videoconferencing, global positioning services (GPS), and multimedia capabilities. See Global Systems for Mobile Communications.

fuzz 1. Blurred effect. A visual artifact, or intentional effect, as in a graphics program. 2. In audio, a tonal distortion usually added to give a rough or rich edge to a sound, as on a guitar. An audio fuzz box is specifically designed to produce this effect.

fV Abbreviation for femtovolt (10^{-15} volt).

FVR flexible vocabulary recognition. A type of speech recognition in which a variety of words, not necessarily just those found in an associated database, can be processed.

FWA See Fixed Wireless Access.

FWIW An abbreviation for "for what it's worth" which is frequently used in conjunction with opinions expressed in email and online public forums. See emoticons, IMHO.

FX See Foreign Exchange.

FYI An abbreviation for "for your information" which is commonly used on business memos, documents, email, and postings on the Internet.

FZA A data compression program developed by D. Carr (Gandalf Data Ltd.), which is derived from Lempel-Ziv (LZ), and which favors high levels of compression over central processing unit (CPU) speed and memory. FZA is based on packet switched networks techniques to compress information into a single frame or across multiple frames. See FZA+, Lempel-Ziv.

FZA+ An updated version of the FZA data compression program. FZA+ was developed by A. Barbir.

G 1. Symbol for conductance. See conductance. 2. Abbreviation for giga-. See giga-. 3. Abbreviation for grid (as in a vacuum tube).

g force (Symbol - *g*) A unit of force of acceleration equal to that which would occur in a falling body acted on by gravity at the Earth's surface, that is, 9.81 meters per second per second.

G Interface In Operation, Administration, Maintenance, and Provisioning (OAM&P), the G Interface is the user-to-computer interface of the Telecommunications Management Network (TMN). The G Interface is intended to promote consistency in the user interface and to reduce errors.

G Series Recommendations A set of ITU-T recommendations which provides guidelines for transmission systems and media, digital systems and networks. These are available as publications from the ITU-T for purchase, and a few may be downloadable from the Net. Some of the related general categories and specific G category recommendations of particular interest are listed on the following pages, organized into general categories. See also I, Q, V, and X Series Recommendations.

G style handset The designation for older telephone handsets which have a round design on the ear- and mouthpieces. They are similar to the newer, squared-off K style handsets, and both G and K style are heavier and more substantial than some of the newer cordless or cell phone handsets which are very flat and small. See K style handset.

G-line A round insulated wire used in microwave transmissions.

G-scan A type of rectangular radar display in which a centralized blip is illuminated and becomes wider or narrower horizontally as the target moves nearer or farther away.

G.lite An International Telecommunication Union (ITU) low-cost, splitterless alternative proposal to ADSL CPE. See Asymmetric Digital Subscriber Line, UAWG.

G/A ground to air communication.

G/G ground to ground communication.

GA go ahead. A common verbal and written communications convention that indicates that the communicator is finished and the listener is welcome to proceed. This is frequently used in question-and-answer style conferences on the Internet, particularly in moderated conferences where a lot of people are queued to communicate, and each one has to wait for a turn.

GAB Group Access Bridging.

gadolinium A light, silvery, crystallizing, ferromagnetic rare earth metal isolated from yttrium in the late 1800s. Gadolinium has been used in the production of phosphors in color cathode ray tubes (CRTs), and for making gadolinium yttrium garnets used in microwave technologies. See europium, yttrium.

gaff The spike which is attached to a utility pole climber's iron. See boomer.

gage Performance indicator. See gauge.

gain Increase in power of a transmission, usually indicated in decibels (dB) when applied to audio gain. Gain is sometimes intentionally created by using various means to boost a signal. Unfortunately, doing so typically also increases noise and interference in analog systems. Gain is descriptive of an antenna's capability to increase its effective radiated signal, relative to a reference like an isotropic antenna, or a center-fed, half-wave dipole antenna.

ITU-T G Series Recommendations

General Categories

Series C	General telecommunications statistics
Series E	Overall network operation, telephone service, service operation, and human factors
Series F	Telecommunication services other than telephone
Series G	Transmission systems and media, digital systems and networks
Series H	Line transmission of nontelephone signals
Series I	Integrated Services Digital Networks (ISDN)
Series J	Transmission of sound program and television signals
Series P	Telephone transmission quality, telephone installations, local line networks
Series Q	Switching and Signaling
Series V	Data communication over the telephone network
Series X	Data networks and open system communication
Series Z	Programming languages

General definitions and vocabulary

G.100	1993	Definitions used in recommendations on general characteristics of international telephone connections and circuits
G.601	1988	Terminology for cables
G.701	1993	Vocabulary of digital transmission and multiplexing, and pulse code modulation (PCM)
G.780	1994	Vocabulary of terms for synchronous digital hierarchy (SDH) networks and equipment
G.810	1996	Definitions and terminology for synchronization networks
G.972	1997	Definition of terms relevant to optical fibre submarine cable systems

Transmission-related

G.101	1996	The transmission plan
G.102	1988	Transmission performance objectives and recommendations
G.113	1996	Transmission impairments
G.114	1996	One-way transmission time
G.117	1996	Transmission aspects of unbalance about Earth
G.120	1988	Transmission characteristics of national networks
G.121	1993	Loudness ratings (LRs of national systems)
G.125	1988	Characteristics of national circuits on carrier systems
G.142	1988	Transmission characteristics of exchanges
G.171	1988	Transmission plan aspects of privately operated networks
G.172	1988	Transmission plan aspects of international conference calls
G.173	1993	Transmission planning aspects of the speech service in digital public land mobile networks
G.174	1994	Transmission performance objectives for terrestrial digital wireless systems using portable terminals to access the PSTN
G.175	1997	Transmission planning for private/public network interconnection of voice traffic
G.221	1988	Overall recommendations relating to carrier-transmission systems
G.180	1993	Characteristics of N + M type direct transmission restoration systems for use on digital and analogue sections, links, or equipment
G.181	1993	Characteristics of 1 + 1 type restoration systems for use on digital transmission links
G.671	1996	Transmission characteristics of passive optical components
G.712	1996	Transmission performance characteristics of pulse code modulation channels
G.773	1993	Protocol suites for Q-interfaces for management of transmission systems
G.801	1988	Digital transmission models
G.961	1993	Digital transmission system on metallic local lines for ISDN basic rate access

Echo-related

G.122	1993	Influence of national systems on stability and talker echo in international connections
G.126	1993	Listener echo in telephone networks
G.131	1996	Control of talker echo
G.164	1988	Echo suppressors
G.165	1993	Echo cancellers
G.167	1993	Acoustic echo controllers
G.168	1997	Digital network echo cancellers

Connection-related noise, distortion, crosstalk, jitter, and wander

G.103 1988 Hypothetical reference connections

G.105 1988 Hypothetical reference connection for crosstalk studies

G.111 1993 Loudness ratings (LRs in an international connection)

G.123 1988 Circuit noise in national networks

G.132 1988 Attenuation distortion

G.133 1988 Group-delay distortion

G.134 1988 Linear crosstalk

G.135 1988 Error on the reconstituted frequency

G.141 1988 Attenuation distortion

G.222 1988 Noise objectives for design of carrier-transmission systems of 2500 km

G.223 1988 Assumptions for the calculation of noise on hypothetical reference circuits for telephony

G.226 1988 Noise on a real link

G.228 1988 Measurement of circuit noise in cable systems using a uniform-spectrum random noise loading

G.229 1988 Unwanted modulation and phase jitter

G.230 1988 Measuring methods for noise produced by modulating equipment and through-connection filters

G.441 1988 Permissible circuit noise on frequency-division multiplex radio-relay systems

G.442 1988 Radio-relay system design objectives for noise at the far end of a hypothetical reference circuit with reference to telegraphy transmission

G.823 1993 The control of jitter and wander within digital networks which are based on the 2048 kbps hierarchy

G.824 1993 The control of jitter and wander within digital networks which are based on the 1544 kbps hierarchy

G.825 1993 The control of jitter and wander within digital networks which are based on the synchronous digital hierarchy (SDH)

Software-related

G.191 1996 Software tools for speech and audio coding standardization

Optical submarine cable systems

G.971 1996 General features of optical fibre submarine cable systems

G.973 1996 Characteristics of repeaterless optical fibre submarine cable systems

G.974 1993 Characteristics of regenerative optical fibre submarine cable systems

G.975 1996 Forward error correction for submarine systems

G.976 1997 Test methods applicable to optical fibre submarine cable systems

Optical - various

G.911 1997 Parameters and calculation methodologies for reliability and availability of fibre optic systems

G.958 1994 Digital line systems based on the synchronous digital hierarchy for use on optical fibre cables

G.981 1994 PDH optical line systems for the local network

G.982 1996 Optical access networks to support services up to the ISDN primary rate or equivalent bit rates

Timing

G.811 1997 Timing characteristics of primary reference clocks

G.812 1988 Timing requirements at the outputs of slave clocks suitable for plesiochronous operation of international digital links

G.813 1996 Timing characteristics of SDH equipment slave clocks (SEC)

G.821 1996 Error performance of an international digital connection operating at a bit rate below the primary rate and forming part of an integrated services digital network

G.822 1988 Controlled slip rate objectives on an international digital connection

G.826 1996 Error performance parameters and objectives for international, constant bit rate digital paths at or above the primary rate

G.827 1996 Availability parameters and objectives for path elements of international constant bit-rate digital paths at or above the primary rate

Connections, cable and cabling

G.322 1988 General characteristics recommended for systems on symmetric pair cables

G.325 1988 General characteristics recommended for systems providing 12 telephone carrier circuits on a symmetric cable pair [12+12 systems]

G.421 1988 Methods of interconnection

G.422 1988 Interconnection at audio-frequencies cable pairs for analogue transmission

G.612 1988 Characteristics of symmetric cable pairs designed for the transmission of systems with bit rates of the order of 6 to 34 Mbps

G.613 1988 Characteristics of symmetric cable pairs usable wholly for the transmission of digital systems with a bit rate of up to 2 Mbps

G.614 1988 Characteristics of symmetric pair star-quad cables designed earlier for analogue transmission systems and being used now for digital system transmission at bit rates of 6 to 34 Mbps

G.621 1988 Characteristics of 0.7/2.9 mm coaxial cable pairs

G.622 1988 Characteristics of 1.2/4.4 mm coaxial cable pairs

G.623 1988 Characteristics of 2.6/9.5 mm coaxial cable pairs

G.631 1988 Types of submarine cable to be used for systems with line frequencies of less than about 45 MHz

G.650 1997 Definition and test methods for the relevant parameters of single-mode fibres

G.651 1993 Characteristics of a 50/125 um multimode grades index optical fibre cable

G.652 1997 Characteristics of a single-mode optical fibre cable

G.653 1997 Characteristics of a dispersion-shifted single-mode optical fibre cable

G.654 1997 Characteristics of a cut-off shifted single-mode optical fibre cable

G.655 1996 Characteristics of a non-zero dispersion shifted single-mode optical fibre cable

G.661 1996 Definition and test methods for relevant generic parameters of optical fibre amplifiers

G.662 1995 Generic characteristics of optical fibre amplifier devices and sub-systems

G.663 1996 Application related aspects of optical fibre amplifier devices and sub-system

G.681 1996 Functional characteristics of interoffice and long-haul line systems using optical amplifiers, including optical multiplexing

Speech/Audio encoding and synthesis

G.115 1996 Mean active speech level for announcements and speech synthesis systems

G.192 1996 A common digital parallel interface for speech standardization activities

G.720 1995 Characterization of low-rate digital voice coder performance with non-voice signals

G.723.1 1996 Dual rate speech coder for multimedia communications transmitting at 5.3 and 6.3 kbps

G.724 1988 Characteristics of a 48-channel low bit rate encoding primary multiplex operating at 1544 kbps

G.725 1988 System aspects for the use of the 7-kHz audio codec within 64 kbps

G.728 1992 Coding of speech at 16 kbps using low-delay code excited linear prediction

G.729 1996 Coding of speech at 8 kbps using conjugate-structure algebraic-code-excited

G.764 1990 Voice packetization

G.802 1988 Interworking between networks based on different digital hierarchies and speech encoding laws

Modulation and multiplexing

G.711 1988 Pulse code modulation (PCM) of voice frequencies

G.726 1990 40, 32, 24, 16 kbps Adaptive Differential Pulse Code Modulation (ADPCM)

G.727 1990 5-, 4-, 3- and 2-bits sample embedded adaptive differential pulse code modulation (ADPCM)

G.731 1988 Primary PCM multiplex equipment for voice frequencies

G.732 1988 Characteristics of primary PCM multiplex equipment operating at 2048 kbps

G.733 1988 Characteristics of primary PCM multiplex equipment operating at 1544 kbps

G.734 1988 Characteristics of synchronous digital multiplex equipment operating at 1544 kbps

G.735 1988 Characteristics of primary PCM multiplex equipment operating at 2048 kbps and offering synchronous digital access at 384 kbps and/or 64 kbps

G.736 1993 Characteristics of a synchronous digital multiplex equipment operating at 2048 kbps

G.737 1988 Characteristics of an external access equipment operating at 2048 kbps offering synchronous digital access at 384 kbps and/or 64 kbps

G.738 1988 Characteristics of primary PCM multiplex equipment operating at 2048 kbps and offering synchronous digital access at 320 kbps and/or 64 kbps

G.739 1988 Characteristics of an external access equipment operating at 2048 kbps offering synchronous digital access at 320 kbps and/ or 64 kbps

G.741 1988 General considerations on second order multiplex equipment

G.742 1988 Second-order digital multiplex equipment operating at 8448 kbps and using positive justification

G.743 1988 Second-order digital multiplex equipment operating at 6312 kbps and using positive justification

G.744 1988 Second-order PCM multiplex equipment operating at 8448 kbps

G.745 1988 Second-order digital multiplex equipment operating at 8448 kbps and using positive/zero/negative justification

G.746 1988 Characteristics of second-order PCM multiplex equipment operating at 6312 kbps

G.747 1988 Second-order digital multiplex equipment operating at 6312 kbps and multiplexing three tributaries at 2048 kbps

G.751 1988 Digital multiplex equipment operating at the third-order bit rate of 34 368 kbps and the fourth-order bit rate of 139 264 kbps and using positive justification

G.752 1988 Characteristics of digital multiplex equipment based on a second-order bit rate of 6312 kbps and using positive justification

G.753 1988 Third-order digital multiplex equipment operating at 34 368 kbps and using positive/zero/negative justification

G.754 1988 Fourth-order digital multiplex equipment operating at 139 264 kbps and using positive/zero/negative justification

G.755 1988 Digital multiplex equipment operating at 139 264 kbps and multiplexing three tributaries at 44 736 kbps

G.791 1988 General considerations on transmultiplexing equipment

G.792 1988 Characteristics common to all transmultiplexing equipment

G.793 1988 Characteristics of 60-channel transmultiplexing equipment

G.794 1988 Characteristics of 24-channel transmultiplexing equipment

G.797 1996 Characteristics of a flexible multiplexer in a plesiochronous digital hierarchy environment

Synchronous Digital Hierarchy (SDH)

G.774 1992 Synchronous Digital Hierarchy (SDH) management information model for the network element view

G.774.01 1994 Synchronous digital hierarchy (SDH) performance monitoring for the network element view

G.774.02 1994 Synchronous digital hierarchy (SDH) configuration of the payload structure for the network element view

G.774.03 1994 Synchronous digital hierarchy (SDH) management of multiplex-section protection for the network element view

G.774.04 1995 Synchronous digital hierarchy (SDH) management of the subnetwork connection protection for the network element view

G774.05 1995 Synchronous digital hierarchy (SDH) management of connection supervision functionality (HCS/LCS for the network element view)

G.774.6 1997 Synchronous Digital Hierarchy (SDH)

G.774.7 1996 Synchronous Digital Hierarchy (SDH) management of lower order path trace and interface labeling for the network element view

G.774.8 Synchronous Digital Hierarchy (SDH) management of radio-relay systems for the network element of view

G.775 1994 Loss of signal (LOS) and alarm indication signal (AIS) defect detection and clearance criteria

G.781 1994 Structure of recommendations on equipment for the synchronous digital hierarchy (SDH)

G.782 1994 Types and general characteristics of synchronous digital hierarchy (SDH) equipment

G.783 1997 Characteristics of synchronous digital hierarchy (SDH) equipment functional blocks

G.784 1994 Synchronous digital hierarchy (SDH) management

G.785 1996 Characteristics of a flexible multiplexer in a synchronous digital hierarchy (SDH) environment

G.803 1997 Architecture of transport networks based on the synchronous digital hierarchy (SDH)

G.831 1996 Management capabilities of transport networks based on the Synchronous Digital Hierarchy (SDH)

G.832 1995 Transport of SDH elements on PDH networks: Frame and multiplexing structures

G.841 1995 Types and characteristics of SDH network protection architectures

G.842 1997 Interworking of SDH network protection architectures

G.957 1995 Optical interfaces for equipment and systems relating to the synchronous digital hierarchy (SDH)

General and miscellaneous

G.151 1988 General performance objectives applicable to all modern international circuits and national extension circuits

G.152 1988 Characteristics appropriate to long-distance circuits of a length not exceeding 2500 km

G.153 1988 Characteristics appropriate to international circuits more than 2500 km in length

G.162 1988 Characteristics of compandors for telephony

G.166 1988 Characteristics of syllabic compandors for telephony on high capacity long-distance systems

G.176 1997 Planning guidelines for the integration of ATM technology into networks supporting voiceband services

G.211 1988 Make-up of a carrier link

G.212 1988 Hypothetical reference circuits for analogue systems

G.213 1988 Interconnection of systems in a main repeater station

G.214 1988 Line stability of cable systems

G.215 1988 Hypothetical reference circuit of 5000 km for analogue systems

G.224 1988 Maximum permissible value for the absolute power level (power referred to one milliwatt of a signaling pulse)

G.225 1988 Recommendations relating to the accuracy of carrier frequencies

G.227 1988 Conventional telephone signal

G.231 1988 Arrangement of carrier equipment

G.232 1988 12-channel terminal equipment

G.233 1988 Recommendations concerning translating equipment

G.241 1988 Pilots on groups, supergroups, etc.

G.242 1988 Through-connection of groups, supergroups, etc.

G.243 1988 Protection of pilots and additional measuring frequencies at points where there is a through-connection

G.411 1988 Use of radio-relay systems for international telephone circuits

G.431 1988 Hypothetical reference circuits for frequency-division multiplex (FDM) radio-relay systems

G.451 1988 Use of radio links in international telephone circuits

G.702 1988 Digital hierarchy bit rates

G.703 1991 Physical/electrical characteristics of hierarchical digital interfaces

G.704 1995 Synchronous frame structures used at 1544, 6312, 2048, 8488, and 44 736 kbps hierarchical levels

G.706 1991 Frame alignment and cyclic redundancy check (CGC) procedures relating to basic frame structures defined in Recommendation G.704

G.707 1996 Network node interface for the synchronous digital hierarchy (SDH)

G.722 1988 7-kHz audio-coding within 64 kbps

G.761 1988 General characteristics of a 60-channel transcoder equipment

G.762 1988 General characteristics of a 48-channel transcoder equipment

G.765 1992 Packet circuit multiplication equipment

G.766 1996 Facsimile demodulation/remodulation for digital circuit multiplication equipment

G.772 1993 Protected monitoring points provided on digital transmission systems

G.795 1988 Characteristics of codecs for FDM assemblies

G.796 1992 Characteristics of a 64 kbps cross-connect equipment with 2048 kbps access ports

G.804 1993 ATM cell mapping into plesiochronous digital hierarchy (PDH)

G.805 1995 Generic functional architecture of transport networks

G.851.1 1996 Management of the transport network

G.852.1 1996 Management of the transport network

G.853.1 1996 Common elements of the information viewpoint for the management of a transport network

G.853.2 1996 Subnetwork connection management information viewpoint

G.854.1 1996 Management of the transport network

G.861 1996 Principles and guidelines for the integration of satellite and radio systems in SDH transport networks

G.901 1988 General considerations on digital sections and digital line systems

G.902 1995 Framework recommendation on functional access networks (AN) architecture and functions, access types, management, and service node aspects

G.921 1988 Digital sections based on the 2048 kbps hierarchy

G.931 1988 Digital line sections at 3152 kbps

G.941 1988 Digital line systems provided by FDM transmission bearers

G.950 1988 General considerations on digital line systems

G.951 1988 Digital line systems based on the 1544 kbps hierarchy on symmetric pair cables

G.952 1988 Digital line systems based on the 2048 bps hierarchy on symmetric pair cables

G.953 1988 Digital line systems based on the 1544 kbps hierarchy on coaxial pair cables

G.954 1988 Digital line systems based on the 2048 kbps hierarchy on coaxial pair cables

G.955 1996 Digital line systems based on the 1544 kbps and the 2048 kbps hierarchy on optical fibre cables

G.960 1993 Access digital section for ISDN basic rate access

G.962 1993 Access digital section for ISDN primary rate at 2048 kbps

G.963 1993 Access digital section for ISDN primary rate at 1544 kbps

G.964 1994 V-Interfaces at the digital local exchange (LE)

G.965 1995 V-Interfaces at the digital local exchange (LE)

Galarno, Roland A Canadian inventor who devised a way, in the mid-1960s, to transfer text into braille by means of telephone relays and a teletype machine. See computerized braille.

Gale, Leonard D. An American chemist who assisted Samuel Morse in his early telegraphic inventions. With Gale's assistance, especially with the design and use of various power cells, Morse was able to come up with a workable design by 1837. During the many problems that plagued Morse's Washington, D.C. to Baltimore line, Gale resigned from the project. See Morse, Samuel F. B.

galena A bluish-gray lead sulphide mineral commonly used as a sensitive radio wave detector in crystal detectors in the early 1900s. Sometimes the galena was thinly coated with other materials to improve its properties.

gallium arsenide A semiconductor substance used to produce electronic components, such as computer chips and solar panels (when combined with germanium). It is sometimes used in place of silicon for high speed devices. It withstands heat and radiation well, making it suitable for orbiting satellite applications.

Galvani, Luigi (1737-1798) An Italian physicist, educator, and physician who experimented with minute levels of electricity in the nerves of frog's legs around 1786. His wife was reportedly the first to notice the muscle twitch in a recently killed frog when a nerve was touched, and Galvani followed up her observation with further experiments. His name has been applied to measuring instruments (galvanometer) and small levels of electricity on skin surfaces (galvanic skin response).

Luigi Galvani did numerous experiments in nonbiological and biological electrical impulses. The galvanometer is named after him.

galvanic Related to, or producing a direct electrical current through chemical rather than electrostatic means. See galvanometer.

galvanic cell A power cell which produces electricity through electrochemical rather than through electrostatic means. See battery.

A historic electrochemical power cell named after Italian physicist Luigi Galvani.

galvanometer Named after Luigi Galvani, an instrument for detecting low levels of electric current through the use of a magnetic needle or coil in a magnetic field. It followed from principles described by Hans Christian Ørstedt and was further developed by Nobili in 1825, who used two needles, each with its own separate coil. This innovation reduced interference from the Earth's magnetic field. In 1858, William Thomson (Lord Kelvin) refined it as the astatic galvanometer. In telecommunications, galvanometers have many uses, including the testing and diagnosis of circuits. See galvanic; Galvani, Luigi.

gamma ferric oxide Ferric oxide is found as dark synthetic pigments and as natural red or black hematite. Gamma ferric oxide has been used since the 1940s to provide configurable magnetic recording surfaces, first on magnetic tapes and later on floppy diskettes.

gamut 1. A range or series. 2. In imagery, the color gamut is the range of colors that can be perceived or reproduced, usually within the context of a particular situation or technology. For example, the gamut of an RGB cathode ray tube (CRT) monitor is quite different from the gamut of printing inks or process color perceptual combinations of dots. It is important to understand the differences in order to adapt the technologies to one another when printing from computer-generated files. See dynamic range.

GAN See global area network.

Ganada Telephone Company, Inc. A Texas-based, independent family-owned local telephone exchange that was established in 1944 with the merger of the Laward and Ganado Telephone exchanges. In those days, operators handled the manual switchboard.

Gandalf The wizard in Tolkien's Lord of the Rings series. The Tolkien books are very popular in the programming community.

gang To group, or aggregate cables, components, objects, or picture elements. To mechanically or electrically combine components or devices so they can be controlled from one source.

Gang of Nine Nine vendor companies who formed a group in 1989 to promote and develop the industry standard architecture (ISA) which is commonly used for computer peripheral device connections on Intel-based microcomputers. Their work resulted in the 32-bit Extended ISA (EISA) standard.

gap Opening, space, small distance between objects or signals, small distance of a less dense material than the surrounding materials.

GAP See Generic Access Profile.

gap, electrical An opening in an electrical connection, which may close to allow a circuit connection, or more commonly, over which a spark will jump to provide a brief connection or discharge through the air, as the gap in a spark plug on an engine. An unplanned gap, through breakage, can interrupt an optical or electrical connection, thus interfering with a transmission.

gap, pickup The distance between a reading/recording head and the medium with which it interacts. This is often a very precise distance, as in hard drives read/write heads and compact disc pickups, where a slight adjustment allows a much greater density of recorded pits in the disc.

gap, time An informational space or time gap inserted to indicate a stopping and/or starting point in recorded material or in a coded transmission.

gap loss The loss in signal strength which occurs when crossing a gap. Even very tiny gaps may substantially affect the quality of a transmission, and gaps are apt to occur at corners and junctions where transmission media change, and in couplers.

garbage 1. In computing, meaningless information, spurious characters, nonsensical output. 2. A meaningless signal or electrical interference. See interference, noise.

garbage collection When software is executed on a computer, memory is dynamically allocated for a number of reasons. In interpreted languages in particular, which are translating commands into machine code instructions as they are encountered, a lot of behind the scenes processing and memory allocation occurs. Sometimes memory starts to fill up and things begin to slow down. In order to 'clean up' memory and make use of memory which is no longer being used, or which could be used more efficiently, a process called 'garbage collection' occurs. This basically consists of a reorganization of variable locations and other information which may cause a temporary, but perceptible system slowdown.

Users of BASIC interpreters are particularly familiar with garbage collection. One way to avoid this annoyance is to overlay new variable values on existing ones, rather than having them dynamically allocated by the system. For example, in earlier versions of Microsoft BASIC, the LSET command could be used.

garbage in/garbage out GIGO. *colloq.* A means of saying that you can't get something good out if you have put something bad into a system. In computer terms, it means that if you supply bad input, you're not going to get good output. In terms of hardware, a weak or poor signal at the origin is not going to result in a good signal at the receiving end. In programming terms, feeding the computer the wrong instructions or the wrong data is not going to result in the software performing in a correct or desired manner.

Garden Valley Telephone Company The oldest telephone cooperative in the United States, chartered in Minnesota in 1912, and still operating after more than 86 years.

garnet A red, more or less transparent semiprecious mineral often used in jewelry and industrial applications. It somewhat resembles ruby, but is generally a little less transparent and a deeper, 'muddier' shade of red. Synthetic garnets can be manufactured from oxides of some of the rare earths. Lasers and some types of computer memory which use magnetic film are made from garnet. See gadolinium, laser.

garter spring In various rotating drums, a device to hold a printing medium, plate, or other material associated with the rotating mechanism.

gas laser A laser in which a gas or vapor is used as the active medium. The gas may be excited by a high-frequency oscillator or direct current. See laser.

gas maser A maser in which a gas or vapor

interacts with microwaves. Gas masers are found in scientific oscillators such as atomic clocks. See atomic clock, maser.

gas plasma display See plasma display panel.

gas tube A type of electron tube which is not completely evacuated. Early electron tubes burned out quickly because of the gas inside the bulb. Later tubes were designed to last longer by being evacuated. In specialized applications, a trade-off is used and a small amount of gas is left in the tube to promote ionization and, hence, enhance current flow.

Gassner An inventor who first created a dry cell in 1888, thus, for the most part, superseding the wet cell. Dry cells were welcomed because they didn't have the problems of leakage associated with wet cells, and they could be made smaller and more portable. See dry cell.

gaston A random noise modulator sometimes used as an antijamming communications transmitting device.

gate In an electrical circuit, in its broadest sense, a junction which selectively controls whether current gets through, when it gets through, or how much of it gets through.

gate, security In a software program, an input or process point that selectively lets users or other processes through, as a security gate, login gate, or system gateway for different types of information or protocols (network traffic, mail, etc.). In the more specific context of programming, a circuit which performs a basic logic operation and provides a single output from that operation. See firewall, gateway, proxy.

gate, telephone In telephone call distribution systems, a physical or virtual trunk through which calls are handled by a group of operators or agents according to some characteristics specified for each gate.

Gates, William Henry, III (1955-) One of the early success stories in the personal computer software industry, Bill Gates began programming seriously in Seattle, Washington, while still in grade school, along with his friend and business partner, Paul Allen. Allen and Gates were both precocious entrepreneurs, successfully winning commercial programming contracts at a young age. Their first business partnership was called Traf-O-Data, founded around 1972.

Later, while Gates was briefly at Harvard and Allen was working for Honeywell in Boston, Allen saw the feature story for the MITS Altair computer in the January 1975 issue of Popular Electronics and got together with Gates to talk about what they could do. They decided to propose to MITS that they provide a BASIC language program for the new microcomputer. Together they quickly developed BASIC while fatefully, another software developer, Gary Kildall, was developing the CP/M operating system. Following the initial success of the Altair proposal, after Allen flew south to demonstrate BASIC to MITS, the two entrepreneurs were approached by International Business Machines about developing an operating system for IBM. This initial contact may have come about in part because Gates' mother and the president of IBM were associated with the United Way organization, if you consider it was unusual in the 1970s for a large company like IBM to contract something this important from a fledgling business run by two young owners.

At the time IBM contracted Microsoft for the BASIC interpreter, the representatives apparently thought they had also purchased rights to an operating system. Gates didn't want to lose the deal and suggested they call Kildall, the owner of Digital Research (DR). IBM went to visit Kildall, who had other commitments, and the DR representative was reluctant to sign IBM's harsh nondisclosure agreement, given that DR's attorney had great reservations about the document.

Evidently IBM was in a hurry, and Gates saw an opportunity; he offered to sign and provide an operating system, even though Microsoft didn't have one that met IBM's needs at that time. Thus ended a gentleman's agreement between Kildall, Gates, and Allen. The significant history of Microsoft began with the signing of the IBM contract. Also significant is the fact that Gates reserved the right to market the results of the development (sold by IBM as PC-DOS) in competition with IBM. Microsoft's version became known as MS-DOS. Since the duo didn't have an operating system, they bought QDOS, developed by Tim Paterson from mid-1970s documentation for CP/M. Microsoft then hired Paterson to quickly develop it into PC-DOS. QDOS was syntactically and functionally extremely similar to Kildall's CP/M, a situation which is still of consequence almost twenty years later, as the current owners of Kildall's early and later operating systems technology are pursuing legal avenues against Microsoft.

Eventually, Allen left Microsoft to pursue other investments and to form his own company, while in 1998 Gates is still head of the organization which has grown from a small group of half a dozen programmers and clerical staff to a campus in Redmond with tens of thousands of employees and assets in the $billions.

Gates has spun off a number of other businesses, including a multimedia company and the Teledesic project which he is codeveloping with Steve McCaw. Gates is the richest man on the continent, one of the richest people in the world, and is one of the most focused, aggressive, and successful business and marketing giants in the software industry. There is a great deal of controversy over his ethics and his methods, resulting in allegations against Microsoft by the U.S. Justice Department; but Gates' tenacity, tactical efforts, competitive drive, and commercial success in the software industry are not in question.

gateway A transmission connection between dissimilar networks which typically performs bandwidth and protocol adjustments and conversions, as needed. Gateways are commonly used between local area networks (LANs) and the Internet (proxy servers), and between land services and satellite services, etc. A gateway differs from a proxy in that a gateway handles requests as though the requests originated from the gateway and not from the original client, thus serving as a server-side portal in a firewall. See firewall, proxy.

Gateway to Gateway Protocol GGP. A network routing protocol used to convey routing information through a distributed shortest path computation.

gating 1. Selectively allowing certain waveforms, frequencies, or portions of waves to pass through a gate point. 2. Performing electronic switching by application of a certain current or waveform. 3. Using logic to control the passage of current or information through a system.

gauge An instrument or other indicator for measuring or testing. An indicator of the thickness or thinness of a substance or object, especially applied to wires and cabling. Position relative to another, as in the distance between the sides of a railroad track. See American Wire Gauge (Brown and Sharpe), Birmington Wire Gauge, Steel Wire Gauge.

gauss A centimeter-gram-second (CGS) unit of magnetic flux density. For example, if one line of magnetic force passes through one square centimeter, the field intensity is said to be one gauss. Named after Karl Friedrich Gauss. See lines of force, magnetic field.

Gauss, Johann Karl Friedrich (1777-1855) A brilliant German mathematician and astronomer who devised the heliotrope, an instrument which could reflect sunlight over long distances, providing a means for making straight lines and calculations related to the Earth's surface. He investigated terrestrial magnetism in cooperation with W. Weber in 1831 and published some significant papers several years later. In 1833, he constructed an electric telegraph, at the same time as such devices were being developed in the United States. The unit for magnetic flux density is named after him.

Gaussian curve, normal curve A Gaussian curve is often called a *bell curve* due to its shape, and a normal curve due to its theoretical distribution. A symmetric mathematical, statistical model and diagrammatic representation that resembles the outline of a bell. The normal curve is based on a combination of population distribution assumptions and experimental observations that place the majority of 'hits' within the center and highest point of the curve (the mean). The changes in the direction of the curve and dispersion of frequencies is often described in terms of *standard deviations*. An interesting demonstration of the shape and distribution of elements in a Gaussian curve can be seen in a number of science museums. It is set up with vertical channels and ping pong balls (or ball bearings), in which the balls are dropped into the top of a device where they roll down into a line of channels that can be viewed through a transparent panel. After many hundreds of balls have fallen down through the mechanism, the shape of the curve will emerge, with the majority of balls in the center, and fewer as you move left or right out to the edges (the 'tails' of the distribution). The normal curve is frequently used as an underlying assumption for the design and scoring of tests. Unfortunately, it is sometimes used as a justification for describing populations where the size of the sample being evaluated is not mathematically sufficient to justify assumptions about its relationship to the normal curve.

Gaussian minimum shift keying GMSK. In wireless communications, a type of wave modulation used in the Global System for Mobile Communications (GSM) system.

Gaussian noise Noise or other electromagnetic interference which conforms to a probability pattern consistent with expectations based on Gaussian statistics. Noise that is somewhat random across a range of frequencies is sometimes known as 'white noise' or Gaussian noise.

GAW Global Atmosphere Watch. A United Nations monitoring and assessment program which watches pollution, ozone, and other aspects of atmosphere composition.

Gb, Gbyte gigabyte. See giga-.

GB gigabit. See giga-.

GBCS See Global Business Communications Systems.

GBIC Gigabit Interface Converter. A small, sometimes swappable gigabit network converter device which attaches to a port connection.

GBR ground-based radar.

GCA 1. game control adapter. 2. ground-controlled approach. A radar-based aircraft landing system.

GCI Ground Control of Interception. A radar-based technique for directing aircraft in the interception of approaching craft.

GCRA See Generic Cell Rate Algorithm.

GCS Government Communication Systems.

GCT See Greenwich Civil Time.

GD&R An abbreviation for "grinning, ducking, and running!" often used in email, on online public discussions and newsgroups to indicate a well-intentioned jab. See emoticons.

GDF group distribution frame. See distribution frame.

GDI See graphics device interface.

geek code A somewhat serious, somewhat tongue-in-cheek, techie encoded message invented in the early 1990s. Geek code is often included in a signature (.sig) file that provides personal information about, and preferences of techie posters to public forums on the Internet.

Geissler tube A type of dual-electrode gas-filled tube which glows when current passes through it.

GEM See Graphics Environment Manager.

gender A designation widely used with respect to cabling connectors to indicate whether they are female 'innies,' or male 'outies,' that is, whether they have holes or extended pins. Most common switchboxes have female connections. Male and female connectors are both found on the backs of computers, and vary from platform to platform and sometimes even from brand to brand. This is inconvenient for users, who often have to pay extra money for gender changers, but cable and connector manufactures don't seem to mind the extra business.

gender bender, gender changer, gender converter A small connecter designed to change the gender of a plug or receptacle in order to enable connection to another plug or receptacle. The male end is the one in which the contact points protrude from the connecter. The female end is the one in which the contacts are depressed. Thus, a simple gender bender typically has two female ends or two male ends. Switchboxes and adaptors sometimes also perform gender bending functions. Extenders (male on one side, female on the other) are sometimes erroneously called gender benders, because they superficially resemble them. Gender benders are usually passive devices which do not influence the signal which passes through the device. See adaptor, extender, switch box.

G

General Magic A California-based commercial venture established in 1989 which is focused on providing innovative, cost-effective computer-Internet-telephony products for mobile applications. Products include Portico, a development name for integrated voice/data communications and information services; and DataRover integrated voice/data hardware devices

General Protection Fault GPF. A common fault condition encountered by users of Microsoft Windows software when applications memory conflicts occur. At the very least, applications should be closed up. Unfortunately, it may also be necessary to reboot the system.

Generalized Trunk Protocol GTP. A telephony protocol which permits complex Channel Associated Signaling (CAS) and Common Channel Signaling (CCS) protocols to be executed in peripheral devices to remove some of the processing load from the main processor.

generations, computers In computer terms, computing systems and computer software have been roughly grouped into an evolutionary hierarchy, with each step in the hierarchy represented by a numeric generation designation. Many of these generations are overlapping, and the later ones tend to become less distinct and more concurrent than the earlier ones. Here is a very general overview of previous computer generations, with some propositions on definitions for future trends and possibilities.

Computer Generations - Hardware

1st - Large, vacuum The early large-scale computer systems were powered and controlled by vacuum-tube technology in the 1950s. They needed elaborate hand-wiring, significant cooling and maintenance, and provided limited computational capabilities and memory.

2nd - Large, transistor The invention of the transistor greatly changed computing. Now much more could be done in much less space and new possibilities opened up in the early 1960s. Memory increased, virtual memory schemes were implemented, programming languages improved, and the idea of 'reusable code' began to spread, championed largely by Grace Hopper. Time-sharing systems were developed.

3rd - Large-scale ICs Integrated circuits provided the next significant step in computing hardware, during the mid-1960s to the mid-1970s. Again, computers became smaller and more powerful, memory increased, languages developed, and computing was entering the mainstream. Intercompatibility and networking began to develop, as did higher level languages that could be learned and used by lay programmers.

4th - Microprocessors Computers suddenly became accessible to homes and businesses in the mid-1970s, when single processor boards could be built, and a variety of general and special purpose languages could be developed by a larger community of programmers for a growing community of users. Very soon, the single-board computers outstripped the capabilities of their room-sized predecessors and computing began to become a mainstream technology.

5th - Parallel processing With the development of less expensive, smaller computers, it also became economical to use multiple processors, and to develop languages and algorithms to take advantage of parallel systems and distributed, networked environments. Networking grew to the point where the global network, the Internet, was reaching into every corner of the world. Most of this evolution occurred in the late 1980s and 1990s.

6th - Parallel entities In the author's opinion, there are two important aspects to sixth generation computers. One is the type of knowledge processing which will be needed to handle the burgeoning information on global networks. Another is the development of high-speed transport systems that are beginning to match and exceed the processing speed of computer central processing units (CPUs). When the communications links between computers match the processing speed of the computers themselves, a fundamental, significant change occurs, in that the system begins to function as a single entity rather than as separate, intercommunicating systems. Where this will lead is yet to be seen. This technology began trickling out in the early 1990s, and will probably continue to build momentum through the turn of the century.

7th - Abio/Bio systems The author has speculated on what may constitute seventh generation systems, and believes that integrated human-computer (or at least bio-electronic) systems may be the next qualitatively different level of computer development that will occur after, or more probably, concurrently with fifth and sixth generation systems. Applied engineering of neural network systems began in the early 1960s, yet when the author proposed the feasibility of human/bionic hybrids in late 1970s, the feedback was disbelief and horror. Even by the mid-1980s, there was strong opposition to the idea, and many said it could never happen. Since then, the use of 'bionic' hearts and limbs and other nonbiological systems has grown and brain implants to provide auditory and visual stimulation have been successful. The author believes that younger generations will accept these electronic/mechanical enhancements as a matter of course. At a conference in the mid-1990s, the author heard about a patented human-electronic interface invented by Howard Davidson, a scientist known for his work with subminiature components. Davidson has engineered a design for 'jacking in' to the brain through a tiny, aesthetic electronic device worn like an earring.

8th - Metacomputers Are there potential computer generations beyond those described above as seventh generation? George Dyson and other authors who have speculated about meta-intelligences, those which arise from the aggregate processing and communications of smaller units which function as a whole on a higher level than it may be possible for us to apprehend or comprehend, have pointed to the future possibilities for eighth generation computing systems, or digital intelligences. Some day the entire world will be networked with high-speed devices and human intelligences may or may not be directly interfaced with these machines. Then, just as individual cells in our body are not 'aware' of the specific thoughts and functioning of our brains, as individual humans we may not be 'aware' of the specific thoughts and functioning of the telecommunications and computer network as a broader entity. This far-flung idea of a vaster digital intelligence or a vaster digital/biological intelligence makes some people uncomfortable, but that doesn't mean it can't happen, or that it isn't already happening.

generations, computer languages Programming languages have evolved to support different systems, different capabilities, and different styles of thinking and programming. There is no cut-and-dry way to describe the evolution of languages, since many were designed to suit specific needs and are not directly comparable. Nevertheless, it can still be valuable to look at some of the trends and general categories in the industry which describe 'generations' of computer languages.

Generations - Computer Languages

1st - Machine code. Low level microcode and machine languages which interacted on the hardware and systems levels. Knowledge of computer architecture was necessary to program machine or microcode. First generation languages were difficult to debug, lengthy, and to most programmers, tedious. Most programming up until the 1960s was done in this way.

2nd - Symbolic low level. In the 1960s and 1970s, there began a transition to symbolic languages (which originated in the 1950s), which used mnemonic symbols. labels, and shortcuts to represent machine code, and which could be more easily written and debugged. Like machine code, however, the program listings were still very long and knowledge of the low level architecture of the processor was still necessary.

3rd - Symbolic high level. Assembly language was difficult for many programmers to learn, as knowledge of the hardware was required and assembly was generally not portable to other systems. New, more portable languages, which could be individually interpreted or compiled on the machine on which the software was running began to arise, some early ones including ALGOL, FORTRAN, and COBOL. These high level languages used syntax which more nearly approximated the English language, and the software could more easily be ported to other systems. Debuggers were developed to further ease the burden of finding errors and removing them.

4th - Symbolic high level. The development of BASIC and various educational scripting and authoring languages made it possible for even new programmers to exercise a moderate amount of control over the execution of computer programs, and these languages were popularly implemented on microcomputers starting in the mid-1970s and 1980s. Many of these languages are interpreted, so that the programmer doesn't have to acquire the technical expertise to compile and link executables.

5th - Symbolic AI. These languages are similar to symbolic high level languages in many aspects of syntax and execution, but they differ in terms of the way they process data and in the kinds of data that they are optimized to handle. Fifth generation languages came predominantly out of research laboratories doing artificial intelligence, neural network, and fuzzy logic work, often with languages like LISP and PROLOG. These languages are good at modeling virtual worlds, making inferences, and modifying themselves while running. They are good at parsing and interconnecting data and concepts in a way that is difficult with 3rd and 4th generation languages.

generator A machine which converts mechanical energy into electrical energy, or one which converts direct current (DC) into alternating current (AC). Dynamo.

Generic Access Profile GAP. A profile is defined by the Open Systems Interconnection (OSI) model as a combination of one or more base standards and association classes, necessary for performing a particular function. The GAP specifies well-defined compatibility levels for DECT products, as an extension of an ETSI-published Public Access Profile (PAP) incorporated into Digital European Cordless Telecommunications. (DECT). See Digital European Cordless Telecommunications.

generic cell rate algorithm GCRA. In ATM networking, an algorithm used to enforce a particular performance level with regard to cell traffic. The GCRA evaluates the cell to determine whether it conforms to the established cell traffic contract. In a network switch, the UPC function will typically incorporate an algorithm such as GCRA to enforce conformance with the specified parameters. See leaky bucket for a fuller explanation and example. See cell rate for related concepts.

generic flow control GFC. In ATM networks, a means of controlling traffic flow. A field in the ATM header can be used to designate flow control parameters. This is evaluated enroute, as appropriate, and is not included in the final delivered communication. See cell rate.

generic flow control field GFC. In ATM networking, traffic flow control is an essential aspect of moving cells from one place to another. In the ATM header, there are priority bits which can be set to inform the end-station that congestion control may be implemented by the switcher. See cell rate.

Generic Security Service Application GSSA. An IETF elective proposed general Internet standard. See RFC 2078.

genlock A device which takes a composite video signal and creates synchronizing pulses so the signal can be combined with signals from other devices such as computers, video cameras, etc. For example, a live feed from a TV set or a signal from a VCR can be fed into a genlock interfaced with a computer. The video image usually is displayed under the computer image and appears on the monitor as a combined image. Genlocks are often used to add computer subtitles to video which is then transmitted as a composite signal to a broadcast medium or saved on another tape. The first widespread distribution of microcomputer genlocks was in 1986 for the Amiga computer, and many local cable companies continue to use them for community events boards.

geographic information system GIS. Any system in which terrain information is gathered, processed, and stored, usually for later retrieval for analysis, long-term comparisons, mapping, navigation, etc. An enormous amount of geographic information is gathered by orbiting satellites, and geocoded and stored on high storage-capacity computer systems. Land-based tracking of networks, utilities, and transportations systems is also carried out with GIS systems. See geoInterface, Landsat.

geographic interface This has two meanings: 1. interfaces designed for users to access and manipulate geographic data, and 2. an evolutionary step in computer user interfaces which models the world in a simulated 3D environment as visual objects or functional constructs.

The second meaning represents a relatively new approach to computer user interfaces. The idea has been around for a while, but the resources to implement it have been cost prohibitive until recently. In the early 1970s, microcomputer user interfaces consisted of dipswitches for input and small blinking lights for output (see Kendak-1 and Altair). These were replaced by simple monochrome, text-based interfaces by 1975 (see SPHERE System). In the early 1980s, developers began producing graphical user interfaces (based largely on 1970s research at Xerox PARC), and it was not long after that the first experimental geographic interfaces began to appear.

Just as a graphical user interface might show a file cabinet icon to indicate a file area, a geographic interface might show an illustration of a room with an illustration of a 'real'-looking file cabinet with drawers that open and close when they are clicked or otherwise activated, and files that slide in and out and display their contents. It's a type of virtual environment that can exist in a simulated 3D space with gloves and goggles, or on a 3D-2D simulated space on a flat monitor. It's computing and graphics intensive, and takes a long time to develop, but it has a lush, tactile appeal that many people like, and will probably continue to gain popularity. With global object modeling standards, there may eventually be a large library to draw upon for this type of interface so individual developers can build on existing objects rather than continually reinventing the wheel. See geoInterface, graphical user interface, object-oriented, Open Systems Interconnection.

geographic north The region on the Earth called the North Pole from which imaginary lines of longitude emanate to meet again on the other side at geographic south. This is the general direction in which the north-seeking needle on a compass points, as magnetic north is near geographic north. See magnetic north.

geoInterface, IBM Geographic Interface International Business Machine's geographic information systems (GIS) interface. GeoInterface was designed to allow users to visualize, analyze, and manipulate IBM Geographic Facilities Information System (GFIS) data on OS/2 and Microsoft Windows systems. It works with IBM's VisualAge object-oriented programming environment in client/server and detached mode systems. Practical uses of this system include tracking and analysis of demographic information, transportations system planning and administration, telephone, cable TV (CATV), and computer networks, utility distribution systems, etc.

Georgia Rural Telephone Museum Located in a former 18,000 foot cotton warehouse, built in 1911, the Georgia Rural Telephone Museum opened officially in 1995. The building was located across the street from the Citizen's Telephone Company. It includes about 2,000 historic telephones and other communications equipment.

geostationary orbit A type of orbit which is timed with the movement of the body it is orbiting so the period is equal to the average rotational period of the orbited body. If, in addition, the orbit is circular, the satellite will appear to be 'not moving' when viewed from the ground, hence the name.

In simpler terms, if you place a satellite into a circular orbit about 35,900 to 42,164 kilometers above the Earth, it will appear to remain in the same place because of its synchronized relationship to the Earth's orbit. This type of orbit has some advantages for communications. The satellite can always be counted on to be available at the same location, and at that high altitude, not many satellites are needed to provide global coverage. The main disadvantage is that powerful sending and receiving stations are needed to send and receive signals from such a high orbit. Geostationary orbits were described by Arthur C. Clarke in the 1940s and 1950s in considerable detail and with remarkable foresight. Also called geosynchronous orbit and fixed satellite orbit.

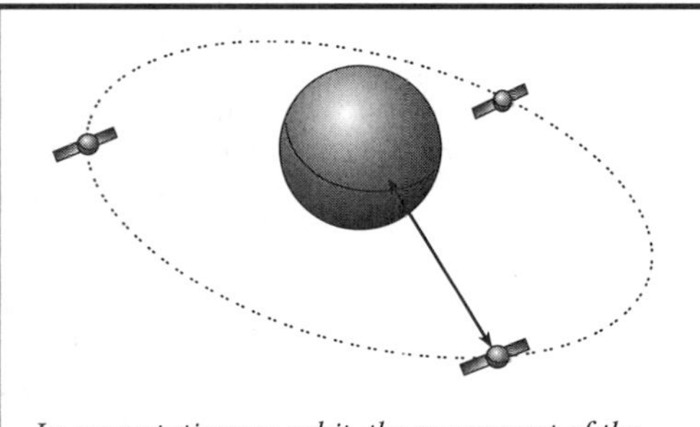

In a geostationary orbit, the movement of the satellite keeps pace with the movement of the Earth, and is thus within the same general visual and communications region at all times.

geosynchronous orbit See geostationary orbit.

GEOTAIL A Japanese research satellite launched in 1992 to study the structure and dynamics of the tail region of the Earth's magnetosphere. The orbit of the satellite was planned so it would cover the magnetotail over a wide range of distances. It contains instruments to measure the magnetic field, the electric field, plasma, energetic particles, and plasma waves. http://www.gtl.isas.ac.jp/

GETS Government Emergency Telecommunications System.

GFC See Generic Flow Control.

GGP See Gateway to Gateway Protocol.

ghost 1. On a monitor with phosphor burn-in from a greatly sustained image, an undesirable pale image that does not disappear when the screen is refreshed. This indicates permanent damage to the monitor (unless the coating is replaced). See screen saver. 2. In audio communications, a quieter repeat, or echo, of a conversation.

ghost, broadcast In broadcast images on a television screen, a slightly offset, pale copy of the desired image caused by secondary transmission of the original signal. Terrain can cause reflections of the direct signal that arrive at the receiver just slightly delayed and make the image appear slightly blurred or double.

ghosted icon A graphical user interface convention in which an icon is displayed in a toned down or paler version of itself to indicate that it is not currently active or available for selection by the user. In other words, if you click on a ghosted icon or ghosted image, nothing is likely to happen except that you feel foolish for having tried to select it anyway.

Ghostscript A PostScript graphics language that is part of the Free Software Foundation's GNU project which is almost fully compatible with Adobe's PostScript. Ghostscript is a great tool for viewing and printing PostScript files.

.gif The standard graphics file name extension used for CompuServe's Graphics Interchange Format (GIF) graphics files. See Graphics Interchange Format.

giga- (Abbreviated G when combined) (*pron.* jig-a) A prefix for 10^9 or 1,000,000,000 in the SI system. One billion. In computing, a giga is 2^{30} or 1,073,741,824 (a multiple of 1024). Giga- has long been used in supercomputing, mainframe and scientific applications, but was relatively unknown in lay language until the mid-1990s when gigabyte (Gbyte) hard drives dropped to consumer price ranges. It used to be a lot of storage space. Ten megabytes used to be a lot of storage. In fact, whole community BBS systems used to run on five megabyte drives in the early 1980s. Now two gigabyte drives are considered to be average. See atto-.

gigabit GB. (*pron.* jig-a-bit) 1,073,741,824 (2^{30}) bits.

Gigabit Ethernet Ethernet networking capabilities which can support half and full duplex transmissions at speeds of 1 Gbps. Objectives for link distances include multimode fiber optic links up to 550 meters, single-mode fiber optic links up to 3 kilometers, and copper-based links up to 25 meters, and Category 5 unshielded twisted pair (UTP) links up to 100 meters. A Gigabit Media Independent Interface (GMII) is also being studied.

Gigabit Ethernet was developed because vendors and users wanted the benefits of a high speed network which could support existing Ethernet frame and protocol characteristics to enhance rather than obsolete existing systems.

Gigabit Ethernet provides a practical way to set up a backbone for interconnecting Ethernet and Fast Ethernet networks, and provides an Ethernet upgrade path as the technology becomes cheaper. The most significant competitor to the faster versions of Ethernet is asynchronous transfer mode (ATM).

Gigabit Ethernet supports the same frame format and size, and carrier sense multiple access with collision detection (DSMA/CD) as Ethernet, and Fast Ethernet. Quality of service (QoS) is not inherent in Gigabit Ethernet, which is primarily a high speed connectivity mechanism, but is incorporated through other standards. RSVP is one way of providing quality through an open standard which can be incorporated into a Gigabit Ethernet system.

Fast Ethernet, the predecessor to Gigabit Ethernet, is an international open standard and widely installed. In 1997, the IEEE approved the P802.3ab study group's proposed 1000Base-T standard for full duplex Gigabit Ethernet signaling over Category 5 networking systems. This lead to the IEEE 802.3z Working Group work towards ratification of a standard for Gigabit Ethernet. See asynchronous transfer mode, Ethernet, Fast Ethernet, Gigabit Ethernet Alliance.

Gigabit Ethernet Alliance GEA. A California-based multivendor open forum established in 1996 to promote the development and acceptance of Gigabit Ethernet technology and to actively support and accelerate the standards process. The GEA supports IEEE activities with regard to the development and ratification of Ethernet standards, particularly the High-Speed Study Group, the IEEE 802.3 Working Group, and the IEEE 802.z Gigabit Ethernet task force. It provides technical resources for implementation and product interoperability. See Gigabit Ethernet.
http://www.gigabit-ethernet.org/

gigabyte GByte, Gb. (*pron.* jig-a-bite) 1,073,741,824 (10^9) bytes.

Gigabyte System Network GSN. See Hippi-6400.

GIGO garbage in, garbage out. An acronym to describe a situation in which output cannot be better than its corresponding input, with the implication that it is the fault and responsibility of the developer or data entry person if the system gives back bad or incomplete information.

GII global information infrastructure. A term used since the mid-1990s by international standards committees with regard to goals, standardization, and development of global interconnected telecommunications systems, including the technology, applications, and related services.

GIIC Global Information Infrastructure Commission.

gilbert A centimeter-gram-second (CGS) unit of magnetomotive force equal to 10 divided by 4p ampere-turn. It is named after William Gilbert.

Gilbert, William (1544-1603) An English physicist and physician who investigated electrostatic charges in various substances. He observed that magnetized iron lost its attractive power when heated to red heat, and published *De magnete* (On the magnet) in 1600. He emphasized the distinctions between the magnetic effect of substances like lodestone, and the attractive properties of amber, a distinction previously promoted by J. Cardan in 1550, but still not widely considered. In his treatise, he used the word *electrica* to describe attractive phenomena. Gilbert established that the Earth is a large magnet, thus explaining the general behavior of compass needles. The gilbert unit of magnetomotive force is named after him. See gilbert, versorium.

GILC See Global Internet Liberty Campaign.

gimbal A mechanism or material which permits an attachment to be freely suspended or inclined in such a way that the suspended attachment remains level, or so the attachment can be inclined in any direction or several directions. Marine compasses and gyroscopes incorporate gimbal mechanisms.

gimp Extremely flexible wire or cable. Wire which can be easily threaded, woven, or spiraled. Gimp is wound up and attached to telephone handsets to allow a length of wire to tighten up like a spring when not in use, so the talker doesn't have to interrupt the phone conversation to reach the refrigerator.

Giorgi System A system of measurement in which the units are: meter, kilogram, second, and ampere (MKSA).

GIP See Global Internet Project.

GIS See geographic information system.

Gisborne, Frederic Newton A Canadian inventor who devised new ways of insulating cable against harsh environments and who, with the financial backing of an American,

Cyrus Field, linked Europe and North America with the first successful transatlantic telegraph cable in 1858.

GITS Government Information Technology Services.

GL See graphics library.

Gladstone-Dale law The refractive index of a substance varies with a change in temperature or volume according to a formula in which the index of refraction (*n*) plus one, over the density (*r*), equals a constant (*k*).

glass house *colloq.* A term to describe the large, glassed-in, controlled environments used to house and protect (and, in some cases, air-condition) large computer installations. These environments still exist, to some extent, in supercomputing systems, but technological advances have decreased the size and fragility of many computers and glass houses are no longer needed for small or medium-sized computing systems. Glass houses or 'clean houses' are still used in chip manufacturing environments to provide a carefully regulated environment where temperature, humidity, and even tiny particles can effect the structure and functioning of certain delicate or microminiature components.

glass insulators A type of portable fabrication ranging in size from about six inches to about 18 inches that was very commonly used for wiring on utility poles. See insulator, utility pole for a chart and more detailed information.

glitch 1. Unexpected small but annoying problem, usually causing a delay or minor informational error. This term is usually applied in instances where repetition of the problem is unlikely or infrequent. 2. Undesirable brief surge or interruption of electrical power.

global area network GAN. A network that is accessible to most or all nations in the world. The Internet is the closest thing we have to a GAN, although it is not yet ubiquitous or accessible by all nations or people.

Global Business Communications Systems GBCS. An AT&T business which was rolled in with the AT&T Laboratory restructuring of Bell Laboratories in 1995-1996, along with the Network Systems Group, AT&T Paradyne, Microelectronics, and Consumer Products. GBCS is moving into the area of multimedia and secure telecommunications services. Products include a Unix-based server that works on a private branch exchange to provide videoconferencing capabilities over data networks like Ethernet and IBM Token-Ring. The server software is called Multimedia Communication Exchange (MMCX) and was implemented first on Unix stations, with the intention of porting it later to PC operating systems.

global directory An internetwork computer database which stores various types of information related to the various networks. It may be user login names and passwords, shared database resources, group member lists, device directories which can be accessed by more than one network, or pointers to various applications or documents common to the various networks. The Internet has a number of global directories of file databases, archives, etc. See Gopher, Archie, Veronica.

Global Internet Liberty Campaign. A human rights group which includes as members the American Civil Liberties Union (ACLU), the Electronic Freedom Foundation (EFF), and others.

Global Internet Project GIP. A private sector organization founded in 1996, consisting of senior level managers representing global software and telecommunications industries with high stakes in Internet development. As part of its activities, GIP encourages the education of world decision-makers in the potential evolution and uses of the Internet.

Global Maritime Distress and Safety System GMDSS. A maritime safety system which incorporates automated distress calls using Digital Selective Calling (DSC). In the late 1970s, maritime experts began to develop systems for updating the safety and distress communications, resulting in the 1979 draft of the International Convention on Maritime Search and Rescue advocating a global search and rescue plan and a Global Maritime Distress and Safety System (GMDSS) as a communications infrastructure for the overall plan. The system is based on a combination of Earth-based and satellite-based radio services, emphasizing ship-to-shore marine signaling. This was to supersede the decades-old Morse code-based system. In addition to the automation of distress signals, it called for the shipboard downloading of maritime safety information as a preventive measure.

In 1996, the Telecommunications Act was written to encompass U.S. marine vessels and ships were required to install GMDSS equipment by 1 Feb. 1999. All vessels subject to Chapter IV of the Safety of Life at Sea (SLOAS) convention must be fitted with GMDSS equipment, with certain stipulated exceptions, as must mobile offshore drilling units (MODUs).

Implementation of GMDSS has not been without problems. It has been criticized for false alarms and general reliability, and many nations have been slow to adopt the system. See COSPAS/SARSAT, NAVTEX.

global mobile personal communications services GMPCS. A phrase coined by the ITU-T to describe mobile communications through low Earth orbit (LEO) satellite systems, but later broadened to include other modes of mobile communications (geostationary FSS, MSS, "Little LEOs" and wideband LEOs). GMPCS was the discussion theme of the World Telecommunication Policy Forum (WTPF), resulting in a set of principles and recommendations described in the "WTAC Report to the Secretary-General [of the ITU-T] on GMPCS" in January 1996.

Global Navigation Satellite System GLONASS, GNSS. A satellite system deployed by the Russian Federation defense department, which shares much in common with the American Global Positioning Service (GPS) in terms of satellite placement and the types of information that are transmitted. The 24 GLONASS system satellites are orbiting in three planes. Unlike the GPS system, GLONASS claims to plan to use the same levels of signals for civilian (CSA) and government use (SA), and civilian use is guaranteed for about the next decade. While the satellites are in place, the system itself was not operational as of early 1998.

The MIT Lincoln Laboratory conducts research on the GLONASS system and reports progress and observations on the project on their Web site. http://vega.atc.ll.mit.edu/glonass/

Global Network Navigator GNN. A Web-based information service which provides listings of and information about new services, sites, and related resources on the Internet.

Global One An international joint venture of Sprint, Deutsche Telekom, and France Telecom.

Global Online Directory GOLD. A commercial product from VocalTec which works in conjunction with their Internet Phone software. Internet Phone lets you plug a microphone into your personal computer and use it as a phone transmitter to communicate with another person with Internet Phone capabilities. The computer speaker provides the equivalent of the phone receiver. Long-distance calls can be placed as though they were local calls through your ISP, without long-distance charges.

Internet Phone connections are full duplex, connecting through the TCP/IP transport protocol. In addition to the features of a conventional phone call, chat lines and other digital enhancements are available.

GOLD is the global directory that stores information about Internet Phone users who can be contacted just as the names of phone subscribers can be accessed through a phone directory.

Global Positioning System GPS. A space- and ground-based 24-hour navigational system originally designed and used by the U.S. military (see Navy Navigation Satellite System), funded and maintained by the U.S. Department of Defense (DOD). It provides the means to monitor, update, and maintain orbiting satellite systems, and to determine a location on or around the Earth through information from these systems.

GPS uses the known positions of satellites as reference points for discerning unknown positions on or above the Earth. There are now over 20 satellites in the system (some are spares), more-or-less evenly spaced, orbiting in 12-hour cycles at an altitude of about 10,898 miles (about 400 miles higher than the original NNSS). A system of sophisticated ground stations with antennas, coordinated by a master control station, administers, deploys, and maintains the satellites and updates them when needed to correct for clock-bias errors.

A variety of types of information can be computed from information from several satellites, including a location, or position of a stationary or moving object, and coordination of time. This information can be incorporated into software applications in vehicle-mounted or handheld positioning receivers. From military operations to recreational navigation out on the ocean in a kayak, to airline navigation, GPS provides a wealth of data with which to determine latitude and longitude, altitude, and velocity. This information can further be combined with maps to record or suggest routes.

GPS satellites transmit timed binary pulses in addition to information constants about the current location of the satellite. The synchronized atomic clocks aboard the satellites permit the transmission of precise timing tags. The combination of the speed of transmitted electromagnetic waves and the atomic clocks installed in GPS satellites provides remarkably accurate timing pulses.

There are thousands of GPS users worldwide. Personal GPS devices can be purchased for as little as $180 to $450, and more sophisticated ones are used in all industries that rely on location information: airlines, shipping firms, ferries, military divisions, etc. For greater details on individual aspects of GPS, see differential GPS, EAGLE, Intelligent Vehicle Highway Systems, GPS Operational Constellation, GPS Navigation Message, local differential GPS, NAVSTAR, Precise Positioning Service, Standard Positioning Service, wide area differential GPS.

global search and replace A function in text editing, word processing, and desktop publishing programs, which allows the user to find and replace a selected entry throughout the entire document, which may even span more than one file. Search-and-replace functions can be constrained in several ways, by limiting the changes to just the highlighted (selected) text to the current page or the current document, or they may be global, to include the entire current document or a multifile document.

Global Software Defined Network GSDN. A high-volume commercial virtual private network service from AT&T which utilizes AT&T's Worldwide Intelligent Network (WIN) to interconnect networks in the U.S. and other countries. GSDN selects an economical route for external calls and provides internal services including order entry, tracking, file transfers, and teleconferencing services. GSDN is aimed at business networks.

Global Standards Collaboration GSC. There is information on this effort on their Web site. http://194.2.180.16/ies/gsc/

Global System for Mobile, Groupe Spéciale Mobile GSM. A digital cellular technology developed jointly by the telecommunications administrations of Europe. The Groupe Spéciale Mobile was founded in the early 1980s, and the Global System for Mobile (GSM) was first publicly announced in 1991, and has since been standardized in Europe and Japan.

GSM was the first fully digital system to provide mobile voice connections, data transfer services, paging, and facsimile at full duplex or half duplex rates up to 9600 bps. GSM operates in two frequency ranges: 890 to 915 MHz for signaling information and 935 to 960 MHz for information transmissions.

GSM is a set of standards specifying a digital mobile communications services infrastructure. It is based on a 900 MHz radio transmission technology and specifies related switching and signaling formats. An 1800 MHz Digital Cordless System (DCS) has also been added. Since mobile systems typically support roaming, and since the multicultural makeup of Europe provides a unique challenge in providing compatible services, interoperability has been emphasized in the GSM specifications.

GSM can be described in three categories: the communications media, the transceiving systems, and the information systems, as shown in the GSM Categories chart.

The GSM subscriber identity module (SIM), also known as a *smartcard,* is a security fea-

GSM General Categories

Category	Notes
Media	GSM works over frequency-modulated (FM) signals using a combination of time division multiple access (TDMA) and frequency division multiple access (FDMA). Peak output power varies with the type of transmitter (mobile station class), ranging from 0.8 to 20 watts. Frequency hopping is used to reduce interference and multipath fading, and encryption increases security. The data rate is 270 Kbps.
Transceiving	There is a base transceiver station (BTS) associated with each cell operating on fixed frequencies unique to its region. Honeycomb-like clusters are handled by base station controllers (BSC), which, in turn, are controlled (routed, switched, handed over) by Mobile Service Switching Centers (MSC).
Information	There are databases associated with GSM which aid in the administration of subscriber information, and those which aid in the administration of security and associated authentication mechanisms. There is also an equipment identity register (EIR) which keeps track of equipment types and configuration, and can block calls on stolen units.

ture which handles encryption and authentication. It includes memory storage which can be used for dialing codes or other information related to the service. The SIM is also a means to download and display call-related information. See Future Public Land Mobile Telecommunication System, Personal Communications Network.

Global Transaction Network GTN. AT&T's extensive 800 service phone network, which was introduced in 1993. This service supports enhanced features, providing more flexible routing and numbering services which can be used, for example, by airline reservation systems.

Globalstar A system of 48 small *bent pipe* communications satellites orbiting at 1400 kilometers (LEO), for providing voice and data services (data files, paging, facsimile). Globalstar was established in 1991 as a joint venture of Loral Space & Communications, Ltd., QUALCOMM, Inc., and a number of corporate partners. Launching began in February 1998. It is expected that half the satellites will be operational by 1998, with the rest slated to come online by 1999.

Globalstar services can be accessed with vehicle-mounted or hand-held devices resembling cellular phones and the system is integrated with cell phone services through dual-modem handsets. Remote users can access the system through Globalstar service providers, with fixed-position and wireline phones. Globalstar is intended to enhance rather than replace existing cellular and other phone services. Services are aimed at international business travellers, commercial vehicle operators, marine craft, field scientists, and others. The competitive aim is low cost for service and accessories.

GLONASS A Russian Global Positioning System similar to the U.S. NAVSTAR system.

GMDSS See Global Maritime Distress and Safety System.

GMPCS See global mobile personal communications by satellite.

GMSK See Gaussian minimum shift keying.

GNN See Global Network Navigator.

GNSS See Global Navigation Satellite System.

GNU Acronym for "GNU's Not Unix!" A Unix work-alike developed under the aegis of Richard Stallman of the Free Software Foundation (FSF). See Free Software Foundation.

GNU as A GNU family of assemblers which is used to write software code for a variety of object file formats. The original GNU assembler for the Digital Equipment Corporation (DEC) VAX system was written by Dean Elsner. Many subsequent programmers and even some commercial vendors have since enhanced and maintained the software. See Free Software Foundation.

GNU C compiler GCC. A C compiler which supports ANSI standard C, C++, and Objective C. The GNU C library includes ANSI C, Unix, and POSIX functions.

GNU Emacs A powerful, extensible, scriptable display editor distributed by Berkeley programmers with BDS, and by many other distributors and commercial vendors. Emacs is so powerful and so well liked by power editor users, many have half-seriously referred to it as an operating system.

The first Emacs was written in 1975 by Richard Stallman. GNU Emacs, which was enhanced by Stallman with true LISP integrated into the editor, was introduced in the mid-1980s. GNU Emacs is widely available on Unix systems.

GNU graphics A set of graphics utilities for plotting scientific data, with support for GNU plot files on various systems and output devices, including PostScript, The X Window System, and Tektronix devices.

GNU's Bulletin A semi-annual newsletter about various GNU projects, produced and distributed by the Free Software Foundation.

go local A command to instruct software to connect to a local connection, usually through a serial null modem interface.

Godwin's law "As a USENET discussion grows longer, the probability of a comparison involving Nazis or Hitler approaches one." The implication is that once the discussion level of a newsgroup drops to this level, its usefulness is pretty much over.

GO-MVIP Global Organization for Multi-Vendor Integration Protocol. GO-MVIP is a non-profit trade association which took over the development and promotion of MVIP in 1994, in order to assure its development and maintenance as a practical, robust common integration standard. GO-MVIP seeks to continue to develop and establish the design specifications for further versions of MVIP. See MVIP. http://www.mvip.org

GOES Geostationary Operational Environmental Satellite.

gold A malleable, metallic chemical element with high conductivity which makes it useful for specialized electrical applications. Gold contacts are often found on sensitive electronics connectors in the computer and video industries. Copper and silver are also good conductors, with copper being the most widely used for electrical installations.

GOLD See Global Online Directory.

gold disk, gold disc 1. The master or final copy of a product (software, music CD, laserdisc, etc.) from which mass production replicas are made. 2. A special limited edition distribution. Collectors' edition. Gold disc music CDs sometimes are marketed as higher quality pressings with special inserts and special tracks that may not be included on a regular copy of the CD.

gold number, custom number, vanity number A phone number specifically selected so that it is easy to remember, particularly if the letters associated with the number spell out a word or other mnemonic. There is typically an extra fee associated with getting a gold number. Sometimes people get lucky, and their number just happens to be easy to remember or to spell something interesting.

golfball printer, daisy wheel A printer with a round, rotating, impact printhead like those found on IBM Selectric typewriters and daisy wheel computer printers. The rotating head is embossed with the character set and impacts with the printing ribbon causing carbon to be deposited on the print medium.

good condition In many rating systems, good condition indicates a product with minimal abrasions from wear and tear, and mechanisms that are in good working order. Good condition does not imply any information about the age of the product or its remaining useful life. Often sandwiched between *fair condition* and *excellent* or *like new* condition.

goodput A generic measurement of network data successfully received, effective throughput; in contrast, discarded cells, or transmitted cells in a congested link, are called badput. See cell rate, throughput.

Goodwin, Hannibal (1822-1900) An American minister and inventor who created celluloid film in 1885 and received a patent for rollable film in 1887. For many years, motion picture films were known as *celluloids* and individual animation frames used to create frame-by-frame animation are still known by the abbreviated form of *cells*.

gopher The command for initiating a Gopher client on a text-based system is "gopher" (all lower case), and "xgopher" is a similar client command that works with The X Window System.

Gopher A document system developed by P. Lindner, M. McCahill, B. Alberti, F. Anklesaria, and D. Torrey at the University of Minnesota in the early 1990s to provide a local campus information server. The Gopher service quickly grew to become a worldwide resource. It is a client/server distributed document delivery system, that is, a means of locating information on the Internet through a simple menu-like text interface (or graphical Gopher client) or of sending information through electronic mail. It is also possible to set 'bookmarks,' that is, Gopher information locations that are frequently used. Links to various Gopher servers together comprise a virtual community known as Gopherspace. The Gopher text menu interface is being superseded by graphical Web interfaces. See Veronica, RFC 1436.

Gopherspace The Gopher document system is composed of many widely distributed document repositories and Internet services in 'cyberspace.' Hence, the Gopher facilities online are called "Gopherspace" by many their users. See Gopher.

GORIZONT A Russian geostationary telecommunications satellite launched in 1996.

GoS grade of service. A phrase to describe service levels, which usually are individually defined on an industry basis. See class of service.

GOSIP Government Open Systems Interconnection Profile. A U.S. government version of the Open Systems Interconnection system which is required in many government data network installations.

Gosling, James Gosling is best known for his contributions to the Java programming language, developed at Sun Microsystems, Inc. He was associated with Bill Joy, Mike Sheradin, and Patrick Naughton on Project Stealth in 1991. Project Stealth's goal was to develop a distributed network in which the various electronic devices could intercommunicate. See Java; Joy, William.

GPF See general protection fault.

GPS See Global Positioning System.

GPS Control Segment A general overall category of the GPS system which comprises a main tracking station, in Falcon Air Force Base, Colorado, and subsidiary tracking stations worldwide as part of the U.S. Department of Defense's Global Positioning System. The tracking stations take the signals from the satellites and incorporate them into orbital models which are further used to compute precise, individual, orbital data and clock corrections. Portions of this orbital *ephemeris* are sent via radio transmissions to GPS receivers. See GPS Space Segment, GPS User Segment.

GPS Navigation Data Satellites in the Global Positioning System (GPS) send out two microwave carrier signals, one of which provides navigation information in the form of a series of time-lagged data frames sent over a specific time period. Subframes are also included for checking data integrity. The satellites are equipped with atomic clocks, and clock data parameters are sent and related to GPS time. Orbits are described by transmitting regularly updated ephemeris data. See Global Positioning System, Universal Coordinated Time.

GPS Operational Constellation The system of over 20 more-or-less evenly spaced, orbiting satellites (some of which are spares), equipped with atomic clocks, in the Global Positioning System (GPS) Space Segment. These satellites orbit the Earth twice a day at about 11,000 miles altitude, transmitting information used in the U.S. Department of Defense's Global Positioning System. The orbital planes are inclined at about 55 degrees in relation to Earth's equatorial plane. From any one point on Earth, it is generally possible to locate between five and eight satellites, four or five of which are typically used to compute location and timing information. The satellites transmit two microwave carrier signals, with L1 frequencies carrying the navigation message (with data describing the orbit and clock parameters) and SPS code signals, and L2 monitoring ionospheric delay of PPS receivers. See Global Positioning System.

GPS receiver/display A fixed or mobile Global Positioning System (GPS) device which interprets GPS information and computes graphics, text, locations, maps, or other displays that provide the user information about position, time, and sometimes velocity. A graphical display of latitude and longitude is common. Receivers vary from room-sized systems to small hand-held units from $180 up to hundreds of thousands of dollars. GPS consoles are used by surveyors, have been combined with map databases to provide car consoles, and have also been incorporated into 'smart cars' that can steer themselves. It is not unrealistic to predict that someday small GPS systems will be designed into wristwatch-style personal locators for travellers, sales representatives, hikers, et al. See Global Positioning System, Intelligent Vehicle Highway System.

GPS Space Segment A general overall category of the GPS system which consists of GPS satellites deployed and administered by the U.S. Department of Defense as part of its Global Positioning System. See GPS Control Segment, GPS User Segment.

GPS User Segment A general overall category of the GPS system which includes GPS receivers and users of the U.S. Department of Defense's Global Positioning System. See GPS Control Segment, GPS Space Segment.

GR generic requirement.

grabber *colloq.* A graphical user interface (GUI) tool that looks like a little hand, and is used to 'grab' or otherwise activate or drag screen elements. Commonly seen on Macintosh computer applications.

graceful close A program or process which shuts down cleanly, with no stray windows, files, buffers, or problems. On many single user computers, shutting off the system may leave stray files or applications in a state that is different from the ideal closing state. On these, there may be a *system shutdown* selection (e.g., Macintosh), which 'cleans up' before shutting off the computer. In layer-oriented networks, *graceful close* refers to a connection that is terminated at the transport layer with no loss of data.

Grade 1 to 5 twisted pair See twisted pair cable.

grade of service GoS. A service level indicator which is evaluated on an industry basis according to the type of service provided. In some industries a hierarchical category scale is applied to various levels or definitions of service. In telecommunications, grade of service is typically described in statistical terms related to the speed and probability of connecting, and the characteristics of the connection, etc. See class of service, quality of service.

gradient Gradual change in elevation, color,

or texture along an axis. Gradual blend or transition. See gradient fill.

gradient fill A common feature of paint programs that allows the user to fill a defined area with graduated tones ranging from one specified color or shade of gray to another. The number of colors in the palette and the two end-tones selected will effect the smoothness and visual appeal of the transition, with more tones generally creating a more pleasing effect. Radial fills can be used to simulate 3D surface areas, as lighter areas appear as highlights.

Graham Act A 1921 act in the United States in which telephone companies are granted exemptions to the provisions of the Sherman Antitrust Act. This resulted in AT&T, more than any other company, in being able to expand and exert further control over the telephone networks. See the Kingsbury Commitment, Modified Final Judgment.

Gramme, Zénobe Théophile (1826-1901) A Belgian engineer who emigrated to France, Gramme developed a direct current (DC) generator, featuring a ring armature in 1869 and 1870. Together with Hippolyte Fontaine, Gramme opened a factory called Societé des Machines Magneto-Electriques Gramme. In 1873, at the Vienna Exposition, it was noticed by a mechanic that an electrical connection from another generator could power the armature of the first generator, thus exhibiting the characteristics of a motor. This was an important historical advancement in industrial and transportation technologies.

A Gramophone developed by Canadian inventor Emile Berliner in 1888. It could play for about 2 or 3 minutes at 70 rpm.

Gramophone, Gram-O-phone A phonograph technology patented by Emile Berliner in 1888, three years after the introduction of the "Graphophone" by Bell and Tainter. The Gramophone used a flat 7" disk with lateral grooves, the earliest of which were single sided. It played at 70 rpm, and the playing time was about 2 or 3 minutes, the same as an Edison cylinder. The model E, introduced in the early 1900s, played at 78 rpm on 7" disks. See Graphophone, phonograph, phonograph record.

grandfather clause A previously existing object, structure, statute, ownership right, or policy which may continue in spite of subsequent restrictions or regulations which would prevent its creation or continuance. A grandfather clause grants a type of pardon, special permission, or immunity. For example, military surplus purchased by a civilian in the 1960s may be regulated in the 1990s such that similar items might not be purchasable by current civilians (such as radiation bunkers). If the ownership is protected by a grandfather clause, when restrictions are imposed or reinstated, current civilian owners may not have to give up the property (but also may not be able to sell it, except perhaps back to the government).

Building codes are often subject to grandfather clauses. If you purchase a house built in 1920, it may not be subject to the same offset, materials, or safety regulations as current structures.

Voting and immigration laws have certain grandfather clauses. Immigrants to the country prior to a certain date do not require the same documents and eligibility requirements as later immigrants.

In telecommunications, phones and various electronic components built or installed before a certain date may not have to meet all current Federal Communications Commission (FCC) regulations.

grandfathered in Instated or installed before certain restrictions or regulations were put in place which would prevent creation, installation, or operation. See grandfather clause.

graphechon A special-purpose 'memory' electron tube used in computer and radar applications. The graphechon can store an electrical charge pattern, similar to the functioning of an iconoscope, and recover the pattern at different scanning rates.

graphic equalizer A component which provides a set of controls for adjusting the tonal qualities at several frequencies in an audio system, usually a music system. The equalizer is not a stand-alone component; it works in conjunction with other components such as receivers, phonographs, tape players, CD players, etc. It frequently has a series of vertical analog sliders for making individual adjustments.

Graphical Kernel System GKS. An official standard for 2D graphics in the mid-1980s, evolved from the Core. A 3D extension was subsequently developed, and GKS-3D became a standard in 1988. See Core, PHIGS.

graphical user interface GUI. A way of facilitating communication between a human and a device, usually a computing machine, by presenting the information in the form of visual metaphors. A graphical user interface works in conjunction with a variety of physical input devices, including speech recognition hardware, mice, keyboards, stylus pens, touchscreens, and joysticks. They provide a means to select and control the various visual elements, which commonly include menus, drag bars, buttons, icons, and window gadgets. Video games like Pong were early electronic adaptations of simple GUIs. Many of the earliest applied GUI ideas in general use today were developed at Xerox PARC and incorporated by Apple Computer into the Macintosh operating system.

graphics accelerator A chip or circuit board integrated into a computer system to relieve the CPU of some of the functions related to the processing and display of graphics. Graphics tend to be computing intensive, and sharing the load can significantly speed the display and refresh of images. Graphics accelerators are often sold as peripheral cards that can be plugged into a slot. See graphics coprocessor.

graphics character A character displayed on a computer screen in terms of its rasterized (or sometimes vectorized) shape. Some systems display characters out of a storage bank which is preset, and may not be configurable. The old TRS-80 Model I computers shipped with upper case letters only. If you wanted upper and lower case characters, you had to install an optional chip in place of the character chip that came with the system. These "hard-wired" characters displayed quickly, but were not very flexible. In subsequent systems, the characters on the screen were displayed as though they were images, even if they were text characters, making it possible to greatly vary the size, shape, and proportions of the letters and symbols.

graphics controller, graphics display processor Specialized computer hardware to improve raster displays by taking some of the load from the CPU. Graphics controllers can speed up scan conversion, and the composition, display and movement of graphics images and primitives. See cathode ray tube, frame buffer.

graphics coprocessor A chip designed to speed computer graphics composition, display, or refresh by sharing the load with the system CPU. Coprocessors are sometimes designed for very specific tasks, such as updating a screen, or storing and displaying graphics primitives, hardware sprites, and the like. Unlike graphics accelerators, which are often sold to consumers as optional system-enhancing peripherals, graphics coprocessors are more commonly sold integrated into the system, often on the motherboard. See graphics accelerator.

graphics device interface GDI. Physical and virtual connections between graphics hardware components and the computer CPU. Since graphics applications tend to be CPU-intensive, it is very common for other graphics hardware (accelerator cards, frame buffers, blitters, etc.) to be incorporated into a system to facilitate the fast creation, display, and refresh of images on a variety of output devices.

graphics engine The part of a computer architecture which supports the graphics functions of the machine, particularly graphics composition, buffering, display, and fast refresh. Graphics engines are typically designed to handle many of the functions in hardware, so there is a minimum of on-the-spot software calculations. Enhanced graphics standards and graphics engines are being developed to support features such as real-time animation; hardware pan, zoom, compression/decompression; instant resolution-switching; and video signal support.

Graphics Environment Manager GEM. An early graphic user interface (GUI) developed by Gary Kildall's Digital Research, the same company which created the popular CP/M text-based operating system in the 1970s. GEM was first demonstrated publicly at the COMDEX computer industry trade show in 1983 and shipped a few months later. The interface greatly resembled the Macintosh interface which Apple had developed after observing development research at the Xerox PARC laboratories. GEM did not become widely distributed, with the exception of providing a front-end to Xerox's Ventura Publisher, a desktop publishing programming that was widely used for documentation page layout on Intel-based microcomputers in the later 1980s.

Graphics Interchange Format GIF. (This ought to be pronounced "gif" given that the G stands for "graphics," but its author apparently uses "jif.") A proprietary raster graphics format introduced by CompuServe, Inc. in 1987. GIF is an 8-bit graphics format developed with the patented Lempel-Ziv-Welch (LZW) compression, whose implementation requires a royalty agreement from Unisys Corporation. The level of compression varies with the type of image and number of colors, but 3 or 4 times compression ratios are common on a typical color image. Due to patent issues, CompuServe agreed in 1994 to secure a license agreement to distribute the LZW technology and issued the Graphics Interchange Format Developer Agreement to provide software developers permissions under CompuServe's software license agreement with Unisys.

GIF is particularly suitable for images which have a small number of distinct colors, as opposed to images which have a great variety of subtle color changes. It also handles line art and grayscale images (through color palette gray matching), and sharp color boundaries better than formats optimized for other characteristics. Because GIF is a 256-color format rather than a 24-bit color format (~1.6 million colors), 24-bit images will be dithered and adjusted, and may not fully satisfy the needs of the user.

GIF will support transparency, which is sometimes desired in order for a background image to be displayed behind the GIF image or through parts of the GIF image as though there were 'holes.' This is often used by Web designers to produce special effects in Web pages, such as buttons with irregularly shaped edges.

GIF is one of the three most common graphics formats supported by World Wide Web browsers, the other two being PNG and JPEG. PNG is an open, non-proprietary format, developed to supersede GIF. In January 1995, CompuServe announced the GIF24 project for designing a replacement format for the original 8-bit GIF, and a month later officially announced that Portable Network Graphics (PNG) would be used as the basis for GIF24.

Support services for GIF users is provided in the form of the CompuServe Graphics Support Forum (GO GRAPHSUPPORT). It is CompuServe's central distribution area for GIF-related information. See Lempel-Ziv-Welch, Portable Network Graphics.

graphics library GL. The writing of graphics routines for computers is time intensive and specialized. For that reason, many companies decide to purchase graphics libraries rather than to write their own. These library routines consist of a collection of commonly used graphics primitives and actions (lines, circles, dots, fills, patterns, etc.) which can be dynamically called from the graphics library, or compiled and linked into the software executables as needed.

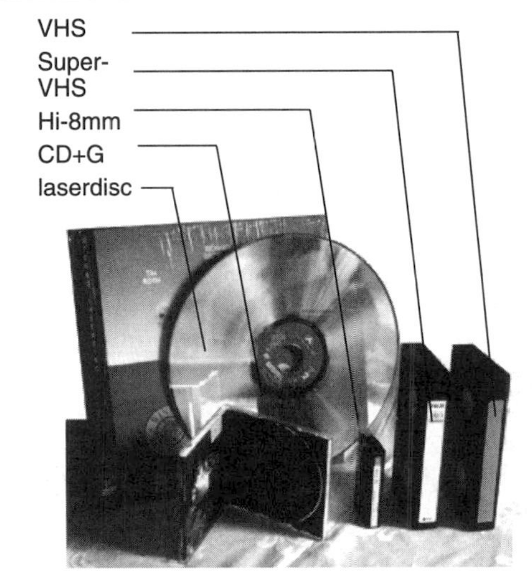

The variety of standardized recording and playboack media for graphics and multimedia products is increasing. Some of the more common formats supporting both images and sound are shown here. It may be that DVD-related formats, which hold more information in less space, will eventually supersede most or all of the above technologies.

graphics mode A setting on dual-display mode systems (typically older IBM-compatibles) that permits the access and display of individually addressable pixels for the rendering of images and *graphics characters*. Some systems distinguish between text and graphics modes and will display only in one mode or the other. With faster processors, the trend is toward the more flexible graphics modes, with graphics characters. This frees the user from having to select a mode and switch between them.

Graphophone A phonograph system patented in the 1880s by Chichester Bell and Charles Tainter, in competition with Thomas Edison. The Graphophone used wax cylinders recorded with vertical grooves. Edison later responded by creating wax cylinders, but was not able to commercially mass-produce them. The products were sold by the Columbia Phonograph Company. See Gramophone, phonograph, phonograph record.

graticule A diagnostic and measurement screen used in conjunction with cathode ray tubes (CRTs). The screen is calibrated and placed on the front of the tube, with the tube image showing through so the relationship of the displayed image to the screen can be observed.

grating A series of narrow slits or grooves specifically designed and oriented so that they will reflect back electromagnetic waves in a spread pattern. Antenna reflectors sometimes incorporate a grating design.

gravity cell, crowfoot cell A type of voltaic wet cell suitable for providing small currents at a constant electromotive force. It derives its name from the way the lower and upper chemical solutions (e.g., copper sulphate over zinc sulphate) align themselves in relation to each other.

Gray, Elisha (1835-1901) A physicist and inventor who developed many early telegraph technologies at about the same time Alexander Graham Bell was working to develop a harmonic telegraph. Gray was mechanically apt and had publicly demonstrated an early version of a harmonic telegraph, a device to send tones over wire, before Bell applied for a patent for his version. Gray filed a caveat for a patent the same day as Bell filed for the patent on what is now considered to be the first telephone, thought of as a speaking telegraph at the time.

In 1867, Gray developed a new telegraph relay instrument. In conjunction with his partner, Enos M. Barton, Gray organized the Western Electric Manufacturing Company in 1869, and expanded by buying out the Ottawa, Illinois Western Union offices.

Late in the year 1873, Gray reports having noticed different vibratory properties in human tissue and described the placing of a galvanometer in the circuit with a microphone for transmitting human speech through wires. This resulted in a patent application that was not accepted until three years later as Gray had to substitute animal tissue for human to satisfy the Patent Office. Gray filed a patent similar to Bell's only hours after Bell, and his company was later purchased by Bell.

In the 1880s, telegraphs that would transmit handwriting were developed and Gray patented a *telautograph* which could lift the pen between letters permitting more natural characters to be transcribed, and sold the rights to a company founded with his name. See telephone history.

Gray, Stephen (1666-1736) An English experimenter who authored an article for *Philosophical Transactions* in 1720, which describes various investigations of attractive properties and light-producing properties of various 'electrics.' He discovered that a substance electrified by friction could pass this property on to another substance. He enlarged on the prior work of Gilbert, demonstrating that Gilbert's 'non-electrics' could conduct electricity from one body to another and could be electrified if insulated with a conductor.

His association with the Royal Society of London indicates that he was likely familiar with the work and writings of F. Hauksbee, and he continued some of the interesting lines of inquiry first investigated by Hauksbee.

In the late 1720s, Gray began a fruitful collaboration with Granville Wheler, a member of the Royal Society. Following this, Gray and Wheler were to discover that substances could be roughly divided into additional substances that readily conducted 'electric virtues' and those which did not. See inductance.

Enos Barton joined the prolific inventor Elisha Gray in establishing the Gray & Barton company.

Gray & Barton A telecommunications company established by Elisha Gray and Enos Barton in 1869 when Elisha Gray bought out George Shawk's interest in the partnership. The physical premises was an electric shop abandoned by the Western Union Telegraph Company. In 1872, this became Western Electric Company, which supplied components to the Western Union Telegraph Company and later became an exclusive manufacturer for the Bell System. See Graybar Electric Company, Inc.; Western Electric Company.

gray market product One of several types of products which are not fully endorsed by the manufacturer. Examples include a product sold by an unauthorized distributor; a product which may be second rate in some way, which normally wouldn't be sold by reputable dealers; a product which is a return item, but which is promoted as being new. In electronics, beware of gray market products from questionable vendors. Find out the history of the company, the warranty terms, and ask if the company deals in gray market products, especially if the price is 'too good to be true.'

gray scale See grayscale.

Gray Telephone Pay Station Company A company formed to commercialize the rotary payphones that were common from about 1930 to the 1960s.

Graybar Electric Company, Inc. A spinoff of the Western Electric Company in 1925, this company handled electrical distribution. The name derives from the original founders of the Gray & Barton company, founded by Elisha Gray and Enos Barton in 1869. In 1928-1929, the employees purchased the company from Western Electric Company, and it is still one of the largest employee-owned companies in the United States. Graybar Electric is still in business after more than sixty years, globally supplying almost a million different electrical and telecommunications products. See Gray & Barton, Western Electric Company.

grayscale Visual information which is represented in shades of gray, that is, with no color. In computing terms, many people confuse the terms *monochrome* and *grayscale*. Monochrome refers to one active color, whether it be white, black, green, or amber. Many older computer monitors were monochrome monitors. Grayscale refers to two or more (usually 8, 16, or 32) shades of gray typically ranging between white and black. Grayscale monitors are less expensive than color, and are very suitable for desktop publishing and other black and white and grayscale print-related applications.

grayscale monitor A monitor capable of displaying a variety of levels of intensity, perceived as shades of gray. Most grayscale monitors can handle from 16 to 64 levels of gray, though 32 is comfortable for viewing tones and details. A grayscale monitor is not the same as a monochrome monitor (sometimes inexactly called a black and white monitor), which has pixels which are all of the same intensity (on or off, and light or dark, depending upon the brightness setting), and which are usually white, green, or amber. See monochrome monitor.

great circle An imaginary circle on the surface of a sphere which is defined as the intersection of the surface and a plane passing through the center of that sphere. On Earth, the great circle is a navigational concept to describe the shortest distance over the Earth's surface (assuming it was flat) between two specified points. Airlines use great circle routes to minimize travel distances, especially over long distances. The *polar aircraft route* from Vancouver, BC on the west coast of North America, which passes over Greenland and Iceland to London or Amsterdam is roughly a great circle route.

greeking A design and layout term which refers to a block of text which is inserted into a design to indicate the presence of text and to serve as a spacesaver until the actual *copy* or content is available. Often an ad layout is designed visually before the content is ready for insertion. A newspaper story may be laid out in the paper before the writer has finished editing the copy. In these circumstances, particularly in electronic layout, 'dummy text' which often consists of Greek or Latin characters is placed in the text box to indicate size, shape, and typefaces, or to help in visualizing and selecting these components.

A design and layout term which refers to a block of text which is inserted into a design to indicate text, and to serve as a spacesaver until the actual copy is made available.

Often an ad layout is designed visually before the content is ready for insertion. A newspaper story may be laid out before the writer has finished editing the copy.

green 1. Young, inexperienced, naive, not hardened or aged. 2. Ecological, environmentally friendly, resource-conserving (as in power saver systems). Green products are represented as being kinder to the environment in terms of resource use, manufacture, materials (low in toxic materials or by-products), or operation than similar products by other manufacturers. See ISO 14000.

Green Book 1. The most technical of the three PostScript reference books produced by Adobe Systems, "PostScript Language Program Design." 2. A standard Smalltalk reference "Smalltalk-80: Bits of History, Words of Advice" by Glen Krasner. See Kay, Alan.

Green Code printer The first commercial typewheel style telegraphic printer which embodied the start-stop synchronization concepts developed by Howard Krum.

green gun In a color cathode ray tube (CRT), using a red-green-blue (RGB) system, the electron gun which is specifically aimed to excite the green phosphors on the inside coated surface of the front of the tube. Sometimes a shadow mask is used to increase the precision of this process, so the red and blue phosphors are not affected, resulting in a crisper color image. See shadow mask.

green machine A physically robust computerized device or system designed and built to military specifications for field work.

green-gain control In a color cathode ray tube (CRT), as in a television screen, a matrix resistor which can be varied to control the intensity of the green signal.

Greene, Harold Judge Greene is a prominent member of the U.S. federal judicial system, best known for the Modified Final Judgment (MFJ) and long divestiture proceedings associated with the breakup of AT&T in the 1980s. Judge Greene has also made important court decisions on Western Electric Company, Inc. (1991). See Modified Final Judgment.

Greenwich Civil Time GCT. See Coordinated Universal Time.

Greenwich Mean Time GMT. A time and geographic reference established by an agreement of 25 countries in 1884. Using astronomical instruments, a local time was established at the Greenwich Meridian in England, after which the world's regions referenced their time in relation to GMT. It is also known as Zulu time. GMT is normally expressed in 24-hour clock notation. The international standard for time for satellite communications and scientific research has since become Coordinated Universal Time, which is based on atomic rather than astronomical clocks. See atomic clock, Coordinated Universal Time.

grep generalized regular expression parser. A very powerful, very useful Unix pattern matching, parsing utility (it takes a whole book to describe its many operating modes and uses) which allows selective search and display of computer data. At its simplest level, grep can help you search for instances of your best friend's name in a directory full of old email messages; at its most sophisticated, it can parse and display complex patterns of instances of information. Grep increases in versatility when used with other commands through a pipe. Egrep (full expressions) and fgrep (fixed strings) are variations of grep. This very simple example of a grep search of an RTF file for the string "CRT" shows grep at its most basic level.

```
abiogen@frodo: /1.2GB/users/abiogen/
RTF $ grep CRT DictG.rtf
\pard\tx20\tx3320\fc0\cf0 CRT
\pard\tx20\tx7560\fc0\cf0  (CRTs)
\b0   In a color cathode ray tube
(CRT), as in a television screen, a
matrix resistor which can be varied
to control the intensity of the green
signal.
\b0   In a color cathode ray tube
(CRT), using a red-green-blue (RGB)
system, the electron gun which is
specifically aimed to excite the
green phosphors on the inside coated
surface of the front of the tube.
Sometimes a shadow mask is used to
increase the precision of this
process, so the red and blue
phosphors are not affected, resulting
in a crisper color image.  See shadow
mask.
```

grid A filtering structure, often used in vacuum tubes, that has the appearance of a grid or small set of blinds, which controls the flow of electrons from a cathode to an anode. Since this structure allows the flow of electrons to be manipulated, it may be called a *control grid.* A large part of electronics involves the harnessing of electrons through cathode rays and grids are an essential control component in many devices.

GRID Global Resource Information Database.

grid, reference A visual guide used in the background of drawing, painting, layout, and CAD programs to help align visual elements and objects. Sometimes a SNAP function will be available in conjunction with the grid, in which the objects will SNAP to designated positions on the grid, to align objects with mathematical precision that cannot be obtained visually.

grid battery Within electron tubes, there is usually an electron flow element called a grid, which selectively controls the movement of electrons from the cathode to the anode. A grid battery supplies a bias voltage to the grid in the electron tube for this purpose. See electron tube, grid bias.

grid bias In an electron tube, a constant potential which is applied between the controlling grid and the cathode to which the electrons are being attracted. The grid bias, or *C bias* is used to establish an operating point. A small cell may be used to supply voltage so the grid becomes more negative than the cathode. See electron tube, grid battery.

grid cap At the top of some electron tubes there is a small 'cap' which attaches to the controlling grid to act as a terminal. Sometimes a spring clip is incorporated into the grid cap to create the electrical connection.

grid modulation In a grid-controlled electron tube with a carrier signal, a voltage can be applied to modulate that signal in order to add information to the signal, as in a radio transmitter. See modulation.

grinder Before the development of electric motors, a term for the people who were needed to turn cranks to operate mechanical equipment, including a variety of telecommunications devices.

gritch *n.* Complaint, gripe.

grok *v.t.* To understand, to get the essential idea of, to comprehend well.

grommet A ring-shaped insulator, usually made of plastic or rubber, which is used as a spacer inside air insulated cables or in panels which require wires to be strung through them, so the inserted wire doesn't touch the materials on the outside edge of the grommet.

groove An indentation in a recording medium, often cylindrical or platter-shaped, which has minute variations which encode information. By creating both horizontal and vertical variations, it is possible to encode two 'tracks' of information, as in a stereo vinyl record. Optical media usually store information in pits rather than grooves. See phonograph.

ground *n.* 1. The surface of the Earth. A large conducting body, such as the Earth, which provides a destination for electrical current. 2. A conductor which makes a connection with the Earth, through which power can drain. 3. A voltage reference point in an electrical circuit. Although it may not actually be touching the ground, it is a reference point whose operation would not be changed if it were grounded to the Earth.

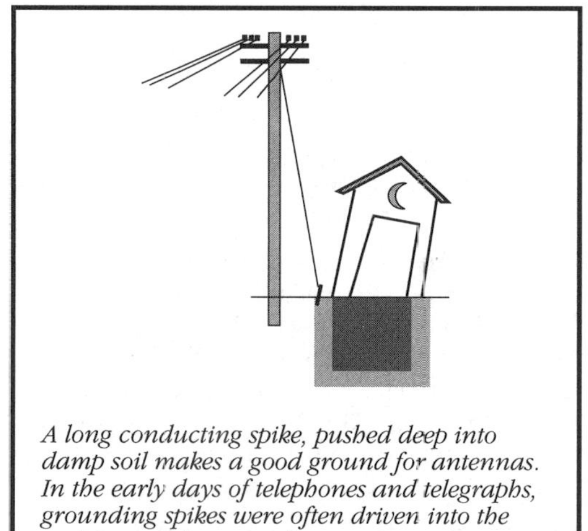

A long conducting spike, pushed deep into damp soil makes a good ground for antennas. In the early days of telephones and telegraphs, grounding spikes were often driven into the moist ground near outhouses.

ground *v.* 1. To put in, or place in contact with, a ground, such as the Earth, or a conductor in contact with a ground. 2. To make an electrical connection with a ground. To provide a path through which an electrical current will drain to the ground, as a lightening rod on a house. A ground is usually established as a safety precaution to direct unwanted or unanticipated electrical charges away from areas where it might cause harm to structures or beings. On an ocean-going vessel, where the ground connection cannot be placed into the Earth or onto a structure connected with the Earth, a device can be grounded on the bed plate of an engine.

ground absorption A loss of transmission energy due to dissipation through the ground. Bounced airwave transmissions are particularly susceptible to ground absorption. Ground absorption may also be greater in regions of soft, uneven terrain.

ground button A button found on some electrical components and power strips to reset components which require a ground start after a power failure. See ground start.

ground clamp A device to connect a grounding conductor to a grounded object. Water pipes are commonly used as grounding objects, as are long metal spikes (often used near phone installations). At the turn of the century, when grounding pipes were not readily available, it was common for a telephone service ground wire to be pushed into the Earth in the damp ground in the vicinity of an outhouse.

ground junction In semiconductor fabrication, a junction can be formed by growing a crystal from a melt. See semiconductor.

ground lead A ground to which other conductors are attached so the ground lead can direct unwanted electrical current to the ground (usually the Earth). Heavy metal spikes are sometimes used as ground leads under external phone connection boxes.

ground noise Residual, usually low level, noise associated with a communications transmission in which no actual information is being transmitted, but there are low-level hisses or hums associated with the transmission devices or media. This type of noise also occurs in many analog audio recording technologies.

ground potential A reference potential associated with the Earth at a particular location. The ground potential at that position is considered the zero potential, and other potentials are referenced against that as a baseline.

ground return 1. A lead to the ground at the end of the circuit, for the return of a signal. 2. A type of circuit which employs the ground as a return to complete the loop. The characteristics of ground circuits were discovered somewhat by accident by early experimenters, particularly when long telegraph cables began to be strung. It was discovered that it was possible to send signals along a single wire, as long as each end of the connection had contact with the ground, so the completion of the circuit happened through the ground rather than through a returning (second) wire.

ground scatter propagation A means of propagating radio waves through a series of hops between the Earth and the ionosphere, rather than following a great circle path. When the signal returns from the ionosphere to the Earth, contact with the terrain scatters it broadly in many directions. See ground wave, ionospheric wave.

ground start In telephony, it is necessary to take control of a line before it can be used. There are two common ways in which to do this, with a *ground start* or a *loop start*. The ground start is the type commonly found in business and other multiple line phones. When you pick up a phone, the plunger is released (off-hook) and the station detects a grounded circuit through the *ring* conductor. This is done so that transfer can be directed to the central office or main switching panel, if desired. See loop start.

ground state A reference descriptor, used to describe the lowest state or energy level of a system, as an atomic system.

ground station An Earth-based station (although the station may actually be located off the ground in a ground-based tower) used for sending, receiving, processing, or relaying communications signals. The term ground station is typically used with reference to services which are partly air-based or space-based, as satellite communications systems. Traditional broadcasting stations are not usually called ground stations, because they bounce signals from ground to ground or from ground to ionosphere to ground, without passing through a space transponder, relay, or other sky-based node. Ground stations may be primary senders/receivers or ground hubs, which relay information or strengthen signals, as in M hop systems. See M hop, satellite.

ground wave A transmitted radio wave that stays close to the ground. Radio waves travel in various directions from the point of transmission, some moving out through the ionosphere, others toward the ground. Ground waves are affected by the surface composition and topography of the surface over which they travel. Very rough or heavily vegetated terrain will interfere with the transmission of ground waves, while smoother surfaces like plains or calm waters may permit transmission for hundreds of miles. See ionospheric wave, radio.

Electronics technicians routinely wear grounding straps when working with components. One-time grounding straps are sometimes included with electronics products that are installed by consumers. This disposable 3M wrist strap was shipped with a computer circuitry upgrade product.

grounding strap A bracelet-like material or component usually worn on the wrist, and common in the electronics assembly and repair industries. The grounding strap prevents discharge from the hands that might damage static-sensitive components. If you must touch electronic components (as when plugging new memory inside a computer), use a grounding strap, or, at the very least, touch a ground such as the power supply, and then install the chips or boards without moving too much or shuffling your feet in the carpet. See ground.

Group 1 to 4 See facsimile formats.

group address A single address which is a logical name for a list of addresses. It may refer to multiple mailing lists, multiple devices, multiple users, or multiple receivers. A group address is used for management simplicity to provide a single reference point for a group of information.

group busy tone In telephone trunks, sometimes the system is at capacity and cannot route any additional calls until volume decreases. In this case, a group busy tone (a fast busy) may be sent out to those attempting to place a call. See fast busy.

group hunting In telephony, the process of searching for available lines in a designated group of trunks. See hunting.

group modulation The process of shifting or collectively modulating a set of signals which have already been individually modulated or multiplexed. This is often done to achieve a block frequency shift. Such frequency shifts may be done to bring a signal into a range which can be handled by the equipment, or may reduce interference between incoming and outgoing signals in a repeater or relay circuit.

groupware Software which runs on a network, so that a number of designated users or all users on the system may use it, either simultaneously or up to a certain number of licensed users at a time. There are two types of groupware software which is accessible and usable by the group, but the data generated by the software is specific and local to the user, and software which is accessible and usable by the group in a type of conference atmosphere, in which the changes made in the software, and perhaps even being viewed by the user, will be seen by other members of the group as soon as the data is updated. Whiteboarding is a type of groupware application. Groupware is popular in corporate cooperative work environments.

Grout, Jonathan An entrepreneur who established one of the first marine reporting telegraphs in North America, which extended from Martha's Vineyard to Boston. The service went into bankruptcy by 1807 although the old signal station still remains.

Grove battery An early, dependable zinc/sulphuric acid and platinum/nitric acid primary cell normally used on closed circuits. The Grove battery provided 1.96 volts per cell. This type of battery was used by Morse in his early telegraph systems.

growler An electromagnetic circuit diagnostic and magnetizing/demagnetizing tool which emits a low growling sound when a short circuit is detected.

GRSU Generic Remote Switch Unit.

grunt *colloq.* A telephone pole and line installation crew member who worked on the ground (as opposed to one who climbed the poles). Grunts were also commonly called groundmen or groundworkers.

GS trunk ground start trunk. See ground start, loop start.

GSA 1. General Services Administration. 2. See Global Standards Collaboration.

GSM See Global System for Mobile Communications.

GSMP General Switch Management Protocol.

GSN Gigabyte System Network. See Hippi-6400.

GSO geostationary orbit.

GSSAP See Generic Security Service Application.

GSTN general switched telephone network. A public switched telephone network (PSTN).

GTE Corporation Formerly General Telephone and Electronics Corporation. A major international telecommunications provider which originally supplied basic local telephone services, but now includes long-distance, wireless, airline services, online directories, Web, and video services. It is building a national private coast-to-coast data network in the U.S. and wireless paging systems overseas. In 1997, GTE acquired BBN Corporation, an end-to-end Internet provider, and Genuity, Inc.

GTN See Global Transaction Network.

GTP 1. general telemetry processor. 2. See Generalized Trunk Protocol.

GTT Global Title Translations.

guard arm A crossbar placed over wires, running in the same direction as the wires, to prevent other wires, debris, people, or animals from contacting the wires, to prevent damage or harm.

guard band 1. A narrow broadcast bandwidth 'safe region' which is interposed between communications channels in order to minimize interference between adjacent channels. Guard bands are particularly prevalent in frequency multiplexed systems, where maximum use of available bandwidth is achieved by dividing the available frequencies into smaller channels. 2. A safety zone in a circuit, or chip, around the active portions of the circuit, to prevent electromagnetic interference with adjacent circuitry.

guard circle The smooth, ungrooved, inner portion of a phonograph record, or other revolving storage medium, which protects the stylus from moving into the center post and being damaged.

guard wire A wire positioned near live wires, as on utility poles, positioned so that if the live conducting wires break or fall, they will come in contact with the guard wire and be grounded, rather than causing danger.

guardian agent A pun on 'guardian angel' in the sense that it is a software tool intended to protect innocent eyes from sites that some person in authority over the user deems unsuitable. With the vast and varied volume of information on the Internet easily accessible through the World Wide Web, some parents and teachers are concerned about the type of Web browsing children might attempt. Software developers have responded to this concern by developing tools to lock out specific 'known' sites or to flag sites with particular characteristics. There is no completely reliable way to screen out all such sites on the Web, particularly since there is no consensus on what people consider objectionable. A child could quite innocently do a biology project search on a major search engine for the word 'beaver.' The hit list that comes up on the search engine isn't going to list only Canada's national animal, it will list every slang sense of the term as well, on commercial sites, home pages, poetry pages, and more. There is a trade-off here. The loss to a user of not being able to access search engines is probably greater than potential harm done by stumbling over an unintended site.

While concern focuses on ways of locking out particular types of sites, a more serious danger on the Net is being somewhat overlooked. Innocent adults and children will sometimes mistake the personality of a person on the Web as being the 'real' person. This is understandable, but naive. It is easier to misrepresent oneself or to cover up hostile or dangerous intentions on the Net than in person. Global communications are here to stay, and it will be necessary to educate users of all ages to cautiously evaluate personal contacts developed through this new medium. It's a new world, and mature and practical strategies have to be developed to enable positive relationships online and to avoid ones which may be emotionally hurtful or physically harmful.

guarding 1. The incorporation of points in a circuit where excess current or leakage are drawn off. 2. The process of maintaining a circuit in its busy state for an interval after it has been released, in order to assure a minimum period of time elapses before the actual disconnect occurs.

Guericke, Otto von The inventor of the air pump in the mid-1600s, which was later improved by experimenters such as R. Boyle, and F. Hauksbee. Guericke did experiments in developing ideas about the rotation of the Earth which were described in 1672 in *de Vacuo Spatio*. In doing so, he created a spinning model to simulate the Earth. Because Guericke noticed that a feather was alternately attracted to and repelled by the spinning globe, this has been credited by some as the first frictional generator, even though Guericke did not specifically design this to create friction for electrical experiments. See Hauksbee, Francis.

GUI See graphical user interface.

guidance system A system which evaluates terrain, sensors, flight information, driving information, or other data pertinent to the object being guided, and reacts accordingly. Guidance systems are used with many airborne objects and transport systems (missiles, radio-controlled models, aircraft, etc.), and underwater systems (torpedos, submarines, diving robots). Some guidance systems incorporate means to hide the guidance control signals so that a target doesn't know the object is coming.

A guidance system typically consists of one or more sensing devices (optical, radar, infrared, sonar), intelligent evaluation of the data from those sensing devices (human, computer, or both), and a controlling system which reacts with input from the information processed from the sensing devices. Dolphins, bats, and some insects have sonar systems that aid them in navigating through water and air, respectively. In dolphins, this perceptual system is so sophisticated that the mammal can actually 'see' the shape of an object inside another object.

In spacecraft, guidance systems are primarily used for trajectory and directional control during the firing of thrusters. Out in space, it's not possible to use hills, trees, and other terrain markers as reference points. Instead, the positions of celestial objects, such as stars, suns, or planets, are used. See attitude and articulation control subsystem, chaff, frequency hopping, intelligent vehicle highway systems.

guided wave An electromagnetic wave whose path is controlled or directed by a structure or process which acts as a conduit, channel, or waveguide. Physical waveguides can be quite sophisticated and are mathematically related to the wavelength of the wave being guided. Reflected waves are a type of guided wave in the sense that the reflection may be carefully organized to channel the wave in the desired direction. For example, a parabolic receiving antenna dish reflects transmission waves into a feedhorn mounted at a specific distance from the dish. See waveguide.

guru Sage, wizard, admired expert.

guru meditation error On the Amiga computer, a whimsical presentation of the error identification number that was displayed on early versions of the operating system if the system crashed. This was later removed in versions 2.0*x* and up, and the OS itself was much less subject to crashes in the updated versions.

gutta-percha A latex substance from Malayan percha trees, which western culture discovered in the 1800s. Perchas are like high-resin rubber trees. Gutta-percha is used to manufacture a tough plastic that was used extensively for insulation materials, general manufacture, and dentistry until the late 1940s, when other types of plastics became more common. Gutta-percha is historically important because it enabled the laying of cable in unfriendly environments, such as ground and deep sea installations.

gutter In page design and layout, a columnar margin between two facing pages of a bound document. Technically, the blank space between columns on a single page is not a gutter, however, not all programmers know this, and some software will label this area as the 'gutter' and will prompt the user for the width of the 'gutter.' It would be nice if programmers and managers would solicit the input of professional specialists when applying terminology in computer programs. In the meantime, users of this dictionary will know the difference.

guy wire A slender support line used to brace and steady an apparatus that might sway or fall. Guy wires are often used in multiples, spaced around whatever they are supporting, and are frequently used for narrow high structures such as transmission towers, aerials, masts, and poles. Guy wires sometimes have small telltales, fine pieces of cloth or plastic attached to the wire to discourage birds from flying into them. Guy wires are useful in areas that are exposed to surf or high winds.

gyrofrequency The frequency at which charged particles naturally rotate under the influence of the Earth's magnetic field. The frequency varies with the type of particle.

gyroscope, gyro A device designed to maintain its axle in a constant vector while rotating. A gyroscope is typically designed to rotate through two axes that are perpendicular to the central structure and to each other. It is called a *gyrocompass* when it is oriented with the axle pointing northward.

gzip GNU zip, a widely used file compression program developed by Jean-Loup Gailly that incorporates Lempel-Ziv coding algorithms with 32-bit CRC. Gzip is widely used on Unix systems and IBM-compatible disk operating systems, especially for compressing and decompressing files to save transmission times for phone line uploads and downloads.

Gzip uses a 'deflate' compression format derived from the freely distributable *zlib* source code distributed by Gailly and Mark Adler. Gzipped files may be unzipped by typing "gunzip [filename]" (the command must be typed all in lower case) at the command line, with relevant optional parameters. On a Unix system, see the "man pages" on gzip for more information on zipping, listing, and unzipping files. See compress, uuencode, zip, RFC 1952.

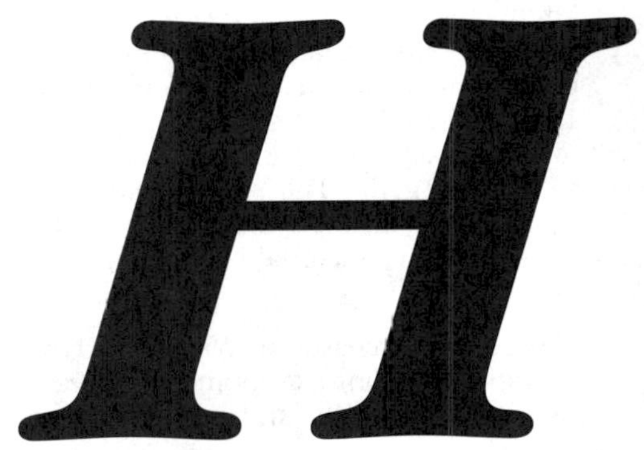

h Abbreviation for hecto-. See hecto-.

H channel An ITU-T-defined transmission channel on packet-switched networks which consists of aggregated B channels (bearer channels), as are used on an ISDN system. See ISDN.

H0 Channel In ATM networking, a 384 Kbps channel consisting of six contiguous DS-0s (64 Kbps) of a T1 transmission line.

hack *v.* 1. To quickly cobble together a program. 2. To create something quickly from available materials, a make-do solution, not necessarily elegant, although it could be, given limited resources. 3. To create a small, quick entertaining showpiece designed to illustrate a cool idea or interesting capability. See Schwabbie.

hacker 1. A person who 'hacks into' a system, that is, gains entry by exploiting the hardware or software architecture through black boxes, stolen or guessed passwords, Trojan horses, design flaws, or back doors. Sometimes called *cracker* to signify someone using these techniques for illegal purposes. See cracker. 2. A person who acquires a sophisticated, in-depth knowledge of a system and applies this knowledge to configuring or programming the system with a high level of expertise or complexity. An elite programmer, engineer, or technician. Two interesting books on this subject are "Hackers" by Stephen Levy and "The Cuckoo's Egg" by Clifford Stoll.

Hacker's Dictionary, The An electronic and print dictionary which evolved from The Jargon File in the early 1980s. The Hacker's Dictionary was an expanded version of The Jargon File with added commentary, published by Harper and Row in 1983, edited by Guy Steele. The co-editors/contributors were Raphael Finkel, Don Woods, Mark Crispin, Richard M. Stallman, and Geoff Goodfellow. After nearly a decade in which it remained essentially unchanged, The Hacker's Dictionary was expanded beyond the artificial intelligence (AI) and hacker cultures to include terms from a broad variety of computers. The 1990s version, called The New Hacker's Dictionary, is maintained by Eric S. Raymond and Guy L. Steele Jr. See Jargon File, The.

hairpin pickup coil A device with a one-turn coil, shaped like a hairpin, used for transferring ultrahigh frequency (UHF) energy.

hairpinning The routing of information or data through a switch in a main facility or network host and sending it out again through another switch or routing device.

HAL-9000 No computer-related dictionary would be complete without mention of the intelligent computer in the science fiction movie classic "2001: A Space Odyssey." HAL stood for "*H*euristically Programming *Al*gorithmic Computer" and apparently the one-letter shift that spells out "IBM" was not intentional, or so say the makers of the film. If not, it's a strong enough coincidence to create an apocryphal legend.

HALE See High Altitude Long Endurance.

half duplex In a circuit, one-directional transmissions. Often half duplex circuits can transmit in either direction, but not simultaneously. Many systems which technically have bidirectional capabilities are operated in half duplex mode to reduce interference and echoes. Modems, satellite voice lines, some cellular radios, and speakerphones are often used in half duplex mode.

half life A property of radioactive decay used as a quantitative measure, of interest to many different branches of science. Radioactive decay happens at widely differing rates for different materials, so *half-life* is not a fixed measure, but one based on our knowledge of the properties of the materials being described. The first half-life of a substance is the interval during which half the radioactive material is left unchanged. The second half-life is the next interval, during which half of the *remaining* radioactive material is unchanged, and so on.

These half life measurements are used by many scientists including astronomers, nuclear physicists, archaeologists, and geologists.

half tap A bridge that is placed across conductors without disturbing the normal functioning of the conductors.

half tap, network In data network communications, a duplicate path established between nodes or systems. A half tap provides redundancy where new cable is being run, as in circuits where fiber optic is replacing copper, but where it's not desirable to disrupt the existing network until the new cabling is functional.

half tap, telephone In telephone communications, a duplicate service installed on the subscriber side of the demarcation point (usually on the customer premises). This may be done in instances where there is a problem with the original circuits, or where a new system is being installed and the old one is left in place until the new one has been tested and is known to be functional.

half wave antenna An antenna designed so that its electrical length is equal to half of the wavelength of the signal being received or transmitted.

halftone A printing technique developed in the late 1800s that takes advantage of human spatial integration to simulate tonal gradations with an ordered series of dots of various sizes and sometimes various shapes. When fine details are seen by humans at a distance, the eye-brain averages and integrates the image so that it appears as levels of gray, even though the dots are of the same color and intensity (usually black against a light surface). Halftone images are commonly used in many types of printing jobs including newspapers and magazines. Observed closely with a magnifying glass, the dots can be seen.

Desktop publishing software typically has settings to control halftone resolutions and screen angles. Newspapers are usually printed with 60 to 85 line screens, magazines at 85 to 150, and art prints at 150 to 200. The first newspaper halftone was printed in 1869 in The Canadian Illustrated News by Canadian publisher G. E. Desbarats.

Hall constant A description of the relationship between current-carrying conductors and magnetic fields. The Hall constant = (transverse electric field) / (magnetic field strength) *x* (current density).

Hall effect If you take a current-carrying semiconductor with a magnetic field perpendicular to the direction of the semiconductor's current, a voltage is created which lies perpendicular to both the current and the magnetic field of flux. It has practical applications in generators and modulators.

Hallwach, Wilhelm A German physicist who discovered the photoemissive properties of certain substances when exposed to light. He demonstrated that photoelectric cells could be used in cameras, a big boost to the evolution of television which was just being developed at that time.

Hallwach's effect In a vacuum, a negatively charged body discharges when exposed to ultraviolet light.

ham operator *colloq.* Amateur radio operator.

ham-in-space program A cooperative technical, educational, communications effort of amateur radio operators worldwide. See AMSAT, Mir, OSCAR.

Hamming code A linear error detection/correction code system named after R. W. Hamming of Bell Laboratories. It can detect single- and double-bit errors in data transmissions and correct single-bit errors. Hamming codes lend themselves to matrix representation. See error correction.

Hammond, John H. An American inventor who developed radio control (RC) systems for vessels in 1912. Many of his patents were later purchased by the U.S. military for use in radio-controlled guidance systems. See frequency hopping.

HAN See home area network. See fiber to the home, home ATM network.

hand off See handoff.

handle A pseudonym, a nickname. These are often very creative, humorous, or obscure. A handle indicates your personality, your interests, or helps preserve anonymity.

handoff, handover 1. The process of passing on a message or transmission to the next leg in a route that takes more than one type of communications medium or more than one transmitting region. A *make-before-break* handover is one in which the transfer to the new leg is carried out in such a way that the user does not perceive a break in communications. 2. The process of a communication being passed through various 'hands,' usually because the user is mobile, as from one zone to another, one station to another, one transmitter to another, or one frequency to another. 3. In cellular communications, the transfer of the call from one cell to the next as the subscriber moves through the various cells. Handoffs often involve frequency shifts. 4. The process of passing a caller to another agent, as from a receptionist to a sales representative or technical support person.

handset A human interface communications transceiver unit, most often associated with telephones. It's the part we pick up and hold to our ears and mouths in order to listen and speak on the phone. Handsets come in a variety of shapes, some of which have names in the telephone industry. The older round handsets familiar on rotary phones are G style, whereas the newer square ones more common on mobile phones and phones with the buttons on the handset are K style handsets.

handsfree A communications unit that does not require the user to hold it in order to be able to communicate with the caller. Headsets and speakerphones are examples of handsfree units in the telephone industry. Some phone systems permit handsfree menu-selection or dialing through voice recognition. Car-mounted cell phones are becoming more prevalent, so the driver can have both hands on the wheel and concentrate on driving, rather than holding the cellular handset. For computer input devices, a voice recognition system can be used along with a headset to create a handsfree unit.

handsfree telephone Any telephone appliance that provides handsfree operation for some or most of its operations, such as a voice operated phone or computer (e.g., for spoken dialing), a speakerphone, a headset, etc. See handsfree.

handshake Communication between two systems to manage synchronization of the transmitted and received signals, often established with ACK or NACK signals, with tones, keywords, or header packets. Handshaking is an essential component of most communications systems and is often incorporated into the transmission protocol itself. Handshaking can be done between people or between machines, or both. The most familiar form of handshake is the verbal spoken "Roger" that allows the other person on a line to know that you've finished talking and they can go on. On public chats on the Internet, where dialogs are typed rather than spoken, "GA" (Go Ahead) serves the same purpose. In verbal communications, this verbal "Roger" handshake is sometimes accompanied by electrical signals that set the half-duplex communications direction to favor the person who is currently talking. In modem communications, handshakes are used to acknowledge a signal, to coordinate baud rates, and to orchestrate the transmission, receipt, and data, so the signals don't override or clobber one another.

handshaking See handshake.

handwriting recognition A software application, often coupled with a scanning device or a stylus that resembles a pen, which interprets written script into computer-readable text. Pen computing uses this type of technology and is of use to those who don't know how to type, or don't want to. Since handwriting is widely variable, most systems need to be trained to recognize an individual's writing and, even then, the results may not be perfect.

Nevertheless, in the shipping industry, scientific field work, and other areas, handwriting recognition is useful, and the technology will eventually improve to the point where anyone's handwriting can be recognized and interpreted by a computer. See Personal Digital Assistant.

hang up *v.* To disconnect from a transmission (two words when it is a verb). On modems, ATH is the Hayes-compatible command for hanging up. On phones, a hangup (one word when it is a noun) occurs when the button is pressed for at least a specific amount of time. In some areas, the callee may not be able to hang up this way if the caller is still on the line. It doesn't work the other way though; if the caller hangs up and the callee is still on the line, the transmission is disconnected. Many Internet Service Providers (ISPs) will automatically hang up (terminate) a computer connection if there is no activity after a certain amount of time, say 10 minutes.

hard copy An image or document which is readable by looking directly at the medium on which it is transcribed, as on a piece of paper, cardboard, stone, or parchment. A soft copy must be accessed with some type of technology in order to be viewed, manipulated, or displayed. Soft copies commonly exist on hard drives, floppy diskettes, tapes, CDs, and other magnetic or optical media.

hard sectored A storage medium, usually magnetic, in which the various boundaries or 'sectors' are physically designated with holes, pits, ridges, or other 'markers' to indicate their extents. Hard sectored media are becoming less common that those which are 'soft sectored' as they are not transportable between different systems.

hard transfer A term for an electronic monetary transaction which involves the actual exchange of funds between individuals or banking institutions. A hard transfer often follows a 'soft transfer.' A paper check is a type of soft transfer. It is a monetary transaction which is not actually finalized until the money is withdrawn from the bank. Similarly, online, there are many monetary transactions which are soft transferred and later 'hard transferred' from the actual bank or other financial institution.

hard tube A type of electron tube which has a high vacuum environment within the sealed glass bulb.

hard wired See hardwired.

hardware The physical circuits and devices associated with systems, especially computerized systems, which are fixed or hard wired and unlikely to be altered by the user. Contrast with software (although the distinction is not actually cut and dry), which is selected and swapped out by the user, modified, or overwritten. See firmware, software.

hardware flow control A capability built into most of the high speed serial card modem combinations that helps to handle flow control. Use of hardware flow control may also require the use of a hardware flow control cable.

hardware interrupt On computing systems, a call to the software to interrupt the current process in order that it may temporarily 'listen to' or interact with a hardware device interfaced with the system. See interrupt, IRQ.

hardwired 1. A circuit that is intended as permanent, or which is not expected to change in the near future, and thus is wired in such as way as to make it efficient to produce or easy to use, rather than making it amenable to change. Contrast this to patch bays and breadboards which are intended to prototype temporary circuits and which are easy to change. Programs or pathways built into computer motherboards are typically hardwired, whereas the various user-added peripherals, especially those which fit into slots or chips designed to be swapped out when better technology comes around are considered to be modular or configurable, and not hardwired. 2. People who are *hardwired* are said to be set in their ways, not amenable to change or open to new ideas. 3. An idea or system that is *hardwired* is one that is entrenched, difficult to change for various reasons, including politics, economics, or complexity.

Harmon, L. D. A Bell Laboratories researcher who initiated a project to simulate the functions of biological nerve cells by means of simple transistors. These could be closely associated with one another in arrays and were applied, for example, in a simulation of mammalian eye nerves. See neural network.

harness Straps, combination connectors, or other means used to consolidate multiple cables so they can be handled more easily as a unit.

Harrison, John A British clockmaker who devised a means, in the 1770s, to create a chronometer which could aid in navigation by determining longitude, even when being bumped around by heavy seas.

Harvard Mark I A historically significant, large, automatic, relay computer constructed by Howard Aiken and IBM engineers in the late 1930s. Although most often remembered as the 'Mark I,' it was also known at the time as the IBM Automatic Sequence Control Calculator. This computer was capable of long computations difficult for humans to undertake, and could run instructions stored on prepunched paper tape. Three programmers worked on the project in the 1940s; the best remembered is Grace Hopper, who joined the project in 1944, after Richard Bloch and Robert Campbell. See Aiken, Howard; Hollerith, Herman; Hopper, Grace.

Harvard Mark III Third in the line of large-scale computing machines developed under the direction of Howard Aiken, the Mark III was delivered in 1951 to the U.S. Naval Surface Weapons Center. It improved on earlier Mark computers and on many competitors by incorporating drum memory with separate drums for instructions and data.

Hauksbee, Francis (died ~1713) An English artisan and experimenter who built on the work of R. Boyle and created an improved double air pump design in the early 1700s, which prevailed for the next century and a half. The availability of air pumps was important not only for commercial purposes, but because the ability to create a good vacuum was invaluable to scientific exploration and study of magnetism and electricity.

In 1705, Hauksbee reported on his experiments with producing light in a mostly evacuated mercury vessel. This led to further experiments, and the observation that lampblack particles would move up and down very rapidly and make an audible sound, when a glass tube that had been rubbed was held above the particles. Following this, he devised a rotating wheel to allow the glass to be rubbed at a great rate, in essence inventing the first friction generator.

Hauksbee was largely without formal education and not highly literate, but his mechanical aptitude and talent for experimentation were highly developed and brought him into association with the Royal Society of London. See barometer; Boyle, Robert; Gray, Stephen; Guericke, Otto von.

Hayes Microcomputer Products Inc. One of the early entrants to the modem market, and a company which set many of the industry's de facto standards for serial communications through modems. See AT command set.

Hayes Standard AT commands See AT commands.

HBA host bus adapter.

HBS See Home Base Station.

HCI 1. See Host Command Interface. 2. human computer interface. 3. See Human Computer Interface standards.

HD See half duplex.

HDB3 See High Density Bipolar Three.

HDD Hard Disk Drive.

HDLC See High Level Data Link Control.

HDSL See high bit-rate digital subscriber line.

HDT Host Digital Terminal.

HDTV See High Definition Television.

head A device for reading, writing, or removing data from a volatile storage medium (usually magnetic). VCRs, hard drives, floppy drives, and tape recorders all have heads which touch, or nearly touch, the surface of the storage medium in order to transmit the information to the logic circuits, or mechanisms that decode the information into human-meaningful form, or to write to the storage medium.

head thrashing If read and/or write heads on storage mechanisms encounter hardware or software problems, especially bad sectors, the mechanism may start to rapidly oscillate, sometimes uncontrollably. This can lead to damage to the head or the data if not terminated in time.

header 1. Identifying text printed in a block at the head of a file or document. Header information frequently includes file format, version, date of creation, author, and typographic information. Header files are common to word processing, desktop publishing, and EDI applications. 2. A commonly used system routine contained in a separate file and referenced during program compilation and linking. System windowing routines and graphics routines are frequently linked in from header files. A header provides modularity and a write-once-use-many solution to many programming tasks. 3. In ATM, the protocol control information which is located at the beginning of a protocol data unit.

H

header area In an EDI file, the area which contains the header information for the document. See EDI, header.

Header Error Control HEC. In ATM transmissions, an error detection mechanism contained in a byte at the end of the 53-byte ATM header. This corrects single bit errors and is efficient over transmissions media with low bit error rates (BERs) like fiber optic cable. In ATM carried over wireless transmissions, the signal is not as clean as a fiber optic signal, and the BER rate can be substantially higher. Satellite transmissions tend to be especially bursty, a situation not handled well with a single bit error mechanism. Some satellite service providers have compensated for this by developing a variety of solutions, including interleaving of cells to isolate the data from burst errors. See ATM Link Enhancer.

headlight antenna A small radar antenna with a beam like a searchlight that can be housed in the wing of an aircraft.

headphone A listening device designed to fit comfortably on the head, with vibrating diaphragms for one or both ears. Designed so the user can listen to private audio input (communications, music, etc.) without distraction from outside noises and without disturbing others.

headset A radio or telephone transceiver unit worn on the head or wrapped around the ear (sometimes referred to more specifically as an earset). Headsets are typically used by professionals who sit and take a lot of calls: receptionists, console attendants, telemarketers, reservation takers; and by warehouse and ground staff, who are on the move, but need to keep in communication with one another. Headsets are becoming a consumer item, with headsets for cellular phones (so drivers can keep both hands on the wheel), and other hands-free applications.

heap memory A type of local memory storage which is dynamically allocated while a program is running. This is usually of more concern to applications programmers than to users, but there are some applications in which heap memory needs to be set prior to running the software in order to provide enough working room for memory-intensive applications. On some systems, heap memory is limited to a maximum of 64 kilobytes.

heat sink A structure for dissipating, or radiating heat away from a heat-generating device such as a semiconductor. Heat sinks often resemble open coils, flat fence rails, or other repeated, spaced elements, usually of metal, that are configured to increase their surface area, and thus their radiating capacity. Some CPUs require surprisingly large heat sinks. Considering that many of these are in competition with similar processors which are smaller and cooler, one wonders if the heat is excessive and unnecessary.

Heaviside layer See Kennelly-Heaviside layer.

English physicist Oliver Heaviside made philosophical and mathematical contributions to our understanding of electrical phenomena.

Heaviside, Oliver (1850-1925) An English physicist who first became a telegrapher, later became increasingly interested in electricity, and began publishing on that subject in 1872. He made thorough studies of Maxwell's equations and then set about simplifying them down to two equations expressed in two variables.

Along with J. J. Thompson, Heaviside theorized about the electromagnetic reactions and mass of electrically charged particles in motion. See Kennelly-Heaviside layer, Maxwell's equations.

HEC See Header Error Control.

hecto- (Symbol - h) An SI unit prefix for 100 or 10^2.

Heisenberg uncertainty principle Proposed by W. Heisenberg in 1927, the uncertainty principle has since become a fundamental principle of physics. (Heisenberg formulated a model of the structure of an atom in the 1930s which has held up well over time.) In studying movement of electronics, Heisenberg proposed mathematically that it is not possible to precisely determine both the *position* and the *velocity* of a material particle at the same time. The uncertainty increases as the size of the particle decreases.

Many researchers have generalized this principle and restated it in various broader contexts, but most commonly it is brought up when describing the results of quantum experimental results. It is said that these are determined in part by the point of view and methods of the researcher. For example, if light is studied as a particle phenomenon, it appears to behave as a particle phenomenon. If it is studied as a wave phenomenon, it appears to behave as a wave phenomenon, at least as far as the observers and measuring instruments are concerned.

In other words, attempts to precisely pin down the location of an electron obscures its energy level, and vice versa, thus challenging the absolute nature of the world suggested by classical physics.

Heisenberg, Werner (1901-1976) A German physicist responsible for deriving a theory of atomic structure and proposing the uncertainty principle in 1927, which has since become widely associated with his name.

Heisenberg built on the work of previous physicists and mathematicians, including Hermann Weyl. For his contributions, he was awarded a Nobel prize in 1932. In the 1940s, he acted as the director to the Kaiser Wilhelm

Institute for Physics. Near the end of the second World War, Heisenberg was captured by American troops and taken to Britain. When he returned to Germany, he helped found the facility that became the Max Planck Institute for Physics. During his later years, he was working to formulate a unified field theory of elementary particles.

Werner Heisenberg was a Nobel laureate who studied the motion and behavior of electrons at the quantum level and challenged many of the accepted notions of classical physics.

held call See hold.

helical antenna, helical beam antenna An antenna designed with a helical (spiral) conductor wound in a circular or polygonal shape. The axis of the helix is usually mounted parallel to the ground. The circumference size of the helix in relation to one wavelength affects the angle of radiation.

heliochrome [sun color] An older word for a color photograph, that is, one that was photographed in color. Originally color photos were in grayscale and pigmented by hand by means of oil pigments. Color photography did not come into common use until the 1960s.

heliograph [sun writing] A visual signaling system employing light signals, which was established around 1865 by H. C. Mance. The heliograph took advantage of the development of glass mirrors in the 1840s to increase the distance over which sunlight could be reflected. It used adjustable mirrors mounted on tripods and could convey messages in Morse code in daylight up to about 100 miles.

In the United States, leaf shutter versions of the heliograph were developed to interrupt the light signals instead of directing the angle of the mirror as was done with the earlier British heliographs.

Most visual signaling systems were superseded by wire telegraphy, but the heliograph survived for several decades, probably because it used Morse, which was then becoming widely accepted, and because it required no external power source.

Since heliograph signals and microwave transmission share some of the same line-of-sight characteristics, heliographs were resurrected to research the placement of microwave relay stations, and the heliograph is still sometimes used for military communications in regional conflicts where other means of communication are scarce.

heliography [sun recording] A type of early photographic process, also called 'sun drawing' which was pioneered by French inventor, Joseph Nicephore Niepce in 1816. Originally Niepce used a camera similar to the camera obscura to temporarily imprint an image of light onto paper coated with silver chloride. It was several years before he developed the process to the point where the image could be permanently preserved. See Daguerre, Louis Jacques Mandé, photography.

helionics The science of the conversion of solar energy to electrical energy.

heliotrope [sun turning] An early surveying instrument which employed the sun's rays to triangulate from mountain prominences. This instrument was developed and used for the highly successful engineering feat of surveying India in the 1800s. It may also have been used for signaling. It was later adapted as a heliograph by H. C. Mance in Britain and used for many decades for daylight signaling of military communications up to 100 miles.

helium-neon laser A type of low power gas laser, commonly used in light shows and monitors. This laser produces warm color tones in the red-orange range.

helix, helical shape A spiral, continuous coil. Many types of springs employ a helix shape. In radio transmissions, a horizontal- and vertical-polarized wave combined as *circular polarization*, is transmitted in a helical fashion so that it can be picked up by both horizontal- and vertical-polarized antennas.

Hellman-Merkle A trapdoor knapsack cryptography system principally designed by Ralph Merkle, with input from Martin Hellman, who was a collaborator with Whitfield Diffie on another cryptography system. The Hellman-Merkle scheme was found to be breakable, and was reported as such in 1982. See Diffie-Hellman.

hello A ubiquitous greeting to indicate a person's acknowledgment of a call (see handshake), especially on a telephone. Since people may have been greeting one another face-to-face with "hello" before the invention of the telephone, it's likely that its use as a standard greeting originated quite naturally. Nevertheless, on early telephones there was probably some discussion as to the best way to alert the callee of the incoming call and of the most natural way to initiate a call.

Hello Girls One of the many colloquial names given to the early female telephone operators. Others include Voice with the Smile, Central, and Call Girls.

Helmholtz, Hermann von (1821-1894) A German physicist who expressed relationships between fundamental phenomena, such as heat and light, by treating them as manifestations of a single 'force,' a concept we now associate with *energy*. Helmholtz further sought to generalize the concepts put forth by James Joule.

help desk A central information and assistance facility found in many large public-service or retail commercial institutions. The help desk is sometimes combined with the reception desk. In the computer industry, the analog to the help desk is the technical support line. Unlike help desks where you can tell if someone is available to help you and where you typically only wait five or ten minutes, tech support lines may be automated so that you don't know if you'll ever be served by a human being, and it's not unusual for large companies to keep you on hold for one to three hours, a situation which shouldn't be considered acceptable.

henry A unit of inductance in a circuit (self-inductance or mutual inductance of two circuits) such that the electromotive force of one volt is produced when the inducing current varies at the rate of one ampere per second. Named after Joseph Henry.

Henry Ford Museum & Greenfield Village The largest indoor/outdoor museum complex in the U.S. providing authentic historical artifacts and educational activities. Among other fields, it features exhibits on the history of transportation and communication, including Thomas Edison's Menlo Park Laboratory. The Henry Ford Museum is located in Michigan State. http://www.hfmgv.org/

Henry magnet A type of experimental magnet designed by Joseph Henry.

Henry, Joseph (1797-1878) A gifted American physicist who began experimenting with magnetism in 1927, Henry produced a high-power industrial electromagnet in 1931. He incorporated his various discoveries into many practical devices, including telegraphs, relays, and electromagnetic motors.

Joseph Henry actively encouraged and assisted other researchers, in addition to carrying on experiments himself, a fact that was not always publicly acknowledged by the many inventors who benefited from his generosity.

Joseph Henry advocated a science and research focus for the great Smithson endowment that eventually became the Smithsonian Institution, and he was appointed its first Secretary in 1878. He also cofounded the U.S. National Academy of Sciences. The henry, a unit of inductance, is named after him (1891).

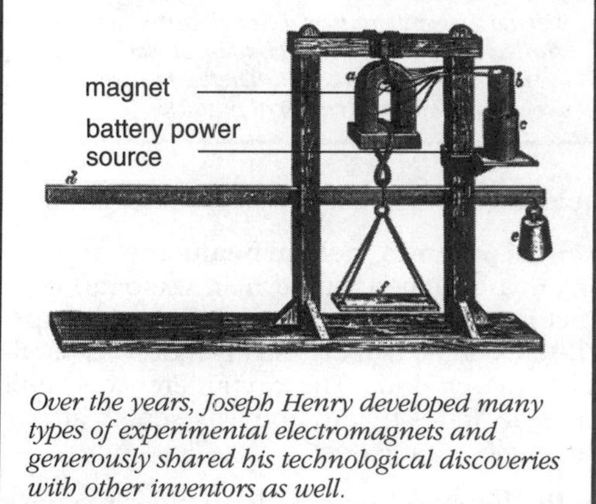

Over the years, Joseph Henry developed many types of experimental electromagnets and generously shared his technological discoveries with other inventors as well.

HEP high energy physics.

hermetic seal A seal that is airtight, leaktight, and which is sometimes used to preserve a gaseous environment inside a sealed device that is different from the gaseous composition outside. Hermetic seals are generally intended to be permanent. Hermetic sealing is used in a variety of industries, including electronics and electrical installation.

Herrold, Charles D. (1875-1948) An American inventor and educator, Herrold is one of the first experimenters to transmit voice over distance starting about 1909. He became known by amateurs for his station SJN broadcasts. When Lee de Forest's transmitter failed, Herrold provided music and news to the 1915

World's Fair about 50 miles away. He was awarded a patent for the Arc Phone in 1915.

hertz Hz. A unit of frequency expressed as one cycle per second, named after H. R. Hertz.

Hertz antenna An antenna system which uses distributed capacitance to determine its resonant frequency, which, in turn, is influenced by the physical length of the antenna. This antenna is used in applications where ground reflection is not a necessary factor for its functioning. Unlike Marconi antennas, a Hertz antenna is not dependent on the ground or the body of a vehicle, as a resonant conductor. This type of antenna is common for television and frequency modulated (FM) broadcasts. See antenna, Marconi antenna.

Hertz, Heinrich Rudolf (1857-1894) A German physicist who demonstrated important properties of electromagnetic radiation, discoveries that later experimenters applied to facilitate the transmission of radiant energy. By 1887 the physical existence of radio waves was established. He also contributed a streamlined reformulation of Maxwell's equations that was widely accepted. The Hertz antenna and hertz unit of frequency are named after him.

Hertzian waves Electromagnetic waves in the range from about 10 kHz to 30,000 GHz. James Clerk-Maxwell had proposed that rapidly vibrating electric currents would emit waves, and Hertz experimentally confirmed this proposition. The waves are named after him. See radio wave.

hetero- Prefix for different, other, not usual. It is often used to describe a mix or variety.

heterodyne *v.t.* To produce a *beat* between two frequencies, which could be of various kinds: audio, optical, or radio. In radio heterodyning, an electrical beat can be selectively created and controlled by heterodyning a received signal current with a steady introduced current. The frequency thus formed can then be further processed by amplification, as in repeater stations or filtering. See beat, beat frequency, heterodyne repeater.

heterodyne repeater A frequency repeating system commonly used in the propagation of radio signals which uses heterodyning to create an intermediate frequency through demodulation, which is amplified, modulated, and retransmitted over the next leg. This technique is less subject to distortion and loss through modulation/demodulation than baseband repeating. Heterodyne repeating may be used in conjunction with baseband repeating if the signal is traveling through several legs and some channels need to be dropped off at the baseband stations. See Armstrong, Edwin H.; baseband repeater.

heuristic problem-solving An exploratory problem solving strategy that employs successive trial and evaluation of the results in such a way that the results can be used in the subsequent trials to 'home in' on a solution. Heuristics are often used in artificial intelligence programs where the result is not known in advance, and where brute force methods are inappropriate due to the large number of possible choices and outcomes. Chess playing programs, for example, use a combination of heuristics to handle novel situations and databases of known moves and strategies. Heuristics are common in robotics, where a robot may have to interact with an unknown or unpredictable environment. See algorithmic problem-solving, brute force problem-solving, neural network.

Hewlett-Packard Company HP. In 1938, Dave Packard and Bill Hewlett, both graduates of Stanford University, began working out of a garage in Palo Alto, California. Their first product, an audio oscillator resulting from Bill Hewlett's research of negative feedback, was designed to test sound equipment. It was a new type of design, utilizing an incandescent bulb to provide variable resistance. This product was followed by a harmonic wave analyzer. Their first big client was Walt Disney Studios, which ordered eight oscillators for the production of the movie Fantasia.

From these beginnings, Hewlett and Packard formed a partnership on New Year's day in 1939, and a toss of a coin decided the company name. HP was officially incorporated in 1947.

Since then Hewlett-Packard has become a well-known supplier of calculators, computers, software, printers, and other accessories to the computing industry. They are known for good quality products (their calculators have been known to survive the 'drop-kick' test) and a corporate culture that seems to produce happier employees than many other companies in the industry.

In 1996, David Packard, successful businessman and philanthropist, died at the age of 83.

Hewlett, William R. (1913-) An inventor and successful business tycoon, Bill Hewlett was co-founder of Hewlett-Packard along with David Packard. An instrumentation engineer,

Bill Hewlett invented an audio oscillator which was used by Disney Studios in the production of Fantasia. He is a past president and director of the Institute of Radio Engineers (now the IEEE), an honorary trustee of the California Academy of Sciences, and has held many other professional and civic leadership positions.

HF, hf 1. hands free. 2. high fidelity. 3. high frequency.

HFC See Hybrid Fiber Coax.

HFU hands free unit.

Hi-band A video standard developed by the Sony Corporation in 1989 to support 500 lines of resolution on a TV screen. This resolution falls between NTSC and High Definition Television (HDTV), and has not been widely adopted.

Hi-OVIS Highly Interactive Optical Visual Information System. A Japanese cable television delivery system employing different types of cable for different parts of the service network.

hibernation A resting state, one of low energy usage and activity. A term often applied to the sleep mode on portable computers which power down during times of low activity to extend battery life. Hibernation has also long been applied to software applications which lie 'dormant' waiting for some event which causes them to become active, or to run at a higher priority level. The event that 'rouses' the program can be many things, including the time of day, input from an interface device such as the mouse or keyboard, activity of users, or data from another program. See sleep mode.

hickey A spot, halo, or other imperfection in the ink or toner of a printout caused by undesired extraneous particles, dried ink, etc.

hierarchical file system An organization structure common to system and user files which arranges files in 'folders' or 'directories' in a descending tree.

hierarchy A group of items, people, or processes ordered according to some type of structure or rank, usually top-down or bottom-up. A hierarchy may or may not be nested. Hierarchies are generally designed to facilitate the location of some item within the hierarchy or are structured in order to simplify the understanding of its contents. People in an organization, and files or discussion groups on a computer or network system, are often assigned positions within hierarchies.

High Altitude Long Endurance HALE. Pilotless platform crafts intended to float above commercial aircraft at about 20,000 meters, which have been proposed as two-way communications links capable of carrying phased antenna arrays. The HALE systems are experimental and a number of systems from helium to jet-engine propulsion have been proposed. The cost relative to traditional satellites makes HALE transceivers attractive to developers. Actual deployment of these systems has to take normal aircraft safety and traffic patterns into consideration.

high ASCII There really isn't such a thing, since ASCII defines the lower 128 characters (0 to 127), but "high ASCII" and "extended ASCII" are often descriptively used to describe characters above decimal values of 127, which are different on each system. See extended ASCII.

high bandwidth A person with high level, wide-ranging intellectual abilities. The author first heard this phrase in the early 1980s among hard-core programming friends in computer users groups and suspects it originated spontaneously in a number of places, and subsequently spread through networks like Fidonet. Other computer-related terms that have been applied metaphorically to human intelligence include *high baud rate, high clock speed,* and *multitasking*.

high bit-rate Digital Subscriber Line HDSL. A digital transmissions technology which can transmit DS-1 or E-1 transmissions for longer distances over the traditional, unshielded twisted pair wire that is widely installed in telephone circuitry. HDSL is typically used in digital loop carrier systems, private branch networks, and cellular antenna systems. Unlike earlier technologies, HDSL transmits over multiple lines without repeaters and uses techniques to pack more information in less bandwidth. It provides rates up to 1.544 Mbps over DS-1 or 2.048 Mbps over E1 in the 80 to 240 kHz bandwidth range. See Digital Subscriber Line, DS-.

High Definition Television HDTV. In 1987, the Federal Communications Commission (FCC) acknowledged that the NTSC television was out of date and formed an Advisory Committee on Advanced Television Service (ACATS) to recommend a revised television standard for the U.S. Most of the improved systems proposed to the FCC were analog or hybrid analog/digital. The FCC stated a preference for a simulcast HDTV system and identified four digital systems. These were extensively analyzed, leading to an ACATS recommendation that digital HDTV should be adopted. In 1993 an alliance was formed in cooperation with ACATS to create an HDTV system, with a specification released in 1994.

high Earth orbit HEO. An orbiting region around the Earth into which certain types of communications satellites are launched. There are advantages and disadvantages to high orbits. The main advantage is that it takes fewer satellites to provide global coverage. Disadvantages include the higher cost of launching, the higher amplification needed for signals to travel the greater distances, and the effects of radiation. The life-spans of high-orbit satellites tend to be around twelve to fifteen years. Most high-orbit satellites travel at about 20,000 to 40,000 kilometers outside earth. High Earth orbit satellites are typically used for geostationary satellites such as the U.S. Global Positioning System (GPS). See low Earth orbit, Global Positioning System, medium Earth orbit.

high fidelity A playback system which reproduces the original so well that it is indistinguishable, or almost indistinguishable, from the original source. This is often accomplished through fast transmission and wide bandwidths. In audio systems, high fidelity is frequently abbreviated *hi-fi*.

high frequency A signal frequency defined as the range from 3 and 30 MHz.

High Level Data Link Control HDLC. An ITU-T standard bit-oriented data link layer communications protocol originally developed by ISO for managing synchronous serial transmissions over a link connection. IN HDLC there are separate bit patterns for control and data representation.

high level language A computer programming language at the user level, designed to be as close to a natural language as possible and generalizable to a variety of platforms. FORTRAN, BASIC, and COBOL are probably the best-known, and most widely used high level languages. High level languages are often interpreted, but may be compiled for the specific platforms on which they will be running. As languages become closer to machine level, they also tend to be more symbolic and, thus, more difficult to read and write. They also increase in platform-dependency. For contrast, see assembly language, low level language, machine language.

high pass filter A filter designed so it doesn't pass waves below a specified cutoff frequency (greater than zero), and the transmission band extends upward indefinitely from that cutoff point. See low pass filter.

High Performance Computing Act An act of the U.S. Congress which was passed in 1991 to facilitate and promote the development and evolution of interconnected computer networks serving educational institutions, research laboratories, and industry.

High Sierra standard A compact disc standard introduced in 1986 by the High Sierra Group (named after the hotel and casino at which the group met). This was subsequently adopted by ECMA (ECMA-119) and ISO (ISO 9660) and released with slight revisions. See ISO 9660.

high speed networking This is a relative phrase and will change as the technology advances, but in the mid-1990s, high speed networking was generally considered to be around or over 100 Mbps. Examples of high speed network technologies include asynchronous transfer mode (ATM), Fast Ethernet, and Fiber Distributed Data Interface (FDDI).

high speed printer Printers with the capacity to feed paper, print on it, and eject (and sometimes collate) it at industrial speeds that exceed consumer printer speeds by a significant margin. The speed that constitutes *high speed* changes, as the technology improves and is set arbitrarily, often by marketing managers. Most consumer printers print at about five or six pages per minute (about 100 to 150 lines per minute). It's reasonable to say that a high speed printer, such as a line or page printer or fast laser printer, has an output of about 500 lines per minute, or about 15 pages per minute or faster.

High Speed Token-Ring HSTR. An enhanced commercial Token-Ring technology developed by International Business Machines (IBM). HSTR can run Token-Ring and Ethernet on one medium, support source routing through data packet headers, and was based on existing standards, including Fibre Channel. Aimed at a market similar to Ethernet, HSTR appears to be primarily supported by those upgrading legacy Token-Ring systems. See Token-Ring.

high usage groups In the telephone industry, high usage groups are trunks between main switching offices that are established as priority routes to handle the majority of transmissions. High usage trunk groups are intended to hand off overflow traffic to alternate trunks. See erlang.

High-performance Network Forum HNF. Developers and promoters of the HIPPI GSN standards. European High-performance Network Forum (EHUG) works in cooperation with HNF and coordinates an annual technical symposium. http://www.hnf.org/

high/low tariff A charge that is selectively made according to a type or level of service. For example, some Internet Service Providers (ISPs) charge different rates for connection time, depending upon the speed at which the subscriber's modem connects. A bulletin board service (BBS) may quote two levels of service depending upon whether or not the user wants to access the games sections. In phone services, two prices may be quoted for a subscriber call, one cost per minute over high density trunks and a different cost per minute over low density trunks.

Hindenburg disaster The terrible crash and flaming destruction of the Hindenburg dirigible in New Jersey in 1937 is etched forever in the minds of those who heard the live radio broadcast of its demise. This historic tragedy was relayed and captured by new technology, radio and tape recordings, in a way that had not been possible prior to the establishment of the broadcast radio industry.

HIPPI High Performance Parallel Interface. HIPPI is a point-to-point high speed data transfer technology created at Los Alamos in the 1980s. HIPPI operates over twisted pair copper cable for distances up to 25 to 50 meters (longer with cascaded switches) and distances up to 300 meters, or 10 kilometers over multimode or single-mode fiber cables.

HIPPI was originally developed for supercomputing applications, but is starting to be adapted to other environments with the dramatic drop in price of the technology, particularly the switches. Transmission speeds include 800 Mbps and 1.6 Gbps (simplex or duplex). HIPPI can be employed with SONET over distances and over satellite transmission links. HIPPI is an ANSI standard with a series of documents spelling out the standard and its various switching and encapsulation characteristics. See HIPPI-6400.

HIPPI-6400, SuperHIPPI High Performance Parallel Interface. Officially it is now known as Gigabyte System Network (GSN). A very high bandwidth, low latency network transmission technology which offers gigabytes-per-second transfer rates, much faster than the capacity of Gigabit Ethernet, ATM, or Fibre Channel. Based on the HIPPI standard, but with enhancements in error correction and lower latency rates, HIPPI-6400 uses a fixed-length cell of 32 bytes. Transfer rates which are as fast or faster than the internal workings of an individual computer on a network will change computing in a significant way. With this development, the network is no longer a bottleneck and individual computers attached to the network can theoretically function as individual parts of the same organism, that is, as a massively parallel computing system. Fast transfer rates with high bandwidth also make it easy to support a diverse variety of protocols, providing flexibility. SuperHIPPI can support applications like uncompressed digital movies and HDTV signals. See HIPPI, Scalable Coherent Interface.

Historic Speedwell Located in New Jersey, this museum features telegraph history through the daily life of the Vail family and their association with Samuel Morse. Morse had a close association with the Vail family as they provided space, materials, and expertise to assist him in fabricating the inventions for which he is known. See Vail, Alfred; Vail, Theodore.

hits 1. Number of system or application accesses of a specified type within a specified time period. 2. The number of attempts of illegal entry to a system within a specified time. This information is used to gauge security needs and adjust procedures, if necessary.

hits, query In a database query, the number of items that match or contain the search criteria. If the initial number of hits brings up too many matches, it is usually advisable to narrow the search, and vice versa. Web search engines often bring up too many hits on the first try, but advanced search options allow the user to specify conditionals and priorities.

hits, Web site On a commercial site, such as a World Wide Web site, the number of times a page or component of a page, is accessed or downloaded in a specified period. For marketing purposes, this information may be further analyzed according to time of day, specific items of interest, number of hits by *different* users, since one user may access a site multiple times, and subsequent ratio of hits to sales.

HNF See High-performance Network Forum.

HNS See Hughes Network Systems.

HOBIS Hotel Billing Information System.

Hoff, Marcian E. (Ted) A pioneer engineer who, in response to a request from a Japanese company for a calculator chip, designed the Intel 4004 in 1971. This highly significant invention was the first commercially successful microprocessor chip, and it launched the microcomputer industry.

hold *v.* 1. To pause, to cause to remain in a particular position or situation. 2. While attempting a computer login, the user may be queued or put *on hold* until fewer users are on the system. This usually manifests as a pause after typing the username or after typing the username and password. On popular archive sites, there may be a hold period or increased lag time while accessing a system. Sometimes it's better to log on later during off-peak hours, or to find a less busy *mirror site*, if one exists. 3. On phone systems where the receptionist is busy with other calls, or the automated system is queuing the caller for the next available operator, the user is usually put on hold, sometimes for extraordinarily unreasonable lengths of time (especially if it's a long-distance call). On systems which don't have music or which don't have a recording informing you that you are in the queue, it is sometimes difficult to know if you are still on hold, or if you have been cut off. On multiline systems, a call can often be put on hold by one individual and picked up by another (line hold). On some systems, the hold call can be continued only on the main console, or by the person originally putting the call on hold (exclusive hold).

Hold Recall An optional telephone feature that alerts you to the fact that someone is on hold. On most multiline systems and some residential phones, the line on hold will be identified by a flashing LED or, on more sophisticated systems, with beeps, or voice messages.

hold time 1. In circuits, the time interval after the clocking of a trigger circuit during which data must remain unaltered. 2. In welding or soldering, the time during which the welded object must be held relatively steady in order for the weld or solder to harden. 3. In telephony, the length of time a caller is kept waiting, and waiting ... .

holding beam A diffused electron stream used to regenerate charges applied to the surface of a storage tube.

holding coil In a communications circuit, an additional coil in a relay for each direction of a transmission, that can be opened or closed independent of the main circuit to enable a single circuit to accommodate alternate two-way communications.

holding gun In an electron storage tube, the source of the electrons that make up the holding stream.

holding time In telecommunications, the entire duration of a call from the time the connection is requested until it is completed and disconnected. The actual time during which the connection is established is only a portion of the holding time, although it's usually the time that is billed. Holding time is important, for example, in tracking sales calls. How long does the sales rep. or telemarketer actually stay on the line trying to connect a call, and handle the entire transaction, as opposed to the amount of time actually spent with the customer? If there is a large discrepancy, another method may need to be tried.

Holding time is also important in computer or telephone circuit planning and management, because the time spent connecting and queuing the caller may affect capacity and efficiency as much as that portion of the call during which the communication takes place.

holding trunk In telephone communications, a queue wherein a call is held until availability is established or an alternate route is found.

holiday factor A concept in service and retail industries that accommodates changes in rates of service use or product purchase during holidays. The transportation, retail, and telecommunications industries tend to have patterns of higher usage during holidays, and need to factor in extra staff, lines, products, etc. to handle the demand. Holidays may also create decreased demand, in which case shutdown or reduction of unneeded systems may result in cost savings.

Hollerith card A sturdy rectangular 80-column punch card designed to store Hollerith code information in the form of punched holes that can later be read and decoded to reassemble the original information. The cards were popularly used to store computer code in the days before tape, diskette, and hard drive storage. (The Hollerith card is a specialized type of punch card.) As the storage capacity of these cards is quite limited, many cards are needed to store a body of information. Punch cards can be fully punched and read in again with mechanical or optical devices, or partially punched to provide a record without the mess and waste of *chad* and read in again with sensing devices. See zero punch.

Hollerith code A 12-level code designed by Herman Hollerith in 1889, which is used on Hollerith cards in which holes are punched at specified intervals in designated rows and columns, with each column corresponding to a character, in order to form a semi-permanent record that can be read and interpreted in the future. This code was widely used in early computing days to store program instructions.

Hollerith, Herman (1860-1929) An American engineer who developed an early tabulating machine in 1884, which he further developed with the concept of punched cards as an information storage medium. Hollerith was subsequently contracted to carry out the storing and reading of U.S. census information. From the success of this invention, he formed the Tabulating Machine Company to market the technology, which eventually became International Business Machines (IBM).

hologram A type of imaging using lasers, based on the recording of an optical interference pattern produced by the interaction of two or more waves from the same source. The effect of viewing a holographic image is a sort of 3D, when moving the head around and focusing on different parts of the image. In most cases, it is not fully a 3D effect, as the image is usually recorded on some 'flat' (with transparent depth) medium such as glass or transparent plastic (although the technology itself is not limited to this form of presentation). The image is typically more ethereal than a photograph, since it is viewed through 'layers' of the transparent medium. Projected holographic images hold great promise for 3D virtual reality, and scientists have been working on holographic memory modules for computers that potentially can store enormous amounts of information in small three-dimensional components. It has been suggested that human memory may share some functional similarities with holograms.

holy wars Interminable subjective online discussions on subjects that have no definitive answer in a general context and have been beaten to death, such as your-computer-versus-my-computer arguments.

home In computer interfaces that distinguish regions of the screen for different functions, one corner is usually designated as the *home* position, that is, a position to where a pointing device may return, or where a cursor might start again after the region has been cleared, or a new window opened.

On English language-based systems, for example, the home position in a text window is usually the top left corner. If the text were being typed in another language that tracks from right to left, such as Hebrew, the home position would be the top right of the window or screen, or bottom right if it is bottom to top reading. In PostScript page layout programming, the home position is generally considered the bottom left corner, as it is with a number of printer graphics languages. Some systems have a *home* key as a one-step shortcut to position the cursor in the home position.

home ATM network HAN. A broadband home network for providing connectivity with a variety of services and devices (computers, television, appliances, etc.). See ATM Forum, fiber to the home, RBB.

home computer A computer configured for basic functionality and ease of use, in a price range low enough to be attractive to general consumers.

home page A World Wide Web concept which indicates the first, primary, or main page of a set of hypertext linked pages on a particular host, or belonging to a particular individual or organization. Web pages are not inherently hierarchical, since any page can link to any other, but humans tend to grasp concepts more easily when information is organized in a top-down or bottom-up manner, and home pages reflect our preference for this type of organizational structure.

A home page serves as a jumping off point, table of contents, or general site information map to help navigate the rest of the links. Commercial sites tend to have home pages that showcase product information and entice the user to explore the rest of the site. Personal sites often show family relations, professional credentials, and personal interests. Educational home pages usually provide information on course offerings, faculty, and facilities. See HTML, hypertext, World Wide Web.

home run A centralized wiring topology, like a star topology, in which cables to individual units or consoles all lead back to the central switching system. Most private branch systems, almost all key systems, and many of the smaller computer network systems, have this type of cabling arrangement. In home run wiring, which has little or no redundancy, a severed line will cut off the end station from all other stations. See topology.

Homebrew Computer Club A historic early

electronic thinktank and tinkerer's organization, founded in California in 1975. Steve Wozniak is one of the most famous of the homebrew members. The Altair was demonstrated here, with homebrewers jumping on the opportunity to write applications for this early computer.

homeostasis A state in which there is a tendency toward balance, stability, or equilibrium. A state in which there is no change.

homing 1. Zeroing in on an intended destination or target. 2. Approaching an intended destination by holding some parameter of navigation constant (with the exception of altitude). 3. In guidance systems, transmitting, receiving, and evaluating signals in order to locate a target. Bats and missiles use homing systems. 4. In telecommunications, homing is the selection of a route through which a call can be set to the next switching center, especially in toll systems, which may pass through several specified switching stations.

homing pigeon One of the earliest means of distance communications, used as least as early as several hundred years BC in ancient Greece. Homing pigeons could be taken on journeys and then set free to fly home with messages. At first, just the appearance of the pigeon represented a prearranged signal. Later, an actual message, written on papyrus, skin, or reed could be attached to the bird's leg. Unfortunately, homing pigeons became extinct following the second World War.

homodyne reception See zero beat reception.

hook switch, switch hook The hook switch was originally designed not just to terminate a connection so the next call could come through, but served also to disconnect from a battery source so it wouldn't be quickly used up, and later an electrical source.

Modern telephones draw current from the line, and don't require a separate battery to operate the basic calling and receiving functions, but the hook switch, the hook on the side of an old traditional box phone, or the buttons (plungers) on top of a traditional rotary desk phone, are still used for disconnecting a call, and sometimes for generating a tone (if they are held down briefly, which doesn't cause immediate disconnection). See hooking signal.

hook switch dialing On older wall box phones and rotary pulse phones, it was possible to dial a number by depressing the hook carefully for each number you wanted to dial. (Depressing the hook switch for too long would disconnect the line.) This is even possible on some of the older pay phones. See hook switch.

hookflash A signal-sending mechanism whereby the hook on an old-style phone, or button plunger on a newer phone is quickly depressed to signal the initiation of a service or operation.

hoot'n'holler, holler down, shout down, squawk box A dedicated, four-wire, open phone circuit connecting speakers or speakerphones at each end of the connection round the clock. It's like a 24-hour public address system using phone lines with full duplex, two-way communication. Other phones on the system can be picked in order to listen to the conversations ongoing on the speaker system.

Hoot'n'holler systems are useful in industrial yards, institutions, and fast-paced financial floors where numbers of free-moving individuals look to centralized sources of information or engage in communal dialog at different locations.

hop *n.* 1. The extent of an individual transmission path between two nodes (with no intermediate nodes). 2. In radio, the extent of a transmission from Earth to ionosphere and back. 3. In frame relay, the extent of an individual trunk line transmission path between two switches. 4. In an IBM Token-Ring network, the extent of an individual transmission path between two bridges. 5. In cellular communications, where the user may be traveling through several transmission zones during the course of a call, a hop is a change in the radio frequency channel.

hop by hop/hop-by-hop routing In contrast to a system that predetermines a route before sending a transmission, hop-by-hop routing creates a route along the transmission path, a step at a time, by using routing information at switchers along the way. There are advantages to both. A predetermined route may be an efficient one, designed to speed the transmission through faster links or perhaps by choosing the shortest path. This is common on small or local systems. On the other hand, on a large system like the Internet, there may be millions of possible routes, too many to store in the routing tables at the source of the transmission. In this case, hop-by-hop routing is a scalable technique that makes use of the best information at each station to progressively build a path for the data.

It has been suggested that ATM implementations of hop-by-hop datagram forwarding on the Internet are no longer adequate to handle traffic volume and proposed improvements have been suggested. See cell switch router, RFC 2098.

hop channel In cellular communications, a radio frequency (RF) channel which is available to continue transmissions for a user with a call in progress who is moving through zones. Available channels are needed to continue uninterrupted transmission while the user is on the move. See cellular, hop, mobile communications.

hop count A sum of the number of hops which make up a route between its source and destination, or between a specified segment of the route. In radio communications, the number of times the wave bounces from the Earth to the ionosphere and back.

In networking, the hop count is the number of segments between individual nodes or routers, a number that is recorded in Internet Protocol (IP) packets on packet-switched data networks. In cellular, the number of times a radio frequency change occurred during the course of a call. Hop counts are one means to gauge the efficiency of a system and to configure or tune it for better performance.

hop off To exit one type of system and complete the route on another. For example, you may initiate a facsimile transmission on the Internet, which then *hops off* to a phone line and a dedicated facsimile machine. Or, you may make a voice call from a telephone which is routed through a voice translation program, and interfaces with the Internet and becomes an email message at the destination. In this case, the *hop off* is from the phone system to the Internet, or, conversely, you can consider it a *hop on* to the Internet, if you are considering the Internet as the main portion of the transmission route.

Hopper, Grace Murray (nee Grace Brewster Murray, 1906-1992) An American mathematician, physicist, and educator, Hopper was the developer of COBOL. She spent many years as a lecturer, research scientist, and programmer for various organizations, including the U.S. Naval Reserve. She is perhaps best known for relating a story in which a technician found a bug inside a Harvard Mark II and solved a problem by removing it. Grace apparently glued the bug into the computer logbook, and in the 1970s announced that she would be contributing it to the National Museum of American History.

Grace Hopper became involved in many of the important computer development projects at the end of World War II. In 1944, she joined Howard Aiken's Harvard Mark I project as its third programmer, and later worked on the Harvard Mark II.

In the early 1950s, when new ideas about programming and reusing existing code began to evolve, Grace Hopper made what was probably her biggest contribution to the field. She became a champion of ideas that led to high-level languages, compiled software, and more efficient coding methods, in spite of the criticisms of many detractors who claimed at the time that such things were impossible. See bug, Harvard Mark I.

Grace Murray Hopper was an American mathematician, physicist, lecturer, and one of the first computer programmers in the days when 'programming' involved rearranging the wires within a vacuum-tube computer system.

hops See hop count.

horizontal blanking interval, horizontal blanking time The period during which a display is suppressed on a cathode ray tube (CRT) to allow the electron gun to return from the right side of the screen to the next display position down and on the left side of the screen. See blanking, cathode ray tube, frame, sweep.

horizontal cross connect The interconnection between a horizontal distribution system and a telecommunications central wiring location such as an equipment or patch panel closet or bay.

horizontal distribution frame The equipment and structural elements that facilitate the interconnection of interfacility cabling configurations, as between subscribers and substations and central offices. The frame technically does not include the wiring, but directs and contains it. Horizontal distribution frames are usually built into flooring or crawl spaces, hence the name. See distribution frame.

horizontal link, inside link In ATM, a link between two logical nodes belonging to the same peer group.

horizontal resolution A description of the amount of information that is contained on a single horizontal line of a rasterized output device such as a monitor or printer. On raster monitors, horizontal resolution is expressed in terms of pixels, usually about 800 to 1024 or more. On black and white laser printers, horizontal resolution on consumer machines ranges from 300 to 1000 dots per inch (dpi), and on prosumer and industrial printers from 1000 to 2700 dpi. Thus, the total would be the number of inches times the dpi. A resolution of about 600 dpi or greater is needed to show clean lines and curves, without staircased artifacts, for common printed documents.

horizontal scan rate A measure of the scan speed of electron beam display devices, usually described in hertz (Hz), as in cathode ray tubes (CRTs) which sweep repetitively from left to right and top to bottom.

The horizontal scan rate describes how many horizontal scan lines per second can be displayed. At a particular scan rate, the number of lines which can be displayed decreases proportionally as the refresh rate increases. Multiscan computer monitors permit a variety of scan rates and resolutions, most ranging from about 40 to 75 Hz. See cathode ray tube.

horizontal segment In wiring distribution systems, the wiring route from individual NAM or IO locations to the riser closets through ceilings or floors, usually up to a maximum of about 250 feet.

horn alert An electronic connection for sounding a horn or loud buzzer to signal an incoming transmission during times when the user might be some distance from the communications device. Horn alerts are used for after hours phone calls or doorbells, for cellular phones in cars, and for a variety of security systems.

horsepower hp. A unit of power designated as equal to raising 33,000 pounds one foot in one minute, which can also be expressed as an English gravitational unit of raising 550 pounds one foot in one second. In the U.S., a unit of power equal to 746 watts. See watt.

host 1. One on which others depend for shelter or sustenance. 2. The main organizer and holder of an event. 3. It's a little difficult to define host as it relates to computer systems, because different groups of computer personnel have given *host* and *client* opposite meanings in the past. For consistency with the English meaning of the word and popular usage, this dictionary defines host as a main server or controlling system, and the client as a subservient system in terms of priority or capabilities. See client.

host carrier In telecommunications, the main carrier through which billing is channeled. In systems where a call goes through various networks or providers, carriers may have arrangements with the host carrier to bill through them to save paperwork and other administrative costs.

host computer A computer in a network which provides primary operations and applications which are run through clients or remote terminals at other locations. A network may have more than one host, and some may be specialized for modem access, email distribution, printing, and other tasks. The term host is related more to function than raw hardware capabilities, but due to resource sharing economics, the host frequently has greater capabilities (more memory, storage, peripherals, etc.) than the clients which are accessing it.

host site 1. A repository, or other archive site which is accessed by remote users through clients programs such as Telnet, FTP, Web browsers, and others. The host site is the one on which the administrative tasks and storage are carried out. 2. A computer bulletin board system, which typically hosts email, chats, games, and file uploads and downloads.

hot 1. Connected; live; ungrounded current-carrying conductor. A term frequently applied to electrical wires. 2. A hot chip is one which either runs at a high temperature and requires cooling, or one which has a fault that causes it to emit more heat than is normal and which is likely to fail soon. See heat sink. 3. Stolen. 4. Topical, popular, desired by a large following. 5. Titillating, arousing. See hot chat.

hot cathode, thermionic cathode A cathode which produces a stream of electrons by means of thermionic emission.

hot chat Realtime conversations on a number of public or private chat forums in which sex is frankly discussed between two or more parties. Private conversations are engaged in by using private messaging commands on public forums or by opening password-protected by-invitation private channels. See Internet Relay Chat.

hot cut, flash cut The transition from one circuit to another while the system is in operation, hopefully without disruption to components or current users. Hot cuts are used when switching from an old wiring system to a new one, or when switching around physical routing paths. On individual computer systems, components are sometimes hot swapped, although it is *never* recommended. Never hot cut a component that is being accessed. It is especially inadvisable to hot cut most types of drives (floppy, CD-ROM, hard drive, etc.). (RAID systems are an exception.) Keyboards and mice are not usually damaged by hot cuts, but make it a habit to power off a system before making hardware configuration changes. See half tap, hot swap, redundant array of inexpensive disks.

hot docking Inserting a component into a docking bay (as in laptop docks or video bays) while the system is powered on. This is generally inadvisable. Whenever possible, power off all components before connecting electrical circuits.

hot key combination 1. A combination of keys, which when pressed simultaneously will perform a specific function or engage a memory resident program, like a printer utility. This is handy for background processes which are frequently needed, but which would be distracting if they were running in the foreground along with other current process. 2. A combination of keys pressed simultaneously to perform a specific operating system function. For example, on an Amiga, *Amiga-Amiga-Ctrl* reboots the machine. On an IBM-compatible running MS-DOS, *Ctrl-Alt-Del* performs a similar function. 3. A combination of keys to access text style attributes and search and replace functions in older word processing programs developed before graphical user interfaces became common.

hot line A private, dedicated phone connection, sometimes indicated by the color of the phone. On a land line, when you pick up the line, it either connects automatically, or does so quickly through the touch of a button or speed dialing. On a wireless service, the system may be configured so the phone can connect only with a specific number. Hot lines are used in security areas as emergency phones in buildings and on roadways, and as dedicated lines to brokerage firms, or important personnel in government or military positions.

hot line service Phone service which expedites an automatic connection through a dedicated private phone. See hot line.

hot links In computer software applications, virtual links that form a connection between information in one document (such as text or images) and another, even if their native formats differ. For example, in a desktop published document, there may be a hot link to text in a word processor and another to an image in a graphics program. Depending upon the system and the software, changing text in the word processor or in the graphics program may immediately effect a change in the corresponding desktop published software, or may effect a change when the page is refreshed or when *update links* is selected from a menu. As systems become more capable (multitasking, faster CPUs, more memory), hot links are more prevalent and updates happen more automatically. See drag and drop.

hot list In computing, a list of frequently used applications programs; directories; or Internet newsgroups, Web sites or archives. A hot list is usually displayed as a text list or pull-down menu from which the user can quickly select the desired destination. See bookmark.

hot spot, hotspot 1. A location on a touch sensitive device that alerts the software to respond in some fashion to user input. 2. A screen location which responds when a cursor is moved into the region, or if the cursor is positioned and a mouse or key clicked to activate the hot spot. 3. A bottleneck, or area of congestion in a network, component, or software routine. 4. An area of a circuit in which some component is generating more heat than would normally be expected, and which may signify a potential problem. 5. A region of a document or image that includes an embedded link so some further action happens if the region is selected or activated. Hyperpage applications use hot spots for various links; graphics programs sometimes use hot spots to activate palettes or specialized drawing menus.

hot standby A backup or background system or program that is operating, but idle and available to take over if failure of the regular system occurs.

Hot Standby Router Protocol HSRP. A protocol that provides resiliency, fault-tolerance, and transparent network topology support for network routers. Standby routers inherit the lead position if the lead router in a group fails.

hot swap The process of connecting or disconnecting an electric circuit, component, or peripheral while the system is powered up. Hot swapping is done to minimize disruption to users of a system. This is highly inadvisable in most circumstances. Some systems are designed to handle hot swaps (some types of video components or redundant hard drive systems), but be sure you know what you are doing before you attempt it. See hot cut.

hot type Cast metal type used in traditional printing presses.

hotel/motel console A private branch system specifically designed for businesses that manage rooms. It provides additional information on the status of the rooms, or status of calls to or from those rooms. Used in convention centers, hotels, motels, private boarding homes, etc.

HotJava An adjunct to Java, the widespread, object-oriented, cross-platform programming language from Sun Microsystems that continues to grow in popularity for use on the Web. HotJava is a Java-enabled Web browser with support for JDK and SSL that is installed on the local computer system and enables Web sites with Java applications to run from a desktop system. Java support enhances a browser's capabilities. See Applets, Java.

hotline See hot line.

Hotline Virtual Private Line Service A commercial Nynex subscriber service which uses public lines specially programmed and configured to operate as though they were private dedicated lines, with the connection activated when picking up the handset. See hot line.

House, Royal E. An American inventor who developed one of the first practical direct paper tape printing telegraphic receivers in 1846. House continued to improve on the original design and patented the improved version in 1852. Two people were required to operate it, as one had to turn a crank to run the mechanism. See telegraph, printing.

housing A protective enclosure commonly used to insulate, protect, or manage wires or electrical connections. Many housings are shaped like boxes, with one side open to provide access. Splice enclosures are a particular type of housing used to connect fiber optic cables between the head end and the node.

howl An irritating, unwanted wailing or screeching sound from acoustic or electric feedback that may occur, for example, when a speaker and microphone from the same transmission are placed too close together. Noise and echo canceling equipment can prevent or reduce howling.

howler, howler tone In telephone communications, a unit which creates a loud sound to signal that a phone has been left off-hook. For example, on some public exchanges, a recording will play first if a phone is left off-hook, "If you'd like to make a call, please hang up and try again ...," followed by a series of raucous beeps that can be heard up to about 15 feet from the phone.

HP See Hewlett-Packard.

HPA See high power amplifier.

HRPT high resolution picture transmission. A specialized image communication for very high resolution images such as those transmitted by imaging satellites.

HSCI High-Speed Communications Interface. A single-port interface from Cisco Systems, which provides full duplex synchronous serial communications.

HSCS high speed circuit switched.

HSD home satellite dish.

HSDA high speed data access.

HSDU High Speed Data Unit.

HSRP See Hot Standby Router Protocol.

HST High Speed Technology. A U.S. Robotics proprietary, high-speed, full duplex signaling and error control transmission protocol. U.S. Robotics manufactures modems complying with this protocol, and some 'dual-standard' modems which are both HST and V.32 bis capable. See Microcom Networking Protocol, modem, V Series Recommendations.

HSV hue, saturation, value. In color imaging, this is a color model which allows settings to be adjusted along these three properties.

HTL high threshold logic.

HTTP See Hypertext Transfer Protocol.

HTTPS Hypertext Transfer Protocol Secure. Security enhanced HTTP. See Hypertext Transfer Protocol.

hub Focal point, center of attachment or activity.

hub, network 1. A connecting point on a network to centralize wiring and connection management. A hub may be passive or active, and is often used in systems with star topologies. 2. A connection box on video or audio systems that permits centralization of cables and easy reconfiguration of devices. Often used in connection with switchers and in many cases, the switcher itself may double as a hub. See bridge, router, switcher.

hub site The location of a hub, which may vary from a small box on a desk or rack to an entire closet or room, depending upon the size of the system. The hub site allows easy cabling and administrative access to a variety of connections. Hubs are often located at main wiring or logical junctions and may connect to external systems.

hue A color of the visible spectrum. Hue does not include white, black, or shades of gray, which are the presence of all colors (white) or absence of color (black) in various intensities (grays). Most people are familiar with hues as the colors of the rainbow: red, orange, yellow, green, blue, indigo, violet. In software, hue may also be called *tone* or *tint*. See intensity, saturation.

Hughes Network Systems A company (actually a group of companies under the Hughes umbrella) which has been involved in satellite communications since the early launches, and has developed a number of associated innovative technologies. One such product is DirecPC, which allows a satellite feed to connect with a personal computer for data communications. See Applications Technology Network Program, DirecPC.

Hughes, David (1831-1900) An English-born American schoolteacher who developed one of the first 'printing' telegraphs, in essence the telegram and later teletype machines. He also invented the carbon microphone in 1877, an important contribution to telephony. See Morse, Samuel F.B.; telegram.

Hughes, David R. A West Point graduate and retired U.S. Army Colonel, Hughes has been acknowledged as a pioneer in internetworking and educational applications in distance learning. He is credited with teaching the first online college credit courses in 1983. He further designed and supported the Big Sky Telegraph network and the Montana state METNET.

Human Computer Interface standards HCI. A series of protocol platform standards from the IETF, including, but not limited to Common Desktop Environment (CDE), Complex Text Layout (CTL), Motif, etc.

hunt, hunting A process through which a call is routed by seeking out the best path or first available path or device.

hunt group In telephone systems, a series of lines set up so calls can be assigned to the next available line in a group if the first accessed line is busy. There are different ways to organize hunt groups, from a straight sequential hunt to random hunts.

Hush-a-Phone decision A landmark communications case in 1956 which challenged the AT&T monopoly of phone line access. The company wanted to attach a mechanical device to a phone set in order to screen out background noise. AT&T argued against it, supported by the Federal Communications Commission (FCC), but the decision was later overturned in the court of appeals, the main arguments including the mechanical rather than electrical nature of the device and the fact that it in no way harmed the phone equipment. See Carterfone decision.

HUT Hopkins Ultraviolet Telescope. See ultraviolet.

HW hardware.

Hybrid Fiber Coax HFC. A transmission system combining fiber optic cable with coaxial cable that can handle simultaneous analog and digital signals. It is less expensive than a full fiber or switched digital video installation, but still provides greater bandwidth than traditional technologies built entirely on copper wire or coaxial. Network technologies such as ATM, SONET, and frame relay can be transmitted over HFC. Discrete wavelet multitone (DWMT) is being proposed as a suitable modulation scheme for existing HFC installations. See discrete wavelet multitone, Hybrid Fiber Coax architecture, SONET.

Hybrid Fiber Coax architecture A hybrid fiber coax technology for carrying video or telephone services, or a combination of both. For video, the bandwidth is typically divided into 'channels' which can further be subdivided into phone lines. It is primarily a 'downstream' technology, which serves broadcast TV very well, but may not be as flexible for interactive TV and phone services. The downstream nature is not inherent in the cable, but rather in the transmission and amplification technology. Typically optical fiber runs from the central office to a node servicing an area neighborhood. From that point, the signal can be converted to be carried via coaxial cables to individual subscribers. At the subscriber point, a device splits the video and telephone signals so they can be directed to the appropriate lines or devices within the premises.

This hybrid system balances some of the speed and bandwidth of a full fiber-based system, with some of the economic advantages of coaxial servicing individual neighborhoods. One disadvantage is that there is not an unlimited amount of bandwidth available for phone 'lines' and phone service must be planned and adjusted as needed. HFC technologies can put cable companies in a position to compete with telephone providers, which may create a shift in future market share. See Hybrid Fiber Coax.

hydraw See octopus.

hydroelectric The generation of electrical power from rapidly moving water or channeled water under pressure.

hydrolysis A process of chemical decomposition which occurs in the presence of moisture. Hydrolysis is of concern in maintaining insulating materials in underwater cable installations.

hydrometer An instrument which measures, by displacement, the specific gravities of liquids. Used, for example, to measure the electrolytes in batteries.

hygroscope An instrument for measuring the amount of moisture in a material. Handheld paper hygroscopes are commonly used in the printing industry to monitor paper moisture balance and the relative humidity of the air in order to adjust printing materials and processes for quality control. See densitometer.

hygroscopic 1. A material with a tendency to absorb and retain moisture. 2. A material which is able to absorb and retain moisture.

hyperfiction A type of computerized fictional narrative which incorporates a multiple-option story line which can be interactively selected by the reader. Hypertext links to words and images within the narrative allow the user to follow different text threads, such as alternate endings, or to find out more about a particular character or the story environment. In other words, different readers have a different experience of reading the story, since they may read different parts of it or read parts of it in a different order from another reader with different preferences.

In hyperfiction, the story essentially exists as a database, with the different scenarios referenced so they can be selectively accessed. When the database is interactively read, it perceptually unfolds for the reader as a story. See hypertext, hyperlink.

hyperlink, hypertext link A logical link between meaningful data organized within a random access database or markup language. Hyperlinks can be hierarchical or 'flat.' They can be one-directional or bidirectional. Although hypertext links are most familiar to users in the form of virtual 'cards' in a computer card catalog or as browser-accessible links on the World Wide Web, a hyperlink in its broadest sense also applies to interconnected visual image links, where the user clicks on an icon or a part of a picture rather than on a word or block of text.

Hyperlinks on the Web have opened up global Internet interactions and cross references to immense, shared information storehouses. There are a number of popular games which are navigated through text or visual links. See browser, hypertext transfer protocol, World Wide Web.

hypertext A means of accessing information through referential links. This idea has been around for a long time, and has had various implementations, with Bush developing a microfilm system and suggesting 'associative indexing' in the 1940s. In the 1960s, D. Engelbart developed the On Line System (NLS) for storing research papers and other information in a hypertext-like manner. A number of other hypertext systems have been developed by various researchers, but the implementation of the concept on computer networks did not become commonly understood and recognized until the distribution of HyperText on the Macintosh computer in the late 1980s.

The most significant implementation of hypertext, which serves as a simple front-end to the Internet in the form of Web pages, is the Hypertext Markup Language. Hypertext tags can be imbedded in Web pages to allow them to connect to any other public page on the Internet. See Hypertext Markup Language.

Hypertext Markup Language HTML. A simple markup language for creating platform-independent hypertext documents. HTML is a generic semantics implementation of Standard Generalized Markup Language (SGML - ISO 8879:1986). HTML is widely used on the Web to represent hypermedia, documentation with inline graphics, database query results, news, and mail.

The formal definition of HTML syntax is described in the HTML Document Type Definition (DTD).

HTML was designed by Tim Berners-Lee at CERN and has been in use since 1990 by the World Wide Web global information initiative. The Document Type Definition (DTD) was written by Dan Connolly in 1992. In 1993 a number of contributors provided enhancements, and the incorporation of NCSA Mosaic software allowed the inclusion of inline graphics. Dave Raggett derived forms material from the HTML+ specification.

In 1994, the HTML Specification was rewritten by Dan Connolly and Karen Olson and edited by the HTML Working Group, with updates by Eric Schieler, Mike Knezovich, and Eric Sink from Spyglass, Inc. Finally, the entire draft was restructured by Roy Fielding. The development and use of Web browsers began to spread.

Since then, the number of users of the Web, interacting through HTML, has climbed to more than 40 million, and many millions have authored personal, institutional, and commercial Web pages using HTML.

HTML has undergone a number of updates and revisions since its initial introduction. See browser, hypertext, RFC 2070 (Internationalization).

Hypertext Transfer Protocol HTTP. An application-level, generic, stateless, object-oriented protocol intended for quick-access, distributed, collaborative, hypermedia systems. HTTP uses typed data representation, allowing system-independent data transfer. HTTP use is widespread, as it has been an intrinsic part of the World Wide Web initiative since 1990, and widely incorporated by Web servers and clients. It also provides a generic means of communication between user agents and proxies/gateways, and Internet protocols for email, search engines, and file servers.

HTTP communications on the Internet are typically over TCP/IP connections, with a default port of TCP 80. A *message* is the basic unit of HTTP communication, which uses a request/response paradigm for serving information. Once a connection is established between the client and the server, the client sends a request, and the server responds with control and error information and, if the request is successful, the requested content.

The syntax of the HTTP URL is as follows:

```
http://<host>:<port>/<path>?<searchpart>
```

See MIME, Secure HTTP.

hysteresis 1. The diminution or retardation of effects upon a body from a force, when the force acting upon the body changes. For example, in a body that is magnetized by a changing magnetizing force (e.g., an electromagnet with a varying current), hysteresis is the amount by which the magnetic values of the body lag (due to friction or viscosity, etc.) behind those of the magnetizing force. 2. The difference in response of a system to a varying force or signal. 3. The difference in the ability of a system or device to respond and change according to a sudden force upon it. To give a simplified example, stomping on a car accelerator or brake does not result in an instantaneous change to a new speed. Hysteresis is the delay effect between the stomping action and the response of the vehicle to the action. Sports car drivers experience less hysteresis than motorhome drivers.

hysteresis curve A diagrammatic representation of a magnetizing force and its related magnetic flux.

In a hysteresis curve for magnetic materials that are subjected to a magnetic influence, then separated from the influence, then magnetized and separated again, it can be seen that materials retain some of their original magnetism after removal of the magnetic influence. This property can be shown to vary among substances by means of a hysteresis curve diagram. Thus, materials with a narrow curve are suitable for the cores of electromagnets in industrial applications those with wide curves can retain their magnetic properties and are used accordingly.

hysteresis device Emergency systems which switch to reserve generators or battery power when voltages drop may fluctuate around the level at which they change from one power system to the other. In this case, a delay mechanism (a hysteresis device) may be deliberately introduced in order to prevent constant fluctuation or *fluttering*, so that the system switches only after a sustained or significant change in power levels.

hystoroscope An instrument for observing, measuring, and recording magnetic characteristics of magnetic materials.

Hz See hertz.

I 1. Symbol for current. 2. A symbol commonly used to designate the 'on' position on a rocker switch, with **O** commonly used for 'off.' 3. Abbreviation for intensity. The I is usually indicated on or near an analog dial on a computer monitor or TV screen, to allow the user to increase or decrease the amount of illumination of the display.

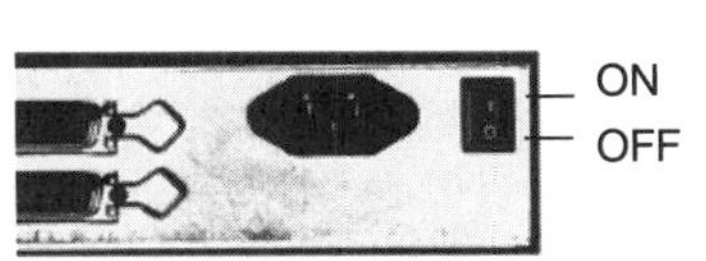

The I and O symbols on the rocker switch on the back of this external hard drive cabinet indicate power on and power off.

I interface, Inter-Service Provider interface An interface between two Cellular Digital Packet Data (CDPD) networks deployed over AMPS. See A interface, E interface.

I & R *abbrev.* installation and repair.

I Series Recommendations A set of ITU-T recommendations which provides guidelines for ISDN. These are available as publications from the ITU-T for purchase and a few may be downloadable without charge from the Net. Some of the related general categories and specific I category recommendations that give a sense of the breadth and scope of the topics are listed here. Since ITU-T specifications and recommendations are widely followed by vendors in the telecommunications industry, those wanting to maximize interoperability with other systems need to be aware of the information disseminated by the ITU-T. See also similar listings under Q, V, and X Series Recommendations that describe other aspects of telecommunications.

ITU-T Recommendations

General Categories

Series C	General telecommunications statistics
Series E	Overall network operation, telephone service, service operation, and human factors
Series F	Telecommunication services other than telephone
Series G	Transmission systems and media, digital systems and networks
Series H	Line transmission of nontelephone signals
Series I	Integrated Services Digital Networks (ISDN)
Series J	Transmission of sound program and television signals
Series P	Telephone transmission quality, telephone installations, local line networks
Series Q	Switching and Signaling
Series V	Data communication over the telephone network
Series X	Data networks and open system communication
Series Z	Programming languages

Vocabulary, Definitions, Descriptions, and General Principles

I.112 1993 Vocabulary of terms for ISDNs

I.113 1997 Vocabulary of terms for broadband aspects of ISDN

I.114 1993 Vocabulary of terms for universal personal telecommunication

I.230 1988 Definition of bearer service categories

I.240 1988 Definition of teleservices

I.250 1988 Definition of supplementary services

I.371.1 1997 Traffic control and congestion control in B-ISDN: conformance definitions for ABT and ABR

I.311 1996 B-ISDN general network aspects

I.313 1997 B-ISDN network requirements

I.324 1991 ISDN network architecture

I.220 1988 Common dynamic description of basic telecommunication services

I.241.8 1995 Teleaction stage one service description

I.210 1993 Principles of telecommunication services supported by an ISDN and the means to describe them

I.310 1993 ISDN - Network functional principles

I.319/Q.1201 1992 Principles of intelligent network architecture

I.330 1988 ISDN numbering and addressing principles

I.334 1988 Principles relating ISDN numbers/ subaddresses to the OSI reference model network layer addresses

I.410 1988 General aspects and principles relating to recommendations on ISDN user-network interfaces

I.510 1993 Definitions and general principles for ISDN interworking

I.601 1988 General maintenance principles of ISDN subscriber access and subscriber installation

I.610 1995 B-ISDN operation and maintenance principles and functions

I.620 1996 Frame relay operation and maintenance principles and functions

I.120 1993 Integrated services digital networks (ISDNs)

I.121 1993 Broadband aspects of ISDN

I.122 1993 Framework for frame mode bearer services

I.130 1988 Method for the characterization of telecommunication services supported by an ISDN and network capabilities of an ISDN

I.140 1993 Attribute technique for the characterization of telecommunication services supported by an ISDN and network capabilities of an ISDN

I.141 1988 ISDN network charging capabilities attributes

I.150 1995 B-ISDN asynchronous transfer mode functional characteristics

I.200 1988 Guidance to the I.200-series of recommendations

I.211 1993 B-ISDN service aspects

I.221 1993 Common specific characteristics of services

I.470 1988 Relationship of terminal functions to ISDN

I.500 1993 General structure of the ISDN interworking recommendations

I.501 1993 Service interworking

I.511 1988 ISDN-to-ISDN layer 1 internetwork interface

I.515 1993 Parameter exchange for ISDN interworking

I.520 1993 General arrangements for network interworking between ISDNs

I.525 1996 Interworking between networks operating at bit rates less than 64 kbps with 64 kbps-based ISND and B-ISDN

I.530 1993 Network interworking between an ISDN and a public switched telephone network (PSTN)

I.555 1993 Frame relaying bearer service interworking

I.570 Public/private ISDN interworking

I.571 Connection of VSAT based private networks to the public ISDN

I.580 1995 General arrangements for interworking between B-ISDN and 64 kbps based ISDN

I.581 1997 General arrangements for B-ISDN interworking

ISDN BRI and PRI

I.420 1988 Basic user-network interface

I.430 1995 Basic user-network interface - Layer 1 specification

I.421 1988 Primary rate user-network interface

I.431 1993 Primary rate user-network interface - Layer 1 specification

Protocols

I.241.8 1995 Teleaction stage one service description

I.320 1993 ISDN protocol reference model

I.321 1993 B-ISDN protocol reference model and its application

ISDN - various

- I.325 1993 Reference configurations for ISDN connection types
- I.327 1993 B-ISDN functional architecture
- I.328/Q.1202 1992 Intelligent Network - Service plane architecture
- I.329/Q.1203 1992 Intelligent Network - Global functional plane architecture
- I.331 1997 The international public telecommunication numbering plan
- I.333 1993 Terminal selection in ISDN
- I.340 1988 ISDN connection types
- I.350 1993 General aspects of quality of service and network performance in digital networks, including ISDNs
- I.351 1997 Relationships among ISDN performance recommendations
- I.352 1993 Network performance objectives for connection processing delays in an ISDN
- I.353 1996 Reference events for defining ISDN and B-ISDN performance parameters
- I.354 1993 Network performance objectives for packet mode communication in an ISDN
- I.355 1995 ISDN 64 kbps connection type availability performance
- I.357 1996 B-ISDN semi-permanent connection availability
- I.364 1995 Support of the broadband connectionless data bearer service by the B-ISDN
- I.370 1991 Congestion management for the ISDN frame relaying bearer service
- I.371 1996 Traffic control and congestion control in B-ISDN
- I.372 1993 Frame relaying bearer service network-to-network interface requirements
- I.373 1993 Network capabilities to support Universal Personal Telecommunication (UPT)
- I.374 1993 Framework recommendation on "Network capabilities to support multimedia services"
- I.376 1995 ISDN network capabilities for the support of the teleaction service
- I.411 1993 ISDN user-network interfaces - references configurations
- I.412 1988 ISDN user-network interfaces - Interface structures and access capabilities
- I.413 1993 B-ISDN user-network interface
- I.414 1997 Overview of recommendations on layer 1 for ISDN and B-ISDN customer accesses

B-ISDN - Physical Layer Specification

- I.432 1993 B-ISDN user-network interface - Physical layer specification
- I.432.1 1996 B-ISDN user-network interface - Physical layer specification: General characteristics
- I.432.2 1996 B-ISDN user-network interface - Physical layer specification: 155 520 kbps and 622 080 kbps operation
- I.432.3 1996 B-ISDN User-Network Interface - Physical layer specification: 1544 kbps and 2048 kbps operation
- I.432.4 1996 B-ISDN User-Network Interface - Physical layer specification: 51 840 kbps operation
- I.432.5 1997 B-ISDN user-network interface - Physical layer specification: 25 600 kbps operation

Multiplexing

- I.460 1988 Multiplexing, rate adaption, and support of existing interfaces
- I.464 1991 Multiplexing, rate adaption, and support of existing interfaces for restricted 64 kbit/s transfer capability

ATM-related

- I.326 1995 Functional architecture of transport networks based on ATM
- I.356 1996 B-ISDN ATM layer cell transfer performance
- I.361 1995 B-ISDN ATM layer specification
- I.363 1993 B-ISDN ATM adaptation layer (AAL) specification
- I.363.1 1996 B-ISDN ATM Adaptation: Type 1 AAL
- I.363.3 1996 B-ISDN ATM Adaptation Layer specification: Type 3/4 AAL
- I.363.5 1996 B-ISDN ATM Adaptation Layer specification: Type 5 AAL
- I.365.1 1993 Frame relaying service specific convergence sublayer (FR-SSCS)
- I.365.2 1995 B-ISDN ATM adaptation layer sublayers: service specific coordination function to provide the connection oriented network service
- I.365.3 1995 B-ISDN ATM adaptation layer sublayers: service specific coordination function to provide the connection-oriented transport service
- I.365.4 1996 B-ISDN ATM adaptation layer sublayers: Service specific convergence sublayer for HDLC applications
- I.731 1996 Types and general characteristics of ATM equipment
- I.732 1996 Functional characteristics of ATM equipment
- I.751 1996 Asynchronous transfer mode management of the network element view

I

I signal In various data transmission schemes, it is common to split a signal and to alter the characteristics of one or both of the data streams so that they can be transmitted together without interfering with one another or creating excessive crosstalk. The signals are then recombined or synchronized at the receiving end.

Streams may also be split according to their different transmissions needs, as in speech, which can be sent on voice grade lines, and graphics, which require better and wider transmissions media.

In various modulation systems, the transmission may be split into two streams, one is the I signal, the other is the Q signal. See quadrature amplitude modulation.

I-beam A common name for a cursor which resembles the capital I, that is, a vertical line with serifs, and sometimes also a small horizontal tick mark in the lower middle to indicate the baseline of the characters below which the descenders will be positioned. This type of cursor is commonly used in painting, word processing, and desktop publishing applications to aid in positioning and alignment of text.

I-frame intra-coded frame. In MPEG animations, a picture which has been encoded into a video frame without reference to past or later frames, using predicted motion compensation algorithms. See B-frame, I-picture.

I-picture intra-coded picture. In MPEG animations, a picture which is to be encoded into a video frame without reference to past or later frames. Once it is encoded, it is considered to be an I-frame.

I-TV See interactive television.

I-way *slang.* An expression for the growing global telecommunications network, derived from a shortening of the phrase "Information Super Highway."

I/O input/output. Generally used in the context of computers as meaning input from users, applications, or processes and output to devices, applications, or processes. See input, input device, output, output device.

I/O bound input/output bound. A processor which is being subjected to a processing load which is in excess of what it was designed to handle, or which causes processes and response time to be uncomfortably slow for the user, is said to be I/O bound. There are number of ways to reduce the incidence of I/O congestion: more efficient algorithms; co-processing chips for computing intensive operations such as graphics, sound, or device management to ease the load on the central processing unit (CPU); faster CPUs; reconfiguration or reorganization of peripheral devices; distributed processing over a network, etc.

I/O device input/output device. A piece of computer hardware which is physically interfaced with a system, and which is electrically and logically configured to engage in two-way communication with the operating system and relevant applications. Many computing devices are input or output only. Joysticks and mice are typically input only; speakers and printers are typically output only. Most monitors are output devices, but touchscreen monitors are used for both input and output. Keyboards are typically input devices, except for those which have small LED displays to send configuration, status, or numeric keypad calculator information to the user. See input device, output device.

IA See intelligent agent.

IAB See Internet Architecture Board.

IAC interapplication communications architecture.

IAHC See Internet International Ad Hoc Committee.

IAM 1. See initial address message. 2. intermediate access memory.

IANA See Internet Assigned Numbers Authority.

IAP See Internet Access Provider.

IAPP See Inter-Access Point Protocol.

IARL See International Amateur Radio League.

IARU See International Amateur Radio Union.

IBM See International Business Machines.

IBM clone See IBM-compatible.

IBM PCjr An Intel 8088-based microcomputer, introduced in the early 1980s. The PCjr was intended as a low-cost home alternative to the IBM Personal Computer XT by International Business Machines (IBM). The list price was $1300 and the computer didn't gain the same acceptance in homes as the XT line gained in business.

IBM Personal Computer AT Advanced Technology. An Intel 80286-based 16-bit microcomputer, introduced in the fall of 1984 by International Business Machines (IBM). The processing speed of the AT was 6 MHz, with 256 kilobytes of memory. It came configured with a 1.2 MByte floppy drive, but the 20 MByte hard disk, graphics adapter, and monitor were optional. A clock/calendar chip was built in.

The base system AT system was listed at $4,000, but the price with an extra 512 kilobytes of RAM, a hard disk, color graphics card, and monochrome monitor brought the total up to $6,700. The hefty price tag was part of the reason competing vendors were able to get a foothold in the industry.

At the same time as the AT computer was released with PC DOS 3.0, IBM announced two additional products: a multitasking, windowing operating system called Topview, which could run existing MS-DOS 1.0 and 2.0 programs; and PC Network, a broadband local area network (LAN), to accompany its Personal Computer line of products.

IBM Personal Computer XT Extended Technology. An Intel 8088-based microcomputer, introduced in 1983 by International Business Machines (IBM). The processing speeds of the various models of XTs ranged from 4.77 to 10 MHz (turbo XTs), with 16-bit data buses. A clock/calendar chip was not standard. Microsoft BASIC was included in ROM, and the computer could use cassettes for program reads and writes. DOS 2.1 was optional, but was needed in order to read and write floppy disk drives.

Although very limited in its graphics and sound capabilities, the XT computer was aimed at and accepted by the business market. Until the introduction of the IBM personal computers, Tandy/Radio Shack had a dominant installed base with almost 80% of the market. It never regained this position in the personal computer market.

As it was being configured and marketed in

IBM Personal Computer XT System Specifications

Category	Item	Notes
Processor:	Intel 8088	4.77 MHz with socket for 8087 math coprocessor
Memory:	40 kilobytes ROM	256 kilobytes RAM up to a maximum of 640 kilobytes
Storage:	10 Mbyte fixed hard disk	one 360-kilobyte double-sided 5 1/4" floppy (additional floppy drives were $425)
Expansion:	5 empty expansion slots	3 slots used by floppy drive, hard disk, and serial adapter
Display:	monochrome/color	monitor optional monochrome ($275) color ($680), color required additional graphics adapter ($250)
Software:	Microsoft BASIC	optional DOS 2.1 ($65)

IBM Personal Computer AT System Specifications

Category	Item	Notes
Processor:	Intel 80286	6 MHz with socket for 8087 math coprocessor
Memory:	40 kilobytes ROM	256 kilobytes RAM up to a maximum of 3 Mbytes
Storage:		one 1.2 Mbyte 5 1/4" floppy
		5 1/4" 360 kilobyte floppy drive optional
		20 Mbyte hard drive optional
Expansion:	5 empty expansion slots	3 slots used by floppy drive, hard disk, and serial adapter
Display:	monochrome/color	monitor and graphics adapter optional
Software:	IBM PC BASIC	PC DOS 3.0, available options included DOS 3.1 and XENIX

1984, a typical 32-pound XT was listed at about $4,000 without monitor, about $5,000 with color monitor, graphics adapter card, and DOS 2.1.

IBM-compatible A de facto marketing term used by various companies to promote a desktop computer incorporating licensed Intel-based International Business Machines (IBM) technology to the extent that most, or virtually all software that was compatible with IBM personal computers would run on the third-party IBM-compatible machines. When IBM originally began licensing their technology to third-party manufacturers, vendors and users made a distinction between *IBM-compatibles* and *IBM clones* to distinguish between machines that were mostly compatible, and those which were virtually compatible or 'identical' in their ability to use various vendors' hardware and software. Since the late 1980s, the distinction between compatibles and clones has disappeared, since there are few machines now which are not completely compatible, and *compatible* has now taken on the connotation that used to be reserved for *clone*. International Business Machines may not be happy with the IBM name being associated with products from other vendors, and likely would prefer that they be called IBM-licensed rather than IBM-compatible, but the term took hold and continues to be widely used.

IBM Token-Ring See Token-Ring.

IBN Institut Belge de Normalisation. A Belgium standards body located in Brussels.

IBS 1. See intelligent battery system. 2. See Intelsat Business Service.

IC 1. integrated circuit. 2. intercom. 3. interexchange carrier. See Inter Exchange Carrier. 4. intermediate cross-connect.

ICAL Internet Community at Large.

ICAPI International Call Control API.

ICB Individual Case Basis. Client requests that are individually handled, as they are not satisfied by standard products or services.

ICCB See Internet Configuration Control Board.

ICCF Industry Carriers Compatibility Forum. An organization which worked on developing an expansion plan for telephony Carrier Identification Codes (CICs) when they began to be scarce in the later 1980s.

ICEA Insulated Cable Engineers Association.

ICI See Interexchange Carrier Interface.

ICM See Integrated Call Management.

ICMP See Internet Control Message Protocol.

ICO Global Communications A planned London-based satellite communications service, spun off from the Inmarsat Project 21. Hughes Electronics has a large interest in the company, and Hughes Telecommunications and Space Company is building the satellites. Other investors include COMSAT Corporation, Beijing Maritime, Singapore Telecom, Deutsche Telecom, and VSNL (India).

The plans include launching 10 satellites plus two spares, into medium Earth orbits (MEO) at 10,000 kilometers using *bent pipe* analog transponders. The satellites will be divided between two orbital planes, inclined 45° relative to the Earth's equator, orbiting once every six hours. Some innovations are planned; the solar wings carry gallium arsenide rather than silicon solar cells and the propulsion system is hydrazine-based. Thermal control is achieved in part with a sun nadir steering system, which orients the panels toward the sun, and the radiating surfaces away from the sun. C- and S-band capabilities will support 4500 simultaneous phone conversations.

Six ICONET satellites are scheduled to come online in the initial stages. They will interface with 12 Earth stations. Start of service is scheduled for the year 2000.

iCOMP Intel Comparative Microprocessor Performance index. A simple means of evaluating and expressing relative microprocessor power, introduced by Intel in 1992. Intel, as a major vendor of microprocessing chips, sought a straightforward way to convey processor information to purchasers. The iCOMP is an index rather than a benchmark in a technical sense, as it narrowly describes instruction execution speed (not clock speed). Benchmarks involve sophisticated and careful evaluation of many performance factors, whereas an index is a basic indicator, in this case, a compilation based on four industry-standard benchmarks, without taking into consideration other aspects of the system architecture, including video display, device addressing, etc.

iCOMP is expressed on a comparative scale, which uses the instruction speed of the 25 MHz 486SX processor as a baseline, assigning it a value of 100, with subsequent processors rated relative to this.

icon 1. Pictorial representation, symbolic image, emblem. 2. In telecommunications documents and applications, a symbolic image, usually small and abbreviated, representing an object, program, state, or task. Visually similar iconic representations are sometimes used to show different aspects or states of the same thing, such as a *ghosted* icon to show something is in use or an iconized version of an application symbol to show something is loaded and available. Icons are used extensively in documentation and graphical user interfaces (GUIs) to represent concepts or contents. Some are specific to an application or platform, but some are common enough to be recognized across a variety of systems, e.g., folder icons to represent directories.

ICONET A satellite communications network being established by ICO Global Communications.

ICONTEC Instituto Colombiano de Normas Técnicas. A Columbian standards body.

ICTA International Computer-Telephony Association.

ID 1. identification, identifier. 2. input device. 3. intermediate device.

IDA 1. integrated data access, integrated digital access. 2. intelligent drive array. See RAID.

IDCMA Independent Data Communications Manufacturers Association.

IDE See integrated development environment.

IDE devices and controllers Integrated Drive Electronics. A control mechanism and format for computer hard disk drive devices developed in 1986 by Compaq and Western Digital. IDE provides data transfer rates of about 1 to 3 Mbytes per second, depending upon other system factors, including the data bus. On common Intel-based microcomputers, the IDE uses an interrupt interface to the operating system.

IDE has been highly competitive with the SCSI standard, another very common drive format. To get the production costs down, and because many Intel-based computers in the early 1980s did not come standard with controllers for extra peripherals, the IDE controller mechanism was incorporated into the drive. Each controller can handle two drives, a 'master' and a 'slave' (compared with seven, including controller, for SCSI).

IDE is more limited than SCSI (fewer devices can be chained, smaller addressable space, IRQs necessary, not compatible with RAID systems, etc.), but it is also less expensive and has become widely established. In order to overcome some of its limitations, a number of 'enhanced' IDE formats now exist.

Most workstations and Motorola-based desktop computers (Suns, SGIs, Amigas, most Apple Macintoshes, NeXTs, and others) include SCSI controllers on the basic machine, making it unnecessary to purchase a separate drive controller to add SCSI peripheral devices to the computer. Some of the newer Macintosh and PowerMac computers support both SCSI and IDE. Most Intel-based desktop computers come with IDE controllers on the basic machine and SCSI controllers can be purchased as options. See SCSI.

IDEA See International Data Encryption Algorithm.

IDEN Integrated Digital Electronic Network.

identifier 1. In database management, a keyword used to locate information, or a category of information. 2. In programming, a variable name, extension, prefix, suffix, or other device to provide a means to easily recognize an element, or distinguish it from others.

IDF intermediate distribution frame. See distribution frame.

IDL See Interface Design Language.

IDLC See Integrated Digital Loop Carrier.

idle In a state of readiness, but not currently activated. Idle is often used as a power-saving measure, and may be a state in which only minimal power is used by the system until full power is needed, as in laptops that power down the monitor and hard drives when they are not in active use.

idle channel code A repeated signal that identifies a channel that is available, but not currently in active use. See idle.

idle channel noise Noise in a communications channel that can be heard, or occurs when no transmissions are active. For example, low level hums can often be heard in phone lines when no one is talking, but are not noticed when talking continues.

idle line cutoff In computer networks, it is not uncommon for Internet Services Providers (ISPs) or network administrators to set the system to log off any clients (machines or applications) which are inactive for longer than a specified period of time (e.g., 15 minutes). This frees up terminals that have been abandoned or frees up modem lines that are no longer in use.

idle signal 1. In networking, a channel which is open and ready, and which may be sending an idle signal, but through which no active or significant transmissions are occurring. 2. Any signal in a circuit intended to signify that no significant transmission is currently in progress. An administrative tool to allow potential users, operators, or operating software to detect available lines and put them into use, or to compile and record usage statistics for further evaluation and tuning of a system. See idle channel code.

IDN Integrated Digital Network.

IDSCP See Initial Defense Communications Satellite Program.

IDTV See Improved Definition Television.

IDU Interface Data Unit. In ATM networking, interface control information transferred to and from the upper layer in one interaction across the layer.

IEC 1. See Inter Exchange Carrier. 2. See International Electrotechnical Commission. 3. International Engineering Consortium.

IEEE Institute of Electrical and Electronic Engineers, Inc. The world's largest electrical, electronics, and computer engineering/computer science technical professional society, founded in 1963 from a merger of the American Institute of Electrical Engineers (AIEE) and the Institute of Radio Engineers (IRE). IEEE serves about a quarter of a million professionals and students in almost 200 countries. IEEE's activities are broad reaching, including standards-setting, publications, conferences, historical preservation and study, conferences, and much more. See American Institute of Electrical Engineers, Institute of Radio Engineers. http://www.ieee.org/

IEEE Canada Institute of Electrical and Electronic Engineers of Canada. The Canadian arm of the well-known IEEE which is organized across the country into groups based on geographic regions. http://www.ieee.ca/

IEEE History Center The historical archive of the IEEE, including about 300 artifacts and a number of oral histories. The IEEE includes among its early members some of the pioneer inventors in the telecommunications field, including Thomas Edison and Nikola Tesla. It works in cooperation with the IEEE library in which the IEEE publications are stored.

IEEE 802.11 Standard for wireless local area networks (LANs) adopted in June 1997.

IEN See Internet Experimental Note.

IETF See Internet Engineering Task Force.

IF intermediate frequency.

IFCM independent flow control message.

IFIP International Federation for Information Processing.

IFRB See International Frequency Registration Board.

IGC intelligent graphics controller.

IGMP See Internet Group Multicast Protocol.

ignition Lighting, kindling, applying a spark so as to inflame or provide sufficient heat or current to set off a chain of events.

IGP See Interior Gateway Protocol.

IGRP See Interior Gateway Routing Protocol.

IGT Ispettorato Generale delle Telcomunicazioni. General Inspectorate of Telecommunications in Italy.

IGY International Geophysical Year.

IIR Interactive Information Response.

IIOP Internet Inter-ORB Protocol. A wire-level communications protocol. See CORBA, Object Request Broker.

IISP 1. Information Infrastructure Standards Panel. 2. Interim Interswitch Signaling Protocol. A basic call routing scheme which does not automatically handle link failures; routing tables established by the network administrator are used instead.

IITC Information Infrastructure Task Force.

IJCAI International Joint Conferences on Artificial Intelligence. An international biennial forum (in odd-numbered years) held since 1969. http://ijcai.org/

ILEC See Incumbent Local Exchange Carrier.

ILLIAC I A historic large-scale computer introduced in 1952 by the University of Illinois. It consisted of vacuum-tube technology and performed 11,000 arithmetical operations per second. See ENIAC, MANIAC.

ILLIAC II The successor to the ILLIAC I, the ILLIAC II was introduced in 1963. It was based on transistor and diode technology and could perform up to 500,000 operations per second.

ILLIAC III The ILLIAC III was introduced in 1966. It was designed to process nonarithmetical data, and so was a departure from ILLIAC II, a special purpose machine.

ILLIAC IV Based on the new semiconductor technology, the ILLIAC IV was introduced in the early 1970s. It was logically designed after the Westinghouse Electric Corporation's

SOLOMON computers developed in the early 1960s. The ILLIAC IV consisted of a battery of 64 processors which could execute from 100 million to 200 million instructions per second. It was significant not only for its speed, but for the ability of its multiple processors to perform simultaneous computations. The services of the ILLIAC IV were made available to other institutions through high-speed phone line timesharing.

ILMI See Interim Link Management Interface.

IM intermodulation distortion. A type of audio distortion that occurs when multiple tones interfere with one another in a way which is not harmonically related to the original tones.

IMA Interactive Multimedia Association. There is info about this organization on the Web. http://www.ima.org/

IMAC See Isochronous Media Access Control.

image antenna A hypothetical antenna, used for mathematical modeling, which is defined as being a mirror-image of an above-ground antenna, located below the ground symmetric to the surface, at the same distance as the actual antenna is above the surface.

imagesetter A professional-level graphics and type imaging machine, an imagesetter is similar to a high quality computer printer. Imagesetters are used in service bureaus and traditional and digital printing houses, to create the image or the color separations which are used to control the ink distribution on the press. Typical resolution on these industrial quality machines is 1200 dpi to 2700 dpi (compared to 300 to 800 for most consumer machines) and they print on paper or film, or both.

While the distinction between consumer printers and imagesetters is blurring, with consumer printers now able to print up to 1200 dpi, there are still technical differences between commercial and consumer machines which are important to design, desktop publishing, and printing professionals.

Imagesetters do more than just print at higher resolutions, they also include more sophisticated and precise algorithms for halftone screens, may include higher quality fonts, may be able to print on special papers and even directly on aluminum, asbestos-based, or other more robust printing plate media. In addition, the distribution of the imaging materials on the printing medium are typically more precise and even. Further, a professional quality imagesetter has better alignment for subsequent printouts.

When printing color separations, especially for four- or five-color process printing, the consistency of the printing from one separation to the next is extremely important to the outcome of the final color printout, especially at resolutions of 175 lines or higher as are used in calendars, posters, and art prints.

In modern digital presses, the trend is to eliminate the separate imagesetter and incorporate the technology into the press itself. In the past, a computer file or traditionally photographed image was taken to a paper or metal plate through an imagesetter and, from there, the physical plate was attached to the press in order to create the printing job. Now it is possible to put a file on a floppy disk or cartridge and have the digital image sent directly to the press without the intermediary steps. It is even possible for a four, five, or six color print job to be printed in one press run, rather than sending each color through the press in a separate pass, and aligning the plates each time. This new technology is revolutionizing the printing industry and eliminating a lot of intermediary steps and jobs in the process.

IMAP See Internet Messaging Access Protocol.

IMASS Intelligent Multiple Access Spectrum Sharing.

IMAX *"I"* - eye + *max*imum. An advanced cinematic system with large film reels to accommodate oversized frames that provide startling detail when displayed on large IMAX projection screens. The system was introduced in 1960 by Canadian inventor W. C. Shaw.

IMHO Abbreviation for "in my humble opinion" commonly used in memos and online correspondence, where facial and tonal expressions are not available to soften the sense of words that may seem harsher than they are intended in print. See emoticons, IMO, IMNSHO.

Immediate Ringing A telephone or private branch system option in which there is no delay between the time of the reception of a call and the ringing of the telephone itself. Favored by those who want to provide quick responses to calls, such as emergency and crisis lines and certain businesses.

Immunity from Suit A legal agreement in which a license holder agrees not to sue the provider of a product or service. Microsoft and certain other large vendors are alleged to be asking for immunity from some of their clients. In the author's opinion, purchasers should avoid signing any licenses that sanction neglect or mismanufacture on the part of the provider, should never sign anything that conflicts with constitutional rights, or which result from coercion. Read your license agreements carefully, *especially the small print*, and question and renegotiate anything that gives you cause for concern.

IMNSHO An abbreviation for "in my not so humble opinion" which is used in memos and online correspondence to emphasize a point somewhat satirically. See emoticons, IMHO.

IMO An abbreviation for "in my opinion" which is used in memos and online correspondence, where facial and tonal expressions are not available to soften the sense of words that may seem harsher in print than they are intended. See emoticons, IMHO.

impact printer Any printer which uses physical movement of a platen, pin, daisywheel, or other printing mechanism to transfer ink or toner to the printing medium. Common impact printers include dot matrix printers, daisywheel printers, and computerized typewriters. Impact printers tend to be noisy, and many offices have purchased sound hoods to fit over their printers to reduce noise levels. They also are usually limited to printing in one color at a time.

Graphics and fonts are limited on impact printers. Daisywheel printers cannot print graphics beyond what might be possible with special daisy heads and a different head is needed to print each typestyle, making it impractical to mix a lot of fonts in a single document. Newer 24-pin dot matrix printers can print a variety of fonts and limited graphics, but usually a lot of effort is needed on the part of the user to exploit these capabilities.

Impact printers were very prevalent in the 1970s and early 1980s, but since then many competing technologies have been developed which are quieter, faster, and more flexible, including laser, ink-jet, thermal wax, and dye sublimation printers. Impact printers are still the most practical solution for documents which require carbon copies, typically invoices, work orders, and other business-related forms. Most impact printers can create up to about 4 to 6 copies before the impression is too faint to be useful. See dot matrix printer.

impact tool See punchdown tool.

IMPATT impact avalanche and transit time.

IMPDU See Initial MAC Protocol Data Unit.

impedance (Symbol - Z) The total opposition, measured in ohms, offered by a circuit to the flow of alternating current (AC) at a given frequency. The ratio, in ohms, of the potential difference across a circuit to the current moving through that circuit. Design and insulating materials can substantially affect the level of impedance in a data cable, with low impedance cables generally costing more, but providing less noise and interference, and sometimes longer transmission distances. See admittance.

impedance bridge A device for measuring in ohms the impedance (combined resistance and reactance) of a portion of a circuit.

impedance compensator An electrical line which effects another circuit in such a way that the combination provides a desired consistent level across a specified frequency range. A compensator is used to minimize fluctuations and distortion.

impedance triangle A diagrammatic model for describing an impedance relationship. Imagine a right triangle with the sides respectively representing resistance and reactance, which change proportional to one another, and the hypotenuse representing impedance, as related to the amount of the resistance and reactance combined.

import 1. Bring in from another source, region, or country. 3. Bring in from a non-native file format, protocol, or transmissions source.

import, file In software applications, to import is to bring in data from another program, file, or transmissions source, usually in a non-native file format. This is usually done through an applications filter or through drag-and-drop capabilities. In drag-and-drop imports, the program will either maintain links to the original imported file or convert the format to one consistent with the program into which it is imported.

import filter Many word processing, desktop publishing, and graphics programs have import 'filters,' plugins, or modules which allow a number of file formats to be brought into an application and then saved in the native format of the application, or exported in the original format, or a new one. This provides better compatibility between programs developed by different vendors. See export; import, file.

import script 1. A script which controls the

assembly of a document by selectively importing information as specified. Often used in spreadsheets and databases. 2. A *very* handy feature of database software in which you can set up a form letter, and then have the script selectively build dozens or hundreds of personalized letters in a few minutes by automatically drawing in names, addresses, and variables from a database to merge with the form letter. Bulk mail companies often use import scripts to personalize letters, contest offerings, and envelopes. 3. In programming, an import script can set up documents or source code by selectively merging modules such as header files, modular routines, Unix "man" pages, etc.

imposition An important stage in printing layout during which the position and order of the elements or pages are planned in order to determine where on the printing sheet they will be located. For example, if you are printing an eight-page 5.5" *x* 8.5" booklet which is going to be saddle-stapled, you will probably be using 8.5" *x* 11" sheets of paper folded in the middle. Thus, the first sheet of paper will be printed with pages one and eight on one side, and pages two and seven on the other side. Setting up the masters so that the pages work out correctly is called imposition. Fortunately, many desktop publishing products now do imposition automatically, assembling the pages in the right order for booklets, or books with specific *signatures* in blocks of 16 or 32 pages. In some types of printing, the positioning of elements is called *stripping* or *stripping in*. See signature.

Improved Definition Television IDTV. A picture broadcast and display system that provides better picture quality than conventional NTSC standards by incorporating field store techniques in the receiving circuitry. For example, the signal can be de-interlaced prior to display to reduce flicker. The originating signal is not changed.

Improved Mobile Telephone Service IMTS. Early mobile phone services were set up on systems based on large antenna transceivers with limited coverage and public operator-assisted broadcast services. The system had little flexibility or privacy, but it served as a forerunner for IMTS, in which the subscriber could place the calls directly, and this in turn developed into current cellular systems where a larger number of smaller, automated transceiver systems allowed broader geographic coverage.

impulse 1. A nonrepetitive pulse so short as to be mathematically insignificant. 2. A very short nonrepetitive pulse which may not seem significant by itself, but which may impede transmission of the affected line or signal. Data transmissions are more sensitive to impulse interference than voice communications. 3. The uncontrolled desire to run out and get the latest techie toy, even though you don't really need it. Cell phones, faster computers, and scanners often fall into this category.

IMSAI 8080 An early 8080A-based microcomputer that used the MITS-developed Altair bus (S-100 bus); it was, in a sense, the first microcomputer *clone*. The 8080A was an enhanced version of the 8008 that was used on the first Altair. The IMSAI was introduced in 1975 by IMSAI Manufacturing with ads that compared it competitively against the Altair. It featured 4 kilobytes of RAM, 22 expansion slots, and, like the Altair, a front panel with LEDs and switches. IMSAI licensed Microsoft BASIC, the same program which Microsoft had first developed to run on the Altair, to the surprise of MITS which had incorrectly assumed it had bought exclusive rights to the Altair BASIC language. See Altair.

IMSI See International Mobile Subscriber Identity.

IMTC The International Multimedia Teleconferencing Consortium. Their Web site includes information. http://www.imtc.org/imtc

IMTS See Improved Mobile Telephone Service.

IMUX See Inverse Multiplexer.

IMW See Intelligent Music Workstation.

in-band A transmissions scheme in which control and data signals are sent together over the same set of wires, or over the same frequencies, sometimes more or less simultaneously and sometimes interspersed with one another.

in-band signaling A type of signaling which is incorporated together with the data that is being transmitted. This is found, for example, in systems which encode signaling codes along with voice transmissions on the same wires (commonly copper twisted pair). In-band signaling has advantages and disadvantages. It doesn't require a separate set of wires to send control signals and thus is less expensive, but it does require more sophisticated handling of data and signals and has a higher potential for slowdown, errors, interference, or fraud.

In-band phone systems are at greater risk for security breaches and unauthorized use of services, because users can send in-band signals over the voice line and control certain telephone functions with illegal control devices such as blue boxes.

Newer out-of-band phone systems, based for example on Signaling System 7 (SS7), make unauthorized use through control signals on the transmissions line impossible, and these types of networks are increasing in prevalence as older equipment is being replaced by newer networks. See ISDN, Signaling System 7.

in-line device A hardware device, commonly a peripheral which can be interposed between two other devices without interfering with the operation of the other devices, or which is intended to interface between two other devices to perform its function (and may or may not change the functioning of the other devices). Daisy-chainable devices are a type of in-line device, though not all in-line devices can be daisy-chained. See daisy chain.

INA Information Networking Architecture.

INC international carrier.

incandescent lamp A common type of illuminating bulb developed by Thomas Edison, originally consisting of a carbonized filament in a glass globe from which the air had been removed. However, the carbon tended to blacken the inside of the bulb and other solutions were sought, with tungsten coming into general use because of its high melting point. Experimentation with the internal environment of the bulb also resulted in the discovery that various gases could alter the glow or extend bulb life.

INCC Internal Network Control Center

incidence angle The angle between the trajectory of an emission or ray of light and the perpendicular to a surface which is in the path of the emission.

incipient failure A failure from degradation of a process or equipment in its early stages.

inclination 1. The angle of a surface or vector in relation to an associated horizontal. 2. A deviation of a surface or vector from horizontal or vertical.

incoherent scattering 1. A behavior of light in some circumstances whereby the phase of the light is random and unpredictable, as in LEDs. 2. A disordered scattering of transmission waves, such as radio, when they encounter a surface and are deflected.

increment 1. *n.* A small change in value. 2. *v.* To add to an existing quantity, as in a software programming loop. Incrementing an integer counter in a procedure is a very common way to keep track of quantities, operations, timing, and events. Although technically a negative value can be incremented, in programming this is usually called *decremented.*

incremental sensitivity A measure of the least amount of change that can be detected by a specific instrument or process.

Incumbent Local Exchange Carrier, Independent Local Exchange Carrier ILEC. Sometimes called dominant carriers, ILECs are comprised of the RBOCs, independent phone companies, GTE, and others. See Competitive Local Exchange Carrier.

Incumbent Local Exchange Carrier duties The Federal Communications Commission (FCC) stipulates a number of duties, in addition to the Local Exchange Carrier duties, in the Telecommunications Act of 1996 as shown in the chart following.

index An organizational tool which provides a key to other types of information, or a larger body of information, which is stored elsewhere. An index is an extremely important aspect of database design, search, and retrieval. It provides a 'hook' or jumping off point, a brief means of indicating the subsequent location of a hierarchy or list. An index in its broadest sense can point to records, further indexes, keywords, locations, sequences, arrays, and much more. An index can be comprised of numbers, symbols, or lexical mneumonics, depending upon the context of the application. Some indexes are seen by the user and set manually; others are transparent to the user and set by the software. Databases, mass storage directory structures, and file hierarchies are typically indexed in one way or another for quick storage and retrieval.

The efficiency of an indexing system depends on the quantity of information that is being indexed, the overall structure of the database, and the types of information that are sought to be stored and retrieved in the system. If a system involves a small amount of data and a complex indexing system, then it is not likely to be efficient. If, on the other hand, there is a large amount of data which can be relatively objectively categorized, an indexed structure is one way to store and utilize the information. See database.

index counter A very common form of feedback that allows a user or technician to monitor usage, or elapsed time or distance. Index

FCC-defined Duties of Incumbent Local Exchange Carriers (LECs)

As defined in the Telecommunications Act of 1996:

"In addition to the duties contained in subsection (b), each incumbent local exchange carrier has the following duties:

'(1) DUTY TO NEGOTIATE- The duty to negotiate in good faith in accordance with section 252 the particular terms and conditions of agreements to fulfill the duties described in paragraphs (1) through (5) of subsection (b) and this subsection. The requesting telecommunications carrier also has the duty to negotiate in good faith the terms and conditions of such agreements.

'(2) INTERCONNECTION- The duty to provide, for the facilities and equipment of any requesting telecommunications carrier, interconnection with the local exchange carrier's network—

'(A) for the transmission and routing of telephone exchange service and exchange access;

'(B) at any technically feasible point within the carrier's network;

'(C) that is at least equal in quality to that provided by the local exchange carrier to itself or to any subsidiary, affiliate, or any other party to which the carrier provides interconnection; and

'(D) on rates, terms, and conditions that are just, reasonable, and nondiscriminatory, in accordance with the terms and conditions of the agreement and the requirements of this section and section 252.

'(3) UNBUNDLED ACCESS- The duty to provide, to any requesting telecommunications carrier for the provision of a telecommunications service, nondiscriminatory access to network elements on an unbundled basis at any technically feasible point on rates, terms, and conditions that are just, reasonable, and nondiscriminatory in accordance with the terms and conditions of the agreement and the requirements of this section and section 252. An incumbent local exchange carrier shall provide such unbundled network elements in a manner that allows requesting carriers to combine such elements in order to provide such telecommunications service.

'(4) RESALE- The duty—

'(A) to offer for resale at wholesale rates any telecommunications service that the carrier provides at retail to subscribers who are not telecommunications carriers; and

'(B) not to prohibit, and not to impose unreasonable or discriminatory conditions or limitations on, the resale of such telecommunications service, except that a State commission may, consistent with regulations prescribed by the Commission under this section, prohibit a reseller that obtains at wholesale rates a telecommunications service that is available at retail only to a category of subscribers from offering such service to a different category of subscribers.

'(5) NOTICE OF CHANGES- The duty to provide reasonable public notice of changes in the information necessary for the transmission and routing of services using that local exchange carrier's facilities or networks, as well as of any other changes that would affect the interoperability of those facilities and networks.

'(6) COLLOCATION- The duty to provide, on rates, terms, and conditions that are just, reasonable, and nondiscriminatory, for physical collocation of equipment necessary for interconnection or access to unbundled network elements at the premises of the local exchange carrier, except that the carrier may provide for virtual collocation if the local exchange carrier demonstrates to the State commission that physical collocation is not practical for technical reasons or because of space limitations."

counters and their electronic counterparts are found on tape drives, VCRs, microwaves, cars (odometers), photocopiers, and almost any appliance in which the location of information or tracking of usage for billing purposes is desired. Counters that give a rough estimate of the number of users who have visited a Web site, or at least the number of accesses to a particular page.

index of refraction The ratio of the speeds of radiating waves or particles in two different materials, as when light passes through a vacuum, air, a mirror, a gem, or a liquid. Index of refraction is of interest to scientists and engineers, of course, but it is also important to computer artists, as it is one of the mathematical values which is entered into ray tracing programs to influence the surface appearance of rendered forms.

indicator light A light which signals a transmission, a fault condition, readiness, or other state that requires attention. Indicator lights are common on appliances, modems, surge suppressors, hard drives, etc.

indirect addressing A common method in computer programming for creating a cross reference to additional related data. Since much of computer data storage cannot be determined in advance, indirect addressing makes it possible to use small segments of memory, or noncontiguous memory, hard drive space, etc. by creating pointers, directories, and other links to the main body of information.

indirect light Light that is not self-emitted, but rather is reflected from another source. For example, the moon does not generate light on its own, but reflects light from the sun.

Indo-European Telegraph Company A company which accomplished the remarkable feat of connecting a wire communications circuit all the way from London to Calcutta in 1884.

indoor antennas Compact antennas which may be used in areas where there are strong broadcast signals. Since broadcast waves are impeded by obstacles and larger antennas tend to pull signals better than smaller ones (there are exceptions based on the frequency range and design of the antenna), indoor antennas are somewhat limited by size and location, yet may improve reception enough to be worthwhile. Television-top rabbit ears and UHF fan dipole antennas are two examples of antennas that can improve indoor reception. See antenna, UHF antennas, VHF antennas.

induced Produced by the influence of an electric current or a magnetic field, usually by proximity.

inductance (Symbol - L) The property of a material which tends to resist change in the flow of electromagnetic current, and thus results in changing lines of force around a changing conductor with alternating current (AC) flowing through it. Thus, the term is used specifically with reference to alternating current, as direct current (DC) does not exhibit the same alternate changes.

An understanding of the properties of inductance was a very important step in the development of induction coils, which could be devised to generate a high-voltage charge, and thus a source of electricity. A basic inductor can be created by winding a conducting wire, such as copper, into a coil. Inductance is typically expressed in henrys. See induction, induction coil, resistance.

inductance bridge A diagnostic circuit configuration instrument which allows comparison of an unknown with a known inductance, similar to the concept used in a Wheatstone bridge. See Wheatstone bridge.

induction Creation of an electric charge or magnetic field in a material by the influence of a proximate electric current or magnetic field. See inductance, induction coil.

induction coil A historic electrical device that played an important role in early electronics inventions. It was a significant provider of high voltage current for many decades, and led to the creation of transformers for converting between alternating and direct current. It also led to various induction-based frequency converters.

A basic induction coil was created in 1836 and described the next year in *The Annals of Electricity* by Nicholas J. Callan. It consisted of a horseshoe-shaped bar of iron, wound with many feet of thick copper wire, and hundreds of feet of thin iron wire. By interrupting the primary circuit with a contact breaker, Callan could induce a charge sufficient to power an arc light. A year after Callan published his findings, an American, Charles Grafton Page, created an induction coil.

induction field In a transmitting antenna, a region associated with the antenna in which changing electromagnetic lines of force are active as current flows through the device.

induction frequency converter A mechanically powered induction device which is connected to a source of fixed frequency current, and which utilizes secondary circuits to deliver a frequency proportionate in speed to the magnetic field. In its most general sense, frequency conversion has become a very important part of communications technology. The conversion of frequencies allows signals to be carried over a variety of media with different transmission characteristics, and further allows signals to be shifted so incoming and outgoing signals are less likely to interfere with one another.

inductive connection, inductive pickup A connection between two devices or objects which does not involve direct electrical contact. The electromagnetic communication between the devices occurs due to electromagnetic 'influence' through proximity to the changing electromagnetic lines of force. Some types of circuit diagnostic tools use inductance to monitor or observe circuits without physically contacting the line. A number of surveillance devices also use this method, in order to avoid detection.

There are regulations to protect privacy governing the unauthorized monitoring of communications through inductive surveillance devices. Fiber optic transmissions are immune to inductive pickup as the optical transmission does not have the same characteristic as electricity of extending beyond the medium through which it is traveling. See wiretap.

inductive coupling The transfer of energy between two circuits which are close together, but not directly electrically connected. Thus, the interaction of the electromagnetic lines of force associated with the interaction of the circuits causes the transfer. The transfer may also occur due to self-inductance of each of the circuits (direct coupling). The transfer of energy may be desirable or undesirable. Unshielded or minimally shielded conducting wires that are too close together may create unwanted noise and interference.

inductive post A conducting bolt, screw, or post associated with a waveguide which provides inductive susceptance to allow tuning of the waveguide. It is usually mounted across the waveguide, parallel to the E field. See E field, waveguide.

inductive tuning In electronic devices such as radio tuners, a means of adjusting the tuning by moving a core in and out of a coil within which it is contained. The core is not in direct contact with the coil, but reacts to the changes in the electromagnetic field associated with the coil by inductance.

Industrial Scientific Medical ISM. A set of electromagnetic frequencies set aside in the 902 to 928 MHz and the 2.4 to 2.484 GHz ranges, which do not require licensing by the Federal Communications Commission (FCC). Spread spectrum technology for local area wireless networks (LAWNs) sometimes uses these ranges for data communications.

Industry Canada A Canadian federal agency responsible for the protection of intellectual property and the allocation of licenses for use of radio frequencies. Formerly the Department of Communications. See Canadian Radio Television and Telecommunications Commission.

Industry Circuit Topography Act ICTA. A Canadian Act intended to protect integrated circuit topographies. See Semiconductor Chip Protection Act.

Industry Standard Architecture ISA. Formerly, a very common input/output bus architecture on International Business Machines and licensed third party computers developed originally on the IBM XT models, and carried through to later models. Originally it was an 8-bit architecture, but was upgraded to 16-bit. The expansion slots inside a computer have to follow a standard format so various manufacturers can create compatible peripheral cards. ISA was one of the common types of slots found in personal computers until the mid-1990s when it was superseded by *Peripheral Component Interconnect* (PCI), *Video Electronic Standard Association* (VESA), *Extended Industry Standard Architecture* (EISA) and others. See Extended Industry Standard Architecture, Peripheral Component Interconnect.

INETPhone A data telephone service which is connected and handled through the Internet, thus substituting the Internet for the long-distance segment of a phone call in a way that is transparent to the users. See RFC 1789.

Infobahn *colloq.* The Information (Super) Highway, based on the German word *bahn*. The Information Superhighway is also colloquially called I-way.

information content provider This is defined in the Telecommunications Act of 1996, and published by the Federal Communications Commission (FCC), as

"... any person or entity that is responsible, in whole or in part, for the creation or development of information provided through the Internet or any other interactive computer service."

Information Network Architecture INA. In the mid-1980s, Bell Communications Research began building its Intelligent Network (IN) to provide a broader range of telephone services and support for data transmission over traditional phone lines. From this grew Advanced Intelligent Networks (AIN), and then Information Network Architecture (INA) with its improved broadband support. There is some discussion as to whether INA will succeed or coexist with AIN, as AIN will meet the needs of many users for some time, considering the lag that exists between the time a new technology is introduced and when it is generally adopted by consumers.

information service This is defined in the Telecommunications Act of 1996 and published by the Federal Communications Commission (FCC), as meaning

"... the offering of a capability for generating, acquiring, storing, transforming, processing, retrieving, utilizing, or making available information via telecommunications, and includes electronic publishing, but does not include any use of any such capability for the management, control, or operation of a telecommunications system or the management of a telecommunications service."

See Federal Communications Commission, Telecommunications Act of 1996.

Information Superhighway A catchphrase promoted by U.S. government representatives, particularly Al Gore of the Clinton administration, and the press, for the domestic and global communications infrastructure. See National Information Infrastructure.

Information Technology Research Centre An Ontario-based Canadian research center.

information theory The pioneer studies in queuing theory, which is related to information theory, were carried out by A. K. Erlang, a Danish engineer. Information theory is a field of inquiry and mathematical modeling which was developed largely through the work of Claude E. Shannon, while he was working at Bell Laboratories in 1948.

Shannon took a theoretical, mathematical look at information, in terms not only of its content and structure, but also its source and purpose. Thus, signals and their frequencies, bandwidths, physical components, and electromagnetic characteristics were set in the broader framework of the information and its human source. This became the basis not only of a broader view of communications, but provided groundwork for more specific measures and descriptors of content and capacity which have real world usefulness. Information theory has aided us in developing more objective system evaluation tools, compression techniques, and practical applications such as voice over IP systems. See erlang, queuing theory.

infrared Electromagnetic radiation with longer wavelengths, between the red part of the visible spectrum and radio waves. Although it cannot be seen by humans, infrared radiation is of commercial importance in remote controls, video game consoles, and fiber optic transmissions.

Infrared serial data links standards are being adapted by a number of manufacturers. Infrared technology can be used to detect differences in heat and, consequently, also movement of bodies emitting heat. Infrared detectors are used in many industries including electronics, construction (structural fault detection, heating, and insulation testing), and medical imaging. Infrared film is used in specialized photographic applications. See Infrared Serial Data Link, snooperscope, ultraviolet.

Infrared Communication Systems Study Committee ICSC. A research committee of the Association of Radio Industries and Businesses (ARIB), organized to study and pro-

IrDA Network Protocol Layers

Layer/Proto.	Notes
IrLMP	A mandatory link management protocol which manages resources and services and higher-level protocols which are made available to other devices. IrLMP sets up and maintains multiple connections.
IrLAP layer	Link establishment, maintenance, and termination. Similar to the half-duplex link control (HDLC) protocol.
physical layer	Provides point-to-point connections and communications between devices with cordless/wireless serial infrared half-duplex links.

mote awareness and use of infrared communications systems. Centered in Tokyo, Japan.

Infrared Data Association IrDA. An organization established in 1993 to support and promote software and hardware standards for cordless/wireless infrared communications links. IrDA is headquartered in California. Just as infrared can be used with remote controls to control various consumer electronics devices, infrared can also be used for data transmission between devices such as laptops, desktop computers, and peripherals. See Infrared Data Association Protocol.

Infrared Data Association Protocol IrDA Protocol. A multilayered networking structure from IrDA for defining hardware and software needs for infrared network communications. The IrDA protocol stack covers physical transfer of information, guidelines for link access, and link management. The layers are briefly described in the IrDA Network Protocol Layers chart.

Infrared Link Access Protocol IrLAP. A serial link access protocol from IrDA which provides three types of connectionless services and six types of connection-oriented services with four types of service primitives. IrLAP provides discovery, address conflict, and unit data services over connectionless services and connect, sniffing, data, status, reset, and disconnect services or connection-oriented services. IrLAP is primary-secondary or primary-multiple station oriented.

The IrLAP layer is intended to facilitate interconnection of computers and peripherals over a directed half-duplex medium provided through the physical layer.

IrLAP stations can be operated in Normal Response Mode (NRM) or Normal Disconnect Mode (NDM) which correspond to connection state and contention state. IrLAP data and control are frame-oriented, with a frame including an address, a control field for determining frame content, and an optional information field.

infrastructure The structural underpinning or base which supports the other layers associated with a system.

ingress 1. Entrance, way in, opening, doorway. 2. In frame relay, frames which are entering toward the frame relay from an access device. The opposite of egress.

initial address message IAM. In Signaling System 7 (SS7) networks, a signaling message sent in the forward direction which initiates seizure of a circuit, and which provides address and routing information for the connection of the requested call. See Signaling System 7.

Initial Defense Communications Satellite Program IDCSP. A project of the U.S. military which first launched three satellites in 1967. They included X-band transponders in the 26 MHz bandwidth, and supported experimental terminals for evaluating images, voices, digital data, and teletype channels using a variety of modulation schemes. The IDCSP were designed to shut down after five years of useful life.

Initial MAC Protocol Data Unit IMPDU. In packet-switched networking, the IMPDU encodes MAC Service Data Unit information. See Media Access Control.

I

Initial Public Offering IPO. A Securities Commission government-regulated mechanism for a company to offer a variety of types of shares (usually common and preferred stock) to the general public. There are a number of categories of public offerings, both state and federal, with levels of restrictions and guidelines depending upon the amount of investment sought. Telecommunications and biotech are two of the 'hot' areas of recent years, and some high-profile stock offerings have been carried out in the technologies industry, one of the most visible being Netscape Communications, developers of Web browsers/servers and other applications.

The investment for a company to 'go public' can be considerable. Being listed on one of the large trading boards can cost millions of dollars. There are many companies which offer public shares which are not listed on the 'big boards,' many as 'pink sheets' known to brokers, but not widely known to the public. This is because there are very strict regulations against 'advertising' stocks. It is permissible to provide information to investors who request it, but not to publicly promote company shares. These regulations are somewhat less stringent at the local level, with small offerings of a $million or half $million state-regulated Direct Public Offerings (DPO) providing some means for offering investor relations information to the public.

DPOs were not widely known or attempted prior to the development of the Internet, because small companies making small stock offerings simply did not have the resources to provide investor information to potential investors and brokers in sufficient quantity to attract investment. Now that the Securities Commission has opened a few doors, to stimulate investment in small companies, which are extremely important to the national economy, trading boards on the Internet are getting a lot of interest from both investors and brokers, and DPOs are becoming feasible for raising capital for seed and startup companies.

ink-jet printing An inexpensive color printing process in which inks from a series of ink 'wells' are fired through a tiny opening called a nozzle. The firing is accomplished through heating the ink chambers to a high temperature so a vapor bubble is formed, which rapidly ejects the ink through the end of the nozzle onto the printing medium, where it cools and adheres. See dye sublimation printing, thermal wax printing.

Inmarsat *In*ternational *Mar*itime *Sat*ellite Organization. Originally an international cooperative agency established in 1979, Inmarsat was then slated for privatization for 1 January 1999. It launched in 1992 and has provided global mobile satellite communications services (voice, data, facsimile), especially maritime services, since 1993. Inmarsat now serves over 80 member countries.

Inmarsat Service Categories

Service	Notes
Inmarsat-A	Analog voice, data, and facsimile services.
Inmarsat-B	Digital voice, data, and facsimile services.
Inmarsat-C	Store-and-forward data, aeronautical voice, and facsimile services. Marine access to email and telex networks, two-way messaging through 5 kilogram terminals.
Inmarsat-M	Briefcase phone.
Inmarsat-Aero	Airline passenger communications supporting voice, facsimile services, with X.25 data services planned.

Inmarsat has a system of four geostationary satellites orbiting at 35,786 kilometers using frequency division multiple access (FDMA). It provides transportation communications and Internet connect services. Five more are scheduled to be launched by the end of the century. Twelve medium Earth orbit (MEO) satellites are also planned.

Customers purchase services from a variety of packages depending upon whether they need phone, facsimile, Internet, emergency services, telemedicine, etc.

The ICONET satellite system is a spin-off of Inmarsat communications services, originally known as Project 21. See ICO Global Communications. http://www.inmarsat.org/

INN 1. See InterNet News. 2. InterNode Network.

INP See Interim Number Portability.

InPerson A consumer-priced SGI-based videoconferencing system supporting video, audio, whiteboarding, and file transfers over analog phone lines and Ethernet networks. Video encoding is accomplished through HDCC compression developed in-house at Silicon Graphics with several audio compression formats.

input Information, in the form of a communication or signal, provided to a person, system, or circuit. Computer software input mechanisms include graphical user interfaces, shell windows, buttons, icons, dialog boxes, etc. Computer hardware input mechanisms include keyboards, mice, trackballs, touchscreens, joysticks, video cameras, and microphones. The input device on a telephone is relatively simple, a small speakerphone or diaphragm (microphone) in the telephone handset.

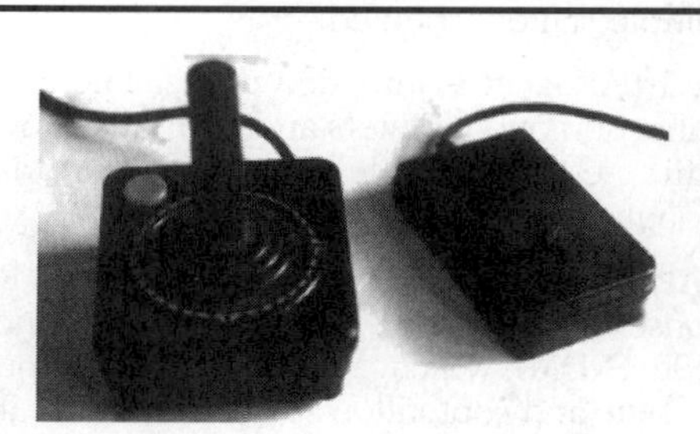

The joystick on the left and the mouse on the right are two common types of input devices for computer games, virtual reality, and day-to-day computing for business, education, and pleasure. Other types of devices include data gloves, pen styluses, microphones, touchscreens, video cameras, and tactile pads.

input device An interfaced device for receiving and transmitting information from an input source (usually human) to a processing system, usually a computing machine or mechanical device. There are a great variety of input devices including keyboards, mice, joysticks, light pens, touch screens, microphones (esp. with speech recognition systems), etc.

Invention of the mouse is attributed to Doug Engelbart in the 1960s. Many of the input devices in common use today were pioneered by Ivan Sutherland in the early 1960s. See individual input devices.

inquiry Systematic seeking of information through queries and observation. A more general term than query, an inquiry may include a series of queries and observations leading hopefully to the desired information. See query.

inside link See horizontal link.

installed base An industry phrase which describes the quantity of products or systems in use extant, or the number of current users of a particular service. The installed base may not indicate the number of products sold. For example, due to software piracy, the number of copies of a software application which have been sold may be 20,000, whereas the installed base may be 220,000. In the other direction, there may be one million disposable widgets sold, but an installed base of only 300,000 if some are discarded, lost, broken, etc. Formulas for establishing installed base statistics are industry specific and somewhat subjective, but still provide useful information for marketing, production, and repair managers.

instant on A consumer electronics term that refers to devices and appliances which come on quickly, usually at the touch of a button, and are immediately ready to use without inspection, training, or configuration. Some VCRs have an *instant on* record mode where popping in the tape makes the component power up automatically, allowing the user to quickly press record.

Many video game consoles are instant on, ready to play as soon as the power is turned on. A few vendors have tried to sell computers with preloaded software with the *instant on* concept, but this is more difficult to achieve. Set top boxes for Internet access are a middle ground between a full computer system and an instant on system, as are dedicated word processing systems.

Turnkey software and hardware products are sometimes designed with an instant on orientation. For example, the Video Toaster by NewTek would boot up directly into the video switcher software, and bypass the operating system startup messages and CLI screen so effectively that many purchasers didn't realize they were using an Amiga computer launched into applications mode. Similarly, library electronic catalog access programs typically boot directly into the catalog user interface, and the user may not even be able to tell which operating system is behind the software.

Institute of Radio Engineers IRE. The IRE was a historic professional organization formed as a result of the merger of the Society of Wireless Telegraph Engineers (SWTE) and The Wireless Institute in 1912, in order to establish and promote an international orientation for the consolidated organization. It served as a standards body, in cooperation with the U.S. federal government, and a professional support group for its members and the radio community at large. See American Institute of Electrical Engineers, IEEE.

Institute for Telecommunication Sciences ITS. The applied research division of the U.S. National Telecommunications and Information Administration (NTIA). The ITS develops, tests, evaluates, and promotes advanced communications networks and domestic standards through its Boulder, Colorado facility.

insulated wire Conductive wire which has been coated, sealed, rubberized, clad, sheathed, or otherwise covered or processed to insulate it from electrical leakage and external interference.

insulation A nonconductive material designed for shielding conductive materials from heat or electrical interference. Insulators are composed of atoms which do not readily give up their electrons and current has difficulty flowing through them. Examples include rubber, glass, and porcelain. Insulating materials sometimes secondarily provide protection from physical damage and spacing or identification marks or colors. Insulation is commonly used to prevent the conduction of heat and electrical currents, and to protect from outside influence. See dielectric.

insulator See insulation.

Insulators on utility poles are a common sight. They used to be constructed of glass, the early ones handmade, but now ceramic insulators are used and glass insulators are collectibles.

insulator, utility pole The fact that glass would make a good insulator for telegraph lines was suggested by E. Cornell, who assisted Samuel Morse in installing the historic 1843 Washington, D.C. to Baltimore line. He originally proposed glass plates and later described a more knob-like design, a larger version of which eventually became standard and widely used on utility poles until the 1970s.

Utility pole glass insulators are thick, threaded, mug- or thermos-sized objects, in a variety of colors, most often blue or green. A number of hand-blown insulators were created in the late 1980s. The oldest commercial mass-produced ones, originating some time in the early 1850s, lacked threads but were colored. Molding processes for creating insulators were patented in the 1870s. The Oakman *beehive* insulator was favored by Western Union for telegraph poles.

Western Union used many thousands of Brookfield and Hemingray insulators over the years. The move to standardize insulators occurred around 1910. Clear insulators were not produced until the 1930s. Ceramic insulators were introduced around 1908 by Locke Insulator, in order to undercut the cost of glass insulators.

Sampling of Insulator Types

Type	Notes
Barclay	Patented by John C. Barclay.
Beaver Falls	A lesser known glass insulator fabricator.
beehive	An early insulator patented by Samuel Oakman in 1884.
bird feeder	Battery rest insulator.
bridle wire	An early AT&T insulator.
Brookfield	A line of various insulators from the Brookfield Glass Company.
double petticoat	Named for the double-tiered skirting.
Hemingray	A line of various insulators from Hemingray.
National corkscrew	Less common type with a characteristic shape.
Storrer	Patented by Storrer in 1906, shipped by Brookfield 1909.

Insulators were developed in many shapes and sizes, in a rainbow of gem-like hues. They provide a legacy of poetically descriptive category names such as slashtops, bat ears, eggs, beehives, and teapots.

Well-known glass insulator manufacturers, like Hemingray, shut down by the mid-1960s.

Historic glass and ceramic insulators are found occasionally in second-hand stores and antique auctions, and older or more interesting ones are favored by collectors and sometimes sell for hundreds of dollars.

Insulator styles are extensive and beyond the scope of this reference, but a few historic insulator types and makes are listed in the Sampling of Insulator Types chart.

INTEGRAL International Gamma Ray Astrophysics Laboratory.

integrated circuit A single electronic component which incorporates what would normally require many traditional electrical circuits. This enables complex, sophisticated capabilities to be incorporated into tiny packages. A computer chip is one particular type of integrated circuit. Credit for the introduction of ICs in 1959 is attributed jointly to Robert N. Noyce, a Dane who joined the Intel Corporation in America, and Gordon Moore. See semiconductor.

Integrated Digital Loop Carrier IDLC. A system designed to integrate Digital Loop Carrier (DLC) systems with existing digital switches as in a SONET network system. A basic installation consists of intelligent remote digital terminals (RDTs) and digital switch elements known as integrated digital terminals (IDTs), interconnected by a digital line. See Digital Loop Carrier.

integrated injection logic IIL. A form of bipolar logic, reduced power circuit intended to provide greater efficiency over TTL chips.

Integrate IS-IS A proprietary routing protocol using one set of routing updates, developed by Digital Equipment Incorporated (DEC). DEC's version is based on the Open Systems Interconnection (OSI) routing protocol called IS-IS. The DEC implementation provides support for a number of other open and proprietary protocols by encapsulating them into Internet Protocol (IP).

integrated messaging, unified messaging A term to describe the combination and consolidation of messaging services such as voice, video, facsimile, email, etc. through a networked computer system.

With a computer phone set, a scanner, and a printer attached to a microcomputer, it is possible to have all the capabilities of these various technologies integrated into one system.

In fact, setting up the system this way provides *more* capabilities than these services have individually, since the computer software can be configured to monitor the calls, store accounting information, transfer data among the various systems, and use files directly, as in directly faxing a document you may be viewing in your word processor, without having to print it and send it through a dedicated facsimile machine. When a facsimile is received, it can be processed to turn it into text and images, or document and PostScript-format files can be sent directly, without any scanning or translation.

By attaching an Internet phone set to the computer, you can have the computer check to see what time it is at the desired destination, dial the call automatically from a database of names, connect the call, alert you when it is connected, keep track of how long you are connected, alert you if you have to attend to other business while making the call, and log the call, if desired, for future reference or statistical or business tracking.

By using an integrated voice, file, email service, you can speak into the headset or a microphone and record your voice in a mail message, send it the same as normal email, which means the recipient can access it whenever he or she is online, and listen to it played on the destination computer as a sound file. This message can easily be combined with text files with binary files as attachments. The NeXTStep operating system has had this flexible type of voice/email/file capability built into its email system since the late 1980s, and Smalltalk object-oriented systems had it even sooner, so it is by no means a new concept. Unfortunately, it is not yet implemented on many commonly used platforms.

integrated model A network traffic routing solution supporting an exchange of routing information between ATM routing and higher level routing. This provides timely external

Overview of Some Common Intel Desktop Computer Central Processing Units

Processor	Data/Int. Bus	Data/Ext. Bus	Address Bus	Clock Speed	Year Introd.	Notes
4004	4/8	4	12	1 MHz	1971	Separate program and data memory. 46 instructions.
4040					1972	Enhanced 4004 with 14 additional instructions, and more space for programming and stack.
8008	8	8	14	2 MHz	1972	Similar to 4040.
8080	8	8	16	2 MHz	1974	Seven 8-bit registers, some of which could be combined into 16-bit register pairs. 256 I/O ports.
8085					1976	An update to the 8080.
8086	8	8	20	5 MHz	1978	Based on the 8080 and 8085. 8-bit 64K I/O.
80286	16	16	24	8 MHz	1982	
80386DX	32	32	32	16 MHz	1985	
80386SX	32	16	24	16 MHz	1988	
80486DX	32	32	32	25 MHz	1989	On-board cache, pipelines, integrated floating point unit.
80486SX	32	32	32	20 MHz	1991	
Pentium	32	64	32	66 MHz	1993	Separate caches. Superscalar.
Pentium Pro					1995	P6, successor to Pentium which converts Pentium instructions.

routing information within the ATM routing and provides transit of external routing information through the ATM routing between external routing domains.

Integrated Services Digital Network See ISDN, Signaling System 7.

Intel One of the best known of the chip manufacturers serving the desktop computer market, rivalled mainly by Motorola. Intel's chips are widely installed in microcomputers worldwide. The Intel 4-bit 108 kilohertz 4004 microprocessor became an important historical impetus in the design of desktop computers, with its successor, the 8008, becoming the world's first commercially significant programmable central processing unit (CPU). The 4004 was developed by Marcian (Ted) Hoff and introduced in November 1971. Three other chips accompanied the 4004, offered as the MCS-4 chip family. The Scelbi computer, first promoted in 1974, and the Altair, which came out as a kit a few months later, incorporated the successor to the MCS-4 family, the MCS-8, based around the 200 kilohertz 8008 (the 8008 was an enhanced version of the 4040) 8-bit microprocessor.

The 4004 was incorporated into many automated systems, including light controls, appliances, calculators, musical instruments, etc.

Gary Kildall developed a programming language for the early Intel processors called PL/M. The 8080 was incorporated into the Altair 8800, as it was in some of the S-100 bus (Altair bus) computers that became competitive with the historic Altair. Since then, the most significant evolution in Intel desktop computer chips is the Pentium series, introduced in the early 1990s.

The Intel Overview table on the preceding page is not comprehensive, but it provides an encapsulated look at some of the highlights in Intel chip development for microcomputer CPUs since the mid-1970s.

Intel Video Interactive IVI. Intel purchased the Digital Video Interactive (DVI) chipset technology and developed it into Indeo 2 and Indeo 3, now known as IVI.

IVI has a number of interesting features, including transparency (e.g., for background overlays), scaling, and the use of an interframe codec for compression, based on relatively new wavelet compression, encoding the images into frequency bands so the image data can be represented at different resolution levels. Data can be password-embedded to protect data. Key frames can be incorporated as reference points for random access. Brightness and contrast settings can be adjusted to adapt to the characteristics of the playback system.

intelligent agent A software application which has been 'trained' to handle tasks dynamically, or which has been trained to recognize certain characteristics of the input, which might be a person's voice, handwriting, or other specialized type of input that may vary from user to user. An intelligent email agent may be configured to screen out 'spam,' unsolicited commercial messages, to sort messages into folders according to sender or priority, or to forward messages to another address if the user is traveling or reading mail at another location.

The difference between a custom agent and an intelligent agent is that the custom agent is explicitly configured by the user, whereas the intelligent agent configures itself on the basis of monitoring the user's habits and interaction history. The agent then establishes actions and parameters based on intelligent analysis of the user's actions and preferences. In other words, a custom agent would require that the user explicitly instruct the email client to put all messages with "Make Money Fast" in the subject line into a 'spam bucket,' a file that contains unsolicited email. An intelligent agent would notice that 15 messages in a row with "Make Money Fast" in the subject line were moved to the other file area, and would subsequently do the transfer automatically on behalf of the user, perhaps prompting the first time it makes this 'decision' in order to confirm that it is carrying out user preferences. See artificial intelligence, expert system.

Intelligent I/O An open standard designed to provide a device-independent device driver architecture. Applied to redundant array of inexpensive disks (RAID) systems, Intelligent I/O can provide faster hard drive access.

Intelligent Music Workstation IMW. A five-year long project which resulted in the 1994 release of a musical software/hardware environment in which commercial products can be integrated as modules. Developed at the Laboratory for Musical Informatics of the Department of Information Sciences of the University of Milan, Italy, funded by the Italian National Research Council.

intelligent transportation systems ITS. Transportation systems which incorporate new computer technologies, such as Global Positioning System (GPS), to improve efficiency. See Intelligent Vehicle Highway Systems.

intelligent vehicle highway systems IVHS. Advanced navigational systems which incorporate computer technologies such as Global Positioning System (GPS) and navigational databases. IVHS vehicles include sensors and compasses to interface with the computer control mechanisms and incorporate dead reckoning, maps, and GPS data to control direction and sometimes velocity. IVH systems can be configured for optimum efficiency and safety and could apply extremely well to specially designed mass transit 'pods' or automated commuter systems. Even regular traffic could benefit from IVH systems. See guidance system.

intelligibility In communications, the degree to which a message can be understood by sound and context. While articulation refers to the specific ability to make out a communication, intelligibility is the ability to make out sentences and phrases based not only on articulation, but on context and inference. Thus, a poorly articulated transmission might still be decipherable in context, especially when enough information is given to figure out the nature of the communication. Intelligibility does not require perfect articulation or good fidelity. If a listener hears "Rog...ov....out" at the end of a CB radio conversation with a lot of noise on the line, it is still intelligible as "Roger, over and out" to an experienced radio operator. See articulation, fidelity.

INTELSAT *In*ternational *Tel*ecommunications *Sat*ellites. The largest commercial not-for-profit satellite communications services provider, founded in 1964. INTELSAT is a cooperative of more than 140 member nations and has 20 communications satellites in geostationary orbit, with further launches planned in 1998. INTELSAT operates as a wholesaler, with subscribers, many of them major broadcasting and telephone companies, paying for services according to their type and duration.

INTELSAT lays claim to having launched the world's first commercial communications satellite in 1965 and the first global communications system in 1969. In 1980, they launched INTELSAT V, the first to use dual-polarization transmissions equipment. In 1995, INTELSAT began providing global Internet access services through its satellite system. See Early Bird. http://www.intelsat.int/

Intelsat Business Service IBS. A commercial telecommunications service based on the INTELSAT satellite communications capabilities. IBS provides almost 10,000 communications channels for a wide variety of services, including voice, facsimile, data, video conferencing, and telex.

Inter Exchange Carrier IEC, IXC. A telephone carrier which is permitted to provide long-distance services between Local Access and Transport Areas (LATAs), but not within a LATA region.

inter- Prefix for between, usually between external and internal systems.

Inter-Access Point Protocol IAPP. A specification developed by Lucent Technologies, Aironet Wireless Communications, and Digital Ocean, IAPP is a means for different vendors to communicate with one another through roaming wireless mobile communications. IAPP describes a backbone-based handover process for mobile stations when implemented in conjunction with the IEEE 802.11 standard.

I

interactive 1. Reciprocal communication, that is, with a back-and-forth, or query-and-answer character. 2. Software which responds to the individual's input, usually in realtime or near realtime, as in multimedia applications. Video games are highly interactive, whereas archive searches over the Internet may be extremely slow (sophisticated searches can take days). Depending upon the circumstances, programs with slow interactivity may be better processed as batch files. Contrast with batch processing.

interactive television A technology in which TV broadcasting becomes a two-way 'dialog' between the user and the broadcaster, enabled by computerization and two-way transmission circuits. Interactive TV has been implemented in a number of ways since the late 1970s, from educational programming to interactive music concerts and on-demand video, but the potential of this technology has only been hinted at so far.

One of the earliest interactive TV networks was the QUBE system from Warner Communications, which was first tested in Columbus, Ohio. Time Warner has been involved in subsequent versions of this technology. Depending upon how it is implemented, interactive TV has been more warmly received by educators than traditional passive-interactive TV. See QUBE.

Interactive Television Association See Association for Interactive Media.

interactive video services IVS. Interactive video in its broadest sense, is public or private image and sound broadcasting through public or private networks that is available upon request by the user. Due to the convergence of broadcast and computer technologies, it is now feasible to provide partial- and full-service interactive video services through a number of transmissions media: twisted copper pair, coaxial cable, fiber optic cable, and wireless. However, with the exception of fiber optic cable, the use of existing technologies, which were designed for other services, means that none of them are ideally configured for IVS, and vendors are hurrying to find ways to deploy services ahead of their competitors. Thus, a variety of technologies are emerging, in spite of the fact that the marketability of these services is not yet fully proven.

Interactive video services potentially include games, movies, and specialized channeling, such as stock quotations and industry-specific news. Some of these have been tried with varying success in different industries and regions, and some companies are devising ways to offer them over the Internet.

Intercarrier Interface ICI. One of the two interface ports of XA-SMDS systems which is used to specify how the carrier switch sends and receives data from an Interexchange Carrier's (IXC's) SMDS network. The other interface is the Subscriber Network Interface (SNI). See Exchange Access SMDS.

Intercast An Intel term for technology which allows a consumer to interface the TV set with a computer hooked up to the Internet, to receive 'push technology' Webcasts or Netcasts, that is, digital broadcasts of information and entertainment that are transmitted over the Web rather than through television broadcast airwaves. The digital information from the Web is displayed in the blanking spaces of the TV signals, so the TV can still receive normal TV broadcasts in addition to displaying Intercasts. See Webcast.

Intercept Service A service in which a call to a changed or disconnected number is routed to a recording or, if a recording is not available, to an Intercept operator. In the case of the latter, the caller will be verbally asked for the destination number and the operator will attempt to complete the call.

Interchange A commercial Internet service from Ziff-Davis, similar to some of the other large Internet Service provisions, but with a slightly more technological slant.

Interchange Carrier IC. A common telecommunications carrier which provides inter- or intra-LATA services through local public exchanges according to regulatory guidelines established by the Federal Communications Commission (FCC) and the Telecommunications Act.

Interchange Format See Rich Text Format.

intercom *abbrev.* intercommunication, intercommunicator. A set of at least two devices, minimally a receiver and transmitter or two transceivers, over which remote communication can take place. Many intercom speakers are wall mounted, like the PA systems in schools or hospitals, and the transmitter may be attached to a handheld microphone or operated through a telephone handset. Baby monitors are a type of mobile intercom, in which one unit is placed near the baby and the other is placed near the parents or babysitter or attached to their clothing so they can move around. Intercoms are often incorporated into phone systems, so that the handset or speakerphone is the transmitter and the receiver is a speakerphone on another console (or on several consoles in broadcast mode).

Wireless intercoms work by broadcasting radio waves. Since broadcast power and frequencies are strictly regulated, intercoms must conform to regulatory guidelines to ensure that they don't interfere with other broadcasts (your neighbor's favorite radio show, for example) or with other appliances. Thus, intercoms must be set within specific frequency ranges and must use low-power output.

interface A hardware connection, or logical connection or translation point. Interfaces are an intrinsic part of interconnected computers, peripherals, and networks. Almost every aspect of data and electrical connections in the telecommunications industry uses a different format or version of a format, and the interface is the point at which all these different hardware and software junctions come together. A cable, peripheral card, card slot, or chip socket are all types of interfaces, as are the images on the monitor and the sounds from a speaker.

Interface Device In frame relay networks, the Interface Device provides a link between an end device and the network through encapsulation. See encapsulation, frame relay Capable Interface Device.

interference Extraneous, unwanted signals which hinder transmission or perception of the desired signal. Types of interference include noise, static, pops, crackles, echo, babble, chatter, crosstalk, cosmic noise, and background noise. Most of these types of interference have individual entries in this dictionary.

interference guard band See guard band.

interferometer A device which detects and displays interference between two or more light wave trains, and can compare wavelengths against reference displacements. This information can be useful in measurement or calibration, for example, to determine angular positions in satellite tracking.

Interim Local Management Interface ILMI. In ATM networking, a service which can be used to obtain an ATM address network prefix. ILMI is not universally implemented and meta-signaling may be used to serve this purpose on some systems.

Interim Number Portability INP. The use of various telephone subscriber services, such as call forwarding, call routing, and call addressing, to allow a call to be redirected to another location, usually on a temporary basis.

interior In ATM networking, interior denotes that an item such as a link, address, or node is inside a PNNI routing domain.

Interior Gateway Protocol IGP. A family of network routing protocols for exchanging information with other routers and switches on the same system. When changes occur in the organization of the network, these changes are communicated to the routers, so the routing table databases can be revised accordingly.

Interior Gateway Routing Protocol IGRP. A Cisco Systems proprietary routing protocol developed for large, heterogenous networks, like the Internet. Enhanced IGRP has also been developed to support TCP/IP, IPX, and AppleTalk.

interlace A system used in video image display to display images in two frame passes, with one pass imaging the odd lines and the next the even lines, in an alternating pattern. Thus, in NTSC, for example, an interlaced screen is imaged in two fields of 262.5 lines (to make up the full 525 scan lines), each field taking 1/60 of a second. Some flicker can be seen on an interlaced display, so noninterlaced monitors, including multisync monitors, have become prevalent on computer systems. Generally, the faster the refresh, the more stable the image. See cathode ray tube, field, frame, interleave, multisync, scan, scanning rate.

interleave *v.t.* 1. To arrange in alternating layers, rows/columns, or time slices. 2. In concurrent programming, a logical means to execute sequences in order to analyze the correctness of concurrent programs. 3. In networking, to transmit pulses through a single path through time-division from more than one source. 4. In graphics file storage and display, a means of arranging the image data so that all odd lines of the image and all even lines of the image are stored, or displayed as a group. 5. In magnetic and magneto-optical data storage, a means and pattern of storing information on a disk so the physical characteristics of the read/write sequence are accommodated without the drive head having to 'backtrack' to find the next section of data. 6. In multimedia applications, a means of 'slicing' up the recording space so that different media (sound, graphics, etc.) are laid down in 'strips' or sections on the tape or disc. 7. A data transmission error-correcting technique in which code symbols are arranged in an interleaved pattern before transmission and reassembled upon receipt.

interleaved video A video display in which a frame is constructed and displayed by alternately scanning all even lines and then all odd lines. This system of display is commonly seen on televisions screens and on some NTSC-compatible computer screens. A certain amount of flicker is usually noticeable on interleaved displays. See interlace.

intermediate frequency IF. In heterodyne receivers, the beat frequency created as a result of the difference between a locally generated signal and the incoming radio signal.

Intermediate Signaling Network Identification In Signaling System 7 (SS7), a capability which allows an application process in the originating network to specify intermediate signaling networks for non-circuit-associated signaling messages, and/or to notify an application process in the destination network about intermediate signaling networks.

intermittent errors Fault conditions that happen occasionally, sometimes without apparent pattern, or which occur from specific causes that happen seldom or irregularly. Difficult to anticipate and diagnose, intermittent problems are often not alleviated until a program has been run hundreds of times or a computer or phone network has negotiated thousands of calls.

internal modem A computer modem installed inside a unit. Sometimes referred to tongue-in-cheek as 'infernal modem' since internal modems can be finicky to install in systems with several peripherals that require IRQs. Internal modems are convenient in that they are out of sight and mind, and don't take up extra space, a real plus on laptop computers. Internal modems usually take the form of small PC boards or very small PCMCIA cards. They have disadvantages as well, as they are often machine or platform specific and can't be reinstalled in a new computer of a different type, as can most external modems. External modems are easier to swap among systems, can be shared by a number of users through a switcher, and usually have status lights which are handy diagnostic tools.

International Alphabet No. 2 An older alphabetic coding system of using equal-duration pulses of negative and positive volts, called marks and spaces in groups of five, to represent character signals. The beginning and end of each character was signaled by a start signal and a stop signal. The use of five elements in two possible polarities results in 2^5 or 32 character encodings. Even for a basic alphabet, this was somewhat limited, and schemes for doubling the number by allowing a code to represent one of two characters, were devised.

This is not unlike what happened later with computer character codes. When International Alphabets developed into ASCII, widely implemented on computers, there were only 128 characters, not sufficient for foreign language, graphics, or mathematical symbols, and so many developers added their own 'extended ASCII' codes (which weren't ASCII at all, and which were not standardized across platforms), which were accessed on many systems by selecting the 'lower bank' or 'higher bank' of characters. See ASCII.

International Amateur Radio Union IARU. A regulatory agency and proponent of world amateur radio activity, established in France in 1925. Amateur radio organizations throughout the world interact with a high degree of cooperation and communications. See American Relay Radio League. http://www.iaru.org/

International Atomic Time, Temps Atomique International TIA. An atomic time scale based on the coordinated efforts of more than 200 atomic clocks from more than 50 centers from around the world, which are maintained in France by the Bureau International des Pods et Mesures. Unlike the Coordinated Universal Time (UTC), which is adjusted occasionally in leap seconds to maintain some coordination with the Earth's axis rotation, TIA is not adjusted, but remains consistent with atomic time scales. Otherwise, TIA and UTC are very similar. See atomic clock, Coordinated Universal Time.

International Business Machines IBM. In the late 1800s, Herman Hollerith, an American engineer, developed the concept of punched cards as a storage medium and applied them to the development of a tabulating machine, an early computer which could be used to store information in categories. This resulted in Hollerith cards, Hollerith code, and a machine which could tabulate the vast amount of census data that was gathered at regular intervals in the United States. The tabulating machine dramatically improved the efficiency of storing and analyzing census data, and Hollerith formed a company called the Tabulating Machine Company. This later merged with several other companies to form the Computer-Tabulating-Recording Company, and the company sold a wide range of industrial products.

Thomas J. Watson, Sr. left NCR to join the company as General Manager in 1914, and remained with the company for over four decades, eventually passing on the position to his son, Thomas J. Watson, Jr.

On Valentine's Day in 1924, the name of the company changed to International Business Machines Corporation. IBM became an enormously influential company in the business and computing market, and funded or partially funded the research and development of several historic room-sized computing machines. IBM's research laboratory has contributed a great legacy of original and fundamental scientific discoveries of interest both inside and outside the computing industry. IBM inventions are awarded more than 1,000 patents per year; in other words, IBM develops as many unique inventions in a single year as the best individual inventors of the 1800s developed in their entire lifetimes.

In 1975, IBM released its first microcomputer, the IBM 5100, but it was not a commercial success, and it was not until five years later that the first of the long IBM PC line was introduced to the public. This time sales were good, particularly in the business market, and IBM and IBM-licensed personal computer technology became the most common platform for desktop computing. See Hollerith, Herman; IBM Personal Computer.

International Data Encryption Algorithm IDEA. A European-designed, 128-bit, single-key encryption algorithm which has been incorporated into Pretty Good Privacy (PGP) partly due to the fact that it doesn't have the same U.S. export restrictions as other encryption algorithms. Use of IDEA is license-free for noncommercial use. See encryption, Pretty Good Privacy.

International Electrotechnical Commission IEC. An International standards-development and recommending body.

International Federation for Information Processing IFIP. An information processing research organization.

International Frequency Regulation Board IFRB. An agency established by the International Telecommunication Union in 1868 to manage the broadcast frequency spectrum. In 1912, the IFRB's Table of Frequency Allocations became mandatory. The frequency allocation Table specified frequency bands for specific uses in order to minimize interference between stations. See International Telegraph Union.

International Internet Association An Internet commercial fee service that provides access to more than 20,000 databases from around the world.

International Mobile Subscriber Identity IMSI. An ITU-T identification number assigned by a wireless carrier to a mobile station to uniquely identify the station locally and internationally.

International Organization for Standardization (International Standards Organization) ISO. An important international standards-setting body which has produced many of the specifications and documents used by telecommunications professionals. ISO is familiar to many through its ISO-9000 series of quality assurance specifications. ISO-9000 standards can be summarized as "Say what you do, then do what you say, and get it certified, if necessary." http://www.iso.ch/

International Radio Consultative Committee CCIR. A standards and regulatory-recommending body founded in 1927, descending from the International Radiotelegraph Conference in 1906, in connection with the International Telegraph Union. This organization was formed in response to public broadcasts over radio waves in the early 1920s.

International Radiotelegraphic Convention One of the early international gatherings, resulting from the growth of telegraphy, held in 1906.

International Switching Center ISC. A gateway exchange whose function is to switch telecommunications traffic between national and international countries.

International Telecommunication Union ITU. A significant global United Nations standards agency descended from the International Telegraph Union. The ITU, headquartered in Geneva, Switzerland, provides publications, promotes communication, sponsors international meetings and conferences, disseminates news, and develops standards and regulations. The ITU oversees a number of subgroups, called sectors, as follows:

Abbrev.	Sector
ITU-R	Radiocommunication Sector
ITU-T	Telecommunication Standardization Sector
ITU-D	Telecommunication Development Sector

For a brief description of the ITU-T history, see International Telegraph Union. For specific publications and recommendations of the ITU-T, see I Series Recommendations, X Series Recommendations, and V Series Recommendations. http://www.itu.ch/

International Telegraph Union ITU. An old and influential organizing and standards-recommending body formed in 1865 when the telecommunications industry was beginning to boom. The ITU was created in response to the need for cooperation and formal agreements related to the installation and use of multinational telegraph systems. Twenty participating countries signed the first International Telegraph Convention.

After the invention of the telephone, the Telegraph Union drew up recommendations for legislation governing international telephony. Radio communications began to develop, and the Telegraph Union convened a preliminary radio conference in 1903 leading to the Radio Regulations and the founding of the International Radio Consultative Committee (CCIR). In 1934, the name was broadened to International Telecommunication Union. It became an agency of the United Nations in October 1947 and the headquarters were transferred from Bern to Geneva in 1948.

The Union later became known as the CCITT, as there were a number of CCIs set up for different areas of communication in the 1920s; the CCIT and the CCIF were amalgamated in 1956.

In 1992, an important conference took place in which the organization was evaluated with the aim of updating it to align with the complex, changing environment of current and future technologies. The organization has recently been renamed International Telecommunication Union (ITU) due to the fact that the fundamental objectives of the original organization remain essentially the same today as they were over 100 years ago, and the convergence of the many media and communications technologies through digital transmission has united many formerly separate areas. [Source: ITU-T Web site history]

In Canada, communication with the ITU is accomplished through the Canadian National Organization for the ITU (CNO/ITU-T) and the Steering Committee on Telecommunications of the CSA (CSA/SCOT). See International Telecommunication Union; Morse, Samuel B.F.

International World Wide Web Conference Committee See World Wide Web Conference Committee.

Internet, the Net A global communications community of more than 60,000 cooperating networks, which evolved in the early 1980s out of ARPANET. The evolution of the Net has been impacted by a wide divergence of technical and lay interests and an equally wide range of commercial and public interests. The vocal promoters of the Net as a universal access communications medium to serve the public good have been joined by commercial interests seeking a way to use the Net to further private and public business interests. The impact of the Internet on our communications and global culture is of the highest significance, and will likely exceed the changes brought about by the industrial revolution.

Speculations about digital intelligence may not be farfetched, and the information glut and impact on personal privacy will be far-reaching. Due to the cooperative communication possible among scientists and interested lay persons, research will move forward at an unprecedented rate. In 1993, the United Nations and the American White House came online, changing the ways in which we access and think about politics. Global doors have opened up to people doing genealogical studies and people are rediscovering friends they haven't seen since elementary school. The phone network is undergoing substantial changes due to competition from long-distance email and chat on the Internet without long-distance phone costs.

The Internet consists of a network of tens of millions of computers linked together through networks small and large. By early 1995, the Internet had more than four million hosts and the term Internet was officially defined by the Federal Networking Council.

This is defined in the Telecommunications Act of 1996 and published by the Federal Communications Commission (FCC) as

"... the international computer network of both Federal and non-Federal interoperable packet switched data networks."

See Telecommunications Act of 1996, RFC 1958.

Internet 2 A consortium of more than 100 academic and nonacademic organizations working to develop a vision and implementation plan for the next generation of the Internet on a content and integration level. For information on the technical successor to the current Internet protocols and physical structures, see IPv6. http://www.internet2.edu/

Internet Access Provider IAP. This is a vendor that provides a connection to the Internet in the form of frame relay, ISDN, a dialup modem, or other physical or virtual connection, and who may or may not provide additional services, such as email, shell accounts, web hosting, etc. Providers with full services available, rather than just an access port to the Internet, are generally called Internet Services Providers (ISPs). See Internet Services Provider.

Internet Activities Board IAB. Established in 1983 to replace the Internet Configuration Control Board. See Internet Architecture Board.

Internet Architecture Board IAB. Formerly the Internet Activities Board established in 1983, the IAB is a coordinating and policy setting board for the Internet Engineering Task Force (IETF) and the Internet Research Task Force (IRTF). All three bodies were combined under the aegis of the Internet Society (ISOC) in the early 1990s and the IAB is now the technical advisor to the Internet Society. See RFC 1358 for a charter of the IAB, and RFC 1160 for a description of its organization and role. See Internet Engineering Task Force (IETF), Request for Comments.
http://www.isi.edu/iab/

Internet Assigned Numbers Authority IANA. An organization which, since the early 1980s, has exercised authority over DNS operations, Internet Protocol (IP) number assignment, Root Name Servers, Request for Comments (RFC) documents, and protocol port number assignments. IANA is the central coordinator for the assignment of unique numbers for Internet protocols and serves as a clearinghouse for this purpose.

IANA also provides registration through a central repository for MIME types, that is, data object types identified by a short ASCII string which can be used to provide rich content types in conjunction with electronic mail.

Jon Postel has almost single-handedly spearheaded this effort, an enormous contribution by an Internet pioneer involved since the days of the ARPANET. IANA is chartered by the Internet Society (ISOC) and located at the Information Sciences Institute (ISI) of the University of Southern California. See domain name, naming authority, name resolution. http://www.iana.org/iana/

Internet Configuration Control Board ICCB. A regulatory board established by the U.S. DARPA in the late 1970s to facilitate the creation of gateways between hosts and the network. Replaced by the Internet Activities Board in 1983. See ARPANET, DARPANET.

Internet Control Message Protocol ICMP. A significant protocol in that it is an IETF-required standard on the Internet for reporting and error messages. Currently the Net is run over IPv4, and migration to IPv6 is planned. ICMP for IPv6 is based on the same definition with some changes and is known as ICMPv6. See IP, RFC 792.

Internet Control Message Protocol for IPv6 ICMPv6. ICMPv6 is a required and integral part of IPv6 which must be fully implemented at every node. It is used for diagnostics and error reporting. ICMPv6 messages are preceded by an IPv6 header and zero or more extension headers, identified by a Next Header value of 58 in the header immediately preceding. ICMPv6 messages are organized into two classes, as shown in the chart.

See RFC 792 for details on Internet Control Message Protocol.

Internet Engineering Steering Group IESG. The executive governing body of the Internet Engineering Task Force (IETF) and technical overseer for the Internet standards process, including final approval. The IESG is a member of the Internet Society (ISOC). See Internet Architecture Board. http://www.ietf.org/iesg.html

Internet Engineering Task Force IETF. The IETF is governed by the Internet Engineering Steering Group (IESG). It is a large, international open community of network research-

Classes and Format of ICPMv6 Messages

Type of message	Identification of type	Message type number
error message	zero in the high-order bit of the message Type field	0 to 127
informational message		128 to 255

ICMPv6 messages have the following format:

```
0                   1                   2                   3
0 1 2 3 4 5 6 7 8 9 0 1 2 3 4 5 6 7 8 9 0 1 2 3 4 5 6 7 8 9 0 1
+-+-+-+-+-+-+-+-+-+-+-+-+-+-+-+-+-+-+-+-+-+-+-+-+-+-+-+-+-+-+-+-+
|     Type      |     Code      |          Checksum             |
+-+-+-+-+-+-+-+-+-+-+-+-+-+-+-+-+-+-+-+-+-+-+-+-+-+-+-+-+-+-+---+
|                                                               |
+                         Message Body                          +
|                                                               |
```

Type	The type of the message. Its value determines the format of the remaining data.
Code	Depends upon the message type. Used to create an additional level of message granularity.
Checksum	Used to detect data corruption in the ICMPv6 message and parts of the IPv6 header.

ers and designers dedicated to the positive evolution of Internet architecture and operations.

The IETF is the primary Internet protocol development and standardization body.

The IETF has worked long and hard on IP Version 6 which will someday fuel the Internet and, in 1997, made some significant changes to support more dynamic addressing schemes. A draft standard for IP Version 6 is scheduled for 1998. In the meantime, some adjustments have been made to lengthen the life of Version 4, which has address space limitations, so that it can continue to be a viable networking solution until vendors begin to implement and support IPv6 toward the end of the century. See Internet Architecture Board, Request for Comments. http://www.ietf.org/home.html

Internet Experimental Note IEN. A document system containing information on Internet specifications and implementations. The IEN is administrated by the Network Information Center (NIC).

Internet Group Multicast Protocol IGMP. An IETF recommended protocol for network transmissions to multiple sites. See RFC 1112.

Internet International Ad Hoc Committee IAHC. The IAHC was a coalition of members of the Internet community cooperating to develop recommendations for the expansion of the Internet Domain Name System (DNS). It published a number of guidelines between 1996 and May 1997, made its Final Report on 4 February 1997, and was dissolved on 1 May 1997. http://www.iahc.org/

Internet Messaging Access Protocol IMAP. An electronic mail protocol descended from Interactive Mail Access Protocol, which is used for electronic mail servers. IMAP is somewhat competitive with Post Office Protocol (POP). See MIME, Post Office Protocol, RFC 1730.

Internet Network Information Center See InterNIC.

InterNet News INN. An NNTP/UUCP USENET newsreading system developed by Rich Salz for Unix systems with socket interfaces. This fast news program was first released in 1992. Later, in 1995, David Barr released a number of unofficial updates. Thereafter, maintenance of INN was taken over by the Internet Software Consortium (ISC). See C News, USENET.

Internet Phone A commercial software/hardware system from VocalTec Ltd. which allows a computer user to place a telephone voice call through the Internet very much the same way that a call is placed through traditional means. The primary difference is that the voice conversation is converted to digital data and channeled through the user's Internet Services Provider (ISP) to the network, rather than through traditional telephone switching offices. The applications software works in conjunction with GOLD, the Global Online Directory that stores information about Internet Phone users who can be contacted online, just as the names of phone subscribers can be accessed through a phone directory. See Global Online Directory.

Internet Protocol IP. A very significant protocol in that it is an IETF-required standard on the Internet along with the Internet Control Message Protocol (ICMP). There are other related IETF protocols which are recommended or elective. IP is very widely used in TCP/IP implementations.

Class	Range	H/O bits	Notes
Class A	0 to 127	0	Larger networks.
Class B	128 to 191	10	Intermediate size networks.
Class C	192 to 233	110	Small networks.
Class D	224 to 239	1110	Specialized, multicast networks.
Class E	240 to 255	1111	Experimental networks.

Internet Protocol provides addressing, segmentation and reassembly, and transport functions in conjunction with a number of associated protocols. Logical IP addresses are used to identify hosts by means of network and node addresses. A number of categories of networks are supported as IP classes.

RFC 768 describes Internet Protocol. RFC 1602 is recommended for its description of the Internet standards process, and RFC 2200 is a useful standards track document for Internet Official Protocol Standards which further describes the standardization process. See IPv6, RFC 950, RFC 919, RFC 922, RFC 2200.

Internet Relay Chat IRC. A worldwide "real-time" 24-hour text-based communications chat link on the Internet developed in the late 1980s by Jarkko Oikarinen. The IRC software is freely distributable through the GNU General Public License from the Free Software Foundation. Many Internet Services Providers provide access to IRC and there are channels in many different languages, although the communication is predominantly English.

IRC is an important meeting ground for people around the world. The form of an IRC chat is somewhat like a group conversation on a teletype machine, except that the output to the screen is very much faster than the transmission and output to a printing teletype. Many celebrities, both in and out of the telecommunications industry, have been known to participate on IRC conversations. To join a chat, you must have access to a provider that provides a port to IRC.

There is a simple command set that you must learn, or you can use a graphical IRC client and point and click your way around IRC. From the text line, a conversation on IRC is joined by typing "#join gardening" or some topic of interest. If it's a common topic, there is probably already a channel allocated to that topic (there are thousands of IRC channels); if it's an uncommon topic, it will be automatically created when the command to join is typed. The channel automatically disappears shortly after the last person leaves, except in the case of registered channels, but comes back as soon as it is re-entered.

Private keyword-protected IRC channels can be created at any time. Courtesy is very important on IRC. If a participant is rude, crude, inflammatory, or off-topic, he or she will be summarily kicked off the channel by an op-

I

Internet Relay Chat Commands Summary

There is a simple set of commands that allows a user to navigate IRC, just as there is a simple set of commands that allows a user to do various file lookup and maintenance functions on a local computers. The IRC commands provide user and status information, as well as chat channel-related functions, as shown in this chart.

The chart on the following page includes a further description of some of the functions of commonly used commands. In interactive mode, commands must be preceded by a slash (/) to be recognized, otherwise they are interpreted as text and broadcast to all the other participants of the same chat channel. Commands can be referenced at any time by typing "/help" with an option for the actual command name.

```
/help
*  choices:
!            :            abort        admin        alias
assign       away         basics       beep         bind
bye          cd           channel      clear        commands
comment      connect      ctcp         date         dcc
deop         describe     die          digraph      dmsg
dquery       echo         encrypt      etiquette    eval
exec         exit         expressions  flush        foreach
help         history      hook         if           ignore
info         input        intro        invite       join
kick         kill         lastlog      leave        links
list         load         lusers       me           menus
mload        mode         motd         msg          names
news         newuser      nick         note         notice
notify       on           oper         parsekey     part
ping         query        quit         quote        rbind
redirect     rehash       restart      rules        save
say          send         sendline     server       set
signoff      sleep        squit        stats        summon
time         timer        topic        trace        type
userhost     users        version      wait         wallops
which        while        who          whois        whowas
window       xecho        xtype
```

erator. If there is no operator present, usually everyone else will leave, and it's ruined for all. Observe courtesy and Netiquette on IRC, and don't talk unless it's something worth saying. The operators or 'ops' are hard-working volunteers who strive to make the IRC an open and fair forum for all.

IRC is a great resource. Companies can have meetings, users' group members can get together for international chats, friends can hang out, nonprofit organizations can keep in touch with members, professionals can seek technical advice, and much more. The Sample IRC Session chart shows an example of entering and exiting a text-based IRC server (graphical clients also exist).

Internet Research Task Force IRTF. An organization which engages in Internetworking research, working closely with the Internet Engineering Task Force (IETF) and the Internet Architecture Board (IAB).

Internet Secretariat An organization which provides administrative assistance to a variety of Internet governing bodies.

Internet Services Provider ISP. A commercial vendor providing access to the Internet and some or all of its services. These services may include email, newsgroup access, World Wide Web access, Internet Relay Chat (IRC),

Sample IRC Session Login, Join Chat Channel, and Log out

```
web1: /s1/abiogen 2% irc -d
* Connecting to port 6667 of server irc.ais.net
* Looking up your hostname...
* Checking Ident
* Found your hostname
* Got Ident response
* ...
* Welcome to the Internet Relay Network abiogen
* If you have not already done so, please read the new user information
with /HELP NEWUSER
* Your host is irc.ais.net, running version 2.8/hybrid-5.2p1
* This server was created Tue Jun 2 1998 at 21: 50:37 CDT
* umodes available oiwszcrkfydn, channel modes available biklmnopstv
* ...
* There are 4488 users and 31563 invisible on 60 servers
* There are 192 operators online
* 15805 channels have been formed
* This server has 3924 clients and 8 servers connected
* Current local  users:  3924  Max: 5207
* Current global users:  36051  Max: 42661
* ...
/join #babylon5
* IRCuser has left channel #earth
* IRCuser (abiogen@web1.calweb.com) has joined channel #babylon5
* Topic for #babylon5: Time is fluid ... like a river with currents,
eddies, backwash.
* #babylon5 MrBawb 899336524
* Users on #babylon5: LadyHawk- @Ivanova @Colen @BobaFet Bro_Theo
@MrBawb CaraD @Dr_Mick @'Dr_Evil' Llorio @JoeyLemur @Delenn @Nu-ghauD
@Spencer @OrenWolf @Wingnut @_Kosh_ @KorMath ETHryAway GleeB @EightBall
@necKro NuFrosty @necKro_II @Pinball @Ramikin @Zathras @Draal @Kosh
@TheOne @LeeThomps @BabCom @Brett IronWing @necK_idle @Sheridan @BB18
@ThemBones @T3GAH_ @Daniel'za Darb @CHeL @BluKnight @Merlyn @ZargDunce
* #babylon5 899133476
<BabCom> Welcome to #babylon5, IRCuser.
-BabCom- Hi IRCuser!  I'm BabCom, the #babylon5 bot.
-BabCom- I do not recognize you.  If you plan to become a regular on
#babylon5,
-BabCom-  '/msg BabCom hello' and I'll add you to the userlist!
....
-BabCom- /exit
```

Internet Relay Chat Commands	
General User Commands	
basics	Very basic introductory information about IRC. A good thing to read the first time you use it.
bye	Drops the user out of IRC. Quit, exit, and signoff do the same.
clear	Clears the current window. Reduces clutter.
date	Displays the current date and time for the local server or a specified server. The time command performs the same function as the date command.
join <channel>	Changes the location to the specified channel. For example, `/join #buglovers` puts the user in the channel with other insectophiles.
info	Provides information about the origins of IRC, its creators, maintainers, slaves, and other perpetrators.
list	Provides a (very, very) long list of channels, and information about the topics and number of participants. Use this command with caution. The * wildcard character may be used to specify the characteristics of the listing, as can a number of useful arguments. -public shows only public channels; -private shows only private channels; -topic shows only channels with a specified topic.
msg <nickname>	Sends a single private message to the specified person. Use query if longer private conversations are desired.
menus	A simple scripting feature for creating custom user menus for an IRC session. This is great for creating mnemonic commands or shortcuts.
newuser	Information about IRC commands and IRC etiquette.
nick	Sets the user's nickname. If the nickname is taken, another must be selected, or the default used.
news	Information about changes, updates, new commands, and other IRC-related functions. It's a good idea to check this once in a while.
query <nickname>	Initiates a private conversion with a specified user. Anything you type now is seen only by that user. The query command with no arguments cancels query mode.
set <variable>	Set various status, logging, and message parameters.
who	Lists users on IRC, or with a wildcard character (*) shows the local channel. A number of arguments can restrict the listing, e.g., -operators lists only operators.
whois <nickname>	Provides more detailed information about the user specified, and his/her 'actual identity.'
Priority Operator Commands	
kick <nickname>	Removes a person from the current IRC channel. Used only when the user contravenes IRC etiquette.
kill <nickname>	Removes a person from IRC. Rarely used, and only with great discretion after the person has blatantly or seriously imposed on others, committed fraud, or disrupted services.
stats	Provides IRC server usage statistical information, such as the number of lines, amount of information carried, authorized users, server uptime, etc.

telnet to other sites, Unix shell accounts, and more. Some providers have flat-rate fees for unlimited access, while others provide unlimited access during off-peak hours, and limited or pay access during times of heavy use. Others charge by connect time. Many distinguish between commercial and personal users, with separate fee scales for each, usually with more mailboxes and longer connect times for business users.

The ISP's link to the Internet may be through a variety of connections usually 56 kbps or higher, up to T1 or even T3 lines. However, if you are dialing up through a regular modem on a phone line, you will not be able to receive and transmit information faster than the slowest point in the link (e.g., the modem speed). There are several large, well-known providers, as well as thousands of small, local service providers. The level of service of many small providers equals or exceeds those of the large companies, so shop around. See Internet Access Provider, National Service Provider.

Internet Society ISOC. A nonprofit international professional organization dedicated to furthering global cooperation and coordination of the evolution of the Internet and its associated technologies. The ISOC oversees and/or works with a variety of other agencies, including the Internet Architecture Board (IAB) and the Internet Engineering Steering Group (IESG). http://www.isoc.org/

Internet standards process The orderly evolution of the Internet is of concern to many networking professionals and the Internet community at large has developed various procedures to facilitate this process. For a technology to become an official or required Internet standard, it must go through a formal discussion, evaluation, and testing process. Through the use of Requests for Comments (RFCs), a protocol must pass through several defined levels of maturity, including proposed standard, draft standard, and standard. The Internet Engineering Task Force must recommend advancement at each stage for the protocol to pass to the next level, and specified waiting periods are imposed. When a protocol has successfully gone through this process, it is assigned a STD number. The process is somewhat recursive in that it is described within itself in RFC 1602. See Internet Engineering Task Force, RFC 1311, RFC 1602.

Internetworking Over NBMA ION. A working group jointly chartered with the Internet and Routing Area of IETF, it is a merger of the IPATM and ROLC groups. The group focuses on issues of internetworking network layer protocols over NBMA subnetwork technologies, including encapsulation, multicasting, address resolution, optimization, and others. See ATM, Frame Relay, SMDS, X.25, ISSLL, ITU, RFC 1932.

InterNIC Internet Network Information Center. This is an authorized central registry for domain names and IP number addresses on the Internet. InterNIC was established in 1993 in cooperation with the National Science Foundation (NSF). To be part of the Internet, you need a unique identifier for your network and the individual host from which you are sending information.

There is a yearly fee (since the mid-1990s) for the registration and maintenance of domain names. The monopolistic nature of InterNIC has been under continued dispute, with various stalled proposals for providing additional domain name extensions and competitive opportunities for other name registries. Established name extensions include .com, .edu, .gov, .net, .int, and .org. See domain, domain name, domain name server, IP address, and the appendix. http://internic.net.

interrogate To query the availability or state of a device or process.

interrupt A system computing resource which causes a suspension of a process, usually to perform another temporary function. On some systems, interrupts were implemented as a means of handling device requests to the CPU and assigned IRQ numbers. This method has a number of significant limitations in that interrupts had to be carefully assigned, and no two devices could use the same interrupt simultaneously; in fact, it was possible to run out of interrupts.

If an IRQ-driven system had several peripherals, it might be necessary to disable one device (e.g., an internal modem) in order to operate another device (e.g., a sound card). This means of managing system resources was not common to all computers, but a significant number of Intel-based consumer machines sold in the 1980s and early 1990s had this form of interrupt-handling.

To overcome the problem of interrupt-handling, a number of vendors developed a system called Plug and Play, which allowed dynamic allocation of interrupts and power-on swapping of devices or device controller cards, provided they support the Plug and Play format (don't just assume a component is Plug and Play; verify it). While this doesn't fully change the underlying concept, it is at least a solution that aids consumers in getting the best use of their machines. See Interrupt Request Numbers chart. See IRQ, Plug and Play.

INTERSPUTNIK The Russian word for satellite is 'sputnik.' The INTERSPUTNIK International Organization of Space Communications system of satellites delivers a variety of programming and data services, including the Voice Of America (VOA), which has formed business relationships with a number of independent Russian radio stations; and Direct Net Telecommunications, which provides international digital voice and data services. See INTELSAT.

Interstate Commerce Act An act established in 1888 to regulate the growing interstate telephone business.

intra- A prefix for inside, within. An intranetwork is a network within a company, home, or other confined locality. In many business contexts, it implies an Internet-compatible internal network, with many of the same functions, such as a Web server, IRC server, email server, etc.

intracellular electrode A device created in 1949 by Ling and Gerard. It consisted of a tiny glass capillary tube with conducting salt, no more than a few tenths of a micron in size. When used in a microprobe, it was possible to measure electrical currents in individual biological neurons. See neural networks.

intrapreneur A person within an organization, usually a large one, who manages, takes risks, proposes and promotes ideas, leads, and generally behaves as an entrepreneur *within* and on behalf of the organization. See entrepreneur.

I

Interrupt Request (IRQ) Numbers and Functions

IRQ #	INT	Notes
0	08h	Reserved for system timer.
1	09h	Reserved for keyboard.
2	0Ah	Reserved for linking (chaining, cascading) upper eight interrupts through interrupt number 9.
3	0Bh	Serial port COM2 and sometimes COM4.
4	0Ch	Serial port COM1 and sometimes COM3.
5	0Dh	Originally assigned to a hard disk controller on 8-bit systems, later 16-bit versions reserved this for a second parallel port (usually designated LPT2). May be available for use by a soundboard, parallel printer, or network interface card (NIC).
6	0Eh	Reserved for floppy diskette controller.
7	0Fh	Reserved for first parallel printer, usually designated LPT1, by some software applications programs (e.g., word processors), but not reserved by the operating system, and thus may be available.
8	70h	Reserved for realtime CMOS clock.
9	71h	Reserved. Used for connection between lower eight an upper eight interrupts. Chained to interrupt #2. In some systems, used for graphics controller.
10	72h	Available. Often used for video display cards.
11	73h	Available. May be used for a third IDE device.
12	74h	Available, although it may be used by a bus mouse (e.g., PS/2 mouse).
13	75h	Reserved for math coprocessor-related functions.
14	76h	Reserved for non-SCSI controllers. Typically used for IDE drives (typical IDE devices include CD-ROM drives, cartridge drives, and hard drives).
15	77h	Available. Sometimes used for SCSI controllers, or a second IDE controller.

inverse multiplexer A multiplexer is a device which takes a circuit, broadcast signal, or given amount of data bandwidth and breaks it up into smaller segments. An inverse multiplexer does the opposite; it takes a number of smaller segments and puts them together to create a larger entity.

An inverse multiplexer is often used in conjunction with computers for high bandwidth applications to coordinate the signals, as in videoconferencing systems that require more than one data line to operate.

As an example, imagine that you have an ISDN data network set up for videoconferencing. Videoconferencing requires fast transmission of high-bandwidth resources: video and sound. Some videoconferencing systems are designed to run over two or three separate ISDN lines. In this case, the inverse multiplexer takes the data from the three sources, coordinates the timing, and sends this information to the computer system, which then displays the images and plays the sound together.

inverter 1. A device or circuit which reverses the polarity of a signal (from positive to negative, or vice versa). 2. A device which changes AC to DC or vice versa. AC to DC converters are very commonly used in digital electronics that draw AC power from a socket. 3. A device or operation that inverts a signal. If the incoming signal is high, the inverted, outgoing signal is low, and vice versa. It is sometimes called a NOT circuit.

Inward Operator Personnel who can assist other operators (e.g., TSPS operators) in making call connections. Normally an Inward Operator does not communicate directly with callers though phone phreakers have been known to do so.

InWATS Inward Wide Area Telephone Services. A subscriber service to receive incoming calls and be billed for them, rather than having the caller billed, somewhat like an automated collect call. This service is provided by a variety of local and interexchange carriers. See OutWATS, WATS.

IOF Inter Office Facility.

IOL InterOperability Lab. Research, development, and vendor verification of interoperability of wireless communications products at the University of New Hampshire.

ION See Internetworking Over NBMA.

ionization 1. The process of dissociating atoms or molecules into ions and/or electrons. 2. The process of rendering a gas to be conducting by causing some of the electrons to detach from its molecules. 3. The process of rendering a solution to be conducting by electrochemical means, assuming the solution is one that contains a compound that can be made conducting.

ionization current A current which results when an applied electric field influences the movement of electrical charges within an ionized medium.

ionoscope A camera tube which incorporates an electron beam and a photoemitting screen where each cell in the screen's mosaic produces a charge. This charge, or electric current, is proportional to the variations of the light intensity in the image that is being captured. The ionoscope produced the television image which was then transmitted to the kinescope for viewing in the days of live broadcasts. Sometimes known by the general use and older trademarked term *iconoscope*. See kinescope.

ionosphere 1. A series of layers of ionized gases enveloping the Earth, the most dense regions of which extend from about 60 to 500 kilometers (this varies with temperature and time of day). 2. The portion of the Earth's outer atmosphere which possesses sufficient ions and electrons to affect the propagation of radio waves. In this region, the sun's ultraviolet rays ionize gases to produce free electrons; without these ionized particles, transmitted radio waves would continue out into space without bouncing back. The deflected path of a radio transmission is effected by the direction of the waves and the density of the ion layers it encounters. See ionosphere sublayers, radio waves.

ionosphere, celestial A region around a celestial body comparable in ionic properties with the earth's ionosphere.

ionospheric sublayers/subregions The Earth's ionosphere has generally been classified into a number of named regions, each of which has properties that make it somewhat distinct from others. These regions are largely hypothetical models, as they may change with the time of day or other factors and don't really form distinct layers as might be implied by the following chart. Nevertheless, the distinctions are useful as a basis for study, even though further refinement and changes are likely.

ionospheric wave Sky wave. A radio wave moving into earth's upper atmosphere. When sky waves are reflected back, at about 2 to 30 MHz frequency ranges, they are known as *short* *waves*. See ionosphere, ground wave, radio, short wave, skip distance.

IP See Internet Protocol.

IP address, Internet Protocol address On a packet network such as the Internet, a number in each packet which is used to identify individual senders and receivers. Under IPv4, this is a 32-bit number, theoretically providing several billion possible addresses, although the actual total is lower due to allocation of subtypes within the system. It is a two-part address identifying the network and the individual device on that network. The IP address is located through an email or domain name lookup. IP addresses can correspond to more than one DNS, although a DNS does not have to have an IP address. The IP system is divided into four classes, roughly according to the size of the network. To be associated with the Internet, a unique network address number must be assigned. See InterNIC IP class.

IP Broadcast over ATM An IP multicast service in development by the IP over ATM Working Group for supporting Internet Protocol broadcast transmissions as a special case of multicast. See RFC 2022, RFC 2226.

IP forwarding The process of receiving a packet, determining how it will be handled, and forwarding it internally or externally. For external forwarding, the interface for sending the packet is also determined, and, if necessary, the media layer encapsulation is modified or replaced for compatibility.

IP Multicast over ATM MLIS Internet Protocol multicasting over Multicast Logical IP Subnetwork (MLIS) using ATM multicast routers. A model developed to work over the Mbone, an emerging multicasting internetwork. Designed for compatibility with multicast routing protocols such as RFC 1112 and RFC 1075.

IP over ATM Internet Protocol over ATM. Implementing ATM involves the coordinated work of many computer professionals and market suppliers of networking products and services. As ATM is a broadly defined format intended to handle a variety of media over a variety of types of systems, there is no one simple explanation for how IP over ATM is accomplished. There are a number of subnet types which need to be supported, including both SVC and PVC-based LANs and WANs. There are also a number of relevant peer models, and end-to-end data transmission models, including Classical IP, TUNIC and others.

See asynchronous transfer mode for general information. See the appendix for diagrams and information about layers. See Internet Protocol, RFC 1577 (Classical IP and ARP over ATM), RFC 1755 (ATM Signaling Support for IP over ATM), 1932.

IP over ATM Working Group Now merged with the ROLC Working Group to form Internetworking Over NBMA (ION). See Internetworking Over NBMA.

IP switching Technology intended to improve transmission speeds and provide consistent bandwidth. In conjunction with a network, IP switching seeks to bring transmission speeds up to the capability of the underlying physical transport medium. It does so by reducing delay in IP routing processing and

Ionospheric Subregions

Name	Approx. location	Notes
D region		A daytime phenomenon and hence not characterized in the same way as some of the regions which exist also at night. Daytime ionospheric activity in this region can impair radio wave propagation.
E region	100 to 120 km	The region which is most distinct in its characteristics and most apt to be classified as a layer.
F1 and F2 regions	150 to 300 km	F2 is always present and is commonly used for radio wave propagation, and has a higher electron density than F1, which is only active in the daytime. The F2 region varies in height, and may sometimes go as high as 400 kilometers in the hottest part of the day.
G region	outer fringes of F	Suggested as a distinct layer by some, but its existence as a definable separate layer is debated.

by making the data transfer mechanism more circuit- than packet-switched.

IPATM See Internetworking over NBMA.

IPCE interprocess communication environment.

IPL Initial Program Load.

IPng IP Next Generation. See IPv6.

IPngWG IPng Working Group. A chartered Internet Engineering Task Force (IETF) group developing the next generation Internet Protocol known as IPv6. Members of the Working Group come from various telecommunications industries, including suppliers of data network hardware, network software, and the telephone industry.

IPO See Initial Public Offering.

IPS Internet Protocol Suite.

IPSec IP Security protocol. Developed in the mid-1990s to resolve some of the issues of conducting secure transactions on the Internet, particularly business-to-business and electronic commerce transactions. A protocol which works at the IP network layer (contrast with Secure Sockets Layer), IPSec provides packet encryption from a choice of encryption algorithms ranging from public-key encryption to secure tunneling. Originally, IPSec worked with an MD5 hashing algorithm, but this was found to be vulnerable to 'collision' attacks, and reinforcement for MD5 and algorithm independence was added in later drafts.

IPSec Working Group A division of the Internet Engineering Task Force (IETF) working on standards specifications for the IP Security protocol (IPSec).

Ipsilon IP switch A commercial switch from Ipsilon, which identifies a stream of IP datagrams for the IP source and destination addresses, and determines if they form part of a longer series. The Ipsilon Flow Management Protocol (IFMP) and General Switch Management Protocol (GSMP) are used in conjunction with specialized hardware to map flow to an underlying network, switching direct IP datagram flows across virtual circuits (VCs). This scheme is most suitable for smaller networks. See IP switching.

IPv4 Internet Protocol Version 4. Developed in the early 1980s, IPv4 is the Internet Protocol for the 1990s, expected to be superseded sometime in the next decade by IPv6. IPv4 features 32-bit addressing, which is suitable for local area networks and widely used there, but no longer sufficient to support the exploding demands on the Internet. See IPv6, RFC 791.

IPv6 Internet Protocol, Version 6. The Internet is a large, complex cooperative network supporting dozens of operating systems and types of computer platforms, tied together with many different circuits, cables, switches and routers. As can be expected in a system this diverse, a flexible, farsighted vision of its future is needed to ensure not only that the technology does not become entrenched and obsolete compared to new technologies that are released, but also that it continue to retain the flexibility to provide universal access, much as is guaranteed by law for North American telephone systems. As such, its evolution is of interest and concern to many, and designers and technical engineers have labored long hours to propose future deployments and to

Extension header	Notes
Hop-by-Hop Option	Unlike other headers, requires examination at each node.
Jumbo Payload Option	Used for packets with payloads longer than 65,535 octets.
	May not be used in conjunction with a Fragment header.
Routing Header (Type 0)	Lists one or more intermediate nodes through which the transmission must pass. Similar to the IPv4 Loose Source and Record Route.
Fragment Header	Used by a source to send a packet larger than would fit on the path MTU, as fragmentation in IPv6 is performed only by source nodes.
Destination Options	A header which is used to carry optional information which is only examined at the packet's destination node.
No Next Header	A value (59) in any IPv6 header or extension header which indicates there is nothing following the header.

develop transition mechanisms to allow the Internet to remain a living 'upgradable' technology.

IPv6 is a significant set of network specifications first recommended by the IPng Area Directors of the Internet Engineering Task Force (IETF) in 1994 and developed into a proposed standard later the same year. The core protocols became an IETF Proposed Standard in 1995.

IPv6 is sometimes called IP Next Generation (IPng). IPv6 was blended from a number of submitted proposals and designed as an evolutionary successor to IPv4, with expanded 128-bit addressing, autoconfiguration, and security features, greater support for extensions and options, traffic flow labeling capability, and simplified header formats.

See 6bone, CATNIP, ICMP, Internet Engineering Task Force, IPv4, X-Bone, TUBA, Simple Internet Transition, SIPP, RFC 1752, RFC 1883, RFC 1885.

IPv6 addresses 128-bit identifiers for interfaces, and sets of interfaces, with each interface belonging to a single node. In most cases, a single interface may be assigned multiple IPv6 addresses from the following types: Anycast, Multicast, or Unicast.

IPv6 extension headers Separate headers are provided in IPv6 for encoding optional Internet-layer information. This information may be placed between the header and the upper-layer header in a packet. These extension headers are identified by distinct Next Header values. In most cases (except for Hop-by-Hop headers), these extension headers are not examined or processed along the delivery path until the packet reaches the node identified in the Destination Address (DA) field of the header. Thus, extensions are processed in the order in which they appear in a packet.

Extension headers are integer multiples of 8 octets, with multioctet fields aligned on natural boundaries. Extension headers in original drafts of IPv6 include Hop-by-Hop, Type 0 Routing, Fragment, Destination, Authentication, and Encapsulating Security payload. If more than one is used in the same packet, there is a sequence which must be followed, both in listing and processing the extension headers. Details can be seen in the extension headers chart. See RFC 1826, RFC 1827.

IPv6 flow A sequence of packets uniquely identified by a source address combined with a non-zero flow label. The packets are sent between a specified source and destination in which the source specifies special handling by the intervening routers. This may be accomplished by resource reservation protocol (RSVP) or by information in the flow packets that may be specified by *extension headers*. There may be multiple flows at one time, in addition to traffic not associated with a flow, and there is no requirement for packets to belong to flows.

IPv6 flow label A 20-bit field in the IPv6 header. Packets not belonging to a flow have a label of zero, otherwise the label is a combination of the source address and a non-zero label, assigned by the flow's source node. Flow labels are chosen uniformly and pseudo-randomly within the range of 1 to FFFFFF hexadecimal, so routers can use them as hashkeys.

IPv6 from IPv4 developments Some of the changes proposed for improving and updating IPv4 which are incorporated into the draft documents for IPv6 include:

- increased address sizes (from 32 to 128 bits) and addressable nodes
- simplified autoconfiguration of addresses
- increased scalability of multicast routing
- new addressing provided through 'anycast' addressing
- simplification of header formats
- improved support for extensions and relaxed limits on length of options
- flow labeling of packets to provide special handling capabilities
- removal of enforcement of packet lifetime maximums
- increased support for security, authentication, data integrity, and confidentiality

IPv6 header format The header format of IPv6, described in the draft RFC document is shown in the chart below.

IPv6 over Ethernet networks IPv6 packets are transmitted over Ethernet in the standard Ethernet frames. The IPv6 header is located in the data field, followed immediately by the payload and any padding octets necessary to meet the minimum required frame size. The default MTU size for IPv6 packets is 1500 octets, a size which may be reduced by a Router Advertisement or by manual configuration of nodes.

IPv6 over Token-Ring networks Frame sizes of IEEE 802.5 networks have variable maximums, depending upon the data signaling rate and the number of nodes on the network ring. Consequently, implementation over Token-Ring must incorporate manual configuration or router advertisements to determine MTU sizes. In a transparent bridging environment, a default MTU of 1500 octets is recommended in the absence of other information to provide compatibility with common 802.5 defaults and Ethernet LANs. In a source route bridging environment, the MTU for the path to a neighbor can be found through a Media Access Control (MAC) level path discovery to access the largest frame (LF) subfield in the routing information field. IPv6 packets are transmitted in LLC/SNAP frames in the data field, along with the payload.

IPv6 security The IPv6 Draft specifies that certain security and authentication protocols and header formats be used in conjunction with IPv6. These are detailed separately as IP Authentication Header (RFC 1826), IP Encap-

Internet Protocol Version 6 Format

```
 0                   1                   2                   3
 0 1 2 3 4 5 6 7 8 9 0 1 2 3 4 5 6 7 8 9 0 1 2 3 4 5 6 7 8 9 0 1
+-+-+-+-+-+-+-+-+-+-+-+-+-+-+-+-+-+-+-+-+-+-+-+-+-+-+-+-+-+-+-+-+
|Version| Traffic Class |           Flow Label                  |
+-+-+-+-+-+-+-+-+-+-+-+-+-+-+-+-+-+-+-+-+-+-+-+-+-+-+-+-+-+-+-+-+
|         Payload Length        |  Next Header  |   Hop Limit   |
+-+-+-+-+-+-+-+-+-+-+-+-+-+-+-+-+-+-+-+-+-+-+-+-+-+-+-+-+-+-+-+-+
|                                                               |
+                                                               +
|                                                               |
+                         Source Address                        +
|                                                               |
+                                                               +
|                                                               |
+-+-+-+-+-+-+-+-+-+-+-+-+-+-+-+-+-+-+-+-+-+-+-+-+-+-+-+-+-+-+-+-+
|                                                               |
+                                                               +
|                                                               |
+                      Destination Address                      +
|                                                               |
+                                                               +
|                                                               |
+-+-+-+-+-+-+-+-+-+-+-+-+-+-+-+-+-+-+-+-+-+-+-+-+-+-+-+-+-+-+-+-+
```

Version	4-bit Internet Protocol version number = 6.
Traffic Class	8-bit traffic class field.
Flow Label	20-bit flow label.
Payload Length	16-bit unsigned integer. Length of the IPv6 payload, i.e., the rest of the packet following this IPv6 header, in octets. Any extension headers present are considered part of the payload, i.e., included in the length count. If this field is zero, it indicates that the payload length is carried in a Jumbo Payload hop-by-hop option.
Next Header	8-bit selector. Identifies the type of header immediately following the IPv6 header. Uses the same values as the IPv4 Protocol field.
Hop Limit	8-bit unsigned integer. Decremented by 1 by each node that forwards the packet. The packet is discarded if Hop Limit is decremented to zero.
Source Address	128-bit address of the originator of the packet.
Destination Address	128-bit address of the intended recipient of the packet (possibly not the end recipient, if a routing header is present).

sulating Security Payload (RFC 1827), and the Security Architecture for the Internet Protocol (RFC 1825).

IPv6 transition IPv6 is a very significant development effort since IPv6 is intended to supplant IPv4, the circulatory system of the Internet. IPv6 is expected to begin commercial implementation by the end of the century. As such, manufacturers and software developers will, in a sense, be overhauling the Net in order to support the updated standard.

In the meantime, the 6bone testbed project has been set up to provide testing of IPv6 and various transition mechanisms. This provides a virtual version of IPv6 which can run on existing IPv4 physical structures. Various mechanisms for providing IPv4/IPv6 interoperability are being developed, including the Simple Internet Transition (SIT) set of protocols. SIT provides a mechanism for upgrade intended not to obsolete IPv4, but rather to gradually phase in IPv6, protecting the connectivity and financial investment of the many IPv4 users.

IR See infrared.

IRAC Interdepartmental Radio Advisory Council.

IRC 1. integrated receiver decoder. A type of satellite receiving device which can be integrated with a multiplexer. This device is used in digital TV broadcasting, especially with MPEG-2 encoded information. 2. See International Record Carrier. 3. See Internet Relay Chat.

IrDA See Infrared Data Association.

IRE See Institute of Radio Engineers.

IREQ *i*nterrupt *req*uest. On interrupt-driven systems, such as widely distributed Intel-based desktop microcomputers, the insertion and use of a PCMCIA card causes an interrupt request signal to be generated to notify the operating system to suspend the current operation and temporarily process the request from the hardware devices attached via the PCMCIA interface. See interrupt, IRQ.

Iridium A series of over 60 low Earth orbit (LEO) communications satellites, sponsored by Motorola. Iridium satellites were scheduled to become operational in 1998. They incorporate FDMA/TDMA techniques and provide voice, data, facsimile and GPS services. Service is estimated to cost about $3 per minute. The name is based on the original estimate that 77 satellites would be needed to blanket the Earth, matching the element Iridium in the periodic table. The number of satellites has since been reduced to 66, but the name has remained. http://www.iridium.com/

IRIS A Macintosh-based videoconferencing system from SAT which provides video capabilities over ISDN lines with JPEG-encoded graphics. See Cameo Personal Video System, Connect 918, CU-SeeMe, MacMICA.

IrLAP See InfraRed Link Access Protocol.

IRQ *i*nterrupt *req*uest. A system of implementing computer processor interrupts which is not common to all computer architectures, but which is characteristic of a large number of Intel-based microcomputers. Many desktop computers can readily accommodate several peripheral devices by just plugging them in and installing a software device driver. However, since Intel interrupt-driven machines are prevalent and some of the most frequent hardware configuration problems encountered by users on these systems are related to IRQ assignments, this section provides extra detail to assist users in configuring their systems. If a system locks up, freezes, or fails to recognize a new device, or a device which was working before a new device is installed, it may be due to an *IRQ conflict*.

When using an application program and an interrupt occurs, a signal is sent by the computer operating system to the processor which says 'pay attention' to the signaling process and temporarily suspend the current process. The IRQ is a number assigned to a specific hardware interrupt. The types of devices for which the system requires hardware interrupts include hard drives, CD-ROM drives, mice, joysticks, keyboards, scanners, modems, floppy diskette controllers, sound cards, and others. There are a limited number of IRQs available and some are reserved for specific tasks.

A peripheral device often comes with a controller card that fits into an expansion slot inside the computer. Sometimes there are small dip switches or jumpers on the controller card or on the device itself (or both), which are set at the factory to a preferred, default, or mandatory IRQ number.

On systems that use a manual IRQ system for hardware devices, it is necessary to assign the interrupts to a corresponding device and a good idea to keep a list of the assignments. On older ISA bus systems, almost the whole process had to be done by hand by the user. With later EISA and Micro Channel buses, there is software assistance for detecting and managing IRQ assignments; and sometimes it is possible to set the IRQs through software, rather than changing dip switches or jumpers.

In earlier systems, interrupts could not be used by more than one device at a time, some were reserved, and only eight were available in total. To complicate matters, some devices had to be associated with a specific interrupt, reducing the number of possible interrupt combinations on a system with several devices. The IRQ may have to be changed in two places, on the computer system and on the controller card or device. To accommodate more devices, more recent machines added a second interrupt controller, increasing the total number of interrupts to 16 (though again, not all were available, as some were reserved, or used for linking).

In general, lower IRQ numbers are higher priority than higher IRQ numbers when two are signaled at the same time, except that IRQs 3 to 8 come after IRQ 15 in priority.

Some peripheral controllers come factory set to a specific interrupt and cannot be changed. Two such cards with the same IRQ requirement cannot be used in the computer at the same time. There are situations where users actually have to physically swap out cards to switch between devices. It is wise to ask about IRQ settings when considering the purchase of 'bargain-priced' peripherals.

The Interrupt Request Numbers chart on page 405 shows various interrupts, including notes about those which are specialized or reserved for certain types of devices, and those which may be available, provided they are not already assigned to another device or already dynamically allocated by a Plug and Play device.

INT refers to the software interrupt association with a command. Two signal lines are used for interrupts, the second being used for parity errors.

Since the interrupt system created administration and configuration problems for users on machines with several devices, some vendors developed the Plug and Play system, which works in conjunction with Windows 95, to ease the burden of setting and tracking interrupts manually. While this doesn't change the architecture of the system and while not all vendors have followed Plug and Play standards, it nevertheless assists users in managing their systems. See interrupt for further information and an IRQ chart. See Plug and Play.

IRR See Internet Routing Registry.

IRSG Internet Research Steering Group.

IRTF See Internet Research Task Force.

ISA 1. See industry standard architecture. 2. Interactive Services Association.

ISD Incremental Service Delivery.

ISDN Integrated Services Digital Network. ISDN represents one of the important technologies developed in recent decades to further the transition of communications networks from analog to digital. ISDN is a set of standards for digital data transmission designed to work over existing copper wires and newer cabling media. It began to spread in the late 1980s, but has not yet received widespread consumer support.

ISDN is a telephone network system defined by the ITU-T (formerly CCITT) which essentially uses the wires and switches of a traditional phone system, but through which service has been upgraded so that it can include end-to-end digital transmission to subscribers. Some systems include packets and frames, as well (see packet switching and frame relay). Nearly all voice switching offices in the U.S. have been converted to digital, but the link to subscribers remains predominantly analog and it has taken some time to work out the logistics of supporting competing switching methods.

ISDN provides voice and data services over *bearer channels* (B channels) and signaling or X.25 packet networking over *delta channels* (D channels). B channels can also be aggregated (brought together) as *H channels*.

ISDN provides an option for those who want faster data transfer than is offered on traditional analog phone lines, but can't afford the higher cost of frame relay or T1 services. ISDN transmission is many times faster (up to about 128 Kbps) than transmission over standard phone services with a 28,800 bps modem. Since the ISDN line doesn't have to modulate the signal from digital to analog before transmission, and then demodulate it back to digital, but rather passes the digital signal through, it's faster. It is also possible to use an ISDN line as though it was up to three lines, sending several different types of transmissions (facsimile, voice call, etc.) at the same time.

A terminal adaptor (TA) is a device commonly used to adapt ISDN B and D channels to common terminal standards such as RS-232 or V.35. A terminal adaptor takes the place of a modem and is provided in much the same way, as a separate component or as an interface card that plugs into a slot.

A network termination (NT1) device is also commonly used in ISDN installations, usually paid for by the subscriber and located at the subscriber's premises.

Not all cities or countries offer ISDN, but its availability is increasing. Many subscriber surcharge services, such as Caller ID, are available through an ISDN line.

ISDN is available in most urban areas with a choice of two levels of service as shown in the ISDN Basic Service Types chart.

ISDN ANSI standards There are a number of important American National Standards (ANSI) of Committee T1 related to ISDN, which are available from ANSI, and which are described in the form of abstracts on the Web. Here is a sampling of those available.

ISDN bonding protocol A protocol which facilitates the use of two ISDN bearer channels (B channels) to transmit a single data stream. The bonding protocol provides dialing, synchronization, and aggregation services for setting up a second call. Both synchronous and asynchronous bonding are supported by various standard and proprietary protocols.

ISDN Caller Line Identification CLI. A feature in which the call address of the caller is sent to the receiving device through the delta channel (D channel). This provides a means for the host router to authenticate the call and to apply any parameters which might be relevant to that particular call.

ISDN ring signal Unlike analog lines, in which an *in-band* ring voltage signal is used to ring the subscriber phone to indicate an incoming call, ISDN uses an *out-of-band* signal, that is, a digital packet on a separate channel, in order to leave established connections undisturbed.

ISDN S/T interface A connecting interface which supports multiple devices on its bus. This provides a connection point for videoconferencing systems, facsimile machines, telephones, and other data communications devices.

ISDN U interface A Network Termination 1 (NT-1), consisting of a two-wire (single pair) installation providing full-duplex data transmissions to support a single device. The NT-1 converts the two-wire interface into a four-wire S/T interface. The NT-1 may be located on the subscriber premises or at the service provider's location, depending on where the service is implemented globally.

ISI Information Sciences Institute. Funded by DARPA to perform research, development, and technology transfer on RSVP.

ISL 1. Inter-Switch Link. A Cisco Systems proprietary protocol which maintains virtual LAN information as traffic moves between switches and routers. 2. ISDN Signaling Link.

ISM 1. See Industrial Scientific Medical. 2. interstellar medium.

ISNI See Intermediate Signaling Network Identification.

ISO See International Organization for Standardization. (In North America, it is sometimes called the International Standards Organization.)

ISO 9000 series A series of quality standards developed in the 1980s for documenting a company's processes and procedures. The ISO 9000 guidelines further track the implementation of what has been documented. Audit, certification, registration, and accreditation programs exist for ISO 9000 and are mandatory in

ISDN Basic Service Types

Service	Abbrev.	Notes
Basic Rate Interface	BRI	Consists of two bidirectional 64 Kbps bearer channels (B channels) for voice and data, and one delta channel (D channel) for signaling or packet networking at 16 Kbps. It requires two conductors through a U Loop, from the carrier to a terminator (NT1) at the customer premises. Except in the rare cases of extremely long phone lines with load coils, most existing phone lines can be used for BRI without significant changes to the actual wire. BRI is aimed at residential and small business users.
Primary Rate Interface	PRI	A service which, in North America, consists of over 20 bearer channels (B channels) and one delta channel (D channel), to serve subscribers with higher capacity needs. In some countries, PRI comprises over 30 bearer channels and one E1 delta channel. PRI requires the same physical configuration as T1, so it is more costly to install than Basic Rate Interface. Multiple support for Primary Rate Interface lines is possible with Non-Facility Associated Signaling (NFAS).

some industries, particularly for manufacturers and parts suppliers.

The full family of ISO standards, including ISO 9001, ISO 9002, and ISO 9003 encompasses more than 20 standards and guidelines, documented with examples in the ISO brochure "Selection and use of ISO 9000" (ISBN 92-67-10267-2). The chemical, electrical, and medical industries were some of the first to adapt ISO 9000 guidelines. http://www.iso.ch/

ISO 9660 A prevalent CD-ROM logical file format standard introduced by the High Sierra Group in 1986. It was subsequently adopted by both ISO and ECMA. It is published as ECMA-119, "Volume and File Structure of CD-ROM for Information Interchange." This compact disc standard aims at portability and future evolution and is widely supported for multiplatform CD-ROM applications. See Yellow Book standard, CD-ROM.

ISO 14000 series A family of standards and guidelines relating to management and the environment. ISO 14000 documents what an organization does to minimize harm to the environment through its activities. Adaptation of ISO 14000 standards can potentially reduce waste management costs, lower distribution costs, and improve a company's corporate image. See ISO 9000 series.

ISOC See Internet Society.

isochronous 1. Uniform signals with embedded timing information, or which depend on an external timing mechanism. 2. In communications, a system in which the transmitter and the receiver use data clocks with the same nominal rate, although not truly synchronous. 3. In data transmission, a process using a specified number of unit intervals between any two significant instants. See asynchronous, constant bit rate, synchronous.

Isochronous Media Access Control IMAC. In Fiber Distributed Data Interconnect (FDDI) networks, a specification for network bridging and access control in an isochronous environment. The IMAC interacts with one or more circuit-switched multiplexers carrying a variety of media that require a constant bit rate and continued connection. Together with the HMUX, the IMAC comprises the HRC element that is FDDI-II. See Media Access Control.

isogonic 1. Exhibiting relative growth or relative scaling, such that individual size relations remain the same. 2. In magnetism, an imaginary line connecting points on the earth with the same magnetic declination (deviation from geographic north). See agonic, magnetic equator.

isotropic 1. Exhibiting consistency in values or characteristics along axes in all directions, often from a central point or point of origin. This concept is used to describe certain physical objects, wave transmissions, and various scientific models. 2. In telecommunications, waves that radiate in all directions at the same time so the leading edge of an unimpeded wave emitted from a single point would form a spherical shape.

isotropic antenna An antenna which radiates in all directions at the same time. Since physical conditions will impede the transmission of waves in some directions, Earth antennas are not isotropic. The phrase is useful for distinguishing this type of antenna from those which direct or concentrate a beam and, also, as a theoretical model. An isotropic antenna is a useful reference point for describing signal variations such as antenna gain.

ISP 1. See Internet Services Provider. 2. Information Services Platform. 3. ISDN Signal Processor.

ISPBX Integrated Services Private Branch Exchange.

ITM See Information Technology Management.

ITR International Telecommunication Regulations.

ITS 1. See Institute for Telecommunication Sciences. 2. See Intelligent Transportation Systems.

ITU, ITU-T See International Telegraph Union, International Telecommunications Union.

IU interface unit

IVDS Interactive Video Data Services. A service considered by the Wireless Telecommunications Bureau (WTC) of the Federal Communications Commission (FCC) to be a means of promoting innovative services and products for telecommunications.

IVHS See Intelligent Vehicle Highway Systems.

IVI See Intel Video Interactive.

IVR interactive voice response. Systems which respond to voice commands or characteristics. Phone systems which can recognize and respond to simple spoken commands are becoming more common, and there are software programs which can interpret spoken commands.

IVS 1. interactive voice service. 2. See interactive video service.

IW interworking.

IWS intelligent workstation.

IXC interexchange carrier. See long-distance carrier.

J Symbol for joule. See joule.

J-hook A piece of J-shaped metal or tough plastic used as an attachment device, often on the end of a cable or elastic fastener.

J-pole antenna A type of simple, long, vertical antenna which is sometimes used in conjunction with short-range, low power frequency modulation (FM) transmitters.

J-scope A type of circular radar screen which shows only the range of objects being detected.

jabber Continuous (erratic or incessant) transmission of inappropriate, corrupted, or meaningless data (garbage), usually beyond the normal protocol interval. A network device which is broadcasting its availability redundantly is said to be jabbering.

Jabber may have limited diagnostic uses, but is mostly fault-related, from a variety of causes, such as software bugs, excessive packet lengths, signal degradation, incompatible systems, hardware defects, or weather-related hardware aberrations or failures. Consequences range from bad data to system lockups.

jabber control, jabber lockup A protective mechanism to inhibit faulty data transmission caused by an overrun of data packets. See jabber.

JACAL A symbolic mathematics system written by Aubrey Jaffer. JACAL runs in Scheme or Common LISP.

jack A contact junction between circuits consisting of a male or female receptacle at one end and usually a line at the other end. Jacks come in a great variety of shapes and sizes. RJ-, RCA, and BNC jacks are commonly used in telephone, video, and networking installations.

jack in To 'attach oneself' to an electronic system. To log on, or interact with a system through worn devices, such as data gloves, body suits, or implanted electronic components, especially those connected to the human nervous system. Often used colloquially to refer to interaction with virtual environments, where there is the illusion of 'being' in the virtual space as opposed to maintaining a separation from it. See avatar, virtual reality.

jack panel A board or panel configured with several jacks, for organizing or centralizing a number of related circuit connections. See cordboard, switchboard.

jacket Sheath which covers a cable to provide protection and sometimes outer insulation, as well as organization and identification. The jacket may bundle a group of related wires or fibers, and can enhance identification through colors, patterns, or materials, e.g., red and green are commonly used to identify tip and ring in phone wires.

Jacquard, Joseph-Marie (1752-1834) An innovative French industrialist who devised a way of automating the storage and retrieval of loom patterns by punching the patterns into cards. This innovation met with strong objections from workers fearful of losing their livelihoods, which is what eventually happened. See Babbage, Charles; Jacquard loom; punch cards.

Jacquard loom A type of automated loom that worked with pattern-encoded punch cards. This loom was devised by Joseph-Marie Jacquard in France in the late 1700s and early 1800s. The holes punched in the cards indicated whether or not a thread was to be woven into the pattern. By associating each card with a color, it was possible to quickly weave complex patterns, and the cards could be modified one by one and recombined, or stored and used in later projects. This caused a revolution in the textile industry. While the idea of encoding information with holes or slots had been used prior to the Jacquard loom, e.g., in music boxes, this large-scale industrial application was significant. It also served as a model for the storage of data and programs in large-scale computers a century and a half later. See punch cards.

JAD joint application design.

jag 1. distortion caused by errors between the transmitter and the recording device, as in a facsimile. 2. distortion or staircasing (aka jaggies) which is an artifact of displaying an image in a resolution which is too low for the amount of information being conveyed. See jaggies.

jaggies *colloq.* In image processing, a descriptive term for aberrations that often occur when the resolution of an image is too complex for the resolution of the device on which it is being displayed or printed. The result is a 'staircased,' 'aliased,' or jagged appearance, especially along edges where there is a sharp contrast or transition from one color or tone to another.

There is not much that can be done with jaggies in monochrome images, but in gray scale or color images, jagged transitions can be smoothed with a technique called antialiasing. This involves selecting intermediate tones between the contrasting elements to visually create the illusion of a transition. Since human perception tends to want to blend such elements, it appears as a smoother line or shape. The technique of antialiasing is often used in the display of images on computer monitors, because the resolution is relatively low on this medium, usually about 75 dpi, as contrasted with print which is usually 300 to 2400 dpi. See antialias.

jam An accidental or deliberate fault condition which hinders or stops the subsequent flow of objects (e.g., printer paper jam) or data (radio, network, or radar signal jam).

On a data network, a jam may occur if more than one device senses idle time or a window, and each device tries to send a packet at the same time, thus causing collisions. A collision may occur unintentionally or be deliberately generated to test the collision-detection response in the network. When testing this on an IEEE 802.3 network, if a station is transmitting and detects a collision event, it should stop the transmission of data, and transmit a 32-bit jam signal to indicate the collision.

JaM The name for one of the predecessors to the PostScript page definition language. JaM was the combined effort of John Warnock and Martin Newell at Xerox PARC in 1978, descended from work on the Design System Language originating at Evans and Sutherland in the 1970s. John Warnock credits John Gaffney with many of the essential ideas. JaM came to be used for various printing, graphics arts, and VLSI design applications. See PostScript.

jam signal A mechanism used in a data network to prevent redundant collisions. When a jam signal is received by more than one device trying to send at the same time, the devices typically wait for a random period and then try again, reducing the probability that they will simultaneously attempt to resend the data and cause further collisions. See babble signal, frequency hopping, jam.

jamming The blocking or obstructing of signals by physical or electronic interference. Jamming is done for reasons of political control or protests, vandalism, competitive obstruction, and national and personal security. Most types of jamming are restricted or prohibited.

JAMSAT Japanese affiliate of AMSAT. See AMSAT.

JAN Joint Army-Navy. The JAN specification is the forerunner of Military Specifications which have been superseded by the MIL designation. See MILNET.

JANET Joint Academic Network. A United Kingdom network established in 1984, which links universities and other academic and research facilities. JANET is funded by the Joint Information Systems Committee (JISC) and developed and managed by UKERNA (formerly by JNT Joint Network Team). The broadband aspect of JANET, capable of transporting video and audio simultaneously with the JANET data, is called SuperJANET (coined in 1989). See JNT.

Japan Amateur Radio League, Inc. JARL. Established in 1926 by a group of 37 radio communication enthusiasts dedicated to promoting the development and use of radio wave technology, JARL's first private license was granted a year later. Since then, the organization has grown and became involved in international activities. In 1985, a reciprocal operating agreement was signed between Japan and the United States. In 1986, JARL began launching a series of amateur radio satellites. JARL has the largest number of amateur radio stations of any country in the world. See JAS-1b, JAS-2. http://www.jarl.or.jp/

Japan Approvals Institute for Telecommunications Equipment JATE. A Japanese regulatory agency, established in 1984, which is roughly equivalent to the Federal Communications Commission (FCC) in the United States. JATE is authorized by the Minister of Posts and Telecommunications, under the provisions of the 1985 Telecommunications Business Law.

JATE grants 'technical conditions compliance approval' and 'technical requirements compliance approval' for public network telecommunications-related equipment. Once approval has been granted, the approval mark must be affixed to the approved equipment.

For non-Japanese manufacturers seeking JATE approval for their products, engaging a good technical translator may expedite the process.

JAR See Java Archive.

Jargon File, The The precursor to The Hacker's Dictionary, the original Jargon File was an electronic dictionary started by Raphael Finkel in 1975 at Stanford University, and has since gone through many revisions with contributions from Mark Crispin, Guy L. Steele Jr., and the ARPANET communities. The dictionary includes jargon, slang, and technospeak from the artificial intelligence (AI) and general computer communities, and is well-known to programmers worldwide. Don Woods later became the Jargon File contact, and in the early 1980s, Richard Stallman made contributions. The early 1980s also saw the first paper publications of The File and the transition to The Hacker's Dictionary. See Hacker's Dictionary, The.

JARL See Japan Amateur Radio League, Inc.

JAS-1b Japan's second amateur radio satellite, launched in 1990, four years after its first satellite was put into orbit. Six years later, in 1996, this was followed by JAS-2. See JAS-2.

JAS-2 Japan's third amateur radio satellite, launched in 1996 using an H-II launch vehicle.

JASON Foundation A nonprofit educational organization founded to administer the JASON Project. The JASON Foundation was established to excite and encourage students to explore science and technology, and to provide professional resources and support to teachers.

JASON Project An effort of the JASON Foundation which was established in 1989 by R. D. Ballard following the discovery of the wreck of the sunken ocean-liner, the RMS Titanic. The enormous interest in the Titanic, especially by thousands of school children, inspired the founders to dedicate themselves to developing interactive telecommunications technologies for participation in global explorations. http://www.jasonproject.org/

JATE See Japan Approvals Institute for Telecommunications Equipment.

Java An object-oriented, platform-independent, threaded programming language which came into being largely because its two earliest contributors were not satisfied with C and C++, and wanted a way to develop programs with less effort and code. Thus, Bill Joy proposed an object environment based on C++ to Sun Microsystems engineers, and James Gosling, author of EMACS, developed a language called Oak. Communications between Patrick Naughton of Sun, Mike Sheridan, James Gosling, and Bill Joy resulted in the Green Project and collaborative work began.

Eventually, in 1995, Java was introduced by Sun Microsystems. Java requires significantly less code than C for many types of applications, is generally considered easier to learn, works well in conjunction with the Web, and has a good chance of becoming a widespread language of choice for software development.

Java support from Sun includes the Java Development Kit (JDK), available for various Sun platforms, Windows NT, and Windows 95 Intel. Independent ports exist for other operating systems, including Linux, NeXTStep, and Amiga. Macintosh support is provided by Apple Computer's Macintosh Runtime for Java (MRJ), and Windows 3.1 support is provided by International Business Machines (IBM).

Java can be used in conjunction with the HotJava Web browser, to allow Java programs to run on a desktop computer. See Java APIs. Java information and specifications are available through the Javasoft Web site.

`http://www.javasoft.com/`

There is a good Java Frequently Asked Questions (FAQ) listing by Elliotte Rusty Harold on the Web.

`http://sunsite.unc.edu/javafaq/javafaq.html`

Java APIs There are a number of important applications programming interfaces associated with Sun Microsystems' Java which provide specifications and procedures for applications development, some of which are shown in the following chart.

Java applet An important component of Java object-oriented programming, an applet is a Java class which is used to extend Java. Applets can intercommunicate within the same virtual machine environment.

Applets are run within the circumscribed context of a Web browser, applet viewer, or other application that supports applets. This provides a measure of extensibility along with a certain amount of security, since the applet can normally only read and write files on the host machine through the application through which it is running. See Java.

Java Archive JAR. A Java open standard platform-independent compressed file format for images and sound, which brings together a set of files into one. In this way, Java applets and their associated components can be bundled and downloaded as a single file. Individual portions of a JAR can be digitally signed and authenticated. JAR archives can be created with the jar utility included with JDK, which functions in a manner similar to many common archive utilities.

Java telephony Applications based on Sun Microsystems Java programming language which enable portable Java applications to set up, manage, redirect, and otherwise administrate telephone calls handled through digital data networks. An important component of Java telephony is the capability of processing speech, that is, both recognizing and generating speech.

JavaBeans A Sun Microsystems Java language object-oriented, platform-independent security model included in JDK. See Java.

JavaScript A cross-platform, scripted open standard programming language familiar to most through the implementation incorporated into Netscape Web browsers. It is only superficially similar to Java, being slower and having a simpler syntax and limited functionality.

JavaTel Platform-independent telephony technology and applications and Java Telephony API support. See Java Telephony API.

JBIG Joint Bi-Level Image Experts Group. A group formed after the JPEG group to concentrate on the task of lossless compression of bilevel, one-bit, monochrome images such

Sampling of Java Applications Programming Interfaces	
Java Media API	Java media applications programming interface.
Java Security API	The Java applications programming interface (API) for building authentication through digital signatures, and other low- and high-level security features into Java programs. Support is provided for key and certificate management, and access control data. This provides a means for Java applets to be 'signed' to ensure authenticity.
Java Speech API	JSAPI. The Java object-oriented open applications programming interface (API) for speech. Specifications for the development of speech recognition and synthesis applications. JSAPI supports speech dictation systems, employing very large vocabularies and grammar-based speech interactive dialog systems (command-and-control). The API provides three basic types of support: resource management, a set of classes and interfaces for a speech recognition system, and a set of classes and interfaces for speech synthesis. Related functions, speech coding and compression, are handled by the Java Media Framework and Codec support.
Java Telephony API	JTAPI. The Java telephone applications programming interface (API) designed to provide portability of telephony applications across applications and across different hardware platforms. JTAPI is a sanctioned specification extension to Java that is used in conjunction with toolkits (such as Lucent's Passageways and Sun's JavaTel), to serve as a guide for the creation of applications. JTAPI was jointly developed by Sun Microsystems, International Business Machines (IBM), Intel Corporation, Lucent Technologies, Novell Corporation, and Nortel Corporation.

as those commonly generated by printers, fax machines, etc.

The JBIG format incorporates discrete levels of detail by successively doubling resolution. The image is divided into strips for processing, each with a horizontal bar and a specified height, with each strip coded and transmitted separately. The order and characteristics of individual strips can be specified by the user. The image can then be progressively decoded, one strip at a time, as received.

Once an image has been segmented according to strips and specified parameters, the resulting bilevel bitmaps are compressed with a Q-coder. Two contexts are defined by JBIG, the base layer, which is the lowest resolution, and the remaining differential layers. These provide contexts for optimization of the compression.

The JBIG format works well with the many common bilevel images that include text and line art. See MPEG, JPEG.

JCL See Job Control Language.

JDBC Java database connectivity. See Java.

JEDEC Joint Electron Device Engineering Council. JEDEC was originally formed as the Joint Electron Tube Engineering Council (JETEC) in 1944. JEDEC is a standards developing body of over 300 member companies representing the electronics industry as part of the Electronic Industries Alliance (EIA). http://www.eia.org/jedec/

JEDI Joint Electronic Document Interchange.

JEIDA Japan Electronic Industry Development Association.

JEMA Japan Electronic Messaging Association.

JFIF A minimalist implementation of the JPEG family of image compression methods. This is often the implementation which is incorporated into Web browsers. See Joint Photographics Experts Group.

jiffy A unit of time equal to 1/60 of a second (North America), or 1/50 of a second elsewhere. Since the proliferation of computers, other definitions of a jiffy have been used, as 1/100 second, or a clock tick in the CPU. The term is most widely used in the film and video editing industries for editing timing purposes. See SMPTE Time Code.

JIPS JANET Internet Protocol (IP) Service. See JANET

jitter 1. Random or periodic signal amplitude or phase instability or degradation. Jitter arises from various causes including poor connections, overly long cables, incompatibilities between software and hardware, or weather. 2. Unstable or erratic display on a television or computer monitor.

jitter, network Jitter refers to a number of problems arising from demultiplexing, incorrect physical connectors or regenerators, and latency times between consecutive packets. In SONET and other high-speed networks, timing is quite important and lack of synchronization can cause fluctuations in the data packets with respect to the reference clock cycle. This type of phase variation can be filtered with adjustment mechanisms. Jitter specifications for SONET networks are described in ANSI T1.105.03-1994 and for computer networks in general, in ANSI T1.102-1993.

JNT See Joint Network Team.

job In computer operations, a process submitted for later execution. In punch card and time-share computing days, a user was required to submit jobs, which were then processed in queues, sometimes according to various priorities, and returned to the user hours, days, sometimes even weeks, later. On current computing systems, a job more often comprises a batch file, low priority process, or background task which is submitted to the operating system (or network server) while the user works in the foreground. Turnaround time may be seconds or minutes.

Job Control Language JCL. A programming language for providing user instructions to a computer operating system, usually in the format of an interpreted scripting language. Although the phrase is now used generically, it was originally developed as a control language by International Business Machines (IBM) for the control of programs on older IBM batch-based computing systems.

Jobs, Steven P. (1955-) An early entrant to the microcomputer industry, Steven Jobs began as an employee of Atari at the age of 17, hired to do video games development. Through the Home Brew Computer Club, he met Stephen Wozniak, an electronic hardware enthusiast, who was working as an engineer for Hewlett-Packard. Wozniak was designing telephone access devices and homebrew computer projects, and Jobs became interested in the business potential of these designs.

By 1976 Jobs had left Atari, and he and Wozniak together created a new company called Apple Computer. They were planning to sell a microcomputer in kit form, a project probably inspired by the Altair, a humble little history-

J

making microcomputer first released as a kit in 1974. Both Steve Jobs and Steve Wozniak had a strong orientation and commitment to educational markets.

Jobs and Wozniak were joined by A. C. (Mike) Markkula, an Intel marketing manager, who provided a business plan and several rounds of investment money to help jumpstart the fledgling entrepreneurs. The Apple computer shipped at approximately the same time as Radio Shack's Model I computer. Normally this would have shut out the young company, since Radio Shack was a large, established corporation with thousands of distribution centers, but the Apple computer had three things the Model I didn't have: color, expandability, and a small-town-to-big-company dynamic duo to promote it. Not only did the technical features appeal to a significant number of computer hobbyists, but they appealed to schools, as teachers recognized the educational value of color programming for children. It also helped that Radio Shack was targeting businesses, with spreadsheets and word processors, while Apple was marketing aggressively to schools.

Despite his youth, Jobs displayed a charismatic personality and marketing flair. Since then these traits have continued to keep him in the headlines for more than twenty years. Apple gained a foothold, and John Sculley was recruited to head the Corporation. Under Sculley's leadership, Apple became a billion dollar company and, as it grew, the two Steves receded into the background due to company growing pains, personal interests, and differences of opinion with the corporation, although not before becoming millionaires while still in their twenties.

Jobs left Apple Computer and founded NeXT, Inc. in 1985. This company designed some of the best computing hardware and software that was available in the 1980s. The elegantly simple hardware, robust operating system, and stunning graphical user interface, straight-forward built-in networking capabilities, Unix underpinnings, and various software utilities, even now, are as good or better than many systems being sold more than ten years later. The NeXT hardware and operating system is aesthetic, well-conceived, and reliable, and business owners, frustrated with the limitations of current business computers, watched with a keen eye when the NeXT computer was released in 1987. Unfortunately, by not cultivating the early interest from the business community and targeting education almost exclusively, Jobs may have missed a narrow window on a large potential market, as the business community subsequently became solidly entrenched in other operating systems based on Intel architectures. The NeXT did well in institutions of higher education, but didn't penetrate other markets to any significant degree, with the result that NeXT was acquired by Apple Computer in 1996-1997. Interest in the NeXT in 1997 was due at least in part to its very good graphical user interface and integration with Internet services, which were now becoming important to consumers. By the mid-1990s, eight years after its introduction, the user learning curve began to appreciate the NeXT design and concept. Jobs' brash statement that the NeXT was the computer for the '90s was more truth than bluster.

During this period, Jobs was involved not only in NeXT development. A year before the NeXT was released, in 1986, he purchased the Computer Divison of Lucasfilm, Ltd., and incorporated it as an independent company called Pixar, co-founded with Edwin E. Catmull as Vice President and CTO. Jobs has long been chairman and CEO of Pixar, a creative software, multimedia, motion picture company which made history with the Academy Award-winning "Toy Story," a computer-generated full-length motion picture, distributed in 1995 by Walt Disney Pictures.

After a few years of quiet creative work, Steve Jobs' name again splashed across headlines in 1997 when Apple bought NeXT. Jobs was back as an executive at Apple, acting in an interim capacity, and speculation about whether he would again head Apple kept reporters on their toes. The management change, and the publicity, created a flurry of activity at Apple, and stocks reacted accordingly. Jobs' return to the limelight showed that public interest in his activities hasn't declined after more than two decades.

Steven Jobs has a philosophical bent, as can be seen from his keynote speeches and interviews with major computing magazines, and it seems clear that his commitment to education and to the harnessing of the creative potential of computers for improving human lives is sincere. It is likely that he will never be far from the creative computing activities that will occur in the future, and will probably, in fact, be the inspirational force for many innovations yet to come. See Apple computer; Wozniak, Stephen.

JOHNNIAC A historic large-scale computer built by Willis Ware, the JOHNNIAC was unveiled in 1954 by the Rand Corporation. Significantly, the first operator of the JOHNNIAC was Keith Uncapher, who became the first chair of the IEEE Computer Group, which is now the renowned IEEE Computer Society. See ILLIAC, MANIAC.

Johnson noise In electronics, the agitation of electrons in conductors through heat creates noise in the circuit. In communications circuits, the amount of noise is related to the receiver bandwidth and source temperature. Johnson noise is sometimes also called thermal noise, and it is characteristically emitted by all objects with temperatures above absolute zero. An understanding of Johnson noise is important to the design and production of antennas and to noise processing and filtering techniques in communications. The phenomenon was discovered and described by J. B. Johnson, a Bell Laboratories researcher.

joint 1. Connection between two or more conductors. This may be a chemical bond, solder joint, or wires touching, clamped, or wound together. 2. A joining part, or space, between two sections, nodes, or articulations. 3. A junction where two or more structural members are combined.

Joint Bi-Level Image Experts Group See JBIG.

joint circuit Shared communication link.

Joint Network Team JNT. An organization established in March 1979 in the UK by recommendations of the Computer Board and Science Research Council (SRC) to study the networking requirements of the academic community and make proposals. The role was transferred to UKERNA 1 April 1994.

Joint Photographic Experts Group JPEG (*pron.* jay-peg). The Joint Photographic Experts Group was founded in 1986 to develop a standard for the compression of still, continuous-tone images. Soon after its formation, its goals were adopted jointly by the International Organization for Standardization (ISO) and the International Telegraph and Telephone Consultative Committee (CCITT), now the ITU-T. Research proposals for such an image compression scheme were solicited internationally, with a deadline of March 1987. By January 1988, the evaluators had narrowed down the suggestions and selected an Adaptive Discrete Cosine Transform method, culminating in a new standard described in ISO 10918-1 Recommendation T.8. Following the publication of the draft standard, work began on improving compression ratios further, and providing scalability.

The JPEG compression format was designed to be used with a wide variety of continuous tone images, without restrictions as to colors, resolution, content, etc. It provides the user trade-off options between compression levels and the quality (lossiness) of the image, and is symmetric, with compression and decompression requiring about the same amount of time and processing power.

The image on the left is a representation of a JPEG file which is only 25K in size. On the right, the same image, stored as a TIFF file, is 150K, a substantial difference considering they don't look very different at this size. However, when the images are compared in detail, the differences can be seen, as the detail enlargements shown below.

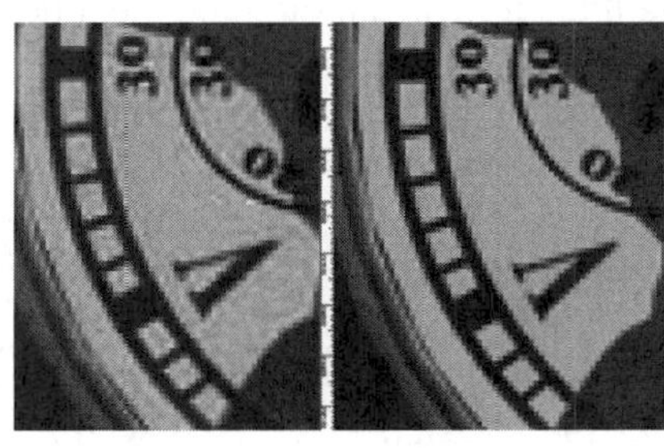

The enlargement on the left shows the 'fuzzy' edges introduced by the lossy JPEG compression process. The TIFF file on the right has cleaner, more consistent areas of color. While the differences may not seem important at this size, they become very evident when blown up to poster or billboard size. Illustrations by the author, copyright 1995 Classic Concepts. Used with permission.

JPEG is not perfect for every type of image. Continuous tone images with many colors generally look good when rendered and compressed with JPEG, in spite of the substantial 'loss' of information and reduction in file size. Crisply rendered images with few colors, sharp boundaries, and thin lines tend to take on a 'fuzzy' or speckled appearance when compressed into JPEG format, and should probably be processed with a different compression format more suitable for that type of image.

JPEG does not support transparency. If transparency is required, another format such as Portable Network Graphics (PNG) or Compuserve Graphics Interchange Format (GIF), should be used.

JPEG is not usually the best format for the storage and rendering of images which are to be printed on a traditional press, as it is a 'lossy' format, that is, it does not retain all the information from the original. The resolutions of printed images on paper are much higher than those of renderings on a computer screen (1800 dpi versus 75 dpi). An image that looks good on the computer may look fuzzy and inadequate on paper. The TIFF format is generally a better choice for images which are to be printed, as the format retains a great deal of information about the image, while still providing reasonably good compression ratios with common compression schemes.

JPEG is actually a family of compression formats and many implementations of it are quite minimal. For example, JFIF is a bare bones version of JPEG which is commonly found on the Web, and SPIFF has been formally defined to be upwardly compatible with JFIF. Variations on the JPEG format can often be identified by looking at the first few bytes in the file header. See JBIG, MPEG, TIFF.

Joint Procurement Consortium JPC. A Bell consortium composed of a number of regional Bell holding companies including Ameritech, BellSouth, Pacific Bell, and SBC Communications, which reviews telecommunications product offerings and makes recommendations. In 1996, the JPC signed contracts with Alcatel for ADSL equipment for use over twisted copper pair networks as an alternative to fiber.

Joint Technical Committee JTC. The JTC is now called JTC 1. It is an "International Standards Organisation/International Electrotechnical Commission" (ISO/IEC) information technology standards body concerned with the specification, design, promotion, and development of systems used for the capture, representation, and processing of information. http://www.jtc1.org/

joint trench A means of aggregating cable installations so more than one department or more than one company can share space within a single conduit or other wiring distribution system. This is done to save money and to limit the number of individual conduits that are installed in public areas. For utility services, there are guidelines and regulations that require that other companies using a joint conduit must be contacted before any street upheaval or digging is undertaken. This is important in order to limit the disruption that inevitably occurs when major line changes or installations are made under or near public streets.

Joint User Service A tariffed, Federal Communications Commission (FCC) system for buying or otherwise sharing telecommunications services by mutual agreement. Local public utility service regulations have restrictions on how certain services may be shared and may require that all associated users be identified. See joint trench.

JOLT Java Open Language Toolkit. See Java.

Jones plug A multicontact polarized receptacle connector.

Josephson effect A quantum effect, and quantum effects are not the easiest to explain, but as an example, imagine a nonsuperconducting material, such as a semiconductor or nonconductor, sandwiched between layers of superconducting material, so that the supercurrent tunnels through the non-superconductor and can variously be effected by magnetic fields. See Josephson junction.

Josephson junction A fast data technology which is sometimes used in place of silicon and which provides a means to do very fast circuit switching. Josephson junctions can be connected together in series, provided their oscillating properties are matched. This is difficult, but has been achieved in devices called Josephson arrays. Josephson junctions have practical applications in many areas, but are particularly of interest to researchers and engineers working with precision voltage metering, microwave electronics, and high-temperature superconductors. Named after British researcher Brian Josephson. See Josephson effect.

Josephson, Brian (1940-) A Welsh-born British physicist who received a Nobel Prize for physics in 1973 for his discovery of the Josephson effect. See Josephson effect, Josephson junction.

Joshi effect When alternating current is passed through a gas dielectric condenser and the gas is continuously irradiated with visible light, there will be a fall or rise in the current.

joule (Symbol -J) An absolute meter-kilogram-second (MKS) unit of work or energy equal to 10^7 ergs. A SI unit of energy equal to 1 kg m^2 s^{-2}. Named after James Joule.

Joule, James (1818-1889) An English physicist who studied the dynamics and efficiency of various types of engines. Joule demonstrated that when mechanical work is used in generating heat, the ratio of work to heat is a constant quantity. A joule, the absolute unit of work or energy, is named after him.

Joy, William (Bill) One of the codevelopers of Sun Microsystems' Java, and attributed with the original idea for the programming language which eventually became Java. In the early 1990s, Joy met with the members of the Stealth Project to develop a language which could create short, powerful programs. See Gosling, James; Java.

joystick A hardware input device that receives and transmits signals containing directional information to a computing device. Commonly used for manipulation of computer software onscreen pointers and selectors. In its most common form, a joystick resembles an aircraft steering control, and may or may not include buttons. It is commonly used for video gaming applications. See mouse, potentiometer, trackball.

JPEG (jay-peg) See Joint Photographic Experts Group.

.jpg A file extension convention commonly used on JPEG-format files. See JPEG.

JPS joint product specification.

JRG GII Joint Rapporteur Group global information infrastructure. A group of rapporteurs and experts from various ITU study groups brought together to further discuss and coordinate global standards-setting tasks.

JTAG Joint Test Action Group. An international body, founded in 1985, JTAG seeks to develop electronics-related test methodologies and related standards. These standards are then recommended to various appropriate standards bodies, such as the IEEE.

JTAPI See Java.

JTC See Joint Technical Committee.

Judge Harold Greene See Greene, Harold.

Jughead Named in association with other Internet tools, including Veronica, Jughead provides a means for Gopher administrators to access and retrieve menu information from the various gopher servers on the Net. They stretched a little to put an acronym to this one, but ended up with "Jonzy's Universal Gopher Hierarchy Excavation and Display."

jukebox A hardware appliance or peripheral which holds and can selectively or randomly access multiple data storage items, generally of the same type, such as audio CDs, tapes or records; computer diskettes; tapes; or cartridges. For audio applications, 20 to 100 items may be accessible, and access may be very quick, whereas for archival purposes, especially with high-capacity storage tapes, there may be 5 to 20 tapes, and access may be slow.

jukebox management High capacity jukebox storage on a network may be accessed by a variety of users with different priority access levels. Jukebox management software detects, sorts, prioritizes, and delivers user data requests. Security and priority levels may vary from industry to industry and installation to installation, so some customization may be needed to enjoy efficient jukebox management.

jumbo group In analog voice phone systems, a hierarchy for multiplexing has been established as a series of standardized increments. See voice group for chart.

jump hunting A means of searching for an available trunk or extension in nonconsecutive order by dialing the first number of the trunk to indicate which trunk to search. Also called nonconsecutive hunting.

jumper 1. A temporary or customized connection used to bypass or reroute a circuit, frequently in the form of a wire, often a short one; common on printed circuit boards. Jumpers are used to test circuits, to correct errors, or make last minute changes in circuit board manufacture. They may also be used to set operating attributes on configurable peripherals such as hard drives (e.g., SCSI settings). Sometimes wire jumpers are equipped with alligator clip heads to facilitate quick connections. 2. A small, paired electrical protruberance on a printed circuit board, intended for configuring a circuit with settings that are infrequent. This type of jumper is common on hard drives, where there may be a row of seven or eight jumpers. The connections are 'jumped' with very small U-shaped electrical tabs with plastic casings to set SCSI ID numbers or other configuration parameters. If the circuit is jumped, the tab is placed over both prongs in the circuit to complete the electrical connection. If not, the tab is removed or placed to one side on a single prong, so as to leave the circuit unconnected.

junction box A wiring box, made of metal or other fire retardant materials, commonly found in homes and businesses. The junction box is usually placed at the junction or demarcation point between internal and external wiring. It often also incorporates fuses or breaker switches to protect against overload and fire hazards.

junctor A circuit extending between frames of a switching unit, terminating in a switching device on each frame, as in an internal network trunk.

JUNET Japan Unix Network. A noncommercial Japanese network dedicated to promote communication among researchers in and outside of Japan.

Junior Wireless Club Limited A pioneer amateur radio organization, formed in 1909.

justify A typesetting term which refers to the regularity of vertical margins related to a line or lines of text. Typically, text is left-justified (aligned on the left), which research indicates is the most readable form, or double-justified (aligned both left and right), which works best with proportional fonts and high resolution printers. Specialized applications (accounting columns, advertising copy, page numbers) are sometimes right-justified.

JVNCnet John Von Neumann Center network. A midlevel, northeastern U.S. regional network.

JWICS Joint Worldwide Intelligence Communications System.

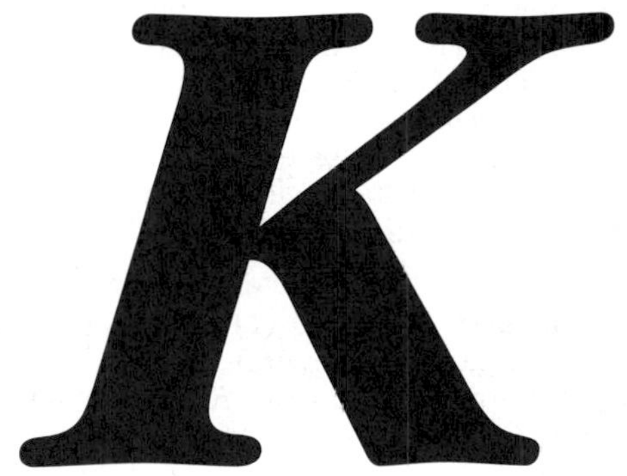

k Abbreviation for kilo-. See kilo-.

K 1. Abbreviation for Kelvin. 2. Symbol for 1024 (usually bits), derived from 2^{10}, 2K = 2048 bits. Also a prefix as in Kbps (kilobits/sec).

K-band A designated portion of the electromagnetic spectrum ranging from 10.9 GHz to 36 GHz. The K-band range is commonly used for small antenna satellite transmissions. See band allocations, Ka-band, Ku-band.

K-carrier A four-wire broadband cable carrier system utilizing frequencies to about 60 kHz.

K-plans See keysheet.

K style handset The designation for the shape of newer telephone handsets which resemble older desk phone G style handsets, except that they have a more squared-off design on the ear- and mouthpieces. They are heavier and more substantial than some of the newer cordless or cell phone handsets which tend to be flat and small. See G style handset.

K56flex modem A 56k modem technology developed by Rockwell Semiconductor Systems and Lucent Technologies. A modem which is competitive with U.S. Robotics' x2 technologies in the absence of an established 56k standard.

kA Abbreviation for kiloampere.

Ka-band The designated portion of the electromagnetic spectrum in the high microwave/millimeter range, approximately 18 GHz to 22 GHz. The Ka-band is used primarily by small antenna satellite transmissions, and is intended to support future applications, for example, mobile voice. A 500 MHz allocation within this spectrum is earmarked for use for non-geostationary fixed satellite orbit services, and there are spectrums for local multipoint distribution services (LMDS), mobile satellite services, and geostationary satellite services. See band allocations for chart.

KA9Q NOS TCP/IP Phil Karn's popular commercial TCP/IP software implementation for packet radio communications. The name of the software comes from his amateur radio callsign. Amateur radio enthusiasts publish a number of amateur radio TCP/IP server gateways on the Web, which connect initially through telnet. Some of these are password-protected, and some can be accessed by anonymous login, somewhat similar to anonymous login sessions on FTP sites.

KADS See Knowledge Analysis and Design System.

Kahle, Brewster Project leader for the Wide Area Information Servers at Thinking Machines Corporation in Massachusetts, and involved with the company since its founding in 1983. Kahle designed the central processing unit (CPU) of the Connection Machine Model 2.

Kahn, Bob In 1974, Bob Kahn co-authored "A Protocol for Packet Network Internetworking" which describes Transmission Control Protocol (TCP). In 1977, Bob Kahn co-demonstrated a gateway system that could interconnect packet radio with the ARPANET. See Cerf, Vince; packet radio.

Kaleida Labs, Inc. A California multimedia development company established as a joint venture between International Business Machines (IBM) and Apple Corporation (including Taligent). Best known, although not well known, for its ScriptX cross-platform hypermedia product. The venture was discontinued in 1995 and the resources rolled into the founding companies.

Kangaroo Network A commercial hardware/software product from Spartacus/Fibronics designed to enable International Business Machines (IBM) mainframes to intercommunicate with other networks using TCP/IP.

K

kanji A symbolic, ideographic language system used to represent Chinese characters which are used in Japanese. It's a challenge to represent the many thousands of Asian characters on computers designed to be programmed and operated in the English language. The Japanese have a number of other ideographic and character symbols in addition to kanji.

Kapor, Mitchell (Mitch) (1950-) The instigator of several historic high profile computer-related organizations, Mitch Kapor founded Lotus Development Corporation in 1982, and was the designer of the well-known Lotus 1-2-3 spreadsheet software. In 1990, he co-founded the Electronic Frontier Foundation (EFF), a nonprofit civil liberties organization. Kapor has chaired the Massachusetts Commission on Computer Technology and Law, and served on the board of the Computer Science and Technology arm of the National Research Council, and the National Information Infrastructure Advisory Council.

Karnaugh map A two-dimensional truth lookup table organized to facilitate combination and reduction of Boolean expressions. See Boolean expression.

Karn's algorithm A mathematical formula used for improving network round-trip time estimations. In layered network architectures, application of the algorithm helps the transport layer protocols distinguish among round-trip time samples. It is used in Transmission Control Protocol (TCP) implementations to separate various types of return transmissions and to establish whether or not to ignore retransmitted signals. It is also applied to backoff timers in Point-to-Point (PPP) tunneling networks. See ATM, Point-to-Point Tunneling Protocol.

Kay, Alan A precocious child and avid reader, Kay was inspired by the work of Seymour Papert at MIT in the 1960s. Kay was committed to the idea that computers should be easy, fun, and accessible, and began developing what was to become the Smalltalk object-oriented programming language. He became a group leader at Xerox PARC in the early 1970s, a period when tremendous innovation in microcomputer technology and user interfaces was stimulated at the lab.

kbps kilobits per second. One thousand bits per second. This is sometimes also written kbits/s.

KBS See knowledge base system.

KDD 1. Knowledge Discovery in Databases. A branch of artificial intelligence applied to databased query, search, and retrieval. 2. Kokusai Denshin Denwa Company, Ltd. A Japanese supplier of international telecom services, equipment, and facilities.

KDKA A historically significant early Westinghouse radio broadcasting station in Pittsburgh, Pennsylvania which used radio waves to report returns of the Harding-Cox Presidential race to the American public on 2 November 1920 about 14 years after the earliest experimental broadcasts. By the following year, KDKA was making regular public broadcasts and more than 500 other broadcasting stations had been established. KDKA originated as amateur call sign 8XK which operated from the garage of Frank Conrad.

Kearney System KS. A parts numbering scheme developed for Western Electric telecommunications equipment, named after the town in New Jersey where the plant was located. The KS system has generally been superseded with vendor-specific and industry standard codes, although Kearney numbers are still found on some pieces of equipment.

keepalive interval Period of time between keepalive messages. The amount of time depends upon the type of network, and the type of activity that is taking place. For example, for a computer process, the interval might be measured in nanoseconds, whereas for a user activity, it might be measured in minutes. See keepalive message, keepalive signal.

keepalive message Messaging between network devices which indicates that a virtual circuit between the two is still active (alive). See keepalive interval, keepalive signal.

keepalive signal A network signal transmitted during times of idleness to keep the circuit from initiating a time-out sequence and terminating the connection due to lack of activity. See keepalive interval, keepalive message.

Kelvin balance Instrument for measuring current.

Kelvin effect When an electric current passes through a single homogeneous, but unequally heated conductor, heat is absorbed or released. The effect is named after William Thompson (Lord Kelvin).

Kelvin scale A temperature scale proposed by William Thompson (Lord Kelvin), which is based on the efficiency of a reversible machine. Zero is designated as the temperature of the sink of the machine working efficiently, that is, complete conversion of heat into work, a situation possible only at absolute zero on a gas temperature scale. Zero degrees Kelvin can be expressed as -273.15 degrees Celsius (C) or -459.67 degrees Fahrenheit (F).

Kelvin, Lord William Thomson (1824-1907). A Scottish physicist and mathematician who made contributions to the field of thermodynamics, and applied his theories to the dynamics and age of the earth and the universe. He utilized field concept to explain electromagnetism and its propagation. The concept of an all-pervasive 'ether' was still prevalent at the time, so he explained a number of his observations in this context. He also developed the siphon recorder. The Kelvin scale and Kelvin effect are named after him.

William Thompson, more familiar as Lord Kelvin, studied electromagnetism. The Kelvin temperature scale, Kelvin receiver, and Kelvin effect are named after him.

Kenbak-1 A discrete logic microcomputer designed by John V. Blankenbaker and introduced in 1971 as the Kenbak-1 Digital Computer. It featured 256 bytes of memory, three programming registers, and five addressing modes. The controlling switches were on the front panel of the machine. It was advertised in the Sept. 1971 issue of Scientific American, three years prior to the introduction of the Altair, for only $750.

The Kenbak-1 was apparently ahead of its time. Unfortunately, only 40 machines sold over the next two years, and the California-based Kenbak Corporation missed a significant business window by a narrow margin. One year after the company closed, the Altair computer kit caught the attention of hobbyist readers of Popular Electronics magazine and sold over 10,000 units.

The Kenbak-1 was not the only commercially unsuccessful computer that preceded the Altair; the Sphere and Micral computers were early entrants to the industry following the Kenbak-1 which also failed to sell in significant numbers. See Altair, Intel MCS-4, Micral, Sphere.

The Kenbak-1 computer was promoted as an educational and hobbyist computer in 1971, three years before the introduction of the Altair. Given its design and price range, it's reasonable to speculate that it had the potential to sell as well as the Altair, which was available assembled or as a kit. Perhaps in 1971, consumers weren't yet ready for the idea of a computer so small it could fit on a desktop.

Kendall effect Distortion in a facsimile record, caused by faulty modulation of the sideband/carrier ratio of the signal.

Kennelly, Arther E. (1861-1939) A British-born American mathematician and engineer who studied mathematical aspects of electrical circuitry. He also studied the properties of the Earth's atmosphere and its effects on radio waves and suggested that an ionized layer above the earth could reflect radio waves, an idea that was soon after independently published by Oliver Heaviside.

Kennelly-Heaviside layer In 1902, A. Kennelly and O. Heaviside proposed, independently of one another, that there is an ionized layer surrounding the Earth which could serve as a reflecting medium that would hold radiation within it. This lead to the discovery of a number of layers surrounding Earth and utilization of the characteristics of some of these layers in long-distance wave transmission. It also led, in the 1920s, to confirming experiments in which radio signals were bounced off this reflecting layer. See Heaviside, Oliver; ionosphere; Kennelly, Arther E.

Kerberos Named for the three-headed dog of Greek mythology who guarded the gates of Hades. Kerberos is a MIT-developed client/server security authentication system based upon symmetric key cryptography.

Kermit An asynchronous, packet-oriented file transfer protocol developed at Columbia University. Kermit supports the 7-bit and 8-bit transfer of text and binary files and is frequently used over local area networks (LANs) and phone lines. It is flexible and configurable. Kermit is not the speediest protocol, as each packet is checked and acknowledged as it is transferred, but it is widespread and well-supported, especially in academic institutions, and, when all other protocols fail, it's often the one which will get the file transfer done.

Kermit is an open protocol, so it is possible to base software development on the protocol, but Columbia's implementation of the Kermit protocol is copyright. See FTP, XModem, YModem, ZModem.

kernel 1. Line within a conductor along which the current-resulting magnetic intensity is zero. 2. Low level of an operating system at which processes and resources (such as memory and drivers) are created, allocated, and managed. Functions and operations at the kernel level form a bridge between hardware and software resources, and are mostly or completely transparent to the user.

kerning In typography, a term which describes spacing between individual letters. Kerning is applied to 'tighten' or loosen the look and feel of a line of type, or to visually correct letter combinations that appear perceptually, or aesthetically to have too much white space between them. For example, if you put the letters T v together, they may seem too far apart because of the overhanging top of the T and the angle of the v. With kerning you can tuck the v under the top bar of the T and it looks better (Tv). When negative letter spacing is applied to the entire line of text, it is called white space reduction.

Tv Tv

The letter pair on the left is unkerned and there is an unaesthetic gap between the T and the v. The pair on the right has been negatively kerned to tuck the v under the top bar of the T to create a stronger relationship between the two shapes.

Kerr effect A phenomenon discovered by Dr. John Kerr in 1875. An electro-optical effect in which certain transparent substances, such as isotropic liquids or gas, become double refracting and optically anisotropic when subjected to a consistent electric field perpendicular to a beam of light.

Kerr magneto-optic effect The change in a light beam from plane polarized to elliptically polarized when it is reflected from the polished end of an electromagnet.

keV Abbreviation for kiloelectronvolt.

Kevlar Dupont tradename for a strong synthetic multipurpose material.

key n. 1. A small, physical security device, often made of metal, inserted into a matching lock receptacle to lock/unlock or activate/inactivate an object or structure. 2. In an image, the overall tone or value of the image, often used to adjust camera settings to balance the amount of light. 3. A switch for opening or closing a circuit. 4. In a database, an organizational means to locate the desired information without searching the entire content of the database. 5. On keyboards, keypads, phone pads, etc. a small, roughly cubic, raised, movable, input attachment which is depressed, usually by a finger, in order to select that particular key. 6. The modern equivalent of the switch on an old phone.

key, encryption A personal or public identifier which helps to establish the owner or recipient of a secure encoded message. A public key cryptographic scheme consists of a 'public key' which is provided openly to anyone who wishes to send an encrypted message and a 'private key' which is used by the recipient to de-encrypt the received message. See Clipper Chip, cryptography, encryption, Pretty Good Privacy.

key, telegraph The signaling device which allows the input of code, usually Morse code, and transmits it to the communications channel. The key superseded the portfule of earlier systems. See telegraph history.

key exchange The transmission or recording of a software key with another party, or swapping among two or more parties. See encryption, PGP, key generation.

key generation The process of creating a software key. Once this has been done, it is expedient to keep track of information related to keys (location, password, etc.) so that key generation does not have to be done again.

key map A table which translates keyboard input values from one configuration to another, commonly used in computer software to accommodate the alphabets of a number of languages. This is useful for translation, alternate typing keyboard setups (e.g., Dvorak), graphics, and music applications.

key pulsing A pulsing system for sending signals through a phone circuit to establish a connection. In older manually operated toll stations, pulsing was sometimes used by operators instead of dialing. Subscriber pulse phones are on the decline, with touchtone phones replacing them.

key service unit, key system unit KSU. The internal electronics and logic which allow the selection of lines and other options in a key telephone system. This may be a small cabinet installed in a closet or some other area where the lines are not cluttering up the environment or causing an obstruction. See key telephone system.

key station Master station from which broadcasts originate.

key telephone system, key system KTS. A multiline telephone system on which individual phones have multiple keys or buttons, which the user presses to select the line over which she or he wishes to communicate. In larger multiline key systems, there may be a main console through which the calls are channeled.

This is not the same as a private branch system, in which there is a separate switching system associated with the phones. In the key system, which is used in many small offices, the switching and selection of lines is done manually by the user. Some larger offices with private branch exchanges will use a hybrid system which also incorporates one or more key systems, sometimes in individual departments. See key service unit, private branch exchange.

keyboard Hardware peripheral for detecting and transmitting user input to a computer system through individually labeled keys. Descended from typewriter keyboards and typically arranged according to the historic "QWERTY" layout which, ironically, was designed to slow down typing in order to prevent key jamming on old manual typewriters. This is unfortunate, because other layouts, such as Dvorak, may be easier to learn and result in faster typing. A variety of keyboards are available for each type of computer system, some with better ergonomics than those which typically come with the system. See keyboard, touch-sensitive.

keyboard buffer Recent input is typically stored in temporary memory in order to prevent loss or corruption, in the event that the system was not yet ready to respond at the time the keys are typed.

keyboard overlay 1. A printed 'skin' or cover, usually of soft molded plastic, which fits over the keyboard and provides additional key designations or modifies existing ones for software programs which require special symbols or keyboard reconfiguration. Useful for custom applications, although they tend to reduce the speed and comfort of typing. 2. A cover to prevent particles and spills from getting under keys and interfering with their operation.

keyboard, touch-sensitive A hardware peripheral for detecting and transmitting user input to a computer system through a flat, touch-sensitive surface. The back side is generally a conductive medium which completes a circuit with very minimal pressure (the touch of a finger). Ideal for general public applications (e.g., kiosks) with simple input demands, where there is a need to prevent clogging with debris; mechanical wear; or vandalism or theft of keys. See keyboard.

keyboarding Typing at a computer keyboard.

keypad Any compact block of functionally related touch-tone telephone, typewriter, calculator, or computer input keys, but typically a group of about 10 to 18 numerical or function keys on a calculator or computer keyboard, physically organized to facilitate touch-typing and fast entry (although some early phone pads were designed in reverse, to slow typing). See numeric keypad.

keysheet An administrative plan for phone extensions which tracks and illustrates the connections and features assigned to that phone. Keysheets are practical in institutional environments with many extensions, particularly if the extension phones have different capabilities and dialing privileges. A keysheet is even more important for keeping track if the phones are also individually programmable. Software exists for developing keysheet connection plans, diagrams, and overlays.

keystoning A visual aberration which occurs when an image is projected on a surface off-plane, that is on a surface which is at an angle to the plane of the surface of the projecting lens. Thus, if a rectangular image from a film or slide projector, for example, were projected on a movable screen which was crooked, the image would be wide on one side and narrow on the other. See barrel distortion.

keystroke Input to a keyboard consisting of a single, or combined key (e.g., Ctrl-C) press. Typing speeds are generally measured in words per minute, but some data entry tasks are measured in keystrokes per minute.

kHz Abbreviation for kilohertz, one thousand hertz.

KIF See Knowledge Interchange Format.

Kilby, Jack St. Clair A Texas Instruments

employee who contributed to manufacturing the first integrated circuit chip, a development that catapulted the miniaturization and speed of electronics into a new level of evolution. See transistor.

Kildall, Gary (1942-1994) American educator and pioneer software developer. Gary Kildall developed CP/M (Control Program/Monitor) over a number of years, beginning in 1973, with contributions from his students, when he was a professor of computer science at a California naval school. Kildall developed CP/M into a very popular, widely used, text-oriented, 8-bit operating system in the late 1970s. Kildall later founded InterGalactic Digital Research, which became Digital Research (DR), to market his software products. Digital Research developed GEM, an early graphical user operating system which predated functional versions of Windows. DR also created DR-DOS, which was competitive with MS-DOS, and claimed by many to be of superior quality.

Kildall is also known for developing PL/M prior to CP/M, the first programming language for the historic Intel 4004 chip, and for co-authoring a floppy controller interface in 1973 with John Torode.

In the ensuing years, Gary Kildall lost one political battle after another with the rapidly expanding Microsoft, and Digital Research never flourished as one might expect for a company so often in the forefront of technology. Digital Research had a history of creating good products, but was overshadowed by its larger, more aggressive competitor. Kildall died at the age of 52.

kill 1. Remove or delete, as a word, line, or file. 2. Abruptly or prematurely terminate a process or broadcast.

kill file An email or newsgroup filter that sends messages from particular people, or on particular topics, to the 'bit bucket,' that is, they are shuffled off to a file that never gets read, or deleted unread.

kilo- (*Abbrev.* - k) Prefix for one thousand (1,000), or 10^3. 10 kilograms = 10,000 grams when used with weights and measures. When used in the context of computer data, more commonly it is capitalized, as in Kbps (kilobits per second) and represents 1024.

kilocharacter One thousand characters. See kilosegment.

kilosegment One thousand segments, with each segment consisting of up to 64 characters. It is used as a billing measure in some systems, such as X.25.

kilovolt-ampere KVA. A unit of apparent power. This is a general measure of power consumption for nonresistive devices such as certain types of lighting and computer components.

kilowatt KW. A SI unit of power required to do work at the rate of 1000 joules per second. See joule, kilowatt-hour, watt.

kilowatt-hour KW-hr. A unit of the energy which is used to perform work as measured over a one hour unit of time. One thousand watt-hours, or 3.6 million joules. This has practical applications as a description of the efficiency of different types of fuel, which can be expressed and compared in terms of kilowatt-hours.

kinescope 1. A cathode ray tube (CRT) in which electrical signals, as from a television receiver, are displayed to a screen. 2. An early term for a motion picture, and probably the inspiration for the term cinemascope. In Britain, the term cinematograph was used to indicate a motion picture or motion picture camera.

kinetograph A device patented in 1889 by Thomas Edison for photographing motion picture sequences. See kinetoscope.

kinetoscope A device patented in 1893 by Thomas Edison for viewing a sequence of pictures, based on the work of earlier experimenters going back as far as 1883. The loop of film images was illuminated from behind and viewed through a rapidly rotating shutter, thus creating a small motion picture film. See kinetograph.

King, Jan The young engineer who coordinated a number of significant amateur radio telecommunications satellite projects, starting with Australis-OSCAR 5 and continuing with the AMSAT satellites. See AMSAT, OSCAR.

Kingsbury Commitment An important event on 13 December 1913 in which the U.S. Attorney General, James McReynolds, informed AT&T of violations of the Sherman Antitrust Act of 1890. AT&T voluntarily gave up controlling interest in the Western Union Telegraph Company, and agreed to stop buying up the independent telephone companies without first obtaining approval from the Interstate Commerce Commission (ICC). AT&T further agreed to provide independent phone companies with access to the long-distance network.

The Kingsbury Commitment derives its name

from Nathan C. Kingsbury, the AT&T vice president who was appointed by Theodore Vail to correspond with the Attorney General. It is sometimes colloquially called the Kingsbury compromise. See Modified Final Judgment.

Kirchhoff, Gustav Robert (1824-1887) A German physicist who conducted pioneer work in spectroscopy and followed up on Ohm's work by providing further information and a more advanced theory of the flow of electricity through conductors. See Kirchhoff's laws.

Kirchhoff's laws Laws for the flow of current first described in 1848 by G. R. Kirchhoff.

1. The current flowing to a given point (node) in a circuit is equal to the current flowing away from that point.

2. In any closed path in a circuit, the algebraic sum of the voltage drops equals the algebraic sum of the electromotive forces in that path.

KIS See Knowbot Information Service.

KISS 1. Keep It Simple Stupid. A tongue-in-cheek, but all too relevant design philosophy.

KISS method A system-independent architecture related to information systems modeling, which is described in terms of object-oriented concepts by its author, Gerald Kristen. KISS concepts are presented in a series of stages, and the model and presentation are sufficiently different from other works in the field of object-oriented (OO) programming that it has not excited a lot of interest in the OO programmers' community.

Kittyhawk A line of very small-sized (less than 2" *x* 3") 20- and 40-MByte, 44-pin IDE hard drives distributed by Hewlett Packard for use in palmtop text and PDA computers.

Kleinrock, Leonard In 1976, authored "Queueing Systems Volume II - Computer Applications," a publication which helped the spread and acceptance of packet switching technology.

Kleist, Ewald Christian von See von Kleist, Ewald Christian.

kludge, kluge Patchwork, improvised, or makeshift hardware or software, which can result from 1. time or material constraints, 2. sloppy workmanship, lack of foresight, 3. communication problems between decision-makers and implementors, or 4. staff changes or design changes during a protracted project.

Kludge usually has a negative connotation, especially with software that tends to dog-down from lack of structure and optimization, while well-conceived, but time-constrained projects are more often called 'quick-and-dirty.' Even well-begun projects can become kludgy after awhile, in which case engineers will generally advise "Time for a ground-up rewrite."

klystron An electron tube which uses electric fields to cause the bunching of electrons. This type of tube is used for generating and amplifying ultra high frequency and microwave frequency waves. The klystron tube was used widely as an oscillator and applied to radar transmitters until it was superseded by the cavity magnetron. Klystron was originally a trade mark. See bunching, cavity magnetron.

KMID key material identifier. A term associated with Message Security Protocol.

KMS See Knowledge Management System.

KNET See Kangaroo Network.

knife switch In old telegraph keys, a type of switch which could short the contacts of the key in a series so the idle line was in a steady mark condition, with current flowing through. This was also called a break switch. Opening the knife switch interrupted the current in all the sounder electromagnets on the line so operators were made aware that someone was about to send a message.

knockout A raised or indented region of a receptacle which can be punched out or otherwise removed to provide access for wires, jacks, or other fittings. Common in general purpose electrical junction boxes.

Knowbot Information Service KIS. A uniform client/server means of interacting with, and displaying, information from a variety of remote directory services typically found on Unix systems, such as Finger, Whois, and others. A query to KIS uses white pages services to these types of systems and displays the results of the search in a consistent format. See Knowbots.

Knowbots In a Knowbot Information Service, programs which carry out the search and retrieval of information from distributed databases as requested by the user. The Knowbots may 'carry' the information or may pass it among one another. See Knowbot Information Service.

knowledge analysis and design system KADS. A structured approach to developing knowledge-based systems.

knowledge base system, knowledge-based system, expert system A computerized system of storing the accumulated 'knowledge' of humans in a system which accesses and manipulates the information using artificial intelligence programming strategies and 'rules' to accomplish information delivery and problem-solving at a sophisticated level.

knowledge engineering Acquisition of knowledge from a human expert or experts and its incorporation into an expert system.

Knowledge Interchange Format KIF. Computer language for the manipulation of knowledge data and interchange of knowledge among disparate programs. Intended not as a user interface, but as an internal representation for knowledge within programs or related sets of programs.

Knowledge Management System KMS. A commercial product from Knowledge Systems, Inc., originally based on research and development of the hypertext management software called ZOG from Carnegie Mellon University.

Knuth, Donald (1938-) Knuth's texts on data structures and algorithms are heavily used, and widely considered by programmers to be the 'bible' of important basic programming structure information. Fundamental search and distribution trees, and much more are in the Knuth texts. It would be difficult to develop sophisticated database software without them. Knuth is also known for authoring the powerful document system called TeX (*pron. tek*) which is one of the few that can handle complex mathematics-related text formatting.

Konexx Modem Koupler A battery-powered commercial modem/modem adapter combination from Unlimited Systems. Konexx allows a modem to be hooked into various types of phone lines and cellular phone systems while traveling.

KQML Knowledge Query and Manipulation Language. Part of the ARPA Knowledge Sharing Effort. A language and protocol for exchanging information and knowledge as part of the larger project to create technology to facilitate development of large-scale, shareable databases.

Krum, Charles and Howard An American father and son team who worked together in the early 1900s to develop and patent a variety of telegraph transmitters and printing machines. One of their early successes was a printer created by interfacing a modified typewriter with a telegraph line, developed at the end of 1908. With a mechanical apparatus ready to use, it became necessary to develop some way to synchronize the pulses and the printing. For this, Howard Crumb applied for a start-stop patent in 1910.

krypton laser A type of gas laser which is primarily krypton that can be used to produce intense red light, or when used with certain optic enhancements, several colors. This is similar to an argon laser, except that it produces a little less light; sometimes argon and krypton are combined. Krypton lasers are typically water-cooled.

KS See Kearney System.

KSR Keyboard Send/Receive. Descended from teletypes, a combination transmitter and receiver which can transmit only from the keyboard, as it does not incorporate storage devices such as punch cards, tapes, floppy disks, or magnetic memory.

KTH Kungliga Tekniska Hogskolan. The Royal Institute of Technology in Stockholm, Sweden.

KTI Key Telephone Interface.

KTS See key telephone system.

Ku-band A range of microwave broadcast frequencies from approximately 11 to 14.5 GHz which is further subdivided into fixed satellite service (FSS) and broadcasting satellite service (BSS). Ku-band is used primarily for data transmission, private networks, and news feeds. Satellites transmitting Ku-band signals tend to be powerful enough for the receiving dish to be small and convenient. Uplinks are in the 14 to 14.5 GHz range and downlinks in the 11.7 to 12.2 GHz range. See band allocations for chart. See broadcasting satellite service, direct broadcast satellite, fixed satellite service.

KV Bell Telephone jargon for key telephones (K = key, V = voice). The jargon was derived from the Universal Service Ordering Code (USOC) which was commonly used until the time of the AT&T divestiture in the mid-1980s.

KVW Bell Telephone jargon for wall-mounted key telephones. See KV.

KWH See kilowatt-hour.

L Symbol for inductance. See inductance.

L-band A portion of the electromagnetic spectrum ranging from 500 MHz to 1500 MHz. Within this range, the frequencies between 950 MHz and 1450 MHz are set aside for mobile communications. Global Positioning Systems (GPS) use the L-band frequencies, as do some of the planet probe systems. See band allocations for chart.

L carrier A common analog frequency division multiplex (FDM) long haul phone services system which was common before digital services began to become more prevalent. Metropolitan areas using FDM more commonly used N carriers.

L1 cache See level 1 cache.

L2 cache See level 2 cache.

L8R An abbreviation for "later" commonly used in online public forums, email, and discussion list messages. It is particularly prevalent in live chat areas on the Internet as a substitute for 'goodbye.'

label 1. A symbol, or group of symbols, often mnemonic, which identifies or describes an item, routine, record, file, application, or process. 2. In programming, a reference point, usually for a procedure, function, or subroutine.

labeling algorithm 1. A mathematical means of calculating the shortest path in network routing. 2. A means of inserting copy protection labels into data, such as compressed video data, so the information can be tagged, or otherwise identified, and can only be written or read under prescribed circumstances. This type of system is being developed to enable vendors to provide digital services to home consumers without fear that the products will be widely pirated and redistributed. It is being experimentally applied to the design of copy-protectable mass storage devices.

LAC Loop Assignment Center.

lacing cord A strong cord, sometimes coated or waxed, used to bundle wires strung along the same path.

ladar laser Doppler radar.

LADS local area data service.

LADT See Local Area Data Transport.

lag To delay, linger, slacken, slow, be retarded, or tarry. Lag occurs in computer applications when the speed of the system is unable to match the speed of the interaction of the user. Lag is characteristic of dialup modem communications, where the speed of the data transmission doesn't match the speed of the computer processor. Lag occurs on data networks when congestion occurs, that is, the number of packets may exceed the ability of the system to handle and transport them. See cell rate, hysteresis, leaky bucket.

Lakeside Programming Group A collaboration of programming friends, which included Bill Gates, Paul Allen, and Ric Weiland. The group created a payroll program in COBOL for a company in Portland, Oregon. They were informally named after the Lakeside private school attended by its members. At the same time, the group was engaged in a programming contract to build scheduling software for a school.

Eventually Gates and Allen formed Traf-O-Data in the early 1970s, to create a traffic analysis program, a partnership which eventually lead to the formation of the Microsoft Corporation.

LAM line adapter module.

LAMA See Local Automatic Message Accounting.

lambda The 11th letter of the Greek alphabet, used as a symbol in a number of mathematical and logical contexts. Lambda symbolizes the null class, von Mangoldt's function, and wavelength.

lambert (Symbol - L) A centimeter-gram-second (CGS) system unit of luminance, equal to the brightness of an ideal diffusing surface which radiates or reflects light at 1 lumen per square centimeter.

laminate *n.* A structure composed of layers, often tightly sandwiched or bonded together. Laminated materials are often used in electronics, from early voltaic piles, which sandwiched moistened materials between layers of metal plates, to magnetic cores and semiconductors, in which layers of various materials are combined according to their electromagnetic properties. An individual layer in a laminated structure is called a *ply*. See semiconductor, voltaic pile.

lamp An illumination apparatus which converts energy into visible light, consisting of a light-producing source and a holder such as a candle and holder, a wick in an oil-holding vessel, a bulb in a handheld, battery-operated container (flashlight), a fluorescent bulb in a fluorescent receptacle, or filamented vacuum bulb in a desk or floor stand. See Edison, Thomas A.

lampblack A sooty, dark carbon dust deposited by a smoking flame, as on the inside of a glass lamp globe. While its presence in lamps is usually undesirable, lampblack has commercial applications in the fabrication of some types of resistors.

LAN See local area network.

LAN adapter A hardware peripheral device or card which connects a computer to a local area network (LAN). Not all computers require LAN adapters; some come equipped with network cards and ports, ready to attach to various connectors, commonly 10Base-T or 10Base-2. However, some require an intermediary device between the network card and the network cable in the form of a LAN adapter.

LAN aware Applications, systems, and devices which can recognize and appropriately respond to an interfaced connection to a local area network (LAN). This involves being able to communicate with other devices on the net, as security permits, locking and unlocking files as needed, using queuing mechanisms appropriately, etc. Some operating systems are designed to be LAN aware. Others run with LAN software installed on top of the operating system.

LAN Emulation See asynchronous transfer mode, LANE.

LAN Manager An early, commercial OS/2-based multiuser network operating system intended to run over TCP/IP or NetBEUI protocols. Microsoft and 3Com's LAN Manager came in two versions, for MS-DOS, Microsoft Windows, and IBM's OS/2, and for Unix/UNIX connections.

LAN Protocols A catchall phrase for a wide variety of protocols developed for local area computer networks, such as AppleTalk, Ethernet, TCP/IP, IPX, and others.

LAN segment In frame relay, a LAN linked to another LAN using the same protocol through a bridge. See bridge, hop, router.

LAN Server A commercial multiuser network operating system from International Business Machines (IBM), based on OS/2 and NetBIOS. LAN Server supports a variety of client computer operating systems, including OS/2, Microsoft DOS and Windows, and Apple Macintosh. It has been superseded by the OS/2 server version.

LAN switch A local area network (LAN) switch, in its simplest sense, is a stand-alone 'box' with connections, which simply directs traffic along one or more pathways. However, with improved technology, LAN switches are incorporating more and more 'intelligent' processing capabilities, and some are almost indistinguishable from routers. Some are available as modular peripheral cards that fit into multiswitch card chassis.

LAN switches can help reduce or more efficiently handle congestion and can improve response time on networks. They have the capability to redirect the World Wide Web queries of users to local caches, and reduce Internet queries; and they can help balance network traffic among servers.

LANCE Local Area Network Controller for Ethernet.

LANDA Local Area Network Dealers Association.

landline, land-line, land line Communications circuits, especially telegraph and telephone, which travel through terrestrial wires and stations. Many mobile units interface with landlines, so that even if a call originates as a wireless call, it may be completed as a landline call to extend distance and free up wireless channels.

Landsat A series of satellites first launched through federal funding in the mid-1960s for remote sensing of the Earth from space.

The Landsat Earth sensing system launches were initiated in 1966 as a response to the announcement of plans for launching civilian Earth Resources Observations Satellites (EROS). As a result, NASA began to plan a satellite launch in order to secure information on Earth resources and provide for national security provisions in space.

This led to the launch of the Earth Resources Technology Satellite (ERTS-1) in 1972, with similar satellites launched in 1975 and 1978. The program was renamed Landsat in 1975. Landsats 4 and 5 were launched in the early 1980s, and jurisdiction was transferred to the U.S. National Oceanographic and Atmospheric Administration (NOAA).

In 1986, during the Reagan administration, jurisdiction was changed to a commercial company, EOSAT, and the primary users became large institutions which could afford expensive satellite data. Complete archiving of data was not always undertaken. EOSAT designed and built Landsat 6, which failed on launch.

In 1992, legislation was passed to return future Landsat missions to the public, and Landsat became part of NASA's Mission to Planet Earth program in 1994. Planning began for the Landsat 7 project.

Landsat satellites are in near-polar orbits, designed to be sun synchronous, that is, the satellites cross the equator at the same local sun time in each orbit. Thus, lighting conditions are kept uniform. The satellites are equipped with telemetry and remote sensing equipment, including cameras and multispectral scanners.

Data collected from early Landsat projects were stored in X-format, that is, band-interleaved by pixel pair (BIP-2). This format was superseded by EDIPS (EROS Digital Image Processing System) with a resolution of 3596 pixels x 2983 scanlines, in band-sequential (BSQ) or band-interleaved-by-line (BIL) formats.

The Landsat digital images are an extremely valuable resource for scientific inquiry, environmental education, mapping, and long-term research and resources planning.

landscape An associative word which refers to the direction of roughly rectangular objects, usually printouts, photographs, or monitors, which are oriented so the short side is vertical and the long side is horizontal. Some monitors can swivel to either landscape or protrait mode at the convenience of the user. The term is widely used in the photography and printing industries to describe the orientation of text or images on a page.

It is important to correctly specify the orientation of a page before sending documents from a computer to a printer. Often an icon will be available as a clickable button for choosing the setting. Most software will default to portrait orientation. Contrast with portrait.

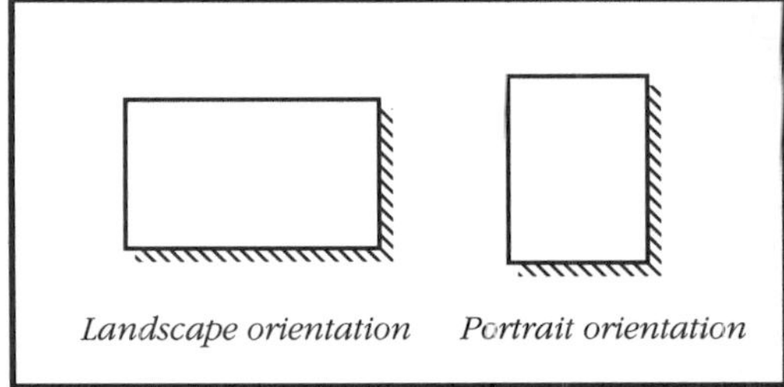

Landscape orientation *Portrait orientation*

LANE LAN Emulation. Local area network (LAN) emulation services and protocols running over asynchronous transfer mode networks. See asynchronous transfer mode and the appendix for greater detail.

language In computer programming, a means of representing instructions, procedures, functions, and data through symbols and syntax which can be interpreted into machine instructions to control the computer. Common high level programming, scripting, and page description languages include Perl, Java, C, C++, PostScript, LISP, Pascal, BASIC, Cobol, and FORTRAN. A common markup language used on the Web is HTML. There are also job control languages, description languages, graphics languages, and low level assembly and machine languages.

LANtastic A commercial peer-to-peer NetBIOS-based network operating system from Artisoft. It supports a variety of client computer operating systems, including Microsoft DOS and Windows, IBM's OS/2, Apple's Macintosh, and various Unix clients.

lap A device which is used for grinding piezoelectric crystals. Since the resonance frequencies of crystals are due in part to their size and shape, the lap provides a means to fine-tune the crystal. See detector, quartz, Y cut.

LAP Link Access Procedure.

laplaciometer An early analog calculator designed for complex mathematical work by J. Atanasoff and some of his graduate students, in the 1930s. The laplaciometer was used to analyze the geometry of surfaces. These developments led to the design and creation of the Atanasoff-Berry Computer (ABC). See Atanasoff-Berry Computer.

Laplink A popular, practical commercial hardware/software networking utility introduced in 1986 for transferring files between computers, especially between laptop computers and office workstations. It is very common for mobile computer users to want to transfer the information from their laptops to their desk computers, and sometimes to transfer in the other direction as well (e.g., sales leads). Laptops typically have smaller hard drives and higher security risks than desktop computers, making it advisable to regularly move the data off the laptop. Transfers can be achieved in a number of ways, through a serial port, over a parallel connection, or through phone lines. In 1995 LapLink Host was added to the product line to provide technical support to remote workers.

Laplink was developed by Traveling Software, Inc., a Washington State company devoted to supporting the needs of mobile users, founded in 1982 by Mark Eppley.

lapping A technique for wrapping electrical tape, foil, or other ribbons around a central core so that the next edge overlaps the previous one, in order to create close contact and a good seal.

laptop computer A low-weight, battery-powered, or combination battery/AC-powered portable computer. Laptops are generally considered to be computers that can fit easily on a lap, an airline tray, train table, or other support surface in common moving conveyances. Laptops range in weight from about three pounds to about seven pounds. Some people distinguish notebook computers as mid-range between laptops and palmtops, at about two to four pounds. Larger portable computers are called 'luggables' or transportables, and the smallest ones are called palmtops and programmable calculators. They weigh from a few ounces up to about two or three pounds.

The range of a laptop battery is usually two to six hours, and may be extended by using powersaver functions or turning off the monitor. Car lighter adapters may be used for powering a laptop or recharging the battery.

Laptops have a variety of types of monitors, designed to be flat, light, and of low power consumption. Various LED, passive matrix, active matrix, and gas plasma monitors are used on laptops. The active matrix screens are brighter and easier to see in dim or bright lighting conditions than screens that depend upon optimal ambient lighting.

Many laptops are equipped with PCMCIA Type I or Type II card slots so the user can add on lower power, compact peripherals such as extra memory cards, fax/modems, network ports, etc.

LARC Livermore Automatic Research Computer. A supercomputer developed for the Lawrence Livermore National Laboratory (LLNL) in the mid-1950s by the Sperry-Rand Corporation.

large scale integration LSI. A term which describes the evolution in electronics from a large number of physically larger components to small chips which perform a group of tasks with a smaller number of physically smaller components. Since the distances between various parts of the logic and physical circuits are greatly reduced in LSI, the processing speeds are also much faster. LSI made possible smaller, more powerful electronic components such as calculators, computers, automated appliances, digital watches, clocks, timers, and much more. See very large scale integration.

laser *acronym* light amplification by stimulated emission of radiation. A device which stimulates photons to produce coherent, nonionizing radiation in the visible spectrum, infrared, and W regions. While lenses and mirrors are commonly used to direct laser beams, the essential components of a laser are a lasing medium, a resonant optical cavity, and a pumping system (optical, mechanical, or electronic).

Lasers were invented in 1958 by Arthur Schawlow, a Bell Laboratories scientist, and C. H. Townes, a consultant to Bell. Numerous other scientists made important contributions to the further development of lasers.

Gems such as ruby and garnet (yttrium aluminum garnet) are used in the production of lasers and many familiar laser implements project a red beam. Gallium arsenide is also used.

A very interesting new type of room-temperature quantum cascade laser technology originated in 1994. QC lasers offer greater control over frequency selection, and have many potential applications in remote sensing and industrial environments.

Unlike light from many other sources, such as incandescent bulbs, a laser beam remains very narrow and straight over long-distances. Many people have seen lasers in the form of business presentation pointers or gun aiming rangefinders on shoot-em-up television. The pointer light usually appears as a small, round, red dot. They make great cat toys, too.

One should never aim a laser pointer at a person or animal, as there is a possibility of harm, especially if the laser shines on a cornea or retina. There may be photochemical or thermal damage from such contact.

Lasers are used for thousands of commercial, industrial, and medical applications. They function as high precision surgical cutting tools in medicine, as imaging components in millions of consumer printers, and as signaling tools in a variety of networking tasks, especially through fiber optic cables.

Laser light is a minimum loss communications tool; in very pure straight fiber optic cable, the loss over distances is low, and the transmission cannot be surveilled in the same way that electrical wires can be monitored through emissions that extend beyond the cable. Laser light communications are not affected by electromagnetic interference in the same way as many other means of transmitting information. See argon laser, cladding, fiber optics, quantum cascade laser, YAG, yttrium.

laser cutting A laser can be used as a high precision cutting tool for surgical procedures and industrial/commercial applications. Rubber stamps, wood blocks, and stencils cut with lasers have very fine, crisp, clean edges.

laser diode A type of semiconductor light-emitting diode (LED) that emits coherent light in response to the application of voltage. See light-emitting diode.

laser fax A combination device which incorporates the printing features of a laser printer with the scanning and transceiver capabilities of a facsimile machine. This is a handy tool for small offices where separate high-capacity devices are not needed and where space is at a premium. Ideally, a laser fax machine should be networkable so users can send and receive faxes without printing each one, and then select to print only the ones which need to be distributed and filed as hard copies. This saves paper and line-ups at the fax machine.

laser printer A printer which uses a computer-directed laser beam to render images. Laser printers typically use a specially treated drum that is influenced by the light of the laser. An 'impression' is made on the drum by aiming very fine, high precision laser light beams at the coating, so the electrical charge is selectively altered. An electrostatic process then attracts the toner to the imaged areas (areas which have been altered by the beam), and heat is used to fuse the toner onto the printing medium, usually paper or card stock.

Laser printers generally print in resolutions ranging from 300 to 1200 dpi, although some can print at higher resolutions with special papers or plates capable of holding a very fine image. Some include Adobe PostScript page description interpreters. Speed of printing and PostScript capabilities are generally enhanced with extra memory.

Laser printing is considered a *dry* printing process, as opposed to offset printing on a press which uses wet inks. It is not advisable to use recharged toner cartridges in laser printers. In most consumer laser printers, the toner cartridge also includes part of the drum mechanism, and the drum mechanism has a limited term of use. Even if the recharge toner is of good quality, it is still possible for the aging drum to stress the printer, perhaps even damaging it. The money you save on toner cartridges may be offset by the potential loss due to damage or reduced lifespan of the printer. See dot matrix printer, dye sublimation, ink-jet printer, thermal wax printer.

laser show A laser show is an entertainment display of differently colored laser lights crossing through the air, sometimes in a darkened room and sometimes falling on a display screen or domed theater surface. If you have seen the various computer screen savers that show lines successively drawing and erasing so that a group of lines appears to move about the screen, then you have a general idea of what a laser light show resembles, except that it's bigger, and three-dimensional.

LAT See Local Area Transport.

LATA See Local Access and Transport Area.

latency Delay or period of dormancy. The speed of acquisition or perception of a thought, object, or communication in relation to the desired speed acquisition or perception.

Latency can result from the intrinsic properties of the communications medium or the communication itself. It can arise from the effects of the time it takes for information to transmit, or from the physical or logical pathways associated with the transmission. It can also result from crowding, congestion, misalignment, mistuning, unanticipated effects or traffic, and a large number of other possible factors. Response time is related to latency, with reduced latency usually being desirable in this context. Every aspect of communications has to concern itself with latency.

In networking, there has been a lot of research and quantification of latency in order to design, evaluate, and tune computer architectures to carry out desired tasks. Here, latency is usually expressed in small units called milliseconds (e.g., latency in ISDN systems is approx. 10 milliseconds). With slower communications pathways, such as slow modems over phone lines, latency may instead be expressed in seconds.

There are many ways to reduce latency: better algorithms, wider bandwidth, better transmissions media, more efficient hardware, different topologies, and new technologies.

Latency is sometimes intended in a dynamic system where connections change and the topology cannot always be anticipated. A queuing system is a means of using latency to good effect, as in email systems, which will hold messages until the intended recipients or routers along the communications path, are available to receive them. Latency may also be used as a signaling system; in other words, delays are taken into consideration and used to convey information. See realtime.

LATNET A Latvian network service, concentrated mainly in Riga, where most of the scientific community is located. LATNET provides services to the RTD community and some businesses. It is operated by the Department of Computer Science at the University of Latvia in cooperation with the Riga Technical University. LATNET utilizes leased lines and radio links, operating with TCP/IP.

launch 1. To start, to set into operation. 2. To start, activate, or begin a computing process, operating system or application. Programs are launched in a variety of ways, such as double-clicking on icons or typing the name of the program on a command line. Programs may also be launched automatically from preprogrammed script files, or transparently from within other programs.

launch, product In management, to begin

Common Layer Hierarchy in Layered Computer Networks

Layer	Notes
application layer	A high level layer at which the user interacts with the network applications programs and utilities. Various types of text or graphical user interfaces may be implemented at this layer. The application may also include remote access mechanisms and information messaging and transfer services.
presentation layer	Data security and data representation during transfer.
session layer	A traffic directing layer that sets up a communication between applications, adjusts synchronization, if needed, and clears the communication when done.
transport layer	A generalized, network-independent means of interlayer communication between the high application-oriented layers and the lower level layers, supporting different types of connections.
network layer	Low level network connection, routing and flow-control functions.
link layer	Low level connectionless or connection-oriented data transfer.
physical layer	Low level electrical connections and interfaces between the computing platforms and the network cables and connections, and the link layer data transmissions.

a new program, project, or marketing plan, sometimes with a lot of fanfare in order to attract the attention of potential customers and the media. New software packages are often *launched* at industry trade shows.

LAWN See local area wireless network and wireless local area network.

laws of electric charges Stated simply: bodies with *unlike* charges will attract one another; bodies with *like* charges will repel one another; bodies with *no* charges will neither attract nor repel one another.

layer architecture Layer architectures are common in computer networks. Asynchronous transfer mode (ATM) is the most broadly distributed layer network architecture.

Defining a number of virtual and physical layers allows communications paths to be organized and administrated so that many different developers and manufacturers can create processes and devices independently of one another, and still apply them to the same system once standards and protocols for the various layers are established. Layers also provide a means to optimize the characteristics of the layer to the type of processes that occur within that layer. The layer architecture is usually described and diagrammed horizontally, from bottom (physical or low level layers) to top (virtual or user interface and applications layers) with variations depending upon the specific organization of the architecture. Layers typically communicate with adjacent layers directly above or below, or may pass through an intervening layer. The Common Layer Hierarchy chart shows a brief overview of some of the common types of layers.

LB See leaky bucket.

LBA See Logical Block Address.

LCD See liquid crystal display.

LCI International An American telecommunications system providing international voice and data services through owned and leased fiber optic networks. LCI is known as the first long-distance provider to bill both business and residential calls in one-second increments, a service known under the Exact Billing service mark.

LCU Lightweight Computer Unit.

LDAP See Lightweight Directory Access Protocol.

LDIP See Long Distance Internet Provider.

LDMS See Local Multipoint Distribution Service.

LE light-emitting.

leader 1. The first segment, or part of a transmission or transmissions medium. 2. The first few centimeters on a magnetic tape (audio, video, etc.), the leader attaches and feeds the tape onto the spool. It is not intended for recording, and may be made of non-magnetic material. 3. A packet, cell, segment, or other leading part of a data transmission, which contains information about data following, without including the data itself. Its function is that of a space or a signal that there will be further information, rather than a component of the information itself. See header.

leadership priority In an ATM network, an organizational function of a logical node by which it is assigned to be the peer group leader.

LEAF Law Enforcement Access Field. A section of classified data that is created in association with a Clipper chip or Capstone, and sent along with the encrypted message. The LEAF includes the session and unit keys concatenated with the sender's serial number and an authentication string. See Clipper chip.

leaf node A type of connecting point in a network which is located at the end of a branch, so there is only one connection between the leaf node and the rest of the network.

leakage In electrical circuits, particularly those which are not well shielded, leakage of the electromagnetic radiation outside the boundaries of the physical medium can occur. This may interfere with other transmissions and devices.

The Federal Communications Commission (FCC) provides guidelines and regulations for shielding various radio and computer devices in order to minimize interference from leakage.

leaking memory See memory leak.

leaky bucket LB. A conformance checking cell flow concept in ATM networking related to congestion, an implementation of the Generic Cell Rate Algorithm (GCRA). Think of the bucket as a point in the network where cells may accumulate, depending upon varying rates of inflow and outflow. If cells are entering and leaving in equilibrium, that is, maintaining a sustainable cell rate (SCR), then the bucket will never be filled. If, however, inflow exceeds outflow, as the network experiences congestion, then the bucket may become full. There are various strategies for dealing with a full bucket, although prevention is advised. If, at the point the bucket becomes full, there are no further incoming cells, then it can be emptied in "bucket depth/ SCR rate" amount of time. Bucket depth, the tolerance to cell bursting, can be set in relation to cell flow and retransmission timing. If however, the incoming cells continue to accumulate, the bucket will 'overflow' and the overflow has to be handled in some manner, with cell discard as one of the options.

There is more than one way to implement a leaky bucket. Cisco Systems, Inc. suggests a means of employing *dual leaky buckets*, so a preconfigured queue depth threshold is set according to an agreed Class of Service (CoS) and Quality of Service (QoS). The first bucket can be configured to provide a service algorithm based on peak cell rate (PCR) and cell delay variation tolerance (CDVT) service parameters. The second bucket is based on sustainable cell rate (SCR) and maximum burst size (MBS). Nonconformant cells can be configured as cell discard, tag, or no change, for each bucket. Dual discard thresholds can be supported to provide a delay mechanism for congestion cell discard rates. See cell rate.

leased line A line whose use is rented over a period of time from the entity which owns and manages the physical connection. Long-distance companies, specialized services, and businesses with direct private lines often lease lines from the local primary telephone carrier rather than installing their own.

Least Cost Routing LCR. A phone service which automatically seeks and selects the line through which to send a call with the least cost. See Automatic Route Selection.

LEC 1. See Local Exchange Carrier. 2. LAN Emulation Client. A LAN software client which keeps address translation and connection information for communication through an ATM network. See LANE, LECS.

LECS LAN Emulation Configuration Server. A LAN software server which maintains configuration information that enables network administrators to control which physical LANs are combined to form VLANs. See LEC.

LED See light-emitting diode.

left-hand rule, Ampère's rule A handy memory aid, once widely used to determine an axis of rotation or direction of magnetic flow in a current. It originally came from Ampère's description of a person swimming in the same direction as the current (in a wire). When the swimmer looks left, it's the same direction that the north-seeking end of a compass will point if it is in the vicinity of the current-carrying wire. Since then, it was decided it was easier to use the left hand and actually look at the thumb and fingers, rather than imagining a swimmer. Extend the thumb and fingers of the left hand so that the fingers are held together and point straight in one direction, with the thumb at a right angle to the fingers, in an 'L' shape. Now curl the fingers around a conductive wire, so that the thumb points in the direction of the current. The direction of the curled fingers is said to indicate the direction of the magnetic field associated with the current.

Some of the confusion associated with left-hand and right-hand rules stems from the fact that pioneer physicists did not originally know in which direction current was flowing in a circuit, between negative and positive terminals. And, in fact, it was not always important to know, as long as the terms of reference were kept consistent in one direction or the other.

Sometimes a distinction is made between current in a motor and current in a generator. By this reasoning, using the left hand, the fingers will show the direction of the current for a conductor in the armature of a motor. Using the same hand relationship for the right hand for a conductor in the armature of a generator. Since the universe appears to be 'right-handed' in its general orientation, some physicists will assert that it makes sense to use the right-hand rule. See right-hand rule.

leg 1. A portion of a trip, broadcast, or transmission. 2. The transmission segment in a network between two physically distinct entities (such as a workstation, switch, router, node, etc.), or between addressable entities, or a combination of the two.

legacy That which is inherited, or which remains from a predecessor.

legacy equipment/software Existing equipment, software, or operating procedures that are becoming dated are called legacy systems. For economic reasons, most legacy systems are maintained and enhanced, rather than being scrapped in favor of new systems. Even when it's *not* economically practical, legacies are sometimes retained because managers are reluctant to let go of the emotional investment tied up in existing systems and procedures. A more practical reason for retaining legacy systems is that staff training costs time and money, and staff members may be reluctant to switch to a new system. Most computer operating systems are legacy systems, incorporating downward compatibility in order to work with older equipment.

Lempel-Ziv A compression format described in an IEEE publication in the late 1970s, named after its developers. It has since been widely used in network and modem technologies, but was not developed for the demands of multimedia networks with widely ranging data characteristics, as these did not even exist at the time. See Ardire-Stratigakis-Hayduk.

Lempel-Ziv-Welch LZW. A 'lossless' data file compression format based on the Lempel-Ziv method, first described by Terry A. Welch in 1984 following the publication of the Lempel-Ziv format in the late 1970s. The developer community has buzzed with a substantial amount of debate over use of LZW, since it was distributed as an 'open standard' for many years before the community at large was informed that LZW was being patented. Many programmers incorporated LZW into their software, thinking it was in the public domain.

A patent for the technology is held by IBM (4,814,746), with a similar one held and enforced by Unisys Corporation (U.S. 4,558,302). Some claim these two patents cover virtually the same technology, or that the Unisys patent is a subset of the IBM technology. A similar patent has also been held since 1989 by British Telecom.

The enforcement of the patent rights caused ripples of unease in the programming community, as several modem file protocols use LZW, many graphics interchange formats incorporate LZW compression, including TIFF, which is widely used in desktop publishing programs, and the Compuserve 8-bit GIF format which is popular on the Web.

In telephone technology, LZW is used in Northern Telecom's Distributed processing Peripheral (DPP) for transmission of compressed data. At one point, Unisys issued a statement exempting freeware authors from paying license fees on the use of LZW in their programs in order to quiet the concerns of software developers who were distributing software without commercial gain. Later, Unisys asserted that *all* software developers would be subject to a minimum royalty ($.10) in order to protect the patent, with exceptions only for charitable institutions. See Ardire-Stratigakis-Hayduk.

LEO See low Earth orbit.

LES LAN Emulation Server. A local area network (LAN) software server which provides Media Access Control (MAC) address-to-ATM address resolution services for LAN Emulation (LANE). See LECS, LEC.

less than A concept often represented symbolically (e.g., <) rather than in words, less than is a very common logical operator in computer programming languages. The inverse is *greater than* (>). While its most obvious use is comparing numeric values, it is also sometimes used for text parsing comparisons, with the comparison based on total ASCII value, or the length of the text word or some other predetermined comparison function.

letterboxing A video display technique that preserves the aspect ratio of a wide screen cinematic production even when it is displayed on a different system such as a TV screen. Wide screen movies are often modified to fit a TV screen ratio, but information at the sides of the image are lost when this is done. In letterboxing, it may appear as if some of the image is gone, because there are larger black areas at the top and bottom of the image, but in fact, it is the letterboxed version that shows the *entire* image and retains the fidelity of the original picture. The difference can be quite dramatic. For example, in the beautiful film Baraka, the non-letterboxed version in some scenes will show two people rather than the original three. See aspect ratio.

letters shift A simple mechanical device on a typewriter or teletypewriter, which puts the machine into the mode for printing letters of the alphabet, or for shifting between upper and lower case letters on those devices which include both.

level 1 cache A small, fast static RAM buffer. On an Intel Pentium central processing unit (CPU) chip, there is a 16 kilobyte cache memory incorporated into the chip.

level 2 cache An external, fast, static RAM buffer. On an Intel Pentium central processing unit (CPU) chip, there is a cache memory incorporated into the processor in addition to the level 1 cache which is included with the CPU. On some of the Pentium chips, the level 2 cache is layered into the CPU (for faster access), and in some, it is a separate section, with a bus allowing it to communicate with the CPU. Level 2 caches may vary in size from 256 kilobytes to 1 MByte.

Leyden jar, Leiden jar A device that concentrates and stores electrical energy, thus it served as an electrical condenser; an early capacitor. The Leyden jar was devised by E.G. von Kleist, a German experimenter, in 1745. It consisted of a nail in a bottle connected to a terminal of an electrical device, with the jar held in von Kleist's hand. He received an unpleasant shock from his experiment. A year later, Cunaeus and Peiter van Musschenbroek created a condenser consisting of a jar mostly coated inside and out with metal foil, with the inner coating in contact with a conducting rod that passed through the stopper (insulator).

The foil would typically cover about two-thirds or a little more of the surface of the jar, and the rod would be inserted through a stopper of cork or rubber. Sometimes a chain was attached to the bottom end of the metal rod. The jar was named after the town of Leyden (Leiden) in The Netherlands. It was subsequently discovered that a Leyden jar charge could be sent through wires over distance. Benjamin Franklin conducted numerous experiments with Leyden jars in his attic laboratory, and they remained prevalent for a another 150 years.

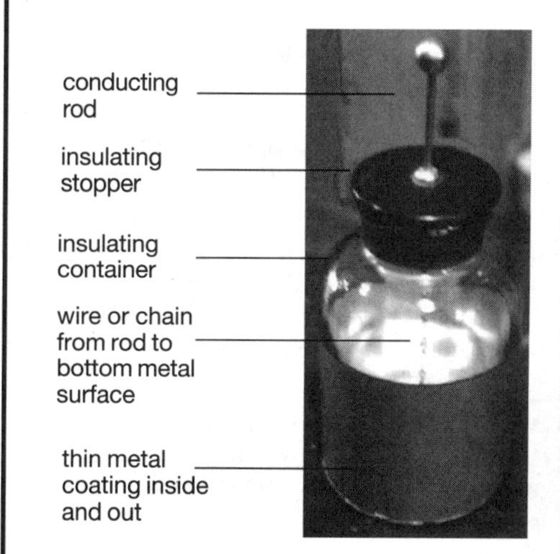

The Leyden jar was an important early 'storage tank' for electrical experimentation, acting as an electrical condenser/capacitor.

LF See line feed.

LFAP See Light-weight Flow Admission Protocol.

LGC Line Group Controller.

LGRS Local Government Radio Service.

Licklider, J. C. R. (Lick) (1915-1990) A computing pioneer who was instrumental in supporting a number of important early developments, including timesharing and the ARPANET. He is best remembered for his inspiration and enthusiasm, and his ability to get the funding and other resources necessary for various computer pioneers to build the stuff of dreams.

LIDB See Line Information Database.

life cycle A period of time describing a product or service cycle, such as the design and development process or the entire life-span from product development through successful sales, to final loss of demand for the product. See shelf life.

life cycle testing The testing cycle of a product from initial testing to product failure. In industrial test environments, products are often exposed to extreme conditions of use or environment in order to determine a baseline life cycle expectancy to estimate what length or type of normal use can be expected from the product. See mean time before failure.

LIFO last in first out. A descriptive term for the order in which data is processed in a queue. For example, if you picture a stack of dinner plates in a plate well in a buffet, when the stack is refilled by the restaurant staff, the last plate on the stack (the one on top) is the first one removed by the next customer. See FIFO, FILO, GIGO, LILO.

On the left, the dot on the "i" collides with the ascender on the "f." Combined as a ligature on the right, the "f" and the "i" have been incorporated into one letter, with more aesthetic contours. Ligatures are typically found in high quality, hand-crafted character sets.

ligature In typography, when setting text, there are instances when two letters placed together will 'collide,' that is, slightly overlap in a way which is unaesthetic, or will leave a gap which detracts from the visual appeal of the composition. In these cases, the typographer may use a *ligature,* a combination of two (or more) letters in which the physical elements of the letters are joined to provide a more appealing or more legible rendition of the combination.

light Radiant energy visible to the human eye having wavelengths in the approximate range of 390 to 750 nanometers, that is, the transition to ultraviolet at one end of the spectrum and infrared at the other end. The phrase *white light* is used to describe light with a mixture of frequencies. The speed of light is 3.00×10^{-8} meters per second, symbolized as *c* in mathematical calculations.

It was discovered that light could be broken up into its component wavelengths with prisms, and this aided researchers in understanding the nature of light and the colors associated with particular frequency ranges. The visible spectrum is specific to human perception; other mammals and insects have broader, shifted, or more specific perception of color ranges. A flower that appears yellow to us, under ultraviolet or infrared light may have other colors which are perhaps 'seen' by pollinating insects. Most dogs and cats are insensitive to colors as we know them (Siamese cats reportedly can perceive color).

Light is the means by which we perceive forms, beings, and their orientation and movement in three-dimensional space. The interaction of the light waves hitting various objects, bouncing back to us through our eyes, and being processed by our brains constitutes the complex phenomenon we call seeing. Some creatures can see beyond the range of humanly visible light. Dolphins use sonar (sound waves) to detect objects which may not be visible to humans, and thus can see 'inside' some objects in a way that is not possible for humans without mechanical aids. See fiber optic, infrared, laser, spectrum, ultraviolet.

light pen A hardware human interface device that translates hand and finger movements into electrical impulses and transmits them to a computing device. These signals are then interpreted into processes by the operating system and applications software.

The light pen resembles a ball point pen with a light detector on one end, where the ink would normally come out of a ballpoint pen, and a 'tail' on the other end, connected to a computer or handheld console. It may or may not include one or more small buttons. On raster displays, the pen detects light on the screen in order for the application to determine the position of the pen. See joystick, mouse, trackball.

light piping Bringing light into an area through fiber optic cables. Laser light will travel along a filament the size of a hair, and filaments can be bundled to provide more light. This is very handy for illuminating hard-to-reach places like small pipes or inside the human body for medical research or procedures.

light-emitting diode LED. An inexpensive semiconductor pn junction structure used in many electronic displays, particularly small ones. The LED lights up when a current is provided. LEDs are common in digital clocks, calculators, microwave readouts, electronic instrument displays, and much more. The LED typically resembles a small illuminated knob with a semiconductor within the knob (which is actually a lens), and leads coming out from the semiconductor/knob arrangement into the device circuitry.

LEDs are now also being used to provide the light rays for certain types of fiber optic transmissions, especially in multimode cables (more precise and more expensive laser lights are used for single mode fiber).

Light-weight Flow Admission Protocol LFAP. A protocol from Cabletron which allows an external Flow Admission Service (FAS) to manage flow admission at the network switch, allowing flexible FAS to be used by a vendor or user without unduly burdening the switch. See RFC 2124.

lightwave communications Optical communications systems using high frequencies. This term helps distinguish optical communications from very short wave microwave communications. Fiber optic cables are used as the physical medium for transmission. This is distinct from *lightwave transmissions* which involve transmission through air or space rather than through a cable as the physical medium. When homodyne or heterodyne detection schemes are used, they are called coherent optical communications systems.

lightwave transmission Optical communications systems based on transmitting a beam through air or space, without using a cable as a physical medium. This is a wide bandwidth, line-of-sight, short distance technology which is relatively inexpensive. It is suitable for building-to-building installations where it is impractical to string wires. This system is subject to loss and is somewhat dependent on weather, and thus is specialized in its practical applications.

Lightweight Directory Access Protocol LDAP. A front-end client/server standard intended to provide a lightweight complement to Directory Access Protocol (DAP), LDAP is based somewhat on ITU-T x.500, with the ability to access X.500 directories. LDAP was developed in the early 1990s at the University of Michigan and submitted as a joint Standards Track RFC with the ISODE Consortium and Performance Systems International.

In LDAP, the protocol elements bypass some of the session/presentation overhead by going directly over Transmission Control Protocol (TCP) or the relevant transport protocol. Many of the protocol data elements are simply encoded as strings. A lightweight best error rate (BER) encoding is used for protocol elements. Extensions to the format such as authentication and server discovery are being discussed and developed. See RFC 1777.

like new A subjective term describing a product which has been opened and stored or used, but is physically unmarred and functionally in good working order. The phrase is not intended to imply that the product will *last* as long as a brand new product in normal use, hence the phrase *like new* instead of *new*. See certified, refurbished, used.

LIM Link Interface Module.

LINC Laboratory Instruments Computer. One of the earliest small computers, developed at the Massachusetts Institute of Technology (MIT) in 1963, the LINC was the inspiration for Digital Equipment Corporation's (DEC's) PDP-8.

line, communications In its most general sense, any path or transmissions link between two or more communicators, including whatever subscriber line, switches, routers, cables, etc. which might comprise the main transmissions pathway. The term line usually implies a physical connection, or series of physical connections including wire or fiber optic cable. Hybrid systems including both wired and wireless connections are sometimes also called lines. Complete wireless connections are usually referred to in terms of *airways* rather than lines.

line, electric A circuit connection physical conductor consisting of shielded or unshielded wire or cable.

line, graphic In computer graphics displays, paint and drawing programs, a visual representation of a narrow, connected path which may or may not be straight. Line tools are very common drawing tools. There are line tools to draw straight disconnected lines, straight connected lines, and curved lines using various freehand drawing or curve algorithms (spline curves are common in vector drawing programs).

line conditioning Improvements and enhancements to a communications line to reduce interference and improve the quality of the signal. Some phone companies offer higher quality line service as an option, which may be of importance to those doing a lot of data communications over a phone line.

line feed In the context of computers, a line feed refers to a movement of a cursor or printer carriage to the subsequent line. In western languages, the subsequent line is down from the previous line. (There are languages which read from bottom to top, but they are unfortunately not well supported by the software on most computer systems.) Line feed does not normally imply a return of the carriage to the start position at the margin (or the beginning of the line on a computer display), although even here some systems make an exception. The return of the carriage is known as a *carriage return*. On some systems, a carriage return or *enter* key includes a line feed, and on some it does not.

From a programming point of view, it's better to differentiate the line feed and carriage return as separate actions so they can be used separately and together in various combinations as needed. In ASCII, character 10 is designated as the line feed.

line finder An evolutionary improvement in step-by-step telephone switching systems that eliminated the need for a separate switch selection for each subscriber line. When the caller picked up the phone, a dial tone was needed before dialing could begin. To accomplish this, a relay would be used to find an available line-finder switch, hunting for the caller's terminal would be initiated, and the caller would be given a dial tone when the line was connected with the switching system.

Line Information Database LID. A national system of telecommunications information databases first deployed in the early 1990s. It is designed so that subscriber and carrier information can be readily accessed and cross-referenced. This information is used for information-querying, validation, and alternate billing administration.

line insulation test LIT. In telephony, a diagnostic test performed from the central office to determine line resistance and line voltages.

line noise Electrical noise in a communications line which interferes with voice communications, or which causes spurious characters to show up in a data transmission. In older modems being used over phone lines, line noise sometimes resulted in strange characters being displayed in the telecommunications software terminal window. Severe line noise can interrupt a data file transfer, or even cause the connection to be disconnected. With newer phone line services and newer, error-correcting modems, this problem is diminishing.

line of sight Many means of communications require an unobstructed straight line of travel for the information to reach the intended destination, or for it to be seen by the appropriate people. Semaphore systems require that the signal arms or flags be visible. Beamed light requires that no impediments block the way or bounce the beam in a direction away from the recipient. Line of sight thus refers to a straight, clear, direct, and otherwise unobstructed travel path.

line powered Any device which receives its power from the main system to which it is attached, or the transmissions medium to which it is attached. For example, most basic phones without extra features do not require a power supply or battery because they are powered by the current in the phone line. Some laptop peripheral devices, to minimize size and weight, derive their power from the laptop itself. See talk battery.

line printer A printing device which prints one line of characters, or a full line at a time. This phrase is used rather generically to mean most impact printers which print a line of characters at a time, though, technically, these are actually character printers. True line printers compose and 'stamp' an entire line of text (they're quite fast and often very noisy), and then compose and impress the next line, composing in the printer memory buffer while the current line is being printed. Line printers tend to be used in institutional and industrial environments where speed is more important than cost or level of quiet.

line status indicator On a telephone, modem, or other appliance which can connect to more than one line at a time, or which can perform a variety of functions on one or more lines, there may be a character display or various lights to inform the user of the status of the line. On multiline telephones, a light usually shows which lines are in use so that the user can avoid barging in on a call inadvertently. On a modem, line status indicators may flash to show whether data is being sent or received, and may indicate whether or not a carrier is present. On network ports, status indicators may indicate loss of frame (LOF) or loss of signal (LOS).

line switching See circuit switching.

Line Terminating Equipment LTE. In SONET networks, an element which originates and/or terminates a line signal. It can originate, access, and/or modify the line overhead, or terminate it, if needed. See SONET, Synchronous Transport Signal.

line tool In image processing programs, a tool which allows straight or curved, joined or separate lines to be drawn. Most paint programs permit the type of line, its thickness, and sometimes its style to be varied according to a number of parameters.

L

line weight In typography, image processing programs, and vector drawing programs, line weight is a setting to control the thickness of a line, sometimes in very precise units such as *points*. In fonts, the line weight can be varied to produce various types of medium, bold, and extra bold effects.

linear transponder A device commonly used in communications satellites and radio relay stations, which takes a small segment of frequencies, amplifies the signal strength across the range of frequencies, and retransmits them at a slightly different frequency range (by shifting or multiplying), so the whole segment is adjusted up or down. This is often done to prevent the transmitted signals from interfering with the received signals. See store and forward repeaters.

lineman, line worker In the early days of the telegraph, when cable was being strung across continents, linemen were assigned to dig holes, cut down trees for poles, set the poles, climb them, and attach the wires, gradually working their way through wilderness, native encampments, and mountain ranges, until the coasts and settlements were interconnected. Once the lines were installed, they would test them, often with portable telegraph keys, and maintain the lines through inclement weather over hostile terrain.

The work was hazardous, no insurance or benefits were available, and linemen who were injured by electrocution, falls, and other hazards, were dependent on the goodwill of their employers for assistance.

Maintaining the increasing number of wires and poles involved the dedication of many 24-hour crews and, until the mid-1960s in North America, much of the work was done by climbing the poles with belts and cleated boots (lineman's climbers), securing in at the point that needed repair and doing the work manually with simple tools. Since that time, power tools, sophisticated testing equipment, and cranes with buckets (cherry pickers) for the line workers have increased in use to the point where it is uncommon to see a worker scale a pole in urban centers.

The line workers now are also responsible for digging, diagnosing, and installing underground transmissions lines, in addition to managing lines on utility poles.

lineman's climbers, line climbers A variety of pole climbing equipment that has been used over the last century to allow installation and repair workers to scale utility poles in various types of weather. These range from climbing irons or spurs strapped on the legs, to cleats on the boots, used in conjunction with a heavy hip belt (sometimes called a 'scare strap') that helped the line worker stay secured and oriented to the pole at a comfortable angle. In urban areas, line climbers have mostly been superseded by mechanized cranes, sometimes called 'cherry pickers,' although climbers are still needed in some circumstances, especially in areas of rough terrain. See lineman.

lines of force, lines of magnetic induction The sphere of magnetic influence of a magnet. For example, in a basic bar magnet, the lines of force can be conceptualized as radiating outward from the poles with no break between the north and south poles. Although diagrams of magnetic lines of force show them as discrete lines in one plane, the actual region of influence is three-dimensional. The *lines* themselves are a descriptive way of conveying the properties of the forces that cause them to crowd each other sideways. Areas where these lines converge, as at the poles, indicate areas of stronger magnetic force. The 'direction' of the lines can be defined as the direction along the north pole if forces other than the magnetic force were hypothetically neutral.

Lines of force can be 'observed' by sprinkling iron particles over a bar magnet and tapping the surface until they form patterns around the magnet. The lines are not fixed in one position; they will change if you sprinkle the filings again, but come about through their interaction, as each piece behaves as a tiny magnet, alternately attracting and repelling near neighbors. Collectively these lines are classed as flux. See flux, magnetic field.

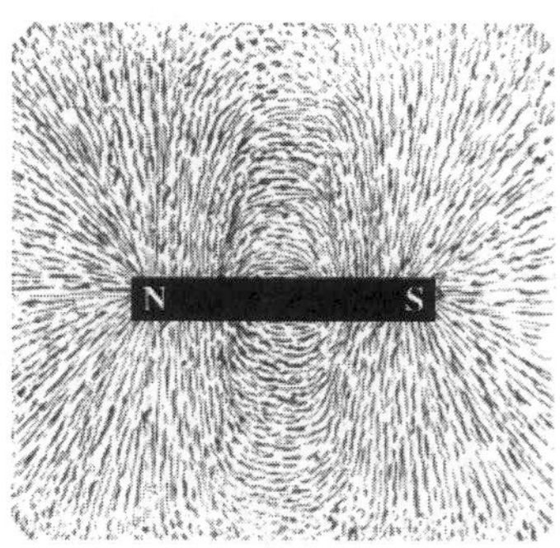

Iron filings sprinkled around a magnet on a light-colored surface or piece of glass will reveal particular types of patterns like those shown in this diagram, depending upon the shape of the magnet. These patterns will change each time the filings are sprinkled, since they are formed not because there are 'lines' emanating from the magnet, but because the magnetic forces associated with the magent cause the particles to interact with the magent and one another in specific ways.

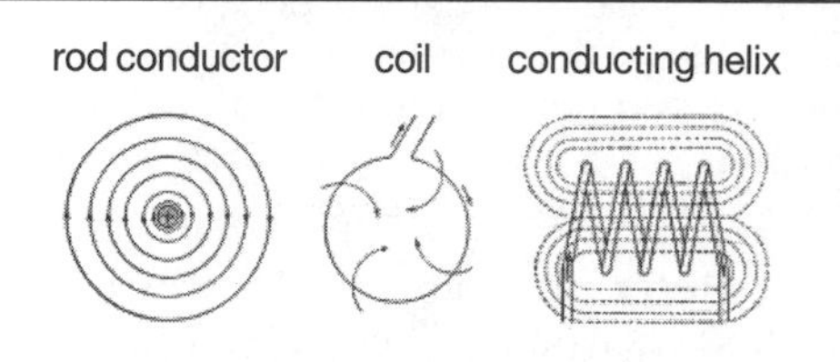

Lines of force are related to the shape and orientation of the objects with which they are associated, and the current flowing through those objects, in the case of electromagnets.

lines per minute LPM. A description of the speed of a recording device such as a line printer, inkjet printer, or teletype-style machine. These devices are sometimes also described in terms of printing speeds of characters per second (cps). The printing speed is not an absolute measure. A printer cannot transcribe the information any faster than it receives it. However, since the bottleneck is more often the printer than the transmission speed, many printers have memory buffers to store incoming information until the printer catches up.

link *n.* 1. In its broadest sense, a communications circuit or channel. 2. A specific leg in a circuit, as between two nodes, or two networks, or two users. In ATM, it is more specifically defined as a logical link, that is, an entity with a specific topological relationship and transport capacity between two specified nodes in different subnetworks. See Ethernet, frame relay, asynchronous transfer mode, tunneling. 3. A communications medium over which nodes can communicate at the link layer. 4. A logical link between software entities, files, or processes, as between an icon and an executable program, or between an alias and a file directory. See alias.

link *v.* 1. To form a logical relationship between software entities, or software and hardware devices. 2. To interconnect hardware devices and/or cables. 3. In software development, while building executable files, to link the compiling source code into appropriate system resources, headers, or other system software that is specified.

link aggregation token See aggregation token.

link attribute In ATM networks, a parameter which is used to assess a link state, to determine whether it is a viable choice for carrying a given connection.

link connection In ATM networks, a connection which can be used to transmit information without the addition of any overhead.

link constraint In ATM networks, a restriction which has been set on the use of that specific link. In other words, restrictions as to whether it may be used as the path for a connection.

Link Control Protocol LCP. In order to support a variety of environments, Point-to-Point Protocol (PPP) provides a Link Control Protocol (LCP). The LCP agrees upon encapsulation format options, handles varying limits on packet sizes, detects common misconfiguration errors, and terminates a link.

The main phases for establishing, configuring, maintaining, and terminating a link are shown in the chart.

The Link Control Protocol packet is encapsulated within the information field of a Point-

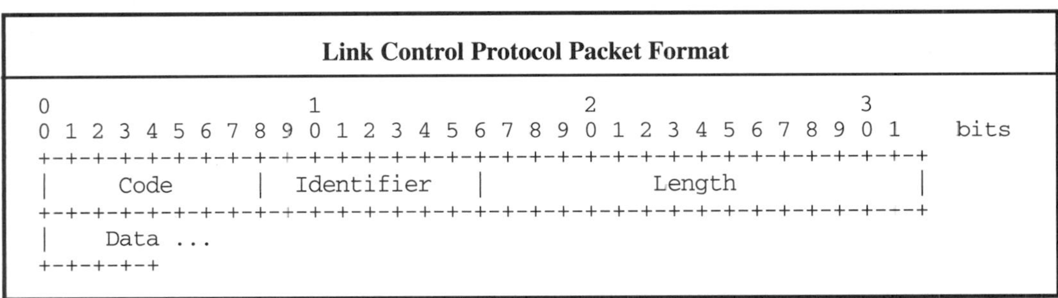

Link Control Protocol Packet Format

```
0                   1                   2                   3
0 1 2 3 4 5 6 7 8 9 0 1 2 3 4 5 6 7 8 9 0 1 2 3 4 5 6 7 8 9 0 1   bits
+-+-+-+-+-+-+-+-+-+-+-+-+-+-+-+-+-+-+-+-+-+-+-+-+-+-+-+-+-+-+-+-+
|     Code      |  Identifier   |            Length             |
+-+-+-+-+-+-+-+-+-+-+-+-+-+-+-+-+-+-+-+-+-+-+-+-+-+-+-+-+-+---+
|    Data ...
+-+-+-+-+
```

Link Control Procedures

Phase	Events	Notes
Phase 1	establishment and configuration	Configuration packets are exchanged, a Configure-ACK packet sent and received, and the Open state entered.
Phase 2	link quality determination	An optional phase in which the link is tested for quality.
Phase 3	network layer protocol configuration	Upon completing optional link quality determination, network layer protocols can be configured and maintained or taken down.
Phase 4	link termination	The link may be terminated at any time, usually at a user's request, but possibly also because of a physical event.

to-Point Protocol data link layer frame. In brief, the packet is configured as shown in the Link Control Protocol Packet Format diagram. See Point-to-Point Protocol, SNA Control Protocol, RFC 1171, RFC 1661.

Link Control Protocol codes Link Control Protocol codes. These link establishment packets are used to establish and configure network links. Link Control Packets are assigned as follows:

Code	Control code assignment
Code 1	Configure-Request
Code 2	Configure-ACK
Code 3	Configure-NAK
Code 4	Configure-Reject
Code 5	Terminate-Request
Code 6	Terminate-ACK
Code 7	Code-Reject
Code 8	Protocol-Reject
Code 9	Echo-Request
Code 10	Echo-Reply
Code 11	Discard-Request

See Link Control Protocol, Point-to-Point Protocol.

link encryption An internetwork security mechanism which, unlike an authentication header (AH), affords some protection from traffic analysis.

link MTU A unit describing the maximum packet size which can be conveyed in one piece over a communications link.

Linux A popular, well-supported, widely implemented, freely distributable, open source Unix-like operating system developed in the early 1990s by Linus B. Torvalds in Helsinki, Finland (he has since moved to California). Torvald has been honored for his contributions to information and telecommunications technology.

Linux is mostly POSIX compliant, and features true 32- or 64-bit multitasking, virtual memory, TCP/IP drivers, shared libraries, protected mode execution, and more.

Linux supports The X Windows system, conforming to the X/Open standard. It also supports the common Internet protocols, including POP, IRC, NFS, Telnet, WAIS, Kerberos, and many more, as a client or a server. Popular Unix shells work with Linux.

Some excellent applications have been ported to Linux, with more being developed all the time. Linux has been ported to many platforms, including PowerPC, Amiga, Macintosh, Intel-based machines, Atari, and others. There are a number of commercially available versions of Linux, including Stac and Red Hat. Linux is a very popular choice for local area network (LAN) and Internet servers. It is reliable, powerful, and is being adapted by many large-scale research and development firms. The Linux Journal is a good trade magazine for Linux developers and users.

Linux International An international organization of developers, vendors, and users of the Linux Operating System and its associated programs. http://www.li.org/

liquid crystal display LCD. A low power display technology comprised of a number of layers, one of which is made of liquid crystal, sandwiched together. The long crystalline molecules in the liquid crystal layer are used to deflect and polarize light. When exposed to an electric field, the crystals orient themselves in the same direction, no polarization occurs, and the light is absorbed; the display remains dark.

Liquid crystal displays were originally developed by RCA's David Sarnoff Research Center and Westinghouse in 1963 and 1964. The first calculator to use LCD technology was introduced by Sharp in 1973.

Simple LCDs include a set of alphanumeric characters and sometimes some symbols. More sophisticated LCDs incorporate raster display technologies, especially when used with computers. LCD panels are popular in phones, calculators, and other low power, mobile devices such as laptops. External light is needed for the user to see the display unless backlighting is added. An *active matrix* LCD includes a transistor at each display point to increase the speed at which the crystals can change state, and can add color. Tektronix further developed 'plasma addressing' which incorporates some of the properties of gas plasma into liquid crystal displays.

liquid detector See electrolytic detector.

LIS 1. Link Interface Shelf. 2. Logical IP Subnetwork.

LISA Laser Interferometer Space Antenna.

LISP *List Processing*. A high level programming language introduced in 1958 by John McCarthy. It is used in many artificial intelligence applications, and as a macro scripting language in applications like AutoCAD from Autodesk Inc. and EMACS. Golden Common LISP is one of the more common implementations of LISP. LISP code is syntactically different from languages such as C, and it has been called Lots of Insignificant Silly Parentheses by subsequent generations of programming students.

list box A list box is a graphical interface device presented to the user for selection from a number of options. The most common type of list box is a directory list. Every time a file is loaded, or saved in a different location, or under a different name, a list box is presented to the user.

List boxes are so important and so commonly used, it's worth mentioning some which are particularly well-conceived, in order to pass on good ideas to software designers.

In 1986 and 1987, Bruce Dawson, author of CygnusEd, and ASDG Incorporated, developers of Art Department Professional, developed some of the best directory list boxes available on any system. They are sizable and modal (background tasks can be carried out while the box is open). The file types that are displayed can be specified by a pattern, with a wildcard capability included. This is very handy for shortening the list and narrowing a search. The Load and Save lists are separate, so there is less chance of accidentally overwriting a file of the same name, and the user doesn't have to constantly switch back and forth between separate Load and Save directories (as often happens when doing file conversions).

The NeXTStep/OpenStep standardized directory list box is quite different from those independently implemented in Amiga applications. It is also well-conceived, which shows that very different ideas can be equally effective.

NeXTStep/OpenStep list boxes show the hierarchical structure of the directories and files in vertical 'panels' in a sizable window. With this design, it is very easy to see the whole structure at a glance and very easy to click to the desired file or directory. In addition, when saving files, it is possible to create a new directory simply by typing in a path. If the directory specified is not found, a dialog box asks if the user wants to create a new directory. If the answer is yes, the system makes a new directory and saves the specified file in that directory in one step. Current versions don't supply pattern-matching or separate Load/Save lists as in the previously-mentioned Amiga applications, but there's no reason why the best features of each system can't be incorporated into future operating systems and software applications.

list server A computer file distribution system used for email, newsgroups, discussion lists, and other types of files which are received from one or many sources, and distributed to one or many 'subscribers' to the service. By using servers to manage the traffic in a centralized manner, it can cut down on administrative overhead, and provide a means to implement security and selective filtering as needed.

LIT See line insulation test.

lithium-tantalate A synthetic crystal first grown in the Bell Laboratories, lithium-tantalate was the first really practical alternative to natural or synthetic quartz for the development of a number of communications-related components such as filters. This substance has practical applications for high bandwidth transmissions. See quartz crystal, quartz crystal filter.

little-endian Stored or transmitted data in which the least significant bit or byte precedes the most significant bit or byte. Many file incompatibilities between computer systems, in which the file formats are otherwise almost identical, are due to platform conventions which store the data in big-endian or little-endian form.

LLC See Logical Link Control.

LLC encapsulation LLC encapsulation is a means to 'envelope' a set of packets so that several protocols can be carried over the same virtual circuit (VC). Type 1 and Type 2 are defined for connectionless and connection services. See Logical Link Control.

LLC/SNAP LLC/SubNetwork Attachment Point. An encapsulation protocol used in Logical Link Control. In an ATM environment, this is the default packet format for Internet Protocol (IP) datagrams. See TULIP, TUNIC, RFC 1483.

LM long-distance marketer.

LMOS Loop Maintenance Operations System.

LMS 1. See Local Measured Service. 2. Local Message Switch. 3. Location and Monitoring Service.

LMSS Land Mobile Satellite Service.

LMU Line Monitor Unit.

LNA See low noise amplifier.

LNW Research Corporation A California company that sold microcomputers assembled and in kit form in the early 1980s. The LNW-80 was a TRS-80 compatible which was arguably better than both the Apple II and the TRS-80 in that it supported a faster CPU, more memory, and higher resolution color graphics. Unfortunately, the LNW computers didn't catch the attention of the public, and the computers never made the impact of the better-known brands.

load coil See loading coil.

Load Number The Canadian counterpart to the U.S. Ringer Equivalence Number (REN) system. As there may be more than one phone device attached to a single line (modem, fax, answering machine, etc.), a system was established to determine and indicate the *ringer load*, that is, the electrical load on the phone line associated with a particular piece of equipment. In this way, the Load Numbers can be summed to show the total load, and make sure the line was not overloaded.

In the REN system, most lines can handle a load of up to 5.0. In Canada, the concept is the same, but the scale uses larger units; in other words, a standard phone might have a load of about 10 to 20 points, with the total load for a single line being about 100 points. Many electronic devices will show the Load Number somewhere on the main circuit board.

loading coil A small electromagnetic induction device which helps prevent attenuation of the signal on a wireline. Loading coils were developed in the early 1900s to improve long-distance transmissions in telephone lines and are still commonly used. By calculating the optimum size and spacing the loading coils carefully along a wireline, it was possible to extend a circuit by several times. With the advent of data communications, loading coils have become a mixed blessing. While they improve transmissions in voice grade lines, they tend to add noise and distortion in the higher frequencies used in data transmissions.

LOC See Loss of Cell.

Local Access and Transport Area LATA. The terms of the 1984 AT&T divestiture resulted in the creation of numerous geographically local telecommunications service areas, of which there are now more than 200 in the U.S. LATAs are determined, to some extent, by population densities. Originally, local Exchange Carriers were not permitted to connect calls across LATA boundaries, as that was the privilege of Inter Exchange Carriers (IXCs). Since that time, the rules have been modified to some extent. See Inter Exchange Carrier, Modified Final Judgment.

local area network LAN. A computer network within a specified geographical space, such as a building or region, or within an institutional entity such as a classroom or department. The network is a means by which the computers are linked to one another and can access a variety of shared resources, typically files, application programs, and peripheral devices such as printers, fax machines, modems, and scanners. Connections between the computers are through wires or wireless signals.

LANs are typically connected directly by telephone wire or coaxial cables, and thus are somewhat constrained in physical size and number of users, due to lower transmission speeds, network topology, signal reduction (see attenuation), and fixed-bandwidth limitations, than wide area networks (WANs).

There are a number of common ways in which LANs are connected, with various commercial, shareware, and freeware products to handle the software tasks associated with networking. One of the most popular server products is Apache, which is robust, freely distributable, and very widely installed. Commercial products include IBM Token-Ring, Microsoft NT, and Novell Netware.

As technology advances and becomes less expensive, direct cabling will probably decrease and wireless solutions become more common, thus reducing the distinction between LANs and WANs. See Token-Ring, virtual LAN.

Local Area Signaling Services LASS. Commercial adjunct processor-based services used to create initial orders. A Bulk Calling Line Identification (BCLID) service provides private branch exchanges with information on calls from outside the PBX group.

Local Area Transport LAT. A proprietary communications protocol for terminal-to-host transmissions, developed as VAX systems by Digital Equipment Corporation (DEC).

local area wireless network LAWN. A local home, business, community organization, campus, or other phone or data network which uses wireless technology to provide the links. The communication is often carried out over radio frequencies (RF) in the frequency modulated (FM) or infrared range. Techniques common to LAWN include spread spectrum, originally developed for government operations, and narrowband. See Industrial Scientific Medical, spread spectrum.

Local Automatic Message Accounting LAMA. An automatic message accounting (AMA) system used in local telephone switching centers in conjunction with number identification information to collect billing data. It also automates the routing of long-distance calls through more than one local office.

local battery Equipment which draws power from a local source, rather than drawing current from the line to which it is attached. Most phones draw current from the phone line sufficient to operate the phone, but if they have extra features (LCD display, speakerphone, etc.), they may require additional power which comes from a battery or local wall socket. Laptop computers use local battery peripherals, such as PCMCIA modems, rather than those which require separate power, e.g., desktop modems, to maximize convenience and portability.

Local Bus Computer processors require a way to communicate with the many devices that make a computer useful: printers, modems, input devices, scanners, etc. The local bus is usually an internal interface that provides a link between the motherboard and various controller cards and other interface connectors. There are many different standards defined for bus transmissions, and the bus speed does not always match the CPU speed, creating a processor bottleneck for some types of processes and activities.

A Local Bus is one of the newer, faster buses which is beginning to supersede other common buses, including EISA, VESA, and PCI.

local call A telephone service phrase referring to calls placed through the local exchange, billed on the subscriber's predetermined regular service plan. These are generally geographically close. There are three common types of billing systems for local calls: 1. unlimited calls for a flat rate monthly fee, 2. flat rate up to a certain number of calls, then a per-call charge beyond that, or 3. a per-charge call which may or may not be scaled according to the total number of calls for the month. The first option is widespread in Canada and the U.S., while many places in Europe use the third option. For contrast, see long-distance.

Telecommunications Act-stipulated Local Exchange Carrier Duties

"Each local exchange carrier has the following duties:

'(1)	RESALE	The duty not to prohibit, and not to impose unreasonable or discriminatory conditions or limitations on, the resale of its telecommunications services.
'(2)	NUMBER PORTABILITY	The duty to provide, to the extent technically feasible, number portability in accordance with requirements prescribed by the Commission.
'(3)	DIALING PARITY	The duty to provide dialing parity to competing providers of telephone exchange service and telephone toll service, and the duty to permit all such providers to have nondiscriminatory access to telephone numbers, operator services, directory assistance, and directory listing, with no unreasonable dialing delays.
'(4)	ACCESS TO RIGHTS-OF-WAY	The duty to afford access to the poles, ducts, conduits, and rights-of-way of such carrier to competing providers of telecommunications services on rates, terms, and conditions that are consistent with section 224.
'(5)	RECIPROCAL COMPENSATION	The duty to establish reciprocal compensation arrangements for the transport and termination of telecommunications."

local differential GPS LDGPS. An implementation of the Global Positioning Service (GPS) designed to improve local accuracy of the data. A single GPS receiver is placed at a known location where it can receive GPS signals. It becomes a reference station which forms a scalar correction for GPS satellites within current view; broadcasting the correction information is provided to local users. Since there is degradation over distance, a series of 'cells' would be needed to apply this system over a large geographical area. See differential GPS, wide area differential GPS.

Local Exchange Carrier LEC. A designation for a local telephone company, now more commonly distinguished as an Incumbent Local Exchange Carrier (LEC) or a Competitive Local Exchange Carrier (CLEC).

This is defined in the Telecommunications Act of 1996, and published by the Federal Communications Commission (FCC), as meaning:

"... any person that is engaged in the provision of telephone exchange service or exchange access. Such term does not include a person insofar as such person is engaged in the provision of a commercial mobile service under section 332(c), except to the extent that the Commission finds that such service should be included in the definition of such term."

See Federal Communications Commission, Telecommunications Act of 1996, United States Telephone Association.

Local Exchange Carrier duties The Federal Communications Commission (FCC) stipulates a number of duties in the Telecommunications Act of 1996 as shown in the chart on the previous page.

local heap See heap memory.

local loop In telephone installations, a physical link through a wire pair connection between the subscriber, which may be an individual, a business, or a private branch system, and the switching office. The local loop used to include the connection right to the subscriber's phone, but now the demarcation point is usually a patch panel or exterior connections box (although to-the-phone is still available for a fee).

Local Management Interface LMI. In frame relay networks, a specification for information exchange between devices. The *frame-relay lmi-type* interface configuration command provides a means to select the type of LMI interface and the *frame relay keepalive* command enables LMI for serial lines. LMI statistics can be displayed with the *show frame-relay lmi* EXEC command.

Local Measured Service LMS. A telephone billing system in which subscribers pay according to the number of calls made or received (or both), rather than according to a flat monthly rate. Measured service is sometimes provided at a flat rate up to a specified number of calls, and then a per-call fee above that number (many banks set up checking charges this way as well). Generally, LMS is of interest to those who make very limited use of the phone or who have a line which is primarily for incoming calls. In some countries, all service is measured and even calls that are unanswered or that result in a busy signal may be billed in some areas.

local multipoint distribution service LMDS. A proposed terrestrial wireless communication service designed to send video over small cells. This would be competitive with urban cable TV services. The Federal Communications Commission (FCC) released a Notice of Proposed Rule Making on the LMDS proposal in December 1992. This led to various FCC proposals to segment the transmissions band, with the effect that the primary spectrum would be limited to the 27.5 to 29.5 GHz range.

Long-term advocates of satellite communications were concerned that this approach would limit future evolution and growth of satellite communications deployment.

LocalTalk A proprietary local area network AppleTalk-compatible protocol developed by Apple Computer and used on Macintosh computers and peripherals. LocalTalk is not a fast protocol, but all Macs come networkable right out of the box, with a simple serial cable, and there's something to be said for ease of use and convenience, particularly in school and work environments. See AppleTalk.

Location Area Identity LAI. A subscriber identity that is allocated on a location basis as part of the Temporary Mobile Subscriber Identity (TMSI) used in a Global System for Mobile Communications (GSM).

lodestone, loadstone A natural magnetic material called magnetite, an oxide of iron. All magnetite can be readily magnetized. This material was used to create early compasses and was called magnes lapis, magnetic stone, or magic stone. It is called lodestone when it comes out of the ground already exhibiting magnetic properties. Lodestone is probably the same stone mentioned by Plato in *Timaeus* as "... the Heraclean stone." since the Heraclean stone was paired in the same sentence with the attractive properties of amber. See amber, magnet.

Lodge, Oliver Joseph (1851-1940) An English physicist who demonstrated that radio waves could carry a signal over distance in 1894. Lodge is also known for his experiments with tuning in radio waves, ideas that were further developed by succeeding scientists.

Lodge-Muirhead detector A simple, early type of self-restoring radio frequency detector which was built by hobbyists and commercial manufacturers in the early 1900s. It employs a small steel revolving wheel, with the outer edge sharpened to a very fine edge, supported between slots on frame posts on either side of a rod passing through the center of the wheel. A small motor supplies the power to quickly turn the wheel. A hard rubber cylinder with a slot cut in the top sits directly under the wheel. Mercury is poured into the slot and makes contact with a binding post threaded in from the outside of the rubber. A thumbscrew is installed in the rubber to raise or lower the mercury level. When the wheel revolves, it makes contact with the mercury and the signal is translated through the mercury, by means of brushes, to the binding post which connects to the rest of the circuit, including a receiver. The battery power to the motor is controlled by means of a potentiometer. See detector.

LOF See Loss of Frame.

log in, log on *v.* To gain access to a console, computer, or network terminal through software, usually by typing certain keystrokes, or selecting a button. Secure systems may require a name, or a name and password. If an authorized name or password is not correctly given, the system, may, after a certain number of failed attempts lock out any subsequent attempts, and, as in banking or government systems, may even notify the system operator of an attempted break-in. Most systems let you keep trying, though, until you get it right. A succession of logins may be necessary in order to gain access to subsections of the system after the first login.

The most effectively whimsical 'failed login' message this author has seen is on NeXTStep, where an incorrect login results in the login window shaking rapidly back and forth for a few moments (as if shaking its head to say no). The verb form of *log in* is two words, as in "I plan to log in now." Otherwise it's one word, as in "My login attempt failed."

log off, log out *v.* To exit a console, computer, or network terminal session, usually by typing: logout, logoff, done, exit, or quit; or by clicking the appropriate button (or simply by hanging up the modem, though this is an untidy way to exit). Some logout routines will display a log summary of the session showing the length of the session, and sometimes even the type of activities. It's important to be absolutely sure that you log out of secure systems, or the next person sitting down at the console may have access to unauthorized files or programs. The verb form of *log off* is two words, as in "I plan to log off now." Otherwise it's one word, as in "My logoff is done, you may have the terminal ..."

log-periodic antenna A periodic antenna is one in which the input impedance varies as the frequency varies. Log-periodic antennas employ a variety of arrangements of active dipoles interconnected to provide broadband, high gain capabilities. They are commonly used for very high frequency (VHF) signals.

logic bomb A software program designed to penetrate a system, present a message (sometimes through graphics or sound), or damage memory or stored data, when some particular logical operation happens. A time bomb is a type of logic bomb which can, for example, wipe all the data off a hard drive when the bomb detects that it is April 1st. Logic bombs are not always malicious, but they are seldom appreciated. They generally fall into the category of practical jokes, which are usually funnier to the perpetrator than to the intended object of the joke.

logical block In storage devices, such as hard drives and magneto-optical devices, the smallest addressable unit. Each block is associated with a unique number, usually starting with 0, and incrementing for each succeeding block. This allows the system to locate data, read and write to the device, partition the drive, etc., in an organized manner.

Logical Block Address LBA. A means for saving and retrieving information by accessing block addresses on a storage medium, rather than by using cylinder-head-sector addressing schemes. The blocks on the storage medium are addressed sequentially, usually starting with zero. SCSI peripherals use this addressing method. Some IDE drives are now beginning to use this method, but it may be necessary to explicitly request LBA mode.

logical drive A drive which is configured separately from the physical configuration of the drives. For example, a computer may have three drives, each with several partitions, but logically, the system may organize them into four logical 'drives' with various partitions from different drives in such a way that the user 'sees' four drives rather than three, of sizes that are set by software. Conversely, a system may have three drives, each with a couple of partitions, which are aggregated into one drive. Thus, the storage space appears to the user as one large drive. Not all operating systems can organize drives in this way. Some of the lower end personal computers have limitations in the configuration of logical drives.

logical link A link between nodes or devices based on an abstract rather than a physical topology. Thus, virtual LANs, 'direct' connections, and other types of paths can be set up in addition to the physical connections.

Logical Link Control LLC. The upper sublayer of the layer 2 Open Systems Interconnection (OSI) protocol. LLC provides data link level transmissions control. It is the default multiplexing layer for Internet Protocol (IP) over AAL5. It was developed by the IEEE 802.2 committee to provide a common access control standard for networking which is independent of packet transmission methods. It includes addressing and error checking capacities.

Logical Storage Unit LSU. A buffer unit which connects to PBX systems to store call information. Also known as a *buffer box* or *poll-safe*.

logical topology In a network, the connections and relationships between computers and various devices may not map in a one-to-one relationship with the physical topology. Thus, logical topologies, organized and managed in software, may be administrated and diagramed separately or as an 'overlay' to the physical connections. In large networks, logical topologies often require tracking and display programs to configure and trouble-shoot the logical connections. From a user's point of view, many aspects of a logical topology can make computing more efficient and enjoyable.

As a simple example, there are workstation networks which allow hard drives to be mapped in such a way that they appear as one giant drive to the user, even though the data may physically be spread over a number of partitions and drives. Thus, the user doesn't have to worry about whether there's enough space, which drive to use, or on which drive or partition that old file was stored. The operating system takes care of all the housekeeping involved in managing the system. On a larger scale, the same can be said of whole systems. The user may be using an application program that's located on a machine in another room, another building, or another state, but it can be logically mapped in so that it appears to be running on the local machine. Commercial software programs with graphical user interfaces exist to help manage logical topologies and, in some cases, various management utilities come with the operating system itself.

logical unit number LUN. The LUN is an identification system used with SCSI devices which allows the computer and controller to distinguish and communicate with up to seven devices including the controller for each SCSI chain. Each device must have a unique ID for the controller to administrate more than one device, when they are chained together.

logiciel Software, in French.

login script, logon script A login script is a file which includes commands and/or variable settings pertaining to the login and initial setup. The two most common functions of a login script are: 1. to set up the system to the specifications of a particular terminal, and/or 2. to set up the preferences for a particular user or set of users. In the first case, the login may include information about the characteristics of the terminal, such as hardware-specific keyboard mapping, graphics card settings, etc., as well as environmental variables, patches, and other initialization parameters. In the second case, the settings may include the user screen size and color preferences, preferred fonts, frequently used applications, permissions, and other characteristics of the user environment. See batch file, JCL, Perl.

Long Distance Internet Provider LDIP. Cooperative services offered by companies that operate under restrictions set by the Modified Final Judgment (MFJ) who wish to provide Internet services.

long haul communication A call which extends beyond the local exchange area.

long key A key held down for longer than a prescribed period which signals an event separate from a short press of the same key. For example, a long key on a computer keyboard might cause the character associated with the key to repeat, or to initiate a key-related process.

long tone In telephony, a long key is one which signals for longer than other keys, in order to communicate through various automated phone menu systems. That is why many phone system menus instruct you to press the 'pound' key (#) after typing in a number, so that a long key signal will be transmitted.

long-distance call, toll call A telephone service referring to any call outside the local service area. Long-distance calls are frequently completed and charged through a carrier other than the local service, and include extra digits in order to identify the desired destination. Long-distance calls are usually billed in a cooperative arrangement through the local carrier, although some services may be billed separately. Sometimes called a *trunk call,* though this phrase is less common.

The first recognized long-distance call is said to have been a one-way message from Alexander Graham Bell's father in Brantford, Ontario, Canada, to Bell in Paris, Ontario in 1876. The first two-way long-distance call was between Bell and Watson in Cambridgeport and Boston.

The first transcontinental phone line went into service in 1915, connecting San Francisco and New York City, with Edison and Watson conducting the first conversation over this line. About a decade later, long-distance radiotelephone service was established across the Atlantic Ocean.

Some of the important inventions which made long-distance communications possible were Pupin's *loading coils*, de Forest's *triode*, Armstrong's *regenerator* circuit, and microwave antennas introduced in the early 1950s. Contrast with local call.

long-distance carrier IXC. Local long-distance providers which are competitive with incumbent local exchange carriers.

longitudinal redundancy check LRC. A data transmission error checking technique incorporating a block check on a group of data. An accumulated Block Check Character (BCC) is compared to the sending BCC; if they match, the block is considered to have been transmitted without errors. See cyclic redundancy check.

Loomis, Mahlon (1826-1886) An American dentist and researcher who was intrigued by the fact that early telegraphs could be run with only one wire, with the earth providing the conductor for the return circuit. He reasoned that if the earth could act as one conductor, then perhaps the air could act as another, especially since Benjamin Franklin's experiments had alerted scientists to the electricity that was in the air. In 1865 or 1866 Loomis devised an experiment in which he raised kites with equal lengths of fine copper wire and demonstrated that a signal could be transmitted from one to the other without direct physical contact. He received a U.S. patent for his improved wireless telegraphic system in 1872.

loop 1. A complete transmissions circuit, or electrical circuit. 2. In telephone systems, a loop comprises the wire transmission path that extends from the central office to the residential or business subscriber and back.

loop, communications hardware A circuit, conduit, or line, which comprises a continuous path with starting and ending points meeting at the same geographical point. The start and end points may or may not be joined. A loop may or may not include nodes. Communications through the loop may be unidirectional or bidirectional. A loop need not be roughly circular, although it sometimes is; often a 'loop' consists of two adjacent lines, one which sends, the other which receives. Some 'loops' send and receive on the same line (especially if it's a wider bandwidth medium such as fiber), so the 'loop' aspect is based more on the nature of the transmission than the configuration of the cable. See local loop, Fiber Distributed Data Interface, Token-Ring.

loop, programming In software, a programming loop is a series of instructions which will repeat until some event or condition occurs to cause the software to drop out of the loop, or to branch to a specified destination. An endless loop is one which, theoretically, goes on forever. In actual practice, an endless loop often indicates a fault condition and is usually externally terminated. See nesting, recursion.

loop antenna A type of radio direction-finding antenna which has one or more complete continuous loops of wire, the ends of which connect to complete the circuit.

loop start In telephony, it is necessary to take control of a line before it can be used. There are two common ways in which to do this, with a *ground start* or a *loop start*. The loop start is the type commonly found in residential and other single line phone lines. When you pick up a phone, the plunger is released (off-hook) and the circuit sends a supervisory signal by bridging the two wires in the phone connection (traditionally called *tip* and *ring*) with direct current (DC). This is done so that the subscriber will get a dial tone and circuit through which to connect the call. The central telephone switching office sends a signal to the phone the caller is trying to reach and rings the number until it goes off-hook when it is picked up by the callee. When the loop is detected, the ringing signal is no longer set. See ground start.

loran *lo*ng *ran*ge navigation. A system of distance navigation in which several radio transmitters (usually land-based) are used to send out pulsed signals from different directions in order to determine the geographic location of the craft using the loran system. Useful for air- and watercraft under some circumstances, but is limited by the availability and distance of loran stations. See Global Positioning System.

Lorimer switch One of the first commercially promoted automatic telephone switches, patented by the Lorimer brothers in 1900 and put into service in 1905. While it had many improvements on its predecessor, the Callender switch, it probably owes some of the impetus for its development to this earlier invention. It was installed in a number of switching systems in Europe, but was never fully reliable. However, the technology was modular and could be extended, an important influence on future telephone switching systems. See panel switch, rotary switch, Strowger switch.

A Lorimer desk phone showing levers which are set to indicate the number to be called.

Lorimer telephone An early telephone design which was powered by a central battery system, and dialed with a series of levers representing units, not unlike an old calculating machine or cash register. Setting levers configured a telephone number.

Lorimer, George William A Canadian employee of inventor, Romaine Callender, George William worked as a telephone operator at the Callender Telephone Exchange Company. He and his brother, James Hoyt Lorimer, later accompanied Callender to New York where Callender was seeking financing to establish a new company after filing a series of patents on telephone switching technology which he was not able to implement in Brantford, Ontario.

In New York, the group succeeded in creating an automatic switching system, after which they returned to Brantford, Ontario. Callender traveled to England to found the Callender Rapid Telephone Company, and the Lorimer brothers founded the Canadian Machine Telephone in Peterborough in 1897. After the death of his brother, James Hoyt, Egbert Lorimer joined George William in marketing their technology. See Lorimer switch; Callender, Romaine; Lorimer, James Hoyt.

Lorimer, James Hoyt The brother of George William Lorimer, James Hoyt originally studied law, but became involved in telephone switching systems research with his brother, George William Lorimer, and George's employer, Romaine Callender. Together the Lorimers founded the Canadian Machine Telephone company in 1897. James Hoyt had a strong mechanical aptitude, and the brothers continued to improve on the Callender switching technology until it was patentable in 1900. James Hoyt met an untimely death after which no significant technological innovation occurred in the partnership, although the products continued to be marketed. See Callender, Romain; Lorimer, George William.

LOS See Loss of Signal.

loss 1. A decrease in power of a transmission signal as it travels toward its destination, usually expressed in decibels (dB). Many factors contribute to loss, such as distance, type of signal, weather, signal modifications through switches and routers, equipment characteristics, etc. Loss through a circuit is cumulative. See amplifier, interference, noise. 2. In a network, a quantitative measure of a reduction in system resources or services arising from undesired factors such as faulty equipment or configuration, vandalism, or incorrect usage.

Loss of Cell LOC. In ATM networking, a performance monitoring function of the PHY (physical) layer in which a maintenance signal is transmitted in the overhead indicating that the receiving end has lost cell delineation.

Loss of Frame LOF. In ATM networking, a performance measure indicating whether frame delineation has been lost. The LOF is transmitted through the physical (PHY) overhead. On some systems, a LOF condition will be signaled on a port with a light-emitting diode (LED), or as a 'yellow alarm.'

Loss of Signal LOS. In ATM networking, a performance measure indicating that the receiver is not getting the expected signal, or that there is simply no signal because nothing is currently connected. The LOF is transmitted through the physical (PHY) overhead. On some systems, a LOS condition will be signaled on a port with a light-emitting diode (LED).

lossless compression A type of data compression technique which does not lose information contained in the image in the compression stage. Some compression algorithms average, sample, or remove image information in order to achieve a high degree of compression, e.g., JPEG. Others retain all the information, e.g., TIFF. See compression. Contrast with lossy compression.

lossy compression A type of data compression technique which selectively or randomly loses information contained in the image in the compression stage. These algorithms average, sample, or remove image information in order to achieve a high degree of compression, e.g., JPEG. New wavelet mathematics is providing some very interesting compression options which provide a high degree of compression with a surprising degree of fidelity to the original image when decompressed and displayed. Others techniques retain all the information, e.g., TIFF. See compression, discrete cosine transform, fractal transform, wavelet. Contrast with lossless compression.

loudspeaker A device designed to amplify sound, especially voice or music. It may be mechanical or electrical. A mechanical loudspeaker may be as simple as a bowl or horn-shaped object that directs sound. An electrical loudspeaker is familiar to most people as the "PA system" (public address system) installed in most public schools. See amplifier, audio, intercom, sound.

Lovelace, Ada Augusta (1815-1851 or 1852) Countess Ada Lovelace (nee Byron) was the daughter of the famed English poet Lord Byron. Ada Lovelace worked with the computer pioneer Charles Babbage, and is regarded as the first computer programmer for her description of how an analytical machine might compute Bernoulli numbers. She proposed the possibility of using computers to compose music or produce graphics. A computer language (ADA) was developed by the U.S. Department of Defense and named after her. See ADA; Babbage, Charles.

Ada Lovelace had an active interest in both the sciences and the arts and speculated on future possibilities for thinking machines.

low Earth orbit LEO. An orbiting region around the Earth into which certain types of communications satellites are launched. There are advantages and disadvantages to low orbits. The main advantage is that it generally requires less power to transmit and receive at lower altitudes; the main disadvantage is that it requires a larger number of satellites to provide full global coverage. Other factors include lower radiation levels and lower launching costs. The life-spans of low-orbit satellites tend to be around five to eight years.

Most low-orbit satellites travel at about 500 to 2,000 kilometers outside Earth. There is a region called the Van Allen radiation belt at the outer regions of low Earth orbit which is generally avoided, between the LEO and medium Earth orbits (MEOs). LEO satellites are primarily used for cell phone and data communications.

Communications designed for lower orbits require a larger number of satellites than those for higher orbits. This necessitates greater coordination, to handle the larger number of systems and to deal with the shorter periods during which each satellite is within range. In contrast, high Earth orbit (HEO) systems can blanket the Earth with only three or four satellites. The trade-off is that higher-placed satellite transmitters require more power to beam the greater distances. See Ellipso, Global Positioning Service, Globalstar, high Earth orbit, Iridium, medium Earth orbit, Orbcomm, Teledesic.

low level formatting In storage devices such as hard drives and cartridge drives, the formatting is the process of arranging the magnetic media on the storage surface to conform to a recognized patterns so the operating system can further organize data on the drive (the next step is usually to high level format [initialize] and partition the drive). Each operating system has its own file formats, the protocols that allow it to create directories, organize files and file pointers, and read and write information from and to the drive. Some operating systems are designed to recognize the file formats of other systems as well. For example, on Macintosh and NeXT systems, if a DOS/Windows disk is inserted in the drive, the Mac or NeXT OS will recognize the foreign drive and read and write data files to the drive (and perform minor conversions as necessary) in the format of the diskette, rather than the native operating system format. This provides the user with a lot of flexibility in terms of data transfer and conversion. This does not mean that executable files from other systems can be run on any platform, but rather that files can be moved about as needed.

Most drives now come low-level formatted from the factory, but if you have serious data loss on a drive, sometimes as a last resort, you may need to reformat the drive to make it usable again.

low level language A computer control or programming language at the machine or assembly level at which individual registers, accumulators, and other aspects of the physical architecture can be directly or nearly directly controlled. Low level languages are rarely used these days except for writing simulators for various types of processors.

It is much more common now to use high level programming languages to create source code, and then engage an intermediary program, called a compiler, to translate the high level language into machine instructions. A certain amount of bit-twiddling can be accomplished in some of the medium level or high level languages, but this is needed only in limited circumstances. Contrast with high level language.

low noise amplifier LNA. A component which amplifies and sometimes converts telecommunications signals, typically from satellite transmissions. In a satellite receiving station, the LNA takes signals from the feed horn, amplifies them, and then converts them or sends them to a separate low noise converter (LNC); from there they are transmitted to the receiver, usually inside a building. See feed horn, low noise converter, parabolic antenna, satellite antennas.

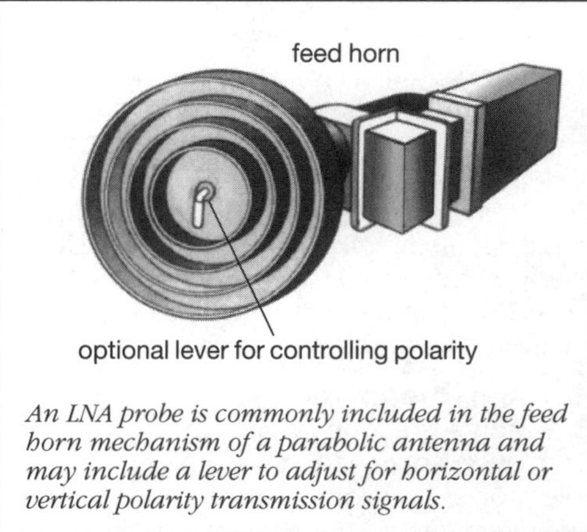

An LNA probe is commonly included in the feed horn mechanism of a parabolic antenna and may include a lever to adjust for horizontal or vertical polarity transmission signals.

low noise amplifier probe LNA probe. A component which works in conjunction with a low noise amplifier to control the signal polarity, which can be set to either horizontal or vertical, in order to accommodate more channels on a single system. The LNA probe is typically built into the feed horn mechanism on parabolic antennas.

low noise block converter LNB. A component which converts amplified signals, usually to a lower frequency to send to a receiver. In telecommunications, it is commonly used with satellites and may be incorporated into the low noise amplifier (LNA). LNBs have a broader range than LNCs, as they are able to convert a range of frequencies (provided they have the same polarization) rather than just a single frequency, as in LNCs. See low noise amplifier, low noise converter, parabolic antenna, satellite antennas.

low noise converter LNC. A component which converts amplified signals, usually to a lower frequency to send to a receiver. In telecommunications, it is commonly used with satellites and may be incorporated into the low noise amplifier (LNA). LNCs work with specific frequencies. See low noise amplifier, low noise block converter, parabolic antenna, satellite antennas.

low pass filter A filter that passes transmissions below a specified cutoff frequency, with little or no loss or distortion, but effectively filters out higher frequencies. See high pass filter.

LPC linear predictive coding.

LRC A type of data error correcting technique. See longitudinal redundancy check.

LRN Location Routing Number.

LRS line repeater station.

LSDU Link layer Service Data Unit.

LSI A term in the semiconductor industry describing capabillities aggregated onto a single chip. See large scale integration.

LSN See Large Scale Networking group.

LSSC lower sideband suppressed carrier. A modulated carrier wave which has had part of the signal 'stripped away' in order to save bandwidth. This lower sideband is rebuilt mathematically at the receiving end to recover the original signal information.

LSSGR LATA Switching System General Requirements.

LSU See Logical Storage Unit.

LTB Last Trunk Busy.

LTC Line Trunk Controller

LTE See Line Terminating Equipment.

LTS Loop Testing System.

Lucent Technologies A company which was created subsequent to the AT&T/Bell divestiture. Lucent was created from the Bell Laboratories research staff, and a number of the electronics, network, and business communications groups, including Systems for Network Operators, Business Communications Systems, Microelectronics, and Consumer Products. See Barton, Enos.

Lucent Technologies Canada Inc. A wholly owned subsidiary of Lucent Technologies, based in Ontario, Lucent Technologies Canada Inc. formed as a result of the restructuring of AT&T after the divestiture. Lucent began in Canada as part of AT&T Canada Inc. in 1984.

lug 1. A projecting attachment point, especially for electrical circuits. See terminal. 2. An attachment added to the end of a wire which provides an eye, or forked end, which allows the wire to more easily be attached to a bolt under a binding screw.

Lugs are designed with with eys or forked ends to facilitate attachment to electrical terminals.

lumen A unit of light (luminous) flux equal to the light produced by one candle intensity on a unit area of a flat surface of uniform distance from the light source.

luminance The luminous (light-emitting) flux reflected or transmitted from a source such as a TV screen or computer monitor. Sometimes called brightness.

luminosity A ratio of light flux to its corresponding radiant flux at a specific wavelength, expressed in lumens per watt.

luminous flux The visible energy (light) produced per unit of time. See lumens.

LUN See logical unit number.

Luneberg lens A type of focusing lens used in antennas to increase gain for ultra high frequency (UHF) transmissions.

lurker A tongue-in-cheek Internet term for those who read the chat channels, newsgroups, and other public discussion areas, without contributing to the discussion or revealing their presence online. Lurkers run the gamut from those who are are new, or simply busy and don't have the expertise or time to contribute, to those who deliberately choose not to reveal their identities in order to protect their privacy (e.g., celebrities). Occasionally there is resentment of lurkers, as some forums have a community take-and-give philosophy, and those who take without giving are perceived as taking unfair advantage of the other participants. However, it's probably good that there is a high proportion of lurkers, for if everyone on the Internet felt compelled to express an opinion on every subject, the discussion groups would become useless through excess volume See handle, newsgroup, pseudonym.

lux A combining word from *lu*minance and

flux. A basic metric unit for expressing illumination (equal to about 10 foot candles). The illumination on a one square meter area on which the flux of one lumen is uniformly distributed.

luxmeter A type of light-measuring instrument that records intensities. Light meters are commonly incorporated into cameras to help to determine aperture and speed settings.

Lynch, Dan Lynch organized the first TCP/IP Implementor's Workshop in 1986, which later developed into Interop in 1988, a large gathering of Internet, network, and other telecommunications professionals. Lynch is also known for his role in the ARPANET transition from NCP to Internet Protocol (IP).

LZS See Lempel-Ziv-Stac.

LZW The abbreviation for a common compression algorithm. See Lempel-Ziv-Welsh.

m 1. Symbol for meter. 2. Abbreviation for milli-. See milli-.

M Abbreviation for mega-. See mega.

M hop A type of pattern that results when communications transmissions are bounced from an Earth station to an airborne receiver, back to an Earth station or intermediary hub, up to an airborne receiver and back down to the final receiving station, thus resembling the letter 'M.' This is a common configuration in hub topology satellite communications. Newer satellites are being designed for intersatellite communication, so the signal goes from an Earth station to a satellite, to another satellite and then down to Earth again, thus forming a shape like three sides of a rectangle rather than the letter M as is shown in the diagram following.

M port In a Fiber Distributed Data Interface (FDDI) network topology, this is an extra port on a concentrator used to attach other nodes in a branching tree topology. M ports can be on both single attachment and dual attachment concentrators. The M port is an addition to the basic attachment to the FDDI network. On a dual attachment station, a redundant link can be created by connecting the A and B ports on different concentrators on the M ports. On a single attachment concentrator, the M port may be connected to the S port. M ports are never connected to one another. The other end of the M port may be attached to a patch panel through a data grade cable.

Ma Bell *colloq.* A term to familiarly describe the Bell telephone system, and later AT&T. The Bell system was so ubiquitous and so well recognized that many came to exclusively refer to the corporation as Ma Bell.

MAC 1. Media Access Control. In a layered network architecture, the lower half of the data-link layer that governs access to the available IEEE and ANSI LAN media. See Media Access Control for a fuller explanation. 2. See multiplexed analog component.

MAC address Media/Medium Access Control address. A network location identifier. See

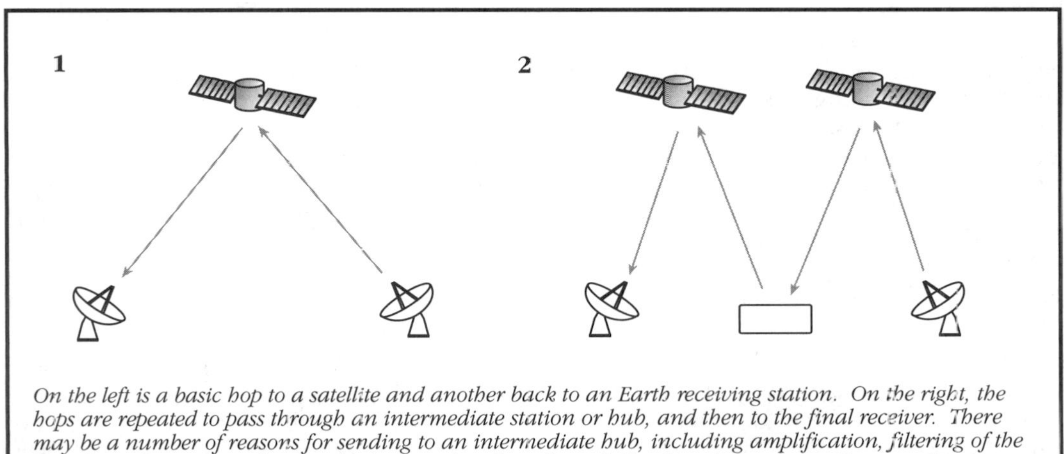

On the left is a basic hop to a satellite and another back to an Earth receiving station. On the right, the hops are repeated to pass through an intermediate station or hub, and then to the final receiver. There may be a number of reasons for sending to an intermediate hub, including amplification, filtering of the broadcast channels, and redistribution to subscribers through Earth stations at more than one location.

Media Access Control.

macadamize To pave using bituminous binding materials. The process was named after John L. McAdam who developed road improvement techniques during the American Revolution. The paving of roads significantly enhanced the speed of communications, since most messages at the time travelled only as fast as the people or horses carrying them.

MACE Macintosh Audio Compression and Expansion.

Mach line, Mach surface The division between regions of supersonic and subsonic flow.

machine dependent Software or peripherals designed to work in conjunction with a specific system or architecture, and not readily usable on other systems (although sometimes it can be done with modifications). Low level routines written to take advantage of a particular chip architecture or peripheral card are machine dependent. Most executables are machine dependent, since they have usually been compiled from a higher level language down into low level machine code for a specific system.

machine language A symbolic computing machine control language that functions at the lowest level possible on a system, in symbols that are readily understood by the computer, but which are inscrutable to most people. Machine language involves the most basic movement and processing of data, in terms that are specific to the computer architecture (usually binary). Thus, *move* and *add* instructions are used frequently in machine language programs. A move instruction transfers data between registers, and an add instruction performs a math operation (add, multiply, subtract, etc.).

Because machine language programs are cryptic, long, difficult to follow, difficult to read, and difficult to debug, assemblers were developed to codify and organize instruction sets so software could be written and debugged more quickly. Later, higher level languages such as FORTRAN, BASIC, LISP, C, Modula, Perl, and Java, etc. were added to make the task of programming easier still, and to adapt programming languages to specific types of tasks.

Some higher level languages are 'compiled' into machine language executables which can be run directly thereafter, and others are 'interpreted' into machine language at the time an instruction is processed. Compiled languages typically run many times faster than interpreted languages because the conversion to machine instructions happens only once, prior to writing the program. This machine language compiled executable is then stored as a file and used as needed.

The Macintosh Plus had a cute, portable all-in-one design, featuring a monochrome monitor, two serial ports, SCSI controller, graphical user interface, mouse, and networking capabilities through Appletalk, as standard features.

Macintosh A family of Motorola 68000-based personal computers developed and distributed by Apple Computer. The first of the Macintosh line was the Lisa computer, introduced in 1983. Its graphical operating system was described as 'radical' by many. There were also vocal detractors that said 'that graphics interface' would never be accepted in an office environment. A graphical interface may not seem unusual now, but at the time, computers were almost exclusively text-based, and this new graphical user interface, which was accessed with a mouse, seemed extraordinary. Many business and computing professionals recoiled in suspicion. In retrospect, now that virtually all computers have graphical user interfaces, the harsh criticisms leveled at the Macintosh graphics appear to have been shortsighted.

The Macintosh graphical operating system was inspired by some of the brilliant research and development that was occurring in the 1970s and early 1980s at the Xerox PARC facility in California. The fledgling founders of Apple Computer were given a tour of the facility, and Steve Jobs came away energetically inspired by what he had seen, determined to add a new line of computers different from the Apple II line.

Unfortunately, while the Lisa was a good machine and most of its characteristics were later incorporated into the more familiar Macs, the initial price tag was high and it didn't sell well. The early Macintosh computers that are familiar to most people started with the Mac in 1984, followed by the Mac Plus and various subsequent Macs (there are now dozens of models). The first major change in the Macintosh line was the changeover to PowerMacs in the 1990s, followed by the evolution to G3s late in 1997. The 1984 release of the 128K 32-bit 4.7 MHz 'little Mac' was accompanied by a significant product in 1985, the Apple Laserwriter. This new laser printer changed the way many people thought about computers. Since 10-pin dot matrix printers with extremely limited fonts were prevalent at the time, few had considered the potential of personal computers as publishing tools. With the Laserwriter and Adobe Systems PostScript fonts, the publishers sat up and took notice and the desktop publishing industry was born, with very substantial repercussions to the traditional layout and printing industry that are still reverberating today. New all-digital presses are completely changing the way printed information is produced.

The Macintosh line soon became the preferred system in print publishing service bureaus, and many of the high-end desktop publishing and graphics programs were available only for the

Overview of Some Early Apple Macintosh Models

Model	Year	Speed	Processor	Notes
Lisa (Mac XL)	1983		68000	First GUI. Didn't sell well.
Macintosh	1984	4.77 MHz	68000	128K
Macintosh Plus	1986		68000	Launched DTP industry.
Macintosh II	1987		68020	First of many successful computers in the Macintosh II line.
Macintosh IIx	1988		68030	
Macintosh IIcx		16 MHz	68030	68882 mathco.
Macintosh SE/30	1989	16 MHz	68000	68882 mathco.
Macintosh SE	1989		68000	
Macintosh IIci	1989	20 MHz	68030	One of the best and most reliable of the Mac II line. 68882 mathco.
Macintosh IIfx	1990	50 MHz	68030	Intended as a graphics machine, it didn't sell well.
Macintosh LC	1990	16 MHz	68020	
Macintosh Classic	1990	8 MHz	68000	A low cost nostalgic release.
Macintosh IIsi	1990	20 MHz	68030	Optional 68882.
Mac. Classic II	1991	16 MHz	68000	Optional 68882.
Macintosh IIvi	1992	32 MHz	68030	Optional 68882.
Macintosh IIvx	1992	32 MHz	68030	With 68882 mathco.
Macintosh LC II		16 MHz	68030	
Mac. Color Classic	1993	33MHz	68030	Optional 68882.
Macintosh LC III		25 MHz	68030	Optional 68882.
Mac. Centris 610	1993	20 MHz	68040	
Mac. Centris 650	1993	25 MHz	68040	
Mac. Quadra 800	1993	33 MHz	68040	
Mac. Centris 660	1993	25 MHz	68040	1 NuBus slot.
Mac. Quad. 840AV	1993	40 MHz	68040	Tower model. DSP. 3 NuBus slots.
PowerMac 6100	1994	Major change from Macintosh line to PowerMacs, subsequently followed by change to G3 and iMac lines in the late 1990s.		

M

Mac. It was not until the early 1990s that some of these important software programs were ported to Intel-based machines, and service bureaus began to use both platforms. The Macintosh Overview chart doesn't include a complete list of Macintoshes and their features, as there are now dozens of models, but some of the Macintosh evolution can be seen through these early desktop models.

MacMICA A Macintosh-based videoconferencing program from Group Technologies, which works over AppleTalk networks. See Cameo Personal Video System, Connect 918, CU-SeeMe, IRIS, Visit Video.

MacMiNT A text-based, Unix-like operating system ported form the Atari ST to the Macintosh which can be used with freely distributable Unix utilities such as GCC, GDB, make, tcsh, perl, etc.

macro A programming routine or script that 'bundles' or combines a number of processes, operations, or other actions. Macros are typically used as time-savers for frequently performed functions. Macros which are scriptable are usually written in text editors, with simple BASIC-like commands. Some macros are 'recordable;' in other words, the user turns on a 'record' feature in the software, then performs a number of operations which are to be used frequently in the same sequence, then turns record off and gives the macro a name. The sequence of events is stored and can be invoked later with a name or hot-key sequence. Batch and job control languages can be used to write macros. Perl is a good programming language for writing powerful macros.

macroblock sampling A compromise technique used in video phone systems to provide a recognizable image in spite of slow transmission media. Standard telephone modems are too slow for full-screen, full-motion video images. By using an averaging system to sample the image, the image information can be sent more quickly. The smaller the blocks, and the more frequent the sampling, the better the image fidelity, but the slower the processing. An alternative to macroblock sampling is wavelet video compression, which may provide better images through frame-by-frame compression.

MAE 1. See Metropolitan Area Ethernet. 2. See Merit Access Exchange.

MAE East The largest Metropolitan Area Exchange, located in Washington D.C. The MAE is a ring system which provides Internet Service Providers (ISPs) with a relay point for exchanging packets with friendly systems through switched and shared Fiber Distributed Data Interface (FDDI) and switched Ethernet communications services. A MAE is, in a sense, a gargantuan wiring closet with thousands of lines of cables, switches, routers, and connections interconnecting many public and private network installations. The MAE system provides access and interconnections, but doesn't provide ISPs with the political connections with the other services on the system. These have to be individually arranged by each ISP. See MAE West.

MAE West A Metropolitan Area Exchange established in 1988 in San Jose, California, providing switched and shared Fiber Distributed Data Interface (FDDI) and switched Ethernet communications services. Originally operated by Metropolitan Fiber Systems, MFS merged with Worldcom Communications in the mid-1990s to provide expanded nationwide services. MAE West interconnects with the Ames Internet Exchange (NASA) and well-known networks like CERFnet, BBN Planet, MCI, and others. See MAE East.

Magellan The name of a prominent Web-based Internet search system.

magic signature A file integrity mechanism built, for example, into PNG graphics files to detect file corruption. The signature is typically incorporated into the file header or the early part of the file image. Different levels of sophistication can be designed into magic signatures, with different byte lengths. See cyclic redundancy check.

magnesium oxide A material which is suitable for insulating against water and heat when compressed around a conducting wire. When used in conjunction with copper wires, it is known as mineral-insulated copper-sheathed (MICS).

magnet 1. A body, person, or situation with attracting properties. It is called charisma when a person has 'magnetic' qualities. 2. A body that produces an external magnetic field which can attract magnetic materials such as iron. Natural magnets are known as lodestone. Magnetic properties were described by the Greeks at least as early as 60 BC. Steel will hold magnetic charges for a long time; iron can be magnetized, but retains the magnetism to a lesser degree than steel, as do nickel and cobalt. Materials that magnetize readily, but lose the property quickly, are useful as cores for various electrical devices. Magnets are used in many industrial applications, generators, monitors, speakers, compasses, and many more. See electromagnet, gauss, lines of force, lodestone, magnetic field, solenoid.

magnetic detector A device designed to receive electromagnetic waves, which was pioneered by Rutherford and further developed by G. Marconi. The magnetic detector comprises a small induction coil with primary and secondary coil windings surrounding a glass tube with a soft iron wire, rather than the usual common soft iron core. The wire is connected to itself in a loop that runs outside of the glass tube and coil windings, and is set to move continuously through the tube. A magnet is installed near the secondary (outer) winding, adjacent to the circulating wire. This fairly sensitive detector was used in conjunction with early telephone receivers, but never came into wide use, and had almost disappeared by the early 1900s. See detector for illustrations.

magnetic disk A circular device coated on the surface with magnetic particles that can be rearranged to encode information. A number of computer storage devices, including floppy disks, cartridges, and hard drives, use magnetic disks. The disk shape is favored because it allows the disk to be rotated very rapidly under a read/write head. By controlling the position of the read/write head, fast access to any portion of the recording surface can be achieved, called random access.

Early magnetic floppy disks were vulnerable to damage because they had an opening for the read/write head where the disk might be inadvertently touched or scratched, and pliable coverings that could be bent, thus damaging the disk. With the commercial introduction of the 3.5" floppy diskette around 1983, these sources of trouble have been removed.

The chief disadvantage of magnetic disks is that they can be damaged by exposure to magnetic sources (be careful to keep them away from monitors and speakers), which gives them a somewhat limited shelf life. See gamma ferric oxide, superparamagnetic.

magnetic equator Similar to, but not coincident with, the Earth's geographic equator; also called the aclinic line as it is the point at which a dip needle is at zero (90 degrees), between its two vertical positions. See agonic.

magnetic field The region of external influence associated with an electromagnetic body in which these forces can be detected or exhibit a measurable influence on magnetic materials or instruments. The intensity of a magnetic field is described in terms of the number of lines of force passing through a specified area, although the field is conceptualized as continous. The influence of a magnetic field can be seen by holding a magnet near small magnetic objects. A magnetic field can be induced in certain materials by running current through them. See flux, gauss, lines of force.

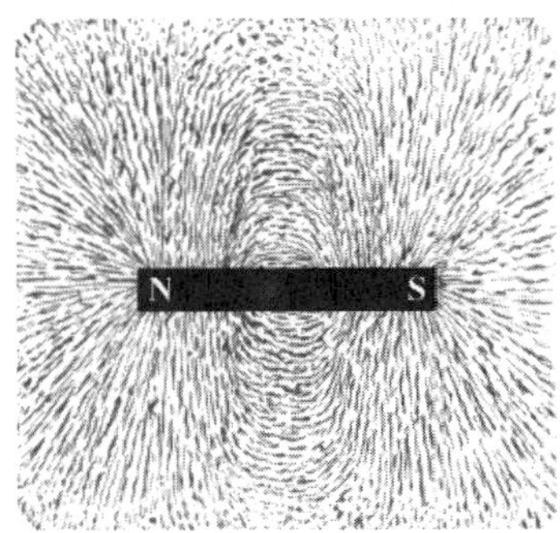

Magnetic lines of force, as illustrated here, vary depending upon the shape and properties of a particular magnet. These lines are not 'fixed' in space, rather they are suggested by the orientation of iron filings and other substances which can be sprinkled over the magnet on a contrasting surface to visually represent the 'influence' of the magnetic field. These lines are conceptualized as extending indefinitely.

magnetic induction The characteristic of certain permeable substances to become magnetized when placed near a magnetic source, without coming in direct contact with that source. Thus, a steel bar does not necessarily have to touch a magnet to be magnetically influenced by that magnet, or to influence other materials in the vicinity of the bar by induction. See lines of force, magnet, magnetic field.

magnetic ink character recognition MICR. A system introduced to the American Bankers' Association in 1956, whose purpose was to recognize printed characters impressed with magnetic oxide ink, to identify bank checks. This permits automation of transactions and works even if ordinary ink, rubber stamp ink, or other pigments are superimposed on the magnetic particles.

magnetic north The northerly direction in the Earth's magnetic field, near, but not corresponding to the north geographic pole, to which north direction-seeking poles of magnets are attracted. Thus, what we call magnetic north is actually Earth's south magnetic pole, located in northern Canada, since the north pole of the compass needle will orient

M

itself toward the south pole of the Earth. See declination.

magnetic storage A medium designed so magnetic materials within it can be aligned and realigned to hold encoded data. Thus, floppy diskettes, hard drives, and audio/data tapes are various forms of magnetic storage. Magnetic storage is inexpensive and very convenient in that it can be easily rewritten; however, it is subject to loss over time through superparamagnetic phenomena, and may be damaged by proximity to equipment with magnetic components, such as monitors. See bulk eraser, superparamagnetic.

magnetic stripe Typically a narrow strip on a portable medium such as a bank card or ID card, which is encoded with information relative to that card. When inserted into a magnetic strip reader, such as a cash machine, the information is used by the system as authorization for access and various transactions.

magnetic tape A narrow, very long magnetic sequential data encoding medium that is familiar to most as audio tapes and computer data backup tapes. In the late 1970s and into the early 1980s, magnetic tape storage was commonly used on microcomputers for storing and retrieving applications programs and data. Due to its slow, sequential nature, tape drives were soon superseded by 8" floppy diskette drives.

magnetite A form of iron ore which is readily magnetized, and is called lodestone when it is already magnetic as it comes out of a mine. See magnet, lodestone.

magneto A magnetoelectric apparatus. In early telecommunications, magnetos were small generators used to create electricity for phone bells. Thus, the subscriber could ring the operator by turning a crank on the phone to provide the necessary power. An alternator is a common application of a magneto that employs permanent magnets (hence the name) to generate ignition current on engines.

magneto phone A historical phone mechanism which employed a crank handle to generate electricity through a magneto. This was used to send a signal from the subscriber to the operator. When the signal reached the operator's switchboard, it rang a bell, notifying the operator that the subscriber wanted to connect a call. The operator would then patch the subscriber's connection to the callee's connection. Magneto phones are no longer needed, as the power to send the signal to notify the switching office that the phone is 'off-hook' is supplied through the phone line itself, and the connection is established automatically by dialing the desired number. See magneto.

magnetometer An instrument for detecting or measuring a magnetic field.

magnetomotive force Descriptive of the relationship of magnetic flux and reluctance through a magnetic circuit, somewhat analogous to electromotive force in an electrical circuit, although the magnetic circuit has a region of influence that differs from an electrical circuit in air. See magnet, reluctance.

magnetron A device that uses a magnetic field acting upon a diode vacuum tube in order to generate microwave frequency power. See cavity magnetron.

MAHO See mobile assisted handoff.

mail bomb A vandalistic or retaliatory transmission sent through email protocols with the intention of disabling an email address or the system on which the address resides or, at the very least, to greatly inconvenience or annoy the recipient. Mail bombs take many forms, but the most common is a repeated message that eventually floods the recipient's email storage space or the storage space on the recipient's service provider's system, depending upon how it is partitioned.

Mail bombs are often sent to people who post absurd messages on public forums, or to originators of junk email (unsolicited email, especially of a commercial nature) to express the extreme displeasure of the recipients at receiving the junk email. A mail bomb may not be a good way to solve the problem, since recipients sometimes retaliate. See flame wars, spam.

mail distributor An agent, script, macro, or filter which takes incoming mail, evaluates the headers or other pertinent information, and distributes the mail accordingly. Thus, a single message might be forwarded to a number of users, or different messages may be funneled to a single user, or groups may be set up to receive certain types of messages. The messages may include certain topics, which are keyed and processed, or may include priority or security information which is handled accordingly. A mail reflector is the simplest type of mail distributor, which passes mail on with a minimum of evaluation and processing of contents (usually only the TO: header).

A mail distributor can be a big time saver when it is used to forward to a mailing list (a large group of recipients) or a discussion list. An address or database entry in the mail distributor can be used to forward a single message to many recipients. This should not be used as a means to distribute junk email, more commonly known on the Internet as 'spam,' as there are regulations against this type of use, and users do not appreciate receiving it (many will boycott companies distributing mail in this way). See discussion list.

mail filter A software utility, or feature of an email client, which automatically evaluates the sender, recipient, subject line, or content of a letter to sort it into designated categories. Mail sent to a specific domain name is often filtered by companies to individual employees' email accounts, junk email messages are often filtered out, and sometimes deleted unread. Some people filter personal and business mail into separate directories before reading the messages. Mail filters are a great convenience and worth the time it takes to initially set them up.

mail gateway Although there are standardized protocols for the distribution of email over networks, not all systems use the same protocols, and not all protocols are implemented exactly the same. Thus, when mail passes from one system to another, if there is a mismatch, there needs to be a way to resolve the differences, or to tunnel or encapsulate the messages so they can reach the recipients. A mail gateway is a system in a computer network that handles mail channeling, or the resolution of protocols.

mail list agent MLA. In SDNS Message Security Protocol (MSP), a mail list agent is one which is addressed by the message originator, and which represents a group of recipients. It provides message distribution services to the participants of that group, on behalf of the message originator.

mail list key MLK. In SDNS Message Security Protocol (MSP), a mail list key is a token held by all the members of a mail list, or by a mail addressable group within the list.

mail reader A software program which permits email to be downloaded from a host system and read offline, so as not to incur connect charges or tie up a phone line. Most mail readers are actually mail readers and writers, and can be used to respond to the received messages or to compose new messages. It may also include filters to preorganize the mail before being read, and a database interface which allows the messages to be organized and stored for later retrieval.

Some mail readers have been enhanced to be used as online news readers as well, for following discussion threads on USENET and for posting to the various lists. Posting is the same as sending email, except that the message will be publicly available and may be read by thousands or millions of readers. Pine, developed by the University of Washington, is one of the most popular mail readers. It is freely distributable, allows flexible processing of mail messages and files, and includes news reading and posting capabilities. See email, USENET.

mail reflector A mail node which is set up to pass messages on according to a provided list. It does only the minimum processing needed to forward the mail to its intended recipients. For information on more sophisticated processing, see mail distributor.

mail server A software system which manages incoming and outgoing electronic mail on a network. Mail servers vary in complexity and features, but most will check the validity of an address; queue, deliver, and store messages (or return them if no valid address is found); forward mail, etc.

Due to overwhelming increases in the quantity of junk email on the Internet, some of the newer mail servers will also check the validity of the sending address, and reject the mail if the sender does not appear to be legitimate. This may result in the loss of some real email messages, for example, if someone is about to change email addresses and close out an old account, they may send you email letting you know the new address, then subsequently close the account before the message reaches its destination. The server may reject the legitimate message. However, some consider the trade-off worthwhile, in order to deflect the thousands, or sometimes tens of thousands of junk mail messages that now flood the systems of most ISPs. See email, mail gateway, mail reader.

mailbox The part of an email client/server software system that comprises addresses and files which store electronic mail.

MAILER-DAEMON, mail daemon A mail management program that works in conjunction with mail servers to process mail that cannot reach the destination address. Since the system may try several times to deliver the message before returning a message to the sender, there are two common messages received from MAILER-DAEMONs, one which is a warning that the message was not delivered and the system will try again later, and one which says the message could not be delivered. The MAILER-DAEMON will usually provide information about the intended recipient and a reason for why the mail was not delivered. A common returned mail response from a MAILER-DAEMON looks like this:

```
Date: Sun, 21 Dec 1997 13:56:41 -0800 (PST)
From: Mail Delivery Subsystem <MAILER-
DAEMON@myISP.com>
To: BigSender@myISP.com
Subject: Returned mail: User unknown

The original message was received at Sun,
21 Dec 1997 13:56:40 -0800 (PST)
from BigSender@localhost

  — The following addresses had permanent
fatal errors —

penpal@goodfriends.education.org

   — Transcript of session follows —

[returned message repeated here]
```

mailto A URL designation for the Internet mailing address of an individual or service. In a Web page, a mailto can be used to make it easy for a person with a Web browser to send a message to the person or organization mentioned. The format to set up the hypertext link is:

```
Click to send email to<A HREF =
"mailto:stan@company.com"> Stan</A>.
```

In the above example, the name Stan will be highlighted in the Web browser to indicate that it can be clicked. When it is clicked, the browser will launch the user's email client, usually inserting the destination address automatically (`stan@company.com`), and allows a message to be written and dispatched to Stan without closing down the browser. It's very convenient. See RFC 822, RFC 1738.

main distribution frame MDF. A central wiring connection point in a larger more complex wiring system that includes more than one distribution frame. The main distribution frame is the one which connects the internal wiring with the external wiring. Within the premises, there may be secondary distribution frames in each department or each floor, depending upon the electrical needs and building configuration. See distribution frame.

main memory In a computer, there are sometimes a variety of types of memory, and there may be more than one memory bank. On some systems, where all the available memory is addressable by the system without significant restrictions, the concept of main memory is not important, as all memory is main memory. However, some systems make a distinction between system memory and expansion memory, and it may not be possible to address all the memory as one contiguous area. These systems treat the 'first' memory, that which is addressed by default, or which is used as first priority as 'main' memory. Extra memory for video display and other specialized uses is not considered main memory as it is not used as general-purpose storage by the system.

mainframe The terms mainframe, miniframe, and workstation are all relative. The most powerful computers in the world are called supercomputers, and the less powerful computers which are above the consumer or workstation price range are called minis or miniframes. Mainframes fall in between these two categories. In general, mainframes are typically the larger, more expensive, more powerful, faster systems with more storage capacity, and the ability to handle many users on a network. Workstations and microcomputers are often used as 'smart terminals' in conjunction with mainframes. Mainframes are used in larger educational institutions, large businesses, and scientific research facilities. Current consumer-priced desktop microcomputers are more powerful than the mainframes available 15 years ago.

Major Trading Area MTA. A service area designation adopted by the Federal Communications Commission (FCC) to administrate Personal Communications Services wireless coverage. There are over 50 MTAs in the U.S., built from Basic Trading Areas (BTAs). Regional designations are somewhat important in administrating wireless services, since the frequency ranges must be reused as efficiently as possible to provide service to as many areas and individuals as possible.

Malus's law When a beam of light that has already been once polarized by reflection hits a second surface at the polarizing angle, the intensity of the beam varies as the square of the cosine of the angle between the two surfaces.

MAN See Metropolitan Area Network.

Management Information Base MIB. A set of modules which contain the definition of a related set of managed object types. In SNMP management systems, it contains the logical names of informational resources on the network.

Management Information Services MIS. Corporate communications professionals whose job is to facilitate the acquisition, flow, use, storage, and retrieval of information within an establishment.

Mance, Henry Christopher (1840-1926) A British engineer who adapted the Indian heliotrope to a heliograph daytime signaling system using mirrors mounted on tripods. The angle of the mirror could convey line-of-sight dots and dashes up to 100 miles. This system was used for military communications for several decades.

Manchester encoding An encoding scheme commonly used for baseband signaling in coaxial cable transmissions, especially 10Base-T network systems. There are variations to the encoding, but a typical differential Manchester employs a voltage transition in the middle of a bit period. A zero is represented with an additional transition at the beginning of a bit period. A one is represented with no transition at the beginning of a bit period.

There is a tradeoff between bandwidth and binary coding, as the coding consumes part of the bandwidth. In Manchester encoded transmissions, the amount of useful bandwidth is about twice the encoding signal.

The Manchester encoding scheme is simple but useful, and can be used as one type of passband signal.

Manchester Mark I An early large-scale computing machine designed and built by Fred Williams, Tom Kilburn, and Max Newman in the late 1940s. It was significant in its ability to store programming information. The earlier prototype for this machine was colloquially known as "Baby."

Mandelbrot, Benoit (1924-) A Polish mathematician who emigrated first to France in 1936, and later to the United States. Mandelbrot extensively researched areas of complex geometry which have come to be known as fractal geometry. At least part of his thinking coincided with, or developed from, the work of G. Julia, who published important mathematical observations on the iteration of rational functions in the early 1900s. One family of fractal images called the Julia set is named after this predecessor.

Mandelbrot's early publications on fractals include "Les objets fractals, forn, hasard et dimension" (1975) and "The Fractal Geometry of Nature" (1982) which created an enormous stir, especially in North America, and fueled much of the fractal imagery since generated on computers.

MANIAC A historic large-scale computer developed in the mid-1950s by the Los Alamos National Laboratory. The construction of its successor, the MANIAC II, inspired professors at Rice to initiate the Rice Computer Project. See Atanasoff-Berry Computer, ENIAC, Rice computer.

MAP See Media Access Project, an important organization representing the public good.

Marconi antenna An antenna that requires the ground, or a large object to which it is mounted (such as a vehicle), to aid in resonance conduction. In other words, it is not a stand-alone antenna like a Hertz antenna. This type of antenna is common for amplitude modulation (AM) broadcasts. See antenna, Hertz antenna.

Marconi detector An adaptation of the Branly detector to which G. Marconi added a vibrating source to quickly 'set' the coherer back to 'zero' or nonconducting status. See detector.

Marconi, Guglielmo (1874-1937) An Italian who as a youth demonstrated wireless telegraphy to his mother in an attic laboratory in 1894, and experimented with radio waves in 1895. With further support and assistance from his mother, Annie Marconi, the 22-year old Marconi filed for a patent and demonstrated radio communications in London the following year, and received a British patent in 1897.

Marconi's first communications were over very short distances, but in 1901 he showed that radio signals could be sent across the ocean between Canada and England, a distance of over 3,000 kilometers. He continued for many years to devise improvements in the technology, and to put them to practical application. In 1909 he was awarded a Nobel prize in physics along with K. Braun. Marconi began broadcasting from Marconi house in 1921 under the famous 2L0 callsign. See Braun, Karl Ferdinand; Tesla, Nikola.

MARECS A European maritime satellite communications service established in the early 1980s; it is similar to the American MARISAT system.

MARISAT Maritime Satellite. First launched in 1976, MARISAT was designed to provide mobile communications services to the United States Navy and other maritime clients. The European MARECS system is similar.

Mark I See Harvard Mark I.

Mark-8 A pioneer Intel 8008-based personal computer kit. The Mark-8 was described in a June 1974 issue of Radio Electronics magazine by Jonathan Titus. See Altair, Intel MCS-4, Micral, Scelbi.

mark-to-space transition, M-S transition In telegraphy, the momentary change when the system reverses polarity, or changes from a closed to an open circuit. At this point, there is a small amount of delay that must be taken into consideration, which can be plotted on a timing wave. The reciprocal is the space-to-mark transition.

MARS Multicast Address Resolution Service. In ATM networking, a protocol used in IP multicasting.

MAS Multiple Address Service.

maser microwave amplification by stimulated emission of radiation. A type of laser technology developed in the late 1950s. See laser.

mask A screen, stencil, or other object which is superimposed between a surface and light, pigments, or other media being put on that surface so only the unmasked portions are seen or affected. When used with light, a mask is known as a beam block.

mask, data In computer programming, a mask is a set of data, flags, or bits which are used as a filter or operator to affect only those bits of data that correspond to the mask template, or which are not included in the mask template.

Massie Oscillaphone A simple type of loose contact electromagnetic wave detector that has long been favored by amateur experimenters and educators. Two carbon blocks (battery carbon can be used) are set up adjacent to one another, about an inch apart, on a non-conducting base such as wood. The top surfaces of the carbon blocks are chiseled or filed so that they have a fine, thin edge. Holes are drilled through their surfaces, near the base, to provide room to insert a screw through each block, with the screwheads on the outside, to secure wires that connect with two binding posts. The top thin surfaces of the carbon blocks are wiped with a woolen cloth, and a light sewing needle or other similar conductor is laid across the top of the two blocks to create a contact between them.

When connected to a circuit including a battery power source, an aerial, ground, and a telephone receiver, an incoming radio wave will interact with the needle-carbon contact, causing the needle to adhere more closely to the blocks, lowering the resistance. This results in an increased flow of current which is translated into sound in the receiver. Further adjustments to the sensitivity of the needle can be attained by placing a small magnet under the needle and adjusting its height. See detector.

master group In analog voice phone systems, a hierarchy for multiplexing has been established as a series of standardized increments. See voice group for chart.

MAU See Media Access Unit.

Mauchly, John W. (1907-1980) An American physicist and engineer who collaborated with J. Eckert to build the historic ENIAC computer. See ENIAC.

MAX See Media Access Exchange.

Maxim, Hiram Percy Founder of the historic Amateur Radio Relay League (ARRL), along with Clarence Tuska, a young fellow radio amateur, who became friends with Maxim after the older radio enthusiast had decided not to purchase radio equipment constructed by Tuska.

Hiram Maxim, W1AW, listening in on a set of earphones in the days when crystal detectors were common radio devices. Maxim encouraged amateur radio participation.

maxwell An electromagnetic unit of magnetic flux to the flux-per-square centimeter equal to magnetic induction of one gauss, or one magnetic line of force. Named after J. Clerk-Maxwell.

Maxwell, James Clerk- (1831-1879) A precocious Scottish physicist who, building on the work of Faraday and Bernoulli, and adding ideas of his own, contributed many important, fundamental theories and equations related to electromagnetism and the nature of particles. He also made mathematical predictions about the composition of Saturn's rings which held up well over time.

James Clerk-Maxwell is remembered for many of his mathematics and physics contributions related to fundamental laws and electromagnetism.

Maxwell's equations A set of fundamental mathematical equations, originated by J. Clerk-Maxwell and further developed by Oliver Heaviside and Heinrich Hertz, for expressing radiation and describing conditions at any point under the influence of varying electromagnetic fields. These concepts and equations are integral to many areas of science, and are of particular interest in understanding and developing transmissions media, antennas, and other basic building blocks in telecommunications. See Heaviside, Oliver; Hertz, Heinrich Rudolph; Maxwell, James Clerk-.

Maxwell's rule Every part of an electric circuit is acted upon by a force which tends to move it in such a direction as to enclose the maximum amount of magnetic flux.

Maxwell's theory of light In 1860, J. Clerk-Maxwell demonstrated that the propagation of light could be regarded as an electromagnetic phenomenon, the wave consisting of an advance of coupled electric and magnetic forces. If an electric field is varied periodically, a periodically varying magnetic field is obtained which in turn generates a varying electrical field and thus the disturbance is passed on in the form of a wave. Maxwell's theory predicted that the speed of light unimpeded was constant.

Maxwell's triangle An ordered representation of color relationships, in the shape of a triangle, developed in the late 1800s by physicist J. Clerk-Maxwell. His premise was that this model would contain all known colors. Red, green, and blue are identified as the three primary colors of light and are located in the three corners of the triangle. The colors progressively blend until, in the center, the combination of all the colors becomes white. A system of color notation was developed by overlaying a grid over the triangle. See color space, Munsell color model.

Mayer, Maria Goeppert (1906-1972) A Polish-born, American physicist who carried out fundamental research in models of the nucleus of atoms. For her independent work, she was awarded a Nobel Prize in physics, along with J. Jensen and E. Wigner.

Mbone multicast backbone. See 6bone, backbone, multicast backbone, X-Bone.

MCC Miscellaneous Common Carrier. See Radio Common Carrier.

McCarthy, John (1927-) A recognized pioneer in the field of artificial intelligence since 1955, McCarthy was one of the first to promote the basic concepts of computer timesharing in the late 1950s. McCarthy is also known as the originator of the LISP interpreted programming language which is popular in artificial intelligence research. See LISP.

McCaw Cellular Communications, Inc. A commercial communications services provider chaired by Craig McCaw, which was sold to AT&T in 1994 and renamed AT&T Wireless Services. McCaw is now collaborating with W. Gates, Motorola, Boeing, and others to develop the Teledesic satellite-based Internet system. See Teledesic.

MCF See Multimedia Communications Forum.

McLennan, John Cunningam A Canadian physicist and educator who explored cosmic rays, gases, and transportation innovations, and was one of the first to propose helium as a substitute for hydrogen in zeppelins and other lighter-than-air ships. McLennan invented a new way to extract helium which dramatically reduced the cost of production and made balloon transport safer and more economical.

MCMS Multimedia Cable Network System.

MCS-4 An early (1970s) Intel chip set. See Intel MCS-4.

MD-IS See Mobile Data Intermediate System.

MDF See main distribution frame.

MDS-xxx A line of commercial digital switching products from Raytheon E-Systems.

MDT mobile data terminals.

mean time between failures MTBF. A performance indicator, the limit of the ratio of the operating time in a device to the number of failures as the number of failures approaches infinity. At the factory, test versions of a product are often subjected to extreme use to estimate in advance what its MTBF rating might be under conditions of actual use.

MECAB See Multiple Exchange Carrier Access Billing.

Media Access Control, Medium Access Control MAC. Functions associated with the lower half of the data-link layer that governs access to the available IEEE and ANSI local area network (LAN) media (or medium, if there is only one). This layer supports multiple downstream and upstream channels.

Media Access Exchange MAX. A system-level network access unit from Ascend Communications, into which peripheral cards can be inserted. A MAX can support multiple host ports or direct network connections, videoconferencing units, and remote LAN connections.

Media Access Project MAP. An important nonprofit, public interest telecommunications law firm that looks out for the First Amendment rights of individuals before the legal system and the Federal Communications Commission (FCC).

Over the years, the broadcast agencies have been provided free use of the airwaves and, in return, have a legal responsibility to provide a portion of programming and resources for the public good. They are bound to uphold these obligations but may neglect this obligation unless citizens support groups like MAP, who take the time to lobby for the interests of the little guy. In recent years more free bandwidth has been allocated to commercial broadcasters, particularly satellite broadcast frequencies.

It is important that citizens support their rights, and that it be impressed upon the government, the FCC, and the broadcasters that these broader free permissions have inherent corresponding responsibilities.
http://www.mediaaccess.org/

Media Access Unit MAU. In Token-Ring local area networks (LANs), a wiring concentrator that connects the end stations. The AU provides an interface between the Token-Ring router interface and the end stations. Also known as access unit (AU).

Media Interface Connector MIC. An eight-pin modular RJ-45-8 plug. This resembles a common RJ-11 phone jack except that it is wider to accommodate connections for eight wires. This is the connector recommended for audio-visual applications by DAVIC specifications.

medium Earth orbit MEO. An orbiting region around the Earth into which certain types of communications satellites are launched, midway between low Earth orbits and the high Earth orbits into which geostationary satellites are typically launched. The life-spans of medium-orbit satellites tend to be around 10 to 12 years. Most medium-orbit satellites travel about 10,000 to 15,000 kilometers outside Earth. There is a region called the Van Allen radiation belt between MEO and low Earth orbits (LEOs) which is generally avoided. MEO satellites are primarily used for a variety of broadcast applications. See high Earth orbit, ICO, low Earth orbit, Teledesic.

Meet Me A commercial FTS2000 conferencing capability initiated by dialing an access number at a prearranged time, or as directed by an attendant. If additional conferees wish to join a conference in progress, this can be done with an Add On conference. See FTS2000.

Meet-Point Trunk Telecommunication trunks configured for two-way traffic in jointly provided Switched Access Services (SAS), to interconnect End Offices and Tandems.

mega- (Abbrev. - M) A SI unit prefix for one million, expressed as 10^6 or 1,000,000. To confuse matters, when used in conjunction with computer-related quantities, if often means 2^{20}, expressed as 1,048,576. The most common of these uses is in descriptions of computer storage capacity as megabytes (MBytes), in which 1 megabyte is 1,048,576 bits. See kilo-.

megger An instrument for measuring values of very high resistance, used, for example, for insulation resistance testing. See Wheatstone bridge.

Melpar model An artificial neuron used at the Wright-Patterson Air Force Base in Ohio in the early 1960s to mimic human reasoning (or at least rodent reasoning). The Melpar model, familiarly called 'Artron' by its inventors, was used as the 'brains' of a maze-running bionic mouse, that physically resembles the input mice used on today's computers. The bionic mouse brain was comprised of 10 Artrons, which was sufficient for a trial-and-error method of learning to run the maze. With a clean slate, the mouse took 45 minutes to complete the maze; eight tries later, it took only 35 seconds. See neural network.

memory In a computing system, a storage area that is dynamically allocated and used by the operating system and various application programs. Most memory in desktop computers is random access memory (RAM), although some programs will also allocate hard drive storage as 'virtual memory.' Memory is one of the most basic elements of a computing system, along with the central processing unit (in fact memory is also incorporated into the CPU), and the input/output (I/O) bus.

Read only memory (ROM) is included in many computers to provide basic operating parameters to a system, particularly on startup. In the earliest microcomputers, a programming language was sometimes included in ROM. Random access memory (RAM) is dynamically allocated by the system and applications programs. RAM is further distinguished as static or dynamic RAM. Most desktop systems include about 8 to 64 Mbytes of RAM, and may be extended up to 64 or 256 Mbytes. RAM typically operates at about 60 to 80 nanoseconds, although this may change, as newer, faster types of memory are developed.

Most types of computer chip memory are volatile, that is, the contents will disappear if the system is not constantly powered and refreshed. However, there are some types of chips that can retain information, such as erasable, programmable read only memory (EPROM) chips.

The price of memory fluctuates dramatically. In 1986, a megabyte of RAM was $600 U.S., by the early 1990s this had dropped to $25, then increased again to $120. By early 1998, the price was down to $4 per megabyte.

Programmers tend to write code that fills available space. This results in applications that require more memory than many consumers have, setting off another round of buying. In 1978, the TRS-80 computer ran with 4 kilobytes of memory, and with 8 kilobytes it could do word processing and spreadsheet applications quite well. By the mid-1980s, the Amiga computer could multitask and run graphics programs concurrently with stereo sound quite comfortably in 4 Mbytes of RAM. Today's systems rarely run efficiently with less than 16, and most vendors recommend 32.

memory leak A memory leak is a characteristic of a software program in which memory is not handled correctly by the application or the operating system (or both), and begins to fill up the system. In other words, as the software is used, more and more memory is allocated without being freed up when it is no longer needed. Eventually, there is no more 'working room' for the program and it may freeze, or on some systems, even effect the operating system (although it shouldn't) and cause spurious problems or system crashes. If the operating system does not clean up the leak on behalf of the application, it will probably be necessary to reboot the machine to clear the problem.

Mercator projection A map projection that preserves angular relationships which is of particular importance in marine navigation.

mercury vapor lamp A lamp that functions by using mercury flowing back and forth through a tube when made horizontal to complete the electric circuit and start the lamp. Ionized mercury vapor is then produced by the heat and current, resulting in light through the length of the tube. The light is very bright, with a greenish glow, and is generally used in industrial applications.

mesh topology A type of 'circular' network backbone topology in which data can travel back along the backbone in the event that a node is unavailable due to a disruption, such as line breakage or failure. The mesh nature of the topology stems from the appearance of the connections between a node and other nodes several nodes away. See topology.

Message Handling System MHS. On a network, MHS provides a means to store and forward messages among MHS users or applications. Unlike traditional telephone networks and the early two-way radio communications, most data networks do not need to establish an end-to-end connection before carrying out communications. Thus, the MHS provides a means to handle the messaging traffic under dynamic circumstances. See X.400 under X Series Recommendations.

Message Security Protocol MSP. A Secure Data Network System (SDNS) protocol for providing X.400 message security. With MSP, a message is given connectionless confidentiality and integrity, data origin authentication, and access control; nonrepudiation with proof of origin; nonrepudiation with proof of delivery.

MSP is a content protocol, in the application layer, and is implemented within originator and recipient MSP user agents. It is an end-to-end protocol which does not employ an intermediate message transfer system. MSP processing is carried out prior to submitting a message and after accepting delivery of a message.

A X.400 messages comprises a content and an envelope. With MSP, a new message content type is defined with a security heading encapsulated around the protected content.

Three types of X.509 digital certificates are supported by MSP. The user's distinguished name and public cryptographic material are bound within an X.509 certificate, which in turn is signed by a certification authority (CA). The CA manages X.509 certificates and Certificate Revocation lists.

message switching A means of switching and multiplexing data packets by storing, queuing, and forwarding the message to the recipient. See circuit switching, packet switching.

Message Transfer System MTS. A general-purpose, application-independent store-and-forward communications service within a Message Handling System (MTS). The MTS uses message transfer agents (MTAs) to relay messages. See Message Handling System.

message unit In packet networking, SNA, a basic unit of data processed by any layer.

Metcalfe, Robert (1940s-) An American engineer and journalist, Metcalfe is the acknowledged creater of Ethernet at Xerox PARC in 1973, along with David Boggs. In 1979 he founded the 3Com Corporation, and since 1990 has been involved with a number of publishing organizations. See Boggs, David; Ethernet.

METEOSAT Meteorology satellite.

Metropolitan Area Network MAN. An urban network of high-speed hosts. See MAE East, MAE West, SMDS.

Metropolitan Fiber Systems MFS. A Competitive Access Provider (CAP) founded in the late 1980s. In the 1990s it established its own backbone, providing national network services. It was subsequently acquired by WorldCom.

Meucci, Antonio An Italian-born Cuban inventor, chemist, stage designer, and engineer, Meucci made many pioneering discoveries in telecommunications concepts and devices, but his findings were not widely communicated to others, and hence his contributions did not significantly impact on subsequent inventions such as telegraphs and telephones, which made telecommunications history. Meucci developed rheostats, electroplating techniques, and experimented with passing electricity through the human body. While studying mild electrical charges, he discovered the 'electrophonic' effect which related to nerve responses to specific applications of current through a wire. By the mid-1800s he had developed several devices for creating a vibrating electric current from spoken acoustical impulses. By using a copper strip and delicate animal membranes as diaphragms, he created one of the earliest telephone-like mechanisms. See telephone history.

MEW Microwave Early Warning system. A phased radar array antenna system used for security warnings and aircraft control.

Meyer code A flag signaling code, employing left and right motions to create characters or syllables, and a forward motion to indicate ends or pauses. This code was in use until it was superseded in the First World War by International Morse code and American Morse code. See semaphore.

MFJ See Modified Final Judgment.

MFS See Metropolitan Fiber Systems.

MFSK See multiple frequency shift keying.

MHEC See Midwestern Higher Education Commission.

mho A practical unit of the measure of conductance, so named because it is ohm spelled backwards. See admittance, ohm.

MHS See Message Handling System.

MIB See Management Information Base.

MICR See magnetic ink character recognition.

Micral The first fully assembled 8008-based microcomputer, the Micral featured 8-bit processing and 2 kilobytes of memory. It was designed in France by François Gernelle. The Micral sold for just under $2,000 and like its predecessor, the Kenbak-1, was not commercially successful in the United States, an important market for microcomputers. It was introduced in May 1973 before the SPHERE, Scelbi-8H, Mark-8, and Altair computers were introduced in the U.S. See Altair, Kenbak-1, Mark-8, MITS, Scelbi-8H, SPHERE.

MicroCal Module A new type of integrated circuit (IC) designed to facilitate self-monitoring in virtually any type of wireless equipment, including base stations, mobile handsets, and subscriber units. The MicroCal Module scans the entire bandwidth, gathering data which is fed back to a central maintenance center. The module, designed by Micronetics Wireless, was awarded a U.S. patent in 1996. Micronetics is working with a number of companies, including Nortel and Motorola, to integrate the module into wireless infrastructure equipment.

Microcom Networking Protocol MNP. A series of proprietary error control and data compression protocols designed for dialup modems which are often incorporated in conjunction with industry standard ITU-T-recommended error control mechanisms.

Name	Notes
MNP-1	Asynchronous mode, half duplex transfer operation.
MNP-2	Simple error correction scheme, asynchronous mode, full duplex operation.
MNP-3	Error correction incorporated, synchronous mode.
MNP-4	Error correction incorporated, increased throughput. Often included with V.42 modems, along with MNP-5 data compression.
MNP-5	Simple data compression scheme. Often included with V.42 modems, along with MNP-4 error control.
MNP-6	Statistical duplexing and Universal Link Negotiation. Full duplex emulation.
MNP-7	Data compression scheme included.
MNP-8	MNP7 for modems which emulate duplex operation.
MNP-9	Data compression scheme included. Incorporates V.32 technology.
MNP-10	Dynamic fall-back and fall-forward adjusts modulation speed with link quality.

For example, MNP-4 works in conjunction with modems that transmit at data rates up to 14,400 bps. MNP-4 is often implemented in conjunction with the V.42 error control protocol standard from the ITU-T. See V.42.

microfiche A somewhat standardized archive system using thin transparent sheets of image-carrying plastic commonly used for the archive of printed matter, especially newspapers, books, journals, etc. Since microfiche information is miniaturized to fit as much data on a sheet as possible, it is typically not human-readable without magnification. Microfiche machines are used for backlighting and magnification. Some photocopiers are designed to enlarge and print microfiche information, although the photocopies do not tend to be very clear. Digital storage techniques are replacing microfiches, but like microfiches, those being used are often low resolution, and do not retain many of the properties of the originals and, unfortunately, also like microfiche archives, the originals are often destroyed, for lack of storage space and funds. Microfiches are common in libraries, post-secondary institutions, and government archives.

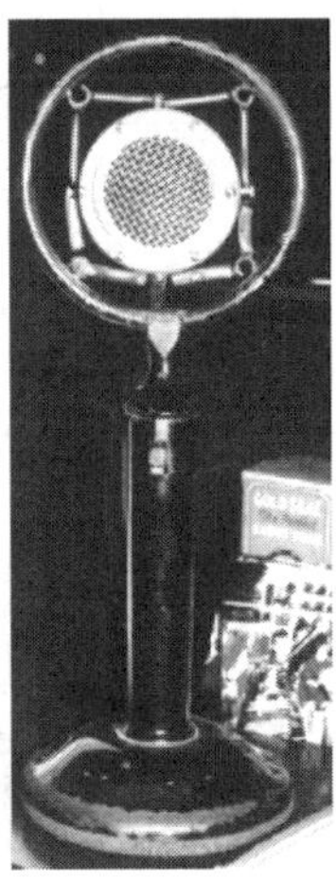

Older microphones were bulky and subject to noise. Newer microphones are used for many functions, including music, public speaking, and videoconferencing over computer networks.

microphone A device for apprehending sounds and transmitting them electrically or acoustically to a receiver or audience. A very simple microphone can be created by wrapping stiff paper into a funnel shape and attaching it to a string or wire, and stretching it to a receiver, another funnel on the other end. If the listener puts an ear near the receiving funnel while the speaker talks into the 'microphone' funnel, the sound, while not loud, can be heard across a room. Add to this electronics to amplify the signal, and you have a basic microphone. Some microphones also include echo acoustics to make the sound of a voice fuller and more resonant. Many singers use this type of microphone to enhance their singing on recordings.

M

Microphones are widely used in camcorders, film cameras, tape recorders, and video recorders. Two microphones are needed for true stereo sound.

Microphones can be used as peripherals with computers for the creation of music and other sound samples, or for videoconferencing. See sampling, videoconferencing.

Microsoft BASIC A BASIC interpreter first released for the Altair computer in 1975. Paul Allen had seen the feature article on building the Altair in the January 1975 issue of Popular Electronics, so he and Gates talked about it in Harvard Square, and conceived the idea of writing a BASIC interpreter for the new kit-based machine. They contacted MITS, made a proposal, and set to work creating a BASIC that could fit into 4K of memory. The entrepreneurs had previous experience in looking at code for interpreters for various languages based on their business activities together through high school, and 8K BASICs were available for the PDP-8. They developed the BASIC in a simulation environment, since it wasn't practical to write it on the Altair itself. Allen created a simulation environment for 8080 programming code and modified a symbolic debugger to understand the 8080 instructions. Gates laid out a design for the BASIC interpreter which was modeled on the BASIC he had encountered on a time-sharing system at Dartmouth, and began coding it, with assistance later from Allen. Monte Davidoff contributed some of the math routines, especially those for floating point operations. On the plane to Albuquerque to demonstrate the software, Allen created a bootstrap loader so the Altair would be able to read the data into memory, using a teletypewriter as an input mechanism. (Gates later streamlined the bootstrap loader.) On the first run at the demonstration at MITS, the BASIC didn't work. On the second try it did. This was a substantial achievement, given the short, hands-off development period and environment.

This first BASIC was later ported to many machines. Not long after the Altair kicked off the microcomputer industry, Microsoft BASIC Level II was bundled with the TRS-80 Model I in ROM in 1976, replacing Level I BASIC, and included with the Commodore PET. Microsoft also contributed some routines to the Integer BASIC designed by Wozniak for the Apple Computer, resulting in AppleSoft BASIC. Later, in 1984, Microsoft BASIC was incorporated into ROM on the IBM Personal Computer XT. In addition to the computer-specific 8-bit operating systems, BASIC was ported to run on the popular CP/M-80 operating system designed by Gary Kildall. At this point, Microsoft BASIC was still a text-based program.

Microsoft BASIC version 2.0, the first graphics-based BASIC for the Macintosh, was not announced until fall 1984, a decade after the text version shipped. In 1985, Microsoft provided a windowing version on floppy diskettes for the Amiga 1000. Later Microsoft BASIC evolved further into Microsoft Visual BASIC which differed chiefly in that graphically entered structures could be used to automatically generate code. See BASIC, Visual BASIC.

Microsoft Incorporated One of the earliest companies supporting the microcomputer market, Microsoft was founded by Paul Allen and Bill Gates in 1975 following their partnership as Traf-O-Data, which they formed around 1972. Although Gates and Allen had worked on programming projects together during high school in Seattle, they formalized Microsoft in 1975 in order to market a version of interpreted BASIC for the Altair computer. The tradename was registered in 1977.

Paul Allen learned of the Altair computer in a Popular Electronics article he saw in Harvard Square. Since the idea for the Altair so closely paralleled a microcomputer hardware idea he had started a while earlier, he contacted Gates to let him know 'someone else is doing it.' They then talked in Harvard Square about writing BASIC for the new machine. They contacted MITS, the makers of the Altair, and Allen, Gates, and Davidoff created a BASIC based on Gates' and Allen's experience with BASIC interpreters at Dartmouth. Six weeks later, Allen flew south and successfully demonstrated the BASIC to MITS in New Mexico.

The entrepreneurs moved their operations first to the Sundowner Motel across the street from MITS, and later to an eighth floor office in Albuquerque, New Mexico. Allen took a position as VP of Software at MITS, while keeping in regular contact with Microsoft and Gates. They hired high school friends to help out. Meanwhile, Gates began enhancing BASIC and porting it to the new platforms that were coming out. When Gates briefly went back to Harvard, Ric Weiland and Marc McDonald formed the core at Microsoft.

Not long after, Marc McDonald designed and coded Stand-alone Disk BASIC, in consultation with Bill Gates. In the late 1970s, Microsoft BASIC was adapted to run on a popular text-oriented operating system called CP/M, developed in various versions by Gary Kildall of

Digital Research between 1973 and 1976.

After three years in New Mexico, Microsoft relocated to Bellevue, Washington, near the co-owners' family members. At this location it was easier to recruit programmers. Microsoft has its own campus now in Redmond, and has grown to be a large, financially successful enterprise.

While it has had occasional forays into hardware development, the primary focus of the company has been software, and a substantial portion of the revenues are derived from operating systems and business-related applications. Paul Allen left the company to invest in a number of other ventures, and formed the Paul Allen Group to oversee his various investments. Allegations of unfair business practices have been leveled at Microsoft a number of times during the late 1980s and the 1990s, and the company began to be scrutinized by the U.S. Justice Department to investigate the numerous allegations. These proceedings are ongoing and will probably not be quickly resolved. Bill Gates continues as CEO of the company as of 1998. See Allen, Paul; Altair; Gates, William; MITS; Traf-O-Data.

microwave Radio wave transmission frequency (1,000+ MHz) generally used for radar and repeaters. Microwave relay systems were in use as early as the 1930s by AT&T. They are now an important aspect of satellite communications. Microwaves are also finding increasing use in connecting local area wireless networks (LAWNs). While they are not used for the primary information-carrying aspects within the network, they are useful for interconnecting line-of-site separated LAWNs, or LANs, as between buildings. This requires a license from the Federal Communications Commission (FCC). See microwave antennas, short wave.

microwave antennas Due to the very short wavelengths used in microwave transmissions, the physical arrangement of microwave antennas is quite different from those for UHF, VHF, and FM broadcasts. Microwave transmissions are directional for both up- and downlinks, quite different from the roughly isotropic, omnidirectional character of traditional television and radio broadcast waves. The common multi-branched Yagi-Uda style aerials and fan dipole antennas are inexpensive and appropriate for VHF and UHF reception, but directional parabolic antennas are the norm for microwave signals.

The diameter of a parabolic dish antenna is a multiple of the length of the microwaves being received and typical dishes range in size from about two feet to about ten feet across, with the curvature of the dish determining the position of the feed horn which focuses the beams.

The first transcontinental microwave communication system began operations in 1951 through a system of relay stations between San Francisco and New York City. Within three years, there were more than 400 additional stations scattered across North America. See antenna, parabolic antenna, UHF antenna, VHF antenna.

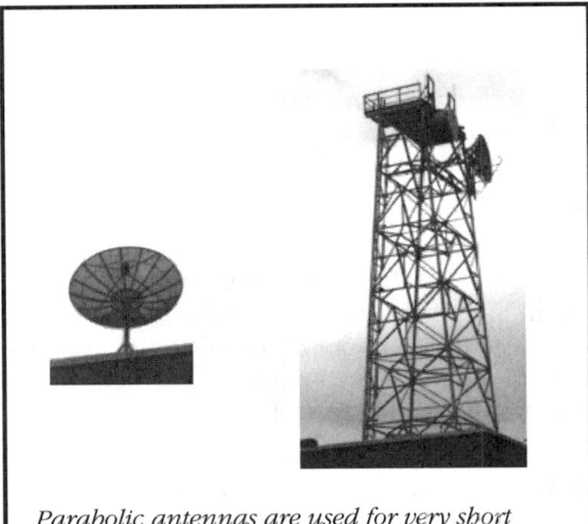

Parabolic antennas are used for very short (microwave) radio waves. The curvature of the dish and the related placement of transmitting or receiving horns are important to the quality of the signal transmitted or received.

M

microwave multi-point distribution system MMDS. MMDS is a system of distributing cable TV programming through microwave communications, more commonly known now as 'wireless cable.' MMDS works in the frequency range of 2.50 to 2.686 GHz, and MMDS service providers are increasing in number. The signals are downlinked from the satellite to the local transceiver, and broadcast from there to subscribers within about a 50-mile radius, depending upon terrain. The subscriber receives the signal on a consumer-priced antenna mounted on or near the home, which is linked through a cable to a 'black box' on or near the TV set. This box provides the decompression of compressed digital signals, and unscrambling of signals intended to prevent unpaid/unauthorized viewing of the programs. The MMDS system is currently in transition from analog to digital technology, which provides opportunities for digital multiplexing through highly linear radio frequency (RF) subsystems, thus providing more television channel choices for viewers.

microwave radar Radar employing microwaves has been extremely important in navigation, tracking, surveillance, guidance, and communications systems. Much of the early research in microwave radar was conducted at the Massachusetts Institute of Technology (MIT) Radiation laboratory in the early 1940s.

Mid-Span Meet An interconnection point between two co-carriers. The Mid-Span Meet is the point up to which the carriers provide cabling and transmissions.

MIDI See Musical Instrument Digital Interface.

MIDI time code MTC. A standard developed to identify timing information associated with a stream of Musical Instrument Digital Interface (MIDI) data. See SMPTE time code

Midwestern Higher Education Commission MHEC. MHEC was founded as an interstate agency in 1991 to promote resource sharing in higher education. As a subgroup, it includes a Telecommunications Committee that takes a regional approach to improving access, services, and costs of telecommunications services.

Mill Street plant This historically significant power plant began providing three-phase alternating current (AC) in 1893. Partly due to the advocacy of T. Edison, most early power plants provided direct current, so the Mill Creek No. 1 Hydroelectric plant was a precedent-setting installation, and many other similar AC power suppliers followed its example.

milli- (Abbrev. - m) A SI unit prefix for one thousandth, 10^{-3} or 0.001. Thus, a milliamp is one thousandth of an ampere.

MILNET Military Net. The ARPANET was a historic network put into operation in 1969. In 1975, ARPANET was transferred to the Defense Communications Agency. Then, in 1983 it was split into MILNET for military usage, and ARPANET, which grew to become the Internet. MILNET is used for nonclassified U.S. military communications. See ARPANET, Internet.

MIMD See multiple instruction/multiple data.

MIME See Multipurpose Internet Mail Extension.

MIN See Mobile Identification Number.

MIND Magnetic Integrator Neuron Duplicator. An artificial 'neuron' developed in the early 1960s at the Aeronutronic Division of the Ford Motor Company. MIND imitates the functions of a human neuron.

Mindset computer A delightfully innovative, ahead-of-its-time 6-MHz Intel 80186 graphics computer with MS-DOS compatibility, released in 1984. The Mindset, developed by Roger Badertscher and Bruce Irvine, former Atari employees, is so hauntingly similar to the early Amiga, with its good graphics, custom coprocessing chips, and MS-DOS compatibility, that the author asked Jay Miner if there was any relationship. His answer was a definite 'no;' he said the machines were independently developed and just happened to converge on the same lines of thinking. Unfortunately, for all its capabilities, the Mindset just never caught the attention of the buying public.

Jay Miner was the gifted hardware engineer who designed microcomputers, games systems, and heart pacemakers. He is best remembered as "Padre, Father of the Amiga."

Miner, Jay (~1930-1994) A gifted design engineer responsible for designing the hardware for the Atari 800 computer, the Amiga computer, and the Lynx color handheld game machine. In 1982, Miner joined Hi Toro to develop the Lorraine computer, which was subsequently sold as the Amiga by Commodore Business Machines. A proponent of open-mindedness and creativity, Miner included his dog's pawprint inside the case of the Amiga 1000. After the Amiga, he created the Atari Lynx, a fast color handheld game machine. Jay Miner was affectionately known as Padré, the Father of the Amiga, to the computing community. Following a serious illness and kidney transplant, Jay Miner devoted his remaining working life to developing medical devices, such as pacemakers, to aid society. Surprisingly, despite the fact that he understood that the creation of the Amiga was a remarkable achievement, Jay Miner didn't anticipate the revolution in the video industry that was launched by his creation. In a computing industry where hardware architectures go out of date in a few months, the viability of the Amiga hardware for more than a decade, particularly for graphics and sound, is a tribute to its efficient and insightful design. See Amiga computer, Commodore Business Machines.

MiniDisk A very compact audio/data magnetic storage medium designed by Sony for sound and computer data storage, popularly used for playing music in portable MiniDisk audio players.

minifloppy A generic term for a number of floppy diskette technologies that store almost 10 times as much data as a regular 3.5 inch floppy, but which are designed by some manufacturers to be downwardly compatible with 1.4 Mbyte drives. The price of storage on these high capacity floppies is substantially cheaper, and they may, in time, supersede current floppies.

minimal shift keying MSK. A type of modulation technique similar to quadrature phase shift keying (QPSK), except that the rectangular pulse in QPSK is a half-cycle sinusoidal pulse in MSK. See modulation, phase shift keying.

minimize button, iconize button Graphical user interfaces on several different operating systems include a small gadget on applications or displays windows which, when clicked, will shrink the window down to an icon. Thus, the program is available and can quickly be retrieved by double-clicking the minimized icon without shutting down the process and having to rerun the program.

Ministry of Posts and Telecommunications MPT. The Japanese radio regulatory administration, the MPT oversees radio communications, based on the Radio Law of 1950. The MPT grants radio station and operator licenses; monitors and inspects stations and radio frequencies; and sets technical standards for radio equipment.

Minitel A French Telecom service that provides free terminals for chat and electronic telephone directory videotext services. It is similar to the German Bundespost's interactive videotext system. See Minitel.

MIP See Multichannel Interface Processor.

MIPG See Multiple-Image Portable Graphics.

MIPS million instructions per second. A measure of processor speed used in system design and cross-system comparisons. MIPS describes the average number of machine instructions that a central processing unit (CPU) performs per unit of time of one second. This is a narrow definition of performance, as many other factors influence overall speed and efficiency. The Digital VAX-11/780 is defined as a baseline at 1 MIP. Most consumer desktop models now deliver about 3 to 10 MIPS. High-end minicomputers and mainframes can range from 10 to 50 MIPS, with supercomputers comprising the top of whatever is state of the art at any particular time. See benchmark.

mirror A highly polished surface, usually of silvered glass or metal, which readily reflects light. Mirrors were used for line-of-sight signaling long before electrical telecommunications methods were available. Hikers still regularly carry them along for emergency signaling in the wilderness. Mirrors are also used in many types of computer devices, especially those which incorporate laser beams, such as laser printers. The mirror serves to direct the beam inside the mechanism onto the appropriate areas, such as a printing drum. See heliograph.

mirror site A computer file archive site that maintains duplicates of files existing on another system. Some mirror sites reflect not only the files on the other system, but the entire file and command structure as well. Mirror sites exist to protect data from loss and also to provide alternate access to popular files, in case the original site is slow to respond or has become inaccessible. FTP sites often have mirror sites, sometimes in other countries.

mirroring A means of providing system backup security by replicating data in different locations. The system can go to the mirror location if the original storage of the data becomes corrupted or the system can be restored with information from the mirror. Redundancy is a very common property of computer systems. Some will mirror whole directory structures and files as a matter of course. Some hard drive systems are set up to constantly mirror information over several devices. While mirroring almost inevitably costs a little more in terms of memory or storage space and in processing time, it is usually worth it. See mirror site, RAID.

MIS See Management Information Services.

MITS Micro Instrumentation and Telemetry Systems. The historic creators of the Altair microcomputer, MITS, under the direction of Ed Roberts, originally sold radio transmitters (telemetry devices) for model planes. These products did quite well and got the company under way, but when the company moved into the area of calculator kits, there was a lot of competition from companies like Texas Instruments, and the Altair was in essence an effort to stave off bankruptcy. MITS developed the MITS 816 in 1972, and later the historic kit for the Altair 8800 in 1974. While the Altair is not the first microcomputer, it is to be credited as being the first commercially successful microcomputer. In spite of the success of the Altair, the company was sold to Pertec, a manufacturer of peripherals. See Altair, Intel MCS-4, Kenbak-1, Mark-8, Micral, Scelbi.

MJ modular jack. Any jack that is designed to interconnect readily with various standardized receptacles in a circuit system. See RJ.

MLA See mail list agent.

MLS microwave landing system.

MMCF See Multimedia Communications Forum.

MMCX See Multimedia Communication Exchange.

MMDS See microwave multi-point distribution system.

MME Mobility Management Entity.

MMF See multimode optical fiber

MMI machine-to-machine interface.

MMIC Monolithic Microwave Integrated Circuit.

MMSP See modular multi-satellite preprocessor.

MMSS Maritime Mobile-Satellite Service.

MMTA See Multimedia Telecommunications Association.

MMU 1. Manned Maneuvering Unit. A human maneuvering unit used in untethered space walks originating from the U.S. space shuttle missions. 2. memory management unit. Computer circuitry often built into central processing chips to handle administration of blocks of storage.

MMX Multimedia Extension. Matrix Math Extension. See Pentium MMX.

mnemonic A memory-jogging device such as an acronym, abbreviation, rhyme, or pun. For example, "I before E, except after C." Programmers often use mnemonic variable names to keep track of code, and will implement mnemonic hot-key shortcuts for the benefit of users.

MNLP Mobile Network Location Protocol.

MNP See Microcom Networking Protocol.

MNRP Mobile Network Registration Protocol.

mobile assisted handoff MAHO. A process in which the handoff of a voice channel by a mobile station is assisted by the base station by providing information on the surrounding radio frequency (RF) signal environment.

Mobile Data Base Station MDBS. In CDPD mobile communications, a system which provides data packet relay functions between the Mobile End System (M-ES) and the Mobile Data Intermediate System (MD-IS). See Cellular Digital Packet Data.

Mobile Data Intermediate System MD-IS. In CDPD mobile communications, a system which provides routing and location management functions, utilizing a Home Domain Directory (HDD) database. The MD-IS communicates with the Mobile End System (M-ES) through the Mobile Data Base Station (MDBS). See Cellular Digital Packet Data.

Mobile End System M-ES. In CDPD mobile communications, the system through which the subscriber accesses wireless network services. M-ESs include modems installed in laptops, palmtops, personal digital assistants (PDAs), etc. See Cellular Digital Packet Data.

Mobile Identification Number MIN. Each wireless phone is assigned an identification number by the carrier. The MIN is not attached to the individual, as the phone may change hands or the individual may change locations.

Mobile IP, Mobile Internet Protocol Mobile data networking through the Internet is coming into demand as the number of laptops and the availability of wireless modem services increases. Since the problems of maintaining contact with a network and network security are concerns on mobile systems, a set of extensions to Internet Protocol (IP) is being developed to handle the special needs of mobile users. Mobile IP uses a dual addressing scheme so that the communications node and the mobile unit can be tracked and administrated. In simple terms, the location of the mobile system becomes a 'forwarding address' to which packets are retransmitted. Security is incorporated to prevent an unauthorized person from intercepting the transmission. See Foreign Agent.

Mobile Network Location Protocol MNLP. In CDPD mobile communications, the MNLP provides a means to track the Mobile End System (M-ES), that is, the laptop modem, cellular phone, or other device that allows the user to link into the network, and to interlink the Home Mobile Data Intermediate System (MD-IS) and the Serving MD-IS. This works in conjunction with a Mobile Network Registration Protocol (MNRP) to verify the user's Network Entity Identifier (NEI), a security ID used to monitor and confine service to authorized users.

Mobile Network Registration Protocol MNRP. See Mobile Network Location Protocol.

mobile phone An audio broadcast system designed to provide mobile communications through hardware interfaces resembling traditional phone handsets. The earliest mobile phone systems were bulky, limited contrivances developed after the turn of the century and first demonstrated in 1919, but they were acknowledged as having an important place in future communications.

Historically similar to broadcast TV, a powerful transmitter was located to provide maximum range, up to perhaps 30 miles, for traveling subscribers. To increase the limited range and channel distribution of the single tower design, cellular networks were developed, which increased available bandwidth by providing many lower power transmitters, closely located to one another, over a wide geographic region.

There are now a number of types of mobile phones, from short-range FM cordless phones with a range of a few hundred feet, to digital PCS and cellular systems with roaming capabilities that range from hundreds to thousands of miles. See cellular phone.

Mobile Subscriber Unit MSU. A main component of a mobile phone system consisting of a portable or transportable control unit and cellular radio transceiver. Convenience, size, transceiver power, and battery life are traded off in the various systems. Larger, more powerful units may be mounted to car batteries, and often split the telephone and the handset into separate units. Smaller handhelds frequently have less range and shorter battery life. See cellular phone, mobile phone.

Mobile Telephone Switching Office MTSO. A main component of a mobile phone service, which performs wireless relaying, switching, and administration tasks similar to those carried out by a wired telephone switching office, except that it must handle the specific technical needs of users who are moving and roaming (changing from one transceiving area to another) with signal monitoring and processing, handoffs, etc. In addition, the MTSO handles the link between the mobile services and connections to wireline services, as many mobile services are actually hybrid technologies, often taking calls from mobile users and connecting them with a wireline destination, and vice versa.

modal In applications programming, a type of user window, dialog, or other input or information display operation which does not suspend access to other processes. For example, suppose the user has selected a Quit function, and the software displays a dialog box that says, "Do you really want to quit? If so, the program will end without saving." Options to Quit or to Cancel will be presented. If the dialog allows the user to go back to the application without responding to the Quit/ Cancel query, the operation is modal. If the user must reply before continuing with using the software, then it is not. While modal (multitasked) operations are preferred in many situations, there are others where a response should be solicited before continuing, especially if it involves the possible loss of data.

mode In some older operating systems, a distinction was made between 'text mode' and 'graphics mode,' but most systems now work in 'graphics mode' with text represented graphically. This system is more flexible.

Two different computer modems: the Global Village on the left has a built-in 9-pin DIN connector, and operates at 33,600 bps, the SuperModem on the right has a standard 25-pin connection supporting V.34 standards. Each can be connected (daisy-chained) with a regular telephone set.

modem modulator/demodulator. 1. A device which modulates and demodulates a signal. Digital data are typically modulated to be carried over analog transmission systems, and broadcast waves are modulated to add information to the carrier band. These are then demodulated again at the receiving end. 2. A computer hardware peripheral specifically designed to convert the digital signals generated by the computer into analog systems that can be carried across an analog transmissions medium such as twisted pair copper wire, and demodulate them back into digital data at the receiving end. Many standards exist for the transmission of this type of data, and the sending and receiving modems must be able to negotiate a common format in order for the signals to be meaningfully received. Current modems commonly transmit at rates of 19,200 bps, 38,400 bps, and higher; most include facsimile transmission capabilities, and some include voice mail capabilities as well. They incorporate a number of error control, data compression, and modulation protocols in order to maximize speed of transmission over lines that people once claimed could never transmit data faster than 600 bps. See error control protocol, data compression protocol, modulation protocol, serial port.

modem pool A set of modems usually servicing a network through which several users can dial out of the system, or through which a number of users can dial in, as to a BBS or Internet Services Provider (ISP). Most higher educational institutions have modem pools through which users can access the system from home or classrooms, or through which they can dial out to community services or extra service providers. Often the modems in a pool will have different characteristics. For example, there may be only a few lines which are high speed lines, due to the higher cost of high speed modems, and a majority which are slower and less expensive.

Some modem pools are extremely large. For example, one of the largest commercial Internet providers has over 100,000 modems in its pool.

A pool is a flexible way to maximize resources. A dozen modems can service a hundred workstations, provided the users do not need constant access to dialup resources. It is also easier for system administrators to do hardware maintenance and to maintain security when the modems are grouped and placed in a secure environment.

modem server A networked workstation application which manages the administrative and access tasks associated with a modem pool, or an intelligent modem hub which manages incoming and outgoing data from more than one user. In large modem pools, a system may be dedicated to assigning user requests for modems, for sending messages to the user (e.g., "All systems are currently busy, please make your request again in 15 minutes."), for evaluating which modems to allocate first (there may be different modems with different capabilities, such as access speed), and for assigning priorities and connect times, when appropriate.

modem standards This is one of the areas where de facto vendor standards and industry standards (e.g., ITU-T V Series Recommendations) have continually leap-frogged one another, and engaged in an uneasy competitive race. The constant consumer demand for faster modems, and the vendor desire to be the first to market with the next generation modem, has caused many vendors to develop their own standards ahead of the global cooperative standards process. For this reason, many modems are dual-standard modems, in order to support both the vendor and generally accepted industry standards. Some modems support either vendor or industry standards, which are often not compatible, and it is important to find out their status before purchasing.

In many cases, the early versions of modems supporting the faster speeds are the ones which are most likely to go out of date quickly. In the early days, many vendors followed Bell and Hayes standards, whereas in recent years, vendors have tended to go with the industry standards once the specifications are finalized and made available. The basic Hayes command set remains, although most vendors implement a superset of the original Hayes commands, which were quite simple and limited.

The Microcom Network Protocol standards for error control and data compression are widely supported modem standards. See Microcom Networking Protocol, V Series Recommendations.

moderated discussion list A public or private online discussion, usually carried out through email or a Web page gateway, in which the messages are screened, and sometimes edited before being posted to everyone on the group. Moderated discussion lists used to be extremely rare. In the 1980s there were thousands of open lists that were kept on topic through voluntary cooperation. Since about the mid-1990s, however, many online discussion lists were ruined by inappropriate use, especially the posting of advertisements for sex sites and get-rich-quick schemes, resulting in some lists being shut down and many others going to moderated status. This is a lot of work for the moderators, who are almost always volunteers, and who have to read everything and decide whether it is appropriate for the list. These unsung heroes deserve a great deal of appreciation for keeping good lists alive, as many have been lost due to abuses of the system. See discussion list, newsgroup.

moderator A person who presides over a meeting or discussion. On the Internet there are a vast number of public discussion groups, and, as the number of participants increases, many of these groups have gone to moderated status. For example, the business groups have become so cluttered with messages promoting get-rich-quick schemes that the only good business groups remaining are those that are moderated. Moderators, usually called *ops*, also help to maintain order and appropriateness on open chat lines. Internet moderators are typically hard-working volunteers who take time to read messages and edit or reject those which are inappropriate to the stated charter of the group. In the early days of the Internet, moderated groups were almost nonexistent. Unfortunately, in the mid-1990s, the need for moderation increased due to repeated abuses of the system. See Net Police, Netiquette, Netizen.

Modified Final Judgment, Modification of Final Judgment MFJ. The name given to a historic seven-year antitrust lawsuit between the U.S. Justice Department and AT&T, which resulted in the breakup of AT&T. It is associated with Judge Harold Greene's decision regarding the 1983 to 1984 (clarification and revision) divestiture of AT&T. Under this judgment, AT&T was permitted to retain ownership of Bell Laboratories and AT&T Technologies (Western Electric), but the Regional Bell Operating Companies (RBOCs) were banned from manufacturing, and Local Access Transport Areas (LATAs) were created rather than retaining the existing local exchange boundaries.

Prior to the MFJ, charges were handled through Division of Revenues, but this was changed to an access charge tariff system. See AT&T, Kingsbury Commitment, Local Exchange Carrier.

modular Composed of separately organized entities, loosely or tightly coordinated or connected to create a larger whole. Modular programming is programming in which the larger application is composed of smaller associated elements such as blocks, objects, primitives, self-contained functions, etc. Object-oriented programming is a type of modular programming. A modular office is one in which the individual components of the facilities can be changed around fairly easily, that is, desks, screens, phones, cables, etc. can be rearranged without undue effort. A modular phone system is one in which handsets or phone sets can be unplugged and moved or rearranged within a building or department. Modular software is software in which a number of separate or related utilities, tools, and functions, can be used together in a number of ways. For example, there may be a variety of functions that do file conversions, image processing, filtering, special effects, etc. which can be used separately or in conjunction with a variety of programs. Some of the more flexible, stand-alone 'plugins' exhibit these properties of modularity. For example, there may be a 'watercolor' plugin which can be used independently to alter the contents of a graphics file, or may work as a plugin in the context of several programs such as an image processing program, a drawing program, etc.

modular multi-satellite preprocessor MMSP. A frame synchronizer designed to provide an interface between a host computer and synchronized mapper telemetry data. The MMSP takes the raw telemetry data, frame aligns and samples it, and transmits the information to the host computer, where it is further processed and the image information extracted from the data.

modulate To change gradually from one state to another. To tune, or adjust. To vary the amplitude, frequency, or phase, typically to add information to a carrier wave. To change the velocity of electrons in an electron beam, as in a cathode ray tube.

modulation A key element in the transmission of information. By changing, or modulating an electrical pulse through a wire, or other conducting medium, or an airborne electromagnetic wave, it is possible to convey information. Similarly, by manipulating its intensity and duration, light can be modulated to send information. Some of the simplest forms of modulation include turning a signal on or off, or varying it between high and low states.

For computer users, one of the most familiar modulating devices is the dialup modem. The modem takes a digital signal from the computer and modulates it to be carried over analog phone lines. At the receiving end, a modem then demodulates the signal, turning it

M

back into digital signals that are transferred to the receiving computer.

There are many modulation techniques used throughout the telecommunications industry, some very simple, and some so sophisticated only computers can control them. The most common types of modulation are amplitude modulation (AM), frequency modulation (FM), and phase modulation (PM). Sometimes different modulation schemes are combined. Each scheme has its own unique characteristics.

Early detractors said frequency modulation was mathematically impossible, but Edwin Armstrong demonstrated, after 10 years of hard research applied to the problem, not only that it could be done, but that it was a great thing. It has since been used in thousands of applications from radio programming to cordless phones and burglar alarm systems. Another important contribution to modulation was the work of John R. Carson, who demonstrated how a portion of a modulated signal could be transmitted, instead of the whole thing, and the original signal 'rebuilt' at the receiving end, thus reducing bandwidth without loss of information. See amplitude modulation; Armstrong, Edwin; frequency modulation; phase shift keying; quadrature amplitude modulation; single sideband.

modulation protocol A data encoding technique used to convert digital data into analog signals. This determines the raw (uncompressed) speed at which the modem can transfer data. Current modems incorporate more than one protocol. See modem.

modulation, light A means of conveying information by manipulating a beam of light. The light can be directly influenced, by turning it on or off, or varying its intensity; or it can be indirectly influenced by interposing shutters, gels, or other objects between the sender and the receiver. Light modulation is used in fiber optic transmissions, with lasers and light-emitting diodes used as common light sources.

moiré 1. In raster-oriented imagery, moiré is a visual artifact that causes an undesirable, distracting pattern which disturbs the intended appearance of the image. 2. In traditional printing on a press, especially process color printing, small dots are often interleaved to simulate the appearance of more colors. If the angles and patterns of these dots are not carefully controlled, a moiré pattern, resembling light through silk, may emerge. Better desktop publishing programs provide print settings to set the angle and type of halftone to match the technology on which the job is being printed. 3. In video images, mixing of high frequencies can create an undesirable, visible, low frequency moiré.

moisture barrier A cover, sheet, bag, or other barrier, usually plastic, intended to retard or prevent moisture from coming in contact with building structures, wires, or electrical components. Moisture barriers are used to prevent rot, condensation, and electrical short circuits.

molding raceway A channel system incorporated into wood, plastic, or metal moldings to hold, protect, and direct interior wiring circuits. Molding raceways are of modular construction with a variety of fittings, so individual sections can be interconnected and holes can be punched where needed. Molding raceways are commonly used on baseboards and wainscots, where they blend naturally with the decor. See raceway.

monochrome monitor A monitor that displays only one color of illuminated phosphor on a contrasting screen. The color is usually white, green, or amber against a dark background, or black against a light background. When the intensity of a monochrome monitor is varied, all the pixels get brighter or dimmer, with no capability of individually setting the intensities. A grayscale monitor, on the other hand, also uses only one color of illuminated phosphor, but each pixel can be individually controlled for its intensity, allowing 16, 32, or 256 shades of gray to be represented. Grayscale monitors make very good work environments for correspondence and desktop publishing, and some are designed in portrait mode (or can swivel between portrait and landscape mode) to more closely recreate the environment of a piece of paper.

monopole A slender self-supporting tower used for attaching wireless antennas/aerials.

monospaced font, fixed width font A font in which each character in the set is of one fixed width, usually equal to the widest character plus a slight bit of space to the right, in order that letters don't touch when strung together. Monospaced fonts lack the aesthetic properties of proportional fonts, because the letter width is not related to the width of the individual characters. Thus, the letter i will have the same width as the letter w with some extra space that may look unappealing. Monospaced fonts are used with printers and typewriters that lack the capability of moving the printhead in small, varying increments. Monospaced type also takes more space on the page, allowing fewer characters, because of the extra white space. Contrast with proportional font.

proportional iiiiii wwwwwww

`fixed width iiiiii` wwwwwww

Moonbounce, Earth-Moon-Earth bounce EME. A means of using the Moon as a passive reflector for communications signals. Due to the great distances involved, very large antennas and strong signals are required, but given these in conjunction with the right weather conditions, Moonbounce transmissions have been demonstrated.

The first Moonbounced signal was recorded in January 1946 in New Jersey, where army engineers used a recently invented FM transmitter and receiver developed by E. H. Armstrong, to send pulses to the Moon, which returned as a slight hum. This was a significant achievement as it showed not only the potential of FM broadcasts, but demonstrated that radio waves could pass through the ionosphere and beyond.

Morse code A system of character encoding using dots and dashes, or long and short sounds or lights, that can be readily sent over distance over many types of transmission media due to its simplicity. International Morse code (Continental Morse code) and American Morse code (Railroad code) have been derived from this.

Morse code is flexible in that it can be sent with tones, clicks, dots and dashes, and lights, in a variety of media. In 1862, two Philadelphia inventors patented a signal light system using a shuttered oil lamp for sending Morse code which was intended to be mounted on the masthead of ships.

International Morse code developed from Austro-Germanic code, a variation on Morse code which is used in radio transmissions partly because American Morse code, while suitable for telegraph communications, was more difficult to interpret over radio waves. In 1851 it became the code of choice for transatlantic cable communications. Basic skill in Morse code has been a requirement of receiving amateur radio licenses for many decades. The code was apparently developed by Morse's collaborator, Alfred Vail, and is named for the inventor of the printing telegraph, Samuel F. B. Morse. See Morse code history.

Morse code history The original paper tape printing telegraph designed by Samuel Morse employed a system of numbers which were then correlated with words, according to a lookup reference. The lookup reference developed by Morse was very large and the system itself somewhat slow and cumbersome; it required the maintenance of a reference and the somewhat arbitrary assignment of nonmnemonic code number sequences to every word. A more simple, direct system was needed.

Alfred Vail was from a family of fabricators and acted for years as assistant to and collaborator with Samuel Morse. Mechanically adept, he built many of the mechanical components designed by Morse. In the process of creating the mechanisms for the Morse printing telegraph key, Vail changed the orientation of the keying mechanism from horizontal to vertical, thus providing a more comfortable hand position. The change also resulted in a stylus which would lift up from the paper, leaving dots and dashes, rather than zigzag-shaped dips on the tape record that Morse's original mechanism produced.

Vail's assistant Baxter reported to Franklin Pope that Vail went to work on simplifying Morse's lookup code system. Vail apparently visited local printers to analyze typesetting cases to determine the frequencies of letter usage. Pope subsequently reported the story in 1888 in "The Century: Illustrated Monthly Magazine." The code he developed eventually evolved into American Morse code, and International Morse Code became a further streamlined variation. [Thanks to Karen Weiss and B. Neal McEwen for unearthing and reporting Vail's apparent unacknowledged contribution to history.]

Morse sounder A type of early telegraph sounding instrument, which used audible clicks to broadcast the incoming message rather than a paper tape printout, which was slow. The sounder incorporated an electromagnet as a pole piece, mounted on a pivoting sounding lever with two stop positions. Releasing the magnet as it was energized produced the clicking sound. The duration of the clicks represented the coded dots and dashes of the Morse code system, and were interpreted aurally by the receiving operator.

Typically the sounder was connected to the sending instrument with only one wire. The viability of the single wire circuit was observed by Steinheil in 1837 in Germany, and independently the following year by Morse in

America. Both discovered that a second wire was not needed to complete the circuit if the two instruments were connected through the ground, using it as the return path for the circuit. This worked even over distance.

Morse, Samuel Finly Breese (1791-1872) An American artist and inventor in the 1800s chiefly known for the code that bears his name. He was a respected artist and one of the founders of the National Academy of the Arts of Design. In the 1820s he became increasingly interested in science and invented electromechanical telegraph devices, some of the first inventions to use electricity for communication.

With advice and assistance from J. Henry and L. Gale, Morse was able to construct a basic working design for the telegraph by 1837.

Samuel Morse demonstrated his invention to the Presidency in 1838 and in 1843 won funding support from the U.S. Congress to construct a telegraph line between Baltimore and Washington. He sent his first public message over this line in May 1844, an event that launched a revolution in communications.

Morse became friends with the Vail family, who were talented fabricators and were able to assist him in constructing practical working models of his ideas. Many of Morse's inventions were built by Alfred Vail, Morse's assistant and collaborator. See Gale, Leonard D.; International Telegraph Union; telegraph; telegraph history; Vail, Alfred.

When spun rapidly, the images inside this cylinder can be viewed through the slits as an animated show of a juggler. In the above picture, the inside is shown for clarity, but normally the rotating device would be viewed straight on so the image passing by the slits is seen. This was one of the early means for creating 'cell animations,' that is animating images by the creation of individual frames that differ slightly from one another.

motion pictures Any images which convey the appearance of motion, whether in real time or by presentation of a fast sequential series of still pictures, especially videos, film reels, and animated computer images. Motion film pictures consist of a series of still images played through a projecter, usually from 20 to 30 frames per second, with 24 or 30 being common, as these are the speeds at which human perception 'merges' successive still frames into a cohesive impression of connected motion.

The development of motion picture photography owes some of its roots to a bet over a dispute as to whether a running horse lifted all four hoofs off the ground. Thomas Edison was one of the first to experiment with displaying a series of still frames in rapid succession in 1889. The first commercial motion picture, backed by the Canadian Pacific Railway, is attributed to Clifford Sutton in the early 1900s. See animation, celluloid, MPEG.

Motorola A significant computer chip designer and manufacturer, and electronic appliances manufacturer, since the 1960s. It is descended from the Galvin Manufacturing Company from the early 1930s. In 1974 it released the MC6800, the first in a long family of chips still being developed a quarter of a century later. One of the first microcomputers developed with the Motorola family of microprocessors was the Altair 680, released late in the fall of 1975. Since that time whole families of computers have been based on the subsequent MC68000 family of chipsets, including Macintosh, Atari, Amiga, Sun, Apollo, SGI, NeXT, and others. Motorola is also well known for products in the mobile data communications industry. The Motorola CPU Sample chart shows a brief summary of some of Motorola's best-known desktop computer microprocessors, prior to the collaboration with IBM to produce the PowerPC chips.

In 1998, Motorola teamed up with the McCaw/Gates Teledesic project to provide Celestri technology to the orbiting satellite network. See Altair 680.

Mountain Bell The familiar name for the Mountain States Telephone and Telegraph Company.

Mountain States Telephone and Telegraph Company An early telephone company, better known as Mountain Bell, which was formed in 1911 from the merger of the Tri-State and Colorado telephone companies, and the purchase of the Rocky Mountain telephone company.

mouse A hardware human interface device

that receives hand and finger movements and transmits them to a computing device. They are then interpreted into actions by the operating system and applications software. The mouse is named for its basic shape, which typically consists of a palm-sized, rounded or squarish object, with one or more buttons under the fingers, and a 'tail,' a cord that electrically connects the mouse with the computer. Mice come in various shapes and sizes: friction mice have a ball on the side that makes contact with a hard surface; optical mice require a grid or special pad. Laptop variations include finger pads and rollerballs, which are not strictly mice, but which employ the same basic movement and input concepts.

The invention of the computer mouse is attributed to Doug Englebart and is variously reported as having been invented around 1959 to 1963. By the late 1960s, Englebart was testing a three-button mouse in conjunction with a 'keyset' that was used in the other hand. During the early 1980s, when the Apple Lisa was being developed, the first of the Macintosh line, there were discussions at Apple as to whether to use a two- or three-button mouse. The testing and rationale supplied by Larry Tesler indicated a one-button mouse was completely appropriate, and the Macintosh line still works very well with this device 15 years later. The majority of competing desktop computers use two-button mice.

mouse blur A visual artifact that makes the pointer look blurred as the mouse is moved too quickly for the refresh rate of the display device to keep up, which is more common on laptop systems that use LCD displays rather than cathode ray tubes (CRTs). Mouse blur is often accompanied by mouse blanking, in which the fast movement of the cursor causes it to momentarily disappear, and reappear when the refresh catches up to its current position.

MP See Multilink Protocol.

MPEG Motion Pictures Experts Group. A series of international standards developed by a joint committee under the aegis of the International Organization for Standardization (ISO), to facilitate the development of digital video and audio. MPEG has received widespread acceptance for the playback of digital animations. Leonardo Chairiglione and Hiroshi Yasuda originated the MPEG development efforts in 1988.

Video and audio technologies typically require a lot of bandwidth and file space, so a large part of the MPEG effort has concentrated on decompression schemes and fast playback algorithms. The compression itself is left up to the discretion of individual vendors.

Individual contributors hold a number of patents to various technologies which have been incorporated into MPEG. These contributors agreed in writing to provide the technology nonexclusively at fair and reasonable royalty rates.

There have been several enhancements to MPEG since its introduction as shown in the MPEG Versions chart. See animation, B-frame, I-frame, JPEG, P-frame.

MPLS See Multiprotocol Label Switching.

M

Sampling of Well-known Motorola Central Processing Units

Processor	Year	Proc.	Data bus	Addr. Bus	Notes
MC6800	1974		8	16	Used in Altair 680.
MC68000	1979	32	16	23	16 32-bit registers. Supervisor and user mode. CISC architecture.
MC68010		32	16	23	Virtual memory.
MC68020	1982	32	16/32	32	256-byte cache. Dynamic bus sizing.
MC68030	1987	32	16/32	32	Paged MMU on processor. 16 byte burst.
MC68040	1990	32	32	32	FPU, cached Harvard buses.
MC68060	1994	32	32	32	Superscalar pipelined. Power-saving.

MPEG Versions

Version	Notes
MPEG-1	"Coding of Moving Pictures and Associated Audio for Digital Storage Media at up to about 1.5 Mbps" ISO/IEC 11172, standardized between 1993 and 1995. An optimized 1.5 Mbps bit stream for compressed video and audio, for compatibility with existing CD and DAT data rates. Non-interlaced color video is typically implemented at 352 *x* 240 (288 in Europe), which is relatively low resolution, as it derives from a CCIR-601 digital television standard. Replay speed is 30 frames per second (25 in Europe), fast enough for natural-looking motion. Sample precision is 8 bits.
MPEG-2	"Generic Coding of Moving Pictures and Associated Audio" ISO/IEC 13818, presented in draft form in 1993. ITU-T recommendation H.262. Similar in structure to MPEG-1, the documentation includes four parts in addition to the categories discussed in MPEG-1. MPEG-2 can address very low bit-rate applications with limited bandwidth needs, and support for surround sound multichannel applications. Video resolution is typically implemented at 720 to 550 *x* 480 not unlike that of many computer monitors, and a frame may be interlaced or progressive.
MPEG-1+	MPEG-1 presented at MPEG-2 resolution. Frames are de-interlaced and compressed.
MPEG-3	Merged into MPEG-2 when it was decided that MPEG-2 syntax could be scaled to support HDTV applications.
MPEG-4	"Very Low Bitrate Audio-Visual Coding" Launched in 1992 to develop new algorithms for providing support for a wider range of applications, and to improve efficiency. New applications include low-bitrate speech coding and interactive mobile communications.

MPOA MultiProtocol Over ATM. A client/server protocol integration effort specified by a working group of the ATM Forum to provide direct connectivity across an ATM network between ATM hosts, legacy devices, and future network-layer protocols from different logical networks. This will enable the production of scalable ATM internetworks. See LANE, IPv4.

MPOA Client An ATM term. A device which implements the client side of one or more of the MPOA client/server protocols, (i.e., is a SCP client and/or an RDP client). An MPOA Client is either an Edge Device Functional Group (EDFG) or a Host Behavior Functional Group (HBFG).

MPOA over ATM sub-working group A group that seeks to solve some of the implementation problems associated with asynchronous transfer mode (ATM). It is integrating LAN Emulation (LANE), Next Hop Resolution Protocol (NHRP), Classical IP, multiprotocol encapsulation, and multicast address resolution in order to provide end-to-end internetworking ATM connectivity. MPOA is a packet-oriented protocol similar to LANE. The group provides courses, support, research, documents, and systems testing services.

MPOA Reference Model MultiProtocol Over ATM Reference Model. A specification approved by the ATM Forum in June 1997 for routing/switching over ATM networks. There are Internet MPOA resources (links, white papers, specifications, etc.) at the ATM Forum's site. http://www.atmforum.com/atmforum/specs/approved.html

MPOE Minimum Point of Entry.

MPP See Multichannel Point-to-Point Protocol.

MPT See Ministry of Posts and Telecommunications.

MRI See magnetic resonance imaging.

MRU maximum receive unit.

MS Mobile Station.

MS-CDEX A set of Microsoft DOS extensions for CD-ROM which allow MS-DOS to recognize the presence of a CD-ROM drive and access it accordingly.

MS-DOS Microsoft disk operating system. MS-DOS originated from a commercial text-oriented operating system developed from Tim Paterson's QDOS, first released by Microsoft in 1981 to accompany IBM's Intel-based microcomputer system. This was a somewhat different move for IBM, as the company had often created its products in-house. But IBM was under pressure to release a successful microcomputer in order to avoid being locked out of the growing market; Radio Shack at one point had almost 80% market share with its TRS-80 line. IBM's move to look outside of its own research and development resources provided a window of opportunity for emerging computer companies. They contracted to purchase an operating system from Microsoft and, in a remarkable turn of events, Microsoft managed to retain the rights to market the product themselves, in competition with IBM, who called their version PC-DOS.

Early versions of MS-DOS were intended for single-user applications, but by version 3.1 network functionality was being added. At about this time, multitasking graphical operating systems were being released by several other vendors, and there was pressure on MS-DOS to provide features found on other systems.

MS-DOS became widespread through the 1980s, only slowly giving away to Microsoft's later graphics-based Windows products, which were developed in the mid-1980s as front-ends to MS-DOS. MS-DOS's text interface is now used much less, but, in syntax and functionality, it was very similar to CP/M, the dominant operating system at the time, developed by Gary Kildall of Digital Research (originally Inter-Galactic Research). MS-DOS became the pivotal product behind the financial success of Microsoft, and it and its successors have been installed on millions of computers worldwide.

To give an introductory idea of the flavor of MS-DOS, the accompanying Outline chart shows some of the most commonly used commands. See MS-DOS history, operating system.

MS-DOS history The Microsoft disk operating system (MS-DOS) was originally Tim Paterson's QDOS. Paterson has reported that he gleaned ideas from a CP/M operating sys-

Outline of Some Commonly Used MS-DOS Commands

Command	Function
CD <> or CHDIR <>	Change the current directory
COMP <>	Compare files
COPY <>	Copy files, duplicate
DEL <> or ERASE <>	Delete/erase a file (use with caution)
DISKCOPY <>	Copy from one floppy disk to another
FORMAT <>	Format a diskette
MKDIR <>	Make a new directory
PATH <>	Define the search path. This is where the system seeks executables and batch files so that the entire command path doesn't have to be typed to run a command or application.
REN <> or RENAME <>	Change the name of a file
TYPE <>	Display the contents of a file (intended for ASCII files)
XCOPY <>	Copy files
CLS	Clear the current screen

tem manual published in the mid-1970s. CP/M was developed by Gary Kildall, a university professor and programmer, founder of Digital Research (originally Inter-Galactic Research). When IBM first approached Microsoft in the early 1980s, to purchase the BASIC programming language for their personal computer, they apparently thought they were also obtaining the rights to CP/M. Bill Gates signed a nondisclosure agreement to work with IBM, and promised to supply BASIC. When IBM found out they hadn't purchased an operating system as well, Gates suggested they call Digital Research, developers of CP/M. Gates didn't want to lose the languages contract for lack of an operating system, and Microsoft didn't have an operating system that would meet IBM's needs. Up until this time, Microsoft had concentrated on languages, Digital Research had concentrated on operating systems, and Kildall had not expected the 'gentleman's agreement' to change.

IBM was in a hurry. They paid a call to Digital Research (DR) at a time when Gary Kildall had another commitment. His wife/business partner was present to talk to the corporate giant, but was uncomfortable with signing the one-sided nondisclosure agreement presented by IBM without consultation with Kildall, especially since DR's attorney didn't like the terms of the agreement. IBM went back to Gates, and, seeing an opportunity, Gates offered to supply an operating system as well as BASIC. IBM accepted. Kildall was probably surprised at not having the opportunity to meet further with IBM to negotiate terms.

Since Microsoft didn't have an operating system that would fulfil the contractual agreement with IBM, they went to a company called Seattle Computer Products (one wonders why they didn't talk to Kildall about licensing CP/M) and bought a simple operating system called QDOS (Quick and Dirty DOS) developed by Tim Paterson. Microsoft contracted Tim Paterson to make a few enhancements in collaboration with Robert O'Rear, and soon after delivered the product to IBM. It was released first as PC-DOS by IBM, and later as MS-DOS by Microsoft. There are some small differences between the two. This was the beginning of a long alliance and the launching of a tiny entrepreneurial effort into a major empire, with MS-DOS as the pivotal product. See Microsoft, MS-DOS, Digital Research.

MSAT Mobile Satellite. A commercial mobile satellite communications service developed jointly by AMSC in the United States and TMI in Canada.

MSB most significant bit, most significant byte. A computer data storage and programming concept that describes the order in which data is organized, stored, or transmitted, or its relative importance. MSB is the leftmost bit or byte, or that which contributes the largest quantity in arithmetic calculations.

MSC Mobile Switching Center. A switch providing coordination and services between mobile network users and external networks.

MSK See minimal shift keying.

MSNF See multisystem networking facility.

MSP See Message Security Protocol.

MSS See multispectral scanner.

MSU 1. microwave sounding unit. 2. See mobile subscriber unit.

MTA 1. Macintosh Telephony Architecture. 2. See Major Trading Area. 3. Message Transfer Agent. 4. Metropolitan Trading Area. A Federal Communications Commission (FCC) designation for a region. This is an administrative process used to facilitate licensing for various communications services.

MTBF See mean time between failures.

MTI Moving Target Indication. In radar, the analysis of Doppler-shifted returning signals to select moving targets from various other nonsignificant or interfering objects (clutter).

MTM Maintenance Trunk Monitor.

MTS 1. member of technical staff. 2. Message Telecommunications Service. An AT&T designation for standard telephone service through direct dialing. 3. See Message Transfer System.

MTSO See Mobile Telephone Switching Office.

MTTR mean time to repair. A computed average of the amount of time it takes to bring a broken or faulty device back into service.

Mu-law A pulse code modulation (PCM) coding and companding standard used in North America. Mu-law is commonly used for encoding speech in 8-bit format by sampling the audio waveforms at 8,000 times per second. See A-law, E carrier, pulse code modulation, quantization, U-law.

MUD Multi-User Dungeon/Dimension. An addictive role-playing game prevalent on the Internet, and on most college campuses since MUDs were invented in the late 1970s. In fact, it was one of the few games that college network administrators tolerated on their systems, because the development and administration of a MUD involved the exercise of quite a few gray cells, and involved more programming and technical expertise than the usual shoot-em-up games that distracted students from their studies. MUDs involve fantasy characters, developed by the players, which interact in various 'rooms' within the MUD. The first MUD has been attributed to R. Bartle and R. Trubshaw at the University of Essex.

MUF maximum usable frequency.

Multi-Vendor Integration Protocol MVIP. MVIP originated in 1989 from an initial consortium of three companies, Natural MicroSystems, Inc., Mitel, and GammaLink, and four more helped the project to coalesce a year later. MVIP has since become one of the two common software/hardware bus standards in computer telephony (along with SCbus). The purpose of the standardization was to bring together the various telephone and computer technologies so they could readily interconnect and intercommunicate. MVIP provides an open, nonproprietary, uniform, yet flexible way of providing telephony components with computer equipment through open software development environments. In other words, phone-related technologies can be accessed and controlled through a desktop computer. The MVIP standard includes the capability to reconfigure 'on the fly' to handle various call functions.

The original single-chassis standard designed for a synchronous environment was MVIP-90 and additional versions followed, including H-MVIP (high capacity MVIP), and MC-MVIP (multi-chassis MVIP). The MVIP Versions chart shows three MVIP formats.

multicarrier modulation MCM. A number of modulation techniques for multiplexed transmission of data by dividing the communications channel into smaller units and evaluating the units individually in terms of speed and suitability for transmission. Widely used to implement Digital Subscriber Line (DSL) services over existing twisted pair copper wires, which can vary widely in their characteristics.

MCM optimizes bandwidth usage for multiple media transmissions and reduces interference from impulsive and narrowband noise. See carrierless amplitude and phase modulation, orthogonal frequency division multiplex, discrete multitone, discrete wavelet multitone.

multicast A type of Internet Protocol (IP) address identifier for a set of interfaces. Frames sent from one end station are received by one or more end stations. In IPv6 multicast addresses supersede broadcast addresses. In ATM networking, the form of the multicast command is: atm multicast <address>. See anycast, unicast, IPv6 addressing, multicast backbone.

multicast backbone Mbone. Technology that extends the Internet Protocol (IP) to support multicasting, developed by Steve Deering at Xerox PARC. The Mbone was adopted by the Internet Engineering Task Force (IETF) in 1992.

Mbone supports two-way transmissions of data between multiple network sites. Thus, with Mbone, a single packet can have multiple destinations and pass through a number of routers before being split up to run through different paths. The packets reach their destinations at about the same time. This leads to greater support for multimedia capabilities over the Internet. Mbone systems are assigned Internet Protocol addresses that are Class D.

Multichannel Interface Processor MIP. A Cisco Systems interface router processor which provides up to two channelized T1 or E1 serial cable connections to a channel service unit (CSU).

Multichannel Point-to-Point Protocol MPP. A protocol from Ascend Communications that is similar to Point-to-Point protocol (PPP), but which supports multiple network channels in inverse multiplexed systems.

multifeed dish, multifocus dish A type of parabolic satellite receiving dish which can be positioned to capture signals from more than one satellite at a time, with the signals being reflected to a series of feedhorns.

multilevel coding A coding scheme or system, usually bandwidth efficient encoding, for carrying more data through a channel in a specified period of time. There are many approaches to improving the efficiency of transmissions.

Multilink Protocol PPP Multilink is an Internet standards-track protocol which enables the splitting, recombining, and sequencing of datagrams across multiple logical data links. Originally designed as a software solution for implementing multiple simultaneous channels over ISDN, the concepts are generalizable to multiple PPP links between two systems. The purpose of multilink implementation is to coordinate multiple independent links (i.e., to aggregate a bundle) between a fixed pair of

Radio Frequency Transmission Schemes		
Format	Abbrev.	Notes
ALOHA		A free-for-all style of transmission; any source transmits at any time, and continues to transmit if there is an acknowledgment. It is not a high-efficiency method, but there are circumstances where it is practical.
Code Division Multiple Access	CDMA	A hybrid scheme which incorporates time/frequency multiplexing to provide spread spectrum modulation. Thus, central channels can be handled without timing synchronization.
Frequency Division Multiple Access	FDMA	A traditional method of channel allocation in which bandwidth is subdivided into frequency bands, with guard bands providing a buffer between channels.
Packet Reservation Multiple Access	PRMA	A type of enhanced TDMA which incorporates aspects of S-ALOHA. Suitable for mobile transmissions.

systems to create a virtual link with greater bandwidth than individual parts used alone. See RFC 1990.

multimedia A catchall term for sensory-rich communications media, that is media that contain images, motion, sound, tactile feedback, etc. Multimedia educational and entertainment products are more common than multimedia business applications. Networks need greater bandwidth and faster processing speeds to handle multimedia traffic. See animation, frame buffer, Mbone, virtual reality.

Multimedia Communications Forum MMCF. A nonprofit international multimedia communications and software research and development organization founded in 1993. MMCF consists of telecommunications service providers, multimedia application and equipment developers, and multimedia users. MMCF promotes market acceptance of networking multimedia products and serves as a clearinghouse for multimedia-related specifications, standards, and recommendations. http://www.mmcf.org/

Multimedia Communication Exchange MMCX. A commercial phone/data server software developed by AT&T's Global Business Communications Systems (GBCS) for providing multimedia services for private phone branch exchanges. See Global Business Communications Systems.

MultiMedia Telecommunications Association MMTA. A professional telephony/computer integration trade association descended from the North American Telephone Association. The MMTA is organized as an open public policy, market development, member support, and educational forum for telecommunications product and service developers and resellers. http://www.mmta.org/

multimeter, multi-range meter A measuring instrument commonly used in the communications industry which has several scales for evaluating an electrical circuit, including voltage, resistance, and current.

multimode optical fiber MMF. A multimode fiber optic transmissions cable usually with a relatively thick core acting as a waveguide to reflect and propagate light at many angles. A thicker core has advantages and disadvantages. More light signals can be sent at one time, but the signal transmission distance is shorter than single mode fiber, due to the interaction of the reflected signals within the core and gradual fading over distance. Thus, it is limited to about two kilometers. Signals are usually transmitted through multimode cables with light-emitting diodes (LEDs) and received at the other end with a photodiode detector. This detector translates the signals back into electrical impulses. See single mode optical fiber.

multipath fading A signal loss characteristic of mobile communications in areas where there are many buildings or uneven terrain which bounces the transmission waves in many directions. The deflected signals may interfere with, or even partly cancel out the primary signal. When a subscriber listens to an analog cell phone conversation or a motorist listens to an AM car radio, the sound volume may decrease. Digital transmission systems provide error-correction techniques to combat the problems of fade, and spreading the spectrum over a broader bandwidth, if feasible, can sometimes help.

multiplayer gaming Many computer games are designed for one player to sit and play against the computer. Some of these games have a mode in which the user can elect to play against an opponent, usually on the same system, with the two players alternately using the input devices, or using two sets of joysticks, for example. These are usually known as single- or dual-player games.

Multiplayer gaming is implemented in a number of ways. There are games in which multiple players can play on the same machine in much the same way as dual-player games, each one taking a turn, but multiplayer more often connotes a game in which multiple players can interact at the same time, with the computer, with each other, or both. This opens up a whole new dimension in gaming, and multiplayer games are very popular due to the 'human' element, and the higher sophistication that they often provide. Most multiplayer games can now be played over networks and, in fact, a large proportion of sales of inexpensive Ethernet cards for home and small business computers happened as a direct result of the popularity of fast-action multiplayer computer games. See MUD.

multiple access A means to increase communications capacity over a limited bandwidth medium. There are many ways to do this; some are specific to the type of technology, and whether it is analog or digital. In computer networks, multitasking and common address schemes and protocols allow multiuser access. Web pages on the Internet are a good example of multiple access technologies, since many thousands of people may be accessing the same site at the same time. In cellular communications and transponders, there are at least three types of basic schemes in use which allow use of limited resources. See the Radio Frequency Transmission Schemes chart for some examples.

Multiple Exchange Carrier Access Billing MECAB. A document prepared under the auspices of the Carrier Liaison Committee of the Alliance for Telecommunications Industry Solutions (ATIS) which establishes the methods for processing orders for access service being provided by two or more local carriers, that is a local exchange carrier (LEC) and a competitive local carrier (CLC).

multiple frequency shift keying MFSK. Frequency shift keying is a modulation technique used in data transmissions such as wireless communications in which binary "1" and binary "0" (zero) are coded on separate frequencies. This scheme can also be adapted to regular phone lines by assigning binary "1" to a tone and binary "0" (zero) to a different tone. In multiple frequency shift keying, a greater use is made of different frequencies to represent information. These multiple frequencies may be transmitted one after the other, or simultaneously. See frequency modulation, frequency shift keying, phase shift keying.

multiple homing 1. In networking, taking an additional service line, which is often a leased line through a larger provider. This is sometimes done by Internet Service Providers (ISPs), as their customer bases increase. Thus, it becomes necessary to maintain and announce an additional set of routes through which addresses can be reached. 2. In phone services, adding phone connections so calls can be routed through more than one switching center to provide redundancy in case of problems with the initial service loop.

Multiple-Image Portable Graphics MNG. A format for using multiple sub-images in a single frame, a multimedia graphics format originally developed as Portable Network Graphics (PNG) for animations. PNF was submitted as Draft 1 by Glenn Randers-Pehrson in June 1996. With Draft 12, in August 1996, PNF came to be known as MNG and, by December 1997, it was up to Draft 42 and considered essentially finished. See Portable Network Frame, Portable Network Graphics.

multiple switchboard A manual switchboard with multiple operators, in which answering jacks are divided up among operators, usually in rectangularly organized banks.

Multiple Virtual Line MVL. A communications system that uses existing analog phone

M

lines to provide higher speed digital communications and simultaneous voice. A mid-range option between cable modems and ISDN modems. An MVL switch must be installed at both the switching center and the subscriber point.

multiplex See multiplexing.

multiplexed analog component MAC. A means of television transmission used with direct broadcast satellite (DBS) programming. The various signal components, audio, luminance (brightness), and chrominance (color) were transmitted independently. The advantage of this, particularly for DBS in European regions, is that the audio can be individually tailored to the language desired.

multiplexing 'Splitting' a circuit in one of a variety of ways so more than one signal can be carried at a time over a single transmission. This is an extremely important way of increasing channel efficiency and capacity in most telecommunications technologies.

Multiplexed phone circuitry began to appear in the late 1930s, and was widely established by the mid-1950s. A historically significant practical application of multiplexing was developed in 1953 by E. H. Armstrong and John Bose, in which they enabled a single FM station to simultaneously transmit more than one signal at a time. This made possible multiple simultaneous broadcasts, including the transmission of stereo broadcasts. Many different multiplexing schemes are included under individual headings in this dictionary.

multiprocessing Using intercommunicating processors to solve computing tasks. The various processors may be housed inside one computer, or may be in separate systems, connected by a network. Multiprocessing is associated with parallel processors with distributed computer systems and, often, with high-end computing tasks such as scientific calculations, rendering, ray tracing, and computer animation. In a multiprocessing system, there is often a centralized processor assigning and distributing tasks, with the other processors feeding back information to the system or following instructions from this central administrator. The software development to optimize these kinds of systems still has room for development, and there are many opportunities to contribute to research and implementations in this area.

Not all tasks benefit from multiprocessing. In other words, it doesn't make everything faster. If the overhead involved in managing and coordinating the different processing tasks is greater than the speedup in the tasks themselves, then it's not a good candidate for multiprocessing.

One of the most exciting aspects of multiprocessing is that the speed of communication between the processors is an intrinsic part of the system. With higher and higher data rates, faster CPUs, and faster data buses, we may someday realize a time when the speed of the data transfer meets or exceeds the speed of its associated processors, in which case the various networked systems will function more like a single organism than a set of separate computers.

Multiprotocol Label Switching MPLS. A routing system which differs from conventional Internet Protocol (IP) forwarding in that the assignment of a particular packet to a particular stream is done only once, as the packet enters the network. This facilitates the use of explicit routing, without each packet carrying the explicit route. The assigned stream is encoded with a short fixed length label which follows it to the next hop. Thus, at subsequent hops, rather than analyzing the packet each time to gather information from the network layer header, the label is used as an index to a table which specifies the next hop, and a new label supplants the old label. MPLS provides explicit class of service (CoS) information, or information for inferring the class of service (CoS) or precedence from the label. Routers that support MPLS are called label switching routers (LSRs).

Multiprotocol Label Switching domain MPLS domain. A contiguous set of nodes which provides MPLS routing and forwarding operations.

Multiprotocol Label Switching edge node MPLS edge node. An MPLS node which connects an MPLS domain with a node outside of the domain.

Multipurpose Internet Mail Extension MIME. Since any type of binary file can be transported over the Internet using appropriate conversion routines, protocols, and compression algorithms, if needed, it is logical that text, graphics, and sound can all be transmitted through binary files. Yet, Internet Mail had provisions for only the transmission of ASCII documents. For this reason, developers sought to give email a richer set of capabilities, and released MIME specifications in 1992 as a general framework and format for the representation of many kinds of data types in Internet mail.

MIME enables a developer to write email software clients which can include pictures of the kids, sounds of the dog barking, and messages to loved ones all in the same message. This provides the user with a multimedia mail interface to the Internet. Since email accounts for at least 75% of the communications of regular consumers on the Net, enhanced email capabilities will probably be appreciated by many.

Content-rich email has been around on several workstation platforms since the late 1980s, but most of these applications were proprietary and could not intercommunicate. Text mail messages for the Internet were defined even earlier, in 1982 (RFC 833). Then, in 1992, a standards track protocol for a generalized Internet extensions format was submitted as a MIME Request for Comments (RFC) from the IETF Network Working Group.

MIME provides ways in which to include multiple objects in a single message, and to represent text in character sets other than just U.S.-ASCII. It supports multiple typefaces, images, and audio. While many email programs are MIME-compliant, not all of them support all the features of MIME, but it is expected that MIME implementations will increase, especially as MIME has been designed to be extensible; additional content types can be defined and supported. See RFC 1341.

multisession A feature of CD-ROM discs and drives which enables the CD-ROM to have information written to it in more than one session with access to everything on the disc. Since historically CDs were write once/read many media, the early CD-ROM drives would not recognize additional information on the disc, if it was written to the disk at a later date from the original information. This wasn't a problem if the consumer was buying a game disk with 600 megabytes of information on the original disc.

Then Kodak developed a type of image storage format called a PhotoCD in which photographs are scanned, digitized, and saved on a CD-ROM. Since the user could bring the CD-ROM back to the processor to have another roll of film digitized and stored on the same disc (assuming there was still space remaining), it became important to create a CD-ROM format and a CD-ROM drive that could read the data from the multiple sessions. Most CD-ROM drives now are multisession.

multispectral scanner MSS. A nonphotographic imaging system, commonly used on remote sensing satellites, that utilizes an oscillating mirror and fiber optic sensor array. As the mirror sweeps, it transmits brightness values a strip at a time. Multiple detectors are placed in an array in order to sense radiant energy in selected spectral bands.

The color that is seen on satellite image posters, that can now be bought in poster and stationery stores, is actually added by computers after the image is geocoded (the strips resolved into a cohesive image). In early days of colorizing satellite images, the color assignments were not as appealing as they are currently. The MSS technology was first developed for the Earth Resources Technology Satellite (ERTS-1) in 1968, a program which later developed into the LANDSAT satellites.

multisync A device which can handle transmissions at a variety of frequencies. This is a common feature of monitors which allows them to display at various resolutions when attached to different graphics display cards and computers. If you are installing a multisync monitor for the first time, it is very important to read the documentation with the graphics card or computer to determine the correct frequency settings for the monitor. If you make the settings too high, you can damage the monitor. Most consumer color multisync monitors run at about 60 MHz.

multitasking Carrying out two or more tasks simultaneously. Many computer systems that claim to be multitasking are actually task switching. In other words, you can switch from one task to another, or one application to another, without having to close the first task or application, but the background task freezes; in other words, it does not continue processing while the foreground task is active. If you can run a sound program in the background or format a floppy while doing something else, like using a paint program or a word processor in the foreground, without the sound stopping or jittering or without having to wait for the floppy drive to finish, then you probably are working on a multitasking system. Most current workstations and some current microcomputers are fully multitasking.

Historically, multitasking microcomputer operating systems have existed since the early 1980s, the most notable examples being some of systems that ran on the Tandy Color Computer (CoCo), Digital Research's multitasking version of CP/M, and the AmigaOS (1985), which had pre-emptive multitasking. In the late 1980s some of the other systems began to implement task switching and multitasking. In 1987, the NeXT computer was introduced with multitasking. Workstation-level computers like the Sun stations, SGIs, etc., have been multitasking for some time. By the early 1990s, most microcomputers had some degree of task switching or multitasking, and full multitasking on desktop systems will probably be standard by the turn of the century.

multithreaded Programming on microcomputers with threads became more common in the 1990s, although the practice has been around longer on mainframes and workstation level computers. A system which is multithreaded can have a number of processes running independently of the others while still using common resources. Even though they function somewhat independently, they can also be programmed to communicate with one another at intervals or as certain events occur. See threads.

Munsell color model A three-dimensional, ordered representation of color relationships for pigments, developed in the early 1900s by an art instructor. For the purposes of illustration, imagine a sphere of no fixed diameter, in which white and black are the north and south poles and hues are represented around the sphere in terms of 'primary colors' of red, yellow, green, and purple and their complements, and are equally spaced around the 'circumference,' with the blended colors occurring as transitions at their junctions. But now, unlike a sphere, imagine chroma as a characteristic that radiates outward from the sphere's central axis, thus influencing the hues that it 'passes through.' Although a complex model, it has the advantage of being somewhat scalable. As new pigments are invented, they can be incorporated into the model without redefining existing colors. See CIE, color space, Maxwell's triangle.

Murray loop test A type of diagnostic procedure which uses resistance through a bridge to locate an 'open' in a length of circuit. It is similar to a Varley loop test, except that instead of having adjustable dials, one arm is eliminated and a variable resistance arm connected in its place, and a third wire is not required. See Varley loop test, Wheatstone bridge.

Museum of Independent Telephony Located in Abilene, Kansas, the home of the United Telephone Company from 1898 to 1966, and one of its former presidents, Carl A. Scupin helped found the Dickinson County Historical Society and Museum. The Museum of Independent Telephony now shares premises with this museum.

Museum of Radio and Technology A nonprofit, volunteer-assisted antique radio technology museum, located in a converted elementary school in Huntington, West Virginia. It includes crystal radio sets, vacuum tube technologies, schematics, vintage books and magazines, and other educational resources and exhibits.

In this music box, tiny spikes have been painstakingly placed in a wooden cylinder to 'store' the song.

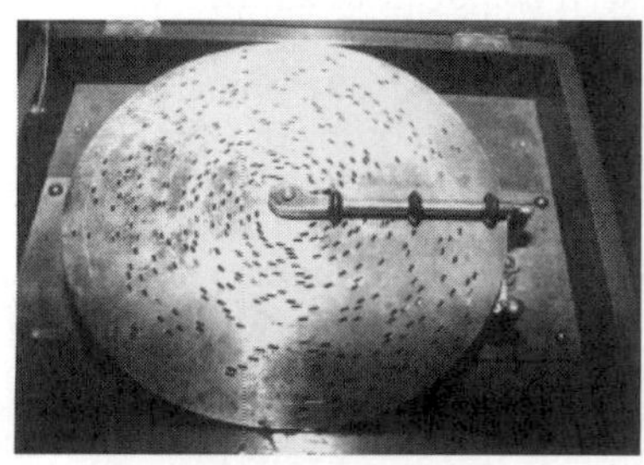

This music box is similar to a gramophone in that it has a revoloving platter; however, the song is stored in small rectangular holes, rather than in a spiral groove. Both music boxes above are included in the Bellingham Antique Radio Museum collection.

Museum of Television and Radio A nonprofit, New York-based preservation and education institution established in 1975 by William S. Paley. Its goal is to collect and preserve historic radio and television programs and make them available for public education and use. In 1991, it was moved to the William S. Paley building. It houses over 60 thousand programs selected for their historic, artistic, and cultural value. The collection adds about 3,000 programs per year. The Museum works in conjunction with the *Museum of Television and Radio in L.A.*, California. Both institutions sponsor seminars and exhibitions.

Museum of Television and Radio in L. A. Also known as the West Coast Museum of Television and Radio, it was established in 1995 in Los Angeles. This museum is named after

Leonard H. Goldenson, considered a pioneer of the broadcasting industry. It works in conjunction with the original *Museum of Television and Radio* in New York.

music box A historic form of entertainment medium which preceded computers in 'storing' data for later playback. Many different means of creating the pattern for the sounds were developed over the years, from nails and holes in boards and cylinders, to paper tapes and piano rolls.

Jacquard looms were another form of technology which incorporated holes in cards to store a weaving pattern. It was not until the 1970s that magnetic storage of software programs for computers superseded the Hollerith punch cards that followed the same conceptual design as the stored songs on these music boxes.

music-on-demand A service that was offered not long after the invention of the gramophone and the telephone. Since switchboards were operated by humans, it was not difficult to set up a system where the caller could request a song and the operator would put on a phonograph record and play the song to the nearby telephone. The system went out of favor as switchboards became automated and radio and stereo systems became prevalent. It has now come back in a slightly updated form as *video-on-demand* in which a user can request a particular movie or other video entertainment through a computer interface or interactive television interface.

A historic music-on-demand system, with a phonograph placed near a telephone receiver to transmit the sound to the listener.

music on hold Background music on a phone line that is heard when the person is put on hold. It can be set up to play music from a radio, cassette tape, or CD player, usually with a simple RCA jack near the phone panel. Recorded music is generally better, as most radio stations play advertising which is not appreciated by most callers. Some people don't care for background music while on hold, but it's probably better than the alternative of not knowing whether or not you've been cut off.

Musical Instrument Digital Interface MIDI. MIDI is a standard protocol for communication of sound information through a number of specified parameters. Functions provided on MIDI-capable instruments are assigned numeric values which can be digitally intercommunicated and remotely or locally controlled. MIDI capabilities are built into many musical editing and sequencing software programs.

MIDI-compatible instruments generally have DIN plugs for interconnecting the various MIDI devices, and usually include MIDI in, MIDI out, and sometimes MIDI through. A simple example of a MIDI setup would be a keyboard connected to the fast serial port or MIDI port of a computer with MIDI-compatible software. There may also be separate speakers, since most computer and keyboard speakers tend to be minimally useful for sound reproduction. MIDI allows the songs from the keyboard to be communicated to the computer and stored and edited. Conversely, compositions created on the computer, sometimes including custom sound patches, can be communicated back to the keyboard.

MIDI is not the only music protocol, but it's definitely the most widespread and best supported. It is built into numerous synthesizers, keyboards, drums, and software music editing and sequencing programs.

Musschenbroek, Pieter van See van Musschenbroek, Pieter.

MVIP supports Unix, OS/2, and Microsoft DOS or Windows. The work was officially taken over in 1994 by the GO-MVIP project. See Go-MVIP. http://www.mvip.org

mute A feature or device which allows sound to be turned off, or substantially lowered in volume. A mute button on a phone can cut out the sound to the mouthpiece so something can be said without being transmitted (handy if you have to call to the dog or ask a coworker a sensitive question), and then released to continue the conversation. A mute on a stringed instrument dampens the vibrations of the bridge so the sound of the instrument is much softer.

MUX See multiplexer.

M

MVIP Multi-Vendor Integration Protocol. See chart below for an overview of MVIP versions and see Multi-Vendor Integration Protocol.

MVL See Multiple Virtual Line.

MVS Multiple Virtual Storage.

MXR 1. mixer. 2. multiplexer.

MZI Mach-Zehnder Interferometer.

Outline of MVIP Versions	
Format	Notes
MVIP-90	Original single-chassis specification consisting of a multiplexed digital telephony bus with 512 *x* 64 Kbps capacity, distributed circuit-switching capability, and digital clocks architecture. MVIP-90 provides for 16 serial lines.
H-MVIP	High Density MVIP. A scalable, high-capacity, downwardly-compatible superset of MVIP-90 which was adopted as a standard by the GO-MVIP technical committee in 1995. H-MVIP provides for 24 serial lines.
MC-MVIP	Multi-Chassis MVIP. A high capacity version which provides several alternate physical layer connections with a common set of software interfaces for interconnecting single-chassis MVIP systems and telephony systems. MC-MVIP provides 1536 *x* 64 Kbps over copper twisted pair (MC1); or FDDI-II for 1536 *x* 64 Kbps over copper or fiber (MC2); or SONET/SDH at 155 Mbps for 4800 *x* 64 Kbps over fiber.

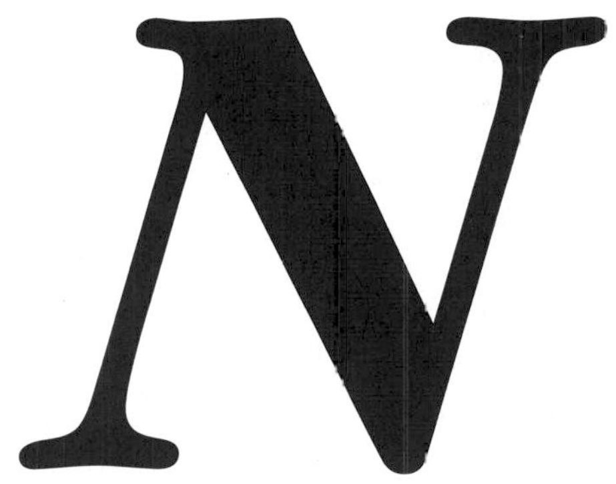

n Abbreviation for nano-. See nano-.

N 1. Symbol for 'north,' on a magnet or compass. The north-seeking end of a compass needle points to a region near the North Pole called magnetic north. 2. Symbol for 'on.' With F as the corresponding symbol for 'off.'

n region In a semiconductor, a region in which the conduction-electron density exceeds the hole density. Usually referenced in relation to the p region, the n materials interact with the p materials at the p-n junction in some types of semiconductors. See p region, p-n junction.

N-scope A type of radar display in which the target appears as a pair of vertical blips coming from a horizontal time base. The direction of the target is inferred by the amplitudes of the related vertical blips. A target distance can also be determined by comparison to a pedestal signal along the base line.

NAB 1. See National Alliance of Business. 2. See National Association of Broadcasters.

NACN See North American Cellular Network.

NADF 1. See North American Directory Plan. 2. North Atlantic Directory Forum.

nadir In satellite imaging, a point on the ground centered vertically below a remote sensing platform.

NAK See negative acknowledge.

name resolution A means of associating an assigned name with its origin, location, or other relevant characteristics. In a network where a name has been used as a mnemonic alias to allow easy recognition of an address, application, or process, there needs to be a mechanism to resolve the name into a form that can be easily recognized and subsequently located by the system. In other words, *mysite.com* has to be translated into a machine-readable address of the location of the host site for *mysite.com*. This is done through name resolution, usually through a lookup table or larger database. Sometimes the name itself will provide some information about its origin or date of establishment, e.g., the name is a set of alphanumeric characters assigned according to a system that can be understood by humans. See naming authority.

naming authority 1. A legislative or organizational body that assigns names, usually as unique identifiers. There are varioU.S. types of naming structures: hierarchical, flat, random, etc. There are many well-known naming authorities: the U.S. Library of Congress; R. R. Bowker (ISBN); IANA (Internet). On the Internet, the various registered domains may assign subauthorities and subnames for local machines. See IANA. 2. In a hierarchical document management system, a tree of entities which provides a unique identifier to each document. This task may be shared by subauthorities.

NAMPS Narrowband Analog Mobile Phone Service. An analog cellular communications technology which provides triple the capacity of an analog cellular voice channel by splitting the channel into 10 kHz bandwidth narrow bands. Narrowband standards were released by the Telecommunications Industry Association (TIA) in 1992 (IS-88, IS-89, IS-90). Digital mobile phone services are gaining ground on traditional analog systems. See AMPS, DAMPS, code division multiple access, time division multiple access.

NANC See North American Numbering Council

nano- (Abbrev. - n) A unit prefix for one billionth (North American system), or 10^{-9}, that is .000 000 001.

NANOG See North American Network Operators Group.

NANP See North American Numbering Plan.

NAP See Network Access Point.

NAPP See National Aerial Photography Program.

narrowband 1. A term which varies in definition depending upon the industry and its bandwidth needs, and on the current state of technology. Narrowband usually represents the lower end of the available capacity or spectrum of a system. In some cases it is used to denote a single band within a multiplexed group of bands being sent more-or-less simultaneously. In traditional telephony, it represents a sub-voice-grade line. In cellular communications, it represents one division of the broadcast spectrum consisting of a channel frequency (CF) of about 30 kHz, usually accomplished through frequency division duplexing (FDD). See AMPS, NAMPS.

Narrowband Analog Mobile Phone Service See NAMPS.

narrowband ISDN ISDN services at basic channel speeds up to 64 kbps, which is fine for voice and some data communications, but only adequate for applications like full-motion video, or video and sound. There are efforts being made to incorporate new standards into broadband ISDN (B-ISDN) which will remove the fixed channel structure limitation of narrowband ISDN.

narrowcasting A type of program delivery that targets specific people and often specific services to those people. If broadcasting is considered to be program delivery to a wide, and sometimes scattered audience, from one to many, then narrowcasting can be seen as one to one or one to few. For example, electronic industries personnel might subscribe to programming on circuit board fabrication. At an even more specific level is 'pointcasting,' that is, program services which target user-selected information, a type of 'electronic clipping service' providing electronic information on specified topics of interest.

NARTE See National Association of Radio and Telecommunications Engineers.

NASA National Aeronautics and Space Administration.

nasa7 A double-precision Systems Performance Evaluation Cooperative (SPEC) benchmark used in scientific and engineering applications. A benchmark tends to be a specific quantitative measure of a particular aspect of computer functioning, and by itself conveys a picture of overall system performance. However, in the specific context for which it is intended, a benchmark can provide valuable information for design engineers, researchers, and manufacturers. Nasa7 generates input data, performs one of seven floating point-intensive kernel routines and compares the results against an expected reference measure. It is used to evaluate performance, memory, I/O operations, and networking factors. See benchmark, Rhealstone, Whetstone.

NASTD National Association of State Telecommunications Directors.

NATA See North American Telecommunications Association.

National Aerial Photography Program NAPP. A program which was initially established in 1980 as the National High Altitude Photography (NHAP) program. In 1987, the height of the satellites was lowered, and the name changed to NAPP. NAPP is administered by the U.S. Geological Survey's National Mapping Division, and exists to coordinate the collection and processing of aerial photos of the 48 contiguous states and Hawaii in a format that meets the requirements of a variety of U.S. federal and state agencies.

National Alliance of Business NAB. An association established to promote innovative, long-term solutions for improving the quality of the workforce, in terms of productivity, education, and security.

National Association of Broadcasters NAB. A well-known American broadcast industry association providing support and education to its members through literature, standards activities, programming, conventions, and seminars. http://www.nab.org/

National Association of Radio and Telecommunications Engineers NARTE. An international professional association which provides support to members along with certification programs.

National Association of Regulatory Utility Commissioners NARUC. A Washington D.C.-based organization serving the needs of United States government utility commissioners. http://www.naruc.org/

National Association of Telecommunications Officers and Advisors NATOA. A professional association which supports and services the telecommunications needs of local governments. NATOA provides education, information, and advocacy for their members. http://www.natoa.org/

National Bell Telephone Company A merger of the Bell Telephone Company and the New England Telephone Company, in a bid to achieve widespread national coverage of services. A court decree dissolved the company only four years later.

National Broadcasting Company NBC. A major broadcast company for many decades, formed in 1926 by David Sarnoff. NBC provides general television programming, entertainment, sports, news, local/interactive, programming transcripts, contests, games, and arts. See Sarnoff, David.

National Bureau Of Standards NBS. A bureau of the United States government which provides testing and standardization services. The NBS had an important role in the development of early computing devices in the 1940s when it undertook the construction of two large-scale computing machines for its internal needs, one to be installed on each coast. This resulted in the building of the Standards Eastern Automatic Computer (SEAC) and the Standards Western Automatic Computer (SWAC).

National Cable Television Association NCTA. A trade association representing the cable broadcast industry founded in 1952. NCTA represents the interests of its members to public policy makers in the U.S. Congress, the judicial system, and the public. NCTA hosts a large annual trade show.

National Communications System NCS. A branch of the U.S. government formed in 1962 during the Cuban Missile Crisis. The recommendation of an interdepartment committee reporting to President John F. Kennedy was to form a single communications system to serve the President, the Department of Defense (DOD), diplomatic and intelligence activities, and civilian leaders. The NCS was officially established in 1963 to link, improve, and extend the communication facilities and components of various federal agencies. It cooperates with various standards bodies, and develops emergency procedures for the American communications infrastructure. The NCS is administrated by the General Services Administration (GSA). http://www.ncs.gov/

National Continental Telephone, Telegraph & Cable Company of America Founded in 1899, this ambitious undertaking was an attempt to gain control of all the independent telephone companies, that is, all those not controlled by Bell. Despite backing by some of America's richest high profile financiers, this project was unsuccessful.

National Coordinating Center NCC. A joint U.S. government-industry organization established by the National Communications System (NCS) to provide for the U.S. government's telecommunications service requirements. The NCC initiates, coordinates, and restores NS/EP telecommunications services. It is one of several divisions of the Office of the Manager of the National Communications System (OMNCS).

National Coordination Office for Computing, Information, and Communications NCO. A U.S. communications coordinating agency. http://www.hpcc.gov/

National Digital Cartographic Data Base NDCDB. A U.S. database of digital cartographic/geographic data files compiled by the U.S. Geological Survey (USGS). The database includes elevation, planemetric, landcover, and landuse data at various scales.

National Electrical Code NEC. A code developed to safeguard public safety and property from hazards associated with the use of electricity. This includes wiring and electrical device construction, materials, installation, and maintenance and is adopted in many parts of the country as law for various building and equipment installations. The Code is developed by the American National Standards Institute (ANSI).

National Exchange Carrier Association NECA. A nonprofit organization established by the Federal Communications Commission (FCC) in 1983 to administrate issues related to service and access charges. The NECA serves the interests of incumbent local exchange carriers (LECs), and administers the universal service fund (USF) which subsidizes certain loop services. http://www.neca.org/

National Geophysical Data Center NGDC. One of three data and information centers of the U.S. National Environmental Satellite, Data and Information Service (NESDIS).

National High Altitude Photography NHAP. Originally established in 1980, with satellites at 40,000 feet, the height was lowered to 20,000 feet, and the program renamed to National Aerial Photography Program in 1987. See National Aerial Photography Program.

National Information Infrastructure NII. The name for the political, administrative, and physical underpinnings of an interconnected collection of public and commercial national narrowband and broadband data networks. One of the biggest stakeholders in the NII is the National Information Infrastructure Advisory Council (NIIAC) established in 1994 through a 1993 executive order. The NIIAC is responsible for advising the government on a national strategy for promoting the development of the NII and the Global Information Infrastructure (GII).

The NII is a physically and regionally diverse system which is considered as a whole mainly on the basis of interconnectivity. It includes small and large networks, wireless and wireline connections, public and private systems, and many sizes and types of organizations and individuals. The NII is also known by the catchphrase "Information Superhighway," although this describes the communications aspect of the NII and could be considered a subset of the NII.

National Institute of Standards and Technology NIST. This standards organization is affiliated with the U.S. Department of Commerce. http://www.nist.gov/

National Internet Services Provider NISP or NSP. An Internet Services Provider of national scope, usually with broader regional access and a variety of connection points. The Internet services provided by local ISPs and NSPs are usually similar. The main difference is that national providers often have dialups in major cities that the user can access with a local call when traveling, thus avoiding long-distance connect charges.

National Laboratory for Applied Network Research NLANR. An organization which researches leading edge networks and supports the evolution of a U.S. national network infrastructure.

National Oceanic and Atmospheric Administration NOAA. A U.S. government agency which sets strategic goals for environmental assessment, prediction, and stewardship and describes and predicts changes in the Earth's environment, to manage coastal and marine resources. http://www.noaa.gov/

National Research and Education Network NREN. A government-funded gigabit-per-second national research backbone proposed in the early 1990s after an initial proposal was presented in 1987 to the Congress by the Federal Coordinating Committee for Science Engineering and Technology (FCCSET). It was intended to support voice and video, and to become a significant means of finding and disseminating information. See National Science Foundation.

National Research Council NRC. A U.S. organization established in 1916 by the National Academy of Sciences to serve the needs of the science and technology community in advising the federal government. It is now the principal operating agency of the National Academy of Sciences and the National Academy of Engineering. The NRC provides services to the government, scientific and engineering communities, and the public. It is administered jointly by the Academies and the Institute of Medicine. http://www.nas.edu/nrc

National Science Foundation NSF An independent U.S. government agency, established in 1950 to promote public welfare through science and engineering research and education projects through various types of educational and financial support. The NSF was established by the National Science Foundation Act of 1950, and provided with additional authority through the Science and Engineering Equal Opportunities Act. It is administrated by the National Science Board appointed by the President with the advice and consent of the United States Senate. See NSFNET. http://www.nsf.gov/

National Spatial Data Infrastructure NSDI. A U.S. Executive Order signed in 1994 under which federal agencies must document, and make accessible through the electronic Clearinghouse network, all new geospatial data collected or produced, either directly or indirectly, using the Federal Geographic Data Committee (FGDC) standard.

National Standards System NSS. A Canadian standards association which is managed by and works in conjunction with the Standards Council of Canada and a committee of volunteers to write standards, and to test and certify products and systems. See Canadian Standards Association, Standards Council of Canada.

National Technical Information Service NTIS. An agency of the U.S. Department of Commerce, through the Technology Administration, the NTIS is the official source for various types and formats of U.S. government-sponsored global scientific, technical, engineering, and business-related information, supplied by many United States government agencies. http://www.ntis.gov/

National Telecommunications and Information Administration NTIA. An Executive Branch agency of the U.S. Department of

Commerce founded in 1978. NTIA is responsible for domestic and international telecommunications policy issues, and is a principal advisor to the President. NTIA works to promote efficient and effective uses of telecommunications information and resources in order to support U.S. competitiveness and job opportunities.

NTIA is descended from a reorganization of the Office of Telecommunications Policy (OTP) and the Office of Telecommunications (OT). It cooperates with the Federal Communications Commission (FCC) in managing broadcast spectrum administration and assignment.

Various endowment and grant programs have been transferred to the NTIA, including the Public Telecommunications Facilities Program from the Department of Health, Education, and Welfare.

The NTIA has a laboratory which conducts applied research that is located in Boulder, Colorado. See Federal Communications Commission, Institute for Telecommunication Sciences. http://www.ntia.doc.gov/ntiahome/

National Television System Committee See NTSC.

native format The structure of a raw application or data file prior to modification for compatibility, emulation, or compression. A format designed to work optimally with a particular architecture. For example, many Amiga computer programs were designed to take advantage of the Amiga's unique hardware features such as the blitter and NTSC output. Color cycling, hardware sprites, video output, and other graphics capabilities that were 'native' on the Amiga were not available on other systems at the time without special hardware add-ons.

Native file formats abound in many applications programs, that is, data formats for saved files that don't work with other applications. Electronic Data Interchange (EDI) was developed as a file-transfer solution for a variety of types of files from otherwise incompatible applications, thus allowing the user to combine the speed and ease of native formats with the practical interchange of data between programs. See ASCII, Electronic Data Interchange.

native mode The common way of running software within the natural operating functions of a specific computer platform, i.e., software specifically designed to run on a system without emulation or modification to mimic or support a different architecture. Since no emulation overhead is needed, native mode software typically runs faster than emulation software. In spite of this, the trend is away from native mode software. Since intercompatibility and especially Internet compatibility are high priorities and since machines are now fast enough that emulators run reasonably well, platform-independent, non-native mode software or software which falls somewhere in between, is becoming more prevalent and practical. See native format. Contrast with emulator, Java, OpenGL, Perl, platform-independent.

NATOA See National Association of Telecommunications Officers and Advisors

natural antenna frequency An antenna's lowest natural resonance frequency when operated without external capacitance or inductance.

natural frequency The frequency at which an otherwise uninfluenced or unimpeded body will oscillate when stimulated to move.

natural language A human language, such as Japanese, Russian, English, French, Indian, Chinese, Arabic, Senegalese, Hawaiian, etc. Natural language processing, both in written and spoken form, is one of the challenges of computing which is ongoing, with many of the routines derived from studies in artificial intelligence.

N

natural language programming Natural language computer programming is the creation of computer instructions that are similar to a human language. The implementation of natural language programming is as yet still somewhat rudimentary and experimental, although success in specific areas has been achieved. Even the BASIC programming language, which was developed to be similar to the English language, is symbolic in its syntax and structure in ways that are different from English. See speech recognition.

natural magnet There are two types of permanent magnets. One is a substance that exhibits and retains magnetic properties without application of a current after it has been 'magnetized' with another magnetic source. The second is a substance which exhibits magnetic properties as it comes out of the ground, without needing to be exposed to magnetic influences for it to become a magnet. The second type of permanent magnet is called a natural magnet. See lodestone.

natural wavelength The wavelength that corresponds to an antenna's natural frequency.

Matching of an antenna's resonant frequency to the characteristics of the wave that is being received (or transmitted) is an important aspect of antenna design.

nautical mile NM. A standard means of describing a distance traveled through water. A knot is one nautical mile per hour, or 6,076 feet (1 knot = 1.15 mph). An international nautical mile is 1,852 meters. The term 'knot' originates from a physical means of estimating the distance. A log line was marked by knots at 47.33 foot intervals. A weighted log chip was attached to one end and thrown overboard to the stern. It would remain somewhat stationary as the boat moved away, and the line was allowed to run for 28 seconds and then hauled back into the boat. The knots that slipped away were tallied to calculate the boat's speed.

navigate 1. To move on or through, with selective consideration of the path taken along the way. 2. To follow a course through the various features of the Internet, to 'surf the Net.' In its broadest sense, the 'path' may be selected in many different ways and combinations, including by geography; topic; name; feature; whim or personal interest; institution; server; etc. Pathways through the Web can be followed through hypertext links embedded in HTML pages. 3. On the World Wide Web, to follow a course through a path of hypertext links set up by the designers of the Web pages themselves.

NAVSTAR A Global Positioning System operated by the U.S. Department of Defense, available to civilian users. See Global Positioning System (GPS), GLONASS, Standard Positioning Service.

NAVTEX An international, automated weather and maritime navigational warning distribution system. NAVTEX sends warnings to ships as they move in an out of areas for which broadcast information is available that may be relevant to marine safety. See Global Maritime Distress and Safety System.

Navy Navigation Satellite System NNSS. A system of satellites moving in polar orbits about 700 miles above Earth, which preceded the Global Positioning System (GPS) systems used today. NNSS Doppler technology could compute group positions on or around the Earth to about one meter accuracy by means of multiple readings. The long time between transits over the same location (about 90 minutes), and the difficulty of determining instantaneous velocity led to the development of the GPS system. See Global Positioning System.

NBC See National Broadcasting Company.

NBFM narrowband frequency modulation.

NBMA nonbroadcast multiple access.

NCAR National Center for Atmospheric Research.

NCC 1. National Communications Committee. 2. National Coordinating Center.

NCCS Network Control Center System.

NCHPC National Consortium for High-Performance Computing.

NCIA native client interface architecture. An SNA applications-access architecture developed by Cisco Systems. NCIA encapsulates SNA traffic on a client computer, preserving the user interface from the native SNA system so that end-user can work in a familiar environment and also have direct TCP/IP access.

NCO See National Coordination Office for Computing, Information, and Communications.

NCOP Network Code Of Practice.

NCS See National Communications System.

NCSA National Center for Supercomputing Applications. A research center at the University of Illinois, best known for the development of NCSA Mosaic, the historic Web information browser that preceded Netscape Navigator.

NCSA Mosaic A well-known Internet information browser and World Wide Web client developed at the National Center for Supercomputing Applications. Mosaic was the predecessor to Netscape Navigator, from Netscape Communications. Navigator is now open source software. See Mosaic, Netscape Navigator.

NCUG National Centrex Users Group.

NDCDB National Digital Cartographic Data Base.

NDIS See Network Driver Interface Specification.

NDSI See National Spatial Data Infrastructure.

NDT No Dial Tone.

NE See network element.

near end crosstalk NEXT. When wires are packed tightly together, and signals are traveling through most or all of the wires, especially in two directions, the signals originating at one end can exceed or interfere with weaker signals coming from the other end, resulting in crosstalk. With much higher speed transmissions media, such as gigabit Ethernet, which involve bidirectional signals in more complex systems of aggregated wires, this can be a severe impediment. One means of compensating for NEXT is to include a NEXT canceler, which detects and adjusts for noise in the circuit. See far end crosstalk.

NECA See National Exchange Carrier Association.

neck The narrow portion of a cathode ray tube (CRT) at the end where electrons are emitted from the cathode.

needle An instrument used to probe the holes in punched metal discs, punch cards, or punch tape, or to create or read the grooves in a phonograph record to play back the sound. A phonograph needle, sometimes called a stylus, is often made of steel or sapphire.

needle chatter, needle talk Undesirable sounds produced by the extraneous vibration of a phonograph needle when it comes in contact with a rotating platter.

needle telegraph See telegraph, needle.

negative acknowledge, negative acknowledgment NAK A commonly used international communications control character which indicates that data was not received, or not received so that it could be understood. This is common to handshaking protocols, in which an acknowledgment is required before the sender can continue. See acknowledge.

negative bias In an electron tube, voltage applied to a control grid to make it hold more of a negative charge than the electron-emitting cathode. Manipulation of the control grid is what makes it possible to control the flow of electrons from the cathode to the anode, and thus to create different types of circuits and effects.

negative glow A luminous glow which can be observed between an electron-emitting cathode and the Faraday dark space in a cold-cathode discharge tube. See Faraday dark space.

negative image An image in which the dark and the light values are reversed, or in which the complements of the colors are displayed instead of the normal colors; also called an inverse image. Photographic negatives contain a negative image. In desktop publishing, negative images are sometimes created so the printout can be processed some way in manufacturing. For example, an image that is being printed on film for subsequent exposure to a printing plate might be printed in negative. Negative images are often used for posterization and other special effects. In monochrome television display systems, a negative image may arise from reversal of the polarity of the signals.

negative plate, negative terminal In a storage battery, the grid and any conductive material directly attached to the negative terminal, that is, the terminal which emits electrons when the circuit is active.

Negroponte, Nicolas Outspoken author, philosopher, and educator, Negroponte is well known for his lectures and back-page editorials in Wired magazine. He is the founder and director of the Massachusetts Institute of Technology's celebrated Media Laboratory, established in the late 1980s. Prior to that, he founded MIT's Architecture Machine Group, a think tank and research lab for discussing new approaches to human-computer interfaces.

neighbors A networking term used to describe nodes which are attached to the same link.

neon gas (Symbol - Ne) An inert gas with many industrial and commercial applications. When ionized, neon glows red. It was popularly used to illuminate signs in the 1940s and 1950s, and is still used for this purpose, along with other gases that emit other colors.

neon lamp A long glass illuminating tube with an electrode at each end, low pressure neon gas inside, which may be angled into interesting shapes. When illuminated, it produces a red-orange light which can be seen in daylight and which can penetrate fog better than most conventional types of lights. Neon has also been used in older tubes in the broadcasting industries, in simple oscillating circuits, and in commercial signs.

Nernst effect A potential difference develops in a heated metal band or strip when it is placed perpendicular to a magnetic field.

NESDIS National Environmental Satellite, Data and Information Service.

nesting In computer programming, a nest is a programming structure in which a block of data, or a function, procedure, or subroutine is logically placed within another structure. Nesting is a means of organizing information, and of organizing the order of processing of data. Nesting is a common means of setting up routines to repeat, sometimes within recursively repeating routines that are several layers deep.

Here is an example of a trivial, inelegant progress and warning subroutine in BASIC-like pseudocode which executes *i*, the outer loop, a specific number of times (3) and produces an audible beep signal each time it begins; then nested within *i* is *j*, which executes up to the value of a variable *n - 1*, which comes from outside the subroutine, but only if *n* has a value of *minimum_var* or more. Further nested within this are two print statements, if *j* is 50 or less, the dashes are not printed, but three dots are always printed (provided *n* has a value of the prestated *minimum_var* or more). If the routine fails the minimum variable test (*minimum_var*), then three signal beeps will be sounded in succession, with no characters printed. At the *minimum_var* value or above, however, three dots followed by three dashes will be printed, a visual SOS warning sign that a value of 50 has been exceeded.

```
Nest_routine (n, minimum_var):
For i = 1 to 3
  Print beep
  For j = 1 to n-1 unless n <
minimum_var
      Print "..."
      Print "–" if j > 50
  Next J
Next I
Return
```

While nests can be many layers deep, it may be a sign of weak program structure if they routinely are more than about three or four levels deep. See recursion.

Net Citizen, Net Denizen See Netizen.

Net Police A generic term for the various individuals who moderate communications on the net for appropriateness, tact, good taste, honesty, and fair use. Although some resent the activities of the Net Police, for the most part, these folks are committed, caring, hard-working volunteers who want to see the broadest possible access to the Internet, and who encourage voluntary compliance with Netiquette in order to try to prevent government regulation of the Internet's open communications forums. See Netiquette.

NET/ROM A packet radio communications protocol which has largely superseded AX.25. It provides support for a wider variety of types of packets with automatic routing. See AX.25.

NETBLT NETwork BLock Transfer Protocol. This is a transport level networking protocol intended for the fast transfer of large quantities of data. It provides flow control and reliability characteristics, with maximum throughput over different types of networks. It runs over Internet Protocol (IP), but need not be limited to IP.

The protocol opens a connection between two clients, transfers data in large data aggregates called 'buffers,' and closes the connection. Each buffer is transferred as a sequence of packets. Enhanced Trivial File Transfer Protocol (ETFTP) is an implementation of NETBLT. See RFC 998.

Netcast Broadcasting through the Internet. See Webcast.

Netgod An individual who is recognized by colleagues, or a large body of Internet participants, as having made significant contributions to the development and functioning of the Internet.

NetHead, Net Junkie Someone who has no life outside the Internet. An Internet addict (yes, they exist). Employers sometimes have to curtail the activities of NetHeads who are accessing the Internet for personal entertainment on company time. Parents sometimes are concerned, as well, that their children will spend too little time outside or with friends and relatives, if they spend too much time on the Net. NetHead need not have a negative connotation; sometimes being a NetHead isn't so bad. The Net has opened up a world of communication to people who are bed-ridden or have trouble getting around. Not only does the Net give them a world to explore, but people on the Net get to know them without any prejudices they may have toward those who have physical limitations to free movement.

Netiquette *N*ewsgroup *etiquette*, *Network* e*tiquette*. An important, well-respected voluntary code of ethics and etiquette on the Internet. Many people have contributed to Netiquette, but it was mainly developed by Rachel Kadel at the Harvard Computer Society, and subsequently maintained by Cindy Alvarez. The whole point of having Netiquette is so that network citizens can enjoy maximum freedom by not abusing the rights and sensibilities of others, so that the Net will remain largely unregulated and unrestricted. This freedom depends upon the cooperation of everyone.

In the early days of BBSs, in the late 1970s, most systems were completely open and not password protected. Gradually the constant vandalism and lack of consideration for others caused passwords to be implemented. Eventually, by the mid-1980s, even this was not sufficient to curtail childish or destructive behavior and many of the sysops gave up trying to maintain the systems.

Many of the same unfortunate patterns of abuse have damaged the USENET newsgroup system, which used to be a fantastic open forum for discussion, with many scientific and cultural leaders participating under their real names in the mid-1980s. Unfortunately, this system is now being abused by bad language, inappropriate remarks, and get-rich-quick come-ons. Consequently, many groups have been forced to close up or go to moderated status, and most celebrities now use assumed names. If members of the Internet community realize that it is completely possible to voluntarily appreciate and respect the rights of others, the Internet can remain an open resource for all.

Read Netiquette. It's a good idea. Its adherents encourge people to choose voluntary self-restraint and freedom over regulation. See emoticon, Frequently Asked Question, Netizen. Also, Arlene H. Rinaldi's "Net User Guidelines and Netiquette" in text format is available at many sites on the Internet, including:

ftp://ftp.lib.berkeley.edu/pub/net.training/FAU/netiquette.txt

Netizen *Net* cit*izen* or *Net* den*izen*—a responsible user of the Internet. Many founders and users of the Internet consider themselves members of a new type of global community that shares and promotes a vision of an open, freely accessible, self-governed communications venue in which participants voluntarily deport themselves with responsibility, integrity, charity, and tolerance toward the many diverse opinions expressed online. A Netizen is one who contributes to the positive evolution of the Net and respects online Netiquette. One could also more broadly say that anyone who uses the Net is a Netizen, although some people online have less polite terms for those who abuse their freedoms and those of others on the Net. See Netiquette.

netmask A symbolic representation of an Internet Protocol (IP) address that identifies which part is the host number and which part is the network number through a bitwise-AND operation. The result of this logical operation is the network number. Netmasks are specified for different classes of addresses, and are used in classless addressing as well. See name resolution.

NETS See Normes Européenne de Télécommunications.

Netscape Communications Originally Mosaic communications, Netscape Communications is the developer and distributor of Netscape Navigator, the best-known browser on the Internet. The company was founded by Mark Andreessen and some very experienced business people from Silicon Graphics Corporation and McCaw Cellular Communications. It had one of the highest profile public offerings in the computer industry. See Andreessen, Mark.

Netscape Navigator The most broadly distributed and used Web browser on the World Wide Web, and the name of its related server software. Descended from Mosaic, the browser was developed by Netscape Communications and widely distributed as shareware until late 1997. At that point, Netscape made the decision, in 1998, to freely distribute the software as open source software and concentrate on marketing their server software. The first beta release was distributed in 1994.

NETscout Applications software from Cisco Systems which provides a graphical user interface (GUI) for network system administrations data collection, reporting, monitoring of statistics, and protocol analysis.

Netware A commercial proprietary networking product from Novell Corporation. Netware is widely distributed for use on local area networks (LANs).

network An interconnected or inter-related system, fabric, or structure. A logical, physical, or electrical grouping in which there is some electromagnetic or biological intercommunication between some or all of the parts. A broadcast network is a physical and communications association of directors, actors, production personnel, and technologies which together cooperate and are used to create and distribute programming to its viewers. A computer network is one in which computers are able to intercommunicate and share resources by means of wireless and/or wired connections and transmissions protocols. A cellular communications network is one in which a cooperative system of wireless communications protocols, geographically spaced transceivers, relay and controlling stations, and transceiving user devices are used to interconnect callers while moving within or among transceiving 'cells.'

network, broadcast *n.* A commercial or amateur radio or television broadcast station. A few examples of well-known broadcast networks include CBC (Canada), BBC (Britain), ABC, NBC, and PBS. Amateurs often run local or special-interest radio, television, or slow-scan television broadcasts. See ANIK.

network, computer *n.* 1. A system comprised of nodes and their associated interconnected paths. 2. A system of interconnected communications lines, channels, or circuits. A small-scale computer network typically consists of a server, a number of computers, some printers, modems, and sometimes scanners, and facsimile machines. Our highway system is a type of network, as is the very effective train system in Europe. See local area network, wide area network.

network, social A social communications system consisting of friends, colleagues, and acquaintances. A great deal of the content of human networks consists of unrecorded oral communications (although with email, video phones, and online public chats, recorded conversations are increasing, a dynamic which is probably of interest to historians and biographers). The term 'networking' is often used in association with business and political alliances formed through social contacts, trade shows, trade association memberships, and referrals.

Network Access Point NAP. A major backbone point which provides service to ISPs and is designated to exchange data with other NAPs. NAP was a development in the mid-1990s which arose from the change in the U.S. Internet from a single, dominant backbone to a shared backbone across four NAPs (California, Illinois, New Jersey, Washington, D.C.). See MAE East, MAE West, Metropolitan Area Ethernet, Public Exchange Point.

network address An identifier for a physical or logical component on a network. Components often have a fixed hardware address, but may also have one or more logical addresses. Logical addresses may change dynamically as the network is altered physically, or as the network software is tuned or protocols changed. Network addresses are typically associated with nodes and stations. See address resolution, domain name, Media Access Control.

network administrator 1. The human in charge of the installation, configuration, customization, security, and lower level operating functions of a computer network. On larger networks, these tasks may be divided among a number of professionals. See SysOp. 2. A software program that handles details of the job of a human network administrator. Activities automated with network administration software include monitoring, archiving, and system checks. See daemon, dragon.

Network Control Protocols NCP. The Point-to-Point Protocol handles assignment and management of Internet Protocol (IP) addresses and other functions through a family of Network Control Protocols (NCPs) which manage the specific needs of their associated network-layer protocols. See Point-to-Point Protocol, RFC 1661.

network drive A drive which is accessible to multiple users on a computer network. On some network systems, users have to specify and access a particular drive to take advantage of the shared storage space. On other systems, this shared arrangement can be set up so that it is transparent to the user and, in fact, a volume may traverse several drives. Network drives are sometimes set up to create data redundancy in case one of the drives or partitions is corrupted. This is a good idea, and since it is in constant effect, there is less chance of loss as with systems that are backed up only at intervals (although it's recommended that companies with important data do both). See redundant array of inexpensive disks.

Network Driver Interface Specification NDIS. A network protocol/driver interface jointly developed by Microsoft Corporation and 3Com Corporation. NDIS provides a standard interface layer that receives information from network transport stacks and network adapter card software drivers. The transport protocols are thus hardware-independent.

network element NE This is defined in the Telecommunications Act of 1996, and published by the Federal Communications Commission (FCC), as:

"... a facility or equipment used in the provision of a telecommunications service. Such term also includes features, functions, and capabilities that are provided by means of such facility or equipment, including subscriber numbers, databases, signaling systems, and information sufficient for billing and collection or used in the transmission, routing, or other provision of a telecommunications service."

See Federal Communications Commission, Telecommunications Act of 1996.

network fax server A workstation equipped with fax/modem hardware and software so multiple users of the network can route a fax in and out through the server. This removes the necessity for having a fax modem (or fax machine) physically attached to each computer in a network. The fax server can then also be located near its associated phone line, or lines. Fax servers also exist which can use Internet connections (T1, frame relay, etc.) rather than phone lines to send and receive messages.

network filter A transducer designed to separate transmission waves on the basis of frequency.

Network Information Service NIS. A client/server protocol developed by Sun Microsystems for distributing system configuration data among networked computers, formerly and informally known as Yellow Pages. NIS is licensed to other Unix vendors.

network interface card, network interface controller NIC. A PC board that provides a means to physically and logically connect a computer to a network. For microcomputers, typically these cards are equipped with BNC and/or RJ-45 sockets facing the outside of the computer and edge card connectors that fit into the expansion slots inside a computer. The cables resemble video cables, or fat phone cables, depending upon the type used. Most systems require a physical *terminator* on the physical endpoints of the network (often if the network isn't working, it's because termination is missing or incorrectly installed). Separate software, not included with the computer operating system, may be required to use the specific card installed. Many workstation-level computers come with network hardware and software built in, and Macintosh users are familiar with the built-in AppleTalk hardware and software. The trend is for microcomputers to use TCP/IP networking over Ethernet.

Network Layer Packet NLP. In High Performance Routing on packet networks, a basic message unit that carries data over the path. See datagram.

Network Management Processor NMP. A network switch processor module which is used to control and monitor the switch.

Network Management Protocol NMP. A set of protocols developed by AT&T to control and exchange information with various network devices.

Network News Transfer Protocol NNTP. A software application developed in the mid-1980s to provide a way to more quickly and efficiently reference, query, and retrieve information from newsgroups through NNTP servers, and to manage listings of newsgroup discussions. NNTP is a network news transport service. Newsgroups may be accessed through the Web from local clients using the NNTP URL scheme as follows:

```
nntp://<host>:<port>/<newsgroup-
name>/<article-number>
```

For global access to newsgroups, the `news:` scheme is preferable. See news, RFC 977, RFC 1738.

Network Operations Center NOC. A centralized around-the-clock facility for monitoring and maintaining a network, which may remotely service smaller centers, such as POPs. NOCs typically provide a number of technical support and accounting services as well. Most large networks (computer, phone, broadcast) have a core staff dedicated to the physical and logistical tasks of keeping the system up and running, well-maintained, and current.

network prefixes Identifiers used to aggregate networks. Networks are divided into *Classes* with the ability to serve up to a certain number of hosts. The prefix identifies the Class, and hence the number of possible hosts.

network printer A shared printing resource on a computer network. Hardware and software connections make it possible for multiple users to share limited printing resources. The software can queue incoming jobs and, if it's good software, it will also analyze the size or importance of a job, assign priorities, and notify the user of when the print job is completed, so that it can be picked up from a remote physical location. Not all printers are equipped with network interface capabilities, but many now come with built-in Ethernet connectors. Some institutions will install the printers in secure rooms to reduce damage and increase security associated with confidential printouts.

Network Reliability Council A CEO-level organization formed in 1992 by the Federal Communications Committee (FCC). See Committee T1.

network server See server.

network service point NSP. Cisco Systems technology which provides native SNA network service point support.

Network Solutions, Inc. NSI. In 1993, this company was awarded the contract for registering Internet domain names with the InterNIC by the National Science Foundation. NSI was acquired by Scientific Applications International Corporation (SAIC) in 1995. See InterNIC.

Network Terminal Number NTN. An identification number assigned to a terminal on a public network by the public network administrator. The ITU-T recommends that public voice/data and digital voice networks also assign NTNs. The NTN is a designation within the Data Network Identification Code (DNIC) for public networks interconnected with X.75. See Data Network Identification Code, X Series Recommendations.

Network Video NV. A freely distributable Sun SPARC-, DEC-, SGI-, HP-, or IBM RS6000-based videoconferencing system developed at Xerox PARC, which supports video, audio, and whiteboarding over Mbone networks.

neural computer A computing system which is theoretically designed to behave like the human brain in terms of performing logical, intelligent problem-solving and inferential 'thinking' activities, and which also may structurally mimic the interconnective structural topology of biological neurons in a centralized nervous system. A neural computer, like the human brain, configures itself through experiential learning, feedback, and internal reorganization over time. Neural computers are not entirely theoretical, except in their most ideal form. There have been many efforts and successes in the design of neural/bionic systems since the early 1960s, with worldwide efforts by major companies to design and implement practical neural computers on a small scale since the 1970s and on a large scale since the late 1980s and early 1990s. A neural computer is a specialized type of supercomputer, since 'supercomputer' implies the state of the art in computing at any one time, and existing neural computers have demonstrated extremely fast processing and problem-solving speeds.

Neural net architectures tend to be highly parallel, with multiple registers, several layers, and a high level of interconnection between nodes. The concepts of neural computers date back to the 1940s, to the work of W. McCulloch, W. Pitts, A. Rosenblueth, and N. Wiener. See artificial intelligence; bionics; neural network; Wiener, Norbert.

neural network In a broad sense, a neural network is a type of network organization that mimics the human nervous system, particularly the brain, in physical structure and connectivity or neural functioning as it relates to thinking, or both. Simulation of neural networks, and modeling of the complex reasoning, generalizations, and inferences that are characteristic of human thinking, have long been of interest to programmers and scientists studying artificial intelligence. While the creation of androids, humanoid thinking robots, is probably some time in the future, there have been some interesting advances in programming resulting from studies of neural network functioning. Software that has the ability to generalize and make choices, and react and further configure itself in response to feedback, is being developed with practical applications in many areas, including robotics. Neural networks can aid machines and humans in navigating unfamiliar environments.

Speculation about neural networks and 'thinking machines' has been around at least since Ada Lovelace proposed in the 1800s that intelligent machines might someday produce art and poetry. In the late 1940s, Norbert Wiener, Arturo Rosenblueth, and their colleagues were discussing concepts related to 'cybernetics,' a term popularized by Wiener in "Cybernetics: or, Control and Communication in the Animal and the Machine." In 1963, in Electronics World, Ken Gilmore described the work on 'bionic computers' that was being carried out at Wright-Patterson Air Force Base in Ohio and modeling of individual neuronal circuits by companies like Bell Laboratories and the Ford Motor Company. In the 1950s and early 1960s there were already many experimental implementations of various aspects of neural networks, including electronic maze-running mice, pattern-recognizing machines, self-organizing machines, and simulations of human vision systems. See artificial intelligence; bionics; Harmon, L. D.; Melpar model; MIND; pattern matching; perceptrons; Sceptron; Wiener, Norbert.

neuroelectricity The very minute level electromagnetic fields generated by the activities of biological neurons. See neural network.

neuron In a biological system, cells which are specialized to code and conduct an electromagnetic impulse are called neurons. A network of interconnecting neurons is called a nervous system, and a network of interconnecting neurons with a main processing center is called a central nervous system, with the main processing center being called the 'brain.'

neutral In stasis, in equilibrium, stable, balanced, normal, unaffected, neither positive or negative, not tending to one side or the other, or one state or another. Neither acid nor base.

neutrodyne In early radios, an amplifying circuit used in tuned receivers. Voltage was fed back by a capacitor to the circuit to neutralize it. See heterodyne, superheterodyne.

New England Museum of Wireless and Steam Located in Rhode Island, this museum preserves the original Massie station, the oldest surviving, originally equipped wireless station.

New Wave Hewlett-Packard's commercial Microsoft Windows-compatible graphical interface desktop computer operating environment.

New York and Mississippi Valley Printing Telegraph Company An early American communications business organized by Hiram Sibley in 1851 which, as it expanded westward, came to be called Western Union, a name suggested by Sibley's associate, Ezra Cornell. Western Union subsequently installed the first transcontinental line in 1861.

newbie A telecommunications green-horn; a new or inexperienced user. We're all newbies of some aspect of computer technology. Newbies are particularly easy to identify on public forums: chat channels, newsgroups, and discussion lists. There's nothing wrong with being a newbie, but new users *must* read the introductory information, charters, and FAQs (Frequently Asked Questions) that are associated with the activity they wish to pursue. It's OK to ask questions on the Net, but the first question should always be "Where can I read the FAQ for this channel/newsgroup/discussion list?" Reading the FAQ will conserve bandwidth, save time, and can spare an individual a great deal of personal or professional embarrassment. See Netiquette.

news, Web access There are a number of ways in which programmers have implemented access to Internet newsgroups through Web interfaces. Traditionally news has been read through Unix command line text interfaces, and many still read the various newsgroups this way. There are also dedicated newsreaders which run on individuals' machines.

When a browser is designed to support the display of newsfeeds, newsgroup articles can be accessed through the Web with two types of Uniform Resource Locators (URLs) as follows:

```
news:<newsgroup-name>
  e.g., news:comp.sys.macintosh
news:<message-id>
```

News URLs are location-dependent. See NNTP, RFC 1036, RFC 1738.

newsgroup A private or public online forum, the largest of which is the USENET system. USENET has more than 35,000 newsgroups, covering every conceivable topic from *alt.religion* to *alt.bondage*. Most newsgroups function on a subscription basis; current software makes it pretty easy to subscribe at the moment at which you would like to read the messages. Not everyone has access to the same USENET newsgroups; it depends partly on what topics your Internet Services Provider has downloaded for its subscribers. Postings on various newsgroup forums can range from one or two messages a day to several thousand a day. A *newsreader* software program can help sort out the topic 'threads.'

In a text-based newsreader, the various newsgroups will be listed alphabetically; in graphical newsreaders, they may be hierarchically organized in menus. The following simple text-based excerpt shows the general format of newsgroup names.

```
alt.humor
alt.humor.best-of-usenet
alt.invest
alt.philosophy.objectivism
comp.sys.be.programmer
comp.sys.mac.advocacy
comp.sys.next.software
comp.theory.info-retrieval
humanities.philosophy.objectivism
misc.business.marketing.moderated
misc.entrepreneurs.moderated
misc.legal
misc.legal.moderated
misc.writing
rec.games.bridge
sci.astro
```

The above names have a hierarchical structure from general to more specific. The general topics listed above include *alt*ernate, *com-p*uter, *humanities*, *misc*ellaneous, *rec*reation, and *sci*ence. Anyone can create a newsgroup, provided there is sufficient community support and interest in doing so. Creation of a new USENET newsgroup requires a body of voters to ferry a proposal through a lengthy submission/acceptance process, one which

N

may take four to seven months. This is necessary as a deterrent to frivolous group creation.

Some newsgroups are moderated. Unfortunately, due to inappropriate postings, open newsgroups are decreasing in number. This puts an unfair burden on moderators, who are generally volunteers, but at least it is a way to keep a forum alive. When you post to a moderated group, the posting is previewed for adherence to the topic, or content, or both. Some newsgroup moderators reserve the right to edit actual postings (although this is rare). Read the charter before you post if you don't wish to have your postings altered. If the message meets the requirements for the group, it is then posted by the moderator. This process can take from a few hours to a few days, sometimes even up to a week and a half.

If you are offended by the topic of a group, don't read the postings. Newsgroups have evolved with a very strong commitment to the tenets of free speech, and their participants vehemently guard their right to express and discuss their views online in the appropriate forum. The following box shows a very brief excerpt from a public newsgroup thread to illustrate the headers and the general format of a message.

The short excerpt from a newsgroup posting shown in the Sample Header and Quotation inset shows the date, the poster, the newsgroup on which the ongoing discussion has been posted (alt.philosophy.objectivism), and the person to whom the poster is responding. When lines are prepended with angle brackets (>) or square brackets ([), it is a convention indicating that the person posting is referencing a previous post, repeated to the right of the brackets. This is generally called a *quote*. It is courtesy on newsgroups to indicate the origin of the quote, and to *not* post the entire previous post, but *only* those parts that are relevant, in order not to waste bandwidth (space on the newsgroup and on people's hard drives). Subjects should be clear, concise, and as complete as space permits. Keep in mind that readers are sifting through hundreds or thousands of posts.

Netiquette has been developed to provide guidelines to the effective and courteous use of the USENET system. It is recommended that you read Netiquette and the charter for each group before posting. And then enjoy; USENET is the closest individuals have ever come to having access to the sum total of human knowledge at any one time. It is a living, breathing 'expert system' where you can seek answers and support on any topic, any time of day or night. See Call for Votes, Netiquette, USENET.

newsreader A software program for accessing, displaying, searching, and posting articles to Internet newsgroups, particularly USENET. You need to have access to the Internet to read the postings on newsgroups. Newsgroups have certain customs and traditions, and you

Sample Header and Quotation from Public USENET Newsgroup Posting

```
Date: 27 Jun 1998 14:42:41 GMT
From: SynthSky <synthsky@aol.com>
Newsgroups: alt.philosophy.objectivism
Subject: Re: The Trouble With Logic

Subject: Re: The Trouble With Logic
From: Verdigris <>
Date: Tue, Jun 23, 1998 21:03 EDT
Message-id: <6mpj8k$37o@newsserver.trl.OZ.AU>

[crc <crc@igg-tx.net> writes: >
[> Verdigris wrote:
[>
[SNIP]

[> Religios and artistic endeavour requires imagination and the trancendence
[> of reason.
As an Objectivist, I absolutely agree with the first one; religion MUST
transcend logic, if religion were logical, it would not be religion, it
would
...
```

should read about newsgroup Netiquette before posting, as well as the Frequently Asked Questions (FAQ) for the particular newsgroup that interests you. It is also wise to read the existing postings for several days before contributing, so that you understand the format and content of typical postings, and so that you don't endlessly repeat a topic which may have been fully discussed. A good newsreader program will allow you to follow *threads*, conversations on a particular topic. Within any one newsgroup there are usually many threads being concurrently discussed. Pine, a poular Unix-based email program developed at the University of Washington, can be used as a newsreader, as can a Web browser. See newsgroup.

newton A unit of force in the meter-kilogram-second (MKS) system of physical units of a size that will influence a body of a mass of one kilogram to accelerate one meter per second per second. Named after Sir Isaac Newton, an English mathematician and physicist of the late 1600s.

Newton, Isaac (1642-1727) An English scientist and mathematician acknowledged as one of the greatest contributors of basic knowledge of our universe through his descriptions of the laws of motion and theory of gravitation (*Philosophiae Naturalis Principia Mathematica* 1687) and the idea that Earthly and celestial events might obey the same laws. He also studied the nature of light (*Optics* 1704) and laid much of the foundation for modern calculus.

Newton MessagePad A commercial personal digital assistant (PDA), a type of handheld computer developed by Apple Computer. It was introduced in 1993. The Newton is designed to 'learn' the user's habits and handwriting input (using a stylus) through 'Newton Intelligence' Apple's catchphrase for artificial intelligence routines built into the software. It is also equipped to interface readily with telecommunications media, including wired and wireless networks.

NEXT See near end crosstalk.

NeXT computer The NeXT computer was unveiled in early October 1988 by Steve Jobs' company NeXT, Inc. It included the first commercial erasable optical drive and incorporated VLSI technology. The programmable digital signal processor (DSP56001) came built in. The operating system was Unix-based, with a gorgeous graphical display interface incorporating Display PostScript. The fonts and graphics are all beautifully rendered in high resolution. The NeXT had some inspired input from Stanford, Carnegie-Mellon, and the University of Michigan. The original NeXT was grayscale, with color added in later versions. The first NeXT was cube-shaped, with later hardware resembling more conventional desktop flat systems, known as NeXT stations.

The first NeXT was based on the Motorola 68030 processor with a built-in 68882 math coprocessor, and it came standard with 8 MBytes of RAM. At the time, most computers had 1 to 4 megabytes of RAM. The original price was $6500 and marketing efforts were aimed at higher education institutions, although business owners expressed early interest due to the networking capabilities of the system.

The NeXT cube was promoted as the 'computer for the '90s' when it was released in 1988. Surprisingly this marketing hype has held true. Many corporations are adopting Unix as their standard, as educational institutions have for years, graphical user interfaces are now ubiquitous, and Display PostScript still provides one of the best WYSIWYG solutions on any system.

This reference was written on a NeXT computer, and even though the basic technology is ten years old, the computer hardware and the operating system, have stood the test of time in essentially their original form. The simple, stunningly aesthetic graphical user interface still beats most systems hands-down, the powerful object-oriented Unix-based operating system and shell connects seamlessly with the Internet, and the multitasking operating system allows dozens of processes to run happily at the same time. In over three years of 24-hour a day operation, this machine hasn't crashed once. That's a strong endorsement, but after using a dozen different types of computers daily for the last eighteen years, the author has seen few systems that can equal it (Sun systems provide similar performance).

The NeXT is an excellent networking computer, connecting easily to the Internet, other NeXT systems, and other types of computers through TCP/IP. It is also an excellent Internet portal, with a full complement of Unix tools, including Telnet, FTP, and others easily downloadable from the Net. OmniWeb, by Lighthouse Design, Ltd., is a powerful Internet graphical browser for NeXTStep that preceded many well-known graphical browsers.

In 1997, Apple Computing, Inc. bought out NeXT, Inc. and is now developing the operating system software under the development name of Rhapsody. See Jobs, Steven.

Next Generation Digital Loop Carrier NGDLC. Developed in the 1980s as an evolution of Digital Loop Carrier systems, NGDLC is based on very large scale integration (VLSI) technology. ISDN was being developed and promoted at about the same time NGDLCs were being implemented, so many were developed to accommodate ISDN. Whereas Digital Loop Carriers were designed to provide services over traditional copper phone lines, NGDLC was designed to work in conjunction with fiber optic cables or fiber/copper hybrid systems. See Digital Loop Carrier.

Next Generation Internet NGI. A U.S. federal multiagency research and development initiative established in 1997. NGI works with industry and academia to develop, test, and demonstrate advanced networking technologies and applications. The following federal networks are being used as testbeds for the NGI initiative:

- NSF's very high performance Backbone Network Service (vBNS)
- NASA's Research and Education Network (NREN)
- DoD's Defense Research and Education Network (DREN)
- DoE's Energy Sciences network (ESnet) (proposed beginning in FY 1999)

NGI is coordinated by the NGI Implementation Team, coordinated by the Large Scale Networking Working Group of the Subcommittee on Computing, Information, and Communications (CIC) research and development of the U.S. White House National Science and Technology Council's Committee on Technology. http://www.ngi.gov/

Next Hop Resolution Protocol NHRP. An internetworking architecture, which runs in addition to routing protocols and provides the information that allows the elimination of multiple Internet Protocol (IP) hops when traversing a NBMA network. Aims at resolving some of the latency and throughput limitations of Classical IP. See Next Hop Server, ROLC, NBMA.

Next Hop Server NHS. In an NHRP networking environment, the Next Hop Server locates an egress point near a given destination and resolves its ATM address, enabling the establishment of a direct ATM connection. See Next Hop Resolution Protocol.

NGDC See National Geophysical Data Center.

NGDLC See Next Generation Digital Loop Carrier.

NGI See Next Generation Internet.

NGSO nongeostationary orbit.

NGSO FSS nongeostationary orbit fixed satellite service.

NHK Nippon Hoso Kyokai.

NHRP See Next Hop Resolution Protocol.

nickel metal hydride NiMH. A rechargeable battery commonly used in portable devices. Hydride is a hydrogen compound.

nickel-cadmium cell NiCd, NiCad. A very common, sealed, rechargeable power cell that works well in low temperatures. The positive electrode is nickel and oxide, and the negative electrode is cadmium, with the plates immersed separately in a potassium-hydroxide electrolyte solution. NiCad batteries have been used in many small, portable telecommunications devices but have the disadvantage of having a 'memory effect,' that is, they will not fully recharge unless they have been first fully discharged, thus reducing the useful time of the battery.

nickname 1. See handle. 2. On Internet Relay Chat (IRC), a name that can be set with `/nick <putnamehere>`. Only one person can have a specific nickname at any one time on IRC.

Nicname See WhoIs.

NIF network interface function.

Night Answer Incoming A type of call forwarding feature of telephone systems, most often used by businesses, which allows an incoming call 'after hours' to be routed to another line, often to phones that can be better heard and answered by maintenance or cleaning staff, night operators, or security personnel.

night chime A feature of a phone system which rings a secondary bell, or loud ringer, during 'after hours' operation. Night chimes are handy for environments where there isn't a full-time receptionist or telephone operator sitting at the console after hours, but the other staff needs to be able to hear the ring without being next to the phone. See Night Answer Incoming.

NIMA National Imagery Mapping Agency.

NIMBUS A satellite program initiated in the early 1960s by the National Aeronautics and Space Administration (NASA), and now operated jointly with the Oceanic Atmospheric Administration (NOAA). NIMBUS is used for research and development by atmospheric and Earth scientists.

NiMH See nickel metal hydride.

NIOD Network Inward/Outward Dialing.

NIOSH National Institute of Occupational Safety and Health.

Nipkow, Paul Gottlieb A German experimenter who developed a rotating dial with a spiral arrangement of holes which he patented in 1884. It was an early electromechanical television system. This was later incorporated into television transmitting and receiving units. See Nipkow disc.

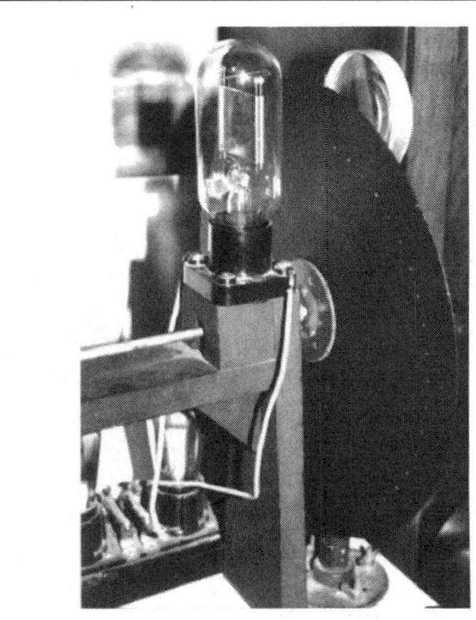

The above Nipkow disc shows the spiral series of holes through which the light is beamed. This historic disc is part of the Bellingham Antique Radio Museum collection.

Nipkow disc A rotating disc with a sequential helical pattern of holes which was used by many early television experimenters to attempt the projection of television images. The perforated disc was rotated in front of the image to be transmitted in order to 'quantize' the signal in a primitive sense, by segmenting the image into lines. The photosensitive material selenium was placed behind the disc to register the dark and light areas of the image. Unfortunately there was no means at the time to amplify the signal for transmission, and the system did not become practical until other technological developments occurred. The disc is named after its inventor, Paul Nipkow. See Baird, John Logie; television history.

Nippon Telephone and Telegraph Corporation NTT. The major Japanese telephone company and largest phone company in the world. In 1997, deregulation allowed NTT to begin operating internationally, and its first international subsidiary, NTT Worldwide Telecommunications Corporation, began servicing overseas corporate customers. See Arcstar.

NIS 1. Network Imaging Server. 2. See Network Information Service.

NISDN 1 See National ISDN-1.

NISP See National Internet Services Provider.

NIST See National Institute of Standards and Technology.

NITF National Image Transfer Format.

NIU network interface unit.

NIUF See North American ISDN Users Forum.

NLA See Network Layer Address.

NLANR National Laboratory for Applied Network Research.

NLPID Network Layer Protocol ID.

NLQ near letter quality. A marketing and print quality designation that was common in the early 1980s, when many computer printers were quite primitive, with only 8 or 9 printing pins. The output from these printers could be quite rough and difficult to read. When 12- and 24-pin printers began becoming more common in the mid-1980s, NLQ was a catchphrase that was used to indicate printouts that were superior to 9-pin output, and almost as good (though not quite) as electric typewriter or daisy-wheel printer output. With the widespread use of laser printers and some of the better ink-jet printers, the use of the phrase has declined.

NMAA National Multimedia Association of American. http://www.nmaa.org/

NMACS Network Monitor and Control System.

NMD See nonintrusive measurement device.

NMP 1. See Network Management Processor. 2. See Network Management Protocol.

NMR See nuclear magnetic resonance.

NNI 1. network node interface. 2. Nederlands Normalisatie-Instituut. A Netherlands standards organization established in 1959, located in Delft.

NNTP See Network News Transfer Protocol.

NNX See NXX.

NO OP, NOP, nop no operation. A low level computer instruction in which no operation occurs; the program passes on to the next instruction. Sometimes used as a timing mechanism (as the computer still has to read and interpret the instructional statement which takes a few cycles), or as a place holder or visual scan aid. In Internet communications, people sometimes respond satirically in chat or mail with "NOP," meaning "Huh?" it didn't register, or wasn't clearly understood.

NOAA See National Oceanic and Atmospheric Administration.

noble gas A rare or inert gas. Examples include argon, krypton, neon, and helium. These gases are useful in illuminated signs and laser technologies. See argon, krypton, neon.

NOC See Network Operations Center.

Emily Noether developed mathematicd group theories that are still widely used by physicists.

node Junction, confluence, meeting point, terminal, intersection. A connection point in a network, which may consist of a router, switcher, dial-up modem, computer, or other interconnecting device supporting the same protocol, or converting to the needed protocol. Together the nodes, equipment and pathways constitute the network topology. See backbone, leaf node.

Noether, Emily (Emmy) (1882-1935) A brilliant German-born mathematician, Emily Noether developed mathematical group theories which underlay many subsequent representations of modern physics. Einstein praised her contributions and offered to write her obituary. Noetherian Rings are named after her.

noise Meaningless, or otherwise unwanted sounds or signals interfering with the desired information or transmission. Noise can arise from bad shielding, wires too close together, overlapping transmissions, weather disturbances, incorrect operation, deliberate human interference, random varying velocity, or faulty or incompatible hardware. See garbage, interference.

noise canceling Techniques and technology to reduce or eliminate noise. This may be background noise or noise on the transmissions pathway. Noise cancellation can be through digital algorithms which analyze the information and screen out calculated noise (a feature now found on digital cellular phones), or may be through conditioning circuits in transmitters or receivers. Noise cancellation is sometimes achieved by adding noise, creating a 'white noise' that may be less objectionable than bursty, intermittent noise.

noise filter An electrical circuit designed to detect or evaluate and exclude extraneous signals passing through a circuit. In digital circuits, fairly sophisticated analysis may take place. In analog circuits, simple exclusions of particular patterns or frequencies may be used.

nonreturn to zero NRZ. A simple binary encoding scheme in which ones and zeros are represented by high and low voltages, and there is no return to a zero level between successive encoded bits, hence the name. Since transitions may or may not occur at each successive bit cell, the NRZ signal has spectral energy and, consequently, a direct current (DC) component, that is a nonzero energy at DC. It is thus one type of baseband signal. See Manchester encoding.

nonintrusive measurement device NMD. A device used to measure various parameters in analog voice transmissions over communications networks. Measured parameters include noise level, speech level, echo path loss, and echo path delay.

nonionizing That which does not cause ionization or change the ion environment around it. A number of transmissions media, including visible light rays and radio waves, do not cause ionization, but can themselves be propagated by ionized particles to which they come in contact.

nonvolatile Circuits or components which retain their information, even if the current is shut off. In computer circuitry, volatile memory is installed in greater quantities than nonvolatile memory. See EPROM, read only memory. Contrast with dynamic random access memory.

nonwireline carrier Also called an A Block carrier, for *a*lternate carrier, that is a competitive phone services carrier which is not the established local phone company (which was usually a Bell carrier, hence B Block carrier).

NORC Network Operators Research Committee.

Norman, Robert A researcher who published a clear statement of the laws governing magnetic attraction and repulsion in 1581. See magnet.

Normes Européenne de Télécommunications NET. An organization which provides compliance testing for commercial telecommunications products for sale throughout the European Union to determine whether they conform to mandatory standards.

Nortel Northern Telecom Limited. A leading global digital network provider providing commercial data, voice, and video services. Nortel is a dominant public switching equipment supplier in Canada descended from Northern Electric. It is also known for manufacturing and distributing radar sets based on magnetron tube technology, particularly during the second world war. Nortel technology is leased by other companies. See Qwest.

North American area codes See the chart of area codes for Canada, U.S. and the U.S. territories.

North American Cellular Network NACN. A commercial provider of international cellular roaming services through their network backbone, serving over 7500 cities worldwide. Supported protocols are System Signaling 7 (SS7), X.24, GSM and IS41.

North American Directory Plan NADP. An X.500-based client/server Directory System for providing global electronic directory and address book capability, distributed by ISOCOR.

North American ISDN Users Forum NIUF. Founded in 1988, the NIUF is coordinated by the National Institute of Standards and Technology (NIST). Management was established three years later through a Cooperative Research and Development Agreement (CRADA) with industry.

NIUF promotes ISDN applications development, implementation, acceptance, and furtherance, and provides services, and opportunities for users and implementors to communicate their needs and goals to one another. http://www.niuf.nist.gov/misc/niuf.html

North American Network Operators Group NANOG. An association of Internet Service Providers which meets several times a year to discuss technical issues regarding the administration and operation of Internet-connected services.

North American Numbering Council NANC. A Federal Advisory Committee established and chartered with the United States Congress in 1995 by the Federal Communications Commission (FCC) to assist in adopting a model for administering the North American Numbering Plan (NANP). This identification scheme is used for many telecommunications networks around the world. NANC advises the FCC and other NANP member governments on general number issues and on issues of portability. See North American Numbering Plan. http://www.fcc.gov/ccb/Nanc/

North American Numbering Plan NANP. A system of assigned codes and conventions introduced in 1947 for routing North American (World Number Zone 1) calls through the various telephone trunks of the public telephone network. In 1995, significant changes were made to the NANP, mainly due to increased demand for area codes, including a change from 1 and 0 as the middle digit to now being 2 through 9. See Area Codes chart.

North American Numbering Plan Administration NANPA. A working group which develops and advises the NANC on processes for selecting a neutral NANP Administrator. It oversees a number of task forces and coordinates with them on issues related to cost recovery for the NANP administration.

North American Telephone Association NATA. Now known as the MultiMedia Telecommunications Association, this is an open public policy, market development, and edu-

Area Codes for the U.S. and Canada

United States

201 New Jersey (Hackensack, Jersey City)
202 Washington D.C.
203 Connecticut (Bridgeport, New Haven, Stanford)
205 Alabama (Birmingham)
206 Washington (Seattle)
207 Maine
208 Idaho
209 California (Fresno, Modesto)
210 Texas (San Antonio)
212 New York (Manhattan)
213 California (downtown Los Angeles)
214 Texas (Dallas)
215 Pennsylvania (Philadelphia)
216 Ohio (Cleveland)
217 Illinois (Champaign/Urbana, Springfield)
218 Minnesota (Duluth)
219 Indiana (Gary, Hammond, Michigan City, South Bend)
228 Mississippi (Biloxi)
240 Maryland (Hagerstown, Rockville)
248 Michigan (Pontiac, Southfield, Troy)
254 Texas (Waco)
281 Texas (Houston metro)
301 Maryland (Hagerstown, Rockville)
302 Delaware
303 Colorado (Aurora, Boulder, Denver, Longmont)
304 West Virginia
305 Florida (Key West, Miami)
307 Wyoming
308 Nebraska (North Platte, Scottsbluff)
309 Illinois (Peoria, Rock Island)
310 Southern California (Beverly Hills)
312 Illinois (Chicago central)
313 Michigan (Dearborn, Detroit)
314 Missouri (St Louis)
315 New York (Syracuse, Utica)
316 Kansas (Dodge City, Wichita)
317 Indiana (Indianapolis)
318 Louisiana (Lake Charles, Shreveport)
319 Iowa (Davenport, Dubuque)
320 Minnesota (St. Cloud)
323 California (LA surrounds)
330 Ohio (Akron, Canton, Youngstown)
334 Alabama (Montgomery)
352 Florida (Gainesville)
360 Washington State (Olympia, Vancouver)
401 Rhode Island
402 Nebraska (Lincoln, Omaha)
404 Georgia (Atlanta)
405 Oklahoma (Enid, Oklahoma City)
406 Montana
407 Florida (Orlando)
408 California (Monterey, San Jose)
409 Texas (Beaumont, Galveston)
410 Maryland (Annapolis, Baltimore)
412 Pennsylvania (Pittsburgh)
413 Massachusetts (Pittsfield, Springfield)
414 Wisconsin (Milwaukee, Racine)
415 California (San Francisco)
417 Missouri (Joplin, Springfield)
419 Ohio (Toledo)
423 Tennessee (Chattanooga, Johnson City, Knoxville)
425 Washington (Bellevue)
435 Texas (Logan, Price, St. George)
440 Ohio (Ashtavula, Lorain)
443 Maryland (Annapolis, Baltimore)
501 Arkansas (Little Rock)
501 Arkansas (Fort Smith)
502 Kentucky (Franfurt, Louisville, Paducah, Shelbyville)
503 Oregon (Astoria, Portland, Salem)
504 Louisiana (Baton Rouge, New Orleans)
505 New Mexico
507 Minnesota (Rochester)
508 Massachusetts (Fall River, Worcester)
509 Washington (Spokane, Walla Walla, Yakima)
510 Northern California (Fremont, Oakland)
512 Texas (Austin, Corpus Christie)
513 Ohio (Cincinnati, Middletown)
515 Iowa (Des Moines)
516 New York (Hempstead, Long Island)
517 Michigan (Bay City, Jackson, Lansing)
518 New York (Albany, Schenectady, Troy)
520 Arizona (except Phoenix metro)
530 California (Chico)
540 Virginia (Roanoke)
541 Oregon (Bend, Corvallis, Eugene, Medford, Pendleton)
561 Florida (West Palm Beach)
562 California (Long Beach)
573 Missouri (Columbia, Jefferson City)
601 Mississippi (Hattiesburg, Jackson)
602 Arizona (Phoenix metro)
603 New Hampshire
605 South Dakota
606 Kentucky (Kentucky, Winchester)
607 New York (Binghamton, Elmira)
608 Wisconsin (Madison)
609 New Jersey (Atlantic City, Cambden, Trenton, Vineland)
610 Pennsylvania (Allentown, Reading)
612 Minnesota (Minneapolis, St. Paul)
614 Ohio (Columbus)
615 Tennessee (Murfreesboro, Nashville)
616 Michigan (Battle Creek, Grand Rapids, Kalamazoo)
617 Massachusetts (Boston central)
618 Illinois (Alton, Cairo, Mt. Vernon)
619 California (San Diego)
630 Illinois (Chicago central suburbs, Elgin, Waukegan)
650 California (Palo Alto)
660 Missouri (Kirksville)
678 Georgia (Atlanta metro, Marietta, Norcross)
701 North Dakota
702 Nevada
703 Virginia (Alexandria, Arlington)
704 North Carolina (Asheville, Charlotte)
706 Georgia (Augusta, Columbus, Rome)
707 California (Eureka)
708 Illinois (Chicago south suburbs)
712 Iowa (Council Bluffs, Sioux City)
713 Texas (Houston)
714 California (Anaheim)
715 Wisconsin (Eau Claire, Wausau)
716 New York (Buffalo, Niagara Falls, Rochester)
717 Pennsylvania (Harrisburg, Scranton, Wilkes-Barre)
718 New York City (Bronx, Brooklyn, Queens, Staten Island)
719 Colorado (Colorado Springs, Leadville, Pueblo)

732	New Jersey (New Brunswick, Piscataway)
734	Michigan (Ann Arbor, Livonia)
740	Ohio (Stubenville)
757	Virginia (Norfolk)
760	California (Barstow, Palm Springs)
765	Indiana (Anderson, Lafayette)
770	Georgia (Atlanta metro, Marietta, Norcross)
773	Chicago (outside central)
781	Massachusetts (Boston suburbs)
785	Kansas (Lawrence, Salina, Topeka)
801	Utah (Ogden, Provo, Salt Lake City)
802	Vermont
803	South Carolina (Columbia)
804	Virginia (Charlottesville, Newport News, Richmond)
805	California (Bakersfield, Santa Barbara)
806	Texas (Amarillo, Lubbock)
808	Hawaii
810	Michigan (Flint)
812	Indiana (Evansville)
813	Florida (Clearwater, St. Petersburg, Tampa)
814	Pennsylvania (Altoona, Erie)
815	Illinois (La Salle, Rockford)
816	Missouri (Kansas City, St. Joseph)
817	Texas (Fort Worth)
818	California (Glendale, Pasadena)
830	Texas (Del Rio, Fredericksburg, New Braunfels)
831	California (Salinas)
843	South Caroline (Charlestown, Florence)
847	Illinois (Chicago north suburbs, Elgin, Waukegan)
850	Florida (Pensicola, Tallahasse)
860	Connecticut (Hartford)
864	South Carolina (Greenville, Spartanburg)
870	Arkansas (Jonesboro)
870	Arkansas (Pine Bluff)
901	Tennessee (Memphis)
903	Texas (Tyler)
904	Florida (Daytona Beach, Jacksonville)
906	Michigan (Marquette, Sault St. Marie)
907	Alaska
908	New Jersey (Elizabeth, Plainfield)
909	California (Riverside, San Bernardino)
910	North Carolina (Fayetteville, Greensboro, Winston-Salem)
912	Georgia (Albany, Savannah)
913	Kansas (Atchison, Kansas City)
914	New York (Peakskill, Poughkeepsie, White Plains, Yonkers)
915	Texas (Abilene, El Paso)
916	California (Sacramento)
917	New York (newer wireless)
918	Oklahoma (Tulsa)
919	North Carolina (Durham, Raleigh)
920	Wisconsin (Fond du lac, Green Bay, Sheboygan)
925	California (Concord)
931	Tennessee (Clarksville, Columbia, Cookeville)
937	Ohio (Dayton/Springfield)
940	Texas (Denton, Wichita Falls)
941	Florida (Fort Meyers, Naples, Sarasota)
949	California (Irving)
954	Florida (Ft. Lauderdale)
956	Texas (Brownsville, Laredo)
970	Colorado (Aspen, Durango, Grand Junction, Steamboat Springs)
972	Texax (Dallas metro)
973	New Jersey (Morristown, Newark, Paterson)
978	Massachusetts (Lowell)

Area Codes for U.S. Territories and Canada

U.S. Territories

242	Bahamas
246	Barbados
264	Anguilla
268	Antigua, Barbuda
284	British Virgin Islands
340	US Virgin Islands
345	Cayman Islands
441	Bermuda
473	Granada
649	Turks, Caicos
664	Montserrat
670	North Mariana Island
671	Guam
758	St. Lucia
767	Dominica
784	St. Vincent, Grenadines
787	Puerto Rico
809	Domincan Republic
868	Trinidad, Tobago
869	St. Kitts, Nevis
876	Jamaica

Canada

204	Manitoba
250	British Columbia (Prince George, Vancouver Island)
306	Saskatchewan
403	Alberta (Calgary, Red Deer), Yukon
416	Ontario (Toronto metro)
418	Quebec (Quebec City)
450	Quebec (Montreal surrounds)
506	New Brunswick
514	Quebec (Montreal metro)
519	Ontario (London)
604	British Columbia (Vancouver)
613	Ontario (Ottawa)
705	Ontario (North Bay, Sault St. Marie)
709	Newfoundland and Labrador
807	Ontario (Thunder Bay)
819	Quebec (Sherbrooke)
867	Northwest Territories, Yukon Territory
870	Alberta (Edmonton)
902	Nova Scotia, Prince Edward Island
905	Ontario (Hamilton, Mississauga, Niagara Falls)

cational forum for telecommunications products and services developers and resellers. http://www.mmta.org/

north geographic pole The point at which the imaginary lines of latitude converge at the north pole. The general direction in which north-seeking compass needles point. See north magnetic pole.

north magnetic pole A point in northern Canada, near the north geographic pole, to which the north-seeking tip of a compass points. Thus, what we call magnetic north is actually the Earth's south pole, if you consider that poles align themselves north on one with south on the other. See north geographic pole.

NOV News Overview.

Novell One of the significant companies providing networking software (Novell Netware) to the business market. Novell is a public company established by a buyout of NDSI by Ray Noorda in 1983.

NOWT Netherlands Observatory for Science and Technology.

NPA 1. National Pricing Agreement. AT&T agreement. 2. See Numbering Plan Area. A three-digit area code. NPAs include special, reserved, and unassigned numbers.

NPR National Public Radio.

NRC 1. See National Research Council. 2. See Network Reliability Council. 3. nonrecurring charge.

NREN See National Research and Education Network.

NRSC National Radio Systems Committee.

NRZ See nonreturn to zero.

NSAI National Standards Authority of Ireland. A standards body for Ireland established in 1961, located in Dublin.

NSAP network service access point.

NSF See National Science Foundation.

NSFNET National Science Foundation Network. A network established by the Office of Advanced Scientific Computing through the National Science Foundation, which is used for the civilian computing operations of the U.S. Department of Defense. See National Science Foundation.

NSInet NASA Science Internet, a network of the National Aeronautics and Space Administration.

NSP 1. See National Internet Services Provider. 2. Native Signal Processing.

NSS See National Standards System.

NSSN National Standards Systems Network. http://www.nssn.org/

NSTAC National Security Telecommunications Advisory Committee.

NT Northern Telecom, Inc.

NTIA See National Telecommunications and Information Administration.

NTN See Network Terminal Number.

NTSC National Television System Committee. An organization formed by the Federal Communications Commission (FCC) which set black and white standards for the emerging television broadcast industry in 1941. By 1953, after the proposal and consideration of several television systems, the FCC adopted a 525-line color standard developed by Radio Corporation of America (RCA), which was downwardly compatible with previous black and white technologies. This is, in part, why luminance and chrominance information are carried separately. This system was accepted by the FCC and is widely used in North America and parts of South America.

In NTSC broadcasts, color, intensity, and synchronization information are combined into a signal and broadcast as 525 scan lines, in two fields of 262.5 lines each (Europe typically uses 625 lines). Only 480 lines are visible, the rest occur during the vertical retrace periods at the end of each field. NTSC is considered to run at 30 frames per second, although, in color television broadcasts, the actual playing rate is approximately 29.97 frames per second. See High Definition TV, PAL, SECAM.

NuBus NuBus is a simple Apple Computer 32-bit backplane card slot standard (ANSI/IEEE P1196) for the connection of peripherals to Apple Macintosh computers. The clock is derived from a 10 MHz reference. NuBus backplane space is limited to 74.55 mm *x* 11.90 mm (even though some models have larger slots). NuBus slots can support up to 13.9 watts of power per card, although more can be used if other slots are not filled.

nuclear magnetic resonance NMR. A technology used to reveal the inside of structures or biological organisms through a series of magnetic scanners or a magnetic field enveloping the body. It is used in addition to, and as an alternative to, X-rays in medical research and diagnostic imaging.

null Empty, having no value. A dummy value, character, symbol, or marker. Null values are sometimes used as delimiters to indicate the beginning and/or end of a value or data stream. Null characters sometimes are used as padding, to even out the size of blocks or to provide extra time for synchronization. Null is a very useful concept, regularly used in programming and network transmissions protocols.

null modem A serial transmissions medium which functions in many ways as a modem, as it uses the same software and protocols, and serial transmissions media, except that there is no modem. In other words, instead of the signal going from a computer to a modem through a phone line to another modem and to the destination computer, the signal goes from the first computer through a serial cable with no modem, connected to the second computer, and back again. The transmit (Tx) and receive (Rx) lines are swapped (usually lines 2 and 3). This provides fast local file transfer capabilities between machines.

null modem cable There are many ways to configure a modem cable, as long as the two computers talking to each other are talking the same language (transmissions protocol), but the most common configuration for a null modem cable is to take a standard RS-232c cable and cross (swap) the transmit and receive lines on one end, that is lines 2 and 3. Or, rather than taking apart a cable, it is usually easier to get a null modem connector that swaps the lines. It looks very similar to an extender or other small coupler. See null modem.

Number Nine Visual Technology Corporation A leading supplier of high-performance visual technology, especially graphics display devices and accelerators, founded in 1982. Perhaps best known for its series of Number Nine graphics boards, this bootstrap company was providing 1024 *x* 768 resolution graphics cards for International Business Machines (IBM) computers in 1983, at a time when CGA (320 *x* 200) was the standard.

number portability NP. A service which enables subscribers to retain a geographic or nongeographic telephone number, they change their location, their services provider, or their type of service.

This is defined with regard to switching services in the Telecommunications Act of 1996, and published by the Federal Communications Commission (FCC), as:

"... the ability of users of telecommunications services to retain, at the same location, existing telecommunications numbers without impairment of quality, reliability, or convenience when switching from one telecommunications carrier to another."

See Federal Communications Commission, Telecommunications Act of 1996.

Numbering Plan Area NPA. A three-digit area code. NPAs include special, reserved, and unassigned numbers. Within each NPA, there are 800 possible NXX Codes (also known as central office codes). NPAs are divided into two general categories:

Category	Notes
Geographic NPA	The Numbering Plan Area code associated with a specific region.
Service Access Code	Nongeographic NPAs associated with specialized services that may be offered over multiple area codes, such as toll free numbers, 900 numbers, etc.

numeric keypad Any compact block of functionally related touchtone telephone, typewriter, calculator, or computer input keys. The most common type of numeric keypad is a group of about 10 to 18 numerical or function keys arranged in a block. These are usually physically organized to facilitate touch-typing or, in some cases, physically organized to slow down typing! On early touchtone phone systems, there was no point in entering numbers quickly as the switching on the network could not be accomplished as quickly as the numbers could be typed, so the digits were reversed to slow down digit entry.

The numeric keypad on a computer keyboard typically consists of 18 keys, with the numerals zero through nine, and symbols usually consisting of period, plus, minus, asterisk (star), and Enter keys. The remaining three keys differ widely, on various computer platforms, but usually include symbols such as the tilde or slash, pipe (vertical bar). See keymap, keypad.

nutating field In radar tracking, an oscillating feed from an antenna which produces an oscillating deflection of the radar beam.

nuvistor A type of electron tube in a ceramic envelope, with cylindrical, closely spaced electrodes.

nV *abbrev.* nanovolt.

NV See Network Video.

NVM nonvolatile memory. Computer memory which retains its information even if there is no current going through it. See EPROM, read only memory. Contrast with dynamic random access memory.

nW *abbrev.* nanowatt.

NXX NXX is also known as Central Office Code, or CO Code. It is an industry abbreviation designating the three digits of a phone number preceding the last four (EXTN). A 10-digit number is expressed with symbolic characters as: NPA-NXX-EXTN. NXX is a reference to the exchange that services that specific area. Each NXX Code contains 10,000 stations numbers. See RNX.

NYNEX Telephone Companies One of the Regional Bell Companies which was formed as a result of the mid-1980s AT&T divestiture.

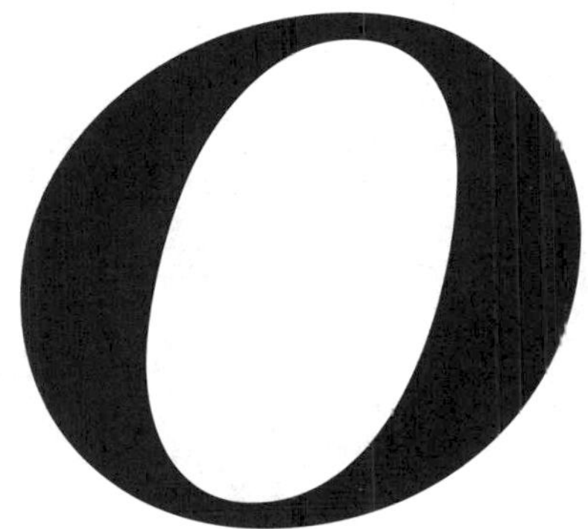

O A symbol used on many consumer electronics devices to indicate "off." On rocker switches, it indicates the side of the rocker which turns an appliance or component off. Its complement is "I" to designate "on."

O&M, O & M operations and maintenance.

O/R Originator/Recipient. A concept associated with the X.400 Message Handling System (MHS). The O/R address is used by the MTS for routing.

The *MTS.ORAddress* is a set of structured and unstructured attribute value pairs which can be ASCII-encoded for interactions with gateways. (Thus, a text string form of O/R address has been specified by ISO.) The standard encoding of the *MTS.ORAddress* is a BNF representation. Each unstructured attribute has a key and a specified encoding. Each structured attribute has the X.400 attribute mapped onto one or more attribute value pairs. For domain-defined attributes, each element of the sequence is mapped onto a triple (a key and two values); each value has the same encoding.

For RFC 822 routing to gateways, O/R address attributes are considered as a hierarchy, and may be specified by the domain. Their order is:

1. country
2. Administrative Management Domain (ADMD)
3. Private Management Domain (PRMD)
4. organization
5. organizational unit (there may be more than one of these)

In cases where there are omitted values, they are skipped. Null values are processed separately from omitted values (a domain cannot be null). As an example:

```
/I=J/S=Doe/GQ=5/@Sales.GreatGoods.COM
```

is an encoded *MTS.ORAddress* which can be broken down right to left as follows:

MTS.CountryName	= "TC"
MTS.AdministrationDomainName	= "BTT"
MTS.OrganizationName	= "GreatGoods"
MTS.OrganizationalUnitNames.value	= "Sales"
MTS.PersonalName.generation-qualifier	= "5"
MTS.PersonalName.surname	= "Doe"
MTS.PersonalName.initials	= "J"

OA See office automation.

OAI See Open Application Interface.

OAM operations, administration, and maintenance. Various related management functions often associated with telephone and computer networks. In telephone networks, there are significant management and accounting tasks associated with maintaining a dynamic environment in which subscribers all request different types and levels of service, and where the subscriber population is very mobile, thus changing their locations on a continual basis. Some systems have computer networks and entire facilities associated with just these aspects of the business. With mobile communications on the rise, these management tasks become even more intricate, and computer systems are used to facilitate the administrative tasks.

OAM Operations And Maintenance. Preventive maintenance information which, in an ATM B-ISDN environment, is included in the transmitted cells.

OAM&P operation, administration, maintenance, and provisioning.

OAM&P ANSI standards There are a number

ANSI Standards OAM&P Abstracts

ANSI Standard	ANSI Document Title
T1.118-1992	G Interface Specification for Use with the Telecommunications Management Network
T1.204-1997	Lower Layer Protocols for Telecommunications Management Network Interfaces, Q3 and X Interfaces
T1.208-1997	Upper Layer Protocols for Telecommunications Management Network, Q3 and X Interfaces
T1.209a-1995	Supplement - Network Tones and Announcements
T1.214-1990	A Generic Network Model for Interfaces Between Operations Systems and Network Elements
T1.215-1994	Fault Management Messages for Interfaces between Operation Systems and Network Elements
T1.221-1995	In-Service, Nonintrusive Measurement Device Voice Service Measurements
T1.224-1992	Protocols for Interfaces between Operations Systems in Different Jurisdictions
T1.226-1992	Management of Functions for Signaling System No. 7 Network Interconnections
T1.226-1992	Management of Functions for Signaling System No. 7 Network Interconnections
T1.227-1995	Extension to Generic Network Model for Interface Between Operations Systems Across Jurisdictional Boundaries to Support Fault Management
T1.228-1995	Services to Interfaces Between Operations Systems Across Jurisdictional Boundaries to Support Fault Management (Trouble Administration)
T1.229-1992	Performance Management Functional Area Services for Interfaces between Operations Systems and Network Elements
T1.233-1993	Security Framework for Telecommunications Management Network (TMN) Interfaces
T1.240-1996	Generic Network Information Model for Interfaces between Operations Systems and Network Elements
T1.243-1995	Baseline Security Requirements for the Telecommunications Management Network
T1.240-1996	Generic Network Information Model for Interfaces between Operations Systems and Network Elements
T1.244-1995	Interface Standards for Personal Communications Services Information Model and Services for Interfaces Between
T1.246-1995	Operations Systems Across Jurisdictional Boundaries to Support Configuration Management - Customer Account Record Exchange
T1.247-1995	Performance Management Functional Area Services and Information Model for Interfaces between Operations Systems and Network Elements
T1.250-1996	Extension to Generic Network Information Model for Interfaces Between Operations Systems and Network Elements to Support Configuration Management - Analog and Narrowband ISDN Customer Service Provisioning
T1.252-1996	Security for the Telecommunications Management Network Directory

of important American National Standards (ANSI) of Committee T1 related to OAM&P, which are available from ANSI and described in the form of abstracts on the Web. See the ANSI Standards OAM&P chart for examples.

OAO orbiting astronomical observatory.

OAS See Organization of American States.

object 1. A thing, article, entity, or unit of information. 2. An individually identifiable part, entity, or component. 3. In programming, an entity, often compartmentalized, that stores or receives data, e.g., a byte, block, register, segment, etc. 4. In The X Windows System, a software concept practically implemented as private data with private and public routines to operate on that data. 5. In typed objects, an entity that interacts as part of a defined operation.

object, drag and drop In drag and drop, or cut-and-paste operations, an object is an entity which is moved as a unit into another environment, window, or application, with a link retained to the original file location. Adjustments to the format of the information may be necessary, and are handled by the drag and drop operation.

object, programming In object-oriented programming, a reusable, modular, 'wrapped up' collection of software characteristics, functions, and parameters at a basic level. For example, a button may be designed with certain visual and operational characteristics and stored for reuse in various applications, so the code for the object isn't constantly reinvented. An object may consist of a collection of other objects to serve some related or higher function. See class.

Object Database Management Group ODMG. An independent standards organization which is now called the Object Data Management Group to reflect the broader efforts of the organization to support Universal Object Storage Specifications (UOSS). See CORBA. http://www.odmg.org/

Object Definition Alliance ODA. A vendor association established by Oracle which aims to promote and develop new interactive TV and other multimedia services and networks that will operate over a variety of platforms. ODA seeks to establish associated technical standards for these products. Vendors include a number of high-profile financial institutions and retailers, and computer and media developers including Time-Warner, Apple Computer, Xerox, and Compaq. See video-on-demand.

object encapsulation A technique for combining related data and functions into an operational 'bundle,' thus simplifying its use within a larger framework. The purpose is not to 'hide' the intrinsic components of an encapsulated object, but to create a common superset of characteristics that work together, and which may be frequently used and reused. This technique is one type of modular approach to programming. See encapsulation, object-oriented programming.

object inheritance A concept in object-oriented programming (OOP) which describes a hierarchical passing on of characteristics down through associated objects.

Object Linking and Embedding OLE. A software system developed by Microsoft Corporation which allows various applications programs which are OLE-compatible to share and exchange information. It is an interoperability system that lowers the distinction between various applications developed by different vendors so users can integrate the applications files and environments, and use them more as a 'suite' of tools than as separate items. It further provides specification guidelines for the interface for accomplishing these tasks. OLE is a very good concept, in principle, and works well a lot of the time. Unfortunately, the various implementations are not yet perfect, as the OLE-compliant programs and OLE software programs that are installed on a system, sometimes will clobber some of the other programs that don't support OLE, causing odd behaviors and situations where software has to be reinstalled, or OLE disabled temporarily. As OLE-capability must be incorporated into each software application by individual developers, there is some variation as to the completeness and dependability of these implementations.

When it works, OLE is a very good means to develop documents that take their various elements from a variety of text, image, sound, and other programs and combine them via links and drag and drop. Spreadsheet totals or statistics can be incorporated into stock offering documents, images can be incorporated into proposals, sounds can be incorporated into multimedia presentations, etc. without having to constantly open and close applications and convert various file formats with external utilities. OLE does more than just provide a means to insert information from one source into another, it further keeps a record of the links so that if the source information in one document is updated, it will also be updated in subsequently linked documents.

O

OLE is used by various applications in Windows operating systems and Macintosh operating systems. See ActiveX.

Object Management Architecture OMA. An architectural framework developed by the Object Management Group (OMG) to lower the complexity and cost of developing new software applications. See CORBA, Object Management Group.

Object Management Architecture Board OMAB. A group established in 1996 by the Object Management Group (OMG) to oversee the OMG Technical Process, including the tracking and revision of technical specifications. See CORBA, Object Management Group.

Object Management Group OMG. A nonprofit organization of over 800 software developers, vendors, and end users whose aim is to establish the widespread use of CORBA through global standard specifications. Headquartered in Massachusetts and established by eight companies in 1989, OMG promotes the theory and practice of object technology for the development of distributed computing systems through a common architectural framework. OMG seeks to establish industry guidelines and object management specifications to further the development of standardized object software, which it hopes will encourage a heterogenous computing environment across platforms and operating systems. See CORBA, Object Management Architecture. http://www.omg.org/

Object Request Broker ORB. The communications center of the Common Object Request Broker Architecture (CORBA) standard developed by the Object Management Group (ORG). It provides an infrastructure for program objects to intercommunicate, independent of the techniques used to implement them and the platform on which the software is running. Compliance with the ORB provides portability over many different systems. The ORB administrates objects, so that an application need only request it by name.

There are now many commercial and freely distributable ORBs. See CORBA, Fnorb, Object Management Group, TAO. There is general information on CORBA at

`http://www.omg.org/`

There is a good list of ORB resources on the Web at

`http://patriot.net/~tvalesky/freecorba.html`

Object Services Management components of the Common Object Request Broker Architecture (CORBA) standard developed by the Object Management Group (OMG). A set of services for facilitating development productivity and consistency of implementation. The Object Services provide generic environments for objects to perform their functions, and interfaces for the creation of objects, control of access to the objects, and administration of the location of objects. See CORBA, Object Management Group.

object-oriented programming OOP. A software development approach that follows a more natural and efficient evolution than many older reinvent-the-wheel approaches to programming. To understand the difference between non-object-oriented programming and object-oriented programming in a simplistic way, imagine a toy shop in which each elf is working in a separate little room, each with a separate set of tools, creating some kind of toy doll. At the end of the day, the creations are brought into a central room and it is discovered that some toys have been duplicated, none have interchangeable parts, and the end result is only a half dozen different toys. That's pretty much how traditional programming has been done. There has been an enormous amount of replication of effort. Every company writes the same sorting algorithms, there are hundreds of half-baked proprietary editors, and file search and retrieval methods are reinvented by thousands of programmers on a daily basis. It isn't very efficient. It isn't even very much fun.

Now picture a toy shop in which some general guidelines are set out for joints and limbs, and in which each toymaker has a magic replicator in which his or her components can be copied an indefinite number of times. Now imagine one of the toymakers is a mechanical wizard, and another is an artist, able to make beautiful embellishments. At the end of the day, instead of having a dozen toys, there are a limitless number of heads and feet, bodies and legs, that can be shared among all the toymakers. Not only that, but there are some particularly intricate mechanical parts and some wonderfully aesthetic ones that can be used by all. Since guidelines were set out, the parts are interchangeable. The elves have created the basis for thousands of toys, rather than just a dozen. Assuming unlimited replication of individual parts is possible, there's no limit to how often each component can be used. That's what object-oriented programming is in the ideal sense. Once you create an *eye object* and give it certain parameters so that the color, shape, and various eye characteristics (contact lenses, eyelashes, ability to track a moving shape, etc.) can be individualized, you don't have to do it again; you can mix and match it with head, nose, and hair objects in thousands of different ways.

Similarly, in programming it is possible to create directory objects, menu objects, window objects, and button objects. That's not to say that object-oriented concepts are limited to 'physical' attributes, the software can also incorporate more abstract user security objects, or sort or fetch objects and functions and behavioral characteristics associated with a type or class of objects.

Object-oriented programming is a modular approach that allows objects to be mixed and matched, or arranged in hierarchies, and customized to suit an individual application. Once created, they can be reused indefinitely. This can save development time and provides the basis for platform-independent software; it also gives a certain level of consistency to the interface, so users don't have a high learning curve for interacting with new applications programs. It further provides the programmer with a number of *levels* of interaction with an object. The developer can use the object in a transparent way, with the definition of the object encapsulated (bound together as an attribute or functional unit), by passing messages and parameters without worrying about how it was coded, or the programmer can take apart the object and use its individual components, or combine it with others to create a larger functional unit. This too is different from traditional programming. In many cases using someone else's non-object-oriented code involves a lot of study and adaptation to make it work in another setting, and it's rarely easy to mix and match parts of the code so the characteristics can be inherited among the different parts. In contrast, program objects can be designed so that their characteristics and behaviors are known, so they can be immediately used without a long ramp-up period or restructuring.

Object-oriented programming languages are still evolving. Smalltalk is one of the first object-oriented programming environments, developed twenty years ago, but many common languages currently used in commercial software development are not, and efforts to create object-oriented versions of traditional languages have not been fully satisfactory. Nevertheless, the trend is toward object orientation, given its obvious advantages of portability and efficiency in many contexts.

For important and interesting information on taking the object-oriented model to global implementation and distribution, see CORBA and Object Management Architecture. See Open Systems Interconnect, Smalltalk.

OBRA See Omnibus Budget Reconciliation Act.

obscenity Obscenity, in its everyday sense, refers to actions or materials which are offensive, repellent, or vile. In a legal sense, it is more specific, as individual interpretations of what is offensive vary dramatically. Questions involving obscenity often conflict with individuals' rights and opinions regarding freedom of speech, and thus are important issues on the global Internet. The Internet has evolved in a strongly tolerant and open advocation of people's rights to their opinions. Much content on the Internet is frank, open, shocking to some, and more explicit than what people typically encounter when going about their daily work or shopping. See Communications Decency Act of 1996, Electronic Frontier Foundation.

OC 1. operator centralization. 2. See optical carrier.

OCC See Other Common Carrier.

Occam's Razor A maxim well known to scientists attributed to William of Occam in the 1300s that it is vain to do with more what can be done with fewer (or less). It has been restated in many ways, in many contexts, but essentially, in science and in human spheres of activity, the idea is that the simplest explanation or one which doesn't require any additional hypotheses, is usually the best, and often correct.

OCP operator control panel.

OCR 1. See optical character recognition. 2. Outgoing Call Restriction.

octal A base eight numbering system utilizing the numerals 0 through 7. See decimal, hexadecimal.

octathorp See octothorpe.

octet A data unit very widely used in digital networks. An octet consists of eight data bits.

octopus, hydra A visually descriptive name for a 25-pair cable common in multiple phone system installations. At the far end, the 25-pair wire is organized into individual connectors (two, four, six, or eight wires) with phone cord connectors. An octopus is useful for stringing a single wire into a location where several phone connections are planned.

octothorpe, octathorp The # symbol, sometimes also called pound, hash, crosshatch, or number sign. It is used as an end signal (or 'long' signal) on some touchtone phone menu systems. It represents a number sign in financial contexts, a suite number in postal addresses, and the *sharp* sign in music notation. See pound.

ODBC See Open Database Connectivity.

ODMG See Object Database Management Group.

ODP open distributed processing.

Odyssey A medium Earth orbit (MEO) satellite communications system which was intended to begin service in 1999 with 12 medium Earth orbit (MEO) satellites. Planned services included voice, data, facsimile, and Global Positioning Service (GPS). The project was discontinued in 1997 and TRW transferred technical expertise to the ICO Global Communications ICONET when it became a leading ICO shareholder later in 1997. TRW Inc. announced that it would turn back the license it had received for the Odyssey program to the U.S. Federal Communications Commission (FCC), in order to make the assigned frequencies available to other communications services. See ICO Global Communications.

OEM original equipment manufacturer. A manufacturer who often supplies to other manufacturers or value-adding distributors.

oersted (Symbol - Oe, Ø) A centimeter-gram-second (CGS) unit of magnetic intensity (field strength) equal to the intensity of a magnetic field in a vacuum in which a unit magnetic pole experiences a mechanical force of one dyne in the direction of the field. It can be expressed as $10^3/4pA\ m^{-1}$. Named after Hans Christian Ørsted (sometimes transcribed as Oersted). See ampere.

Oersted (Ørsted), Hans Christian (1777-1851) A Danish physicist and educator who demonstrated the effects of current on a magnetic needle to a class of physics students around 1819. He reported on the magnetic effects of electric currents, information that brought together magnetism and electricity as never before, the results of which led to a change in scientific thinking and the development of electric telegraphs in Europe and America. The oersted unit of magnetic intensity is named after him.

OFDM See orthogonal frequency division multiplex.

off-hook On a phone set, the state of having the plunger or switch-hook in the active or 'up' position so the circuit is connected. The term comes from the old wall phones on which the earpiece (receiver) was taken off a curved hook when in use (and when the battery power was engaged). When the phone is first taken off-hook, it alerts the switching exchange to the fact that the caller wants to use a line. The switching exchange returns a dial tone to the caller to let them know the line is available and they can dial through. See on-hook.

off hours The times outside of normal operating or working hours. Telephone and Internet services are often discounted in off hours or off-peak hours.

off-peak hours Hours of low usage. In telephony, the hours between 11:00 pm and 7:00 am are designated as off-peak in many areas, and calling rates are lower. The concept also applies to transportation systems, and fewer buses, trains, or subway cars may be in service during these hours.

off the shelf Products and services that are ready to use without any customization. Products which can be readily purchased by anyone walking into a store or ordering from a catalog, and run with little or no configuration. Essentially the same as shrinkwrapped products.

office automation A catchall term for procedures and systems that are designed to streamline or increase the efficiency of the operations of a business, often through the installment of technology which may or may not displace human workers. In some respects office automation has freed people from drudge work; it is no longer necessary to have rooms full of 'human calculators' sitting and working out sums by hand, but the introduction of technology has also introduced greater needs for training, storage of information, information retrieval, and other time-consuming activities that don't necessarily improve quality of life or shorten the work day.

Office of Science and Technology Policy OSTP. The Science Policy coordinating group for the Federal Government Executive Branch. The OSTP is led by Presidentially appointed directors, and is organized into four divisions: environment, national security and international affairs, science, and technology. The OSTP provides expert advice to the President of the United States in matters of science and technology.

http://www.whitehouse.gov/WH/EOP/OSTP/html/OSTP_Info.html

Office of Telecommunications Along with

the Office of Telecommunications Policy, this organization was rolled into the U.S. National Telecommunications and Information Administration in 1978 as a result of a reorganization.

Office of Telecommunications Policy OTP. The OTP was rolled into the U.S. National Telecommunications and Information Administration in 1978 as a result of a reorganization.

OGT outgoing trunk.

ohm A practical unit in the meter-kilogram-second (MKS) system equal to the resistance of a circuit in which a potential difference of one volt produces a one ampere current. Thus, if the values of two of these three are known, the third can be calculated. Named after Georg Simon Ohm. In 1908 the International Congress established the International ohm as the resistance offered to an unvarying current by a column of mercury at zero degrees centigrade, 106.3 centimeters long, of a constant cross-sectional area of 1 square millimeter, and weighing 14.4521 grams. In the U.S. in 1950, Congress defined the ohm as equal to one thousand million units (10^9) of resistance. See ampere, electromotive force, Ohm's law, resistance, volt.

Ohm, Georg Simon (1787-1854) A German physicist who, in 1820, investigated the conducting properties of various materials. He described the flow of electricity through a conductor and discovered the relationships between current, resistance, and electromotive force, information that greatly influenced subsequent theory and application in electricity. See Ohm's law.

Ohm's law In any specific direct current electrical circuit, the strength of the current is directly proportional to the potential difference in the circuit and inversely proportional to the resistance. Thus, current (in amperes) equals electromotive force (in volts) divided by resistance (in ohms), or $I = E/R$. See ampere, ohm, resistance, volt.

OHR See optical handwriting recognition.

OLE See Object Linking and Embedding.

OLIU Optical Line Interface Unit.

OLNS Originating Line Number Screening.

OM Operational Measurement

OMA See Object Management Architecture.

OMAT Operational Measurement and Analysis Tool.

OMG See Object Management Group.

Omnibus Budget Reconciliation Act OBRA. OBRA is a 1993 U.S. Congress amendment to the Communications Act of 1932 which preempts state jurisdiction in such a way that individual states no longer regulate rates and entry by companies offering wireless services. The federally controlled spectrum was transferred to the Federal Communications Commission (FCC). It further organized wireless into two categories: commercial mobile radio services (CMRS) including cellular radio services and personal communications services (PCS); and private mobile radio services (PMRS), including public safety and government services. See Telecommunications Act of 1932.

omnidirectional Effective in all directions, radiating in all directions, or receiving from all directions. Functional in many directions without preference to any one. An omnidirectional antenna is one which is designed to send or receive signals in a maximum number of directions. A theoretical isotropic antenna is fully omnidirectional and is often used as a reference for comparing antenna patterns or effectiveness. An omnidirectional speaker directs sound in all directions. Since this is structurally difficult to achieve, the speaker is usually a collection of speakers pointing in many directions, housed in one cabinet.

O

omnidirectional antenna An antenna designed to transceive signals through a wide range of directions. Since an antenna's capabilities are determined by shape and location, it is rarely completely omnidirectional, but broad omnidirectionality is achieved by maintaining equal field strength through the horizontal plane, and radiating in or out through the vertical plane. See isotropic, omnidirectional.

omnidirectional microphone A microphone designed to capture sound from all around its location. This is actually less common than directional microphones. Tape players, camcorders, digitizing sound sample microphones, phoneset microphones, and others have directional microphones to zero in on the crucial input, so they can screen out extraneous noises and conversations. Omnidirectional microphones can be said to capture sound 'environments.'

on-hook On a phone set, the state of having the plunger or switch-hook in the inactive, depressed or 'down' position to interrupt the circuit, so it is not active while the phone is not being used. The term comes from the old wall phones on which the earpiece (receiver) was cradled on a curved hook when not in use (to conserve battery power). See off-hook.

on-line See online.

on/off keying A type of modulation scheme similar to frequency shift keying (FSK), except that no signal is used for binary "0" (zero). See frequency shift keying, phase shift keying.

on-ramp *colloq.* Access to a main communications link, such as highway, phone trunk, or networking service. The on-ramp is the link between the user's system and the main system. An Internet Access Provider (IAP) can be considered an Internet on-ramp.

ONA See Open Network Architecture.

ONAC Operations Network Administration Center.

ONAL Off Network Access Line.

online 1. Having access to a system which is at least minimally functioning, and which is largely automated. Interacting with a proceduralized system. A user is said to be *online* when he or she logs into a computer or a network, or accesses an automated phone system. 2. To bring a system *online* is to connect or power it up so that it is at least minimally functioning. 3. To bring an employee *online* refers to fitting the person into an organizational structure within an established system of priorities and procedures.

OOP See object-oriented programming.

OPAC Outside Plant Access Cabinet.

open 1. Unbounded, having no barriers or extents, unconcealed, exposed, uncovered. 2. An open circuit is one which is not currently connected, usually because there is no power coming to it (as when it is turned off). A circuit breaker or blown fuse may create an open circuit. 3. An open transmission channel is one that is not currently in use, and may be available. An open channel may also imply one that is unsecured, where others can hear any communication that occurs.

Open 56k Forum A consortium of telecommunications vendors promoting K56flex modem technology.

open air transmission A type of transmission which either depends on air for the propagation of the signals or which is commonly broadcast through the air. Radio, shortwave, and microwave transmissions are primarily open air systems.

open applications interface OAI. An interface built into a system and documented in such a way that third party vendors can develop equipment and software applications that tie into that system.

Open Collaboration Environment OCE. An environment created by Apple Computer which provides a means for third-party developers to create telecommunications applications that interface with the Macintosh operating system (MacOS). Thus, developers can create Internet phone, facsimile, network, and other telecommunications-related products for Macintosh owners.

Open Database Connectivity ODC. Microsoft's telephony software open application processing interface (API), which is part of a system that provides interoperability between Microsoft business-oriented database, spreadsheet, and word processing software, which is especially useful for digital telephony applications. The interface itself is independent of the application that provides the formatted data. In this way, call records and statistics can be stored and manipulated with popular software applications, providing computer-telephone integration and advanced call-recording capabilities.

Open Group, The Formerly the Open Software Foundation, The Open Group is an organization which aids in the development and implementation of secure and reliable network infrastructures. The Open Brand is a registration mark (X) awarded by The Open Group to products which conform to the standard specifications. http://www.opengroup.org/

Open Network Architecture ONA. A system being developed to encourage third-party vendors to supply public phone network products and services. Under the Federal Communication Commission's (FCC's) ONA, the telephone companies must provide the same service guarantees and levels to outside vendor's products that use the phone lines, as they use themselves.

open office An administrative and physical structure in which low walls or no walls are favored over high walls, movable walls are favored over fixed walls, and work stations are generally within view of more administrators and employees than in other office designs. Open office concepts are designed to promote flexibility and communication.

open skies *colloq.* Regulatory policies that are liberal enough to allow private use. Prior to 1972, the Federal Communications Commission (FCC) did not permit private American satellites to be launched for commercial communications. It then opened the doors on private domestic satellite launchings and operations, a move that created an opportunity for new competitive services to be established. MCI is one of the companies that got its start partly through recognizing and taking advantage of the opportunities presented by these openings.

Open Software Foundation OSF. This has now become the Open Group. See Open Group. http://www.opengroup.org/

open-space cutout In telephone wiring, a protective grounding mechanism, often used in conjunction with fuses and heat coils to guard against possible danger to people and equipment from large power fluctuations. If voltage is too high, the wire grounds by arcing across a small air gap between carbon blocks mounted on an insulator such as porcelain.

open system An open computer system is one which has few security barriers. Passwords may not be needed or individual users' directories may be open to all users. In many ways the Unix operating system and the Internet global network have been developed with an effort to keep them open and accessible, and there are some people who advocate that all systems should be that way.

Open Systems Interconnection OSI. An important layered architecture specification released as a standard by the International Organization for Standardization (ISO). OSI is designed to facilitate communications development between computer equipment and network software. Many vendors have opted to support this standard. Essentially, the communication is mapped onto seven layers as shown in the Overview of OSI Layers chart.

operand A quantity or information which is being manipulated. For example, in mathematics, if you divide 200 by 10, then 200 is the operand. In computer algorithms, data, or the address of the data to be operated upon, are passed to an instruction in order for the instruction to act upon the data.

operating environment This has two meanings, depending upon the context. It is used in a limiting context to describe an operating system which isn't fully integrated or fully multitasking. It may be task-switching, or it may have a graphical user interface on top of a text-based operating system which is not fully implemented as multitasking. Many vendors have claimed multitasking operating systems which were not really so. In its second, broader context, it refers to the environment surrounding an operating system, that is the system, the software that runs it, the peripherals and applications which support it, etc. An operating environment, in this broad sense, may encompass more than one operating system.

operating system OS. The most important software on a computer is that which lies between the user applications and the hardware. It's not possible to control the CPU, to manage memory, to access a disk drive, to send images to a monitor, or data through a network connection or modem without an operating system. The operating system handles interrupts, timing, the movement of data from one register to another, and all the nitty-gritty operations that are typically not seen or understood on a technical level by most users.

Microcomputer operating systems began to be

Overview of OSI Layers

Layer	Notes
applications layer	The user software application layer. Interapplication communication, file access, emulation, etc.
presentation layer	Character sets, text handling, and other prompt, input, and output details.
session layer	Bidirectional intranet communications.
transport layer	A logical data link between the user and the lower level network service that may be required.
network layer	Routing, flow control, and other virtual routing factors above the physical layer.
data link layer	Basic transmission and packet handling, error detection and correction. Subdivided into two layers: the Logical Link Control (LLC) and the Media Access Control (MAC). Ethernet is one of the common protocols implemented at this layer.
physical layer	Cables, connections, devices, buses, processors, signaling, etc.

developed in the 1970s. The early ones were text-based. One of the first widely used, popular operating systems was CP/M, designed by Gary Kildall. CP/M was the forerunner of QDOS, and hence, MS-DOS, and the syntax and commands are very similar. LDOS, TRS-DOS, UltraDOS, and other TRS-80 operating systems shared many common properties with CP/M. Kildall also later designed a multitasking operating system and a graphical user environment (GEM).

In the early 1980s, Apple created proprietary operating systems for their Apple and Macintosh lines of computers, featuring the first widely distributed graphical operating system descended from pioneering work at Xerox PARC. This concept was so successful, it has since been adapted by virtually all subsequent vendors, including Atari, Commodore Amiga, Apollo, Sun Microsystems, Microsoft, SGI, The X Windows System, and NeXT. It's difficult to find a computer now that doesn't have a graphical user interface on top of, or in conjunction with a text operating system, or which is fundamentally a graphical operating system. The various versions of Windows are popular on consumer-level Intel-based systems.

A number of multitasking systems were developed in the late 1970s and early 1980s, but the first widely distributed commercially successful preemptive multitasking operating system on a microcomputer was AmigaOS, introduced in 1985. Most of the workstation level computers had multitasking operating systems by the early or mid-1980s, and the other vendors began to follow this lead in the late 1980s, most notably OS/2 (Operating System 2), originally developed jointly by IBM and Microsoft. Many other operating systems released around this time were task-switching rather than being fully multitasking. Microsoft Windows has since become widespread on personal computers, with Windows NT, originating out of the OS/2 collaboration, used on many server systems.

Unix is one of the most robust, earliest, and most important operating systems. A high proportion of institutional computing operations, much scientific research, and many Internet hosts run on Unix systems. Unix is freely distributable, powerful, flexible, dependable, well-supported, and runs on most computers. Linux is a popular implementation of Unix (as is BSD), available from a number of commercial and free distribution sources. Along with Apache, a freely distributable server software, Unix/Apache systems are used by thousands of Internet Services Providers to provide gateways to the Internet through the Web.

An operating system runs a computer, and computers are increasingly being delegated control tasks beyond that which humans can handle alone or in cooperation with one another. Navigational aids on aircraft are a good example. Fighter jets traveling at hundreds of miles per hour move too quickly for the human nervous system to react in time to control every aspect of the plane's behavior, so computerized systems handle many functions on the pilot's behalf. If you extrapolate that type of control to appliances, houses, security systems, currency exchange, intelligent vehicle systems, and every aspect of human society which will someday be controlled by computers through a 24-hour Internet connection to all the other computers in the world, imagine the importance of carefully choosing an operating system. Buying the cheapest system or the most popular system, or allowing the choices to be made for us by a profit-based corporate entity is a more far-reaching decision than selecting a President or other political leader. Why? Because a President who does a bad job can be voted out. An operating system which does a bad job cannot be disabled if it has been delegated important tasks such as running medical equipment or environmental controls. Once a fighter plane has been designed with a dependency on computer control, you can't just turn off the software in mid-flight. Similarly, if five or ten years from now a family's financial transactions, prescriptions, education, and travel arrangements are all handled by the computer operating system through the Internet, it may no longer be possible to just 'turn off' the computer. That bears some thought and some forethought. See BSD, Linux, Microsoft Windows, OS/2, Solaris, SunOS, Unix, UNIX.

operational load 1. The full power requirement of a facility, which may be expressed in terms of averages or maximums. 2. The administrative and personnel requirements of an organization when in full operation.

operator-assisted call Any phone call in which the caller contacts the operator to handle some part of the transaction or connection, rather than direct dialing. There are usually surcharges associated with operator-assisted calls.

operator console, computer In computer networks, a console is a computer terminal which allows access to management and administrative functions that monitor and control the network. Common operations carried out on the console include user password assignments, new user account allocations, virtual configuration of devices and the network topology, server configuration, etc. Often the operator console is in a separate room for security reasons and, at the very least, is password protected to prevent unauthorized users, or well-meaning, but uninformed users from accessing lower level system functions that could interrupt the functioning of the network.

operator console, phone In telephony, the main console of a multiline phone system. The operator console often is programmable and may include a hands-free earphone set.

Operator Service Providers OSP. Previously called Alternate Operator Services. A competitive provider of operator-assisted long-distance calls, especially third-party billing, collect, etc., which usually leases the services of existing phone networks. Large hotels sometimes provide AOS services to hotel guests for a premium. See splashing

OPRE Operations Order Review.

Ops Abbreviation for operations, operators, operator services, and system operators.

OPS off-premises station.

optical amplifier A type of device used as a cable repeater on fiber optic transmission lines, which functions without converting the optical signal. Early versions were based on semiconductor lasers. Most are now erbium-doped fiber amplifiers (EDFAs), that is, directly doped silica fibers, with the principal parameters defined by atomic composition.

optical bypass In a Fiber Distributed Data Interface (FDDI) token-passing network, port adaptors can be equipped with optical bypass switches to avoid segmentation which might occur if there is a failure in the system and a station is temporarily eliminated. Normally optical signals pass through the bypass switch uninterrupted, but if a station fails and is eliminated from the ring, the optical bypass can reroute the signal back onto a ring before the signal reaches the failed station, thus providing fault tolerance for the system.

The optical bypass switch is attached to the FDDI port adaptors, between the token-passing ring and the attachment station. See A port, dual attachment station.

optical carrier OC. A series of transport levels defined in conjunction with SONET.

optical character recognition OCR. A software process, or combination of hardware scanning devices and software, which evaluates marks on a page and determines whether they have the predefined characteristics of text, symbols, and images, depending upon the software.

Most OCR programs use a combination of intelligent algorithms and character tables to process marks. These marks are typically text and character symbols, although some programs will also recognize mathematical and logical symbols if these are added to the program dictionary or the user dictionary. Some programs can automatically discern columns of text and images, and divide the page up appropriately, handling the regions separately. Once the document has been 'recognized,' the characters and symbols are converted to a common format, such as ASCII or one of the many flavors of extended ASCII, and stored as a file that can be further edited with a text editor, word processor, or desktop publishing program. In its strictest sense, OCR just recognizes the characters, and software which handles images as well is called *optical document recognition*. However, most OCR applications these days have some document processing and image recognition functions, so the phrase is used broadly here.

Most of the early work on pattern matching algorithms used in character recognition was done in the late 1950s and the early 1960s. By 1963, IBM had a system that could recognize Roman and Cyrillic characters at the rate of about 50 words per minute, the speed of a moderately competent typist, and faster than most typists can type information with many numerals. See optical document recognition, scanner.

optical computer A type of processing hardware based on photons rather than electrons. This is a more experimental technology than is used with traditional computers, but it has possible advantages particularly in speed and resistance to interference over current technologies, and may become more prevalent in the future.

optical connectors Connectors specially designed to couple fiber optic cable junctions so they interfere as little as possible with the path of the optical beams passing through the connectors. The connectors are usually used at points where the fibers connect with routing or switching circuitry, or with the optical interface to the system itself. Optical connections have to be well-engineered, as they must handle very precise beams and paths, and often must maintain the proximity and orientation of a bundle of optical fibers. Connectors for blown fiber installations are easier to install and maintain than a number of other types of fiber attachments. See blown fiber.

optical detector A substance or circuit which detects and converts electromagnetic waves in the form of optical waves. Solar panels contain substances which allow the conversion of light into electricity, which are widely used in the satellite industry to provide power for telemetry adjustment of orbits. See solar panel.

optical disc Any of a number of technologies which are used to store digital data which is subsequently accessed by light, usually a laser beam pickup. The information on an optical disc is commonly stored in a series of pits, that is, indentations in a metal disc coated with a plastic substrate. The placement, size, and proximity of these pits is defined in part for the specification for the type of optical technology. Newer technologies have been developed to increase the information capacity of a disc, as in digital videodiscs (DVDs) that are slightly thicker, to bring the disc closer to the laser pickup, with slightly smaller, tighter pits. Compact discs, videodiscs, PhotoCDs, and DVDs are popular forms of audio and visual storage media. Optical discs are also used for computer data storage, particularly for backups.

optical document recognition ODR. A software process or combination of hardware scanning devices and software, which evaluates the various elements on a page to determine whether they are distinguishable as text, images, symbols, or lines, and identifies them accordingly. The ODR in its simplest sense does not interpret the text into characters; it merely identifies page layout elements: images, columns, page numbers, text, etc. However, in commercial products, it is commonly combined with optical character recognition (OCR) capabilities in order to optically identify and interpret those areas which are text into characters that can be edited with a word processor, text editor, or desktop publishing program. ODR is a somewhat more complex process than OCR; together they comprise powerful tools which are a part of image document management and processing.

ODR is particularly valuable for converting archived image information (e.g., microfiche) into documents that can be desktop published, or which can be archived in databases for faster and more efficient search and retrieval. See image document management, optical character recognition, scanner.

optical fiber A flexible, light-conducting, filamentous plastic or glass medium used for optical signal transmissions. Optical fiber is typically from 2 to 125 micrometers thick, and is capable of carrying a variety of high-speed, wide bandwidth transmissions with relatively low loss, when correctly installed. Unlike wire, fiber is not subject to electromagnetic interference or most of the types of radiant eavesdropping techniques that can be used on wire. This has made it very popular for backbone, hazardous area, and multimedia installations. It also does not present the same potential fire hazard as wire electrical cables.

Two common types of optical fiber include step-index fibers and graded-index fibers. Step-index fibers have two layers: a lower refractive outer *cladding* layer and an inner core with a high refractive index. Graded-index fibers also

SONET Carrier Transport Levels

Carrier	Rate (Mbps)	DS-3	DS-1	DS-0
OC-1	51.84	1	28	672
OC-3	155.52	3	84	2016
OC-9	466.56	9	252	6048
OC-12	622.08	12	336	8064
OC-18	933.12	18	504	12096
OC-24	1244.16	24	672	16128
OC-36	1866.24	36	1008	24192
OC-48	2488.32	48	1344	32256
OC-96	4976.64	96	2688	64512
OC-192	9953.28	192	5376	129024

are less refractive toward the outer edge, but rather than being two layers sandwiched together, as in the step-index fibers, the refractivity in the material overall decreases gradually in relation to its distance from the innermost point of the cylinder.

Fiber can be bundled without the electrical interference common to bundled wires; a group of fibers works together to provide greater capacity. Single-mode fiber transmissions, in which the signal can only follow one path through the filament, can travel greater distances without repeaters. See multimode optical fiber, single mode optical fiber.

optical fiber cable See fiber optic cable.

optical fiber ribbon A way of bundling the fibers so that they are laid side-by-side to form a flat ribbon or strip. This is convenient if they are to be installed in narrow areas such as walls or under carpeting. This way of arranging fibers is less common than bundling them into a cylindrical shape.

optical handwriting recognition OHR A specialized form of character recognition designed to separate joined shapes and recognize variations of particular letters, in addition to other generalized optical character recognition (OCR) functions. Most OHR systems have to be 'trained' to recognize a particular style of handwriting, since there is so much variation in the letter forms in the way different people write.

optical scanner See optical character recognition, scanner.

OQPSK offset quadrature phase shift keying. See quadrature phase shift keying.

ORB See Object Request Broker.

ORBCOMM Orbital Communications.

orbit The path described by a body that is in more-or-less stable balance with the gravity of the body being orbited so that it continues in that path for a significant period of time (usually at least a few hours or days, although orbits of artificial satellites can last for years in a stable orbit). The Earth is in orbit around the Sun, and the Moon is in orbit around the Earth.

When the balance is lost and the orbit becomes smaller as the orbiting body is drawn inward, it is said that the orbit is *decaying*. There are many types of orbits (orbits at different heights, with differently shaped paths) used in telecommunications with artificial satellites.

Early communications satellite orbits tended to be circular or low and somewhat flatly elliptical (and tended to decay quickly). Communications were hindered by the necessity of locating the orbiting satellite and keeping it in range before it passed around to the other side of the Earth. Later satellites were put into higher, more stable orbits, which were often geostationary, that is, the orbit was synchronized with the movement of the Earth so that the location of the satellite was roughly above the same location at all times. The amount by which an orbit deviates from a circle is referred to as its *eccentricity*. See geostationary, satellite.

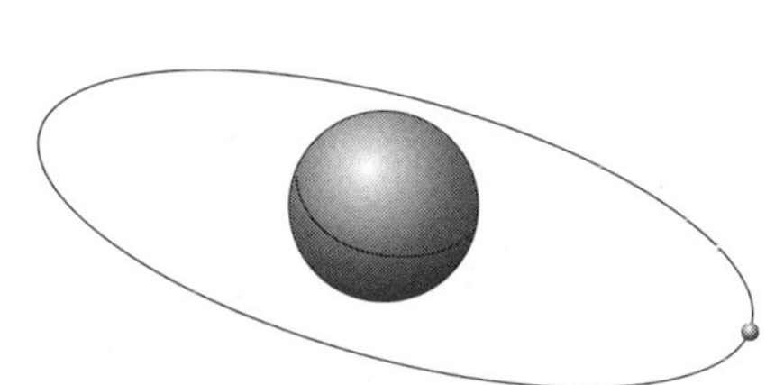

Satellite orbits are circular or more flatly elliptical, and may be further controlled in order to spend a greater part of the orbit over ocean or land masses, depending upon the function of the satellite. Orbits are often described in terms of their distance from the Earth, as low, medium, and high Earth orbits. Geostationary orbits are a type of high Earth orbit in which the satellite remains in the same position relative to the Earth.

Orbital Sciences Corporation Commercial developers of global satellite communications services. Orbital Sciences is a space and information systems company which designs, manufactures, and markets space-related infrastructures and products. See OrbLink.

OrbLink A global broadband commercial satellite communications network being developed by Orbital Sciences Corporation for deployment around 2002. The designers intend to use the newest technology to construct broadband services that can be offered at lower prices than existing services. OrbLink is based on seven medium Earth orbit (MEO) satellites orbiting at 9,000 kilometers in an equatorial orbit, transmitting in extremely high frequency radio bands. Services will include digital voice, data videoconferencing, computer networking, imaging, and other broadband applications.

Pending Federal Communications Commission (FCC) approval, intersatellite communications will be at 65.0 to 71.0 GHz at speeds up to 15 Gbps, supporting high-capacity intercontinental trunking. Two-way digital connections will be between 37.5 to 38.5 GHz and 47.7 and 48.7 GHz bands up to about 1.5 Mbps.

The seven satellites, plus one spare, are based on Orbital's STARBus, a small, lightweight, geostationary technology, acquired through a purchase of CTA Incorporated's space system business. Each satellite will support 100 spot beams. As an economic note on the dynamics of new telecommunications technologies, the total cost of building and deploying the orbital network, according to Orbital Sciences, compares to the cost of installing two transatlantic fiber cables, which provide only about 10% of the trunking capacity of the proposed satellite system.

Orckit Communications A developer and manufacturer of high-speed local loop communications systems and participant in a number of network standards working groups. Orckit partners with Fujitsu Network Communications, Inc.

order entry The inputting, usually by voice or keyboard, of a customer request for a product or service. On the World Wide Web, Web *forms* are often available for customers to put in their own order entry. These Web applications are often in the form of *shopping carts* in which the customer browses various Web pages and enters each product desired into the shopping cart (order batch). The order is typically preprocessed by a CGI and then sent to the appropriate order fulfillment personnel for shipping and billing. There are other automated systems which allow customers to carry out a complete ordering transaction over a touchtone phone by using the keypad to enter codes and digits. See Automatic Call Distribution.

Organization of American States OAS. The OAS Web site provides information on the group. http://www.oas.org/

Organizationally Unique Identifier OUI. A globally unique Ethernet address identifier for LANs and MANs managed and assigned by the IEEE Registration Authority online or by phone. The OUI is assigned as a three-octet field in the SubNetwork Attachment Point (SNAP) header, identifying an organization which further creates unique six octet numbers. Together they constitute a distinct Media Access Control (MAC) address or Ethernet address. See Ethernet.

originate/answer On a computer data modem, when the user wants to dial out to connect to another computer, a bulletin board, or an Internet access point, the commands for controlling the phone line and dialing the desired number come from the originating modem. "Originate mode" sets up a sequence of events which checks for a dial tone, dials, and handshakes with the receiving modem to establish the connection rate and protocol (or hangs up if the line is busy, or is dropped). The receiving modem is set to "answer mode" so it detects an incoming call, answers it, and participates in the rate and protocol negotiation. Most of this is automatically handled through a terminal software program, but it may be necessary to set originate or answer through menu selections, or direct commands through the software to the terminal program.

originate restriction A security or specialized use restriction on a phone line which causes it to work only for incoming calls. Outgoing calls are blocked. This restriction is sometimes set on phones adjacent to public areas to prevent people from monopolizing or misusing a phone line. Sometimes the originate restriction applies only to long-distance calls or calls outside of a private branch exchange.

In some circumstances, the local phone company will partially disconnect a line by setting an originate restriction if the subscriber is behind in the payment of the phone bill. After paying the bill, it is usually necessary to request restoration of full service, as it is seldom done automatically.

originator Initiator, beginner, inventor, introducer, founder. The person, entity, device, or station that first communicated a message or started an action or process.

Orion A broadband data satellite service provider aiming at international common carriers and individual companies.

ortho-correction In satellite imaging, a correctional adjustment for distortion resulting from terrain.

orthogonal frequency division multiplex OFDM. A multicarrier modulation system which is similar to discrete multitone in that it utilizes Fourier transforms of data blocks. OFDM is suitable for Digital Subscriber Line (DSL) services. See Digital Subscriber Line, discrete multitone.

OS/2 Operating System/2. International Business Machines' 32-bit preemptive multitasking text and object-oriented graphical operating system which was targeted for Intel-based microcomputers in the late 1980s. It was originally being developed for IBM by both IBM and Microsoft Corporation, and version 1.0 was released in 1987 to succeed MS-DOS. When an upgrade to OS/2 was well under way, Microsoft pulled out to concentrate on their own operating system in competition with IBM.

Many of the same concepts that were part of OS/2 were incorporated into Microsoft Windows NT, which was marketed as a direct competitor to OS/2.

In 1991, IBM released OS/2 version 2.0. Version 2.1 added support for multimedia and Windows 3.1 applications. OS/2 had some commercial success in the early and mid-1990s, but by 1996, through aggressive advertising and bundling programs, Windows was better known and more widespread in North America. In spite of this, there are many strong supporters of OS/2, including a worldwide network of Team OS/2 Groups. It is still a popular choice in Western Europe and Canada. Team OS/2 information for OS/2 users is available at their Web site. See Team OS/2.

OS/2 SMP OS/2 Symmetric Multiprocessing. This version of IBM's OS/2 supports systems with multiple processors, making it suitable for Internet services, graphics, and corporate applications, particularly those which operate as various types of resource servers in a networked environment. See OS/2.

OS/2 Warp Operating System/2 Warp. By version 3 of IBM's OS/2 operating system, their operating system product was called OS/2 Warp. The Warp version added increased support for various peripheral devices and reduced memory requirements. It was succeeded by OS/2 Warp Connect which offered networking through full TCP/IP capabilities. OS/2 Warp version 4 was aimed at corporate users. This version included increased networking features, speech-to-text speech recognition software, and built-in support for Sun Microsystems' Java. See International Business Machines, Java, OS/2.

OS/2 Warp Server IBM's Warp Connect system integrated with their local area network (LAN) server 4.0. This version of OS/2 was designed for handling file and device service sharing on networks. See OS/2.

OSCAR Orbiting Satellite Carrying Amateur Radio. A series of orbiting satellites originally developed in the homes and garages of a group of amateur radio enthusiasts. In 1962, the OSCAR Association was incorporated as Project OSCAR, Inc. The early OSCAR satellite projects began in 1961 and continue today in a much expanded and more sophisticated form.

Early OSCARs used fairly simple beacon transmitters with non-rechargeable batteries, so they were only useful for a few weeks, but they showed what might be accomplished with relatively simple materials and a lot of cooperative effort. Solar cells and telemetry equipment were added to later versions in order to extend useful life and provide greater control over positioning. Relays were then added, with the aim of eventually providing two-way communications.

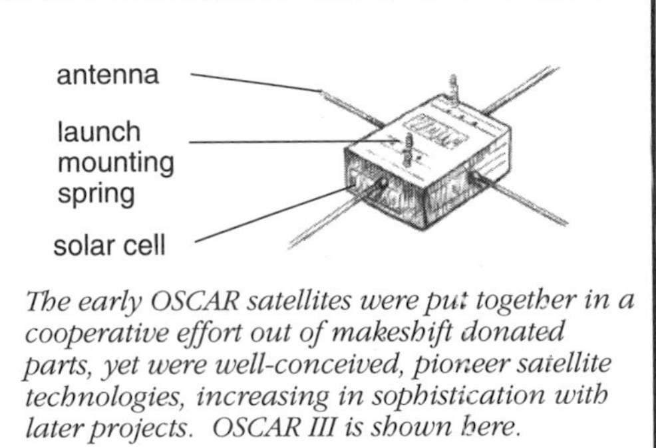

The early OSCAR satellites were put together in a cooperative effort out of makeshift donated parts, yet were well-conceived, pioneer satellite technologies, increasing in sophistication with later projects. OSCAR III is shown here.

OSCAR-AMSAT projects became increasingly sophisticated and, by the time the OSCAR 6, 7, and 8 were in orbit, telemedicine and search and rescue satellite communications were demonstrated to be feasible.

The deployment mechanisms of the early OSCARs were of particular interest to scientists researching satellite installation. The building, launching, and especially the tenuous securement of domestic and international regulatory permissions to launch and operate, were a great achievement for amateur enthusiasts, and there are still benefits accruing from the hard work and voluntary contributions of radio amateurs. See Overview of OSCAR Projects chart. See AMSAT.

OSCE Organization for Security and Co-operation in Europe.

oscillation 1. Variation, fluctuation, continuing periodic reversal. Although oscillation in the general sense does not imply a regular oscillation, many waves, materials, and circuits studied or constructed by scientists exhibit fairly regular, predictable oscillating properties. See quartz. 2. The cyclic alternation of electrical properties in a circuit.

oscillator An electronic device designed to generate a low-current alternating current (AC) power at a particular frequency according to the values of certain constants in its circuits. In microcomputers, an oscillator can be used to provide a reference frequency for clocking. An oscillator is also useful for generating test signals. See oscilloscope, quartz.

oscilloscope A device designed to provide a visual representation of variations in electrical quantities as a function of time, displayed in the form of pulses or waves on a monitor. The size and form of the waves are traditionally 'tuned' for optimum viewing with knobs, as on an old radio. Oscilloscopes are sometimes interfaced with computers to provide a means of directly adjusting and analyzing the oscilloscope signals through software. Oscilloscopes are useful for diagnosis and testing of electronic circuits.

OSF See Open Software Foundation (now the Open Group).

OSI See Open Systems Interconnection.

OSI Transport Protocol OSI TP. The ISO-recommended communications protocol used by X/Open.

OSN operations system network.

OST Office of Science and Technology. A U.K. government group founded in 1992 to coordinate science and technology issues across government departments.

OSTA Optical Storage Technology Association. http://www.osta.org/

OSTP See Office of Science and Technology Policy.

OT See Office of Telecommunications.

OTA Office of Technology Assessment (U.S.).

OTAR over the air rekey.

OTGR Operations Technology Generic Requirements.

OTH over the horizon.

OTOH An abbreviation for "on the other hand" commonly used in email and on online public forums. See AFAIK, IMHO.

OTP See Office of Telecommunications Policy.

OUI See Organizationally Unique Identifier.

out-band/out-of-band signaling Control signaling that is carried separate from the informational portion of a message. See Signaling System 7.

out-of-range alert In wireless communications, a beep or light that alerts the user that the handset is at the edge of its range and the user shouldn't move further from the source of the transmission.

outage Loss of power, service interruption. See blackout, brownout.

outlet 1. Exit, opening for egress, vent. 2. Plug receptacle in a circuit, usually for electricity or connectivity to data transmissions. 3. Source of goods, supplier.

outline font, algorithmic font, vector font A character set which is defined by mathematical algorithms that describe the shape of the letters with graphics primitives such as lines, arcs, ellipses, spline curves, etc. That way, when printed or displayed on a monitor, they will be drawn at the best possible resolution offered by the display system. Unlike bitmap fonts which are hand drawn as raster images that cannot be significantly reduced or enlarged, vector fonts look good at sizes ranging from 4 points to 100 points and much larger. Outline fonts are resolution- and platform-independent, provided an interpreter is available on the system for the particular format that is being used. Since outline fonts are widely supported on many platforms, this is usually not a limitation.

The first outline fonts widely used on desktop computers were defined with the PostScript page description language in the mid-1980s. These were released for use with desktop publishing programs on the Macintosh computer, and were incorporated into the Apple LaserWriter printer introduced in 1985, which almost single-handedly launched the desktop publishing revolution. The introduction of outline fonts at a price that consumers could afford allowed individual publishers, writers, and small presses to release documents, books, pamphlets, and other printed matter that were cost prohibitive prior to the mid-1980s.

Adobe continued to develop PostScript fonts after they were first released. 'Hinting' was added, a process that adjusts small fonts to reproduce better on lower resolution devices (e.g., 300 dpi printers). Type I, Type II, and recently, Type III fonts were specified. Type I and Type II are still widely in use. PostScript fonts have been ported to many computer platforms and printers, and continue to be well-supported, particularly in the service bureau and printing industries.

TrueType fonts are another type of vector font commonly used in computer applications. Unlike bitmap fonts, TrueType fonts provide professional-looking output on common printers, such as the HP LaserJet series, without the need of a PostScript interpreter. Thus, it's a lower cost option to PostScript. The main difference between Adobe PostScript and Microsoft TrueType fonts is that PostScript is a full page description language, so it can define not only the shape of the font, but the behavior of the letters, and their positioning on the page (swirls, curves, graphic inserts, clipping, etc.). TrueType fonts were intended more specifically

Overview of Early OSCAR Satellite Projects

Satellite	Launch Date	Tech. Details	Notes
Phase I Satellites - experimental, low orbit, short lifespan.			
OSCAR I	12 Dec. 1961	10 lb., beacon, 22-day orbit. Non-rechargeable batteries.	Initiated by a U.S. west coast group. U.S. Air Force launched.
OSCAR II	2 Jun. 1962	Better coatings and temperature control.	Similar to OSCAR I, but incorporating improvements.
OSCAR *	Not launched	Phase-coherent keying.	Similar structurally to previous.
OSCAR III	9 Mar. 1965	First relay transponder. Solar backup.	Tracking and telemetry equipment. Approx. 3,000 mi. range. 18 day transponder use.
OSCAR IV	21 Dec. 1965	High altitude, transponder. Solar, beacon, no telemetry.	Unplanned varying elliptical orbit. Two-way communication achieved. Link between Russia and U.S.
OSCAR 5	23 Jan. 1970	Controllable, magnetic attitude stabilization. No solar or transponder.	Seven analog telemetry channels. Australis-OSCAR 5 (AO-5). Built in Australia. First NASA-launched OSCAR.
Phase II Satellites - developmental, low orbit, operational, longer lifespan.			
OSCAR 6	15 Oct. 1972	Telemetry, command, transponder. Solar. Store-and-forward system.	Twenty-four telemetry channels. Two-way communications. Falsing and beacon problems. Life span exceeded four years. Educational materials printed.
OSCAR 7	15 Nov. 1974	Two transponders, linear frequency translation. Telemetry, radio teletype. Beacons. Up to 4,500 miles low altitude.	Many countries contributed various technologies and parts. AMSAT-OSCAR 7 (AO-7). Relayed with OSCAR 6! Almost seven-year lifespan.
OSCAR 8	5 Mar. 1978	10-meter antenna. Two transponders (Modes A & J) that could operate simultaneously.	ARRL operated. Cooperatively built by Project OSCAR, AMSAT and JAMSAT. Lasted five years.
Phase III Satellites - operational, high elliptical orbit, longer lifespan. - See AMSAT.			

to define typefaces, rather than all the graphic elements of a page, and thus are more limited and more commonly used for less embellished text effects in word processing.

outline font attribute While an 'outline font' generally means a vector font, an 'outlined font' or font manipulated with a outline *text attribute* means one which is rendered as an outline, rather than as a filled entity. Outlining is a common feature in desktop publishing programs and some word processors. The effect is usually used for emphasis or variety, particularly in headlines. It should be used with discretion. Too much outlined text or outlined text that is too small can interfere with legibility.

ABCDEFGabc12345

ABCEDFGabc12345

output 1. That which results from, or comes out of, a process or system. 2. The combined signal and content information of a transmission. 3. The result of a computer process, e.g., the output of a word processing session might be a printed document, Web page, or a facsimile transmission.

output device A device that facilitates the communication or transmission of information, usually in another form or format. In most cases, an output device is a 'human interface' in the sense that it facilitates the translation and/or movement of information between nonhuman-readable forms and human-readable forms, or between single-copy modes and multiple-distribution modes intended for a wider audience.

outsourcing Outsourcing is the process of assigning production or management tasks to an external consultant or organization. Outsourcing is practical when special expertise is needed, or the project is short, and it is not cost effective to hire new permanent staff to handle a project that may take only a few days or months. Specialized design projects, advertising, documentation, and cyclic/seasonal projects are often outsourced. Network administration is often outsourced by small companies, whereas a company with a larger or more complex network would probably have an in-house system administrator. Telephone answering services are a common form of outsourcing used by small businesses and home businesses. Utilizing an answering service is less expensive than hiring a receptionist, a good solution for small companies that don't receive a lot of incoming calls.

OutWATS Outward Wide Area Telephone Services. A WATS service for outgoing calls, which is available at bulk-use discounts. See InWATS, WATS.

overflow 1. Traffic or data in excess of what is typically found on a system, or in excess of what the system is capable of handling. Some systems have additional or alternate circuits, lines, systems, or operators to handle overflow, while others may be slowed down in terms of speed of service, or may cease to function. 2. In telephone circuits, overflow traffic may be diverted to another trunk line. See erlang.

overflow, data In programming, an overflow occurs when an operation generates a result for which there is insufficient address or storage space.

overhead The portion of a task, data block, or operation which provides 'management' information pertaining to the task, data, or operation, which is not part of its integral content. For example, the overhead in a graphics file may consist of a header containing size and palette information, which is not part of the image itself. The overhead in a parallel processing operation may be the time and processing it takes to handle the logistics of farming out the tasks and recombining the results of the processes. In networks, overhead exists in the form of protocol information, timing information, error data, security bits, routing, priority, and more. Given the amount of overhead in networks, it's a marvel that they can work so effectively.

overhead transparency, foil A transparent medium that is receptive to photocopy toner or various inks which is used in conjunction with a bright light and projector to project information on a large surface such as a screen or plain wall. Overhead transparencies are often used for presentations, especially to illustrate lectures. Overhead transparency films come in a variety of compositions; some can be photocopied in black and white, some in color. Don't use regular transparency paper in your laser printer or photocopier, as the plastic may melt and destroy the internal mechanisms. Cardboard frames can be purchased to sturdy the transparencies, which are somewhat flimsy and otherwise hard to hold and organize.

overlay 1. *n.* A keyboard template or sheath. See keyboard overlay. 2. *v.* A programming technique in which a limited amount of storage is extended by reusing portions which are not immediately or subsequently required, or by initiating less commonly used routines only on demand. In telephone applications, overlays may be used to bring various tasks into memory as needed. Some versions of BASIC have commands (e.g., LSET) which allow a variable in RAM to be overwritten with a subsequent variable in order to prevent eventual slow-downs from *garbage collection,* that is, from the reorganization of storage to accommodate more information.

overlay area code A telephone area code assigned as a parallel code in an existing service area. These are commonly assigned to mobile services, like cellular and pager services, so that the area code is separate from the geographic code assigned to that region. These are not yet prevalent, but are expected to increase as mobile services are more widely distributed. See North American area codes for a chart of telephone and mobile service area codes.

overlay network A protocol or application-specific subnetwork, managed and configured independently of its underlying infrastructure, and interconnected by Internet Protocol (IP) encapsulation tunnels over production networks. Recent protocols are supported on overlay networks, including Mbone (multicast IP) and 6bone (IPv6).

overlay, video In video editing, it is common to *overlay* two video signals, or to overlay a computer signal over a video signal, or vice versa. Newscasts will often overlay a human weather forecaster over a computer-generated weather map. In cinema action shots, a stunt worker in a barrel may be overlaid on an image of the Niagara Falls. See chromakey.

override 1. To overlap, to neutralize, to take over, to dominate. A stronger signal, such as an emergency signal, can override a regular transmission. A boss can override the decision of a subordinate; a priority transmission can override current transmissions. An operator can override a current phone conversation. Some private branch phone systems are configured so that someone in authority has the option of overriding other conversations, a power that should be used with discretion.

overrun To overwhelm, to swarm, to go above or beyond an edge or capacity, to overflow. A cost overrun happens when someone goes over budget or some other allotted quantity. A data overrun can happen when the receiving system isn't fast enough or smart enough to handle the incoming transmission. A printer overrun can happen if the print mechanism continues to function after the paper runs out (some facsimile machines still do this). Overruns often result in discard or loss of information. See cell rate, leaky bucket.

oversampling A process of redundant sampling used in some multiplexing schemes.

overscan, full scan An image on a monitor which extends to the maximum outer extents of the cathode ray tube (CRT). Overscan on computers may be achieved by increasing the resolution of the display or by adjusting position and size controls associated with the display device. Overscan display modes are common in video applications, where the signal is not being optimized for the computer monitor, but for the video recording medium to which it is being output. Overscan may also be a screen option on some systems that are adapted for desktop video and usually adds about 10 to 30 pixels to each edge of the displayable resolution. (Thus, a 320 *x* 480 interlaced image might become 360 *x* 525 in overscan mode, for example.)

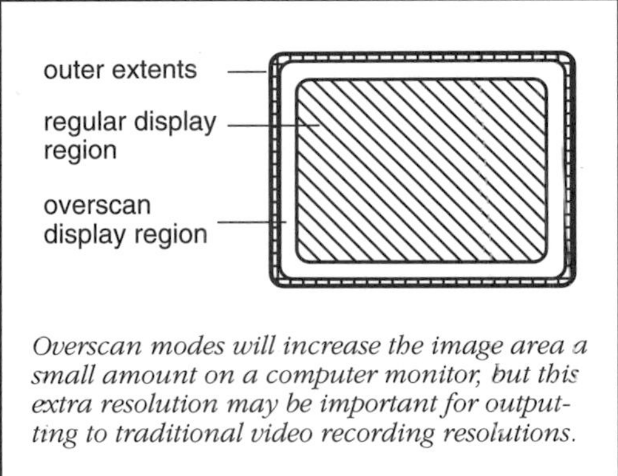

Overscan modes will increase the image area a small amount on a computer monitor, but this extra resolution may be important for outputting to traditional video recording resolutions.

Flat screen monitors are becoming more widely available, but in the past, cathode ray tubes had a significant curvature at the outer edges which would distort the image (like looking through a lens) at the outer edges. In order to minimize distortion, the image is usually not displayed to its fullest extent, but rather to the point on the front of the tube at which the curve begins. The edge of the monitor casing is usually designed by the manufacturer to fall approximately at the same point or slightly outside the point at which the overscan image falls. See cathode ray tube.

overtime period In a pay-per-time-connected service, the time that elapses after the paid-up period has been exceeded. When using a payphone, the time after the first insertion of the coins has run out is overtime, and the operator may request additional funds or terminate the call.

The same general idea applies to per-pay network access, time-sharing, or any other system in which a set amount of time is billed periodically, or is prepaid, with the option for the user to exceed the usage period, as long as additional fees are paid, often at a higher rate.

OWT Operator Work Time.

oxidation The process of combining with oxygen, often resulting in a significant change in the material oxidized that may degrade it or otherwise influence its integrity or usefulness for a particular purpose. Oxidation is a particular concern in external wiring installations or those exposed to water or chemicals. See corrosion.

oxymoron A combination of contradictory, incongruous words. Puns sometimes employ oxymoronic implications which may or may not be true. Satirical examples include: common sense, military intelligence, smart ass, casual dress, friendly fire, and authentic reproduction.

p 1. Symbol for pico-. 2. Abbreviation for power.

P 1. Symbol for permeance. 2. Symbol for peta-.

p region A region in a semiconductor in which the conduction-electron density characteristics result in positive 'holes.' This is usually discussed and utilized in relation to its interaction with n materials in the n region, where conduction-electron density exceeds hole density. See n region, p-n junction.

P-frame predictive-coded frame. In MPEG animations, a picture which has been encoded into a video frame according to *past* frames in the sequence, using predicted motion compensation algorithms. See B-frame, I-frame.

P-picture predictive-coded picture. In MPEG animations, a picture which is to be encoded according to *past* frames in the sequence using predicted motion compensation algorithms. Once it is encoded, it is considered to be a P-frame.

pA Abbreviation for picoampere.

PA See public address system.

PABX See Private Automatic Branch Exchange.

PACA Priority Access and Channel Assignment.

pack To compress characters or data together to conserve space. In the old 4 kilobyte computers from the 1970s that used BASIC as the programming language, 'string packing' was often employed to save precious memory. String packing involved running as many characters and commands together in as brief a way as possible on each numbered line of the source code. Packing is most commonly used now to reduce the size of medium and large binary files, especially sound and image files, to lower the amount of time it takes to transmit the files over a network, as when downloading from an ftp site. Database entries often have a lot of 'empty' space in them and so may be packed to reduce the storage size of files.

Packard, David (1912-1996) Founder, along with William Hewlett, of the Hewlett-Packard computer company, one of the well-respected pioneering companies of the computing industry. The company had its humble beginnings in the Packard garage in Palo Alto, California, and has grown into a multinational company with over 100,000 employees. Packard also cofounded the American Electronics Association, and was a member of the President's Council of Advisors on Science and Technology for four years. In 1996, David Packard, successful businessman and philanthropist, died at the age of 83. See Hewlett, William R.; Hewlett-Packard.

packet 1. A generic term for a unit of data formed as a bundle with a certain specified organization, according to a protocol. Other designations for network units and bundles include *cell* and *frame*. Although packet formats vary, they most typically include a header, an information payload, and a trailer. The header may contain a number of pieces of information, including priority, source, destination, length of packet, etc. The payload is the message or information being sent, and may be split over a number of packets. The trailer may include flags, signals, and error detection or correction data. When a series of related packets is transmitted over a network, they may not all take the same route, and so disassembly, routing, and assembly procedures may be applied to transmitted packets, and instructions to coordinate this process may or may not be included in some of the packets.

Sometimes packet-switched networks are connected to non-packet-switched networks, in which case tunneling takes place, or conversion through a packet assembler/disassembler, to accommodate the differences in formats.

packet assembler/disassembler PAD. In packet-based systems, information is converted into data units known as packets, and then transmitted. At the receiving end, these packets are apprehended and disassembled to turn them back into the information contained in the original content.

packet filtering The evaluation of packet structure or contents in order to selectively reject or accept passage of the packet through a network junction. See firewall.

packet radio Packet radio is a combination of computer equipment and radio transmissions used to exchange messages. Microcomputers and terminal node controllers (TNCs) are commonly used in packet radio systems. The computer is cabled to a radio transceiver at each end of the communication. Because computers have store and forward, or other types of scheduling capabilities, the operator doesn't have to be present when the message is sent or received. In radio, this is called *time-shifted* communications. The system could be configured to send at a time when interference is less likely to be encountered, or when a favorable time of day occurs at the sending or receiving end.

Packet radio transmission speeds are fast enough that various types of propagation can be used, including meteor-scatter. Due to the nature of packet transmissions and its built-in error-correcting mechanisms, packet transmissions are reliable. Packet radio uses a number of protocols and favors the Open Systems Interconnection (OSI) reference model. Common protocols in use include NET/ROM, AX.25, TCP/IP, and ROSE.

packet reservation multiple access PRMA. An enhanced time division multiple access (TDMA), which incorporates aspects of S-ALOHA. It is suitable for mobile transmissions. See ALOHA, time division multiple access.

packet sniffer A diagnostic and snooping mechanism for examining the contents of network packets during transmission. See packet tracing.

packet switched radio See packet radio.

packet switching A computer communications technology developed in the early 1960s that bundles up information into discrete data *packets* which can be sent out in separate paths, like breaking up the cars on a train sending them on separate tracks, and putting them all back together again at the destination. In the 1960s, computing was becoming more accessible, generating greater interest in its use and spurring the manufacture of various types of systems. Practical packet-switched implementations began to appear in the 1970s, and separate server computers to handle various specialized purposes such as accounting, opened the doors to the development of various types of distributed computing architectures.

The rise of ARPANET greatly influenced the development and acceptance of packet switching. With hosts springing up in distant locations and specialization and the variety of computing tasks increasing, packet switching was a practical way to facilitate intercomputer communications.

Gradually layered architectures emerged, separating user functions and applications from lower level operating functions. This enabled information carried in packets to be communicated through many different types of systems, while still retaining the unique operating features and user interfaces of each system.

Historically, telephone networks were built around circuit-switching. This meant that a dedicated path through the switching system had to be established (and was tied up) for the duration of the call. In a large global network where many institutions are online all the time, this is not a practical solution. A better way for large systems is to route information through whatever path is most practical at the time (since some systems may be inaccessible or offline without notice), to divide the packets up, if necessary, if routes change while the data is enroute, and to resend any portions of the message that don't make it through. It works 24 hours a day, and will continue to try to send the data in a dynamically changing environment, even if intermediate hosts or the receiving party are temporarily offline. This essential flexibility is at the heart of packet-switching architectures and is incorporated into huge cooperative systems like the Internet.

See circuit switching, Open Systems Interconnection, Systems Network Architecture, X.25.

packet switching network A communications network in which a channel is occupied only for the time during which the packet, a unit of data, is transmitted, a common distributed data network format. See frame relay.

packet tracing See packet sniffer.

page description language PDL. A means of providing commands to a system for the placement and formatting of page elements, such as text and graphics. Adobe PostScript is widely-used, powerful page description language, and HTML is a very basic page description language extensively used to format information for viewing with a Web browser. Various printers include page description languages which are usually somewhere between PostScript and HTML in complexity.

PageNet A registered trademark service of Paging Network, Inc., one of the largest international wireless messaging companies. PageNet services include a range of one- and two-directional voice, text, and numeric paging services.

pager 1. A general broadcasting loudspeaker connected to a phone or microphone, usually in a business, or educational or health care institution. See public address system. 2. A portable, wireless handheld device which can emit an audible, short verbal message or short alphanumeric message. These are often used by emergency workers, sales representatives, and business professionals. See paging.

paging Alerting a recipient that there is a message or item awaiting his or her attention. Public address systems can be used to page employees or clientele when packages are ready, when there is a phone call, or when lost children or items have been located or turned in. Pagers commonly known as *beepers* are portable wireless devices that will make an audible beeping sound to signal that a message or call is waiting, or that the user has to go to a certain location if paged. Portable wireless alphanumeric pagers can display a short message or telephone number to notify the user of a situation or phone message. Pagers are commonly used by professionals in the field, emergency workers, and industrial yard workers. See public address system, Short Message Service.

paging system PS. A system which allows a message to be broadcast broadly to anyone within range of the speaker, usually to attract the attention of a particular person or party, to give them instructions, or to ask them to pick up a message. Paging systems are common in hospitals, schools, and shopping malls. See public address system.

pair A pair of associated wires, often twisted together to facilitate electrical conductance and/or to reduce noise. Most phone networks are based on decades-old circuits of twisted pair copper wires.

pair assignment The assigning of a specific current, transmission, or function to a twisted pair wire. These are often designated with a code or color, in order to make interconnections quicker and less error-prone.

PAL 1. See phase alternate line. 2. programmable array logic.

palette The colors which collectively comprise the available options for creating an image. In paint pigments, the palette usually includes a number of primary colors, a few difficult to mix intermediary pigments (such as some shades of mauve and brown), and white and black. In computer palettes, several choices are usually available, depending partly upon the number of colors in which the image will be displayed, and also depending upon whether the output will be RGB, CMYK or some other system.

palette board The physical board or virtual menu which holds the available pigments or colors. Thus, a painter's palette is usually some type of wooden or plastic tray, and a computer artist's palette is usually a window, menu, or dialog box. In some software, the term palette is more loosely used to mean any window that comes up with a range of drawing-related options (patterns, line drawing tools, etc.).

Paley, William S. An American experimenter and business tycoon, who purchased and developed the Columbia Phonograph Broadcasting System (1927) into the Columbia Broadcasting System (CBS) in 1928. Under his leadership, the company grew and added new products and services to its line. In 1983 Paley retired from CBS, only to return three years later to work with Lawrence Tisch. In 1995 CBS was bought by Westinghouse.

In 1975, Paley established the Museum of Television and Radio in New York, an educational resource and archive of historical and culturally important broadcasts. The William S. Paley Foundation, Inc. has been established in his honor.

Palo Alto Research Center PARC. One of several Xerox research installations, PARC was founded in 1970 in the Stanford University Industrial Park. It is the site of many remarkable pioneer developments in the field of computers and telecommunications. The PARC was a hotbed in the 1970s for many original developments in object-oriented programming and computer interface design. Both Apple and Microsoft toured the facility in their early days and were inspired by their experiences there, particularly demonstrations of the Alto computer running Smalltalk applications. See Kay, Alan; Smalltalk.

PAM 1. payload assist module. A shuttle satellite deployment mechanism. The satellite in this context is considered the 'payload.' 2. See port adapter module. 3. See pulse amplitude modulation.

panel switch A commercially successful electromechanical telephone switching system developed in the AT&T labs in 1921, based on Lorimer one-step selection concepts. It incorporated mechanical selectors to connect calls.

At the time the panel switch was introduced, independents were widely using the step-by-step switch developed a year earlier. The panel switch technology allowed customers to dial their own calls, albeit with a lot of noise in the early versions. The panel switch was widely used in the United States until the 1950s when it was superseded by the crossbar switch which had been developed in the late 1930s. See crossbar switch, Lorimer switch, rotary switch, step-by-step switch.

panoramic camera A camera configured to take very wide, cinematic shots to provide the same 'feel' that humans experience when scanning the horizon. The first patent for the panoramic camera was granted in 1888 to Canadian inventor, John Connon. Connon's camera could swivel a full 360 degrees in one exposure by advancing the film in synchronization with the movement of the camera, resulting in a long, two-and-a-half-foot, unspliced image. In spite of its innovative aspects and the fidelity of the images it produced, it unfortunately never caught on, with most photographers using fish eye lenses or taking a series of shots and then combining them later.

panoramic receiver A device used in radio communications which provides continuous monitoring of a specified band of frequencies. On a computer monitor, signals are displayed in graph form, with vertical blips moving horizontally along the X axis and amplitude graphed on the Y axis.

Pantone Matching System PMS. A model for standardizing the selection of pigments and reproduction of color printed materials. Pantone colors are used almost universally by printing professionals as one of the means to display, select, and specify the contents of colored inks used on presses. In the last few years, Pantone matching has also been added to many better quality computer desktop publishing and graphics programs. Pantone systems describe both solid (spot) color printing and process color printing for glossy and matte papers. The swatches are invaluable for design work. They should be shielded from light sources, and replaced every year or so.

PAP 1. packet-level procedure. 2. See Public Access Profile.

paper tape An information storage medium. Paper that is designed to have specific areas of the tape encoded and punched or electrostatically recorded onto the tape, for subsequent reading by a paper tape reader, or other interpretive device, such as a computer, stock ticker machine, player piano, or music box. This means of information encoding and storage was used to program early computers and had many characteristics in common with computer punch cards.

Early telegraph receivers used paper tape systems designed by inventors such as Bain and Morse. Later teletypewriter systems used tapes to save transmission time and money by being composed offline and sent only when complete. This also provided a way to correct significant errors before transmission, since a bad tape could always be repunched. Paper tapes have been superseded by tape drives, hard drives, floppy diskettes, magneto-optical discs, cartridges, and memory cards. See Bain, Alexander; Morse, Samuel F. B.

paper tape punch A device designed to receive or interpret coded information and translate it into physical locations on a paper tape and punch them accordingly.

paper tape reader A device which detects and translates the encoded holes in punched paper tape as the tape moves through the machine. The machine may be an interface to a display device, or may be self-contained. Older paper tape readers required that the holes be completely punched out and were usually read by optical means. Later machines could read semi-perforated or *chadless* tape, usually by means of physical sensors. See Hollerith code, paper tape punch, punch card.

papertape, electrostatic A type of chemically sensitive paper tape which is stimulated by electricity such that a mark appears on the paper. Used in early telegraph receivers.

PAR 1. Positive Acknowledgement Retransmit. 2. Precision Approach Radar.

parabola A plane curve that is frequently studied and described in various disciplines including physics, geometry, and art. Parabolic curves are observed in the motion of objects and are used in the manufacture of reflectors and antennas. See parabolic antenna, parabolic reflector.

parabolic antenna An antenna designed with a characteristic parabolic dish shape that captures a directional beam and focuses it through a feed horn. This shape is appropriate for very short, directional transmission waves, such as microwaves, and the diameter of the antenna is designed to correspond with a multiple of the length of the wavelength being received. Parabolic antennas may be made from a variety of materials: solid metal, mesh metal, fiberglass. This style of antenna is commonly used for microwave satellite transmissions. See antenna, feed horn, microwave antenna, low noise amplifier.

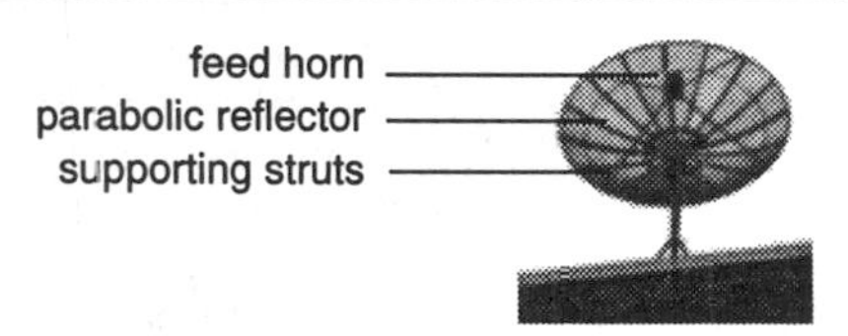

This roof mounted parabolic antenna is about eight feet across and employs a mesh parabola to reduce weight and wind resistance.

parabolic reflector An antenna, or other reflector, which utilizes the characteristics of the shape of a parabola to concentrate and direct reflecting beams. See parabola, parabolic antenna.

paradigm A clear or typical example, a standard, ideal, or archetype.

paradigm shift A fundamental, significant change in the way something is perceived or understood, particularly if it has been taken for granted, or assumed to be 'true' for a long time, or by a majority of the population. In other words, the situation or thing itself has not changed, but our way of understanding it has.

A general paradigm shift occurred when humans, most of whom believed that the Earth was the center of the solar system and even the universe, acknowledged that the Earth revolves around the sun. The discovery that matter at the atomic level (quantum mechanics) did not behave according to accepted models of classical mechanics represented a paradigm shift in physics. Paradigm shifts often take a long time, sometimes decades or centuries (although transition periods are collapsing as education and television become widespread), and those who first propose new ideas and ways of looking at things are often pilloried or persecuted (even beaten to death or hanged) for their assertions. The suggestion that computers could be taught to be 'intelligent,' or to play games intelligently, was met with almost universal contempt in the 1960s and 1970s, but in 1997, a computer beat a grandmaster chess player, an event that adds credence to the argument that intelligent computers could be developed, and may someday surpass humans in specific or generalized intelligence, or develop 'machine intelligence' of which humans are not capable.

parallel port An interface port on a computing system that permits the connection of parallel devices for the simultaneous transfer of data across multiple transmission wires. Most microcomputers are now standardized to 25-pin parallel D connectors, communicating with Centronics-compatible parallel protocols (although there are individual makers who use slight variations of the standard). Due to the increased speed of transmission over serial communications, parallel ports are commonly used for outputting to printers and other types of peripherals like cartridge drives. See serial port.

parallel processing Carrying out two or more tasks, more or less concurrently, usually with the intention of carrying out the processing at a faster speed, or otherwise more efficiently. See concurrent programming.

parameter A property which records, embodies, or determines a characteristic of an object or system. In communications, parameters affect many characteristics such as size, shape, speed, timing intervals, addresses, identities, etc.

parametric amplifier A type of low noise, radio-frequency amplifier which employs high-frequency alternating current (AC) for power. Used with microwave frequency electron beam devices.

parametric design The process of using general parameters, rather than individual measures, to automate computer-aided design and drafting (CAD) and computer-aided manufacturing (CAM). Parametric design incorporates a form of expert system and is particularly valuable in situations where many small variations on a basic design (bolts, boxes, modem covers, PC boards, telephone handsets, etc.) are needed. In these cases, a computer program can be used to automate the design process, with the parameters given, to turn out the many variations thousands of times faster than a CAD operator could draw each one by hand. One of the early patents for an applied parametric design computer program was awarded to Synthesis, in Washington State, in the 1980s. It is available online, along with other patents since the mid-1970s on the U.S. Patent and Trademark Web site. See CAD, expert system. http://patents.uspto.gov/

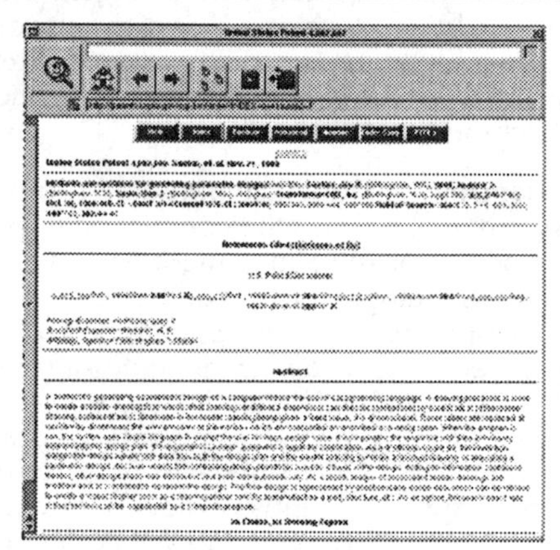

The U.S. Patent and Trademark Office is endeavoring to make patent and trademark information available online to the public. Patent abstracts and descriptions are currently available, and diagrams are scheduled to come online following the text information.

parametric equalizer A component device used in sound systems to selectively manipulate selected frequencies in order to adjust the sound, usually to suit the taste of the listener.

parasite An organism or process which feeds off another without providing a return. In technology, the term can refer to a process, or a mechanical or electrical device that monitors or uses transmissions clandestinely, or without the usual compensation to the provider of the transmission. Small wire-tap devices are sometimes called parasites, especially if they draw their power from the line being tapped.

PARC, Xerox PARC See Palo Alto Research Center.

parity Equality, state of being the same, equivalent, matching.

parity bit A bit which is included in a transmission for error checking or status purposes. In telecommunications over a modem, most protocols allow the use of a parity bit appended to a data stream of a specified length, the parity bit set to zero or one, depending upon the preceding data. Parity values calculated and stored as the sent bits are checked against parity values calculated from the received bits. See parity checking.

parity checking A simple means of checking data integrity after a transmission by comparing the calculated value of the parity at the receiving end with the value calculated and stored at the sending end. Parity checking is very commonly used in file transfer through modems over phone lines.

First the transmitting and receiving ends negotiate a common protocol, for example, ZModem, then the parity setting is selected as odd or even (or none). Assume a parity setting of even for this example. Parity is calculated prior to sending, by tallying the ones or zeros which are in a group of bits (usually seven), and then assigning a parity value of *zero* if there is an even number of one bits, and a parity value of *one* if there is an odd number of one bits, (or the converse, by looking at zero bits for odd parity). The sender transmits the data and its associated parity bit. The receiver calculates the parity of the received bits and checks to see if there is a match with the transmitted parity bit. If not, there is a problem.

The system is not foolproof; a match does not guarantee that the data was correctly transmitted, as the parity bit itself may have become altered along with the data, but there are mechanisms in most software to evaluate the frequency of parity errors, so that the user may be alerted and the transmission aborted, restarted from an earlier point or resumed later, depending upon the protocol.

park drive In hard drives, 'parking' the drive is a means to secure any moving mechanisms which may be damaged by being jiggled in transit. Some hard drives park automatically when not in use, and some use mechanisms which do not cause damage if the unit is transported (e.g., drives in laptops). Older drives were often equipped with software-parking, and it was quite important to run the software command to park the drive before moving the system or removing the drive. This system is now uncommon. Mobile computers are equipped with self-parking drives.

park phone In telephony, parking is the process of putting a line through to a particular phone so that it can be picked up at another station, or to put a line on 'soft hold' so the conversation can be continued from another phone.

park timeout In telephony, a time limit on a parked line after which it hangs up the line if the call is not resumed on another line (or the same line).

Parkinson's law C. Northcote Parkinson wrote in the 1950s that work expands to fill the time available for its completion. (For those who are perfectionists, and believe that if a job is worth doing, it's worth doing right, this is doubly true.)

partition Subset, class, section, or division.

partition, drive On hard drives, a usually-contiguous section of a disk which is individually initialized and handled by the operating system as a distinct unit. Some systems can format the individual partitions in a variety of formats, i.e., you could have a 1 Gigabyte hard drive with a NeXTStep 400 megabyte volume on one partition, a 400 megabyte Linux volume on another, and a 200 megabyte Macintosh volume on a third, all recognized by the OS and readable/writable without any unusual technical expertise or demands on the user.

On many microcomputer operating systems, disk volumes and files cannot cross partitions, but many Unix and workstation operating systems can handle volumes that cross partitions transparently to the user, e.g., two 500 megabyte hard drives used together might appear to the user as a 1 Gigabyte virtual drive. There are many schools of thought as to whether a hard drive needs to be partitioned. A few operating systems can only handle up to four partitions, each with up to 2 Gigabytes of space and, consequently, a larger hard drive must be sectioned into smaller pieces in order to be handled by the operating system. Others don't have this limitation on the number of partitions, and can manage larger-sized partitions. In terms of disk management, in the case of problems, it may be easier to rebuild partitions or handle data recovery procedures, if there are several partitions rather than just one. Redundant array drives are another way of handling error recovery. Often a small 200 to 500 megabyte partition will be set aside as a 'swap drive' and not used for other purposes. See RAID.

partition, memory In computer memory, a linked or contiguous section, separate from other sections, that is allocated for a specific purpose or process, such as video display or frame buffering.

partly perforated tape See chadless tape, paper tape.

party 1. One of the individuals in a transaction. A common legal term used to stipulate an individual or organizational entity. To be *party to* a transaction is to listen in or participate. In telecommunications, the transaction might be a telephone call, a conversation, or a computer communication. 2. Two or more individuals engaged in a group activity, especially a social activity (e.g., celebrating after completing a very long writing project).

party line In telephony, a line that is shared by two or more subscribers, so if one or more subscribers pick up the line and listen when someone else is engaged in the call, they can hear the conversation, and can't make further calls until the current conversation is disconnected. Party lines were very common on older shared phone circuits until the 1960s; they are now uncommon in North America. On ISDN lines and frame relay networks, a sort of party line system exists, but is rarely a hindrance to the user, unless too many subscribers are assigned to the line.

party line, following Following the *party line* is a phrase from politics that indicates acceptance and promotion of the administration's point of view. The administration might be a political party, a business entity, or other institution. It is sometimes used as a derogatory phrase for ambitious compliance, or for a person who doesn't think for him or herself, but promotes the current popular point of view.

pascal An SI unit of pressure equal to one newton per square meter.

Pascal A programming language descended from ALGOL, developed by Niklaus Wirth in 1970. Pascal became especially popular in the 1980s for teaching programming concepts and techniques. A structured, typed language, Pascal is somewhat similar to Modula II, and fits somewhere between C and higher level languages like BASIC and FORTRAN. It is less cryptic than C, but also less preferred by programmers in commercial development environments, yet is generally preferred over the less structured BASIC in educational environments. See Modula II, C.

Pascal, Blaise (1623-1662) A talented French inventor and mathematician, Pascal devised one of the earliest calculators, a "Pascaline", while still in his teens. It was a numerical base ten, movable dial, wheel calculator designed to assist his father in carrying out his duties as a tax collector. Pascal appears to have come up with the design independently, and probably was not aware of the earlier

P

calculator developed by Schickard at about the time of Pascal's birth. Pascal also did research in fluid dynamics. See Schickard, Wilhelm.

pass through *v.* To move through a component device or leg of a network without significantly altering the characteristics of that which has just been passed through, or without being altered by that which is passed through. See passthrough device, tunneling.

passband The range of transmissions frequencies which can pass through a filter without a significant decrease in amplitude (attenuation). A passband filter allows selective screening out of irrelevant or undesired frequencies in order to create a device for a specific purpose, or to simplify its operation.

passband A signal which poses no spectral energy at direct currents (DC), unlike a baseband signal. A Manchestor-encoded signal is one example of a passband signal.

passthrough device 1. A device which is chained between two other devices, and which passes data through without changing it. For example, an external memory module might be attached to a computer, with an external hard drive attached to the memory module. The memory module passes through the hard drive signals in such a way that the hard drive works just as though it were directly attached to the computer. See daisy chain. 2. A device which provides access to another and passes back the signals transmitted by the other. Sometimes used as a diagnostic tool.

password A word or combination of characters which, when provided by a person or entity wishing to gain entry to a system or situation, is checked against certain characteristics, or a list of those who are authorized to have access. If a match is found, entry is permitted. Password protection systems are very common on computers and networks. It is very unwise to tape passwords to monitors or desks where anyone can see them. It is also unwise to use common words as passwords; a moderately long password with a combination of letters and symbols is safer. See anonymous FTP, back door, back porch.

patch *v.* To connect one circuit with another, usually through an intermediate line. For example, on old telephone switchboards, the operator would *patch* through a call by taking a jack that was connected at the other end to the main switchboard, and plug it into the phone receptacle for the individual getting the call. A patch is a temporary connection, one subject to frequent change, or which is being used for diagnostic purposes.

patch, software *n.* In software, a patch is a piece of code that is inserted into the original code to override some of the original programming, or to add capabilities or data which weren't in the original code, and perhaps should have been. A patch is distinguished from an upgrade in that it typically is intended to correct oversights or errors, whereas an upgrade is usually of greater scope, intended to enhance or extend the capabilities of the program. In many products, the two are combined.

patch, sound In electronic music, a sampled segment of sound stored digitally. The sound is measured and recorded, that is, 'quantized,' at rapid intervals in order to create a digital impression of the analog sound wave. For the most part, the more frequent the sampling, up to the limits of human perception, the more true to the original the sample tends to sound (the capabilities of the playback mechanism contribute as well). Sound patches can be generated by, and used with many commercial sound synthesizers and computer synthesizer software. MIDI is a common protocol used in the music industry for communicating digitized sound between MIDI-compatible instruments and software programs. Speech and music sound patches are often used to enhance multimedia CD-ROM educational and entertainment products. More recently messages composed from speech patches are becoming common on the Web. See quantize, sampling.

patch bay, patch board A hardware panel designed with connections which can be easily changed. In other words, it is set up so that temporary circuits, or those which are frequently changed, can easily be rewired. Patch bays are often equipped with wheels, and usually have receptacles or terminals for easy insertion and removal of patch cords and/or wires. Patch boards are useful for prototyping, monitoring, and testing new circuit layouts. See patch panel.

patch cord A short length of wire or cable used to connect circuits. The connectors at either end vary, but are often RCA jacks or BNC connectors. Patch cords are commonly used with patch bays, patch panels, and electronics components. Videographers and musicians often refer to video and audio connecting cables as patch cords, since video equipment connections are frequently reconfigured.

patch panel A hardware device, often wall-mounted, that facilitates the connection and reconfiguration of temporary circuits. A patch panel may resemble a distribution frame, in that it has a grid of openings or connectors through which circuits can be routed. It commonly has mounted receptacles to match the types of jacks used in that particular circuit.

patent A registration process formally established in the United States in April 1790 which provides a record of the ownership, development, and date and method of creation of unique products and processes. The first American patent was granted on 31 July 1790. By 1802, applications had increased to the point where a separate Patent Office was set up, and more rigorous scrutiny was established by 1836.

In the United States, the documents are processed and stored in a central government repository that is open to the public and which is intended to further technological progress by the encouragement of the dissemination of ideas. Japanese patents have been available over networks for some time now, and recent U.S. patents are now searchable on the Web through the U.S. Patent and Trademark Office site. The Clinton Administration announced on 25 June 1998 that over 20 million pages of patent and trademark information will be provided free to the public on the Internet by year's end. This is being supplied through the Commerce Department's large database of text and images. The collection will include the full text of two million patents dating from 1976, 800,000 trademarks, and 300,000 pending registrations dating from the 1800s. Images will follow, with low and medium resolution images available for printing with Web browsers, and high quality print copies available by order.

Patent applications must follow very specific format and content guidelines laid out by the patent office. Patent registration grants exclusive intellectual and certain commercialization protections to the inventor for a term of 17 years in the U.S. (international patents are similar). In cases of others coming up with the same idea simultaneously or previously, without knowledge that the idea has been patented, preference for the idea now goes to the inventor who first is granted the patent. This is a change from historical procedures in which an earlier inventor, if she or he had documents to prove the case, could have a patent from a later inventor overturned.

Many people incorrectly think that the patent process exists to explicitly prevent others from infringing on patents, but it is the responsibility of the patent owners, not the patent office, to police the use and abuse of patented ideas. The patent does, however, define the nature and extent of the legal protection available to the inventor through the justice system. Granting of a patent does not include granting of a right to manufacture a product incorporating the idea, since other patents for other aspects of the invention may exist.

The most important aspect of the patent and the submission of patent applications is the *Claims* section, in which the inventor lays out, in point form, the characteristics which make the invention *unique* and *nonobvious*. Some or all of these claims may be accepted by the patent office, and the document is critiqued and rejected or returned to the applicant for revisions. Since uniqueness is often evaluated in a historical context, the *Prior Art* section, in which the historical antecedents and current similar inventions, must be described by the applicant thoroughly and succinctly. The invention must also be more than a half-baked idea, since the patent application must include a clear description of how to build or otherwise recreate the invention itself, without undue difficulty to a layperson or someone appropriately skilled in the area of specialization appropriate to a specialized invention.

Hardware patents usually fall under the products category and software patents under the process category. Note that patents, copyrights, and other legal registration procedures may grant ownership to the *employer* of the invention rather than the inventor, if the employee undertook the invention in the course of his or her normal work hours or duties.

One of the most famous patent clerks in history was Albert Einstein, who worked as a junior clerk in the Swiss Patent Office when he was unable to find work as a teacher or research scientist. While employed there, he wrote some of his most startling and insightful treatises on relativity. See copyright, trademark. http://www.uspto.gov/

Paterson, Tim Paterson developed a historic disk operating system for Seattle Computer Products in the late 1970s. The product was derived from Gary Kildall's CP/M operating system, which was the most successful and well known at the time, with over 1/2 million copies distributed. Paterson created a basic operating system called QDOS (Quick and Dirty Operating System) which he has stated was derived in part from the program interface described in a CPM manual from the mid-1970s. Microsoft bought it, fixed it up a little, and provided it to IBM soon after. IBM released it initially as PC DOS 1.0. Meanwhile, Seattle Computer Products retained the 'rights' to QDOS. Microsoft subsequently bought out all QDOS distribution rights for $50,000. The Microsoft financial empire essentially sprung from this transaction as the product was developed into MS-DOS and, eventually, after many facelifts and enhancements, evolved into Windows. See Digital Research; Kildall, Gary; Microsoft Corporation.

path A route, track, directional identifier, runway, conduit, or other end-to-end, hop-to-hop, or as-you-go means of delineating the track followed by a person, process, transmission, or data unit while traveling from one point, node, or endpoint to another. A file path is one which indicates the hierarchical organization and location of a specific file or grouping of files. A transmissions path is the specific or general direction of radiant energy travel.

path information unit PIU. In packet networking, a message unit which consists of a transmission header (TM) or a transmission header combined with a following basic information unit (BIU) or segment. See datagram.

Path Terminating Element See SONET path terminating element.

pattern matching, pattern recognition The process of comparing text, symbols, images, or other elements to determine whether they are the same, similar, or mathematically equal. The process of pattern matching is widely used in database search and analysis mechanisms, and its cousin, pattern recognition, is common to artificial intelligence applications including expert systems, robotics, and others.

Pattern recognition was in its infancy in the late 1950s and early 1960s, when computing systems were expensive, cumbersome, and programmed with punch cards. Nevertheless, early researchers at the time, sensing its potential, developed equipment and algorithms which could read a few handwritten letters, if they were plainly written. See Perceptrons.

pay phone, pay telephone See payphone.

payload The user information, and sometimes accounting and network administration information, carried in the upper layers in a layered architecture, within a cell, frame, packet, or other network data transmission unit. Separate from, but associated with the payload, there is frequently signaling, header, error checking, and other data which relate more to the type and manner of transmission than to the information content from the user or process sending the transmission.

Coin payphones have in many ways remained unchanged from their historical antecedents. Most now employ a single slot, and credit card/ calling card slots are being added as the phones are gradually replaced.

payphone, paystation phone Any self-contained public or private telephone unit which requires a per-call or per minute fee, usually directly transacted with the phone, although some human-operated stations exist. The first pay telephones were attended by operators who collected the fees for the calls.

payphone history One of the early coin box patents was issued in 1885, and William Gray installed a public coin phone in Connecticut in 1889. He later founded the Gray Telephone Pay Station Company with George A. Long, and together they filed additional patents for payphone technology, establishing a strong hold on the market.

The first payphones were actually human-attended *pay stations*. The Social Telegraphy Association charged 15 cents per call in 1878. In 1880, pay stations began to be established in New York, and a pay station was announced by the Connecticut Telephone Company for 10 cents per call, which is interesting since the reduction in price of the technology and wider distribution of payphones over the next century meant that calls in the 1960s were still only 10 cents each. In Europe, payphones were beginning to appear at the turn of the century.

Since the early attended stations were not as profitable as automated stations, there was a lot of economic incentive to find an automated way to collect the money. This also made it easier to provide service 24 hours a day.

While there were many incremental improvements, like the change from three slots to one in the mid-1960s, payphones did not change significantly until the 1980s, when calling card and credit card mechanisms were added to selected stations. By the 1990s, in the United States, Canada, and Western Europe, card-reading payphones were common.

payphone postpay Payphone calls which are paid after completion, usually with a calling card or credit card.

payphone, private Also known as COCOT, this is a customer-owned coin-operated phone, as might be found in a hotel lobby or tavern. COCOTs may provide only limited access to long-distance carriers.

PBX Private Branch Exchange. See Private Automatic Branch Exchange.

PC 1. personal computer. While many people use PC to specifically refer to International Business Machines (IBM) and third-party licensed hardware, in fact, PC correctly refers not only to Intel/IBM computers, but to any personal computer or microcomputer priced in a consumer or small business price range. 2. printed circuit. See printed circuit board. 3. program counter. 4. protocol control.

PC card See PCMCIA card.

PC Pursuit A commercial Telnet dialup service that was very popular in the 1980s for accessing Internet Service Providers in distant cities. PC Pursuit allowed users to connect with any PC Pursuit-supported city for a low monthly charge, and dial out a call from there to anywhere in the local dialing area.

PC, IBM/Intel *colloq.* In general and marketing terms, PC is understood as a subset of personal computers, consisting of Intel-based International Business Machines hardware, or third-party licensed hardware, running the IBM OS/2 software, or, more commonly running the Microsoft Windows graphical operating environment in conjunction with MS-DOS. For information on specific IBM desktop computers, see the listings under IBM Personal Computers.

PC-AT See IBM Personal Computer AT.

PC-XT See IBM Personal Computer XT.

PCA 1. point of closest approach. In a satellite communications system, a point on a segment of the orbit or ground track when the satellite is closest to a specific ground station. 2. Premises Cabling Association. 3. protective connecting arrangement. Commercial connecting rental agreement, required by AT&T/Bell prior to divestiture for telecommunications devices which where not AT&T/Bell, that were connected to the AT&T/Bell system. See Carterfone decision.

PCB 1. power control box. 2. process control block. 3. See printed circuit board. 4. protocol control block.

PCCA See Portable Computer and Communications Association.

PCI 1. See Peripheral Connect Interface. 2. Protocol Control Information.

PCIA Personal Communications Industry Association. Formerly known as Telocator, PCIA is a national association representing the mobile communications industry.

PCjr See IBM PCjr.

PCL See printer control language.

PCM See pulse code modulation.

PCMCIA Personal Computer Memory Card Interface Association. A professional association of electronics peripherals and semiconductor manufacturers and software engineers. See PCMCIA card, PCMCIA standards.

PCMCIA card, PC card A standardized computer peripheral card format, which is not much bigger than a fat wallet card which is commonly used in portable computing applications. PCMCIA cards (since the mid-1990s called PC cards because it's easier to say) are microminiaturized devices with a thin edge connector, including memory cards, hard drive cards, fax/modem cards, network interface hookups, and more. They are used in radio phones, laptop and palmtop computers, digital cameras and camcorders, and various other portable electronic devices. The most common cards are called Type I or Type II (Type III is less common, and Type IV is vendor-specific). Most laptop peripherals use Type II cards, and it pays to have at least one Type II slot on a portable computer. Hard drives and radio devices tend to use the thicker Type III cards.

PCMCIA standards A set of 8-bit bus standards which bares the same name as the organization which developed the standards, the Personal Computer Memory Card Interface

Association. PCMCIA standards were developed and tested in the late 1980s, and released for general use in 1991. While there is fairly good adherence to the standards, compatibility is not absolute. It's advisable to try cards before buying them, or to get them with a good return policy. The set of standards includes Type I, Type II, Type III, and Type IV. See PCMCIA card.

PCMIA Personal Computer Manufacturer Interface Adaptor.

PCN See Personal Communication Network.

PCP 1. Private Carrier Paging.

PCR 1. See peak cell rate. 2. Program Clock Reference. In MPEG-2 encoding, a timestamp inserted into the transport stream to enable recovery.

PCS 1. See Personal Communications Service. 2. personal communications software.

PDA See Personal Digital Assistant.

PDF See Portal Document Format.

PDL See page description language.

PDP 1. See plasma display panel. 2. See power distribution panel.

PDU Protocol Data Unit. A unit of datum consisting of control information and user data which is exchanged between peer layers. See asynchronous transfer mode, LLC, RFC 1042.

PDUS Primary Data User Station. The combination of a ground station and a satellite image processing system.

PE processing element.

peak cell rate PCR. In ATM networking, a traffic flow measure that describes the upper cell rate limit, which may not be exceeded by the sender. See cell rate.

peek/poke Higher level language instructions, generally found in BASIC, which allow the programmer to examine (peek) the contents of a specified area of memory or insert (poke) data into a specified area of memory. Often used by BASIC programmers to manipulated data directly on early microcomputers to speed up interpreted BASIC programs, or to carry out operations which were not directly supported by the BASIC instruction set. Peek is a safe operation, but poke should always be handled with deference and care, since *poking* information directly into memory may interfere with subsequent running of the software, or may immediately overwrite something crucial and crash the system.

peer entities In layer-oriented network models, entities within the same layer, usually diagramed and visualized as horizontally related.

peer model A networking model built with the assumption that internetwork layer addresses can be mapped onto ATM addresses and vice versa, and reachability information between ATM routing and internetwork layer routing can be exchanged. See integrated model.

peering The voluntary exchange of routing announcements in order to effectively establish data paths among providers.

PEG Regulated Public, Educational or Government access. See cable access.

Pel See picture element.

pendulum A suspended object which, when stimulated into an unimpeded swing, will oscillate in regular, predictable patterns related to the plane of the oscillation and the movement of the Earth.

The most celebrated experiments with pendulums are those of Foucault in the 1800s in which he studied the behavior of large pendulums on very long lines, over 200 ft. long. These were suspended so as to inscribe a pattern in sand at the lowest point of the pendulum. As Foucault's immense pendulum moved, the pattern changed, becoming the first tangible rather than theoretical evidence of the movement of the Earth.

penetration Gaining access to a system, circuit, facility, or operation, usually for security reasons or unlawful access. Physical penetration of circuits or networks can be done through means of taps or black boxes. Logical penetration can be done through password-guessing, Trojan horses, viruses, and back doors. Bodily penetration can be done through overriding electronic security measures, entering as an impostor, or using insider privileged access in an unethical manner. See back door, Trojan horse.

penetration tap 1. Any means by which a conductor is accessed by piercing the outer layers of shielding and grounding and connecting to the current circuit, with the intention of not disrupting current transmissions. 2. A network connection technique which enables devices to be attached to the network cable without interrupting current network operation. The tap is carried out with a sharp tool which can pierce the outer and inner ground shielding of the network cable, such as a coaxial cable commonly used in Ethernet implementations.

penetration testing Testing a system for the integrity of its security. This is sometimes done by internal staff, by the contractors installing the security measures, or by outside experts hired to try to penetrate the system. In the telephone and computing worlds, known 'hackers' are sometimes hired to try to penetrate a system to try to identify security holes before the system is opened up to employees, or to the public, depending upon its nature. In 1998 it was found that cash cards, which were generally considered reasonably safe from decryption and unauthorized use, could be penetrated by measuring their electrical emanations and properties, a finding that calls into question the use of cash cards in place of traditional means of currency exchange.

Pentium An Intel Corporation 80586-based central processing unit (CPU), designed to succeed the 80486, introduced in 1993. Originally released at 66 MHz clock speed, other versions came out, including a 100 MHz version with a 16-bit cache and a 64-bit memory interface, and eight 32-bit general-purpose processing registers. The name is derived from the '5' in the processor line 80x86 due to a court ruling that a number cannot be trademarked.

Pentium II An Intel Corporation central processing unit (CPU) similar to the Pentium Pro. Unlike the Pentium Pro, which incorporates the level 2 (L2) cache into the chip with the CPU, the Pentium II operates with a cache inserted in a slot on the motherboard, thereby increasing the amount of time it takes for the two to communicate. It also incorporates MMX circuitry intended to improve graphics and multimedia-related operations.

Pentium MMX Pentium Multimedia Extension, Pentium Matrix Math Extension. The MMX is essentially a Pentium Pro chip enhanced with a number of new data types and floating point instructions that enhance computing-intensive operations such as graphics. Since applications are becoming increasingly visual in nature, with more graphical user interfaces, image processing, rendering and raytracing, videoconferencing, realtime games, and virtual reality applications, support for commonly executed graphics and math-intensive computing processes on the chip is intended to support these growing areas of interest. Also, by incorporating capabilities similar to those supplied by direct memory access (DSP), Intel can reduce its reliance on the DSP technologies of other vendors

The Pentium MMX incorporates what Intel calls Single Instruction, Multiple Data (SIMD) techniques, to allow several processes to be carried out with a single instruction. See Pentium II, reduced instruction set computing.

Pentium Pro, P6 An Intel Corporation 80686 central processing unit (CPU) in the Pentium brand name line, introduced in 1995 as a successor to the Pentium processor. The Pentium Pro originally shipped as a 133-MHz CPU and shares a number of commonalities with the Pentium, including a 64-bit memory interface. It is a two-part chip in the sense that it has a CPU and a level 1 memory cache, plus a level 2 (L2) memory cache layered into the CPU rather than residing separately on the motherboard. It is a hybrid chip with an underlying RISC structure, but also includes a CISC-RISC translator for downward compatibility. The clock speed of the first version was 133-MHz, with other versions following.

PEP 1. Packetized Ensemble Protocol. A high speed, proprietary, full duplex transmission protocol from Telebit. It has error-correcting mechanisms and is said to handle line noise well. It is no longer in general use. 2. See Public Exchange Point.

Percent Local Usage PLU. A measure of telecommunications usage by time. PLU is a ratio of the local minutes to the sum of local and intraLATA long-distance minutes between exchange carriers, sent over Local Interconnection Trunks. Switched access and transiting calls are not included.

Perceptrons Self-organizing, pattern recognition systems built in the early 1960s at Cornell University. These systems were rudimentary, barely managing to recognize simple letters, yet studies and experiments of this kind led to the optical character recognition and handwriting recognition systems we now take for granted.

At the same time, at the Massachusetts Institute of Technology (MIT), researchers were developing pattern-matching systems for medical diagnosis, with a system designed to screen for cancer cells through a microscope. See neural networks, pattern matching.

Peregrinas, Peter An earlier experimenter in magnetism, Peregrinas designated the magnetic *poles*. He was a student of Roger Bacon, an English scientist.

perforator A tool to make a hole, to penetrate a substance, to punch an opening. Common three-hole punches are perforators. Electronic perforators are widely used to turn electronic signals into code pattern holes in punch cards and paper tapes. See chad, Hollerith card, punch card.

performance category See category of performance.

perigee The point in an orbit nearest the gravitational center of the body being orbited. See apogee, periapsis.

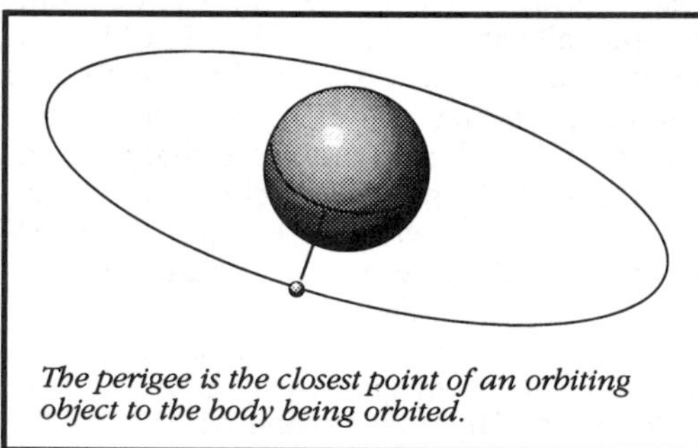

The perigee is the closest point of an orbiting object to the body being orbited.

period 1. Cycle, interval of time, portion of time encompassing a distinct culture (historical period). 2. Geologic time division that is part of an era, and longer than an epoch. 3. The time interval between two consecutive orbits of a satellite through a specific point (usually the perigee) in the orbit. 4. In electronics, one interval of a regular, repeating event.

periodic postings Most public discussion lists on the Internet have a number of areas of general interest, guidelines for postings, and frequently answered questions that are collected and organized by newsgroup and discussion list administrators or volunteers. These are disseminated as periodic postings to all readers in order to bring new participants up to speed quickly and to keep repeated postings of common questions to a minimum.

Peripheral Component Interconnect, Peripheral Connect Interface PCI. A very popular local bus standard developed by Intel in the early 1990s which supports 32/64 bit data and was compatible with the new Pentium processors coming out at the time. It was designed with a newer chipset, to improve on the ISAs and VLBs that were then common, and to include bus mastering, use of the system bus. Since PCI's development, PCI slots have become common in Apple Macintosh and Intel-based IBM-licensed machines, along with upgraded versions of the VESA VL bus. The PCI Mezzanine Card (IEEE P1386.1) was designed to work with the PCI specification.

peripheral device 1. A piece of equipment which is not a main component of a system, but which, when connected to that system, enhances its functionality, speed, or storage capabilities. Peripheral devices generally cannot perform useful functions unless connected to the main system. Monitors, speakers, keyboards, scanners, printers, etc. are examples of peripheral devices. CD-ROM drives are an exception in that some are now designed to play audio CDs even if not connected to a computer. 2. In the telephone industry peripherals may also be called outboard processors, applications processors, or adjunct processors.

Peripheral Technology Group An international commercial distributor of computer connectivity products.

Perl Practical extraction and reporting language. A powerful, flexible, general-purpose, interpreted scripting language (originally spelled with a lower-case "p") developed by Larry Wall in 1986, and now extensively used for platform-independent scripting on multiple platforms on the Internet. The syntactical structure of Perl is quite remarkable (perhaps owing to Larry Wall's expertise as a linguist), and useful, powerful routines can be written in a few lines or sometimes even in a few characters. An important tool for shell scripting, Common Gateway Interface (CGI) development, and much more. When combined with Penguin, it may be a serious contender with Java for object-oriented, Web-related interface design and Automation. The Perl Journal gives practical assistance to Perl programmers.

Permanent Number Portability PNP. A means by which a telecommunications subscriber can transparently maintain an existing telephone number, even if changing to a different service provider in the same locality.

permanent virtual connection, permanent virtual circuit PVC. A logical communications channel, which may differ from the physical topology over which it is laid, which is established with the intent that it stay the same for some time. In an ATM environment, there are two types of PVCs: permanent virtual path connections (PVPCs) and permanent virtual channel connections (PVCCs). PVCs provide manually-configured connections between end systems. The addressing information, Virtual Path Identifier/Virtual Channel Identifier (VPI/VCI), must be put into both devices for connectivity. 2. In frame relay networks, a PVC is a logical link, with network management-defined endpoints and Class of Service (CoS). The link consists of: an originating element address and data link control identifier and a terminating element address and termination data link control identifier. See RFC 1577, VC, SVC, SPVC.

permeability The porosity or penetrability of a substance. The degree to which liquids or gases can pass through a substance. Contrast with reluctance.

permeability, magnetic The property of a magnetizable material that determines the degree to which it will modify the magnetic flux in a region it occupies within a magnetic field. See magnetic field.

persistence 1. Perseverance, endurance, running the course, keeping on or with, tending to continue. What it takes to write a reference of this magnitude. The telecommunications industry changes and grows so quickly, writing a telecommunications reference is like trying to gas up a car that is driving away at full speed. 2. The tendency to continue a signal, echo, electrical charge, or data transmission, after the actual communication has ceased or the message part has been received.

persistence of vision A phrase that describes the way in which human visual perception 'holds' an image for a brief moment, about a tenth of a second, even if the objects in the visual field have changed or moved. Thus, humans can only scan or perceive still images up to a speed of about 24 to 60 frames per second. Faster than that and they are no longer seen as still images, but as a series of moving or related images, especially if the forms in the images are closely related to the previous ones. Researchers Muensterberg and Wertheimer demonstrated in the early 1900s that this was a property of brain processing and perception more than a physical property of the retina. These characteristics of visual perception have greatly influenced the design and development of moving visual communications technologies. See frame, scan lines.

Personal Communication Network PCN. See Global System for Mobile Communications for the background and technology base for PCN. PCN was developed, starting in the late 1980s, as a modified form of GSM operating in the 1800-MHz frequency band (GSM is 900 MHz). It has smaller cell sizes, requires lower power, and is optimized to handle higher density traffic than GSM, but otherwise is essentially the same. The PCN standard was finalized in 1991. It is primarily used in the United Kingdom. See Global System for Mobile Communications.

Personal Communications Service. PCS. A low-power, higher frequency, standards-based, wireless mobile communications system, operating in the 1800- and 1900-MHz range, implemented in the mid-1990s. Most PCS systems are 100% digital. In contrast to cellular, which is limited to A and B carriers, PCS operates across six (A to F). In other words, cellular can be thought of as a subset of PCS in its broadest sense.

Three operational categories of PCS have been defined by the Federal Communication Commission (FCC) as shown in the PCS Categories chart.

Category	Notes
narrowband PCS	PCS operating in a limited bandwidth in the 900-MHz spectrum and not suited to high speed data communications, although low-bandwidth short text messages would work. Best suited to in-building and near outside premises use, pagers, and cordless phones.
broadband PCS	PCS in the 1.9-GHz spectrum range for better quality voice communications and higher duplex-mode data communications.
unlicensed PCS	PCS in the 1910- to 1930-MHz range, suitable for in-house and in-company systems, and small independent service providers. Limited to low power signals.

In PCS, particular channels are assigned to specific cells, with provision for reuse. A channel is associated with one uplink and one downlink frequency. A specific number of channels is assigned to an operator's authorized frequency block. PCS service can be installed as a centralized or distributed architecture, and supports both *time* and *code division multiple access* (TDMA, CDMA). Designed to broaden market distribution of wireless services, the system may have more limited range than traditional cellular, but the cheaper connect times and handsets may be appealing to consumers. Industry watchers are predicting there may be up to 15 million subscribers by the year 2000. See cellular phone, DCS, GSM, AMPS, DAMPS.

personal computer PC. A compact, relatively low-cost computer system designed for home, school, small business, and prosumer (high-end consumer) use. The first PC, in the sense that we know it, with a keyboard and CRT monitor, was probably the SPHERE computer released in 1975, but it didn't sell well. Subsequently, the Radio Shack TRS-80 series, followed closely by the Apple computers and the Commodore PET were all commercially successful.

At the time of the introduction of personal computers in the mid- and late-1970s, the cost of a workstation-level computer was typically $40,000 and more, so the price tag of about $2,000 - $6,000 for a personal computer with useful peripherals (printer, modem, etc.) was revolutionary in terms of availability to individuals. In the early 1980s, when networks that could interconnect individual PCs began to proliferate and CPUs became more powerful, the distinction between personal computers and higher end systems began to blur, a progression that continues to this day, with personal computers of the 1990s being more powerful than minicomputers of a decade ago. The development of PC networks also opened up hybrid systems, with PCs sharing the computing power of mainframes and mainframes using PCs as I/O devices.

The term PC has been generically applied to systems used by individuals for personal, educational, and business purposes, and so does not fit the term 'personal' in its strictest sense. Some people use PC to refer only to IBM-compatibles, which is not really a correct use of the term, and has probably proliferated because "IBM-compatible" is such a mouthful. The distinction between a PC and a workstation is not as cut and dry as many people think. By the time you add a graphics card, sound card, CD-ROM drive, more memory, and network interface card to a personal computer, its cost is comparable to many off-the-shelf workstation-level computers. See listings for individual personal computers. See workstation.

Personal Digital Assistant PDA. A handheld computerized device that is optimized for common time-scheduling and note-taking activities that many business and personal users particularly desire. These include calendars, account keepers, note-takers, calculators, alarm signals, modem connections, databases, etc. Some PDAs support handwriting recognition through a penlike interface, others have small text keypad input screens, and some have both. PDAs were introduced in the late 1980s, with pen-recognition PDAs coming out in the early 1990s. Most PDAs work on batteries or AC power with a converter. Some work only with batteries. Battery life ranges from two to five hours on most systems, depending upon usage.

Apple ClockWorker is an interesting evolution in PDA technology. This little 300 MHz RISC chip with 30 MBytes of RAM and 70 Mbyte memory chip outruns many full-sized desktop computers. Even more surprising is that it is powered by a clockwork mechanism developed in the U.K. Twelve turns of the AppleKey are said to provide up to three hours of continuous use. The idea is not entirely new. Analog wound watches have existed for decades, but this is an interesting adaptation to computer technology, and it is being tested in full-sized notebook computers.

Personal Identification Number PIN. A system of alphanumeric characters, usually numerals, which identifies a particular user or holder of an identification card. PINs commonly used for credit cards, bank cards, ID cards, calling cards, and other forms of wallet-sized identification to access security doors, ATMs, phones, and vending machines.

Personal Information Manager A collected group of applications, usually including an address book, time management tools, database, appointment, notes, and other similar applications or print materials used to keep track of schedules and contacts.

peta- A prefix for an SI unit quantity of 10^{-15}, or 0.000 000 000 000 001. It's a very, very tiny quantity.

petticoat insulator A type of early glass utility pole insulator, also the double petticoat insulator, developed around 1910. See insulator, utility pole.

PGP See Pretty Good Privacy.

PGP Inc. A company jointly established by Philip Zimmermann, the developer of Pretty Good Privacy, and Jonathan Seybold. See Pretty Good Privacy; Zimmermann, Philip.

PGP/MIME Pretty Good Privacy/Multipurpose Internet Mail Extensions. An IETF working group Internet messaging standard for the transmission of secure network communications. A variety of content types have been provided for MIME, and more continue to be added. Unlike S/MIME, PGP/MIME does not use public keys distributed through X.509 digital certificates. PGP can generate ASCII armor (required) or binary output for the encryption of data. The trend is for the signed portion of the message and the message body to be treated separately. PGP/MIME can support 128-bit encryption, although not all implementations will use the full 128 bits. See S/MIME, RFC 1847, RFC 1848, RFC 2015.

phantom circuit In telephony, a means of devising an additional circuit by utilizing resources from existing circuits on either side. Thus, three circuits can be configured to prevent cross-talk and used simultaneously with only four line conductors. The use of phantom circuits has, for the most part, been superseded by a variety of multiplexing techniques. See Carty, John J.

phantom group In telephony, a phantom circuit and the balanced circuits which flank it, from which it draws some of its circuitry, is called a phantom group.

phase alternate line PAL. A color television broadcast and display standard widely used in the United Kingdom and a number of European, South American, and Asian countries. The name originates from the fact that the color signal phase is inverted on alternate lines. The format was introduced in the early 1960s. It displays at 25 frames per second and can support up to 625 scan lines (not all are seen on the screen; some at the bottom may be obscured). It provides a better picture than the NTSC format prevalent in North America and is not compatible with NTSC or SECAM. PAL-M is a variation on PAL which supports 525 lines.

phase jitter A particular type of undesirable aberration in which analog signals are abnormally shortened or lengthened. See jitter.

phase shift keying PSK. A type of modulation scheme which distinguishes between a binary "1" (one) and a binary "0" (zero), by changing the phase of the transmitted signal 180 degrees if the next input unit is a binary "0" (zero). If it is binary "1" (one), then a phase shift is not executed. See frequency modulation, frequency shift keying, on/off keying, quadrature phase shift keying.

PHIGS Programmer's Hierarchical Interactive Graphics System. An official standard for 3D graphics from the late 1980s. The PHIGS+ extension added sophisticated rendering of realistic looking objects on raster displays. Simple PHIGS (SPHIGS) is a powerful, display-independent subset of PHIGS which incorporates some PHIGS+ features.

Phillips code A type of telegraphic code designed to meet the needs of press communications.

Philmore Manufacturing Company Originally founded in 1921 as the Ajax Products Company by Philip Schwartz and Morris L. Granat, it was incorporated under the name Philmore in 1925. Philmore was the predominant manufacturer of crystal detectors and crystal detector parts used in early radio wave detection, continuing for almost sixty years.

phoneme A unit of speech, considered to be the smallest distinguishable unit, which may vary from language to language and among dialects of a particular language. Phonemes are of interest to programmers for speech recognition and speech generation applications. See speech recognition.

phonograph A device for reproducing sounds stored in grooves on a solid medium by means of a stylus or needle. Early phonograph recording media took many different forms, from the Edison cylinders to flat vinyl discs, which came to be known as *records*, and later as *LPs* (long-playing records) and *45s*. The vibration of the stylus within the groove as the recorded medium rotated was translated into an acoustic signal and amplified through a horn on early machines, and was translated into an electrical signal and amplified electronically on later machines.

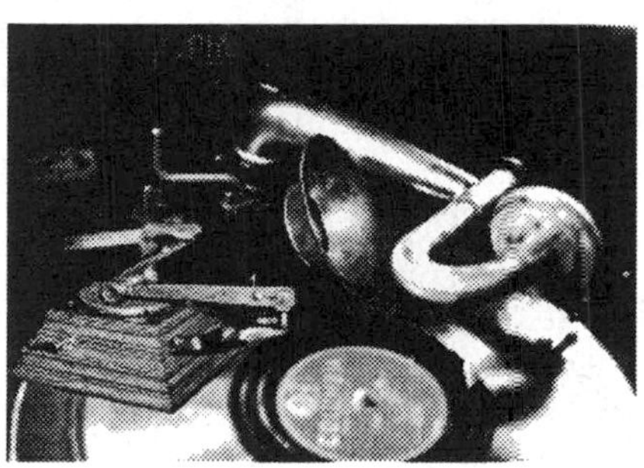

Two early phonographs are shown here, a GramoPhone and a Phonola, both of which relied on acoustics for conveying the sound, whereas modern electronics provide electronic amplification through speakers. Old phonographs used sharp steel needles and heavy platters or cylinders that were recorded at 70 or 78 revolutions per minute.

One of the earliest recording devices was the *phonautograph*, built in 1857 by Léon Scott.

P

It incorporated a sensitive diaphragm to move a stylus and thus record information. In 1876 and 1877 Edison did some historic experiments with recording voice on paper and later on tinfoil. Edison took the basic mechanism of a phonautograph and wrapped fine metal foil around the recording cylinder. He then spoke into a microphone as the cylinder was revolving and played back the recording by substituting a 'reading' stylus for the cutting stylus. It worked, and Edison named it the *Phonograph*.

Later, a flat rotating disc was found to be more practical than a rotating cylinder. One of the earliest examples was patented by Emile Berliner in 1888 as the Gramophone, a name which became generically associated with the technology. Basic monaural record players, essentially the same in design as modern records, emerged, although cylinders and discs competed for a little over a decade before Edison gave in and starting selling flat discs. The phonograph became very popular in the early part of the century until live radio broadcasts claimed part of the listening market.

Changes from that point on focused on recording quality and playing speed, with the industry eventually standardizing on 33 1/3 revolutions per minute (rpm) in the 1970s. Phonograph records were largely superseded by compact discs in the 1980s.

Stereo sound was eventually added by taking advantage not only of the up and down movement of the stylus, but also the left and right movement, thus allowing additional information from two sound sources to be recorded and played back, resulting in richer, more life-like stereo sound. See compact disc, phonograph records, telegraphone.

This Edison "Amberol Record" was one of the first "long play" cylinders, with a duration of four minutes, enough for two songs. The Edison name and image was licensed and the phonograph cylinder made and sold by the National Phonograph Company in 1909.

phonograph records A medium on which sound is stored for playback on phonograph players. Various materials have been introduced as the recording and distribution media for phonograph records, from paper and tinfoil, to rubber and vinyl.

The original tinfoil cylinders had a playing time of about 2 or 3 minutes. Eight years later, wax cylinders were introduced by Bell and Tainter, and in 1888, Berliner produced the first commercially successful phonograph that is essentially the same as those used today. It was a flat 7" rubber vulcanite disk, with lateral grooves. The earliest copies were one-sided and rotated at 70 rpm. Their capacity was about 2 minutes, roughly equivalent to the earlier Edison cylinders. By 1897, shellac records were replacing rubber vulcanite, and by 1904 double-sided records were being produced in Europe.

Edison continued to provide innovations in recording and playing technology for many years, in hot competition with other inventors and historic recording companies. Materials were tried, recording times increased, and the market for audio recordings boomed.

In the 1920s, electrical recordings were developed at the Bell Laboratories, making it possible for symphonic music to be recorded. Acetate coatings were developed around the same time. Long play recordings were introduced in the late 1940s, with 45s becoming popular about a year later. Records were widely used for music distribution until compact discs began to outsell records in the 1980s.

Phonograph technology led to the related development of tape recorders and video cassette tapes. Even although the shape of these media was different, the basic recording and playing principles were similar, as was their use for consumer audio. Consumers also welcomed the increased playing times and tapes became popular and practical for car stereo systems. Phonograph use began to decline in the 1980s. Records were easily scratched, would hiss and pop on all but the very finest stereos, and didn't have the features of the compact discs that superseded them. While records still remain as collectors' items, the golden age of the phonograph is over, having enjoyed about 100 years of popularity. See compact disc, Graphophone, Gramophone, phonograph, telegraphone.

Photo CD Kodak Digital Science Photo CD System. An image storage and retrieval format developed by Kodak and introduced in 1992. PhotoCD is a means to store digitized still images in various resolutions on a compact disc so it can be read back from CD-ROM drives. It is used by many stock photo suppliers and graphic design professionals.

Conventional 35mm film shot with a traditional camera can be taken to photofinishers supporting PhotoCD and developed into both pictures and digital images. At the lab, the file is scanned with a high resolution drum scanner and saved onto PhotoCD discs. If there is room, additional pictures can be added to the disc later, and read back with a multisession CD-ROM XA drive and an appropriate software driver (including Apple QuickTime Photo CD extension, SGI's IRIX, Sun's Solaris, IBM's OS2/WARP, AmigaOS 3.1, IBM AIX, etc.).

A Photo CD disc can hold about 100 images, that is, about three or four rolls of film. The images are stored in Photo YCC color encoding, with multiple resolution levels. Resolutions include: 2048 *x* 3072, 1024 *x* 1536, 512 *x* 768, 256 *x* 384, 128 *x* 192. The Photo CD Pro format also includes 4096 *x* 6144. See compact disc.

photocopy A dry transfer replication, sometimes also called a xerograph, after Xerox, the company that popularized the technology. C. F. Carlson was awarded a patent for the technology in 1942 and failed to sell it to some of the larger businesss-oriented companies. But a small company called Xerox took a chance on the technology. See the Carlson patent diagram.

photoelectric cell A type of sensing device which is activated by light and which is widely used in security systems, automatic lighting systems (e.g., street lights), automatic doors, etc. A photoelectric cell can be made by coating cesium on one of the electrodes in a vacuum tube. This technology was used in early television cameras.

photography The art and science of registering light from objects in a scene and storing them in the form of an image. Later it became possible to produce multiples of these images by a number of means. Most photography involves the capturing of three-dimensional imagery in a two-dimensional format. Light is usually recorded from the visible spectrum, but there are cameras and films designed to record heat and infrared radiation which show images in a form different from the way humans perceive them, and electron microscopes record the movement of a beam of electrons.

Traditional photography was developed in the early 1800s by a number of inventors includ-

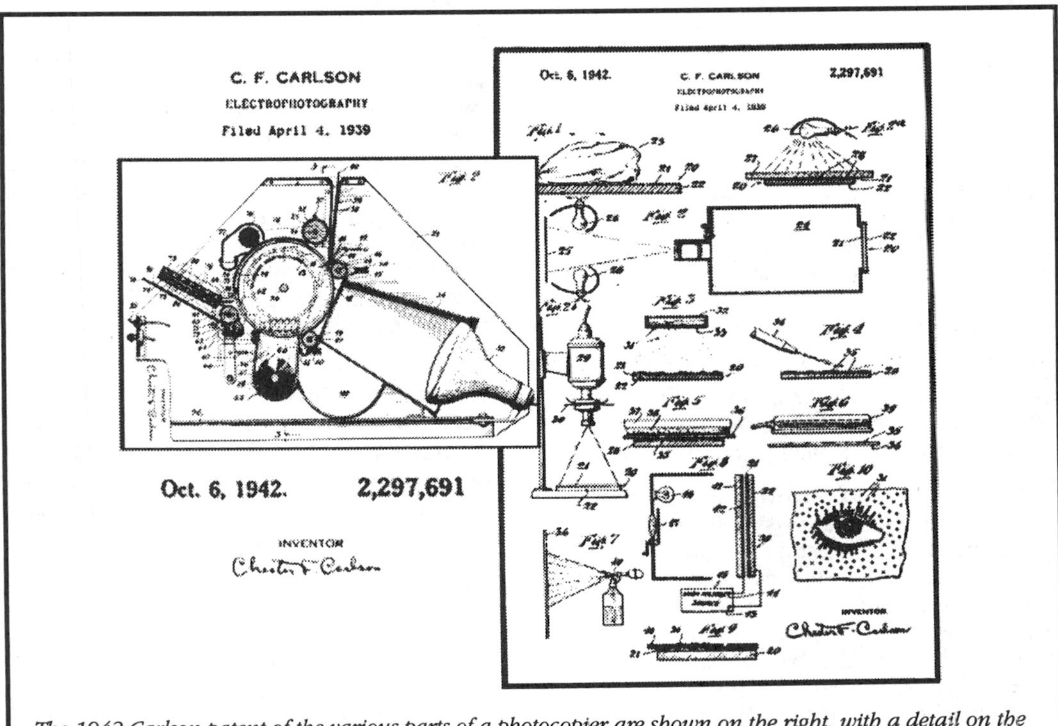

The 1942 Carlson patent of the various parts of a photocopier are shown on the right, with a detail on the left of the drum mechanism. Large companies were not willing to risk purchasing the new technology.

ing Joseph Nicephone Niepce, a French inventor, who developed a process called heliography or sun drawing, on paper coated with silver chloride. Other pioneers included Daguerre (originator of the daguerrotype), Herschel, Talbot, and Archer. One of the earliest photos was captured with silver chloride by Thomas Wedgewood in 1802. More than a hundred and fifty years passed before 3D photography, in the form of holographs, became practical. Newer digital cameras can immediately relay an image to a computer network so the image can be viewed almost instantly at great distances from the actual scene of the event. See Daguerre, Louis Jacques Mandé; heliography.

photophone A device which transmitted voice by means of light waves, invented by Alexander Graham Bell in 1880.

photovoltaic A specialized semiconductor device which converts light into electrical energy. A basic photovoltaic cell is a slice of semiconductor material. These are combined into an integrated array called a *panel*, many of which are commonly called *solar panels*. A number of photovoltaic technologies exist, with monocrystalline silicon and polycrystalline silicon, both of which are wafers cut from silicon, being common. Photovoltaic panels are important for powering orbiting satellite transceivers and their various telemetry devices.

Physical Layer Convergence Procedure PLCP. A networking convergence procedure commonly used in Distributed Queue Dual Bus (DQDB).

physical layer signaling PLS. In layer-oriented network architectures, a way to transfer information from the physical interface components to the communications channel. In some standards, the PLS is treated as a specific signaling sublayer of the physical layer.

pica In typography, a unit of measurement applied to graphic elements and to linear measures of type size and composition. 6 picas = 1" and 12 points = 1 pica. See em, point size.

Pickard, Greenleaf Whittier A scientist who followed up on the observations of Karl Ferdinand Braun by testing hundreds of natural and synthetic substances for their conductivity. Braun had observed that some substances conducted radiant energy in one direction, inhibiting the backward return of a wave. Pickard researched these properties more closely, examining substances that exhibited this property, and to what degree. In the course of his research, he found more than two hundred substances which, in conjunction with a metal contact, could detect radio waves.

This led to the development of an early form of radio component called a crystal detector, which was first marketed in 1906 and patented by Pickard in 1908. See crystal detector.

pico- A prefix for an SI unit quantity of 10^{-12}, or 0.000 000 000 001. It's a very tiny quantity.

picocell A limited-reach wireless services base station with low output, intended to serve a very limited area, such as a single building or industrial yard.

PICS See Platform for Internet Content Selection.

pictographs Images dating back about 30,000 years painted by humans on stones with natural dyes which have retained their shapes and some of their colors to this day. The first known examples of publishing, records of human thoughts and events. In a poetic sense, pictographs have *tele* (far) *communicated* over great distances of time.

picture element PEL. The smallest individual unit that can be addressed and controlled on a computer video display.

Picturephone A pioneer videoconferencing system developed by AT&T Bell Laboratories. It has been decades since the development of the first prototype Picturephone system, and it was not practical to introduce it to the public until 1970 and, even then, the technology, while appealing, was too expensive and cumbersome for consumer distribution. Since then the Picturephone concept has been refined to take advantage of advances in technology, and competing products have been developed to provide a number of practical videoconferencing options to businesses and individual consumers.

Picturephone Meeting Service PMS. An AT&T service that combined TV and voice transmissions to provide a means to carry out audiographics conferences. It was a little before its time and generally expensive, and there is now a lot of competition for videoconferencing technologies. See audiographics, Picturephone.

PictureTel Live 100 A commercial, standards-based microcomputer videoconferencing system designed to work with Microsoft Windows 95. It provides six transmission speeds up to a maximum of 384 kilobytes per second. Three ISDN telephone lines are required to operate the system.

piezoelectricity A form of electricity or electromagnetic polarity that arises from pressure, especially in crystalline substances. For example, compressing or twisting a quartz crystal causes its ends to assume opposite charges. The charge itself is not different from other charges; the name is descriptive of the manner in which it arises. A basic condenser can be constructed using a quartz crystal slice sandwiched between thin plates of metal foil. The vibrational interval of a quartz crystal is so constant, it is used for very fine clocks and to stabilize broadcast waves. See lap, quartz, Y-cut.

pike poles Long poles with spikes on the end which aid in erecting tall poles, masts, antennas, utility poles, and other tall narrow extensions which are frequently used to hold transmission wires. These are sometimes called boom poles, and their users boomers.

PIN See Personal Identification Number.

pincushion distortion A type of visual aberration in which the outer corners of an image are stretched outward and the centers between the corners curve in. The opposite of pincushion distortion is barrel distortion. See keystoning.

ping *p*acket *in*ternet *g*roper. In data communications, a software utility which employs an echo to detect the presence of another system and any delay which might be occurring in the connection. Often used as a diagnostic tool in conjunction with *traceroute*. Here is some sample output from ping, interrupted after three status outputs to the screen.

```
PING othercomputer: 56 data bytes
64 bytes from 192.42.172.20: icmp_seq=0.
time=14. ms
64 bytes from 192.42.172.20: icmp_seq=1.
time=2. ms
64 bytes from 192.42.172.20: icmp_seq=2.
time=1. ms
^C
—othercomputer PING Statistics—
3 packets transmitted, 3 packets received,
0% packet loss
round-trip (ms)  min/avg/max = 1/5/14
```

pink noise A slightly altered pure tone. This is sometimes used by 'blue boxers,' attempting to gain unauthorized use of phone lines for long-distance calls, to avoid detection up until the connection with the toll network. Once it reaches the network, however, only a pure 2600 Hz tone will be processed by the system. See blue box.

pipelining 1. In data networks, the transmission of multiple frames without checking for acknowledgment of individual frames at the time of receipt (this may be done later by a variety of means). 2. A technique used in certain central processing units to aggregate processor instructions into a set of overlapping processing steps.

pits Minute indentations in the plastic medium of an optical recording disc which encode the information, and are read by laser pickup mechanisms.

PIU See path information unit.

pixel picture element. A unit representing the smallest resolvable area on a monitor or broadcast display. Typically used to describe individual picture elements in raster displays. See addressable graphics, pixmap, raster display.

plan position indicator A type of radar display which shows target bearing in terms of a rotating intensity trace which is synchronized with the rotation of the aerial.

PLAR See Private Line Automatic Ringdown.

plasma display panel A type of display technology which consists of outer layers sandwiched around an inner layer of tiny neon bulbs imbedded in glass. The bulbs can be selectively lit by controlling voltages through matrix addressing. Unlike excited phosphors in CRTs, the bulbs remain lit until explicitly turned off, and thus do not need to be refreshed. Plasma panels are sturdy and flat, and are used in a variety of field applications. See liquid crystal display.

Platform for Internet Content Selection PICS. An effort of the World Wide Web Consortium since 1995, PICS seeks to create a means to determine and voluntarily declare the content of information on Web pages. This provides a way to implement filtering and to control searches. The first officially recommended PICS document was offered in 1996 as PICSRules1.1, and a number of Web browsers can now detect PICS ratings. PICS contains a language for defining Web page profiles and a number of rating systems have been developed, including ARC (Ararat Software), Net Shepherd CRC, SafeSurf Internet Rating System, and Voluntary Content Rating (VCR).

platform-independent Software or hardware which follows a standard, or is self-contained, so it can run without modification or without significant modification, on a variety of computers. With faster processors, it is possible for more programs and languages to be designed as platform-independent, thus increasing the user's choice of systems. There exist standards that are designed to promote consistency in the design of software objects so that computer applications can be platform-independent as well. See CORBA, HTML, Java, Open Systems Interconnection, Perl, Unix.

PLC Power Line Carrier.

PLCP See Physical Layer Convergence Procedure.

PL/M Programming Language/Microprocessor. The first programming language for the first commercially distributed microprocessor, the Intel 4004, designed by a consultant to Intel, Gary Kildall, who later founded Inter-Galactic Digital Research and developed CP/M. PL/M was developed to run on an International Business Machines (IBM) 360 computer to generate the code which was burned into the ROM of the 4004. It was later modified by Kildall to support the successor to the 4004, the 8008.

PLMR Private Land Mobile Radio.

PLS See physical layer signaling.

PLU See Percent Local Usage.

Plug and Play PnP. A format designed by a number of commercial vendors to overcome some of the problems associated with interrupt-driven computers based on IRQ assignment systems. Not all computer architectures use this means of handling interrupts, but many widely purchased Intel-based computers do. In the days when consumers had only one modem and one printer as peripherals, the IRQ-assignment design of these microcomputers was not seen as a particularly severe limitation. Unfortunately, as users added mice, sound cards, extra printers, scanner cards, network cards, and other device controllers to their systems, it became a significant problem to keep track of and set up IRQ numbers for interrupts so that there were no system conflicts, especially since some devices were designed to be associated with specific interrupts. As a result, some peripherals wouldn't work together on the same machine, and the user could run out of interrupts. At this point the architectural structure became a significant hindrance, and vendors developed Plug and Play to overcome some of the problems.

Plug and Play is a system that provides dynamic arbitration of system interrupts, and which sometimes also permits hot swapping of peripheral components or cards. To fully take advantage of Plug and Play, the user needs a Plug and Play compatible operating system, a Plug and Play BIOS, and Plug and Play compatible peripheral cards that don't have overlapping IRQ requirements.

Plutarch An ancient Greek philosopher who attempted to explain the static phenomena associated with lodestone, a natural magnet, and the attractive properties associated with the rubbing of amber. Although it took centuries for them to be understood, these early speculations led to our understanding of electricity. He also made an important observation about the differences between lodestone and amber, in that lodestone appeared to attract only certain substances, primarily iron, whereas amber attracted a multitude of objects, as long as they were very small and light.

PM performance monitoring.

PMA physical medium attachment. A device which connects physically with a network.

PMARS Police Mutual Aid Radio System.

PMDF A store-and-forward system from Innosoft International, Inc. for distributing electronic mail. PMDF is a mail transport agent (MTA) which directs messages to the appropriate network transport and ensures reliable delivery over that transport. PMDF is implemented for DEC Unix and VMS systems, and Sun Solaris and SunOS systems.

PMR 1. See poor man's routing. 2. private mobile radio.

PMS 1. See Pantone Matching System. 2. See Picturephone Meeting Service.

PMT 1. Photo Multiplier Tube. Light control technology commonly used in higher end drum scanners. See charged coupled device. 2. Photo Mechanical Transfer. A camera-ready layout tool used in the printing industry.

pn junction, p-n junction Within some types of semiconductors, a region of transition between p materials (*p*ositive holes) and n materials (*n*egative electrons). See n region, p region.

PNG See Portable Network Graphics.

PNG Development Group A group initiated in January 1995 by the release of the first draft of the Portable Bitmap Format (PBF), which evolved into Portable Network Graphics (PNG). Thomas Boutell, Scott Elliott, and Tom Lane, and many other early contributors developed an open source, freely distributable graphics compression format successor to Compuserve's proprietary Graphics Interchange Format (GIF). In an unprecedented development effort, the format went from first draft to final draft in only two months, a remarkable example to the standards-development community. See Portable Network Graphics.

PNM public network management.

PnP See Plug and Play.

PNP See Permanent Number Portability.

PODP Public Office Dialing Plan.

point of interconnection POI. The physical point of interconnection between various telecommunications networks. This may also represent a demarcation point for service and responsibility for connections.

Point of Presence POP. In data networks, a node consisting of a server, network connection, router, and one or more hosts, but not including a network operations center or network information center. A POP is sometimes serviced remotely from a Network Operations Center (NOC).

point of presence, telephony POP. In telephone networks, the central location of an Inter Exchange Carrier (IXC), the point at which the IXCs long-distance lines connect with the various local phone companies' lines. The point of connection for long-distance calls or for cellular calls.

point size A typographic unit of measure equal to ~.0139", approx. 1/72" (72.27). Point size is sometimes used to describe the sizes of graphic elements, and almost universally used to describe the height of a typestyle and distances between lines of type. It *may* include a small amount of space above and below the ascenders and descenders as it was originally based on the physical size of a wooden or metal block on which the typeface was mounted, but *does not* include extra space added between the lines with *leading*. Type sizes from 10 to 12 *points* are common for business or personal correspondence. Headlines in papers are frequently as large as 72 points. Point sizes below 9 are difficult for many people to read and do not render well on printers with resolutions of 200 dpi or lower. It is *never* correct to refer to the size of a font on a computer monitor in terms of point size if you are talking about pixel resolution or size relative to the settings of the monitor. A 10 point font can vary from 8 scanlines to 20 scanlines (or more), depending upon the monitor settings and the degree of zoom. A 20 *scanline* font is *not* the same as a 20 *point* font. It is best to use *point size* in relation to fixed media, such as the size of the font on a piece of paper once it has been printed. See ascender, descender, leading, pica, typeface.

point-to-multipoint A system in which a transmission originates from a single source and is transmitted to multiple destinations. Most broadcast media are point-to-multipoint.

point-to-point A system in which a transmission originates at a single source and is transmitted to a single destination. Many personal and business transactions, such as email addressed to a single recipient, are point to point communications. See Point-to-Point Protocol.

Point-to-Point Protocol PPP. A standard method for transporting multiprotocol datagrams over point-to-point links. PPP is intended for facilitating connections through a wide variety of hosts, bridges, and routers. PPP provides a method for encapsulating datagrams; a Link Control Protocol (LCP) for establishing, configuring, and testing the data-link connections; and a family of Network Control Protocols (NCPs) for establishing and configuring different network-layer protocols.

PPP is intended for simple, bidirectional, full duplex links transporting packets between peers. With PPP, different network-layer protocols can be simultaneously multiplexed over the same link.

PPP encapsulation requires framing to indicate the beginning and ending of the encapsulation. PPP consists of a one- or two-octet protocol field, followed by an information field of zero or more octets, followed by an arbitrary number of octets to pad up to the Maximum Receive Unit (MRU) which includes the Information field and Padding, as shown in the PPP Framing chart.

In order to establish communications, each end of the link must first send LCP packets to configure and test the data link. Once the link is established, peer authentication may be optionally carried out. PPP must then send NCP packets to select and configure one or more network-layer protocols. Once this has been done, datagrams may be transmitted. The link remains open until the LCP or NCP packets close it, or until some external event terminates the link.

See Link Control Protocol, RFC 1171, RFC 1220 (Extensions for Bridging), RFC 1661.

Point-to-Point Tunneling Protocol PPTP. A networking protocol proposed as a draft in July 1997 by a group of commercial networking vendors including Ascend Communications, U.S. Robotics & 3Com, Copper Mountain Networks, Microsoft Corporation, and ECI Telematics. PPTP allows Point-to-Point Protocol (PPP) to be tunneled through an Internet Protocol (IP) network, that is, it provides a means of carrying PPP traffic so users can benefit from secure remote access over virtual private networks (VPNs).

PPTP seeks to increase the flexibility of IP management.

Dial-in users of a common PNS can retain a single IP address. Users of other protocols such as Appletalk, could be tunneled through an IP-only system. In ISDN systems, where multilink PPP is typically used to aggregate B channels, a single PPTP Network Server (PNS) can handle the bundle, instead of grouping it at a single Network Access Server (NAS), thus making it possible to spread the bundle across multiple PPTP access concentrators (PACs).

PPTP specifies a client/server call-control and management protocol, allowing the server to control access for dial-in circuit-switched calls originated from a public switched telephone network (PSTN) or ISDN. It can also initiate outbound, circuit-switched connections. Flow- and congestion-controlled encapsulated datagram packet services are provided by an enhanced Generic Routing Encapsulation (GRE). PPTP provides for the decoupling of a number of Network Access Server functions in order to gain flexibility. See Point-to-Point Protocol.

pointcasting The broadcasting of specialized information or custom-selected information to a particular subscriber. Pointcasting is occurring over the Net, where it is possible to sign up for subscriptions for individualized information and have them electronically forwarded, somewhat like a traditional clipping service. See broadcasting, narrowcasting.

polar keying A transmission technique, commonly used in telegraphy, which employs two current states (as opposed to just turning the current on and off), i.e., current flowing in opposite directions, positive or negative voltages, to indicate *mark* and *space* signals. See telegraph, needle. The idea of using polarities led to later duplex and multiplex systems, and alternate polarities are now incorporated into digital transmissions systems.

polar mount A common type of antenna mount for parabolic antennas that is installed with the elevation mount pointed at the North Star (hence the name). Once installed, the hour axis, on which the dish swivels, is adjusted to correspond to the arc of the satellite orbit, and can be interfaced with a computer control device to simplify positioning. This type of mount is used with antennas that require orientation toward highly directional beams. See azimuth-elevation mount, parabolic antenna, microwave antenna.

polar relay A type of relay used in older communications switching systems such as telegraph systems, which uses a permanent magnet to center the armature, and a split magnetic circuit so the relay can be polarized to operate in one direction or the other. A vibrating circuit is sometimes added to a polar relay to increase its sensitivity and reduce its response time.

polarization *n.* 1. A movement or division in opposite directions or into opposite parts. 2. When radiation, especially light, vibrates perpendicular (normal) to the main electromagnetic beam (which may be straight or elliptical). In photography and sunglasses, polarization is selectively controlled to influence the light reflections that pass through the lens. 3. In an atom, the slight displacement of the positive charge of a dielectric, when influenced by an electric field. 4. The orientation of

Point-to-Point Protocol Framing

```
+----------------+----------------+----------------------+-
|    Protocol    |  Information   | Padding up to MRU    |
| 8 or 16 bits   |   0 or more    |    ...               |
+----------------+----------------+----------------------+-
```

A basic, unembellished frame as described in RFC 1171 looks like this.

```
+----------+----------+----------+----------+----------+----------+----------+
|   Flag   | Address  | Control  | Protocol | Informa- |   FCS    |   Flag   |
| 01111110 | 11111111 | 00000011 | 16 bits  |   tion   | 16 bits  | 01111110 |
+----------+----------+----------+----------+----------+----------+----------+
```

A Morse code-based rapid telegraphy system, theoretically capable of transmitting 88,000 words an hour was developed in the late 1800s by Hungarian inventors, Pollak and Viràg.

molecules in a magnetic material, when aligned in the direction of magnetic lines of force. 5. In antenna transceivers, a way to process a transmission wave's characteristics and direction. Horizontal and vertical polarization provide ways to reduce certain types of interference and facilitate the directing of a beam. Vertical and horizontal polarization can be combined to create circular polarization. This creates a helical wave that can be received by either vertically- or horizontally-polarized antennas.

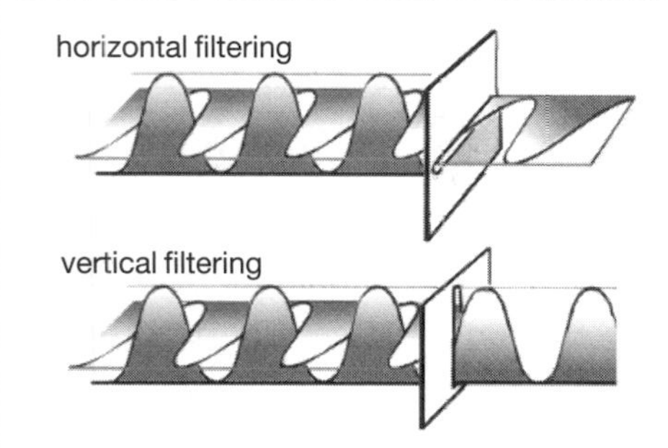

By filtering an incoming wave, it can be separated into its horizontal or vertical polarized components. This has two advantages: it permits selection of incoming information and the transmission of more information in the same bandwidth.

policy-based management In networking, an administrative adaptation to the management of networks that are too large to evaluate and maintain on a device-by-device basis. Policy-based management typically entails the development of a set of rules and guidelines for implementing network-wide decisions in areas such as security, quality of service (QoS), reliability, etc.

poll-safe See Logical Storage Unit.

Pollak and Viràg telegraph A rapid telegraphy system developed in the late 1800s by two Hungarian inventors, Pollak, a telegraph agent, and Viràg, a patent examiner. The system employed Morse code transferred to a perforated tape which passed around an electrically connected wheel. The perforations were organized in two lines, the top for dashes. Metal brushes connected to a galvanic battery registered the pulses as the brushes passed across the holes and briefly closed the electrical circuit. The negative and positive pulses caused a mirror to swing right or left, reflecting incandescent light onto light sensitive paper. Subsequent development of the paper message took about five minutes, imaging the Morse code. In 1900, the inventors communicated a message of 220 words in nine seconds (a rate of 88,000 words per hour) between Berlin and OfenPest.

polyvinyl chloride PVC. A weather-resistant polymerized vinyl compound, plastic or resin, widely used in piping and as insulation in low-voltage electrical installations. PVC is not suitable in high temperature environments, as the material will give out dangerous gases or will burn when subjected to heat.

Pony Express A colorful contributor to American history, the Pony Express service was organized by William Hepburn Russell to provide postal and parcel delivery among growing urban centers and farflung pioneer towns and villages. It was an ambitious undertaking in a large, undeveloped country with only about 15 million residents spread across a vast distance of 3,000 miles east to west. In good conditions, a rider could cover two-thirds of that distance, with a relay of horses, in less than ten days. The service was organized in shifts, with stations dotting the landscape at about ten- to fifteen-mile intervals, where riders would change their horses, and the riders themselves would switch off after about five to seven changes of horses.

The Pony Express began in April 1860 and would have only run until July 1861 when news spread of the creation of the Western Union transcontinental telegraph line. The Express was induced to continue for the four extra months it took to complete the line, and ceased operating in October 1861. It achieved such legendary status in books and movies that many people are unaware of the briefness of its history.

poor man's routing A means of routing packets according to a route defined at the source, without using network layer routing algorithms. It can be a simple, appropriate means of accomplishing the transmission on a small network with known characteristics, or on a larger network which is known to be stable, and through which an efficient route has been precharted and considered to have a high likelihood of being available.

POP 1. See point of presence. 2. See Post Office Protocol. 3. Abbreviation for *pop*ulation. One population unit, in other words, one person. In wireless services, carriers are rated partly according to POPs.

Popow, Aleksandr Stepanowitsch (1859-1906) A Russian physicist and educator, who improved on the discoveries of Oliver Lodge and created a receiver with an antenna wire for better reception in 1895 or 1896. Around the same time, he sent a wireless message from an ocean vessel to his laboratory in St. Petersburg. See telegraph history.

populated In electronics, a printed circuit board (PC board) that has had components added to the traces, as opposed to a bare board.

popup As one word, it is a noun; as two words, it is a verb, as in "I saw it pop up a moment ago." A memory-resident program that remains out of sight until needed and then pops up in front of the current application (and then disappears again when not needed). Making the program memory-resident is a convenient way to have it load and display very quickly, hence the name. See popup menu, TSR.

popup menu A type of computer menu frequently used in graphical user interfaces in situations where the user has a number of possible selections, but those selections should be displayed only on an as-needed basis, so as not to obscure the rest of the screen. A popup menu or dialog box is usually selected through a keyboard or pointing device. On dedicated computer devices, popups sometimes are activated by pressing a specific button or hot key sequence.

port 1. Point of ingress or egress, or both. Data input/output point. 2. Entrance or exit to a network, firewall, or gateway; a transmissions interface. 3. Data input or output path to peripheral devices such as modems and printers.

port, access Access point to a virtual port, such as a 'chat room' or a physical port, as a connecting point, switch or card interface. Port management, security, protocol conversion, and traffic management are essential aspects of networks, especially mixed networks and those with shared resources. See port adaptor.

port adapter In a Fiber Distributed Data Interface (FDDI) token-passing network there are two physical ports, designated PHY A and PHY B. Each of these ports is connected to both the primary and the secondary ring, to act as a receiver for one and a transmitter for the other.

The port adaptors are typically configured as peripheral cards that fit into slots in an interface processor. In many systems, there are several slots available, and it may be necessary to insert a blank card in open slots for ventilation, and to conform to emissions requirements for computer-related devices. Port adaptors may be equipped with optical bypass switches to avoid segmentation which might occur if there is a failure in the system and a station is temporarily eliminated.

FDDI ports can be directly connected to either single mode or multimode fiber optic media, providing half duplex transmissions. LEDs are commonly used on port adaptors as status indicators. Optical bypass switches may in turn be attached to the port adaptors. See Fiber Distributed Data Interface, optical bypass.

port adapter module PAM. A Cisco Systems ATM module with ports that can be configured as redundant links for use by ATM routing protocols. Port status is indicated by LEDs, and each port can be configured to a variety of clocking options. Aggregate output traffic rates can be controlled to accommodate slow receivers of public network connections with peak rate tariffs.

port sharing A networked system in which two or more devices or two or more virtual connections (VCs) share a single port. Common ports are printer and modem ports, and it is not uncommon for many computers to share a limited number of printers and modems. Port sharing is usually accomplished through a combination of software and hardware. In manual systems, the port may be shared by using an A/B switch device. For example, two computers may be sharing one modem and, since most modems cannot handle two ongoing transmissions, the switcher is set to one computer or the other. Multiple printers on medium or large systems may need more sophisticated management to handle the traffic, in which case an automatic switcher may handle the queues and access to the various printers or other ports. Not all ports connect to physical devices in the sense of a modem or printer. Some ports are virtual communications ports for chat lines, or other networked services, in which the handling of the port access is generally handled through software.

port-sharing device A device which selectively limits or administrates shared access to a virtual or physical port. A switcher, from simple A/B switch to an intelligent logical switching device, is a common means with which to share resources on a networked systems. See port sharing.

portability 1. An attribute of a hardware device that makes it easy to carry around. Pen computers, palmtops, and laptops are light and battery-powered to maximize their portability. 2. An attribute of software that makes it easy to move from one platform to another. Hardware-independence. Languages like HTML, Sun's Java, and Perl were designed to be portable. Unix software is portable to a wide variety of systems. 3. An attribute of a virtual address that allows the user to access a phone or computer network from a variety of locations. See mobile computing, cellular.

portable Easy to carry around, particularly devices which have desktop analogs, as a desktop computer and a smaller portable computer, a desktop or wall phone, and a battery-powered, small portable phone. Portable is considered the category of devices that are easiest to use while mobile. Luggables and transportables are the next level up, as they are a little heavier and usually require more power consumption.

Portable Computer and Communications Association PCCA. A nonprofit organization established to provide a forum for the various industries to cooperate in the evolution of interoperable mobile computing and communications standards and implementations. http://www.outlook.com/pcca

Portable Document Format PDF. An Adobe Systems Inc. proprietary document encoding and display format which is used on many Web sites to distribute read-mostly document files. PDF files can be created and exported from a number of desktop publishing and layout programs, or with Adobe Systems Acrobat which is designed for this purpose. Freely distributable PDF readers are widely available for download on the Internet. PDF files typically carry a *.pdf* file extension.

Portable Network Frame PNF. A format for using multiple sub-images in a single frame, a multimedia graphics format intended as the foundation for the Multiple-image Network Graphics format (MNG) which is essentially Portable Network Graphics (PNG) for animations. PNF was submitted as Draft 1 by Glenn Randers-Pehrson in June 1996. With Draft 12, in August 1996, PNF came to be known as MNG. See Multiple-image Network Graphics, Portable Network Graphics.

Portable Network Graphics PNG. (*pron. "ping"*) A graphics standard developed during the first two months of 1995, with the Web in mind, which is recommended by the World Wide Web Consortium (W3C, WWWC). It received official 'done' status in May 1996. A standard for lossless, variable-transparency, cross-platform, 'truecolor' graphics which includes, in the file, information about the authoring platform so the viewer can adjust the image accordingly.

PNG was developed as a result of an unexpected patent announcement by Unisys Corporation in the mid-1990s. Unisys claimed intellectual property ownership over the Lempel-Ziv-Welch compression algorithm, which was integral to many graphics compression schemes, the most visible of which was CompuServe, Inc.'s GIF format. GIF was well supported and widely used on the Web. Programmers were taken aback and reacted with concern, and soon there was a move to create an alternate format, not just a simple alternate, in fact, but a better format, which would provide an evolutionary improvement over GIF. At about the same time, CompuServe announced that they would begin development of a GIF successor, GIF24, followed by an official announcement a month later that PNG would be used as the basis for GIF24.

The Internet community of programmers, some of whose contributers included members of The World Wide Web Consortium (W3C) and Compuserve, began work on the development of PNG. The current consensus is that PNG is a good format and it is quickly being adopted by the Web community. The PNG specification was transcribed by Thomas Boutell and Tom Lane and released as RFC 2083 in 1997.

The features of PNG include an open software standard, good lossless compression ratios, 8-bit color palette support, 16-bit grayscale support, 24- and 48-bit 'truecolor' support, alpha blending transparency for supporting different 'degrees' of transparency, gamma correction, two-dimensional interlacing, text chunk support, multiple CRCs for error checking without viewing, security signature, full online references, and source code availability. PNG was not designed for animation support; it was intended for single images. MNG was subsequently developed to support animation. See Graphics Interchange Format, Lempel-Ziv-Welch, magic signature, PNG Development Group, Portable Network Frame, Virtual Reality Modeling Language, RFC 2083.
http://www.cdrom.com/pub/png/png.html

portrait A descriptive word that refers to the direction of roughly rectangular objects, usually printouts, photographs, or monitors, which are oriented so the long side is vertical and the short side is horizontal. The term is widely used in the photography and printing industries to describe the orientation of text or images on a page. Some monitors have been designed to pivot 180 degrees so they can be used in both portrait and landscape modes. Most monitors are designed to be used in landscape mode, but portrait mode monitors are handy for WYSIWYG desktop publishing of printed page-oriented layouts. Contrast with landscape.

portrule A historic telegraphic device with a metal, toothed bar acting as symbolic contact points to define digits or code symbols. This device was later superseded by the simpler telegraph key.

POSIX 1003.0 Portable Operating System Interface UNIX. An open systems standards architectural framework, also known as IEEE 10030.0

Post Roads Act A regulation of the telecommunications industry in the United States began in 1866 with the Post Roads Act in which authority was granted to the Postmaster General to oversee rates for government telegrams and to assign rights of way through public lands. By 1934, after passing through some intermediate bodies, including the U.S. Department of Commerce, telecommunications became the primary responsibility of the Federal Communications Commission (FCC).

Post Telephone & Telegraph administration. Telecommunications operating bodies around the world which are individually controlled by their regional governments.

Postal Telegraph Company A historic communications provider, founded in 1881. Postal was formed to market some of the new telegraph technologies, particularly those of Elisha Gray. Postal pioneered the use of a new type of copper wire with a steel core, known later as Copperweld. In 1883, controlling interest in Postal was acquired by John W. Mackay, co-founder of Commercial Cable Company. It reorganized after filing for bankruptcy in 1884. Control of the company then passed through the hands of Albert B. Chandler to Mackay's son Clarence H. Mackay. It merged with IT&T in 1928, reorganized in 1939, and finally was merged into Western Union in 1943.

Postel, Jonathan B. (1943-1998) No dictionary about the Internet would be complete without expressing appreciation to Jonathan Postel. From the early ARPANET to the current Internet, he contributed three decades of low-key, passionate, dedicated service and volumes of fundamental information as a developer, advisor, protocol prototype implementer, and RFC Editor. Like the underlying thread that runs through a tapestry, Postel held quietly to a vision, avoiding the fanfare and business opportunities that constantly presented themselves to pioneers of the computing industry. He chose instead to concentrate on the structure and orderly evolution of this most important communications medium, for the benefit of all. Over the years, Postel worked for a number of educational institutions and high technology companies including the Network Measurement Center at UCLA (ARPANET) and SRI International with Doug Engelbart. See IANA, Request for Comments.

postmaster The person responsible for configuring and maintaining a network mail server, often including administrating users, setting up mailboxes, distribution lists, aliases, filters, etc. Many of the postmaster functions are actually handled by the computer software, such as dragons and mailer daemons.

PostScript, Adobe PostScript A powerful, high-level, device-independent page description language and document format widely used in desktop publishing and electronic document design. PostScript, from Adobe Systems, Inc., is used for bitmap and scalable images, scalable fonts, computer monitor display systems, and much more.

PostScript originated as the Design System language in 1976 at Evans & Sutherland Computer Corporation, a company renowned for its pioneering flight simulator programs. John Gaffney is credited by John Warnock as being the inspiration behind many of PostScript's major design components. In 1978, when John Warnock joined Martin Newell at Xerox PARC, they reimplemented Design System as JaM (after their first names, John and Mark). At this time, the language was used for experimental applications in VLSI design, printing, and graphic arts, resulting in Xerox's printing protocol called Interpress.

John Warnock joined forces with Charles (Chuck) Geschke in 1982 to form Adobe Systems Incorporated. He and Geschke further developed JaM in collaboration with Doug Brotz, Bill Paxton, and Ed Taft, and PostScript was born of the effort.

One of the important developments that helped introduce the PostScript language and make consumers aware of its capabilities, was the release of the Apple LaserWriter PostScript-capable printer in 1984. Although PostScript doesn't seem as remarkable now, in 1984 most people had 9-pin dot matrix printers and bitmap fonts that provided output that even the most undiscriminating viewer would admit was 'crude at best.' Suddenly, with PostScript, a prosumer-priced laser printer could print legible, beautiful text down to 4 points in size in some of the finest fonts in the world. This launched the desktop publishing revolution, one which is still having a far-reaching impact on the publishing market, especially among small presses, self-publishing individuals, and genealogists.

PostScript fonts are some of the best computer fonts in the world. Adobe Systems maintains a large library for sale to consumers and service bureaus. PostScript fonts are scalable, in order to print out at the best resolution of the output device. Many people think that PostScript fonts and other scalable fonts are essentially the same, but most other scalable fonts are not integrated into a page description language, and so are not as flexible and powerful as PostScript fonts. Since PostScript is a programming language, fonts can be swirled, stretched, and individually rendered so that each letter differs from the previous in some essential way. The possibilities have even now not been fully exploited.

PostScript is commonly used to distribute documents on the Web, as is Adobe Acrobat format, a second-cousin to PostScript for displaying text and graphics. It is also possible to send high-quality PostScript documents through email, by sending the file as an email *attachment*. This is a means by which people can send professional-looking text and graphics résumés, business documents, manuals, and much more over the Internet, or they can be linked to a Web page for instant download. See vector fonts.

potentiometer 1. An instrument for measuring electromotive forces. 2. A device used to regulate a current by varying the resistances at either end. It can also perform the functions of a rheostat, which is more limited. Potentiometers are commonly incorporated into dials and computer input devices like joysticks. See rheostat.

POTS Plain Old Telephone Service. The basic analog phone service which has been available from local phone companies and used in homes for years and years. No ISDN, no surcharge services such as Caller ID, Call Waiting, etc. See loop start.

potting To embed within an insulating or protective material or layer, usually for the purpose of reducing electrical interference or fire hazards. Potting is sometimes required in cases where higher voltage computer components may be interfaced with lower voltage phone lines.

Poulsen arc A device enclosed in a gas atmosphere with a strong magnetic field that created an electric arc which could generate high frequency radio waves. It was found that larger versions of the Poulsen arc could generate even greater arcs. The rights to market this technology were purchased in 1909 by C. Elwell, who formed the Federal Telegraph Company to design and build industrial arc transmitters which were used in the newly developing broadcast industry. Eventually the technology was superseded by vacuum tube transmitters. See arc converter; Poulsen, Valdemar.

Poulsen, Valdemar (1867-1942) A Danish scientist and inventor who devised a way to use an electric arc to generate continuous waves at high frequencies by placing the arc in a controlled atmosphere within a strong magnetic field. He collaborated with P. O. Pedersen in inventing wireless telegraphy technology. In 1898, Poulsen recorded electronic waves on a thin conducting wire, a pioneer electromagnetic tape recorder that was called the *telegrafon* or *telegraphone*.

pound 1. An alternate term for the number, or crosshatch, sign #, sometimes called an octothorpe. Programmers often refer to it as a *hash* sign. On pushbutton phones using automated menu systems, the instructions will sometimes tell you to input a number followed by the *pound key*. The pound key signals the system that you have finished typing in the number. 2. In music notation, the symbol for a *sharp*, a note raised a half tone.

power, electrical Expressed in watts, the product of the electromotive force (in volts) and the current (in amperes). Thus, $P = E\ I$. In terms of resistance, according to Ohm's law, this can be expressed as $P = I^2\ R$. See ampere, Ohm's law, ohm, resistance, volt, watt.

power down To initiate or perform a sequence of operations in order to shut down a system. For example, a power down on a computer may involve closing files, asking the user to save data, logging out the user, etc. before actually terminating the power to the system. Power down sequences are designed to clean up systems, remove unwanted files, and prevent the accidental loss of data.

power hole digger A machine for digging deep narrow holes for utility poles, starting with the early telegraph lines. It was introduced in North America around 1915, although line workers using long spades still dug the holes in undeveloped regions for several decades after the introduction of the power hole digger. See pike pole.

Power Macintosh computer The successor to the original Macintosh line, the Power Mac is based on the IBM POWER RISC chip architecture. These are gradually being succeeded by even faster Macintoshes built with G3 chips. See G3, Macintosh computer, POWER.

POWER, Power PC Performance Optimization With Enhanced RISC. A complex processor, one of the first superscalar processors, initially implemented by IBM with three integrated circuits (branch, integer, floating point). This technology was further developed as a microprocessor by IBM, Motorola, and Apple Computing in the early 1990s. The idea was to create a successor to both the Motorola 68000 line and the Intel 80x86 line; the PowerPC was the result of this collaboration. The first version was the PowerPC 601, released in 1993, derived strongly from the IBM POWER specification. Since then, a series has been released, including the 603, 604, and G3, and the chips are incorporated most familiarly into the PowerMacs and Macintosh G3s.

power save mode See sleep mode.

power up sequence The operational bootstrap and test sequence that a computer goes through when first powered on. This usually includes loading very low level routines, often from read only memory (ROM), which then make it possible to load other routines and operating system capabilities from a hard disk, floppy, cartridge, or CD-ROM drive. It is very common for a computer to run through a hardware systems check in the power up sequence to test memory, sound, graphics, and other basic input/output devices. Device drivers and external device checks may also be performed, in addition to locating and interfacing with a network, if applicable.

If there are a lot of devices attached to a computer, it is often important to power them up in the right order. If you turn on the computer before turning on external hard drives or CD-ROM drives, the computer may not recognize the device or know that it can be accessed. As a general rule, turn on peripherals before turning on the computer. Give a hard drive a moment to 'spin up,' that is get the drive revolutions up to speed before turning on the computer. Similarly, with a device such as a scanner or printer which may also have test sequences, count to five before you turn on the computer. If a system is being powered up right after being shut down, it is important to wait 30 seconds or so before turning it back on. Some of the electronic components in a computer will retain current after the system is shut down, and a sudden surge of additional current may stress the circuitry. Give the current a few moments to drain off, then turn the system back on.

PPDN Public Packet Data Network.

PPI 1. pixels per inch. See resolution. 2. See plan position indicator.

PPP See Point-to-Point Protocol.

PPS 1. packets per second. A means of quantifying network traffic by tallying the number of packets transmitted through a given point in a given amount of time. 2. Path Protection Switched. See SONET. 3. See Precise Positioning Service. 4. pulses per second.

PPSN Public Packet Switched Network.

PPTP See Point-to-Point Tunneling Protocol.

PRAM programmable random access memory. A chip that is sometimes used on computers to save configuration settings that are semi-permanent, such as monitor settings. The chip retains its information by being refreshed with power from a battery, usually a small lithium cell. This battery may have to be replaced every six years or so.

PRB Private Radio Bureau.

Precise Positioning Service PPS. One of the precise location data signals transmitted from Global Positioning System (GPS) satellites. This signal is for military and general government use, requires a specially equipped receiver, and is encrypted. Horizontal and vertical accuracy are about 22 to 28 meters, and time accuracy is 100 nanoseconds. See Global Positioning System, Standard Positioning Service (SPS).

Preece, William A British researcher who experimented in the late 1800s with conductivity methods for sending wireless communications. He was able to send a message a distance of five miles. See conductivity method.

prepaid phone card A credit card-like monetary storage medium which is 'charged' or credited with a certain amount of prepaid phone access. This phone card can then be inserted in a card-compatible phone and will automatically allow access to calls up to the amount that is charged on the card. Copying machine cards are somewhat similar. Unfortunately, many phone cards are not rechargeable, which is an unfortunate waste of resources since a phone card can last many years, as do copy cards. This is mainly due to the way the accounting is done. The value of the card is not simply embedded in the card as it is on a copy card, rather it is handled through accounting software at the switching center. The best reason to get a phone card is to avoid inserting coins into a telephone, especially for a long-distance call during which you might be interrupted by an operator to add more change (which you might not have handy). The denominations on a phone card vary from region to region, but amounts such as $4.95, $9.95, and $19.95 are common.

There are a variety of vendors providing phone cards; the service is not necessarily directly provided by the local phone company. This accounts for the different designs on the cards, the different ways in which they are promoted, and the different denominations that are available.

Presentation Manager PM. International Business Machines' (IBM's) standard toolkit and graphical user interface for OS/2, a multitasking operating system for desktop Intel-based computers. PM provides a means to develop applications that conform to the windows-based interface characteristics of OS/2.

Presentation Time Stamp PTS. In MPEG-2 encoding, a timestamp that is encoded into the elementary packet stream. This is used for synchronization of different streams by comparing it against the System Time Clock (STC). A video decoder synchronizes the MPEG video data with the STC, with the assumption that the audio decoder follows suit. If the synchronization is within acceptable parameters, the decoded picture is displayed, otherwise, it is repeated, the STC is readjusted, or the next B or P frame skipped over to maintain synchronization.

preset A setting that is configured in advance so it can quickly be accessed later. Thus, in video and audio editing, cuts and dubs are sometimes preset; in computer programs, times or online activities may be preset. Pushbutton radios can be preset to instantly tune to a desired frequency.

PREST Centre for Policy Research in Engineering, Science and Technology. U.K. group.

Pretty Good Privacy PGP. A powerful high-security encryption scheme developed in the early 1990s by Philip Zimmermann, based on the Blowfish encryption technology.

PGP provides both privacy and authentication of transmitted messages. Only the person intended to see the message can read it. An intercepted message cannot be deciphered. Authentication provides assurance of the authenticity of the sender and that the message has not been changed. PGP is freely distributed to U.S. and Canadian citizens for non-commercial use by the Masachusetts Institute of Technology (MIT) in cooperation with Zimmermann and with RSA Data Security, Inc., which licenses patents to the public-key encryption technology incorporated into PGP. PGPfone has now been developed to provide for secure online digital phone conversations. In 1998, Network Associates acquired PGP technology. See Blowfish; International Data Encryption Algorithm; PGP Inc.; Zimmermann, Philip. There is a useful PGP FAQ on the Web. http://cryptography.org/getpgp.htm

preventive maintenance Regular inspection, testing, adjusting, and maintenance of equipment in order to prevent problems before they cause damage or affect service. Computers (especially floppy disk drives and the area around the fan) tend to collect dust after a year or two of use, which should be *gently* brushed out or vacuumed with a low-power, fine-nozzled vacuum, making sure the equipment is turned off and probably even unplugged. A good time to do this is when you are swapping out or installing new memory. Monitors should be turned off when not in use, and should have a screen saver active when in use. Batteries in phones that have extra features (like LCD readouts) should be replaced regularly, *before* you lose all the phone numbers programmed into the system. See screen saver.

Price Cap Regulation A means by which local monopolistic phone companies are regulated so that rates remain the same for a specified period. Unlike Profit Cap Regulation, which did not carry large incentives to pare back staff or adopt more cost effective, newer technologies, it was intended that Price Cap Regulation would provide incentives for innovation. Many companies changed from Profit Cap Regulation to Price Cap Regulation in the mid-1990s.

Primary Rate Interface PRI. One of two major categories of ISDN services, PRI caters to higher-end customers and businesses. See Basic Rate Interface, ISDN.

primitive A representation of a basic 'unit' in computing. As examples, a graphics primitive may be a circle, line, or square; an audio primitive may be a phoneme or sound; a programming primitive may be a routine, procedure, or object. A primitive is some type of basic building block, usually one which is frequently used or reused. In networking, a primitive is an abstract representation across a layer service access point, where information is exchanged between a user and provider.

principal Main, central, overriding; the highest authority or administrator.

principle A general, or fundamental concept, statement, or truth. A basis for decision-making, actions, or operations.

print server, printer server A system that handles the logistics of requests to one or more printers on a network. Frequently there will be a variety of types of printers shared among users. These might include plotters, laser printers, dot matrix printers, high-speed page printers, and specialized color dye sublimation or thermal wax printers. The print server handles queuing; messages to users if a printer is not in service; alternate routing if the printers have been reorganized; scheduling, if some types of jobs (e.g., big ones) are to be run at night or after hours; and prioritizing, if some users have higher precedence. The print server can also be used to send messages to maintenance personnel if trouble is suspected, as with paper jams, empty trays, etc. Some printers have sufficient processing power to send status and error messages to the server, which in turn may be relayed to the user or the appropriate service center.

print spooler An application which manages and schedules a printing job to a printing device in such a way that the computer is not tied up, waiting for the print job to finish. For example, suppose you send a large plotting job to a plotter from a CAD program. If it is a single-tasking system and cannot handle the print job in the background, and if there isn't a large buffer in the plotter itself, it might take 10 to 40 minutes for the plot to finish, and the computer would be unusable for that period of time. In order to reduce wait time on this type of system, print spoolers were developed so a plot could be printed to disk rather than to the plotting device, a process which might take two minutes instead of twenty. The plot can then be spooled to the printer during a lunch break, after hours, or when the plotter is not tied up by another user. With the spread of multitasking systems and printers with large buffers, the use of spoolers is diminishing.

On larger networks, there may be a print server to handle printing tasks as a type of 'smart spooler.' In other words, the print job might be sent to a file, or sent directly to the print server, and scheduled and spooled from there rather than from the originating machine.

printed circuit board PCB. A board on which electronic circuits are mounted, with the circuit connections etched, foiled, or blasted onto the surface of the board, usually on the side opposite from the majority of the components. The etched electrical pathways, called *traces,* provide flat, convenient electrical contacts without wires. This is a very practical, lightweight way to do away with wires and allow mass production of PCBs.

The conventional wisdom is that printed circuit boards were first invented in the 1940s, but the Bellingham Antique Radio Museum has a radio in its collection that dates from the late 1920s which is designed with a copper, blasted circuit on the underside of the board as shown in the following photos. Thus, the technology was introduced almost twenty years before its purported invention.

Copper traces blasted onto the underside of a 1928 cabinet radio discovered by Jonathan Winter (the wires were added later).

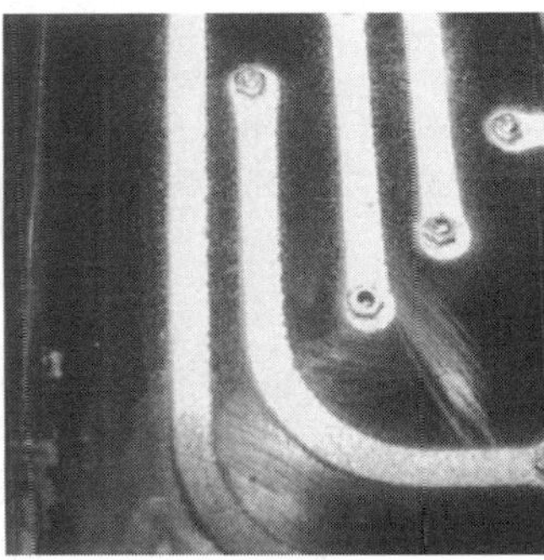

Detail close-up of copper blasted traces on the underside. Small bolts and nuts were used to provide the electrical connection to the vacuum tube components on the top side. This historic example of one of the first printed circuit boards in existence is part of the Bellingham Antique Radio Museum collection.

printer A device for transcribing information onto a medium which can be read directly or otherwise understood directly, or with a minimum of manipulation (as in mirror writing), by someone familiar with the communications medium (writing, illustrations, Morse code, seismographic charts, etc.). The printing medium is often portable, as paper, card stock, or metal plates.

printer control character A character which has a specific control effect on the action of a printer. The effects include line feeds, page feeds, carriage returns, mode changes, font changes, page length control, and other features that might be specific to the printer. Control characters can be sent to a printer before sending the document, in order to set up the printing parameters, or may be imbedded in the document itself to set typefaces, font sizes, text attributes, space, line feeds, etc.

Printer control characters are handled transparently by word processing programs, which send the appropriate characters through a printer driver without explicit programming by the user. In cases where a user is imbedding printer control characters in the document, a hex editor, or ASCII editor with hex capabilities, is often used, as the control characters cannot always be entered from the keyboard, as they lie beyond the range of the alphabet. Printer control characters are often not displayable, or may display as unusual symbols.

P

printer control language A language designed to utilize the capabilities of a particular type of printer, or one conforming to a common standard such as Hewlett Packard Graphics Language (HPGL), which is widely used on plotters, for example.

printer driver A program which provides information on the physical, operational, and control characteristics of a printer. This may include relevant control codes, available font shapes and sizes, paper feed controls, etc. Typically an operating system or applications program will interact with a printer through a printer driver stored as a computer file. There may be many printer drivers available, and if the relevant one is not online, often a substitute can be found from a maker with a similar printer. Many different laser printers will function with the same commands as Hewlett Packard or Apple laser printers, and many impact and ink-jet printers will work with Hewlett Packard or Epson printer drivers.

printer emulation Any software which formats and outputs text to a device as though it were outputting to a printer. This way, for example, a word processed program with all the text formatting, margins, and images can be sent through a facsimile transmission without first being printed and scanned. This saves paper and avoids problems such as slippage through the fax machine.

printer font A font that is stored in the printer's memory or on a hard drive directly attached to the printer. Dot matrix printers often have a few memory-resident font styles optimized for printing at low resolutions. High end PostScript printers often have the capability of storing a variety of high quality Adobe PostScript fonts on a hard drive, for quick access and printing, so the system doesn't have to download the fonts from the computer to the printer.

printer server See print server.

printing Printing is the production and reproduction of characters, symbols, and images in order to communicate or distribute them to others. There is evidence that movable type, in the form of clay discs, may have been used in Crete as early as 1500 BC, and type was used in southeast Asia by the 11th century. In the mid-1400s Johann Gutenberg introduced movable type to the west. Type was set by hand right up until the early 1800s, when the first typesetting machines began to appear.

Mass market microcomputer printers were introduced in the 1970s, and the home and prosumer desktop publishing markets became widespread in the mid- and late-1980s. Dry printing processes (e.g., photocopying with toner) became prevalent in the 1970s, and have replaced a large proportion of wet printing processes (e.g., ink). A modern desktop laser printer costing $1,000 is easier to use and more flexible than the letter presses of the 1960s that cost $30,000 or more.

Color printing processes are numerous, from traditional ink presses which use *spot color* while running the paper through the press several times, once for each color, to *process color*, which uses cyan, magenta, yellow, and black dots to simulate a wide range of colors. Computer printers use thermal wax, dye sublimation, ink-jet, and colored ribbons to create printouts that rival commercial color photocopies and photographs, and may someday supersede them.

printing plates Printing plates were originally used to insert images into layouts which included movable type. The plates were then mounted on traditional presses for use with ink reproduction processes. Today, printing plates often contain an image of the entire page, both type and images. The earliest printing plates were probably clay tablets and, later, hand-carved wood blocks, which evolved into copper and steel engravings in the 1400s. In most cases the metal engravings provided finer detail and longer edition life than wood blocks, although M.C. Escher has created wood engravings that rival some of the finest metal engravings. Metal and asbestos plates can now be run through certain phototypesetting printers in the $7,000 range to create plates for use on traditional ink presses. These same desktop typesetting machines can produce dry process laser printouts at about 1200 dpi.

priority A level of access or usage which ranks higher than others. For example, on a computer system, operating system functions usually take precedence over user or network requests. On a server, such as a print server or network server, certain types of tasks may be handled first. In graphical user interfaces, a window which is clicked to the front may be considered a higher priority process, and may be given a greater proportion of processing time, than windows in the background. On most network systems, system administration functions take priority over user functions.

Private Automatic Branch Exchange, Private Branch Exchange PABX, PBX. A private telephone exchange, usually located in a business or educational institution which can handle switching and other functions automatically. In an automated exchange, an operator is not needed to handle outgoing calls, for example, as they can be connected by first dialing "9" to access an outside line. PABX was derived from Private Branch Exchange (PBX) which originally was attended by a switchboard operator. Since private exchanges are almost all automated, the terms PABX and PBX are now used interchangeably. See Centrex.

private carrier A privately owned, commercial public messenger service or telecommunications service provider that may or may not be in competition with a dominant commercial carrier or government-funded service.

private line In the early days of phone service, a private line was a line that went from one business or person to another, without necessarily going through a public phone exchange, or from one floor or room to another. It was not uncommon in the early part of the century for hundreds of wires to crisscross over a street between one building and another. As more people were connected through public wiring systems that could handle multiple connections, the meaning of the phrase changed and, until about the 1960s, a private telephone line came to mean one which was not a party line, that is, the phone line was dedicated to only one user, and there was no possibility of a neighbor on the same exchange listening in to the conversation or tying up the connection so another subscriber couldn't dial out. As private lines became the norm and party lines began to disappear, private line began to take on a different meaning, similar to its early meaning, referring to a directly connected line between two businesses, or between a home and a business, or between different departments in a business or institutional complex.

Private Line Automatic Ringdown PLAR. A means of interconnecting two lines to form a 'hotline' direct connection.

privileges 1. On a telephone system, particular functions and services available on particular consoles or to particular individuals in a company, sometimes through keying in an access code. 2. On a data network system, access to specific applications, processes, devices, or data files. More specifically, file privileges are a record of the actions an individual or group member can take on a file, typically *read*, *write*, *execute*, or *delete*.

PRMA packet reservation multiple access.

probe A detection or measuring device, often with a narrow tip, which is used to assess temperature, wind, humidity, current, voltage, amperage, polarity, or other properties of air or electrical circuits. A probe is often used in conjunction with an analog or digital readout displaying the results of the probe. On a computer system, a probe is any software process which seeks out specified information, or which detects certain actions or types of data. The results of the problem are typically reported to another program which analyzes the information and acts accordingly. A probe can be used to locate a particular site, user, address, archive, or other type of information. It may be used as a diagnostic tool to troubleshoot or configure a network.

process A software activity consisting of carrying out a set of predetermined or situation-influenced logical instructions, which may be low-level background processes integral to the operating system, router, or intelligent switcher, or higher level processes related to the running of an applications program.

process switching On a network, packet processing at process level speeds, without the use of a route cache, as is used in fast switching. A Cisco Systems distinction.

Prodigy One of the earlier commercial online services established by IBM and Sears Roebuck. Prodigy employed proprietary software to provide a graphical user interface access to the Internet. A lot of parents signed on to Prodigy to give their children educational access to online services. Like most of the largest commercial services, Prodigy earned revenues through placing ads on viewer pages. This practise is uncommon on the smaller, independent Internet Services Providers. Now that freely distributable Web browsers are available, proprietary software like that provided by Prodigy is less common. However, the ads remain, as many Web sites are subsidized by banner ad revenues.

Profit Cap Regulation, Rate of Return Regulation Prior to the mid-1990s, the predominant means by which local monopolistic phone companies were regulated. Excess profits were required to be passed back to consumers as, for example, a rate reduction. By the mid-1990s, many companies changed to Price Cap Regulation. See Price Cap Regulation.

Program Clock Reference A synchronization reference clock used, for example, in MPEG decoding. The PCR can be used to synchronize the Station Time Clock (STC). An MPEG-2 video decoder chip can be designed to include an internal counter which can be used as a Station Time Clock, which can in turn be accessed by other components through an interface.

program counter In general terms, a display or internal reference that keeps track of a location in a presentation (e.g., video or laserdisc program, TV broadcast, computer animation, etc.), sometimes to provide information to the viewer and sometimes as a reference point for searches or editing. In computer software execution, a program counter is a reference that monitors the location in a program that is currently being accessed. This is handy when debugging, testing a program, or tracing a logical path.

programmable Any device which can be controlled or altered through logical instructions without having to reconfigure the physical connections.

programming language Instructions used by a programmer to control computer operations. Programming languages are roughly divided into low, medium, and high level languages. Low level languages are those which most directly translate into machine instructions and interact most directly with the hardware architecture of the system. Machine language and assembly language are considered low level programming languages. Machine language programs are typically written in binary, with ones and zeros. Assembly language is similar to machine language, except that instructions are more symbolically represented, and routines can be written to pass control to different parts of the program. Machine language and assembly language are more difficult and time consuming for some to learn, and more difficult to trace and debug than higher level languages.

Medium level languages include those which are reasonably powerful, somewhat cryptic in their instruction sets and syntax, but comprehensible enough that some of the commands resemble written English. C is a common language which is somewhat of a medium level language. It is powerful, but requires a good understanding of memory allocation, pointers, and arrays, and takes some time to learn and to apply. C is a compiled language, which means that the code is compiled down to machine code in advance, before the program is run.

Higher level languages like BASIC and various authoring systems, were designed to be easy to learn and to use, with commands and syntax that are fairly close to written English. They often are run as interpreted languages (although compilers may exist), and so do not require knowledge of how to configure and run a compiler before they can be used. Interpreted languages are translated to machine code as the program is being run, and thus will execute more slowly than a program which is precompiled. Interpreted languages tend to be more limited, but also more portable than lower level languages.

programming overlay, configuration overlay A covering made from membrane, plastic, cardboard or another material that provides information on a keyboard, keypad, or graphics tablet setup. Overlays were especially prevalent before graphical user interfaces, when complicated, difficult-to-remember control code combinations were used to run word processors and graphics programs.

progressive scanning A method of displaying broadcast video signals in which each frame is transmitted one after the other, rather than dividing the frame into sets of two fields interlaced. See interlaced.

Project Athena See Athena project.

promiscuous mode In data networks, promiscuous mode is an open mode in which the network interface controller (NIC) passes all the frames which it receives, regardless of the destination address, to high level layers in the network. This is usually only done in diagnostic situations, or by users gaining unauthorized access to information from the system. In normal operations, frames are evaluated and selectively passed along if the destination address maps to that device.

prompt *n.* 1. A mechanism for gaining the attention of the user to indicate that the system is ready for input or that input is required before continuing. 2. A prompt on a computer system may be in the form of a cursor, a dialog box, a flashing area, or an audible tone or spoken message. 3. A prompt on an automated phone system may be a spoken question or suggestion to which the user can respond by typing in codes or, in some cases, by clearly speaking numbers or words.

PROMPT Institute for Prospective Technological Studies. Founded in 1989 as a result of reorganization of the Joint Research Centre to monitor and analyze new sciences and technologies. Monitoring is carried out by by the European Science and Technology Observatory (ESTO).

proof of concept A strategy for communicating an idea which is not readily accepted when communicated through verbal means alone. Proof of concept usually involves producing a prototype which is partially or mostly functioning, at least enough to show that the idea can work. Many new inventions or ideas are disbelieved until they are actually demonstrated. Edwin Armstrong spent years going against the stated assumption that frequency modulation was mathematically impossible, but eventually succeeded in his attempts. When it was shown that it *was* possible, resources were made available to develop and implement the technology. Proof of concept demonstrations are created to attract interest, support, or research and investment dollars.

propagate 1. Pass along, continue, extend. 2. To travel through a material or space, to cause to spread out over a greater area.

propeller head *slang* A highly technical person lacking in social graces.

proportional font A font in which each character is designed for maximum legibility and aesthetic beauty so the width of individual characters varies with the needs of the character and its relation to other characters. Most high end printing uses proportional fonts. See ligature. Contrast with monospaced font.

Prospero The Prospero Directory Service. A client/server directory resource on the Internet, accessible through the `prospero://...../` Uniform Resource Locator (URL).

prosumer A combination word derived from *professional* and *consumer* intended to indicate a level of literacy or competency somewhere between a skilled layperson and a technical professional. In other words, the reading level or skill level for the product or service requires more knowledge than possessed by an average layperson but not as much as might be required by a technician or design engineer in that industry. Computer technology has brought many new products into the marketplace that were previously used only by professionals with years or decades of experience. The printing, desktop publishing, and desktop video industries are good examples of sectors that now include a large number of prosumer products.

protocol A code of etiquette and procedure.

protocol, networking In networking, a procedure for organizing and exchanging data transmissions which may include specific rules and/or formats. Protocols for various Internet transmission and application protocols are well documented in online RFCs (Requests for Comments), some of which are standards or draft standards. If you need technical information on protocols, it is strongly suggested that you consult the RFCs. Many of the important protocols are briefly described in this dictionary under individual entries, but especially important on the Internet are Internet Protocol, Point-to-Point Protocol, and Transmission Control Protocol. See the appendix for a list of RFCs related to Internet protocols.

proxy An agent, deputy, or authority acting on behalf of another.

proxy, computer A software intermediary or agent that acts on behalf of clients and can act as a server or client. Proxy systems are frequently placed in points of a network where there are connections between LANs, or between a LAN and the Internet, or between a LAN and another external system such as an Internet Services Provider (ISP) or phone network. Proxies act as protocol managers and security administrators, and handle requests by servicing them or passing them through to other services. In cases where they are passed through, the proxy may interpret and modify a request before sending it on. Conceptually, a proxy server differs from a gateway in that requests are passed through as though from the original client, whereas a gateway handles the request as though it were originating from the gateway, thus forming a barrier between the server of the request and the client. A system can be configured as a security firewall to allow selective passing of messages through its portal. See firewall, gateway.

PRS Police Radio Service.

PRSM Post Release Software Manager.

PS See paging system.

PSA Protected Service Area.

PSDN packet-switched data network. See packet switching.

PSDS Public Switched Digital Service.

PSE packet-switched exchange.

pseudocode 1. A software program or process listing written in somewhat plain language (not a programming language per se) which delineates the steps and algorithms in a process in order to outline or draft it so it can be transcribed into a programming language. Many programmers find pseudocode more practical and useful than flow charts for drafting software program flow. 2. P-code. A type of code which is compiled down to an intermediary stage without being specifically tied to one computer architecture. There are a number of high level languages which can be compiled down to P-code, so they can execute faster (and sometimes to protect proprietary programming algorithms); they can then be distributed for a variety of computers. Each computer on which the P-code is run will need software to execute the program to do any further platform-specific translation of machine instructions which might be needed.

PSI 1. packet switching interface. 2. Policy Studies Institute. U.K. government consortium.

PSK See phase shift keying.

PSTN See Public Switched Telephone Network.

PSU Packet Switch Unit.

PSWAC See Public Safety Wireless Advisory Committee.

PSWN Public Safety Wireless Network

PTE See SONET path terminating element.

PTFE polytetrafluoroethylene. A synthetic material useful for insulating wires in environments where they might be subjected to heat.

PTI Payload Type Identifier. An ATM cell header descriptor which indicates the type of payload in the cells, such as user or management. See cell rate.

PTN Public Telecommunications Network.

PTO public telecommunication operators.

PTS 1. Personal Telecommunications System. 2. See Presentation Time Stamp. 3. Public Telecommunications System.

PTT See Post Telephone & Telegraph administration.

Public Access Profile PAP. A profile is defined by the Open Systems Interconnection (OSI) model as a combination of one or more base standards and association classes, necessary for performing a particular function. The PAP is an ETSI-published profile incorporated into Digital European Cordless Telecommunications (DECT) as a test specification. See Digital European Cordless Telecommunications.

public address system PA, PA system. A system designed to receive and transmit amplified sound, especially voice or music, to a wide audience. It may be acoustical or electrical. An acoustical PA may amplify through a horn-shaped object that directs sound. An electrical PA is familiar to most people as the microphone, circuits, and speakers installed in most public schools and hospitals. See amplifier, intercom, loudspeaker, sound.

public key In public key encryption schemes, this is the 'key' given to the public to act as one of the tools to create a message addressed to the person to whom the key corresponds. The recipient then uses his or her private key to open and read the encrypted message. It is important to create a private key password that can be remembered for a long, long time, otherwise it isn't very useful to distribute it to the public, as it will not be possible to decipher encrypted messages if the password is forgotten.

public key encryption An encryption scheme, which often also incorporates an authentication scheme in which public keys are distributed for encryption of messages to the person owning the key, and private keys are established for decrypting messages. Sometimes the encrypted message is differentiated into two components, a signature and a message. It is possible in some systems to encrypt the signature to provide authentication without encrypting the message. Pretty Good Privacy is a public key encryption technology which has been incorporated into various Internet applications, for example, in email through MIME. See Blowfish, Diffie-Hellman, PGP/MIME, Pretty Good Privacy.

Public Safety Wireless Advisory Committee PSWAC. A committee established in 1995 by the Federal Communications Commission (FCC) and the National Telecommunications and Information Administration (NTIA). The PSWAC provides advice on the various public safety agencies' wireless communications requirements through to the year 2010. This includes identification of emerging technologies and recommendations as to their utility and role, and emphasizes the importance of semiconductor technologies.

Public Service Commission PSC. A state regulatory authority for communications. See Public Utility Commission.

Public Switched Telephone Network PSTN. The national telephone infrastructure consisting of the RBOCs, the IXCs, and LECs. The PSTN is intended to further the goals of universal access to telephone service described in the 1934 Communications Act.

Public Switched Telephone Network history PSTNs have long been regulated in the U.S., beginning with the Post Roads Act in 1866. The Bell System and AT&T are an intrinsic part of the history of the PSTN, their dominance, research contributions, and many voluntary and required reorganizations have formed a colorful history in which the rewards of a competitive market, and the demands of independents to further fair competition, have often been difficult to sort out and maintain.

AT&T was divested from Western Union in 1913 and mandated to provide access to independent carriers across its long-distance network. Further divestiture occurred in 1975, and a Modified Final Judgment, approved by Judge Harold Greene, came into effect on 1 January 1984. The Justice Department formed 22 Bell Operating Companies (BOC) from AT&T, which were organized into seven regional Bell holding companies. Until 1991, when the Supreme Court granted wider privileges, the BOCs were not permitted to provide electronic data services.

Service divisions that resulted from the restructuring fall into the following general categories, which are described more fully under individual headings in this dictionary: Local Access and Transport Areas (LATAs), Local Exchange Carriers (LECs), Independent Telephone Companies (ITCs), Interexchange Carriers (IXCs), and Other Common Carriers (OCC).

Public Utility Commission PUC. State regulatory commission with jurisdiction over phone companies. See Public Service Commission.

publish *v.t.* To consolidate, organize, and present information in one or more communications media such as print, video, or Web pages. While *publish* in its broadest sense can mean to publish music or voice, and certainly music and voice are often included in multimedia publications, publish nevertheless tends to be used in conjunction with visual media, or media which are predominantly visual, with *record* more often used to describe the publishing of sound media. The concept of publishing has broadened with the new media-rich technologies. In fact, the creation of software is called software publishing, even though software itself is diverse and sometimes quite different in form from traditional publications. Nevertheless, it can be broadly stated that visual information brought together and organized on some media, in order to provide information and/or entertainment, has been published.

PUC See Public Utility Commission.

pull box In wiring installations, a box that is inserted into a long cabling run, especially at major junctions that allows easier access and working room for changing existing cables, removing them, or running new ones.

pulling eye A round open device incorporated into lines and other objects which are intended to be threaded through conduits or similar tight spaces. The eye is used to attach a line to thread the line. See birdie.

pulling glass *colloq.* Installing fiber optic cable, especially by pulling it through with various implements for this purpose, as opposed to air-blowing. See pulling eye.

pulse 1. A rhythmic beating, throbbing, vibrating, or burst of electricity, sound, or light. 2. A briefly transmitted electromagnetic wave or modulation. 3. In telephony, a brief, timed signal sent out by a pulse-dialing phone to indicate the desired destination. 4. In radar, a brief burst of microwaves.

pulse amplitude modulation PAM. A common means of converting a continuous stream of information into a series of samples with assigned discrete values used in analog to digital conversions. PAM takes into consideration the fact that for many media, it is not necessary for the entire communication to be transmitted for it to be understood. For example, in a voice conversation, if you were to take samples of the speech at frequent intervals and transmit those to a human listener, enough of the information is retained that most people can still understand what was said, even if subtle parts of the original message are not included. The more frequent the samples, the closer to the original, up to the limits of human perception. For example, moving images played by a film movie projector are typically displayed at the rate of about 24 to 30 frames per second. Higher sampling rates (more frames per second) do not substantially improve the quality, since humans can't see the individual frames, and the whole is perceived as motion, rather than as a series of still images.

PAM is a baseband transmission multiplexing scheme, used, for example, in Digital Subscriber Line (DSL) transmissions. PAM uses the amplitude of the information being modulated to determine the amplitude modulation characteristics of the transmission. PAM's advantage is that it uses lower frequency bands, which are less subject to attenuation and crosstalk, two areas that have caused concern in DSL installations over existing copper wires. See carrierless amplitude and phase modulation, pulse code modulation.

pulse code modulation PCM. A means of sampling a signal, subdividing it, and assigning values to the individual parts. Since this can be done in a number of ways, not all PCM transmissions are compatible, and different schemes are used in Europe and North America. PCM is a very common means of converting analog to digital signals and is widely used in telecommunications. Sampling rates may vary, but 6,000 to 10,000 times per second is typical. Eight-bit sampling is sufficient for voice communications so PCM is widely used in telephone systems (in contrast, at least 16-bit is preferred for quality music sampling).

Modulation allows information to be carried on a signal, and digital conversion allows a host of processing algorithms, compression routines, and selective regenerative techniques to be applied to, or used in conjunction with, the data.

Pulse code modulation is sometimes used in multiple modulation schemes. For example, PCM subcarriers may be used to frequency modulate another carrier.

Pulse code modulation first came into practical commercial use in the early 1960s when it was implemented as a T1 system by Bell Laboratories. T1 connections are now important fast-access data network technologies and are used in local area network, wide area network, and other corporate and educational systems. See differential pulse code modulation.

pulse compression In radar, a technique of using long pulses to increase the energy of the received signal while still retaining the resolution of short pulses.

pulse dialing A means of transmitting phone numbers through a phone line by converting the length of the signal that occurs with the rotation of the rotary dial into electrical pulses. Pulse dialing phones are becoming rare and pulsed signals will not work with online automated menu systems. The other common means of dialing is through tones, with each number assigned a particular tonal frequency. See tone dialing.

pulse dispersion The gradual spread of pulses as they travel over some transmissions medium.

pulse duration modulation PDM. A type of pulse modulation in which the length of time of the modulation is controlled to impart information. See pulse code modulation.

pulse generator A device for creating pulses of various specific amplitudes, shapes, repetition rates, and durations. Thus, a pulse generator can be used in a variety of pulse modulation schemes.

pulse inverter A wideband, low-distortion waveform-modifying device which enables an output wave to be the inverted form of the input wave.

pulse modulation A simple means of modulating a signal by applying pulses of current (often just on or off) without changing the frequency of the signal. See pulse code modulation, pulse width modulator.

pulse repetition frequency PRF. The number of pulses which occur in a unit of time, such as a second or minute.

pulse width modulator PWM. A digital circuit that can be configured to produce a pulse with desired characteristics (period, cycles, etc.). Thus, it can be used for a variety of purposes, including the control of the speed of mechanisms or to modulate transmission signals.

pulsing See pulse dialing.

PUMA Product Upgrade Manager.

punch *v.* 1. To prod, poke, or perforate. 2. To perforate small regular holes in paper, card, fine metal, or other similar flat surface in order to create a semi-permanent record of a code. Early gramophone cylinders, player pianos, patterned looms, music boxes, and punch card reading computers employed this technique to encode and store information intended to be read back later. See Hollerith card. 3. To apply pressure to an enclosure in order to pop out a small section, to enable the threading of circuit lines. 4. To apply pressure with a punchdown tool to a wire and terminal block. See punch down, punchdown tool.

punch card Any sturdy paper or card stock in which parts of the card are punched out according to a code that can later be reread and decoded to produce the original meaning. Punched codes in various media have been around for centuries, with early ones incorporated into music boxes. The Jacquard loom incorporated punch cards to store loom patterns in the early 1800s, revolutionizing the textile industry and causing a substantial uprising among human weavers, whose jobs were obsoleted by the automation.

Punch cards can be used to store many sorts of codes and variations of the concept are used in music boxes, player piano rolls, older computing devices, etc. See Hollerith card, zone punch.

punch down *v.* (The verb form is two words, the noun form often combined into one word.) To apply pressure, usually with a specialized punchdown tool, to wires looped around a terminal block to strip insulation from the end of the wire, and insert the conductive surface between the prongs of the terminal in order to make a solid, clean connection. See punchdown block, punchdown tool.

punchdown block, terminal block, cross-connect block A multiterminal block designed to facilitate electrical cross connections, especially those which may change from time to time or where future expansion is planned. Very common in telephone installations where a multiline wire is threaded into a building and then, at the punchdown block, is split into pairs to wire several individual phones in different locations. See punchdown tool.

punchdown tool A handheld wire installation device that resembles a screwdriver with an extremely short, notched shaft. The handle may be spring-loaded. The ends are designed to fit particular sizes of terminals and they are used to quickly connect wires to a punchdown block. The wire is looped over the prongs of one of the terminals in the block, and the punchdown tool is pushed against them to spread the prongs and snap the wire securely into place. Some punchdown tools will both punch and cut. See punchdown block.

Pupin, Michael Idvorski (1858-1935) An Eastern European-born American inventor and educator. He studied low pressure vacuum tube discharges, X-rays, and invented an electrical resonator and inductive loading coils, paving the way for long-distance wired transmissions.

Pure ALOHA See ALOHA.

purge *v.* To rid of, or remove. In computing, there are many instances where the operating system or applications software retains information, files, or backup files until explicitly instructed to remove them. For example, on many workstations, you can set the number of backup versions of a file you wish the system to automatically retain. Thus, each file may be saved under a slightly modified name or with the same name, but a different time stamp. *Purging* the directory may delete all the multiple copies of each file and retain only the most recent. Similarly, many desktop publishing and word processing programs save a history of edits and additions within the file, so that the information is available for recovery, if needed. This may result in *very* large files. These programs often will *purge* the file edit history if the user selects *Save As* instead of *Save*, resulting in a smaller, cleaner file, but one which is more difficult to recover if problems arise.

Many database programs retain information such as customer names and addresses, even if the customer entry has been deleted. In this way, it can be recovered if the entry is needed in the future, but doesn't clutter up directories or slow down searches in the meantime. There may be a menu item called *Purge* which allows you to permanently remove these records, or in some programs, you can set the system to *Purge* any deleted records that have remained inactive for longer than a specified amount of time.

Purkinje effect The human visual system is more sensitive to blue light in conditions of lower illumination and more sensitive to yellow light in conditions of higher illumination.

purple wire A color designation used by IBM to indicate wires that have been incorporated during testing and debugging to accommodate errors (purple wires may actually be yellow). See blue wire, red wire, yellow wire.

push-pull circuit A circuit consisting of elements operating in a phase relationship which is rotated 180 degrees one from the other. This makes it possible to cancel or filter certain elements and to amplify or magnify others. This type of circuit is used in oscillators, amplifiers, and transformers.

pushbutton dial A means of creating a signal on a transmissions system through a set of buttons, usually located on a keypad, as on a computer/telephony system or on a phone. The dial causes a tone to be generated at a specific frequency, which is then transmitted and interpreted as a numbered location of the destination receiver. A few units have both rotary and pushbutton capabilities. See touchtone. Contrast with rotary dial.

PVC 1. See permanent virtual connection. 2. See polyvinyl chloride.

PVCC Permanent Virtual Channel Connection.

PVN See private virtual network.

pW Abbreviation for picowatt.

PWB printed wire board.

PWM See pulse width modulator.

PWR An abbreviation for power sometimes seen in technical manuals or on components.

pylon 1. A tower or other tall supporting structure for stringing wire over a wide span. 2. A broadcast transmissions tower.

pylon antenna A vertical standing antenna consisting of slotted sheet-metal cylinders sometimes combined in sections, one atop the other to achieve greater height. The name comes from the cylindrical shape and orientation which resemble the log pylons on a dock.

pyrheliometer An instrument for measuring infrared radiation.

pyroelectricity Electromagnetic charge created through a change in temperature (usually heating). Pyroelectricity refers to the means of generating the charge, not the nature of the charge itself, which is the same as others. Crystals have valuable oscillating characteristics and are commonly used in timing mechanisms and radio electronics. Some of them have interesting electrical activity when exposed to heat.

pyromagnetic effect The combined effect of heat and magnetism in a material or circuit.

pyrometer An electronic instrument used for determining temperature in situations that are hotter than those typically measured by a traditional mercury thermometer. Temperature can be measured in a number of ways by electrical resistance, or by optical or other radiant energy emissions.

pyrone detector A type of radio wave crystal detecting device using iron pyrite and other conducting materials, with rectification occuring between the materials. See detector.

Pythagoras' theorem, Pythagorean theorem A mathematical rule that states that in a right angled triangle, the sum of the squares of the sides is equal to the square of the hypotenuse (the hypotenuse being the longest side). This theorem is widely used in mathematics for calculating distances and a multitude of other applications. It is named for the Pythagoreans, a philosophical group closely connected with Pythagoras. It is believed that Pythagoras, or one of his close disciples, developed this theorem.

q 1. Symbol for quantum value. 2. Symbol for electrical quantity in coulombs. See coulomb.

Q 1. Abbreviation for quality factor. See Q factor. 2. Abbreviation for queue. See queue. 3. A merit indicator for a capacitor or inductor equal to the reactance divided by the resistance.

Q address A storage location for data, from which the information can be accessed and retrieved.

Q antenna A type of dipole antenna in which the feed line impedance is made to match the radiometer center impedance by the interposition of a vertical section, consisting of parallel bars between the two.

Q bit A qualifier bit in the first octet of an X.25 packet header. This bit indicates the existence of control information, allowing the data terminal equipment (DTE) to signal that it wishes to transmit data on more than one level.

Q channel 1. In NTSC color television broadcasting, a frequency band in which green-magenta color information is transmitted. 2. In ISDN Basic Rate Interface (BRI) S/T interface implementations, an 800 bps maintenance channel.

Q demodulation Demodulation of an incoming broadcast signal in a color television receiver to combine the chrominance signal and the color-burst oscillator signal in order to recover the Q signal.

Q factor (Symbol - Q) Quality factor. 1. In electronic circuits, a means of describing the desired characteristics of a system. The terms of the Q factor vary depending upon what is being described (capacitance, inductance, etc.). Generally, a higher number is used to indicate a more efficiently operating component. 2. A measure of frequency selectivity, or the sharpness of resonance in a resonant vibratory system which has one degree of mechanical or electrical freedom.

Q multiplier A circuit used to enhance the selectivity of a component by feeding the signal back through the resonant network. This was used in early superheterodyne receivers, but various types of filters have, for the most part, superseded it.

Q output The reference output of an electronic flip-flop state, which may be one or zero.

Q Series Recommendations A set of ITU-T recommendations which provides guidelines for switching and signaling. These are available as publications from the ITU-T for purchase and a few may be downloadable from the Net. Some of the related general categories and specific Q category recommendations are included in charts on the following pages to give a sense of the breadth and scope of the topics listed here. See also I, V, and X Series Recommendations.

Q signal 1. In various data transmission schemes, it is common to split a signal and to alter the characteristics of one or both of the two data streams so that they can be transmitted together without excessive interference or crosstalk. A Q signal is one of two common streams; the other is the I signal, into which data is commonly split in various modulation systems. See quadrature amplitude modulation. 2. A telegraph code shorthand signal consisting of two or three letters prefaced by a "Q." See QBF.

Q

General Categories of ITU-T Documents

Series C	General telecommunications statistics
Series E	Overall network operation, telephone service, service operation, and human factors
Series F	Telecommunication services other than telephone
Series G	Transmission systems and media, digital systems and networks
Series H	Line transmission of nontelephone signals
Series I	Integrated Services Digital Networks (ISDN)
Series J	Transmission of sound program and television signals
Series P	Telephone transmission quality, telephone installations, local line networks
Series Q	Switching and Signaling
Series V	Data communication over the telephone network
Series X	Data networks and open system communication
Series Z	Programming languages

General

Q.9	Vocabulary of switching and signaling terms
Q.1300	Telecommunication applications for switches and computers (TASC) - General overview
Q.1302	Telecommunication applications for switches and computers (TASC) - TASC functional services
Q.1303	Telecommunication applications for switches and computers (TASC) - TASC management: Architecture, methodology and requirements
Q.1290	Glossary of terms used in the definition of intelligent networks
Q.1201/I.312	Principles of intelligent network architecture
Q.1202/I.328	Intelligent Network - Service plane architecture
Q.1203/I.329	Intelligent Network - Global functional plane architecture

Automatic and semi-automatic switching

Q.4	Automatic switching functions for use in national networks
Q.5	Advantages of semi-automatic service in the international telephone service
Q.6	Advantages of international automatic working

Signaling systems

Q.7	Signaling systems to be used for international automatic and semi-automatic telephone working
Q.8	Signaling systems to be used for international manual and automatic working on analogue leased circuits
Q.48	Demand assignment signaling systems
Q.50	Signaling between circuit multiplication equipment (CME) and international switching centers (ISC)
Q.698	Interworking of signaling system No. 7 ISUP, TUP and signaling system No. 6 using arrow diagrams
Q.700	Introduction to CCITT Signaling System No. 7
Q.701	Functional description of the message transfer part (MTP) of Signaling System No. 7
Q.721	Signaling System No. 7 Functional description of the Signaling System No. 7 Telephone User Part (TUP)

In-band and out-band

Q.20	Comparative advantages of “in-band” and “out-band” systems
Q.21	Systems recommended for out-band signaling
Q.22	Frequencies to be used for in-band signaling
Q.25	Splitting arrangements and signal recognition times in “in-band” signaling systems
Q.25	Splitting arrangements and signal recognition times in “in-band” signaling systems

Phone features and signals

Q.23	Technical features of push-button telephone sets
Q.24	Multifrequency push-button signal reception
Q.27	Transmission of the answer signal
Q.28	Determination of the moment of the called subscriber’s answer in the automatic service
Q.35	Technical characteristics of tones for the telephone service
Q.109	Transmission of the answer signal in international exchanges

Network access

Q.26	Direct access to the international network from the national network

Quality of transmissions; interference and noise

Q.29	Causes of noise and ways of reducing noise in telephone exchanges
Q.30	Improving the reliability of contacts in speech circuits
Q.31	Noise in a national 4-wire automatic exchange
Q.32	Reduction of the risk of instability by switching means
Q.33	Protection against effects of faulty transmission on groups of circuits
Q.44	Attenuation distortion

ISDN and B-ISDN

Q.71	ISDN Circuit mode switched bearer services
Q.761	Functional description of the ISDN user part of Signaling System No. 7
Q.762	General function of messages and signals of the ISDN user part of Signaling System No. 7
Q.763	Formats and codes of the ISDN user part of Signaling System No. 7
Q.764	Signaling System No. 7 ISDN user part signaling procedures
Q.767	Application of the ISDN user part of CCITT Signaling System No. 7 for international ISDN interconnections
Q.768	Signaling interface between an international switching centre (ISC) and an ISDN satellite subnetwork
Q.850	Usage of cause and location in the digital subscriber signaling system no 1 and the Signaling System No. 7 ISDN user part
Q.922	ISDN data link layer specification for frame mode bearer services
Q.923	Specification of a synchronization and coordination function for the provision of the OSI connection-mode network service in an ISDN environment
Q.2010	Broadband integrated services digital network overview - signaling capability set, release 1
Q.2100	B-ISDN signaling ATM adaptation layer (SAAL) overview description
Q.2110	B-ISDN ATM adaptation layer - service specified connection oriented protocol (SSCOP)
Q.2119	B-ISDN ATM adaptation layer - Convergence function for SSCOP above the frame relay core service
Q.2120	B-ISDN meta-signaling protocol
Q.2130	B-ISDN signaling ATM adaptation layer - service specific coordination function for support of signaling at the user network interface (SSFC At UNI)
Q.2140	B-ISDN ATM adaptation layer - service specific coordination function for signaling at the network node interface (SSCF at NNI)
Q.2144	B-ISDN Signaling ATM adaptation layer (SAAL) - Layer management for the SAAL at the network node interface (NNI)

Wireless connections

Q.14	Means to control the number of satellite links in an international telephone connection
Q.1001	General aspects of public land mobile networks
Q.1032	Signaling requirements relating to routing of calls to mobile subscribers
Q.1100	Interworking with Standard A INMARSAT system - Structure of the Recommendations on the INMARSAT mobile satellite systems
Q.1101	General requirements for the interworking of the terrestrial telephone network and INMARSAT Standard A system
Q.1111	Interfaces between the INMARSAT standard B system and the international public switched telephone network/ISDN
Q.1112	Procedures for interworking between INMARSAT standard-B system and the international public switched telephone network/ISDN
Q.1151	Interfaces for interworking between the INMARSAT aeronautical mobile-satellite system and the international public switched telephone network/ ISDN
Q.1152	Procedures for interworking between INMARSAT aeronautical mobile satellite system and the international public switched telephone network/ ISDN

Q

Q

Q spoiling A technique used with lasers in which a more powerful burst or pulse is attained by inhibiting the action of the laser for a few moments, to allow an increase in the number of ions, and then Q switching to allow the extra burst of light to be emitted.

Q-band A microwave frequency spectrum ranging from 36 to 46 GHz, between the Ka-band and the V-band. Frequencies in this range tend to be used for radar and small aperture satellite transmissions. See band allocations for a chart of designated frequencies.

Q.SIG A global common channel signaling protocol (CCS) based on the ISDN signaling protocol, used in the digital transmission of voice over digital networks such as ATM. In addition to the features in the ISDN signaling protocol, Q.SIG includes private branch exchange (PBX) features so a network of PBXs can interact as a distributed system. CCS systems are more prevalent in Europe than in the United States. See voice over ATM.

QA 1. quality assurance. 2. queued arbitrated. In DQDB, an information field segment used to transfer slots when they arrive through a nonisochronous transfer.

QAM See quadrature amplitude modulation.

QBE See query by example.

QBF, fox message QBF = "quick brown fox." The Q signal code to send a test sentence that includes all the letters of the English alphabet, which is commonly used to verify whether all letters in a device or coding system are present and/or working correctly. Familiar to most as "THE QUICK BROWN FOX JUMPS OVER THE LAZY DOG" (which is then repeated as all lowercase, if needed, on systems that support or need both). See Q signal, Z code.

QC quality control.

QC laser See quantum cascade laser.

QCIF See Quarter Common Intermediate Format.

QD See queuing delay.

QD-DOS, QDOS A historic microcomputer operating system (Quick and Dirty Operating System) developed by Tim Paterson, which was derived from a mid-1970s manual describing Gary Kildall's CP/M, and extremely similar in syntax and functionality. At that time, International Business Machines (IBM) was looking for an operating system for its line of microcomputers. IBM contacted Microsoft about contracting their (computer language) products, thinking they had also purchased the rights to CP/M. When they found that Microsoft didn't have an operating system, they went to visit Digital Research (originally Inter-Galactic Research), but the DR representative was reluctant to sign IBM's nondisclosure agreement on DR's behalf, especially when the attorney didn't like the terms of the contract.

IBM went back to Microsoft and DR thought it would have a further opportunity to talk terms with IBM, especially since Microsoft didn't have an operating system that could meet IBM's needs at the time, as they had been concentrating their efforts on developing computer languages. Microsoft, however, promised one to IBM in a very short time period, and delivered on the contract by purchasing the code for QDOS from Seattle Computing, the company for which Paterson was working. They provided it to IBM who released it as PC-DOS. Microsoft subsequently also purchased the distribution rights for QDOS, at the price of $50,000 and later released a slightly altered version of PC-DOS as MS-DOS (Microsoft Disk Operating System). Microsoft managed to contractually stipulate that they could retain the rights to independently sell the product they had developed for IBM, in competition with IBM. Thus, QDOS, derived from CP/M became IBM's product, rather than CP/M itself, and evolved into MS-DOS, and eventually Windows. See CP/M, Microsoft Corporation, MS-DOS, Digital Research.

QDU See quantizing distortion units.

QFC See Quantum Flow Control.

QFM See quadrature frequency modulation.

QIC See Quarter Inch Cartridge Drive Standards.

QL See query language.

QLLC See Qualified Logical Link Control.

QMS Queue Management System. See queue management.

QoR Query on Release.

QoS See quality of service.

QPSK 1. See quadrature phase shift keying. 2. See quaternary phase shift keying.

QRP A designation for low power amateur frequency radio transmissions. Low power transmitters and receivers are an interesting subgroup of hobbyist radio, when used with respect for the privacy of individuals and within regulatory guidelines. Regulations for short distance, low power transmissions are more lenient than other types of broadcasts. QRP transmitters can be used for short distance broadcasting, home security systems, door intercoms, climbing communicators, baby and child monitors, and other short-range projects.

QRP ARCI The QRP Amateur Radio Club International is a nonprofit organization dedicated to amateur design, construction, and use of QRP (low power) transmitters. See Amateur Radio Relay League, QRP.
http://www.qrparci.org/

QSAM see quadrature sideband amplitude modulation.

QSDG See Quality of Service Development Group.

QTAM See Queued Telecommunications Access Method.

QTC Quick Time Conference.

quad- Prefix for four.

quad antenna A type of array antenna which is similar in principle to a Yagi-Uda antenna, except that it uses full-wavelength loops in the place of half-wavelength straight elements, thus providing greater gain over a similar Yagi-Uda antenna. A two-element quad antenna is called a *quagi*.

quad wiring Wiring bundles consisting of four individually sheathed, untwisted wires brought together (aggregated) within a single cover. Quad wiring is often used for the internal wiring of two-line analog phones, with the lines inside generally color-coded green and red (tip and ring) for the first line, and black and yellow for the second line. This type of wiring is not recommended for data transmission installations. Quad fiber cables consist of four individual fiber cables bundled together within a single cover.

quadrature A state in which cyclic events are 90 degrees out of phase. In signal transmission quadrature, phasing is a common technique used to distinguish information in signals. It is also used to vary a signal so crosstalk between two closely associated transmissions is reduced.

quadrature amplitude modulation QAM. A modulation technique employing variations in signal amplitude. This is a two-dimensional coding scheme which can be transmitted in a narrower spectrum. It is a combination of amplitude and phase modulation. The QAM spectrum derives from the spectrum of the baseband signals as they apply to the quadrature channels.

QAM is similar to nonreturn to zero baseband transmission and multiphase phase shift keying (PSK), except that QAM does not have a constant envelope as in PSK.

QAM requires lower sampling frequencies and spectral width can be optimized by keeping the baud rate lower, thus reducing the potential for crosstalk. See modulation.

quadrature phase shift keying QPSK. A type of phase shift keying modulation scheme in which four signals are used, each shifted by 90 degrees, with each phase representing two data bits per symbol, in order to carry twice as much information as binary phase shift keying. QPSK can also be used to carry bit timing. Even more sophisticated systems exist, which employ differential encoding of symbol phases. See frequency modulation, frequency shift keying, on/off keying, modulation, quadrature sideband amplitude modulation.

quadrature sideband amplitude modulation QSAM. A modulation encoding technique in which different signal amplitude states are used to represent data.

quadruplex circuit A circuit which is carrying two bidirectional transmissions simultaneously to make a total of four.

Q

Qualified Logical Link Control QLLC. A data link control protocol from International Business Machines (IBM) which works with the IBM SNA systems to allow them to operate over X.25 packet switched data networks.

quality 1. Meeting subjective and/or objective standards of excellence in operation, manufacture, aesthetics, or a combination of these. 2. In manufacturing, quality is more narrowly defined as conformance to high objective standards of appropriateness, functionality, and longevity within the context of related products. 3. In service industries, quality is generally determined by adherence to operating and ethical standards of the industry and degree of customer satisfaction. See quality assurance.

quality assurance Systematic actions which seek to assure satisfactory levels of manufacture, service, functionality, and longevity.

quality factor See Q factor.

quality of service QoS. This has a general meaning across many industries and somewhat more specific meanings in the context of telecommunications networks. In general, quality of service is a descriptor and reference of performance for the provision of services on a network. It includes parameters and values such as data rates, acceptable delays, losses, and errors, etc.

As part of the QoS requirements for an ATM network, four class of service (CoS) traffic types have been specified.

CoS	Characteristics
Class A	Connection-oriented, constant bit rate (CBR), with a strong timing relationship between source and destination. Constant bit rate video and PCM encoded voice are included in this category.
Class B	Connection-oriented, bit rate may vary, with a strong timing relationship between source and destination.
Class C	Connection-oriented, bit rate varies, no timing relationship between source and destination. TCP/IP and X.25 are included in this category.
Class D	Connectionless, bit rate varies, no timing relationship between source and destination. Connectionless packet data is included in this category.

There are many types of data, and how they are perceived in part determines how their quality is evaluated. Consequently, QoS requirements vary with the type of data. See cell rate, class of service.

Quality of Service Development Group QSDG. A Telecommunication Standardization Sector group of the International Telecommunications Union established in 1984 to help develop practical implementations of international telecommunication quality of service (QoS) standards. It is funded primarily by administrations and ROAs.

quantization A process in which a continuous range of values, such as an incoming analog signal, is subdivided into ranges, with a discrete value assigned to each subset. This is a means of converting analog data to digital data, and is used in musical sound sampling, modem communications, voice over data networks, radio wave modulation, and many other aspects of telecommunications.

Generally the frequency of the sampling influences the quality and fidelity of the outgoing quantized signal, within certain limits set by the capabilities of the equipment and the characteristics of the human perceptual system. Higher sampling rates tend to produce closer approximations to the original signal, but also require greater transmission speeds and bandwidth.

Quantization is used in a number of modulation schemes, including pulse code modulation (PCM) which is commonly used in voice communications. See modulation, patches, pulse code modulation, sampling, quantization error.

quantization, vector A vector version of scalar quantization, designed to reduce the volume of data files or the bit rates of data transfers. Vector quantization has practical applications for image and speech coding.

quantization error A number of aspects can introduce error into a quantized signal, including the amount of noise and interference accompanying the signal, the signal range, or amplitude as it relates to the capabilities of the quantizing mechanism, the strength and complexity of the signal being quantized, and the mathematics used to carry out the conversion. Quantization error is sometimes assessed after a digital signal is reconverted to analog format, and the end signal is compared to the original, with the differences assessed subjectively (as in music systems) or evaluated with various measuring instruments.

quantize To convert a continuous range of values into discrete, nonoverlapping values or *steps*. This is an important means to convert analog to digital values.

quantizing distortion units QDU. A measure of the degree of degradation in a voice channel that occurs as a result of format and signal conversions (e.g., analog to digital to analog). This is described in the ITU-T G Series Recommendation G.113 (transmission impairments).

quantometer An instrument for the measurement of magnetic flux.

quantum (plural - quanta, symbol - q) A name for a relatively recently discovered and investigated phenomenon related to the movement of electrons. Quantum theory was first stated by physicist Max Planck in 1900. A quantum is a discrete quantity of electromagnetic energy, the smallest possible amount of energy at any given frequency v.

Quantum phenomena are of great interest to physicists, and researchers are now investigating ways of enlisting quantum behaviors in the manufacture and use of various industrial products such as lasers, and in operations associated with digital logic, with some surprising and provocative success. See quantum cascade laser.

quantum cascade laser QC laser. A new type of 'NanoLaser' developed by Frederico Capasso and Jerome Faist at Bell Laboratories in 1994. The QC laser has a number of advantages over diode lasers, including higher optical power and finer linewidth. QC lasers can be used in a wide variety of applications, including medical diagnostics, radar heterodyne detectors, and remote sensing applications, particularly environmental monitoring in toxic environments.

The wavelength of the laser is determined by quantum confinement. Thus, it can be used selectively over a wide range of the infrared spectrum by varying the layer thicknesses and spacing of the different materials used in its manufacture. This differs from other technologies in that the output wavelength is not dependent on the chemical composition of the semiconductors, but on their thickness and positioning. These layers, created with a molecular beam epitaxy (MBE) materials-growth process, are sometimes only a few atoms thick. It also functions at higher temperatures than diode lasers, making it practical for room temperature use.

The QC laser is a continuously tunable, single-mode, distributed-feedback device. To understand how a QC laser works, imagine an electric current stimulating a number of electrons to cascade over a series of steppes (a terraced organization), squeezed through quantum wells in successive layers, dropping off energy in the form of photons (light pulses) as they contact and travel through each steppe. At each steppe, the electrons perform a quantum jump between well-defined energy levels. The photons which are emitted as a result of their activity reflect back and forth in an amplification process that stimulates other quantum jumps and emissions, and results in a high output. See Capasso, Frederico.

Quantum Flow Control QFC. In ATM networks, a congestion avoidance scheme proposed for use on available bit rate (ABR) connections. For example, in a network in which VCI tunneling is implemented, the ATM device will send only after receiving explicit credit from a receiving ATM device at the other end of the connection. If tunneling is not used, buffer allocation and a credit manager must be included. If the buffer allocation is exceeded, noncomplying cells will be discarded.

quantum mechanics The study of atomic structure and behaviors using various measuring instruments and techniques. See Heisenberg uncertainty principle, quantum.

quantum noise When using a detector to investigate quantum characteristics in electromagnetic phenomena, there may be noise from random variations or fluctuations in the average rate of incidence of quantum interactions with the detector. These may be expressed in terms of photons.

Quarter Common Intermediate Format QCIF, Quarter CIF. A standard for the transmission of video frames in the ITU-T H.261 standard. QCIF consists of 144 lines of luminance and 176 pixels per line (144 *x* 176 CIF format is optionally supported by H.261). This relatively low resolution creates an image that has a soft focus, indefinite appearance, but has the advantage of using fewer system resources and less bandwidth. In fact, the standard was developed with the needs of circuit switched networks in mind. For small windows, simple images, and small display devices, it has practical applications, and it is widely favored for videoconferencing, especially on ISDN networks. H.261 is usually implemented in conjunction with other related standards. See Common Intermediate Format.

Quarter Inch Cartridge Drive Standards QIC. An international association, established in 1987, to promote the acceptance and use of quarter-inch readable/writable data cartridge drives and media. These types of storage media are commonly used for computer backup, secondary storage, and temporary storage for files that need to be transported.

More than a hundred QIC standards have been developed since 1988. The list on the following pages describes a few of the standards of interest, to provide a sampling of their scope and contents. QIC-40, QIC-80, QIC-3101, and QIC-3020 have been particularly prevalent in the tape cartridge field, although they are now being superseded by higher capacity formats. A complete list and fuller description of each standard are available on the QIC Web site. http://www2.qic.org/

Q

Q

QIC Standard	Date	Notes
Interface-related		
QIC-02	19-Apr-88	1/4-Inch Cartridge Tape Drive Intelligent Interface
QIC-59	28-Mar-85	1/4-Inch Cartridge Tape Drive Enhanced Basic Interface
QIC-36	14-Sep-84	1/4-Inch Cartridge Tape Drive Basic Interface
QIC-103	3-Feb-86	Basic Interface for QIC-100-MC 1/4-Inch Cartridge Tape Drive
QIC-104	12-Feb-87	Implementation of Small Computer System Interface (SCSI) for QIC-Compatible Sequential Storage Devices
QIC-106	11-Feb-87	Single-Channel Magnetic Head for Use with QIC-40-MC Floppy-Interface Minicartridge Tape Drives
QIC-107	5-Jun-86	Basic Drive Interface for Flexible-Disk-Controller-Compatible 1/4-Inch (6.35 mm) Mini Data Cartridge Tape Drives.
QIC-115	3-Feb-88	Basic Drive Interface for PS/2 Flexible-Disk-Controller-Compatible 1/4-Inch (6.35 mm) Minicartridge Tape Drives
QIC-117	28-Aug-96	Command Set Interface Specification for Flexible Disk Controller Based Mini Data Cartridge Tape Drives
QIC-121	14-Dec-95	Implementation of Small Computer System Interface (SCSI-2) for QIC-Compatible Sequential Storage Devices
Information Interchange-related		
QIC-24-DC	22-Apr-83	Serial Recorded Magnetic Tape Cartridge for Information Interchange (9 tracks, 10,000 FTPI, GCR, 60 Mbytes)
QIC-40-MC	2-Sep-92	Flexible-Disk-Controller-Compatible Recording Format for Information Interchange (20 tracks, 10,000 BPI, MFM, 40 Mbytes)
QIC-80-MC	20-Mar-96	Flexible-Disk-Controller-Compatible Recording Format for Information Interchange (28 tracks, 14,700 BPI, MFM, 80 Mbytes)
QIC-136	23-Apr-97	Unrecorded Magnetic Tape Cartridge for Information Interchange (0.25 in, 45 000 and 50 800 ftpi, 900 Oe)
QIC-137	20-Mar-96	Unrecorded Magnetic Tape Cartridge for Information Interchange (0.25 in, 38 750 ftpi, 900 Oe)
QIC-139	23-Apr-97	Unrecorded Magnetic Tape Cartridge for Information Interchange (0.25 in, 50 800 ftpi, 900 Oe)
QIC-140	26-Feb-92	Read-Only Unrecorded Magnetic Tape Minicartridge for Information Interchange [For Program Distribution] (0.25 in, 10 000-14 700 ftpi, 550 Oe)
QIC-141	26-Feb-92	Read-Only Unrecorded Magnetic Tape Cartridge for Information Interchange [For Program Distribution] (0.25 in, 10 000-12 500 ftpi, 550 Oe)
QIC-142	26-Feb-92	Unrecorded Magnetic Tape Minicartridge for Information Interchange (0.25 in, 550 Oe, “Pegasus”)
QIC-156	20-Mar-96	Unrecorded Magnetic Tape Minicartridge for Information Interchange (0.25 in, 45 000 ftpi, 900 Oe)
QIC-159	20-June-96	Unrecorded Magnetic Tape Minicartridge for Information Interchange (0.315 in, 400 ft, 10 000-14 700 ftpi, 550 Oe)
QIC-160	14-Dec-94	Unrecorded Magnetic Tape Minicartridge for Information Interchange (0.25 in, 425 ft, 10 000-14 700 ftpi, 550 Oe)
QIC-161	13-Dec-95	Unrecorded Magnetic Tape Minicartridge for Information Interchange (0.315 in, 750 ft, 14 700 ftpi, 550 Oe, Travan)
QIC-162	20-Mar-96	Unrecorded Magnetic Tape Minicartridge for Information Interchange (0.315 in, 750 ft, 44 000 ftpi, 900 Oe, Travan)
QIC-163	20-Jun-96	Unrecorded Magnetic Tape Minicartridge for Information Interchange (0.315 in, 425 ft, 76 300 ftpi, 1,800 Oe, MP++)
QIC-180	27-Aug-97	Unrecorded, Servo Bursts, Track I.D., Magnetic Tape Minicartridge for Information Interchange (0.315 in, 740 ft, 79 800 ftpi, 1650 Oe)
QIC-525-DC	10-Mar-94	Serial Recorded Magnetic Tape Cartridge for Information Interchange (26 tracks, 20,000 FTPI, GCR, ECC, 525 MB)
QIC-1000-DC	10-Mar-94	Serial Recorded Magnetic Tape Cartridge for Information Interchange (30 tracks, 45,000 FTPI, GCR, ECC, 1010 MB)
QIC-1350-DC	4-Mar-93	Serial Recorded Magnetic Tape Cartridge for Information Interchange (30 tracks, 38,750 FTPI, RLL 1,7, ECC, 1350 MB)
QIC-2100-DC	4-Mar-93	Serial Recorded Magnetic Tape Cartridge for Information Interchange (30 tracks, 50,800 FTPI, RLL 1,7, ECC, 2.1 Gbytes with 875 Feet of 900 Oe Tape)
QIC-3010-MC	27-Aug-97	Serial Recorded Magnetic Tape Minicartridge for Information Interchange (40 or 50 tracks, 22,125 BPI, MFM, 340 MBytes with 400 feet of 900 Oe 0.25-inch tape, 425 MBytes with 0.315-inch tape)

QIC-3020-MC 20-Mar-96 Serial Recorded Magnetic Tape Minicartridge for Information Interchange (40 or 50 tracks, 42,000 BPI, MFM, 680 MBytes with 400 feet of 900 Oe 0.25-inch tape, 833 MBytes with 0.315-inch tape)

QIC-3030-MC 2-Sep-92 Serial Recorded Magnetic Tape Minicartridge for Information Interchange (40 tracks, 50,800 FTPI, GCR 4,5, ECC, 555 Mbytes with 275-foot 900 Oe Tape)

QIC-3040-MC 20-Jun-96 Serial Recorded Magnetic Tape Minicartridge for Information Interchange (42 or 52 tracks, 50,800 FTPI, GCR 0,2 4,5, ECC, 840 MBytes with 400 feet of 900 Oe 0.25-inch tape, 1 Gbytes with 0.315-inch tape)

QIC-3050-MC 14-Jun-95 Serial Recorded Magnetic Tape Minicartridge for Information Interchange (40 tracks, 38,750 FTPI, RLL 1,7, ECC, 750 MBytes with 295 feet of 900 Oe tape)

QIC-3060-MC 4-Mar-93 Serial Recorded Magnetic Tape Minicartridge for Information Interchange (38 tracks, 50,800 FTPI, RLL 1,7, ECC, 875 MBytes with 295 feet of 900 Oe tape) *(inactive)*

QIC-3080-MC 15-Dec-94 Serial Recorded Magnetic Tape Cartridge for Information Interchange, Fifteen track, 0.250 in (6.35 mm), 10,000 bpi (394 bpmm) Streaming Mode Group Code Recording

QIC-126 15-Jun-94 Magnetic Recording Head for Use in 1 GBQIC-1000-DC Cartridge Drives

QIC-131 2-Dec-92 Magnetic Recording Head for Use in 555 MBQIC-3030-MC Minicartridge Drives

QIC-133 27-Aug-97 Magnetic Head for Use in QIC-3010-MC and QIC-3020-MC Recording Format

QIC-134 20-June-96 Magnetic Recording Head for Use in 4 GBQIC-3070-MC Minicartridge and 13 GBQIC-5010-DC Cartridge Drives

QIC-151 15-Jun-94 Magnetic Recording Head for Use in 750 MBQIC-3050-MC Minicartridge Drives

QIC-152 20-Mar-96 Preformatted Magnetic Tape Minicartridge for Information Interchange (0.25 in, 45 000 ftpi, 900 Oe)

QIC-153 20-Mar-96 Unrecorded Magnetic Tape Minicartridge for Information Interchange (0.25 in, 70 000 ftpi, 1400 Oe)

QIC-158 20-June-96 Magnetic Head for Use with 1.6 GBQIC-3080-MC Recording Format

QIC-171 11-Dec-96 Magnetic Head for Use with 4 GBQIC-3095-MC Recording Format

QIC-177 27-Aug-97 Magnetic Head for Use with QIC-3220-MC Recording Format

QIC-178 27-Aug-97 Magnetic Head for Use with QIC-3210-MC Recording Format

QIC-179 27-Aug-97 Magnetic Head for Use with QIC-4GB-DC Recording Format

QIC-181 27-Aug-97 Magnetic Head for Use with QIC-5210-DC Recording Format

Format-related

QIC-112 27-Oct-87 ECC Format

QIC-113 15-Jun-95 Host Interchange Format

QIC-123 1-Sep-94 Registry of Data Algorithm Identifiers for 1/4-Inch Cartridge Tape Drives

QIC-147 13-Dec-95 Cleaning Cartridge Recognition

QIC-174 21-Mar-96 CD-ROM Compatible Tape Format for Installable File System

QIC-176 28-Aug-96 Cleaning Cartridge Recognition with Light Prism

QIC-CRF1 11-Dec-96 Common Recording Format with Multichannel Capability for Use with RLL 1,7 Encoded Recording Formats

QIC-CRF2 15-Dec-94 Common Recording Format for Use with RLL 1,7 Encoded Recording Formats with Data Capacities of Less than 8 GB

QIC-CRF3 21-Mar-96 Common Recording Format for Use with Flexible Diskette Encoded Recording Formats

QIC 90-14 24-Apr-97 Hole Pattern assignments - 5.25 Inch Data Cartridge Identification Table

QIC 91-16 24-Apr-97 Hole Pattern assignments - 3.5 Inch Minicartridge Identification Table

QIC 95-101 27-Sept-97 Medium Types and Density Codes

Format-related - Compression

QIC-122 6-Feb-91 Data Compression Format for 1/4-Inch Data Cartridge Tape Drives

QIC-130 3-Sep-92 DCLZ Data Compression Format

QIC-154 10-Mar-94 Adaptive Lossless Data Compression (ALDC)

Format-related - Protocols, Command Sets

QIC-157 13-Dec-95 Common SCSI/ATAPI Command Set for Streaming Tape

QIC-172 20-Mar-96 Common SCSI/ATAPI Command Set for Floppy Tape

QIC-146 2-Dec-92 Autoloader SCSI Gateway Protocol: Serial Bus Implementation

quarter wave The distance, or elapsed time, in a conducting line or through a conducting space, which is 90 degrees to a wave disturbance. This information in used in the design of antennas and in the quadrature transmission of signals, particularly in modulation schemes. See quadrature.

quartz A silicon dioxide mineral found, or synthesized, in crystal form and in crystalline masses which is widely used in scientific research and telecommunications due to its oscillating qualities. Quartz is transparent, harder than glass, and varies in its oscillating frequencies depending upon its size and shape. Quartz crystal watches are extremely accurate, and quartz arc lamps are used for sterilization, due to the way ultraviolet light passes through the crystal. See piezoelectricity.

quartz crystal A quartz crystal is a piece of quartz cut to a precise size for a specific purpose. Quartz has remarkable constancy in its vibratory qualities which makes it suitable for extremely precise time devices. These vibratory qualities can be controlled by manipulating the shape and size of the crystal. Early radio sets were called 'crystal detectors' as they employed various crystals (galena and carborundum were popular) to detect (rectify) and to channel a radio wave. Quartz is commonly used in oscillators and filters. Quartz crystals are used to provide timing in watches and to stabilize broadcast waves. See quartz, quartz crystal filter.

quartz crystal filter The properties of quartz crystals make them useful for a variety of applications which require highly selective electrical circuitry, and hence they are used in the creation of various types of filters. Synthetic quartz crystals, developed in the 1950s, furthered the manufacture of quartz filters for use as electronic components. There were, in fact, few other materials that offered the advantages of natural or synthetic quartz until the development of lithium-tantalate crystals in the Bell Laboratories. See lithium-tantalate, quartz, quartz crystal.

quaternary phase shift keying QPSK. A modulation technique which is used to encode digital information to be transmitted over wire or fiber networks. It is a subset of phase shift keying (PSK), and is essentially a four-level version of phase modulation (PM). QPSK divides the bit stream into two streams, and sends them alternately to in-phase and out-of-phase modulators, where they are subsequently demodulated at the receiving end.

QUBE An Interactive TV information utility. Warner instituted the QUBE interactive educational TV network in the late 1970s. The first interactive television concert, broadcast live over the QUBE system in 1978, featured Todd Rundgren, pioneer multimedia recording artist.

quench To bring to a sudden halt, to cool rapidly, to quickly extinguish a flame, spark, or gas emission.

quench oscillator In some super-regenerator circuits, a type of ultrasonic oscillator which serves to quench, or rapidly reduce, the regeneration when it has almost increased to the point of oscillation.

quenched spark gap Early wireless transmitters used spark gaps in their spark transmitters, with several types of gaps: open gaps, rotary gaps, and quenched gaps, each with different strengths and weaknesses. Quenched gaps employed a racklike series of metal plates separated by thin layers of mica, resulting in a very small spark which is quickly quenched and does not tend to overheat as do open gaps. Due to improvements in technology and the need for regular cleaning to keep quenched gap transmitters working optimally, they were eventually superseded by continuous wave (CW) transmitters.

query 1. Request for data, in which the content of the data itself is the desired result. Common in database applications. 2. Request for data which provides information about the state (operating parameters, mode, security, etc.), or functioning (availability, readiness, status, responsiveness, etc.) of a system. Usually at a low operating level, and generally transparent to the user.

query by example QBE. An idea introduced in the 1970s whereby a user interacts with a front end to a database by supplying examples of the type of information that the user wants to retrieve. There are circumstances where this is more practical than querying by keywords or algorithms. A number of popular database programs provide this capability.

query language A programming language intended to facilitate search and retrieval of information, usually from a database. Query languages are frequently in the form of interpreted scripting languages or graphical report generators, with commands that are similar to common English words, to make them easier to program by those without programming backgrounds.

queue A stream of items or tasks waiting to be processed or executed, such as calls to an operating system, a network, or a phone system. Queues are used to maximize the use of existing resources, especially on shared systems. It's expensive to put a printer on every computer in a network and, since printing doesn't happen as often as data input/output, it's not efficient either. By allocating one printer to every few workstations, user print requests can be handled efficiently by the network, with simultaneous requests administrated through a set of parameters. This also can improve resource choice. By sharing printers, it may be possible to offer a variety of types of printers and paper sizes, which is more practical and economical than trying to purchase several printers for each computer. See queuing.

queue administration Queues are widely used to manage resource-sharing on a network. Whether the resource is a printer or modem, an applications program, data file, or gateway to the Internet or Web, computer systems create, manage, authorize, and prioritize access to these resources and services through queues which are usually transparent to the user.

On phone networks, queue administration may involve putting a caller on hold, checking to see if and when agents are ready to take the call, playing periodic messages to the caller, and assigning the call to the appropriate agent.

On computer networks, queue administration may involve logging in users as they sign on to the system, checking for the existence of devices when a resource request occurs (e.g., a printing job), determining if there are others in the queue, and where to slot the new request (the size of the print job, or relative priority of the user requesting the job may be taken into consideration), and may even change the queuing arrangement dynamically if another printer comes online or a print request is canceled before the job is run.

Queued Telecommunications Access Method QTAM. An International Business Machines (IBM) communications control protocol which handles some applications processing tasks. QTAM is used in a number of telecommunications applications, including message switching, data processing, etc.

Queuing Delay QD. In cell-based transmissions, a delay imposed on a cell due to the current inability of the cell to be passed on to the next element or function (because of congestion or errors). Depending upon the system and priorities, significant delays may have several results; the buffered cell data may be returned or destroyed.

queuing theory Queuing, in its broadest sense, involves an understanding of mathematics, statistics, modeling, data flow, and human behavior in order that machines may be configured, tuned, and operated so as to carry out worthwhile tasks and processes in an efficient and orderly manner. In a more specific sense, as applied to networking, queuing focuses on organization, priorities, delay, and loss, and, as such, is related to standards for quality of service (QoS).

Researchers in queuing theory regularly come from fields such as probability mathematics, complex systems theory, and simulation research.

One of the earliest developers of queuing concepts was Danish telephone engineer A. K. Erlang, who studied and described telephone traffic in its mathematical context and practical applications. Another significant contributor to the body of knowledge in queuing theory is Leonard Kleinrock, who was involved in the early development of the ARPANET and who authored "Information Flow in Large Communication Nets" in 1961. He subsequently wrote "Communication Nets" in 1964, which provides design and queuing theory for building packet networks, in spite of a common sentiment at the time that packet switching wouldn't work.

QUICC Quad Integrated Communications Controller. A single-chip integrated microprocessor from Motorola designed for embedded telecommunications and internetworking applications.

quick-break fuse A type of fuse which breaks a circuit very quickly, if a surge or other anomalous electrical condition occurs. Quick-break fuses are especially useful with electronics components, which are sensitive to electrical fluctuations and prone to damage.

QuickDraw A widely used proprietary computer drawing and display specification from Apple Computer Inc. QuickDraw provides a means for displaying images on the screen and processing PostScript files so they can be printed on nonPostScript-equipped printers.

quicksilver *colloq.* mercury.

QuickTime A proprietary cross-platform computer display and animation program from Apple Computer Inc. that runs on Macintosh, PowerMac, Windows 95, and Windows NT. QuickTime allows some interesting applications to be developed and distributed, including frame-based animation, whiteboarding, video clips, teleconferencing applications, virtual reality environments, games, and more.

Most recently, QuickTime has been enhanced to support streaming media in Internet browsers, support for more than 30 different audio and video file formats, and modules for saving digital video (DV) camcorder formats for the development of digital video.

QuickTime 3D Apple Computer's 3D QuickTime cross-platform 3D rendering software.

QuickTime Conference QTC. Designed on Apple Computer's QuickTime compression technology, QuickTime Conference supports videoconferencing in a window on the computer screen. Electronic whiteboarding is also supported, so participants can communicate and collaborate on shared drawing, text, or other projects. The software can be used to deliver Web events using QuickTime Live! software. See Simple Multicast Routing Protocol.

QuickTime VR An Apple Computer extension to QuickTime which adds cross-platform virtual reality capabilities through a movie-like presentation of images. The user can move through the scene, pan the surroundings, interact with objects, and much more. QuickTime Authoring Studio can be used to create virtual reality scenarios for display in QuickTime VR.

quiet tuning In radio receivers, a tuning characteristic in which the signal is kept quiet, that is, not broadcast to the listener, except when the tuner is getting a clean, clear signal of a specified frequency on the incoming carrier wave. In other words, if it isn't a good signal, the receiver mutes the sound to save the listener from the distraction of weak or noisy stations.

QWERTY A ubiquitous computer and typewriter keyboard configuration designation, named after the six top left lettered keys. Although each computer keyboard has a variety of individual symbol and function keys, most follow common QWERTY configurations for numbers, letters, and common punctuation marks. This configuration was originally designed to slow down typing in order to help prevent jamming on the old manual typewriters (if you've ever used one, you know how easily they jam).

Other keyboard layouts have since been proposed which consider ergonomics and physical properties, the most recognized being the keyboard designed by August Dvorak. The Dvorak keyboard was developed on the basis of studying finger motion and lettering combinations which were easier and more efficient to execute, and incorporating them into new keyboard character arrangements. A number of variations of this by other people have also been called Dvorak keyboards, even when they differ from that developed by A. Dvorak.

Unlike typewriters, computers make it relatively easy to remap the key positions and alternate keyboards can be designed to put the letters anywhere the user desires. In spite of this, the QWERTY keyboard has remained prevalent, and manufacturers and teaching institutions are reluctant to change to other systems, whether or not they have advantages.

Qwest Communications A telecommunications company creating fiber optic networks in over 100 U.S. and Mexican cities. Commercial services provided by Qwest include dedicated business Internet access, Internet faxing, Internet phone (Q.talk), and video.

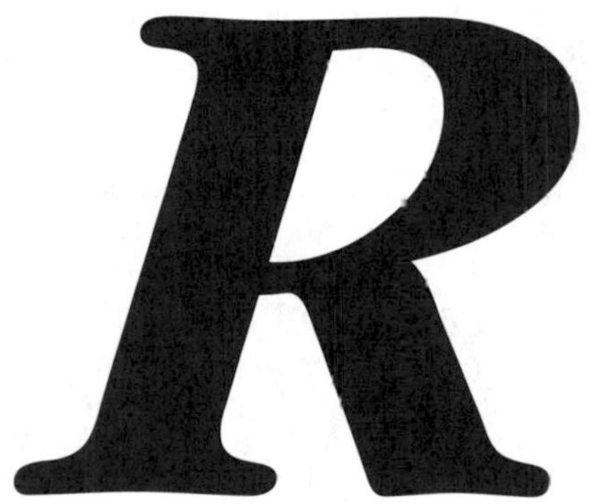

r Abbreviation for roentgen (røntgen). See roentgen.

R 1. Symbol for range. 2. Symbol for resistance. See resistance.

R interface In ISDN, a number of reference points have been specified as R, S, T, and U interfaces. The R interface is a generic reference point at boundaries between nonISDN station equipment (TE2) and an ISDN terminal adapter (TA). RS-232c and analog phone line tip and ring are examples of R interfaces.

R/T See realtime.

R & D, R&D research and development.

R and E, R & E See Research and Education.

RA real audio. A streaming protocol which allows realtime audio or audio files to be served to a Web client and played.

RA, RA number See return authorization.

RAC See Radio Amateurs of Canada.

RACE 1. random access computer equipment. 2. Research into Advanced Communications in Europe. This organization is a predecessor of Advanced Communications Technologies and Services (ACTS).

RACES Radio Amateur Civil Emergency Service.

raceway A duct or channel system designed to hold, protect, and direct interior wiring circuits. Raceways are typically plastic or metal modular construction, with a variety of fittings so individual sections can be interconnected and holes can be punched where needed. Raceways can be mounted on or in walls or floors. See molding raceway.

RACF resource access control facility.

rack, tray A support structure designed for the easy insertion, removal, and configuration of modular component systems. Racks are frequently equipped with rollers, although large ones may be attached to a wall for better support. They are generally assembled from rigid metal strips, interconnected to produce a strong open structure so the components can be quickly slid in and out of the individual bays from the front, and cabled to one another at the back. Racks are commonly used in the broadcast TV and video editing industries; they can also be found in telephone switching installations and on large computer networks with a variety of storage media. See distribution frame.

rack mountable A component designed to specifications so it will fit easily and securely into a rack of a standard size for components from that industry. See rack.

rad radiation absorbed dose. A quantification of radiation energy that describes how much radiation is delivered to 1 gram of a substance by 100 ergs of energy. Radiation absorbed by body tissue is measured in roentgens.

RAD 1. rapid application design. 2. recorded announcement device. 3. remote antenna driver.

radar *ra*dio *d*etection *a*nd *r*anging. In its basic form, radar is a means of detecting distant or unseen objects by emitting radio waves in the high frequencies and measuring the reflected response. As such it can operate at night, during fog, and in situations where something is too distant to be seen by unaided eyes. Radar works on the principle that radio waves will deflect off of solid or sufficiently dense objects in a way that can be controlled so the returning signal can be analyzed for the presence of the objects, their general shape and size, and their distance.

The earliest use of radio frequencies for bouncing signals was in the 1920s and 1930s, where it was used to determine the presence of marine vessels and aircraft. Radar typically operates in ultra high frequencies (UHF) and microwave frequencies. See Taylor, A.H.

radar detector A device designed to detect the presence of radar signals. These are used in military applications and are sold for civilian use in the form of car-dash devices to detect police speed-detection radar systems. The use of radar detectors is regulated and prohibited in some areas.

radar screen A small display device, usually round or rectangular, which shows target signals as illuminated dots or blips. There may be grids and other alignment and location marks superimposed over the illuminated blips on the screen to aid in tracking and location.

radar systems Devices incorporating radio waves to detect the presence and characteristics of distant or otherwise 'unseen' objects. Although radio echoes were observed in the 1920s, and put into practical use in the 1930s, developments in radar guidance, detection, and identification systems did not flourish until the second World War. See cavity magnetron.

radial acceleration Acceleration in a circular trajectory, characteristic of a spinning solid or liquid substance. Radial acceleration is used in centrifugal separators to isolate particular particles or substances. The radial acceleration characteristics of various spiraling entities are of interest to astronomers. In optical media, radial acceleration is one of the characteristics which is measured to determine conformance with expected properties or standards, along with axial acceleration and radial runout.

radiant energy Transmitted electromagnetic energy such as heat, light, or radio waves. Radiant energy is typically measured in calories, ergs, or joules.

radio An appliance or other device designed for the transmission and/or receipt of radio wave communications. There are many types of radio technology: amplitude modulation (AM), frequency modulation (FM), shortwave, cellular, short-range (cordless phones, wireless intercoms), etc. With increased demand for wireless communications, harnessing and using radio waves efficiently has become extremely important in both scientific and commercial research. More details about radio communications can be found under individual listings in this dictionary. See crystal detector, detector.

Radio The publication name of a widely-distributed Soviet electronics journal which, in June 1957, announced the Soviet Union's plans to soon launch a satellite ("sputnik" in Russian), and provided details of the planned launch date, modulation techniques, and frequencies to be used. Sputnik I did in fact launch at the end of that year. See Sputnik I.

Radio Act of 1912 With increasing interest in radio broadcasting and demand on airwaves, the U.S. Congress passed an act which granted the U.S. Department of Commerce the authority to regulate amateur broadcasting in order to prevent interference with government stations and to increase maritime safety, largely due to the sinking of the Titanic. See Titanic.

Radio Act of 1927 As a response to the enormous rising demand for broadcast channels in the early part of the century, a conference was held to sort out the chaos. As of the Radio Act of 1912, the U.S. Department of Commerce took control of radio broadcasting. Zenith Radio Corporation applied for a license to operate at a frequency that was being used by other stations as well, and so was granted a license to broadcast at a different frequency. Zenith changed frequencies to one that had already been granted, instead of using the one that had been licensed. In the process of investigating the violation, it was found that the Department of Commerce didn't have sufficient jurisdiction to stop the actions of the broadcaster, and one of the consequences was the creation of the Federal Radio Commission (FRC) in 1927. This was later to become the Federal Communications Commission (FCC) through the Communications Act of 1934. See Communications Act of 1934, Federal Communications Commission.

Radio Amateurs of Canada RAC. RAC provides liaison, coordinating functions and policy decisions for the benefit of Canadian amateur radio organizations and individual amateur radio operators. http://www.rac.ca/

radio astronomy The art and science of electromagnetic waves, especially those distant from the Earth, used in radio communications, including their characteristics, control, and practical applications. See sky maps.

radio broadcasting Commercial radio broadcasting began in the early 1900s, arising out of the experimental broadcasts of inventor R. Fessenden in 1906. There were many amateur broadcasts between 1906 and 1920, including the regularly scheduled shows by

Charles 'Doc' Herrold, at the Herrold College of Wireless and Engineering in California, and the pre-KDKA broadcasts from the garage of F. Conrad in 1919. KDKA itself is acknowledged as the first commercial station, hitting the airwaves in 1920.

Commercial broadcasting in Europe was underway by 1913, and the Eiffel Tower still stands as a historic reminder of the lofty ambitions of the broadcast pioneers. It was built for the Paris World's Fair in the 1800s and there have been several attempts to remove it since then, but its usefulness as a giant antenna is one of the reasons it was preserved. Lee de Forest participated in one of the first transcontinental broadcasts from the world's largest radio tower. The Radio Corporation of America (RCA), founded in 1920, is one of the most well known and influential of the early radio pioneers, and much of its history is related to the activities of David Sarnoff. Sarnoff was also instrumental in the forming of the National Broadcasting Corporation (NBC) in 1926. The following year the Columbia Broadcasting System (CBS) was formed (originally Columbia Phonograph Broadcasting until William S. Paley bought out the company in 1928).

From 1921 to 1922 the number of commercial stations in the U.S. increased from five to over 500. In the early 1930s, record companies became nervous about competition from radio stations and began restricting the open broadcasting of audio recordings. From that point on, royalties and other means of enforcing payment for broadcasts were instituted. By the late 1930s the wonderful music from bands and orchestras around the world could be heard through the magic of radio, and listeners who had never been to a theater to hear a live performance enjoyed the new form of entertainment. The advent of radio meant the eventual death of vaudeville, but some of the vaudevillian actors, perhaps best exemplified by George Burns and Gracie Allen, made a successful transition to radio, and eventually to TV programming.

By the early 1940s, frequency modulated (FM) broadcasting, made possible by the tireless efforts of inventor Edwin Armstrong, was beginning to catch on and, while it didn't supersede AM, it provided clean, clear transmissions that were favored by public broadcast and classical music stations. By the 1950s, radio had competition from TV broadcast stations, but unlike many technologies, it didn't lose its practicality and appeal. Radio stations in North America still outnumber TV stations, and radio sets continue to be in demand.

The next major milestone in radio broadcasting came with the exploration of space. In 1969, American astronauts sent sound and images from the Moon to Earth. And soon communications satellites were being launched into orbit in the 1970s and 1980s. This provided a means to develop mobile communications, and linked computers and radios as never before. Many of the pioneer communications efforts and new technologies were contributed by amateur radio enthusiasts, most notably through the OSCAR and AMSAT satellite programs.

With digital electronics, laptops, and cell phones, the importance of radio continued to grow, as wireless communications were integrated into increasingly mobile lifestyles. One of the significant recent events in radio broadcasting is the introduction of digital broadcasting, pioneered by Sweden in 1996. See AMSAT; ANIK; CKAC; Emergency Alert System; KDKA; OSCAR; Radio Corporation of America; Sarnoff, David.

radio broadcasting regulations Many different sets of guidelines and regulations have been developed to manage radio broadcasting. Some of these were intended to curtail unfair business practices, such as more powerful transmitting stations deliberately drowning out less powerful ones, and some were implemented to organize and coordinate the use of limited 'airspace,' that is, the limited availability of broadcast frequencies. Others were put into effect in wartime to shut down broadcasting almost entirely, curbing the broadcast pirates, but also curbing responsible amateurs. In 1963 the Emergency Broadcasting System (EBS) was established, recently replaced in 1997 by the Emergency Alert System (EAS).

Several Radio Acts and later Telecommunications Acts have controlled American broadcasting over the decades. The jurisdiction has changed hands a number of times, from the U.S. Secretary of Commerce to the Federal Radio Commission (FRC) in 1927, to the Federal Communications Commission (FCC) in the mid-1930s. The FCC has retained its wide-ranging licensing and regulatory powers up to the present time. See Emergency Alert System, Federal Communications Commission, Radio Act of 1912.

radio button A physical button on a component, or iconic button in a software program, which permits selection of only one option from a group of mutually exclusive selections. Selecting any one option automatically deselects the previous option. The name derives from the action of pushbutton radio sets in which buttons can be pretuned to selected stations, and then pushed for the desired station, one at a time. Software radio buttons are often seen on input forms on Web pages.

Radio Common Carrier RCC. Service providers of mobile telephone and paging services employing radio technology, as opposed to land line transmissions.

radio control The control of various models, vessels, and other moving objects by means of radio waves. J. Hammond received a patent for radio control technology in 1912.

Radio Corporation of America RCA. An offshoot of General Electric founded in 1919 as a result of a merger with the Marconi Wireless Telegraph Company of America. In 1920, RCA made a significant agreement with WSA, AT&T, GE, Westinghouse, and others, to be the exclusive distributor of radio receiving sets and crystal detectors. In 1921, David Sarnoff joined the company as its general manager, and later moved up in the Corporation, becoming vice president in 1926. Sarnoff was a colorful part of its history for many decades. See Armstrong, Edwin Howard; Sarnoff, David.

radio facsimile The transmission of the contents of pages including text and images by means of radio signals. Radio facsimiles were pioneered in the 1800s, and this early form of facsimile machine was in use at least as early as 1943. See facsimile.

radio frequency RF. Radiant electromagnetic waves that range from about 3 to 10 kilohertz at the lowest end to just about 300 gigahertz at the high end, a position that falls between the audio frequencies and the boundary of the visible spectrum where infrared is found. Radio frequencies are widely used for radio and television broadcasting, and for various types of wireless communication. The frequency range has been administratively subdivided into a number of categories in order that limited airwaves can be assigned and licensed in an efficient way. In the U.S., this responsibility is managed by the Federal Communications Commission (FCC); in Canada it's managed by the Canadian Radio Television and Telecommunications Commission (CRTC). See band allocations for chart.

radio wave An electromagnetic wave, commonly used to carry audio transmissions, in a frequency spectrum that ranges from 10 KHz to 200 GHz. Transmission waves such as radio waves are further classified into subcategories according to various properties; examples include ionospheric waves (sky waves), ground waves, short waves, and others. The characteristics of various transmissions media, chiefly the Earth's ionosphere, are exploited to facilitate in aiming and propagating these waves. Frequency divisions of radio waves according to wavelength (higher frequencies have shorter wavelengths) have been designated as shown in the chart under band allocations. Sounds and other signals are converted to radiating waves for transmission, then converted back at the receiving end. See antenna, ionospheric wave, ground wave, radio, short wave.

radiogram 1. A telegram sent through radiotelegraphy, also called "radiograph." 2. Combined radio receiver and phonograph.

radiometeorograph See radiosonde.

radiophone A device that transmits sound through radio waves. Although the term is less common, radio phones are everywhere; they just have more individual names now due to their specialization (cordless phones, cell phones, etc.).

radiosonde, radiometeorograph A miniature, automatic radio transmitter usually sent aloft on an aircraft or meteorological balloon, to transmit back meteorological information, such as temperature, humidity, pressure, etc.

radiotelegraphy Transmission of telegraph signals through radio waves. The carrier wave was modulated to carry Morse code. The two main types were continuous-wave (CW), in which the carrier wave itself was interrupted to form the coded symbols, and interrupted continuous-wave (ICW) in which the carrier was modulated at a fixed frequency.

radiotelephony The art and science of communicating through radio waves, often by means of various types of radiophones.

radome A radar 'dome,' a housing around a radar antenna which protects it without interfering with the signals. Radomes are especially important in radar antennas that are exposed to the elements, as in an airplane.

RADSL Rate Adaptive Digital Subscriber Line. See chart under Digital Subscriber Line.

ragged In desktop publishing, an uneven margin, usually on the right side of the line. This occurs when proportional text is not justified (lined up). Ragged text is generally easier to read than double-justified text, particularly if the column widths are somewhat narrow.

RAID See redundant array of inexpensive disks.

raised floor distribution A type of structure designed to accommodate a horizontal distribution frame for the attachment and management of wiring installations. It is typically designed so that the floor covering can be pulled aside or lifted to gain access for changes or additions. See distribution frame.

RAM See random access memory.

RAM disk An area of chip memory allocated and managed as though it were a disk drive. Unlike a disk drive, RAM is volatile; it requires a continuous source of power to retain its information and will lose the stored data if the system is turned off. A RAM disk is a means of disk caching that was popular when many systems had only floppy drives and no hard drives. It provided a fast way to access data without doing disk seeks or having to swap out disks. RAM disks are now less prevalent.

RAM Mobile Data An open architecture, nationwide commercial data communications service offered by Ericsson, BellSouth, and RAM Broadcasting. It is similar to ARDIS, a packet data service offered by Motorola. Base stations are used for relaying messages to users or to other stations.

RAMDAC random access memory digital-to-analog converter. A graphics adapter display circuit which converts the computer digital information for representing the screen image into analog signals which a cathode ray tube (CRT) display monitor can use.

random access memory RAM. A type of computer memory in which data in any part of memory can be accessed in any order, that is, it is not restricted to reading and writing data sequentially as in serial data, tapes, etc. This makes it a very fast access device, and RAM is almost universally incorporated into computing systems. In the mid-1970s, microcomputers typically had 4 kilobytes of RAM and the price per kilobyte was about $100. Since then, the amount of memory installed in microcomputers has increased as prices have decreased. While there have been some dramatic interim fluctuations in the prices, there has also been an overall trend, as illustrated in this general summary of the quantity/price changes over two decades.

Time Period	Typical Quantity	Approx. Price/Mbyte
mid 1970s	4 kilobytes	$100,000
late 1970s	8 kilobytes	$2,000
early 1980s	128 kilobytes	$1,000
mid-1980s	256 kilobytes	$700
late 1980s	4 to 8 Mbytes	$400
early 1990s	8 Mbytes	$250
mid 1990s	8 to 16 Mbytes	$200
early 1998	16 to 32 Mbytes	$4
late 1998	32 to 64 Mbytes	$1

range 1. The extent, distance, or scope represented, or traversed. 2. In a Global Positioning System (GPS), a fixed distance between two points, such as the distance between a GPS receiver and a satellite. 3. In mobile communications, the maximum distance of a transmission sufficiently clear to be useful.

Rapid City IP switch See Accelar routing switch.

rare earth Some rare earth elements are commonly used as doping agents which can aid in the propagation of signals when added to transmissions media, such as optical fiber, during manufacture. Rare earth doping is also being applied toward the design and manufacture of electrically pumped lasers that employ electronic circuitry. Erbium, gadolinium, europium, and samarium are examples of rare earth elements.

rare earth doping A means of using small amounts of rare earth substances to alter the transmission-carrying capacity of a medium such as a fiber optic waveguide. Doping allows a signal to be amplified by the stimulation of the rare earth substances, thus increasing the transmissions capability and the distance that the signal can transmit. Since fiber optic cable is not a long-distance carrier in the same sense as other media, anything which increases the distance is a great boon to fiber cable manufacture. Transoceanic cable applications can particularly benefit from this technology. Erbium is one of the rare earths used in this process. Samarium is another rare earth used to dope lasers. See doping.

RARP See Reverse Address Resolution Protocol.

RAS remote access server.

RASC Radio Amateur Satellite Corporation.

raser Acronym for radio amplification by stimulated emission of radiation.

raster A sequence of adjacent scanning lines on a cathode ray tube (CRT) displayed quickly enough and closely enough together that they are perceived as a fairly uniform coverage of the display surface of the tube. The full coverage of the screen is called a 'frame.' Most television broadcasts and computer monitor images start the raster at the top left corner, with each line sweeping horizontally left to right down the tube, and ending in the bottom right corner. There may be two sets of interleaved rasters displaying concurrently. Color raster systems typically employ three beams: red, green, and blue (RGB) the primary colors of light (the primary colors of pigment are red, yellow, and blue). Unlike a vector display, in which a straight line rendered at a angle appears reasonably straight (depending upon the resolution of the monitor), raster displays may have artifacts which cause the image to appear jagged or *staircased.* Antialiasing can perceptually decrease this effect. See antialiased, bitmap, interlaced, vector display, refresh.

raster fill The filling in of spaces between raster lines on a CRT to provide an image that appears brighter or sharper. See raster.

raster image processor RIP. A device to accelerate the process of data conversion, such as scan conversion on a monitor, or vector to raster conversion on a high-end printer.

raster line A single line sweep (usually horizontal) of the electron beam on a cathode ray tube. The time during which the image is rendered by exciting the phosphors on the inside front of the CRT. When the beam travels back to start the next raster line, it is suppressed in a process called blanking. See blanking, frame, raster.

rastering, rasterizing The process by which an image is converted to data, usually as a stream of bits. Rastering is a common process in document transfer, and is often accompanied by compression and decompression of the data in order to minimize transmission time.

rate The cost per object unit or unit of time of an equipment lease or service. Phone services are typically billed at a flat rate per month with individual surcharges for connect time for long-distance calls or cellular calls. Internet Service Providers (ISPs) typically charge a flat rate per month, although some add surcharges for popular services like email, file storage, and Web access.

rate averaging An economic method for providing uniform, simpler pricing options for equipment or services which normally might vary widely in their costs of installation and operation to different groups of consumers. For example, phone companies have fairly uniform rates over a wide variety of terrains, services, and population densities. Postal services also employ rate averaging, in other words, a letter to the next town requires the same postage as a letter to the most distant part of the country from the sender's locality.

rate decrease factor RDF. In ATM networking, an available bit rate (ABR) flow control service parameter which controls the decrease in the transmission rate of cells when it is needed. See cell rate.

rate increase factor RIF. In ATM networking, an available bit rate (ABR) flow control service parameter which controls the increase in the transmission rate on receiving an RM-cell. See cell rate.

rate period In telephone service, a segment of time designated as a specific period in order to assign billing charges. Rate periods are determined by evaluation of phone call traffic volume, cultural customs, and time of day, and then usually established semi-permanently so that subscribers become familiar with peak and off-peak rate periods. Rate periods vary from country to country. In the U.S., for example, the least expensive rate period on weekdays is from 23:00 to 08:00, and there are cheaper rate periods available on weekends. Companies often schedule fax transmissions to be sent out automatically after midnight to take advantage of the cheaper rate period. See rate period specific.

rate period specific When telephone calls which cross rate periods are billed at a higher or lower rate when the period changes, they are called *rate period specific.* International calls originating in the U.S. are usually not rate period specific, and the call is billed according to the rate period during which it was initiated. See rate period.

Rate Quote System A computerized telephone rate/quote system which can be accessed by TSPS operators.

rated voltage A designation of the voltage at which an electrical component is set to operate, or, if put in a variable voltage environment, the safest maximum voltage at which it can be used for extended periods without risk of hazard or component burnout.

rat's nest Mess; poor configuration; snarled, complicated arrangement of wires, machines, processes, or code statements.

RATS Radio Amateur Telecommunications Society. RATS broadcasts to a Java-enabled site on the Internet on a 145.790 MHz channel.

Rayleigh disc An instrument for the fundamental measurement of particle velocity by means of acoustical radiometry.

Rayleigh fading Fading, or loss of signal strength, as a result of interaction with the various objects or particles which are part of the environment of the transmission. This phenomenon is often found in mobile communications in which the interaction of the radio signals with the surrounding terrain causes signal fading. A number of techniques are being developed to reduce the incidence of fading. For example, in systems where long delays are acceptable, fade can be reduced by interleaving. Named after J. W. Strutt (Lord Rayleigh).

Rayleigh scattering Scattering of radiant waves by contact or interaction with minute suspended particles such as dust or moisture, or by impurities in a transmissions medium such as fiber optic cable. Named after J. W. Strutt (Lord Rayleigh).

Rayleigh, Lord (1842-1919) John William Strutt, an English physicist and mathematician who made fundamental mathematical contributions to the field of physics, including atomic physics, acoustics, and optics. In 1870, he published "On the Light from the Sky - Its Polarization and Colour," which presented his ideas and calculations based on observations of the scattering of light and the relationship of the scattered radiation to wavelengths. In 1904 he was awarded a Nobel Prize for his discovery of argon. Rayleigh scattering is named after him.

RB reverse battery.

RBOC See Regional Bell Operating Company.

RBS See robbed-bit signaling.

RCA See Radio Corporation of America.

RCC See Radio Common Carrier.

RD See routing domain.

RDF 1. radio direction finding. In radar, a British name for a tracking system based on locating the source of unidentified or foreign radio signals. 2. See rate decrease factor.

rdist remote file distribution program. A program to distribute and maintain file copies on multiple hosts on a network. See DHCP.

RDT 1. recall dial tone. 2. remote digital terminal.

RDY ready.

read A command commonly used in software application menus to provide the user with the ability to load data from permanent or semipermanent storage such as a floppy diskette, hard drive, cartridge drive, tape, RAM disk, or other medium. Files on a drive can be set with protections to be read only, or read/write, or write only so that they can't be read. Similarly a disk can usually be set to write enabled or write protected mode. Most optical storage media are read only, and cannot be rewritten or written without special, more expensive devices than are used by most consumers.

read only memory ROM. A nonvolatile, random-access data storage unit which is preconfigurable, and not changeable by the user by normal means. ROM chips are commonly used for kernel level operating instructions or other information for the low-level functioning of a system which needs to be quickly accessed, and transparent to the user. See CPU, RAM, PROM, EPROM, kernel.

RealAudio A commercial on-demand, realtime audio player for multimedia-capable computers from Progressive Networks.

Real Time Streaming Protocol RTSP. An IETF data format draft standard.

realtime The term realtime is used somewhat differently by two groups of people: marketing personnel and users; and technical systems designers and operators. First, a more general description of realtime.

Realtime is a description of computer processes that occur at a speed which corresponds with human perception of the speed of events in 'real life,' and in immediate response to requests. In other words, a ray tracing program that takes two hours to render and display a frame of an animation is not realtime, as there is a delay during which the viewer must wait for the image to be constructed and displayed. In contrast, a fast action video game, in which the motions are displayed at 20 or 30 frames per second so that they are perceived as natural motion, and in which the joystick, mouse, or other inputs from the user have immediate effect upon the game, is considered to be a realtime game. Realtime flight simulators are used to train pilots, and realtime rendering programs exist on some fast, high-end platforms.

In telephony, realtime processing involves handling calls as they are received. If callers are put on hold or experience delays in automated menu processing systems, the system is not providing realtime service.

Realtime effects and processing, especially if they involve graphics, typically require fast, wide data buses, fast CPUs, and efficient mathematical algorithms for handling input, calculations, and display. In spite of the resources needed, humans seem to have a compelling interest in creating realtime scenarios, and striving for real and fantasy simulations that mimic or outstrip the pace of life. This creates economic incentives for creating realtime simulations, especially in the entertainment industry, with audiences eager for these scenarios. Indeed, many of the advances in computer technology have been pioneered, fueled, and financed by the games industry.

In a more technical systems implementation sense, realtime occurs in a computing system when computations are processed not only as expected, and with logical correctness, but within certain predetermined or expected timing frames, and with a certain guaranteed minimum level of usefulness of the service. In this sense of the word, speed is not so much the issue, as the appropriateness of the response time to the task at hand. Some realtime systems rely on sensor and other feedback mechanisms, and may be used not only in consumer computing operations, but in industrial robotics or remote sensing applications. Realtime functionality is likely to be important in future space probes and the vehicles that deploy them, as well as in intelligent vehicle autonavigation systems.

realtime capacity The capability of a system to handle calls, inputs, requests, or other stimuli, as they are received. In configuring and tuning various types of networks, realtime capacity is one of the criteria many systems use as a reference point for smooth operations.

realtime diagnostics Tests which allow measuring, diagnostic, or display instruments to monitor and report events as they are occurring. Most electrical instruments work in realtime, reporting circuit status at the moment the instrument is applied to the circuit. This is not so easily done with sophisticated computer systems, where it is difficult to track everything that is happening on the system at any one time. More often software "monitors" (statistical display programs) for specific processes are used, which include the representation of statistics for load, CPU speed and processes, congestion, failed packet ratios, quality of service (QoS), etc.

Realtime Transport Protocol RTP. An IETF data format that provides higher video priorities to facilitate realtime multimedia transport over Internet Protocol (IP).

reassembly An important aspect of network communications in which an Internet Protocol (IP) datagram or other type of data unit, which has been split up at the source or enroute and may have been transmitted in sections at different times and/or through different routes, is reassembled at the receiving end. The process of disassembly and reassembly allows packets to be transported through a large, dynamic network environment, like the Internet, which changes topologically in unpredictable ways. Reassembly and synchronization is also important in applications like videoconferencing, where more than one line may be used to transmit the various audio and video signals that make up the communication.

reboot To cause a system to return to its initial operating status, as it was at the beginning of a system startup, usually without turning off the power. This typically clears memory, closes all applications and files, sets initial test sequences and starting parameters for timing, sound, video, etc. and reinitializes devices. The term is derived from 'boot,' which comes from 'bootstrapping.'

If the power is turned off to reboot a system, it is called a *cold* boot. You should always count to 20 before flipping the power switch on again. Electronic components are sensitive to sudden power surges, and there is always some residual power in some of the chips that needs to drain off when electronics devices are turned off.

Most reboots are *warm* boots, in which the power to the system is not interrupted. Rebooting is seldom necessary in stable operating systems, which can operate 24 hours per day for years without crashing, hanging, or fragmenting memory. However, some operating systems do not handle error conditions or memory management well and may hang, freeze, or crash, in which case a reboot may be necessary in order to continue using the system.

receive only device A device which can receive data but not send it. Technically, there are very few receive only devices in computer networking, since most devices employ *handshaking* to negotiate a transmission. For example, a computer printer may seem to be a receive only device, but a printer has to be able to tell the computer when it is ready to receive, when it is busy printing and can't receive more data, and when it is available again for other jobs or other users on a network. This involves two-way communication. It may even signal the sender about its capabilities and configuration parameters. Most receive only devices are passive devices or broadcasting devices such as simple PA speakers, buzzers, lights, etc.

receiver 1. A device for receiving signals, impulses, or data transmissions. 2. A device which captures, and sometimes converts electromagnetic waves or signals into a form which is meaningful to humans. Receivers are often combined with tuners, to specify the frequency desired, and amplifiers, to increase the power of the signal. See telephone receiver.

receiving perforator See reperforator.

Recognized Private Operating Agency RPOA. An ITU-T designation for telephone companies providing internetworking services.

rectification 1. A condition in which current flowing through a material or circuit in one direction encounters greater resistance than current flowing through in the opposite direction. 2. The one-directional processing of an alternating current (AC).

rectifier 1. A material or circuit that offers greater resistance to an electrical current flowing in one direction than in its opposite direction. 2. A device for converting alternating current (AC) to direct current (DC). Rectifiers are commonly used on power transformers for electronics devices with power requirements different from the power coming directly from an electrical source. Vacuum tubes were used as rectifiers in early radios, with selenium rectifiers beginning to supersede them in the mid-1940s.

recursion 1. Returning, moving back upon. 2. A repetitive succession of elements or operations which affects the preceding elements or operations in a like (although not necessarily identical) manner according to a finite rule or formula. Recursive formulas are commonly used in programming, with fractal display programs being a popular, visually appealing example.

red alarm In telephone transmissions systems, a critical failure alert signal which occurs if an incoming signal is lost or corrupted. This is implemented in various T3, T1, or SONET network systems.

Red Book 1. The original Compact Disc audio specification, developed in the late 1970s and early 1980s. Audio sectors, tracks, and channels are specified, along with other physical parameters. This was followed, in the mid-1980s, by the Yellow Book, which specified CD-ROM parameters. 2. In telephony, books in the ITU-T (formerly CCITT) 1984 series of recommendations. 3. The Adobe PostScript Language Reference manual.

red gun In a color cathode ray tube (CRT), using a red-green-blue (RGB) system, the electron gun which is specifically aimed to excite the red phosphors on the inside coated surface of the front of the tube. Sometimes a shadow mask is used to increase the precision of this process, so the green and blue phosphors are not affected, resulting in a crisper color image. See shadow mask.

red wire A color designation used by International Business Machines (IBM) to indicate wires used to establish a hardware patch to accommodate a code change. See blue wire, purple wire, yellow wire.

Reduced Instruction Set Computing RISC. A type of programming and system architecture which uses a set of simpler instructions performing single, discrete functions to carry out an operation than would be used in a comparable operation by a Complex Instruction Set Computing (CISC) design. Most of the newer computers tend to incorporate RISC architectures, although not all the support circuitry enables the full capabilities of RISC architecture to be used.

Unlike CISC commands, RISC commands are of the same size, which means that less time is required for subsequent processing of the instructions, because individual evaluation of the commands for size and conversion to microcode is not required. When RISC software is compiled, it is evaluated to determine which operations are not dependent on the operation or results of others, and slates them for simultaneous execution.

Due to the reduced instruction set and processing that takes place, the circuitry on RISC chips is simpler than on most CISC chips, resulting in a smaller physical size and, usually, lower heat output.

Not all chips are strictly RISC or CISC. For example, in the Intel line of processors, the Pentium chips are a transitional architecture which maintains some downward compatibility with the earlier CISC architectures, while still incorporating some of the advantages of RISC architectures. The chips tend to be larger and hotter than straight RISC chips, but meet a market demand through a transition period.

redundancy Replication, duplication, superfluity, repetition. Redundancy is important in computing because the loss of data, whether stored or in the process of transport, can have serious consequences to human safety, economics, or business transactions. See redundant array of inexpensive disks.

redundant array of inexpensive disks RAID. A data storage, retrieval, and protection system using multiple disk storage devices, a system commonly used in networks. RAID consists of multiple hard drive storage devices linked together to provide data mirroring or data striping and parity-checking across disks in order to record the information redundantly. Duplication or data mirroring is primarily a function of software, whereas parity-checking requires a controller and is associated more closely with hardware. Many RAID systems are SCSI-based.

A basic low-end RAID center may consist of four drives, each with 2.1 gigabytes of storage, sometimes set up in a rack, talking through a centralized controller system, usually through a server.

Although a certain amount of storage is inevitably lost due to duplication of data, the big advantage of RAID systems is that they provide pretty good protection against data loss if a drive goes down. There is less protection if several drives go down, but, since this happens rarely and since companies are reluctant to back up data sufficiently often on systems like tape drives, the RAID alternative works well in practice. Many RAID systems are 'hot-swappable' which means that an individual drive can be pulled out and replaced while the system is online, thus not necessitating a system shutdown or inconveniencing current users on the system.

Specifications released in 1988 in the RAID paper proposed five levels. Since that time, changes and enhancements have occurred; the levels are not cut and dry since configuring various parameters, such as stripe size, creates overlapping characteristics between the different levels. Hybrid systems also exist. Generally, however, to provide an introductory understanding, the RAID levels can be summarized as shown in the RAID Levels chart.

In addition to redundancy and parity checking, a RAID system may have some intelligent monitoring incorporated into the system, which does periodic checks and analysis and reports anomalies to the controller. The controller can then signal a warning which allows the device administrator to check for potential problems, or swap out a drive before it fails. See dynamic sector repair, SMART.

re-engineer To step back from a system or process, take a new look at it, and redesign it, sometimes from the ground up, with the intention of making it more efficient and cost effective. Software often has to be re-engineered, as legacy systems tend to be slower and less efficient over time, due in part to the way they are upgraded and, in part, because of technological improvements and changes in hardware which are accommodated in a variety of ways. Market pressures also cause many software programs to be released before their time, in which case, they may be re-engineered before the next release. Work environments in companies that are growing or downsizing quickly often have to be re-engineered as the ways of organizing facilities and staff that are appropriate to a small company are not necessarily appropriate to a large company.

reed relay switch A type of electronic telephone switch developed in the 1960s. Reed relay switches began to supersede crossbar switches, which were prevalent at the time, and some of the step-by-step switches still in use. Electronic switches opened up possibilities for many new types of caller services, such as Caller ID, Call Waiting, etc.

reel-to-reel tape Magnetic recording tape wound onto separate round reels which are usually about 4" to 8" in diameter. Although most tape is now distributed on cassettes rather than reels, reel-to-reel is still used in some professional recording studios, especially if eight or 16 tracks are required for sound mixing and dubbing. Gradually these reel-to-reel sound recorders are being superseded by digital recording media. See cassette tape.

reference clock A clock that is considered very accurate or stable, which is used as a reference for other clocks or processes, as computer processes. Quartz crystal clocks are considered very accurate due to their vibrational properties and are often used in computers and watches. The speeds of various computer processes are described in *clock cycles.* Atomic clocks are used to establish Coordinated Universal Time and are used in satellite positioning systems which require accurate clock references.

In multimedia editing environments, a reference clock is used to provide 'house sync,' that is, long-term synchronization of various audio or other signals, which resolve to the reference clock rather than to the time code signal. See atomic clock, chase trigger, Coordinated Universal Time, quartz crystal, time code.

Referral Whois See RWhois.

reflector/director elements On antennas, two or more protruberances from the main rod that are usually narrow and regularly spaced in an array. Reflector and director rods help improve gain and directivity. See Yagi antenna.

refresh rate, scan rate The rate per unit of time, at which information or an image, is recreated in order for it to remain current or visible. This phrase is frequently applied to broadcast and computer display technologies, especially cathode ray tube (CRT) displays in which the action of electron beams on the phosphors is very limited and must be reinitiated (refreshed) in order for the image to continue to be visible. Refresh is a general concept which applies to many different types of situations in computing, from individual phosphor refreshes to graphical user interface element refreshes.

The refresh rates of the phosphors on the inside coating at the front of a cathode ray tube will effect the clarity and amount of flicker seen on the screen. Monochrome or gray scale monitors have longer persistence; that is, the image from the excited phosphors is visible longer, and thus do not need to be refreshed as often as color images.

Refresh of the entire CRT image is described as the number of times per second the frame is redrawn. Refresh rates slower than about 20 to 40 frames per second are perceived as flickering to the human eye, especially if the image involves fast action. Still or slow-mov-

RAID Level Specifications in Brief

Level	Notes
Level 0	Striping, no redundancy or error correction. This can provide faster access, but does not protect data from loss.
Level 1	Disk mirroring. Complete redundancy. Provides data protection.
Level 2	Byte striping, dedicates at least one drive for parity information. Uses Thinking Machines, Inc.'s proprietary setup which is not commonly used.
Level 3	Generally used instead of level 2. Block striping will improve performance if data is written in large blocks, and simultaneous reads are used. Distributed parity information (originally required a dedicated parity disk, a stipulation that was removed in 1994). In other words, when appropriately implemented, both better performance and data protection can be achieved.
Level 4	Similar to level 3, but larger data blocks are striped across disks, each drive is not necessarily involved in each access.
Level 5	Block striping, parity information distributed across drives. At least three drives are required for a minimum implementation. Each drive is not necessarily involved in each access. Parity information is also striped across disks. Provides data protection and, in many cases, will improve performance. This is a popular implementation of RAID.
Level 6	Not consistently specified or implemented.
Level 7	Similar to level 4, with larger data blocks striped across disks. Uses Storage Technology, Inc.'s proprietary caching mechanism and operating system.

ing images do not need to be refreshed as often.

The refresh rate of a computer program image is a combination of operating system and applications programming, and is not simply dependent on the hardware attributes of the system. In order to optimize speed on a computer display, the OS or programmer may choose to refresh only a section which has just been manipulated or changed. If the software does not keep track of what is transpiring on the screen or if several processes are active at once, it may seem that the display is slow to update or refresh a new window, gadget, or element drawn in a paint program.

In general, quicker refreshes are desired over slower ones, but the cost of more computing power and faster hardware puts some economic constraints on the refresh rates of various systems.

refurbished equipment Equipment which is used and has been serviced and tested by a technician to bring it back into original operating condition. If further work is done and substantial numbers of parts replaced or upgraded, it may also be referred to as remanufactured equipment.

Refurbished equipment, or *refurbs,* are usually cleaned up and made to appear new or nearly new. Refurbished items are typically sold at a discount of about 15% to 30% over the price of new ones. Sometimes that's worth it. If the refurbished item comes with a good warranty and is the type of product that doesn't wear out readily, like switch boxes, modems, peripheral cards, or motherboards, it may be worth it to buy a refurbished product. However, limited life products like monitors and various types of printers, which are susceptible to mechanical wear and tear, should be considered only if the discount is significant.

regenerate To restore, bring back to original condition, recreate, duplicate. In electronics, signal regeneration is an important issue. Transmissions typically suffer from loss and interference over distances, and any means that can be used to maintain a signal or regenerate parts or all of a signal that is experiencing loss or change in some form, is usually desired.

There are many physical and digital schemes for regenerating systems. In some digital systems, regeneration may involve putting a signal back into its original form at the receiving end. In a sense, single sideband transmissions are a type of regenerated transmission, since only a portion of the signal is sent. The opposite sideband and the carrier signal are mathematically constructed, and the original signal thus reconstructed at the receiving end. See regenerative repeater, relay, repeater, single sideband.

regenerative repeater A type of repeater used in communications that are characterized by uniformity of length and signal, to correct the timing of the signal and retransmit the 'cleaned up' impulses. These are common in older teletype communications. See repeater.

Regional Bell Operating Company RBOC. One of a number of companies which were formed by the divestiture of AT&T, which originated from the original Bell Telephone Company through a long, colorful history of mergers and splits. In the mid-1980s Judge Harold Greene broke AT&T into seven RBOCs. For more detail, see divestiture, Consent Decree of 1982, Modified Final Judgment.

register A repository for data, a storage area which may or may not also be used for data manipulation. There are many areas of telecommunications where registers are used. Many computer chip architectures have registers for holding information that is about to be moved or manipulated. Palette configurations for computer displays may be saved in color registers. Data modems have registers for setting various parameters, with Hayes AT command 'S' registers being common. Flags, configuration settings, etc. are stored in registers.

registration marks, register marks In printing and desktop publishing, small marks, often resembling very fine targets, which are placed around an image to be printed in more than one run through the press, or in more than one color, in order to align the various layers so that they print perfectly on top of one another. Registration is especially important in process color printing, since the dots that make up the multiple colors will not create the correct patterns, and the illusion of many colors, unless the alignment is nearly perfect. Note that there is no room for registration marks if the extents of an image reach the edge of the paper on which they are laid out. If you are desktop publishing an 8.5 *x* 11 image with *bleed*, or with registration marks for color or multiple press passes, you will have to select a paper size *larger* than 8.5 *x* 11 (*letter extra* is a common setting) for there to be room for registration and other printers' marks. See crop marks.

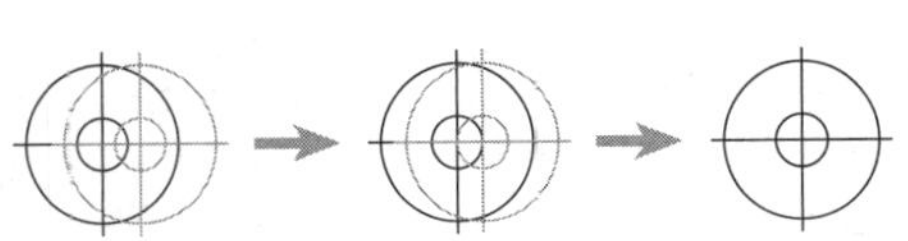

By lining up registration marks, which resemble targets, color separations and other types of separate press runs can be 'registered,' that is, lined up or composited with all the elements and colors in the right places.

Registration Number As part of the Federal Communication Commission's (FCC's) jurisdiction over equipment which may emit radiant waves that interfere with other equipment, appliances, radios, etc. There is a process of submission, evaluation, and certification which warrants that the equipment has passed FCC requirements. This Registration Number is not related to quality, suitability for a particular use, or other usability issues; it simply confirms that the equipment falls within acceptable emission standards.

Reis, Johann Philip (1834-1874) A German inventor who did substantial pioneer investigations in transmitting tones and possibly also voice, over wires. Reis accomplished this with various transmitters and other equipment that he developed and publicly demonstrated in Frankfurt to the Physical Society in 1861. No directly verifiable evidence indicates whether voice was transmitted at the 1861 demonstration, but Reis' subsequent work indicates that he recognized the potential for voice communications and concentrated many of his efforts in that direction, eventually developing a telephone design that was not unlike telephones actually put into production in the United States some years later.

Subsequent inventors, excited about the breakthrough, made improvements on Reis's early crude mechanisms, while Reis himself continued to study and improve the technology until his early death in 1874. See telephone history.

relay *n.* 1. A means of passing on signals, objects, or communications to another 'node' after which it is further transmitted or transported. 2. An electromagnetic device in a circuit for providing automatic control, which is activated by varying electrical impulses. A relay is usually combined with switches to control when they open and close, and widely used to automate older telephone switching centers. Thus, it was important to design relays for durability, since they had to open and close circuits many millions of times. Because the relay is essentially a simple mechanism, it can be greatly varied by adjusting contact springs and windings, thus producing a large variety of types of relays. Multicontact relays were developed in order for numerous switching contacts to operate simultaneously. See crossbar switch.

reliability An expression of the dependability of a system under actual conditions of use. See availability, mean time between failures (MTBF).

Reliable SAP Update Protocol RSUP. A bandwidth-saving protocol developed by Cisco Systems for propagating services information. RSUP enables routers to reliably transmit standard Novell SAP packets only when a change in advertised services is detected by the routers. Network information can be transported in conjunction with, or independently of, the Enhanced IGRP routing function for IPX.

reluctance Opposition, or resistance in a magnetic circuit against the creation of magnetic flux. Similar to the concept of resistance in an electrical circuit. See resistance. Contrast with permeability.

remailer Any online electronic mail transit station which changes or prepends the header in such a way that the originating information is changed or obscured, or which intercepts mail and then forwards it on to its destination. Sometimes these remailers are LAN servers which have been configured so that the header changes when incoming mail is served out to the local recipients. This is unfortunate in that the recipients cannot automatically 'reply' to the sender and must manually type in the return email address in order to respond to their correspondents. This is not a recommended way of configuring a mail server and should only be done in circumstances where a specific reason warrants it.

Remailers are sometimes used irresponsibly. There are a number of get-rich-quick and commercial products promoters who use remailers to obscure the origin of their postings. This is because there is a lot of legitimate opposition to unsolicited commercial messages online.

Anonymous remailers are mail transit points which deliberately obscure the identity of the poster in order to ensure his or her privacy. See anonymous remailer, spam.

remanufactured equipment See refurbished equipment.

remapping On a computer system,

remapping is moving data, often in the form of blocks, arrays, or tables, from one area of storage to another. Memory remapping, address remapping, file location remapping, and keyboard remapping are some common examples. Remapping is sometimes used as a means of 'double-buffering' computer graphics screens. By building a screen in a background while the current screen is being viewed by the user, and then displaying it by remapping the entire image to the video display area can result in fast, clean transitions. See frame buffer.

remote access An important aspect of networking in which access to computing services, devices, and information can be gained through a remote device on the network, usually a computer terminal or phone line.

On a phone line, remote access to an answering machine can enable a user to dial up the answering machine from a different phone, punch in some codes to see if there are messages available, retrieve those messages remotely, and even change the message on the answering machine through the phone line.

Remote access does not imply the level of operations that can be accomplished, only that the device can be accessed in some basic way. Remote access terminals vary greatly in their ability to interact with a server or other user functions. For example, on a basic text-oriented 'dumb terminal' connected to the main computer with a serial line, the user may only be able to execute simple text commands and won't be able to display graphics or run sophisticated applications locally.

On the other end of the spectrum, some systems provide full access to remote applications, especially if they are connected with a fast transmissions protocol over fiber. In other words, there may be a graphical database program available on the server that the user can run on a smart terminal just as though the terminal was the main computer. Not all operating systems can do this. The X Window System is designed to provide this type of capability in conjunction with various Unix systems. It has also been upgraded to provide similar services over the Internet. See X Window System 11.

remote access PBX A private branch telephone exchange which can be accessed from an outside line with appropriate authorization codes. Once 'logged on' to the internal branch system, various features can be used such as voice mail messages, long-distance calls connected and billed through the PBX, etc.

remote batch processing A means of submitting a computing job remotely to a processing system and receiving it back as or when the job is processed. This is rarely done at the consumer level, but it is still common for high-end mathematical calculations, scientific research, and other intensive computing applications which may require large amounts of computing time or more sophisticated computing resources. In the earlier days of computing, remote batch processing, especially with punch cards, was the only type of service available, and it could take hours or days to receive the results of a simple calculation.

remote call forwarding A service in which a phone number is located in the central office of one exchange and any calls made to that number are forwarded to a line in another exchange. This may be of value to businesses which want to maintain a local 'presence' without the expense of a local office, so that customers can call a local number instead of long-distance.

remote control *n.* A device to allow control of another device without making direct physical contact. The control of the device may be through indirect physical means (through a remote controller and cable), through a network (computer-controlled vending machines in another part of a building), or through various wireless methods (infrared, FM, audible sound control, etc.). Remote control of computers on a network can be done through various telecommunications products, specialized remote applications and file serving software, or through operating systems which support this capability.

remote diagnostics Systems diagnostics which can be run from a remote location. It is common for higher-end routing and switching devices on a computer network to be controlled through software at a main administrative location. This software typically permits the running of test and diagnostic routines and may show graphical diagrams of problems or potential problems or bottlenecks. On phone systems, diagnostic checks can sometimes be carried out with devices that generate specific tones or signals, which can initiate processes at the other end of a phone line.

remote job entry RJE. Entry of computer commands from a remote terminal, that is, a terminal at another location. See remote programming.

Remote Operations Service Element ROSE. An application layer service that provides the capability to perform interactive remote operations through a request/reply mode. ROSE is a generic information exchange technique which is not application-specific. It is defined as ISO 9072-1, and as an X Series Recommendation by the ITU (X.219). See X Series Recommendations.

remote procedure call RPC. A means of making a request to a remote system so that it appears to the user as though the request is being fulfilled on the local machine. In other words, a user may open a word processing program and load in a file. The file may actually reside on a computer in another room or another city, but the user is unaware of any difference in using the file from the remote system or using a file from the local system. In other words, the RPC is transparent to the user. Another example would be the use of a terminal communications program which accesses a modem on another computer as though it were physically attached to the local machine. A number of conventions for making requests to a remote system, and fulfilling those requests, have been developed.

remote programming The capability whereby a system can be programmed from a remote location, either through a data network or phone lines, usually after input of appropriate authorization codes. Remote programming allows a field worker, or telecommuter, to administrate a system without being physically present. In computing, remote programming is often done by BBS operators who want to check and manage their systems when out of town. By dialing their own BBSs and logging in as the Sysop, they can validate new users, check mail, configure the bulletin board, and accomplish various maintenance tasks through a phone connection.

Remote programming is often implemented in high-end corporate and industrial software programs. The software is set up with security mechanisms so an authorized programmer working for the software vendor can dial into the customer's machine and do routine maintenance, software tune-ups, diagnostics, and configuration without having to travel to the customer's site, and during nonbusiness hours, if needed. This type of service is usually provided through a separate service contract for a specified period, or is billed hourly, as needed.

remote site A facilities or equipment location which is distant from the one presently occupied, or from which certain maintenance or administration tasks may be carried out. A sales representative with a laptop, a scientist with an intercom radio doing field research, a computer terminal in an annex building, all constitute remote sites if their systems are not the main operations or administration systems, and they are in some way directly or indirectly communicating with or through the main system.

removable media Storage cartridges, drives, or diskettes which can be 'swapped out' and replaced with another. This provides a less expensive, portable option to numerous fixed storage devices. In the mid-1990s various cartridge drives became very popular, as it was possible to store from 100 to 1,000 Mbytes on a cartridge not much bigger than a floppy. The problem was that every cartridge drive had a different format and the formats were not intercompatible. More recently, super diskettes have been introduced, which use normal floppy-sized disks that can store 100 MBytes, but the drives are still downwardly compatible with 1.44 floppies, so that it's not necessary to have several devices attached to the computer. It is not clear, as of this writing, which of these technologies will prevail or whether another new one will leapfrog them before one or the other is firmly established.

REN See Ringer Equivalence Number.

repeat dialing This is both a function of some phones and telephone/computer software programs, and a service of some phone companies in which a number which is found to be busy can be repeatedly dialed until the connection goes through, without the user having to dial the number again. Repeat dialing is very commonly used in telecommunications software programs to dial up BBS numbers which are frequently busy.

repeater A device for receiving signals and retransmitting those signals in order to propagate or amplify the signal. Repeaters are commonly used in technologies with signal attenuation and fade. Repeaters are used in both digital and analog systems and are often spaced at intervals over paths that cover long-distances. In digital systems, it is possible to reconstruct the informational content of the signal; in analog systems, more often the signal is amplified, which means that accumulated noise and degradation are still limiting factors. Radio broadcast repeaters and microwave repeaters are examples of common implementations. In networks, there are a number of devices that assist in the conversion and propagation of signals (bridges, routers, etc.). A repeater is simpler than most of these devices, serving only to continue the signal and extend its range, or to clean it in its most basic sense, rather than to change the informational content of the data. See amplifier, doping, regenerative repeater.

reperforator An instrument which translates received signals into a geometrically coded series of locations that are punched or otherwise impressed onto a paper tape. Early telegraph systems and most of the early computing devices used reperforators. These were then read with optical or tactile sensing devices to turn the code back into human-readable form. See chad.

reperforator/transmitter RT. A teletypewriter device which includes both a reperforator for punching received codes on paper tape, and a tape transmitting unit for sending the codes to a tape punching mechanism.

repetitive pattern suppression RPS. A means of data optimization which compresses digital communications by removing repetitive patterns and reproducing them at the receiving end.

replication A process commonly used on computer systems for security, redundancy, distributed access, or other backups. In large companies and on the Internet, whole file archives are often replicated or 'mirrored' in order to provide access at a reasonable speed to a larger number of uses. Replication of data is a means to protect data in case of a serious problem with a system or storage device. RAID systems are a means of replicating data to preserve data in case of a fault. Replication in the form of regular backups is recommended for all computer data which is important and which needs to be preserved. In radio communications, transmissions are sometimes repeated to improve the chances of a message getting through. The replication may be of individual small units of transmission or may be repetition of a short message or signal. See ALOHA.

reprography Copying and replicating.

Republic, S. S. In 1909, due to a collision, the S. S. Republic sank, yet all hands were saved but two. The lifesaving role played by wireless equipment and operators led to legislation requiring other vessels to use wireless emergency equipment. Similarly, when the Titanic sank, legislation was further instigated to require greater diligence in round-the-clock wireless monitoring. See Titanic.

Request for Comments RFC. A formatted, open communications forum for technical experts which accepts, edits, numbers, publishes, and disseminates Internet-related documents including protocols, draft and official standards, notices, opinions, and research. Known as RFCs, these electronic documents form a body of more than 2,200 contributions that provide a remarkable overview of the evolution of the Net, its structure, functioning, and philosophies.

There are a number of categories of RFCs. Some of them are tiered and cannot be submitted without passing through previous categories in a specified order, with specified waiting periods for comments and revisions from the RFC community.

Code	Category
(no code)	Unclassified
I	Informational
H	Historical
E	Experimental
PS	Proposed Standard
DS	Draft Standard
S	Standard

RFCs are not changed once they have been submitted, assigned a number, and distributed. Any changes regarding an RFC must be submitted as a new RFC.

There are many excellent RFC repositories on the Internet, with good indexes and abstracts. Anyone can submit an RFC provided it is topical and follows the official format and procedures. This dictionary includes references to specific RFC numbers where the author felt the technical origin of the information would be of interest, and there is a list of significant RFCs in the appendix. For information on submitting RFCs, see RFC 1543.

reroute A temporary or permanent change in a data path. Rerouting frequently occurs in large, dynamic networks like the Internet. Systems where rerouting is common usually use a hop-by-hop method of routing in order to accommodate changes and to create new paths as needed. In electronics, rerouting of a circuit may be accomplished by a patch or shunt, a wire which bypasses the original path. On Fiber Distributed Data Interface (FDDI) networks, using dual rings, rerouting to the second, backup ring is carried out if a problem is detected on the primary ring.

RES Residential Enhanced Service.

Resale and Shared Use decision A decision of the Federal Communications Commission (FCC) to allow competition in value-added networks.

resampling The process of subsequent sampling of data, as image or sound data, and reencoding it. Resampling usually occurs when the original sample was not of the resolution level or compression rate desired. Resampling may also occur in order to update or refresh information that may be changing, as in Internet videocam shots, videoconferencing, etc. See sampling.

Research and Education R&E. Those served by the National Science Foundation's (NSF) networking efforts.

reseller A company which purchases a block of services (e.g., long-distance services) or numbers (e.g., cellular phone numbers) and resells them directly to customers. The reseller does not own and usually does not maintain the physical structures or services, but usually provides information and technical support to the resale customers. See agent, aggregator.

Residential Broadband RBB. A broad term for higher capacity/higher speed broadcast and networking services to residences. Larger providers such as NTT and AT&T are already engaged in upgrading cables and services to provider a greater range of programming choices to home subscribers. New switching technologies are an important technological and economic component of these upgraded services. See fiber to the home.

Residential Broadband working group. RBWG. A working group of the ATM Forum Technical Committee which promotes a single set of global specifications to maximize interoperability of products from various vendors. The Residential Broadband group provides documents and recommendations regarding enhanced services to residential users of ATM-related technologies.

resinous electricity A term coined by Dufay to denote the type of electrostatic charge produced on sealing wax when rubbed with flannel (or amber rubbed with wool). Later, in the 1700s, Benjamin Franklin proposed *negative*, a term that superseded vitreous. See electrostatic, static electricity, vitreous electricity.

resistance 1. Opposition, counteracting force or retarding force against. 2. In electricity, opposition to the flow of current, usually expressed in ohms. There is great variability in the resistance of various materials. Materials with low resistance, such as silver and copper, make good conductors. Resistance in a particular material may change with temperature, moisture, or the presence or absence of current. Resistance is defined as the reciprocal of conductance. See reluctance, resistor, ohm. Contrast with conductance.

resistor A component or system which provides resistance to electrical current. Some materials are naturally resistant, and this property may be further exploited by the way a circuit is configured (e.g., longer wires, more loops, etc.). Controlled current is useful in a number of circumstances and can be used to protect sensitive components or to provide operational control. Electronic resistors use standard color coding schemes to identify the degree of resistance they provide. Applying Ohm's law, it can be stated that the combined resistance of any two resistors connected in parallel can be expressed by dividing their product by their sum. See Ohm's law, resistance.

resolution Resolution is somewhat technology-specific, since it is often based not just on the size or discrete value of an individual unit, but also on the total area occupied by a block or line of units, and the units will vary depending upon the type of medium or technology being described.

resonance 1. The enrichment of a sound by supplementary vibration. For example, the body of an acoustical stringed instrument is designed to increase the resonance of the string vibration by transmitting the sound through the bridge, the sound holes, and the body. 2. A greater amplitude vibration arising from smaller periodic vibrations with the same, or nearly the same, period as the natural vibration period of the system. This can arise in both electrical and mechanical systems. 3. The enhancement of an event by creating excitation within the system, as in particle reactions. See magnetic resonance imaging. 4. In a circuit, a balanced condition between inductive- and capacitive-reactance components.

resonance curve A diagrammatic representation of the relationship of various frequencies at or near resonance to a tuned circuit.

resonant frequency The frequency at which

R

a maximum amplitude response occurs in a given object or system when acted upon by a constant amplitude sinusoidal force. In an antenna, for example, the transmitting current is greatest when the impedance level is lowest at a given frequency and power.

resonator A device which increases and/or directs sound, as in a musical instrument, music box, or telegraphic sounder. In gramophones, a horn which amplifies and directs sound to the listener. In telegraphic systems, a box which holds a sounder and directs the audible clicks to the ear of the operator. In microwave communications, a hollow, metallic container in which microwaves are produced. In crystal detectors, a piezoelectric crystal which oscillates when stimulated by radio waves.

Resource Management RM. In an ATM system, cells contain information for managing bandwidth, buffers, and flow control aspects such as loads, traffic congestion, etc. These resource management (RM) cells are thus associated with the administration of the data transmissions. They are passed along the path through various switches to monitor and control congestion by adjusting the various cell rates (current, explicit, minimum) as needed. See cell rate.

response time 1. The time it takes to react appropriately to a given situation or signal. In business, response time to customer inquiries is often critical in making sales and getting repeat business. The response time to a phone call usually needs to be four rings or less in a business environment, or eight rings or less in a residential environment, or the caller may give up and terminate the call before it is answered. In computer operations, response time of an input device, or the software, is important in terms of productivity and user satisfaction. 2. A command called *ping* can be used in network testing and management to determine a response time, or whether a host is even available to respond. See ping.

restart 1. Initiate again, begin again. Put back into service or operation, to power up again. 2. In computer operating systems, to reinstate operations without having to power down and power up the entire system. Many computers have a key sequence, menu selection, or restart button which will reinitialize the operating parameters without a full power up sequence. In some systems (e.g., Macintosh), the restart option will also do a clean shutdown of the system in order to ensure that files and applications are closed, and important processes finished so no data corruption occurs from a sudden shutdown. See reboot.

restocking fee A fee charged by many companies to partly compensate for the time and expense incurred by the company when the consumer returns an item for a refund or exchange.

restore 1. To put back to its previous or original state, to renew, revive, to return to good operating order. 2. To return to its original position, location, or owner, to re-establish, to reinstitute service. 3. To get back data which has been erased or damaged, either by rebuilding or relinking pointers, file tables, and directory tables on the storage medium, or by accessing a backup archive of the data as it was last saved. 4. In programming, to return the value of a variable to a previous value, which may be a default or original value. 5. To recharge or refresh information in a memory circuit with continuous or periodic current. Lithium batteries are sometimes used for this purpose. 6. A gadget on the edge of the windows of most operating systems with graphical user interfaces (GUIs) which allows the window to be automatically sized to its original size without the user having to remember the setting, or do it manually.

Restricted Numeric Exchange See RNX.

retransmission consent In television cable broadcasting, consent is a local TV station's right to negotiate fees associated with program provision. It is common for a local station to purchase a variety of television programming from various distributors and broadcasters. They then resell these programs in various packages to local cable subscribers. See cable access, Cable Act.

retrofit *n.* To equip a device or system with new parts or capabilities that were not available, or perhaps not requested, at the time of initial purchase and installation. For example, it is very common for computer parts retailers to offer accelerator cards, storage device controllers, faster CPUs, and other enhancements to users trying to upgrade or extend the life of their systems.

retry Another attempt to perform the same operation if the previous attempt failed. This is an intrinsic part of most computing processes. If a process fails to read or write data from or to a storage location, it will retry a certain number of times before alerting the user that there is a problem. If a modem fails to connect, it may redial a specified number of times before signaling an error condition. If a mail server tries to send email to a recipient who cannot be found, it may try a number of times before bouncing the email back to the sender. For many operating systems, the retry parameters are set by the programmers and are transparent to the user, and not changeable. For individual applications programs, there may be settings in the user preferences which allow retries for various operations to be controlled.

return CR. A common designation for the carriage return key on a keyboard (sometimes called the *enter* key) or carriage return function in a program such as a word processing program. On most systems, the carriage return incorporates not just a *return* (which returns the cursor to the far left or right of the screen, depending upon the language), but also a *new line* (which drops the cursor down

Resolution of Various Computer-Related Peripheral Devices

Device	Notes
Scanners	The resolution of scanning devices describes the smallest discernible or recordable unit into which the image can be digitally encoded, usually dots per inch (dpi). In most current scanners, this is an optical resolution of about 1200 to 2400 dpi, and interpolated resolutions may be higher.
Printers	The resolution of printing devices is usually described in terms of dots per inch (dpi), with 300 to 800 dpi being common, particularly in consumer laser and ink-jet printers. Professional typesetting machines range from 1000 dpi to 2700 dpi and require special papers, metal plates, or film to retain very fine details.
Monitors	In common computer display devices, resolution is a measure of a block or other combination of the minimum possible discrete sizes or values of a physical system. For example, the resolution of most consumer desktop computer monitors is described in terms of a line of horizontal dots of a particular size, designated by dot pitch, the size of the individual dots that are created on the screen in response to the excitement of the phosphors on the inside coating of the screen. These monitors usually range from about .30 mm dot pitch to .27 mm dot pitch, with lower numbers signifying smaller, finer dots, and thus higher resolution, because it is possible to fit more of them across a specified area. The resolution in this case is really dependent on two factors, not only the size of the individual dots, but the distance between them. As a rule of thumb, generally the smaller the dots, the closer they are positioned, and the more that can be fit horizontally across the screen. Since most CRT display devices employ scanning to create a picture image as a 'frame,' the vertical resolution is better described in scan lines than in individual pixels, but since most computer monitors these days are addressable in terms of pixels within individual scan lines, vertical resolution is generally described in the same terms as horizontal resolution.
	Common screen resolutions are 640 *x* 480, 800 *x* 600, 1024 *x* 768, 1280 *x* 768, 1024 *x* 1024, 1280 *x* 1024, 1150 *x* 786, 1600 *x* 1280, and others. Most current computer operating systems can display in the higher resolutions, but the graphics display card that comes with the computer may only support the lower resolutions, and the monitor, if it is an RGB monitor, may not work with higher resolutions, which require faster refresh rates. To get the higher resolution modes, it may be necessary to purchase a multisync monitor and a high resolution graphics card or, if the monitor already supports higher resolutions, to swap out the graphics card for one with better capabilities.
Rangers	The resolution of a ranging system, one which attempts to discern objects in three dimensions, is expressed in terms of its ability to distinguish one object from another. This is difficult to express precisely, as color, movement, shape, and other factors affect this measurement. Nevertheless, ranging resolution can be used to distinguish high-end from low-end systems.
Positioners	The resolution of positioning systems is a description of how small a unit of distance can be resolved, recorded, and reported by the system. For example, the resolution of the Global Positioning System (GPS) for civilians is lower than the resolution for military and other government use. See Global Positioning System.

to the next line).

return authorization RA. Permission from a manufacturer or vendor for a consumer or dealer to return a product, usually because it is defective, damaged, or does not meet the advertised specifications. Few vendors will process a return without prior authorization. Authorization is usually identified with a return authorization (RA) number, sometimes called a return material authorization (RMA). The RA is used for internal database tracking and inventory control.

return loss A measure of the ratio of incoming to outgoing power, usually expressed in decibels, at a specified reference point. Return loss is a diagnostic means of evaluating various factors such as loss, quality, echo, etc.

return material authorization RMA. See return authorization.

Reverse Address Resolution Protocol RARP. A client/server tool for reporting an Internet Protocol (IP) address to its client. The protocol is intended for workstations to dynamically determine protocol addresses when only the hardware address is known. The Ethernet address is mapped to the logical IP address. Used in TCP/IP. See RFC 903.

reverse bias See back bias.

reverse engineering A process of working backward from a finished product or process to determine the steps that were taken to construct it. Thus, a clock can be taken apart to see what makes it tick, and a software program can be taken apart to see what algorithms and conditional relationships were used in its construction. Software developers are generally nervous about having their software reverse engineered, because it is difficult to prove reverse engineering in cases of theft of intellectual property.

reversing error, reversal error A condition in which current divided through two circuits, such as through a component and a measuring instrument, will vary due to the deflection of the measuring instrument for the same current passing in both directions. See shunting error.

RF See radio frequency.

RFC See Request for Comments.

RFD Request For Discussion. Similar to a Request for Comments (RFC) in that it is a means on the Internet to solicit and generate discussion on a specified topic.

RFE Radio Free Europe.

RFI 1. radio frequency interference. Electrical noise resulting from some wire or attachment acting as an antenna. 2. Request for Information. A solicitation and notification of interest in receiving feedback or information on a specified topic, product, or process, without implying that the requester necessarily wishes to purchase or use that for which the information is solicited.

RG-58U A type of thin-wire cable used in 10Base-2 data communications cabling installations.

Rhealstone A type of computer processing system benchmark used in realtime multitasking systems. Run times for a set of operations (task switching, interrupt latency, etc.) are independently measured. See benchmark, Whetstone, Dhrystone.

rheostat A device with one fixed terminal and a movable contact, used to regulate a current by varying the resistances. Similar to a potentiometer, except that a potentiometer can connect to both ends of the resistance-varying element. See potentiometer.

Rhumkorff coil An induction coil which can be used to induce high voltages, first constructed in 1850 by Heinrich Rhumkorff. This technology was developed into ignition coils. See coil, induction.

Rhumkorff, Heinrich A German physicist of the 1800s who first constructed the induction coil.

Riad computer A type of IBM System/360 series-compatible computer developed in Russia.

ribbon cable A cable design in which the wires are encased so they are aligned side by side, in close proximity, with insulating material separating the wires and holding the whole structure together like a long strip of ribbon. Typical ribbon cables carry between nine and 60 wires. Ribbon cables are commonly used for parallel wiring connections, disk drives, and other data transfer applications, although some types of ribbon cables carry AC power. Ribbon cables are somewhat more fragile than other types of more heavily shielded cables and kinks in the cables can snap the conductors, if they are very fine. Wide ribbon cables can be difficult to attach to other types of connectors. Nevertheless, they are convenient to use in a number of places, because they are flat and flexible, such as inside computers and under floor coverings.

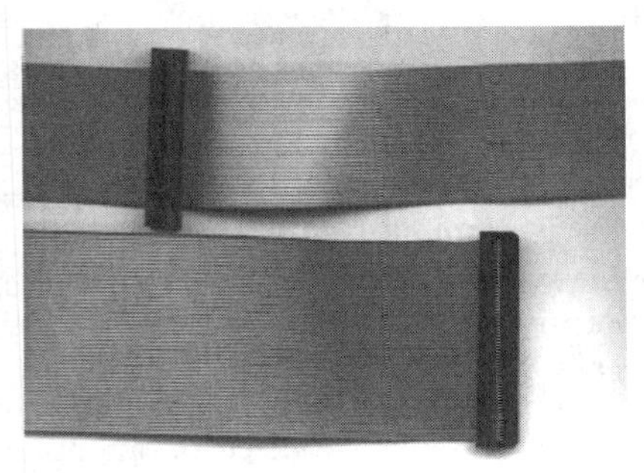

Examples of two ribbon cables, the top is 34 pins, the bottom 50 pins. These are commonly used inside computers where space is limited and narrow. Several devices, such as hard drives and CD-ROM drives, can be daisy-chained on one length of cable.

Rice computer The Rice Computer Project was inspired by the MANIAC II at Los Alamos. Three professors, Zevi Salsburg, John Kilpatrick, and Larry Biedenham, initiated the project, which culminated in the development of a computer for research, and research into the development of computers. Joe Bighorse, as head technician, implemented most of the hardware design. The computer came online in 1959 and was fully functional by 1961. For almost a decade it was the primary computing machine on the Rice campus. Architecturally, the Rice 1 (R1) was descended from the Brookhaven computer and the MANIAC II. It was essentially a tagged-architecture computer, using 54-bit vacuum tubes plus two tag bits and seven error-correcting bits. It implemented indirect addressing capabilities and memory was stored in cathode ray tubes (CRTs). After 1963, transistor-based logic was added.

Rich Text Format RTF. Also called Interchange Format. A document encoding file format developed by Microsoft that retains simple text formatting codes and basic text attributes (typestyle, size, **bold**, *italic*, <u>underline</u>, etc.). A widespread format that can be exported and imported between word processing, OCR, and desktop publishing programs. It is a good intermediary format to use when moving text with attributes from one system to another. Many people use ASCII to export/import text, and despair because the formatting is lost. You may wish to try RTF. Although there are different flavors of RTF, it usually works and can save hours of reformatting. Here is a very basic example of Rich Text Format showing the syntax for various formatting parameters.

```
{\rtf0\ansi
{\fonttbl\f0\fswiss Helvetica;}
\paperw9880
\paperh3440
\margl120
\margr120
\pard\tx20\tx7120\f0\b\i0\ulnone\fs20\fc0\cf0
Rich Text Format
\b0  RTF.  Also called Interchange
Format.  A document encoding file
format developed by Microsoft that
retains simple text formatting
codes, and basic text attributes
(typestyle, size,
\b bold
\b0 ,
\i italic
\i0 ,
\ul underline
\ulnone , etc.).  A widespread
format that can be exported and
imported between most word
processing, OCR, and desktop
publishing programs.  It is a good
intermediary format to use when
moving text with attributes from
one system to another.  Many people
use ASCII to export/import text,
and despair because the formatting
is lost.  You may wish to try RTF.
Although there are different flavors
of RTF, it usually works, and can
save hours of reformatting.  Here
is a very basic example of Rich
Text Format, showing the syntax for
various formatting parameters.\
}
```

RIF See rate increase factor.

RIFF Resource Interchange File Format. A platform-independent multimedia specification developed in the early 1990s by a group of vendors including Microsoft Corporation.

right-hand rule, Fleming's rule A handy memory aid widely used in mathematics and physics to determine an axis of rotation or direction of magnetic flow in a current. Extend the thumb and fingers of the right hand so that the fingers are held together and point straight in one direction, with the thumb at a right angle to the fingers, in an 'L' shape. Now curl the fingers around a conductive wire, so that the thumb points in the direction of the current. The direction of the curled fingers then indicates the direction of the magnetic

R

field associated with the current. Using the same hand relationship, point the thumb in the direction of wire motion and the fingers will show the direction of the magnetic lines of force and the direction of the current, for a conductor in the armature of a *generator.* For the direction of current for a conductor in the armature in a *motor,* see left-hand rule.

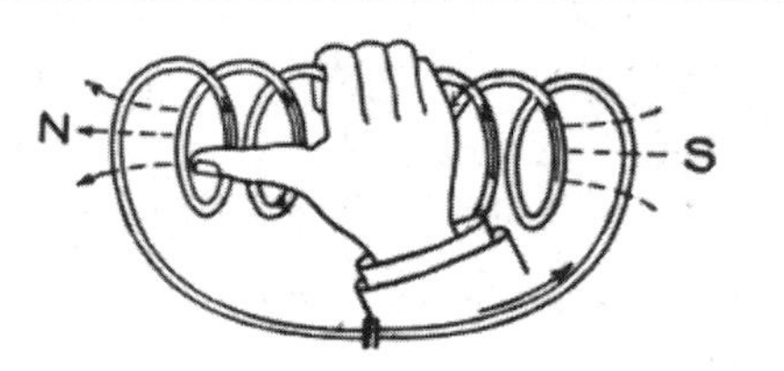

The 'right-hand rule' is a visual mnemonic device for remembering the relationship of the direction of the current in an electromagnetic field.

ring 1. The sound made by a phone or other communications device to indicate an incoming call or imminent announcement (as on a PA system). The frequency of the tone and its cadence vary from country to country. In North America, phones typically ring once every six seconds. 2. Traditionally the red wire in a two-wire telephone circuit. The name originates from the configuration of a manual phone jack in an old telephone switchboard in which the large plug was divided into two sections, with an internal wire electrically connected to the *tip* of the plug and another wire to the *ring* around the plug partway up the jack, nearer the insulated cord. The ring is traditionally around –48 volts, with the negative charge used to help to prevent corrosion. See tip, tip and ring.

ring-armature receiver A telephone receiver which began to be widely installed in the 1950s; it followed the development and widespread use of the bipolar receiver. It differed substantially from earlier receivers in its details, incorporating a lighter, more efficient dome-shaped diaphragm, driven piston-like by magnetic fields across the armature ring. See bipolar receiver.

ring-around-the-rosy A theoretical hazard situation on the phone system, in which a circular tandem connection exists, somewhat like an endless loop in a software routine.

ring, network A type of circular network topology. See Token-Ring.

ring topology A network topology in which each station in the network is connected in a closed loop so that no termination is required. In this architecture, data packets are passed around the loop through each intervening node until they reach the destination machine. Ring topologies are fragile in the sense that failure of one node or system affects the entire network. See star topology, topology.

ringback A usually undocumented, self-ringing telephone test number that can be used to verify the number of a particular phone. In some regions, when a two- or three-digit number, followed by the expected phone number is dialed, a new dial tone is provided, which can then be used for ringing tests. Ringback functions are used by service technicians after a phone has been installed, to verify that the system is functioning correctly. They are also sometimes used for confirmation by subscribers. In some regions, the ringback service connects the caller to a recording which identifies the exchange in which the phone is located. See ringing tone.

ringdown See ringing current.

ringer, bell The mechanical or digital sound generator that indicates the presence of an incoming call. The interval and type of tones generated vary from phone to phone, and even more so, from country to country. With the proliferation of digital devices, it is likely that ringers will eventually be configurable. You'll be able to load in a sound patch and have the phone sound like anything you desire: a cat, a parrot, a jet airplane, your spouse's voice, etc. In fact, paired with a Caller ID system, there is no reason why the ringer couldn't switch to a device that says out loud, "Hank is calling, want to talk to him?"

Ringer Equivalence Number REN. When telephone equipment is purchased and several devices are placed on the same line, there is a need for a way to designate and organize the ringer load on the line so as not to exceed the available current. In the United States, a certification number has been developed for telephone products which indicates that they meet certain specified requirements and guidelines. This REN helps installers and consumers to set up various pieces of equipment (phones, answering machines, fax machines) so they will not interfere with one another when connected.

Various devices on the same line may have different REN values. If a certain maximum number of RENs is exceeded, the various devices cannot be guaranteed to work correctly and may not ring.

For data modems, the Ringer Equivalence Number is usually printed on the main chip in the center of the internal modem board, along with the Federal Communications Commission (FCC) registration number. It may also be listed on a label on the back of an electronic device. In most regions, the sum of the RENs on one line should not exceed 5.0. In Canada, the concept of Load Number is essentially the same as the REN.

ringing The production of an audible signal at the receiving end by means of a mild AC or DC current to indicate signaling or the presence of an incoming call. In telephony, a ringing current is sent from the central office to the subscriber or from a local console or branch to a local phone device.

ringing current, ringdown The current on a telephone system used to transmit ringing signals and ringing tones. This varies with the type of switching system and the distance over which it has to be carried. In older phone signaling systems, for example, the ringing signal was carried by 20 hertz AC current for local distances, and 135 to 1000 hertz AC current for long-distance calls. See talk battery.

ringing key In a telephone switching system, a means by which the subscriber's telephone ringing is initiated by a key at the central office to indicate a call. Signaling current was sent as alternating current from a central office to the subscriber, a system that didn't work well for long-distance calls due to the loss of the signal over distance.

ringing signal Any signal transmitted over a telephone to initiate ringing on the receiving phone to indicate that a call is coming through. Various schemes for sending this signal have been used over the years to make it possible to send the ringing signal over distances without the current interfering with actual call transmissions. See ringing key.

ringing tone, ringback A tone generated in the caller's line to indicate a call is being routed to a receiving phone. The ring at the receiving end is initiated by a ringing signal sent from the central office or private branch switching system. This doesn't absolutely guarantee that the callee's phone is ringing. If the bell is defective or the routing of the call was in error (either through incorrect switching or because of dialing the wrong number), the callee may never hear the ring. See ringback, ringing signal.

ringing voltage The amount of voltage applied in an analog telephone switching system to cause the called phone or other phone device to ring, usually about 88 volts. There is a limit to the number of devices which can be rung when attached to a single line. Too many, and there will be interference with the ringing circuitry. For more information, see Ringer Equivalence Number.

RIP 1. See Routing Information Protocol. 2. See raster image processor.

RISC See Reduced Instruction Set Computing.

Ritchie, Dennis M. (1941-) An American Bell Laboratories researcher who codeveloped Unix in collaboration with Ken Thompson. He is well known for his development of the C programming language, along with Richard Kernighan. See C, Unix.

Ritchie, Foster An inventor, and early assistant of Elisha Gray, who designed a writing telegraph in 1900 based on principles different from those originally patented by Gray a few years before. Known generally as *telautographs,* these devices could transcribe handwriting across short distances and were in use for several decades.

RJ The Universal Service Ordering Code (USOC) for Federal Communications Commission (FCC) registered jacks for connecting to a public network. USOC was developed in the 1970s by AT&T and the telecommunications industry is widely standardized on this system. Each type of jack has a number of wiring configurations, depending upon the number of wires that are connected. Thus RJ-25 wiring uses the same jack as RJ-11 except that eight wires are connected instead of two. In many cases, not all the wires are used. Wires are often connected in pairs.

In order to promote pair continuity, when plugging into receptacles with varying numbers of active wires, the wires are usually connected in pairs beginning in the center and working out.

In the following Common Wiring Jacks chart, C represents flush or surfaced mounted, W represents wall mounted, and X is complex line.

RJE See remote job entry.

RLCM Remote Line Concentrating Module.

RLE See run length encoding.

RLP Radio Link Protocol.

RM See Resource Management.

RMA, RA returned merchandise authorization.

A coding system used to control and track returned merchandise. See return authorization.

rms See root mean square.

rn A full-screen, configurable news reader developed by Larry Wall, the author of the Perl programming language, and released in 1984. Wayne Davison developed a superset of rn known as *trn.*

RNC radio network controller.

RNX Restricted Numeric Exchange. A local telephone exchange in which calls are restricted to the network on which they originate. Suppose a number is 555-1111 when dialed from *outside* the local exchange. The first three digits are the NXX digits. From *inside* the restricted exchange, you might instead dial 105-1111, with 105 being the RNX digits. The call then rings through to the number without passing through circuits outside the local exchange. See NXX.

roaming Using wireless telecommunications services while moving around, either on foot or in a vehicle. The logistics of designing and managing roaming subscribers is quite complex and requires sophisticated software. See cellular phone. See Inter-Access Point Protocol.

robbed bit A commandeered bit in a transmission for something other than its usual purpose. This technique may be used to acquire extra bits for signaling information, especially if the signals are only occasionally needed. See robbed-bit signaling.

robbed-bit signaling RBS. In data communications, a means of taking one bit from a data path to provide signaling information. For example, in voice communications over inband T1 systems, a bit may be robbed to indicate the hook condition of the line.

Roberts, H. Edward Founder of MITS, which developed and distributed the first commercially successful computer kit, the Altair 8800, that launched the microcomputer industry.

Sample of Common Wiring Jacks Described in the Universal Service Ordering Code (USOC)

Jack	Wiring	Notes
RJ-11		Can accommodate up to six wires, though typically only two or four are connected as one or two pairs. A very common type of single-line phone jack used for telephones, modems, and fax machines.
	RJ-11C/W	One pair of wires connected, as for a single line phone. Traditionally the green and red wires are connected as tip and ring for the first line. The connection is bridged.
	RJ-14C/W	Two pairs of wires connected, as for a two-line connection, e.g., line 1 might be a phone or answering machine and line 2 a modem or fax machine. Traditionally green and red are assigned as tip and ring for the first line, and black and yellow as tip and ring for the second line. The connections are bridged.
	RJ-25C/W	All three pairs configured. Thus, line 1 might be a phone, line 2 a modem, and line 3 a fax machine.
RJ-45		Can accommodate up to eight wires and is common for multiple line phones (up to four lines) and for data communications, especially Ethernet and Token-Ring.
	RJ-48C/X	Four wires are typically connected as two pairs to provide 1.54 Mbps digital data network services.
	RJ48S	Four wires are typically connected as two pairs to provide local digital data network services.
	RJ61X	Four pairs of wires connected to accommodate four phone devices. Three lines are bridged.
	10Base-T	Twisted Pair. Pairs two and three are connected.

Roberts codesigned the computer when the market for calculators began to slip. Roberts left MITS in 1977 and is now a physician. See Altair.

robot In its simplest form, a robot is a mechanical apparatus which automatically moves or senses according to a set program, or an adaptive program. In its most complex form, it is a sophisticated electromechanical logical device that can interact with its environment in ways which are ascribed to human 'intelligence.' That is, it is adaptive, and responds in ways that are appropriate to its task or to its benefit. Robot arms are used in many production-line tasks, whereas many humanlike robots are portrayed in science fiction stories and films. A humanlike robot (in form and functioning) is called an android.

A robot is generally associated with hardware and physical forms, but with the development of software artificial intelligence techniques, online avatars that seem to have a humanlike presence on public forums and chat lines have also been called robots, or more commonly 'bots or bots. A bot can be online 24 hours a day, can monitor processes, log user activities and interaction, and more. Bots have been banned from some of the chat rooms due to badly programmed bots violating chat Netiquette. However, system operators sometimes have authorized bots running on Internet Relay Chat (IRC) to perform many useful functions.

Bots are also associated with search engines. Just as IRC bots do housekeeping tasks on the chat channels, search bots do useful Web page search and retrieve jobs, like a crew of gofers working around the clock. See mailer daemon.

rocket camera A creative early 1900s invention for taking photographs at high altitude (close to 800 meters) by attaching a camera to a stabilizing rod and equipping it with a small parachute.

Rocky Mountain Bell Telephone Company An early Bell telephone exchange, established in 1883, with financial assistance from American Bell Company. It rapidly acquired the Ogden Telephone Exchange Company, the Montana Telephone & Telegraphy Company, the Idaho Telephone & Telegraph Company, and the Park City exchange. This pattern of acquisitions continued for a number of years.

Rocky Mountain Telephone Company An early telephone exchange established in Salt Lake City in 1800.

roentgen, røntgen (Symbol - R) An international unit of X-radiation or gamma radiation which is equal to the amount of radiation which produces one electrostatic unit (esu) of charge in one cubic centimeter of dry air at zero degrees centigrade and standard atmospheric pressure ionization of either sign.

Rogers receiver An early batteryless radio receiver first introduced in 1925 at the Canadian National Exhibition in Toronto.

ROLC Routing Over Large Clouds. The ROLC Working Group has now merged with the IPoverATM Working Group to form Internetworking over NBMA (ION). See frame relay, Internetworking over NBMA, RFC 1735.

Roll About A commercial, self-contained videoconferencing unit which includes the monitor, camera, microphone, and other components on a stand that can be moved from one office to another as needed.

ROM See read only memory.

Rømer, Ole (Olaf) Christensen; Rümer, Ole Christensen (1644-1710) A Danish astronomer, physicist, and scientific instrument-maker who, in 1675, demonstrated the velocity of light as 11 minutes per astronomical unit (AU) based on observations of the planet Jupiter and its moon Io. In the 1660s, he was entrusted with the editing of the great scientist Tycho Brahe's manuscripts. He was appointed in France by Louis XIV to tutor the Dauphin in astronomy, and also was appointed as the *Astronomer Royal* at the Danish court of Christian V. He invented a new type of thermometer and communicated some of his ideas to D. Fahrenheit in the early 1700s. In Denmark, he introduced a new system for numbers and weights in which the concepts of weight and length were brought together.

Røntgen, Wilhelm Konrad (1845-1923) A German physicist and educator who developed the vacuum tube (1895), the fluoroscope, and pioneer experimental studies in X-ray emissions in industrial radiography and medical radiology. He was the first recipient of the Nobel prize in physics. The roentgen unit of X-radiation is named after him. See roentgen, X-rays.

root mean square (*Abbrev.* rms) The effective value of a quantity in a periodic circuit, measured through the duration of one period.

ROSAT Røntgen Satellite. A research satellite which has expanded our knowledge of the universe and past events in our galaxy, by aiding us in discovering local hot X-ray plasma.

ROSE See Remote Operations Service Element.

Rosenblueth, Arturo A scientist who collaborated with Norbert Wiener in investigations in artificial intelligence and self-organizing systems, much of which was documented in Wiener's book on cybernetics.

rotary dial A circular dial mechanism typically activated by placing a finger (or pencil end) in one of a series of punched out holes, and turning (dialing). The mechanism springs back to its original position after each selection. The alphanumeric selections are displayed under each associated hole. Turning the dial activates the telephone carrier's electrical loop for specified intervals that form a simple code to identify the number dialed.

Dials were not always circular. Some of the earliest dials consisted of levers, resembling the front of a small slot machine.

Rotary dials are normally associated with pulse dialing signals. Rotary phones are steadily being superseded by pushbutton phones, especially since automated systems require touchtone phones in order to use their services. See pulse dialing, tone dialing, keypad, touchtone phone.

rotary switch A commercially successful electromechanical telephone switching system developed in the AT&T labs in the early 1900s, based on Lorimer one-step selection concepts and incorporating a permanently rotating motor. These were installed in Europe as a result of an International Telegraph and Telephone buyout of Western Electric's International division in 1925. See Lorimer switch, panel switch, rotary switch.

ROTF, ROTFL Abbreviations for "rolling on the floor" and "rolling on the floor, laughing" both of which are often used in chat rooms and email correspondence. See emoticons, AFK, IMHO.

route 1. *n.* Path taken by data or other transmissions. See traceroute. 2. *v.* To delineate a communications path. This may be fixed before the transmission or may be dynamic according to availability and load levels evaluated enroute. See hop-by-hop.

route flap, router flap A fault condition in which changes in routes propagating across a network (usually from losing one or more nodes) exceed the capacity of a router's processor and memory to cope with the change, and consequently impacts its ability to route. Routers are generally selected to well exceed the number of routing paths expected to be needed to prevent this serious problem.

router 1. A device or mechanism for selecting a path. 2. In a simple network, an interface device which selects a path for the transmission packets. In layered networks, the router typically functions at layer two or layer three, depending upon the degree of automation and 'intelligence' built into the router. There used to be somewhat of a distinction between routers and switches, but switches are becoming so sophisticated that the distinction is disappearing. Routers frequently include routing databases, in addition to algorithms to dynamically select routes. See bridge, switcher.

router, ATM In ATM networks, a router delivers and receives Internet Protocol (IP) packets to and from other systems, and relays IP packets among systems. Routers vary in sophistication, with some able to contribute significant network management functions, such as priority and load balancing. They can be protocol-dependent or protocol-independent. Also called an intermediate system. See LIS.

router, frame relay A frame relay-capable router has the ability to encapsulate local area network (LAN) frames in frame relay-format frames and feed them to a frame relay switch, as well as receiving frame relay frames, and stripping the frame relay frame to restore the information to its original form, passing it onto the end device. With improved technologies, the distinction between routers and bridges is lessening. Even switches now have many of the capabilities of routers. See bridge, gateway, switch.

routing Selecting or establishing a path by which to transmit signals or information.

routing aggregation An administrative tool for organizing and optimizing the use and availability of routes to deal with the continually rising demand on networks. Users are encouraged to return unused addresses, and old addresses are assigned prefixes so multiple routes can be aggregated into one.

Routing Arbiter Database RADB. A routing database established by the Routing Arbiter project. One of several databases in the Internet Routing Registry.

Routing Arbiter Project A National Science Foundation funded project given the task of coordinating routing for the new NSFNet architecture in cooperation with a number of educational and private business concerns.

Route Servers were to be installed at connection points to reduce the need for peering. Due to NSFNet legacy database information, a number of large providers have shunned the Routing Arbiter and the project has changed to a service available through some of the Public Exchange Points. See peering.

routing code 1. In telephone communications, the area code. 2. In U.S. postal communications, the last four digits of the ZIP+4 code. 3. In networks, the data parsed by the router or switcher to establish a path to the intended destination.

routing computations, routing algorithms Mathematical schemes to compute efficient routes through a network. The algorithms may be straightforward if the topology and size of the network is known and is relatively stable. The situation is more complicated on the Internet, which is extensive, encompasses many different types of configurations, and which changes constantly as networks are added or changed.

routing domain RD. In ATM networking, a collection of systems which have been grouped topologically within one routing system.

Routing Information Protocol RIP. A very common routing protocol from a family of protocols known as the Interior Gateway Protocols (IGPs). RIP evaluates the path between two points in terms of *hops* between the points. See RFC 1058.

Route Servers Specialized servers from the Routing Arbiter project intended to hold clearinghouse routing information in a Routing Information Base at each network interconnection point in order to eliminate, or at least reduce, the need for peering. These servers are not intended to actually transmit the traffic, but serve to handle the flow of information concerning pathways.

routing table, data network In data networks, a table detailing paths to specific Internet Protocol (IP) addresses. With the explosive growth of the net, the number of paths, and hence the size of the tables, can become very large. Primary routers sometimes list more than 25,000 routes and the routers themselves must be designed to keep ahead of capacity. Discussions are ongoing as to the benefits and problems in various assignments of routing paths, with provider-based routing being favored due to greater ease of implementation, and geographic-based routing proposed because it has less of a tendency to concentrate power in the hands of a few large providers.

routing table, telephone network In telephone communications, routing tables serve to record and provide information for the processing of incoming calls. Thus, calls may not just go to a particular caller or workstation, they may be directed to automated voice services, voicemail, queued holds, recordings, etc. They may require the capability of stepping back through the route as well, depending upon the sophistication of the system and the selections available to the user.

routing update Information provided by a router as to network configuration, and, in some cases, costs associated with use of particular routes. Routing updates can be scheduled to be automatically sent out at specified intervals, and are typically broadcast if significant network configuration changes have been made.

row The grouping, between the left and right edges, of more-or-less horizontally aligned elements arranged within a grid or tabular format. Commonly used to reference screen locations or positions within a spreadsheet. See column.

RPC See remote procedure call.

RPG Report Program Generator. A computer programming language for processing and displaying large data files.

RPM Remote Packet Module.

RPOA See Recognized Private Operating Agency.

RPS See repetitive pattern suppression.

RPW remotely piloted vehicle.

RQS See Rate Quote System.

RS recommended standard. See RS-232 for an example using this prefix.

RS-1 Along with RS-2, the first USSR amateur satellites, launched in October 1978.

RS-232 Recommended Standard 232. A decades-old standard for serial transmissions widely adopted in desktop computers and other devices, commonly used for communicating with modems, remote terminals, and printers. The RS-232 specifies the electrical and physical characteristics of the connection. The most common implementation is RS-232c, which was developed by the Electrical Industries Association (EIA). It defines a means of connecting data terminal equipment (DTE) with data circuit-terminating equipment (DCE).

Most systems support RS-232c through a 25-

pin D connector, although minimally, nine pins are needed to implement the specification, and 9-pin D connectors are sometimes used. A few systems specify more than 11 or 12 pins (pin 12 is not part of the spec. but some vendors add a low voltage power signal to one pin). The basic RS-232 pinouts are as follows:

Pin	Abbrev.	Function
2	TxD	transmit data
3	RxD	receive data
4	RTS	request to send
5	CTS	clear to send
6	DSR	dataset ready (signals that the device is on)
7	GND	signal ground
8	DCD	data carrier detect (signal that carrier is on)
15		transmit clock
17		receive clock
20	DTR	data terminal ready
24		auxiliary clock (provided by some vendors for local connections)

Pins 2 and 3 are the most important, as these are the ones that send and receive data. A null modem cable can be made from an RS-232 cable by swapping pins 2 and 3 on one end of the connection. This allows two locally connected computers to 'network' through the cable with suitable communications programs running on each system. The initial RS-232 standard was superseded in 1987 by the standard defined by the Electronic Industries Alliance (EIA) as EIA-232-D, which was followed in 1991 by the EIA/TIA-232-E. See EIA Interface Standards for a list of common standards.

RSA Rural Service Area.

RSC Remote Switching Center.

RSGB Radio Society of Great Britain.

RSUP See Reliable SAP Update Protocol

RSVP Internet Reservation Protocol. An extensible, scalable protocol designed in the mid-90s to provide efficient, robust ways to set up Internet-integrated service reservations, RSVP became an Internet standard in 1997. It has primarily been promoted by commercial interests, as it makes it possible to establish 'priority' connections through reserved bandwidth, a feature of interest to large competitive business network users.

RSVP is appropriate for multicast applications, although it supports unicast as well. RSVP interfaces existing routing protocols rather than performing its own routing. The RSVP is used by a host to request a specific Quality of Service (QoS) from the network. RSVP attempts to make a resource reservation for the data stream at each node through which it passes. RSVP communicates with two local decision modules: admission control and policy control, to determine whether the node has sufficient resources to supply the QoS, and whether the user has administrative permission to make the reservation. One of the difficulties in implementing RSVP has been assessing fees for connections across more than one network. Some opponents of the system fear the establishment of 'elite' Internet users based on economics rather than on quality of information or services offered. RSVP development has continued since 1995 as RSVP2. See STII.

RT 1. realtime, real time 2. remote terminal. 3. reorder tone.

RTC runtime control.

RTCP Real Time Conferencing Protocol.

RTF See Rich Text Format.

RTFM An abbreviation for "read the freaking manual" used on public forums on the Internet, when a user asks a question which has been asked and answered hundreds of times and is clearly answered in the appropriate documentation, or FAQ. An exhortation for the user to not be lazy, to look it up *before* using up people's valuable time in asking an obvious question. See Frequently Asked Question.

RTP 1. realtime protocol. 2. routing table protocol.

RTS Request to Send. Flow control, typically used in serial communications, which is an output for DTE devices and an input for DCE devices. See TxD, RxD, CTS, DSR, DCD, DTR, RS-232.

RTSP See Real Time Streaming Protocol.

RTV realtime video. Video that appears as natural movement, usually with at least 20 frames per second, and which further may be a live broadcast as opposed to playback from stored information.

RU 1. In packet networking request unit, response unit, request/response unit. See basic information unit. 2. receive unit.

rubber bandwidth *jargon* A communications channel whose bandwidth can be dynamically altered, that is, it can be changed without terminating and reinitiating the transmission. This colorful phrase apparently originates from Ascend Communications, a supplier of networking-related products, to describe characteristics of an inverse multiplexing system.

run To initiate a software program, or linked suite of programs which form an application.

run length encoding RLE. A lossless data compression technique that works well with data that includes repeated sequences. The repeated sequences (white spaces in a document, a single background color in an image, etc.) are replaced with a code that indicates what follows is a string of the same character of a particular length. If run length encoding is used on data with little or no redundancy, the encoded file may be *longer* than the original.

Rundgren, Todd A multimedia recording artist who has managed to stay at the forefront of emerging interactive entertainment technologies, synthesizing the new capabilities in media into video and sound. Rundgren began programming microcomputers in the late 1970s, adapting Macintoshes, Amiga Video Toasters, and other systems to many new creative venues, producing new types of music albums, computer-generated rock videos, and interactive TV entertainment concerts. Since the mid-1990s, Rundgren has been President and CEO of Waking Dreams, which develops, licenses, and distributes products and services which originate from creative and undervalued ideas.

runtime 1. The time during which a software routine is actually processed. 2. The duration during which a software program executes. See runtime.

Rural Service Area RSA.

rural telephone company This is defined in the Telecommunications Act of 1996 and published by the Federal Communications Commission (FCC) as:

> "... a local exchange carrier operating entity to the extent that such entity—
>
> (A) provides common carrier service to any local exchange carrier study area that does not include either—
>
> (i) any incorporated place of 10,000 inhabitants or more, or any part thereof, based on the most recently available population statistics of the Bureau of the Census; or
>
> (ii) any territory, incorporated or unincorporated, included in an urbanized area, as defined by the Bureau of the Census as of August 10, 1993;
>
> (B) provides telephone exchange service, including exchange access, to fewer than 50,000 access lines;
>
> (C) provides telephone exchange service to any local exchange carrier study area with fewer than 100,000 access lines; or
>
> (D) has less than 15 percent of its access lines in communities of more than 50,000 on the date of enactment of the Telecommunications Act of 1996."

See Federal Communications Commission, Telecommunications Act of 1996.

Rural Utilities Service RUS. A U.S. Department of Agriculture agency that provides technical and funding support for rural utilities infrastructure projects involving electricity, water, and telecommunications.

RURL Relative Uniform Resource Locator. See RFC 1808.

Rutherford, Ernest (1871-1937) A New Zealand-born British physicist who contributed substantially to our knowledge of atomic physics. Rutherford researched at the Cavendish lab studying ionizing gases and following up much of the work of the Curies and Philipp Lenard. He collaborated with Hans Geiger, developer of the Geiger counter, and influenced Paul Villard's studies of gamma rays.

RWhois, rwhois Referral Whois. RWhois gives users a means to look up information on the Internet. It is a primarily hierarchical, client/server distributed system for the discovery, retrieval, and maintenance of directory information on computer networks. RWhois allows for the deterministic routing of queries based on hierarchical tags, referring the user closer to the source of the information. The RWhois specification defines both a directory architecture and a directory access protocol, as they are intrinsically related.

RWhois extends and enhances its predecessor, Whois, in a hierarchical, scalable way in order to meet the increased demands on Whois from the explosive growth of the Internet. The protocol and its architecture are structurally derived from the Domain Name System (DNS), and concepts from the X.500 Protocol and Simple Mail Transport Protocol (SMTP) have been incorporated into the specification. To use RWhois from a command line shell that supports the utility, type "rwhois" all in lower case. See InterNIC, Whois, RFC 2167.

RxD receive data. Data channel, typically used in serial communications, which is an input for DTE devices and an output for DCE devices. See TxD, RTS, DSR, DTR, RS-232.

S-band A broadcasting frequency spectrum ranging from 2310 to 2360 MHz. S-band is typically used for radar and some types of mobile services. It is terrain-sensitive. See band allocations for a chart.

S-HTTP See Secure HTTP.

S/W software.

SA 1. See Service Agent. 2. source address.

SAA 1. Standards Association of Australia. 2. Supplemental Alert Adapter. A connection device for interfacing alerting devices to analog multiline phones. 3. See Systems Application Architecture.

Saco River Telegraph & Telephone Company Maine's oldest independent phone company, established in 1889.

SAFE 1. Security and Freedom through Encryption.

SAFE Act Security and Freedom through Encryption Act. See Clipper Chip, Pretty Good Privacy.

sagan A tongue-in-cheek tribute to Carl Sagan, indicating a very, very large amount. "Billions and billions" as he would say with infectious enthusiasm in his popular TV series when referring to the many stars in the cosmos. Sagan's premature death was mourned by many amateur astronomers who got their first taste of the wonders of the galaxy and beyond through Sagan's show.

Sagan, Carl E. (1934-1996) An American astronomer, writer, educator, and inspirational host of the popular "Cosmos" television series on the U.S. Public Broadcast System (PBS). Sagan was the director of the Laboratory for Planetary Studies and the David Duncan Professor of Astronomy and Space Sciences at Cornell University.

SAGE Semi-Automatic Ground Environment. A U.S. government security digital communications, detection, and craft control network.

SAIL See Stanford Artificial Intelligence Laboratory.

SAM security accounts manager.

samarium A silver rare earth metal discovered through spectroscopy in samarskite in the late 1800s. Samarium is used for doping calcium fluoride crystals for use in lasers. It is one of the rare earth metals used in carbon-arc lights. See doping.

sampling Recording a signal by quantizing it at intervals in order to capture its basic properties, usually also accompanied by saving the samples in a digital or abbreviated form. In digital sound sampling, for example, the sound of a musical instrument or a voice can be digitally sampled a certain number of times per second in order to be able to play back the sound so it retains and conveys the character of the original, although not necessarily all the information or format of the original.

Generally, the more frequent the sampling, the better the fidelity of the playback. Some sounds, like the sound of a concerto played on a violin, are more complex than others, like a doorbell ringing, and require more frequent samples and a greater frequency range to retain the perceptual quality of the original.

In videoconferencing over slow transmission lines, the image is usually sampled rather than played in real-time, that is, a new still image is 'grabbed' and transmitted every few seconds or every few minutes. Generally, as in sound sampling, more frequent samples provide greater fidelity to the original. In video sampling, rates of at least 20 to 30 frames per second are perceived by humans as natural motion. See animation, audiographics.

sampling rate The number of captures of an input, such as light or sound waves, per unit of time. Sampling rates for images are generally expressed in frames per second, with 24 or more appearing natural to the viewer. Sampling rates for audio are generally expressed in kilohertz (kHz); an instrument might be sampled at 50 kHz, that is 50,000 bits per second. Higher sampling rates generally require more sophisticated equipment, higher processing speeds, and faster transmission speeds, especially if the signals are being sent over a network.

SAN See satellite access node.

Sandbox A Sun Microsystems Java security block. Since Java applets are freely shared through public sites on the Internet, there is always reason to question whether they contain hidden viruses or other destructive or annoying capabilities. The Sandbox is a means of restricting doubtful applets to a confined area, that is, quarantined, so they can't affect other data on the disk or other Sandboxes.

sanity check The process of sitting back and reflecting on one's current priorities, responsibilities, and level of stress. Anyone in the computer industry probably needs to do a sanity check once in a while to evaluate whether the technology is improving the quality of his or her life, or whether it is just eating up time and money and forcing everyone to do more in less time. (Probably descended from the phrase *systems check* in which a step-by-step run-through of a system's current operating status is made to see if there are any issues that require attention or repair.)

SANZ Standards Association of New Zealand.

SAP 1. Service Access Point. An interface point in a network, often associated with a specific layer. 2. See Service Advertising Protocol.

SAPI Service Access Point Identifier.

SAR 1. See segmentation and reassembly. 2. synthetic aperture radar.

Sarnoff, David (1891-1971) A Russian emigrant to America, Sarnoff was an ambitious, energetic radio operator and Marconi station manager at the Radio Corporation of America (RCA) who is widely reported as having intercepted the messages of the Carpathia when the Titanic struck an iceberg, and relayed the messages to relatives and friends of the passengers on board the sinking ship. While it is likely that Sarnoff did play a part in relaying the messages, the RCA promotional information of this event included a doctored picture of Sarnoff at the console, which lends some doubt to the claim that Sarnoff handled the post single-handedly for 72 hours, as widely promoted.

Sarnoff is known for the large part he played in the early history of the Radio Corporation of America (RCA), since becoming its general manager in 1921. He is also remembered for his association with the inventor E. Armstrong in the late 1920s which ended abruptly in 1935 with the removal of Armstrong's test equipment from the premises, and the development of television.

In 1926, Sarnoff was instrumental in founding the National Broadcasting Company (NBC).

The IEEE now honors Sarnoff's contributions with the David Sarnoff Award in Electronics. See Radio Corporation of America.

SART See search and rescue radar transponder.

SAS 1. single address space. 2. simple attachment scheme. 3. Survivable Adaptive Systems.

SASL See Simple Security and Authentication Layer.

SASMO Syrian Arab Organization for Standardization and Metrology.

SASO Saudi Arabian Standards Organization.

sat-, -sat A prefix/suffix often prepended/appended to satellite-related names and technologies.

SATCOM Satellite Communications.

satellite, artificial A manufactured object launched to orbit the Earth, Moon, or other celestial body. There are currently many communications and Global Positioning System (GPS) satellites in orbit around Earth which send and receive signals to and from ground stations and transportation vehicles (cars, trains, boats, planes, etc.). Satellites now provide the main means for wireless long-distance communications. The first artificial satellite was Sputnik I in 1957, followed by the first geostationary satellite in 1963. See global positioning systems, satellite antennas.

satellite, natural A celestial body in orbit around another.

satellite access node SAN. A terrestrial satellite link, usually consisting of an Earth station or Earth station hub.

satellite antennas Satellite antennas were

originally launched into orbit for military monitoring and communications, space research, and cable TV broadcasting, but increasing numbers serve individual parabolic home receivers and data communications providers. GPS satellites orbit at about 18,000 kilometers (11,000 miles) and broadcast satellites at about 36,000 kilometers (22,300 miles) altitude.

In its basic form, a broadcast satellite system consists of a broadcasting station sending signals through an uplink dish aimed directly at a geostationary satellite antenna in synchronized orbit with the Earth. The signal subsequently is sent from the satellite to a downlink dish (parabolic antenna) attached variously to business complexes or rebroadcast stations, and subsequently directed to subscribers through through coaxial cable. Some stations broadcast directly, through scrambled signals, to apartment blocks or individual households. The lead from the downlink dish feeds into the user's television or computer system. See antenna, C-band, feed horn, Global Positioning System, Ka-band, Ku-band, microwave antenna, parabolic antenna.

satellite broadcast frequencies The various ranges of frequencies over which satellite antenna transmissions take place. These are dependent on many factors, including the type of transmission, the type of satellite, and regulatory guidelines and restrictions. Broadcast stations typically operate in the C-band, with uplinks at about 6,000 MHz and downlinks at about 4,000 MHz to rebroadcast stations with powerful antennas. The frequency levels are tied to the size of the receiving dishes, with higher frequencies being more difficult to accommodate technologically, but having the advantage of much smaller receiving dishes. Higher frequencies can broadcast to smaller receivers, making it possible for some frequencies to be broadcast directly to smaller consumer dishes. See C-band, Ka-band, Ku-band.

satellite closet A centralized wiring closet for interconnection of cables and equipment. In a number of satellite installations, the programming is beamed from the satellite to a central service provider with a satellite receiving dish and, from there, delivered by wire or cable to subscribers, necessitating local loop hookups. See distribution frame.

satellite communications A wide variety of radio, television, telephone, data, and other broadcast and two-way wireless communications provided by transmission via orbiting satellites to centralized distribution providers or individual subscriber satellite dishes.

The age of satellite communications began in the late 1950s, with the launch of Sputnik I, although it was described with remarkable insight by Arthur C. Clarke in the 1940s and 1950s in various articles and books.

The early satellites did not last long, from a few weeks to a few months, and power consumption and radiation problems had to be solved before widespread use became practical.

In less than four decades from their modest beginnings, satellite communications have developed rapidly and there are now hundreds of satellites of different designs orbiting the Earth at various distances. Their lifespans now range from about five to 15 years, and most are powered by solar panels with battery backup.

It was not long after the first satellites were launched that they were used by commercial and amateur radio stations. The first television broadcast station to use satellites was the Canadian Broadcasting Corporation, transmitting through ANIK in 1972. Direct broadcast to consumers, rather than through intermediary stations, did not really become prevalent until the 1990s, when broadcasts of higher frequencies became practical, and smaller, more convenient satellite dishes were manufactured.

Satellites are launched in conjunction with spacecraft or shuttle crafts into elliptical or geostationary orbits from about 500 km to about 36,000 km above the Earth and are able to either passively transmit data back to Earth (these are becoming rare), or actively regenerate or otherwise amplify the signal and retransmit, usually at a different frequency to avoid interference of uplink and downlink signals. Satellites are general purpose with many transponders, or specialized for data, voice, broadcast, etc. Broadcast satellites tend to be unidirectional, while data and voice satellites, as for mobile systems, are bidirectional. See AMSAT; Clarke, Arthur C; Global Positioning System; direct broadcast satellite, OSCAR, Syncom, Telstar, and the many listings under satellite services.

satellite communications control SCC. The Earth-based station facilities and equipment which control satellite transmissions, including signaling functions, access control, error correction, signal conditioning and noise reduction, etc.

satellite constellation A group of satellites in a cluster. A group of satellites is commonly used for Global Positioning System (GPS) applications, for example, where data from three or more related satellites are mathematically manipulated to yield precise positioning information of an Earth location.

satellite link A system of transmitters and receivers communicating with a satellite, usually through an active transponder which will amplify and shift the received communications to another frequency before retransmitting on the downlink. Uplinks and downlinks are often managed separately. For example, television broadcasts are primarily in one direction (though interactive TV applications are increasing), while phone and computer data communications are typically in two directions. See geostationary, satellite.

satellite scanner See scanner.

satellite scatter This has two opposite meanings, as scatter can be the undesirable diffusion and weakening of a signal, or, conversely, it can be a deliberate manipulation of the environment to enhance communications. In the latter, it was discovered that ionization of airborne particles could open up 'communications windows' which otherwise were not available. Thus, burns from launched spacecraft or deliberate 'seeding' of high regions with elements like barium, could provide possibilities for detecting or sending transmissions through these temporary holes. There have also been experiments with heating the ionosphere with high powered waves to form a type of 'aurora' which can facilitate transmissions.

Due to their transient nature and the strength of the signals needed, these are not major sources of communications, but it's valuable to understand the nature of the various phenomena, and provides occasional glimpses into frequencies emanating from space.

satellite services Profit and not-for-profit organizations which provide various types of satellite-based communications services. See American Mobile Satellite Corporation, AMSAT, ARIES, Astrolink, Constellation Communications, Inc., CyberStar, ECCO, Ellipso, ICO Global Communications, INMARSAT, Globalstar, OrbLink, Skynet, Spaceway, Teledesic.

Satellite Work Centers A telework organization similar to a branch office, placed in a residential or rural village area by a business entity, and made commercially viable by the implementation of new communications technologies. See ADVANCE Project, Shared Facility Centers, telework.

saturation To add or adjust such that no more can be absorbed, or contained. In color applications, saturation refers to color purity. Undithered colors on a computer monitor or printing inks made from primary pigments, tend to be highly saturated.

save 1. In computer applications, a selection that permits the storage of information for later retrieval. 2. In phone hardware, a key sequence or button, that enables the user to store a number for later quick retrieval. See speed dialing.

SBE Society of Broadcast Engineers.

SBR spaceborne radar.

scalable Adjustable, able to increase or decrease in size, capacity, or other relevant characteristics, without significant degradation in quality of service or functioning.

Scalable fonts and images, usually defined as vectors, have the capability of being adaptable to lower and higher screen and printer resolutions, displaying at the best possible resolution for that particular device due to the internal algorithmic nature of the font definition. Scalable images are sometimes called *resolution-independent* images.

Scalable networks allow the system to accommodate to changing conditions. In static environments, scalability is not a critical factor, and nonscalable systems tend to be less expensive. In dynamic environments, such as the Internet or large WAN implementations, scalability can be a crucial factor, especially over time, contributing to the flexibility and usability of a system.

Many aspects of networks need to be scalable. The system software should be scalable to adapt to smaller and larger numbers of users, sometimes on a minute-to-minute basis. Physical storage mediums need to be scalable to accommodate less or more storage as needed. Routing protocols need to be scalable to accommodate changing topologies and numbers of workstations.

Scalable Coherent Interface SCI. A high bandwidth, scalable, media-independent network transmission technology developed in the late 1980s that operates up to about 1 Gbps. It is an ANSI/ISO/IEEE standard (1596-1992). SCI supports parallel distributed multiprocessing and cache-coherent interconnection. SCI fits into the upper mid-range in throughput. It is faster than ATM, Fibre Channel, and Ethernet, but slower than HIPPI-6400, and does not have HIPPI-6400's retransmission capabilities.

Initial implementations of SCI tend to be high-end commercial/industrial and military supercomputing applications.

scalable typeface A character set defined algorithmically with vectors so that it can be displayed to the highest resolution of any particular display device. In other words, if you have a 75 dpi monitor, the typeface will display as well as the monitor can handle, but will also display as well as a high-end typesetting machine can handle, say 2400 dpi, that is, with smooth curves and fine details readily visible. In contrast, a nonscalable typeface is usually a raster screen font (composed of a specific number of dots arranged in a grid). It may look fine on the computer monitor, but print it at 600 dpi or 1200 dpi or higher, and it looks the same is it does on the screen: grainy, chunky, staircased, and unappealing. The Adobe PostScript page definition language is used to define high quality scalable fonts and images, in addition to being able to render images, charts, etc. Microsoft's TrueType is a common format for scalable fonts. See PostScript, TrueType.

A

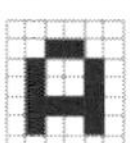

The raster font on the right is very limited, as it is composed of a set number of pixels and cannot be effectively reduced or enlarged. In contrast, the vector font on the left can be scaled effectively from 4 points to 100 or 500 points or more as it is defined mathematically based on the relationships of its shapes and proportions, and is not tied to a fixed number of dots.

scaling *v.t.* 1. Sizing, adjusting to size. May be proportional scaling, or selective scaling in one or more axes. Scaling is a common operation in image processing. See cropping. 2. Adjusting to capacity, or number of members. In programming there are often scalable ways of designing algorithms. For example, an operating system may have a fixed number of windows which can be open at one time (e.g., maximum of 200), or it may be scalable, in which the maximum number of windows is limited only by the system resources, and becomes greater as greater resources are added (e.g., memory, storage space, CPU speed, etc.). Scalable systems are more flexible and less likely to go out of date, but are often more resource-intensive and sometimes more difficult to program. In network transmissions with a variety of protocols, scaling may occur to accommodate differences in bandwidth, data, or speed capacities of the various systems through which the transmission may travel.

scan converter A device for converting a video signal. With computers it is common to take the RGB signal that normally leads to the computer monitor and feed the signal through a scan converter so it can be recorded on a video tape.

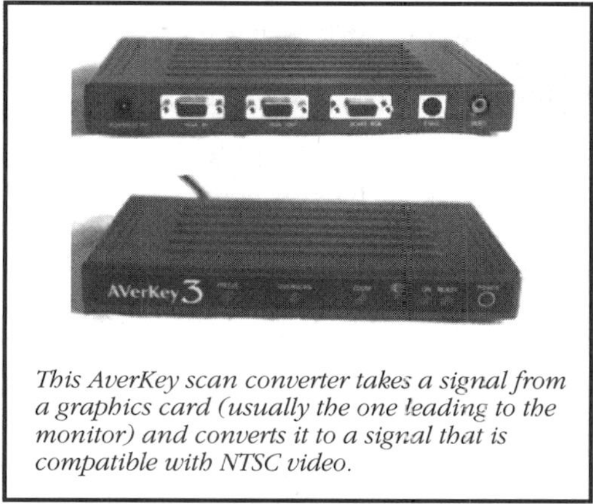

This AverKey scan converter takes a signal from a graphics card (usually the one leading to the monitor) and converts it to a signal that is compatible with NTSC video.

scan line, scanning line On a display monitor, a narrow more-or-less continuous line illuminated by the movement of the electron beam across the inside surface of the tube. In television broadcasting and raster monitors, these are typically horizontal. On vector monitors, the scan line can be traced in any direction. See raster, vector.

scanner A device that samples objects, information, or processes, and quantifies and records the results in some form that can be further processed or analyzed. Scanners, like digitizers, frequently sample analog information and convert it to digital, or convert a digital sample to a waveform for transmission. Typically the scanner does not process the information; its job is to *capture* the information, and often it can store that information in a variety of selectable formats. Once the information is captured, it is then sent 'live' to a processing application or it is stored for later conversion or further processing. Scanners are used in a multitude of imaging and sensing applications, including those described in the Scanning Technologies chart.

scanning rate 1. The speed, in units per time period, at which a scanning device captures information. Many common technologies are expressed in terms of inches per second. Generally, for moving images, faster scan rates are associated with higher accuracy or fidelity in conveying motion. In still images, faster scan rates may compromise the resolution or fidelity of an image. 2. In cathode ray tubes (CRTs), the scanning rate is the speed at which the electron beam sweeps the screen to refresh a full 'frame.' This is usually about 30 frames per second. See cathode ray tube, frame, interlace. 3. For technologies that use a beam for scanning (antennas, radar), the rate is more often expressed as a specified number of sweeps over a unit of time.

scanning rate, optical For optical scanning of images, such as those used with computer scanners and facsimile machines, scanning 'rate' is often not as important as resolution. In other words, the fineness of the scan is described rather than the speed at which it is scanned, with resolutions of 1200 to 4800 dpi interpolated being common on desktop scanners. See scanner.

scare-straps *colloq.* Telephone line worker safety belts.

scatter *v.* To separate, to distribute widely or randomly, to disperse, to diffuse in various directions.

scatter, transmission To diffuse or spread out in such as way as to lose the strength or directionality of a transmission signal. In most communication transmissions, this is undesirable. 3. To enhance communications through exploiting deliberate, controlled scatter. See satellite scatter.

SCC 1. Specialized Common Carrier. A carrier competing with the dominant carrier in niche markets. 2. Standards Council of Canada.

SCCP See Signaling Connection Control Part.

SCE See Service Creation Environment.

Scelbi-8H *Sc*ientific, *el*ectronic, and *bi*ological. A forerunner to personal computers, developed between 1972-1974 and released in 1974, it followed the Kendak-1 which was first advertised in 1971. The Scelbi-8H was the first publicly advertised Intel 8008-based per-

Scanning Technologies for a Variety of Devices

Application	Notes
scannning antenna	Parts of a moving antenna which cause the directional scanning of the antenna beam.
facsimile scanner	A component of a facsimile machine which converts a sampled digital image to a waveform for transmission over phone lines. See ITU-T, facsimile, TIFF.
image processing	A computer input device which passes a beam over a 2D or 3D object or image, and converts it to digital data, usually a 2D raster image. See digitizer, optical character recognition, TWAIN.
scanning radio	A radio receiver that scans a range of transmission frequencies automatically, so the user can locate and listen to conversations occurring at the time of the scan. Often used to find emergency or cellular phone conversations.
remote sensing	In satellite remote sensing, a scanner employing an oscillating mirror which captures images in strips, was first proposed in 1968 by Hughes Aircraft for use in orbiting satellites, and first deployed in 1972 in the first Earth Resources Technology Satellite. This technology revolutionized our understanding and recording of Earth's topological and geological features. The program became Landsat in 1975.
robot vision	In robotics, a scanner samples visual input and processes the information in a way that provides data which is useful for tracking, navigating, or object sensing. Robot scanner interfaces vary widely, but often are small video cameras, light detection devices, or pattern sampling devices mounted on the robot itself. Robot scanners may be simple, to detect and record light or dark areas, or complex, to capture sophisticated patterns, which are further processed as faces or recognized as objects.
scanning software	A software program that searches data for new entries of a particular kind, such as newly uploaded files, email, user logins, etc., or which scans processes, such as network load, states, CPU usage, etc.

sonal computer kit, developed by the Scelbi Computer Consulting Company in Milford, Connecticut. It featured one kilobyte of programmable memory and was priced at under $600. It was initially advertised as the Scelbi-8H in QST magazine in March 1974. The Scelbi was not a big seller and was soon overshadowed by the Altair, which was featured on the cover of Popular Electronics less than a year later. See Altair, Intel MCS-4, Kendak-1, Mark-8, Micral, SPHERE.

Sceptron spectral comparative pattern recognizer. An intelligent pattern-recognition system developed in the early 1960s by Robert D. Hawkins of Sperry Gyroscope. The Sceptron employed quartz or glass fibers, a photocell, electrical current, and mechanical motion to recognize a spoken word and optionally print it on a little illuminated display. When exposed to audio stimulation, the Sceptron 'learns' the sound and can recognize it in the future. Other types of signals, translated into the audio frequency, can also be recognized. Sceptron was able to pick out a learned word from a stream of human speech.

The fiber optic array was the most unique aspect of Sceptron's design, loosely packing about 700 fibers into 1/4 cubic inch. When stimulated by mechanical excitation, the fibers, all of the same diameter but different lengths, would vibrate, each at its own natural frequency. A piezoelectric or electromechanical driver was provided to convert electrical signals into mechanical motion. A photocell then detected the motion of the fibers, registering the movement or lack of movement, as a dark or light spot. Sceptrons were configured in pairs, one with a reference static mask, one as a memory mask balanced together in a bridge circuit. See neural networks, pattern matching.

Scheduled Transfer ST. A media-independent upper-layer protocol, which was originally developed as part of the HIPPI-6400 network transmission standard.

schematic A diagrammatic representation of an electrical circuit, floorplan, network, or other interconnected system. Electronics drawings have conventions and symbols for various types of components and connections.

Scheme A dialect of the LISP programming language, Scheme was developed by Guy Lewis Steele Jr. and Gerald Jay Sussman. It is dynamically typed, statically scoped, and tail-recursive, incorporating first class procedures. See LISP.

Schrott effect In a cathode ray tube (CRT), a random variation that occurs in the emission of electrons.

SCI See Scalable Coherent Interface.

SCO An AT&T specialized network maintenance organization providing a single point of contact for resolving customer network faults.

SCO Unix The Santa Cruz Operation's adaptation of Unix. See Unix, UNIX.

scope 1. colloq. See oscilloscope.

score *v.* In the printing industry, to create a crease or indentation in the printing medium, usually heavy paper or card stock, so the medium can be easily and cleanly folded along the score line. Scoring is used in folded pamphlets, covers, and cards.

SCP 1. Satellite Communications Processor. 2. Service Control Point. See Signaling System 7.

SCR 1. silicon-controlled rectifier. 2. sustainable cell rate. See cell rate for chart. 3. System Clock Reference. A synchronization time reference used, for example, in MPEG decoding.

scrambler A device which rearranges or distorts a data communication or broadcast transmission to provide a measure of security. A descrambler is required at the end of the transmission to convert the information back into comprehensible form. Scrambling is a type of encryption, although the term *encryption* tends to be associated more with sophisticated data encryption schemes, and scrambling is associated more with the simple rearrangement of a few parameters, such as inverting audio frequencies. Black market descramblers are common, especially for television broadcast signals, since scrambling schemes are easier to decode than sophisticated key encryption algorithms. Scrambling and encryption are sometimes combined to maximize security. Scramblers are most commonly used to protect pay services like cable channels, but they are sometimes used to disrupt the transmissions of others, such as communications in war zones or traffic radar detectors.

screen font A font specifically designed for optimal readability on a limited resolution computer screen (usually about 75 dpi). Taking a vector or raster font and scaling it to the type size desired on a computer screen does not always result in legible and aesthetic characters, especially at small sizes. For this reason, some operating systems will instead display a font especially designed to overcome some of the limitations. Thus, a system may include two types of fonts, screen fonts, and printer fonts. Ideally, printer fonts are vector fonts designed to print at the highest resolution of the printing device. If you want a printout that matches the screen fonts (e.g., a screen capture for a software manual), it is best to use a capture utility designed for this purpose.

Scribner, Charles Ezra (1857 or 1858-1926) An American inventor and engineer, Scribner was chief engineer at Western Electric, after joining the firm at the age of 18. He had already patented a telegraph receiver at that age, and eventually held over 440 patents, more than any other single man in electrical history. On of his most significant inventions was the multiple telephone switchboard. Scribner founded the Western Electric engineering department, which later evolved into Bell Telephone Laboratories.

scroll bar A graphical user interface (GUI) device which serves two purposes: to indicate that there is more information beyond what can be seen in the current window or box; and to allow the user to scroll up and down (or across) the contents of the box by clicking on and dragging the scroll bar or by clicking on arrow indicators at either end of the scroll bar. A well-designed scroll bar may have a third function, to give a proportional idea of how much information is being viewed or which is hidden. Displaying a scaling drag area, to match the size proportional to the total of the information contained in the listing, helps the viewer to perceive whether there is a little or a lot hidden from view.

Some scroll bars are designed so the scrolling speed accelerates the longer the listing and the longer the user holds down the scroll bar. This type of selectively accelerated scroll can be very handy and is perceived by most users as natural and comfortable.

SCSI See Small Computer System Interface.

SCT Secretaria de Comunicaciones y Transportes. This state agency was given jurisdiction over communications for Mexico in the 1938 Law of General Means of Communications. It has played an important role in regulation and public services in Mexico's telecommunications history. In 1992, telecommunications responsibilities were transferred to the Telecomunicaciones de Mexico.

SCTE See Society of Cable Telecommunications Engineers, Inc.

SDL See Specification and Design Language.

SDLC See Synchronous Data Link Control.

SDLLC SDLC Logical Link Control. A Cisco Systems IOS feature which can provide translation between Synchronous Data Link Control (SDLC) and IEEE 802.2 type 2.

SDN See Software Defined Network.

SDNS See Secure Data Network System.

SEAC Standards Eastern Automatic Computer. A historic large-scale computer, undertaken by the U.S. National Bureau of Standards for its internal use in the 1940s. It was developed for installation on the east coast. A similar system was built at the same time for the west coast.

SEAL Simple and Efficient Adaptation Layer. See asynchronous transfer mode in the appendix for information about ATM adaptation layers.

seal *n.* 1. A part designed to tightly close, or close off a part, adjoining parts, or a container. Seals are sometimes used to make parts air- or water-tight. 2. A security/tamper device designed to indicate whether the parts adjoining or underlying the seal have been opened or altered. It is common for technical components such as hard drives, internal parts to computers, and warranteed parts to be sealed by a plastic or metallic adhesive seal, or a dot of colored paint or resin. Breaking the seal may void a warranty.

search and rescue radar transponder SART. As part of the new automated marine warning and safety measures implemented through the Global Maritime Distress and Safety System (GMDSS), ships are equipped with radiophone and/or radioteletype equipment in order to receive broadcasts of safety information provided over high frequency bands. One or more search and rescue radar transponders is required to be mounted in the vessel, which aids in locating the marine craft if it is distressed. The SART system creates a series of blips on the searching ship's radar scope within a range that is under 10 miles. See Global Maritime Distress and Safety System, NAVTEX.

search engine A software application designed to search and retrieve information from a database according to user-specified parameters or keywords. See WAIS.

search engine, Web A software application combined with a Web site to provide search and retrieval of a great variety of information, such as newsgroup postings, personal or business names and addresses, individual Web pages, or Web site topics. Many include advanced search features with logical operations for more specific or complex searches. Considering the millions of sites on the Web, Web search engines are indispensable. See Appendix C for a list of Web search engine sites.

SECAM sequential couleur avec memoire, sequential color with memory. A composite color television standard developed in France, and used also in the French colonies and western regions of the former USSR. It supports up to 625 scanlines at 50 cycles per seconds at a frame rate of 25 per second at 4.42 MHz, with an inverted signal which makes it incompatible with PAL, the other common format in Europe, and NTSC, the format used in North America.

second 1. A brief unit of time defined as 1/60th of a minute in reference to a solar day. 2. A unit of time, designated in 1967 by the General Conference of Weights and Measures, as the duration of 9,192,631,770 periods of the radiation corresponding to the transition between the two hyperfine levels of the ground state of the cesium-133 atom. See atomic clock.

Section Terminating Equipment STE. In SONET networking, the STE may be a terminating network element or a regenerator between any two adjacent network elements (such as line repeaters or lightwave terminals). It can originate, access, and/or modify the section overhead, or terminate it if needed. See SONET, Synchronous Transport Signal.

Secure Data Network System SDNS. A system which incorporates the SDNS Message Security Protocol (MSP) which was developed as a cooperative project between government and industry participants, sponsored by the U.S. National Security Agency (NSA). MSP was specified to be integrated into the X.400 Message Handling System (MHS) environment.

Secure HTTP S-HTTP. A security-enhanced version of Hypertext Transfer Protocol (HTTP) introduced in the mid-1990s. This version is designed to support multiple public-key algorithms, although a 1997 revision designates the Diffie-Hellman algorithm as the default.

Secure Multipurpose Internet Mail Extension See S/MIME.

Security Multiparts specification A specification for secure messages which separates the content data from the signature and formats them as multiple parts of a MIME communication. See S/MIME, RFC 1847.

Seebeck, Thomas Johann A German researcher who described thermoelectrical effects in 1823, after observing a connection between electricity and heat. The Seebeck effect is named after him.

seek time A quantified description of the time it takes to locate specified information. In software, for example, this could be expressed as the average time in milliseconds or clock cycles it takes for a specific tool to locate queried information from a database of a given size. In hard storage devices, it could be expressed as the average time it takes for the read head to position itself on the track where the information lies. Industry definitions of seek times exist for specific types and sizes of devices.

segmentation and reassembly A common process in packet-based networking of dividing up the packets so they can be individually processed or routed, and reassembling them at the receiving end to recreate the original message or transmission.

seize To take control of a circuit or system so it cannot be used by others. Computer files are sometimes seized and locked so that the data cannot be inadvertently modified simultaneously by more than one user. This helps protect data integrity. Transmission circuits may be seized to prevent interference on the line or to preserve privacy.

Selective Sequence Control Computer SSEC. The successor to the Automatic Sequence Control Calculator, better known as the Harvard Mark I, instigated by International Business Machines (IBM) in 1948. By this time a number of different designers and manufacturers were getting involved in the development and marketing of large-scale computing machines and IBM was motivated by the competition. Machines built around this time represent a transitional evolution from calculator/tabulators to computing machines in the sense that we understand them.

selenium A substance which was first isolated in the early 1800s, selenium's light-sensitive properties were noted by British scientists in 1873, which subsequently led to many of the important early experiments in television transmissions using selenium. As the technology evolved, selenium came to be used in television cameras.

semaphore A visual signaling system employing movable apparatus like arms or flags. Individual symbols, words, or instructions are made to correspond to distinct positions of the arms or flags. While electronic communications have superseded most semaphore systems, they are still sometimes preferred in situations where electronic messages might be overheard. See Chappé, Claude.

France had an extensive system of semaphore signaling before the telegraph was invented. Weather and constant monitoring were two disadvantages of the semaphore system.

semaphore, programming 1. An access or exclusion indicator, such as a variable flag. Semaphores are useful for controlling file locks to preserve data integrity. In other words, they can be used to prevent multiple users from accessing a file simultaneously and changing data in a way that could disrupt the information or corrupt the data. 2. A low level integer variable having only nonzero values; a primitive which can be used for synchronization in concurrent processing implementations.

semiconductor A material widely used in electronics due to its relative balance of electrical conducting and insulating properties (hence the name *semi*conductor). Semiconductor materials are typically crystalline in structure, and their properties of enabling or impeding the flow of current are used in designing solid state electronic circuitry.

Materials which are commonly used to create semiconductor components include silicon, germanium, and gallium arsenide. Doping, the addition of other elements, is used in the creation of semiconductors to further control and enhance their properties. Current flow in semiconductors is commonly controlled by electricity, but may also be controlled by the influence of light or magnetic fields. Semiconductors are important materials used in the manufacture of integrated circuits. See integrated circuit.

sensor glove, data glove A human interface device which fits over the hand and uses electronics to translate hand movements into signals which can be interpreted by a software program. Data gloves are used for many purposes: video games, virtual reality environments, special effects in movies, experimental computer interfaces, and scientific studies in human movement and perception. They are often combined with special types of monitors mounted in helmets, on glasses, or on tiny headsets, in order to facilitate natural movement and the illusion of a 3D virtual environment.

sequential A nonoverlapping succession or series, in chronological or data order, with no significant intervening time or data. See concurrent, consecutive, parallel, serial.

serial communication A means of transferring data one element at a time, often through a single wire or trace in a circuit. While it may not seem very fast or efficient, serial communication is easily implemented and very commonly used in computing systems. The RS-232 standard is the most common specification for the physical/pin connections for serial communications. See modem, parallel, RS-232.

serial interface card A printed circuit card which fits into a slot in a computer or other computerized device, or piggybacks on a motherboard to provide standardized electrical connections for the synchronous serial transmission of digital data. The connection on the card is typically a 25-pin D connector. On consumer desktop computer systems, most serial interface cards support data rates up to about 28,800 bps or 38,400 bps. A serial interface card is a common way to connect remote computer terminals and data modems to a computer. See RS-232.

Serial Line Interface Protocol, Serial Line IP SLIP. Originating with an early 1980s 3COM UNET TCP/IP implementation, SLIP became a de facto standard encapsulation protocol for serial lines, used for point-to-point communications with TCP/IP. SLIP has now been superseded by Point-to-Point Protocol (PPP). See Point-to-Point Protocol, RFC 1055.

SERN See Software Engineering Research Network.

server A system which provides services to other computers connected to it through a network. A server may store and administrate software applications, security measures, access to peripherals or external systems, etc. The server does not necessarily have to be an enhanced system, as servers can be specialized as print servers, mail servers, etc. (and there may be several servers on a system), but servers performing the bulk of centralized or generalized tasks often have more memory, processing speed, and storage than other systems on the network.

The software is probably the most important aspect of a good server. Good network software is robust, configurable, and usually fully multitasking. There are many well-tuned network workstation options that are reliable and do not crash, except in the most unusual of circumstances. Shop around when selecting your network server software; if you pay a few hundred or a few thousand extra dollars initially, you may recoup it in only six months in terms of downtime, software reinstallation, and administrative costs that accrue on unreliable systems.

server agent In server/client systems, software that handles the major processing or protocols and serves a request from a client as a Web server, mail server, or FTP server.

Service Advertising Protocol See Service Location Protocol.

Service Agent A network utility which, when queried, provides information about a network service (printer, modem, etc.) such as its URL.

service bureau A center providing services that usually require specialized equipment and/or expertise. Copy centers, private postal centers, data entry services, and facsimile centers are examples of common service bureaus. Service bureaus commonly arise when there is a service occasionally required by many businesses or individuals, but those businesses or individuals don't have the equipment or the expertise to do the job themselves. See outsourcing.

Service Control Point SCP. A point which provides access to the database in an Intelligent Network (IN), which is connected to a Service Management System (SMS), and which accesses Internet Protocol (IP) as needed. SCPs enable advanced services by processing the format or content of transmitted information. See Intelligent Network.

Service Location Protocol SLP. An intelligent resource discovery and registration protocol developed in the mid-1990s. Described as a 'quieter' alternative to Service Advertising Protocol (SAP), SLP includes extended attributes information to reduce network traffic queries. Thus, a printer may be described in terms of its capabilities (such as duplex printing, PostScript-capable, tabloid paper) and found transparently, without the user querying for its IP address. See Service Advertising Protocol, Service Agent, SLIP, RFC 2165.

Service Management System SMS. An interactive computer system dedicated to coordinating national 800 numbers, from International Business Machines (IBM).

Service Profile Identifier When hooking up ISDN BRI services, the carrier provides the user with a SPID for each number being installed, typically two. The SPID points to a memory location in the carrier's central office where ISDN parameters, including which services are enabled for a particular subscriber, are stored. As not all phone carriers have automatic SPID detection, some newer modems can determine what type of ISDN service is connected and configure the SPIDs accordingly. When connected to carriers with automatic SPID detection, they can configure themselves whether or not a computer is attached to the modem. Modems with these capabilities help compensate for some of the problems traditionally associated with the installation of ISDN services. It is recommended that the subscriber keep a record of SPID numbers filed away somewhere, as it's easier to look up a lost SPID than to have to get it again from the phone carrier.

service quality Standards of service established by businesses that include such things as service without outages, available lines without lag or busy signals, technical support availability, good data integrity, etc. This is not the same as quality of service (QoS), which has a more specific meaning.

Service Specific Convergence Sublayer See SSCS.

Service Switching Point SSP. A point which provides local access and an ISDN interface for a Signaling Transfer Point (STP), which, in turn, provides packet switching for message-based signaling protocols in an Intelligent Network (IN). See Intelligent Network.

services-on-demand SoD. Services provided to an audience on a request basis, rather than on a scheduled broadcast basis. The concept is not new, in fact, it has been available for media services for over 100 years, but new digital technologies are providing automated services, thus making available cost-effective SoD delivery options which were not previ-

S

ously possible. See audio-on-demand, video-on-demand.

set-top box *colloq.* A media device which sits on top of a TV set or within a home entertainment component cabinet to hook into the system in some way. Set-top boxes provide a variety of capabilities, including conversion of cable TV signals, provision of WebTV services, etc. Some set-top boxes are proprietary units offered through lease or purchase by a service provider or vendor.

SETI Search for ExtraTerrestrial Intelligence. An interesting, federally funded scientific project in which arrays of radiotelescopes are used to search for signs of intelligence in other parts of the universe. The movie "Contact" provides an idea of a SETI-like project.

The rationale of SETI is that you can send signals farther and faster using radiowaves traveling through space than you can send spacecraft (that's not to say that SETI supporters don't also support spacecraft missions), and it's worthwhile to send out signals in the hope that other life forms may intercept them, or humans may intercept the transmissions of other life forms. Unfortunately, to date, no signs of sentient communications have been detected, but SETI concepts have resulted in the discovery of some interesting radiowave signals from distant celestial objects.

The Columbus Optical (COSETI) Observatory, a pioneering observatory in Ohio near Columbus, searches for extraterrestrial intelligence in the optical spectrum. http://www.seti.org/ http://www.coseti.org/

SF 1. single frequency. 2. See SuperFrame.

SFTP See Simple File Transfer Protocol.

SGML Standard Generalized Markup Language. A markup standard adapted by the International Organization or Standardization (ISO) in 1986 which is not itself a language, but is designed for specifying the content and structure of a document or document language, with the assumption that the actual output or display of the document may vary according to the output device. SGML allows the development of cross-platform applications and documents, and a document can be processed by an SGML compiler by referencing a document tag definition (DTD). HyperText Markup Language, widely used on the World Wide Web, is a descendant of SGML that incorporates some of its capabilities.

Yuri Rubinsky (1952-1996) was one of the pioneers who did much to enthusiastically promote the use of SGML through educational programs.

SGMP See Simple Gateway Monitoring Protocol.

SGRAM synchronous graphics random access memory. A type of memory optimized for use in memory-hungry graphics applications, particularly 3D rendering and ray tracing.

shadow mask A type of cathode ray tube (CRT) color display technology which incorporates a thin, performated metal plate mounted close to the front of the inside of the tube to create a mask through which red, green, or blue (RGB) phosphors can be selectively excited. See cathode ray tube.

Shannon, Claude E. (1916-) A theorist who contributed significantly to the study and understanding of information theory. The history of communications emphasizes the inventors, the programmers, and the hobbyists who have developed the mechanisms and operations of information systems. Fewer people have taken a broad look at what information is, how it relates to the technology (e.g., channel capacity), and what the process of conveying information entails from a more abstract, theoretical, statistical, and broadly practical viewpoint. Shannon, while working at Bell Laboratories, is credited with bringing together fundamental theories of information in 1948.

Shared Facility Centers A telework organization similar to a branch office, but co-owned or partially community- or freelance professionals-funded, situated in a residential or rural village area, and made commercially viable by the implementation of new communications technologies. See ADVANCE Project, telework.

ShareView 3000 A Macintosh-based videoconferencing system from Creative Labs which supports audio, video, whiteboarding, application and document sharing, and file transfers over analog phone lines. An IBM-licensed PC-version called ShareVision PC3000 is also available. See Cameo Personal Video System, Connect 918, MacMICA, IRIS, VISIT Video.

Shaw, Wiliam Chester An Ontario-born Canadian who invented the IMAX motion picture system in the 1960s. This large-format film technology provides startling realism and detail. See IMAX.

shear In image processing software, to shear is to alter the shape of an object to displace it sequentially in one direction or the other, through one axis. For example, if you take a rectangle and shear it along the X axis (horizontally), either the top or the bottom stays in one place and the opposite top or bottom follows the pointing tool so it becomes a parallelogram that somewhat resembles a diamond. Shearing a straight vertical bar through the X axis would result in a longer straight bar at an angle (actually, it may have jaggies, depending upon the visual resolution, but mathematically it remains straight). Shearing the same bar through the Y axis would simply lengthen the bar. A good shear tool will allow you to select the axis of shear, although many permit shearing only through X or Y.

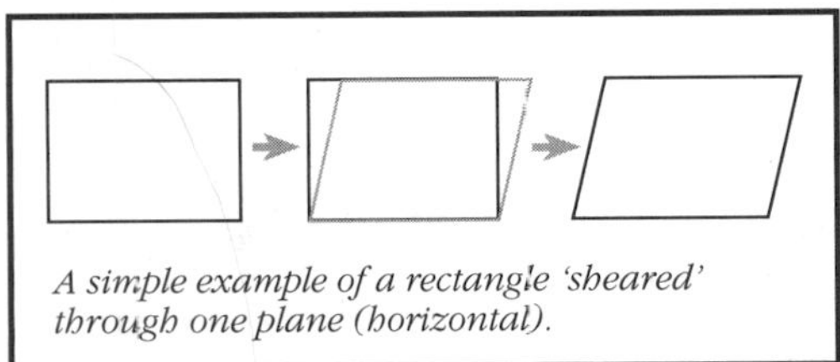

A simple example of a rectangle 'sheared' through one plane (horizontal).

sheath *n.* A close-fitting protective covering, usually tubular, often made of plastic. Sheaths can be used to bundle wires, to insulate, to protect from moisture or wear, or to provide identifying colors or symbols. They are commonly used on conducting wires and fiber cables. See conduit.

shelf life The useful life of products in terms of physical functioning, public demand, or profitability. Most electronic products currently have a very short shelf life in terms of the compatibility and hence viability of the technology. Unfortunately, the short shelf life is not related to physical deterioration; the products themselves are perfectly usable and obsolete only because of lack of market demand. The shelf life of computers has changed from years to months and individual components, such as modems and high capacity storage devices, may be only weeks. This results in massive quantities of unsellable products on warehouse shelves and companies scrambling to unload inventory of perfectly good products.

shell, command shell A computer user interface input and display environment which translates user commands into operating system instructions.

Sherman Antitrust Act An important 1890 U.S. act which was established to prevent the establishments of monopolies which could hinder U.S. trade and competition.

SHF super high frequency. About 3 to 30 Ghz, used for satellite transmissions.

ship to shore telephone See marine telephone.

shock, electric A sudden, often hazardous, electrical stimulation to a living body which may greatly affect nerves and cause convulsive contractions through muscles, possibly endangering the heart muscle. It may also cause severe burning, confusion, and unconsciousness.

Light electric shocks are uncomfortable, but not always dangerous and are sometimes used as perimeter boundaries for livestock or secure areas. Light electric shocks are also used in animal experiments for studying the nervous system and are occasionally used in riot control and law enforcement.

It is unwise to open up or attempt to repair cathode ray tubes (CRTs), which may store a considerable charge, without careful preparation and knowledge of safety procedures.

Electric shocks must be taken seriously, and, if severe, may require the contacting of emergency services or the application of cardiopulmonary resuscitation (CPR). Never touch someone who is experiencing shock from an electrical source until the electricity is turned off or the source of the contact knocked away with a nonconducting material. Consult emergency first aid sources for information.

Shockley, William Bradford (1910-) An English-born American physicist who worked in the Bell Telephone Laboratories from 1936. He discovered the rectifying properties of impure germanium crystals at a time when vacuum tube rectifiers had replaced the old galena and carborundum crystal detectors. This led Shockley to explore the various impurities in germanium and he found there was electron drift toward the positive or negative pole under controlled conditions. By combining these 'solid-state' rectifiers, the transistor was born and vacuum tubes superseded. One of the most significant consequences of transistors at the time was miniaturization of communications devices and room-sized computing machines.

Shoemaker detector A type of electrolytic detector that incorporates a battery into itself and, consequently, requires no outside power source. It consists of a glass tube with a platinum-sealed point, with a zinc strip, rather than the platinum point, coming in contact with a mild sulphuric acid solution. Shoemaker detectors were used commercially in wireless telephone receivers in the early 1900s. See electrolytic detector.

short See short circuit.

short circuit An unintended or harmful cross connection, of low resistance, of electrical circuits. Short circuits can occur from: an excess of solder, an incorrectly connected wire, conductive debris (such as a screw falling into a circuit box), worn out insulation in bundled wire, water, or physical bumping of electric conductors, etc. The result is often a sudden flow of current in the wrong direction or of the wrong magnitude, which can potentially damage components. Some systems are configured to shut down or blow a fuse or breaker, in the event of an excess of current or other abnormal electrical activity. See burst, spike.

short haul A short travel or installation distance. The actual length depends upon the situation or medium employed. A short haul for a SCSI cable is about three feet or less, above six feet serious signal degradation occurs, and special hardware is needed for distances over twelve feet (see Fibre Channel). A short haul for other media may be several yards or thousands of miles.

short haul modem A software/hardware combination used for short distance communications up to a couple of dozen miles usually over a copper single-channel line. See baseband modem.

Short Message Service SMS. A global, wireless, low bandwidth, two-way service first distributed in Europe in the early 1990s and later in North America. SMS provides the capability of transmitting alphanumeric messages between mobile systems and external systems that support paging, email, and voice mail. The handsets used in these services can send or receive at any time, regardless of whether a data or voice call is in progress. SMS is appropriate for applications like stock quotes, paging, short fax and email messages, online quick banking, etc.

Short Message Service Center SMSC. A relay and administrative center for Short Message Service (SMS) which provides store and forward services. This is somewhat like an enhanced alphanumeric paging system with two-way service and guaranteed delivery. See Short Message Service.

shortwave, shortwave Long-range radio transmission frequencies in approximately the 1.6 MHz to 30 MHz range, above the commercial broadcast bands. Shortwave signals are easier to apprehend at night, due to lowered atmospheric noise and the fact that many shortwave broadcasters prefer to send in the evening hours. Coordinated Universal Time (UTC) is often used as the reference time for broadcasts. The Internet has sites that list broadcast times and frequencies for various shortwave stations around the world. See microwave, ionospheric wave, radio.

shrinkwrapped products Products that come ready to use 'off the shelf' that require little or no configuration or customization. Shrinkwrap software is generally that which is intended for a mass market. Contrast to custom software written to a customer's specifications, written for a small market segment, or provided by consultants.

SHT Short Hold Time.

Shugart A historic disk development company. Shugart floppy disk drives were used on some of the earliest microcomputers.

shunt *n.* A switch, pipe, detour sign, or other diverting mechanism.

shunt, electrical In electrical circuits, a means to divert some or all of the current. A shunt is sometimes used to divert part of a current in order to prevent damage to sensitive measuring instruments. Temporary shunts are sometimes established with jumper wires or alligator clip connections.

shunt circuit, bypass circuit, detour circuit A circuit configuration through which a specific portion of the current is redirected or subdivided. Often used for diagnostic purposes, temporary arrangements, or circuits in which variable conditions are accommodated or where the original current can be more effectively used by dividing it. Shunts are sometimes incorporated into the internal workings of diagnostic instruments.

shunting error A condition in which current divided through two circuits, as through a component and a measuring instrument, will vary depending upon the frequency. See reversing error.

SIA Securities Industries Association.

Sibley, Hiram (1807-1888) Sibley founded the New York and Mississippi Valley Printing Telegraph Company which took on the name of Western Union, suggested by Ezra Cornell in the mid-1850s when it began westward expansion. Sibley remained president during the expansion and Western Union installed the first transcontinental cable in 1861. After the failure of Western Union's first atlantic cable, Sibley traveled to Russia to investigate the installation of a Siberian-Alaskan communications line, and the Russians offered to sell Alaska to Western Union. Sibley turned down the offer, but alerted the U.S. government of the opportunity. Along with his colleague, Ezra Cornell, Sibley helped to found Cornell University. See Western Union.

side circuit In telephone installations and other circuits where additional endstations are desired, but where resources are limited, a side circuit is a means to build an additional circuit using the resources of two adjacent circuits. See phantom circuit.

side lobe An undesirable condition in which a directional antenna loses power due to dispersion of the signals or wave patterns out of the sides of the antenna.

sideband The frequencies on either side of the main frequency or carrier band in a communications signal. These frequencies are within the *modulation envelope* of a transmission wave, but were originally not used because of problems with noise. Later, as technology improved and the demand for airspace continued to grow, sideband transmissions became more interesting, and it was found that one sideband could be transmitted, sometimes even without the carrier wave, and the original wave mathematically 'rebuilt' at the receiving end. The advantages included lower power requirements for the transmission and a narrower wave overall, leaving more room for other transmissions.

sidetone In a telephone receiver, transmitting currents are directed into the receiver to make it possible for the speaker to hear his or her own voice, as a form of feedback mechanism. This has to be carefully controlled so that it doesn't become excessive, and various anti-sidetone circuits are applied to minimize feedback and reduce transmission of acoustical noise.

Siemans Telecom Networks A provider of telecommunications services and network equipment emphasizing robust, secure technologies to regional Bell operating companies and independent telephone and holding companies.

Siemens, Werner (1816-1892) An American inventor who, along with his brother, William, developed the dynamo, a device to convert mechanical energy into electrical energy without the use of permanent magnets. In the 1870s, he demonstrated that the velocity of electrical conductivity through a wire could approximately equal that of light.

sign in, sign on See login.

sign language Any system of gestures or movements that conveys information. In its broadest sense, sign language includes handheld semaphore systems and, in its most specific sense, includes a number of standardized systems for using arms, hands, and facial expressions to convey meaning. Sign language in the form of hand and arm or flag semaphores is still standard in airports where noise and distance prevent spoken communications with ground personnel. See American Sign Language, semaphore.

Signal Transfer Point STP. A point which provides access to a database and packet switching for message-based signaling protocols for the Service Control Point (SCP) in an Intelligent Network (IN). The Intelligent Network is based around Signaling System 7. See Intelligent Network, Service Switching Point.

Signaling System 6 SS6. An early out-of-band signaling system developed in the 1960s, which is gradually being superseded by Signaling System 7. SS6 was the first system to incorporate packet switching into public switched telephone networks (PSTNs). See Signaling System 7.

Signaling System 7, Signaling System No. 7 SS7. SS7 is a common channel network signaling system, descended from Signaling System 6, which is displacing telephone in-band multifrequency signaling systems. SS7, introduced in the 1980s, is more flexible and powerful than earlier systems, making it possible to implement broadband digital services far in advance of basic voice circuits.

Unlike earlier phone signaling systems, which operated through many semi-independent switching centers, SS7 brings together the communications channels into a more integrated whole.

SS7 is being gradually integrated into ATM/T1 and PCS/UPT networks.

SS7 is an *out-of-band* signaling system, which means that the signals which control the system are transported on a different line from the data itself. This makes SS7 far more secure than traditional *in-band* analog lines.

Signaling System 7 ANSI standards There

are a number of important American National Standards (ANSI) of Committee T1 related to SS7, which are available from ANSI and which are described in the form of abstracts on the Web. The Signaling System 7 chart shows a sampling of those available.

signature In printing, a grouping of pages in order to organize multiple pages for binding. Common sizes for signature groups are eight or 16 pages. See fascicle, imposition.

silence compression A technique used in voice over data networks applications which involves removing the pauses and spaces that typically occur in many conversations. This reduces transmission time. Two common techniques typically used together include voice activity detection (VAD), which distinguishes speech from the surrounding background noise, and comfort noise generation (CNG), which creates a low type of static that gives humans a certain comfort level and trust that

Signaling System 7 ANSI Standards

ANSI Standard	Title	Subtitle
T1.110-1992	Signaling System No. 7	General Information
T1.111-1996	Signaling System No. 7	Message Transfer Part
T1.111.1-1996		Chap. 1 - Functional Description
T1.111.2-1996		Chap. 2 - Signaling Data Link
T1.111.3-1996		Chap. 3 - Transfer Functions & Procedures
T1.111.4-1996		Chap. 4 - Transfer Functions & Procedures
T1.111.5-1996		Chap. 5 - National & International Networks
T1.111.6-1996		Chap. 6 - Signaling Performance Requirements
T1.111.7-1996		Chap. 7 - Testing & Maintenance
T1.111.8-1996		Chap. 8 - Signaling Codes Number Scheme
T1.111a-1994		Numbering of Signaling Point Codes
T1.112-1996	Signaling System No. 7	Signaling Connection Control Part Func. Descript.
T1.113-1992	Signaling System No. 7	ISDN User Part
T1.114-1996	Signaling System No. 7	Transaction Capability Application Part (TCAP)
T1.116-1996	Signaling System No. 7	Operations, Maintenance, and Admin. Part
T1.116-1996	Signaling System No. 7	Intermediate Signaling Network Identification
T1.116-1996	Signaling System No. 7	Operations, Maintenance, and Admin. Part
T1.118-1992	Signaling System No. 7	Intermediate Signaling Network Identification
T1.118-1992	Signaling System No. 7	Intermediate Signaling Network Identification
T1.118-1992	Signaling System No. 7	Intermediate Signaling Network Identification
T1.226-1992	OAM&P	Management of Functions for Signaling System No. 7 Network Interconnections
T1.134-1993	Signaling System No. 7	MTP Levels 2 and 3 Compatibility Testing
T1.135-1992	Signaling System No. 7	SCCP Class O Compatibility Testing
T1.136-1993	Signaling System No. 7	ISDN User Part Compatibility Testing
T1.118-1992	Signaling System No. 7	Intermediate Signaling Network Identification

the line is still active and the call hasn't been cut off.

silent discard In packet networking, the discard of a packet without further processing. The system may log the event and may even store the contents of the discarded packet for later evaluation.

silicon An abundant nonmetallic, tetravalent element, widely used in semiconductor technology. Silica, silicon dioxide, occurs in many common forms, including sand, quartz, and opals.

Silicon Bay Bellingham, a port town an hour and a half north of the Silicon Forest (Redmond/Seattle), has a number of independent software development corporations, many supporting the computer graphics and CAD markets.

Silicon Bayou The region of high technology companies in and around New Orleans.

silicon detector An early type of radio wave detector which is similar in some aspects to electrolytic detectors. Silicon is used in place of the electrolyte, making contact with a platinum wire, and the thumbscrew contact with the silicon can be finely adjusted by filing the end of the thumbscrew to a fine point, using a spring with the thumbscrew to assure even pressure. The interaction of the thumbscrew and the silicon sets up a thermoelectric reaction which can be translated into audible waves in the receiver. See detector, electrolytic detector.

Silicon Forest The Microsoft campus and surrounding high technology region in the Redmond/Seattle area has been dubbed the Silicon Forest due to the surrounding stands of trees which include Douglas Fir, Cedar, Sitka Spruce, and Western Hemlock.

Silicon Graphics Incorporated SGI. A computer company known for innovative software and hardware workstation-level computers, especially those with good graphics and sound. SGI was founded in 1981 by James Clark who later became affiliated with Netscape Communications Corporation.

Silicon Valley A region of California with a high density of high technology companies, many of which pioneered computer technology. The economy, educational institutions, research labs, and climate were all factors that contributed to the growth of technology companies in Silicon Valley.

SIMM single inline memory module.

Simple File Transfer Protocol SFTP. A simple file transfer protocol that fills the need for a specification that is easier to implement than File Transfer Protocol (FTP). It provides file transfer capabilities combined with user access control, listing of directories, traversing directories, file renaming, and file deleting. In other words, it incorporates the most common and necessary functions of FTP. See RFC 913.

Simple Gateway Monitoring Protocol SGMP. Developed in the mid-1980s and demonstrated in 1987, SGMP later evolved into Simple Network Management Protocol (SNMP).

Simple Internet Transition SIT. A set of Internet protocol mechanisms for hosts and routers designed to smooth the transition between IPv4 and IPv6, its successor. SIT eases the transition by supporting incremental upgrades of hosts through upgrading of the DNS server with support of existing addresses. SIT employs a number of mechanisms to achieve interoperability and compatibility including:

- embedding of IPv4 addresses within IPv6 addresses
- encapsulation of IPv6 packets in IPv4 headers for transmission through IPv4 legacy routers
- dual IPv4/IPv6 protocol stacks model for hosts and routers
- header translation for IPv6 only routing topologies

simple line code SLC. A means of transmission through four-level baseband signaling that filters the baseband and restores it at the receiving end.

Simple Mail Transfer Protocol SMTP. A transmission subsystem-independent electronic mail protocol which establishes and negotiates communications between sender and receiver (or multiple receivers) across transport service environments. Transmissions may be direct, depending upon the transport service, or may pass through relay servers.

When a user mail request is generated, the sender-SMTP establishes a two-way transmission channel to the intermediate or ultimate destination-SMTP. SMTP commands are then sent between the two ends. Once a transmission channel is established, a lock-step negotiation of the transmission and identification of the recipient or recipients is carried out, and the mail data sent, with a terminating sequence to indicate the end. When successfully received, the recipient sends an OK reply. See electronic mail, email, RFC 821.

Simple Multicast Routing Protocol SMRP. A routing protocol from Apple Computing, Inc. which is used for AppleTalk network data from applications such as their QuickTime Conference, which in turn is used for videoconferencing, electronic whiteboarding, etc.

Simple Network Management Protocol SNMP. SNMP evolved from, but is not backwardly-compatible with, the Simple Gateway Monitoring Procol (SGMP). Essentially, SNMP communicates management information between network management stations and the agents in the network elements (NEs).

SNMP was designed for TCP/IP-based network environments and manages nodes on the Internet. SNMP was originally designed as an interim solution with the intention that it follow generally along Open Systems Interconnection (OSI) guidelines. Over time, these were found to be more different than originally envisioned.

Along with MIB and SMI, SNMP has been designated by the IAM as a full Standard Protocol with "recommended" status. The SNMP Extensions working group was formed to evaluate and further develop the SNMP definition, with the mandate of retaining its simplicity. See RFC 1157.

Simple Raster Graphics Package SRGP. A low level graphics package which incorporates features from a variety of graphics systems (such as GKS and PHIGS standards, The X Window System, Apple QuickDraw). SGRP typically functions as an intermediate layer between the applications program and the display device.

Simple Security and Authentication Layer A Network Working Group proposed standard for providing a quick method of negotiating an authentication mechanism, even if the client has minimal knowledge of the system. See RFC 2222.

Simple Server Redundancy Protocol SSRP. A network protocol which provides resiliency for LANE services on ATM-based local area networks (LANs).

simultaneous voice/data SVD. A number of analog and digital techniques and standards which permit limited use of simultaneous voice and data through regular phone lines with computer voice/data modems. These might be considered medium level applications, since they do not support full realtime videoconferencing, but they allow whiteboarding and switching between voice and data as needed (alternate voice/data (AVD)). SVD is accomplished through multiplexing. In analog SVD, voice is multiplexed with data in digital SVD, data and digitally compressed voice are multiplexed into a digital data stream.

The ITU-T has established standards, draft standards, and specifications related to SVD. These are periodically reviewed and updated to reflect improvements in modem technology. V.61 has been specified for 14,400 bps standard for analog SVD, and V.70 for 28,800/ 33,600 bps for digital SVD.

simulator A software program, or software/ hardware combination which models, reconstructs or 'mimics' an environment or situation, which may be real or imagined. Simulators are used in many areas of scientific research to enact scenarios; to test, confirm, or investigate hypotheses; to compare or contrast the effects of various changes to a system; or to monitor the evolution of a system. Simulators are also popular in the entertainment industry. Flight simulators have been developed into interactive, environmental video games with helmets, moving seats, and more, to provide a strong emotional/intellectual/tactile experience. Virtual reality simulators go a step farther, creating 3D effects which appear to inhabit the space around the user, sometimes so convincingly that the user will duck to get out of the way of a virtual image.

Sinclair ZX81 The successor to the ZX80, the ZX81 personal computer was introduced in spring 1982 and sold for under $200 (without monitor; it could be hooked up to a television set). It sported 8 kilobytes of extended BASIC, math functions to 8 decimal places, a built-in printer interface, 1 kilobyte of memory expandable to 16 kilobytes, and a 32-column *x* 24-line display. It was also available as a kit for under $100.

sine wave A fundamental waveform present in almost all vibratory motion, which can be represented as a *sine curve* with periodic oscillations in which the amplitude of displacement at each point in the wave is proportional to the sine of the phase angle of its displacement. In telecommunications, the sine wave is important in many representations, but especially in alternating current (AC) circuitry and in representing sound. See oscilloscope.

sine galvanometer An early current-detecting instrument in which the coil is rotated until the reading needle again registers zero. This type is subject to interference from the Earth's magnetic field. See galvanometer.

SINGARS Single Channel Ground and Airborne Radio Systems. A tactical radio system. See Enhanced Trivial FTP.

single line repeater A mechanism for allowing two-way communication on a single line by permitting the transmission to be alternately broken in one direction in order to initiate or resume communication in the other direction. This is accomplished by an additional holding coil on each relay which can open or close independent of whether the main circuit is open. See half-duplex.

single mode optical fiber A single mode fiber optic transmissions cable has a relatively thin core acting as a waveguide such that light is reflected and propagated at a consistent angle. A thinner core has advantages and disadvantages over multimode fiber. Signals cannot be sent at a variety of angles, but distortion is minimized and transmissions can occur over longer cable runs. Thus, where multimode fiber in data network installations are limited to about two kilometers, single mode fiber can transmit to about 15 kilometers. For other types of transmissions, longer distances are possible, sometimes up to 200 kilometers. Signals are usually transmitted through single mode cables with laser diodes, in order to get the precise alignment needed for the fine filaments, and received at the other end with a photodiode detector. This detector translates the signals back into electrical impulses. See multimode optical fiber.

single sideband Transmissions created by manipulating frequencies that are selected from one side of the *modulation envelope* of a transmission wave to recreate the original baseband transmission. Much of the credit for the development of single sideband technology, which is essential to frequency division multiplexing, belongs to John R. Carson, a mathematician with AT&T and later Bell Laboratories who mathematically demonstrated the relationship between the information in the sideband signals and the original baseband.

Sideband frequencies were not originally used because of problems with noise. Later, as technology improved and the demand for airspace grew, sideband transmissions became more interesting, and Carson demonstrated in 1915 that one sideband could be suppressed from the transmission and the other could even be transmitted without the carrier wave. Due to its predictable characteristics, the original baseband wave could then be mathematically 'rebuilt' at the receiving end. In a sense, this was a type of 'wave compression' accomplished by removing extraneous and redundant information. The significant advantages included lower power requirements for the transmission and a narrower wave overall (i.e., requiring less bandwidth), leaving more room for other transmissions. See frequency division multiplexing.

single sign-on SSO. A network security and management strategy to help reduce the number of passwords needed to access a variety of software and hardware resources on a network.

single slot Architecture for coin operated devices that uses one slot for the various denominations of acceptable coins. Thus, the coins are processed by the mechanism after it is inserted by the user, rather than the user selecting an appropriate slot. These are becoming more prevalent and are now common in coin-operated phone installations.

Single UNIX Specification Developed within the Common Applications Environment by the X/Open Company, the Single UNIX Specification is a collection of documents which includes interface definitions, interfaces, headers, commands, utilities, networking services, and X/Open Curses. This specification is distinct from the AT&T licensed source-code commercial product and is intended as a single stable UNIX specification for which portable applications can be built. It provides vendors a means to provide a 'branded' product and assumes voluntary conformation to the specification. Basic components within the Specification are shown in the Single UNIX Specification chart. See Unix, UNIX.

single use batteries Batteries which are not intended to be recharged after the power is drained, such as common alkaline batteries. There are, in fact, rechargers in the $20 to $30 range that can safely recharge these 'nonrechargeable' batteries up to about six or eight times. Not only does this save money, but is easier on the environment as there are toxic substances in batteries, especially the older ones. Batteries should be recycled rather than thrown in the trash to reduce environmental contamination.

single wire circuit A transmission path used in early telegraph lines and still used for telephone service in some rural areas. The single wire circuit relied on the conductive characteristics of the Earth to ground the circuit and complete the return path.

sink 1. A device to drain energy from a system. Heat sinks are common on devices or components which 'run hot' and need to be cooled for safety and to maintain operating temperatures. 2. A point where energy from a number of sources is directed, and then drained away. 3. A point in a communications system where information is directed.

Sioussat, Helen J. Sioussat was the Director of the Talks and Public Affairs Department of CBS radio from 1937 to 1958. Her extensive correspondence with many of the radio and television broadcast pioneers has been preserved in the Library of American Broadcasting at the University of Maryland Libraries. See Broadcast Pioneers Library.

SIPP 1. Simple Internet Protocol Plus. One of three candidate protocol proposals eventually blended into IPv6 by the Internet Engineering Task Force (IETF).

SIS Standardiseringen i Sverige. The Swedish standards institute located in Stockholm.

SIT See Simple Internet Transition.

SITA See Société Internationale de Télécommunications Aéronautiques.

site license A legal arrangement granting specific use or distribution permission of a copyright product to a specified location, firm, or other entity. A site license is a common method for specifying and controlling software use and distribution within a firm, particularly if the firm wishes to install the software on a network for access by multiple users or on several user machines within the organization. Typically, software companies will offer site licenses with the first copy and installation of the product priced at one level, and discount subsequent installations. This is common in educational institutions. For example, the first copy might cost $1,000 and permit installation on up to five machines, with subsequent installations, in groups of five, at $200 each. Network licenses typically specify how many users may simultaneously access the software, and the software itself may monitor and control access to itself. Distribution of any sort, other than as specified by the license is, in most cases, a criminal offense. See piracy.

skew *n.* 1. In computer imaging, a function which allows a selected area to be stretched through one plane at an angle, e.g., a square can be skewed to form a diamond. 2. In statistics, the distortion from an expected or true value. 3. In parallel data transmissions, the deviation in arrival of data from each individual transmissions path.

skin 1. Outer protective layer. A skin is often used to isolate conductive materials and/or to provide insulation and, sometimes, identification through the use of colored or marked skins.

skin antenna An antenna used on aircraft,

Single UNIX Specification Components

Components	Notes
XPG4 System Calls and Libraries	Internationalized, covering POSIX.1 and POSIX.2 callable interfaces, the ISO C library and Multibyte Support Extension addendum, the Single UNIX Specification extension including STREAMS, the Shared Memory calls, application internationalization interfaces, and other application interfaces.
XPG4 Commands and Utilities V2	Covering the POSIX.2 Shell and Utilities and a large number of additional commands and development utilities.
XPG4 Internationalized Terminal Interfaces	Including the new extensions to support color and multibyte characters.
XPG4 C Language	
XPG4 Sockets	See sockets.
XPG4 Transport Interfaces (XTI)	

in which a region of the metal craft is delineated and isolated on its edges by insulating materials.

skin effect In electricity, a situation in which the current tends to pass through the outer portions, rather than through the core of a conductor. This is due to the magnetic field that arises in the wire which prevents penetration to the core of the wire. It may increase the effective resistance in long wires and interfere with transmissions in the high frequencies used in broadcast transmissions.

skinning Stripping an outer protective layer. This is commonly done with wires to expose the conductive material within in order to make a connection.

skip distance The distance traveled by a reflected radio wave from the transmitter to the point at which it reaches the Earth's surface or the receiving antenna. This distance is affected by the frequency of the wave, the angle at which it passes into the ionosphere, and various atmospheric characteristics and conditions. See ionospheric wave, radio.

skip selection In computer software applications, a selection that halts the current process, or lets it finish in the background, and allows the user to continue to the next menu or activity without waiting. In automated voice or tone systems, especially menu-driven touchtone phones, a key press that allows you to continue to the next selection, menu, or local phone number without waiting for completion of the current message.

skip zone See zone of silence.

SKIPJACK The name of an encryption algorithm which is the basis of the Escrowed Encryption Standard (EES) incorporated into the Clipper chip. See Clipper chip.

skunkworks *colloq.* A facility in which clandestine or time-pressured activities take place in an environment which is closed off to increase security. Government operations, sensitive research, and high technology design often operate in environments that are without sunlight or adequate ventilation, and in which the participants are working long hours without much free time for personal hygiene.

sky maps Charts of the electromagnetic radiation in the radio frequencies emanating through space and around Earth. Much pioneer work in this area was conducted in the 1930s and 1940s by Grote Reber, an amateur radio operator, using a home-built 32-foot parabolic antenna. Cosmic frequencies can sometimes be detected when the ionosphere is temporarily affected by the burn of a spacecraft or deliberate seeding with elements such as barium. See radio astronomy.

sky wave See ionospheric wave.

Skybridge A medium Earth orbit (MEO) satellite system from Alcatel.

Skynet A U.S. domestic communications satellite service purchased in 1997 from AT&T by Loral Space & Communications.

slamming A reprehensible trade practice in which a long-distance supplier switches a person's long-distance service without his or her explicit permission. In the early 1990s, some companies did this by phoning potential subscribers and having them verify their name and address over the phone and then signing them up without actually asking for consent. Since that time more stringent customer consent is required before a change in the service can be initiated, and the customer usually must initiate the request, or the company making the change must obtain written authorization or outside verification.

SLAR side-looking airborne radar. A self-illuminating (through microwaves) electronic image-creation system derived from a radar beam transmitted perpendicular to the ground track during acquisition from an aircraft. Thus, the signal hits the terrain at a rather flattened angle and the view of the terrain is vertical, revealing fine surface features useful in interpretation of the data. The imaging is provided in strips or mosaics, as is true for many satellite imaging systems. SLAR imagery is used by geologists, Earth resource scientists, cartographers, engineers, and others. SLAR encompasses real-aperture radar, and synthetic-aperture radar (SAR). SLAR is not used for very precise topographic mapping, as the resolution is only up to about 30 meters.

slave A subsidiary structure, system, process, or device which takes direction or data from a master. Many computer peripherals are slave devices. In programming, slave processes are sometimes used to gather and report information to a master controlling process. In communications circuits, slave consoles, subsidiary switching centers, and substations are often used to supply low-density populations or workstations some distance from the main controller or switching center.

SLC See simple line code.

sleep mode, power save mode Since portable appliances typically run on batteries and batteries wear out and must be replaced or recharged, vendors have come up with ways to extend the time between replacement or recharge. One solution is to provide a manual or automatic *sleep mode* in which the appliance is put on a standby setting which uses less power until it is needed again. This is very common on laptop computers. Since screen displays, modems, and hard drives require more power, but may not be needed while the user is talking or waiting, the laptop will often power down into sleep mode until the user touches a key or moves the mouse.

slide contact A small sliding ball or tab attached to a thin rod which acts as a contact mechanism on a tuning coil. Tuning coils were used in early radio sets to select a frequency. A radio might come with several tuning coils for selecting various frequencies, as desired. See tuning coil.

SLIP See Serial Line Interface Protocol.

SLM System Load Module.

slot 1. In programming, a time or data 'opening' into which other processes or data can be inserted. 2. A physical opening for connectors or wires/cables which is typically narrow and rectangular. The slots on the back of a computer allow external connection access to peripheral cards such as serial, graphics, or network interface cards (NICs). See slot types. 3. In building structures, an opening that may be built into a wall or floor in order to enable cables to be fed through the building.

slot types Most computers and switching stations have slots into which electronics peripheral cards can be inserted. In order for third-party suppliers to be able to develop options for consumers, a number of standards have been adopted for the shape and electrical configuration of these slots. Most of these slots are long narrow edge card configurations, with two to six slots in the typical desktop computer. Many computers will accommodate two different card formats. The software drivers for the cards which are inserted into these slots are sometimes supplied on diskettes, to be loaded on the system, and are sometimes supplied in hardware, on chips on the actual card. Some of the more common card slot types include PCI, ISA, ESA, ZORRO, and PCMCIA.

Slotted ALOHA See ALOHA.

slotting In setting up a network, the assignment of a circuit to available channel capacity.

slow scan television, slow scan TV SSTV. A type of black and white TV signal which can function within a narrow spectrum, similar to single-sideband transmissions for voice. SSTV has been used since the late 1950s by amateur television and radio operators to send series of images over radio frequencies. SSTV can be viewed on a television set with a scan converter or on a computer monitor with the appropriate interface.

In the U.S., SSTV uses frequencies ranging from about 3.845 to 145.5 MHz to transmit a series of images which can be captured through a dedicated system or through a computer linkup. Interface circuits for setting this up are in the hobbyist price range. Hicolor mode can provide color images up to 320 *x* 240 in thousands of colors. Even higher resolution 640 *x* 480 24-bit images (millions of colors) can be transmitted, but they take seven or eight minutes, compared to low resolution black and white images that take only seven or eight seconds.

Radio broadcasting is regulated throughout the world; those interested in SSTV technology will have to be licensed, usually for voice grade channels, by local regulatory authorities.

A related technology is *amateur TV* (ATV) which refers to *fast scan* amateur television. See amateur television.

SLP See Service Location Protocol.

SLR 1. send loudness rating. 2. single lens reflex.

Small Computer System Interface SCSI. A standardized interface specification which provides a means for the central processing unit (CPU) and main circuitry on the motherboard to communicate with computer devices that are interfaced to the system. This requires standardization of electrical circuitry and data protocols because peripheral devices are manufactured by many different companies. One of the most common of these formats is SCSI, which is widely used to interconnect hard drives, scanners, cartridge drives, digitizers, CD-ROM drives, and more.

The SCSI standard is approved by the American National Standards Institute (ANSI), and several enhanced versions have appeared (variously called SCSI-2, extended SCSI, SCSI-3, wide-SCSI, etc.)

SCSI typically consists of a SCSI controller on a motherboard or a peripheral card, which is terminated, and usually designated as zero or six, depending upon the system and one or more peripheral devices, set to SCSI ID number zero through five or one through six, depending upon which one is reserved for the motherboard, and terminated at the end of the last device. The devices can be hooked up end-to-end, that is, 'daisy-chained.' Each SCSI controller can chain up to seven devices, with the motherboard or main controller counting as one. The cable for SCSI devices is either a 50-pin edge connector or a 25-pin D connector (or a hybrid cable with an edge connector at one end and pin connector at the other). SCSI-3 cables are wider.

There can only be one device assigned to each SCSI ID. Conflicts or lack of termination will cause failure to recognize a device or spurious errors. Many systems expect CD-ROM devices to be set to ID 3, although there is no inherent reason why ID 3 has to be assigned to only this type of device. Scanners often default to SCSI ID 4. The ID number will determine the priority setting for loading the device, thus boot disks are usually assigned a number closest to the number of the controller. In other words, if the controller on a motherboard is zero, then the boot hard drive should probably be set to one and a relatively low-use tape drive to five or six.

SCSI ID settings are sometimes on the outside of the device, with a thumb-turn switch or DIP switch, and sometimes on the inside, with DIP switches or jumpers.

Termination is accomplished either by placing a physical terminator in one of the cable connection slots, by setting DIP switches, or by setting jumpers inside the device. Automatic termination is available on some devices, which means that if the device senses that it is that last device in the chain, it will terminate automatically. These types of automatic terminators are sometimes specific to the slot. There will be two slots on the back of most SCSI devices so they can be chained. Take care to follow instructions for which one to connect if the device is last in the chain and intended to terminate automatically.

Most SCSI devices can only work with cables up to about six feet in length, and three feet or shorter is generally recommended. Newer Fibre Channel Standard technologies can support longer connection lengths, allowing SCSI devices to be centralized in an operations room or wiring closet.

SCSI controllers are standard in many consumer and workstation computer systems, including Macintosh, Amiga, server-level IBM-licensed desktop computers, NeXT, Sun, SGI, some HP systems, and DEC. Most of these systems include both an internal SCSI controller (for up to six hard drives and internal CD-ROM drives, etc.) and an external SCSI controller (for up to six scanner, printer, external CD-ROM, external hard drive devices, etc.). Thus, a total of 12 devices can easily be daisy-chained to these systems without any modifications to the operating system or the hardware, other than perhaps adding a software device driver and cabling. In the author's experience, SCSI is a good format. The inexpensive eight-year old Motorola 68040-based computer used for the illustrations for this dictionary has two SCSI connectors (internal and external) with eight SCSI/SCSI-2 devices attached (scanner, tape drive, cartridge drive, six-disc CD-ROM drive, and four different kinds of hard drives). These are chained to the two controllers and worked together the first time they were connected without any compatibility problems.

SCSI drives are incorporated in mirroring and redundancy combination working drive/backup systems such as *redundant array of inexpensive disks* (RAID) systems. These drives can be conveniently hot-swapped in and out of the system if a drive fails and needs to be replaced, with the information rebuilt by the controller and software when the new drive is installed.

For consumer desktop Intel-based, IBM-licensed computers that come standard with IDE drives, a SCSI controller card can be added to the system to accommodate SCSI devices. However, on this type of system, it is important to determine whether appropriate device drivers are available for the peripheral, and that there is no contention with the IDE drive, and also that any appropriate IRQ issues are settled.

Two common varieties of SCSI hard disk drives showing the various connectors, components, and jumpers for setting the SCSI ID number.

small vocabulary In speech recognition, it has been found that software can be designed to recognize a variety of voices, without special training of the system, if the total vocabulary of the recognition is kept small. These 'small vocabulary' systems work well in specific environments such as stock buy/sell systems. While definitions of 'small' vary, recent systems of this type typically recognize 200 or fewer words.

Smallhouse, Charles "Chuck" An amateur radio enthusiast who contributed substantially to the first three OSCAR satellites' design and construction. See OSCAR.

Smalltalk An object-oriented computer exploration and development language developed through the Xerox Corporation in the 1970s. It was evaluated by four Xerox-selected companies in 1980, before being broadly distributed. By the mid-1980s, commercial versions of Smalltalk-80 were being released for a variety of platforms including International Business Machines (IBM) licensed personal computers and Apple II systems. Smalltalk has been favored by developers working in object-oriented programming environments and artificial intelligence applications. See Palo Alto Research Center.

SMART Self-Monitoring, Analysis, and Reporting Technology. A preventive system implemented in data protection schemes such as RAID which uses predictive failure analysis to anticipate possible failures. Impending problems are communicated to the controller which signals a warning so that faulty drives may be examined or replaced prior to any failure which might occur. See redundant array of inexpensive disks.

SMASH Project A project dedicated to developing mass storage devices for multimedia applications for home use. This is intended to promote commercial offerings of video services to the home, with part of the goal of SMASH to provide a labeling algorithm system in the storage system to provide vendor copy protection. Thus, data on the storage device can be set so that it can only be stored or copied once.

SMASH seeks to develop realtime labeling methods for compressed video. Common schemes for this include spatial or discrete cosine transform (DCT). The SMASH Project also introduces two new realtime labeling techniques that can be used in conjunction with MPEG-1 or MPEG-2 format video information. See watermark.

SMAS switched maintenance access system.

SMATV Satellite Master Antenna Television. A satellite communications distribution system designed to send transmissions to hotels, motels, apartments, etc. Since these are sent mainly to commercial establishments, they are often used as marketing leaders or as pay-per-view revenue-generators.

SMDS See Switched Multi-Megabit Digital Service.

SME 1. Security Management Entity. 2. Small- and medium-sized enterprises.

smear 1. Descriptive term for a television signal display distortion in which the image is blurred and appears stretched in the horizontal direction. 2. Low-level frequency distortion in an audio signal. 3. In digital imagery, distortion of details resulting from sampling frequencies or compression algorithm compromises, so transitions which normally would be sharp and crisp in the original image exhibit blurring or smear.

SMI See Structure of Management Information.

S/MIME Secure Multipurpose Internet Mail Extension. An IETF working group (inherited from the S/MIME Consortium) Internet messaging standard for the transmission of secure network communications. Unlike PGP/MIME, S/MIME public keys are distributed via X.509 digital certificates. S/MIME can support 128-bit encryption, although not all implementations will use the full 128 bits. See PGP/MIME.

smoke signal An early form of distance communications signal code that involved burning materials that produced a quantity of thick smoke at a high vantage point, and controlling the smoke with a blanket used to alternately suppress and liberate the puffs of smoke. Extensively employed by native North American plains dwellers, and later adapted by some of the early white settlers during times of conflict.

SMPTE See Society of Motion Picture and Television Engineers.

SMPTE Registration Authority SMPTE RA. A format and specification authority for technologies related to the motion picture and television industries. For example, the SMPTE RA is approved by IEC and ISO for the registration of MPEG-related format identifiers. See Society of Motion Picture and Television Engineers.

SMPTE time code A standard developed by the Society of Motion Picture and Television Engineers which provides synchronization for information recorded on audio and visual video tapes. SMPTE time code digitally encodes hours, minutes, seconds, and frames.

SMPTE time code is recorded onto audio tracks and video tracks as follows: in audio as Longitudinal time code (LTC); in video as Vertical interleave time code (VITC).

A time code 'word' consists of 80 'bits' (zero or one) per video frame, with 2400 'bits' per second corresponding to 30 frames per second for North American TV. In Europe, 2000 'bits' per second corresponds to the standard of 25 frames per second. See drop frame, Society of Motion Picture and Television Engineers.

SMR See Specialized Mobile Radio.

SMRP See Simple Multicast Routing Protocol.

SMS 1. See Service Management System. 2. See Short Message Service.

SMTP See Simple Mail Transfer Protocol.

SNA See Systems Network Architecture.

SNA Control Protocol SNACP. A protocol which handles the configuration and enable/disable functions at the ends of a point-to-point link. Subdivided into two protocols that independently negotiate SNA with or without LLC 802.2. Similar to Link Control Protocol. See See RFC 2043.

SNACP See SNA Control Protocol.

SNAFU An acronym for "situation normal—all fouled up" originating in World War II, or perhaps earlier, this expression hasn't lost its currency and is frequently heard in industries struggling to implement, use, and maintain constantly changing technology.

snail mail *slang* A tongue-in-cheek reference to physical messages (letters) being delivered very slowly (at a snail's pace) by the postal service, in contrast to electronic mail, which is often received within minutes of being sent. See email.

snake A cabling aid consisting of a flexible, long, thin cord of metal or plastic used to feed wire and cable through conduit or through structures (ceilings, walls, attics, etc.) where space is tight, or access is limited. See birdie.

SNAP See SubNetwork Access Protocol.

sneak currents Low level undesired currents which seep into circuits and may, if continued long enough, cause damage. Sneak currents are those which do not cause immediate harm and are not sufficient to trigger safety mechanisms such as normal fuses and breakers. Sneak currents can result from causes such as worn sheaths and insulators, incorrect wiring, temporary contact due to settling, etc. See sneak fuse.

sneak fuse A special low-level current detection fuse specifically designed to trigger if sneak currents are detected. See sneak currents.

SNI See Subscriber Network Interface.

sniffer A network traffic monitoring tool used for diagnostics and snooping. The sniffer is a useful system administration tool which can help monitor and log peak traffic times, network load, and possible problems. This information can help 'tune' a system to operate efficiently. See packet sniffer.

SNMP See Simple Network Management Protocol.

snooper, snooperscope, night scope A device designed to enhance night vision by sending out and intercepting an infrared beam. The incoming beam is interpreted into an image that shows objects not visible to the human eye.

snow An undesired aberration in a broadcast or display of a video image in which there are many randomly distributed speckles, often white. Snow can result from transmission problems, such as a weak or drifting signal or from display device problems, as in a cathode ray tube (CRT).

Société Internationale de Télécommunications Aéronautiques SITA. An international airline reservations and telegraphic transmissions service backbone network established in 1983.

Society of Cable Telecommunications Engineers Inc. SCTE. A U.S. national nonprofit professional organization founded in 1969. The society includes over 13,000 members from around the world, representing a broad spectrum of cable professionals. The society provides education, certification, and standards development. http://www.scte.org/

Society of Motion Picture and Television Engineers SMPTE. An international organization founded in 1916, originally as the Society of Motion Picture Engineers. The T was added in 1950 to encompass the emergence of the television industry. The society includes over 8,500 members in 72 countries, including engineers, technical directors, and production/post-production professionals dedicated to advancing the theory and application of motion-picture technologies. SMPTE contributes to standards development, encourages consensus-based recommended practices (RPs), and industry engineering guidelines (EGs).

In 1957, the society was awarded an Oscar for its contributions to the advancement of the motion picture industry. It has also received three Emmy awards for various recording and video systems and standards.

SMPTE is best known for developing SMPTE time code methods, which are used for video editing. When video tape began to be widely used for recording and editing, a way was needed to synchronize edits, to locate specific places on the tape, and to dub sound to match the video sequences. The SMPTE began in 1969 to develop a standard for digitally encoding time information in terms of hours, minutes, seconds, and frames onto audio or video tape. See MIDI time code, SMPTE time code. http://www.smpte.org/

socket 1. A means of providing unique identification to which or from which information is transmitted on a network. RFC 147 specifies a socket as a 32-bit number; even sockets identify receiving sockets, odd sockets identify sending sockets. Each socket is identified with a process running at a known host.

SOCKS An access and security technology designed to provide a framework for TCP and UDP client/server applications to conveniently, transparently, and securely utilize and traverse a network firewall. There have been a number of versions of SOCKS, with RFC 1928 representing version 5. Version 5 adds UDP and authentication capabilities, and extends addressing to accommodate the future needs of IPv6.

The protocol fits between the application layer and the transport layer and does not provide ICMP message-forwarding services. Traversing a firewall securely depends upon the various authentication and encapsulation methods selected and used in the negotiation between the SOCKS client and server. See firewall, gateway, proxy, RFC 1928.

sodium vapor lamp A lamp that glows a warm golden color, from the passage of electricity through metallic vapors in a cylinder encased in a glass tube. Sodium vapor lamps have been used as street and bridge lamps. See mercury vapor lamp.

soft copy A stored image, document, or file which is recorded on a medium which must be accessed with some type of technology in order to be viewed, manipulated, or displayed. Soft copies commonly exist on hard drives, floppy diskettes, tapes, CDs, and other magnetic or optical media. A hard copy is readable by looking directly at the medium on which it is transcribed, as on a piece of paper, cardboard, stone, or parchment.

soft transfer A term for an electronic monetary transaction which precedes the actual exchange of funds between individuals or banking institutions. A paper check is a type of soft transfer. It is a monetary transaction which is not finalized until the money is withdrawn from the bank. Similarly, online, there are many monetary transactions which are soft transferred and later 'hard transferred' from the actual bank or other financial institution, such as a credit union.

SoftCard An early commercial product from Microsoft, from an idea suggested by cofounder Paul Allen. The SoftCard was an internal peripheral card equipped with a Z80 processor, which ran CP/M-80 from Digital Research. This card, when installed in an Apple II computer, allowed its user to install and run CP/M-compatible applications programs.

softlifting A term that became prevalent on the Internet and in trade magazines, when commercial Web browsers became widespread, although it had previously been mentioned somewhat in the context of other types of applications. Softlifting is a process of having a piece of software gather information that is stored on a person's computer or gathering information about the computer configuration, and reporting that information back to the manufacturer. The reporting is usually without the knowledge of the user, and often happens when the user dials up to a public network or when they dial up to 'register' their software on the vendor's bulletin board system. This is considered by the consumer and Internet communities as a reprehensible, blatantly unethical activity, and developers who attempt this covert information-gathering put themselves at risk of substantial criticism and possible legal reprisals.

software Computer instructions which are stored on a medium which is reasonably portable and accessible by users. Actually, the distinction between hardware and software is much less clear than many people realize. It may seem reasonable to designate everything inside the computer as hardware and everything that holds information that can be inserted into external storage read/write devices as software, but that's not really the best distinction. Floppy disks and computer chips are tightly integrated hardware/software combinations whether external or internal, so the matter is really one of accessibility coupled with structure. Since computer instructions stored on disks are easily read, written, and moved, they are thought of as software. Since computer instructions on computer chips are not easily read or written and not easy for a lay person to access or move, they are considered as part of the hardware.

The lowest level software functions are programmed into the computer chips themselves. At the hardware operations level, this software acts to start up the system, test it, bootstrap the device drivers to come online, and initiate the operating system to accept user input and output, and to otherwise communicate with the central processor. Some of these operating instructions may be read, in turn, from hard drives, CD-ROMs, cartridges, or other storage media. High level software interacts with the user through application programs.

Software is created with a variety of programming, editing, debugging, compiling, interpreting, and linking tools in a great assortment of languages, which are general purpose or optimized for specific types of programming. See programming.

Software Engineering Research Network SERN. An engineering and research joint venture of the Department of Computer Science and the Department of Electrical and Computer Engineering at the University of Calgary, Alberta. It is sponsored by the Government of Alberta, the University of Calgary, Motorola, Computing Devices, and Northern Telecom.

Sol A very early microcomputer descended from the Altair and based on the Altair bus (S-100 bus).

solar cell In the 1940s, Bell Telephone Laboratories developed a storage cell from thin strips of silicon which had the characteristic of developing a charge in the presence of light. Since the silicon is not directly depleted in this process, solar cells are not subject to the limited life spans of traditional wet and dry cells. Solar cells have since been developed and refined in many ways and are used in many aspects of electronics. See photovoltaic.

Solaris A popular 32-bit operating system from Sun Microsystems, which is commercially distributed, as is their SunOS operating system and, more recently, OpenStep. Solaris is multiprocessing, multithreaded, and network-friendly (using NFS), based on an open systems architecture. Many large Internet Services Providers, university systems, and enterprise local area networks (LANs) run on Solaris. Solaris is available for various Sun SPARC, Intel-based, and Motorola-based systems.

solenoid A long, cylindrical, current-carrying coil with properties similar to a bar magnet into which an iron bar will be drawn when current is applied to the coil. Solenoids are commonly used in circuit breakers which have replaced traditional fuses. See electromagnet.

SONET *S*ynchronous *O*ptical *Net*work. SONET is a set of ANSI telecommunications standards which specify a modular family of rates and formats for synchronous optical networks.

SONET provides a standard operating environment for managing high bandwidth services, and incorporates multiplexing, service mappings, and standardized interfaces, so commercial vendors can develop interconnecting technologies.

SONET has been adopted by the ITU-T as the basis for the Synchronous Digital Hierarchy (SDH) transport system, and is a subset of this system. SONET is based on STS-1 which is suitable for T3, and SDH is based on STM-1, suitable for E4 transmissions.

Communication between nodes, to permit control, provisioning, administration, and security, is accomplished through the Synchronous Transport Signal (STS) transmitting at a line rate of 51.84 Mbps. The STS is comprised of payload information and signaling and protocol overhead. Since the two ends of a SONET transmission may vary in format and speed, data is converted to the STS format, transmitted, and, when received, converted into the appropriate user format. OAM&P is integrated into SONET. See detailed information under the following SONET listings.

SONET ANSI standards There are a number of important American National Standards (ANSI) of Committee T1 related to SONET that are available from ANSI and are described in the form of abstracts on the Web. The ANSI Standards chart shows a sampling.

SONET frame The frame length is 8000 fps or 125 µsec. SONET uses Synchronous Transport Signal level 1 (STS-1) as its basic signal rate of 51.84 Mbps. SONET frames are organized in a row by column structure totaling 810 bytes. Transport overhead is contained in the first three columns and is subdivided to include section and line overhead. The remaining columns, from four to 90, are used for the Synchronous Payload Envelope (SPE).

The STS-*N* frame consists of frame-aligned, byte-interleaved *N* STS-1 signals.

The STS-*Nc* frame consists of concatenated STS-1 signals to form a multiplexed, switched signal that can be transported together. This is done to accommodate broadband services such as ISDN.

Here is an overview of some of the bit rate speeds for the Synchronous Transport Signal levels and how they compare to European equivalents. Keep in mind that this chart only indicates bit rates, as the frame formatting for each system differs even further.

U.S.	Europe	Bit rate
STS-1	—	51.84 Mbps
STS-3	STM-1	155.52 Mbps
STS-12	STM-4	622.08 Mbps
STS-24	STM-8	1244.16 Mbps
STS-48	STM-16	2488.32 Mbps
STS-192	STM-64	9953.28 Mbps

SONET multiplexing SONET signals can be multiplexed to make efficient use of network capacity. There are a number of ways to accomplish this, as shown in the chart.

SONET optical interface layers SONET includes a hierarchy of interface layers. Each one builds on the previous; from high to low, they are: path layer, line layer, section layer, and photonic layer. Individual layers communicate to peers on the same layer and to adjacent layers above and below.

SONET path overhead In SONET, path overhead is transported with the payload until the signal is demultiplexed at the receiving end. The path overhead supports four classes:

Class	Functions	Notes
Class A	Payload independent functions.	Required by all payload types.
Class B	Mapping dependent functions.	Required by some payload types.
Class C	Application-specific functions.	
Class D	This is reserved for future use.	

SONET path terminating element PTE. The PTE is an element which multiplexes and demultiplexes the Synchronous Transport Signal (STS) payload and processes the path overhead as needed to originate or access it. If necessary, the PTE can also modify or terminate it. See SONET, Synchronous Transport Signal.

SONET timing In SONET networking, synchronization is accomplished by referencing a high accuracy clock and information from its slaves, so synchronization characters between equipment nodes are not used. Due to the high data rates carried by SONET, it is important to maintain clock accuracy. The three major timing modes supported are: external timing based on a clock, generated free run/holdover timing from an internal clock, OC-N signal line timing.

SOP standard operating procedure.

sound Radiant mechanical energy produced by vibration, which requires a physical medium for its transmission (such as air), and is detected by hearing, accomplished through physical sound-detection, perception, and interpretation by the nervous system. Compared

SONET Multiplexing Techniques

Technique	Type of interleaving	Notes
Interleaving	Interlaces individual bytes.	Reduces overhead at receiving end.
Single-stage interleaving	Direct byte interleaving.	STS-N signal created directly.
Two-stage interleaving	Direct byte interleaving.	Accommodates European ITU-T rate.

ANSII Standards Related to SONET		
ANSI Standard	Title	Subtitle
T1.105-1995	Synchronous Optical Network	Basic Description including Multiplex Structure, Rates, and Formats
T1.105.01-1995	Synchronous Optical Network	Automatic Protection
T1.105.02-1995	Synchronous Optical Network	Payload Mappings
T1.105.03-1995	Synchronous Optical Network	Jitter at Network Interfaces
T1.105.03a-1995	Synchronous Optical Network	Jitter at Network Interfaces - DS1 Jitter
T1.105.04-1995	Synchronous Optical Network	Data Communication Channel Protocol and Architectures
T1.105.05-1994	Synchronous Optical Network	Tandem Connection Maintenance
T1.105.06-1996	Synchronous Optical Network	Physical Layer Specifications
T1.105.07-1996	Synchronous Optical Network	Sub-STS-1 Interface Rates and Formats
T1.105.09-1996	Synchronous Optical Network	Network Element Timing and Synchronization
T1.119-1994	Synchronous Optical Network	Operations, Administration, Maintenance, and Provisioning - Communications
T1.119.01-1995	Synchronous Optical Network	OAM&P Communications Protection Switching Fragment

to light and heat, sound waves move very slowly. Human sound perception through hearing covers a frequency range from about 20 hertz to about 20 kilohertz, although lower, and sometimes higher frequencies are felt, although not heard, through vibrations in the body. Other creatures perceive broader, narrower, or more specific frequencies, and sound is a ubiquitous means of species communication. We know enough about the nature of

SONET Interface Layers		
Layer	Peer	Notes
Path	Services & path overhead mapping	Transport services between path terminating equipment (PTE). Mapping signals to line layer format.
		Conversion between STS and OC signals.
Line	SPE & line mapping	Transport of path layer payload and overhead across physical medium. Synchronization and multiplexing.
Section	STS-N & section overhead mapping	Transport of STS-N across physical medium.
Photonic	Optical conversion	Transport of data across physical medium.

sound waves to record, reproduce, and modify them, and to propagate them over great distances. Humans can project unamplified voice through the air for a few dozen or hundred yards, depending upon atmospheric conditions. Whale songs will resonate for thousands of miles through water, although whale communication distances have been drastically reduced by interference from industrial shipping noise.

sound spectrograph An instrument for measuring the structure of speech and displaying it visually, developed in the early 1940s by Bell Laboratory researchers. This opened the door to more objective, quantitative measures of speech, information that is of interest not only to speech therapists, physicians, and educators, but also to developers of communications technologies.

sounder A sound amplification device incorporated into a communications receiver, usually a telegraph receiver, to make the code clicks audible to a human operator. Sounders were invented when it was noticed that telegraph operators had learned to interpret Morse code clicks and transcribe them manually faster than a paper tape could print the messages. A *mainline sounder* was an adaptation that allowed variable adjustments without a relay. See resonator.

space division multiple access SDMA. One of two common optical multiplexing techniques which utilizes an angle diversity receiver, that is, multiple receiving elements receiving from different directional angles. See wavelength-division multiple access.

space-charge field In electronics, an electric field created outside the physical surface of a conductor or semiconductor.

space-to-mark transition, S-M transition In telegraphy, the momentary change when the system reverses polarity, or changes from an open to a closed circuit. At this point, there is a small amount of delay that must be taken into consideration, which can be plotted on a timing wave. The reciprocal is the mark-to-space transition.

Spaceway A commercial constellation of geostationary communications satellites from Hughes Communications. Spaceway was formed from the merger of the Hughes Galaxy Satellite Services and the PanAmSat Corporation. Hughes Electronics is a subsidiary of General Motors Corporation. Spaceway is intended to be a global broadband communications system with service planned for 2001.

spade lug, spade tip A small, flat, notched (somewhat U-shaped) conductive connector that is attached to the end of a conductive wire in order to easily secure it to an electrical terminal by sliding the end around the mounting screw and applying pressure via a bolt, or by soldering. Spade lugs are still common inside small residential phone wire junction boxes, but in large installations, punchdown blocks and modular components are more prevalent. See lug.

spaghetti code *jargon* Software code that is disorganized and logically 'tangled.' Code often turns into spaghetti code, because it can be difficult to solve a problem and create an efficient design until more aspects of the concept are explored and known. Consequently, half way through a project, many programmers exhort, "Time for a rewrite!" The code may have become bulky, obscure, too hard to follow or maintain to be practical, or it may use slow, inefficient routines where better ones have been discovered or developed along the way. Spaghetti code often develops when software is created by committee or individual programmers in the project are intent on following personal methods for its development that don't mesh with the other portions of the code. It is often observed in the software industry that code tends to decrease in efficiency and elegance of design in proportion to additional people added to a project.

spam *slang* A term widely used on the Internet to describe annoying, unsolicited, irrelevant, illegal, or worthless communications, usually in the form of email or public postings. It's generally said that the word originated as a tongue-in-cheek reference to a Hormel meat product called Spam, which is frequently pilloried and satirized in the media. Whether or not that is the case, the *spam* on the Internet, especially in the form of unsolicited email bulk promotion of get-rich-quick schemes and sex sites, has become a big problem due to the intrusive way in which the spammers violate the space and privacy of recipients. Not as often acknowledged is the fact that spam causes substantial expense to ISPs, and general annoyance and expense to users who pay for email messages individually or for extra storage space for mail messages on their online accounts.

Junk email creates ill will and a negative impression. It is unwise for legitimate businesses to send unsolicited commercial messages to email addresses, as many recipients of spam will boycott those businesses indefinitely. It is also unwise to send exaggerated and unsupported claims, as there are laws governing false advertising that will probably be more stringently enforced as the number of complaints to authorities increases. See spamming.

spamming *slang* Posting or emailing irrelevant, annoying, illegal, or unsolicited opinions or promotional materials, sometimes through anonymous mailers or with false return email addresses. In 1997, many ISPs were forced to install anti-spamming software on their systems to stem the floodtide of these types of messages. Many ISPs will now also reject an email message which does not have a legitimate sending address in the mail header. On the USENET newserver, also in 1997, many formerly useful discussion groups became worthless (especially business-oriented groups) due to the large volume of spam, and some recovered only by going to moderated status (a strain on the many volunteers who have to check all those messages before reposting them). See cross-posting, spam.

SPAN Switched Port Analyzer. A Cisco Systems network switch feature which extends the monitoring abilities of existing network analyzers into a switched Ethernet environment. SPAN takes the traffic at one switched segment and mirrors it onto a predefined SPAN port. A network analyzer which is attached to the SPAN port can monitor traffic on other compatible switched ports.

spanning tree algorithm STA. A standard technique described in IEEE 802.1 which is incorporated into bridges in computer networks, for example, in Fiber Distributed Data Interface networks, where it is incorporated into bridges that connect the primary and secondary rings. The spanning tree logic can prevent duplicate bridging and allows the backup ring hub to handle bridging if the primary ring hub fails. See Fiber Distributed Data Interface, Token-Ring.

Spanning Tree Protocol STR. A protocol based on IEEE 802.1d that provides resiliency through system and link redundancy that is especially suitable for virtual local area networks (VLANs).

spark A short, bright electrical discharge. Sparks may be intentional, as those generated by spark plugs in an engine, or those created to start a fire in tinder. Unintentional sparks may be dangerous and may occur as a result of incorrect electrical connections (shorts, crossed wires), inadequate insulation, or contact with unintended conductive substances such as water.

spark coil A device incorporating an inductive magnetic core surrounded by helical windings of conductive materials used to generate a spark. The spark coil was typically used in conjunction with a condenser and vibrator (or interrupter) for telecommunications applications. Spark coils are still used to ignite internal combustion engines, but for many electronics applications, transformers began to replace spark coils in the early 1900s. See armature, coil, dynamo, induction, winding.

spark gap The distance across which a spark jumps between electrodes. Adjusting this gap effects the behavior of the spark. If the gap is too large, the spark may not jump the gap.

SPARS code Society of Professional Audio Recording Studios. A three-letter code found in compact discs which indicates the analog or digital nature of a portion of the recording history. For example, ADD indicates that the original recording was analog, the mixing was digital, and the mastering stage was digital. See compact disc.

SPCAS See SPC Allocation Service.

SPEC See Standard Performance Evaluation Corporation.

Specialized Common Carrier decision A decision by the Federal Communications Commission (FCC) in 1971 to permit competition with AT&T in the provision of specialized voice and data services.

S

Specialized Mobile Radio SMR. A national, analog, two-way radio dispatch system favored by commercially dispatched passenger and cargo fleets. The system is converting to digital which makes it competitive in some areas with cellular services. While SMR spectrum is more limited than cellular, it has an advantage for some applications in that it has a longer range.

Specification and Design Language SDL. An ITU-T-defined language for the description and specification of the behavior of telecommunications systems. XDL is an extension of this language.

spectrograph An instrument for spreading light into its spectral components. By studying the brightness of the spectrum at each wavelength, it is possible to study the composition and characteristics of substances through their light-emitting properties and patterns. This is widely used by astronomers studying our solar system.

spectroscopy A technique used by scientists to study the composition and/or characteristics of a substance based on its light-emitting properties. In astronomy, spectroscopy has enabled more detailed study of stars, beyond just distance and brightness.

spectrum In general, a continuous sequence or range of some property or radiant energy. Many phenomena are described in terms of their characteristics or position within the spectrum of radiant energy, including electromagnetic, radio, sound, visible light, specific colors, etc. See band allocations, radio, sound, visible spectrum.

speech recognition The process of receiving, interpreting, and parsing spoken words. On computers, this is often accomplished with a microphone input device, an analog-to-digital peripheral card, and a software program that works independently or in conjunction with other programs such as word processors. It may also include *noise-canceling* features that help to separate the voice of the speaker from ambient room noise such as the hum of computers or low conversations in the distance.

Speech recognition is distinct from voice recognition in that voice recognition is the processing of the particular characteristics of a specific voice so that it can be recognized, such as in a security identification system. Voice recognition does not involve making sense of the content of the message, as does speech recognition. Speech recognition systems typically require a minimum sampling rate of about 3,000 samples per second in order to reliably recognize words. In many systems 8,000 samples per second is used.

Speech recognition can be used to dictate text, give commands, and send information over a communications system in digital or altered form. Since speech recognition is a complex process, most current systems are specialized to recognize a specific limited vocabulary as spoken by a number of speakers in a specified language or a general (or specific) vocabulary as spoken by one particular speaker. More sophisticated systems can recognize and react to sentences and grammatical structures. Many speech recognition programs have 'training' algorithms (*speaker adaptive* algorithms) included so the software can gradually adapt to the idiosyncrasies of a particular speaker's means of expression and pronunciation of certain words.

In the mid-1990s, speech recognition computer software began to be reliable and inexpensive enough to be of interest to small businesses and individual consumers, and its use will probably spread through a variety of applications, perhaps adding another means of input to standard software such as word processors and electronic mail programs. See phonemes, voice recognition.

speech synthesis The reproduction of audible human communication, through the use of computers. There are many different ways to create synthesized voice. Sound samples of human voices uttering certain words, sounds, and syllables can be recorded as separate entities, stored digitally, and then combined and played back to create words and phrases. Other schemes, such as pure digital recreation of voice-like sounds are also available, but tend to have a distinctly mechanical quality to them.

The naturalness of the sound of synthesized voice is highly variable. Just as a graphical rendering program in the hands of a good artist results in better pictures than in the hands of a nonartist, speech programming and recreation done by a person with a good ear for sounds yield better results than when done by someone without a sense of the rhythm, timbre, and pitch of human voice communications. There is still much that can be done, within the capabilities of the current technology, to improve speech synthesis.

The most famous synthesized voice in the world is probably that of Stephen Hawking, world acclaimed physicist who 'talks' through words and phrases programmed into a computer keyboard installed on his wheelchair. When Mr. Hawking is finished composing the message, it is replayed to the listener through a speech synthesizer.

Synthesized voices are used in multimedia applications, on storybook CD-ROMs, in automated telephone mail order and banking systems, etc. See phoneme, speech recognition, voice recognition.

speed dialing A means of keying in a shorter code to represent a longer one in order to speed up the dialing of long phone numbers. See abbreviated dialing.

SPF shortest path first.

SPHERE System A microcomputer system from SPHERE Corporation that was sold as a complete unit with a black and white CRT display, keyboard, 4 kilobytes of memory, serial port, and realtime clock. It was capable of running an extended BASIC compiler. A color CRT display was even available as an option. It was shipping in the fall of 1975, before the Tandy TRS-80 and Apple computers. The Utah company ran full page ads in Popular Electronics offering the system for only $860, or $350 for just the logic board, yet for reasons unknown, the system didn't catch on.

The SPHERE came as an assembled unit, the first microcomputer to come standard with a monitor and keyboard, or it could be purchased as a kit; either was under $900 in 1975.

SPHIGS Simple PHIGS. See PHIGS.

SPI 1. Security Parameters Index. 2. Service Provider Interface.

SPID See Service Profile Identifier.

SPIE The International Society for Optical Engineering. SPIE is a nonprofit Washington State-based professional society dedicated to advancing research, engineering, and applications in optics, photonics, imaging, and electronics. SPIE produces educational publications, sponsors conferences, and workshops, and now also provides Web resources in conjunction with the Institute of Physics (IOP). http://www.SPIE.org/ http://optics.org/

splashing When using competitive operator services for long-distance calls, if the caller places a call from San Francisco to Portland and the alternate service is based in Los Angeles, the call is said to be *splashed* if the billing is determined by the distance from Los Angeles to Portland. See Operator Service Providers.

splice *v.* 1. To unite or combine separate lines, usually by weaving together the individual strands. 2. In electrical splices, care is usually taken to match the data lines so as not to cross one type of data channel with another, and bare wires are generally covered with an insulator such as a cap, electrical tape, or plastic shrink sheath to prevent short circuits or shock.

spoofing 1. Deceiving, covering up the identity of, conveying an impression of something else. A means of gaining unauthorized access by deceit, impersonation, or other misrepresentation. 2. In routing network traffic, a means of rerouting, or otherwise changing the destination of a transmission by mimicking the destination through responses, signals, or other identification.

spool An acronym for "simultaneous peripheral operations on line." A technique for improving efficiency by accommodating the different operating speeds of a number of types of peripherals and processes. While the concept is still relevant, spooling programs are becoming rarer, since they are mainly required on nonmultitasking systems. On multitasking systems, it is not necessary to wait for a print or a plot to finish, for example, in order to continue word processing or drawing. On single-tasking or task-switching systems, however, and with peripherals which don't support a *queuing* system, wait time can be a problem. For example, plotters tend to print rather slowly. If the user must sit and wait for 25 minutes while a plot is being created, productivity is lost. By sending the plot to a software spooling process (or a hardware spooling buffer) and plotting it at a convenient time, activities can overlap. Print spooler.

spread spectrum A communications technique for broadcasting in which more bandwidth is used than would normally be needed to transmit the signal. At the sending end, specific types of 'noise' are incorporated into the signal and the signal is spread over a wider frequency range. At the receiving end, a lower bandwidth version of the original signal is recovered. Why use more bandwidth than is necessary? Spread spectrum was originally designed to improve security for sensitive communications, particularly within the military. A spread spectrum signal is harder to detect, harder to jam, and requires specialized receiving equipment.

Spread spectrum is generally divided into frequency hops, a system developed over 50 years ago by Hedy Lamarr, and direct sequencing which utilizes the noise-like characteristics of pseudorandom sequences to control a phase modulator.

Spread spectrum broadcasts are not only used in covert or private communications, they are sometimes now also used as a means to make best use of increasingly limited broadcast space due to the increased demand. In the United States, unlicensed spread spectrum transmitters are permitted to broadcast within specified frequency ranges. As an example, the 900-MHz cordless telephones and a number of wireless local area networks (LANs), operate in the allowed spread spectrum frequencies.

Pseudonoise, or direct sequence spread spectrum, is a technique used in local area wireless networks (LAWNs). Redundant data bits called *chips* are incorporated into the transmissions and the receiver must have knowledge of the spreading code to remove the added chips and decipher the incoming message. The insertion of chips provides a means to provide more frequencies within a given area, with a tradeoff in speed. The throughput of this type of system is about 2 to 8 MHz.

The Federal Communications Commission (FCC) regulates the amount of time that may be spent on any one channel. For example, in the ISM band of 902 to 928 MHz, the frequency hop on a particular channel must not exceed 0.4 seconds once every 20 seconds. Similarly, in the 2.4- to 2.484-GHz range, the interval is not more than once every 30 seconds. Recommendations are also being drafted by the IEEE. See frequency hopping.

spreadsheet A columnar ledger sheet used for listing items and, often, financial information. The advantage of electronic spreadsheets is that they can be configured to dynamically change cells which are related to information in other columns (such as calculations, totals, etc.), if the information in the original column is changed. This is extremely convenient for testing budget scenarios, for example. Electronic spreadsheets are great time savers and tend to be much less prone to arithmetic errors than hand-calculated equivalents.

SPS See Standard Positioning Service.

spurs Foot spikes which are worn by line workers who must manually climb utility poles for maintenance, testing, and repair.

Sputnik I The world's first artificial satellite, launched by the Russian Federation on 4 October 1957, studied the ionosphere, and heralded the space age. It transmitted in a frequency range just above global frequencies for standard time signals, so that those listening in, which included Earth stations all over the world, and a large number of amateur enthusiasts, would be able to monitor and report the status of the satellite. The project was announced in print by the Russian Federation several months before its launch so that results of communications from the satellite could be reported.

Sputnik 2 The world's first successful launch and retrieval of a living creature, Laika, a dog, was carried out by the Russian Federation in Sputnik 2 on 3 November 1957. This craft also studied space radiation and used a slow-scan TV camera to relay images to the ground.

SRA See system reliability architecture.

SRAPI Speech recognition API. A speech recognition and translation applications programming interface (API) developed by the SRAPI Committee, a Utah nonprofit corporation consisting of some well-known vendors of various audio and multimedia products. http://www.srapi.com/

SRGP See Simple Raster Graphics Package.

SS7 See Signaling System 7.

SSB See single sideband.

SSCS Service Specific Convergence Sublayer. A component of the ATM adaptation layer (AL) which coordinates protocol of the above layer with the requirements of the next lower layer, the Common Part Convergence Sublayer (CPCS). See asynchronous transfer mode.

SSEC See Selective Sequence Control Computer.

SSL See Secure Socket Layer.

SSO single sign-on.

SSP See Service Switching Point.

SSRP See Simple Server Redundancy Protocol.

SSTO Single-Stage to Orbit.

SSTV See slow scan television.

ST 1. See Scheduled Transfer. 2. signaling terminal. 3. straight-tipped. 4. systems test.

STII Stream Protocol Version II. A connection-oriented Internet protocol. A connection must be set up between sender and receiver(s) before datagrams are transmitted, and flow quality is accomplished through datagram scheduling by the routers. In contrast, a great deal of Internet traffic is connection-oriented, in that the receiving system doesn't have to be determined to be online before a transmission is sent. See RFC 1819.

STA 1. See spanning tree algorithm. 2. Science and Technology Agency. Japan's organization to support research and development, and to plan and coordinate national science and technology. Founded in 1956.

stand alone, stand-alone, standalone A system or device which provides a self-contained service or function, or which works independently of other major components. See turnkey system.

standard cell A fragile, special-purpose cell which provides very small amounts of current (1.019 volts) for short periods.

Standard Performance Evaluation Corporation SPEC. A nonprofit organization which supports the establishment and maintenance of standardized, relevant benchmark computer performance evaluation tools that can be applied to various high-performance systems. http://www.specbench.org/

Standard Positioning Service SPS. One of the precise location data signals transmitted from Global Positioning System (GPS) satellites. This signal is available without charge for private and commercial use, and is not encrypted. It provides information about the functioning of the satellite and its approximate location. Combined with information from three or four other satellites, the user can pinpoint a location with horizontal and vertical accuracies up to about 100 meters and 340 nanoseconds of time, depending upon the quality and accuracy of the receiving equipment. See Global Positioning System, Precise Positioning Service.

Standards Council of Canada SCC. The Standards Council of Canada is a federal Crown corporation charged with the responsibility of promoting efficient and effective voluntary standards in Canada for the health, welfare, and economy of Canada. The SCC works in cooperation with, and manages, the National Standards System. The SCC coordinates input from the SCC to the International Organization for Standardization (ISO) and the International Electrotechnical Commission (IEC). See Canadian Standards Association, Telecommunications Standards Advisory Council of Canada. http://www.scc.ca/

standards reference model A communications framework for describing and classifying information processing standards by the Institute of Electrical and Electronics Engineers, Inc. (IEEE). Within the reference model, general sets of requirements are developed. Generally the reference model divides standards into two categories:

Category	Notes
application program interface	Standards which influence application portability by describing and defining interoperability between applications software and the computer operating system. Consideration is given to systems, communications, information, and human-computer interactions.
platform external interface	Standards which influence system portability and interoperability, that is, the behavior of information processes which interact with their external environment. Data portability and user interfaces are important considerations.

Conformance to the above standardization concepts provides a framework for vendors to develop and distribute products and services compatible with those of other vendors, and provides users with some assurance that products purchased from different vendors will work together.

standby processor A spare or secondary processor in a computer system, which is ready to take over if there are problems with the first, or if extra computing power is needed on an irregular basis. The standby processor is usually a 'hot standby' device, that is, it can come online without turning the system off or interrupting its functioning to a substantial degree. Sometimes the second processor is not idle, but is used for less intensive computing operations, while still remaining ready if there is a need for it to come online as a substitute for the main processor. In some cases, the standby processor does low-level maintenance work on updating its databases and file structures, so that if the primary processor goes offline, the file information is known by the standby processor.

Standby processors are most commonly used in high-end systems that require a high degree of reliability. Examples would include medical or navigational applications where people's lives might be in danger if there were a processor failure.

standby time 1. A power-saving feature built into many consumer electronics. Camcorders, calculators, laptop computers, and various other devices that rely on limited battery power, will often have timing mechanisms that monitor idle time, that is, time during which the device receives no input from the user. When the idle time expires, the device is powered down or put in a minimal power-consuming mode, in order to save battery life. 2. The amount of time a fully charged battery-powered unit can remain on before the battery runs out. This applies to many devices, including cell phones, cordless phones, short-range radios, laptop computers, etc. Standby time is often used as a marketing statistic to characterize a system and generally refers to idle time, rather than time during which the item is being used. The use time (or talk time on a cell phone) is typically less if many features are being used.

standing wave A phrase which is used to describe physical relationships and movements in real life as well as diagrammatic representations on display systems such as computers and oscilloscopes. Very large standing wave patterns are described by astronomers when talking about the movement, relationships, and symmetry of galaxies.

standing wave ratio SWR. A diagnostic measurement of a standing wave which is commonly used in transmitter and transmission line testing and maintenance. When cables are affected by moisture, wear, and loose connections, an impedance mismatch may occur which can be detected through the amplitude ratio of a standing wave. The transmitted wave is used as the signal source for measurement by an inline SWR meter.

Stanford Artificial Intelligence Laboratory SAIL. This research laboratory is the source of many pioneer ideas and developments in artificial intelligence and general concepts related to computers. SAIL contributed to the development of the LISP programming language. The SAIL facility also developed the SAIL programming language, an ALGOL-like language created in the early 1970s. The SAIL facility was closed down in 1990.

Stanley, William An inventor who created the first practical alternating current generator, after it was pioneered by Elihu Thomson. Thus, the first Stanley alternating current (AC) distribution system came into being in 1885, three years after Thomas Edison opened the first direct current (DC) power utility company. In conjunction with Elihu Thomson and Sebastian de Ferranti, Stanley also developed the transformer. See Alexanderson alternator.

star topology A common type of network topology in which remote systems and nodes are connected point-to-point to a central system, and not to one another. Unlike some topologies with redundant connections, if the central system on a star network fails, the entire network is unable to intercommunicate. One advantage is that problems are easier to isolate, another is centralized administration and security. A star system provides the option of physically isolating the server from unauthorized access. Star topologies are used in many phone and data networks. See hub, topology.

start pulse In a start-stop teletypewriter system, a mechanism to release the receiving line relay and permit the receiving arm to move.

start-stop synchronization A method developed by Howard L. Krum for use with permutation code telegraph systems.

static *n.* 1. In radio transceivers or phone systems, static (also called *atmospheric*) is noise resulting from weather phenomena and related atmospheric electrical charges. Proper grounding can help reduce static interference. See interference, noise.

static declaration In programming, a static property, as might be associated with a variable, is one which is established prior to execution of a program.

static electricity An electrical charge at rest, familiar to many people as *static* electricity. Static electricity associated with rubbing amber (*amber* in Greek is *elektron*) was known to the Greeks by at least 600 B.C. when Thales recorded that a 'fossilized vegetable rosin' (amber), when rubbed with silk, acquired the property to attract very light objects to itself.

static generator An early experimental device used to investigate static electricity and its effects. Static generators still are valuable as educational tools. See Van de Graaf generator.

static object A software term for an embedded object that cannot be edited (other than to perhaps change its position in the document). Graphics are often embedded as static objects in word processing documents, and can thereafter only be deleted from the original document, changed by an external program and reinserted into the document. Contrast with linked object.

static RAM SRAM. A type of fast electronic random access memory chip which is often used in computers in conjunction with dynamic

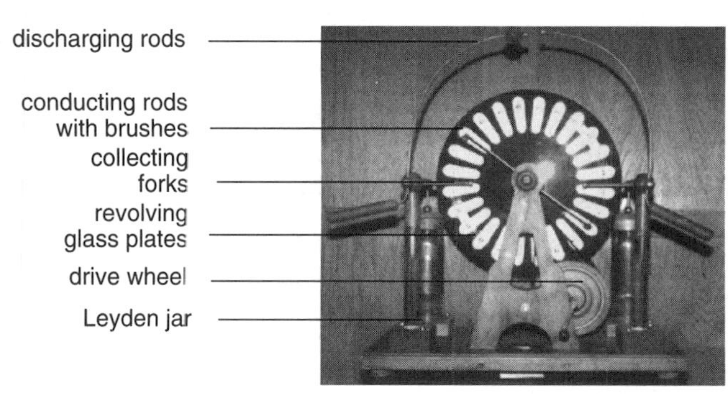

This excellent example of a static generating machine mounted on a base with Leyden jar condensors on either side is on exhibit in the Bellingham Antique Radio Museum.

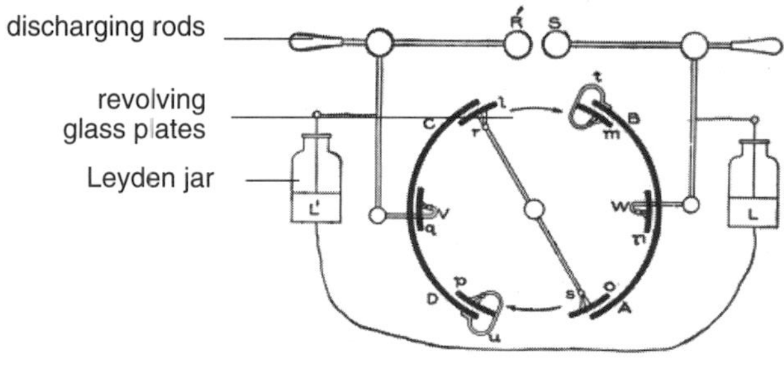

This schematic diagram shows the basic components of a static generator which functions essentially the same as the one shown in the photo above. A Leyden jar is positioned at each side, with the revolving plates in the center. The discharge rods are shown across the top.

RAM (DRAM) as they have different characteristics. Unlike DRAM chips, SRAMs do not need to be refreshed while in operation, thus providing fast access. As with many other types of memory chips, they require power to retain their information. See dynamic RAM.

static route A data transmissions path that is fixed and stored in a table or other form of database in a network router or high-end switcher (the distinction between routers and switchers is not as great as it used to be). Static routing is often faster than dynamic routing, but is not suitable for all types of installations. Static routing works well in small systems or those in which the routes are fixed and known, whereas dynamic routing is suitable for large, changing, distributed networks.

station A phone, computer, or other telecommunications device service office, console, or workstation.

station battery A battery used in early telephone switching stations which commonly provided 48 volts of direct current (DC).

station clock A centralized timing clock which provides a local reference for broadcast or other telecommunications functions carried on at that station. Timing information may be actively transmitted from the station clock to other equipment. In broadcast stations in North America, the idiosyncrasies of black and white TV caused discrepancies in playback speeds of different media, even though they might use the same 'time code' basis. Thus, the playing time of a broadcast may be considered 'compensated' when it has to be adjusted to match the station clock.

In networks, internal clocks and station clocks are used to provide timing information.

In very precise timing situations, such as in astronomical observatories, cesium-beam station clocks may be used for timing purposes.

stationnaires French semaphore operators in the late 1700s and 1800s.

STE 1. See Section Terminating Equipment. 2. Station Terminal Equipment. 3. Spanning Tree Explorer.

steady mark condition With telegraph keys arranged in series, an operator, when not sending, would close a switch to short the key contacts in order to leave the series unbroken. This put the idle line into a *steady mark* condition.

Steinheil, K. A. A Bavarian researcher who experimented with an Earth conductivity method to send wireless communications over distance in the 1830s. See conductivity method.

Steinmetz, Charles Proteus (1865-1923) A German-born American electrical engineer acknowledged for his genius in the investigations of lightning, alternating current phenomena, magnetism, and other discoveries which lead to the development of safer power distribution systems and better motors, generators, and electrical appliances.

step-by-step switch An early electro-mechanical automatic telephone switching system developed in 1920 and quickly favored by independents. It used rotating blades for setting switching connections. In competition with this, AT&T developed the first commercial panel switch in 1921. This type of switch was widely used until the mid-1970s, when crossbar switches superseded most panel switches and many step-by-step switches. Step-by-step switches must be modified to support touchtone dialing. See Callender switch, crossbar switch, Lorimer switch, panel switch.

Stibitz Complex Number Calculator The first relay calculator, released in 1939, developed at Bell Laboratories.

Stibitz, George A researcher at Bell Telephone Laboratories who developed relay arithmetic devices in the mid-1930s. This led to Stibitz's subsequent development, in collaboration with Samuel B. Williams, of an electromagnetic relay calculator, the Complex Number Calculator, which could manipulate complex numbers. By 1940, this machine had become the Bell Labs Model 1 and was incorporated into the telephone network in such a way that it could be remotely accessed, setting an early precedence for future network systems.

STL 1. Standard Telegraph Level. 2. Studio-to-Transmitter Link.

stock ticker A type of early stock reporting machine, somewhat resembling an anniversary clock with mechanical parts within a glass globe on a round base. This machine used a paper tape and telegraphic line hookup to provide fast reporting of stock activities with alphanumeric characters. In many ways the stock ticking machine is the ancestor to asynchronous telecommunications.

Stoll, Clifford Author of "A Cuckoo's Egg" (1989), an account of computer espionage by foreign infiltrators as experienced by Stoll, who was determined to track down the source of a tiny, but puzzling accounting error, and found much more.

Stoner, Don An electronics experimenter and amateur radio enthusiast who, in 1957, suggested the amateur construction of a relay satellite capable of two-way communications. This idea was in advance of its time, preceding the widespread use of electronic transistors and the construction of government two-way communications satellites and probably inspiring the development of the OSCAR satellites.

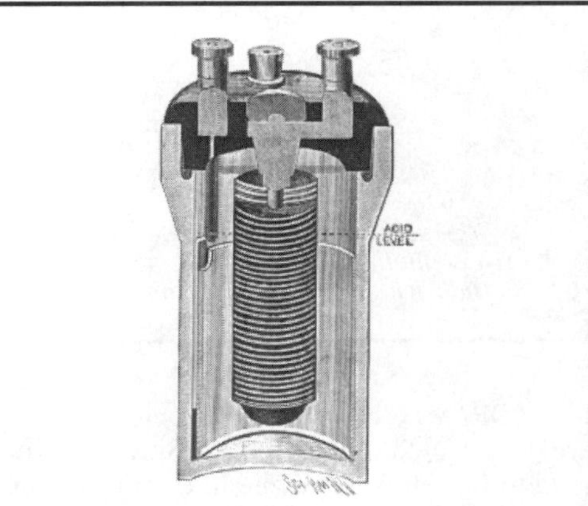

This zinc-lead storage battery was introduced by the United States Battery Company in 1900. It featured high voltage in a small size. The amalgamated coiled plate is zinc, fitting within a glass container. The lead is formed into suspended spongy, superposed peroxide plates. Scientific American, April 7, 1900.

storage cell, accumulator A secondary source of electricity, since it does not provide power immediately, but rather is charged up and then used. Car batteries are typically lead-acid storage cells that derive their power from a generator when the car is running, then store the power for later starting of the vehicle or operation of its electrical system when the motor is not running. See solar cell.

store-and-forward A technique for temporarily holding information until the conditions are right for transmitting the data to the receiver. This method is very common on data networks, where a router, local network, or individual machine may be offline or down. The data may be held indefinitely and transmitted when conditions are right (the right time, when traffic is lower, when the recipient is online, etc.), or may be bounced back to the sender after a certain interval or number of tries. It may even be abandoned, depending upon its nature and priority level.

store-and-forward repeaters Transmissions devices which store and forward information when conditions are favorable. In radio receiving and transmitting stations, both Earth and satellite stations, the conditions for transmitting a received signal may not be optimal right away, due to weather, political unrest, high traffic levels, or the movement of a satellite out of transmissions range. The message is thus not sent until conditions improve or the satellite comes into a favorable position in orbit, and then sent.

Storey, G. J. A scientist who first used the term *electron*, in 1891, to specifically describe an electric charge.

Storrer insulator A type of early utility pole insulator patented by L. W. Storrer in 1906, and first shipped by the Brookfield Glass Company in 1909. See insulator, utility pole.

STP 1. shielded twisted pair. 2. Signal Transfer Point. See Signaling System 7. 3. Spanning Tree Protocol.

STPC 6800 An early Motorola MC6800-based computer kit from Southwest Technical Products Corporation. It featured 2,048 bytes of static memory, a serial interface, case, and cover. The STPC was available in the fall of 1975, and sold for $450 without a monitor or keyboard. See Altair 680, SPHERE System.

strand A single long thread of uninsulated wire. When two or more of these strands are combined or twisted around one another in the same bundle, it is called *stranded wire*. Wire is stranded for a number of reasons; it can make it more flexible and it may alter the electrical characteristics of the wire for some particular purpose.

StrataCom A commercial supplier of ATM-related telephone networking products, particularly frame relay switching systems and network management control software based on open, standards-based interfaces for integrating with other vendors' products.

street price The price you might pay for a product for which you have nominally shopped around, as opposed to *suggested list* or *suggest retail,* which is the price the wholesaler or manufacturer has designated for the product and which is usually higher, due to dealer discounts. This suggested list price is often imprinted on the product packaging and overlaid with a lower dealer price. In other words, through discounts, volume deals, and competitive pricing, dealers often offer a street price which is lower than the list price. The street price may also be higher than list price. For example, a ticket for a popular rock concert sold by street hawkers an hour before showtime may have a street price that is two to ten times the list, or original ticket price.

stripping, image assembly In traditional page layout and printing, the process of positioning page composites, in the form of negatives or positives, on a flat in preparation for creating the printing plate.

strobe *n.* 1. High speed intermittent illumination. 2. Older term for an electronic flash. 3. In asynchronous communication, input of parallel data to a register or counter. 4. A momentary intensified sweep of a beam on, for instance, a scope. 5. On a computer bus, strobe lines indicate when data is being transferred.

Stroud, William A Canadian printer and inventor who developed new ways to print entire projects as a single run through a printing press, rather than as separate sections, as had been done prior to this time. His work resulted in a new type of belt press and he unveiled the Stroud, Bridgeman press in 1959.

S

Strowger, Almon B. An American mortician and inventor who created the first commercial automatic telephone switching system, a step-by-step switch patented in 1889, and dial-switch patented in 1891. This system allowed a subscriber to dial-connect a local call without going through a human operator. The first Strowger exchange was established in Indiana in 1892.

Strowger co-founded Automatic Electric in 1901, the largest telephone equipment manufacturer servicing Bell's competitors, the independent telephone companies. This was a successful fit, since Bell was creating its own switching technology, such as the panel switch, in competition with Strowger's technology, and the Strowger switch was somewhat unmanageable in large installations, a limitation that was only a minor problem when the majority of Automatic Electric's customer base was small independent telephone companies. See

Callender, Romaine; Lorimer, George and James; Strowger switch.

Strowger switch The first automatic telephone switch put into commercial service, in Indiana, patented in 1889 by Almon B. Strowger. Thus, direct dialing was born, and a human switchboard operator was no longer needed for connecting local calls. This also promoted a small revolution in phone design, since now dials were needed for callers to dial their own calls. The Strowger technology was further developed and put into service by the Automatic Electric company, co-founded by Strowger and directed by Alexander E. Keith. Surprisingly, the Bell system did not adopt the Strowger system until twenty years after its introduction. See Callender switch, Lorimer switch, panel switch, step-by-step switch.

Structure of Management Information A standard for mechanisms used for naming and describing objects for the purpose of network management. See RFC 1155.

STU Secure Telephone Unit. A telephone designed to include cryptographic protection for voice, data, and fax transmissions.

STU-3 Secure Telephone Unit 3. A secure telephone unit used for government communications. See STU.

STUN serial tunnel.

Sturgeon, William Credited with producing the first electromagnet in 1823.

SU subscriber unit. The device or system at the end of a circuit. This may be a phone, handset, or computer terminal.

Subnetwork Access Protocol SNAP. An evolution of the Logical Link Control (LLC) method, with backward compatibility with Ethernet, which facilitates communication of entities at a given network layer. SNAP was developed by IEEE to support multiple standard, public and private Network Layer protocols. SNAP expanded 8-bit SAP space to 40-bit (5 byte) protocol ID, and uses the first five bytes in the LLC Protocol Data Unit (PDU). SNAP supports more upper layer protocols than previous methods. It also allows Ethernet protocol type numbers to be used in IEEE 802 frames, to provide easy translation between Ethernet and IEEE 802 frames.

subscriber loop The circuit between the telephone company's central office and the subscriber station. In earlier times, the 'subscriber station' extended all the way to the phone, but more recently this demarcation point has been changed to the service box on the outside or inside of the premises to which the interior wiring usually attaches. It's still possible to get service right to the telephone, it just costs more.

Subscriber Network Interface SNI. One of the two interface ports of XA-SMDS systems which is used to connect an end user to the SMDS network. The other interface is the Intercarrier Interface (ICI). See Exchange Access SMDS.

SUMAC SuperHIPPI Media Access Controller. See HIPPI-6400.

Sun Microsystems Computer Company SMCC. A California hardware and software manufacturer established in 1982, Sun computer systems are commonly found in higher educational institutions, scientific research and medical imaging applications, and as servers for local area networks (LANs) in corporations, educational institutions, and Internet Services Provider (ISP) premises. Many of Sun's products are aimed at telecommunications applications for both voice and data. Sun's products cover a wide range from desktop systems to high-end research and supercomputing systems. The SunOS and Solaris operating systems are well known.

In 1996, Sun acquired Integrated Micro Products (IMP), including their fault-tolerant computer specifically targeted to the telecommunications industry, and Cray Research's high-end server system.

Sun's JavaSoft is the developer of the well-known Java object-oriented platform-independent general purpose programming language. JavaSoft collaborated with Lucent Technologies to develop a Java telephony application programming interface as part of a series of Java Media APIs intended to provide an open framework for Java applications development.

Sun Microsystems Inc. and Motorola Inc.'s Multimedia Group joined forces to develop products for cable operators to deliver high-speed data communications and Internet access to the home through Motorola's CyberSURFR (tm) cable modem.

Sun's XTL Teleservices for Solaris is a set of telephony software services and open application programming interfaces that extend Solaris LIVE, an integrated multimedia environment.

SunXTL A teleservices product delivery vehicle developed by Sun Microsystems, known as Sun XTL Teleservices Platform for Solaris.

SunXTL provides Teleservices development support for applications intended to run on personal workstations. The types of teleservices which can be implemented with this technology include integrated voice mail, answering machine, automated dialing, faxing, etc. Because these are generated within the computing environment, they can be integrated with input and output from word processors, address books, databases, and spreadsheets.

SunXTL is a foundation library for telecommunications-related applications, which includes call control functions, data stream access methods, and data flow control.

SunXTL API A SunXTL Teleservices object-oriented applications programming interface which allows the development of personal desktop applications with C++, including on-screen phone graphical user interfaces, remote workstation access, personal voice mail, etc. which work through telephony hardware peripherals.

SunXTL Call Objects The SunXTL API provides developers with C++ *XtlCall* objects to control various aspects of a telephone call, including querying the call state and the numbers associated with the call, the call's current status, and its data type or media class. It can also request a change in call state. The XtlCall objects also have callback methods for the asynchronous notification of state changes.

SunXTL Provider Configuration Database The SunXTL Teleservices configuration database is a repository for installed providers. The database provides information on each provider and how to invoke it, and lists its characteristics and capabilities. The database describes telephony resources such as available bandwidth, number of available lines, types of voice services available, etc. A graphical user interface (GUI) tool *xtltool* is provided for browsing and editing the Provider Configuration Database.

SunXTL Provider Interface A SunXTL Teleservices open interface which provides a means for third-party developers and Independent Hardware Vendors (IHVs) to use the Provider Library to ensure compatibility and compliance with basic system protocols. This message set can be extended to add user-specific features. The provider interface fits between the server and/or datastream multiplexer and the various drivers.

SunXTL Provider Library A SunXTL Teleservices library which works in conjunction with the Provider Interface to keep the Provider information distinct from system services. The library provides interfaces to various data streams and services, including the provider database and various server functions.

SunXTL Server The SunXTL Teleservices server provides administration, message passing, and security to networked personal workstations running SunXTL Teleservices applications.

SunXTL System Services The SunXTL Teleservices System Services provide an intermediary between the application view of a call object and the provider's implementation of the call. Interprocess message passing, object identification and creation, call ownership, and security are handled by the server.

super group In analog voice phone systems, a hierarchy for multiplexing has been established as a series of standardized increments. See voice group for chart.

super server A high end server which consists of a number of computers networked together with communications links which are as fast, nearly as fast, or faster than the processing speed of any one individual computer, so the collection functions as a fast, integrated, distributed 'unified' entity. With very fast transmissions media and protocols like HIPPI and SuperHIPPI, the distinction between individual machines becomes less critical, and the processing algorithms for carrying out the tasks are more crucial to the concept of the system as an 'organism.' A super server can also be a single machine with multiple CPUs, set up to function together to handle higher end processing requests at faster speeds, or of greater complexity than might be achieved with a typical one-CPU system. A number of interesting distributed processing 'supercomputing' applications have been configured at several U.S. research labs using Linux running on personal computers communicating through fast network links.

Super Speed Calling A telephony subscriber option, which is essentially the same as Speed Calling in that it allows an abbreviated set of characters to be dialed to invoke a longer number. The distinction is more of a marketing distinction to describe enhanced systems where a name can be entered, which is easier to remember, rather than just a number (usually four digits or characters). See abbreviated dialing.

super video graphics array SVGA. A graphics standard common on International Busi-

S

ness Machines (IBM) and licensed third party computers, supporting a variety of palettes and resolutions, including 800 *x* 600; 1,024 *x* 768; 1,280 *x* 1,024; 1,600 *x* 1,200; 1,024 *x* 768 (16 or 256 colors). See video graphics array.

supercomputing A term applied to high-end computing applications provided on the best hardware/software available at any particular state of the technology.

Supercomputers originated sometime in the 1950s, when the viability of computers as a commercial item became apparent. Some of the earliest supercomputers were designed by International Business Machines (IBM) and shipped in the early 1960s.

The supercomputers of thirty years ago had fewer capabilities and were slower than many hand-held calculators of today, (they were also much larger in physical size). The definition of supercomputing is thus a relative one, since most desktop computer systems now are faster and more powerful than mainframes which were running multiuser networks in many universities and colleges fifteen years ago. Supercomputers tend to be characterized by faster (or multiple) CPUs, wider data buses, faster network links, and larger, faster-access storage devices than are available as consumer products. They also may be run with more sophisticated distributed processing algorithms, although the writing of parallel applications is an art and there is still a lot of research and discovery yet to be done in this field.

Supercomputers tend to be used in scientific research and military applications.

superframe In its generic sense, superframe is used to describe a period in time during which a specified number of downstream and upstream frames are transmitted. Thus, the transmission time of a superframe will be related to the bit rate. The concept of the superframe is used in the context of frame timing and alignment.

SuperFrame standard SF. A 1969 improvement to the original 1962 DS-1 standard for a frame format for 1.544 Mbps transmissions (2.048 in Europe with 30 channels) which improves the signal-to-noise ratio and combines 12 frames into one SuperFrame. Frames 6 and 12 are used for robbed-bit signaling. This has since been superseded by Extended SuperFrame, which provides increased error detection and removes the need to take down an entire line for servicing. See Extended SuperFrame.

superheterodyne receiver An early improvement in radio receivers designed to be more sensitive than radio frequency receivers of the time. The superheterodyne receiver incorporated a signal detector working in conjunction with a local oscillator to mix the signals, producing an intermediate frequency which was then amplified and passed on to a second detector, and from there to the earpiece. The superheterodyne circuit was invented by "The Father of FM," Edwin Howard Armstrong. See heterodyne.

SuperHIPPI See HIPPI-6400.

SuperHouse A trademark of BellSouth, to signify a house which is designed with information services resources built in (conduit, wiring, etc. to support computing and Internet applications). This is often also referred to as a "smart house" and, in fact, SmartHouse has been trademarked by the National Society of Home Builders.

SuperJANET See JANET.

superparamagnetic Phenomena which contribute to magnetization and signal decay of magnetically recorded information over time, thus limiting the useful lifespan of magnetic recording media. The density of recording information is related to the superparamagnetic effects, as well, resulting in a practical superparamagnetic limit. Studies for arranging magnetic data in particle array systems to study superparamagnetic effects to develop practical schemes and optimize recording density are being carried out at the IBM Research labs.

superpose 1. To place or lay over, with or without contact with that overlaid. 2. To overlay upon another such that all like parts of the overlay coincide with the overlaid.

superposition principle A principle which can be applied to networked electrical circuits to solve current values in individual branches of the network. See Kirchoff's laws.

superstation A television broadcast station whose signal reaches a very wide audience by being retransmitted beyond what would be possible by standard airwaves. The extended viewing area is often reached by a satellite transmission which is further extended through cable.

supertrunk A high-end data transmissions cable system which carries multiple high bandwidth services such as several video channels.

support The provision of additional products, peripherals, or technical expertise to support a customer system or service. In the software industry, technical support for installing and using the software is routinely provided free, or for a fee, from the developer. A small amount of support may also be available from the dealer, depending upon the complexity of the system. Peripheral support for computing products includes repair and replacement of accelerators, upgrade chips, memory expansion interfaces or chips, graphics display controllers, and hard drive controllers.

surf *colloq.* A common term for riding a wave, a technology, a trend, or other force or medium. *Crowd-surfing* happens at rock concerts when someone is thrown across the top of the crowd and carried from hand-to-hand above people's heads. Traditional *wave surfing* is catching a large ocean or river wave with a surfboard, and riding it as far (and well) as can be accomplished. *Channel surfing* describes a television watcher who uses a remote control device to skim programs from channel to channel, particularly during commercials. This is done with the hope of finding better programming (or as thumb exercises for couch potatoes). *Surfing the Net* means to travel, in the virtual sense, through the myriad resources and sites on the Internet, especially through the Web, a graphical interface to the Internet. See browser, Internet, World Wide Web.

surge Large, sudden changes in a circuit current or voltage. See burst.

surge protector, surge suppressor A device which conditions or filters a current to provide a constant level of power. It is placed where it can provide protection to subsequent devices or components in the system. Ground lines are sometimes used to drain off the surge.

For computer systems, a surge protector resembles a fat extension cord and can provide protection from power fluctuations which might damage the electronics. Some limited protection from lightening storms may be possible with a surge protector, but a direct hit will likely damage both a simple consumer surge protector and the system to that it is connected. Surge protectors are generally a good idea, especially with laptops which get plugged into a lot of circuits of dubious nature (motel sockets, ferryboat sockets, etc.).

SUT 1. System Under Test. See ATM. 2. Service User Table. A telephone use authorization term.

Sutton, Clifford A Canadian film maker attributed with creating the first commercial motion picture in the early 1900s. See motion pictures.

SVC See switched virtual connection.

SVD simultaneous voice and data.

SVGA See Super Video Graphics Array.

SWAC Standards Western Automatic Computer. A historic large-scale computer, undertaken by the U.S. National Bureau of Standards

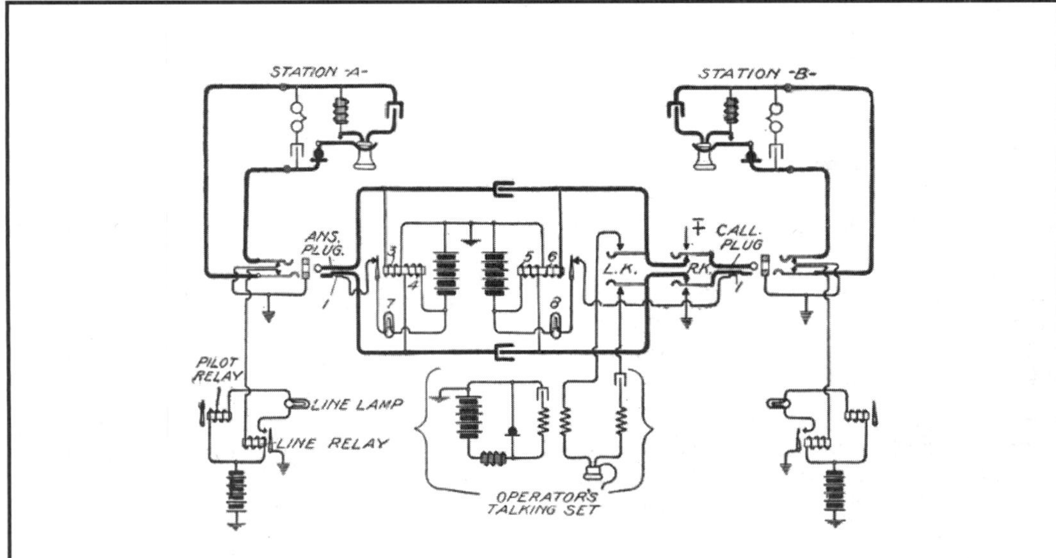

This historic schematic diagram shows the layout of a basic telephone common battery system with two subscriber lines in contact with one another through a manually operated cordboard switchboard.

for its internal use, in the 1940s. It was developed for installation on the west coast. A similar system was built at the same time for the east coast.

Swiss army knife An indispensable geek tool for anyone who happens to need a screwdriver (or pair of scissors, or corkscrew) to open up a hard drive, CD-ROM, pizza box, etc.

switch *n.* 1. A mechanical or electrical device that breaks or completes a path in a circuit, or changes the path. See switcher. 2. An electronic circuit designed to carry out a logic operation. New switches have capabilities that were once found only in routers, and some can do switching at the application level (fourth layer). 3. In software, a means to direct a routine; a branch.

switch hook See hook switch.

switchboard In its most general sense, any device into which a number of incoming and outgoing circuits are routed, where the routing of the individual circuit connections can be changed manually, mechanically, or electronically.

Human-operated telephone *cordboards* are probably the most picturesque of the various historic switching boards. The earliest ones required that a foot-pedal be pumped to generate the power to ring the subscriber's phone. A manual telephone cordboard was often built into a wall, and was hand-connected with simple jacks and cables. In early telephone history, young men were hired to staff switchboards, as women were not permitted to work in most clerical positions, but the arrangement had problems. Some of the male employees were rude to callers, chewed tobacco while talking, and used excessive profanity. As a result, in the mid-1880s, women were hired, and eventually replaced men entirely until the late 1970s when a few male operators re-entered the field.

Most male and female switchboard operators are now being replaced by automated switchboard systems with voice recognition and touchtone menu dialing functions.

The large panels of switch connections in early cordboards and switchboards have been replaced by multiline phone consoles in many businesses, although some large phone installations or central switching services still have wall panels. Even here, human receptionists are becoming rare, with many small businesses adopting computer voicemail services instead.

switchboard cable This has two meanings. It was originally a patch cable which was used in old manual cordboard telephone systems to patch two circuits together to create the end-to-end connection for a phone call. Now that manual switchboards have been replaced by automated switching systems, the switchboard cable is considered to be the one which connects a central office switchboard with an associated automated system, such as a computer.

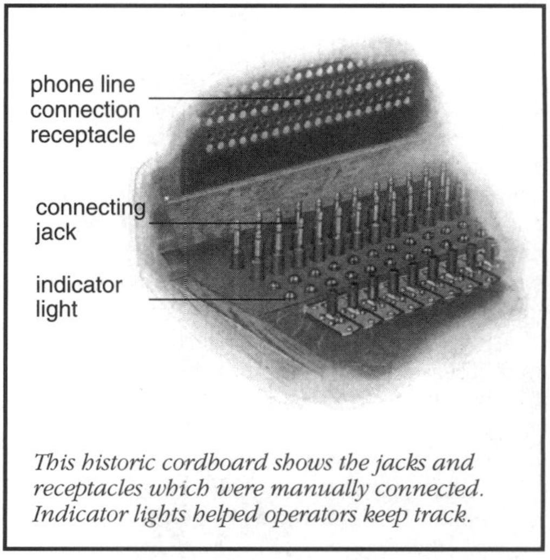

This historic cordboard shows the jacks and receptacles which were manually connected. Indicator lights helped operators keep track.

Switched 56 The name of a 56 Kbps switched network voice/data service provided by some local telephone companies which allows calls among several points through one pair or two pair copper wires. Switched 56 can be used for voice, file transfers, Internet access, facsimiles, connections to other local area networks, and videoconferencing. This service is gradually being superseded by ISDN services. See DS-0.

Switched Multi-Megabit Digital Service SMDS. A high-speed wide area networking (WAN), connectionless, packet-switched cell relay transport service based on IEEE 802.6, offered by telephone service providers. It provides capabilities to interconnect LANs (Ethernet, Token-Ring, etc.) and WANs through public switched telephone networks (PSTNs).

SMDS can be integrated with transmission technologies such as ISDN, DS-x, and frame relay with associated bandwidths ranging from 56 kbps or 64 kbps to 34 Mbps or more. It works with asynchronous, synchronous, and isochronous data and can be used over optical fiber. It can provide congestion control to protocols such as frame relay which don't have congestion control as an intrinsic part of their specification or implementation. See cyclic reservation multiple access.

switched virtual connection, switched virtual circuit SVC. A generic term for a logical communications connection. In ATM systems, there are two types of SVCs: switched virtual path connection (SVPC) and switched virtual channel connection (SVCC). SVC provides on-demand connections between communicating end systems. Using signaling software, the Virtual Path Identifier/Virtual Channel Identifier (VPI/VCI) information will be dynamically allocated to the participating end systems.

Switched Voice Service SVS. The standard service offered with FTS2000. See FTS2000.

switcher An audio/video component which provides easy reconfiguration of several circuits. In a sense, the button on your receiver that lets you select between CD, phono, or tape is a switcher, although the meaning is more often ascribed to a separate component with a number of connectors, inputs and outputs. See A/B switchbox, switch.

switching The three types of switching commonly used in telecommunications networks include message switching, circuit switching, and packet switching. Message switching is a system where the entire message is relayed, intact, through a variety of nodes or service points, from the sender to the recipient. Circuit switching is commonly used in end-to-end direct communications, as in telephone connections. Packet switching involves the segmentation, routing, and reassembly of communications so different parts of the messages may be transmitted at different times and through different routes. While this may not sound very efficient, it is an excellent way to manage data in a large, dynamic, distributed environment, and provides many possibilities for data sharing, representation, authentication, and filtering as desired. The Internet is built on packet switching concepts. See circuit switching, packet switching.

switching simulation The simulation on a computer, usually with graphical output, of the circuitry and operations of a switching system. Hayward and Bader did early computer simulations of telephone switching networks in 1955.

SWR See standing wave ratio.

SXGA super-extended graphics array.

symbol A character, icon, or other agreed-on abbreviated representation, useful in representing objects, quantities, languages, arithmetic and logic operations, rules, layout schemes, qualities, sounds, ideas, and others. Examples of symbols include street signs, logos, musical notes, electronics diagram elements, computer buttons and icons, words, punctuation marks, and arithmetic operators.

symbolic Representative of something else, or of some greater meaning, usually in an abbreviated form. See symbol.

symbolic code A computer code which represents programs in source language, with symbolic names and addresses, in contrast to machine code, which has hardware-specific names and addresses. Use of mnemonic symbols in higher level languages aids in programming and debugging, and symbolic names and addresses can be used to increase portability to other platforms. See assembly language.

symbolic debugger An essential software programming and debugging tool used to control and monitor the application under development. A debugger allows you to step through execution of the code, set breakpoints and temporary branches, to determine and change the contents of variables, and view the design and functionality of software-in-progress. A symbolic debugger eases the task of finding code segments (many current software applications have tens of thousands, and even hundreds of thousands of lines of code) by jumping to a symbol, such as a label name, so that cryptic assembly listings don't have to be searched.

S

symbolic language A computer programming language which uses mnemonic symbols, rather than machine code or actual hardware names and addresses, in order to make the process of creating the code more comfortable for humans and easier to read and debug. See symbolic code.

symbolic logic A written symbol language developed to express logical and mathematical concepts and arguments in a way that is more specific to itself and less ambiguous than natural human languages.

synchronize 1. To cause to occur at the same instant in time. 2. To precisely match two waves or two functions.

synchronous Signals with the same timing reference and the same frequency. See isochronous, asynchronous.

Synchronous Data Link Control SDLC. In IBM Systems Network Architecture (SNA) systems, SDLC is a bit-oriented, link-level protocol which provides a means of moving data between network Addressable Units (NAUs). The Data Link Control layer lies between the higher layers and the Physical Control layer and communications links, and passes information through.

SDLC is a subset of the High-level Data Link Control (HDLC) protocol.

SDLC is packet-oriented, with each frame comprising a header, information, and trailer. It transparently provides flow control, multipoint addressing, error detection, and multimessage capabilities. See Systems Network Architecture.

Synchronous Optical Network See SONET.

Syncom 3 The first geostationary satellite, designed for telecommunications use, launched on 19 August 1964 by the United States.

syntax 1. The rules of structure or grammar for a language, natural or computer. 2. In programming, the words and symbols which are valid within the structure and scope of a computer language. Debuggers and compilers will provide a general syntax error alert to the developer when it encounters unrecognized words/symbols, or, sometimes, of word/symbol positions or combinations. (In other words, it will flag spelling errors as syntax errors, for example.)

syntax error An error in a programming statement which indicates an unrecognized word, symbol, or structure. Syntax errors are usually flagged and displayed by debuggers and compilers so that the error can be corrected. See syntax.

synthesized voice A mechanically or electronically generated speaking system. Synthesized voices are now used on phone systems, computers, and certain publica address systems. The voice may be constructed from recordings of natural human voices, pieced together electronically from sound samples, or may be entirely synthetic. See speech synthesis.

Sysop system operator. The Sysop, or systems administrator, sometimes also called the *super user*, due to his or her higher access privileges and power over other accounts on a system. The system operator is a technical expert with high security privileges, managing a bulletin board system or computer network. On small systems, like BBSs, these tasks may be performed by one person. On medium-sized systems, sometimes assisting sysops are assigned intermediary privileges, between those of the super user and regular system users. Larger systems often split the installation, security, administration, file management, diagnostic, and tuning responsibilities of a network among a number of system operators, or may even have an entire facility devoted to the administration of the network.

system disk A disk which includes 'boot' information, that is, low-level operating information, from which a computer system can be started. This may be a floppy drive, hard disk, CD-ROM, or other disk with system files which are somewhat transparent to the user. Without certain system files, a computer cannot configure itself to recognize peripherals, available memory, monitor types, etc. See bootstrap, operating system.

system integrator A commercial vendor offering a variety of network design and implementation services according to the configuration needs of various customers.

system reliability architecture SRA. Systems which are designed as fault tolerant, reliable, and which function even while undergoing maintenance checks and procedures. SRA implies systems which incorporate redundancy, the ability to *hot swap* components, fast recovery from power failures, and online upgrading of software.

Systems Network Architecture SNA. One of the first significant layered architectures, developed in the 1970s by the International Business Machines (IBM) Corporation. It shares many common overall concepts with the Open Systems Interconnection (OSI) model which evolved at about the same time, however, the two are not directly compatible.

Layered architectures like SNA were developed when computers became smaller and less expensive, resulting in an increase in mass production and a greater variety of hardware configurations and operating systems. Thus, interconnectivity and specialization challenges were posed and new market opened up. Layered architectures were a practical way to resolve these needs.

SNA provides a cohesive way for users to communicate between systems, for transmitting and receiving, by specifying the operating relationships of various components of the different systems. To achieve this, various communications peripherals, i.e., adapters, modems, data encryption devices, etc. are designed to be consistent with the implementation of the SNA specification. See Token-Ring.

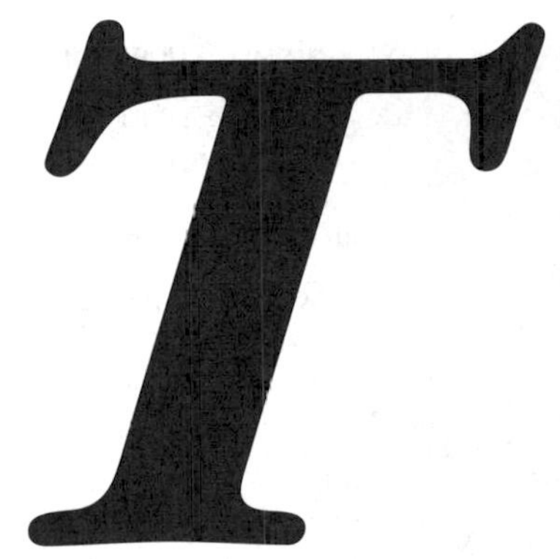

T carrier Any of a number of multiplexed carrier systems, commonly known as T1, T2, and T3, widely used for fast, wide bandwidth data transmissions. T carriers are capable of transmitting full duplex digital signals. See the chart below, and see listing for DS-.

T1, T-1 A communications system which can be carried over ordinary twin cable pairs, or fiber optic, which provides significant speed and bandwidth improvements over earlier technologies. The T1 time division multiplexing (TDM) pulse code modulation (PCM) system is capable of carrying multiple simultaneous conversations (24 over twin cable pairs), and began to be incorporated into central office trunk switching technologies in the 1960s with more significant, widespread implementation to subscribers in 1982.

The capacity of T1 was originally stated as 1.544 Mbps (U.S., Canada, Japan), although European ITU-T standard implementations are faster, 2.048 Mbps, and upper limits tend to change as more efficient techniques are incorporated to improve the throughput of a system as a whole. It is a low loss transmissions system when delivered over fiber optic cable, but is subject to cross-talk in long metal wire installations.

Pulse code modulation was pioneered in the 1940s and 1950s, but it was not until the early 1960s that some practical success was achieved. Proponents wanted the improved transmissions technologies to be compatible with existing switching systems, potentially saving billions of dollars by using, rather than replacing, central office switching circuitry.

While high capacity 22-gauge is preferable for T1 transmissions, the system is not limited to this and can work with a number of paper or plastic insulated cable pairs, or staggered twist cable which has been in use in telephone systems for decades. A single cable can handle

T Carrier Transmissions Systems Overview

Type	Signal	Bandwidth	Typical cable	Notes
	DS-0	64 Kbps		
T1	DS-1	1.544 Mbps	19, 22, or 24 gauge	Originally developed for digital voice, also used for data communications. European E1 standard is similar.
T1C	DS-1C	3.152 Mbps	19, 22, or 24 gauge pairs	
T2	DS-2	6.312 Mbps	low capacitance	
T3	DS-3	44.736 Mbps	fiber optic or microwave	
T4	DS-4	274.760 Mbps	fiber optic or microwave	High-speed, high bandwidth applications.

up to almost 5,000 channels.

Due to its cost, T1 is still primarily used in large installations, phone trunks, government installations, campus backbones, and medium and large enterprise networks.

T connector A generic name for any type of cable connector which is shaped roughly like the letter "T." This is a common shape for splitters, Ethernet cabling, certain types of terminators, and adaptors.

T-Bone A freely distributable Java-based distributed network broadcasting system which permits interconnection of JavaBeans on different systems. It can be used for remote monitoring and control, 'push' channels on the Internet, stock ticking feeds, and online discussions. See Java, JavaBeans.

TAB tone above band.

table Collection of ordered data, often stored in arrays or printed in columnar form.

tablet Hardware peripheral input device, usually rectangular, with an associated mouse or stylus. Graphics tablets are frequently used with engineering and CAD/CAM software.

TABS Telemetry Asynchronous Block Serial Protocol. A medium-fast AT&T network protocol.

Tabulating Machine Company Herman Hollerith, who developed Hollerith punch cards to store census data from the 1890 U.S. census, founded the Tabulating Machine Company in 1896 to market his products. Eventually, the company was sold out to the Computer-Tabulating-Recording company, which, in turn, became International Business Machines (IBM) in the 1920s.

TACD Telephone Area Code Directory.

TACS total access communications systems. A designation for the analog cellular radio system in the United Kingdom, similar to the U.S. AMPS system.

Tag Distribution Protocol A protocol which builds routing databases for the handling of tagged datagrams that are accessed by Tag Switches and Tag Edge Routers.

Tag Image File Format TIFF. A very widely used platform- and application-independent, lossless, color, raster image file format. The TIFF format is used in faxes, image processing programs, scanner files, and many graphics creation programs. It is well supported by service bureaus and the printing and graphics design industries. Files are often identified by the *.TIF* or *.tiff* file extensions.

Creation of the format took into consideration the needs of the desktop-publishing industry and other related graphics applications, with the goal of making image information broadly interchangeable. TIFF was created to be extensible, so that it may accommodate future needs.

TIFF has gone through a number of major revisions but, in general, fields are identified with unique tags so that various applications can elect to include or exclude particular fields depending upon their needs and capabilities. A TIFF file consists of three main parts: an image file header, a directory of fields, and the file data. See facsimile, scanner, TWAIN.

Tag Switching A Cisco proprietary method of using Tag Edge routers and assigning software *tags* to each IP datagram in a sequence in order to identify router paths. With the goal of reducing the time needed for each router to send datagrams across the network to speed throughput, the Tag Edge routers append a special string of bits, the *tag*, to the datagrams before they are transmitted across the backbone. This tag provides routing information to other routers (Tag Switches) so they are freed from table lookups and processing. It is similar to IP switching, except that nonstandard tag bits are appended. See IP switching, Tag Distribution Protocol.

TAI See International Atomic Time.

tail circuit A final segment in a connection between a central switching location and the subscriber.

talking battery In early telephone central offices, a 24 or 48 volt battery was used for supplying the power for a phone conversation. Later, starting around 1893, these were replaced by *common batteries* at the central office which supplied a 'talking battery' to each subscriber. See battery.

tandem Two, dual, pair. Acting together, in conjunction with, partnership.

Tandem Connection Tandem connections are used at the discretion of network carriers, and software with TC are mostly inter-office network applications rather than general public subscriber network applications. In SONET, the Tandem Connection layer is optional.

Tandem Connection Overhead TCO. In SONET, an optional overhead layer between the Line and Path layers as defined in ANSI T1.105. The layer deals with the reliable network transport of Path layer payload and its associated overhead.

TANE The Telephone Association of New England. A regional association providing information, education, and support to its members. http://www.tane.org/

Tandy Color Computer TRS-80C, "CoCo". A Tandy Corporation computer introduced in spring 1981, aimed at the home market, following up on the TRS-80 Models I and II, which targeted business users. Called CoCo for short by most of its users, the computer featured a Motorola 6809E microprocessor, 4 kilobytes of RAM expandable to 64, color graphics (256 *x* 192), two joystick ports, a cassette interface, and an RS-232 serial interface, for a list price of $399.

tangent galvanometer An early current-detecting instrument employing a card to record the degree of deflection. This type of galvanometer is subject to interference from the Earth's magnetic field. See galvanometer.

TAO Project TAO is part of the Satori project being carried out at Washington University. It is being develeoped by the Distributed Object Computing group, funded partly by the DARPA Quorum program. It is a high-performance, realtime Object Requester Broker (ORB) designed to provide end-to-end network quality of service (QoS) guarantees to applications by integrating CORBA middleware with operating system input/output subsystems, communications protocols, and network interfaces. TAO is freely distributable to researchers and developers within ACE copyright restrictions. See CORBA.

tape recorder A serial data recording instrument, commonly used for music, dictation, talking books, and computer data backup, which is descended from the Danish inventor Valdemar Poulsen's *telegraphone* patented in 1890. The telegraphone was initially used for write-once/read-many dictation recordings. Through various improvements by different contributors, the magnetic tape recorder as we know it was developed in the early 1930s. One of the important related inventions was contributed by Fritz Pfleumer of Germany, who patented a way of applying magnetic powder to paper or film.

By the late 1940s, experimenters were scrambling to find a way to record video images on tape, and the early systems were demonstrated by the mid-1950s.

Tape recorders were used experimentally to record live music concerts between 1936 and 1948, and the music recording industry boomed. By the 1960s, audio tapes were common in homes, businesses, and automobiles. See telegraphone.

TAPI See Telephony Application Programming Interface.

tariff Scheduled rate or charge between a carrier and its subscribers, usually published, and sometimes regulated by government agencies.

TAS Telecommunication Authority of Singapore.

TAXI Transparent Asynchronous Transmitter/Receiver Interface.

Taylor, A.H. A member of the U.S. Naval Research Laboratory who observed in 1922 that a radio echo from a steamer could potentially be used to locate a vessel. This observation was not put into practical use until some years later. See radar.

Tc See Committed Rate Measurement Interval.

TCAP See Transaction Capability Application Part.

TCI Telecommunications, Inc.

TCIF Telecommunications Industry Forum.

TCM trellis code modulation. See trellis coding.

TCO See Tandem Connection Overhead.

TCP/IP Transmission Control Protocol/Internet Protocol. An internetworking transmissions protocol combination developed in the 1970s on the ARPANET to enable the intercommunication of various types of computers across wide area networks (WANs). It was widely adapted by educational institutions by the 1980s and by corporations by the 1990s. Although it appeared for a time that Open Systems Interconnection might overtake TCP/IP, TCP/IP has now become an international standard which has been implemented on most microcomputers since the mid-1990s.

TDM See time division multiplexing.

TDMA See time division multiple access.

TDRSS Tracking and Data Relay Satellite System.

te wave transverse electric wave.

Team OS/2 A strong, independent, international support and advocacy group for IBM's OS/2 (Operating System/2). It provides education, demonstrations, resources, Web links, and other services for OS/2 users and the general community. OS/2 users have a strong enthusiasm for this operating system, and many members independently advocate that it is superior to other commercial Intel-based personal computer operating systems. However, the official position of Team OS/2 is that OS/2 is strong enough to stand on its own merits without comparisons to other systems. Team OS/2 was founded by Dave Whittle shortly before the release of OS/2 version 2. The Team OS/2 Web site has further information on the formation and goals of the organization, and lists of other OS/2-related links on the Web. See OS/2. http://www.teamos2.org/

Technical and Office Protocols TOP. A protocol development effort to support the needs of engineering and office environments. TOP was initiated by Boeing, and is now part of the Manufacturing Automation Protocol/ Technical and Office Protocols (MAP/TOP) users' group. TOP was designed to conform to the Open Systems Interconnection (OSI) model.

telautograph A type of early telegraph machine invented in the late 1800s, which could transmit handwriting over short distances. The earliest models used a pen writing a continuous line, and did not leave breaks between letters or words. Subsequent improvements were made by E. Gray, F. Ritchie, and others, which allowed the pen to be lifted off the paper when desired. These devices were used for several decades. Modern versions of the telautograph, using electronics, are now known as *telewriters*, and they have only recently been superseded by facsimile machines.

Telco, TelCo Abbreviation for telephone company, a local or regional telephone carrier.

telebusiness British counterpart of telemarketing, teleresearch, and telesales.

telecom *abbrev.* telecommunications.

Telecom Developers A telephony industry trade show, the precursor to the Computer Telephony Conference and Exposition, held regularly in the spring.

Telecommunication Standardization Advisory Group TSAG. A division of the International Telecommunications Union which interprets global standardization concepts and goals into practical implementations.

telecommunications 1. Meaningful wired/ cabled or wireless transmission and receipt of signals over distance. 2. Broadcast, telegraph, phone, and computer network communications, frequently with a 'give-and-take' quality or by choice of the receiving party, carried through a variety of media, including wires, fibers, air, etc. 3. The term is sometimes used to indicate a broader scope of communications *telephony,* to include video, for example (although telephony's meaning is not quite as narrow as thought by some). 4. This is defined in the Telecommunications Act of 1996, and published by the Federal Communications Commission (FCC), as:

"... the transmission, between or among points specified by the user, of information of the user's choosing, without change in the form or content of the information as sent and received."

See Federal Communications Commission, Post Roads Act, Telecommunications Act of 1996.

Telecommunications Act of 1996 This is the first substantial overhaul of telecommunications regulations since 1934, signed into law by President Clinton on 8 February 1996. The goal and intent of the law is to enable open access to the communications business and to permit any business to compete with any other telecommunications business. The chief impact of this act is on phone and broadcast services. Regulatory responsibility is largely shifted away from state courts and regulatory agencies to the Federal Communications Commission (FCC), but much of the administrative workload remains with the state authorities.

Some of the changes include the lifting of some long-standing restrictions, with the Regional Bell Operating Companies (RBOCs) now permitted to provide interstate long distance services. Telephone companies can now provide cable television services and cable companies can now provide local telephone services.

The FCC and individual states are responsible for implementing the terms, and the FCC has published an implementation schedule for this important act regarding the various issues of interconnection, universal service, access, assignment of broadcast licenses, etc. See Above 890 decision, Commercial Space Launch Act of 1984, Communications Act of 1934, Federal Communications Commission.

Telecommunications and Customer Service Foundation TCSF. A Canadian-based association formed to promote excellence in customer service in the telecommunications industry.

Telecommunications and Information Infrastructure Assistance A United States Department of Commerce grant program established in 1994 to assist local government and nonprofit organizations in funding projects which contribute to the design and development of the national information infrastructure (NII).

telecommunications broker An entity (person or business) which assists in negotiating contracts for communications services on the part of a user, or one which purchases specialized or bulk telecommunications services with the intent of reselling these services to consumers, sometimes at discount rates. See broker.

telecommunications carrier This is defined in the Telecommunications Act of 1996 and published by the Federal Communications Commission (FCC), as:

"... any provider of telecommunications services, except that such term does not include aggregators of telecommunications services (as defined in section 226). A telecommunications carrier shall be treated as a common carrier under this Act only to the extent that it is engaged in providing telecommunications services, except that the Commission shall determine whether the provision of fixed and mobile satellite service shall be treated as common carriage."

See Federal Communications Commission, Telecommunications Act of 1996, telecommunications carrier duties.

telecommunications carrier duties The Federal Communications Commission (FCC) stipulates a number of duties in the Telecommunications Act of 1996 as follows:

"Each telecommunications carrier has the duty—

'(1) to interconnect directly or indirectly with the facilities and equipment of other telecommunications carriers; and

'(2) not to install network features, functions, or capabilities that do not comply with the guidelines and standards established pursuant to section 255 or 256."

telecommunications closet A wiring panel or room or other centralized, secured or separated administrations center for equipment junctions and/or demarcation points. Larger systems may have a series of panels, punch-down blocks, racks or other furnishings to secure and organize the wiring system. See wiring closet.

Telecommunications Development Bureau BDT. An agency established as a result of the Plenipotentiary Conference in Nice, France in 1989 to set up technical assistance in developing countries for coordinating, standardizing, and regulating telecommunications in third world countries. France is the location for many regulatory and standardization bodies. BDT activities began in 1990.

telecommunications equipment This is defined in the Telecommunications Act of 1996, and published by the Federal Communications Commission (FCC), as:

"... equipment, other than customer premises equipment, used by a carrier to provide telecommunications services, and includes software integral to such equipment (including upgrades)."

See Federal Communications Commission, Telecommunications Act of 1996.

telecommunications facilities The integrated structures and equipment which enable telecommunications to be conducted and managed. This may include secure rooms for servers or patch bays, consoles, PBX systems, satellites, telephones, facsimile machines, modems, wires and cable, video cameras, radio transceivers, etc.

Telecommunications Industry Association TIA. TIA began in 1924 as a small group of communications suppliers. Later, the group became a committee of the U.S. Independent Telephone Association (USTSA). This group split off from the USTSA in 1979 to become a separate, affiliated association, and TIA was formed in 1988 through a merger of USTSA and the Information and Telecommunications Technologies Group of EIA

A national trade organization representing about 1,000 member companies which provide communications and information technology products, materials, and services. TIA provides a forum for discussing industry information and issues, organizes industry trade conventions, and serves as a voice for manufacturers and suppliers of communications products for matters of public policy and international commerce. http://www.tiaonline.org/

Telecommunications Information Network Architecture TINA. A common architecture for building and managing communications services developed by the Telecommunications Information Network Architecture Consortium in the early 1990s. This architecture logically separates the physical infrastructure and the applications from the need to communicate directly with each another. Control and management functions are integrated. Control and management functions can be

T

placed on the network independent of geography through a single Distributed Processing Environment (DPE). See Telecommunications Information Network Architecture Consortium.

Telecommunications Information Network Architecture Consortium TINA-C. An international association of over 40 telecommunications operators and manufacturers who first came together at a TINA Workshop in 1990 and formed the consortium to cooperatively define a common architecture (TINA) to be promoted as a global standard for building and managing telecommunications services. This work draws heavily on the work of other organizations and standards bodies in order to take advantage of ongoing studies and developments, to expedite the progress of the TINA project, and to promote the harmonious cooperation of various groups with similar goals. http://www.tinac.com.

telecommunications lines Physical lines, usually metal wire or fiber optic cable, over which communications are transmitted, usually by electrical impulses or light. Contrast to wireless communications.

Telecommunications Management Network TMN. A global network management model for Network Elements (NE) and Operation System (OS) and the interconnections between them. Global standardization provides greater incentives for common interface development. Discussions of O&M aspects of intelligent transmission terminals began, and TMN was first formally defined in 1988, with the recommendation for M.3010 (Principles for TMN) published in 1989, in addition to others over the next three years. OSI Management, originating in ISO, was adopted as a framework for TMN to provide transaction-oriented capabilities for operations, administration, maintenance, and provisioning (OAM&P). Elements of a TMN interface consist of various definitions, models, and profiles, including architectural definition of TMN entities, OAM&P functionality, management application and information models, resource information models, communication protocols, conformance requirements, and profiles.

Telecommunications Reform Act An act by government opening up local and long distance markets to competition. The act included a highly controversial provision called the Communications Decency Act (CDA) which was, after a great deal of discussion and input from the Internet community, declared unconstitutional.

Telecommunications Regulatory Email Grapevine TREG. An informal organization that carries on regular online discussions about real world issues associated with taking products and services through the various regulatory processes. This self-help group answers queries and shares experiences, archiving the information on the Web.

telecommunications service This is defined in the Telecommunications Act of 1996 and published by the Federal Communications Commission (FCC), as:

"... the offering of telecommunications for a fee directly to the public, or to such classes of users as to be effectively available directly to the public, regardless of the facilities used."

See Federal Communications Commission, Telecommunications Act of 1996.

Telecommunications Standards Advisory Council of Canada TSACC. A Canadian industry-government alliance formed in 1991 to develop strategies for Canadian and international standardization in information technology and telecommunications. Information on telecommunications technologies is provided on their Web site. http://www.tsacc.ic.gc.ca/

Telecommunications Technology Association TTA. Established by the Korean Ministry of Communication in 1988, beginning operations in 1992. http://www.tta.or.kr/

telecommuting Virtual 'commuting' to the work site, that is, communicating through various telecommunications methods instead of physically traveling to the work site. A number of factors have contributed to the increasing desire for, and availability of, telecommuting jobs: increasing congestion in cities causing higher housing costs and less availability of housing; increased traffic congestion; more families with two working parents who don't want to leave children unattended; improved telecommunications services, with faster and better transmission, more hookup services through phone lines, and videoconferencing options.

Telecommuting is not for everyone; many people prefer to work under direction or to work in close physical proximity to coworkers, but there are also many who work better undisturbed and will use the time saved by not having to commute to produce a higher quality product. There are also increasing numbers of businesses willing to provide telework options so they can recruit highly skilled workers from diverse regions. See telework, virtual office.

telecomputer, computerTV A TV broadcast system-computer integrated system that allows a user to control the choices of which programs to watch, and/or which provide menu options for viewing, such as split screen for more than one show, digital effects, sound options, integration of TV and phone (e.g., on-screen Caller ID on the TV when the phone rings), email and Web access, shopping from home, etc. This is an example of the convergence of the computer and broadcast industries. Standards for ATM for the home are being promoted so that standardized commercial consumer systems can be developed which allow these many technologies to link and work together. See Broadband Residential, fiber to the home, Home Area Network, WebTV.

teleconference A telephone conference where three or more participants share in a conversation. Conference call buttons or codes are available on some local multiline systems, and operators can set up conference calls across public lines for participants who are distant from one another. See videoconference.

telecopier See facsimile machine.

teledensity A measure of the number of telephone lines per 100 POPs (individual people) which is used to assess service distribution, economic compromises, revenues, etc.

Teledesic A privately owned constellation of literally hundreds of satellites orbiting at 700 kilometers (LEO) designed to provide switched broadband bidirectional network services, including Internet access, data, voice, videoconferencing, and interactive multimedia. It is designed to operate at up to 64 Mbps for downlink and up to 2 Mbps for uplink. The top transmissions speed is more than 2,000 times faster than standard modems operating over wired phone lines. Connection is through small parabolic antennas.

The Teledesic group approached the Federal Communications Commission (FCC) in 1994 for a 500-MHz frequency allocation within the Ka-band for this service.

In May 1998, Motorola Inc. joined the venture as the prime contractor, bringing in its Celestri technology, along with Boeing Company and Matra Marconi Space, a European satellite manufacturer.

Teledesic LLC is a McCaw/Gates company scheduled to begin launching in 2001 and to be in service by 2002.

Telefunken A German radio station founded in 1903 soon after Marconi's wireless demonstrations in London, England excited the imaginations of radio experimenters and future broadcasters.

telegaming Gaming over a distance communications medium (telephone, computer network, postal service). Telegaming has been with us for a long time. For centuries, people have played long distance chess and backgammon games by messenger and, more recently, by mail or phone. Currently it implies an unbroken connection, since that is now possible through computer networks and games like *chess* and *go* are routinely played on the Internet. Video arcade games are played on local networks, usually on an Ethernet link, although the term telegaming doesn't apply as well to an activity in which the participants can see or hear one another in the same or next room.

telegenic Having characteristics that appeal to television audiences, such as charisma, talent, humor, relevance (news), or other qualities favored by broadcast networks and viewing audiences.

telegram Originally *telegramme* (France, 1793). A printed record of a telegraphic communication. Early telegraph signals were transcribed on paper tape as wiggly lines; later, audible signals were interpreted by human operators and written down by hand; and, finally, devices that could interpret the signals into text and impress them on paper as telegrams were devised. For decades the telegram was delivered into the hands of the intended receiver or at least brought to the doorstep. Courier services and facsimile machines are superseding telegram services. See telegraph system, teletypewriter.

telegraph fire alarm That telegraph signals could be used to report fires through signal boxes was realized not long after the invention of the telegraph, and many of the larger communities installed this type of safety system by the early 1900s. The Boston Fire Alarm system was one of the first, following a published description of its feasibility by William F. Channing in 1845. Later, with the help of a telegraph engineer, Moses G. Farmer, Channing supervised the 1851 city funding and 1852 construction of the first fire alarm telegraph in the world. Originally based on manual crank boxes, painted black, the mechanisms were later changed to pull switches, and eventually dials. By 1881, the fire boxes were changed to red.

telegraph history The telegraph was a system of equipment and coding that enabled communication over distance, originally through signal towers, and later by wires powered by high-intensity batteries.

As with many technologies, the telegraph was invented in a number of places at about the same time, and many of the early models were never practically or commercially implemented. Lesage had created a *frictional telegraph* as early as 1774, and A. Ampère and P. Barlow proposed early designs as well. In Germany, Samuel T. von Sommering created a 35-wire telegraph based on electrochemical concepts.

One of the first practical implementations of the telegraph was in 1837 by C. Wheatstone and W. Cooke in England. The telegraph in America owed much of its design and development to Samuel Morse and Alfred Vail. Morse's original telegraph caveat (an intention to file a patent) described a mechanism with a horizontally moved key which made corresponding zigzag marks on a moving paper tape to represent numbers, which were then looked up to find the corresponding words in a reference dictionary prepared by Morse. Vail improved on the mechanics of the key, making it move up and down instead of side-to-side, thus forming dots and dashes with breaks in between on the paper. As this system was simpler and more direct than doing a dictionary lookup, it evolved into the system now known as Morse (Vail) code. Their telegraphic invention was demonstrated to the Presidency in 1838. Morse subsequently won funding from Congress to construct a telegraph long distance line, carried out the project with assistance from Ezra Cornell, and began to spread telegraphy throughout America in the mid-1800s. Both Wheatstone and Morse received advice and encouragement on the development of telegraphic instruments from Joseph Henry in the 1830s. Morse, unfortunately, didn't duly credit Henry's assistance.

In its simplest form, the telegraph consists of a sender (a keying device), a receiver (with a sounder or printer), and a simple code for conveying characters. Early telegraph receiving machines used paper tapes to record messages (Morse's telegraph created a wiggly line), but operators began to recognize the slightly audible incoming clicks and could copy messages faster than a paper tape could print them, and machines were soon equipped with *sounders* and *resonators* to amplify and direct these clicks. Not surprisingly, many inventors sought ways to translate the signals into let-

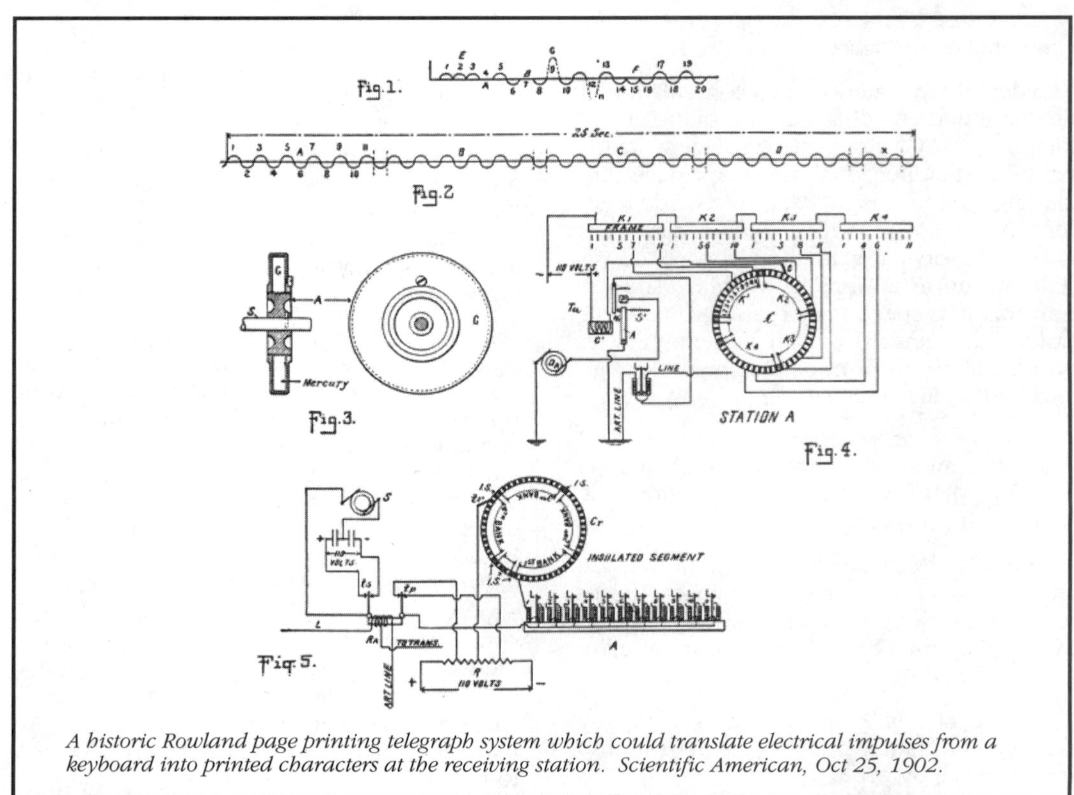

A historic Rowland page printing telegraph system which could translate electrical impulses from a keyboard into printed characters at the receiving station. Scientific American, Oct 25, 1902.

ters that could be recorded directly, as in a telegram or teletype-style printout. One of the first to succeed was David Hughes, a schoolteacher, in 1856.

In America, messages were sent by shutting current on and off, while in Britain, Wheatstone introduced *polar keying,* a means of using polarity to convey signals. The concept of polarity is still used today in high-speed data transmissions.

In 1866, M. Loomis demonstrated that signals could be sent from one airborne kite to another, when each was strung with fine copper wire of the same length, without direct physical contact. This later lead to his 1872 U.S. patent for a wireless improved telegraphic system, although it was some time before his discoveries were put into practical use.

By the 1880s, scientific investigations and demonstrations had confirmed the viability of wired and wireless telegraphy. The end of the century then became a time of creative application of the concepts and evolutionary improvements in speed and practicality.

In 1895 and 1896, in Russia, A. S. Popow was conducting experiments with wireless telegraphy and succeeded in sending a shipboard message to his laboratory in St. Petersburg. Unfortunately, due to the secrecy surrounding Russian naval technology and inventions in general, Popow's discoveries were not communicated to the rest of the world, and he did not receive credit for his early experiments.

In the late 1800s, *telautographs* which could transcribe handwriting were created by a number of inventors including E. Gray and F. Ritchie. While these remained in use for a number of decades, they didn't originally work over long transmission lines and were superseded by telewriters and, eventually, facsimile machines.

In 1886, Amos Dolbear, a Tufts University scientist and writer, was awarded a patent for a wireless telegraph based on induction.

In 1889, F. G. Creed invented a High Speed Automatic Printing Telegraph System. By 1898, his Creed Printer could transmit 60 words per minute and his technology was widely sold in many countries. He broadened his enterprise in 1923 by demonstrating marine wireless printed telegraphy, a system eventually used for marine safety.

The telegraph had a revolutionary impact on communications, changing forever the concept of distance. It networked not only the predominantly rural early settlers of North America, but spurred the installation of the first transatlantic cable, providing 'instant' (by 1800s standards) communication with Europe. Prior to the oceanic cable, messages typically took two months or longer to travel in ships from one continent to the other. News, business, warfare, and family contacts were dramatically affected by the availability of fast long distance communications.

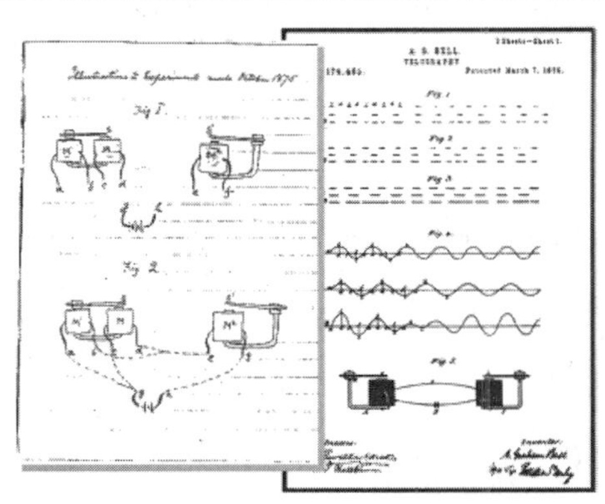

One of the early Bell telegraph patent documents. There were many inventors at the time independently making similar discoveries, and there was substantial competition to be the first to patent and commercialize the new communication technologies.

See Creed, Frederick George; heliotrope; Dolbear, Amos; Morse, Samuel F.B.; Popow, Aleksandr Stepanowitsch; telegram; telegraph system; telephone; Wheatstone, Charles.

T

telegraph key A mechanical switch on early telegraph systems that enabled a circuit to be opened and closed in order to generate transmissions through a signal such as Morse code.

The original telegraph key was a fairly straightforward device for opening and closing a circuit at intervals that corresponded to the dots and dashes in Morse code.

telegraph signals For telegraph signals through wires, two main methods were used: *polar* transmission, in which the polarity was changed to reverse the current; and *neutral*, or *open/close* transmission, in which open current (space) is interspersed with no current (mark).

telegraph system An apparatus for sending and/or receiving information over distance, coded in some fashion, usually in Morse code dots and dashes. A basic telegraph circuit consists of a key to translate finger or other mechanical pressure into signals, a relay that is sensitive to the very small current that may be coming through the wire, and a receiving device which can express the message by means of audible tones, paper tape code, or printed letters.

Telegraph systems have co-existed with, rather than been superseded by, telephone systems for a number of reasons, including the expense and time delays of setting up long distance toll calls to some areas, and the importance, in some situations, of creating a written record in the form of a telegram. With electronic telephony advancing and facsimile machines proliferating, the telegraph is becoming more historically interesting than practical. See telegraph history; telegraph, needle.

telegraph, needle A type of five-needle telegraph devised by Charles Wheatstone and put into service in England in 1837. Faulty equipment lead to the gradual realization, by telegraph operators, that two needles were sufficient and, eventually, only one needle and one dial were used to efficiently convey messages. The needle telegraph also represents the development of *polar keying*, which employed positive and negative voltages for indicating *mark* and *space* signals. See polar keying.

An automatic telegraph sender. The wheels shown at the top represented characters that could be selected and placed in order to spell out a message. This example is from the Bellingham Antique Radio Museum collection.

telegraph, printing Early telegraph paper-

Overview of Telphone Development Phases

Type	Time period	Notes
original invention	late 1800s	Proof of concept, the first discernible, intelligible human voices can be heard over distances.
hand crank phones	late 1800s, early 1900s	Phones were large, to accommodate a battery, and had to be cranked to send a ringing current. Hand crank phones were still in use in rural areas, including some of the San Juan Islands in the 1960s.
dial phones	early 1900s to 1980s	Common batteries and automatic switching systems made it possible to create smaller, line-powered phones and rotary dials so the subscriber could direct dial a local call, and later, long-distance calls.
touchtone phones	late 1970s to present	Phones that sent tones rather than pulses through the line, which were interpreted according to pitch. This made automated menu-controlled systems possible.
digital phones	early 1990s to present	Interface speakers or headsets which attach directly to a computer as peripherals to allow the user to talk into a digitizing program which samples the sound and transmits it over public data networks.

tape and manually-operated sounding systems did not satisfy the needs of inventors and users who wanted quick, automated written messages. Thus, the development of printing telegraphs was of interest to many. One of the first successful systems was developed by A. Vail in 1837, employing a type wheel. Later D. Hughes developed a practical working type wheel system in 1855 which became established in Europe, but didn't catch on well in America, where Morse systems were in use. Improvements to printing telegraphs continued and, in 1846, R. E. House developed a printer that printed telegraphically transmitted letters directly. Further improvements to House's system resulted in a patent in 1852.

The necessity of noise-free transmissions and technical expertise to maintain the equipment prevented printing telegraphs from coming into widespread use until decades later. See teletypewriter.

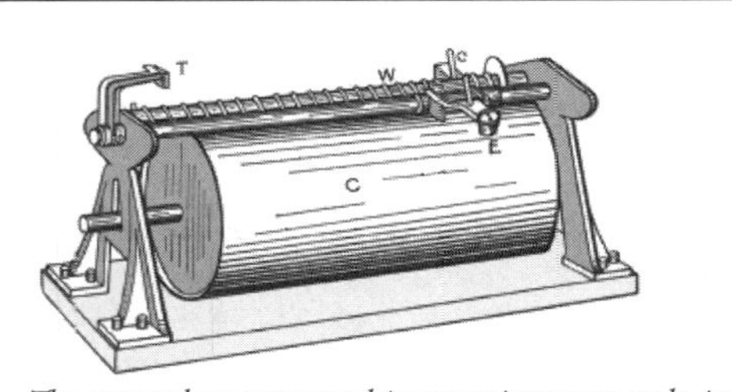

The tegraphone spurred innovations not only in telegraphy, but in sound recording as well.

telegraphese A terse, abbreviated mode of messaging (or speaking) which has the character of a telegram. Since telegrams were often charged by the letter or by the word, a compact style of communication emerged in order to keep the cost as low as was practical.

telegraphone, telegrafon This is not only a type of telegraph instrument, but more important, was an early electromagnetic tape recorder, designed in 1898 by Danish inventor Valdemar Poulsen. Poulsen succeeded in recording electronic waves on a thin wire of steel, and improved on the technology enough to receive a U.S. patent in 1890. This developed into dictating machines sold through the American Telegraphone Company. See tape recorder.

TeleLink Project The full name is TeleLink 'Training For Europe' Project. This is a European Community (EC), Euroform-funded project which seeks to promote and develop telework training opportunities and qualification guidelines. This includes qualification level certification (currently at the vocational level) for teleworkers and a system of TeleLink centers around Europe. See ADVANCE Project, telework.

telemarketing The promotion of products and services through telephone calls to individual premises. There are various regulations governing when telemarketers may call, whom they may call (e.g., calls to a person at their place of business must be stopped if the callee requests it), and what they must say to identify themselves and their affiliations. There are also restrictions on where they may obtain names, and how they must dial the call. Many scams have been perpetrated through telemarketing schemes, and it is important for the callee to get sufficient information to ascertain that the offering is legitimate. If you don't

Except for the very earliest experimental models, the essential mechanical design of the dial telephone hasn't significantly changed in about sixty years, but a number of cosmetic updates have occurred over time.

wish further calls from the source, you should request that your name be taken off their list. See war dialer

telemarketing broadcasts The promotion of products and services through mass market advertising usually providing a 1-800 or 1-900 number for the interested buyer to call. Automated systems for taking the caller's name and billing information through touchtone selections are becoming prevalent.

telemedicine Medical information and services and medical education provided over distance through telephone, radio, facsimile, videoconferencing, and the Internet. Information such as medical imaging results can readily be transferred as data, since much of it is digital in nature. Teaching and other communications among medical professionals and their patients are possible through newer technologies.

telemetry, telemetering The art and science of gathering information at one location, usually in terms of some quantity, and transmitting that information to another location for storage, analysis, or evaluation. Weather balloon data gathering and transmission through a radiosonde to a weather station for interpretation is one example of telemetry. The transmitting of information from space probes is another. Telemetry equipment is typically included on artificial satellites to aid in the control and orientation of the satellites.

Teletext A commercial computer service offered by NBC, which was discontinued in 1985. Many of these early computer services came and went, but they are coming back in updated forms now that there is a large user base drawn to the Web.

telephone A communications apparatus designed primarily to convey human voice communications. In its simplest form, a telephone consists of a transistor, that converts sound into electrical impulses, and a receiver, which converts them back again into sound. Additional technology is used to amplify and direct the communication between these two basic devices. The design of the telephone set has gone through five overlapping phases in its development. See the Telephone Phases chart. See telephone history.

telephone amplifier A device to amplify sounds at the receiving end of a call. This can be incorporated into the handset, headset, or speakerphone, or may be an add-on to provide even more amplification for the hard of hearing. Most handset telephone amplifiers draw current from the phone line, but many speakerphones and add-on amplifiers require a separate power source. The amplifier is often adjustable through a dial or slider on the side of the phone.

telephone answering machine An electronic or mechanical device for answering calls and often for recording them digitally or on tape. Telephone answering machines based on reel-to-reel mechanisms have been available since the early 1960s, but small cassette and digital answering machines did not become common until the late 1970s and early 1980s.

Most households now have answering machines to respond to calls, take messages, or to screen calls. Many of these will include information on the time and date of the call, and some will record the identity of the caller, if Caller ID is activated on the subscriber line. Computer voicemail applications can also be hooked to a phone line through a data/fax/voice modem to allow the software to function as a full-featured answering machine with multiple mailboxes.

telephone answering service 1. A service offered by commercial vendors in which a human operator or voice-automated system will answer the subscriber's phone line when it is call forwarded, or when the answering service number is called directly and forwards the message to the subscriber. This service is widely used by small businesses, freelancers, and real estate agents. Sometimes these services are combined with paging. 2. A service offered by local phone companies in which a human operator or voice-automated system will take calls and forward messages to the subscriber, or through which the subscriber can use a touchtone phone to retrieve messages.

The earliest telephone looked more like a pinhole camera than current familiar desktop phones and mobile handset phones.

telephone central office See central office.

telephone circuit An electrical connection consisting minimally of a transmitter, receiver, amplifier, and connecting wires, and more commonly comprising a system of two-way audio and signaling connections between local exchanges and subscriber lines and telephones.

telephone exchange Switching center for telephone circuits. See central office, private branch exchange.

telephone history The telephone was a significant evolutionary development, occurring a few decades after the invention of the telegraph. While the telegraph revolutionized telecommunications by making communications over great distances possible, the telephone personalized it, and many inventors were excited by the potential of sending tones, or even voice, over phone lines.

An innovative optic telephone, based on the stimulation through a diaphragm of a flame from an acetylene burner. The impulses were then further transmitted optically through a light-sensitive selenium cell and reflector. The optic telephone was developed by Ernst Ruhmer, and was used for 'long distance' communications. Scientific American, Nov. 1, 1902.

In the early 1800s, a German inventor, Philip Reis, observed that a magnetized iron bar could be made to emit sound, and he first demonstrated the transmission of tones through wire in Frankfurt in 1861. In a letter, Reis reported he could transmit words, although we have no direct way of verifying this. Around the time of Reis's death, an American physicist, Elisha Gray, was making numerous experiments in telegraphy and developed early concepts for harmonic telegraphy, the transmission of tones, and telephony. In the mid-1800s, Italian-born Antonia Meucci was experimenting with wires attached to animal membranes to transfer sound through current, but news of his discoveries did not become widely known outside Cuba.

Here Alexander Graham Bell demonstrates his telephone invention. The inset shows one of his early sketches of the invention, from the famous Bell notebooks. Bell achieved great financial success from commercializing his discoveries.

The precursors to the telephone and later variations appear to have been invented more-or-less independently by Elisha Gray and Alexander Graham Bell, but Bell filed his 'telephone' patent (it was actually a precursor to the telephone, a harmonic telegraph) a few hours before Gray filed a caveat (intention to file within three months). In a 1911 lecture on the origins of the membrane telephone, Bell described how he worked out the idea in discussions with his father while on a family visit in Canada in the summer of 1874, two years before it was successfully implemented. Bell and Watson reported that Bell first spoke intelligibly over wires in March 1876. The transmission succeeded by use of a liquid medium, something not mentioned in Bell's patent. This voice capability was not publicly demonstrated until some time later, which seems odd given the magnitude of the reported achievement. Ironically, Bell had been discouraged by investors from trying to make a talking telegraph and was prodded to concentrate on a harmonic telegraph instead.

Gray had publicly demonstrated rudimentary telephone-related technology before the Bell patent was filed, and later successfully earned a number of telephone-related patents. He designed a telephone in the 1870s not unlike the second-generation switch-hook phones that employed separate ear and mouth pieces which came into use in later years.

The first commercial telephone exchange was established in Connecticut, U.S. and became operational in 1878. It was followed the same year by the second commercial exchange in Ontario, Canada.

The Bell patents formed the basis of the early Bell System in the United States, a company that has influenced the development of communications, and thus the course of history, in countless important ways. The Bell Telephone Company of Canada was incorporated in 1880.

The most interesting evolutionary step in telephone technology, besides the growth of wireless communications, is probably the videophone, descended from early picture telephones such as the Picturephone. The Bell Labs were transmitting pictures in the late 1920s and demonstrated the early technology to the Institute of Radio Engineers in 1956, but it was not until 1964 that a practical experimental system was completed and the Picturephone exhibited cooperatively by Bell and the American Telephone and Telegraph Company (AT&T) at the New York World's Fair.

Currently there are many companies scrambling to be the first to get a cheap, publicly accepted version of a picture telephone or as they are known now, audiographics systems, videophones, or videoconferencing systems. With the growth of the Internet and the drop in price of small video cameras, this may become a common microcomputer computer peripheral by 1999.

Another significant change in telephony has been the sending of voice over computer networks by means of a specialized handset attached to a computer. This permits the connection of long distance calls world round without any long distance toll fees. The technology threatens to dramatically change the established economic structure of the telephone system, and it is difficult to predict whether the same revenue-generating model that has worked for about 100 years will be viable in the future, given the current rate of change. In fact, some of the long distance carriers, worried by this threat to their survival, have lobbied for this type of transmission to be blocked. See Bell, Alexander Graham; Callender switch; Gray, Elisha; Meucci, Atonio; Reis, Philip; telegraph history.

telephone land-line density A measure of the number of installed phone lines per 100 people.

telephone pickup Any of several devices for connecting into an ongoing telephone conversation, usually for monitoring purposes.

Telephone Pioneers of America TPA. A nonprofit organization founded in 1911, with chapters throughout the United States and Canada. Originally consisting of telephone pioneers with 25 years of service or more, with Theodore N. Vail as its first president, membership later opened up to a wider group, now numbering almost 100,000, as fewer pioneers remained from the original group.

TPA engages in a number of community-oriented activities, with a particular focus on education. A somewhat analogous organization for non-Bell employees is the Independent Pioneers. http://www.telephone-pioneers.org/

telephone receiver The portion of a handset, headset, or speakerphone which converts electrical impulses into sound. On a handset, the receiver is the part that you hold up to your ear. Inside a basic traditional receiver is a magnet, with coils wound around the poles connected in series and a light, thin, vibrating

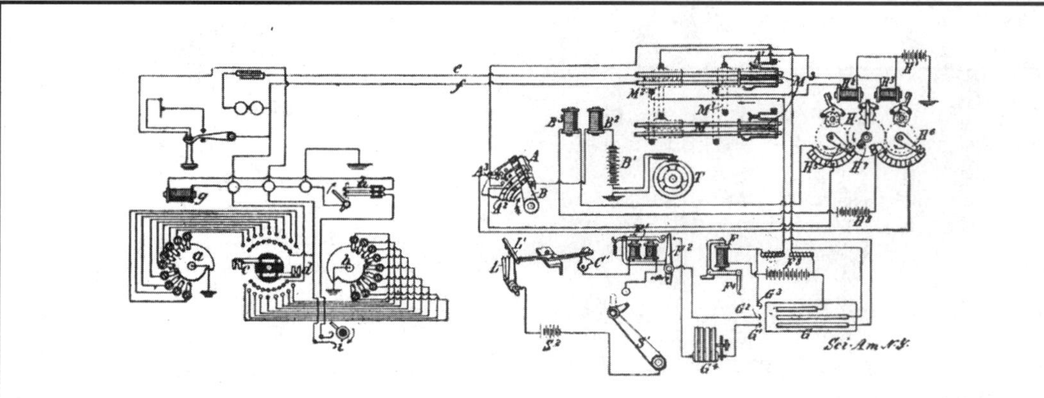

A schematic for a historic semiautomatic telephone switching system (it still required a human operator to turn a spring-loaded knob to send the dial pulses through the wire). Scientific American, Oct. 11, 1902.

diaphragm mounted very close to the magnet poles. When current passes through the coils, the diaphragm vibrates, producing sound by moving the air next to it. Early receivers used a bar magnet, which later was replaced by a horseshoe magnet. See telephone transmitter.

telephone repeater An amplification device employed on telephone circuits to rebuild and maintain signals across distances, which otherwise would be subject to loss.

telephone signaling Any device that indicates an incoming call, usually a bell, but may also be a light or moving indicator.

telephone switchboard A centralized distribution point for managing telephone calls. Early switchboards consisted of a human operator answering calls, and plugging a large physical jack into the receptacle of the person to whom the call was being patched. The first commercial switchboard in North America went into operation in Connecticut in 1878. Switches were mechanized in the mid-1900s, although it was not uncommon for human switchboard operators to staff manual switchboards in rural areas and private branches until the 1950s. Although mechanical switching stations still exist, updated switchboards function electronically.

telephone tag Colloquial phrase for two parties attempting to contact one another by phone, not reaching the other person, and having to leave messages with an answering machine, operator, or voice mail system. Do this back and forth a few times, and you're playing telephone tag.

telephone transmitter The portion of a handset, headset, or speakerphone which converts sound into electrical impulses. On a traditional handset, the receiver circuit connects to the part that you hold next to your mouth. Inside the mouthpiece is a movable diaphragm with an attached carbon electrode, behind which another carbon electrode is fastened securely inside the housing. Between the electrodes are carbon granules (it's possible to build a simple phone transmitter using the core of a carbon pencil laid across two conducting surfaces connected to wires and a diaphragm). When a current is applied, resistance decreases, as a result of the carbon granules compressing more closely together. Thus the current increases and attracts the diaphragm more strongly. The diaphragm vibrates to produce an electrical impulse that corresponds to the movement of air caused by the speaker's voice. An induction coil may also be used to increase the voltage to compensate for signal loss through the transmissions medium. See Blake transmitter, coherer, telephone, telephone receiver.

telephone user interface TUI. The use of telephone equipment, usually a handset or headset or telephone line attached to a peripheral card, to interact with computer software. Instead of using a keyboard and mouse as the input devices, voice or touchtones over the handset or phone line are used to control the actions of the computer. For example, you may have a computer set up like an answering machine to answer calls, respond to callers, and log time, date, and caller messages. Then, from a remote location, you may call the line attached to the computer, and by speaking or pressing touchtone buttons, have the computer send back information about the calls or to replay the calls themselves.

telephony The transmission of audio communications over distance, that is, over a greater distance than these communications could be transmitted without technological aid. The breadth of the term has increased, typically now including visual communications that accompany sound communications (as in audiographics and videoconferencing), although it is preferred that the more general term *telecommunication* be used for audio/visual transmissions. Most telephony occurs over wires, but wireless services transmitted by radio waves and satellite links are increasing.

Telephony, in its simplest sense, is not a high bandwidth application with each conversation requiring only a narrow channel, but because of its continuous bidirectional nature, bandwidth needs increase as the number of simultaneous calls increases. Traditional telephony media, such as copper wires, are no longer strictly used for oral communications; they now service a large number of data transmission services such as Internet connectivity, facsimile transmission, and more. Due to increased demands for lines with greater speed and accuracy than are needed for simple voice transmissions, fiber and coaxial technologies are being used to upgrade data lines and, consequently, the phone lines themselves. See HFC, telephone, telegraph.

Telephony Application Programming Interface TAPI. A set of guidelines developed jointly by Intel Corporation and Microsoft Corporation for connecting microcomputers directly to phone lines in local area networks.

Telephony Services Application Program-

ming Interface TSAPI. A set of guidelines developed by a group of developers including Novell, Inc. and AT&T for interconnecting corporate telephone systems into the data network server in medium and large business networks.

telephoto, telephotography Visual information conveyed through conventional photographs or digital photographs from data received remotely. Journalists, geographers, navigators, and others use telephotos to send or receive visual information from remote sources through wired or wireless communications, and to print them in various resolutions through photographic, laser, or other means. Satellite photos of the Earth's surface are extremely popular examples of telephotos. Many of the images now printed in national newspapers are telephotos sent through wireless modems by journalists using digital cameras and laptops.

Teleport Communications Group TCG. At one time, a national competitive local telecommunications provider with fiber optic SONET networks in over 50 large markets, acquired in early 1998 by AT&T.

teleprinter 1. Teletypewriter. 2. A Western Union trade name for printing telegraph terminals. See telex.

TelePrompt Project A European Community (EC) project funded by a consortium of academic and commercial groups designed to develop and further technology-based distance learning resources for European teleworkers. The term *teleworking* in Europe is roughly equivalent to the term *telecommuting* in North America.

teleran An aerial navigational guidance system employing information received through television waves and radar transmitted to aircraft by ground stations.

telesales A British term for telemarketing. See telemarketing.

Teletype A name trademarked by Teletype Corporation for a variety of teleprinting devices used in communications. See teletypewriter.

Teletype Corporation An early printing telegraph company, the Morkrum-Kleinschmidt Corporation, which was acquired in 1930 by the Bell System and renamed Teletype Corporation.

teletypesetter A machine for remotely controlling typesetting machines. When these were originally put into service, teletype machines relied on a five-unit code that was insufficient to transmit all the characters needed by a similar teletypesetting machine. Thus, a six-unit signal code was developed for teletypesetters to increase the size of the character set from 32 to 64.

teletypewriter TTY. A printing apparatus which, in its common form, resembles a typewriter on a pedestal with continuous feed or tractor feed paper so that it can print unattended. Sometimes it is used to send and receive signals over phone lines and is used for transmitting messages or computer data in text form.

The teletypewriter superseded key and sound telegraph systems because it could operate unattended and achieve transmission speeds of 60 to 100 words per minute. The earliest teletype-style printers and start-stop synchronization methods were developed by Charles and Howard Krum. See Baudot code; Krum, Charles and Howard; telegraph, printing; Teletype; telex.

teletypewriter code A five-unit code that employs elements of uniform length. Start and stop pulses are used to distinguish each character in the transmissions. See Baudot code.

teletypewriter exchange service Any commercial service which provides teletypewriter communications sending and receiving services through a switching exchange. Similar in concept to a long distance telephone exchange. TWX is one such service of the Bell System, established in 1931, subsequently owned by AT&T. See Telex.

television TV. A system of sending and receiving broadcast moving images (even if the object in the transmission isn't moving), usually in conjunction with sound, although some closed-circuit television systems don't include sound circuitry. Television broadcasts can be transmitted through air or over cables, with cable TV (CATV) increasing in popularity. Air transmissions are captured with a television antenna designed for a portion of the broadcast spectrum (although three-in-one antennas exist for UHF, VHF, and FM signals). At the receiving end, a television set (tuner and monitor combined), or a VCR tuner and monitor are typically used to display the broadcast.

television broadcast band The various frequencies which are assigned and regulated for television broadcast transmissions. Due to the proliferation of programming and the increased availability of access through satellite transmissions, there is constant pressure to increase available frequencies and channels, and there are now hundreds of programming channels available. See band allocations for a chart.

television camera A lens-equipped, optical-sensing pickup device designed to capture moving images and transmit or pass them on to receiving, editing, and broadcast equipment. The type of signal generated by the camera varies according to the receiving or editing equipment, and varies from country to country. Television cameras have traditionally been expensive, large, heavy, analog high resolution apparatuses. This is all changing, with small hand-held digital and analog personal cameras beginning to rival the quality of traditional TV cameras for only a fraction of the price. See NTSC, PAL, SECAM.

television history Television, perhaps more than any other of the major communications technologies, arose in fits and starts in the late 1800s with many geographically diverse announcements of success and few demonstrated working systems. One of the important discoveries in the history of television was the photoconductive characteristics of selenium, which responded to the amount of light hitting the surface. A French researcher, M. Senlacq, suggested in 1878 that selenium might be used to register the shapes of dark and light areas on documents. British researcher, Shelford Bidwell, was able to successfully transmit silhouettes by 1881, and the now famous German inventor, Paul Nipkow, after whom the Nipkow disc is named, patented an electromechanical television system in 1884. But the transmission of moving images and shades of gray in high enough resolutions to be practical eluded the early inventors.

Although patents for television-related technologies began to appear in the late 1800s, it was not until the 1920s that television transmission and reception as we know it was demonstrated by inventors such as John L. Baird in the west and Kenjito Takayanagi in Asia. Baird's first significant success was in 1926, the same year Tekayanagi transmitted Japanese script with a cathode ray tube.

In the U.S., a precocious 15-year-old, Philo T. Farnsworth, described an idea for a television to his schoolmates and reportedly showed a sketch to his teacher in 1922. In 1927 he succeeded in building a working model.

Experimental television stations sprang up in the late 1920s and, by the mid-1930s, regular public broadcasting began to develop. In Europe, television images were being transmitted by 1931.

Television sets were available by the late 1930s, but it was a while before the technology became cheap enough for home use. By the late 1940s, there were at least 20 broadcast stations in North America, with hundreds of hopefuls clamoring for the limited licenses.

Black and white televisions came into widespread use in the 1950s in North America and color television was common about 15 years later. By the mid-1980s, melon-sized portable televisions became inexpensive and wrist-sized consumer TVs had been developed.

Commercial sponsorship provides much of the funding for television in North America, thus controlling, to some extent, the type of programming which is available, influenced by majority consumer demand or perceived viewer preferences. In many other countries, television is funded and controlled by local governments.

The next major step in television broadcasting was the launching of communications satellites such as the Telstar 2 in 1962 which permitted intercontinental communication. Commercial application of satellite television broadcasting was pioneered by the Canadian Broadcasting Corporation through the ANIK satellite in 1972, followed in the late 1970s by Turner and the Public Broadcasting System (PBS) in the U.S.

In North America, satellite TV broadcasts can now be received by consumers on small parabolic dishes through monthly subscription services, with hundreds of potential stations available. Television is widely used for mass-media entertainment, education, distance monitoring, and local security monitoring. The influence of television on world culture is very substantial, with a preponderance of the programming originating in the United States, widely promoting American values, styles of dress, and cultural priorities to all parts of the globe. See Baird, John Logie; Farnsworth, Philo T.; Nipkow, Paul Gottlieb; Nipkow disc; Takayanagi, Kenjito; television; television camera; Zworykin, Vladimir.

television relay A station designed to pass on a television broadcast signal to the next station so the signal is protected from loss. The relayed signal is not intended for reception by viewers until it reaches the destination station.

television signal The coding of images can be accomplished in a number of ways, and there are several standards, each of which is preferred in a different part of the world.

T

Common formats related to the broadcast and display of moving image signals are shown in the Common Broadcast Formats chart.

telework Work at home or at satellite locations made possible through computer and telecommunications technologies. In 1988, Jack M. Nilles proposed a broad definition of telework as "... all work-related substitutions of telecommunications and related information technologies for travel," thus, employer/ employee interactions across distance through new technologies. This term is more common in Europe and is roughly equivalent to the term *telecommuting* in North America. See ADVANCE Project, European Community Telework Forum, TelePrompt Project.

telex *tele*printer *ex*change. Generic term for a communications service developed near the end of the second world war that uses teletypewriters to transmit through wire lines and automatic exchanges to produce a written message at the destination. In Europe, this technology used audio frequencies over phone lines. See Baudot code, Telex, Western Union.

Telex A global message service established in the United States by Western Union in the early 1960s. This was competitive with AT&T's TWX service.

Telkes, Maria A physicist who did pioneer work in the development of solar energy in the early part of the 20th century. Solar energy has subsequently become an extremely important power source for orbiting communications satellites.

Telnet Protocol A widely supported 8-bit byte-oriented network protocol for remote terminal access, originating from the days of the ARPANET. Telnet allows the user to log on to another system through a TCP/IP network, and perform file functions and other activities. Telnet is the protocol 'telnet' spelled all in lowercase is the command to launch a remote utility that uses the Telnet Protocol. The form of the Telnet command is:

```
telnet [IP_address|host_name][port]
```

(with the command entered in lower case). A sample Telnet session looks like this:

Common Broadcast Formats

Abbrev.	Name	Notes
NTSC	National Televisions Systems Committee	The North American standard since the 1950s. 525 vertical lines. NTSC uses negative video modulation and FM sound.
PAL	Phase Alternate Line	The predominant standard in the United Kingdom and parts of Western Europe since the early 1960s. 625 vertical lines. There are a number of variations of the PAL system, including PAL-B, PAL-H, PAL-M, etc. PAL uses negative video modulation and FM sound.
SECAM	Sequential Color and Memory	Developed in France and used in North Africa, Russia, and parts of Europe since the early 1960s. 625 vertical lines. There are a number of variations of the SECAM format, including SECAM-B, SECAM-H, etc.
HDTV	High Definition Television	Introduced in Japan and proposed as a global standard, but not readily adapted by American and other manufacturers, some of whom would prefer to enhance current standards rather than adopt a new one. 1125 vertical lines at 60 frames per second. HDTV is supported by some Internet 'push' channels and can be viewed with an interface peripheral and a computer with a fast connection.
C-MAC	Multiplexed Analog Components	Developed in the U.K. and recommended by the EBU as a European standard.

```
abiogen@mycomputer: /1.2GB/users/
abiogen $ telnet remotecomputer

Trying 192.42.172.20... Connected to
remotecomputer.
Escape character is '^]'.


NeXT Mach (remotecomputer) (ttyp0)

login: myname
Password:

bash$ ls -la
...
bash$ logout
Connection closed by foreign host.
abiogen@mycomputer: /1.2GB/users/
abiogen $
```

See RFC 318, RFC 854, RFC 855 to RFC 861 (various options).

TELSTAR 1 A historically significant low-altitude communications satellite that broadcast microwave transmissions and tracked satellites in the 1960s. This AT&T endeavor is claimed to be the first active communications satellite, launched 10 July 1962 by the United States, although some RCA engineers launched a transmissions satellite earlier. It is the first transponder-equipped satellite. Prior to this, satellites were passive transmitters, but the use of transponders for amplifying the signals was preferred from this time on, and some satellites now include as many as ten transponders. The TELSTAR had some early problems that were fixed in 1962; it ceased functioning in 1963. By 1964, two more TELSTAR satellites had been successfully launched.

tem wave transverse electromagnetic wave.

template 1. A pattern, guide, table, or mold used to provide the basic configuration, format, or design for creating a new version, or multiple versions of a project with few or no changes. A template is intended to save time by automating the creation of new versions. A word processing template can be used to set up documents which are reissued frequently with only minor changes (e.g., form letters).

Temporary Mobile Station Identifier TMSI. A dynamically assigned mobile station identifier (MSID).

TENET See Texas Educational Network.

tensile strength A descriptor for the greatest amount of longitudinal stress that can be borne by a particular material before it will rip apart. The units used to describe this property vary from industry to industry. It is an important factor in many manufacturing and industrial applications.

tera- A prefix for an SI unit quantity of 10^{13}, or 1,000,000,000,000. It's a trillion, a very large quantity, but considering there are now hard drives with terabytes of storage space, it's not as big as it used to be. It comes from the Greek root *terat* or *teras* meaning 'monster.'

terminal 1. An endpoint, extremity. 2. A conducting device, often a small metal post or receptacle, provided for facilitating a good electrical connection. 3. A device or system which provides remote access to a central computer. 4. An endpoint in a communications line, or one which can be, but is not necessarily, extended to other circuits.

Terminal Adapter TA. A device available in various configurations from a number of vendors, which provides protocol adaptation and interfacing with an ISDN line. A TA enables a variety of consumer electronic products such as computers, fax machines, etc. to connect to the ISDN service.

terminating office In a transmission such as a phone call or telegraph message, the terminating office is the switching center which is the final one that connects directly to the subscriber line or other receiver of the communications. In Internet dialup communications, the local ISP would be considered the terminating office.

TESC Technology Subcommittee.

tesla A meter-kilogram-second unit of magnetic flux density equivalent to one weber per square meter. Named after Nikola Tesla.

Tesla coil An air-core transformer for creating high-voltage discharges at very high frequencies.

Tesla, Nikola (1856-1943) A Croatian-born American engineer and inventor who developed the alternating current induction motor, an essential part of alternating current distribution systems. Tesla began his research in Hungary, and then emigrated to the United States in 1884.

In America, Nikola Tesla and Thomas Edison came into regular contact with one another, not always with happy consequences, and an enmity grew between the two men. When it was proposed in 1912 that the Nobel prize be awarded jointly to Edison and Tesla, Tesla eschewed any association with Edison, and the prize went to a Swedish scientist instead.

Tesla's inventive mind turned power generated devices into interesting applications such as aircraft power systems and robotic submarines. He earned more than 700 patents in his lifetime, even though many of his ideas and inventions were not patented. He was somewhat temperamental and eccentric. One of his most practical contributions was the adaptation of alternating current into everyday applications. His colleague, George Westinghouse, further implemented many of Tesla's ideas. The tesla unit of magnetic flux is named after him.

test board A switching panel used for making temporary connections in conjunction with the panel or equipment being tested. By diverting some of the signals through the test panel, problems can sometimes be more easily isolated or identified. A test clip can also be used for the purpose of making quick temporary connections. See breadboard, shunt.

test pattern Any pattern generated for a particular transmission medium that indicates the integrity of the various characteristics of its signal, which may include resolution, signal strength, stability, linearity, contrast, brightness, colors, sound range and quality, etc. 2. In video editing, a series of bands of specific colors. 3. In television broadcasting and television set calibration and diagnosis, a pattern (known to some as 'the Indian head pattern') which includes particular lines and line widths, ellipses, tonal gradations, and numerical values that allow the diagnostician to determine problems and make adjustments. This test pattern was frequently used in the 1950s and 1960s by local stations as a visual signal to viewers to indicate that there was no programming currently in progress, although this use has greatly declined due to the multitude of programming now available.

tetrode A four-element vacuum tube. The three-element tube, called a triode, was developed by Lee de Forest. This no doubt inspired experimenters to try other configurations. The four-element tube followed, consisting of a filament, plate, and two grids rather than one. The second grid, the tetrode or screen grid, was positioned between the first grid and the plate (anode).

Texas Educational Network TENET. A Texas education communications infrastructure dedicated to fostering educational innovation and excellence among educators and students. TENET developed through the collaboration of the Department of Information Resources, the Texas Education Agency, and the University of Texas. TENET provides various resources, including publications, discussion forums, and professional development seminars and facilities. http://www.tenet.edu/tenet-info/main.html

TFT See thin film transistor.

TFTP See Trivial File Transfer Protocol.

TH transmission header. In packet networking, a header that includes control information, and may be followed by a basic information unit (BIU) or segment. Used for network routing and flow.

theremin An electronic musical instrument incorporating radio frequency oscillators in which two similar frequencies were combined to provide a lower, human-audible frequency. This was done by combining a reference frequency with a variable frequency. The theremin was played by interposing a hand to vary the capacitance between two projecting electrodes, thus controlling the pitch and volume. It was first constructed in 1920 and became popular in the late 1920s.

The process of mixing signals of slightly different frequencies is called heterodyning and was incorporated into many radios over the next couple of decades. A transistor version of the Theremin still exists, and Mystery Science Theater 3000 fans are familiar with its eerie sounds.

It was named after its inventor Leon Theremin, who originally called his invention an "aetherphone." See heterdyning, Theremin, Leon.

Theremin, Leon (1896-1993) A Russian engineer and inventor who devised electronic musical instruments, most notably the "aetherphone" (theremin) while a student at the University of Petrograd. He traveled to America in 1927 to play a concert, and two years later licensed the Radio Corporation of America (RCA) to manufacture a "thereminvox." While in the U.S., Theremin also experimented with 'multimedia' concerts, combining light shows and dance with the theremin music, later returning to do research at the University of Moscow. See theremin.

thermal noise Random noise arising from heat generated by the motion of charged particles. Thermal noise in electrical circuits is undesirable if it interferes with transmission.

thermal circuit breaker A breaker mechanism that trips when heat generated by excessive current expands the conductor. See circuit breaker.

thermal wax printing A printing process by which various colors of wax are heated, sprayed, and adhered on the printing medium where they subsequently cool to form the desired image. The colored wax is usually supplied in the form of a long continuous film roll with blocks of sequential colors (cyan, magenta, and yellow) or four colors (CMY and black). The block of each color covers the width and length of the printing area, usually about the size of the paper being printed or slightly smaller. Paper is then passed through the printer in three or four passes, depending upon the number of colors, in close contact with the waxed surface, and tiny heated print filaments 'hammer' a dot of color onto the paper. As the printout advances, the color wax film is advanced until all blocks of color have passed across the surface of the paper. The three or four color process combines to produce the effect of many intermediate colors. This is not as economical a means of printing as ink-jet, since ink-jet is an 'on-demand' printing technology, which ejects only as much ink as is needed, while thermal wax sheets feed through the same amount of film whether there is one dot on the page or a million.

Thermal wax printing has become a medium price-range solution for desktop color printers. While the process is not as photographic as dye sublimation printing, it's pretty good and costs about 70% less per printout. One of the advantages of thermal wax printers is that they can print on many different types of paper, even on very fine sheets of aluminum. Many other color print processes rely on specialized papers, sometimes very specialized high-gloss paper. See dye sublimation printing, ink-jet printing.

thermion An electrically charged particle (a positive or negative ion) emitted from a heat source. See thermoelectron.

thermionic emission The emission of electrically charged particles under the influence of heat. Thermionic emissions are characteristic of *hot cathode* ray tubes. Cathodes without thermionic emissions are called *cold cathodes*.

thermionic valve See vacuum tube.

thermistor An electrical resistor comprised of a semiconductor with a high, nonlinear temperature coefficient. The resistance of the semiconductor varies sufficiently in relation to the temperature to make it useful in a number of applications. See thermostat.

thermocouple, thermal junction A device which measures temperature at the junction of a pair of joined wires employing dissimilar materials, with the difference in potential proportional to the temperature, determined by an instrument connected to the other ends of the wires. See potentiometer.

thermocouple wire A wire used with a thermocouple which is made of iron or particular alloys calibrated to the appropriate specifications.

thermodynamics The art and science of heat-related phenomena, their properties and relationships.

thermoelectron An electron (negative thermion) emitted from a heat source. See thermion.

thermography A printing process in which nondrying inks are treated to simulate a raised, engraved surface. After passing through the press, the ink is dusted with a compound which, after the excess is removed, is exposed to heat that causes it to fuse with the ink to form a raised surface.

thermoplastic A material with industrial significance due to the fact that it can be heated and reshaped and rehardened by cooling. It has various uses including insulating and information recording. Contrast with thermoset.

thermoset A resin or plastic material which can be shaped and cured, but once this has been done, cannot be reshaped and cured again, as with thermoplastic. Contrast with thermoplastic.

thermostat 1. A sensing and regulating device triggered by temperature which is useful in turning machines on or off, for controlling fire safety devices such as alarms and sprinklers, and for regulating heating and cooling systems. 2. A device which regulates temperature, by measuring it and controlling heating equipment (or heating and cooling equipment) in order to maintain the temperature at the setting which is selected on the thermostat. This is usually accomplished by triggering the heating circuit when the temperature falls a certain amount below or above the desired setting. Programmable thermostats, which can be programmed for specific temperatures at certain times of the day, are becoming increasingly common. Temperature regulation (cooling) in large supercomputing implementations is important. See thermistor.

THF See tremendously high frequency.

thin In typography, a unit of measure related to type size indicating a distance equal to one third or one quarter of an em space. Thus, a thin in a 12-point font is 4 or 3 points wide. See em, en, point size.

thin film transistor TFT. A technology used in display devices which creates a correspondence between a transistor and pixel on the screen so pixels can be independently controlled. Used in color (RGB) active matrix LCD panels. This technology has been applied to portable display projectors and similar devices.

Thompson, Joseph John (1856-1940) An English experimenter who investigated electricity and X-Rays. He was awarded a Nobel prize in physics in 1906 for gaseous conductivity of electricity.

Thompson, Ken (1943-) Principal developer, in 1969, of the Unix operating system, along with Dennis M. Ritchie, which is quite a distinction considering its widespread use and utility. It has since evolved through extensive support by the programming community and exists in a variety of forms, although all bear similar features. Thompson also authored B, which was a predecessor to C. See Unix, UNIX.

Thompson, William See Lord Kelvin.

Thomson, Elihu Inventor of one of the first alternating current (AC) generators in 1878. At the time, the predominant form of power was direct current (DC). This was a significant achievement, because it permitted the transmission of much higher voltages, necessary to cross some of the distances desired. Improvements to the concept were soon developed by William Stanley. Thomson also experimented, in 1892, with electric arcs. He collaborated with Sebastian de Ferranti and William Stanley in the development of the transformer.

thread 1. In piping, a helical indentation used to match and secure separate sections. 2. Continuing element, theme, or train of thought. 3. In programming, a flexible process organization mechanism by which individual processes can use common resources, but continue to operate unimpeded by other threads, if needed, in order to improve program efficiency, or to increase simultaneous access to various system or applications resources. Common in object-oriented systems.

thread, discussion In online newsgroups, a topic of conversation characterized by the same (or similar) subject line, and theme and direction of discussion. Threads are a very convenient way to follow one line of thought through the myriad opinions that are discussed in the general context of a newsgroup. Good newsreading software will organize threads into groups and subgroups much the same way computer directories (folders) are organized on the computer operating system. Thus, the user can selectively open and read, or close and ignore, a thread.

three finger salute *slang* A descriptive phrase for rebooting the operating system (without powering down the system) with three designated keys held down simultaneously for most OSs that run on Intel-based IBM-licensed systems (Ctrl-Alt-Delete).

throughput Production; output; nonredundant information or items of relevance moving through a system. Throughput is used in industrial and computing industries to describe the efficiency of a system or end-result of a communication (how much information got through).

The measurement of throughput is quite specific to the system and information or objects being transferred, so there are few generalized standards for time intervals or total data against which to compare the throughput (end result). Nevertheless, relative measures of throughput, as compared to another manufacturer, another type of machine, or when processed in a different manner, can be very useful in tuning a production line system. Relative measures of data throughput in different parts of a network, or over different data protocols or operating systems similarly can be used to improve the configuration and efficiency of a computer network.

TI-99/4 A Texas Instruments home computer introduced early in 1980. It featured 16 kilobytes of RAM, sound capabilities, 16-color graphics on a 13-inch color monitor, extended TI BASIC, and cartridge-like 'solid state' program modules for a list price of $1,150 U.S.

TIA See Telecommunications Industry Association.

ticket In telecommunications, a record of a transaction or paid toll, fare, or fee. The ticket indicates either that the transaction has been confirmed and it's ok to bill the client, or that the transaction and billing have both taken place (as in many credit card transactions). Tickets traditionally were on paper, but electronic tickets are becoming prevalent, with online transactions sometimes going directly through to the credit card company from the vendor without any slips or other paper confirmations.

tickler In computer applications, a program designed to hibernate until a certain time or until certain events take place, and then become active to remind the user of something timely or important, such as appointments, anniversaries, events, etc. These applications have variously been called ticklers, reminders, and naggers.

tickler, electronic In electronics, a feedback or regeneration device consisting of two small coils connected in an electron tube, one to the anode (in series), the other to the grid-circuit.

tie *n.* Fastener, electrical strap, bundler. A strip, usually of plastic or Velcro to hold wires away from one another, or to bundle them together, or fix them in place, sometimes to a post or other secure structure.

TIES Telecom/Information Equipment and Services. A government-to-government program which provides U.S. and Russian support for the expansion of international commerce in high technology. This is a subgroup under the U.S.-Russia Business Development Committee (BDC).

TIFF See Tag Image File Format.

TIIAP See Telecommunications and Information Infrastructure Assistance.

tiling 1. In printing, a technique for printing a large image on pieces of paper that are small, relative to the size of the image. Commonly used for billboards, banners, and wall-sized murals. Most computer printers have options for tiling, in order to print large images on letter sized paper. 2. In digital image display, a visual artifact common to heavily compressed images which causes a blocky, mosaic-like appearance to otherwise smooth lines and transitions. See DCT, JPEG.

TIME Time Server Protocol. An IETF elective standard Internet protocol. See RFC 868.

time code A system of encoding timing information on a recording medium, usually along with the information that is being stored. This technique is commonly used with audio/visual recordings that will later be edited, dubbed, or otherwise manipulated or played within strict time constraints. Time code is typically stored as hours, minutes, seconds, and frames.

Time code was developed in the late 1960s when analog recording tapes became prevalent, in many cases replacing film. The system was developed because video tape lacked the sprockets which previously had been used on film to synchronize sound and images. In the 1990s, another transition is being made from analog video tapes to digital recording technologies, and the time code techniques used for analog video and audio encounter certain problems when applied to digital recording technologies. See chase trigger, MIDI time code, reference clock, SMPTE time code.

time division multiple access TDMA. A digital technology designed to overcome some limitations of analog cellular mobile communications. Time slot assignments allow several calls to occupy one bandwidth, thus increasing capacity for various wireless technologies. E-TDMA (Extended TDMA) provides even more time slots. TDMA is widely supported by AT&T Wireless Services. It is similar to code division multiple access (CDMA).

There are a number of TDMA implementations, with three primary ones: European TDMA (GSM), Japanese TDMA (PHS/PDC), and North American TDMA (IS-136). See AMPS, DAMPS, cellular phone, time division multiplexing.

time division multiplexing TDM. A technique for combining a number of signals into a single signal by allocating a time slot in the combined signal with a multiplexer. At the receiving end, a demultiplexer is used to separate the interleaved signal back into its original signals. Some of the early developments of this technique were accomplished by J. M. E. Baudot in the 1870s. In current usage, TDM allows a variety of types of communications, audio and video, to be transmitted at the same time in one interleaved signal.

time signals From around the mid-1800s, before time zones were established, to the present day, people have sought to devise ways to determine the time and synchronize their activities. The first time signals were drums or bells which were regularly sounded based on the sun's position in local communities. Later, in the 1860s, the U.S. Naval Observatory used the telegraph to transmit time signals, and soon Western Union was sending standard time signals, a tradition they continued for a century. Telegraph time signals were similar to current Coordinated Universal Time signals, in that audible clicks were used coming up to the hour, just as tones now signal the upcoming minute. See Coordinated Universal Time, Greenwich Mean Time.

Time T An ITU-T designation for 2359 hours Coordinated Universal Time (UTC) on 31 December 1996.

timing Configuration of a system so successive repetitions are controlled for the desired interval (which may be desired to be variable), or so certain events begin and/or end at designated times or according to certain events. Timing is important in magnetic storage mechanisms, motors, signal amplitude sequences in electronics, broadcast equipment configurations. Timing is also important on networks, where, for example, video and audio signals may be sent separately or on separate lines, but have to be coordinated at the end to provide services like videoconferencing. Constant oscillators are often used in conjunction with very precise timing devices. See atomic clock, quartz, SMPTE time code.

timing signal 1. A signal generated according to an accepted standard of time, usually for the purpose of providing a precise or objective baseline against which to measure events. 2. A signal generated by measuring some repetitive event which is then compared to some standard or clock. 3. A signal simultaneously recorded with data to provide a measure or standard against which the data can be analyzed. 4. A regularly emitted signal against which other time-related events can be synchronized.

TINA Telecommunications Information Networking Architecture. A networking telecommunications software architecture intended to be developed into a global standard. See Telecommunications Information Network Architecture Consortium.

TINA-C See Telecommunications Information Network Architecture Consortium.

tinned wire Wire that has been treated with tin to provide insulation and/or to facilitate soldering. Common on copper wire.

tinsel A fine, very long thread or strip of metal sometimes wound around electrical conductors between the insulator and the main core or wire. Because of its properties, tinsel is used in cables that need to be tightly wound or very flexible (such as phone handset cords).

tint Lighter or darker values of a particular color; hue. Tints are created by successive additions of white or black pigments, or by successively increasing or decreasing values of red, green, and blue (RGB) together at the same time. Greater amounts of each color of RGB will produce lighter tints and lesser amounts will produce darker tints.

tip 1. The line or connection attached to the positive side of a circuit or battery. 2. In two-wire telephone wiring, the tip is traditionally the green wire attached to the positive side of the circuit at the central switching office. The name originates from the configuration of a manual phone jack in an old telephone switchboard in which the large plug was divided into two sections, with an internal wire electrically connected to the *tip* of the plug, and another wire to the *ring* around the plug partway up the jack nearer the insulated cord. See ring, tip and ring.

tip and ring Historically, the tip and ring designations derive from the configuration of a phone jack from a manual switchboard, called a cordboard. The *tip* was the positive circuit connected to the tip of the jack, and the *ring* was negative located slightly away from the tip, encircling the jack, sometimes called 'sleeve.' Later the tip and ring became standardized to correspond with the green and red color-coded wires traditionally used to install phone line services. Often telephone wire is composed of four wires, with red, green, black, and yellow sleeves. Since dual lines have become more common in small businesses and in some homes, the black and yellow lines are used for tip and ring, respectively, for the second line. While these codes are standardized in North America, there are variations in other countries and in larger installations with multiple phone lines.

tip jack, pup jack One of the simplest types of connectors, a tip jack has a single, usually round, contact point plug that fits into a matching single-hole plug.

Titanic, RMS The famous, ill-fated 'unsinkable' ship that sank in 1912, at the cost of hundreds of lives while crossing through ice fields north of Canada. The radio operator of the Titanic sent distress calls, but two of the closest ships didn't receive the communications, as 24-hour watch programs and radio regulations had not yet been established (this changed after the sinking of the Titanic). Fortunately, however, some of the sea-goers were saved by the Carpathia, who came to their rescue after receiving the radio distress call.

The sinking of the Republic, an earlier ocean-going ship, had a strong influence on legislation requiring wireless communications systems to be installed on ocean-going vessels. Due to the wireless distress calls, all but two of the hands on the Republic were saved. The sinking of the Titanic resulted in further legislation associated with keeping those communication lines open and monitored 24 hours a day. See the JASON project, MARECS, MARISAT.

title Commercial product, usually shrink-wrapped. A term often applied to off-the-shelf entertainment and educational software products, especially multimedia CD-ROM programs.

TJF test jack frame.

TL tie line.

TLF trunk link frame.

TLP Transmission Level Point.

TM 1. terminal multiplexer. 2. traffic management. A term associated with network transmission cell traffic flow, monitoring, and control. See cell rate.

TMA Telecommunication Managers Association.

TMC traffic management center.

TMGB Telecommunications Main Grounding Busbar

TMN See Telecommunications Management Network.

TMS 1000 A one-chip 4-bit microcomputer introduced by Texas Instruments in 1972. Arguably the second microcomputer ever released, following the release of the Intel MCS-4 chip set, it never gained popularity.

TMSI See Temporary Mobile Station Identifier.

TOGAF See The Open Group Architectural Framework.

toggle Two-state process or switch. Flip-flop. An electrical current that alternates in intensity at two distinct levels, or which has two states: on or off. A traditional light switch is a type of toggle, unlike a dimmer switch which has many possible settings. In software applications, a button or icon that has two states or positions. In computer programming, a flag that is either set or not set, on or off. Toggles are very commonly used to keep track of settings.

token On a Token-Ring network, a status/priority information block used in coordination of traffic on the network. A token consists of a 24-bit frame the operates at the Media Access Control (MAC) level. It is continually passed around the ring in one direction. It consists of a start delimiter (SD), an access control (AC) field, and an end delimiter (ED). Most of the information for controlling events is contained in the AC. It is further subdivided into a 3-bit priority field, a token bit (zero indicates it is a token), a monitor bit, and a 3-bit reservation field that lets the sta-

Token-Ring Basic Frame Components

Abbrev.	Item	Notes
SD	Starting Delimiter	Indicates the beginning of a frame.
AC	Access Control	Contains the Priority, Token, Monitor, and Reservation bits. In a frame, a workstation can only change the Reservation bits in the access control field. Only the active monitor can change the "M" bit, and only the workstation or device changes the Token bit.
FC	Frame Control	Indicates the type of frame: data frame or maintenance frame. A maintenance frame is used by the protocol to manage the ring.
DA	Destination Address	Physical or NIC address of the receiving workstation or device.
SA	Source Address	Physical or NIC address of the sending workstation or device.
	Information	Layer control, routing control, and data
FCS	Frame Check Sum	Error checking at the destination.
ED	Ending Delimiter	Indicates the end of a frame. If one of multiple frames or the last frame in a transmission, it's an I bit; if an error occurred, it's an E bit.
FS	Frame Status	Indicates if a frame has been recognized and copied by the destination station.

tion get in the priority 'queue' for future transmission of frames. A token is reissued after use by a station of the suitable priority, and continues on its way. See token-passing, Token-Ring frame, Token-Ring network.

token latency In a token-passing scheme, the time it takes for the token to make it all the way around a token-passing local area network. See token-passing, Token-Ring.

token-passing A process on a Token-Ring network by which status/priority 'tokens' are used as a mechanism to coordinate traffic around the unidirectional ring. A token is a 24-bit frame the operates at the media access control (MAC) level. It consists of a start delimiter (SD), an access control (AC) field, and an end delimiter (ED). The token-passing continues around the ring where each station checks the priority before adding frames to the traffic on the ring. Once a token has been used, assuming the proper priority, it is sent out again by the transmitting station to continue its journey around the ring through each station, carrying out the same sequence of events, according to priority levels.

Token-Ring frame A Token-Ring frame in the IEEE 802.5 specification consists of the components specified in the Token-Ring Basic Frame Components chart.

Token-Ring Interface Processor TRIP. A Cisco Systems high-speed interface processor which two or four Token-Ring ports which can be independently set to speeds of 4 or 16 Mbps.

Token-Ring network TR network. A local area network developed by International Business Machines (IBM) in the mid-1980s, employing a star or ring topology, that is, with nodes connected directly to one central hub, or in a continuous loop not requiring termination. The token-passing is a scheme for data transmission between the stations which prevents collisions from different workstations sending messages at the same time. A workstation cannot transmit until it receives 'permission,' that is, a token of the proper priority. The token is passed from station to station around the ring in one direction. At each station, the priority is checked before a frame is transmitted on the ring.

Token-Ring uses a source routing system that moves information among stations based upon information in data packet headers, thus utilizing inexpensive bridges, a system different from Ethernet LANs. Speeds are up to about 16 Mbps, a limitation that has been addressed in High Speed Token-Ring, and throughput is about 60 or 70 percent, somewhat higher than Ethernet. Frames hold about 4,000 bytes. Token-Ring LANs are sometimes combined with Ethernet LANs. They typically run over copper twisted pair cables, although some are now being implemented with fiber.

Since IBM's introduction, Token-Ring has been developed into a Media Access Control (MAC) level standard protocol by the IEEE (802.5).

A Fiber Distributed Data Interface network is one which employs various Token-Ring token-passing concepts using dual rings to provide redundancy and fault tolerance. See Ethernet, Fiber Distributed Data Interface, High Speed Token-Ring, ring topology, star topology, topology.

TokenTalk An Apple Computer Macintosh-based implementation of a Token-Ring local area network. See AppleTalk, Token-Ring network.

toll call Any call to a location outside the local service area, so called because it is billed at a rate above and beyond the local subscriber service. A long distance call.

toll denial Denial of service outside the local service area. Toll denial may be part of a private exchange in order to limit calls to local calls, except as authorized. Toll denial may also be set up on an individual subscriber line (e.g., a subscriber who is behind in payments on long distance charges), or a line in a college dorm, or other location where potential toll abuse or fraud may occur. It may still be possible to make toll calls by going through an operator or entering authorization codes. See toll diversion, toll restriction.

toll saver A feature of answering machines and some computer software programs that lets you call into an answering machine from a long distance location and know whether there are any messages *before* the line is connected. The system is based upon the number of rings, usually four or two. With toll saver enabled, if there are no messages, the answering system will ring four times before answering. If there are messages, it will answer on two rings. That way, if the caller hears three rings, there are no messages, and he or she can hang up before the machine answers on the fourth ring and save the long distance charge.

toll terminal A phone system set up only for long distance calls (toll calls). Toll terminals may be set up to expedite long distance calls, or to compartmentalize billing or reporting on long distance calls. A toll terminal may be in a secure location, accessible only to authorized toll callers.

tone dialing A system of audio tones called *dual tone multifrequency* (DTMF), used to generate distinct signals with which phone numbers and symbols can be transmitted. Frequencies are selected in such a way that dual tones are not harmonically related. Each tone is actually a combination of two tones, a high tone and a low tone, which are decoded when they are sent down the line to the switching office. The high tone is usually slightly louder, or at least as loud, as the low tone. The tones range in frequency from about 697 Hz to 1633 Hz.

Two advantages of tone over pulse dialing are speed and flexibility. As direct dial long distance services became available, numbers became longer, and it takes a long time to dial a number on a pulse phone, partly because of the mechanical act of rotating the dial and partly because the dial has to return to the base position before dialing the next number. Touchtone systems also provide more options. With a combination of digital processing and touchtone signals, automated menu systems can be accessed through the phone. These are now widely used by banks, mail order houses, and others. See pulse dialing, DTMF, touchtone dialing.

tone generator Any device or software application which generates tones. These may be at a particular frequency or may vary. Tones can be generated in the audible frequency range for humans, or higher ranges for electronic detection. See buzzer.

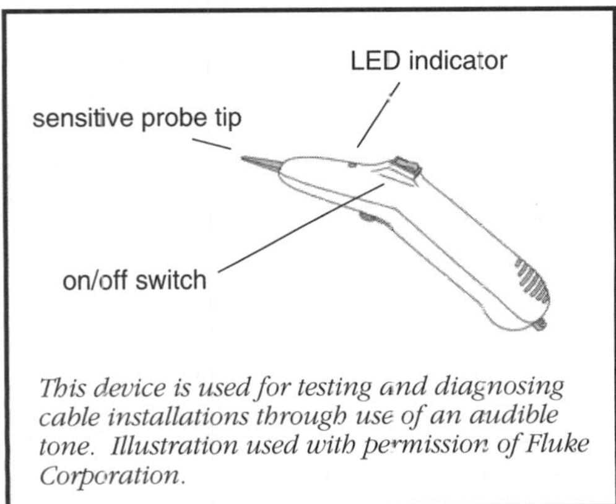

This device is used for testing and diagnosing cable installations through use of an audible tone. Illustration used with permission of Fluke Corporation.

tone probe A diagnostic device which is used to detect and amplify audio tones, usually in the 500 to 5,000 Hz frequency range, although it depends on the application. Cable technicians use these to trace cable links, for example. The tip of the probe is unshielded in order to selectively position it to detect a signal. Filtering may be incorporated to eliminate noise. Some probes have earphone jacks so they can be used with stereo headphones. A tone probe may be used in conjunction with a tone sender.

tone receiver unit TRU. The electronics in a telephone receiver which detect and interpret touchtone codes.

toolbar In computer graphical user interfaces, a type of selection menu that consists of buttons, usually iconized, grouped together. Commonly, toolbars consist of about four to 20 buttons grouped in one or two rows or columns. Toolbar buttons are often single-clicked to activate them, or double-clicked to expose more options. They are especially common in painting and drawing programs, or along the tops and bottoms of word processing or desktop publishing software. Toolbars may or may not be movable.

TOP See Technical Office Protocol.

top-down An organizing or processing hierarchy that distributes itself downward, usually in a branching pattern. Top-down often implies higher priority or more generalized functions or items at the top of the hierarchy. Thus, a top-down outline lists more important concepts first, a top-down personnel chart usually shows executive managers at the top, a top-down phone system starts with priority-listed agents first, etc.

Top Level Aggregate TLA. An IPv6 prefix, a coveted commodity to the Internet community, the proposed assignment of Top Level Aggregates to privileged companies caused controversy. To quell the objections, TLAs were significantly increased and TLA requirements were removed from the IPv6 specification. There are many Internet developers who are concerned with preventing the development of a VIP system on the Net. See IPv6, Next Level Aggregate.

topology, network topology A schematic representation and configuration of the geometric and electrical connections and relationships of a network and its various routing components. Depending upon the number of servers, terminals, routers, and switches, there are a variety of possible configurations that are practical, including token rings, stars, and others. See backbone, mesh topology, node, star topology.

topology management In networks, the configuration, tracking, and management of connecting devices, particularly switches. The software used to manage them will often show graphical images of the various devices and their interconnections in the system.

Torricelli, Evangelista (1608-1647) An Italian physicist and mathematician who invented the torricellian tube, now known as a mercury barometer. Barometers later became important in weather forecasting and in altitude-measuring instruments, particularly for aeronautics and ionospheric experimentation.

touchscreen A specialized computer monitor which is activated by contact with a finger or pointing instrument. The idea was that it was more natural for people to point than to use a mouse of keyboard. Unfortunately, it didn't take into consideration the fact that holding an arm up for any length of time is uncomfortable, so touchscreens haven't become popular for extended or repetitive work. However, they are suitable for occasional input, as in kiosks and directory systems, and are suitable for some types of childhood education.

Townes, Charles H. The Nobel Prize-winning scientist who co-invented the laser along with Arthur Schawlow, in 1958, while working at Bell Laboratories. See Capasso, Frederico; laser.

TPA See Telephone Pioneers of America.

TPDU Transport Protocol Data Unit.

TPI tracks per inch.

TPWG Technology Policy Working Group.

TQM total quality management. A management philosophy and means of putting it into action to develop and maintain quality principles in commerce, service, and manufacturing. Quality assurance and ISO-900*x* certification are two kinds of tools in the TQM arsenal. See ISO 9000.

TR transmit/receive.

traceroute A Unix shell utility written by Van Jacobson at Lawrence Berkeley Labs which seeks out and displays the path of a transmission packet as it travels from host to host, detailing the hops in the path. Traceroute sends an IP datagram to the destination host, then it iterates through each router, decrementing TTL, discarding the datagram, and sending back ICMP messages until the destination host is reached. Traceroute is extremely useful as a network diagnostic and optimizing tool and often used in conjunction with ping. See ping. The Traceroute Example inset below shows a sample of traceroute output showing the IP numbers, hops, and packets.

trackball A hardware peripheral device which receives tactile directional input and transmits it to a computing device. The information transmitted is similar to that of a mouse or stylus, and is often used in conjunction with graphical user interfaces (GUIs). Unlike a mouse or stylus, a trackball is generally fixed in place, with physically separate buttons. Trackballs are common in video arcade games and on laptop keyboards to increase portability. See joystick, mouse.

tractor feeder A sprocketed paper alignment device used mostly on impact printers. It pro-

Traceroute Example

```
$ traceroute abiogenesis.com
traceroute to abiogenesis.com (207.173.142.184), 30 hops max, 38 byte packets

1  chapman.nas.com (198.182.207.6)  322 ms  252 ms  251 ms
2  orthanc.nas.com (198.182.207.1)  251 ms  247 ms  275 ms
3  milo-s4.wa.com (204.57.232.1)  277 ms  257 ms  296 ms
4  dilbert-fe4-0.wa.com (192.135.191.254)  530 ms  303 ms  273 ms
5  dogbert-f4-0.nwnexus.net (206.63.0.254)  263 ms  282 ms  286 ms
6  borderx2-hssi2-0.Seattle.mci.net (204.70.203.117)  306 ms  277 ms  373 ms
7  core2-fddi-1.Seattle.mci.net (204.70.203.65)  288 ms  511 ms *
8  bordercore1.Denver.mci.net (166.48.92.1)  299 ms  323 ms  293 ms
9  electric-lightwave.Denver.mci.net (166.48.93.254)  294 ms  351 ms  313 ms
10 H3-0.1.scrlib01.eli.net (207.0.56.162)  329 ms  394 ms  591 ms
11 gw2-CALWEB-DOM.eli.net  (208.131.46.46)   730 ms  gw1-CALWEB-DOM.eli.net
(208.131.46.30)  507 ms gw2-CALWEB-DOM.eli.net (208.131.46.46)  357 ms
12 abiogenesis.com (207.173.142.184)  362 ms  358 ms  350 ms
```

vides more precise control of positioning than most roller feeders by preventing slippage. The tractor feed resembles a pair of short regular series of inverted cleats which fit through corresponding holes in tractor feed paper. Tractor feed mechanisms are especially useful for multipage documents (invoices, checks, etc.) on dot matrix printers.

trade secret Information, data, process, or procedure which would lose its commercial value by being revealed to outsiders. Nondisclosure agreements (NDAs), policy statements, and inservice training are mechanisms which restrict external communication of trade secrets. If trade secrets are not specifically identified by an employer, it may be more difficult to stop or prosecute an offender who has willfully or inadvertently revealed them. See patent, copyright, nondisclosure agreement, trademark.

trademark A legal designation for the right of an association to the use of a mark in trade. It can consist of a word, phrase, or symbol sufficiently unique to be distinguished from others in the same general industry. Provided it is not already owned by another entity, a trademark becomes valid as soon as a company uses it in trade according to certain stipulations, provided the company continues using it. You cannot come up with a trademark idea, not use it, and claim it later if someone else uses it. Trademarks may be registered federally for a fee, although this is not required. The motivation for trademark registration is to provide prima facie evidence in the event of a legal dispute. Unregistered trademarks must be identified with a ™ symbol, and registered trademarks must be identified with a ® symbol. Policing of trademark violations is not a responsibility of the federal government or of the agency that registers a trademark; it must be done by the company seeking to protect the trademark. Trade names are similar to trademarks and are registered at the state level. There can be more than one company using the same trade name, if the line of business is sufficiently different to prevent confusion.

The Commerce Department's large database of patent text and images, 800,000 trademarks and 300,000 pending registrations dating from the 1800s are being uploaded to the Web during 1998 and 1999 and thereafter. See copyright, patent. http://www.supot.gov/

Traf-O-Data The first business partnership of Paul Allen and William R. Gates, founded around 1972, growing out of the business experiences of the less formally organized Lakeside Programming Group. Allen and Gates worked together on a variety of software projects including an automobile traffic flow system. At about the same time, they prototyped an early microcomputer based upon the recently released Intel 8008, with the participation of a hardware designer, Paul Gilbert. They made some tentative attempts to sell this early machine, but it didn't work during an early demonstration.

At about the time Gates graduated from high school, Allen encouraged him to join him to start a company to build and sell computers based upon the Intel 8080 chip set, but Gates wasn't fired up about the idea of a hardware company, and his parents were encouraging him to continue his post-secondary education. The majority of Gates and Allen's subsequent efforts were software-related, most significantly, a BASIC interpreter for the Altair, and a disk operating system for International Business Machines (IBM), which resulted in the founding of Microsoft Corporation. The subsequent success of the company was due to sales of software, primarily operating systems and business applications, and Apple Computer instead became the small business success story in computer hardware.

traffic A term often used on large communications systems to describe communications signals, data, cells, or packets which comprise the information and signaling associated with the transmissions. The term is sometimes also used to describe the overall flow and pattern of traffic in the context of the system it is on, as in traffic congestion, traffic flow, traffic monitoring, traffic engineering, traffic routing, etc. In telephony and various systems that are largely analog based, the term may be used to describe the flow of information in the context of calls, as in call attempts, call connects, call volume, etc. In digital systems, it is often used in a more specific sense to indicate numbers of packets or cells. See leaky bucket.

traffic capacity A measure of the capability of a system to carry a certain maximum number of calls, cells, frames, or packets, depending on how capacity is measured or data carried on that particular system. The maximum capacity may not be a fixed number, as on systems where multiple channels may be aggregated to carry certain broad bandwidth types of information. On other systems, where one wire or two wires can carry one and only one communication, traffic capacity is more likely to be expressed as a set amount. *Traffic capacity* is used in systems design and marketing to provide buyers a general guideline as to the capability of a system. For example, a rural phone switching system in a small community may only require a capacity of 20,000 call seconds (usually expressed in increments of *hundred call seconds*) per hour during peak calling times, providing a measure of whether a system can meet or exceed those needs. See traffic concentration.

traffic concentration A measure of communications traffic that indicates how different peak traffic periods are compared to traffic on the whole. In telephony, the service day is usually broken down into hours for traffic monitoring to determine high and low traffic periods. Traffic concentration is typically expressed as a ratio of the traffic (number of calls) during the busiest hour to the total traffic during a 24-hour period. This may further be calculated over a period of weekdays, weekends, and/or seven-day weeks to provide statistical averages.

traffic engineering In more traditional communications systems, traffic engineering is the estimation and application of the type, quantity, and configuration of devices and equipment which are required to meet the needs of a predicted number of users of the system being designed. Experience, trial and error, probability theory, similar system comparison, and insight are all brought into play in designing a system to meet current and predicted future needs. This system works quite well with traditional telephone switching systems and local area networks.

As systems become more complex, however, and distributed digital data systems more prevalent, traffic engineering, once the initial equipment is set in place, becomes an even more esoteric process, with much of the configuration and selection being carried out by computer software, not just through physical connections. Virtual networks are now laid over physical networks, resulting in several levels of traffic engineering. When these networks are interfaced with a global network like the Internet, then prediction of the number of potential users becomes more difficult and less important than incorporating flexibility and scalability into the system to handle unpredictable traffic loads and activities. At this level, traffic engineering becomes a collaborative activity between programmers and traffic engineers, with careful evaluation of data and message priorities, scheduling and incorporating deliberate delays, for example, so some types of traffic can be set to transmit during off-peak hours. On data networks, many traffic decisions are now built into the software on servers, gateways, routers, and some of the high-end switchers.

traffic load The sum total of the traffic on a particular system, or portion of a system such as a specific trunk, leg, or hop, measured at a particular point in time, or during a particular specified range of time.

traffic monitor A mechanism for keeping track of traffic on a system. On data networks, some of the software included for system administrators allow the monitoring of various processes, and typically display them as sampled or realtime graphical charts. Thus, CPU usage, number of users, number of transmissions, level of traffic, etc. can be visually assessed, and statistics derived from computerized traffic and analysis can be stored and evaluated to make changes as needed. The results of the monitoring are sometimes directly incorporated into other software on the system which makes adjustments to priorities, number of users permitted online at one time, and other parameters that can be changed to increase or decrease the traffic to optimize use of the system.

Sometimes traffic monitoring is a very simple mechanism. One of the simplest and most familiar applications is on modems, where little blinking lights give some indication as to how much data is traveling through the modem, and when. If a user is downloading a large file from a Web site and those lights stop blinking, it's possible that the connection has been dropped or there is a glitch in the system. In ATM systems, traffic monitoring is an essential aspect of preventing network congestions and bottlenecks, and some cells will be flagged as to their discharge eligibility. See cell rate, leaky bucket.

traffic overflow A situation where demand exceeds capacity. When overflow occurs, the traffic is either routed to another trunk or leg, or it is rejected and a signal sent to the user in the form of a signal (like a fast busy) or a text message (as a broadcast on a computer system). In some data networks, particular packets may be discarded if there is traffic overflow, or they may be routed back to the sender.

traffic path The physical or virtual pathway taken from the sender to the receiver. This may be fixed, as in direct wire communications and smaller data networks, or it may be flexible, as in switched systems and many larger data networks. On very small systems, where the setup is known and doesn't often change, a fixed traffic path may be the fastest and most practical implementation. In dynamic data networks, the communication may not be quite as fast, but the system has an advantage in that it can tolerate and adjust to unpredictable changes.

A path in a phone system usually passes from the subscriber to a *demarcation point*, usually a junction box between the internal and external wiring, through one or more *switching centers* which are connected with pathways called *trunks*, or sometimes through a wireless link during part or all of the transmission. In a digital computer network, the may data pass from the sender through various *servers*, *gateways*, *switches*, and *routers*, and each section in the path is called a *hop* or a *leg*, with the various terminal points generically called *nodes*. A data network transmission may also be wireless for part or all of the path.

traffic policing In ATM networking, a mechanism which detects and controls cell traffic according to specified parameters. See cell rate.

traffic recorder A means to record traffic on a specified transmissions channel in order to monitor capacity, load, efficiency, etc. It may or may not be paired with software that helps analyze the traffic.

traffic reporting In networks, information on traffic flow gathered and organized by external analyzing devices and/or internal monitor agents. This information about packet volume, distribution, collisions, and errors, may be reported in the form of tables, ASCII graphs and charts, or images, and is essential for configuration, tuning, and troubleshooting.

traffic shaping In ATM networking, a mechanism which shapes or modifies bursty traffic characteristics in order to create the desired traffic. See cell rate.

train *v.* To instruct, teach, indoctrinate, or drill. Training is an important component of any computing system that must recognize input beyond point and click or keyboard instructions. In pen computing, handwriting must be interpreted into commands, and it is usually necessary to 'train' the system, by successive trials and feedback, to recognize an individual's style of writing. Some camcorders have eye-controlled systems which need to be trained to track the direction of focus of the user's eye. In voice recognition systems, software is trained to recognize an individual's mode of speaking. OCR systems can be trained to improve recognition of unfamiliar or unusual character sets. While software training systems are not perfect, they have evolved to the point where they do much useful work, and improvements in technology and software algorithms indicate that some day natural methods of computer input may supersede keyboard and mouse entry for many applications.

training An automatic feature of some hardware and software systems to evaluate the characteristics of incoming signals (timing, delay, etc.) or information (handwriting, voice, etc) and improve performance or recognition through successive adjustments and feedback. See artificial intelligence, expert system, robotics.

transaction 1. An agreement, or exchange of information or goods, between two or more parties or entities. 2. A business transaction, which may be subject to various recording requirements (contracts, taxation, audit records, etc.). 3. An entry into a database or spreadsheet.

transaction, network Any of a number of situations in which information or signals are exchanged, passed on, recorded, or evaluated. Examples of network transactions include protocol determinations and conversions, security level evaluation and processing, routing, transmission between nodes, error processing, etc.

Transaction Capability Application Part TCAP. In the SONET specification, there are several chapters regarding the function, formats, and definitions for TCAP provided in the ANSI T1.114-1996 document.

transaction tracking A system of recording each instance of a transaction, and sometimes also the actual transaction or its outcome. In database or spreadsheet applications, for example, each transaction may be recorded as entered, in order to prevent loss of data due to a series of transactions not having been saved at regular intervals. This type of 'live' recording of transactions frees the user from worrying about 'saves' and ensures that, under most fault situations, no more than the most recent entry may be lost. See ticket.

transatlantic cable The invention of the telegraph communications system provided a strong motivation for laying a cable to bridge the gap between North America and Europe. The first cables did not succeed, but improvements in insulation, and trial and error attempts resulted in a successful installation in August 1858, initiated by F. N. Gisborne and financed by Cyrus Field. Another was installed between Ireland and Canada in 1894. See Gisborne, Frederic Newton; gutta-percha.

transceiver Device which transmits and receives within a single unit, often sharing circuitry to reduce size and weight. See transmitter, receiver.

transcontinental telegraph The first transcontinental telegraph line was initiated by Hiram Sibley, with encouragement from Ezra Cornell, in the mid-1800s. Sibley, the founder of what was to become Western Union, made a reasonable estimate that it would take two years and about $1 million to complete the project.

The telegraph line had a significant impact on the new Pony Express service, which had been in operation for less than two years when it shut down in October 1861, after inducements to stay in service at least until the telegraph line was completed.

Not surprisingly, problems other than weather plagued the construction of the line. Buffalo discovered that telegraph poles made good scratching posts, sometimes bringing down the poles in their enthusiasm. Native bands sometimes made off with the wires, as the lines stretched through their treaty lands, and within days or hours, exquisitely woven copper wire bracelets would appear in local trading markets. Remarkably, despite the great distance, harsh conditions, and the small size of the work crew, only about 50 line workers, the transcontinental telegraph was completed in October 1861. It had taken only four months at a fraction of the projected cost, one of the most stunning achievements in western engineering.

transducer A general term for a device which converts one form of energy to another, a process that is used throughout communications. When sound waves from a telephone conversation come in contact with a telephone mouthpiece diaphragm, the diaphragm causes small polished carbon granules to cohere and the energy is converted to electrical impulses that are transmitted along the phone line. When the mechanical movements of a phonograph stylus are turned into electrical impulses, and then, at the speaker, converted again to audible sound waves, the signal has gone through (at least) two transducers.

transformer An electrical device for changing the qualities of a current by mutual induction. Transformers did not come into wide use until the early 1900s, gradually superseding spark coils for providing power for communications and electronic components. They were similar to spark coils in that they had a core surrounded by conductive windings. However, the core used soft iron sheets rather than a bar. Like a spark coil, two sets of windings, one within the other, were commonly used. The core could be closed or open. Transformers required alternating current and direct current was still prevalent at that time, but the use of alternating current allowed the elimination of a vibrator, as the natural alternations of the current caused inductive discharge. This was more compact and practical than a spark coil.

Since many modern electronic appliances (modems, printers, answering machines, model trains) have electrical requirements different from that which comes out of a household socket (110 or 220 alternating current), the power cord may be equipped with a transformer which modifies the current to the needs of the device being powered (nine or 12 volts is common). It is important to use the correct transformer; if the voltage is too high, it will likely 'blow' your components.

Transient Mobile Unit A mobile communications term for a unit that communicates through a foreign base station.

transistor A small device developed in the late 1940s which provided a means to amplify signals with very low power consumption and very little heat. The importance of the transistor to the development and evolution of electronics cannot be overstated. A whole new world of tiny components opened up, including portable radios, hearing aids, computers, satellites, and much more. See transistor history.

transistor history The invention of the transistor in 1947 is widely attributed to William Shockley, John Bardeen, and Walter H. Brattain of Bell Laboratories, although Ralph Brown is sometimes also mentioned. Starting in 1951, when they began to be commercially produced, transistors replaced large, power-hungry, cumbersome vacuum tubes, enabling electronics to be smaller and less expensive, and to run cooler and faster. (There are still some high-frequency applications where the use of vacuum tubes is practical.)

The development of the transistor was foreshadowed by the 1926 patent application of Julius Edgar Lilenfeld, who had devised a way to control the flow of current in a solid conducting body by establishing a third potential between the two terminals. Later, in the 1970s, many types of transistors were succeeded by semiconductors. See de Forest, Lee; Kilby, Jack St. Clair; Pickard, Greenleaf.

transistor radio A radio developed in the 1950s, based on small semiconductor transistors instead of larger electron tubes. The smaller size and power consumption of transistors made it possible to design handheld portable radios, which became popular and widespread in the early 1960s. Portable radios were not new, since the early crystal detector sets required no outside power and could be carried around in a small case. However, practical, amplified, battery-driven portable radios did not become widespread until the development of small, low-cost transistor components.

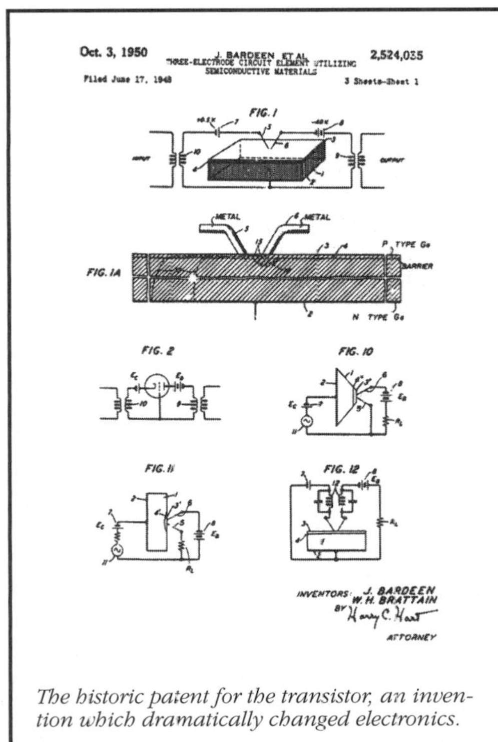

The historic patent for the transistor, an invention which dramatically changed electronics.

transistor-transistor logic TTL. A logic circuit design similar to diode transistor logic (DTL), but with multiple emitter transistors. Sometimes called multi-emitter transistor logic.

transit In network communications, if a provider wants to send data to a destination but does not have the needed routing information, the provider may arrange temporary, permanent, fee or nonfee access (transit) to the destination through a second provider which has the necessary information or access. See peering.

transliterate To spell or represent the characters of one alphabet in the closest possible corresponding characters of another alphabet. Transliteration does not imply translation of the actual meaning of words composed of those characters. Western European languages transliterate reasonably well. It is harder to transliterate between cyrillic and roman characters, and is decidedly a challenge to transliterate between pictographic or symbolic alphabets, such as Asian languages and sequential, phonetic alphabets such as western European languages. The difficulties in transliteration on computing systems have led to many alternate keyboards, character mappings, and input systems. See Unicode.

transmission The sending of information through electrical signals. It is very common for information to be added to a signal through various forms of signal modulation, often with a 'carrier wave.' See carrier wave, modulation.

Transmission Control Protocol TCP. A widely used Internet and local area network (LAN) connection-oriented packet-switching transmission protocol descended from previous DARPA editions. Together with Internet Protocol, the TCIP/IP combination is a means for the transport of host-to-host information over layer-oriented network architectures. See Internet Protocol, RFC 793.

transmission medium Any material through which transmission is facilitated either due to its inherent characteristics or through inherent characteristics enhanced by technology. Common transmission media include air, light, wire, coaxial cable, fiber optics, etc. Broadcast transmissions primarily are sent through air, fiber, and coaxial cable. Computer transmissions are typically sent through copper wire or coaxial cable, although the use of fiber is increasing. Various media vary greatly in the amount of information (bandwidth) they can carry at any one time. See individual media for more detailed information.

transmitter 1. That which transmits, or sends through some means such as chemical, optical, or electrical signals. 2. A device which sends out a signal, as a transmitting antenna, telegraph instrument, or modem. A transmitter may also include various mechanisms to amplify, compress, or modulate, or encode a signal. See telephone transmitter.

transparent A transparent technology is one in which the inner workings are not apparent to the user. For example, in computer operating systems with graphical user interfaces, the user sees the applications through point-and-click icons, text windows, and resizable gadgets and dialogs. The conversion of the information into operating instructions, the device drivers, queuing mechanisms, buffers, priority and security mechanisms, and binary arithmetic are essentially in the background, and thus 'invisible' or *transparent* during typical interactions.

transponder, radio In radio communications, a transceiver that transmits information automatically, on receipt of an appropriate interrogation signal.

transponder, satellite In satellite broadcasting, a device which receives and retransmits electromagnetic signals. Broadcast satellites employ this technology with multiple transponders. With compression, the capacity of a transponder can be significantly increased.

transportable cellular phone A transportable phone consisting of a handset, antenna, and battery, usually bundled together in a carrying bag (sometimes known as 'bag phones'). It's a little heavier than a self-contained, handheld cellular phone. This type of phone can operate on up to 3 watts of power. It can be operated independently of a car battery and is typically used for field work that requires a phone with greater mobility, and sometimes a larger antenna and longer life battery. They may include fax and modem features, and may be used in conjunction with a laptop. Journalists, scientists, and business people who need something a little more full-featured than a simple handset system use transportable systems.

trap To confine, narrow in on, circumscribe, or surround.

trap door A hidden device or software mechanism which allows entrance to an application, environment, or system only to those aware of its presence. On computer systems, trap doors are often paired with passwords or puzzles to further hinder access. Trap doors are notorious for being inserted by programmers into software applications, for later entry, sometimes for clandestine purposes. They have legitimate uses as well, providing access to authorized administrators without signaling their presence to general users or unauthorized invaders. See Trojan horse.

trapping In printing, a technique for assuring that adjacent inks meet, or slightly overlap so there won't be an undesirable, paper-colored gap between the inks if the registration of the press is slightly off. When a user creates desktop publishing page layouts that are intended for printing at high resolutions, there are usually settings that must be adjusted to maximize trapping to ensure the quality of the final product. Typically, small, light-colored objects like fonts are *trapped* (slightly overlapped) over dark ones, so that the detail in the small objects is not lost. See choking.

traveling user TU. In a Secure Data Network System (SDNS), a traveling user is one who is visiting a Message Security Protocol-equipped (MSP-equipped) facility other than the usual one where the user reads and sends messages. In networks in general, a TU is someone who may interconnect or interact with a network from a variety of facilities while traveling. This type of network access is increasing as mobile systems become more prevalent.

traveling-wave tube TWT. A tube in which a stream of electrons interacts in a more-or-less synchronized manner with a directed electromagnetic wave so energy transfers from the stream to the wave. This type of mechanism is used in communications satellites. The TWT was designed by inventor R. Kompfner, and later improved by Kompfner and J. R. Pierce at Bell Laboratories.

tree structure A common structure in computer programming and file organization. A tree branches from a main trunk to its various branches, just as the roots successively branch into subdivisions as you move away from the main trunk. Data branching structures are found in various database data storage schemes, in file directory structures, in fractal images, and more. Physical branching structures are found in network topologies; for example, branches may come off a backbone, and branch further in individual local area networks (LANs). Physical branching also occurs in phone circuits, with a main switching station supplying local private branches, which further subdivide to service individual lines within the local branch. Programmers use various types of tree structures including binary trees, B trees, B* trees, etc.

TREG See Telecommunications Regulatory Email Grapevine.

trellis coding A source coding technique used in a variety of contexts, from high-speed modems to MPEG decoding, to produce a sequence of bits from an incoming stream that conforms to certain desired characteristics.

TRI transistor resistor logic.

TRIBES Tri-Band Earth Station. A commercial satellite tracking station from California Microwave, Inc. (CMI), operating on C-, X-, and Ku-Band frequencies, aimed at government applications. TRIBES-Lite is a downsized version.

tributary Secondary, or subsidiary peripheral, signal, or process. Subsidiary peripherals receive their control data from servers or devices higher up in a hierarchical network. Subsidiaries may be aggregated to create a combination medium or signal. See tree structure.

triode An electron tube with three primary elements, an electron-emitting cathode, an electron-attracting anode, and a grid superimposed so that it can be used to control the flow of electrons. The invention of the triode by Lee de Forest, who developed the Audion, is one of the most significant developments in the history of electronics. See Audion.

TRIP See Token-Ring Interface Processor.

Trivial File Transfer Protocol TFTP. A simplified, lock-step version of File Transfer Protocol (FTP) for file transferring over networks which is easily implemented and provides just the most basic of features. A multicast option introduced in RFC 2090 suggests a way to use multicast packets to allow multiple clients to concurrently receive the same file. See Enhanced Trivial FTP, Simple File Transfer Protocol, RFC 1350.

trn A versatile, full-screen news reading program which was developed by Wayne Davison, as a superset of rn, written by Larry Wall. This incarnation adds threads through a hierarchical database.

Trojan horse In software programming, a front-end that appears to be a normal interface to the user, but is really a layer on top of the interface designed to capture information to be used later for penetration of the system. For example, a programmer might create software that looks like a normal login prompt. The user sits down at the terminal, types in a name and password, and gets an error message back (from the Trojan horse) that the password was entered incorrectly, try again. The Trojan horse has already recorded the username and password and gone away. Now the user gets the normal login prompt without being aware of the deception, types the password again, and logs in successfully. The writer of the Trojan horse has grabbed a name and password, with a low chance of detection by the user, and can now access the system by normal means that are difficult to detect as unauthorized access. Named after the invasion of Troy. See trap door.

troposphere The lower layer of Earth's atmosphere, which contains clouds and most of the air, varying from a height of about 10,000 meters at the poles to about 18,000 meters at the equator. In radio transmissions, some frequencies can be bounced off the troposphere, that is, bent back to earth through super-refraction. See ionosphere, tropospheric scatter, tropospheric scatter transmission, tropospheric wave.

tropospheric scatter The dispersion and propagation of waves resulting from the varied and discontinuous physical properties of the troposphere. This can be predicted and controlled sufficiently to be useful in communications. See tropospheric scatter transmission.

tropospheric scatter transmission A method of electromagnetic wave propagation, employing frequency modulation, that exploits the irregular propagation properties of the troposphere. Tropospheric scatter transmission is a way to propagate, for example, microwave transmissions for thousands of miles, in segments up to about 500 miles per hop.

tropospheric wave The troposphere includes a diversity of moisture, heat, and other properties which can result in electromagnetic waves that undergo abrupt changes sufficient to create tropospheric waves distinguishable from the original tropospheric scatter transmission waves.

trouble unit A systems diagnostic or descriptive measure to indicate the expected performance of a circuit over a given period of time.

TRS-80 computer A mass market microcomputer series introduced by Tandy/Radio Shack in the mid-1970s, at one time said to have 80% of the microcomputer market share. It ran quite effectively at 1.8 Megahertz with 4 kilobytes (*not* Mbytes) of RAM. The same generation as the first Apple computer and the Commodore Pet, it featured the Z80 chip, a monochrome screen, built-in BASIC programming language, and ran a variety of applications including Scripsit and Electric Pencil word processing, VisiCalc spreadsheet, and the Scott Adams adventure game series.

For telecommunications, it could be equipped with a PC board that interfaced with the old 'suction-cup' style acoustic modems, and later with direct-connect modems to dial into local BBSs and precursors to the Internet as we know it today.

Leo Cristofferson and David Lien contributed

high quality educational recreations for the TRS-80, and Louis Rosenfelder wrote one of the best books on BASIC applications programming. A TRS-80 enhanced clone series called the LNW-80 designed by a California company called LNW Research never got the market share it deserved. Arguably superior to both the TRS-80 and the Apple, it featured a faster CPU, higher memory capacity, and color screen.

The TRS-80 series ran quite a variety of operating systems, including TRS-DOS, LDOS, CP/M, and others. (Consumers may not realize that they aren't tied to one operating system; a variety of OSs can run on current hardware platforms, just as they did on far more limited systems in the '70s.)

Various models of TRS-80 were shipped over the next decade, the most well-known being the TRS-80 Model I, Model II, Model III, and the historic laptop, the TRS-80 Model 100, which caught the attention of journalists and traveling sales reps in the early 1980s and is still used by some of them 16 years later!

In spite of its strong start, good distribution, and commercial success, the TRS-80 series did not survive the introduction of International Business Machines (IBM) computers to the marketplace. While Radio Shack computers were targeted at both business and personal users, business users were nevertheless conservative and slow to adopt the new technology. They wanted warrantees, assurances, service contracts, and familiarity with the company that was providing the equipment. Despite the very limited capabilities of the early IBM computers, the IBM name and service guarantees made them the computers of choice, especially with Apple targeting the education and home markets. The Tandy series of computers continued under the Tandy name, but their visibility had dropped substantially by the late 1980s, and Intel-based machines cornered most of the business market. See Altair, Apple computer.

TRU See tone receiver unit.

true bearing A bearing given with relation to geographic north, rather than magnetic north.

true north The Earth's geographic north. See magnetic north.

Truetype A system of widely used scalable vector-format fonts developed by Microsoft Incorporated. See vector fonts.

truncation 1. Cutting or chopping off at an end, sometimes abruptly. 2. The quick or abrupt termination of an operation or process. 3. The removal of characters from the end of a word or numeral, as in reducing a four-decimal numeral to two decimals without altering the numerals remaining (i.e., not rounding). Truncation applies to either the leading or trailing end, but more often than not, the part truncated tends to be the trailing end.

trunk A communications link between two switches or distribution points. A generic term that applies to many technologies, but is most often used in connection with telephone and network lines. Trunks can be set up to be bidirectional (left to right to left) or unidirectional (right to left or left to right). See path, route.

trunk exchange A specialized telephone exchange facility dedicated to interconnecting trunk lines.

trunk group A group of trunks which share essentially the same electrical characteristics and often the same physical characteristics, which connect to the same switching endpoints or connections. Multiple trunks are common in areas where one trunk would not be sufficient to carry the traffic. If the main trunk is busy, traffic may be manually or automatically switched to the next one in the group, and so on. See trunk hunting.

trunk hunting A call management system that seeks an available communications trunk over which to route a call. Economy is achieved by hunting the most frequently used trunks first. See hunting.

TSACC See Telecommunications Standards Advisory Council of Canada.

TSAG See Telecommunication Standardization Advisory Group.

TSAP See Transport Service Access Point.

TSO time share operation.

TSRM Telecommunication Standards Reference Manual.

TTA See Telecommunications Technology Association.

TTAB transparent tone above band.

TTC See Telecommunications Technology Committee.

TTIB transparent tone in band.

TTL See transistor-transistor logic.

TTS text-to-speech. A type of speech synthesizer.

TU See traveling user.

TUANZ Telecommunications Users Association of New Zealand.

TUBA TCP and UDP with Bigger Address. One of three candidate protocol proposals eventually blended into IPv6 by the Internet Engineering Task Force (IETF). See IPv6.

TUG Telecommunication User Group.

TUI See telephone user interface.

tumbling A type of cell phone fraud that involves successively switching the electronic serial number for each call too quickly for the cell operator to detect the user.

tungsten A heavy metallic element with properties similar to chromium and molybdenum used in electrical installations, filaments, and contact points, and for hardening alloys.

tuning To adjust to resonate at a particular wavelength, as setting an instrument to a specific pitch or setting a radio antenna or tuner to receive a particular frequency of radiant energy. See tuning coil.

Radio tuning coils resembled large spools. They often came in sets, designed for different frequencies, and could plug into the circuit by means of two or more prongs.

Since tuning coils could be purchased in sets, they often included a base to keep the coils in order and protected from damage. These excellent examples are from the Bellingham Antique Radio Museum collection.

tuning coil A winding coil specifically configured to pick up certain frequencies of radiant energy, particularly radio waves. Early tuning coils consisted of nothing more complicated that a coil of fine conductive wire wound around a wooden or robber core, with the coil in circuit with a connecting pin or pins. Tuning coils for consumer sets tended to range from the size of a sewing spool to about the size of a human hand. Often they were stored like thread in banks or rows on little wooden shelves, so that the appropriate frequency coil could quickly be selected and inserted in a connector on the radio. The wire on each coil would be slightly different, to pull in a different frequency range, with different thicknesses and spacing between successive windings.

Sometimes a small sliding tab or knob attached to a bar would be placed along the edge of the winding, in order to contact with a specific portion of the winding to provide further fine control. This was called a slide contact.

Other tuning coils used a type of intricate basket weaving in various patterns supported by a slender frame in such a way that a lot of wire could be wound into a small space and no cylindrical spool was used. See basket winding, coil.

tunnel *n.* 1. A hollow tube, conduit, or passageway through an obstruction. 2. In software, an intermediary program that provides a temporary relay between connections without interpreting the communication. A tunnel often provides a temporary *portal* for passing data through a system such as a proxy, and ceases to exist when the ends of the connection are closed.

tunneling 1. Encapsulating a network transmission in an IP packet for secure transmission over a network. See virtual private network. 2. To temporarily reroute a network transmission packet in order to utilize routers which would not normally be able to route the transmission to the original destination due to not having the needed destination entry.

TUR Traffic Usage Recorder.

Turing, Alan Mathison (1912-1954) A British mathematician who traveled to Princeton in 1936, where he studied as a graduate student, and wrote "On Computable Numbers ... ." He proposed some provocative ideas in 'ordinal logics' and created a cipher machine while at Princeton, but Turing is best remembered for his description of a hypothetical device which could handle logical operations and manipulate symbols on infinite paper tape. This *Universal Turing Machine* is described as a *finite state* machine due to the finite set of instructions from which individual actions were derived. This provided the roots for thinking about computers in terms of algorithms and equations, and for positing devices that could be used for all possible tasks. Many of the conceptual roots of computer science, particularly general purpose machines and reusable code, were developed through his research. See Turing machine.

Turing machine A hypothetical model devised by Alan M. Turing, which could handle logical operations and manipulate symbols on infinite paper tape. This *Turing machine* is described as a *finite state* machine due to the finite set of instructions from which individual actions were derived at that point. Many extrapolations from and implementations of these basic concepts have contributed to the development of computing devices. See Turing test.

Turing test In 1950, Alan Turing published "Computing Machinery and Intelligence" in *Mind*, a philosophical journal. In it much of his thinking of the last few years was put into print, providing an effective inspiration to many future scientists and providing some of the concepts that developed into the field of artificial intelligence (AI). The Turing test was a provocative assertion that computers would, in the succeeding 50 years, be able to pass for a human under certain test conditions. With the year 2000 drawing near, the imaginative efforts to support this prediction have led to a wide variety of interesting programs, and there are even prizes offered for innovative software that can pass the Turing test.

turnaround time The time of a transaction, especially one that passes from one hand to others and back, as in sending something to another department or external service bureau. For example, the turnaround time for printing a brochure may be two weeks from the time it leaves the hands of the in-house layout production staff until it returns printed, folded, in boxes of 500 each. 2. In network transmissions, the time it takes to send a transmission and receive an acknowledgment that transmission was received, or can continue. 3. In detecting or *pinging* another system, the time it takes for the signal to reach the other system, and report back statistics on the connection. The phrase *turnaround time* is sometimes applied loosely to the time it takes for a human to send out a signal and receive an acknowledgment, and sometimes more precisely to the number of clock cycles or actual measured time it takes for the signal to leave the sending site, reach the receiving site, and report back to the sending site, used for system diagnostics and tuning. Or even more specifically, in handshaking or half-duplex applications, turnaround time is the interval that occurs when the system switches to communication in the other direction. In terms of half-duplex satellite phone conversations, this turnaround time is perceived as a blip, lag, or break in the line by the callers, and, if it is long, can be quite distracting to the conversation. See hysteresis.

turnkey system A self-contained system that can be purchased, installed, or relocated as a unit or as a package. Turnkey systems typically arise from two situations: 1. The technology is complex and the vendor combines options in such a way that the system meets a need but need not be configured or technically understood by the user. The user simply purchases it, turns it on, and uses it. For example, many video outlets bundled Video Toaster systems as turnkey solutions for desktop video applications. 2. There are many options available for configuring a system, and the vendor best knows how to combine and configure the individual components to meet the needs of the purchaser. Private branch telephone systems and multiline telephone systems with lots of options are often bundled and purchased this way.

turnstile antenna An antenna comprised of two dipole antennas perpendicular to one another with axes intersecting at their midpoints.

turntable 1. A round, rotating platter, frequently with a central shaft or outer rim to hold objects in place, commonly used for playing audio and visual media, especially vinyl phonograph records. 2. A round, rotatable, plate-like surface designed to give easy access to objects placed on it by turning. A Lazy Susan. Turntables are used in loading platforms, cupboards, microwave ovens, and Chinese food restaurant tables.

turtle The common name for a programmable turtle-shaped robot designed by William D. Hillis, a graduate student at MIT. Hillis' work arose from the writings and development of the LOGO programming language by Seymour Papert at the MIT Artificial Intelligence Laboratory in the 1960s. LOGO was a LISP-like language intended for robot control. A great deal of excitement was inspired by Papert and Hillis' inventions when microcomputers became available to schools and homes. Finally, there was a mechanism for teaching LOGO and downloading the programs to the little turtles which were commercially introduced in 1978. The Terrapin Turtle was acknowledged by many to be an excellent educational resource for teaching robotics, guidance, artificial intelligence, computer programming, and much more, in a fun, interactive context.

The Hillis/Papert 'turtle' was a delightful, educational, interactive, hands-on teaching tool for helping children learn the essential characteristics and relationships of programming concepts by translating them into movement.

TVM See time-varying media.

TVRO television receive-only

TWAIN An image-oriented communications protocol standard widely used in computer scanning devices and digital cameras (which are, in essence, portable scanners). Most high-end graphics programs can handle TWAIN-compliant devices, as can many optical character recognition systems. TWAIN is supported and promoted by a number of graphics industry vendors, including Eastman Kodak, Hewlett-Packard, UMAX, Adobe Systems, and others. See Tag Image File Format.

tweak freak A technical user who is exaggeratedly interested in the inner workings, whys, and wherefores of the tiniest technical details in a system. This characteristic is an asset when diagnostics and fine-tuning are needed, and a liability in general social situations.

twinning Systems configured in parallel to create redundancy, or alternate means of access or transmission. Twinning is sometimes done to install a new system while an old one is still being used, or to provide a backup in the case of emergencies. In newer technologies, twinning can provide backups or transition systems until the old system is little used or phased out, such as twinning word processors with typewriters or physical facsimile machines with software facsimile programs.

twisted pair cable Twin strands of intertwined, insulated copper wire, used for many decades in the telephone industry. The twists are organized to help reduce interference in the line. Thicker cables tend to provide cleaner transmissions, at a higher cost. Although the theoretic data transmission limit of twisted pair has been underestimated many times, and improved with new ideas and data protocol schemes, it is generally accepted that their practical capacity under normal operations is about 56 kps. Twisted pair is now also commonly used in Ethernet LAN connections. Twisted pair cable is sold commercially in several grades of transmission performance. See category of performance.

two electrode vacuum tube A historically important, but ultimately not very useful early vacuum tube. The two electrode tube consisted of a simple filament (cathode) and an electron-attracting plate (anode). Since there is no controlling mechanism, it wasn't of much practical use, but it was a history-making invention in the sense that it led to the three element tube with a controlling grid that was to subsequently open the door to the entire electronics industry. See Audion; de Forest, Lee.

two-phase coding A means of increasing data transmissions by splitting the signal into two orthogonal channels, one which is in phase and one which is a quadrature signal. They are then transmitted simultaneously with this 90 degree phase offset, each operating at half the data rate of the originating signal. Thus, a signal might be transmitted at the same data rate in half the bandwidth, but for the trade-off of converting it from a baseband signal to a passband signal when splitting it into two channels. The increased potential for noise, when two signals are transmitted simultaneously is also a consideration.

two tone keying A means of using two tones, one for *mark* and one for *space*, to modulate a telegraph signal so as to create two channels transmitting in the same direction.

two-way trunk A network trunk which operates in both directions. In the case of telephone service trunk lines, it refers to one which can be seized from either end of the connection, as opposed to one-way trunks, which may be set up to send only or receive only.

TWS two-way simultaneous. A mode in which a router optimizes communications over a full-duplex serial line.

TWT See traveling-wave tube.

TWX *T*eletype-*W*riter E*x*change. A Bell system printing telegraph service which could operate over the existing long distance network, established in the 1930s. In 1970, Western Union purchased the TWX service from AT&T and merged it into its own Telex service. See Western Union.

Tx, TX transmit.

TxD transmit data. A data channel, typically used in serial communications, which is an output for DTE devices and input for DCE devices. See DTR, DSR, RTS, RxD, RS-232.

type bar A bar, usually in an impact printer such as a line or page printer, which is embossed with the printable characters and used to impress the characters onto the printing medium, usually through a carbon ribbon. A similar concept is used on pressure label makers, which use dialable characters to select a line of characters which are then impressed on the printing medium.

typeahead 1. A capability of systems to store or *buffer* commands that are entered by the user, even if the current instructions have not fully completed. For example, a fast typist using a slow word processor might be able to type faster than the word processor can display the characters. Typeahead allows the characters to be stored and displayed as the software catches up. 2. Better phone menu systems using touchtones have a typeahead capability which allows a user to type further menu selections before the current instructions have been completed.

typeface, typestyle, type family A collection of character sets with an overall aesthetic relationship, which may include a number of styles (bold, oblique, italic, condensed, display, etc.) and sizes. For example, Helvetica and Times are common typefaces. Typefaces are commonly used in printing, and computer printers are capable of rendering a wide variety of fonts. See font, PostScript, Truetype.

typesetting and composition The art and process of laying out a page with type in order to transcribe information into written form and to present it visually so that the composition enhances the reader's ability to understand and appreciate the information. In most cases, the type is then transferred to a printing press for the creation of multiple impressions. The selection of type, its organization, visual relationships, and aesthetics are all considered toward these ends.

The automation of typesetting began in the late 1800s when the ability to provide electricity and control machines and the flow of information made it possible to create typesetting machines. Early typesetters, also called type-compositors, that were developed around the late 1880s were of two types: those which employed movable type and those in which the type materials were cast line-by-line in mold-beds of assembled matrices. Each of these types of machines were operated through keyboards, like those of typewriters, thus removing the traditional 'hands-on' positioning of type.

U interface In ISDN data communications over phone wires, the telephone company typically has to install a number of devices to create the all-digital circuit connection necessary to send and receive digital voice and data transmissions. One of these devices, installed in U.S. phone circuitry, is the U interface.

The U interface is full-duplex device that works over a single pair (2-wire) cable which interfaces with the line coming from the telephone switching office. The U interface then links to a small interface device called a Network Termination 1 (NT1) device. The NT1 converts the 2-wire U interface into a 4-wire S/T interface which, in turn, can support multiple devices in a single bus loop configuration, such as a telephone, computer, or facsimile machine. See ISDN, V interface.

U-law A pulse code modulation (PCM) coding and companding data standard used in audio systems on computer multimedia peripheral cards. This takes some of the load of specialized applications off of the central processing unit (CPU). It is often implemented in addition to A-law companding. See A-law.

U reference point A demarcation point in ISDN services installed in North America, where the local loop connects with the NT1 device. See U interface.

UA See User Agent.

UART See universal asynchronous receiver-transmitter.

UAV unstaffed aerial vehicle.

UAWG Universal ADSL Working Group. A commercial consortium formed to promote an easy-to-deploy, fast version of Digital Subscriber Line (xDSL) based on ANSI T1.413. Since traditional ADSL installations require a splitter to be wired to the subscribers' premises and a custom modem installed in their computer, there have been a number of initiatives to simplify the installation process and, hence, the cost, and to allow the subscribers a choice of modem hardware. See G.lite.

UBR See unspecified bit rate.

UCC Uniform Commercial Code.

UCF See UNIX Computing Forum.

Uda, Shintaro Designer of the Yagi-Uda antenna, a sensitive, directional antenna which worked in the higher frequency ranges and became the model for many of the antennas that came later and are still in use. See Yagi-Uda antenna.

UDI Unrestricted Digital Information.

UDP See User Datagram Protocol.

UECT See Universal Encoding Conversion Technology.

UHF See ultra high frequency.

UI 1. Unix International. A consortium of computer software and hardware vendors promoting the development and implementation of Unix, and of related and other open software standards. See Unix, UNIX. 2. See user interface.

UL See Underwriters Laboratories Inc.

ultra high frequency UHF. A designation for a range within the radio frequency spectrum commonly used for broadcast communications, which ranges from 300 MHz to 3000 MHz.

ultra high frequency (UHF) antenna A category of antennas which are designed to take advantage of the particular characteristics of ultra high frequency (UHF) waves. Because of the wavelength differences between UHF and

Sampling of UL Standards Related to Telecommunications	
UL No.	Descriptions of UL Numbers from Underwriters Laboratory
UL 1409	Low-Voltage Video Products Without Cathode-Ray-Tube Displays: Antenna signal amplifiers, CATV adapters and digital converters, channel balancers and processors, distribution amplifiers, commercial TV cameras, disc players, electronic viewfinders, internal distribution amplifiers, laser disc players, modulators, picture tube degaussers, power packs, power supply-battery chargers, satellite receivers, satellite receiver dish controllers, single-channel converters, teletext and television decoders, television descramblers, tuner adapters and power supplies, UHF amplifiers, tuners, and converters; VHF amplifiers and tuners, video printers, video-production, -processing, -receiving, and -recording equipment, and video tape recorders.
UL 1418	Cathode-Ray Tubes: Bonded frame, laminated, prestressed picture tubes (CRT), rebuilt picture tubes, picture tubes for business equipment, dental, and medical equipment.
UL 1410	Television Receivers and High-Voltage Video Products: Household and commercial television receivers and monitors, and health care facility television equipment.
UL 1412	Fusing Resistors and Temperature-Limited Resistors for Radio- and Television-Type Appliances: For use in appliances that do not involve potentials greater than 2500 volts peak.
UL 1414	Across-the-Line, Antenna-Coupling, and Line-by-Pass Capacitors for Radio- and Television-Type Appliances: For nominal 125- and 250-volt, 50- to 60-hertz circuits, includes double protection capacitors rated 1.0 B5F maximum.
UL 1419	Professional Video and Audio Equipment: Video tape recorders, audio/video editing equipment, audio/video receiving and processing equipment, signal transmission equipment, television cameras, video digitizers, video monitors, metering equipment, and similar equipment.
UL 1492	Audio-Video Products and Accessories: Audio and video products intended for use on supply circuits. Audio products and accessories intended for household use and involved with the reproduction or processing of audio signals. Video products that are intended for household or commercial use, that receive signals in ways such as off the air, through a CATV/MATV cable system, from a video-recorded medium, and from image-producing units. Auxiliary products and accessories intended for use with audio or video products wherein the auxiliary and accessory products are separate and do not perform the desired function, but are used in addition to or as a supplement to products mentioned above. Cellular telephones and similar transceiving devices used on a vehicle, boat, or the like, where the telephone interconnects to the telephone network through a radio transmitter and receiver. Portable audio or video products of the types described above that are intended for use with a vehicular, marine, or any other battery circuit as the power supply means.
UL 6500	Audio/Video and Musical Instrument Apparatus for Household, Commercial, and Similar General Use: This standard applies to the following apparatus that is to be connected to the supply mains, either directly or indirectly, intended for domestic and commercial and similar general indoor use and not subject to dripping or splashing: radio receiving apparatus for sound or vision; amplifiers; independent load transducers and source transducers; motor-driven apparatus which comprise one or more of the above-mentioned apparatus or can be used only in combination with one or more of them, such as radio-gramophones and tape recorders; other apparatus obviously provided to be used in combination with the above-mentioned apparatus,

UL No.	Descriptions of UL Numbers from Underwriters Laboratory
	Sampling of UL Standards Related to Telecommunications, cont.
	such as antenna amplifiers, supply apparatus and cable-connected remote control devices; battery eliminators; electronic musical instruments; electronic accessories such as rhythm generators, self-contained tone generators, music tuners and the like for use with electronic or nonelectronic musical instruments; video apparatus intended for entertainment purposes in health-care facility locations; cellular telephones and similar transceiving devices used on a vehicle, boat, or similar location where the telephone interconnects to the telephone network through a radio transmitter and receiver; portable audio or video apparatus that are intended for use with a vehicle, marine, or any other battery circuit as the power supply means.
UL 1685	Vertical-Tray Fire-Propagation and Smoke-Release Test for Electrical and Optical-Fiber Cables: Limits for each fire test to make the tests equally acceptable for the purpose of quantifying the smoke. The cable manufacturer is to specify, for testing each "-LS" (limited-smoke) cable construction, either the UL vertical-tray flame exposure or the FT4/IEEE 1202 type of flame exposure. The same test need not be specified for all constructions.
UL 1577	Optical Isolators: Optically isolated switches and insulation systems, photocouplers.
UL 1651	Optical Fiber Cable: Single and multiple optical-fiber cables for control, signaling, and communications as described in Article 770 and other applicable parts of the National Electrical Code.
UL 1690	Data-Processing Cable: Electrical cables consisting of one or more current-carrying copper, aluminum, or copper-clad aluminum conductors with or without either or both (1) grounding conductor(s) and (2) one or more optical-fiber members, all under an overall jacket. These electrical and composite electrical/optical-fiber cables are intended for use under the raised floor of a computer room (optical and electrical functions associated in the case of a hybrid cable) in accordance with Article 645 and other applicable parts of the National Electrical Code.
UL 2024	Optical Fiber Cable Raceway: Covers the following types of optical fiber cable raceways and fittings designed for use with optical fiber cables in accordance with Article 770 of the National Electrical Code: *Plenum.* Evaluated for installation in ducts, plenums, or other spaces used for environmental air in accordance with the National Electrical Code as well as general purpose applications; *Riser.* Evaluated for installation in risers in accordance with the National Electrical Code as well as general purpose applications; *General Use.* Evaluated for general purpose applications only.
UL 1459	Telephone Equipment: Cordless telephones, key systems private branch exchange equipment, telephone answering devices, dialers, and telephone sets.
UL 1863	Communication Circuit Accessories: Telecommunications equipment such as jack and plug assemblies, quick connect assemblies, telephone wall plates, cross connect enclosures, network interfaces, and connector boxes.
UL 1950	Practical Application Guidelines On-Line Service (PAGOS). A reference service providing information for understanding and applying the requirements of UL Standards for Safety. Of interest is the UL 1950 Standard for Safety of Information Technology Equipment, Including Electrical Business Equipment.

U

very high frequency (VHF) waves, and the relationship of the rods on the antenna to the length of the wave, it is possible to make UHF antennas relatively small with more branching elements, compared to VHF antennas. However, as UHF television broadcast signals are generally weaker than those from VHF, there is a greater potential for loss, and they must be designed and installed with greater care to be effective. See antenna, combination antennas, VHF antennas.

ultraviolet Electromagnetic radiation with shorter wavelengths, between the violet part of the visible spectrum and X rays. Although it cannot be seen by humans, ultraviolet radiation is of commercial significance because it can degrade many types of materials and pigments. Commercially, it is used in a variety of lamps, such as arc lamps, and can be used to remove data from erasable/programmable computer chips.

In astronomy, ultraviolet sensing devices use ultraviolet radiation focused through a spectrograph to study the characteristics of celestial objects. Telescopes and some satellites are equipped with this capability. The Hubble Space Telescope and the International Ultraviolet Explorer satellite enable study of objects using ultraviolet light which is near the visible spectrum. The Johns Hopkins Ultraviolet Telescope extends the range to the *far ultraviolet*, that is, the region farther from the visible spectrum. See infrared.

UMA See upper memory area.

umbrella antenna An antenna that resembles an umbrella in that the lines extend out and down from a central pole.

UMTS Universal Mobile Telecommunications Systems.

unattended call A situation that occurs when, for example, an automatic dialing system dials a line, then tries to pass the call to the first available human agent, but no agent is available. Consequently, the call is abandoned. This type of calling occurs in the telemarketing industry and the call is terminated in order not to irritate a potential customer. Unattended calls are also used by collection agencies and the system hangs up if no agent is available or if the call is answered by an answering machine, thus not impinging on the agent's time.

unattended systems Devices or systems which function without a human operator or without significant human attention except for installation and routine maintenance and upgrades. Unattended systems have become prevalent since the late 1970s, when computer automation became inexpensive enough to incorpo-

UL Conformity Assessment Services Related to Telecommunications

Service	Brief description
Listing Service	A UL Listing Mark indicates that representative samples have been tested and evaluated according to nationally recognized safety standards.
Classification Service	A UL Classification mark indicates that products have been evaluated for certain properties under specified conditions.
Component Recognition	A service for factory-installed components in complete products.
Certificate Service	A service for completely installed systems.
Field Engineering Service	A service for installed products without UL Listing Marks or UL Classification Marks.
Testing Environ. Products	Evaluation of innovative environmentally friendly products.
LAN Cable Performance	Safety evaluations and evaluation of LAN cable according to industry performance specifications, including TIA/EIA standards.
Energy Efficiency	Electrical appliances are certified according to U.S. or Canadian standards for energy efficiency through the UL Energy Verification.
SDS Verification Testing	Verification of input/output products to Honeywell's Smart Distributed System (SDS) for compatibility of components to an industrial control communications network.

rate into a wide variety of components and machines. Computer bulletin board systems were some of the first information-rich systems to function 24 hours a day, and phone systems now are frequently automated with menu selections and voice mail options. Recently, faxback systems allow users to request product information or technical support in the form of fax documents. The system logs the phone call, gets the customer's document selections from a list of options, then dials back the user's fax machine and transmits the requested documents. Unattended systems generally function 24 hours per day at a significant cost savings over human operators. Many businesses are willing to give up personalized service in favor of the economy offered by automated systems. See Auto Attendant.

unbalanced line A transmission line with two conductors (such as coax or a telephone circuit) with unequal voltages with respect to the ground. In phone circuits, this is generally an undesirable condition.

unblanking The portion of the sweep in a cathode ray tube (CRT) where the beam is turned on, with pulses from the generator. See blanking.

unbundled Products or services which are sold separately. For example, a company may release a graphics card/monitor combination as a package deal, and later unbundle the items, that is, allow them to be sold separately in order to clear the products, get a higher return, or respond to market demand for one product over the other. Contrast with bundled.

Underwriters Laboratories, Inc. UL. UL is a not-for-profit organization established in 1894, which provides conformity, safety, and quality assessment services and publications to a variety of organizations, including manufacturers. In addition, UL provides educational materials, input to international safety systems, and assistance to various regional authorities.

UL publishes a catalog of its standards and the standards themselves in print, on microfilm, CD-ROM, and diskettes. The UL also sponsors a UL Standards Electronic BBS (accessed directly, or through the Web). The majority of the UL published standards have been approved as American National Standards by the American National Standards Institute (ANSI). UL has a number of publications of interest to professionals in the communications industry, including WireTalk for the wire and cable industry. http://www.ul.com/about/wtalk/index.html

UL provides ISO 9000 standards quality registration through its accredited RvA, Registrar Accreditation Board (RAB) and other international quality affiliations. UL provides information on new international environmental management standards through ISO 14001. http://www.ul.com/

The full UL catalog is available on the Web, but the UL Standards chart shows some telecommunications-related UL Standards which may be of interest and which provide an idea of the types of use and safety issues concerned.

Underwriters Laboratory Inc. assessment The UL provides a number of conformity assessment services for product certification. These include listing, classification, field engineering, and various types of safety and performance testing. The UL Conformity Assessment chart shows some of the services relevant to telecommunications.

Underwriters Laboratory Inc. Mark UL Mark. UL provides a number of listing marks to indicate that products or systems have been evaluated by UL and conform to certain specifications. Those shown in the UL Listing Marks chart are relevant to telecommunications.

unerase A system command or feature of a software application, which permits the user to restore files which have been erased from a hard drive or similar hard storage device. Note, if other *disk write* activities have taken place, it may not be possible to unerase the files. Subsequent writes may overwrite portions of the data or the file pointers indicating where the data is stored. In most cases, if you accidentally delete a file, you must unerase before doing anything else.

UNI User Network Interface, User-to-Network Interface. As specified by the ATM Forum, an ATM network switch which interfaces user equipment to private or public ATM network equipment, or connects between Customer Premises Equipment (CPE) and public network equipment. See PMP, PCR, OCD, SCR.

unicast A type of Internet Protocol (IP) address identifier for a set of interfaces. Unicast involves the transmission of a single Protocol Data Unit (PDU) to a single destination (unlike multicast, where it may go to multiple destinations). The format of the ATM subinterface unicast command is: atm smds-address *<address>*. See anycast, IPv6 addressing, multicast.

Unicode A character-encoding standard designed to support text-encoding in data files. Unicode, Inc. was originally a collaboration between Apple and Xerox, who produced the

U

original specification. They were later joined by Adobe, Aldus, Borland, IBM, Microsoft, NeXT, Novell, Sun, and others. Unicode has been rolled in with an ISO specification to become a subset of ISO 10646. Unicode is loosely based upon the widely supported ASCII standard, but in a greatly extended form to support major world languages, including those which are not represented with Roman characters, including Cyrillic, Greek, Arabic, Hebrew, Japanese Kana, Chinese bopomofo, Korean hangul, and others. Symbols, punctuation marks, mathematical symbols, and technical symbols are also supported. Unicode uses a 16-bit character set, supporting over 65,000 characters. Each character is assigned a unique 16-bit value. No special modes or control or escape sequences are needed.

Unicode comprises the first 65,536 code points of the ISO 10646 standard; the rest are reserved for future use. See Unicode Consortium.

Unicode Consortium A nonprofit association founded in 1991 to promote and support the acceptance and implementation of Unicode. The Consortium publishes a pamphlet on the Unicode specification. The Unicode Technical Committee, descended from Unicode, Inc., now functions as part of the Consortium to actively maintain the standard. See Unicode.

unidirectional Moving, responding, or transmitting in one direction, or in only one direction at a time.

unified memory architecture A system on which the video display drivers are integrated into the motherboard, and system random access memory (RAM) is used to buffer graphics displays, rather than having them as separate systems. Some systems use this very effectively, providing graphics coprocessor chips, and allowing greater video graphics memory and more control over memory for programmers, applications, and users. On other systems, this type of integration slows down the graphics rendering and overloads the CPU. This is not the fault of the concept, but rather with the way some manufacturers have implemented it.

unified messaging See integrated messaging.

Uniform Resource Locator URL. A compact string representation for a resource available on the Internet. URLs have been in use since 1990 as Universal Resource Identifiers in WWW. A URL is a means to locate resources, by providing an abstract identification of its location. Generally, a URL follows this format:

```
<scheme>:<scheme-specific-part>
```

Examples:

```
http://www.abiogenesis.com/telecomdict
ftp://www.peanut.org/
```

UL Listing Marks Related to Telecommunications

Listing Mark	Brief description
UL Listing Mark	Commonly seen, and indicates that samples of the product conform to UL safety requirements according to UL published standards for safety.
C-UL Listing Mark	Canadian market products evaluated according to Canadian safety requirements.
Classification Mark	Products evaluated for specific properties under specified conditions. These usually consist of industrial and building materials and equipment.
C-UL Classification Mark	Classification Mark products intended for the Canadian market.
Recognized Component Mark	Specific to components used in products sold as complete units, and thus not usually seen from the outside. There is also a Canadian version.
International emc-Mark	Products which conform to electromagnetic compatibility requirements of Europe and/or U.S. and/or Japan and/or Australia.
Field Evaluated Product Mark	A product which is evaluated in the field rather than in a laboratory.
Facility Registration Mark	A facility which has passed UL quality assurance standards, specifically ISO 9000-series and ISO 14001 (environmental).

Scheme names consist of a sequence of lower-case characters from a to z, numerals 0 to 9 and the characters "+" (plus), "." (period), and "-" (hyphen). It is recommended that upper case be treated as lower case in resolving a URL.

A number of specific schemes for particular protocols are standardized or commonly used and there is a process for registering new ones. Common schemes (typed in lower case when used in a URL) are shown in the Common Schemes chart. See RFC 1630, 1738.

uniform line A line which has essentially identical electrical characteristics throughout the transmission path.

uniformity The capability of a broadcast or other communications medium to deliver a steady and consistent signal within the desired range.

uninterruptible power supply UPS. A safety and steady-service device which protects equipment and data by guaranteeing a sufficient and steady source of electrical power in the event that other power sources are interrupted or lost.

UPS systems may take their current from an alternating power supply while the system is up, may store charges from this source, and then switch to an alternate source, such as a direct current storage battery or separate alternate current generator in the event of power disruptions to the normal supply.

UPSs are used on computers, phones, lighting systems, and in emergency centers.

Power outages can create severe problems on computer systems, particularly to network servers, queues, backup file systems, and applications which are reading or writing to storage media at the time of a power outage. UPS systems can prevent loss of files which are in the process of being saved and prevent possible corruption to the medium on which they are being written.

unipolar Having only one pole, direction, or polarity.

United States Electric Illuminating Company A month after Thomas Edison's famous Pearl Street electrical company began providing power for incandescent lighting, this competing company became South Carolina's first central power station in 1882.

United States Geological Survey USGS. The USGS carries out fundamental and applied research in geological surveying, cartographic data collection, storage, search, retrieval, and manipulation. It is responsible for assessing natural ecological events, energy, land, water, and mineral researches. The USGS conducts the National Mapping Program, and publishes thousands of reports and maps each year.

United States Telephone Association USTA. An organization founded in the 1800s which promotes the well-being of the industry, and provides technical and standards assistance, discussion forums, and publications for its members. USTA represents more than 1,200 local exchange carriers (LECs).

U

Common Uniform Resource Locator Schemes

Scheme	Name	Type
ftp	File Transfer Protocol	Files and directories
http	Hypertext Transfer Protocol	Internet resources, Web pages
gopher	Gopher Protocol	File directories in Gopherspace
mailto	Electronic mail address	Internet electronic mail address
news	USENET news	Newsgroups and individual articles
nntp	USENET news using NNTP access	Alternate means of accessing news
telnet	Reference to interactive sessions	Interactive telnet remote logon sessions
wais	Wide Area Information Servers	WAIS databases, searches, documents
file	Host-specific file names	Accessible files from various hosts
prospero	Prospero Directory Service	Resources on the Prospero service

USTA provides representation before Congress and various regulatory bodies, training courses, technical bulletins, conferences, and media relations.

USTA arose from the National Telephone Association, established in 1897. This organizational strength provided a voice for independents in the dominant Bell marketplace. The National Telephone Association later became the United States Independent Telephone Association (USITA). The Kingsbury Commitment, an important step toward cooperation between Bell and the Independents, entered into in 1913, may have averted government take-over of the telephone industry arising from charges of the monopolistic control exerted by Bell at that time.

After the mid 1980s, divestiture of AT&T, USITA became the United States Telephone Association (USTA). http://www.usta.org/

United Telephone Company A historic phone company founded in 1898 by Cleyson L. Brown in Abilene, Kansas, and later expanded to other communities. It operated there until 1966, and then moved to Shawneed Mission, Kansas, where it forms the local division of Sprint Corporation. See Museum of Independent Telephony.

UNIVAC Universal Automatic Computer. A historic, large, general-purpose electronic computing system in active use in the 1950s. It was designed and built in the mid-1940s by the Eckert-Mauchly Electronic Control Corporation, but taken over before its completion by Remington-Rand. X-1 was a programming language used on this system. It is significant for its use in the tabulation of 1952 presidential election returns. Due to the media exposure, UNIVAC became so well known that the name became a generic term for large computing devices.

universal asynchronous receiver-transmitter UART. UART chips and UART circuitry perform a conversion function within a computer. When a computer software program generates data that travels from the computer to the serial card in a peripheral slot, or through a serial device to an external modem, the parallel data generated by the computer is converted by the UART into serial data that is then transmitted through the modem. The same process occurs in reverse at the receiving end. This is *not* the same process as that performed by the modem, which is to modulate and demodulate a signal which is converted from digital to analog to transmit through the phone line, and back again. The UART does its job before the modulation/demodulation process occurs in the modem. A UART chip may be in the computer, or it may be in the modem itself.

Universal Encoding Conversion Technology UECT. A Digital Equipment Corporation (DEC) proprietary software system for converting documents to and from Unicode. UECT has been incorporated into the AltaVista search engine, one of the significant search tools on the Web from Digital Equipment Corporation.

universal mailbox A centralized computer point of access for a variety of types of messages, including email, digitally encoded voice messages, facsimiles, etc., so the user can look at one listing to determine what to read and when to read it, and to simplify the filing and cross-management of document databases. See integrated messaging.

universal payphone A payphone with a wide scope of payment options including coin, calling card, credit card, collect, etc.

Universal Serial Bus USB. An open serial data bus standard developed by a consortium of prominent computer products and telecommunications services providers which allows peripherals to be attached to a computer through a single peripheral attached to the motherboard, with other devices chaining or attached in a star topology. Commercial USBs are designed to support a large number of devices, sometimes up to 64 (the host computer is considered a device). A USB will sometimes also provide additional power to devices which might require it.

Universal Service Order Code USOC. An identification system for tariff services and equipment introduced in the 1970s by AT&T, and later adopted by the Federal Communications Commission (FCC). Since divestiture, the code is even less 'universal' than before, with individual Bell operating companies developing billing systems somewhat independent of one another.

Universal Time See Coordinated Universal Time.

Universal Transverse Mercator projection UTM. A map projection technique which preserves angular relationships and scale. UTMs are used in many planimetric and topographic maps. A UTM consists of a series of identical projections, each six degrees of longitude oriented to a meridian, taken from around the world's mid-latitudes.

Universal Wireless Communications UWC. A wireless communications collaborative program initiated by wireless operators and vendors in 1995. The program is built on the TIA IS-136 Time Division Multiple Access (TDMA) radio frequency standards, along with IS-41 Wireless Intelligent Network (WIN) standards.

Universal Wireless Communications Consortium UWCC. A Washington State LLC, established to support carriers and vendors of IS-136 TDMA/IS-41 WIN standards. The UWCC sponsors a number of working forums, including the Global TDMA Forum (GTF), the Global WIN Forum (GWF), and the Global Operators Forum (GOF). See Universal Wireless Communications. http://www.uwcc.org/

Unix A widespread, powerful operating system, originally developed in 1969 by Ken Thompson at AT&T Bell Laboratories. The trademarked version of Unix is spelled all in caps as UNIX, whereas Unix spelled in upper and lower case is used generically in the computer industry to refer to all the many freely distributable flavors of Unix that have been implemented by different groups. UNIX has gone through a number of hands, from AT&T, to Novell Inc., to the X/Open Company Limited. See UNIX.

UNIX UNIX is a powerful, widespread, cross-platform, Internet-friendly, multitasking, multiuser operating system. When spelled in all capitals, UNIX is a registered trademark, licensed exclusively through the X/Open Company Limited. See Single UNIX Specification, Unix.

UNIX Computing Forum UCF. A comments and feedback forum through the Santa Cruz Operation, Inc. (SCO), which provides UNIX server operating systems and related products.

Unlicensed Personal Communications Services, Unlicensed PCS UPCS. There are a number of low-range communications systems which can be used without broadcast licensing. These are commonly used for applications such as cordless phones, intercoms, monitors, etc. Some are incorporated into short range wireless local area network (LAN) data and phone systems. Specific frequency ranges have been assigned to UPCS services by the Federal Communications System (FCC). UPCS are permitted within the 1890 to 1930 MHz frequency ranges and are further subdivided for use with asynchronous (1910 to 1920 MHz) and isochronous (1890 to 1910 and 1920 to 1930) communications. See band allocations.

unlisted phone number A service requiring a fee, in which a phone listing is not published in printed directories or available through directory assistance. Some carriers also provide unpublished service, which is excluded from printed directories, but may be listed with directory assistance, as a partway privacy measure. People pay to prevent their numbers from being listed for a variety of reasons: to avoid crank calls, undesired telephone solicitations, harassment from ex-spouses, etc. Some carriers make it possible for callers to leave a message for an unlisted number, which the caller may or may not return at his or her option. This is useful for emergency calls. See unpublished phone number.

unmatched calls Calls which do not have a corresponding match in a Service User Table (SUT). Call matching is a way of determining whether the call is authorized and should be permitted to ring through. If there is no match to the destination number in any of the relevant lookup tables, such as the Authorization Code Table (ACT) or Calling Card Table, the card will likely be rejected or may be redirected to someone in authority.

unpublished phone number A service, usually requiring a fee, in which a phone listing is not published in printed directories, but may or may not (depending upon the carrier) be available through directory assistance. Thus, if listed with directory assistance, it is a midway solution between a listed number and an unlisted number. Some people choose to have unpublished phone numbers to avoid crank calls and undesired telephone solicitations. Some carriers will allow you to exclude your address from your published listing, without charging extra. For further privacy, see unlisted phone number.

unshielded Unprotected from emitting or receiving electromagnetic interference or broadcast signal interference. Most cables are shielded with plastic and/or metal foil, but since this increases the weight and cost of the cable, there are still circumstances where low shielded or unshielded cables are used. In video applications, well-shielded cables are recommended. Monitors should be well-shielded to protect users from radiation exposure, and computers shielded to prevent interference with nearby broadcast devices, such as radios. Improper or insufficient shielding may result in Federal Communications Commission (FCC) rejection in the manufacture of new products.

unshielded twisted pair UTP. A very common type of cable consisting of one or more pairs of twisted copper wires bound together. Frequently used for phone wire installations intended to carry faster data rates.

unspecified bit rate UBR. An unguaranteed ATM networking service type in which the network makes a best efforts attempt to meet the sender's bandwidth requirements. See available bit rate, cell rate.

unsupervised transfer, blind transfer A phone call transfer in which the recipient is not advised as to the identity of the caller. This is common on automated systems in which the caller can select an extension by way of the keypad.

unused A product which may have been opened, or taken home and returned, but which has not been used. It may have slight abrasions and, if sold, may carry a warranty that differs from a new warranty.

UPC Usage Parameter Control. A mechanism by which a network monitors and controls traffic and guarantees service for legitimate uses. See traffic policing, traffic shaping.

UPCS See Unlicensed Personal Communications Services, Unlicensed PCS.

uplink In broadcast communications, the uplink is the leg from an Earth station to a satellite. From the satellite back to the Earth is a downlink. The distinction is made partly because of the different technologies used in satellite and Earth stations, but also because uplink and downlink services can often be purchased separately.

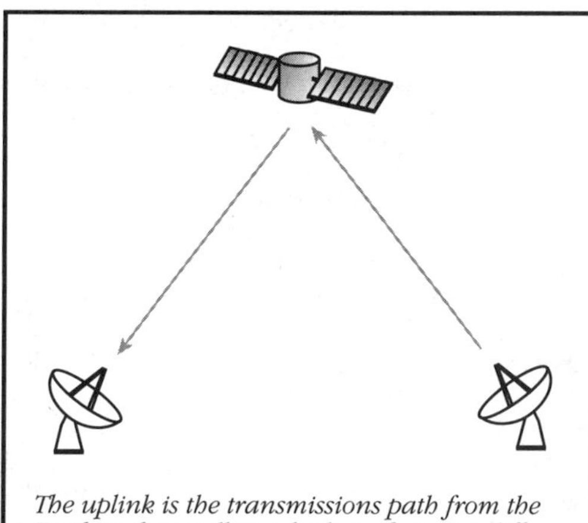

The uplink is the transmissions path from the Earth to the satellite, which is often at a different frequency from the downlink in order to reduce interference between incoming and outgoing signals.

upload, send, transfer To transmit a broadcast or transfer data from the current device to another one, usually at a different location or desk. Computer data is often uploaded from a personal computer to the Internet or to a mainframe. Information from a laptop may be uploaded to a desk computer. Telecommunications software, Web browsers, and FTP are common ways in which people upload files. Broadcasts may be uploaded to a satellite link. See upstream. Contrast with download.

upper memory area UMA. A section of memory on Intel-based International Business Machines (IBM) and licensed third party computers commonly used to buffer video data which can be accessed and read by a video graphics display card.

upstream Generally, the transmission going in a direction away from the reference point. Thus, the stream of data from a personal computer to a mainframe would be considered upstream. Sometimes the designation implies from a smaller or less powerful system to a larger or more powerful system, so its use is not completely standardized. In cable networks, the transmission from the transmitting station to the cable television headend is the upstream direction. See upload. Contrast with downstream.

uptime An uninterrupted interval during which a system or process has been in active service. The active, functional time between failure or maintenance periods. Contrast with downtime.

upwardly compatible A device or program intended to work with later upgrades or revisions. Upwardly compatible may also mean compatible with a larger or more complex version. For example, a handheld device bar code device may be designed to be upwardly compatible with a desktop computer. Upward compatibility in terms of later versions is much more difficult to achieve that downward compatibility, since future changes or improvements cannot always be anticipated. Contrast with downwardly compatible.

URI Uniform Resource Identifier. See URL, URN, RFC 1630, RFC 1738, RFC 1808.

URL See Uniform Resource Locators. See RFC 1738.

URM user request manager.

URN Uniform Resource Name. See RFC 1737.

U.S. West One of the regional companies created when AT&T was divested in the mid-1980s. It is comprised of Mountain Telephone, Pacific Northwest Bell, Northwestern Bell, and other related firms servicing the "Fourth Corner."

USB See Universal Serial Bus.

USDC U.S. Digital Cellular. A telephone standard which uses frequency division multiple access (FDMA) and time division multiple access (TDMA) techniques in the 824 to 894 MHz range.

USDLA United States Distance Learning Association.

used A term describing a product which has been opened and used, with no implications as to the quality, age, or remaining useful life of the product. Used equipment is generally represented as being in working condition, as far as is known. See certified, fair, like new, refurbished.

USENET Created in late 1979, shortly after the release of a Unix V7 which supported UUCP, USENET is best known for its more than 35,000 public newsgroups that flourish to this day. USENET was developed by Tom Truscott and Jim Ellis at Duke University, and Steve Bellovin at the University of North Carolina. The first two-site installation was described in January 1980, at the Usenix conference and, after modifications by Steve Daniel and Tom Truscott, it became known as A News.

As soon as it caught on, A News volume began a steady rise and, in 1981 Mark Horton, from UC Berkeley and Matt Glickman, enhanced the software and made it more able to cope with the increasing volume of information. This 1982 version was known as B News.

Two years later, administration of the software was taken over by Rick Adams from the Center for Seismic Studies. Moderated groups capability was added, in addition to compression, a new naming structure, and control messages. A rewrite by Geoff Collyer and Henry Spencer of the University of Toronto was released as C News in 1987.

In 1992, Rich Salz released InterNetNews (INN), a program optimized for NNTP hosts, but with support for UUCP. INN was designed for socket-oriented Unix hosts. Enhancements and bug fixes to INN were released by David Barr, beginning in 1995. Maintenance of INN was taken over by the Internet Software Consortium.

UUCP gave way to TCP/IP, and TCP/IP's greater compatibility across platforms was a means to provide wider access to newsgroups. The Network News Transfer Protocol (NNTP) was also developed and, in 1986, a means to use this for news articles was released. [This information is courtesy of those who compiled the "USENET Software: History and Sources FAQ."] See newsgroup, RFC 822/RFC1123, RFC 977, RFC 1036, RFC 1153.

user Sometimes called *end user*. Although often used to indicate a nontechnical consumer of a product or service, a user also generically refers to anyone interacting with that product

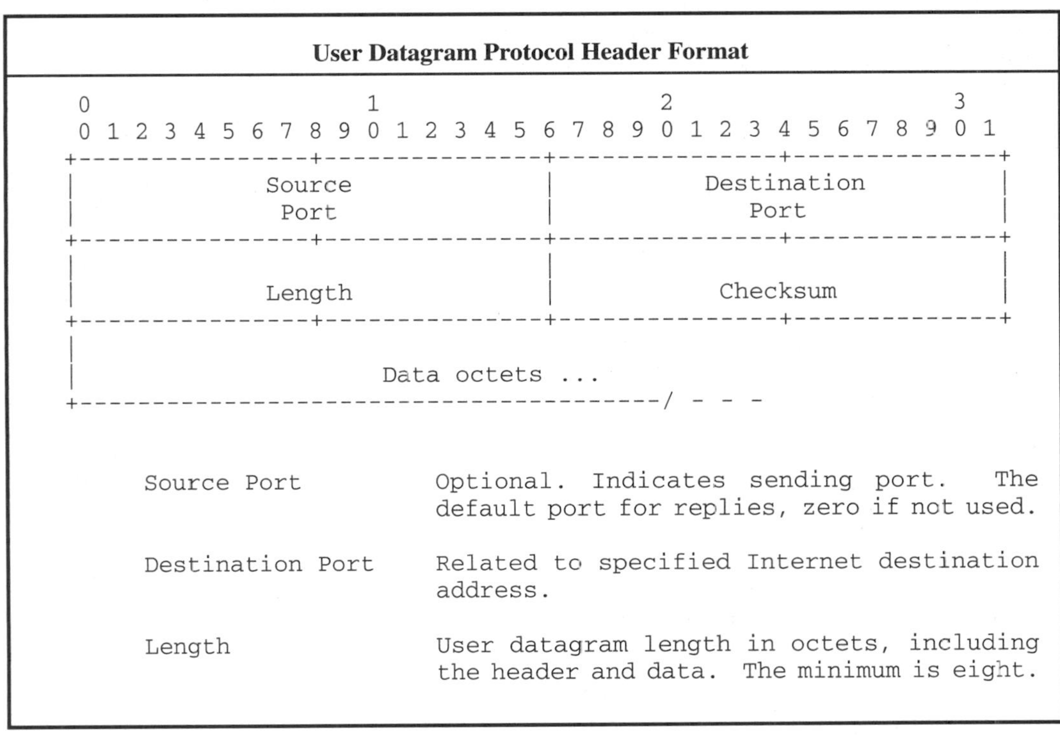

User Datagram Protocol Header Format

```
 0                   1                   2                   3
 0 1 2 3 4 5 6 7 8 9 0 1 2 3 4 5 6 7 8 9 0 1 2 3 4 5 6 7 8 9 0 1
+---------------+---------------+---------------+---------------+
|             Source            |          Destination          |
|              Port             |              Port             |
+---------------+---------------+---------------+---------------+
|                               |                               |
|             Length            |            Checksum           |
+---------------+---------------+---------------+---------------+
|
|                     Data octets ...
+-------------------------------------------/ - - -

     Source Port         Optional. Indicates sending port.  The
                         default port for replies, zero if not used.

     Destination Port    Related to specified Internet destination
                         address.

     Length              User datagram length in octets, including
                         the header and data.  The minimum is eight.
```

or service, as opposed to developing or distributing it.

user account An account assigned for a specific individual on a computer network or on a multiuser machine. A user account is a security system which is configured by the system administrator. The sophistication of the security can range from a simple name prompt at the time of login, to name and password logins at various levels of access, and different protections attached to directories, processes, and programs.

User Agent UA. A network service used by clients to find available services on behalf of the user. See Directory Agent, Service Agent, Service Location Protocol.

User Datagram Protocol UDP. An IETF-recommended protocol for the Internet which provides a datagram mode for Internet Protocol-based (IP-based) packet-switched network communications. UDP is primarily used with the Internet Name Server and Trivial File Transfer. The format of UDP header is shown in the User Data Protocol Header chart. For more details about UDP, see See RFC 768.

user event In programming, a type of input event which is signaled through an input device such as a mouse, joystick, keyboard, or touchscreen, and interpreted into a response by the operating system or applications program. User events typically include button, window, or menu selections/adjustments; and movement of icons, windows, or objects. The most challenging types of user events tend to occur in fast action video games and realtime graphics input processing.

User Glossary Working Group UGWG. A group within the User Services Area of the Internet Engineering Task Force (IETF) which has created an Internet Users' Glossary. See RFC 1392.

user group, user's group, users' group An organization of users of a particular product or service. A support group. With the introduction of computers to our world, society took a technological leap that was difficult for any one individual to understand or bridge. In order to facilitate the use and understanding of complex systems, programming languages, and technologies, many users' groups sprang to life, beginning in the mid-1950s, to provide mutual support and assistance in sharing information and meeting technological challenges.

The proliferation of users' groups is important not only for the support they provided to members, but also because this venue provided a forum for computer hobbyists, amateur radio groups, and many other amateur and professional enthusiasts to brainstorm ideas and contribute to the developing fields. The development of computer technology was no longer in the hands of large educational institutions and corporations. Individuals and small companies, particularly in the 1970s, had a window of opportunity during which they were able to make highly significant contributions to the development of the industry.

user interface UI. The communications link through which a person interacts with a machine. On computers, the UI, in its broadest sense, includes the various symbolic text, images, sound, and other sensory cues and gadgets presented to the user, with which the user interacts. This is commonly done through peripheral devices such as touchscreens, keyboards, mice, joysticks, microphones, data gloves, and others not yet invented. It is considered the highest layer of the computer system structure, with the machine instructions for physical operations comprising the lowest layer.

The user interface is the single most important aspect of computing and should never be undervalued. Computers were designed to serve the needs of people and, if people are having to adopt unhealthy or uncomfortable habits to interact with computers, or if computers take time away from people rather than freeing them from repetitive or drudgery tasks, then human needs are not being adequately served by the technology. User interfaces and software applications should be designed with the goal that the purpose of the technology is to improve the quality of life.

The design of user interfaces is an art. It demands common sense, a knowledge of ergonomics, psychology, philosophy, electronics, aesthetics, and a large dose of sympathy for a broad range of users. As such, user interfaces have developed in fits and starts, with many software programs providing very poor support for users, forcing the user to conform to the idiosyncrasies of the machine (or the programmer who wrote the software), rather than the other way around. See user interface history.

user interface history The earliest telecommunications interfaces consisted of physical semaphores (smoke, flags, and arms) and telegraph keys sending coded messages that needed to be decoded and transcribed when received. The received message was usually presented as a long paper tape inscribed with wiggly lines, dots and dashes, or punched holes. Computers right up until the 1950s used a similar model. This kind of user interface wasn't very friendly, so inventors, even in the earliest days of telecommunications technology, sought ways to encode the alphabet, so that letters could be directly sent and received (eventually resulting in teletypewriters) without the operators having to do the translation. But the basic methods prevailed for decades, mainly because they could be used anywhere, with the simplest of equipment. Telegraph key codes are still a requirement of attaining amateur radio licenses.

With the development of personal computers, user interfaces took a leap. The Altair microcomputer, sold originally as a hobby kit in 1984, had no monitor, mouse, or keyboard. It was programmed by means of flipping little dip switches; if you made a mistake, you had to start again. Yet within two years microcomputers as we know them, inspired partly by high-end systems with better resources than the Altair, came into being in the form of the TRS-80 and Apple computer, and keyboards and monitors became standard. Almost every change since then has been an evolutionary refinement or logical addition rather than a revolutionary change. Even the lifelike and startling three-dimensional virtual reality world represents, for the most part, an evolutionary development, albeit an exciting one.

Early computer user interfaces consisted primarily of monochrome screens displaying limited text, often with no lower case letters, and large rectangular graphic blocks. While ingenious computing pioneers wrung astonishing surprises from this primitive technology, it was obvious that improvements were needed in order for a computer to be more fun, versatile, and consumer-friendly.

UserID User Identification. A unique computer account designation used to gain access to a secure or monitored system. A UserID is frequently paired with a password for system access. Historically many systems accepted only eight characters for the UserID, and, for backward compatibility, this limitation persists on many systems today. On networks using the most common mail systems, the UserID typically forms the first part of an email address.

USGS See United States Geological Survey.

USITA Formerly United States Independent Telephone Association. See United States Telephone Association.

USKA Union Schweizerischer Kurzwellen-Amateure Union (Union of Swiss Shortwave Amateurs). A member organization associated with the International Amateur Radio Union. http://www.uska.ch/

USOC See Universal Service Order Code.

USOP User Service Order Profile.

USTA See United States Telephone Association.

UT1 A time reference based on Earth's axis rotation. It is related to Coordinated Universal Time in that UTC was set to synchronize with UT1 on 0000 hours on 1 January 1958.

UTAM Since some of the frequencies being used by incumbent carriers have now been designated for USDC services (1890 to 1930 MHz), companies are having to change their operating equipment and software to operate instead in the 2.0 GHz microwave C-band. UTAM Inc. is an open industry resource for assisting in frequency relocation. See band allocations.

UTC See Coordinated Universal Time.

UTDR Universal Trunk Data Record.

A typical utility pole with crossbars bearing telephone and power lines, ceramic insulators, coils, and protective vessels.

utility pole A sturdy tall pole installed in the ground (or *on* the ground in mountainous areas, supported by rocks and guy wires). The pole is used to support utility wires for power and telecommunications and may have crossbars and insulators. Utility poles are raised with the aid of long poles with spikes on the end

U

called *pike poles*, or with industrial machines designed for the job. Most poles are made from logs, although some areas have metal poles and, at one time in history, it was thought that metal poles would soon replace all the wooden poles, a prediction that didn't hold true. In some areas, especially avalanche areas, it was necessary to reinstall poles once or twice a year, a costly, time-consuming business, so various alternatives were tried including just laying the wire along the ground. Unfortunately, rodents like to chew through the wires, causing almost as much interruption to service as the avalanches. Transmissions in inclement regions are now often sent with microwave transceiving systems rather than with wires, a system which requires less maintenance.

UTM See Universal Transverse Mercator Projection.

UTP See unshielded twisted pair.

UTR Universal Tone Receiver.

UTS Universal Telephone Service.

UUCP UNIX-to-UNIX Copy. UUCP was developed in the mid-1970s by AT&T Bell Laboratories and distributed with UNIX in 1977. This was quickly adopted by many educational and research institutions for disseminating mail. Two years later, USENET, the global public news forum, was established using UUCP. By the mid-1980s national networks in other countries were being set up with UUCP; it was becoming clear that UUCP was an important catalyst for intercommunications and the development of distributed networks. In 1987, UUNET was founded to provide commercial UUCP and USENET access. See Unix, UNIX, USENET, UUNET.

UUNET A Unix-based network provider and backbone (long-haul network). UUNET provides Internet name serving, connectivity, MX forwarding, and news feeds. The formation of UUNET was probably in part due to the Acceptable Use Policy (AUP) focus on research and education enforced by the National Science Foundation's NSFNET. This stimulated commercial establishment of computer networks.

In the 1980s, UUCP over long-distance dialup lines was the primary means by which providers, institutions, and individuals received their messages. However, Internet connectivity has become ubiquitous and the situation has changed. In the 1990s, UUNET is the only remaining significant network that uses UUCP transport for USENET messages. UUNET Canada, Inc. is located in Toronto; UUNET Technologies Inc. is located in Virginia.

UV See ultraviolet.

UWB ultra wideband.

UWC See Universal Wireless Communications.

UWCC See Universal Wireless Communications Consortium.

v 1. Symbol for volt. See volt.

V 1. Symbol for vacuum tube. See electron tube, vacuum tube. 2. Symbol for voltmeter.

V & H Coordinates, V H Coordinates Vertical & Horizontal grid coordinates. Imaginary coordinate points on a virtual grid which are used to determine straight-line mileage between two specified points, with each exchange's location represented by a pair of V & H coordinates. This is used for various products and services which are charged on a distance or mileage basis, as are long distance calls.

The V & H system is based upon a 'flattened Earth' system from a Donald Elliptical Projection, developed by Jay Donald of AT&T in the mid-1950s. The basic idea is to create a triangular distance calculation over a flattened surface.

V & H Tape Vertical and Horizontal Coordinates Tape. A recorded tape provided primarily to assist with billing, it includes NXX types, major and minor V & H coordinates (latitude- and longitude-like regional designations), LATA Codes, and other information related to long distance accounting and service areas. It is made available for purchase through Bellcore.

V interface In ISDN data communications over phone wires, the telephone company typically has to install a number of devices to create the all-digital circuit connection necessary to send and receive digital voice and data transmissions. One of these devices, which connects the telephone service carrier to the local loop, is the V interface. See ISDN, U interface.

V Series Recommendations A set of ITU-T recommendations which provides guidelines for interconnecting networks and network devices. These are widely implemented in computer modems. The V Series specifications are available as publications from the ITU-T for purchase, and a few may be downloadable from the Net. Some of the related general categories and specific V category recommendations of particular interest are listed here. See also I, Q, and X Series Recommendations.

The listing on the following pages includes a number of V Series Recommendations of particular interest. Note that while a larger number often denotes a more recent standard, there are exceptions, and the categories are intermingled resulting in the number schemes series being interleaved.

V5 A telephony standard adopted by the European Telecommunications Standard Institute (ETSI) for use in digital local exchanges in Europe, Australia, South America, and the Far East. Exchanges around the world are gradually being upgraded to V5.

V-band A frequency band commonly used in radar applications. See the band allocations listing for a chart and details of frequencies.

V.Fast A vendor-developed format which was brought out while the ITU-T standards bodies were developing the V.34 protocol. V.Fast is similar to V.34, but not the same, and consequently some V.Fast modems are not compatible with V.34, while others are hybrid and support both. See V Series Recommendations listing under V.34.

ITU-T V Series Recommendations

General Categories

Series C	General telecommunications statistics
Series E	Overall network operation, telephone service, service operation, and human factors
Series F	Telecommunication services other than telephone
Series G	Transmission systems and media, digital systems and networks
Series H	Line transmission of nontelephone signals
Series I	Integrated Services Digital Networks (ISDN)
Series J	Transmission of sound program and television signals
Series L	Construction, installation and protection of cable and other elements of outside plant
Series P	Telephone transmission quality, telephone installations, local line networks
Series Q	Switching and Signaling
Series R	Telegraph transmission
Series S	Telegraph services terminal equipment
Series U	Telegraph switching
Series V	Data communication over the telephone network
Series X	Data networks and open system communication
Series Z	Programming languages

Duplex Modem ITU-T-Recommended Transmission Standards

V.21	A 1984 dialup modem standard supporting data rates up to 300 bps. Most modems in the late 1970s and early 1980s were acoustical couplers, with direct connect modems just beginning to catch on in the mid-1980s.
V.22	A 1988 dialup modem standard supporting data rates up to 1200 bps with fallback to 600 bps. It is interesting to note that in the 1980s, many technical people insisted it was impossible to support speeds faster than 1200 bps over standard phone lines, and that this was probably the fastest speed at which modems would ever transmit data.
V.22 bis	A 1988 update to the V.22 standard supporting data rates up to 2400 bps through frequency division techniques, with link negotiation fallback to 1200 bps, and fallback to V.22.
V.23	Dialup modems supporting data rates up to 1200 bps with fallback to 600 bps and a 75 bps *back channel* or *reverse channel.*
V.26	A 1984 standard for modem supporting data rates up t o 2400 bps on 4-wire leased telephone lines.
V.26 bis	A 1984 update to V.26 supporting 2400 and 1200 bps over public phone dialup lines.
V.26 ter	A 1988 update to V.26 and V.26 bis which supports both echo cancellation (public phone lines), and point-to-point two-wire leased phone lines.
V.27	A 1988 standard for modem data rates up to 4800 bps with manual equalizer for use with leased telephone lines.
V.27 bis	A 1984 update to V.27 which supports data rates of 4800 and 2400 bps with automatic equalizer for use with leased telephone lines.
V.27 ter	A 1984 update to V.27 and V.27 bis which supports data rates of 4800 and 2400 bps over public phone dialup lines.
V.29	A 1988 standard for modem data rates of 9600 bps for 4-wire leased telephone lines.
V.32	A standard for public dialup and 2-wire leased line modems with rates up to 9600 bps with fallback to 48,800 bps. This standard was approved in 1984. When modems incorporating V.32 were first introduced in the late 1980s, it was not unusual for a single modem to cost between $600 and $2,000. Modems of this speed can now be found for $10, as they have been superseded by much faster data rates of 33,600 bps. Many of the V.32 modems were dual-standard modems which supported V.32 in addition to various proprietary protocols from individual manufacturers. See V.32 bis.
V.32 bis	A 1991 standard for public dialup and 2-wire leased line modems which

supports rates up to 14,400 bps with fallback to 12,000, 9600, 7200, and 4800 bps. Like V.32, a number of these modems are dual-standard modems with proprietary protocols included, and they are generally all backwardly compatible with V.32. Many of them support MNP-2 to MNP-4 error control and MNP-5 data compression, in addition to V.42/V.42 bis error control and data compression.

V.32 ter An update to V.32 bis designed by AT&T, which supports data rates up to 19,200, and fallback to 16,800 and V.32 bis. This format is freely distributable and has become a de facto standard. The corresponding ITU-T approved format for 19,200 is V.34.

V.33 A 1988 standard for modem data rates of 14,400 bps for 4-wire leased telephone lines.

V.34 A 1996 standard for dialup modems and fax/modems with rates up to 33,600 bps in full duplex, and up to 28,800 bps in half duplex for facsimile transmissions. (These are sometimes also called VFast, or VFC, but those are actually slightly different interim protocols developed by vendors who were anxious, in the mid-1990s, to get products to market while the work on the V.34 standard was being completed. Some modems became V.34/VFast hybrids.)

V.34 modems are designed to adapt to the line by probing the connection and adjusting according to quality and capacity. They interact with the telephone circuit through handshaking. Optional control data can be sent through an auxiliary signaling channel. V.34 modems are not as subject to noise as earlier modems, due to the implementation of multidimensional trellis coding.

V.34 bis A 1996 standard which supports the higher data rates of 56,000 bps and 31,200 bps.

Digital Modems, Digital/Analog Hybrids, Wideband, & Parallel Transmission Modems

V.19 A 1984 standard for modems for parallel data transmission over telephone signaling frequencies.

V.36 A 1988 standard for wideband synchronous transmission modems using 60 to 108 kilohertz group band circuits.

V.37 A 1988 standard for wideband synchronous transmission modems for signaling rates higher than 72,000 bps, using 60 to 108 kilohertz group band circuits.

V.38 A 1996 standard for wideband data circuit-terminating equipment for rates of 48,000, 56,000, and 64,000 bps for use on digital point-to-point leased circuits.

V.70 A 1986 standard for digital simultaneous voice and data (DSVD) modems in which a data transmission and a digitally encoded voice transmission can be sent at the same time over a single dialup phone line. DSVD are typically downwardly compatible with standard dialup modems based on more recent high-speed technologies (e.g., V.34). This type of technology lends itself well to teleconferencing applications, and a number of vendors have incorporated V.70 to this end.

V.75 A 1996 standard for digital simultaneous voice/data (DSVD) transmission terminal control procedures. See V.70.

V.76 A 1996 standard for multiplexing using V.42 LAPM-based procedures. V.76 multiplexing has been incorporated into V.70 systems, although it is not limited to V.70.

V.80 A standard for the application interface for communications through data terminal equipment (DTE) of a synchronous H.324 bit stream, such as video, for asynchronous transmission over public switched telephone networks. V.80 works in conjunction with a number of other related technologies and standards. H.324 is an ITU-T approved standard which provides the foundation for combining video, voice, and data communications on a single analog phone line using a 28,800 bps data rate connection. This opens the door to a variety of practical, reasonably priced stand-alone and computer-based videoconferencing products that display up to 15 video frames per second. H.263 and G.723 are video and voice compression schemes used in H.324.

V.90 A 1998 standard for digital to/from digital or digital to/from analog connections, supporting download data rates up to 56,000 bps with upload data rates up to 33,600. During development it was known as V.pcm and some vendors call it PCM. This standard reflects the gradual conversion of data communications to digital format. This standard has its own Web site at http://www.v90.com/

Half-Duplex Modem ITU-T-Recommended Transmission Standards

V.17 A two-wire scheme for facsimile machines and fax/modems used in conjunction with extended Group 3 facsimile standards for image transfer at rates of 12,000 bps and 14,400 bps.

V.27 A modulation scheme for dialup facsimile machines and fax/modems used in conjunction with Group 3 facsimile standards for image transfer at rates of 2400 bps and 4800 bps.

V.29 A modulation scheme for dialup facsimile machines and fax/modems used in conjunction with Group 3 facsimile standards for image transfer at rates of 7200 bps and 9600 bps.

ISDN- and Digital Communications-Related

V.100 A 1984 standard for interconnection between public data networks (PDNs) and public switched telephone networks (PSTN).

V.110 A 1996 standard for ISDN support of data terminal equipment (DTE) with V-series interfaces. Listed also as I series 463.

V.120 A 1996 standard for ISDN support of data terminal equipment (DTE) with V-series interfaces with provision for statistical multiplexing. Listed also as I series 465.

V.130 A 1995 standard for an ISDN terminal adaptor framework.

V.230 A 1988 specification for a general data communication interface layer 1.

Error Control and Data Compression Protocols

V.41 A 1972 code-independent error control standard.

V.42 A 1996 error control standard which greatly enhances the functioning of modems over standard phone lines. Phone lines tend to be noisy, slow, and somewhat unreliable for high speed data communications. In the past, if there were errors on the line, modems would react by exhibiting line noise, losing the connection, or aborting the current operation. With new error control capabilities built in, some of these problems are overcome through filtering and selective retransmission.

V.42 includes two error control protocols for dialup modems. These are link access procedure for modems (LAP-M) and Microcom Networking Protocol (MNP-4), such that connections with a variety of modems are supported.

V.42 uses a filtering process somewhat like the error correction schemes incorporated into a number of file transfer protocols, such as XModem and ZModem.

V.42 bis A 1990 standard for data circuit terminating equipment (DCE).

Miscellaneous Modem/Transmission/Network-Related Standards and Protocols

V.7 A 1988 guide to terms concerning data communications over telephone networks.

V.8 A 1994 guide to procedures for starting sessions of data transmissions and setting up connections parameters over general switched telephone networks.

V.8 bis A 1996 guide to procedures for the identification and selection of common modes of operation between data circuit-terminating equipment (DCE) and between data terminal equipment (DTE) over general switched telephone networks, and on leased point-to-point telephone-type circuits.

V.18 Interoperability guidelines for communications services for the hearing impaired.

V.24 A 1996 set of definitions for interchange circuits between data terminal equipment (DTE) and data circuit-terminating equipment (DCE) which specify the characteristics of interfaces, including pinout circuitry. This is similar to the RS-232 specifications.

V.25 A 1996 standard for dialup automatic answering equipment and automatic calling equipment.

V.25 bis A 1996 command set designed for synchronous communications through serial ports and connections.

V.54 A modem diagnostics standard implemented in high speed modems.

vaccine A whimsical name for a virus protection program that resides on a computer and signals the user if there are anomalies or known viruses present, and allows the virus to be disabled or deleted. As quickly as virus detectors and vaccines are written and disseminated, virus creators come up with new ways to create mischief or outright destruction on people's computer systems. With the Internet foreshadowing a not-too-distant day when every computer is linked to the net, perhaps 24 hours per day, more opportunities for vandalism exist, and education about virus detection and protection needs to be made known to the computing public.

vacuum An enclosed space in which the gases are at pressures below atmospheric pressure. Since many vacuums are actually near vacuums, categories of vacuums from low to ultra-high depending upon the pressure, have been described. The discovery and creation of vacuums has aided scientists in making important discoveries and provides environments in which various phenomena take place, or fail to take place, due to the absence of gases. For example, a filament in a bulb can burn much longer in a vacuum.

vacuum column In some magnetic tape drivers, there may be a vacuum mechanism to control the tape loop. This provides a lower air pressure 'suction' next to the tape and the drive mechanism.

vacuum gauge An instrument for measuring the degree of vacuum in an enclosed space. There are a number of types of vacuum gauges, including manometers, ionization gauges, and thermal conductivity gauges.

vacuum tube A ubiquitous, essential, and versatile electron tube that was common until about the mid-1960s, after which it was superseded by various early electronic transistors, and later more sophisticated semiconductor technologies.

The vacuum tube in its basic form consists of an electron-emitting filament (cathode) and a metal plate (anode) to which the electrons are attracted. Various types of control grids might be interposed between the cathode and the anode (or in other locations), with the whole thing sealed in a glass vacuum tube to control the internal environment and to prolong the life of the filament. The first vacuum tubes adapted for radio broadcasting equipment were developed just after the turn of the century in the early 1900s.

While most small vacuum tubes have passed out of use, cathode ray tubes (CRTs) are still widely used in television sets and desktop computer monitors. There are still situations in which vacuum tubes can be a better solution than the solid-state components which are now commonly used in electronics. For high power-level, high-frequency applications, vacuum tubes are sometimes more efficient and still merit consideration. See cathode ray tube, electron tube.

Vacuum tubes fulfilled thousands of roles in electronics for over fifty years, before the invention of the transistor and modern semiconductor technologies. Even now they are appropriate for certain high-frequency applications. The most significant step in the evolution of vacuum tubes was the triode, which included a controlling grid to harness electron energy. These excellent historic examples are from the Bellingham Antique Radio Museum.

vacuum tube amplifier An important development stemming from Lee de Forest's invention of the audion, developed by Harry DeForest. The vacuum tube amplifier was incorporated into telephone repeating units, which extended communications distances.

VAD See voice activity detection.

Vail, Alfred (1807-1859) An American scientist and inventor, and associate of Samuel Morse, from whom Morse adapted a number of ideas related to the building of telegraph systems. Vail was very mechanically apt and continued over the years to make technical improvements to the technology. In 1837 Vail made an agreement with Morse to turn over the rights for his inventions in return for a share of the commercial rights. In 1848, due to the mounting workload, and the lack of sufficient appreciation and compensation, Vail decided to terminate this relationship with Morse. Morse code may be one of Vail's biggest contributions. When Morse was designing a system for coding messages, he initially created a complex numeral-letter relationship that required a time-consuming dictionary lookup to decode the words. Vail apparently came up with the simpler system, after altering the mechanics of a telegraph instrument so it moved vertically rather than horizontally, to allow the instrument to leave spaces (thus yielding dots and dashes) when recording a message. See Morse code, telegraph history.

Vail, Theodore N. (1845-1920) Theodore Vail was from the same Vail family that had a close association with Samuel Morse at the time of his invention of the telegraph, and Vail learned telegraph code from the elder Morse. Vail became a telegraph operator, and later devoted most of his life to the management and promotion of universal telephone services.

Vail became company general manager of the first Bell company in 1878 in Boston. He was the first president of the Telephone Pioneers of America and a cofounder of Junior Achievement Inc. (1919). He was instrumental in continually expanding the company's offices and services, and directed the building of the first transcontinental telephone line, completed in 1914 and officially opened in 1915.

After what has been regarded as a brilliant career, where he maintained his commitment to quality and service, Vail resigned in 1887 due to his disgust at the narrowly commercial vision of the Bell financiers.

In 1907, he was induced to return to AT&T as president, at a time when Bell was rapidly buying independents. Bell also purchased Western Union stock, and installed Vail as president of what had once been Bell's chief rival. With the death of J. P. Morgan in 1913, Vail voluntarily took steps to reduce AT&T's monopolistic buyouts and control of the long distance networks, in order that reorganization could be done from the inside rather than being imposed by regulatory authorities. Vail retired as president in 1919. See Kingsbury agreement.

value In imagery, the relation of a color to white or black, or levels of gray. Sometimes called lightness. In color, the intensity of the color.

value-added Services offered, usually for a fee, in addition to standard product or subscriber services. Options.

VAN value added network.

Van Allen radiation belt A region of space surrounding Earth in which there is high-intensity particle radiation which is sufficiently destructive that it is avoided for communications satellite orbits. This region is roughly between the low Earth orbit (LEO), starting at about 1000 kilometers, and medium Earth orbit (MEO). It is named after James A. Van Allen.

Van de Graaff generator A device for creating electrostatic effects by charging insulated electrodes to high energy potentials. Van de Graaff generators are popular exhibits in science museums. They typically are configured as melon-sized metallic spheres atop a central pole, and visitors can place their hands on the globe and watch their hair stand on end. Named after Robert J. Van de Graaff, an American physicist of the early 1900s.

van Musschenbroek, Peiter (1692-1761) A Dutch educator and experimenter who created the Leyden jar condenser (named after the region) in 1745 independently of other inventors, including E. G. von Kleist. It was a simple, but practical capacitor, and effective enough that it could cause harm to a person not careful in its handling. Van Musschenbroek also described a way to carry out experiments with the jar so that a person would not be mortally shocked. Many subsequent experimenters, including Benjamin Franklin, devised variations on the basic Leyden jar.

vaporware A derogatory term for software which has been announced prior to completion, or prior to distribution. The negative aspects of the term stem from two sources, truth in advertising and the fact that many announced software products never hit the shelves. Because the software industry is new and confusing to a great portion of the public, unsupported product announcements have not been as strongly condemned and regulated as in other industries, a situation which may change as the buying public becomes more familiar with the new technologies and tired of unsubstantiated promises.

variable bandwidth Bandwidth which can be tailored, usually on an on-demand basis, to the capacity needs of the current transmissions. The ability to adjust capacity allows the system to more efficiently allocate resources and provides a mechanism for setting up accounting systems which bill on an as-used basis.

variable bit rate VBR. A data transmission commonly represented by irregular groups of bits or cell payloads, followed by unused bits or payloads. VBR traffic is generated by most media other than voice. In an ATM environment, a VBR service can be realtime or non-realtime, and is guaranteed sufficient bandwidth and quality of service (QoS). See cell rate for a chart.

variometer An instrument used for measuring magnetic declinations, particularly of the Earth.

Varley loop test A type of diagnostic procedure which uses resistance through a bridge to locate a fault in a length of circuit. Variable resistance is connected in series with the resistance of the broken or defective line. Similar to the Murray loop test, but with a third wire, and more commonly used. See Murray loop test; Wheatstone bridge.

Varley, Cromwell Fleetwood A British researcher and technician who investigated ionization, and was involved in early Atlantic telegraph cable installation. He was hired by Western Union to evaluate the telegraph system in the U.S. Varley standardized many of the lines and systems, and diagnostic techniques. The Varley loop test is name after him.

VAX *V*irtual *A*ccess E*x*tension. A series of minicomputers from Digital Equipment Corporation (DEC) which followed the PDP-*x* series. VAX machines running the VMS and Unix operating systems were widely installed in educational institutions and corporations, often using personal computers as remote terminals.

VBE VESA BIOS extensions. A high resolution VESA BIOS standard which can be implemented in hardware or software to provide control of video graphics display on a computer, typically an Intel-based International Business Machines (IBM) or third party licensed computer.

vBNS very high speed Backbone Network Service. A research network established in 1995 by the National Science Foundation.

VBR See variable bit rate.

VBX Visual BASIC Extension.

VC virtual connection, virtual circuit. A generic term for a logical communications medium which is established on request. A connection typically includes a concatenation of channels forming an end-to-end path. A circuit refers to transmission in both directions. Three types of VCs include permanent (PVC), smart/soft permanent (SPVC), and switched (SVC). In an ATM environment, data to be transmitted by a VC is segmented into 53 octet quantities called cells. This consists of 5 octets of header, and 48 octets of data.

VCC virtual channel connection. A generic term to describe a logical connection. In an ATM environment, a virtual channel refers to the unidirectional transport of ATM cells associated by a common unique identifier value. Virtual circuits (VCs) can be combined to form a virtual channel.

VCI Virtual Channel Identifier. A value in the header of each ATM connection cell which identifies that connection. See VC, VCC, virtual channel.

VCL Visual Component Library. A library used for applications development for Borland Delphi products.

vector Any quantity having both magnitude and direction.

vector display A cathode ray tube (CRT) vector display is one in which the sweep of a beam follows a vector (line or stroke) and illuminates (and refreshes) the *specific* part of the display that is needed to render the desired colors and shapes. In other words, the beam doesn't follow the sawtooth scan characteristic of raster displays. This is in contrast to a television screen, in which a beam constantly sweeps the screen to form a *frame* in which the beam travels across the entire display area on a constant, repetitive basis.

In a vector monitor, a vector generator takes the coordinates supplied by the processor and converts them into analog voltages that are used to control the direction of the beam as it excites the phosphors coated on the inside of the CRT. Vector graphics appear very 'crisp' and clean, but refreshing large areas or the entire screen is generally slow and impractical for many applications.

Vector monitors were prevalent during the 1960s and '70s, but have largely been replaced by mass market raster monitors. Tempest was an early video game that employed vector graphics, whereas Space Invaders used raster graphics.

vector font A set of textual characters or symbols defined by vector algorithms (usually lines, spline curves, and arcs) rather than by relative positioning of dots within a grid (raster or bitmap font). A vector font looks smooth and appealing at almost any size, except very tiny sizes, and displays at the highest resolution available to the output device on which it is being displayed or printed. In general, the higher the resolution, the smoother the lines and more attractive the overall look of the font. Contrast with raster font.

vector quantization See quantization, vector.

verifying punch A punch card perforator which also verifies the punches to each card to see that the perforations are correct, automatically replacing those which are defective.

Veronica Named tongue-in-cheek for a comic-book character reference to a related tool that queries FTP sites named Archie. What Veronica does for Gopher information is similar to what Archie does for FTP sites. Veronica is an Internet keyword query tool which searches Gopher sites, sometimes called *Gopherspace*, and typically displays the results of the search in the format of a Gopher menu. Users are linked transparently to the source, and may not see the location of the source unless they explicitly ask. Veronica was introduced in 1992 by S. Foster and F. Barrie of the University of Nevada (Reno).

Computer programmers love to come up with acronyms, and Veronica is no exception. It's a stretch, but it has been said that Veronica stands for Very Easy Rodent-Oriented Netwide Index to Computerized Archives. See Archie, Anarchie, Gopher, Jughead, WAIS.

Versatile Interface Processor VIP. A Cisco Systems router interface card which provides multilayer switching.

versorium A device for detecting electrical properties of various materials, invented by William Gilbert in the late 1500s. The versorium resembles a compass needle in that it is a horizontal movable needle balanced on a small support stem, but differs in that the needle itself is not made of magnetic material, like lodestone, but rather of wood or a nonmagnetic metal. Gilbert used this sensitive device for evaluating attractive properties of different materials when altered by rubbing. Descendants of this instrument are now called *electroscopes*. See Gilbert, William.

vertigo A perception by an individual that the environment is moving around the individual, or the individual is moving in relation to the environment when no such physical motion actually exists. For example, watching a big-screen movie with realistic action may cause a viewer to feel as though he is moving, when in fact he is sitting in a theater seat.

very high frequency VHF. Electromagnetic waves in the approximate range of 50 MHz to 300 MHz, part of which is allocated for amateur use (50 to 54 MHz and 144 to 146 MHz) with some regional variations.

very large scale integration VLSI. In the semiconductor industry, VLSIs are integrated circuits (ICs) which combine hundreds of thousands of logic and/or memory elements into one very small chip. This type of circuitry has brought about a revolution in the cost and manufacture of computers. VLSI has enabled the manufacture of palm-sized computers which are more powerful than room-sized computers from a few decades ago, which were dependent on vacuum tubes and wires for their circuitry.

Very Small Aperture Terminal VSAT. Very small commercial terminals for two-way satellite transmissions in the United States, and one-way communications in countries with restrictions. VSATs are generally organized in a star topology, with the Earth station acting as a central node in the network. This Earth station operates with a large satellite dish and a commercial quality transceiver.

In some VSAT implementations, the signal from the transmitting Earth station to the satellite is amplified and redirected to a hub Earth station. Since all transmissions pass through this hub, two hops are needed for intercommunication between satellites. (This results in a bounce pattern known as an M hop.) Some newer implementations, modulation and amplification systems are included onboard the satellite, so that an interim transmission to a hub is not required.

VSAT systems are appropriate for centralized business and institutional networks. Commercial VSATs typically communicate in the C- and Ku-band frequencies.

VESA See Video Electronics Standards Association.

VF access voice frequency access.

VFast See V.Fast, V Series Recommendations.

VFC Version Fast Class. A vendor-developed interim format for modem-based serial communications which was commercially implemented while the ITU-T was working out the V.Fast standard. See V.Fast, V Series Recommendations.

VGA See Video Graphics Array.

vgrep visual grep. A variant of a very powerful, useful Unix command. See grep.

VHF See very high frequency.

VHF antennas A category of antennas which is designed to take advantage of the particular characteristics of very high frequency (VHF) waves. Because of the wavelength differences between VHF and ultra high frequency (UHF) waves, and the relationship of the rods on the antenna to the length of the wave, VHF antennas tend to be larger and more varied in their shapes than UHF antennas, and can be installed with less precision and still be relatively effective. They are not as broad, however, as a single UHF antenna can cover the entire UHF band, but a VHF antenna is usually optimized for a particular range or stations. See antenna, combination antenna, fan dipole antenna, UHF antenna.

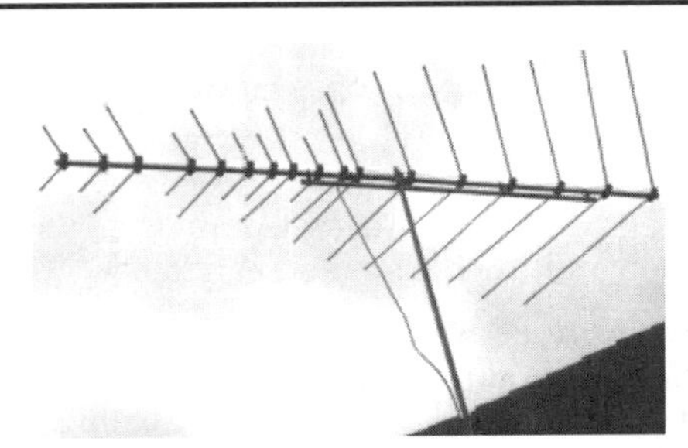

This common style of rooftop antenna is designed primarily for capturing VHF television broadcast frequencies. The length and orientation of the reflectors and directors, and the general orientation of the entire structure will influence the strength of the signal that is transferred through the feedline to the television tuner.

VHS Video Home Systems. A widely used video format developed by JVC that is compatible with millions of home user systems. VHS and Beta formats were released at about the same time. Beta was acknowledged as being superior, but was also a bit more expensive, so VHS won the marketing wars. It is slowly being superseded by S-VHS, 8mm, and Hi-8mm formats, in addition to a number of digital formats, including DVD. S-VHS systems are downwardly compatible with VHS tapes, that is, you can play a VHS tape in an S-VHS system (but not the other way around).

VIA See Virtual Interface Architecture.

Vibroplex Trade name of a type of semi-automatic telegraph key introduced in the later 1800s, more commonly called a *bug key*. This particular type of bug key was patented in 1904 by Horace G. Martin. By using a vibrating point for automatically generating dots and dashes, it relieved telegraph operators from physical and mental strain.

video capture board See frame grabber.

video chipset A logic circuit in a computing device which handles the processing, and sometimes acceleration, for video graphics display. Various means of configuring this circuit, and integrating it with the system, control the speed, resolution, and palette set of the display.

Video Electronics Standards Association VESA. An industry standards body established in 1990, which develops various peripheral standards for Intel-based microcomputers. This organization is responsible for defining Super VGA (SVGA) and the VESA local bus for peripheral device interfaces with personal computers. See super video graphics array, VESA VL.

video floppy A 2" digital image storage floppy released in the mid-1980s to hold video images of 360 lines of resolution. Various methods are now used to store digitized images, including flash memory, 3.5" floppies, and various proprietary cards.

Video Graphics Array VGA. A graphics standard common on Intel-based International Business Machines (IBM) and licensed third party computers, supporting 640 *x* 480 (16 colors) and 320 *x* 200 (256 colors). It has been superseded by super video graphics array (SVGA). See super video graphics array.

video switcher A generic phrase for a wide variety of types of passive routing boxes for video signals. Home systems sometimes have simple switchers to select between a VCR and a laserdisc player. Professional switchers may have banks of connectors, sliders, and settings.

This simple consumer switcher provides switching for composite video and audio inputs and outputs through standard RCA receptacles. It can switch between four different devices.

V

video tape A magnetic recording medium resembling common audio tape, which is designed to store both images and sound. The most common formats for video tape are VHS, S-VHS, S-VHSc (compact S-VHS), Beta, 8mm, and Hi-8mm. S-VHS and Hi-8mm are sufficiently good for many professional applications, although higher quality formats are preferred for commercial broadcast quality tapes. Many tapes can store audio in two places, intermixed with the images or on a separate track along the side of the tape, for high fidelity recordings which come close to the quality of CD. Recording times range from 20 minutes to several hours depending upon the type of tape and the quality settings.

All television programming used to be live. The performances were saved only in the minds of those who watched them. Then, in the mid-1950s, taped broadcasts became practical and the live aspect of television changed forever. Broadcasts could now be archived, played as reruns, broadcast during convenient times for a specific timezone, or sent overseas to other markets. Station managers could re-air programs without providing royalties to the actors, thus reducing costs (actors weren't happy about this). A whole series could be shot in a period of weeks and then aired over a period of months, freeing up the actors and production staff to work on the next project.

VHS taped entertainment has been widely available through video rental stores since the early 1990s.

In the 1980s, less expensive camcorders (camera/recording combinations) were utilized by consumers to tape special events, weddings, birthdays, graduations, and amateur movies. By the mid-1990s digital camcorders began to appear and, by the late 1990s, the price dropped to the point where they became consumer items, with film use declining.

video tape recorder, video cassette recorder VTR, VCR. A recording and playback device specifically designed to record simultaneous motion images and sound. The input is usually through video patch cords from microphones and cameras for live recording, or from camcorders and CDs, phonographs, and tapes for re-recording or editing. VCRs have been common consumer items since the mid-1980s; prior to that, they were generally used in the television and video editing industries.

One of the earliest patent applications for a video tape recorder was submitted in 1927. One of the early 'portable' video tape recorders, based on new transistor technology, looked just like a large reel to reel audio recorder. It was introduced by Ampex in 1963. This large desktop model weighed about the same as a medium-large TV set, but was nevertheless only one-twentieth the size of previous floor-standing models. It used a single-head helical scanning mechanism and could record 64 minutes of programming on standard 8" reels. VCRs are much smaller now, and use convenient cassettes rather than reels. They have also been improved to support high fidelity sound and higher resolution images.

video-on-demand VoD. A commercial interactive video system in which the user can request a specific video to be played at a particular time, unlike traditional TV programming where the station determines which programs are to be broadcast, and when. A number of these programs have been tried in various regions with mixed success. It's difficult to institute a pay service in competition with hundreds of 'free' channels on TV, which are primarily financed by advertising sponsors. The most successful video-on-demand systems appear to be those installed in motels and hotels which cater to business people attending professional conferences. Thus, one could say tongue-in-cheek that success depends in part on what the market will bare. See audio-on-demand, services-on-demand.

videoconferencing The transmission of coordinated motion images and sound through computer networks. This is an exciting area with many systems vying for the front row seat. Technologies to transmit speech and images have been around since AT&T's Picturephone system, which was developed many years before it was introduced to the public in 1970s. However, full motion video as found in videoconferencing, or still frame video and sound as found in audiographics, didn't reach practical speeds and consumer price ranges until the mid-1990s. Even then, they were mostly of interest to educational institutions and corporations. By 1997, however, consumer systems were beginning to be practical, especially with the proliferation of tiny monochrome and color video cameras, similar to those found in security systems. See audiographics.

vidicon A television with a photoconducting pickup sensor.

VidModem A patented signal-processing technology from Objective Communications Inc. which can accommodate simultaneous two-way video, voice, and data over standard copper wires. VidModem uses FM signals and compression to transmit a 24 MHz FM signal through the 20 MHz bandwidth that is supported on phone lines.

Vines A commercial virtual network based on Unix system V, from Banyan Systems.

VIP See virtual IP.

Virtual Interface Architecture VIA. An association of vendors who seek to describe and promote a generic systems-area network in order to facilitate the development of software for various X86- and RISC-based computers and their interconnections. VIA was established in 1996 as a small vendor consortium, and has grown to over 50 companies. See Scheduled Transfer.

virtual IP, virtual Internet Protocol VIP. A function which enables the creation of logically separated switched IP workgroups across the switch ports of a Cisco switch running Virtual Networking Services (VNS) software.

virtual LAN virtual local area network. A local area network in which the internal mapping is organized other than on the geography (physical relationship) of the stations. This allows the system to be segmented into manageable groups. Network software is used to administrate bandwidth and load, and to maintain a correspondence between the virtual LAN and the physical LAN. Newer versions of software will even allow configuration and connections to be established through software with graphical user interfaces that display the equipment itself, as graphics, with lines to indicate the various connections. See local area network.

virtual office A company, or department which is loosely connected physically, but which is communications-linked through various business telecommunications options such as cellular phones, videoconferencing systems, satellite modems, the Internet, etc. Some of the participants may be working at home or traveling. Some corporations mistakenly consider the 'virtual office' to be a new concept, but publishers and their associated writers have successfully employed this business model for decades. See telecommuting telework.

virtual path connection VPC. A path connection established on ATM networks along with an associated quality of service (QoS) category, which defines traffic performance parameters.

virtual private network VPN. A secure encrypted connection across a public network which allows organizations to utilize a public network as a virtual, private communications tool. Through a process called tunneling, the packet is encapsulated and transmitted. A VPN is a cost-saving measure for businesses that don't want the expense of setting up an internally funded secure network, and yet desire interconnectivity between remote branches and departments accessible through the relatively inexpensive services of an ISP. VPNs provide a cost-effective alternative to laying cables, leasing lines, or subscribing to frame relay services. The disadvantage to VPNs over public networks is the response time.

virtual reality VR. A phrase to describe electronically generated environments which interact with human senses to provide the illusion of the 'real world' or to provide a fantasy world experience that cannot be achieved in the 'real world.' Sensory headsets, goggles, helmets, implants, gloves, shoes, body suits, computers, monitors, chambers, and a whole host of visual/tactile/auditory two- and three-dimensional inputs are used to create virtual reality worlds. See Virtual Reality Modeling Language.

Virtual Reality Modeling Language VRML (*pron.* ver-mul). VRML was originally released in 1994 by Tony Parisi and Mark Pesce. Initially dubbed Virtual Reality markup language by Dave Raggett, VRML is a file-format standard, built in part from Silicon Graphics Inc. (SGI) Open Inventor File Format, which was made freely distributable later in 1994.

VRML provides a means for the creation of 3D multimedia and shareable virtual environments. Its inventors describe it as a 3D Web browser. It can be used in geographical, architectural, and industrial modeling; simulations; education; and games.

In August 1996, when the version 2.0 specification of VRML was released, JPEG and PNG were specified as the two image formats required for conformance with the specification.

VRML plug-ins are available for a number of browsers. The files tend to be very large, but there are unique opportunities, too, like taking a virtual ride on Mars Pathfinder, for example, an experience that's worth the download time. VRML 97 was approved in January 1997 as International Standard ISO/IEC 14772-1. See Joint Photographics Group Experts, Portable Network Graphics, virtual reality, VRML.

Virtual Tributary VT. In SONET networking, a sub-STS-1 signal designed for switching and transporting data. A VT Group (VTG) is defined as 12 columns, which can be formed

V

by interleaved multiplexing, and a group may contain only one type of VT. VTs operate in two modes: locked and floating. The VT types are as follows:

VT Signal	Rate	Digital Signal	Rate
VT-1.5	1.728 Mbps	DS-1	1.544 Mbps
VT-2	2.304 Mbps	CEPT-1	2.048 Mbps
VT-3	3.456 Mbps	DS-1C	3.152 Mbps
VT-6	6.912 Mbps	DS-2	6.312 Mbps

Virtual Trunking Protocol VTP. A virtual LAN (VLAN) autoconfiguration protocol from Cisco Systems.

visible spectrum The region of light waves that is perceived by humans as color, ranging from approximately 380 to 700 nanometers, or 3800 to 7000 angstroms. Technology cannot reproduce all of the colors of the visible spectrum, but then humans cannot always distinguish between very closely related colors either. For practical purposes, the approximately 17 million colors displayable on better quality computer monitors and the approximately 10 million colors that can be printed with pigments on a press are sufficient for most personal and commercial needs. Directly outside the visible spectrum are the infrared and ultraviolet wavelengths.

VisiCalc Visible Calculator. A historic early computer spreadsheet program, introduced in 1979, which was developed by Dan Bricklin and Bob Frankston for the Apple computer.

VISIT Video A Macintosh- and IBM-licensed PC-based videoconferencing system from Northern Telecom Inc. which supports video, whiteboarding, and file transfers over Switched 56 or ISDN. An extra transmissions line is needed for audio. See Cameo Personal Video System, Connect 918, MacMICA, IRIS, ShareView 3000.

Visual BASIC A basic Microsoft Windows programming application development product with a graphical user programming interface from Microsoft Corporation. Suitable for prototyping, although extensive use of dynamic linked libraries (DLL) may be needed for extensive applications development.

visual ringer A small lamp on a phone console, usually a light-emitting diode (LED), which lights up when the phone rings. This is convenient in a noisy environment or for those who are hearing impaired. It is also common on multiline phones, to indicate which of the multiple lines are currently ringing.

vitreous electricity A term coined by Dufay to denote the type of electrostatic charge produced on glass when rubbed with silk. Benjamin Franklin later proposed *positive*, a term that superseded vitreous. See electrostatic, resinous electricity.

VLAN See virtual LAN.

VLSI See very large scale integration.

VMI V Series Modem Interface. A standard software front end and software layer that provides an entry and exit point to modem functions implemented through a variety of modem standards, for DSP Software Engineering modem products.

VNS virtual network system, virtual network service.

vocoder *voice coder*. A late 1930s invention which provided a means for analyzing the pitch and energy content of speech waves. This technology led to the development of a device designed to transmit speech over distance without the waveform. The transmission was expressed at the receiving end with a synthetic speaking machine. This general concept has evolved into linear predictive encoders.

VoFR voice over frame relay. See frame relay, voice over.

voice activity detection VAD. A capability of digital voice communications systems to distinguish between information, such as speech, and the silences in between the speech elements. Typically, voice conversations consist of only about 40% talking, with the rest being pauses, silence, or low level background noise. By transmitting information only when it is meaningful and filtering out the silent moments, it is possible to create a significant savings in the amount of data which needs to be transmitted. See silence suppression.

voice-activated system A system such as a computer, phone, door, etc. which responds to the sound of a voice, which might be any voice or a specific voice. Voice-activated systems are calibrated to separate out the patterns and frequencies common to human voices from general background noise or other sounds. Technically, there's no reason why it couldn't also be configured to recognize the bark of a dog. This should not be confused with a speech-recognition system which recognizes actual words, not just a general or particular voice. Sometimes the two are combined.

voice grade channel A transmission circuit which is sufficiently fast (usually up to about 56 Kbps) and suitable for transmitting clear voice conversations within frequencies of between 300 and 3300 Hz. It is not typically suitable for other faster, higher bandwidth uses. Voice is a relatively low bandwidth application and does fine over copper wires, but others such as data transfer and video images require more.

voice group In analog voice phone systems, a hierarchy for multiplexing has been established as a series of standardized increments. These are organized as follows:

Group Name	Composition	Number of Voice Channels
group		12 voice channels
supergroup	5 groups	60 voice channels
mastergroup	10 supergroups	600 voice channels
jumbo group	6 mastergroups	3600 voice channels

voice over ATM A growing area of interest, voice over ATM involves the digital transmission of voice conversations (which traditionally have been carried over analog phone lines) over asynchronous transfer mode (ATM) networks. Typically this involves taking a synchronous voice signal, segmenting it into cells, each with its own header, and interleaving the cells into the network with cells from other sources, eventually delivering the cell packets to their destination where they are converted back into a synchronous data stream.

Since various queuing delays on the network will affect the transmittal of the cells, the receiving buffer must have timing capabilities to organize the arriving cells so as to not leave gaps in the synchronous output signal. Delays of greater than 50 milliseconds of the conversation roundtrip must be avoided in order to prevent echo on the line. ATM networks make use of echo cancellers to reduce echo delay problems. Delays of greater than 250 milliseconds must also be avoided, as they result in perceptual discomfort on the part of the participants in the conversation.

In order to maximize bandwidth over a public network, in which thousands of phone conversations coexist, compression techniques are commonly used to reduce transmission time and resources. See echo canceller, jitter, silence suppression, voice activity detection.

voice over frame relay VOFR. See frame relay, voice over.

voice over IP VoIP. Voice over IP involves digitizing conversations and other human vocalizations so they can be transmitted over data networks. This usually involves compression of the sound, as voice applications tend to be somewhat bandwidth intensive (though not as much as music and other types of sounds). Commercial VoIP offerings usually include familiar phone services like Caller ID, and newer ones like follow-me services that allow forwarding to cell phones or pagers. Some systems are designed to use the public switched telephone network as a fallback if there are problems with transmission over the data network.

Voice over IP Forum VoIPF. A group within the International Multimedia Teleconferencing Consortium (IMTC) which promotes and recommends voice over Internet Protocol (IP) technologies. See voice over IP.

voice over networks There are now a variety of ways in which wide bandwidth data networks can be used to send telephone voice calls. The call can be initiated through a regular phone line that connects to a private or public network, or through a computer voice system hooked directly to a network. Thus, Internet Service Providers (ISPs) are emerging as collaborators and competitors with traditional copper line phone service carriers.

Voice with the Smile One of the many colloquial names given to the early female telephone operators. Others include Hello Girls, Central, and Call Girls.

voicemail A type of data communication in which a voice message is digitally recorded, usually through a small microphone interfaced with a computer, and sent through an email or voicemail client as an attachment or message. In order to hear the message, the receiver must have the capability to replay the message on the destination system. This is usually done either directly through the voicemail client, or, if sent as an email attachment, it can be played with a separate player utility that is compatible with the type of sound file in which the message is stored.

voicemail, electronic A system for intercepting an incoming phone call, playing a prerecorded digital message, and recording a message left by the caller. Many voice mail systems support multiple messages, multiple mailboxes, and menu hierarchies accessed through touchtones entered from the caller's phone keypad, and may also allow a facsimile message to be transmitted manually, since many voice modems and voicemail systems support data and facsimile communications as well.

Voice mail systems are not used just as fancy answering machines; they are also employed in faxback systems, technical support systems, and for providing product information and purchase options to callers. Because electronic voice mail applications are digital, they can be programmed to provide a wide variety of services, according to the needs and imagination of the programmer and user.

volt (Symbol *v* or *e* for voltage) An SI unit of electrical potential. When a difference of electrical potential occurs between materials or portions of materials where there is a pathway between them, electrons seek a direction of flow that balances that potential. A volt is a unit of electromotive force (EMF) equal to that needed to produce a one ampere current through a one ohm resistance. In any given circuit, voltage, current, and resistance are related, so any one of those values can be computed if the other two are known. The unit is named after Alessandro Volta. See ampere, ohm, Ohm's law, resistance.

Volta, Conte Alessandro Giuseppe Antonio Anastasi (1745-1827) A physicist who pursued many of the ideas proposed by Luigi Galvani by studying the varying electrical properties of different materials. He questioned Galvani's explanation of 'animal electricity' and proposed that the reaction of the muscle to stimulation of a nerve was due to unequal temperatures, and set up more rigorous experiments to determine what was happening. He showed how electricity could be generated by chemical action, which became known as *galvanic electricity*. This was the forerunner of the electrolytic cell. In 1800 he described his invention of the voltaic pile.

Volta devised a condensing electroscope to respond to very sensitive charges, and with it, was able to demonstrate contact charges (though some were actually chemical interactions). The volt, a unit of electromotive force, is named after him. See Faraday, Michael; volt; voltaic pile.

voltaic pile Alessandro Volta developed a system of layers of metal plates and paper or briny cloth, which exhibited a difference in potential between the top and bottom, which could be varied with the materials used and the number and organization of the layers. Volta attributed this difference to 'contact' electricity, though we know now chemical factors play a role. Volta later modified the pile design to create what he called a *crown of cups*. The metal plates were placed in separate cups containing liquid, some distance apart. The plates were the poles or electrodes. Each cup is now known as a *voltaic cell* and a pair is known as a *voltaic battery*.

Two historic voltaic piles show the layers of materials piled within supporting rods.

voltmeter, voltameter A galvanometer or other instrument such as an ammeter, connected in series with a resistor, calibrated to indicate electric pressure from electromotive force, or voltage differences in potential at different points of an electrical circuit. The voltmeter is connected in parallel across the circuit being tested and must have a higher resistance than that of the circuit being measured. In the past, sometimes also called a coulomb-meter or coulometer. See volt.

von Bunsen, Robert Wilhelm A German chemist in the 1800s who did numerous experiments with wet cells and made improvements on the early inventions.

von Guericke, Otto (1600s) An early experimenter who devised a machine that amplified and demonstrated the properties of negative and positive electromagnetic forces. Von Guericke used a spinning large sphere, molded out of sulphur, to investigate theories related to the spinning and magnetism of the earth. He noted also that holding certain substances up to the sphere would produce a spark. He discovered basic principles of air pumps and demonstrated characteristics of vacuums with his Magdeburg hemispheres in 1663. A university in Magdeburg, Germany is named after him.

von Kleist, Ewald Christian (1715-1759) A German physicist who discovered in 1745 that an electrical charge could be held in a glass vial with a nail or piece of brass wire inserted. A similar jar was developed independently by P. van Musschenbroek, known as the Leyden jar. See Leyden jar.

von Neumann, John (1903-1957) A Hungarian-born American mathematician who worked in advanced mathematics, game theory, and quantum mechanics. He produced a body of work in the mid-1900s that significantly influenced the design and evolution of computing machinery, including practical implementation ideas, conditional control, self-modifying code, program storage, and much more. Von Neumann collaborated with Mauchly and Eckert on the EDVAC. See EDVAC.

Many of von Neumann's contributions to game theory have applications and consequences for practical applications outside the realm of pure mathematics. These can be applied in the design and operation of 'thinking' machines.

von Neumann machine A classification of computing systems, based on the work of John von Neumann, which includes single-instruction, single-data computation, which requires the repeated access and fetching of instructions and data.

VOR VHF omnidirectional range.

VP virtual path. A generic term to describe a logical connection consisting of combined *virtual channels*. In an ATM environment, it refers to the unidirectional transport of ATM cells belonging to virtual channels with the same endpoints, associated by a common identifier value. Related abbreviations include VPCI (Virtual Path Connection Identifier), VPI (Virtual Path identifier), VPL (Virtual Path Link), and VPT (Virtual Path Terminator). See virtual path connection, virtual private network.

VPC See virtual path connection.

VPN See virtual private network.

VQ vector quantization.

VRAM Video RAM. Memory chips designed to enhance graphics display. Bisynchronous input/output. Related is SVRAM, Synchronous VRAM which reads only in or out at one time. See WRAM.

VRML Virtual Reality Modeling Language. A programming language for developing 3D interactive image environments. This is a popular goal of video gaming and simulation developers. VRML allows you to create a sequence of images that can be presented in a World Wide Web environment in combination with a VRML client/browser. There are standalone and Web browser-compatible VRML clients available from several vendors. See Virtual Reality Modeling Language for historical background.

VRML Review Board VRB. Originally the VRML Architecture Group (VAG), founded in 1995, the VRML Review Board participates in and oversees Virtual Reality Modeling Language development, documentation, and formal specifications. http://vag.vrml.org/

VSAT See Very Small Aperture Terminal.

VSX Verification Suite for X/open. See X Windows System.

VT-100 A data terminal, and terminal emulator, originally developed by Digital Equipment Corporation (DEC), which has become an industry standard and is widely used in telecommunications. The VT-100 emulation setting in telecommunications programs works with almost any remote system and is probably the one to select if you get strange characters or formatting in your software. Web browsers are quickly overtaking VT-100 as a front-end to online sessions, but VT-100 is still a valuable standby when connecting to remote systems in text mode. For the most part, it has been superseded by VT-220 and other newer standards, but it is still a good fallback if compatibility is a problem.

VTAM Virtual Telecommunications Access Method. A data communications access method used in International Business Machines' (IBM's) Systems Network Architecture (SNA). See Systems Network Architecture.

VTP See Virtual Trunking Protocol.

VTS Vehicular Technology Society.

Vulcan Street plant A historic site on which was built the world's first hydroelectric central power supplier. It provided direct current, as did many of the early power stations. This was a subject of hot debate in the early days of power stations, with Thomas Edison supporting direct current against strong opposition by Nikola Tesla and Westinghouse, who felt alternating current was a better choice. Alternating current first gained a foothold in Europe, where the high cost of batteries spurred inventors to look for other solutions. See Mill Creek plant

W 1. Symbol for watt. See watt. 2. The USOC FCC code for wall mount jack. 3. Symbol for work.

W-DCS See wideband digital cross-connect system.

W2XBS The Radio Corporation of America's (RCA's) first television broadcasting station, located in the city which was the hotbed of broadcasting for decades, New York. W2XBS was established in 1928 and gave the popular cartoon character "Felix the Cat" the exposure that made him a 'star.'

W3 See World Wide Web.

W3C See World Wide Web Consortium.

WAAS See Wide Area Augmentation System.

WABI Windows Application Binary Interface. Software from Sun Microsystems which enables Microsoft Windows applications to run on the Solaris desktop system, thus users can have access to the large library of software available for the Windows operating environment and run them, along with Solaris applications, on computers installed with Solaris.

WABIserver Windows Application Binary Interface server. A Windows application server providing integrated Windows/Solaris options to SPARCstation users running Sun Microsystems' Solaris, Solaris Intel, or an X terminal system.

WACK *w*ait *ack*nowledgment. A signal sent by the receiving station which indicates that there needs to be a wait or delay before transmitting a positive acknowledgment (ACK).

wade insulator A very early blown glass, unthreaded telegraph pole insulator. See insulator, utility pole.

wafer A fine thin disk, usually cut from a larger piece of the substance. Many materials are cut into wafers including semiconductor materials, quartz crystals, and synthetic gems used in optical systems. Silicon is one of the most common materials used in semiconductors. Many electronic chips are layers of wafers and photovoltaic panels are arrays of wafers.

Since many wafers used in electronics and other industries are extremely thin, production methods are very specialized. A traditional metal saw is not appropriate, especially since the part cut away and lost by the saw blade, the *kerf*, would be larger than the width of the wafer itself.

WAIS See Wide Area Information Server.

Wait on Busy U.K. term for a Call Waiting type of optional subscriber service in which a caller encountering a busy signal can wait until the call in progress is over and be automatically connected.

wait state In computer programming and processing, a time during which the processor waits. This may be explicitly established or may be dependent on other events. Wait states are introduced for many reasons, for timing, synchronization, to reduce power demands, etc.

walk time Propagation delay in a Token-Ring network. Walk time plus service time combines to form scan time, the mean interval between the arrival of tokens at any given station.

wall outlet A socket, phone connector receptacle, cable connector, or other conductive receptacle mounted on the wall for easy access, often positioned above the baseboard or at shoulder height.

Wall, Larry Software author of *rn*, a popular

newsreading system, and *Perl*, a significant interpreted scripting language widely used on the Internet. Larry Wall has also authored several bestselling programming books, most notably books on the Perl programming language. See Perl.

Walmsley antenna An antenna comprising an array of rectangular loops mounted vertically in parallel with one another.

WAN See Wide Area Network.

wander Timing deviation or drift. In networking, especially high speed networks in which synchronization is important, wander and jitter can contribute to signal degradation. Physical factors such as connectors, regenerators, or temperature variants can contribute to wander due to propagation delay. Over longer distances, this effect can become magnified, with the pulse position gradually shifting.

In SONET networks, wander has a more specific meaning; it consists of a phase variation tracked and passed on by a phase locked loop. This is managed by tracking the incoming signal and passing it through a filter, to extract timing data.

Wang Global A firm which has formed an alliance with Microsoft to provide local area network (LAN) services.

war dialer An automated dialing system that sequentially dials a new number for each succeeding call, sometimes taken from a computer database. War dialers are used by those dialing to a large number of phone-access BBS systems, by collection agencies, telemarketers, and teleresearchers. There are restrictions in some areas on the use of war dialers for commercial solicitations.

war room A strategy and decision making room, often related to critical big business or government activities. The war room may be a closed, secure environment with no equipment other than perhaps tables and chairs, or it may include sophisticated electronics for monitoring and communications. See skunkworks.

warble tone A tone which resembles a bird's warble in that it fluctuates in tone periodically, sometimes quickly. Warble tones are often used as signal tones, as on public address systems. A warble tone is usually one of the options included with multitone generators, which also feature sirens, steady tones, and timed pulses for various public or employee alert needs. Warble tones are used diagnostically in conjunction with an integrating detector device, to measure crosstalk in a transmissions line.

WARC See World Administrative Radio Conference.

warm boot See reboot.

warranty A promise on the part of a manufacturer, retailer, or service provider that the goods or services provided will meet stated terms of manufacturing quality, use, or lifespan. Most warranties are limited to manufacturing defects, and refunds, replacements, or damages up to the original price or replacement price of the product. Terms of warranties in the computer industry tend to be about three months, although longer warranties on equipment are now beginning to be honored, ranging from one to five years. Few warranties will cover abuse, loss, or damage from natural disasters.

WASI Wide Area Service Identifier.

watch, watchpoint A means of monitoring program functioning, usually within a software debugger, while the application is being executed. Watch commands are often used in conjunction with trace and break commands and watchpoints are set much the same way as tracepoints.

water bore A device which uses a highly-pressurized jet of water to bore holes which can be used for the insertion of underground conduit and cabling.

watermark In the paper printing industry, a symbol or shape embedded in the paper, which is usually so subtle that it is seen only if the paper is held up to the light or held up to particular frequencies of light. The watermark is typically paler or more transparent than the surrounding medium. Watermarks are often used as identification signatures on quality papers and as identification security marks on currency, bonds, stocks, etc.

The term has been used analogously in computer applications to indicate a subtle background image which is inserted behind the active layer on a computer screen, that is, behind a text document, Web page, or other application. In addition, the term is being used to describe a type of copy protection or data security mechanism, in which a label is inserted into the data to identify it, without interfering with the quality of the presentation of the data. See SMASH Project.

WATM Wireless ATM. A number of initiatives are under way to provide better support for wireless services over asynchronous transfer mode (ATM) networks. One of the proposals is for a Radio Access Layer (RAL) by Olivetti Research, which has developed a prototype wireless ATM local area network (LAN).

WATS Wide Area Telephony Service, Wide Area Telecommunications Service. A discounted long-distance service. This service originated from AT&T, but the name became generic and is now broadly used. WATS services can be incoming, outgoing, or both; WATS lines can be installed with incoming and outgoing services handled over the same line (although this may limit the service). As with many discount services, the savings are dependent upon the pattern of usage. If WATS lines are used more often and longer, the savings may be negligible or nullified.

Watson, Thomas A. (born 1854) A machinist and assistant to Alexander Graham Bell, Watson filed for a telephone patent for a two-bell ringer in 1878. He also designed the Watson board, a very early and not entirely practical, telephone switchboard.

Watson, Thomas J. (1874-1956) Watson became president of the Computing-Tabulating-Recording Company in 1914. This later became International Business Machines (IBM), a significant company in computing history. After four decades, he passed the position on to his son, Thomas J. Watson, Jr. (1914-).

Watson, William (1715-1789) An English experimenter who demonstrated in 1746 that electrical current could be sent through a wire about 3 kilometers long, using the Earth as a return conductor, a technique later applied in many technologies including early two-way telegraph systems.

watt W. An absolute meter-kilogram-second (MKS) unit to describe electrical power, which is equal to the amount of work done at the rate of one absolute joule per second. Described a different way, a watt is the electrical power expended when 1 ampere of direct current (DC) passes through a resistance of 1 ohm. For large units of power, the kilowatt (1,000 watts) is typically used. See Ohm's law; power, electrical; Watt, James.

Watt, James (1736-1819) A Scottish inventor who pioneered the steam engine, after whom the watt is named.

WATTC World Administrative Telegraph and Telephone Conference.

wattmeter An instrument for measuring electrical power in watts. The wattmeter is similar to a dynamometer, which measures force or power in that it employs a moving coil and a field coil, however, the windings on the coils differ from the dynamometer. See dynamometer.

wave A periodically oscillating or undulating process, or physical or electromagnetic phenomenon.

WAVE A commercial product from MPR Teltech Ltd. which permits the simultaneous realtime connection of up to eight different sites through ATM switches for broadcast TV quality videoconferencing.

wave audio See waveform audio.

wave division multiplexing WDM. A means of using separate channels grouped around distinct wavelengths to increase the capacity of a fiber optic transmission system. Proposed methods for multiple wave division are known as dense wave division multiplexing (DWDM).

wave filter A device, as a transducer, which separates out waves on the basis of frequency. Some loss occurs during this process, depending upon the method used and the characteristics of the wave. See wave trap.

wave length See wavelength.

wave packet A short burst or pulse of waves.

wave trap A device which is usually placed between the receiver and the incoming waves that excludes unwanted waves, especially undesired frequencies or interference waves. Like the receiver itself, a wave trap is often tunable to optimize control over incoming waves. See wave filter.

waveform The shape, or spatial characteristics, of an electromagnetic wave. 2. A graphical representation of the spatial characteristics of an electromagnetic wave, as on a scope or 2D or 3D coordinate system illustration or modem. Waveforms are typically graphed according to amplitude across time. Waveforms with certain recognizable shapes, when graphed, have been given names to distinguish them from one another.

waveform audio A digital representation of sound waves, often created by sampling through a pulse code modulation (PCM) technique. Waveform editing on microcomputers became well supported in the mid-1980s. The early Macintosh supported audio waveforms, and many pioneer computer musicians used Macintosh music sequencing and editing software to create electronic music compositions. The Atari ST was released with a basic midi

W

device built in and the Amiga in 1985 came with built-in multiple channel 8-bit stereo sound. By the late 1980s, 16-bit third party sound cards were available for most of these computers. Todd Rundgren, musician, composer, and multimedia designer, began using Macintoshes and Amigas to create digital sound videos and CDs in the mid and late 1980s.

There are many file formats for storing sound waves, and sound files can be played from Web pages on the Internet with browsers that support sound (assuming the computer has a sound card; most types of computers came with built-in sound cards by 1986). The `.wav` extension is commonly used to designate audio files of a particular format on the Internet.

waveform editor An applications program on a computer, or specialized electronic device, which allows the user to display, evaluate, and alter the characteristics of a wave. A computer display often uses a graphics system to represent the wave, as in traditional oscilloscopes. The dials on the simulated computer scope are often represented as buttons on the screen or may be input from a joystick or specialized peripheral.

Waveform editors are used to alter the characteristics of music patches, voice, or speech files. Adjustments can alter the volume, tone, harmonics, echo, and other characteristics, and when converted to digital form on the computer, the adjusted files can be stored and replayed later, or cut and pasted to create songs or speeches. Digital sound editing with a waveform editor can be combined with digital video sequences to synchronize the sound and video and to 'put words into people's mouths.' See waveform audio.

waveform monitor An oscilloscope or oscilloscope-like computer applications program which surveys an input signal and displays its characteristics on a screen. Typically there are dials or graphical user interface gadgets and buttons to adjust the displayed wave.

waveguide A device for confining and channeling the propagation of electromagnetic waves, often through a hollow round tube, hollow rectangular tube, coaxial cable, or fiber optic cable. The interior environment of the waveguide will vary with the type of wave being channeled, since it must allow sufficient room relative to the characteristics of the wave so as not to change or diminish the signal. Thus, waveguides are more practical for high frequency waves such as microwaves.

waveguide dispersion The process by which an electromagnetic wave becomes distorted as it passes through a waveguide. Since the dimensions and shape of a waveguide interact with the phase and velocity characteristics of a wave, the waveguide's geometric properties may cause dispersion of the guided signal.

waveguide laser A gas laser which incorporates a tube as a waveguide to channel the direction of the laser beam.

waveguide lens A device used with microwaves, in which the waveguide elements act as lenses to produce the required wave phase changes through refraction.

waveguide phase shifter A device which takes the phase of the incoming waves and adjusts them in terms of their output current or voltage.

waveguide propagation A type of long-range communication that makes use of the atmospheric wave guiding channels which arise between the ionospheric D region and the Earth's surface. See ionospheric sublayers.

waveguide scattering Scattering of an electromagnetic wave which occurs due to the geometric characteristics of the waveguide structure, as in an antenna or fiber optic cable, not due to the materials of which the waveguide is constructed.

WaveLAN A wireless local area network (LAN) from Lucent Technologies.

wavelength The distance, when measured from any point on a wave, to the corresponding point in the phase of the next wave in a related type or series of waves, such as sound or light waves. A wave's length is sometimes described in units of distance and sometimes in terms of the time it takes, from the phase of a wave to the corresponding phase of the next related wave, to travel through the same point.

Waves can differ dramatically in length, as can be seen, for example, from the various categories of designated radio bands, which range from a single millimeter (EHF band) to tens of kilometers (VLF band) when expressed in distance, or from 30 GHz to 10 kHz when expressed in frequency or cycles per second. See band.

wavelength division multiple access WDMA. One of two common optical multiplexing techniques in which each transmitter transmits at different wavelengths within a narrow spectrum, and the receiver extracts the desired wavelengths with a bandpass optical filter. See space-division multiple access.

wavelength division multiplexing WDM. A means of multiplexing different wavelengths through the same strand of fiber to greatly increase the capacity of data transmission over fiber optic cables. Optical signals at different frequencies do not interfere with one another. The technology permits a substantial amount of data to travel over even one strand, and when the strands are bundled, it permits transmission in the terabits per second range, ample for high bandwidth applications like video, data, and simultaneous voice. See frequency division multiplexing.

wavelength shifter A device or process which takes an incoming series of waves and shifts their frequencies so the outgoing waves are related to the incoming waves, but in a different range. It is very common in satellite communications for the incoming signals to be shifted so they don't interfere with subsequent outgoing signals. In photocells, wavelengths may be shifted by means of compounds so that the length of the outgoing waves is related but greater.

wavelet analysis Wavelet analysis involves looking at time and frequency. A prototype function (*analyzing wavelet* or *mother wavelet*) is used as a starting point for dilations and translations by creating a high-frequency reference and a low-frequency reference, which are analyzed, in turn, for aspects of time and frequency. See wavelet theory.

wavelet filter compression An analysis low/high-pass filter technique used in wavelet compression by selective quantization, in which images are decomposed into frequency bands. Wavelet encoding combined with vector quantization (VQ) has been shown to be a good means of compressing image data, and several schemes for accomplishing this have been developed. See wavelet.

wavelet packets Calculated linear combinations of wavelets which retain many of the properties of the parent wavelets from which they are derived.

wavelet theory A set of mathematical concepts and techniques related to the representation and manipulation of oscillating wave forms *according to scale*. Wavelet analysis provides a means to use approximating functions contained within finite domains and are well-suited to representing data with sharp discontinuities.

Wavelet theories and algorithms are being applied to audio and image compression with some practical and interesting results. When wavelet concepts are used in image compression, they share some characteristics with discrete cosine transform (DCT), although the functions used are more complex than cosines.

Wavelet compression is sometimes used in conjunction with other methods, such as vector quantization, to provide low-loss, high-compression ratios. They have also been used in turbulence studies, human vision, radar systems, astronomy research, and fractal imaging. Wavelets are used in conjunction with, and sometimes instead of, Fourier transform methods, depending upon the application. Unlike Fourier transforms, wavelet transforms are not limited to sines and cosines, and can comprise an infinite set. See discrete cosine transform, Fourier transform, wavelet transform.

wavelet transform, discrete wavelet transform DWT. A linear mathematical technique which is a subset of wavelet packet transform. DWT is used to generate a data structure with segments of various lengths. In a sense, a transform is a means of rotating a function so it can be visualized and analyzed with a different set of tools, those tools being not only mathematical algorithms, in this case localized frequency wavelet functions, but also conceptual models. To simplify calculations, the DWT is factored into a product with a few sparse matrixes using self-similarity. Wavelet transforms differ from Fourier transforms in that they are localized in space, and an infinite possible number of basis functions can be applied to them, unlike Fourier transforms which use only sines and cosines, but they also share some common basic properties. See Fourier transform, wavelet, wavelet packets, wavelet theory.

wavelet types Due to the variety of possible types of wavelet transforms, some wavelets have been grouped into families, on the basis of *vanishing moments*, and subclasses, on the basis of the number of coefficients and level of iteration, and many more await to be developed and discovered. The Daubechies wavelet family has a fractal structure. Others are Symmlet, Coiflet, and the simpler Haar family, often used to introduce wavelets in educational contexts.

wax master An original physical master intended as a prototype for making one or more copies, usually with a more durable material. Since wax is easily shaped, stretched, changed, and otherwise manipulated, it is a good medium for creating prototypes of production parts. Wax is used to design jewelry, sculpture, certain types of audio recordings, prototype components, and more. Once sufficient copies of the original are made, the wax is sometimes reused for other projects.

way station An intermediate office in a communications line. An intermediate phone in a way circuit, one which is not the main console.

way wire, way circuit A party line circuit that connects a number of subsidiary stations to a main switching or relay station. See party line.

WBC See wideband channel.

WBEM See Web-Based Enterprise Management.

WDL Windows Driver Library.

WDM See wave division multiplexing.

WDMA See wavelength-division multiple access.

Web *colloq.* See World Wide Web.

Web address See Uniform Resource Locator.

Web browser A display and hypertext client used as a front-end to Web-related services on the Internet. For a fuller explanation, see browser, Web.

Web Crawler One of the significant commercial search engines on the Internet, known for its quick simplicity and very cute little spider mascot. http://www.webcrawler.com/

Web hosting A service in which Internet Service Providers (ISPs) enable a business or individual to store Web pages on the ISPs computer system, instead of on the user's computer. Since an ISP's machines are typically connected to the Internet all the time and since a certain amount of Web server setup is needed to make a Web site function properly, it is very practical to have the ISP manage the administration of the Web server. The design, management, and updates of the individual Web page are left up to the user or can be handled by the ISP for a fee.

An ISP can also arrange for the registration of a *domain name* for the Web site, with the user paying the fee to the ISP, who passes it on to the registration authority, or with the user paying the registration authority directly, and the ISP handling the setup of the necessary computer configuration. See domain name.

Web master See Webmaster.

Web search engine The World Wide Web is an enormous repository of information and it changes all the time, so there's no practical way to find a particular page without a little help. Enterprising programmers quickly realized that tools were needed to make it easier to locate information on the Web. As a result, they developed 'search engines,' which are application programs in Web format designed to facilitate the location and retrieval of information according to user-specified parameters or categories. Most search engines provide a text window for the user to type in a keyword, after which the user clicks a "Search" button, resulting in the display of a listing of relevant Web pages to which the user can jump by clicking on the highlighted hypertext link.

It is not unusual for a search keyword to result in hundreds, thousands, or even millions of 'hits,' that is, pages that include the specified keyword. It's obviously not practical to try to visit several thousand sites, so most search engines allow the user to narrow the search by adding more keywords and providing operators such as *AND* and *OR* to focus the search. This can reduce the resulting *hits* to a more manageable number.

Search engines also provide means for businesses and individuals to get their Web pages listed so others can find them. There is usually a button at the top or bottom of the page for this purpose.

There are thousands of search engines on the Web. Many sites have their own local search facilities, but there are also a dozen or so very prominent Web search tools which are commonly used. The applications listed in the appendix all perform general searches of the Web, with the exception of DejaNews, which searches newsgroups. Many also have 'specialties' to set them apart from the others. Note that these search engines catalog millions of pages and they are not necessarily up to date. The best search engines seem to have a lag of about three weeks to three months from the time a site is submitted until it is added to the database. See the appendix for a list of major search engines.

Web server A client/server model system on a multi user network that serves requests for HTML-based Web pages which are part of the World Wide Web. A Web browser is a type of client software that communicates with the server through HyperText Transfer Protocol (HTTP), and displays the information received from the server in the form of a Web page containing a variety of text, graphics, and sound. Most Internet Services Providers have Web servers. See HyperText Transfer Protocol, server, World Wide Web.

Web Service Provider WSP. A commercial provider of computer network access to the World Wide Web or to a local Web-based network. Most, although not all, Internet Service Providers (ISPs) provide Web access and often include it in their monthly subscription cost. They may or may not provide a diskette with a Web client, called a browser. If not, Web browsers are widely available for download, or through friends, as many of them are freely distributable or shareware. Some service providers use proprietary servers and the user can only use the browser provided for this service; others allow users to select the browsers of their choice. See Web browser, Web server.

Web site On the World Wide Web, which is a mechanism for viewing many portions of the Internet, there are businesses and individuals who have organized data files in such a way as to provide a virtual community, storefront, library, educational resource, or other form of informational/educational/entertainment source for the public or authorized users. When a Web user accesses the site with a computer program called a *browser*, she or he is presented with various communications media (text, graphics, sound, etc.) which are available on the site.

A Web site is comprised of a series of related files, managed by a client/server Web system which serves the Web files, most of which are called Web *pages*, to processes that request them over the network. These files typically consist of text and graphics organized into hypertext relational links through a markup language called HTML. There may also be sound, forms, Sun Microsystems Java applets, and various scripts that provide additional functionality to the basic HTML layouts.

Web sites are as varied as the people who design them. To mention just a few, there are commercial sites promoting products and services; scheduling sites, providing listings of television, radio, and other broadcast venues; weather sites, that allow lookup of weather conditions almost anywhere in the world; educational sites, through which *distance education* is becoming more and more available; personal sites that journalize the day-to-day activities of individuals; genealogical sites that detail family histories; and chat sites in which opinions are readily offered and debated. See World Wide Web.

Web streaming A capability of a Web browser through addition of a plugin, or which is incorporated directly into more recent browsers, which allows audio or video playback as the data is transmitted to the browser. In older browsers, if the user clicked on a sound or video file, the browser downloaded the sound or video to the local computer from where it was played after the download. More recently, through Web streaming utilities, the audio or video is played as it is received in 'realtime,' albeit sometimes at a lower quality level and with variations due to the connection speed to the Internet.

A number of commercial plugins and applications are available to take advantage of Web streaming, ranging from $20 to over $3,000. A freely distributable version, called RealVideo, is available from Progressive Networks.

Web TV See WebTV.

Web-Based Enterprise Management WBEM. A distributed management system based on Web technology. WBEM uses a midlevel manager approach, employing HTTP and Web browsers, to provide access to management data and reporting mechanisms. There have been a number of efforts to standardize distributed management, with mixed results so far. WBEM is supported by HMMS, an open standard of the Desktop Management Task Force, and the HyperMedia Management Schema (HMMS), under the aegis of the Internet Engineering Task Force (IETF).

WBEM provides a means to extend the system through HMMS and utilizes HyperMedia management Protocol (HMMP) to link HMMS to run over HTTP.

Weblock *colloq.* The slowdown that may occur when trying to access extremely popular sites on the World Wide Web. The Web version of a traffic gridlock, where everything temporarily slows to a halt.

Webcast, Netcast A broadcast using the World Wide Web as the communications medium. While traditional Web browsing involves clicking on desired pages, text, and images, and thus 'pulling' the information to the user, Webcasting involves a method of 'pushing' the information at the user in the sense that the user's screen is continuously updated without the user having to click the screen or manually request an update (like watching television). Since early browsers did not inherently function this way, it was usually necessary for the consumer to first download additional software to his or her computer. Subsequent browsers, such as Netscape Netcaster, directly incorporate the 'push' capabilities and can be used in conjunction with Channel Finder, a directory of Webcasting channels.

Webcasting is a significant development. It means the advent of 'digital TV' through a computer, with access to a multitude of 'channels' on the Internet. When this catches on, TV will probably never be the same, particularly since very specialized information, such as stock and finance channels, art and music channels, etc. can be made available through the Web without the major sponsorship which is necessary to participate on the big broadcast television networks.

HDTV-quality push reception can be accessed with a TV set, interfaced with a computer peripheral. See Intercast.

Webmaster A professional who is responsible for a World Wide Web site. A Webmaster's duties typically include coding, installation, and maintenance of the site and *may* also include page layout, content decisions, and the production of graphics and text. Many individuals have responded to the demand for Webmasters by learning a little bit of layout and HTML and hiring themselves out as professional, with mixed results. Proper creation, presentation, and maintenance of a good site, especially a commercially viable site, involves a large number of professional skills including marketing know-how; writing, editing and proofreading; graphics design and production; Web statistics monitoring, analysis, and reporting. Choose your Webmaster carefully. In fact, you may do best to team up a user-interface and CGI-savvy programmer with a market-savvy artist and writer. HTML skills alone are *not* sufficient to create good database, point-of-purchase, or shopping cart software, and these may be essential to a business presence on the Web.

WebNFS Web Network File System. A proposed specification for a client/server protocol as an extension to recent versions of the Network File System (NFS). NFS is an RPC-based, platform/transport independent protocol and WebNFS is intended to provide semantic extensions to NFS. It serves as a means to make file handles faster and easier to obtain, and may improve transit of firewalls and system scalability. WebNFS clients assume the availability of a WebNFS server registered on port 2049. See Network File System, RFC 1094, RFC 2054, RFC 2055.

Webspace Derived from 'cyberspace,' Webspace refers to the sum total of the hardware, software, people, transmissions, and interactions which comprises the World Wide Web.

WebTV A simple consumer-oriented Web access device that uses a TV as the display. It was purchased in August 1997 from the California-based WebTV company by Microsoft Corporation. Sony Electronics, Philips Consumer Electronics, and others have products based on the WebTV technology.

Webzine An electronic publication available on the Web, either free or through subscription. Webzines may be pure electronic publications or may be electronic versions of printed publications, with or without enhancements possible on the Web. Many of the major publications are providing back issues of print publications as Webzines in order to attract readers to both the print and Web versions.

With electronic access increasing and trees declining, this type of publishing may, in time, supersede many of the print publications. At the present time, it is difficult for lesser known Webzines to be profitable. There is so much free content on the Web, people are reluctant to pay, and there are as yet no really easy ways to pay small amounts for such a product. This may change. See e-zine.

WECO See Western Electric Company.

Wehnelt interrupter A type of early electronic apparatus used in laboratories to provide interruptions ranging from 100 to 1,000 per second. The positive side of a circuit was connected to a platinum electrode conducting through a well-insulated primary winding coil. This in turn was connected to a lead plate in a dilute solution of sulphuric acid. A tube for circulating water was included to provide cooling.

WELL, The The Whole Earth 'Lectronic Link. This Web site is somewhat like a bohemian meeting ground for intellectuals and artists. Established in 1985 by Stewart Brand and Larry Brilliant, the WELL's discussion community patrons include many artists, educators, writers, and high-level technical professionals. http://www.well.com/

West Coast Computer Faire A historic meeting place for computer hobbyists, many of them pioneers in the microcomputer field. Established in 1978 by Jim Warren, it was then the largest computer show in the world.

West Ford satellites A set of satellites launched by the United States, beginning in 1961. The pioneering West Ford project launched surfaces into orbit which contained millions of slender copper dipoles to provide a reflective 'blanket' around Earth. The first launch attempt was unsuccessful due to a failed ejection, but the second launch, in 1963, demonstrated that communication could be achieved, at least with high powered ground stations. At this time, concerns were expressed over side effects from certain types of satellites and space debris, and since the feasibility of active relays was successfully demonstrated, the project was discontinued. See ECHO satellites.

WESTAR The name of a family of Western Union communications satellites, WESTAR I was the first U.S. domestic communications satellite, launched in 1972. A WESTAR satellite was launched with the Challenger space shuttle mission in February 1984. Unfortunately, the Payload Assist Modules (PAMs) didn't function correctly; the WESTAR was launched into a lower orbit than had been planned, and it had to be retrieved by the Discovery some months later.

Western Electric Company WECO. Originally established as the Gray & Barton company in 1869 by Elisha Gray and Enos Barton. In 1872 it became known as Western Electric Company and, in thirty years, grew to be one of the world's largest manufacturers. The Graybar Electric Company, Inc. was spun off from WECO in 1925 and is still doing business as an employee-owned distributor of telecommunications products.

In 1881, Western Electric Manufacturing Company became Western Electric and acquired exclusive rights to manufacture and provide Bell Equipment.

In 1915, Western Electric took over Western Electric Company of Illinois and, in 1925, Western Electric Research laboratories were consolidated with part of AT&T's engineering department to form Bell Telephone Laboratories, Inc.

See Gray, Elisha; Gray & Barton; telephone history.

Western Union, Western Union Telegraph Company Western Union was originally organized as the New York and Mississippi Valley Printing Telegraph Company in 1851 by Hiram Sibley, and named Western Union Telegraph Company in 1956 by his business associate Ezra Cornell (co-founder of Cornell University), when it merged with Cornell's New York & Western Union Telegraph Company. Cornell was the largest share-holder for over fifteen years.

Western Union installed the first North American transcontinental line in 1861. Sibley traveled to Russia and was offered an opportunity to purchase Alaska on behalf of Western Union, but turned it down, passing on the information and the historic opportunity to the U.S. government.

In 1868, Western Union hired C. F. Varley to evaluate the U.S. telegraph system with the result that he introduced many standards for coordinating the various lines and systems.

Western Union acquired the rights to new technology, called duplex telegraphy, in the early 1870s. This enabled two messages to be sent concurrently, one in each direction, over the same wire.

In 1873, Western Union entered the international market by a takeover of the International Ocean Telegraph Company.

In 1877, Western Union went into competition with the Bell system, acquiring patents or licensing rights from Edison, Gray, and others, and forming the American Speaking Telephone Company. As Western Union had thousands of miles of telegraph cable strung across the country, this posed a very real challenge to the Bell system, and they reacted with a patent infringement lawsuit which was upheld two years later in the Supreme court.

Western Union had a short alliance with the Bell system from 1908, when AT&T gained control of the company, to 1913, when it voluntarily sold it off again to forestall government breakup of the company.

In 1943, the Postal Telegraph Company, founded in 1881, was merged into Western Union.

In 1950, Samuel Morse's original telegraph instrument was presented by Western Union to the Smithsonian Institution. Other versions of the Morse telegraph are replicas.

In 1970, Western Union acquired TWX (from AT&T) and merged it with its own Telex system. Western Union International was acquired by MCI International in 1982.

In the mid-1980s, the world was changing rapidly. Communication through the Internet was catching on, overnight couriers were in heavy competition with the telegraph industry, and divestiture changed the competitive atmosphere of the phone industry. As a result of this and other changes, the 138-year reign of the Western Union Telegraph Company came to an end in 1989. See Western Union Telegraph Company Collection.

Western Union Telegraph Company Collection This extensive historical collection was presented by the Western Union Telegraph Company to the National Museum of American History as a gift in 1971. It consists primarily of manuscripts, telegraphs, and photographs previously housed in the Western Union Museum. A tremendous amount of telegraph history is contained in these archives. See Western Union.

Western Union Telegraph Museum A historic repository of Western Union documents and photographs established in 1912 by H. W. Drake, an in-house engineer. By 1930, the collection had its own room within Western Union and historical instruments were being added until the late 1960s, when the materials were warehoused. In 1971, the collection was transferred to the National Museum of American History, then the National Museum of History and Technology. See Western Union.

Westinghouse, George (1846-1914) An American inventor who pioneered many aspects of alternating current (AC) power, generators, and the railway air brake (1868). He also made practical applications from ideas derived from or shared with Nikola Tesla, and had more than one disagreement with Thomas Edison. His light and power system was used in the 1893 Chicago World's Fair, the same year he won the historic contract to develop alternating hydroelectric current from the Niagara Falls.

WestNet One of several National Science Foundation funded regional TCP/IP networks, WestNet is located in Salt Lake City and serves the states of Arizona, Colorado, New Mexico, Utah, and Wyoming.

wet cell A basic electricity-providing apparatus employing a positive and negative terminal, each separately in contact with a liquid electrolyte medium (battery acid is an electrolyte). Wet cells are inconvenient in that the electrolyte tends to evaporate and may spill, so other types of cells have been devised, chief among these, the dry cell, available since the early 1900s. A type of wet cell called an air cell became widely used in phone applications. See air cell, Bunsen cell, dry cell.

wetting Adhering a uniform, smooth film of solder to a surface, usually a base metal.

wetting agent 1. A substance which, when applied to a surface, prevents it from repelling wetting liquids to enhance the adhesion of liquids to the surface. 2. An agent that facilitates the smooth, even spread of liquid over a surface.

WFS Woodstock File Server.

WFWG Windows For Workgroups. See Windows.

WGDTB Working Group on Digital Television Broadcasting.

WGIH Working Group on Information Highway. A working group of the TSACC.

WGS World Geodetic System. A very extensive geologic data set, which is updated every decade or so to take advantage of improved data which can be acquired with improved technologies.

Wheatstone, Charles (1802-1875) An English physicist who researched acoustics and invented the concertina. He also experimented with electromagnetic and solar clocks, and developed a speaking machine based on earlier work by W. Kempelen. Like Samuel Morse, he received encouragement and assistance from Joseph Henry, and invented a telegraph that predates the one designed by Morse. In collaboration with W. F. Cooke, Wheatstone's telegraph was operating in 1837. In 1840, he proposed an underwater cable between England and France. The Wheatstone bridge that bears his name was developed by Samuel Christie and described by Wheatstone in 1843. See polar keying; telegraph, needle.

Wheatstone bridge, resistance bridge A device employing a galvanometer and a group of interconnected resistors for measuring resistance against a comparative standard. This tool can be used for determining faults in a length of wire, so the entire wire doesn't have to be dug up or pulled out. By creating a balanced bridge through various loop tests, the approximate location of the fault point can be determined. See megger.

Wheler, Granville An English experimenter who collaborated with Stephen Gray in discovering conductors and nonconductors, and demonstrated in the late 1720s that an electrical charge could be conducted through a thread of more than 600 feet in length. In fact, it was found it could be conducted simultaneously over multiple threads. See Gray, Stephen.

Whetstone When microprocessors were slow and rudimentary in design, and operating systems and software applications were limited, it was easier to run a few tests to evaluate and compare the relative speeds of various systems. Thus, a number of benchmark tests were devised to measure performance. The Whetstone test was developed in the 1970s by B. Wichmann with an Algol 60 compiler. It is named after the English town where it originated. The Whetstone monitored the number of floating point operations that could be carried out by a process in one second. Floating point operations are common in processor-intensive computing applications such as graphics and scientific work. See benchmark, Dhrystone, Rhealstone.

Whirlwind A historic, large-scale computing machine developed at MIT in the late 1940s and early 1950s. This is credited as being the first realtime-processing digital computer. It is also significant for incorporating random access memory in a matrix core memory.

White Book 1. Any book in the set of 1992 ITU-T recommendations. 2. A document which specifies MPEG video and audio file structure, coding, and indexing. See MPEG, Red Book.

white noise Human-audible signals which consist of a spectrum of frequencies more-or-less evenly distributed across the range so that no one tone predominates, as in background noise. White noise is sometimes used to create ambience in sound systems and it is used in audio experiments.

white pages 1. In most English-speaking countries, the directory portion of residential, or residential and business listings, in the local telephone directory. 2. On the Internet, directories of individuals' physical and electronic addresses usually accessible to the public through the World Wide Web. Web white pages can be searched through keywords and are more flexible and powerful than standard printed phone listings. You don't have to be using the Internet to be listed in these directories; many are compiled from telephone listings and other public sources of information. See ego surfing, yellow pages.

white room, clean room A controlled environment in which dust, smoke, bacteria, and moisture are eliminated or regulated in order to reduce interference with and contamination of the functioning of the environment. Controlled environments are used in component production, research labs, medical facilities, etc.

white transmission 1. In an amplitude modulated transmission of an image, a black transmission means that the greatest divergence in amplitude in the signal represents the black tones, and the narrowest divergence represents the lightest tones (or no tone at all). In white transmission, the opposite is true. 2. In a frequency modulated transmission, a black transmission means that the lowest frequency corresponds to black and the highest frequency corresponds to white, or no tone, and in a white transmission the opposite relationship is used. The concept applies to image scanners, facsimile machines, photocopiers, etc.

whiteboarding Communicating through means of text and graphics drawn on an erasable wall board or large sheets of paper.

whiteboarding, electronic 1. An electronic software application in which input from various keyboards and pointers is displayed or projected on a large white screen that resembles an erasable whiteboard. 2. A dedicated computer network display system that allows remote two-way communication of ideas through text and graphics. 3. The conceptual analog of a whiteboard communication, carried over a computer-based videoconferencing system. The author first saw this demonstrated by SGI with a videoconferenced paint program at a trade show around 1991, in which both participants, in different locations, contributed to the same illustration and text while another window simultaneously showed their faces as they talked to one another. As it was demonstrated, the whiteboarding concept was very broad and could conceivably include collaborative use of any type of software application by two or more participants. This is a powerful concept that goes several steps beyond mere conversation on a videoconferencing system and shows great promise. See audiographics, telecommuting, videoconferencing.

whois An Internet username directory service application that responds to a name query by accessing a central database and returning a listing of users found to match the name or a portion of the name. Whois is based on the WhoIs Protocol (NICNAME) which is an elective proposed Draft Standard of the IETF. See finger, rwhois, RFC 812, RFC 954.

wicking *v.* 1. The process of drawing a liquid out of a substance or along a path. Wicking is occurring when solder runs along a wire or up underneath an insulating sheath. Diapers and hiking socks are designed to wick moisture away from the skin to prevent irrita-

W

tion. Oil lamps draw oil up through the wick as they burn.

Wide Area Augmentation System WAAS. A satellite-related capability of the Federal Aviation Administration (FAA) which includes GPS-based aircraft navigation capabilities.

wide area differential GPS WDGPS. An implementation of the Global Positioning System (GPS) which includes a network of reference stations that act as data collection sites to receive and preprocess GPS satellite signals. The information is forwarded to a central processing hub which creates correction vectors for each satellite, including clock corrections. WDGPS systems can provide more accurate local positioning information for a variety of industries, including surveying and navigation, particularly aviation, where precise positioning for takeoffs and landings are important. See differential GPS, local differential GPS, Global Positioning Service.

Wide Area Information Server WAIS. Developed in the early 1990s by Brewster Kahle, WAIS is a powerful search and retrieval system widely used on databases on the Internet. The WAIS URL scheme designates WAIS databases, searches, and individual documents available from a WAIS database. WAIS can be accessed on the Web through the following URL schemes:

WAIS database

```
wais://<host>:<port>/<database>
```

Specific WAIS search

```
wais://<host>:<port>/<database>?<search>
```

Specific WAIS document

```
wais://<host>:<port>/<database>/<wtype>/<wpath>
```

There are freely distributable and commercial versions of WAIS available. With the burgeoning information on the Net, WAIS is a tool of some significance. See RFC 1625.

Wide Area Information Server gateway WAIS gateway. A computer which is used as a 'go-between' between incompatible networks or applications to translate WAIS data.

wide area network WAN. Unlike LANs, which tend to be directly cabled and thus limited in scope, Wide Area Networks can connect users over broad geographical regions through the use of long-distance transmission technologies, such as telephone and satellite services. WANs are often used to connect LANs with a variety of architectures and protocols. A WAN accomplishes the transmission of various formats through *routers*, or alternately, through *bridges*, which are protocol-independent in order to connect WANs and LANs. See bridge, Local Area Network, router.

wide band channel WBC. In FDDI-II isochronous networks, WBC is the circuit-switching capability. Any bandwidth not allocated to WBCs can be used for other data, such as statistical information about data traffic. See Fiber Distributed Data Interface.

wide characters Character codes consisting of two bytes (16 bits), rather than the traditional one byte (8 bits), in order to accommodate a much larger number of characters from different languages. See Unicode.

wide open receiver A receiver which is receiving a range of frequencies simultaneously. CB radios and various emergency systems are sometimes set to receive a range of transmissions at one time.

wideband 1. A band wider than that which is necessary for transmitting voice, sometimes called medium-capacity band, in the 64-Kbps to 1.5-Mbps range. 2. A range between narrow-band and broadband, typically between about 1.5 Mbps and 45 Mbps. 3. A band with a broad range of frequencies, often multiplexed. See broadband, narrowband.

wideband digital cross-connect system W-DCS. A digital cross-connect system which accepts a variety of optical signals, and is used to terminate SONET and DS-3 signals. In other systems, it may also cross-connect DS-3/DS-1. W-DCS can be used as a network management mechanism. Switching is carried out at the VT level. See broadband digital cross-connect system.

wideband modem A modem designed with a bandwidth (frequency spectrum) capability greater than that of common consumer modems designed to work over basic voice channels.

widescreen TV A home theater TV set which supports and can display a video transmission with a 16:9 horizontal to vertical picture ratio, as is found in movie theatres. These are the same proportions which are supported by 'letter-boxed' laserdiscs and videos, those in which none of the original movie imagery is cut off of the sides when it is displayed. The 16:9 ratio is also supported by some of the better consumer camcorders.

Wiener, Norbert (1894-1964) An American mathematician who collaborated with Arturo Rosenblueth and a group of scientists from various disciplines in developing many fundamental concepts of artificial intelligence. He authored "Cybernetics: or, Control and Communication in the Animal and the Machine" in 1948 and updated it in 1961 to include ideas about self-reproducing machines and self-organizing systems. He also contributed to the fields of stochastic processes and quantum theory. See neural network.

wildcard A symbol which takes the place of and represents a series of characters, or an unknown quantity or any quantity, usually in a numeric or alphanumeric context. For example, the asterisk (*) is a wildcard character frequently used in computing applications, especially file management, for representing unknown characters or any characters. Thus, the UNIX shell command `rm myfile.*` would *re*move any filename in the current directory beginning with "myfile." and ending in any extension (or no extension).

WiLL See wireless local loop listing 2.

Willis Graham Act of 1921 An act which not only recognized AT&T's monopoly in the phone industry, but legitimized it as well. However, the U.S. federal government 28 years later filed antitrust proceedings against both AT&T and Western Electric. This long process was not settled until 1956 with a consent decree.

WIN See Wireless Intelligent Network.

Win-OS/2 Microsoft and International Business Machines (IBM) were originally collaborating on an operating system for IBM's microcomputers which IBM released as OS/2, but at some point during the project, Microsoft stepped back and began concentrating on developing their own Windows products. The windowing component of OS/2 and a number of the concepts were incorporated into Microsoft's Windows product line.

WIN95 See Windows 95.

Winchester disk A random access hard disk drive marketed in 1980 by Shugart Associates, founded by Alan Shugart, formerly of International Business Machines (IBM). Considering that computer users were predominantly using cassette and reel-to-reel tapes to sequentially write and read data up until this time, the hard disk drive was a welcome computer peripheral and is still one of the foremost storage media.

winding *n.* 1. A conductive path, coupled inductively to a magnetic core or cell, usually made of metal wire. Helical coil windings are used in simple armatures, inductors and transformers. 2. A structure to enhance the transmitting or receiving of electromagnetic waves.

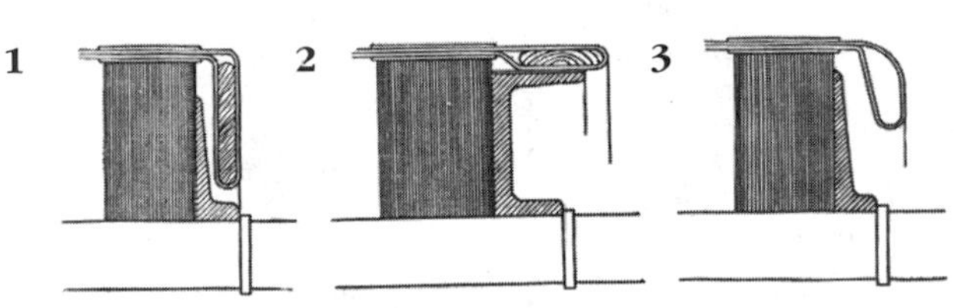

Three types of armature windings are shown here: 1. evolute-wound, 2. barrel-wound, and 3. bastard-wound (a variation on barrel-wound is also shown bottom right).

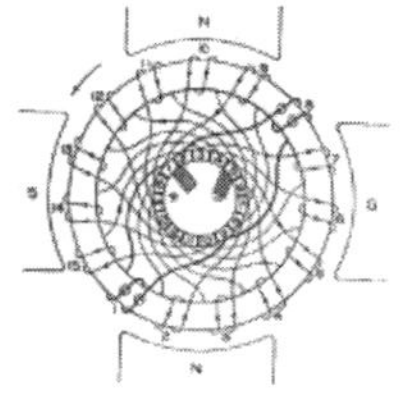

This schematic shows one type of ring armature, a wave-wound armature connected in series. The spacing of the wire windings is called 'pitch.' (Cyclopeda of Applied Electricity, Chicago American School of Correspondence, 1908)

This is the armature from a Westinghouse generator. It is a bastard barrel winding, a type of winding whose end connections are inward and cylindrical, thus shortening the length of the armature parrallel to the shaft compared with barrel winding.

A type of compact antenna mechanism. Various types of windings can be found around small spool-like cores in old radio sets. These were used for frequency selection, employing different thicknesses of wire and winding patterns. Depending upon the purpose of the winding, the wire may be left open or may be sealed in paraffin, rubber, or some other protective material.

Winding can be somewhat tedious and exacting. For this reason, windings are sometimes done in sections and winding machines may be used to create the coil. Since the amount of wire that is wound is sometimes critical, the spool on which the wire is wound may be weighed before and after the winding, to check that the desired amount has been used.

Boiling the coil in linseed oil was one of the techniques used in early fabrication to drive out moisture and provide a tight insulating layer. Paraffin and rubber were sometimes also used. See basket winding, coil.

Historic Basic Armature Winding Types

Type	Characteristics
ring	The two common types are spiral wound (single closed helix) and series-connected wave-wound (see the preceding diagram).
drum	The two common types are lap and wave, which are similar on bipolar machines, but different on multipolar machines.
wave	The winding passes along the conductor to the back of the armature once through each conductor.
multiplex	Two or more independent windings on a single armature. These are more commonly used on generators intended to supply large currents with small voltages.

winding machine Any machine improvised or designed to facilitate wire windings around various cores for the creation of armatures, antennas, frequency tuners, spark coils, and other apparatus that utilize wire windings. Since even, tight windings are important in many electronic applications and tedious to wind by hand, a spinning bobbin, lathe, or specialized winding machine may be employed to ease the task.

window 1. An opening, entrance, time interval, or opportunity. 2. An opening or transparent material that permits light to penetrate, or permits a viewer to see beyond the structure in which the window is installed. 3. A graphical user interface structure developed at Xerox PARC in the 1970s, and incorporated into the Alto computer that first came into widespread use on Macintosh computers in 1984 (it was used also on the Apple Lisa in 1983, but the Lisa never caught on). It contains a group of related information or functions. Most computer interfaces now use similar conventions for sizing, scrolling, opening, closing, and iconizing a window. See graphical user interface. 4. In networking, an opportunity, space, or transmissions lull, in which information can be sent, processed, or otherwise efficiently handled. System tuning, capacity monitoring, and flow control, are all ways of taking advantage of windows.

windowing A description for a means of organizing and interacting with graphical user interface (GUI) structures called *windows*. A window is essentially a *portal* into a portion of the computer's data, visually represented within a bounded entity which can usually be moved around, sized, iconized, or placed in priority overlapping with other windows. There are text windows, graphics windows, sound 'windows,' and others. When these are moved about a computer screen in order to make them comprehensible and easily accessible, popped to the front, or pushed to the back to bring a relevant window to the front, it's called 'windowing.'

Windows A Microsoft Corporation graphical operating environment which works in conjunction with Microsoft's text-oriented MS-DOS (Microsoft Disk Operating System), primarily on Intel platforms, although a number of Windows emulators for other systems are available.

Windows is descended from concepts developed in the 1970s at the Xerox PARC research lab, which were first widely incorporated into microcomputers by Apple Computers, Inc. in the early 1980s. Apple shipped the Lisa computer with a graphical user interface in 1983, shortly before the announcement of Windows 1.0, which didn't actually ship until late in 1985. Current versions of Microsoft Windows use essentially the same basic concepts as their Xerox-inspired predecessors, including icon and gadget point-and-click interaction, sizable windows, visually-accessed directory structures, and window overlap priority for the active window. The following Windows-related entries describe some of the history and development of the various versions.

Windows - early versions Originally called Interface Manager, the first version of Windows was developed between 1981 and 1985. Windows 1.0 was officially announced in 1983, not long after the release of Apple's Lisa computer, although it was not commercially distributed until two years later, beginning late in 1985, around the same time the Atari ST and Amiga 1000 shipped with full graphical user interfaces. While these competitors never unseated MS-DOS and Windows as the prevalent operating systems, they did give consumers and developers food for comparison and many of the ideas pioneered on competitive machines were incorporated into later versions of Windows.

This could be seen two years later, in 1987, when an improved version of Windows, capable of having more than one application open at a time with overlapping windows, was released as Windows 2.0. Icons were not yet being used in this version. It was renamed Windows/286 when Windows/386 was released later the same year. The early versions of Windows were limited to 640 kilobytes of address space and 16 color palettes.

Windows 3.0/3.1 Windows was greatly enhanced and overhauled between 1987 and 1990 to provide an improved user interface, the capability to support more than 16 colors, and memory addressing beyond 640 kilobytes. The result was released as Windows 3.0.

Two years later, in 1992, Windows 3.1 shipped with significant enhancements over 3.0, with scalable fonts, object linking and embedding, and better multimedia capabilities (Multimedia Windows was absorbed into this product). Windows 3.1 became a widely distributed version of the Microsoft graphical user operating environment, popular in the mid-1990s. Windows 3.11 was released as a free upgrade in 1994 to correct some of the networking functions of Windows 3.1. Windows 3.1 has, for the most part, been superseded by Windows 95, although a sizeable number of corporations still use it.

Windows 95 Windows 95 was the highly promoted successor to Windows 3.1 and, to some extent, Windows for Workgroups. Windows 95 first shipped in 1995. It did not have the robustness, multiprocessor support, or security features of Windows NT, but it also cost less, and thus became a popular consumer and small business alternative to the more extensive, more expensive Windows NT operating system.

Windows 98 The successor to Windows 95.

Windows for Pen Computing 3.1 Although not well-known, this version of Windows, specifically designed for pen computers requiring handwriting recognition for the execution of commands, shipped in the spring of 1992.

Windows for Workgroups 3.1 This version of Windows, designed for integrated networking and sharing of resources, shipped in 1992, the same year as Windows 3.1. It should be noted that the Macintosh operating system provided file sharing network support off-the-shelf eight years earlier, so Workgroups was not a forerunner in microcomputer networking, but it was a welcome enhancement to corporate users running Windows systems and included network mail capabilities. Windows for Workgroups 3.11 was enhanced with a number of features, the most important being the addition of 32-bit file access.

Windows NT Windows New Technology. A 32-bit multitasking, multithreaded, networking operating system first released in 1993, Windows NT shares many basic features and user interface concepts with Windows 95, but it is more polished, more reliable, more secure, and includes symmetric multiprocessor support. Bill Gates has publicly stated that Windows NT originated as "OS/2 3.0," with development work beginning in 1987 based on the IBM/Microsoft collaboration on developing OS/2, with D. Cutler heading the project that became Windows NT since 1989.

Windows NT is favored in development environments, server applications, and corporate networks. Windows NT was designed to run on processors other than just the Intel chips, providing portability to other platforms not well-supported by earlier versions of Windows, except in the form of third-party emulators. It is available in workstation and server versions.

Windows Internet Name Service WINS. A name resolution client/server application which resolves the internal names applied to networked Windows computers to corresponding Internet Protocol (IP) addresses.

Windows NT Advanced Server NTAS. An extended version of Windows NT specifically aimed at server applications.

Windows Open Services Architecture WOSA. A Microsoft Corporation distributed, semi-open, client/server-based architecture, first announced in 1992. WOSA is built on a three-level model consisting of service providers, applications programming interfaces for each category of services, and the applications themselves. While the service provider interface (SPI) level is essentially open, the applications programming interfaces (APIs) and service provider interfaces (SPIs) are proprietary to Microsoft.

There are a variety of WOSA elements including:

Type	Notes
Messaging API (MAPI)	Provides electronic messaging access, including a variety of services such as voice mail, electronic mail, facsimile transmissions, etc. Address databases are supported by Extended MAPI.
Speech API (SAPI)	Provides a means for a speech engine to run on a Windows client or server, thus enabling a telephony application, for example, to utilize speech capabilities.
Telephony API (TAPI)	Telephony support for Windows NT systems to enable use of voice, data, and video over existing networks. The API provides generic means for connecting between multiple machines and accessing media streams being transmitted among them. TAPI supports standards-based H.323 conferencing and IP multicast conferencing.

The Open Group Architectural Framework (TOGAF) has information and opinions on WOSA and WOSA-related open architectures at the following Web site:

http://www.opengroup.org/public/arch/wosa.htm

WINF See Wireless Information Networks Forum, Inc.

wink In telephony, a handshaking signal indicating the transition between on-hook and off-hook states. A wink signal can indicate that the central office is ready to receive the dialed number. A wink is a fairly generic type of signal that can be used in many contexts and can be created in a number of ways, depending upon the type of communications protocol and whether the line is analog or digital. In an analog system, the wink can be indicated by a change or interruption in tone or a change in polarity. On a digital line, a wink may be indicated with signaling bits.

WINS See Windows Internet Name Service.

Winsock Windows sockets. Winsock provides a standardized applications programming interface (API) for Windows- and OS/2-equipped computers to interact with the Internet. It originated at an Interop conference discussion group in 1991. Winsock 2 is a further development effort of the Winsock Standard Group. In addition to the capabilities of the original version, Winsock 2 incorporates multiprotocol and multicast capabilities, quality of service (QoS), and a layered provider architecture. Microsoft's Windows Open Services Architecture (WOSA) incorporates Winsock into its APIs.

Winsocks were designed on the same concepts as UNIX-based Berkeley sockets, with extensions to accommodate Windows-specific functions; programming them is somewhat similar to programming Berkeley sockets.

Winsock functions by means of a dynamic link library file called *winsock.dll* which comes in 16-bit and 32-bit versions, and is loaded into memory. The .dll file format is compatible with Windows 3.x and newer versions. Winsock 32-bit drivers are included with Windows 95 and Windows NT. There are also variations in the .dll files to provide compatibility with various TCP/IP stacks. In a sense, the .dll provides a communications layer between the TCP/IP stack and the Winsock applications programs developed to communicate through the sockets.

Winsock Standard Group An organization of network software developers and vendors, established in the fall of 1991, which meets several times a year to enhance, refine, and extend the Winsock specifications. Winsock 2 specifications were developed and published by this group.

Wintel systems Windows and Intel. A term which refers to Intel-based computers running Microsoft Windows operating software.

WIP See Women Inventors Project.

Wire Center A physical facility of a telecommunications provider, to which outside subscriber lines lead.

wire concentrator Any wire path or conduit, which serves to bring together a number of wires running roughly through the same installation path.

wire pair Two separate conductors following the same transmissions path in close proximity to one another. It is called twisted pair when the wires are twisted around one another in order to reduce noise. Copper twisted pair wire is very common in the telephone industry.

wire speed The raw speed at which data bits are transmitted over wire. This is not the same as the amount of information that can be transmitted. Various factors will hinder the transmission when all the devices and wires are interconnected, including noise, interference, connection devices, etc. Conversely, a variety of compression and multiplexing techniques can be applied to the data to increase the informational-carrying capacity of a wire.

wire stripper A tool which quickly removes the protective sheath from wires so they can be easily attached to mounting posts or to one another, to make good electrical contact. Punchdown tools for installing telephone line often have a wire stripping blade included.

wire tap *n.* A covert listening device which is associated with a communications circuit (usually voice) in order to listen in on, and record, transmissions (conversations). The device need not necessarily be physically attached to the line being tapped. Other than high level encryption through a digital line, there are no sure-fire ways to prevent a wire tap, especially on an analog line. Strictly regulated, wire taps require legal authorization and are permitted only in specific, carefully evaluated circumstances, usually by law enforcement officials. See bug, trace.

wireless A system which performs a function, or transmits information fully or predominantly, without wires. An older term for a radio set, since radio signals were intercepted as waves. Now the term is used more broadly for telecommunications technologies that work without wires; cellular phones; intercoms; and video security systems that work over FM frequencies.

WIRELESS The largest annual wireless industry trade show, sponsored by the Cellular Telecommunications Industry Association (CTIA).

Wireless Access Controller WAC. A central point in a private branch or public switched network which provides a communications link to a host network or base station. Wireless systems require constant monitoring of signal strength and administration of user registration, hand-off, and roaming provided by the WAC so the user is not aware of the changing connections.

wireless cable Wireless cable doesn't utilize physical cables or at least, is not predominantly based on wired connections, but is named that way to provide marketing familiarity for subscribers who are used to associating 'cable services' with television broadcast services. Thus, *wireless cable* refers to television broadcast services and, to some extent, interactive TV services delivered through the airwaves without physical cables (usually through microwave transmissions).

Wireless Consortium WC. An association and long-term independent facility established in 1996 through the University of New Hampshire InterOperability lab. It represents the cooperative interests of wireless products and services vendors wishing to develop and test IEEE 802-11-conformant products.

Wireless Data Forum WDF. Formerly the Cellular Digital Packet Data Forum, the WDF is a nonprofit organization established to promote the acceptance and development of wireless data products and services, particularly those employing the Commercial Mobile Radio Services (CMRS) spectrum assigned by the Federal Communications Commission (FCC). This consensus-building organization holds its meetings in a variety of locations around North America. http://www.wirelessdata.org/

Wireless Information Networks Forum, Inc. WINF, WINForum. A trade association of manufacturers of unlicensed communication systems, including personal communications services (PCS). WINForum works with the Federal Communications Commission (FCC) in the allocation of spectrum with the Unlicensed National Information Infrastructure (U-NII). See band allocations.

Wireless Intelligent Network WIN. International Standard IS-41. Actual implementation of this standard is typically done in conjunction with time division multiple access (TDMA) technology. WIN is a network of real-time databases and transaction processing systems, which support the processing of the messages in a wireless transmission and which manages roaming. See time division multiple access technology.

Wireless LAN Interoperability Forum WLI Forum. A trade organization of wireless product and service organizations established to promote and support the growth of wireless LAN. The WLI Forum has developed and published an open interface specification for interoperability not intended to conflict with IEEE 802.11. The specification is based upon the RangeLAN2 radio frequency wireless LAN technology developed by Proxim, Inc. in 1994. It uses a multichannel spread spectrum frequency hopping architecture in the 2.4 GHz range at a data rate of 1.6 Mbps per channel.

Wireless LAN Research Laboratory WLRL. A research laboratory established through the Center for Wireless Information Network Studies at the Worcester Polytechnic Institute (WPI) in Massachusetts. The lab serves as a center for the development and specification and testing of hardware and software performance, compatibility, and interoperability.

wireless local area network WLAN. A data network which is connected through radio wave and/or infrared transmissions rather than through cables and wires. Inter-building connections may be through line-of-site microwave connections.

The IEEE is working on standards for wireless LANs through the IEEE 802.11 Working Group. The basic structure of the standard specifies three physical layers (two for radio, one for infrared) and one Media Access Control (MAC) sublayer to provide authentication, association, integration, and distribution services. Two configurations for wireless stations have been generalized as independent (directly communicating stations) and infra-structure (communicating through inter-access points).

There are actually a few campuses which have wireless LANs up and running, and the technology is gradually being adapted by corporations as well. There are advantages to both wired and wireless communications. Signals can be kept separate more easily in wired communications, but wireless is sometimes the only practical means for individuals or groups on the move to keep in communication with the network. There will probably be many situations where hybrid systems serve the greatest needs. See Inter-Access Point Protocol.

wireless local area network physical layer WLAN physical layer. There are three physical layer (PHY) specifications for the IEEE 802.11 wireless local area network, two for radio and one for infrared transmissions. The radio PHYs specify spread spectrum frequency hopping and direct sequence spread spectrum. The infrared PHY uses pulse position modulation with 16 positions.

wireless local loop 1. WLL. A local telecommunications services distribution system which uses low power radio waves instead of wires. This is particularly cost effective in rugged terrain or sparsely populated areas, where it is not practical to lay cables. Wireless services may also be easier to install in some areas from a legislative/jurisdictional point of view with regard to Federal Communications Commission (FCC) licensing to local carriers. 2. WiLL. A Motorola, Inc. fixed wireless system designed to provide an alternative to traditional landline systems. WiLL systems are provided in two configurations: as public switched telephone network direct connect (PSTN-based) or as mobile telephone switching office/mobile switching center (MTSO/MSC-based) networks. The WiLL System Controller (WiSC) provides performance monitoring and radio channel control functions.

wireless packet switching A means of sending data over wireless networks which is similar in concept to packet switched wired networks. Like wired packet transmissions, a communication is subdivided into small data packages called *packets*, the packets are routed over the network, sometimes different portions

Wireless Local Area Network Physical Layers

Type	Scheme	Modulation	Notes
radio	spread spectrum/	Gaussian frequency shift keying	1 Mbps with 2-level GFSK
	frequency hopping		2 Mbps with 4-level GFSK
radio	spread spectrum/	differential binary phase shift	1 Mbps
	direct sequence	differential quadrature phase	2 Mbps
infrared		pulse position modulation	1 Mbps with 16 PPM
			2 Mbps with 4 PPM

of the message taking different routes, and reassembled at the receiving end. Routing will vary depending upon the current load on the routing stations, and the distance and number of 'hops' over which the message is transmitted. The message may even travel over wires for part of its journey. There are a variety of protocols for accomplishing this. See packet switching.

wireless service provider An authorized wireless communications supplier, usually providing cellular phone and paging services.

Wireless Speciality Apparatus Company WSA. A pioneer radio parts manufacturer founded in 1906 by Greenleaf Whittier Pickard and Philip Farnsworth. By the mid-1940s, WSA had become one of the largest radio engineering manufacturers in the world, but they are chiefly remembered for their production of rugged, well-constructed radio wave crystal detectors in the early part of the century. Later, in conjunction with a number of other significant manufacturers (GE, Westinghouse, AT&T, etc.), WSA made an agreement for Radio Corporation of America (RCA) to be an exclusive distributor of radio receiving sets.

wireless switching center WSC. A terminal center for switching wireless communications with other trunks or service. It's not uncommon for wireless centers to interface with public switched telephone networks (PSTNs) and various landline facilities in order to route a call that originated as a wireless call through traditional routes. This makes use of the best features of the various technologies, extends the range, and improves the transmissions clarity of wireless-originated calls.

wireless telephone service A variety of forms of wireless voice transmissions have been around for decades, but not in significant numbers, or they had significant limitations such as distance. The service now primarily means cellular telephone transmissions. These first became established in 1983 in the U.S. Almost half the wireless systems in use were purchased to be used in emergency situations and the rest are divided among business and personal users.

wireline 1. Communications circuits which are comprised of physical wires and cables, usually copper twisted pair as in traditional telegraph and telephone lines, or fiber optic. 2. In cellular communications, not all aspects of the transmission are necessarily carried through wireless radio waves. Sometimes the system interfaces with a wired system for part of the transmission. A wireline cellular license granted to the local telephone company from the Federal Communications Commission (FCC) is designated as a B Block. It provides permission to operate at FCC-specified frequencies.

wiretapping See wire tap.

wiring closet In medium- and larger-sized wiring installations, whether for phone, computer networks, or other services, a wiring closet will often be designated on the premises as a central administrative and physical connections facility. This provides consolidated access to equipment and connections, and makes it possible to restrict access to unauthorized users. This is typically the location at which wires enter the premises from external sources or from the building distribution system, and may or may not be the demarcation point for the services, depending upon the type of service and the particular contract. There may be several wiring closets on a premises or campus, which may be cross-connected.

wiring grid The wiring architecture of a department, section, or building. This may be illustrated with a diagram of locations, connections, and circuits. The wiring grid may be monitored on a computer system, with consoles for determining and configuring routes and pathways, and for diagnostic testing.

WITS Wireless Interface Telephone System.

Wizard 1. An elite professional, as a programmer, who appears to perform miracles with his or her solutions to problems, or ability to troubleshoot or configure a system. 2. A shortcut computer software applications helper which may consist of hotkeys or other means to quickly perform often-repeated tasks. Some programming Wizards are user-scriptable. 3. A computer software applications informational helper, which may pop up in a particular context or be requested by the user.

WLAN See local area wireless network and wireless local area network.

WLANA An organization established to promote awareness and acceptance of wireless local area network technologies (also called local area wireless networks (LAWN)). WLANA provides education and support for the technology through standards committees and other venues.

WLI Forum See Wireless LAN Interoperability Forum.

WLL See Wireless Local Loop.

WLRL See Wireless LAN Research Laboratory.

WOM write-only memory.

WOMBAT waste of money, brains, and time. One of the wittier of the acronyms seen in email and online public forums on the Internet.

Women Inventors Project WIP. A Canadian project which highlights women who have contributed to technology in a variety of industries. The results of a collaboration of this project with the Canadian National Museum of Science and Technology is available as an educational traveling exhibit for loan periods of 6 to 8 weeks.

WORA Write Once Run Anywhere. A slogan usually used in context with Java, the portable programming language from Sun Microsystems, but which can apply to any software which is designed to be portable, that is, platform- and/or application-independent.

word In computer programming, a contiguous group of a specified number of bits (basic information units) constituting an information entity which is transferred or processed as a unit. The length of a word varies from system to system, with smaller systems usually processing a word as 8 bits, and larger systems as 32 bits. Some supercomputing systems handle a word as 64 bits. Commonly, a word is a multiple of 8 bits, as a byte is 8 bits, and is a basic unit of information between a *byte* and a *longword.* The lack of standardization results from the fact that systems continue to become more powerful, with CPUs and buses handling larger and larger information units at a time; larger systems typically lead smaller systems in this evolution.

In networking, the term 'octet' is often used to mean eight bits in order to prevent the ambiguity associated with 'byte' and 'word.'

word spotting In speech recognition, the selective recognition of a predetermined sound. This can be used to glean specific information or to filter extraneous noise or information. Word spotting is demonstrated in Star Trek episodes on the turbolift, where turbolift commands are recognized, but general conversation is ignored. The same concept applies to automated door and phone systems, and software programs configured to recognize only relevant material.

words per minute WPM. A description of the speed of a communication over time. Commonly used to measure speed in typing, printing, code sending, and some data communications.

work order WO. A widely used administrative tool, especially in the trades, to request and record work to be done. Usually in the form of a document, a work order may be assigned a number and/or a priority, and provides a sequential record of the people, parts, and procedures involved in carrying out a requested job. Work orders help track customer requests, work and completion cycles, and sometimes include details as to who did the job and how. The work order may also include billing information or may be used as a reference for creating a separate invoice.

work station See workstation.

workflow The sequence and path of work in an organization, environment, or department. Workflow management is especially important in production line settings where actions must take place in a particular order and location to keep the line up and running. In service-oriented environments, it refers to the efficient deployment of staff and resources, timed, allocated, and adjusted to process calls and requests as they come in. The term workflow applies in computer technology to the sequence and path of data transport and its management.

workgroup Individuals within an organization who share certain tools, software applications, or tasks, typically on a network. Priority and access levels within a specified workgroup are often similar. A person may belong to more than one workgroup, and workgroup access to resources may overlap from one group to another. Software companies have been targeting as workgroups people within an organization who share communications or applications needs without necessarily being in the same department or physical location. Workgroup applications include electronic mail, word processors, spreadsheets, scheduling programs, and others.

workgroup manager A supervisory individual who has authority to assign login, priority, and security parameters to individual computer network users organized into workgroups. The manager may use a software utility to set up the system, or may pass on information about the organizational structure to the system administrator, who then sets up the computer configuration.

workload 1. The capacity or capability of an employee or system, usually evaluated within a specified time period. 2. In networks, workload may be measured by the number of transmissions or the quantity of information, that can be processed within a specified interval.

workstation In telecommunications, a networked or networkable computer which fits somewhere between the low-end consumer market and the high-end computing market (scientific research, military, high-end medical imaging). Some people refer to all microcomputers as workstations, some refer only to microcomputers with enhanced processors, storage, and memory as workstations, and some refer to microcomputers with decent processing speed and robust operating systems suitable for professional work as workstations. In general, the least and most expensive computers are not called workstations.

World Administrative Radio Conference. WARC. A global space telecommunications conference that convened regularly. Founded in June 1971 by the ITU to review regulations relating to radio astronomy and extraterrestrial communication. At the 1979 general conference, significant access to bands between 1 and 10 GHz were provided for amateur radio satellite programs, bandwidths that the operators had previously been restricted from using.

World Intellectual Property Organization WIPO. A Geneva-based global organization comprising over 160 member countries which promotes, supports, and educates about intellectual property protection through international treaties. http://www.wipo.org/

World Numbering Zone See World Zone.

World Report on the Development of Telecommunications Information and economic reports on global telecommunciations developments. This is a publication of the International Telecommunications Union (ITU).

World Telecommunication Policy Forum WTPF. An organization established as a result of the Kyoto Plenipotentiary Conference wherein members can discuss matters of telecommunication policy and regulation, and provide reports and opinions to the ITU. The first meeting was in 1996, with a discussion theme of 'global mobile personal communications by satellite' (GMPCS).

World Telecommunications Advisory Council WTAC. A global private and public sectors organization established in 1992, which studies telecommunications implementations and advises the Secretary-General of the ITU on matters of policy and strategy. Originally composed of top level managers, the organization now also includes consultants, government officials, and entrepreneurs. The WTAC has provided a number of publications and sponsorships, including a booklet entitled "Telecommunications: Visions of the Future," WorldTel sponsorship, and a global mobile personal communication systems (GMPCS) symposium.

World Wide Web Web, WWW, W3. The W3 was first proposed in March 1989 by Tim Berners-Lee (CERN), who created the first Web software a few months later which he publicly introduced in 1991. The Web is the single most significant implementation on the Internet which has spurred global access and acceptance not only by technical professionals, but by computer novices and laypersons. In a sense, it constitutes a simple graphical user environment for the Web (it is actually much more than that, but that is how 80% of new users first perceive it) and, as such, its use will probably greatly change and grow in future revisions of Web browsers, especially now that Netscape, the primary browser, is freely distributable with source code.

A browser actually gives a thin window on the Web, as it consists of more than just a browser front-end. It is built on a set of protocols and conventions that make access and transfer of information widely available. The first international WWW conference was held in 1993. See browser, HTML.

World Wide Web Conference Committee, International World Wide Web Conference Committee IW3C2. Founded in 1994 by NCSA and CERN, the IW3C2 was incorporated as a nonprofit Swiss Confederation association in 1996. IW3C2 organizes international telecommunications conferences focusing on the World Wide Web, and makes the information resulting from these conferences open to the widest possible audience. IW2C2 promotes and assists in the global evolution of the Web through a broad-based international body. Currently, the IW3C2 selects one academic institution per year to host the IW3C2 conferences. http://www.iw3c2.org/

World Wide Web Consortium WWWC, W3C. An industry consortium of more than 150 organizations, founded in 1994 to promote the development and use of common standards for the evolution of the World Wide Web (WWW). The W3C develops and maintains information repositories for programmers and users; reference code for implementation of recommended standards; prototypes, and sample applications. http://www.w3.org/

World Zone, World Numbering Zone Geographic divisions devised to facilitate the global linking of national telephone services, a system developed in the 1960s. Country codes of one to three digits in length were assigned to specified regions. For example, World Zone 1 represents most of North America and some Caribbean countries, 2 is Africa, 3 and 4 represent parts of Europe, etc. In an international long-distance call, the country code is the first part of the phone number dialed.

WORM write once read many, write once read multiple. Although this commonly refers to optical storage media which are not as readily written as magnetic media, it generically means any storage medium which is typically written (easily) only once, and is usually read, or can be read, many times by the user. Most CD-ROMs fit into this category, although read/write optical systems are coming down in price to the point where they may become consumer items. Kodak PhotoCD discs demonstrate that consumer products on CDs can be written in several sessions. See compact disc.

WORM, The Internet A virus released onto the Internet in 1988 which affected about 10% of the hosts on the Internet. The law and network administrators are not tolerant of potentially destructive activities on the Net, and this instance and its originator were not treated lightly. See Computer Emergency Response Team, virus.

WOSA See Windows Open Services Architecture.

wow An undesirable audio distortion from various causes including uneven rotation of a turntable or tape reel.

Wozniak, Stephen (Woz) (1950-) Steve Wozniak was a shy, inventive child, given to tinkering with electronics and entering projects in science fairs. When he first became friends with Steve Jobs, neither one realized they would soon be making history and millions of dollars.

Wozniak was constantly working on hardware projects in his parents' home and was naturally drawn to the hobbyists who began to meet and talk shop at the Homebrew Computer Club. Together with Bill Fernandez in 1971, Steve Wozniak built a simple computer from cobbled together parts called the Cream Soda Computer.

From 1973 to 1976, Wozniak worked in electronics at Hewlett-Packard. After meeting John Draper, Woz began to collaborate with Steve Jobs in developing blue boxes, phone devices that provided unauthorized access to long-distance services. After this ran its course, he was inspired by the homebrew members and a Popular Electronics feature of the Altair computer kit to develop the Apple I computer. This lead to formation of Apple Computer in 1976, a change of operations to Steve's home, and the subsequent creation of the historic Apple II computer.

Wozniak's engineering feat consisted not only of building one of the first microcomputers, but of doing it with elegance and flair by substantially reducing the number of chips needed to perform various functions in the computer. Meanwhile Jobs sought out marketing and venture capital experts, and added a crucial link that had been missed by other electronics buffs building small computers and computer kits: The Apple II succeeded in part because it was the first easy-to-use 'fun' computer (in spite of the fact that they first considered marketing to businesses), and the Apple turned out to be a great tool for schools and home users. Thereafter they put their focus on education, a direction that appealed to the humanitarian spirit in both of them.

Wozniak stayed with Apple Computer until 1985 when he left to form a wireless electronics company called CL-9 (Cloud Nine). Through all his ventures, he has participated in many educational and philanthropic activities. He is the founder and current president of Unuson Corporation in which he enjoys his time playing and teaching guitar.

WRAM Windows Random Access Memory. Similar to VRAM, but faster and less expensive, commonly used to enhance graphics performance. Offers bisynchronous input/output. See VRAM.

wrap A redundancy mechanism used in International Business Machines (IBM) Token-Ring local area networks (LANs). Token-Ring cabling is set up with two trunks, one for normal use and one for a backup. If something happens to the main trunk, a disconnected TCU connection creates a signal which causes the path to 'wrap' onto the backup trunk in order to maintain the interconnections through the ring. See Fiber Distributed Data Interface.

wrap connector A diagnostic tool which interconnects cables or controllers to verify the circuit.

wrap plug A diagnostic device which, when plugged into a computer port, causes the data to be looped around back to the port.

wrap test A diagnostic test in which a signal is looped back through a device, usually to see if the input signal and the output signal match.

wrap-around A circumstance where visual or textual information goes off the edge of the display device, usually the bottom or right hand side, and the information that would be clipped off by the edge is 'wrapped around' to the opposite side. For example, if a window drops off the bottom of a computer monitor, the bottom part will be displayed at the top of the monitor. Wrap-around is usually undesired and is a result of programming bugs. However, there are circumstances, such as graphical scenes in which a landscape is intentionally wrapped around, usually from side to side, to give the illusion of an infinite landscape.

WRC World Radiocommunication Conference.

write To record data, as to a storage device, such as RAM, a hard drive, or a floppy drive. Many computer data storage media use a means of rearranging magnetic particles in order to form patterns which are subsequently recognized and interpreted by software. Since this can be done many times, these media can be re-used. There are also media which can only be written once. See WORM, write protect.

write head A storage write mechanism which is designed to put data on the writable medium, which may be tape, a diskette, an optical disk, a vinyl record, etc. Write heads are sometimes separate from read heads, and are sometimes incorporated into a combined read/write head.

write protection A device or method for protecting a medium from accidentally being overwritten with other data or erased. Most computer diskettes, audio, and video tapes have simple physical write protect mechanisms which usually involve covering up a small hole with a label or a plastic tab. This allows the drive or tape mechanism to recognize that it should not alter the diskette or tape if the user accidentally tries to save something to the diskette (or erase it) or record on the tape. With computer storage media, software write protect methods exist, as well.

WRS Worldwide Reference System. A global indexing scheme designed for Landsat.

WRT An abbreviation for "with respect to" used in memos and online communications. See FYI.

WSA See Wireless Specialty Apparatus Company.

WSC See wireless switching center.

WTAC See World Telecommunications Advisory Council.

WTB Wireless Telecommunications Bureau. An organization of the Federal Communications Commission (FCC).

WTO World Trade Organisation. An international organization established in 1994.

WTPF See World Telecommunication Policy Forum.

WWW See World Wide Web.

WWWC See World Wide Web Consortium.

WYPFIWYG (*pron.* wip-fee-wig) An abbreviation for "what you pay for is what you get." It applies as much in computer technology as anywhere else. If you buy a cheap system, you may pay more later for adding extra memory, sound cards, graphics cards, hard drive controllers, CD-ROM drives, etc. The same applies to network server software. A cheap system may cost substantially more in the long run in terms of administrative time, productivity, and lost access due to downtime.

WYSIWYG (*pron.* wiz-ee-wig) An abbreviation for "what you see is what you get." A desktop publishing term that refers to a computer display of a document or image which looks on the screen the way it would look on the intended output device, usually a printer. Display PostScript in NeXTStep (Rhapsody) comes close to WYSIWYG, since many high-end printers and publishing programs employ Adobe Systems PostScript. Most professional-level desktop publishing software provides a measure of WYSIWYG if the system is run with a high resolution display (e.g., 1024 *x* 768 or better) and scalable fonts.

Some word processing programs provide WYSIWYG, but many do not (kerning, leading, and even page breaks may change when printed). WYSIWYG capability depends partly on having the system correctly configured. If the fonts on the computer are not available on the printer or are not downloaded to the printer, there will be substitutions (often Courier). Obviously, a 72 or 75 dots per inch (dpi) computer display cannot visually match a 1200 dpi printout and, in fact, the fonts used for the screen display are often optimized for the monitor, separate from printer fonts. In some ways, WYSIWYG on low-priced consumer systems is really more of a goal than a reality at present; in the meantime, good software, fonts, and a little imagination and familiarity with the various types of output can provide good results even without a WYSIWYG system.

WZ1 World Zone 1. See World Zone.

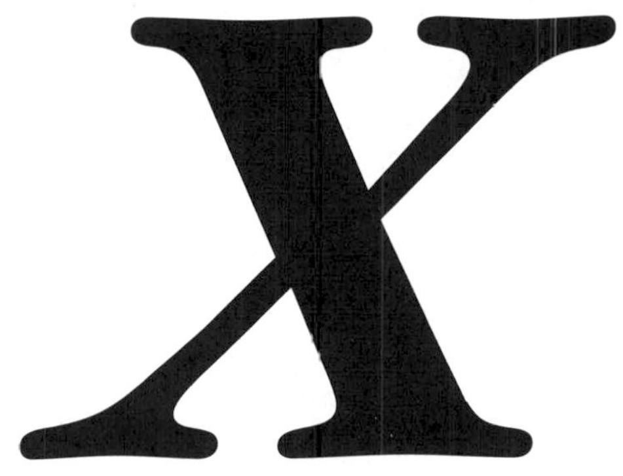

X, x 1. Symbol for an unknown, situational, or arbitrary quantity. Commonly used in mathematics and software programming, written both lower- and uppercase, and often italicized. The symbol is also often used in product identification to indicate a family of products, e.g., *x*DSL for the family of Digital Subscriber Line services. 2. Abbreviation for cross. 3. Abbreviation for exchange. 4. Abbreviation for trans- (prefix). 5. The USOC Federal Communications Commission (FCC) code for complex multiline or series jack. 6. *colloq.* See X Window System.

X axis, x axis A common convention for coordinate systems is to designate the horizontal axis as the X axis. When graphing processes which occur over time, the X axis is often used for the time variable.

X Consortium A group which continued development and management of the X Window System, now part of the Open Group. See X Window System.

X cut A type of cut used with piezoelectric crystals. Crystals are used in radio wave detection and timing applications, and their piezoelectric properties are partly determined by their shape and size. An X cut creates a crystal plate with the plane perpendicular to the crystal's X axis. See crystal, detector, quartz, piezoelectric, X-ray goniometer, Y cut, Y bar.

X Protocol A low-level client/server standard communications protocol which handles window manipulation routines for the graphical user interface (GUI) X Window Systems. See X Window System.

X Series Recommendations A set of ITU-T recommendations which provides guidelines for interconnecting networks and network devices. These are available as publications from the ITU-T for purchase and a few may be downloadable from the Net. Some of the related general categories and specific X category recommendations of particular interest are listed here. See also G, I, Q, and V Series Recommendations.

General Categories	
Series C	General telecommunications statistics
Series E	Overall network operation, telephone service, service operation, and human factors
Series F	Telecommunication services other than telephone
Series G	Transmission systems and media, digital systems and networks
Series H	Line transmission of nontelephone signals
Series I	Integrated Services Digital Networks (ISDN)
Series J	Transmission of sound program and television signals
Series P	Telephone transmission quality, telephone installations, local line networks
Series Q	Switching and Signaling
Series V	Data communication over the telephone network.
Series X	Data networks and open system communication
Series Z	Programming languages

Specific X Series Recommendations of Particular Interest	
X.25	Definitions of the procedures for exchanging data between user devices (DTEs) and network nodes in a public switched packet data network (PSPDN) in order to provide a common interface across a variety of systems. X.25 is a layered packet transmissions protocol commonly used in wide area networks (WANs). A version of X.25 specifically designed for packet radio has been developed as AX.25.
X.400	An international ISO/ITU-T series of standards for electronic messaging architecture for the exchange of data between computer systems. X.400 was published by the ITU-T in 1984. The standard was jointly rewritten by ISO and ITU-T in 1988. X.400 does not stipulate the formatting of data. It provides guidelines for internetworking various messaging systems, addressing individual messages, and describing message contents. Within X.400 there are also substandards and recommendations to X.400, some of which are: X.402 which describes the overall architecture, X.420 for transferring email, X.435 which defines the electronic movement of Electronic Data Interchange (EDI), and X.440 for voice messaging. See Electronic Data Interchange.
X.445	Asynchronous Protocol Specification (APS). A commercially promoted multiple media client/server extension of the X.400 standard which facilitates the exchange of digital data over public phone networks rather than X.25 standard leased lines.
X.500	A directory service protocol for building distributed global directories. It was developed in response to a need to design directories which would not experience the same problems and bottlenecks that were developing with many of the large databases being accessed by thousands or millions of users on the Internet. X.500 employs decentralized maintenance, searching capabilities for complex queries, homogenous global namespace, and a structured standards-based information framework.

Additional X Series Recommendations of Interest	
X.1	International user classes of service in, and categories of access to, public data networks and integrated services digital networks (ISDNs). Includes information on access to leased or switched circuits by data terminal equipment (DTEs) in various modes, access by facsimile terminals, and access to frame relay systems.
X.6	Multicast service definition. Service definitions and capabilities of a multicast service providing a common model for the description of service elements. Interface specifications and protocol elements are not specified by X.6.
X.31	Support of packet mode terminal equipment by an ISDN. Service and signaling procedures definitions operated at the S/T-reference point of an ISDN for subscribing packet mode terminal equipment and terminal adapter functionalities to support existing X.25 terminals at the R-reference point of the ISDN.
X.75	Packet-switched signaling system between public networks providing data transmission services. A description of packet-switching signaling systems among public data networks.
X.76	Network-to-network Interface between Public Data Networks providing the Frame Relay data transmission service. A description of interface interconnections between frame relay networks and public data networks. Layer, data transfer, and signaling information are provided.
X.77	Internetworking between PSPDNs via B-ISDN. Definitions of procedures for internetworking which includes reference configurations, protocol stacks, and signaling procedures.
X.121	International number plan for public data networks. A description of the design, characteristics, and applications of the numbering plan for public data networks. The International Number Plan was developed to facilitate the linking of public data networks with the worldwide system. It describes country

identification (assigned by the ITU-T), regional/local network identification, and a mechanism for integrating with other numbering plans. Guidance for efficient allocation of numbers is included, along with eligibility criteria and procedures. See Data Network Identification Codes.

X.122 Numbering plan interworking for E.164 and X.121 number plans.

Information on interworking with other numbering plans.

Brief Listing of Further X Series Recommendations

X.2 International data transmission services and optional user facilities in public data networks and ISDNs. X.3 defines a set of parameters which are used for regulating basic functions such as terminal characteristics, flow control, data forwarding, etc.

X.3 Packet assembly/disassembly facility (PAD) in a public data network.

X.4 General structure of signals of international alphabet No.5 code for character oriented data transmission over public data networks.

X.5 Facsimile Packet Assembly/Disassembly facility (FPAD) in a public data network.

X.7 Technical characteristics of data transmission services.

X.8 Multiaspect pad (MAP) framework and service definition.

X.10 Categories of access for data terminal equipment (DTE) to public data transmission services.

X.20 Interface between data terminal equipment (DTE) and data circuit-terminating equipment (DCE) for start-stop transmission services on public data networks.

X.20 bis Use on public data networks of data terminal equipment (DTE) which is designed for interfacing to asynchronous duplex V-series modems.

X.21 Interface between data terminal equipment and data circuit-terminating equipment for synchronous operation on public data networks. A digital signaling interface which includes specifications for physical interface elements, character alignment, and data transfer.

X.21 bis Use on public data networks of data terminal equipment (DTE) which is designed for interfacing to synchronous V-series modems.

X.22 Multiplex DTE/DCE interface for user classes 3-6.

X.24 List of definitions for interchange circuits between data terminal equipment (DTE) and data circuit-terminating equipment (DCE) on public data networks.

X.25 Interface between Data Terminal Equipment (DTE) and Data Circuit-terminating Equipment (DCE) for terminals operating in the packet mode and connected to public data networks by dedicated circuit.

X.28 DTE/DCE interface for a start-stop mode data terminal equipment accessing the packet assembly/disassembly facility (PAD) in a public data network situated in the same country.

X.29 Procedures for the exchange of control information and user data between a packet assembly/disassembly (PAD) facility and a packet mode DTE or another PAD.

X.30 Support of X.21, X.21 bis, and X.20 bis based data terminal equipment (DTEs) by an integrated services digital network (ISDN).

X.32 Interface between Data Terminal Equipment (DTE) and Data Circuit-terminating Equipment (DCE) for terminals operating in the packet mode and accessing a packet switched public data network through a public network.

X.33 Access to packet switched data transmission services via frame relaying data transmission services.

X.34 Access to packet switched data transmission services via B-ISDN.

X.35 Interface between a PSPDN and a private PSDN which is based on X.25 procedures and enhancements to define a gateway function that is provided in the PSPDN.

X.36 Interface between data terminal equipment (DTE) and data circuit-terminating equipment (DCE) for public data networks providing frame relay data transmission service by dedicated circuit.

X.37 Encapsulation in X.25 packets of various protocols including frame relay.

X.38 G3 facsimile equipment/DCE interface for G3 facsimile equipment accessing the Facsimile Packet Assembly/Disassembly facility (FPAD) in a public data network situated in the same country.

X.39 Procedures for the exchange of control information and user data between a Facsimile Packet Assembly/Disassembly (FPAD) facility and a packet mode Data Terminal Equipment (DTE) or another FPAD.

Brief Listing of Further X Series Recommendations, cont.

X.42 Procedures and methods for accessing a Public Data Network from a DTE operating under control of a generalized polling protocol..

X.45 Interface between Data Terminal Equipment (DTE) and Data Circuit-terminating Equipment (DCE) for terminals operating in the packet mode and connected to public data networks, designed for efficiency at higher speeds.

X.48 Procedures for the provision of a basic multicast service for Data Terminal Equipment (DTEs) using 25.

X.49 Procedures for the provision of an extended multicast service for Data Terminal Equipment (DTEs) using 25.

X.50 Fundamental parameters of a multiplexing scheme for the international interface between synchronous data networks.

X.50bis Fundamental parameters of a 48-kbps user data signaling rate transmission scheme for the international interface between synchronous data networks.

X.51 Fundamental parameters of a multiplexing scheme for the international interface between synchronous data networks using 10-bit envelope structure.

X.51bis Fundamental parameters of a 48-kbps user data signaling rate transmission scheme for the international interface between synchronous data networks using 10-bit envelope structure.

X.52 Method of encoding anisochronous signals into a synchronous user bearer.

X.53 Numbering of channels on international multiplex links at 64 kbps.

X.54 Allocation of channels on international multiplex links at 64 kbps.

X.55 Interface between synchronous data networks using a 6 + 2 envelope structure and single channel per carrier (SCPC) satellite channels.

X.56 Interface between synchronous data networks using an 8 + 2 envelope structure and single channel per carrier (SCPC) satellite channels.

X.57 Method of transmitting a single lower speed data channel on a 64 kbps data stream.

X.58 Fundamental parameters of a multiplexing scheme for the international interface between synchronous nonswitched data networks using no envelope structure.

X.60 Common channel signaling for circuit switched data applications.

X.61 Signaling System No. 7 - Data user part.

X.70 Terminal and transit control signaling system for start-stop services on international circuits between anisochronous data networks.

X.71 Decentralized terminal and transit control signaling system on international circuits between synchronous data networks.

X.80 Interworking of interexchange signaling systems for circuit switched data services.

X.81 Interworking between an ISDN circuit-switched and a circuit-switched public data network (CSPDN).

X.82 Detailed arrangements for interworking between CSPDNs and PSPDNs based on Recommendation T.70.

X.92 Hypothetical reference connections for public synchronous data networks.

X.96 Call progress signals in public data networks.

X.110 International routing principles and routing plan for public data networks.

X.115 Definition of address translation capability in public data networks.

X.116 Address translation registration and resolution protocol.

X.123 Mapping between escape codes and TOA/NPI for E.164/X.121 numbering plan interworking during the transition period.

X.130 Call processing delays in public data networks when providing international synchronous circuit-switched data services.

X.131 Call blocking in public data networks when providing international synchronous circuit-switched data services.

X.134 Portion boundaries and packet-layer reference events: Basis for defining packet-switched performance parameters.

X.135 Speed of service (delay and

	throughput) performance values for public data networks when providing international packet-switched services.
X.136	Accuracy and dependability performance values for public data networks when providing international packet-switched services.
X.137	Availability performance values for public data networks when providing international packet-switched services.
X.138	Measurement of performance values for public data networks when providing international packet-switched services.
X.139	Echo, drop, generator, and test DTEs for measurement of performance values in public data networks when providing international packet switched services.
X.140	General quality of service parameters for communication via public data networks.
X.141	General principles for the detection and correction of errors in public data networks.
X.144	User information transfer performance parameters for data networks providing international frame relay PVC service.
X.145	Performance for data networks providing international frame relay SVC service.
X.150	Principles of maintenance testing for public data networks using data terminal equipment (DTE) and data circuit-terminating equipment (DCE) test loops.
X.160	Architecture for customer network management service for public data networks.
X.161	Definition of customer network management services for public data networks.
X.162	Definition of management information for customer network management service for public data networks to be used with the CNMc interface.
X.163	Definition of management information for customer network management service for public data networks to be used with the CNMe interface.
X.180	Administrative arrangements for international closed user groups (CUGs).
X.181	Administrative arrangements for the provision of international permanent virtual circuits (PVCs).

X Window System, The X Window System, X Windows, X, X.11 Hardware-independent foundation software for the development of graphical user interfaces (GUIs) based upon a client/server model. The X Window System is a nonproprietary, distributed, multitasking, network-transparent protocol which has been implemented on many different Unix-based systems. Originally used as a graphics display protocol for text-based UNIX platforms, developers are recognizing and exploiting its ability to enable popular OSs to run on a UNIX workstation or, conversely, to run UNIX applications on popular hardware platforms, and to run applications from within Web browsers. Development tools such as Motif facilitate the quick design of X GUIs.

The X Protocol is an X Windows System client/server protocol and the X server is a client/server process which controls a display device on The X Window System.

X Windows code for noncommercial purposes is freely downloadable from The Open Group Web site. As of version X11R6.4, commercial users must pay a license fee, which will be used by The Open Group to continue support for these development efforts. See Athena project. http://www.opengroup.org/

X Window System 11 Release 6.x (X11R6.x) A substantial initiative by The Open Group to enable the X Window System to be used to create and access interactive World Wide Web applications through The X Window System and a downloadable plugin. Applications linked to the Web using X11R6.x can be found, accessed, and executed with the same Web browsing utilities used to access current static HTML documents. This may become a very significant means of networking through the Internet. See The X Window System.

X Window System history X was originally developed by Robert Scheifler and Ron Newman from the Massachusetts Institute of Technology (MIT) and Jim Gettys from Digital Equipment Corporation (DEC) to provide a user interface for the Athena Project. It has been further developed by The X Consortium and is now trademarked and managed by The Open Group. See Athena project, X Window System.

X Window System User Interface Toolkit XUI. A suite of utilities which facilitates the implementation of user interfaces for hardware platforms installed with the X Window System. See X Protocol, X Window System.

X-band An assigned spectrum in the microwave frequencies used by military satellites.

See band allocations for a chart of assigned frequencies.

X-Bone A system designed to facilitate and automate the rapid deployment and management of multiple overlay networks. X-Bone is an overlay technology combined with teleconferencing-style coordination and management tools. X-Bone provides a virtual networking infrastructure which is configurable. While X-Bone is intended to be implemented with networks running more advanced systems such as IPv6, some of the automatic tunneling services can be deployed to a limited extent on IPv4 systems. See 6bone, Mbone, overlay network, X-Bone xd.

X-Bone xd An X-Bone directory tool which performs a number of tasks including the coordination of resource sharing at the local site, the support of local daemons through authentication, configuration, and creation of IP-encapsulation tunnels between daemons, and the provision of a user interface and API for users or programs wishing to manually parameterize and override overlays. See X-Bone.

X-dimension of recorded spot In facsimile transmissions, a means of describing variation density in terms of the minimum density. The largest center-to-center space between recorded spots is measured in the direction of the recorded line. This can also be assessed perpendicular to the recorded line as the Y-dimension of recorded spot. The same principles can be applied to assess the scanning spot.

X-ray A radiant energy within the spectrum of high energy, invisible, ionizing electromagnetic radiation which ranges about 0.08 nanometers in wavelength, between ultraviolet light and gamma rays. X-rays were somewhat naively and irresponsibly used in early radio signal experiments, and for extended imaging inside objects and humans. These practices are now used with great care due to the damaging influence of X-rays on living cells. X-rays are used in many medical, industrial, and fabrication applications. See X-ray goniometer; Roentgen, Wilhelm Konrad.

X-ray goniometer An instrument for determining the position of the axes in a quartz crystal. X-rays are aimed at the atomic planes of the crystal and the reflected rays are evaluated. Since crystals are physically manipulated to alter their oscillating properties and often cut in very thin slices, it is important to know the orientation of the crystalline structure before cutting. See quartz, X cut, Y cut.

X-ray spectrometer An instrument which is used, by means of reflected rays and evaluation of the resulting diffraction angles, to study the characteristics and composition of materials, including crystals. See X-ray goniometer.

X/Open A global, independent organization of computer manufacturers, founded in 1984. X/Open seeks to promote an open, multivendor Common Applications Environment (CAE) to enhance application portability. This is a good concept, as it allows software developed by different vendors to run on a variety of platforms, leaving the choice of equipment up to the individual purchaser. See Common Applications Environment, Open Systems Interconnection.

X/Open Federated Naming A naming mechanism from the X/Open group for developers to access network naming services and to provide integration with industry-accepted naming services such as X.500, Domain Name Service (DNS), DCE, and others.

X/Open Portability Guide XPG. A guide which documents the X/Open common applications environment system.

XA extended architecture.

XA-SMDS See Exchange Access SMDS.

Xanadu, Project Xanadu A unified interconnected document storage/access/retrieval/publishing environment. Designed over a period of decades, Nelson, the author, has entitled this a 'docuverse.' The Xanadu scheme is a write-once/keep-always system with a hypertext-like file structure.

XAPIA X.400 Application Program Interface Association. See X Series Recommendations.

Xaw The Athena Widget set. A set of widgets which is distributed with The X Window System, which began as Project Athena. See Athena, X Window System.

Xbar *abbrev.* crossbar.

XBase, Xbase A generic designation for applications which read and/or write dBase-compatible files.

XC *abbrev.* cross connect.

XCA extended communication adapter.

Xcoral A multiwindow text editor for The X Window System, which can be used in conjunction with a mouse.

XCVR *abbrev.* transceiver.

xd See X-Bone xd.

XDF extended distance feature.

XDL An object-oriented extension to the ITU-T-defined Specification and Design Language

(SDL) for telecommunications systems. See Specification and Design Language.

XDMA Xing Distributed Media Architecture. A commercial streaming media architecture for delivery of live and on-demand audio-video from Xing Technology Corporation. It is built around standards such as TCP/IP and MPEG, and supports multicasting to multiple simultaneous users over local area networks (LANs) and wide area networks (WANs). XDMA can be implemented over ISDN networks for services such as news and distance learning.

XDMCP X Display Manager Control Protocol. A protocol used to communicate between X terminals and UNIX workstations.

XDR See External Data Representation.

xDSL Generic term for a variety of digital subscriber line technologies, which include ADSL, EDSL, and HDSL. See digital subscriber line and individual listings for further information.

XENIX A Unix implementation best known as being from the Santa Cruz Operation, Inc. (SCO), it was originally codeveloped by International Business Machines (IBM) and Microsoft as XENIX-11 for Intel machines. Microsoft had licensed UNIX from Bell Laboratories and developed it into XENIX for Digital Equipment's PDP-11, the predecessor to the VAX line. XENIX-11 was popular on the 80286/80386 line in the late 1980s. SCO is now marketing UnixWare 7.

xerographic printer A printer which uses the same basic electrostatic mechanisms and techniques as a xerographic copier. The information is imaged onto a drum with lasers, the printing medium is passed across the drum, and picks up the dry transfer toner, which then fuses the toner to the printing medium. See laser printer

Xerox Corporation One of the first companies to see the commercial benefits of new photocopying technology developed by Carlson in the early 1940s. When still a relatively new, small company, Xerox took a chance on the new photocopying invention that had been passed up by other companies. Xerox is now known throughout the world for its technology, especially in the replication industry, and many people refer to all photocopies as "xeroxes." See photocopy for further information and an illustration of Carlson's patent.

Xerox Network Services XNS. A multilayer, distributed file network architecture developed by the Xerox Corporation which is somewhat similar to TCP/IP. Unlike many networks from other vendors, XNS permits a user to use files and devices from a remote machine as if they were on a local machine. XNS functions compatibly with the third and fourth layers of the Open Systems Interconnection model (OSI).

Xerox PARC Xerox Corporation's Palo Alto Research Center. This research center provided enormous impetus to early computer companies and software developers (e.g., Apple Computer Inc., Microsoft Incorporated), especially those developing object-oriented systems and graphical user interfaces (GUIs). Inventors of mice and various laser printing technologies, developers of Smalltalk, the PARC researchers came out with one good idea after another throughout the 1970s and early 1980s, yet surprisingly few of these saw widespread commercial application through Xerox itself. Charles Simonyi, one of the early founding members of PARC, was demonstrator of the Alto computer that inspired so many of those fortunate enough to see it in those early days. Later Simonyi was hired by Microsoft to move the company into graphical applications.

XFN See X/Open Federated Naming.

XFR *abbrev.* transfer.

XFS X11 File System.

XGA 1. See extended graphics adapter. 2. See extended graphics array.

XID exchange identification. In data networking, XIDs are request and response packets which are exchanged prior to communications between a router and a network host. XID is used for device discovery, address conflict, resolution, and sniffing. The XID packet includes the parameters of the serial device, and a connection can only be negotiated if this configuration is recognized by the host.

XIP An abbreviation for "execute *in place*." This is a way to access memory and execute code on PCMCIA cards without having to load them into system memory first. See PCMCIA.

XIWT See Cross-Industry Working Team.

Xlib X Library. A program interface for the X Windows System.

Xmission *abbrev.* transmission.

Xmit *abbrev.* transmit.

XML See Extensible Markup Language.

XModem A widely used error-correcting file transfer data transmissions protocol developed by Ward Christensen in the late 1970s. XModem utilizes 128-byte packets, so files of various lengths will be padded to adjust the packet length and may be longer than the original file. The filename is not sent with the transmission. YModem, a successor to XModem, with support for longer data packets and file attributes, was developed by Chuck Forsberg.

XModem is often used with computer modems to transfer files to and from bulletin board systems over traditional phone lines. XModem is not fast, but it has error correction and it's reasonably reliable. It's well supported, an important consideration since both ends of the connection have to agree on a protocol. Many service bureaus use XModem for file transfers, however, check if they have ZModem, which is faster and capable of restarting an interrupted transmission from where it left off. See Kermit, YModem, ZModem.

XModem-1K, XModem-K A variant of XModem which manages data in 1K (1024 byte) packets. See XModem.

XModem-CRC XModem with 16-bit cyclic redundancy checking (CRC) error detection mechanisms, instead of checksum. XModem-CRC can communicate with XModem versions which use checksum for error correction.

XMP X/Open Management Protocol.

XMS See Extended Memory Specification.

XMT *abbrev.* transmit.

XNMS MlCOM's commercial IBM-licensed Intel-based desktop computer packet data network (PDN) network management system software.

XNS See Xerox Network Services.

XO *abbrev.* crystal oscillator. See crystal detector, quartz.

XON/XOFF transmission on/transmission off. Common flow control signals used between two communicating devices or software programs, typically through modems. Since many transmissions media are inherently slow, there may be a delay between receiving a block of data and resuming transmissions. XON/XOFF signals allow the communicators to signal when to stop sending data and when to resume, in order to prevent loss or corruption.

XON/XOFF is also known as software flow control. In newer high-speed modems, flow control may be handled by hardware, often in conjunction with specific types of cables.

Flow control signals are not limited to modem communications. If a user is working on a terminal which understands XON/XOFF commands, usually signified by Ctrl-S (stop) and Ctrl-Q (resume), then it is possible to suspend a listing or other activity and resume when it is convenient. This is a common way of preventing a screenful of text information from scrolling by so fast that it cannot be read in text windows which lack scroll bars for back-referencing.

Xover *abbrev.* crossover.

XPAD external packet assembler/disassembler.

XPG See X/Open Portability Guide

XPM Extended Peripheral Module.

Xponder transponder.

XRB transmit reference burst.

xref *abbrev.* cross reference.

XRemote A serial transmission protocol for The X Window System.

XRF Extended Recovery Facility.

XSG X.25 Service Group.

XSI X/Open System Interface Specification.

XSMP X Session Manager Protocol.

XT 1. *abbrev.* crosstalk. 2. See IBM Personal Computer XT.

Xtal 1. *abbrev.* crystal.

Xtal Set Society A society founded in 1991 for building and experimenting with radio electronics, particularly crystal radio sets.

Xtalk *abbrev.* crosstalk.

xterm A popular terminal emulator for The X Window System which has been ported to several other operating systems. Xterm lets you can have more than one terminal window active at a time through a single modem, each with its own input/output process running independently of the others. In other words, you can be reading Internet newsgroups in one window, interacting in an Internet Relay Chat (IRC) channel in another, and downloading binary files in a third at the same time, which is very convenient.

XTI X/Open Transport Interface. See X Window System.

XTL See SunXTL.

XTL API See SunXTL.

XTL Provider Interface See SunXTL.

XTL Provider Library See SunXTL.

XTP Express Transfer Protocol.

XUI See X User Interface Toolkit.

XWS See X Window System.

XY cut A means of angle-cutting a piezoelectric crystal so its electrical characteristics are between those of an X cut and a Y cut. See quartz, X-ray goniometer, X cut, Y cut.

Y 1. Symbol for admittance. The ease with which alternating current (AC) flows through a circuit, as opposed to impedance. See impedance. 2. Symbol for yttrium. See YAG, YIG. 3. A general purpose programming language distributed from the University of Arizona in the early 1980s which is semantically similar to C, but without C pointers and structures.

Y antenna A single-wire antenna with leads connected in a Y shape, with the top part of the Y corresponding to the transmission line. Since the top of the Y is closed, causing it to resemble the Greek "D" (*delta*), the Y antenna is sometimes also known as a *delta matched* antenna. This style of antenna is commonly used for very high frequency (VHF) and frequency modulated (FM) signals.

Y axis, y axis A reference baseline or vector within a coordinate system, most often associated with rectangular or Cartesian coordinates. The Y axis is oriented vertically by convention, perpendicular to a horizontal X axis in a two-dimensional system, and perpendicular to the Z and X axes in a three-dimensional system. See Cartesian coordinates, X axis, Z axis.

Y bar A type of cut used with piezoelectric crystals in which the plane of the long direction is parallel to the crystal's Y axis. See Y cut.

Y cable A cable which splits from a single line or bundle, into two usually equivalent lines or bundles. A splitter. Y cables are frequently used in audio applications to split a mono signal into two jacks (not the same as real stereo) to connect systems with different inputs and outputs, or to combine a stereo signal onto one jack. Y cables are also used to split power sources, as when adding an extra drive to a computer system. Depending upon the application, the Y cable may or may not cause a degradation of the transmission once the signal is split. A Y cable may or may not be combined with other connectors or converters. See converter.

Y cables are commonly used in audio and video applications. Shown here are two Y cables with RCA connectors, male connectors on the left and female on the right.

Y cut A type of cut used with piezoelectric crystals. Crystals are used in radio wave detection and timing applications, and their piezoelectric properties are partly determined by their shape and size. A Y cut creates a crystal plate with the plane perpendicular to the crystal's Y axis. See crystal, detector, quartz, piezoelectric, X cut, X-ray goniometer, Y bar.

Y signal A monochrome signal luminance transmission. When combined with a color signal, luminance provides brightness to the image. See Y/C.

Y-dimension of recorded spot In facsimile transmissions, the X-dimension of recorded spot is a means of describing variation density in terms of the minimum density. The largest center-to-center space between recorded spots is measured in the direction of the recorded line. When it is assessed perpendicular to the recorded line, it is the Y-dimension of recorded spot. The same principles can be applied to assess the scanning spot.

Y/C A color image information encoding scheme in which the chroma (color) signals (C) are separated from the luminance (brightness) signals (Y). A Y/C splitter cable allows the two signals to be handled separately.

YAG yttrium-aluminum-garnet. A substance used in laser electronics. See YIG.

Yagi, Hidetsugu A Japanese researcher and educator who worked with his assistant, Shintaro Uda, to develop and describe a new, more sensitive directional antenna structure. See Yagi-Uda antenna.

Yagi-Uda antenna, Yagi antenna A narrow bandwidth, directional antenna that resembles a driven dipole antenna with branches, usually along one plane. The Yagi-Uda antenna improves gain through the use of reflectors and directors (the branches extending out from the main rods) and works in high frequency ranges. Rooftop television aerials commonly use Yagi-Uda configurations. This antenna was designed in the mid-1920s by Shintaro Uda, and described in a paper by Hidetsugu Yagi in 1928.

Yahoo A significant, high-profile, extensive search and information site originally developed at Stanford University and established on the World Wide Web in the mid-1990s. Yahoo provides general search capabilities as well as a large database of sites organized according to popular categories and topics. http://www.yahoo.com.

Year 2000, Y2K A designation for the electronic changeover to the new century, a circumstance which was not anticipated and accounted for by all programmers when designing hardware, operating systems, spreadsheets, databases, backup software, schedulers, and the like. Many software programs and hardware clocks accommodate years only up to 1999 and cannot 'roll over' to 2000. A significant number take into consideration only the last two or three digits of the year, and thus cannot resolve a "00" that follows a "99," for example, causing potential file and data problems for backup systems as well as various application programs.

Further problems may be encountered with software that doesn't take into consideration the extra day that is added to February every four years in a leap year. As the year 2000 is a leap year, there may be a combination of problems with data storage and retrieval resulting from the failure to accommodate this aspect of the changeover as well.

Some operating systems were designed without taking the year 2000 changeover into consideration; others, like the Macintosh OS were intended from the beginning to be Year 2000 compliant.

For users who are concerned about Year 2000 compliance, it is recommended that all important data be thoroughly backed up before the rollover and software programs used with temporary files for a few days to identify and correct any potential problems. Do not try to reinstall backed up data until potential problems are understood and alleviated.

Yellow Book CD-ROM A CD-ROM authored and written according to *Yellow Book standards* in ISO 9660 format, a format very common in computer multimedia applications.

Yellow Book standards. Standard for the physical format of a CD-ROM disk (as opposed to the logical format). See CD-ROM, ISO 9660.

Yellow Book, Jargon This is a common name for the illustrated printed publication "The New Hacker's Dictionary" which is descended from the infamous Jargon File. See Jargon File.

Yellow Box An Apple Computer Inc. designation for an object-oriented, OS-independent developer platform. See Blue Box, Rhapsody.

yellow pages *colloq.* 1. A common name for a number of business/advertising sections in various phone directories printed on yellow paper to distinguish the section from residential listings. Many online electronic business directories are called the Yellow Pages. 2. A Sun Microsystems client/server protocol for system configuration data distribution now known as Network Information Service (NIS). See Network Information Service.

Yellow Pages A trademark of British Telecommunications.

yellow wire 1. A color designation used by IBM to indicate wires used to re-establish a broken connection in traces or flat cables (ribbon cables). See blue wire, purple wire, red wire. 2. A color commonly used for 'ring' on the second phone in four-wire phone installations (two wires for each phone). The corresponding 'tip' wire is usually black.

Yerk An object-oriented programming language somewhat like a very extended Forth, originally sold as a commercial product of Kriya Systems until the late 1980s. It is named after the Yerkes Observatory and is now maintained at the University of Chicago. See Forth.

Yes A Novell product-compatibility certification program.

Yes, yes key A pushbutton shortcut key on some appliances or keyboards to provide an affirmative response transmission without typing in the whole word 'yes.'

YIG yttrium-iron-garnet. A crystal used in the manufacture of YIG devices (amplifiers, filters, multiplexors, etc.) which, in conjunction with a variable magnetic field, is used to tune wideband microwave circuits.

YIG filter A wide bandwidth filter in which a YIG crystal is positioned within the field of a permanent magnet associated with a solenoid and tuned to the center of the frequency band.

YIQ A color model originating from hardware characteristics which is used in color television transmission and for some computer monitors. Y is luminosity; I and Q provide chroma signals separate from the luminosity in order to provide backward-compatibility with black and white standards. On black and white (grayscale) televisions, only the Y component of the signal is displayed. See Y signal.

YMMV An abbreviation for "your mileage may vary" which is frequently seen online in email and on public discussion lists to indicate that the information provided is very generalized and must be adjusted to a particular situation. In other words, the user may get a different transfer rate, different processing speed, different number of replies, different price quote, etc., according to circumstances and context.

YModem A data transfer protocol commonly used with modems developed by Chuck Forsberg as a successor to XModem. Typically faster than XModem or Kermit, though not as well supported on BBSs and at services bureaus as XModem and ZModem. YModem comes in two flavors, batch and nonstop (YModem-G). The batch version allows multiple files to be sent in a transmission session and wildcard characters for filenames are supported. When errors in transmission are frequent, YModem is able to fall back automatically to smaller packets, thus accommodating a poor line connection, but is not compatible with the error correction of XModem. YModem supports the various XModem characteristics, including 128-bye and 1,024 byte frames and both types of error checking (checksum and CRC), and so may often be used in conjunction with XModem. YModem was eventually succeeded by ZModem, also by Chuck Forsberg. See XModem, ZModem.

YModem-G A non-stop streaming variation of the batch-oriented YModem developed by Chuck Forsberg. YModem-G does not wait for receiver acknowledgments and does not have built-in error correction, instead it relies on an error-correcting modem to supply the error logic. It's appropriate in situations where a good clean connection is available and speed is desired. If a problem with a bad block is encountered, the entire transfer is aborted. Unlike ZModem, it cannot resume a transmission in a second session at the point at which errors and termination of the initial session occurred. See YModem, ZModem.

yoke 1. A clamp or frame which unites or holds two parts or assemblies firmly together. 2. A coil assembly installed over the neck of a cathode ray tube (CRT) to deflect the electron beam as currents pass through it. 3. A ferromagnetic assembly, without windings, which forms a permanent connection between two magnetic cores.

YP See Yellow Pages.

yttrium Yttrium oxide is used in conjunction with Europium to create the red phosphors for color cathode ray tubes. When combined with aluminum and garnet in YAG devices, it is used in laser technology. See europium, gadolinium, yttrium.

Yukawa, Hideki (1907-) A Japanese physicist who proposed a new nuclear force-field theory and a massive nuclear particle, the meson. The existence of the meson was verified by Cecil Powell a couple of decades later. Yukawa furthered the nuclear process of 'K capture,' the absorption of an innermost encircling electron. This theory, too, was subsequently confirmed. Yukawa received a Nobel prize in Physics in 1949 for his significant contributions to our understanding of quantum mechanics.

yurt A sturdy, circular, temporary, mobile shelter of Asian origin. Yurts range from simple tent-like shelters to lavishly furnished mountain-top ski resort hideaways. Yurts can be used on work sites where shelter for equipment and workers is needed during construction or installation work. In the author's town, little yurts have been popping up around the various road construction sites under which new fiber optic cables are being installed.

YUV A color encoding scheme designed to accommodate the characteristics of human visual systems. Human visual perception is less sensitive to color variations than to intensity variations (particularly in individuals who are 'color blind'). Thus, YUV uses full bandwidth to encode luminance (Y) and half bandwidth to encode chroma (UV). See Y/C.

yV Y-matrix of a vacuum tube.

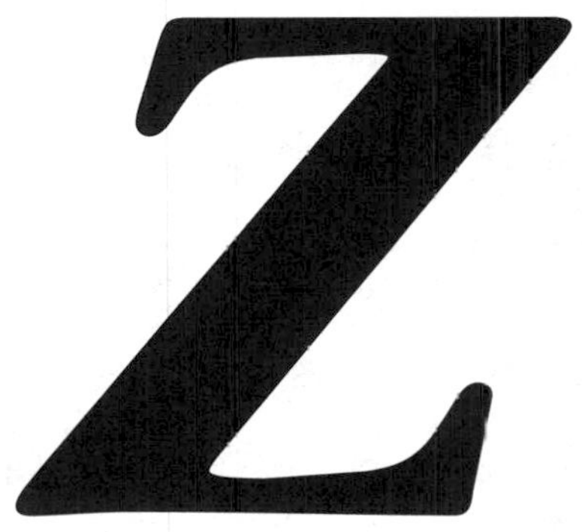

Z 1. Symbol for impedance. See impedance. 2. Abbreviation for Zulu time. See Zulu time. 3. Abbreviation for Zebra time. See Zebra time. 4. The name of a formal specification language for describing and modeling computing systems, based on axiomatic set theory and predicate calculus. Developed at Oxford University in the early 1980s.

Z axis A reference baseline or vector within a coordinate system, most often associated by convention with rectangular or Cartesian coordinates. The Z axis is oriented perpendicular to the X and Y axes in a three-dimensional system. See Cartesian coordinates, X axis, Y axis.

Z axis modulation The varying of the intensity of an electron stream in a cathode ray tube (CRT) by manipulating the cathode or control grid.

Z code In telegraphy, a system of shortcut codes which are related to short phrases, to save transmissions time. Z codes were those prefixed with "Z," a rarely used letter, to reduce the chance of confusing them with the content of a message. For example, ZFB meant "Your signal is *F*ailing *B*adly."

Z force The pressure sensitivity of a touch-activated devices, such as a touchscreen monitor or touch-sensitive pad, as are often used in kiosks.

Z-1 An early home-brewed binary relay computer, developed in Germany in the mid-1930s by Konrad Zuse. While this is one of the pioneering computer constructions, it was largely unknown outside of Germany due to the second World War and, as such, did not significantly influence the industry abroad.

z-fold, zig-zag fold, fanfold A term for a type of fold often used in continuous-feed forms and other computer printouts. The name is derived from the shape of three sheets of paper separated by perforations. The design allows long lengths of paper to pass through a printer and refold itself on the other side to form a neat pile. In the printing industry, the z-fold is sometimes used for company brochures and handouts, and typically doesn't require perforations unless a folded section is designed as a tear-off return sheet.

Z-marker See zone marker.

Z++ Just as C++ is seen as an object-oriented derivative of C, Z++ is an object-oriented extension of Z, a formal specification language for specifying computing systems. See Z.

Z3 A historic programmable, general purpose computer, developed by Konrad Zuse. The Z3 was released in 1941.

Z80 The Zilog Z80 8-bit computer microprocessor was released in 1976 and quickly was incorporated into many control and robotics applications, and into a number of popular microcomputers such as the Tandy Radio Shack Model I and LNW-80 computers. While it was capable of clock rates up to 2.5 MHz, implementations of 1.4 to 2.4 were common.

The Z80 evolved from the Intel 8080 with which it was more-or-less compatible. The Z80 has a simple register structure, including index registers and an accumulator, and is capable of 16-bit addressing through 8-bit double register pairs, something not found on most of the other microprocessing chips that were incorporated into 8-bit microcomputers at the time.

Many hobbyists acquired their first *machine language* and *assembly language* programming skills on the Z80 chip. The original Z80 was followed by faster versions, such as the

Z80A, Z80B, and others, and has been used for two decades in many control applications (e.g., robotics, telemetry, etc.) due to its simple efficiency, low cost, and practical instruction set.

zap *slang* To eradicate data, to burn out a circuit. It may or may not be intentional, and can result from power fluctuations such as those caused by lightening (which is probably where the term originated). See kill.

ZBTSI See zero byte time slot interchange.

Zebra time, Z time The same as Greenwich Mean Time. See Greenwich Mean Time, Z.

Zeeman effect The effect observed on the structure of gas spectrum lines, when subjected to the influence of a moderately strong magnetic field. Named after Zeeman's 1896 observation.

Zen An Asian Buddhist system of belief, which encourages meditation and self-discipline, and the attainment of enlightenment through direct intuitive experience. Some aspects of Zen are unfamiliar to Westerners who are raised with different philosophies and customs, but it is also of interest to many and frequently referenced in philosophical writings about mathematics, physics, and telecommunications.

Zen mail A tongue-in-cheek descriptive phrase for computer communications which erroneously (or deliberately) arrive with no content in the body of the message. See Zen.

zener current In an intense electric field, a current through an insulator sufficient to excite an electron from the valence band to the conduction band.

zener diode An electronic device which may behave like a rectifier below a certain voltage, but exhibits a sudden increase in current-carrying capacity above a specific voltage level (zener value). Zener diodes are used in voltage regulators and power supplies.

zener effect A reverse-current breakdown effect that occurs at a semiconductor or insulator junction in the presence of a high electric field. See zener current, zener diode.

Zenith Corporation One of the early entrants to the radio industry, Zenith was founded in 1931 by a radio amateur and soon became a major manufacturer.

zeppelin antenna A horizontally oriented antenna, which is a multiple of a half wavelength in length with a two-wire transmission coming into one end, which is also a multiple of a half wavelength.

zepto- A standard metric prefix of the Système International (SI). The zepto unit is used in scientific measurement requiring very small numbers and represents 1000^{-7} in decimal. See zetta-.

zero balancing A telephone service accounting technique in which a specific dollar quantity is distributed over a large category of calls. The total base price of all the calls within the category is used to calculate an adjustment percentage so the full dollar amount produces a zero balance (rounding errors are not permitted).

zero beat A condition during which two combined frequencies match, and consequently do not create a *beat.*

zero beat reception, homodyne reception In radio transmission reception, a system that uses locally generated voltage at the receiving end of the transmission, which is the same frequency as the original carrier and combines it with the incoming signal. See beat reception.

zero bias 1. In a cathode ray tube (CRT), the absence of any difference in potential between the cathode and its control grid. 2. In teletypewriter transmission circuits, zero bias is a state during which the length of the received signal matches the length of the transmitted signal.

zero bit insertion, bit stuffing In data communications, the process of inserting a zero bit after a series of one bits in order to specify a distinct break or change. Thus, the beginning or ending of a frame is not misconstrued. This is sometimes used in place of control signals.

zero compression A data compression technique in which nonsignificant leading zeros are removed.

zero fill A data manipulation technique in which zeros are inserted into a file or transmission without affecting the meaning of the data. See zero bit insertion.

zero insertion force ZIF. A type of socket used in integrated circuits which allows a chip to be inserted without undue pressure. A lever or screw is then pushed or turned to lock the component securely in place so that it is not dislodged due to bumping or transit. This type of socket costs more than standard pressure sockets and tends to be used in specialized systems such as test systems or with specific chips which are larger or more expensive.

zero level A level established in order to have a reference from which to judge further states or activities of sounds and signals, for observation, calibration, or testing. The definition of zero level is technology-specific.

zero potential The potential of the Earth, used as a reference measure.

zero power peripheral A device which requires very little power and which consequently can draw that power from the primary device to which it is attached, or from the circuit with which it is associated. Some modems and most telephones take their power from the phone line, unless they have extra features (e.g., speakerphone), which require additional power. External floppy drives and pointing devices often draw power from the laptops to which they are attached. Zero power peripherals are favored in mobile communications as they are less bulky than standard peripherals with power supplies, and easier to attach, since extra electrical outlets are not required.

zero punch A punch located specifically in the third row from the top of a punch card. See Hollerith card, punch card.

zero shift, zero drift A descriptive measure of the amount of shift or drift which has occurred from the original setting or calibration point, at a subsequent point in time. See zero stability.

zero stability The ability of an instrument to retain its original state or settings over time, that is, to withstand zero shift. Zero stability is generally considered a desirable characteristic. See zero shift.

zero stuffing See zero bit insertion.

zero suppression The elimination of zeros that are not meaningful. Zero suppression is often used to increase the readability of information with leading zeros, for formatting or transmission purposes, which would otherwise be distracting or confusing to a human reader. In these cases, the zeros are often replaced with a blank (a space character on printouts). Tables and columnar data (like financial statements) are usually printed with zero suppression.

zero transmission level reference point For an arbitrarily selected point in a circuit, a level reading which is subsequently used against which to measure transmission levels in other points of the circuit or at the same point at another time. In telephone transmissions, the reference point is frequently selected at the location of the source of the transmission.

zero usage customer A listed subscriber who has not used the network to which he or she has access.

zerofill, zeroize To insert the zero character into unused storage locations. This is done for a variety of reasons, for formatting, for creating space savers, and sometimes as a delay mechanism to match up transmission speeds with output speeds of slower output devices (printers, facsimile machines, etc.).

zetta- A standard metric prefix of the SystÜme International (SI). The zetta unit is used in scientific measurement requiring very large numbers and represents 1000^7 in decimal. See zepto-.

ZIF See zero insertion force.

Zimmermann, Philip R. A software engineer and cryptography specialist, best known for developing the Pretty Good Privacy (PGP) encryption scheme, based on the Blowfish technology, Zimmermann is the founder of PGP, Inc. He is a software engineer with a long history of experience in cryptography, data security, data communications, and real-time embedded systems development.

Zimmermann has been honored with numerous humanitarian awards, due to his contribution to the safeguarding of personal privacy, including the 1996 Norbert Wiener Award, the 1995 Pioneer Award from the Electronic Frontier Foundation, and many more. Zimmermann is a member of the International Association of Cryptologic Research, the Association for Computing Machinery (ACM), and others. See Blowfish, PGP Inc., Pretty Good Privacy.

zinc Zn. A malleable, bluish metallic element with a relatively high electropotential, useful in plating, dipping, and galvanizing, to prevent corrosion in other metals. Combined with copper, it forms brass, used for piping, construction accessories, and household items.

zip *v.* To bundle up or compress a file or set of files, usually for transferring or archiving. The term is now commonly used for any such processed (zipped) files, regardless of the archive utility used, but originally referred to programs called gzip and PKZIP. See arc, compress, gzip, PKZIP, uuencode.

ZIP Zone Information Protocol.

zip code A series of characters added to a mail address which specifies the destination in order to expedite delivery. In Canada and the U.K. (called *postal code*), it consists of numbers and letters. In the U.S. it consists of five digits, a hyphen, and four additional digits more recently introduced as a routing code. Zip codes can be optically scanned and converted into bar codes for processing by automated mail systems.

zipped See zip, gzip, PKZIP.

Ziv See Lempel-Ziv.

ZModem A fast, flexible, error-correcting, full-duplex file transfer data transmission protocol similar to XModem and YModem, but with updates and enhancements. YModem and ZModem were written by Chuck Forsberg.

ZModem is well supported by various telecommunications programs, and many BBSs and service bureaus use it.

ZModem includes fall-back and dynamic adjustment of packet size, which is important if the connection is a line with fluctuating characteristics. One of ZModem's most valuable features is its ability to resume a file transfer that has been aborted. With almost all earlier desktop serial file transfer programs, if the line was interrupted or the file transfer aborted 99% of the way through, the program would start again from the beginning when reconnected, rather than resuming from where it left off. ZModem can detect a partial file, establish a new starting point at the end of the partial file, and continue transmitting until the transfer is completed or interrupted. If you've ever spent a couple of hours transferring a seven megabyte file and lost it when the transmission was almost complete, you will appreciate ZModem's "resume" feature. See XModem, YModem.

ZOG A high-performance frame- and link-based hypertext system designed for local area networks (LANs) developed at Carnegie-Mellon University which has evolved into a commercial product called Knowledge Management System (KMS).

zone 1. An area, usually contiguous, which may or may not be self-contained; or has some common characteristics within itself; or which is distinguished as different, in some way, from the surrounding area; or which is assigned on the basis of size, population, or some other economic or practical characteristic (as in shipping zones). 2. A section of computer storage which has been allocated for a particular purpose. 3. In telephone communications, a specified area outside the local exchange. 4. In cellular phone communications a zone consisting of adjacent cells is called a cluster. 5. A physical or virtual region of access within a network, sometimes delineated by physical structures such as a workstation or router. See firewall.

zone, optical recognition A region manually specified by the user, or automatically detected and enclosed by the optical recognition software, which corresponds to a particularly type of information. For example, some optical document recognition sytems, and some of the better optical character recognition systems, can distinguish columns and page numbers, images and formulas, and scan each separately from the others. Some software allows zones to be set up in a template in advance and order priority assigned to the zones. This way a long document, such as a book in which most of the pages are identical in format, can be scanned without resetting the zones each time.

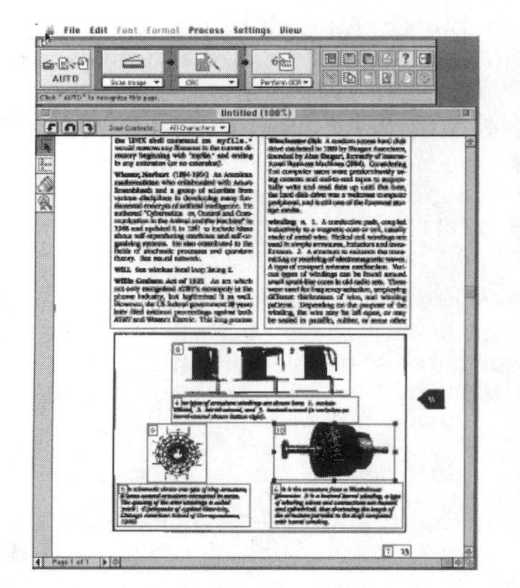

Many optical recognition programs will automatically determine zones or allow them to be manually configured. This allows the flow of text, and separation of graphics and text to be handled more efficiently by the software, as in this Caere Omnipage example.

zone, punch card One of three specific positions on the top of a punch card. See punch card, zone punch.

zone blanking Turning off a cathode ray tube (CRT) at a point in the sweep of an antenna.

zone cabling A cabling architecture designed for open office systems in which various physical zones are designated and cabled in such a way that office desks and equipment can be moved around without ever being too far from the various necessary outlets and connectors.

zone marker, Z-marker A vertically radiating beacon of light that signals a zone above a radio station transceiver.

zone method A wire installation ceiling distribution system, in which the space above rooms is organized into sections or zones. Cables are centralized in each zone, with main arteries running between zones or to the central power source. See distribution frame.

zone of authority The set of names managed by, or under the authority of, a specific name server.

zone of silence, skip zone In radio transmissions, a geographic region that does not receive normal radio signals, frequently due to abrupt changes in terrain, (e.g., mountains).

zone paging The capability of an intercom or phone system to selectively page certain groups of speakers. See public address system.

zone punch In a punch card, a punch located in one of the upper three rows (the section which usually has less text displayed on the card). See Hollerith card, punch card. Contrast with digit punch.

zone time A system in which the Earth is divided longitudinally into 24 time zones of about 15 degrees each starting in Greenwich, England. It was developed in the late 1800s by a Canadian, S. Fleming, to establish a standard time. See Greenwich Mean Time.

zoning, stepping In microwave transmissions, displacement of portions of the surface of the microwave reflector in order to prevent changes in the phase front in the near field.

zoom To continuously reduce or enlarge an image, as on a monitor or in a viewfinder. The capability of zooming is usually provided to improve visibility of details (zoom in), or to provide a 'big picture, wide angle' view (zoom out) of a diagram, object, image, or scene. See zoom factor.

zoom factor The degree to which an image can be scaled, that is, decreased or enlarged. The X and Y axes may or may not be capable of sizing independent of one another. The enlargement zoom factor on consumer camcorders often ranges up to 20 or 40 times the normal viewing factor. In some cases, the zoom factor on still cameras and camcorders may be digitally enhanced, that is, the zoom up to 20 times may be an optical zoom and, beyond that, it may be a digital zoom, which may show some pixelation at higher zoom factors. See zoom.

zoom lens An apparatus that provides the ability to reduce or enlarge the apparent size of an image in order to frame that image with the desired scope or to enhance detail. Commonly used on video and film cameras, and sometimes on telescopes and binoculars.

zsh, Z shell A Unix command interpreter shell, similar to ksh, developed by Paul Falstad. Zsh is said to be similar to a bash shell, but faster and with more features. Zsh is not a Posix-compliant implementation.

Zulu time Greenwich Mean Time, Coordinated Universal Time.

Zworykin, Vladimir Kosma (1889-1982) A Russian-born American physicist and electrical engineer who emigrated to America in 1919 and worked for Westinghouse Electric Company in the 1920s. Zworykin developed the idea of controlling the passage of beams in an electron tube with magnets to devise a cathode ray tube (CRT), an idea that he patented in 1928 and which led to his development of the *iconoscope*, a practical television camera. In 1929 he demonstrated a cathode-ray display, the same basic concept as current computer and television monitors. The same year, Zworykin became the director of research for the Radio Corporation of America (RCA).

Numerals

4B/5B Fiber transmissions cable that is commonly used in asynchronous transfer mode (ATM) and Fiber Distributed Data Interface (FDDI) networks. This 4-byte/5-byte multifiber cable can support transmission speeds up to about 100 Mbps.

6bone An IETF-supported international collaboration testbed providing policies and procedures for the evolution of Internet Protocol (IP). The name 6bone is derived from *backbone* a major 'artery' of the Internet, IP version 6. 6bone is designed to be used in the development, deployment, and evolution of the Internet Protocol Version 6 (IPv6) which is intended to succeed the current Internet protocol IPv4.

This testbed and transition project is essential in that the Internet is not one machine and one agency running it, but a global collaboration of computing devices managed and owned by many different personal, commercial, and governmental entities. The 6bone provides not only a means to test the many features and concepts of the new systems, but also a means for developing and deploying a transition infrastructure.

The 6bone is a virtual network which is layered on portions of the current physical structure of the IPv4-based Internet. IPv4 routers are not designed to accommodate IPv6 packets. By layering IPv6 on the existing structure, the routing of IPv6 packets can be accomplished prior to the implementation of enhanced physical structures, particularly routers, designed to take advantages of the features of IPv6.

To understand the 6bone virtual structure, imagine various workstation-class computers, such as those commonly used as servers in various communities and institutions. Provide these machines with operating system support for IPv6 so that they have direct support for the IPv6 packets. Now provide a means through the Internet for these machines to interconnect and communicate with one another through virtual point-to-point links called *tunnels*, thus managing the links on behalf of physical routers until IPv6 support is more widespread. Eventually, as the Internet is upgraded to IPv6, this interim system will be replaced by agreement with direct physical and virtual IPv6 support.

The 6bone Web site is sponsored by the Berkeley Lab Networking & Telecommunications Department. See IPv4, IPv6, MBone, X-Bone. http://www-6bone.lbl.gov/6bone/

8B/10B Fiber transmissions cable that is appropriate for high speed networks. This 8-byte/10-byte multifiber cable can support transmission speeds up to about 149.76 Mbps.

10Base-2 A physical transmissions medium that supports up to 10 Mbps of baseband usually carried over coaxial cable. The Manchester scheme of binary coding is typically used with 10Base-T.

10Base-T A physical transmissions medium that supports up to 10 Mbps of baseband transmissions over twisted pair (T). The Manchester scheme of binary coding is typically used with 10Base-T.

73 A telegraph numeric 'shortcut' code dating back to at least the mid-1800s which indicated various sentimental, amorous, and fraternal greetings, depending upon the time, place, and operator. See Z code.

100VG-AnyLAN A commercial LAN from IBM. It is similar to Fast Ethernet, and capable of carrying Ethernet and Token-Ring transmissions simultaneously. This technology has not found widespread commercial acceptance. See High Speed Token-Ring.

800 calling A service in which calls are billed to the receiver. 800 numbers are widely used by businesses to encourage potential buyers to contact the company through a toll-free 800 number (or 888 number) without concern about long distances charges. These services are sometimes used internally, through an unpublicized 800 number for traveling employees to contact their main office or branch.

802.3z A Gigabit Ethernet technology.

888 calling A service similar to 800 toll-free calling which was developed as an extension to 800 numbers when 800 numbers assignments were nearing capacity.

900/976 calling A set of numbers which are billed to the caller through a rate determined by the callee. 900 services are used by information brokers, public opinion pollers, advice counsellors, astrologers, other prognosticators, and by vendors offering phone sex. Charges for 900 calls range from $1.95 to $4.95 per minute.

These calls are not only somewhat expensive, but some of the less scrupulous 900 vendors will keep the caller on the line longer by asking him or her many personal questions at the beginning of the call. It's not unusual for the average call to be around $25 and often they are much more. Because of the abuse or overuse of 900 numbers, subscribers demanded a way to disable 900 calling and this is now provided by the phone companies. This is mainly to curtail calls by children, 900-addicts, and 900 calls by unauthorized callers using a phone without permission.

911 calling A service designed to expedite connections to telephone emergency services such as medical services, law enforcement, and fire departments. By dialing only three easy-to-remember digits, subscribers can more easily get help when needed. This concept was first introduced in the 1970s. The calls are connected quickly to a Public Safety Answering Point (PSAP) where a trained emergency dispatcher records the call, determines the origin and nature of the call, responds to the caller, and dispatches services as needed.

1000Base-T An IEEE standard approved in 1997, developed by the P802.3ab study group. This standard defines a full duplex Gigabit Ethernet signaling system for category 5 (Cat 5) network systems. Unlike 100Base-TX transceivers, which use only two pairs of wires, one in each direction, 1000Base-T transmits on all four pairs simultaneously from both directions of each pair. This creates a more complex system and a greater potential for crosstalk. See far end crosstalk, near end crosstalk.

4004 An early 4-bit central processing unit (CPU) from Intel as part of the MCS-4 chipset released in 1969. See Intel, MCS-4.

8008 An early 8-bit central processing unit (CPU) from Intel, released as a successor to the 4004 as part of the MCS-8 chipset in 1972. The historic Altair computer was based on this processor. See Altair, Intel.

8080 An 8-bit central processing unit (CPU) released by Intel in 1974. RAM addressing was limited to 64 kilobytes. It was incorporated into a number of early microcomputers including the first model released by International Business Machines (IBM) in 1980.

8086 A successor to the 8080, the 8086 was an Intel 16-bit central processing unit (CPU). It could address 1 Mbyte of RAM. This chip was quickly incorporated into new versions of the IBM computers and was also used by manufacturers licensing IBM technology in competition with IBM. See Intel and Motorola for charts of other numbered microprocessors.

DATA & TELECOMMUNICATIONS DICTIONARY

Appendices

Telecommunications Timeline

Essential Concepts and Early Engineering

~2970 BC The great step pyramids are designed and engineered in Egypt, possibly by Imhotep.

~1650 BC Ahmose transcribes Egyptian mathematics which include fractions.

~670 BC Thales engages in abstract and deductive mathematics and investigates magnetism.

~500 BC Pythagoras and the Pythagoreans make important contributions to mathematics and the study of sound frequency relationships.

~260 BC Archimedes establishes many important, basic principles of physics.

~800 The concept of zero is used in mathematics in Asia.

1200s Fibonacci authors "Liber Abaci" (Book of the Abacus) in which he promotes the use of Arabic numerals and positional notation.

1400s Gutenberg develops movable type, thus ringing in the age of mass-produced books and the publishing industry.

Copernicus studies the celestial movements and relationships, and develops concepts that challenge accepted notions of the Earth in the cosmos.

15/1600s Galileo makes important observations of the laws of physics, especially of gravity and bodies in motion.

1600s Robert Boyle studies physical properties and the compression of air, resulting in Boyle's law, sometimes called Marriotte's law due to parallel studies in France.

late 1600s Sir Isaac Newton makes important observations about basic physical laws, now widely known as *Newtonian* physics or *classical* physics. These highly significant discoveries form the basis of modern physics.

1676 O. Røhmer calculates the velocity of light as a constant 227,000 kilometers/second.

1724 Fahrenheit reports his new temperature scale to the Royal Society.

1900 Max Planck states quantum theory.

1905 Einstein publishes a paper on the special theory of relativity.

Physics and Basic Electronics Technology

1738 Bernoulli writes on the relationship of pressure and the velocity of fluids, now known as Bernoulli's principle, used in vacuum technology.

mid-1700s Benjamin Franklin conducts numerous experiments with electricity, inspires other scientists, and coins many terms associated with the emerging science.

1700s Luigi Galvani studies electromagnetism and its effects in living tissue.

1775 Alessandro Volta invents the electrophorus, the basis of subsequent electrical condensers, replacing the Leyden jar.

1800 Alessandro Volta invents the voltaic pile, a pioneer wet cell.

1800s Ritter discovers ultraviolet light.

1807 J. B. J. Fourier announces Fourier's theorem, which forms the basis for Fourier transforms, now widely used as analytical tools in mathematics.

Humphry Davy uses battery power to separate out and discover potassium and, about a decade later, invents an arc lamp (arc lamps were also invented by others).

1819 H. C. Ørsted demonstrates the relationship of electricity and magnetism.

1820 A. M. Ampère studies the mathematical characteristics of electromagnetism and announces the *right-hand rule*.

1826 J. N. Niepce develops a primitive type of photography.

1829 L. J. Daguerre develops one of the first permanent photographs - the *daguerreotype*.

1832 Michael Faraday publishes Faraday's laws of electrolysis.

1877 The microphone is invented by David Hughes.

1878	The first telephone exchange is established in London.
1879	American Telephone & Telegraph is founded based on the technology in the Bell patents, later to be known as AT&T.
1891	Almon B. Strowger invents the automatic telephone switching system so subscribers can dial the desired number, rather than calling a human operator.
1904	Fleming releases the two-element Fleming tube, which leads to de Forest's development of the three-element tube.
	Lee de Forest invents the Audio leading to the evolution of three or more element vacuum tubes which revolutionizes the electronics industry. R. A. Fessenden broadcasts the first voice and music broadcasts, using an Alexanderson alternator to supply the power.
1910	The Computing-Tabulating-Recording Company is founded, later to become International Business Machines (IBM).
1911	Wireless communications are established between the United States and Japan.
1915	The Radio Corporation of America (RCA) is founded.
1920	The historic KDKA radio stations begins broadcasting, having moved out of Frank Conrad's garage.
1921	John Logie Baird carries out pioneer television broadcasting experiments.
1927	Scheduled television broadcasts are begun by station WGY in New York state.
1928	Bell Laboratories develops pioneer color television technologies.
1957	The Russian Sputnik communications satellite is launched into orbit.
1972	The Canadian ANIK satellite becomes the first domestic television broadcast communications satellite.

Communications Applications

1774	Lesage created the frictional telegraph.
1836	C. Wheatstone and W. Cooke's first practical implementation of the telegraph.
1837	Samuel Morse received approval and funding from the U.S. Congress to build a Washington, D.C. to Baltimore telegraph line.
1838	Samuel Morse and Alfred Vail demonstrates a telegraph to the President.
1843	Samuel Morse transmits historic "What God has wrought?" message which launches the age of the practical, commercial telegraph.
1848	The laying of the first transatlantic telegraph cable. It only lasted a few days.
1856	David Hughes translates telegraph signals into letters.
1858	Introduction of the Pony Express.
1861	The first transcontinental telegraph line is built in a record four months. At the line's completion in October, the Pony Express ceased operations.
	P. Reis demonstrates the transmission of tones through wire.
1866	The laying of the first successful installation of a transatlantic telegraph cable. This revolutionizes communications. Previously, sea voyages of two or three months were necessary to transmit overseas messages.
	A. Loomis demonstrates essential basics of wireless radio wave transmissions.
1872	Loomis patents wireless telegraphic technology.
	James Clerk-Maxwell publishes an important paper on electromagnetic wave theories.
1874	Elisha Gray submits a caveat to the U.S. patent office after Alexander Graham Bell submits a patent for the 'harmonic telegraph,' the precursor to the telephone.
1876	Bell reports having spoken intelligibly over wires to his assistant, Watson.
1878	Public telephones make their commercial debut in Connecticut.
1886	A. Dolbear receives patent for induction-based wireless telegraphy.
1880	Bell Telephone is incorporated in Canada.
1889	Creed high-speed automatic telegraph system is developed.

1895-96	A. S. Popow demonstrates wireless telegraphy from ship to shore in Russia.
1920s	Bell Laboratories begins transmitting picture phone images.
	Various types of facsimile transmissions are implemented by different inventors.
1947	Bell Laboratories scientists invent the transistor.
1956	Bell demonstrates Picturephone technology to the Institute of Radio Engineers.
1967	The concept and design of the ARPANET is born.
1969	The ARPANET is put into operation.
1980	T. Berners-Lee develops Enquire hyptertext system.
1983	The ARPANET is split into Milnet and ARPANET, the precursor to the Internet.
1989	T. Berners-Lee develops a Web browser.
1990	The ARPANET is officially discontinued, as it has evolved into the Internet.

Tabulating Machines, Computers, and Microcomputers

B.C.	In many parts of the world, from the Americas to Asia, counting systems arise. The Native Americans use grains, Eastern cultures use sand, clay, counting sticks, and beads, eventually evolving many different forms of the abacus.
1600s	By this time, in Western Europe, the slide rule as we know it is being used, evolving from a number of different sliding tabulating systems.
1623	Wilhelm Schickard invents the first calculating machine.
1642	Blaise Pascal independently invents a different type of calculating machine.
1673	René Grillet develops a general adding machine.
1674	Leibnitz develops a calculating machine.
1786	J. H. Muller publishes ideas for automatic 'difference engines.'
1822	Charles Babbage develops important historic models for 'different engines.'
1834	Charles Babbage develops the concept of an 'analytical engine.' Technology has not yet developed to the point where his ideas can be fully carried out, but the design concepts are sound.
1853	The Scheutzes undertake the construction of an automatic 'different engine.'
1928	Punch card equipment is put into use for storing information from calculators in Germany.
1930	V. Bush's idea for a difference analyzer is constructed at the Massachusetts Institute of Technology (MIT).
1936	Konrad Zuse applies for a patent for 'mechanical memory.'
1937	John Atanasoff conceives the Atanasoff-Berry Computer (ABC).
1937	Howard Aiken teams up with IBM to produce the Harvard Mark 1, which became operational six years later.
1943	The ENIAC project is initiated and is partly operational a year later and fully operational by 1945.
1946	Eckert and Mauchly form the Electronic Control Company.
1951	The UNIVAC is put into service at the U.S. Census Bureau.
1971	The Kendak-1 computer is advertised in the September issue of Scientific American.
1972	Bill Gates and Paul Allen partner for various business ventures, eventually forming Microsoft Corporation.
1973	The Alto 'minicomputer' blazed trails at Xerox PARC. It later serves as inspiration to Apple Computer and Microsoft, who eventually incorporate the graphical interface ideas into the Macintosh and Windows lines of products.
1974	The Altair 8800 is released and subsequently featured in January 1975 issue of Popular Electronics.
1975	The Altair 680 MC6800 computer is released.
1976	Radio Shack ships the TRS-80 Model I. Apple Computer is formed by Steven Jobs and Stephen Wozniak.
1980	International Business Machines releases its Personal Computer line of products.

Asynchronous Transfer Mode (ATM)

asynchronous transfer mode ATM. ATM is a highly significant protocol due to its flexibility and widespread use for Internet connectivity. It is a high-speed, cell-based, connection-oriented, packet transmission protocol for handling data with varying burst and bit rates. ATM evolved from standardization efforts by the CCITT (now the ITU-T) for broadband ISDN (B-ISDN) in the mid-1980s. It was originally related to Synchronous Digital Hierarchy (SDH) standards.

ATM allows integration of local area network (LAN) and wide area network (WAN) environments under a single protocol, with reduced encapsulation. It does not require a specific physical transport, and thus can be integrated with current physical networks. It provides virtual connection (VC) switching and multiplexing for broadband ISDN, to enable the uniform transmission of voice, data, video and other multimedia communications.

Two methods for carrying multiprotocol connectionless traffic over ATM are routed and bridged Protocol Data Units (PDUs). Routed PDUs allow the multiplexing of multiple protocols over a single ATM virtual circuit through LLC Encapsulation. Bridged PDUs carry out implicit higher-layer protocol multiplexing through virtual circuits (VCs).

ATM employs fixed length cells consisting of an information field and a header. The information field is transparent through the transmission. The U.S.A. and Japan proposed the use of 64-byte cells, and Europe proposed 32-bye cells. As a consequence of the discrepancy, 48-byte cells are favored by many as a compromise.

Charts and simplified diagrams on the following pages show an ATM system through user input and reception of a variety of media, including voice, video, and data. The data are inserted and extracted by the ATM adaptation layer (AAL) into a logical package called a *payload* which makes up part of the ATM cell. The ATM layer, in turn, adds or removes a five-byte header to this payload, and the physical layer converts the information into the appropriate format for transmission, which may extend over large areas, and pass through other networks switches, and routers. The physical layer is comprised of two sublayers, the physical medium (PM) sublayer and the transmission convergence (TC) sublayer. See dictionary entries for Ethernet, frame relay, HIPPI, TCP/IP.

ATM Cell Header and Payload Format

```
|<------------------- Header ----------------------------->|<Payload>|
+----------+---------+----------------+-------------------+----/----+
|VCI Label | control | header checksum | option. adaptation | payload |
| 3 bytes  | 1 byte  |     1 byte      |   layer 4 bytes    | 44 or 48|
+----------+---------+----------------+-------------------+----/----+
```

ATM Cell at the User Network Interface (UNI)

Item	Abbrev.	Num. bits	Notes
cell loss priority	CLP	1 bit	Cell loss priority of '1' is subject to discard, without violating agreed upon quality of service (QoS). If CLP is '0,' resources are allocated.
generic flow control	GFC	4 bits	Point-to-point and point-to-multipoint. The field appears at the user network interface.

ATM adaptation layer AAL. In ATM, there is a set of ITU-T-recommended, service-dependent layer types which interface the user to the ATM layer. The AAL is the top of three layers in the ATM protocol reference model. Higher layer services are translated through one or more ATM cells. AAL0 to AAL5 perform a variety of connection, synchronization, segmentation and assembly functions for adapting different classes of applications to ATM. Within the AAL, information is mapped between the PDUs and ATM cells. Upon creation of a virtual connection (VC), a specific AAL is associated with that connection. See the following diagrams for the relationships of the adaptation layers to the ATM format.

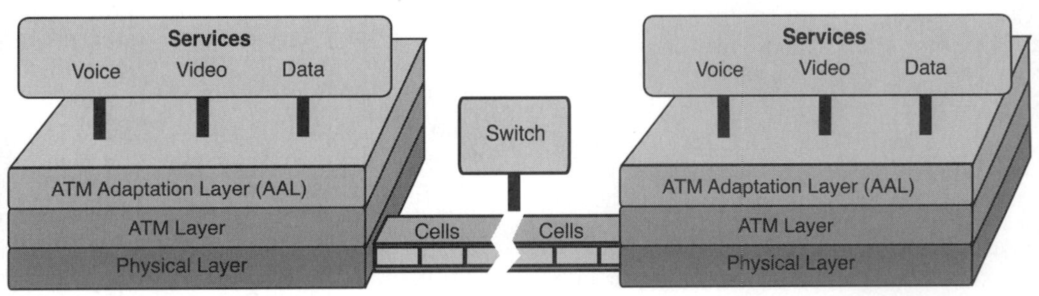

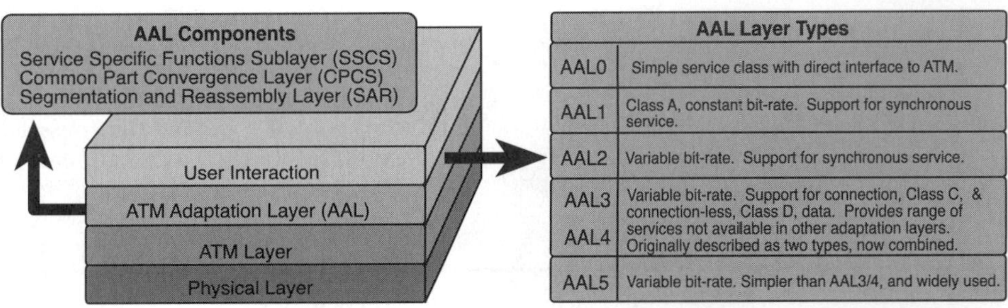

AAL Layer Types	
AAL0	Simple service class with direct interface to ATM.
AAL1	Class A, constant bit-rate. Support for synchronous service.
AAL2	Variable bit-rate. Support for synchronous service.
AAL3 AAL4	Variable bit-rate. Support for connection, Class C, & connection-less, Class D, data. Provides range of services not available in other adaptation layers. Originally described as two types, now combined.
AAL5	Variable bit-rate. Simpler than AAL3/4, and widely used.

ATM adaptation layer 0 AAL0. A layer implementation intended to provide a direct connection between the user and the ATM. It is limited in that it provides no service guarantee mechanisms. It is recent and rarely used, except in proprietary, stand-alone systems. Nevertheless, some standard commercial drivers support AAL0.

ATM adaptation layer 1 AAL1. This is a constant rate service level. It is useful for time-sensitive applications such as voice, video, and circuit emulation.

ATM adaptation layer 2 AAL2. This is a variable rate service. It is rarely used.

ATM adaptation layer 3/4 AAL3/4. This is a variable rate service. It is the most comprehensive of the adaptation layers, and was originally specified as separate AAL3 and AAL4 for connectionless and connection communications.

ATM adaptation layer 5 AAL5. This is a variable rate service similar to AAL3/4. It is sometimes called SEAL for Simple and Efficient Adaptation Layer. It is widely used, especially in TCP/IP implementations. This is a non-assured service, and retransmission must be accomplished by higher-level protocols. It specifies a packet with a maximum size of 64K minus 1 octets.

ATM adaptation layer 6 AAL6. This is a recent addition, designed to accommodate demand for some of the recent multimedia, high-bandwidth applications.

For further information related to ATM adaptation layers, see RFC 1483, RFC 1577, RFC 1626.

ATM cell The ATM cell is the basic unit of information transmitted through an ATM network. An ATM cell has a fixed length of 53 bytes, consisting of a 48-byte *payload* (the information being transmitted) and a 5-byte header (addressing information). Interpretation of the signals from different types of media into a fixed length unit of data makes it possible to accommodate different types of transmissions over one type of network.

There are a number of important traffic flow control, congestion management, and error-related concepts related to ATM, including those listed in the ATM Cell Rate Concepts chart.

ATM Cell Rate Concepts

Abbrev.	Name	Notes
ACR	allowed cell rate	A traffic management parameter dynamically managed by congestion control mechanisms. ACR varies between the minimum cell rate (MCR) and the peak cell rate (PCR).
CCR	current cell rate	Aids in the calculation of ER and may not be changed by the network elements (NEs). CCR is set by the source to the available cell rate (ACR) when generating a forward RM-cell.
CDF	cutoff decrease factor	Controls the decrease in the allowed cell rate (ACR) associated with the cell rate margin (CRM).
CIV	cell interarrival variation	Changes in arrival times of cells nearing the receiver. If the cells are carrying information which must be synchronized, as in constant bit rate (CBR) traffic, then latency and other delays that cause interarrival variation can interfere with the output.
GCRA	generic cell rate algorithm	A conformance enforcing algorithm which evaluates arriving cells. See leaky bucket.
ICR	initial cell rate	A traffic flow available bit rate (ABR) service parameter. The ICR is the rate at which the source should be sending the data.
MCR	minimum cell rate	Available bit rate (ABR) service traffic descriptor. The MCR is the transmission rate in cells per second at which the source may always send.
PCR	peak cell rate	The PCR is the transmission rate in cells per second which may never be exceeded, which characterizes the constant bit rate (CBR).
RDF	rate decrease factor	An available bit rate (ABR) flow control service parameter which controls the decrease in the transmission rate of cells when it is needed. See cell rate.
SCR	sustainable cell rate	The upper measure of a computed average rate of cell transmission over time.
UBR	unspecified bit rate	An unguaranteed service type in which the network makes a best efforts attempt to meet bandwidth requirements.
VBR	variable bit rate	The type of irregular traffic generated by most non-voice media. Guaranteed sufficient bandwidth and QoS.

ATM models Because of the great variety of needs in the networking community, many types and implementations of ATM networks have been developed. Information on some of the more common and emerging models is shown in the ATM Models chart. For further details on specific models, see dictionary entries under ATM Transition Model, Classical IP Model, Conventional Model, Integrated Model, Peer Model.

ATM Models

Model	Notes
Classical IP over ATM	A model for allowing compatible, interoperable implementations for transmitting IP datagrams and ATM address Resolution Protocol (ATMARP) requests and replies over ATM adaptation layer 5 (AAL5). LLC/SNAP encapsulation of IP packets. IP address resolution to ATM addresses via an ATMARP service within the LIS. One IP subnet is used for many hosts and routers. Each virtual connection (VC) directly connects two IP members within the same LIST. TCP/IP applications. See RFC 1577.
IP Broadcast over ATM	An IP multicast service in development by the IP over ATM Working Group for supporting Internet Protocol (IP) broadcast transmissions as a special case of multicast. See RFC 2022, RFC 2226.
IP Multicast over ATM MLIS	Internet Protocol (IP) multicasting over Multicast Logical IP Subnetwork (MLIS) using ATM multicast routers. A model developed to work over the Mbone, an emerging multicasting internetwork. It is designed for compatibility with multicast routing protocols such as RFC 1112 and RFC 1075.
LANE	Local area network (LAN) Emulation. Protocol-independent applications which aid in the transition from legacy internetworks to ATM.
Native ATM API	ATM-specific applications which take advantage of its quality of service (QoS) capabilities.

World Wide Web
Major Search Engines

Name	URL	Notes
AltaVista	http://www.altavista.com/	Extensive searching, advanced search parameters, priority ranking. First introduced by Digital Equipment in 1995.
DejaNews	http://www.dejanews.com/	A huge archive of the posts to various USENET newsgroups. A remarkable record of the public conversations online, searchable by keywords or author.
Excite	http://www.excite.com/	General search, weather, stocks.
i-Explorer	http://www.i-explorer.com/	Search in popular, general interest categories.
InfoSeek	http://guide.infoseek.com/	Web pages, newsgroups, and individuals.
Inktomi	http://inktomi.berkeley.edu/	Fast distributed searchable database from the University of California at Berkeley.
LinkStar	http://www.linkstar.com/	Business directory search.
Lycos	http://www.lycos.com/	General searching, maps, and personal names from Carnegie Mellon University.
Magellan	http://www.mckinley.com/	Sites reviewed and rated by the McKinley Group, Inc.
Sleuth	http://wwwisleuth.com/	The Internet Sleuth lets you search over 3,000 Internet databases. Selections can be found through general categories and subcategories.
Starting Point	http://www.stpt.com/	Searches the Web and other Internet resources (selectable), includes advanced search capabilities.
Switchboard	http://www.switchboard.com/	Personal and business listings of names, addresses, and email addresses.
Webcrawler	http://www.webcrawler.com/	Quick, to-the-point listings.
Yahoo	http://www.yahoo.com/	Search engine and hundreds of topics organized under categories of interest.

For further information from the publisher: http://www.crcpress.com/

For further information from the author: http://www.abiogenesis.com/Books/

Internet Domain Name Extensions

Primarily U.S.A.

.us	United States	U.S., not commonly used.
.um	United States	Outlying islands
.com	commercial	Typically business.
.net	network	Net-related.
.org	organization	Nonprofit, not-for-profit, charitable.
.edu	education	Schools, colleges, universities, other educational facilities.
.gov	U.S. government	Local, state, and federal government agencies.
.mil	U.S.A. military	Military agencies, bases.
.arpa	ARPANET	Advanced Projects Research Agency

Countries other than the U.S.A.

.int	International
.nt	Neutral Zone
.ca	Canada
.mx	United Mexican States
.gp	Guadeloupe (French)

United Kingdom, Western Europe

.gb	Great Britain
.ie	Ireland
.uk	United Kingdom
.at	Austria
.ba	Bosnia-Herzegovina
.be	Belgium
.cz	Czech Republic
.dk	Denmark
.fo	Faroe Islands
.fi	Finland
.de	Federal Republic of Germany
.fr	France
.fx	France
.gi	Gibraltar
.gr	Greece
.gl	Greenland (Denmark)
.hu	Hungary
.is	Iceland
.it	Italian Republic
.li	Principality of Liechtenstein
.lu	Grand Ducy of Luxembourg
.nl	Netherlands
.no	Norway
.pt	Portuguese Republic
.sm	San Marino (Italy)
.si	Slovenia
.es	Spain
.se	Sweden
.ch	Switzerland
.va	Vatican City State
.yu	Yugoslavia

Southeast Europe and Middle East

.bh	Bahrain
ir	Iran
.iq	Iraq
.il	Israel
.jo	Hashemite Kingdom of Jordon
.lb	Lebanon
.tr	Turkey

Eastern Europe and Middle Asia

.af	Afghanistan
.al	Albania
.am	Armenia
.az	Azerbaidjan
.bg	Bulgaria
.by	Bielorussia
.hr	Croatia
.ee	Estonia
.kz	Kazakhstan
.kg	Kirgistan
.lv	Latvia
.lt	Lithuania
.md	Moldavia
.pl	Poland
.ro	Romania
.ru	Russian Federation
.rw	Rwanda
.sk	Slovakia
.si	Slovenia
.tj	Tadjikistan
.tm	Turkmenistan
.ua	Ukraine
.us	Union of Soviet Socialist Republics
.uz	Uzbekistan

Mediterranean and Caribbean

.aw Aruba
.cy Cypress
.mt Malta

African Continent

.ad Andorra
.eg Arab Republic of Egypt
.ae United Arab Emirates
.dz Algeria
.ai Anguilla
.bd Bangladesh
.bj Benin
.bw Botswana
.bi Burundi
.cm Cameroon
.cf Central African Republic
.cg Congo
.gq Equatorial Guinea
.et Ethiopia
.ga Gabon
.gm Gambia
.gh Ghana
.gn Guinea
.gw Guinea Bissau
.gy Guyana
.gf Guyana (French)
.ci Ivory Coast
.ke Kenya
.kw Kuwait
.lr Liberia
.ly Libya
.mg Madagascar
.mr Mauritania
.mc Monaco
.ma Morocco
.mz Mozambique
.na Namibia
.ne Niger
.ng Nigeria
.sa Saudi Arabia
.sn Senegal
.so Somalia
.za South Africa
.sd Sudan
.sr Suriname
.sz Swaziland
.tz Tanzania
.tg Togo
.to Tonga
.tn Tunisia
.ug Uganda
.eh Western Sahara
.zr Zaire
.zm Zambia
.zw Zimbabwe

.an Antilles (Netherland)
.ky Cayman Islands
.mo Macau
.mt Malta
.lc Saint Lucia

Asia and the South Pacific

.io British Indian Ocean Territories
.in India
.bt Buthan
.kh Cambodia
.hk Hong Kong (Xianggang)
.jp Japan
.kp Korea (North)
.kr Korea (South)
.la Laos
.my Malaysia
.u Mauritius
.mn Mongolia
.np Nepal
.pk Pakistan
.cn People's Republic of China
.sg Singapore
.kr South Korea
.Ik Sri Lanka
.tw Taiwan
.th Thailand
.vn Vietnam

.am American Samoa
.au Australia
.bn Brunei Darussalam
.ck Cook Islands
.fj Fiji
.gu Guam
.id Indonesia
.my Malaysia
.mh Marshall Islands
.fm Micronesia
.nz New Zealand

.nf	Norfolk Island
.pg	Papua New Guinea
.ph	Philippines
.pf	Polynesia (French)
.ws	Samoa
.sb	Solomon Islands

West Indies, Antilles

.ag	Antigua and Barbuda
.bs	Bahamas
.bb	Barbados
.bm	Bermuda
.ky	Cayman Islands
.cu	Cuba
.dm	Dominica
.do	Dominican Republic
.gd	Grenada
.ht	Haiti
.jm	Jamaica
.mq	Martinique (French)
.pr	Puerto Rico (U.S.)
.lc	Saint Lucia
.tt	Trinidad and Tobago
.vg	Virgin Islands (British)
.vi	Virgin Islands (U.S.)

Central and South America

.ar	Argentine Republic
.bz	Belize
.br	Brazil
.bo	Bolivia
.cl	Chile
.co	Colombia
.cr	Costa Rica
.ec	Ecuador
.fk	Falkland Islands (Malvinas)
.gt	Guatemala
.hn	Honduras
.mo	Macau
.ni	Nicaragua
.pa	Panama
.py	Paraguay
.pe	Peru
.uy	Uruguay
.ve	Venezuela
.do	Dominican Republic

Miscellaneous

.aq	Antarctica

Request for Comments Documents (RFCs)

The Request for Comments documents are an essential resource and history of the structure, format, and evolution of the Internet. There are over 2,000 of these documents and, unfortunately, not sufficient space here to list abstracts or even all the RFCs. Nevertheless, it is valuable to have a quick lookup to find a particular title, and many of the earlier titles have been superseded or obsoleted, so even an incomplete listing can be a useful reference. The reader is encourage to consult the many excellent RFC repositories on the Internet archived in various formats including ASCII, editable PostScript, Adobe PDF, and HTML formats.

General

RFC 1920	INTERNET OFFICIAL PROTOCOL STANDARDS
RFC 1925	The Twelve Networking Truths
RFC 1935	What is the Internet, Anyway?
RFC 1941	Frequently Asked Questions for Schools
RFC 1958	Architectural Principles of the Internet
RFC 1999	Request for Comments Summary RFC Numbers 1900-1999
RFC 2000	INTERNET OFFICIAL PROTOCOL STANDARDS

Various Protocols

RFC 768	User Datagram Protocol (UDP)
RFC 791	Internet Protocol (IP)
RFC 792	Internet Control Message Protocol (ICMP)
RFC 793	Transmission Control Protocol (TCP)
RFC 821	Simple Mail Transfer Protocol (SMTP)
RFC 826	Ethernet Address Resolution Protocol
RFC 903	Reverse Address Resolution Protocol
RFC 951	Bootstrap Protocol (BOOTP)
RFC 959	File Transfer Protocol (FTP)
RFC 977	Network News Transfer Protocol (NNTP)
RFC 1058	Routing Information Protocol
RFC 1350	The TFTP Protocol (Revision 2)
RFC 1813	NFS Version 3 Protocol Specification
RFC 1831	RPC: Remote Procedure Call Protocol Specification Version 2
RFC 1934	Ascend's Multilink Protocol Plus (MP+)
RFC 1987	Ipsilon's General Switch Management Protocol Specification Version 1.1
RFC 1986	Experiments with a Simple File Transfer Protocol for Radio Links using Enhanced Trivial File Transfer Protocol (ETFTP)

Compression Protocols

RFC 1950	ZLIB Compressed Data Format Specification Version 3.3
RFC 1951	DEFLATE Compressed Data Format Specification Version 1.3
RFC 1952	GZIP file format specification Version 4.3
RFC 1962	The PPP Compression Control Protocol (CCP)
RFC 1967	PPP LZS-DCP Compression Protocol (LZS-DCP)
RFC 1974	PPP Stac LZS Compression Protocol
RFC 1977	PPP BSD Compression Protocol
RFC 1978	PPP Predictor Compression Protocol

Telnet

RFC 854	Telnet Protocol specification
RFC 855	Telnet option specifications
RFC 856	Telnet binary transmission
RFC 857	Telnet echo option
RFC 858	Telnet Suppress Go Ahead option
RFC 859	Telnet status option
RFC 860	Telnet timing mark option
RFC 861	Telnet extended options: List option

Whois

RFC 1913	Architecture of the Whois++ Index Service
RFC 1914	How to Interact with a Whois++ Mesh

SNMP

RFC 1157	A Simple Network Management Protocol (SNMP)
RFC 1445	Administrative Model for Version 2 of the Simple Network Management Protocol (SNMPv2)
RFC 1446	Security Protocols for Version 2 of the Simple Network Management Protocol (SNMPv2)
RFC 1447	Party MIB for Version 2 of the Simple Network Management Protocol (SNMPv2)
RFC 1901	Introduction to Community-based SNMPv2
RFC 1902	Structure of Management Information for Version 2 of the SNMPv2 Working Group
RFC 1903	Textual Conventions for Version 2 of the Simple Network Management Protocol
RFC 1904	Conformance Statements for Version 2 of the Simple Network Management Protocol
RFC 1905	Protocol Operations for Version 2 of the Simple Network Management Protocol
RFC 1906	Transport Mappings for Version 2 of the Simple Network Management Protocol
RFC 1907	Management Information Base for Version 2 of the Simple Network Management Protocol

PPP

RFC 1332	The PPP Internet Protocol Control Protocol (IPCP)
RFC 1334	PPP Authentication Protocols
RFC 1661	The Point-to-Point Protocol (PPP)
RFC 1877	PPP Internet Protocol Control Protocol Extensions for Name Server Addresses
RFC 1915	Variance for The PPP Connection Control Protocol and The PPP Encryption Control Protocol
RFC 1962	The PPP Compression Control Protocol (CCP)
RFC 1963	PPP Serial Data Transport Protocol (SDTP)
RFC 1967	PPP LZS-DCP Compression Protocol (LZS-DCP)
RFC 1968	The PPP Encryption Control Protocol (ECP)
RFC 1969	The PPP DES Encryption Protocol (DESE)
RFC 1973	PPP in Frame Relay
RFC 1975	PPP Magnalink Variable Resource Compression
RFC 1976	PPP for Data Compression in Data Circuit-Terminating Equipment (DCE)
RFC 1979	PPP Deflate Protocol
RFC 1990	The PPP Multilink Protocol (Obsoletes RFC 1717)
RFC 1993	PPP Gandalf FZA Compression Protocol
RFC 1994	PPP Challenge Handshake Authentication Protocol (CHAP)
RFC 1989	PPP Link Quality Monitoring

BGP, BGP-4

RFC 1657 Definitions of Managed Objects for the Fourth Version of the Border Gateway Protocol (BGP-4) using SMIv2
RFC 1771 A Border Gateway Protocol 4 (BGP-4)
RFC 1772 Application of the Border Gateway Protocol in the Internet
RFC 1773 Experience with the BGP-4 protocol
RFC 1774 BGP-4 Protocol Analysis
RFC 1965 Autonomous System Confederations for BGP
RFC 1966 BGP Route Reflection an alternative to full mesh IBGP
RFC 1997 BGP Communities Attribute
RFC 1998 An Application of the BGP Community Attribute in Multi-home Routing

POP

RFC 1725 Post Office Protocol Version 3
RFC 1939 Post Office Protocol Version 3
RFC 1957 Some Observations on Implementations of the Post Office Protocol (POP3)
RFC 1833 Binding Protocols for ONC RPC Version 2

Format and Transmission Standards

RFC 822 Standard for the format of ARPA Internet text messages
RFC 850 Standard for interchange of USENET messages
RFC 894 Standard for the transmission of IP datagrams over Ethernet networks
RFC 1042 Standard for the transmission of IP datagrams over IEEE 802 networks
RFC 1832 XDR: External Data Representation Standard
RFC 1922 Chinese Character Encoding for Internet Messages
RFC 1947 Greek Character Encoding for Electronic Mail Messages

IP General

RFC 1932 IP over ATM: A Framework Document
RFC 1954 Transmission of Flow Labeled IPv4 on ATM Data Links Ipsilon Version 1.0

IPv6 (IPNG)

RFC 1924 A Compact Representation of IPv6 Addresses
RFC 1933 Transition Mechanisms for IPv6 Hosts and Routers
RFC 1955 New Scheme for Internet Routing and Addressing (ENCAPS) for IPNG
RFC 1970 Neighbor Discovery for IP Version 6 (IPv6)
RFC 1971 IPv6 Stateless Address Autoconfiguration
RFC 1972 A Method for the Transmission of IPv6 Packets over Ethernet Networks
RFC 1981 Path MTU Discovery for IP version 6SNMPv2
RFC 1902 Structure of Management Information for Version 2 of the Simple Network Management Protocol
RFC 1903 Textual Conventions for Version 2 of the Simple Network Management Protocol
RFC 1904 Conformance Statements for Version 2 of the Simple Network Management Protocol
RFC 1905 Protocol Operations for Version 2 of the Simple Network Management Protocol
RFC 1906 Transport Mappings for Version 2 of the Simple Network Management Protocol
RFC 1907 Management Information Base for Version 2 of the Simple Network Management Protocol

MIME

RFC 1521	MIME (Multipurpose Internet Mail Extensions) Part One: Mechanisms for Specifying and Describing the Format of Internet Message Bodies
RFC 1522	MIME (Multipurpose Internet Mail Extensions) Part Two: Message Header Extensions for Non-ASCII Text
RFC 1927	Suggested Additional MIME Types for Associating Documents

HTTP/HTML

RFC 1942	HTML Tables
RFC 1945	Hypertext Transfer Protocol—HTTP/1.0
RFC 1866	Hypertext Markup Language - 2.0
RFC 1980	A Proposed Extension to HTML : Client-Side Image Maps

Datagrams and Subnets

RFC 919	Broadcasting Internet datagrams
RFC 922	Broadcasting Internet datagrams in the presence of subnets
RFC 950	Internet standard subnetting procedure

LDAP

RFC 1959	An LDAP URL Format
RFC 1960	A String Representation of LDAP Search Filters

DNS

RFC 1995	Incremental Zone Transfer in DNS
RFC 1996	A Mechanism for Prompt Notification of Zone Changes (DNS)

Domain Names, Routing, Mail Routing, and Miscellaneous

RFC 896	Congestion control in IP/TCP internetworks
RFC 974	Mail routing and the domain system
RFC 1034	Domain names - concepts and facilities
RFC 1035	Domain names - implementation and specification
RFC 1072	TCP extensions for long-delay paths
RFC 1112	Host extensions for IP multicasting
RFC 1122	Requirements for Internet hosts - communication layers
RFC 1123	Requirements for Internet hosts - application and support
RFC 1144	Compressing TCP/IP headers for low-speed serial links
RFC 1155	Structure and Identification of Management Information for TCP/IP-based Internets
RFC 1212	Concise MIB Definitions
RFC 1213	Management Information Base for Network Management of TCP/IP-based internets: MIB-II
RFC 1451	Manager to Manager Management Information Base
RFC 1850	OSPF Version 2 Management Information Base
RFC 1321	The MD5 Message-Digest Algorithm
RFC 1323	TCP Extensions for High Performance
RFC 1441	Introduction to Version 2 of the Internet-standard Network Management Framework

RFC 1452	Coexistence between Version 1 and Version 2 of the Internet-standard Network Management Framework
RFC 1519	Classless Inter-Domain Routing (CIDR): an Address Assignment and Aggregation Strategy
RFC 1583	OSPF Version 2
RFC 1630	Universal Resource Identifiers in WWW: A Unifying Syntax for the Expression of Names and Addresses of Objects on the Network as used in the World-Wide Web
RFC 1700	Assigned Numbers
RFC 1737	Functional Requirements for Uniform Resource Names
RFC 1738	Uniform Resource Locators (URL)
RFC 1808	Relative Uniform Resource Locators
RFC 1900	Renumbering Needs Work
RFC 1940	Source Demand Routing: Packet Format and Forwarding Specification (Version 1)
RFC 1956	Registration in the MIL Domain
RFC 1812	Requirements for IP Version 4 Routers
RFC 1911	Voice Profile for Internet Mail
RFC 1917	An Appeal to the Internet Community to Return Unused IP Networks (Prefixes) to the IANA
RFC 1918	Address Allocation for Private Internets
RFC 1919	Classical versus Transparent IP Proxies
RFC 1985	SMTP Service Extension for Remote Message Queue Starting

Security, Encryption, Authentication

RFC 1421	Privacy Enhancement for Internet Electronic Mail: Part I: Message Encryption and Authentication Procedures
RFC 1422	Privacy Enhancement for Internet Electronic Mail: Part II: Certificate-Based Key Management
RFC 1423	Privacy Enhancement for Internet Electronic Mail: Part III: Algorithms, Modes, and Identifiers
RFC 1424	Privacy Enhancement for Internet Electronic Mail: Part IV: Key Certification and Related Services
RFC 1510	The Kerberos Network Authentication Service (V5)
RFC 1929	Username/Password Authentication for SOCKS V5
RFC 1948	Defending Against Sequence Number Attacks
RFC 1961	GSS-API Authentication Method for SOCKS Version 5
RFC 1964	The Kerberos Version 5 GSS-API Mechanism
RFC 1968	The PPP Encryption Control Protocol (ECP)
RFC 1969	The PPP DES Encryption Protocol (DESE)
RFC 1938	A One-Time Password System
RFC 1984	IAB and IESG Statement on Cryptographic Technology and the Internet
RFC 1991	PGP Message Exchange Formats

Bibliography

Author's Note: I consulted many historic books and hundreds of pre-1900s journals, searching for early and original sources on the telecommunications industry, and am indebted to the radio pioneers who recorded not only their knowledge, but their excitement and wonder at the birth of a new era. I would like to credit the historians who searched tirelessly for the truth about the origins of the early discoveries, for I found in the course of my research that the threads of invention are not so clear and clean as they are depicted, and I strove in this reference, wherever possible, to give the originators of the new technologies credit where it was due. Jonathan Seagrave provided a valuable sounding board for some of my tougher questions about physics, and thanks go to him for his assistance.

Books

Asimov, Isaac; "Asimov's Biographical Encyclopedia of Science and Technology," Doubleday & Company, Inc., 1964.

Baylin, Frank; Gale, B.; Long, R.; "Home Satellite TV Installation and Troubleshooting Manual," Baylin Publications, 1997.

Coe, Lewis; "The Telegraph: A History of Morse's Invention and Its Predecessors in the United States," McFarland & Company, Inc., 1993.

Crocker, Francis B.; Esty, William; "Cyclopedia of Applied Electricity: Volume II," American School of Correspondence, 1908.

Crocker, Francis B.; Esty, William; "Cyclopedia of Applied Electricity: Volume IV," American School of Correspondence, 1908.

Dibner, Bern; "The Atlantic Cable," Blaisdell Publishing Company, 1964.

Dyson, George B.; "Darwin Among the Machines: Evolution of Global Intelligence," Addison-Wesley Publishing Company, Inc, 1997.

Evans, Alvis J.; "Antennas: Selection and Installation," Tandy Corporation, 1989.

Everitt, W. L., Editor; "Fundamentals of Radio," Prentice-Hall, Inc., 1943.

Fike, John L.; Friend, George E.; "Understanding Telephone Electronics," Tandy Corporation, 1983.

Fisher, David E.; Fisher, Marshall Jon; "Tube: The Invention of Television," Counterpoint, 1996.

Foley; van Dam; Feiner; Hughes; "Computer Graphics: Principles and Practice," Addison-Wesley Publishing Company, 1996.

Graham, Frank D.; "Audels Radiomans Guide [sic]," Theo. Audel & Co., 1935.

Heyn, Ernest V.; "Fire of Genius: Inventors of the Past Century," Anchor Press/Doubleday, 1976.

Hodge, Winston William; "Interactive Television: A Comprehensive Guide for Multimedia Technologists," McGraw-Hill, Inc., 1995.

Horowitz, Paul; Hill, Winnfield; "The Art of Electronics," Cambridge University Press, 1994.

Jordan, Nelson, Osterbrock, Pumphrey, Smeby, Everitt; "Fundamentals of RADIO," Prentice-Hall, Inc., 1943.

Kloeffler, Royce G.; Sitz, Earl L.; "Basic Theory in Electrical Engineering," The Macmillan Company, 1955.

Laughter, Victor H.; "Operator's Wireless Telegraph and Telephone Hand-Book," Frederick J. Drake & Co., 1909.

Long, Mark; "The Down to Earth Guide to Satellite TV," Quantum Publishing, Inc., 1985.

Mabon, Prescott C.; "Mission Communications: The Story of Bell Laboratories," Bell Telephone Laboratories, Inc., 1975.

Meadowcroft, William H., "A-B-C of Electricity," Harper & Brothers Publishers, 1915.

Miller, Kempster B; Patterson, George W., et al.; "Cyclopedia of Telephony and Telegraphy, Volume II: Manual Switchboards, Automatic Systems, Power Plants," American Technical Society, 1912.

Miller, Kempster B; Patterson, George W., et al.; "Cyclopedia of Telephony and Telegraphy, Volume IV: Telegraphy, Transmission, Electricity," American Technical Society, 1914.

Myers & Crosby, Supervisors; "Principles of Electricity Applied to Telephone and Telegraph Work," American Telephone and Telegraph Company, 1953.

Pierce, John R.; "Signals, The Telephone and Beyond," W. H. Freeman and Company, 1981.

Roller, Duane; Roller, Duane H. D.; "The Development of the Concept of Electric Charge," Harvard Case Histories in Experimental Science, Harvard University Press, 1954.

Thompson, J. Arthur, Editor; "The Outline of Science, First Volume," G. P. Putnam's Sons, The Knickerbocker Press, 1922.

Walker, J. B., Editor; "The Cosmopolitan: A Monthly Illustrated Magazine, Vol. XVI. November, 1893-April, 1894," The Cosmopolitan Press, 1894.

Weaver, Jefferson Hane; "The World of Physics: A Small Library of the Literature of Physics from Antiquity to the Present," Simon & Shuster, 1987.

Wellman, William R., "Elementary Electricity," D. Van Nostrand Company, Inc., 1959.

Journals

Byte Magazine, The Small Systems Journal
Numerous issues from 1980 to 1987

Computer
Numerous issues from 1970 to 1980

Microcomputing, Wayne Green Publisher
June 1982

Popular Electronics
Numerous issues from 1974 to 1976

Popular Mechanics
Numerous issues from 1909 to 1924

Scientific American
Numerous issues from 1896 to 1921

Acronyms & Abbreviations

A

A/D analog to digital
AA Automated Attendant
AAAC all aluminum alloy cable
AAAI American Association for Artificial Intelligence
AAAS American Association for the Advancement of Science
AABS Automated Attendant Billing System
AAC 1. abbreviated address calling 2. Aeronautical Administrative Communications
AACS attitude and articulation control subsystem
AAL ATM adaptation layer.
AAMOF "as a matter of fact"
AAP Applications Access Point
AAPI Audio Applications Programming Interface
AAPT American Association of Physics Teachers
AARP AppleTalk Address Resolution Protocol
AAS authorized application specialist
ABIST autonomous built-in self test
ABM asynchronous balanced mode
ABME asynchronous balanced mode extended
ABR 1. available bit rate, cell rate 2. autobaud rate
abs absolute value
ABS Alternate Billing Services
ABT Advanced Broadcast Television.
ABX Advanced Branch Exchange.
AC 1. Authentication Center 2. alternating current
AC/WPBX Advanced Cordless/Wireless Private Branch Exchange
ACA 1. American Communications Association 2. Australian Communications Authority 3. Automatic Circuit Assurance
ACAR aluminum conductor alloy-reinforced
ACARD 1. Advisory Council for Applied Research and Development 2. Acquisition Card Program
ACB 1. Annoyance Call Bureau 2. Architecture Control Board 3. ATM Cell Bus 4. automatic callback
ACCS Automatic Calling Card Service.
ACD Automatic Call Distribution
ACF Advanced Communication Function
ACIA asynchronous communications interface adapter
ACK acknowledge.
ACL 1. access control list 2. Applications Connectivity Link 3. Association for Computational Linguistics
ACM 1. Association for Computing Machinery 2. Automatic Call Manager 3. Address Complete Message
ACO 1. Additional Call Offering 2. alarm cutoff
ACOnet Austrian Academic Computer Network
ACOST Advisory Council on Science and Technology
ACP activity concentration point
ACR 1. allowed cell rate 2. attenuation to crosstalk ratio
ACS 1. Advanced Communication System 2. automatic call sequencer
ACSE Association Control Service Element
ACTS 1. Advanced Communications Technologies and Services 2. Advanced Communications Technology Satellite 3. Association of Competitive Telecommunications Suppliers 4. Automatic Coin Telephone Service
ACUTA The Association of College and University Telecommunications Administrators
AD administrative domain
ADACC Automatic Directory Assistance Call Completion
ADAS Automated Directory Assistance Service
ADB Apple Desktop bus
ADC, A/DC 1. analog-to-digital converter 2. automated data collection, automatic data collection
ADCA Automatic Data Collection Association
ADCCP Advanced Data Communication Control Procedures
ADCIS Association for the Development of Computer-based Instruction
ADCU Association of Data Communications Users
ADF 1. automatic direction finder 2. automatic document feeder
ADIO analog/digital input/output
ADK application-definable keys
ADM 1. adaptive-delta modulation 2. add/drop multiplexer
ADN Advanced Digital Network
ADONIS Article Delivery Over Network Information Systems
ADP automated data processing
ADPCM adaptive differential pulse code modulation
ADQ Average Delay in Queue
ADR 1. achievable data rate 2. aggregate data rate 3. analog to digital recording 4. ASTRA Digital Radio
ADRMP autodialing recorded message player
ADRT approximate discrete Radon transform
ADS 1. advanced digital system 2. AudioGram Delivery Services 3. automated data systems
ADSL asymmetric digital subscriber line
ADSP AppleTalk Data Stream Protocol
ADSTAR Automated Document Storage And Retrieval
ADSU ATM Data Service Unit
ADT abstract data type
ADTV Advanced Definition Television
ADU asynchronous data unit
AE 1. acoustic emission 2. Application Entity
.ae Internet domain name extension for the United Arab Emirates
AEA 1. American Electronics Association 2. American Engineering Association
AEC acoustic echo canceller
AECS Plan Aeronautical Emergency Communications System Plan
AECT Association for Educational Communications and Technology
AEEM Aerospace Engineering and Engineering Mechanics
AEGIS Advanced Electronic Guidance and Instrumentation System
AEP AppleTalk Echo Protocol
AES 1. Application Environment Standard, Application Environment Service 2. atomic emission spectroscopy 3. Audio Engineering Society
AESS Aerospace and Electronics Systems Society
AEW 1. aircraft early warning. 2. airborne early warning
AF audio frequency
AFAIK "as far as I know"
AFAST Advanced Flyaway Satellite Terminal
AFC 1. Advanced Fibre Communications 2. automatic frequency control
AFCEA Armed Forces Communications and Electronics Association
AFE 1. analog front end 2. antiferroelectric
AFI Authority and Format Identifier
AFIPS American Federation of Information Processing Societies. A national organization of data processing societies which organizes the National Computer Conference (NCC).
AFK "away from keyboard"
AFM 1. Adobe Font Manager. 2. Adobe Font Metrics. 3. antiferromagnetism.
AFMR antiferromagnetic resonance.
AFNOR Association Français Normal

AFOSR Air Force Office of Scientific Research
AFP AppleTalk Filing Protocol
AFS Andrew File System.
AFT Automatic Fine Tuning
AFTRA American Federation of Television and Radio Artists
AFV audio-follow-video
AGC 1. AudioGraphic Conferencing 2. automatic gain control
AGP Accelerated Graphics Port
AGT AudioGraphics Terminal
AGU 1. address-generation unit 2. Automatic Ground Unit
Ah ampere-hour
AHT Average Handle Time
AI 1. Airborne Interception radar 2. artificial intelligence
AIA 1. Aerospace Industries Association 2. Application Interface Adapter
AIEE American Institute of Electrical Engineers
AIFF Audio Interchange File Format
AIIM Association for Information and Image Management
AIM 1. amplitude intensity modulation 2. Ascend Inverse Multiplexing protocol 3. Association for Interactive Media 4. ATM inverse multiplexer
AIN Advanced Intelligent Network
AIOD Automatic Identified Outward Dialing
AIP ATM Interface Processor
AIR 1. additive increase rate 2. Airborne Imaging Radar 3. All India Radio
AIS 1. alarm indication signal 2. Automatic Intercept System 3. Association for Information Systems 4. Automated Information System
AIST Agency of Industrial Science and Technology
AISTEL Associazione Italiana Sviluppo delle Telcomunicazioni
AIT 1. assembly, integration, and testing 2. Atomic International Time (more correctly known as TIA) International Atomic Time 3. Automatic Identification Technology
AITS Australian Information Technology Society
AIX Advanced Interactive Executive
AJ anti-jam
AJP American Journal of Physics
AKA also known as
Al aluminum
AL Adaptation Layer
ALA American Library Association
ALAP AppleTalk Link Access Protocol
ALBO automatic line buildout
ALC 1. automatic level control 2. automatic light control
ALDC adaptive lossless data compression
ALE 1. Application Logic Element 2. Atlanta Linux Enthusiasts 3. Automatic Link Establishment
ALI 1. ATM Line Interface 2. automatic location identification, automatic location information
ALIT Automatic Line Insulation Testing
ALLC Association for Literary and Linguistic Computing
ALM 1. airline miles 2. AppWare loadable module 3. automated loan machine
ALPS automatic loop protection switching
ALU arithmetic and logic unit
AM 1. access module 2. active messages 3. active monitor 4. amplitude modulation
AM/VSB amplitude modulated vestigial sideband
AMDM ATM multiplexer/demultiplexer
Ameslan American Sign Language
AMHS Automated Message-Handling System
AMI 1. alternate mark inversion 2. analog microwave link
AMIS Audio Messaging Interchange Specification
AML 1. Actual Measured Loss 2. ARC Macro Language
AMLCD active matrix liquid crystal display
AMN Abstract Machine Notation
AMPS Analog/Advanced Mobile Phone Service
AMR anisotropic magneto-resistance
AMS 1. Account Management System 2. American Meteorological Society 3. Attendant Management System
AMSAT Amateur Satellite program
AMSC American Mobile Satellite Corporation
AMSS Aeronautical Mobile Satellite Service
AMTOR amateur teleprinting over radio
AMTS Automated Maritime Telecommunications System
AN access network
anamux analog multiplexer
ANC All Number Calling
ANI Automatic Number Identification
ANT Access Network Termination
AO acousto-optic
AOCS attitude and orbit control system
AOL America OnLine
AOM acousto-optic modulator
AOR Atlantic Ocean Region
AOS 1. Alternate Operator Services 2. Area of Service
AOSSVR Auxiliary Operator Services System Voice Response
AOTF acousto-optic tunable filter
AP 1. action potential 2. aiming point 3. application program 4. Applications Processor 5. array processor 6. Associated Press
APAD asynchronous packet assembler/disassembler
APAN Asia-Pacific Advanced Network Consortium
APaRT Automated Packet Recognition/Translation
APC 1. adaptive-predictive coding 2. advanced process control 3. Association for Progressive Communications
APC adaptive predictive coding
APCC The American Public Communications Council
APD avalanche photodiode
APDU Application Protocol Data Unit
API application programming interface
APIC advanced programmable interrupt controller
APL A Programming Language
app application.
APPA American Public Power Association
APPC advanced program-to-program communications
APPN Advanced Peer-to-Peer Networking
APS 1. Advanced Photo System 2. Automatic Protection Switching
APTS Association of Television Stations
AQL acceptable quality level
Ar argon
AR Automatic Recall
ARA AppleTalk Remote Access
ARB all-routes broadcast
ARC Ames Research Center
ARD 1. advanced research and development 2. Automatic Ring Down
ARE All Routes Explorer
ARES Amateur Radio Emergency Service
ARF Alternative Regulatory Framework
ARI Automatic Room Identification
ARIB Association of Radio Industries and Businesses
ARIES Angle-Resolved Ion and Electron Spectroscopy
ARINC Aeronautical Radio, Inc.
ARISE Advanced Radio Interferometry between Space and Earth
ARISTOTELES Applications and Research Involving Space Technologies/Techniques Observing The Earth's Fields from Low Earth Orbiting Satellites
ARL Association for Research Libraries
AROS Amateur Radio Observation Service
ARP Address Resolution Protocol
ARPA Advanced Research Projects Agency

ARQ automatic retransmit request
ARRL American Relay Radio League
ARRN Amateur Radio Repeater Network
AS autonomous system
AS&C Alarm Surveillance and Control
ASA Acoustical Society of America
ASAI Adjunct Switch Application Interface
ASAPI Advanced Speech API
ASCA Advanced Satellite for Cosmology & Astrophysics
ASCII (as-kee) American Standard Code for Information Interchange
ASDSP application-specific digital signal processor
ASE Application Service Element
ASH Ardire-Stratigakis-Hayduk
ASI 1. Adaptive Speed Leveling 2. Advanced Study Institute 3. artificial sensing instrument 4. Astronomical Society of India
ASIC application-specific integrated circuit
ASK amplitude-shift keying, modulation
ASN 1. Abstract Syntax Notation 2. Autonomous System Number
ASN.1 Abstract Syntax Notation 1
ASP 1. Abstract Service Primitive 2. Adjunct Service Point 3. administrative service provider 4. analog signal processing. 5. AppleTalk Session Protocol 6. ATM switch processor 7. Attached Support Processor 8. Association of Shareware Professionals
ASPI Advanced SCSI Programming Interface
ASQ Automated Status Query
ASR 1. Access Service Request 2. Airport Surveillance Radar 3. Automatic Send/Receive 4. Automatic Speech Recognition
ASSP 1. acoustics speech and signal processing 2. application-specific standard product
ASSTA Australian Speech Science and Technology Association
ASTC Australian Science and Technology Council
ASTER Advanced Spaceborne Thermal Emission Reflectance Radiometer
ASTRAL Alliance for Strategic Token-Ring Advancement and Leadership
ASU application-specific unit
ASV Air-to-Surface-Vessel radar
AT 1. Access Tandem 2. AudioTex
AT, PC/AT Advanced Technology
ATB all trunks busy
ATCP AppleTalk Control Protocol
ATCRBS air traffic control radar beacon system
ATD 1. asynchronous time division 2. Attention Dial 3. advanced technology demonstration
ATDRSS Advanced Tracking and Data Relay Satellite System
ATG address translation gateway
ATIS Alliance for Telecommunications Industry Solutions
ATM 1. asynchronous transfer mode 2. Automated Teller Machine
ATS Applications Technology Satellite program
ATT Automatic Toll Ticketing
ATV 1. advanced TV 2. amateur television
AUI Attachment Unit Interface
AUP Acceptable Use Policy
AUU ATM User-to-User
AUUG Australian Unix User Group
AUXBC auxiliary broadcasting
AVC automatic volume control
AVD alternate voice data
AVHRR advanced very high resolution radiometer
AVIOS American Voice Input/Output Society
AVRS automated voice response system
AVSSCS Audio/Visual Service Specific Convergence Sublayer
AWA Antique Wireless Association
AWC Association for Women in Computing
AWG American Wire Gauge

B

B8ZS binary/bipolar eight-zero substitution
B-CDMA Broadband Code Division Multiple Access
B-DCS broadband digital cross-connect system
B-ICI B-ISDN InterCarrier Interface
B-ICI SAAL B-ICI Signaling ATM Adaptation Layer
B-ISDN Broadband ISDN
B-LT broadband line termination
B-MAC Broadcast Master Antenna Control
B-NT broadband network termination
B-TE broadband terminal equipment
BABT British Approvals Board for Telecommunications
BACP Bandwidth Allocation Control Protocol
BAFTA British Academy of Film and Television Arts
BAN 1. base area network 2. Billing Account Number
BB 1. baseband 2. broadband
BBC 1. Broadband Bearer Capability 2. British Broadcasting Corporation
BBG Basic Business Group
BBN 1. Bolt, Beranek, and Newman, Inc. 2. BBN Planet
BBS bulletin board system
BBT broadband technology
Bc Committed Burst Size
BC 1. backward compatible 2. beam coupling 3. binary code 4. broadcast
BCC 1. Bellcore Client Company 2. block check character
BCD binary coded decimal
BCNU "be seein' you"
BCOB Broadband Connection Oriented Bearer
BCRS Bell Canada Relay Service
BCS 1. basic control system 2. Batch Change Supplement 3. Boston Computer Society 4. British Computer Society
BDCS Broadband Digital Cross-Connect System
BDF block data format
BDT Telecommunications Development Bureau
Be 1. Excess Burst Size 2. Be, Inc.
BE 1. base embossed 2. Bose-Einstein
BEC Bose-Einstein condensation
BECN Backward Explicit Congestion Notification
BEP Back End Processor
BER 1. Basic Encoding Rules 2. bit error rate
BETRS Basic Exchange Telecommunications Radio Service
BFI Bad Frame Indicator
BFO beat frequency oscillator
BFOC bayonet fiber optic connector
BFT binary file transfer
BG, BGND background
BGP Border Gateway Protocol
BHLI Broadband High Layer Information
BIB Backward Indicator Bit
BIBO bounded input, bounded output
B-ICI Broadband Inter-Carrier Interface
B-ICI SAAL Broadband Inter-Carrier Interface Signaling ATM Adaptation Layer
BICEP bit-interleaved parity
BIDDS base information digital distribution system
BIG broadband integrated gateway
BIND Berkeley Internet Name Domain
BIS border immediate system
BISDN B-ISDN
BISSI Broadband Inter-Switching System Interface
BIST built-in self-test
BITS Base Information Transport System
BIU basic information unit
BL bit line
BLSR Bidirectional Line Switched Ring
BLU basic link unit

BM burst modem
BMEWS Ballistic Missile Early Warning System
BNC bayonet nut connection, bayonet navy connector
BNCC base network control center
BO branch office
board printed circuit board
BOC Bell Operating Company
BOF 1. Birds of a Feather 2. Business Operations Framework
BOM 1. Beginning of Message
BOP Bit Oriented Protocol
BP 1. bandpass 2. base pointer 3. bypass
BPAD Bisynchronous Packet Assembler/Disassembler
BPDU Bridge Protocol Data Unit
BPI bytes per inch
BPM Beam Position Monitor
BPON Broadband Passive Optical Network
BPS bits per second
BPSK binary phase shift keying
BR 1. beacon receiver 2. Radiocommunications Bureau
BRB "be right back"
BRCS Business and Residence Customer Services
BRI Basic Rate Interface
BS 1. back scatter 2. band signaling 3. base station 4. beam splitter
BSAM Basic Sequential Access Method
BSCC Bellsouth Cellular Corporation
BSD Berkeley Software Distribution
BSE 1. back-scattered electrons 2. Basic Service Element 3. Basic Switching Element
BSF bit scan forward
BSI British Standards Institution
BSL British Sign Language
BSMS Broadcast Short Message Service
BSMTP Batch Simple Message Transfer Protocol
BSP 1. Bell System Practice 2. Byte Stream Protocol
BSR 1. bit scan rate 2. bit scan reverse
BSS 1. Base Station System 2. broadcasting satellite service 3. Business Support System
BSVC Broadcast Switched Virtual Connections
BT 1. British Telecom 2. Burst Tolerance
BTA 1. Basic Trading Area 2. Broadband Telecommunications Architecture 3. broadband terminal adaptor
BTag Beginning Tag
BTBT band-to-band tunneling
BTE 1. Boltzmann Transport Equation 2. broadband terminal equipment
BTI British Telecom International
BTL Bell Telephone Laboratories
BTM Broadband Transport Manager
BTN Billing Telephone Number, Billed Telephone Number
BTRL British Telecom Research Laboratories
BTS 1. Base Transceiver Station 2. bit test and set
BTU 1. basic transmission unit 2. British Thermal Unit
BTW "by the way"
BW bandwidth
BZT Bundesamt für Zulassungen in der Telekommunikation

C

C/A Code civilian code, S-code
C64 Commodore 64
CA Call Appearance
CAB Canadian Association of Broadcasters
CACM Communications of the Association for Computing Machinery
CAD 1. computer-aided dispatch 2. computer-aided design/drafting
CADS Code Abuse Detection System
CAE 1. Common Applications Environment 2. computer-aided engineering
CAFA computer-aided financial analysis
CAI 1. computer assisted instruction 2. common air interface
CAL computer-aided learning, computer-assisted learning
CALC customer access line charge
CALS Continuous Acquisition and Life-Cycle Support
CAM 1. carrier module 2. Call Accounting Manager 3. Call Applications Manager 4. computer-aided manufacturing 5. computer-assisted makeup, composition and makeup
CAMA Centralized Automatic Message Accounting
CAN Control Area Network
CAP 1. carrierless amplitude and phase modulation 2. Cellular Array Processor 3. Competitive Access Provider
CAR computer-assisted retrieval
CARAB Canadian Amateur Radio Advisory Board
CAS 1. Centralized Attendant Service 2. channel associated signaling. 3. Communications Applications Specification
CASE computer-aided software engineering, computer-assisted software engineering
CAST computer-aided software testing, computer-assisted software testing
CATNIP Common Architecture for Next Generation Internet Protocol
CATS Consortium for Audiographics Teleconferencing Standards
CATV 1. Cable Television 2. Community Antenna Television
CAU controlled access unit.
CAV Constant Angular Velocity
CB citizen's band (radio)
CBC Canadian Broadcasting Corporation
CBDS Connectionless Broadband Data Service
CBM Commodore Business Machines
CBR constant bit rate
CBS Columbia Broadcasting System
CBSC Canadian Broadcast Standards Council
CBT 1. Canadian Business Telecommunications Alliance 2. Computer-Based Training
CBTA Canadian Business Telecommunications Alliance
CBX Computerized Branch Exchange
CC carbon copy
CCB Common Carrier Bureau.
CCC 1. clear channel capability 2. Communications Competition Coalition
CCD charge coupled device
CCIA Computer and Communications Industry Association
CCIR Comité Consultatif International des Radiocommunications
CCIRN Coordinating Committee for Intercontinental Research Networks
CCIS Common Channel Interoffice Signaling
CCITT Comité Consultatif Internationale de Télégraphique et Téléphonique
CCP Compression Control Protocol
CCR current cell rate
CCS Common Channel Signaling
CCS/SS7 Common Channel Signal/Signaling System 7
CCT 1. Calling Card Table unmatched call 2. Consultative Committee Telecommunications
CCTA Central Computer and Telecommunications Agency
CCTV Closed Circuit TV
CCU 1. camera control unit 2. communications control unit
CD 1. carrier detect 2. compact disc 3. count down

CD-V compact disc video
CD-WO compact disc write once
CDA Communications Decency Act
CDCS Continuous Dynamic Channel Selection
CDDI Copper Distributed Data Interface
CDE Common Desktop Environment
CDLC Cellular Data Link Control
CDMA Code Division Multiple Access
CDMP Cellular Digital Messaging Protocol
CDO community dial office
CDP 1. Cisco Discovery Protocol 2. Customized Dial Plan
CDPD Cellular Digital Packet Data
CDR Call Detail Record
CDT credit allocation
CDV 1. cell delay variation 2. Compressed Digital Video
CDVT cell delay variation tolerance
CE Connection endpoint
CEO Chief Executive Officer
CEPT Conférence Européenne des Administrations des Postes et Télécommunications
CER cell error ratio
CERB Centralized Emergency Reporting Bureau
CERFnet California Education and Research Federation Network
CERN Centre European des Recherche Nucleaire
CERT Computer Emergency Response Team
CEST Centre for the Exploitation of Science and Technology
CEV controlled environmental vault
CFB Call Forward Busy
CFDA Call Forward Don't Answer
CFGDA Call Forward Group Don't Answer
CFP Channel Frame Processor
CFR Confirmation to Receive
CFUC Call Forwarding UnConditional
CFV Call for Votes
CFW Call Forward
CGA 1. Carrier Group Alarm 2. Color Graphics Adapter
CGI Common Gateway Interface
CGM Computer Graphics Metafile
CGSA Cellular Geographic Service Area
CI 1. congestion indicator 2. Certified Integrator
CIAJ Communications Industry Association of Japan
CIC Carrier Identification Code
CIDR Classless Inter-Dǒmain Routing
CIE The Commission Internationale de L'Eclairage (Internal Commission of Illumination)
CIF 1. cells in flight 2. Cells in Frames 3. Common Intermediate Format 4. cost, insurance, freight
CIID Card Issuer Identifier Code
CIIG Canadian ISDN Interest Group
CIP Carrier Identification Parameter
CIR Committed Information Rate
CISC Complex Instruction Set Computing
CISCC Collocation Interconnection Service Cross Connection
CITEL Inter-American Telecommunications Commission
CITR Canadian Institute for Telecommunications Research
CIV cell interarrival variation
CIVDL Collaboration for Interactive Visual Distance Learning
CIX Commercial Internet Exchange
CJC Canadian Journal of Communication
CLEC Competitive Local Exchange Carrier
CLP 1. Cell Loss Priority 2. Connectionless Transport Protocol
CLR Cell Loss Ratio
CLTS Connectionless Transport Service
CMC connection management controller
CMDS Centralized Message Distribution System
CMOS Complementary Metal Oxide Semiconductor
CMR cell misinsertion rate
CMRS/PMRS Commercial and Private Mobile Radio Service
CMTS Cellular Mobile Telephone System
CMY cyan, magenta, yellow
CN Complementary network
CNA 1. Centralized Network Administration 2. Cooperative Network Architecture
CND 1. Calling Number Delivery 2. Calling Number Display
CNET Centre National d'Études de Télécommunication
CNIS Calling Number Identification Services
CNR 1. Complex Node Representation 2. customer not ready
CNRI Corporation for National Research Initiatives
Co Symbol for cobalt
CO 1. cash order 2. central office 3. commanding officer
COAM Customer Owned And Maintained Equipment
COB Close of Business
COBOL Common Business-Oriented Language
COBRAS Cosmic Background Radiation Anisotropy Satellite
COCOT customer-owned coin-operated telephone
COD connection-oriented data
COM 1. Component Object Model 2. continuation of message
COMETT Community Action Programme in Education and Training for Technology
COPT Coin Operated Pay Telephone
CoS class of service
COS 1. compatible for open systems 2. Corporation for Open Systems International
COSETI Columbus Optical SETI
COSINE Cooperation for Open Systems Interconnection Networking in Europe
COTS 1. commercial off the shelf 2. Connection Transport Service
CP/M Control Program/Monitor, Control Program for Microcomputers
CPAS Cellular Priority Access Service
CPCS Common Part Convergence Sublayer
CPD Call Processing Data
CPE Customer Premises Equipment, Customer Provided Equipment
CPI common part indicator
CPL commercial private line
CPN Calling Party Number
cps characters per second
CPU central processing unit
CQ come quick
CQD come quick, distress
CR 1. carriage return 2. call reference 3. connection request 4. customer record
CREN Corporation for Research and Educational Networking
CRF 1. cell relay function 2. connection-related function
CRIS Customer Record Information System
CRM cell rate margin
CRMA Cyclic Reservation Multiple Access Protocol
CRS cell relay service
CRT Cathode Ray Tube
CRTC Canadian Radio Television and Telecommunications Commission
CS communications satellite
CSA 1. Callpath Services Architecture 2. Canadian Standards Association
CSC customer service center
CSMA Carrier Sense Multiple Access
CSNET Computer+Science Network
CSPP Computer Systems Policy Project
CSR 1. cell switch router 2. customer service record

CST Computer Supported Telephony
CSTA Computer Supported Telephony Application
CSU Channel Service Unit
CSU/DSU Channel Service Unit/Data Service Unit
CSUA Canadian Satellite Users Association
CT 1. Call Type 2. Cordless Telephone 3. Conformance Test
CT3IP Channelized T3 Interface Processor
CTCA Canadian Telecommunications Consultants Association
CTD 1. cell transfer delay 2. Continuity Tone Detector
CTI 1. Call Technologies, Inc. 2. Computer Telephony Integration 3. Critical Technologies Institute
CTIA 1. Cellular Telecommunications Industry Association 2. Computer Technology Industry Association
CTL Complex Text Layout
CTRL Control (also written *Ctrl* and *ctrl*)
CTS 1. clear to send 2. Communication Transport System 3. Conformance Testing Services
CTSS Compatible Time-Sharing System
CTTC coax to the curb
CTX Centrex

D

DA 1. desk accessory 2. destination address 3. Directory Agent 4. Directory Assistance
DAA Data Access Arrangement
DAB 1. digital audio broadcasting 2. dynamically allocatable bandwidth
DACS Digital Access and Cross-connect System
DAF Destination Address Field
DAL Dedicated Access Line
DAMA Demand Assigned Multiple Access
DAP Directory Access Protocol
DAQ Delivered Audio Quality
DAR digital audio radio
DARPA Defense Advanced Research Projects Agency
DARS Digital Audio Radio Service
DASD direct access storage device
DAT digital audio tape
DBMS Database Management System
DBS direct broadcast satellite
DBT Deutsche Bundespost Telecom
DC 1. In telephone communications, Delayed Call 2. direct current 3. disconnect conform
DCA 1. Defense Communications Agency 2. Document Content Architecture 3. Dynamic Channels Allocation
DCC 1. data communications channel 2. Data Country Code
DCD Data Carrier Detect
DCE data communications equipment
DCM dynamically controllable magnetic
DCP Digital Communications Protocol
DCS 1. digital communications system 2. digital cross-connect system 3. distributed computing system
DCT discrete cosine transform
DCTI desktop computer telephony integration
DDB digital databank
DDCMP Digital Data Communications Message Protocol
DDD Direct Distance Dialing
DDE dynamic data exchange
DDN Defense Data Network
DDP distributed data processing
DDS 1. digital data service 2. digital data storage 3. distributed data system
DE Discard Eligibility
DEW line Distant Early Warning line
DFA doped fiber amplifier
DGPS Differential Global Positioning System
DGPT Department General of Posts and Telecommunications, Viet Nam
DGT Dirección General de Telecommunicaciones
DHCP Dynamic Host Configuration Protocol
DID Direct Inward Dialing
DII defense information infrastructure
DIM document image management
DIMS Document Image Management System
DIN 1. Deutsches Institute för Normung 2. dual inline
DINA Distributed Intelligence Network Architecture
DL 1. distant learning 2. distribution list
DLC 1. Data Link Control 2. Digital Loop Carrier
DLCI Data Link Connection Identifier
DLL 1. data link layer 2. Dynamic Link Library
DLR Design Layout Record
DLS, DLSw Data Link Switching
DLSW DataLink Switching Workgroup
DM delta modulation
DMA 1. direct memory access 2. Document Management Alliance
DMD differential mode delay
DMI 1. Desktop Management Interface 2. Digital Multiplexed Interface
DMS Digital Multiplex System
DMSP Defense Meteorological Satellite Program
DMT discrete multitone
DMTF Desktop Management Task Force
DN Directory Number
DNA 1. Digital Network Architecture 2. distributed network administration
DNC 1. distributed networking computing 2. dynamic network controller
DNCF Directory Number Call Forwarding
DNIC Data Network Identification Code
DNIS Dialed Number Identification Service
DNR Dynamic Network Reconfiguration
DNS Domain Name System
DP 1. data processing 2. demarcation point 3. Dial Pulse
DPA 1. Demand Protocol Architecture 2. digital port adapter
DPBX digital private branch exchange
DPCM differential pulse code modulation
DPE distributed processing environment
DPLB Digital Private Line Billing
DPNSS Digital Private Network Signaling System
DPO Direct Public Offering
DPP Distributed Processing Peripheral
DPX DataPath loop extension
DQDB Distributed Queue Dual Bus
DRAM 1. digital recorder, announce mode 2. dynamic RAM
DS 1. Dansk Standardiseringsrad 2. digital system 3. Distributed Single Layer Test Method
DSA 1. data service adapter 2. Digital Signature Algorithm 3. Directory System Agent
DSAT Digital Supervisory Audio Tones
DSC Digital Selective Calling
DSCS Defense Satellite Communications System
DSE Distributed Single Layer Embedded (Test Method)
DSH double-superheterodyne
DSI digital speed interpolation
DSL Digital Subscriber Line
DSLAM DSL access multiplexer
DSP 1. digital signal processor 2. Display System Protocol
DSR data set ready
DSRC Dedicated Short Range Communications
DSRR digital short range radio
DSS 1. Digital Signature Standard 2. Direct Station Select 3. direct satellite system
DSTO Defence Science and Technology Organisation
DSU Digital Service Unit

DT Deutsche Telekom
DTE Data Terminal Equipment, End Device
DTL 1. Designated Transit List 2. diode transistor logic
DTMF dual tone multifrequency
DTP desktop publishing
DTR Data Terminal Ready
DTRS Digital Trunked Radio System
DTSR Dial Tone Speed Recording
DTT Digital Trunk Testing
DTU Digital Test Unit
DVBG Digital Video Broadcasting Group
DVD digital videodisc
DVST direct view storage tube
DXI data exchange interface

E

EA Equal Access
EACEM European Association of Consumer Electronics Manufacturers
EAGLE Extended Area Global Positioning System (GPS) Location Enhancement
EARN European Academic Research Network
EARP Ethernet Address Resolution Protocol
EAS 1. Emergency Alert System 2. Extended Area Service
EBS Emergency Broadcast System
EBU European Broadcasting Union
EC 1. exchange carrier 2. European Community, European Common Market, European Union (EU)
ECC 1. elliptic curve cryptography 2. Emergency Communications Center
ECHO European Commission Host Organization
ECI emitter coupled logic
ECMA European Computer Manufacturers Association
ECP 1. Encryption Control Protocol 2. Enhanced Call Processing
ECPA Electronic Communications Privacy Act
ECSA Exchange Carriers Standards Association
ECTF 1. Enterprise Computer Telephony Forum 2. European Community Telework Forum
ECTUA European Council of Telecommunications Users Association
ED Electronic Directory
EDA electronic design automation
EDAC error detection and correction
EDACS Enhanced Digital Access Communications System
EDDA European Digital Dealers Association
EDF erbium-doped fiber
EDFA erbium-doped fiber amplifier
EDGAR Electronic Data Gathering Archiving and Retrieval
EDM electronic document management
EDO RAM extended data-out random access memory
EDP electronic data processing
EDSAC Electronic Delay Storage Automatic Computer
EDTV Enhanced-definition TV
EEC European Economic Community
EEI external environment interface
EEMA European Electronic Messaging Association
EEPROM electronically erasable programmable read only memory
EF entrance facility
EF&I engineer, furnish, and install
EFI&T engineer, furnish, install, and test
EFCI explicit forward congestion indicator
EFF Electronic Frontier Foundation
EFS error free seconds
EFT Electronic Funds Transfer
EFTA European Free Trade Association
EFTPOS Electronic Funds Transfer Point of Sale
EGA Enhanced Graphics Adapter
EGNOS European Geostationary Navigation Overlay Service
EGP Exterior Gateway Protocol
EIA Electronic Industries Alliance
EIG Electronic Information Group
EIR Equipment Identity Register
EIRPAC Eire packet network
EIS Expanded Interconnection Service
EISA Extended Industry Standard Architecture
EIU Ethernet Interface Unit
EKE electronic key exchange
EKTS Electronic Key Telephone Service
ELIU electrical line interface unit
ELOT Hellenic Organization for Standardization
ELSU Ethernet LAN Service Unit
EMA Electronic Messaging Association
EMP electromagnetic pulse
EMR Exchange Message Record
EMS Expanded Memory Specification
EMT electrical metal tubing
EMU European Monetary Unit
ENN Emergency News Network
ENOS Enterprise Network Operating System
ENS Emergency Number Services
EO 1. end office. 2. erasable optical.
EOB end of block
EOF end of file
EOM end of message
EOS Earth Observing System
EOT 1. end of transmission 2. end of tape
EOTC European Organization for Testing and Certification
EPD Early Packet Discard
EPF Electronic Payments Forum
EPP Enhanced Parallel Port
EPROM erasable programmable read-only memory
EPS Encapsulated PostScript
EPSCS Enhanced Private Switched Communications Service
ERP effective radiated power
ERS European Remote Sensing Satellite
ERTS-1 Earth Resources Technology Satellite
ESA 1. emergency stand-alone 2. European Space Agency
ESC, Esc escape
ESCA European Speech Communication Association
ESD electrostatic discharge
ESF Extended SuperFrame
ESI 1. Enhanced Serial Interface 2. End System Identifier
ESMR Enhanced Specialized Mobile Radio
ESMTP Extended Simple Mail Transport Protocol
ESN 1. electronic serial number 2. electronic switched network 3. emergency services number
ESnet Energy Sciences Network
ESP 1. Encapsulating Security Payload 2. Enhanced Serial Port
ESPA European Selective Paging Association
ESPAN Enhanced Switch Port Analyzer
ESS electronic switching system
ESTO European Science and Technology Observatory
ETACS Extended TACS
ETF European Teleconferencing Federation
ETFTP Enhanced Trivial File Transfer Protocol
ETNO European Public Telecommunications Network Operations Association
ETS European Telecommunication Standard
ETSI European Telecommunications Standards Institute
EUROTELDEV European (Regional) Telecommunication Development
EUV extreme ultraviolet
eV electron volt
EW Electronic Warfare
EWOS European Workshop for Open Systems
EWP electronic white pages
ExCa Exchangeable Card Architecture

EXTN extension
EYP electronic yellow pages

F

FAM fast access memory
fanfold z-fold
FANP Flow Attribute Notification Protocol
FAQ Frequently Asked Question
FB Framing bit
FBT Fused Biconic Tape
FBus Frame Transport Bus
FC 1. feedback control 2. frame control
FCA Fibre Channel Association
FCC Federal Communications Commission
FCLC Fibre Channel Loop Community
FCS 1. Federation of Communications Services 2. Frame Check Sequence 3. Fraud Control System
FCSI Fibre Channel Systems Initiative
FDD floppy disk drive
FDDI Fiber Distributed Data Interface
FDM frequency division multiplexing
FDMA frequency division multiple access
FEC forward error correction
FECN Forward Explicit Congestion Notification
FEP Front End Processor
FER Frame Error Rate
FEXT far end crosstalk
FF form feed
FGDC Federal Geographic Data Committee
FID Field Identifier
FIF Fractal Image Format
FIFO first in, first out
FM 1. fault management 2. frequency modulation
FMAS Facility Maintenance and Administration System
FMV Fair Market Value
FNC Federal Networking Council
FNR fixed network reconfiguration
FNS Fiber Network Systems
FOA 1. fiber optic amplifier 2. Fiber Optic Association, Inc. 3. First Office Application
FOC Firm Order Confirmation
FOD fax-on-demand
foil overhead transparency
FOT Fiber Optic Terminal
FOTS fiber optic transmission system
FPLMTS Future Public Land Mobile Telecommunication System
fps frames per second
FPU floating point unit
FRA fixed radio access
FRAD frame relay access device
FRM focus-rotation mount
FRND frame relay network device
FRSE frame relay switching equipment
FRTE frame relay terminal equipment
FSAN Full Services Access Network
FSK frequency shift keying
FSO Foreign Service Office
FSP File Service Protocol
FSS fixed satellite service
FSTC Financial Services Technology Consortium
FTA Federal Telecommunications Act
FTIP Fiber Transport Inside Plant
FTNS Fixed Telecommunications Network Service
FTP File Transfer Protocol
FTS 1. file transfer support 2. Federal Telecommunications System
FTTC fiber to the curb
FTTH fiber to the home
FTTL fiber in the loop
fV femtovolt
FVR flexible vocabulary recognition
FWA Fixed Wireless Access
FWIW "for what it's worth"
FYI "for your information"
FX Foreign Exchange

G

G/A ground to air communication
G/G ground to ground communication
GA go ahead
GAB Group Access Bridging
GIGO garbage in, garbage out
GII global information infrastructure
GIIC Global Information Infrastructure Commission
GILC Global Internet Liberty Campaign
GIP Global Internet Project
GIS geographic information system
GITS Government Information Technology Services
GL graphics library

H

HALE High Altitude Long Endurance
HAN home area network
HBA host bus adapter
HBS Home Base Station
HCI 1. Host Command Interface 2. human computer interface 3. Human Computer Interface standards
HD half duplex
HDB3 High Density Bipolar Three
HDD Hard Disk Drive
HDLC High Level Data Link Control
HDSL high bit-rate digital subscriber line
HDT Host Digital Terminal
HDTV High Definition Television
HEC Header Error Control
HEP high energy physics
HF, hf 1. hands free 2. high fidelity 3. high frequency
HFC Hybrid Fiber Coax
HFU hands free unit
Hi-OVIS Highly Interactive Optical Visual Information System
HNF High-performance Network Forum
HNS Hughes Network Systems
HOBIS Hotel Billing Information System
HP Hewlett-Packard
HPA high power amplifier
HRPT high resolution picture transmission
HSCI High-Speed Communications Interface
HSCS high speed circuit switched
HSD home satellite dish
HSDA high speed data access
HSDU High Speed Data Unit
HSRP Hot Standby Router Protocol
HST High Speed Technology
HSV hue, saturation, value
HTL high threshold logic
HTTP Hypertext Transfer Protocol
HTTPS Hypertext Transfer Protocol Secure
HUT Hopkins Ultraviolet Telescope
HW hardware
Hz hertz

I

IBN Institut Belge de Normalisation
IBS 1. intelligent battery system 2. Intelsat Business Service
IC 1. integrated circuit 2. intercom 3. interexchange carrier 4. intermediate cross-connect
ICAL Internet Community at Large
ICAPI International Call Control API
ICB Individual Case Basis

ICCB Internet Configuration Control Board
ICCF Industry Carriers Compatibility Forum
ICEA Insulated Cable Engineers Association
ICI Interexchange Carrier Interface
ICM Integrated Call Management
ICMP Internet Control Message Protocol
ICO Global Communications
ICONTEC Instituto Colombiano de Normas Técnicas
ICTA International Computer-Telephony Association
ID 1. identification, identifier 2. input device 3. intermediate device
IDA 1. integrated data access, integrated digital access 2. intelligent drive array
IDLC Integrated Digital Loop Carrier
IDCMA Independent Data Communications Manufacturers Association
IDE integrated development environment
IDEA International Data Encryption Algorithm
IDEN Integrated Digital Electronic Network
IDF intermediate distribution frame
IDL Interface Design Language
IDLC Integrated Digital Loop Carrier
IDN Integrated Digital Network
IDSCP Initial Defense Communications Satellite Program
IDTV Improved Definition Television
IDU Interface Data Unit
IEC 1. Inter Exchange Carrier 2. International Electrotechnical Commission 3. International Engineering Consortium
IEN Internet Experimental Note
IETF Internet Engineering Task Force
IF intermediate frequency
IFCM independent flow control message
IFIP International Federation for Information Processing
IFRB International Frequency Registration Board
IGC intelligent graphics controller
IGMP Internet Group Multicast Protocol
IGP Interior Gateway Protocol
IGRP Interior Gateway Routing Protocol
IGT Ispettorato Generale delle Telcomunicazioni
IGY International Geophysical Year
IIR Interactive Information Response
IIOP Internet Inter-ORB Protocol
IISP 1. Information Infrastructure Standards Panel 2. Interim Interswitch Signaling Protocol
IITC Information Infrastructure Task Force
IJCAI International Joint Conferences on Artificial Intelligence
ILEC Incumbent Local Exchange Carrier
ILMI Interim Link Management Interface
IM intermodulation distortion
IMA Interactive Multimedia Association
IMAC Isochronous Media Access Control
IMAP Internet Messaging Access Protocol
IMASS Intelligent Multiple Access Spectrum Sharing
IMAX "I" - eye + maximum
IMHO "in my humble opinion"
IMNSHO "in my not so humble opinion"
IMO "in my opinion"
IMPATT impact avalanche and transit time
IMPDU Initial MAC Protocol Data Unit
IMSI International Mobile Subscriber Identity
IMTC The International Multimedia Teleconferencing Consortium
IMTS Improved Mobile Telephone Service
IMUX Inverse Multiplexer
IMW Intelligent Music Workstation
INA Information Networking Architecture
INC international carrier
INCC Internal Network Control Center
IOF Inter Office Facility
IOL InterOperability Lab
ION Internetworking Over NBMA
IP Internet Protocol
IPATM Internetworking over NBMA
IPCE interprocess communication environment
IPL Initial Program Load
IPng IP Next Generation
IPO Initial Public Offering
IPS Internet Protocol Suite
IPSec IP Security protocol
IR infrared
IRAC Interdepartmental Radio Advisory Council
IRC 1. integrated receiver decoder 2. Internet Relay Chat
IrDA Infrared Data Association
IRE Institute of Radio Engineers
IREQ interrupt request
IRR Internet Routing Registry
IRSG Internet Research Steering Group
IRTF Internet Research Task Force
ISA 1. industry standard architecture 2. Interactive Services Association
ISD Incremental Service Delivery
ISDN Integrated Services Digital Network
ISI Information Sciences Institute
ISL Inter-Switch Link
ISM 1. Industrial Scientific Medical 2. interstellar medium
ISNI Intermediate Signaling Network Identification
ISO International Organization for Standardization
ISOC Internet Society
ISP 1. Internet Services Provider 2. Information Services Platform 3. ISDN Signal Processor
ISPBX Integrated Services Private Branch Exchange
ITM Information Technology Management
ITR International Telecommunication Regulations
ITS 1. Institute for Telecommunication Sciences 2. Intelligent Transportation Systems
IVDS Interactive Video Data Services
IVHS Intelligent Vehicle Highway Systems
IVI Intel Video Interactive
IVR interactive voice response
IVS 1. interactive voice service 2. interactive video service
IW interworking
IWS intelligent workstation
IXC interexchange carrier

J

JAD joint application design
JAMSAT Japanese affiliate of AMSAT
JAN Joint Army-Navy
JANET Joint Academic Network
JAR Java Archive
JARL Japan Amateur Radio League, Inc.
JATE Japan Approvals Institute for Telecommunications Equipment
JCL Job Control Language
JDBC Java database connectivity
JEDEC Joint Electron Device Engineering Council
JEDI Joint Electronic Document Interchange
JEIDA Japan Electronic Industry Development Association
JEMA Japan Electronic Messaging Association
JIPS JANET Internet Protocol (IP) Service
JPS joint product specification
JRG GII Joint Rapporteur Group global information infrastructure
JTAG Joint Test Action Group
JTAPI Java
JTC Joint Technical Committee
JUNET Japan Unix Network
JVNCnet John Von Neumann Center network
JWICS Joint Worldwide Intelligence Communications System

K

kbps kilobits per second. One thousand bits per second. This is sometimes also written kbits/s.
KBS knowledge base system
KDD 1. Knowledge Discovery in Databases 2. Kokusai Denshin Denwa Company, Ltd.
keV kiloelectronvolt
KIS Knowbot Information Service
KISS Keep It Simple Stupid
KMID key material identifier
KMS Knowledge Management System
KNET Kangaroo Network
KQML Knowledge Query and Manipulation Language
KS Kearney System
KSR Keyboard Send/Receive
KTH Kungliga Tekniska Hogskolan
KTI Key Telephone Interface
KTS key telephone system
KWH kilowatt-hour

L

LAC Loop Assignment Center
ladar laser Doppler radar
LADS local area data service
LADT Local Area Data Transport
LAM line adapter module
LAMA Local Automatic Message Accounting
LAN local area network
LANCE Local Area Network Controller for Ethernet
LANDA Local Area Network Dealers Association
LB leaky bucket
LBA Logical Block Address
LCD liquid crystal display
LCU Lightweight Computer Unit
LDAP Lightweight Directory Access Protocol
LDIP Long Distance Internet Provider
LDMS Local Multipoint Distribution Service
LE light-emitting
LEAF Law Enforcement Access Field
LEC 1. Local Exchange Carrier 2. LAN Emulation Client
LECS LAN Emulation Configuration Server
LED light-emitting diode
LIT line insulation test
LLC Logical Link Control
LM long distance marketer
LMOS Loop Maintenance Operations System
LMS 1. Local Measured Service 2. Local Message Switch 3. Location and Monitoring Service
LMSS Land Mobile Satellite Service
LMU Line Monitor Unit
LNA low noise amplifier
LOC Loss of Cell
LPC linear predictive coding
LRC longitudinal redundancy check
LRN Location Routing Number
LRS line repeater station
LSDU Link layer Service Data Unit
LSI large scale integration
LSN Large Scale Networking group
LSSC lower sideband suppressed carrier
LSSGR LATA Switching System General Requirements
LSU Logical Storage Unit
LTB Last Trunk Busy
LTC Line Trunk Controller
LTE Line Terminating Equipment
LTS Loop Testing System
LUN logical unit number
LZS Lempel-Ziv-Stac
LZW Lempel-Ziv-Welsh

M

MAC 1. Media Access Contro 2. multiplexed analog component
MACE Macintosh Audio Compression and Expansion
MARISAT Maritime Satellite
MARS Multicast Address Resolution Service. In ATM networking, a protocol used in IP multicasting
MAS Multiple Address Service
MCF Multimedia Communications Forum
MCMS Multimedia Cable Network System
MD-IS Mobile Data Intermediate System
MDF main distribution frame
MDT mobile data terminals
MFJ Modified Final Judgment
MFS Metropolitan Fiber Systems
MFSK multiple frequency shift keying
MHEC Midwestern Higher Education Commission
MHS Message Handling System
MIB Management Information Base
MICR magnetic ink character recognition
MILNET Military Net
MIME Multipurpose Internet Mail Extension
MIN Mobile Identification Number
MIND Magnetic Integrator Neuron Duplicator
MIP Multichannel Interface Processor
MIPG Multiple-Image Portable Graphics
MIPS million instructions per second
MIS Management Information Services
MITS Micro Instrumentation and Telemetry Systems
MJ modular jack
MLA mail list agent
MLS microwave landing system
MMCF Multimedia Communications Forum
MMCX Multimedia Communication Exchange
MMDS microwave multi-point distribution system
MME Mobility Management Entity
MMF multimode optical fiber
MMI machine-to-machine interface
MMIC Monolithic Microwave Integrated Circuit
MMSP modular multi-satellite preprocessor
MMSS Maritime Mobile-Satellite Service
MMTA Multimedia Telecommunications Association
MMU 1. Manned Maneuvering Unit 2. memory management unit
MMX Multimedia Extension. Matrix Math Extension
MNLP Mobile Network Location Protocol
MNP Microcom Networking Protocol
MNRP Mobile Network Registration Protocol
MPLS Multiprotocol Label Switching
MPOA MultiProtocol Over ATM
MPOE Minimum Point of Entry
MPP Multichannel Point-to-Point Protocol
MPT Ministry of Posts and Telecommunications
MRI magnetic resonance imaging
MRU maximum receive unit
MS Mobile Station
MSAT Mobile Satellite
MSB most significant bit, most significant byte
MSC Mobile Switching Center
MSK minimal shift keying
MSNF multisystem networking facility
MSP Message Security Protocol
MSS multispectral scanner
MSU 1. microwave sounding unit 2. mobile subscriber unit
MTA 1. Macintosh Telephony Architecture 2. Major Trading Area 3. Message Transfer Agent 4. Metropolitan Trading Area
MTBF mean time between failures
MTI Moving Target Indication
MTM Maintenance Trunk Monitor

MTS 1. member of technical staff 2. Message Telecommunications Service 3. Message Transfer System
MTSO Mobile Telephone Switching Office
MTTR mean time to repair
MUX multiplexer
MVIP Multi-Vendor Integration Protocol
MVL Multiple Virtual Line
MVS Multiple Virtual Storage
MXR 1. mixer. 2. multiplexer
MZI Mach-Zehnder Interferometer

N

NAB 1. National Alliance of Business 2. National Association of Broadcasters
NACN North American Cellular Network
NADF 1. North American Directory Plan 2. North Atlantic Directory Forum
NAK negative acknowledge
NAMPS Narrowband Analog Mobile Phone Service
NANC North American Numbering Council
NANOG North American Network Operators Group
NANP North American Numbering Plan
NAP Network Access Point
NAPP National Aerial Photography Program
NARTE National Association of Radio and Telecommunications Engineers
NASA National Aeronautics and Space Administration
NASTD National Association of State Telecommunications Directors
NATA North American Telecommunications Association
NATOA National Association of Telecommunications Officers and Advisors
NBC National Broadcast Corporation
NBFM narrowband frequency modulation
NBMA non-broadcast multiple access
NCAR National Center for Atmospheric Research
NCC 1. National Communications Committee 2. National Coordinating Center
NCCS Network Control Center System
NCHPC National Consortium for High-Performance Computing
NCIA native client interface architecture
NCO National Coordination Office for Computing, Information, and Communications
NCOP Network Code Of Practice
NCS National Communications System
NCSA National Center for Supercomputing Applications
NCUG National Centrex Users Group
NDCDB National Digital Cartographic Data Base
NDIS Network Driver Interface Specification
NDSI National Spatial Data Infrastructure
NDT No Dial Tone
NE network element
NECA National Exchange Carrier Association
NGDC National Geophysical Data Center
NGDLC Next Generation Digital Loop Carrier
NGI Next Generation Internet
NGSO non-geostationary orbit
NGSO FSS non-geostationary orbit fixed satellite service
NHK Nippon Hoso Kyokai
NHRP Next Hop Resolution Protocol
NIMA National Imagery Mapping Agency
NiMH nickel metal hydride
NIOD Network Inward/Outward Dialing
NIOSH National Institute of Occupational Safety and Health
NIS 1. Network Imaging Server 2. Network Information Service
NISDN 1 National ISDN-1
NISP National Internet Services Provider
NIST National Institute of Standards and Technology
NITF National Image Transfer Format
NIU network interface unit
NIUF North American ISDN Users Forum
NLA Network Layer Address
NLANR National Laboratory for Applied Network Research
NLPID Network Layer Protocol ID
NLQ near letter quality
NMAA National Multimedia Association of American
NMACS Network Monitor and Control System
NMP 1. Network Management Processor 2. Network Management Protocol
NMR nuclear magnetic resonance
NNI 1. network node interface 2. Nederlands Normalisatie-Instituut
NNTP Network News Transfer Protocol
NNX NXX
NO OP, NOP, nop no operation
NOV News Overview.
NOWT Netherlands Observatory for Science and Technology
NPA 1. National Pricing Agreement AT&T agreement 2. Numbering Plan Area
NPR National Public Radio
NRC 1. National Research Council 2. Network Reliability Council 3. non-recurring charge
NREN National Research and Education Network
NRSC National Radio Systems Committee
NRZ non-return to zero
NSAI National Standards Authority of Ireland
NSAP network service access point
NSF National Science Foundation
NSFNET National Science Foundation Network
NSP 1. National Internet Services Provider 2. Native Signal Processing
NSS National Standards System
NSSN National Standards Systems Network
NSTAC National Security Telecommunications Advisory Committee
NT Northern Telecom, Inc.
NTIA National Telecommunications and Information Administration
NTN Network Terminal Number
NTSC National Television System Committee
nV abbrev. nanovolt
NV Network Video
NVM nonvolatile memory

O

OA office automation
OAI Open Application Interface
OAM operations, administration, and maintenance
OAM Operations And Maintenance
OAM&P operation, administration, maintenance and provisioning.
OAO orbiting astronomical observatory
OAS Organization of American States
OBRA Omnibus Budget Reconciliation Act
OC 1. operator centralization 2. optical carrier
OCC Other Common Carrier
OCP operator control panel
OCR 1. optical character recognition 2. Outgoing Call Restriction
ODBC Open Database Connectivity
ODMG Object Database Management Group
ODP open distributed processing
OHR optical handwriting recognition
OLE Object Linking and Embedding
OLIU Optical Line Interface Unit
OLNS Originating Line Number Screening
OM Operational Measurement
OMA Object Management Architecture
OMAT Operational Measurement and Analysis Tool
OMG Object Management Group
ONA Open Network Architecture

ONAC Operations Network Administration Center.
ONAL Off Network Access Line
OOP object-oriented programming
OPAC Outside Plant Access Cabinet
OQPSK offset quadrature phase shift keying
ORB Object Request Broker
ORBCOMM Orbital Communications
OSF Open Software Foundation (now the Open Group)
OSI Open Systems Interconnection
OSI TP Open Systems Interconnection Transport Protocol
OST Office of Science and Technology
OSTA Optical Storage Technology Association
OSTP Office of Science and Technology Policy
OT Office of Telecommunications
OTA Office of Technology Assessment (U.S.)
OTAR over the air rekey
OTGR Operations Technology Generic Requirements
OTH over the horizon
OTOH "on the other hand"
OTP Office of Telecommunications Policy
OUI Organizationally Unique Identifier

P

pA Abbreviation for picoampere
PA public address system
PABX Private Automatic Branch Exchange
PACA Priority Access and Channel Assignment
PAP 1. packet-level procedure 2. Public Access Profile
PAR 1. Positive Acknowledgement Retransmit 2. Precision Approach Radar
PCA 1. point of closest approach 2. Premises Cabling Association 3. protective connecting arrangement
PCB 1. power control box. 2. process control block. 3. printed circuit board 4. protocol control block
PCCA Portable Computer and Communications Association
PCI 1. Peripheral Connect Interface 2. Protocol Control Information
PCIA Personal Communications Industry Association
PCL printer control language
PCM pulse code modulation
PCMCIA Personal Computer Memory Card Interface Association
PCMIA Personal Computer Manufacturer Interface Adaptor
PCN Personal Communication Network
PCP Private Carrier Paging
PCR 1. peak cell rate 2. Program Clock Reference
PCS 1. Personal Communications Service 2. personal communications software
PDA Personal Digital Assistant
PDF Portal Document Format
PDL page description language
PDP 1. plasma display panel 2. power distribution panel
PDU Protocol Data Unit
PDUS Primary Data User Station
PLAR Private Line Automatic Ringdown
PLC Power Line Carrier
PLCP Physical Layer Convergence Procedure
PL/M Programming Language/Microprocessor
PLMR Private Land Mobile Radio
PLS physical layer signaling
PLU Percent Local Usage
PM performance monitoring
PMA physical medium attachment. A device which connects physically with a network
PMARS Police Mutual Aid Radio System
PMR 1. poor man's routing 2. private mobile radio
PMS 1. Pantone Matching System 2. Picturephone Meeting Service
PMT 1. Photo Multiplier Tube 2. Photo Mechanical Transfer
PNM public network management
PnP Plug and Play
PNP Permanent Number Portability
PODP Public Office Dialing Plan
PPDN Public Packet Data Network
PPI 1. pixels per inch 2. plan position indicator
PPP Point-to-Point Protocol
PPS 1. packets per second 2. Path Protection Switched 3. Precise Positioning Service 4. pulses per second
PPSN Public Packet Switched Network
PPTP Point-to-Point Tunneling Protocol
PRAM programmable random access memory
PRB Private Radio Bureau
PSI 1. packet switching interface 2. Policy Studies Institute. U.K. government consortium
PSK phase shift keying
PSTN Public Switched Telephone Network
PSU Packet Switch Unit
PSWAC Public Safety Wireless Advisory Committee
PSWN Public Safety Wireless Network
PTE SONET path terminating element
PTFE polytetrafluoroethylene
PTI Payload Type Identifier
PTN Public Telecommunications Network
PTO public telecommunication operators
PTS 1. Personal Telecommunications System 2. Presentation Time Stamp 3. Public Telecommunications System
PTT Post Telephone & Telegraph administration
PVC 1. permanent virtual connection 2. polyvinyl chloride
PVCC Permanent Virtual Channel Connection
PVN private virtual network
pW picowatt
PWB printed wire board
PWM pulse width modulator
PWR power

Q

QDU quantizing distortion units
QFC Quantum Flow Control
QFM quadrature frequency modulation
QIC Quarter Inch Cartridge Drive Standards
QL query language
QLLC Qualified Logical Link Control
QMS Queue Management System
QoR Query on Release
QoS quality of service
QPSK 1. quadrature phase shift keying 2. quaternary phase shift keying
QSAM quadrature sideband amplitude modulation
QSDG Quality of Service Development Group
QTAM Queued Telecommunications Access Method
QTC Quick Time Conference

R

R/T realtime
R&D research and development
R & E Research and Education
RA real audio
RA, RA number return authorization
RAC Radio Amateurs of Canada
RACE 1. random access computer equipment 2. Research into Advanced Communications in Europe
RACES Radio Amateur Civil Emergency Service
RACF resource access control facility
RARP Reverse Address Resolution Protocol
RAS remote access server
RASC Radio Amateur Satellite Corporation
raser radio amplification by stimulated emission of radiation

RB reverse battery
RBOC Regional Bell Operating Company
RBS robbed-bit signaling
RCA Radio Corporation of America
RCC Radio Common Carrier
RD routing domain
RDF 1. radio direction finding 2. rate decrease factor
RDT 1. recall dial tone 2. remote digital terminal
RDY ready
REN Ringer Equivalent Number
RES Residential Enhanced Service
RF radio frequency
RFC Request for Comments
RFD Request For Discussion
RFE Radio Free Europe
RFI radio frequency interference
RIF rate increase factor
RIFF Resource Interchange File Format
RJE remote job entry
RLCM Remote Line Concentrating Module
RLE run length encoding
RLP Radio Link Protocol
RM Resource Management
RMA, RA returned merchandise authorization, return authorization
rms root mean square
RNC radio network controller
RNX Restricted Numeric Exchange
ROLC Routing Over Large Clouds
ROM read only memory
ROSE Remote Operations Service Element
ROTFL "rolling on the floor, laughing"
RPC remote procedure call
RPG Report Program Generator
RPM Remote Packet Module
RPOA Recognized Private Operating Agency
RPS repetitive pattern suppression
RPW remotely piloted vehicle
RQS Rate Quote System
RS recommended standard
RSA Rural Service Area
RSC Remote Switching Center
RSGB Radio Society of Great Britain
RSUP Reliable SAP Update Protocol
RSVP Internet Reservation Protocol
RT 1. realtime, real time 2. remote terminal 3. reorder tone
RTC runtime control
RTCP Real Time Conferencing Protocol
RTF Rich Text Format
RTFM "read the freaking manual"
RTP 1. realtime protocol 2. routing table protocol
RTS Request to Send
RTSP Real Time Streaming Protocol
RTV realtime video
RU 1. In packet networking request unit, response unit, request/response unit

S

S-HTTP Secure HTTP
S/W software
SA 1. Service Agent 2. source address
SAA 1. Standards Association of Australia 2. Supplemental Alert Adapter 3. Systems Application Architecture
SAFE Security and Freedom through Encryption
SAFE Act Security and Freedom through Encryption Act
SAGE Semi-Automatic Ground Environment
SAIL Stanford Artificial Intelligence Laboratory
SAM security accounts manager
SAN satellite access node.
SANZ Standards Association of New Zealand
SAP Service Access Point
SAPI Service Access Point Identifier
SAR 1. search and replace 2. segmentation and reassembly
SART search and rescue radar transponder
SAS 1. single address space 2. simple attachment scheme 3. Survivable Adaptive Systems
SASL Simple Security and Authentication Layer
SASMO Syrian Arab Organization for Standardization and Metrology
SASO Saudi Arabian Standards Organization
SATCOM Satellite Communications
SCC Specialized Common Carrier
SCCP Signaling Connection Control Part
SCE Service Creation Environment
SCI Scalable Coherent Interface
SCP 1. Satellite Communications Processor 2. Service Control Point
SCR 1. silicon-controlled rectifier 2. sustainable cell rate 3. System Clock Reference
SCSI Small Computer System Interface
SCT Secretaria de Comunicaciones y Transportes
SCTE Society of Cable Telecommunications Engineers, Inc.
SDL Specification and Design Language
SDLC Synchronous Data Link Control
SDN Software Defined Network
SDNS Secure Data Network System
SEAC Standards Eastern Automatic Computer
SEAL Simple and Efficient Adaptation Layer
SETI Search for ExtraTerrestrial Intelligence
SIMM single inline memory module
SIPP Simple Internet Protocol Plus
SIS Standardiseringen i Sverige
SIT Simple Internet Transition
SITA Société Internationale de Télécommunications Aéronautiques
SLAR side-looking airborne radar
SLC simple line code
SLIP Serial Line Interface Protocol
SLM System Load Module
SLP Service Location Protocol
SLR 1. send loudness rating 2. single lens reflex
SMAS switched maintenance access system
SMATV Satellite Master Antenna Television
SMDS Switched Multi-Megabit Digital Service
SME 1. Security Management Entity 2. Small- and medium-sized enterprises
SMI Structure of Management Information
S/MIME Secure Multipurpose Internet Mail Extension
SMPTE Society of Motion Picture and Television Engineers
SMR Specialized Mobile Radio
SMRP Simple Multicast Routing Protocol
SMS 1. Service Management System 2. Short Message Service
SMTP Simple Mail Transfer Protocol
SNA Systems Network Architecture
SNACP SNA Control Protocol
SPCAS SPC Allocation Service
SPEC Standard Performance Evaluation Corporation
SPHIGS Simple PHIGS
SPI 1. Security Parameters Index 2. Service Provider Interface
SPID Service Profile Identifier
SPIE The International Society for Optical Engineering
SRA system reliability architecture
SRAPI Speech recognition API
SRGP Simple Raster Graphics Package
SS7 Signaling System 7
SSB single sideband
SSCS Service Specific Convergence Sublayer
SSEC Selective Sequence Control Computer
SSL Secure Socket Layer

SSO single sign-on
SSP Service Switching Point
SSRP Simple Server Redundancy Protocol
SSTO Single-Stage to Orbit
SSTV slow scan television
ST 1. Scheduled Transfer 2. signaling terminal 3. straight-tipped 4. systems test
STA 1. spanning tree algorithm 2. Science and Technology Agency
STU Secure Telephone Unit
STU-3 Secure Telephone Unit 3
STUN serial tunnel
SUMAC SuperHIPPI Media Access Controller
SUT 1. System Under Test 2. Service User Table
SVC switched virtual connection
SVD simultaneous voice and data
SVGA Super Video Graphics Array
SWAC Standards Western Automatic Computer
SWR standing wave ratio
SXGA super-extended graphics array

T

TAI International Atomic Time
TAS Telecommunication Authority of Singapore
TAXI Transparent Asynchronous Transmitter/Receiver Interface
Tc Committed Rate Measurement Interval
TCAP Transaction Capability Application Part
TCI Telecommunications, Inc.
TCIF Telecommunications Industry Forum
TCM trellis code modulation
TCO Tandem Connection Overhead
TCP/IP Transmission Control Protocol/Internet Protocol
TDM time division multiplexing
TDMA time division multiple access
TDRSS Tracking and Data Relay Satellite System
TENET Texas Educational Network
TFT thin film transistor
TFTP Trivial File Transfer Protocol
THF tremendously high frequency
TIES Telecom/Information Equipment and Services
TIFF Tag Image File Format
TIIAP Telecommunications and Information Infrastructure Assistance
TJF test jack frame
TL tie line
TLF trunk link frame
TLP Transmission Level Point
TM 1. terminal multiplexer 2. traffic management
TMA Telecommunication Managers Association
TMC traffic management center
TMGB Telecommunications Main Grounding Busbar
TMN Telecommunications Management Network
TMSI Temporary Mobile Station Identifier
TOGAF The Open Group Architectural Framework
TPA Telephone Pioneers of America
TPDU Transport Protocol Data Unit
TPI tracks per inch
TPWG Technology Policy Working Group
TQM total quality management
TR transmit/receive
TSACC Telecommunications Standards Advisory Council of Canada
TSAG Telecommunication Standardization Advisory Group
TSAP Transport Service Access Point
TSO time share operation
TSRM Telecommunication Standards Reference Manual
TTA Telecommunications Technology Association
TTAB transparent tone above band
TTC Telecommunications Technology Committee
TTIB transparent tone in band
TTL transistor-transistor logic
TTS text-to-speech
TU traveling user
TUANZ Telecommunications Users Association of New Zealand
TUBA TCP and UDP with Bigger Address
TUG Telecommunication User Group
TUI telephone user interface
TUR Traffic Usage Recorder
TVM time-varying media
TVRO television receive-only
TWS two-way simultaneous
TWT traveling-wave tube
TWX Teletype-Writer Exchange
Tx, TX transmit

U

UA User Agent
UART universal asynchronous receiver-transmitter
UAV unstaffed aerial vehicle
UAWG Universal ADSL Working Group
UCC Uniform Commercial Code
UCF UNIX Computing Forum
UDI Unrestricted Digital Information
UDP User Datagram Protocol
UECT Universal Encoding Conversion Technology
UHF ultra high frequency
URI Uniform Resource Identifier
URL Uniform Resource Locators
URM user request manager
URN Uniform Resource Name
USB Universal Serial Bus
USDC U.S. Digital Cellular
USDLA United States Distance Learning Association
USGS United States Geological Survey
USITA Formerly United States Independent Telephone Association
USKA Union Schweizerischer Kurzwellen-Amateure Union (Union of Swiss Shortwave Amateurs)
USOC Universal Service Order Code
USOP User Service Order Profile
USTA United States Telephone Association
UTC Coordinated Universal Time
UTDR Universal Trunk Data Record
UV ultraviolet
UWB ultra wideband
UWC Universal Wireless Communications
UWCC Universal Wireless Communications Consortium

V

VOR VHF omnidirectional range
VP virtual path
VPC virtual path connection
VPN virtual private network
VQ vector quantization
VRAM Video RAM
VRML Virtual Reality Modelling Language
VSAT Very Small Aperture Terminal
VSX Verification Suite for X/open
VTP Virtual Trunking Protocol
VTS Vehicular Technology Society

W

W-DCS wideband digital cross-connect system
W3 World Wide Web
W3C World Wide Web Consortium
WAAS Wide Area Augmentation System
WABI Windows Application Binary Interface
WACK wait acknowledgment
WAN Wide Area Network
WARC World Administrative Radio Conference

WATTC World Administrative Telegraph and Telephone Conference
WBC wideband channel
WBEM Web-Based Enterprise Management
WDL Windows Driver Library
WDM wave division multiplexing
WDMA wavelength-division multiple access
WFS Woodstock File Server
WFWG Windows For Workgroups
WGDTB Working Group on Digital Television Broadcasting
WGIH Working Group on Information Highway
WGS World Geodetic System
WIN Wireless Intelligent Network
WIN95 Windows 95
WINF Wireless Information Networks Forum, Inc.
WITS Wireless Interface Telephone System
WLAN wireless local area network
WLI Forum Wireless LAN Interoperability Forum
WLL Wireless Local Loop
WLRL Wireless LAN Research Laboratory
WOM write-only memory
WOMBAT waste of money, brains, and time
WRC World Radiocommunication Conference
WRS Worldwide Reference System
WRT "with respect to"
WSA Wireless Specialty Apparatus Company
WSC wireless switching center
WTAC World Telecommunications Advisory Council
WTB Wireless Telecommunications Bureau
WTO World Trade Organisation
WTPF World Telecommunication Policy Forum
WWW World Wide Web
WWWC World Wide Web Consortium
WYPFIWYG "what you pay for is what you get"
WYSIWYG "what you see is what you get"
WZ1 World Zone1

X

XA extended architecture
Xbar crossbar
XC cross connect
XCA extended communication adapter
XCVR transceiver
xd X-Bone xd
XDF extended distance feature
XDMA Xing Distributed Media Architecture
XDMCP X Display Manager Control Protocol
XDR External Data Representation
XFN X/Open Federated Naming
XFR transfer
XFS X11 File System
XGA 1. extended graphics adapter 2. extended graphics array
XID exchange identification
XIP execute in place
XIWT Cross-Industry Working Team
Xlib X Library
Xmission transmission
Xmit transmit
XML Extensible Markup Language
XMP X/Open Management Protocol
XMS Extended Memory Specification
XMT transmit
XNS Xerox Network Services
XO crystal oscillator
XPAD external packet assembler/disassembler
XPG X/Open Portability Guide
XPM Extended Peripheral Module
Xponder transponder
XRB transmit reference burst
xref cross reference
XRF Extended Recovery Facility
XSG X.25 Service Group
XSI X/Open System Interface Specification
XSMP X Session Manager Protocol
XT 1. crosstalk 2. IBM Personal Computer XT
Xtal crystal
Xtalk crosstalk
XTI X/Open Transport Interface. X Window System
XTL SunXTL
XTP Express Transfer Protocol
XUI X User Interface Toolkit
XWS X Window System

Y

YAG yttrium-aluminum-garnet
YIG yttrium-iron-garnet
YMMV "your mileage may vary"
YP Yellow Pages

Z

ZIP Zone Information Protocol
Ziv Lempel-Ziv